Zookeeping

Zookeeping

An Introduction to the Science and Technology

Edited by Mark D. Irwin, John B. Stoner, and Aaron M. Cobaugh

THE UNIVERSITY OF CHICAGO PRESS • CHICAGO AND LONDON

With the official endorsement of the following organizations:

The University of Chicago Press, Chicago 60637
The University of Chicago Press, Ltd., London
© 2013 by The University of Chicago
All rights reserved. Published 2013.
Printed in the United States of America
29 28 27 26 25 24 23   7 8 9 10

ISBN-13: 978-0-226-92531-8 (cloth)
ISBN-13: 978-0-226-92532-5 (e-book)

Library of Congress Cataloging-in-Publication Data
Zookeeping : an introduction to the science and technology / edited by
    Mark D. Irwin, John B. Stoner, and Aaron M. Cobaugh.
        pages cm
    Includes bibliographical references and index.
    ISBN 978-0-226-92531-8 (cloth : alk. paper) —
ISBN 978-0-226-92532-5 (e-book) 1. Zoos—Management. 2. Animal
handling. I. Irwin, Mark D. II. Stoner, John B. III. Cobaugh, Aaron M.
QL76.Z736 2013
636.08899—dc23
                                              2013016610

♾ This paper meets the requirements of ANSI/NISO Z39.48-1992
(Permanence of Paper).

Contents

Part Ten

Government and Legislation

Preface

The profession of zoo or aquarium keeper is unique, and everyone in society is familiar with the term "zookeeper." The profession is regularly featured (sometimes inaccurately) in movies, on television, and in books, but how does one become a zookeeper? It varies with the zoo and the region. Keepers need to have a working, practical knowledge of animal care, welfare, and husbandry; an appreciation of the issues facing endangered species and the environment; an understanding of biological principles; the ability to interact with the public; and a broad perspective of the captive wild animal community. Increasingly, zookeepers need to network professionally and synchronize their efforts globally. Indeed, as wild populations of endangered species dwindle and become restricted to smaller areas, the role of zoos and aquariums and their keepers will become more important. Zoos and aquariums exist to connect animals with the public; and, as direct ambassadors for the zoo or aquarium's animals, keepers are often viewed by the public as the facility's most credible source of animal information. They are often an underused resource. When used fully, a professional, well-rounded, and properly trained keeper will be one of a facility's greatest assets.

There is surprisingly little consistency in how keepers are trained and credentialed, and few foundational textbooks have been targeted to entry-level keeper training. Compare this to other professions, such as electrician or nurse, and the difference is stark. Training is essential to the realization of a keeper's potential and, when done effectively, it will contribute to improved animal welfare, strengthened public communications and relations, efficient use of resources, fewer accidents in the workplace, and greater success in conservation activities. This textbook will be of value to colleges and experienced zoo and aquarium personnel who need a broad keeper-training resource, and it is targeted to future and new keepers who seek an orientation to the career. It will also be a reference for other zoo or aquarium staff who desire an understanding of keeper work, and to zoo and aquarium visitors seeking detailed insight into the profession of keeper.

The idea for this book started to develop many years ago, when I was a newly hired animal care seasonal employee at Toronto Zoo. My supervisor gave me a copy of the *Metro Toronto Zoo Manual of Zoo Keeping* (which was already 25 years old), and I was captivated. On several occasions I visited John Stoner, then animal care manager, in his office and discussed the zoo field and keeper training manuals. It would be many years later that we finally began work on this book. In 2008 my employer, the State University of New York's (SUNY) Jefferson Community College (JCC) granted me a four-month sabbatical to begin development of a textbook that could be used for training student keepers. We wanted to develop a practical book that would provide a foundational orientation for potential and new zoo professionals, as well as a bridge between a general science background and a more advanced zoo- and aquarium-focused book such as Kleiman's *Wild Mammals in Captivity: Principles and Techniques for Zoo Management* (also known as *WMIC*). It didn't take long for us to realize that the book would need to be a larger team effort. John and I brainstormed possible formats and topics. As the project grew, we enlisted Aaron Cobaugh from SUNY Niagara County Community College as a third volume editor to compliment our knowledge base, and solicited input from numerous other colleagues.

We decided to develop a contributed work, along the lines of *WMIC* and the *Fowler's Zoo and Wild Animal Medicine* series. While this approach added tremendous complexity to the project, we felt that it would result in the most effective result. We solicited a diverse group of expert contributing authors from different backgrounds, regions of the world, professional organizations, and so on. Authors were asked to provide readers with an orientation to their topic, focusing to what information a new keeper should know or have immediately accessible when starting general keeper work in an accredited zoo (although we want this book to be of value to all keepers, not just those at accredited facilities). We also solicited artistic support through the American Association of Zoo Keepers. There is a lot of talent within this field. With

66 major chapters, 73 contributing authors, and 7 contributing artists, this book is an all-encompassing overview of basic zoo and aquarium animal keeping.

The multiauthor approach held inherent challenges. We worked extensively with authors to sculpt chapters that would fit both in content and in style with other parts of the book. Chapters were being written concurrently, so aside from our communications, authors had to develop their chapters without the advantage of knowing what others were writing. Whenever possible, we advised them on issues of content overlap and prose style, and made suggestions for topics on which there might not have been agreement within the field or between regions. Authors were asked to include information based on their experience and, when appropriate, to provide their recommendations for best practices. Of course, such recommendations may not be applicable everywhere and will need to be considered in the context of each facility's situation, including regional association standards and local laws. The writing style may vary somewhat between authors, some overlap between certain topics may still exist, and chapters may differ in breadth and depth, but we believe that the finished volume's content is much stronger for having such an expert group of contributors. Most chapters were prepared to stand alone, but many are complimented by others in the book, and readers are encouraged to read related chapters together for the best understanding. For example, readers of chapter 15, on physical restraint and handling, should also review chapter 14, on stress and distress. When possible, chapters have been categorized and grouped with others that cover related topics. By the nature of the content, each major section of the book has a somewhat different flavor.

To develop from scratch a synchronized book whose parts would fit together, significant coordination and cooperation was necessary. Writing conventions were developed to promote consistency throughout. Even deciding on the term to be used for our primary audience involved some deliberation. Zookeeper, zoo technician, aquarist, biologist, handler, trainer, animal caretaker, and animal manager are all terms that have been used; the term often varies from one facility to the next. Ultimately, in this foundational text we want the term "keeper" to refer to those who care for the animals in a zoo or aquarium (including facilities such as safari parks and sanctuaries), and we do not want to distinguish them on the basis of the type of animals they work with, the distribution of their work tasks, and so on. Similarly, we chose to use "zoo and aquarium" as the general term for the facilities in which the keepers work. Since this is an introductory text, we also asked authors to follow the convention of leading with common terms, which are then followed by scientific or technical terms in parentheses.

It has been a pleasure working with each contributor. It seems as though we have been working on this project for a very long time, but from start to finish, it will have taken less than five years—not bad for a comprehensive, first-edition textbook! Everyone involved has been very cooperative and understanding of the challenges and frustrations that accompany such a large project. Conservation and management of endangered species can only be successful when done cooperatively, ultimately on a global scale. We are proud to be in a field with so many passionate and dedicated colleagues willing to contribute their expertise to advance the profession of keeper, improve the quality of zoo animal care, and therefore support global conservation. Keepers are often motivated by a love of animals and/or a passion for conservation, but they may be unaware of their role in the larger picture. We hope that new keepers will strive to contribute in their own way to their regional zookeeper and zoo associations, and on a global scale to the International Congress of Zookeepers (ICZ) and the World Association of Zoos and Aquariums (WAZA).

As issues relating to science-based animal management, the environment, legal liability, and animal welfare become more prominent, effective zoo operation will increasingly depend on having effective teams of skilled, trained, professional keepers. We hope that *Zookeeping* will help to advance the profession of keeper, improve consistency whenever appropriate, and promote best practices for the keeping and care of zoo and aquarium animals. Feedback and suggestions from readers for a possible second edition are welcome.

Mark D. Irwin

Part One

Professional Zookeeping

1

The Profession of Zookeeper

Ken Kawata

INTRODUCTION: PROFILE OF ZOOS CONTINUES TO EVOLVE

THE VOICE OF *SILENT SPRING*

The modern zoo, conceived and born in Europe, has been in existence for two centuries. In the United States it began in the mid-nineteenth century and still represents a relatively young institution, continuing to evolve with time. The first aquarium did not appear until 1853 in the London Zoo; an independent aquarium (not a part of a zoo) did not open until 1859, in Boston (Solski 1975, 362–97). In more recent decades, zoos and aquariums have gone through periods of metamorphosis. "The low point against which zoos today measure their progress occurred between the 1950s and the 1970s, as cities entered fiscal crisis and zoo facilities deteriorated from years of neglect. In the first four decades of the twentieth century nobody complained about zoos; even with their animals behind bars, they were civic gems. But in the 1960s the environmental movement began to change public attitudes about wildlife and expectations for how zoo animals ought to be displayed and cared for" (Hanson 2002, 164).

In her 1962 book *Silent Spring*, Rachel Carson, a marine biologist, chronicled the frightening effects of pesticides, and the public responded with a tremendous outcry. US Senator Gaylord Nelson is often referred to as the "father" of Earth Day, the first of which was held on 22 April 1970. "Soon 'ecosystem' and 'endangered species' became household words. Such a social change was bound to affect the zoo world, because a zoo is not an isolated island" (Kawata 2003, 114). While zoos were given the mission to be survival centers for endangered species, zoos in any given country are often affected by the dynamics of economic status of the time. In the United States, for example, zoos were facing a financial crisis that had begun in the 1970s: "The present precarious financing of zoos makes it difficult for individual zoos to commit themselves to establishing and maintaining long-term programs. These programs involve commitment of major economic resources needed to maintain the larger groups of individuals of each species selected" (Donaldson 1983). Through a variety of challenges, zoos have grown to adjust themselves for institutional survival and to balance the internal needs and the requirements of our time.

This chapter will outline the basic role of the keeper as an integral part of zoos and aquariums. After studying this chapter, the reader will understand

- how zoos and aquariums have evolved in recent decades into a multidisciplinary entity, with diverse demands cast upon them;
- how the keeper's duties have also increased with time while the core of the keeper's work—caring for the animal collection—remains unchanged;
- that in becoming a professional keeper, one's inner transformation is an important step, to comprehend the reality of wild animals' nature and the human aspects of the workplace;
- that communication skills are essential for the keeper to help establish an efficient environment and work harmoniously with colleagues and members of the public;
- that learning is a lifelong process for the keeper's growth and development, particularly in working with groups engaged in diversified activities within a zoo, some of which are seemingly unrelated to animal care;
- that it is vital for the keeper to make contributions through daily work toward organizational goals, to advance science as well as to strengthen a positive image of zoos and aquariums as cultural and public institutions.

DIVERSE REQUIREMENTS FOR ZOOS

Along the way, with the necessity to satisfy increasingly diverse requirements for fund-raising, education and conservation activities, the modern zoo has become a multidisciplinary entity, bringing together expertise from various fields to the daily operation. Such a change has been noticeable in the evolution of zoos' staff structures. The basic core structure of the zoo staff, however, remains relatively unchanged. After all, it is a business organization with groups of workers and those who supervise them in daily operations, and that pattern stays the same across the world. Here are the generalized profiles of key positions that directly affect the work of a keeper. Among individual zoos the definitions and content of work will vary.

DIRECTOR

Depending on the zoo's size and the preference of its governing body, this executive-level position may have titles such as president, chief executive officer (CEO), executive director, or (in smaller zoos) superintendent. Today's zoos have grown into complex organizations, and that is reflected in the requirements for this position. The director's job is to manage the system as a whole, which is in many ways comparable to the job of a director in any other business organization, even though a zoo undoubtedly has very unique missions. Responsibilities for this position include planning, organizing, budgeting, administering, and directing programs and activities of the zoo. As the zoo's representative, the director must be able to deal efficiently with labor forces, elected officials, members of governing bodies, civic groups, and news media. He or she is also expected to participate in activities of the national and international zoo and conservation organizations as a member. (In some zoos the director has assistants and associate or deputy directors, as noted in box 1.1.)

CURATOR

Typically, this is a middle-management position for directing a group of workers and their activities. In some zoos there are general (or senior) curator positions for overseeing subordinate supervisors and the entire collection of animals, while other zoos have more specialized positions such as curators of mammals, birds, and reptiles. Also, there are curatorial positions for non–animal care disciplines such as education. There may also be assistant curatorial positions in larger zoos, often for specific groups of animals or exhibit areas. In all, a curator's duties can be categorized in three major areas, not in order of importance. *Managerial* duties include hiring keepers, drafting work policies and guidelines, proposing budgets, and labor negotiation. *Technical* duties include planning breeding programs, coordinating research activities, and assuming a regional and international studbook keepership. *Operational* activities include supervising keepers, coordinating shipment of animals, and modifying animal diets. The amount of work in each of the above categories varies widely from one zoo to the next.

ANIMAL CARE SUPERVISOR

The animal care supervisor occupies a position in the forefront of animal husbandry. In addition to overseeing keepers, his or her responsibilities may include immediate care of animals (e.g., cleaning exhibits and feeding). Individual zoos give this position varying titles such as headkeeper, senior keeper, or keeper II or III. In larger zoos with more keepers there may also be a layer of subordinate supervisory positions. Regardless of titles, the responsibilities focus on daily care and safe maintenance of assigned exhibits and holding facilities, according to established protocols and under the direction of the curatorial staff. Duties include coordinating activities with personnel of other divisions, such as veterinarians and maintenance workers, and submitting various written reports. They also often include providing information on animals and zoos to the public during various events, and to other divisions within the zoo, such as education and public relations.

For the purpose of historical perspective, a quick review of the zoos in the United States in 1930 reveals a rather simple, linear structure. A manager in charge, called the park superintendent or headkeeper, was the zoo's only chief representative, with an occasional second man in command (a zoo was a man's domain then; Doolittle 1932, 90–100). Basically, this was a system for overseeing workers who cared for the animal collection and maintained the grounds, buildings, equipment, and machinery. By 1970, upper-level staff had become more structured, as some zoos listed a specialized curatorial system (Truett, 1970). Looking in from the outside, the public may assume zoo work to be simple: there are keepers, and people who sell ice cream and peanuts, as well as the director, who appears on television every now and then. For sick animals there is a veterinarian, and that is the extent of the zoo staff. A quick look at a zoo's organizational structure, however, reveals a different picture. There are many job titles unrelated to daily care of animals, especially in larger zoos. A modified, generalized, and generic outline of the senior staff is shown in box 1.1 (AZA 2010a, 33, 37, 40, 52, 54, 56, 80, 88, 104).

A zoo's animal care division often represents a diversified group; but not all zoos have a complex system. In fact, the majority of the world's zoos are small, located in smaller cities. For example, Erie Zoo in Pennsylvania—with 21 permanent employees and 90 additional temporary part-time employees during peak season—has a much more scaled-down organization, as shown in figure 1.1.

Although all zoos may appear outwardly similar, privately owned and managed institutions differ considerably in terms of operation from those managed by federal, state, and municipal governments and nonprofit groups. Thus, zoos can be grouped into the public sector and the private sector. In this context, even the term "zoological society" means different kinds of operations. In a government-operated zoo, the term refers to a support group, but not the zoo's governing body. (In some zoos such a society may have a different label, such as "friends of the zoo," but its functions are basically the same.) The society's role is to assist the zoo by raising funds, purchasing animals, funding departments, and conducting programs such as publication of the zoo's periodicals. In the zoo's organizational chart, the society is placed "off to the side"; its role is to aid the chief executive officer. In other zoos, however, the function of a "zoological society" is quite different: it runs the zoo. The zoo may or may not receive assistance from a tax levy, but the zoological society, an officially chartered organization, is the zoo's operator. Its board controls the budget and manages the zoo staff. This board, often called the board of trustees, is a group of persons with managerial or investigative powers. Its members are usually community leaders from business, government, education,

Box 1.1. Senior Staff Organizational Structure

The number of senior (or upper-level) staff members varies depending on the size and scope of work of individual zoos. As a general rule, larger zoos have more diverse and specialized programs that require more staff and a wider variety of positions. Senior staff members may be categorized arbitrarily into subgroups such as administrative, scientific, and operational. Below is a description of a composite model of a large zoo (with at least several dozens of employees), patterned after several major American zoos. Senior staff work under the top executive, who hold one of various job titles, depending on the zoo: president, chief executive officer (CEO), executive director, director, or superintendent. Under this position, larger zoos have another layer of executive officers, as noted below.

Sub-executive group. Under the top executive in many zoos are subordinate executive officers, such as chief operating officers (COOs) and vice presidents. The latter position may be divided into senior vice presidents, and assigned into specific areas such as vice (senior) presidents of conservation science, government affairs, and guest services operations. There also exist subordinate directorial positions with such titles as deputy, assistant, or associate director, with responsibilities in areas including education, food service, administration and technology, and exhibits and park management. (Note: the title "executive assistant" in a staff roster usually refers to the secretary to the CEO, not to an executive per se.)

Administrative group. This group may be defined as the executive branch of the organization. Its duties are closely connected to the management of funds, personnel, policy making, and dealing with the zoo's governing body. Job titles in this category include vice presidents (or directors) of finance, development, marketing, membership, merchandising, public relations, strategic initiatives, and human resources.

Scientific group. This group's responsibilities include management of live collections (both animals and plants), ex situ and in situ conservation, education, and the planning and administering of various programs using the collections. Job titles in this category include vice presidents (or directors) of veterinary services (or animal health), conservation science, animal care, audience research, education and international training, and graphics and publications. (The curatorial positions, which are directly involved in the programs under the direction of this group, are reviewed in this chapter.)

Operational group. Maintenance of buildings and grounds—monitoring of water and energy supplies, sewer lines, and safety and security, the nuts and bolts of the zoo—falls into the hands of this group. Officers in this category carry job titles such as vice president of plant and facilities management, director of facilities and construction, and exhibits and facilities manager (modified from AZA 2010a).

Erie Zoological Society

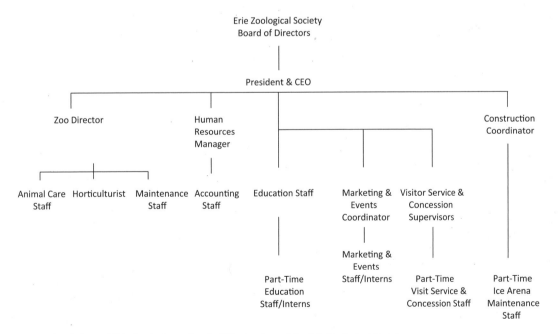

Figure 1.1. Organizational chart of Erie Zoo in Pennsylvania (US). Courtesy of the Erie Zoological Society.

Wildlife World Zoo and Aquarium Organizational Chart

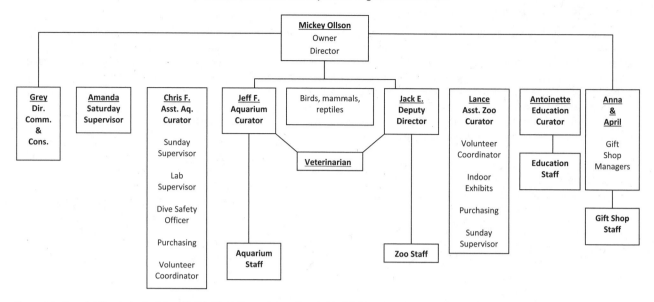

Figure 1.2. Organizational chart of the Wildlife World Zoo and Aquarium in Litchfield Park, Arizona (US). Courtesy of the Wildlife World Zoo and Aquarium.

and other fields, and are appointed to serve for a specific term with no monetary compensation.

Zoos in the private sector receive no financial assistance from the tax levy and must rely on revenue-generating activities for institutional survival. One of the successful private institutions that has emerged in recent years is Wildlife World Zoo and Aquarium in Litchfield Park, Arizona; figure 1.2 shows its organizational chart. Its owner/director (pers. comm.) has stated that the number of full-time employees stays between 45 and 58 (these do not include workers in the food concessions, which are operated on a contractual basis). Of these zoo employees, 10 are keepers; at the aquarium there are nine keepers in addition to two people who handle interpretive aspects such as the touch (or contact) tank. The rest of the employees are assigned to animal ride operations, gift shops, maintenance, and office duties.

GETTING STARTED AS A KEEPER

The keeper's job is considered an entry-level position. The requirements and expectations for this position may differ between the public and private sector, yet these differences, in historical perspective, appear rather insignificant. The basic requirements and work processes for a keeper, which will be reviewed in depth in this book, have remained largely the same throughout the years. In the beginning, after various selection protocols ranging from a competitive civil service examination to a more informal job interview, a fortunate applicant arrives at a new workplace, in most cases with a fresh-minted college degree. It might be noted at this point that realistically, four years of higher education have hardly been adequate to fully prepare a person for the zoo and aquarium profession, either technically or mentally. It is commendable

to be ambitious and upwardly mobile, but it must be realized that a keeper's real education is beyond the scope of university curricula, and beyond the realm of immediate on-the-job training; it begins on the first day of work, and continues throughout the keeper's career. Also, an inner and personal transformation is vital for a keeper to grow and develop as a professional.

Such a transformation takes a couple of directions. First, from childhood our perception of *an animal* has been mostly nurtured by household pets such as dogs, which have been domesticated for millennia to serve our needs. By contrast, wild animals, which include a vast array of different species, have evolved and survived over millions of years in nature's harsh environments. In the mind of the new keeper, a transformation must occur in understanding that the zoo animal is not a pet but a potentially dangerous *wild* animal. (There are certain things that only experience can teach, often the hard way.) Also, in a public zoo and aquarium the animal belongs to the owner institution or the taxpayers, and it should be cared for in accordance with the prescribed protocol, not the keeper's personal methods.

This leads to another fundamental area that requires a transformation from existing within a family, college campus, or other entity to that of a structured zoo or aquarium workplace. In this system, higher education is no automatic pass for promotion or entitlement for professionalism. There is a linear chain of command that places a keeper at the organization's entry level. As an employee, the keeper is required to adhere to established policies and procedures at every step of work life, and bureaucratic hurdles and politics, internal and external, are not uncommon. It takes time to build up credentials and seniority, and promotional opportunities for upper-level positions, such as curatorship, are by no means abundant in most places. In short, patience is a virtue, and the learning curve is steep. What is needed is to anchor down instead of grappling for instant gratification; to make a commitment for two to three years and dig in intellectually *beneath* the surface of the everyday work routine, be it animal care or other activities. Experienced and seasoned keepers can tell you that the sense of fulfillment and rewards will come, albeit slowly. Encouragingly, the image of keepers has gone through considerable change over the decades. However, the process of change has represented a gradual path.

THE CHANGING IMAGE OF ZOOKEEPERS

A GRADUAL PROCESS OF METAMORPHOSIS

In European zoos, the position of keeper belongs to the blue-collar workforce. For example, "zookeepers in Germany are rarely college or even *Gymnasium*—in British terms, 'A-level graduates'; most have only a *Realschule* education ending in the equivalent of O-levels" (Reichenbach 2003, 495). A-level graduates in Britain are roughly equivalent to German pupils who have 13 years of school and the right to attend a German university; O-level graduates have the equivalent 9 or 10 years of school and are not qualified to attend university (Reichenbach pers. comm. 2004). In the United States only a few decades ago, the typical image of zookeepers was of a mostly Caucasian, blue-collar workforce: middle-aged municipal-worker-type males with no college education. Some keepers were functionally illiterate even in the 1970s.

The American keeper was soon to experience a metamorphosis of sorts. The shift began under the surface in the 1960s, depending on the region and individual institutional cultures. This was also the time when women began to join the zookeeping workforce. To cite an example, Pat Sass began her zoo career in June 1961 as a volunteer in the Lincoln Park Zoo children's zoo in Chicago. She was hired as a part time zoo leader in the following year; at that time women were not allowed to become keepers, and the zoo leader was a position for women working in the children's zoo only. In 1965 she became a full-time employee in the children's zoo, paid less money than male employees. In 1972 women were finally allowed to take the civil service exam for the position of animal keeper, and Sass became one of the first female animal keepers at Lincoln Park Zoo (Anon. 1999, 382). Notice that she started in the children's zoo, not in the main part of the zoo where the "ABC animals" such as elephants, big felids, bears, and great apes were housed; only the infants of great apes and big felids were kept in the children's zoo. (Women are no longer a rarity in the zookeeping workforce, but some ethnic groups are still poorly represented in the zoo world, including at the keeper rank.)

In 1985 the Professional Standards Committee of the American Association of Zoo Keepers (AAZK) made a survey of keeper hiring criteria in zoos after making 282 contacts. Of these, 169—or 60% of employers—responded (McCoy 1985). The result shows varying requirements in many categories, such as ability to follow written and oral directions, physical strength and stamina, ability to operate mechanical and power equipment, and willingness to accept regular week-

end, holiday, and after-hours assignments. It is interesting to note in this report that for all job titles in the zookeeping workforce (e.g., assistant, trainee, keeper I, and keeper II), 107 zoos required a high school diploma while 63 required either a two-year or a four-year college degree. Thus, it appears that by the mid-1980s in many zoos in the United States, higher education was becoming a job requirement for keepers.

Today, more than half of keeper job applications received by zoos are from young persons with college educations. An experience at the Staten Island Zoo in New York reflected this trend. Between January 2004 and June 2005 a total of 184 resumes were received for keeper positions. Of these, 32 (or 17%) were from males and 152 (83%) were from females (the remaining two resumes had no clear indications of gender, but appeared to be from one of each). Nine of the applications were from foreign countries (Australia, Canada, Mexico, Peru, and Spain). As for the applicants' educational backgrounds, 64% had at least a two-year degree. One foreign applicant had a veterinary degree; three (one from a foreign country) had PhDs; 11 had Master's degrees. Seventy-nine (excluding those with advanced degrees), or 43% of all applicants, had four-year degrees while 23 had two-year degrees. Of those who lacked four-year degrees, 59 said they expected to receive one within a year.

One of the reasons for this increasingly educated applicant pool, aside from the fact that zoos appear to be a promising field, is simply that there are so many college graduates in the US job market. "In the fall of 2005, more than 1.5 million students enrolled in America's four-year colleges or universities, a number equal to 50 percent of high school graduates that year," notes Charles Murray. The number of bachelor's degrees awarded in 2005 amounted to 35% of all 23-year-olds (Murray 2008, 67–68), indicating a large pool of job seekers. If we consider the situation closely, college campuses tend to encourage unrealistic expectations about the job market; those who arrive at the zoo gates fresh from campus often find a chasm between reality and their aspiration. Some new keepers may expect to spend most of their time "studying animals." There is a need for those with imagined expectations to make a realistic adjustment if they are to stay in the field, grow, and develop. The reality of zoos dictates that someone has to clean up after animals and feed them every day, and that responsibility falls on the shoulders of the keeper, regardless of his of her educational background.

As mentioned before, changes for keepers arrived slowly. A review of keeper job descriptions and duty outlines in operation manuals from the 1970s through the end of the last century reveals hardly any functions other than the custodial care of the animal collection. For instance, knowledge, skills, and abilities required for a keeper in a large municipal zoo included ability to move objects weighing up to 22.6 kg (50 lbs.), seeing well enough to read standard English text, ability to stand and walk, and graduation from an accredited four-year high school (anon. 1997). Similarly, requirements for zookeeper 1 at a small municipal zoo included a driver's license; ability to walk, climb, balance, stoop, talk, and hear; and ability to lift and carry equipment and supplies such as feed bags weighing up to 45.3 kg (100 lbs.). Specific education requirements were not included (anon. 1995).

As for current job descriptions for the zookeeper positions, the AZA website offers a wide range of examples (AZA 2010b). One from the Brookfield Zoo, in a Chicago suburb, includes a comprehensive description, given here in part.

Responsibilities
Prepare and distribute food as directed. Observe assigned animals closely, routinely, and objectively report to supervisor on their behavior, health, and welfare. Keep animal records as assigned. Observe condition of assigned area(s), animal enclosures, and equipment. Monitor and clean public space(s), plants, and interpretives as assigned. Report maintenance needs. Perform minor maintenance tasks as assigned. Clean animal enclosures, pools, and service areas. Study specific reference materials about animals in assigned area. Read other zoological and husbandry texts as assigned. Perform routines of assigned areas and specialized procedures as directed. Monitor visitors to insure animal and human safety and protection of park property. Communicate with and assist guests. Take initiative to facilitate guests' appreciation of animals and enjoyment of their zoo visit. Demonstrate knowledge of and implement zoo-wide and area emergency procedures as directed. Other related duties as assigned.

Requirements
Bachelor's degree in relevant biological field or equivalent combination of training and/or experience required. Animal care facility, farm, kennel, or equivalent animal experience required. Incumbent must possess an understanding of practical principles of animal behavior and the ability to work with live animals safely. Good interpersonal skills. The ability to interact in a courteous and professional manner. Knowledge of natural history, zoology, or animal husbandry preferred. Must be capable of dealing with emergencies calmly and efficiently. Must learn and integrate information, knowledge, and direction quickly. Good decision-making skills. Must be alert and make sound independent judgments.

Interestingly, the job description also states, "Spanish Fluency a plus but not required," reflecting our time of societal diversity. As for specialized skills, it states: "For the Living Coast and Seven Seas keeper positions, SCUBA Certification required within one year of hire." This is an example of a more generalized zookeeper position. There also are job openings for specialized work, typical of which are elephant care positions, which appear often. Here is an example from Tennessee:

The Memphis Zoo has an opening for a Keeper in our Elephant area. This position requires a minimum of two years of experience in pachyderm care, including knowledge of training concepts and philosophies needed to work with pachyderms and a varied collection of hoofed animals and birds, and a college degree in zoology, biology or a related field (or the equivalent combination of education and experience). Significant experience with operant conditioning is required. Ours is a protected contact program with a herd of three female elephants, including one who is a recent addition from a free contact environment. Responsibilities include providing a healthy, attractive environment for the animals in a specified area, all aspects of daily animal husbandry, exhibit maintenance, observation, enrichment, and encouragement of public interest in the animals. Must have excellent communication skills and a willingness to interact positively with the public and Zoo staff.

FACING THE CHALLENGES

The evolution of the keeper's role, however, is almost always accompanied by perceived and often increased expectations from within the zoo or aquarium institution. This is especially true since zoos are facing the increasingly diverse demands of our time. As the "torchbearers" of the zoo or aquarium's essential component, animal care, keepers are also required to take up a broader range of activities beyond the traditional care for the animal collection. In order to meet the challenges of our time, keepers must develop diversified skills in broader areas by arming themselves with knowledge, both academic and practical, for duties such as assisting researchers, public speaking, and helping children to learn. Once out of college, the "education" is not over! True, there will be no more midterms or finals, yet in their work life the process of education has just begun; exams now come in various, subtle, and often unexpected forms. And for those who wish to advance as professionals, learning is a lifelong process.

LEARNING: LIFEBLOOD FOR KEEPERS

Marlin Perkins was one of the most prominent mentors in American zoos during the latter half of the twentieth century. For all his experience, knowledge, and wisdom, he stated in his memoir: "When I realized I would like to stay at the zoo, I decided I should acquire still more education. I enrolled in night school at Washington University and for several years took courses in advanced zoology, Latin, and psychology. . . . This self-education process has continued all the rest of my life. Besides my reference library, so important to me in my zoo work, I have accumulated volumes on archeology and anthropology, travel, and many other subjects" (Perkins 1982, 44). Even in the 1970s, learning was mandated for keepers in some of the more traditional municipal zoos. According to one work manual, "in talking about modern zoo philosophy, it is obvious that zoo employment involves not only a working process but a learning process as well. A zoo man never reaches a point where he knows everything" (anon. 1972, 2). Another manual emphasized: ". . . it is expected that you will take an active interest in your animals and attempt to learn as much about them as possible" (anon. 1973, 6).

Learning can mean different things to different people. For daily care of animals, mastering practical skills is essential. Veteran reptile man Peter Brazaitis recalls: "I could immediately recognize a new keeper who might have reams of academic credentials and knowledge about specific animals but no actual work skills. There really is a way to hold a broom and dustpan that allows you to collect all of the dirt on a floor. There is a way to rinse a mop to get rid of the dirt it has already absorbed, so that you don't simply spread it back on the floor with each successive sweep. There is a method for unpacking live frogs from a cardboard shipping box so that they can't leap out en masse in every direction as soon as you lift the first flap." And the list goes on (Brazaitis 2003, 34).

THE ART OF MENTORING

Another vital tool for learning is mentoring. Before the arrival of waves of young persons with higher education, there used to be more active transgenerational conveyance of experiences through mentorship. There existed a body of knowledge, techniques, skills, thoughts and wisdom that had accumulated over the years and were transferred from one generation to the next. In those days, for aspiring young men mentoring was the basis for developing skills and critical thinking; a firm sense of camaraderie served as ligament to bind keepers together, and they took pride in their work. Josef Lindholm III (the author of chapter 4 in this volume) gives a small example of how transmission of information took place in a rather casual style, in his case in the San Francisco Zoo, where the feeding of large felids became a beloved local public activity. Lindholm remembers: "I have very distinct memories from my childhood of unkempt middle-aged men trundling wheelbarrows full of horse meat past roaring and snarling cats. In 1972 I asked one of these men, just after the lion feeding, about the Baikal seals and saigas that used to be at San Francisco. John Alcaraz, actually Principal Ape Keeper, befriended me (12 at the time), and among other things showed me where the zoo library was. That's where I was introduced to the *International Zoo Yearbook* (IZY) and *International Zoo News* (IZN). I also remember Jack Castor, portly, mustached, glasses, who took care of that cat house for around fifty years. At any rate, this does make one think" (Lindholm pers. comm. 2008). It is probably safe to assume that these old-timers, who likely had never received higher education, did not realize their influence on a young boy with an impressionable mind.

Lindholm's mention of his introduction to the zoo library leads us to another essential mode of learning: reading. Even in this era of the dot-com, iPod, and Blackberry, familiarity with print media remains fundamental to being a professional. There is an unfathomable amount of knowledge and wisdom in the print world, the result of human experience accumulated over millennia. The above-mentioned *IZY* and *IZN* have been around for half a century, and *Animal Keepers' Forum*, published by AAZK, must also be added to the basic list. These volumes are such an immense reservoir of knowledge; one must simply dive into it, look around and dig in. Immeasurable pleasure exists there for those who are eager to learn and grow.

Aside from learning from animals and books, keepers must learn interpersonal skills. A longtime zoo curator who had started as a keeper once said: "Right from the beginning in this field I never thought zoo work was animal work. It's people work." She was quite insightful. A careful look at eight hours of a zookeeper's work raises the question: How much time does a keeper spend with animals? If one employs an ethologist's method of studying time budgeting and categorizing work activities, the result could be surprising. Throughout the day a keeper is constantly dealing with fellow keepers, supervisors, coworkers in other departments, and members of the public. In between comes time for cleaning animal quarters on and off exhibit, preparing for feeding, handling tools and equipment, and moving from one work station to the next.

The public's perception—"It must be a cushy job, playing with animals all day and getting paid!"—could not be less accurate. Precious little of a keeper's time is spent in direct physical contact with animals. This is where interpersonal skills are indispensable to an efficient zoo worker. Those who continuously have problems getting along with other human beings cannot make themselves efficient zookeepers, no matter how excellent an innate rapport they possess with animals.

WORKING WITHIN THE ZOO STRUCTURE

We have briefly reviewed some of the essential requirements for a professional keeper, including the willingness and ability to learn as well as the ability to work with others harmoniously. These ingredients are vital to becoming a productive member of a progressive zoo staff. After all, a zoo is a workplace, and within a zoo's structure a keeper is required to work with persons from a variety of divisions, depending on the zoo's size or specific institutional culture; these people include curators, veterinary staff, registrars, maintenance crew, and horticulturists.

INTRADIVISIONAL ACTIVITIES

Keepers maintain close contact with workers who are also charged with caring for live collections, including horticulturists. Communication is the glue that holds people together, and it begins with a close-knit work group. If coworkers do not or, worse, choose not to understand the importance of communication, the consequences could be quite serious. In his book *My Wild Kingdom*, Marlin Perkins, the aforementioned zoo man, shared an episode he heard about, which had occurred in a New York zoo.

Tom McLeary, the zoo's animal foreman, had a hand-reared male white-tailed deer under his care. One year McLeary took a month's vacation, and a relief keeper was assigned to this area. This new man was afraid of deer, and brought some loaves of bread with him when he entered the deer yard. "As the deer came up to him, he would throw a slice of bread way off and the deer would run after it. Then he would do a little sweeping up and when the deer came back, he would throw another slice of bread to keep the deer away. When Tom returned, the replacement keeper didn't say he had been feeding the deer in this manner. Tom went in as usual with his wheelbarrow, rake, and broom and started his work in the yard. The buck came over to Tom." The deer, expecting to be fed, charged him. "He hit Tom squarely in the backside, knocking him flat, and gored him in the back with his antlers. Pulling his antlers out of Tom's body, the deer started at Tom's head.

"Lying on the ground, Tom realized he had to defend himself—it was early morning and there were no visitors in the zoo and no other keepers nearby. He also realized that he must grab the antlers, but must grab only on one side, so as to throw the deer off balance. If he grabbed both antlers the deer, who was stronger, would lift him off the ground. So he grabbed one antler and held it to the ground. . . . The antler had gone through his buttocks clear into his coelomic cavity and, of course, the wound became infected. He nearly died and spent nine months recuperating after he left the hospital. Tom explained that he hadn't known the relief keeper had been paying off the deer. How different the story would have ended had there been good communication between those two" (Perkins 1982, 85–86).

As Perkins noted, closer communication and more cordiality are in order within the work group. To take another aspect of communication, consider a large indoor community exhibit that houses mammals, birds, reptiles, fish, and invertebrates. The exhibit may require keepers of different taxonomic groups of animals to work together in the same space. Peter Brazaitis, the aforementioned reptile expert, talks about a classic example of friction between groups as "the problem of the animosity between some very intense 'bird people' and 'reptile people,'" citing a young senior bird keeper he called Ames, who "was particularly passionate in his desire to see me dead. Ames was indeed an excellent bird person: his gentleness, great knowledge, infinite patience, and close attention to detail made him a master of his work. The root of the problem was that I was a 'reptile man' and he was a 'bird man'" (Brazaitis 2003, 321).

In such an exhibit there often are live plants, and requirements for the health of animals and plants could be on a collision course. Plants need light to produce food for themselves, but the light intensity in nearly all indoor zoo exhibits is insufficient for the plants. For zoo horticulturists, it is a challenge to find plants that will survive and grow under diminished lighting. In such conditions the plants need less water. However, most exhibits must be hosed often to remove animal waste, which can lead to too much water for the plants. Roots cannot survive in a waterlogged soil, since they need oxygen; as the roots die, so do the stems and leaves. Another problem is that in captivity, animals tear plants, eat their foliage, destroy their bark, or uproot them; keepers and maintenance workers cause damage, often unintentionally. Heating and ventilation systems could also pose a threat to plants, making it difficult to overcome losses caused by animals and other factors (Rives 2001, 1–2). At the least, keepers should be sensitive enough to comprehend the challenges that horticulturists face daily.

INTERDEPARTMENTAL ACTIVITIES

Outside of the realm of live collections there are other divisions, such as maintenance and groundskeeping. These are more traditional, age-old parts of the zoo, as basic as air and water to a zoo's institutional well-being, and coordination with these groups ensures smooth and safe operation of the zoo. Personnel in these groups may not realize the danger of wild animals and, conversely, the hazard for animals caused by seemingly minor items. Letting an electrician or a repair crew go into animals' living quarters requires extreme caution. Pulling a wrong door at the wrong time may place a plumber with an African buffalo, which could have fatal consequences for the human. A nail or screw carelessly left behind by a carpenter in a hoofstock yard could cause pica (ingestion of unnatural objects), which could lead to a painful death for the animal.

Another department closely related to the animal collection is the research department, which is more commonly established in larger zoos. Research, which is vitally important for a zoo as a cultural institution, may be viewed a luxury in zoos with limited resources. Smaller zoos are fortunate to have even a part-time scientist on the staff.

Zoo and aquarium animals are basically for public exhibition, and although some of them offer precious research opportunities, they have their limitations: "Rare and endangered animal species, by definition, often present very few opportunities to learn much about them. Daily management priorities and animal well-being also take precedence over strictly controlled research methodologies; therefore, some creative logistical and statistical strategies are often employed to make the most of the limited access to the rare and endangered" (Wharton 2007, 181). Keepers can offer important assistance to researchers: "Keepers often do not have advanced training in research techniques, but they can contribute to science or participate in scientific activities. In fact, much zoo and aquarium research would not be possible without the logistical assistance, knowledge and direct participation of keeper staff" (Hutchins 1997, 34–35). It is encouraging to note that as early as 1979, an internal document at Smithsonian National Zoo in Washington, DC, emphasized the role of keepers in research: "Keepers often work with researchers and undertake research projects of their own. . . . Keepers also write and present papers to professional organizations about particular aspects of their work" (anon. 1979).

SERVING THE PUBLIC

Zoos are no longer places to merely exhibit animals, as they were perceived decades ago. Today's zoos are diversified operations with a variety of functions. The work of keepers is vital, yet they are one of the work groups that serve zoo visitors, and they do so in various ways. A group frequently encountered by keepers, formerly known as "concessions" but sometimes now known as "visitor (or guest) relations," operates gift shops, snack bars, and restaurants. Other groups are relatively new in the zoo field, and are more common in larger zoos with more diversified functions. They are in the front offices, often isolated from the daily care of animals. Since their work paths rarely cross those of the keepers, the keepers often do not know them on a daily basis and may not realize the importance of their work. One of these departments is called "development" and is engaged in fund-raising: "There are two main purposes for an organization's fund-raising activities: to raise money for a project or operation and to involve the community in the organization's vision and missions" (LaRue 2002, 141).

Other groups also enhance and ensure the zoo's financial health. Their work takes various directions, although they all have the common goal of promoting the zoo. First, zoos must increase the number of visitors who pass through their gates, and the press, both local and national, plays a pivotal role in how citizens perceive them. As public organizations zoos and aquariums are subject to public scrutiny, and knowing how to deal with news media can create a positive image for them. In the forefront of this endeavor is the public relations department, which focuses on building a bridge between the zoo or aquarium and the news media. These personnel organize press releases and press kits which contain facts and figures, the basis for news stories, and they stay in touch with the local press corps. Similar tasks fall to the marketing department, whose work includes, but is not limited to, supporting advertising sales and corporate sponsorship for the zoo. This department may also be involved in drafting applications for grants from government agencies and private sources, making presentations to corporate representatives, or initiating fund-raising programs such as "adopt-an-animal" (or exhibit).

There exists yet another type of promotional work for the zoo: special events, which involves planning, promoting, and conducting activities and events for the public. As discussed above, revenue-generating measures have become important in recent decades, and zoo calendars are filled with such programs as breakfast with animals, keeper chats, elephant's birthdays, golf outings, and the now-popular sleepovers for children. Rented space is available for weddings in many zoos and aquariums, and for after-hours events for corporations and other organizations.

These divisions—development, public relations, marketing, special events, and at times education—share overlapping areas of responsibility, depending on the individual zoo or aquarium's institutional culture and priorities. Yet personnel in these departments must all have one ability in common: to communicate effectively, both verbally and in writing.

These departments once comprised the support groups of a zoo. In more recent decades they have increasingly been given decision-making status in the mainstream of the zoo's administration. "Phoenix (Zoo) for instance in making its collection plan allowed the marketing and PR departments to join in the group to decide what animals the zoo should have, keep or acquire. Many are now allowing non-animal staff to make these kinds of decisions" (Marvin Jones pers. comm. 1996). It is self-evident that without a collection of animals a zoo or aquarium will have no structure; it is appropriate for keepers, as persons who maintain the animal collection, to express their concern about the welfare and well-being of their charges, thus also making others aware of the needs of animals. For instance, pulling out a neonate or newly arrived specimen for a press debut prematurely will likely jeopardize the health of the very animal the zoo is trying to promote. Such decisions must be made in a reasonable manner by those who know the animals.

Another oft-discussed behind-the-scenes issue is accuracy of the information disseminated by the zoo or aquarium in the form of press releases, brochures, or verbal announcements. Most if not all information inaccuracies originate from within the organization, not from outside. For instance, some of us have probably read or heard accounts stating that the *domesticated* form of the Bactrian camel is highly endangered, or that penguins are from the Arctic. Neither statement is true. Persons in the animal management sector have an obligation to encourage those who formulate such texts to be more accurate concerning facts about animals, zoos, and aquariums, to avoid glaring and embarrassing public errors. This is yet another example of a broader role that keepers can play.

Zoos and aquariums today face many a challenge. Directly or indirectly, keepers as a critical part of the work force can make tremendous contributions to help resolve issues. A zoo can fill its grounds with endangered species, conduct outstanding conservation programs, gain praise from its peers, and attain a leading position in its field. But if its grounds are littered with trash, if its bathrooms are filthy, if there are long waits at its restaurant and the food is of low quality, if its animals hide from the public in "naturalistic" exhibits, and, more importantly, if employees, particularly keepers, are discourteous, then the most important component for institutional survival—public support—is in jeopardy. All zoo employees must strive to be dependable, well-spoken, and trouble-free. In this way keepers can help to strengthen the zoo's positive image and cement public support from the ground level upward.

ACKNOWLEDGMENTS

Thanks go to Jen LaPaglia, human resources, Erie Zoological Society, Erie, Pennsylvania; Mickey Ollson, owner/director of Wildlife World Zoo, Litchfield, Arizona; and Elizabeth Frank, former large mammal curator of Milwaukee County Zoo, Wisconsin, for providing information for this chapter.

REFERENCES

Anon. 1972. *The Keepers' Responsibilities of Mammal Care.* Fort Worth, TX: Fort Worth Zoological Park.
———. 1973. *Manual of Operations.* Tulsa, OK: Tulsa Zoological Park.
———. 1979. Supplemental Qualification Statement: Animal Caretaker. Form approved OMB no. 3205–0038. Washington: US Office of Personnel Management.
———. 1995. Job announcement #J9501. City of Duluth, MN.
———. 1997. Job announcement, zookeeper, code 53–55–21 (issued in 1970, revised in 1997). Detroit: Human Resources Department.
———. 1999. "1999 AAZK and Animal Keepers' Forum Awards." *Animal Keepers' Forum* 26:10.
AZA (Association of Zoos and Aquariums). 2010a. *AZA Member Directory 2010.* Silver Springs, Maryland.
AZA. 2010b. Website job announcements. AZA.org. February 25.
Brazaitis, Peter. 2003. *You Belong in a Zoo! Tales from a Lifetime Spent with Cobras, Crocs, and Other Creatures.* New York: Villard.
Donaldson, William F. 1983. "Putting the Endangered on the Ark." *Philadelphia Inquirer.* August 17.
Doolittle, Will O., ed. 1932. *Zoological Parks and Aquariums.* Tulsa: American Association of Zoological Parks and Aquariums.
Hanson, Elizabeth. 2002. *Animal Attractions: Nature on Display in American Zoos.* Princeton, NJ: Princeton University Press.
Hutchins, Michael. 1997. "Keepers and Conservation." *Communiqué* 37.
Kawata, Ken. 2003. *New York's Biggest Little Zoo: A History of the Staten Island Zoo.* Dubuque, IA: Kendall/Hunt.
LaRue, Michael D. 2002. *This Place Is a Zoo! How to Manage the Unmanageable Organization.* Lincoln, NB: iUniverse.

McCoy, Janet. 1985. Unpublished internal memorandum. American Association of Zoo Keepers Professional Standards Committee, Washington Park Zoo, Portland, OR.

Murray, Charles. 2008. *Real Education.* New York: Crown Forum.

Perkins, Marlin. 1982. *My Wild Kingdom.* New York: E. P. Dutton.

Reichenbach, Herman. 2003. Book reviews. *International Zoo News* 329.

Rives, James. 2001. "Plants and Animals: Inseparable Relationship." *Animaland* 55:2.

Solski, Leszek. 2006. "Public Aquariums 1853–1914." *Der Zoologische Garten* 75, no. 5–6: 362–97.

Truett, Bob. 1970. *Zoos & Aquariums in the Americas.* Washington: American Association of Zoological Parks and Aquariums.

Wharton, Dan. 2007. "Research by Zoos." In *Zoos in the 21st Century: Catalysts for Conservation.* Alexandra Zimmermann, Mathew Hutchwell, Leslie A. Dickie, and Chris West, eds. Cambridge: Cambridge University Press.

2

Professionalism and Career Development

Jacqueline J. Blessington

INTRODUCTION

Keepers are generally employed by an institution to perform a job. It is then up to the keeper (and potentially the institution) to determine whether that employment is a profession or merely a job, and whether it then develops into a career.

Job \jäb\ *n* 1: A piece of work 2: something that has to be done: TASK; *also:* a specific duty, role, or function 3: a regular remunerative (compensation or payment) position (*Merriam-Webster* 2006, 586)
Profession \prə-ˈfe-shən\ *n* 1: an open declaration or avowal (acknowledgment) of a belief or opinion 2: a calling requiring specialized knowledge and often long academic preparation 3: the whole body of persons engaged in a calling (*Merriam-Webster* 2006, 829)
Career \kə-ˈrir\ *n* 1: COURSE, PASSAGE; *also:* speed in a course 2: an occupation or profession followed as a life's work (*Merriam-Webster* 2006, 151)

In 2008 a survey was conducted by Jeff Thompson, PhD, of Brigham Young University and Stuart Bunderson, PhD, of Washington University in Saint Louis on behalf of the American Association of Zoo Keepers (AAZK). The purpose of this survey was to (1) assist the AAZK in understanding the demographics, attitudes, and opinions of its membership and (2) promote academic research on high-commitment professionals. The strongest point conveyed by respondents was that as animal care professionals (i.e., keepers) they are passionate about what they do. The following are responses from three different components of the survey: (1) 83.8% of respondents agreed with the statement "I have a meaningful job that makes a difference"; (2) 68.2% of respondents agreed with the statement "My profession is an important part of who I am"; and (3) 87.9% of respondents agreed with the statement "Working with animals feels like my calling in life" (Thompson and Bunderson 2008).

From that survey and additional interviews conducted in March 2009, a paper was published titled "The Call of the Wild: Zookeepers, Callings and the Double-Edged Sword of Deeply Meaningful Work" (Bunderson and Thompson 2009). From this article, "we found that a neoclassical calling is both binding and ennobling. On one hand, zookeepers with a sense of calling strongly identified with and found broader meaning and significance in their works and occupation. On the other hand, they were more likely to see their work as a moral duty, to sacrifice pay, personal time and comfort from their work and to hold their zoo to a higher standard . . . ." One might infer from the generalized survey results that a majority of keepers believe that their "jobs" are in fact a profession.

This chapter is intended to aid the reader in acquiring a job in the zoo and aquarium field and to provide information on how he or she can then develop a professional career within that field. After studying this chapter, the reader will also understand

- how to get started in a zookeeping career
- basic principles for writing cover letters, resumes, curriculum vitaes, and job applications
- how to perform well at an interview
- the importance of professional image and mindset in a job
- career tracks for the animal care provider.

GETTING YOUR FOOT IN THE DOOR

Most if not all keepers, at some point in their career, are asked the question, "What did you have to do to get such a cool job?" The answer is generally related to two things: education and experience. High school students are encouraged to take courses related to biology, zoology, or anything animal-associated. Once students move on to an advanced degree they can choose to attend a major university and study biology, zoology, animal ecology, fisheries and wildlife, or other animal sciences. On the other hand, they may choose to attend one of the specialty "zookeeping" courses or programs at various universities that offer semester credits or degrees (see appendix 3 for a listing of schools).

Experience is the second major requirement for getting a job in the keeping field. High school students are encouraged to participate in junior keeper or explorer programs that various institutions provide. Many institutions provide internship programs for college students. During the summer months, interns volunteer to work side by side with keepers to get hands-on experience. Some of the experience is paid work, but most of it is not. Many organizations and in-the-field research projects also advertise for interns to help collect data. Most institutions also provide seasonal or part-time paid positions. These positions generally only pay minimum wage and do not provide benefits, but do allow individuals to get the much-needed paid experience that is required for full-time work. Throughout the global community, beginning a career by volunteering through structured programs shows one's commitment to an institution, and may open the door to a permanent position.

An applicant considering a job outside his or her country of origin will need a valid passport. Many countries will also require a working visa and/or "green card." One should research government websites of the country of interest to identify the relevant requirements and restrictions.

JOB SEARCHES

Many resources are available for the keeper seeking a job or a student seeking an internship. The vast majority of opportunities can be found on the internet. Websites such as www.aza.org, www.aazk.org, and www.caza.ca will target primarily North American institutions, while job opportunities in Australia will be outlined at www.australasianzookeeping.org, and www.zooaquarium.org.au, and jobs in the United Kingdom and Ireland at www.biaza.org.uk.

Institutional websites may list employment opportunities and details about who to contact for a job. They may also include applications and the process for filling vacancies. Institutions run by government entities may post job openings on city, county, state, provincial, or territorial websites. They may also post openings at conferences, workshops, and seminars. Because the zoo and aquarium field is a relatively close-knit community, many job openings are still communicated via word of mouth.

THE JOB APPLICATION PROCESS

RESUMÉ VERSUS CURRICULUM VITAE

The primary differences between a resumé and curriculum vitae (CV) are in the length of the document, the information included, and the purpose for which each is going to be used. A resumé contains full name and contact information (address, phone number, and e-mail). It should also include a summary of skills, work history, and education/academic background. In the United States the resumé is expected to be a maximum of two pages in length, whereas in Australia it is expected to be a minimum of two pages in length. Don't pad a resumé; if a prospective employer confirms details of the employment history and finds the resumé is inaccurate, the applicant may be dismissed as a candidate.

A CV includes the basics of a resumé plus a summary of other relevant information such as teaching and research experience, publications, presentations, honors and awards, professional memberships, and licenses. Employers in some countries may expect to receive a CV in lieu of a resume. In the United States, employers rarely ask for a CV in reference to a zookeeping job. However, AZA may request one when an individuals applies for positions such as species coordinator, regional studbook keeper, or professional development course instructor.

Generally, work history is listed first and in chronological order. However, listing the most recent job first may be more appropriate, depending on its relevance to the job being sought and the area custom. The listing of each job should include the employer, start and end dates, job title(s), and main duties or responsibilities. More detail should be given on more relevant jobs, and information should be kept minimal on those that are less relevant. If work history is limited, the applicant should highlight his or her education and training; these qualifications should be listed with most the recent certifications given first. A list of animal taxa worked with should also be included if it is pertinent to the position.

The list of skills and experience should relate to the job description, as the institution may be looking for specific expertise. For example, if the job involves working on an education or presentation team, examples of participation in presentations (e.g., keeper chats) and development of training programs (e.g., crating and weighing for a barn owl) should be included. It is also advisable to include any relevant temporary work, volunteer experience, and internships. Gaps in the employment history should be avoided; if time between jobs was filled by care of family, job seeking, and so on, that information should be included.

Hobbies, interests, and other achievements may be included if they pertain to the job, and should be made specific and interesting. For example, listing the hobby of training horses may be applicable to a position working hoofstock; volunteer work at a library may show the applicant's interest in research and resources. Other special skills may also be included, such as foreign language proficiency or first aid/first responder certification.

THE COVER LETTER

A cover letter is used to formally introduce a resumé for employer review. It is an opportunity to quickly introduce the applicant and gain the hiring manager's attention. The cover letter is also where one might explain any gaps in employment; one should be sure to include what was learned during or from that gap time. A cover letter should be simple yet specific to each job and institution. There are several types of cover letters. Following are three common styles:

1. The application or invited letter is used when applying for a position in response to a specific job advertisement or posting.
2. The prospecting or uninvited letter is used when

sending a resumé cold (unsolicited) to an employer seeking information on possible or future job openings.

3. The networking letter is used when requesting leads or information on job opportunities within the industry. Often these letters are sent to individuals the job seeker knows personally.

All three types of cover letters should include the applicant's contact information, the date on which the letter was written, the employer's contact information, and an appropriate salutation. If it is being sent to a contact person at the institution, that person's name should be included in the letter's address; otherwise the letter should be addressed to the hiring manager or to the human resources department.

The application cover letter must let the employer know which position is being applied for (with its reference number, if one is listed) and where the job posting was found. It should also explain what the applicant has to offer the institution, and how that person's qualifications meet the requirements of the job. The letter of interest is slightly different, since the applicant needs to sell his or her strengths without having a specific job description to tailor it to. All three types of letters should incorporate information from the resumé or CV, but should only be three to five paragraphs long. The letter should conclude by thanking the institution for consideration for the position, indicating interest in further communication, and providing contact information.

THE JOB APPLICATION

Even after a cover letter and resumé have been submitted, most employers will request that applicants fill out an employment application. This application differs slightly from the previous documents in that it generally requests more detailed personal and work history. It may also require the applicant's signature as verification of the information provided.

Employers may ask for an application to be filled out online or in hard copy. Online forms can be convenient if the applicant can save the information as it is being filled out. One tip is to print a blank form and fill it out completely, and then transcribe it online. Another is to write down all of the questions and formulate answers for them before completing the online application. If there is no option to save the document while working, one should be sure to have all of the information ready and the time to fill it out before starting. When submitting a paper application, it is a good idea to make a practice copy first. One should write as neatly and legibly as possible, using block letters (as cursive can be difficult to read) and black ink (as it is most visible). As with all documents provided to a prospective employer, the applicant should proofread, proofread, proofread. All information should be complete and free of errors. No questions on the form should be left blank.

ADDITIONAL TIPS FOR COMPLETING JOB APPLICATIONS

- Many applications have a space at the end for "additional information or personal statements." This section should be used to highlight why the job is desired, what skills the applicant has, and how those skills relate to the job, along with examples.
- Once the application is complete, the applicant should not forget to sign it.
- A record of the application should be kept as a photocopy or printout.
- A mailed application should be sent in plenty of time before the closing date.

RECOMMENDATION LETTERS AND REFERENCES

In general, recommendation letters and references provide information on personality, character, work ethic, and/or academic achievement that may or may not be found in the job application. The person writing the recommendation or giving the reference should know the individual well enough to speak about these particular traits. An applicant should secure recommendations or references from work or internship supervisors (former or current), professors or school administrators, coworkers, or anyone else familiar with his of her work ethic history. At least one reference should be work-related, but people should not be chosen simply for their titles. There are no guarantees to obtaining letters of reference. In some cases, the person who is asked for a letter of reference would be support the job applicant with their own professional reputation and may determine that it is not in their best interest and decline to write the letter. The applicant should not take it personally, but rather seek out multiple references. If there is a gap in employment, the applicant should seek the recommendation from someone who is familiar with his or her character and experience and would be considered responsible and professional; one should avoid using family and neighbors.

The applicant should communicate to the person writing the recommendation letter specifically what it is to be used for. This will help that person tailor the recommendation. If the reference is needed to stress strong work ethic and leadership qualities, the applicant should not hesitate to indicate that. A good letter of recommendation takes time to write, so the letter writer should be given enough time to properly complete the task, as well as a specific deadline if necessary. One should try to obtain three to five reference letters, so as to have a few to choose from. Copies of the letters should be kept for future references, and their authors should be acknowledged with a thank-you.

Contact details for references on the CV or resumé can be included, or one can simply include the sentence "References available on request." If the contact details are listed, they should also include each person's title and/or relationship to the applicant (e.g., "John Buck, lead supervisor" or "Jane Doe, editor and personal friend").

THE INTERVIEW

The interview is the next step in the job seeking process, and probably the most stressful. Keeper interviews are typically held either face-to-face or over the phone. Occasionally a prospective employer might extend an invitation for a dining interview. The following tips might help alleviate some anxieties.

PREPARATION FOR THE INTERVIEW

Many interviewers will ask standard questions like "What are your strengths and weaknesses?" "Why do you want to work at this institution?" "How would you describe yourself?" or "Do you consider yourself a team player or do you prefer to work independently?" There are also questions typical to the zoo and aquarium industry, such as "What is flight-versus-fight response?" Responses to these and other standard questions should be prepared in advance, and should include actual examples of how one might describe oneself and one's skills. The answers should be rehearsed, either in front of a mirror or with a friend or family member. Actually saying the responses out loud will help the applicant build confidence and pave the path to a solid interview.

The applicant should bring copies of the resumé (enough for all interviewers) to the interview, along with an application and a list of three references with contact information. Copies of publications relevant to the applicant and job are also suggested. A pen and paper are handy for taking notes.

RESEARCH

The applicant should research the institution he or she is applying to. One should visit its website, read its mission statement, and review the institution's history, products, services, and management. It is to the applicant's advantage to know the background on these points: the size and diversity of the institution's collection; the natural history of some of the species; the institution's layout, staffing numbers, organization structure; and how the job being applied for fits into that structure. Understanding the institution's education and conservation programs can be of benefit in the interview. The applicant should use social media to research the institution, and become a fan of the company on Facebook. One should also spend time tapping into personal networks to see who might be able to help give one an interview edge over other candidates.

TIMING

The applicant should be on time for the interview—or, better yet, 10 to 15 minutes early. This allows the applicant time to compose: to comb hair, use the restroom, turn cell phone to silent, and so on. One should be familiar with the institution's location, including where the interview will take place, and drive there ahead of time to determine exact locations and travel duration.

STAYING CALM

A standard greeting and parting gesture is to shake hands, and to look the individual in the eyes while doing so. This shows self-confidence as well as a positive attitude. One should try to remain poised and relaxed without becoming too lackadaisical, to not slouch when sitting, and to pay attention and make sure to listen to the entire question before answering. If the question is unclear or vague, one should ask for clarification and take time, if needed, to think. Use of the notepad and pen to take notes will make it easier to keep track of the interview.

ETIQUETTE

Some general points of etiquette apply during an interview:

- Don't swear or use slang words and use caution with the over use of "um" and "like."
- Don't smoke, chew gum, or eat, but do keep a glass of water nearby if needed.
- Don't lie or exaggerate; an applicant may be dismissed if the employer finds false information.
- Don't be arrogant, disrespectful, or argumentative with the interviewer.
- Don't read from notes, a resume, or CV; be familiar with information that has been submitted.
- Don't criticize former employers, colleagues, or institutions. Doing so may imply that the applicant is a troublemaker or a gossip. There are numerous connections within the zoo and aquarium industry; the interviewer may be familiar with the criticized individual.
- Keep answers focused on what the applicant can do for the employer and institution, not the other way around.

CONNECTING

Interviews may be conducted by an individual or team of panelists. One should try to connect with the group as a whole as well as with the individuals. When one interviewer asks a question, one should respond directly to them, but also make eye contact with the rest of the group. Group dynamics should be considered, as it is possible to get a feel for an institution based on how all of the interviewers respond to each other.

CONFIDENCE

When answering questions, one should try to incorporate researched information about the institution, and to relate one's career accomplishments to what the interviewer is seeking. Employers often use questions that have straightforward answers, but many also implement behavioral interview questions. These focus on past performance versus future performance. One should have specific examples of achievements ready to use in response to typical behavioral interview questions such as "Tell me about a major accomplishment / challenging situation / difficult coworker." Interviewers look for specific answers to understand how situations were handled, and whether the skills used by the applicant are what they are looking for. One should be honest and positive; there are no right or wrong answers.

QUESTIONS

As the interview comes to a close, it is common for an interviewer to ask whether the applicant has any questions. The following are some common questions that may be asked by the interviewee: "What are the opportunities for promotion and advancement?" "Will there be opportunities for greater responsibility and experience?" "What is the institution's management/operating style?" "Does the zoo or aquarium have a strategic plan?" The applicant can also ask general questions, such as "Why do you like working here?" "Why do people want to stay here?" "How open are managers to differing viewpoints?" Generally speaking, one should not ask about salary during the first interview, but should instead wait until one has been offered the job. Responses to specific questions might help determine in advance if the applicant thinks the job is right for them.

PHONE INTERVIEW

Most phone interviews in the zoo and aquarium industry are held because the institution does not or cannot provide expense coverage for a keeper's time and travel. Additionally, the candidate may be unable or unwilling to pay for expenses to travel to the institution. Generally, it is acceptable to perform a "first-round" interview over the phone for an out-of-town candidate. It is highly encouraged, though, that the candidate perform an in-person interview if participating as a finalist. Preparation for a phone interview is basically the same as that for an in-person interview. One should compile a list of strengths and weaknesses, and answers to typical interview questions. All of the previous etiquette tips also apply, and here are a few additional tips for the phone interview:

- Have a copy of the resumé, CV, and/or application available for reference.
- Turn off call waiting, so the interview isn't interrupted.
- If the interview time isn't convenient, ask for it to be rescheduled and suggest alternative times.
- Clear the room of all distractions including children, pets, stereo, and television; then close the door.
- Consider using a landline rather than a cell phone if service or reception is questionable.
- Talking on the phone isn't as easy as it seems, so practice interviewing. Record a mock interview, and then listen to it.
- Smile, as it will change the voice tone and thus project a positive image to the listener.
- Speak slowly and enunciate clearly.
- Don't interrupt the interviewer; wait until it is clear that they are finished speaking before answering.
- Stand during the interview, as it inspires self-confidence that can be conveyed in one's voice.

THANK-YOU LETTER

Taking the time to say thank you after a job interview is not as common as it used to be. However, not only is it good etiquette, but it also reinforces the applicant's expression of interest in the position. Alternatively, if the applicant is no longer interested in the position, the thank-you letter is an appropriate venue in which to respectfully withdraw the application. The letter can also be used to address any concerns or issues that may have come up and were not addressed or resolved during the interview.

If possible, each interviewer should be sent a thank-you note. If that is not feasible, such a note can simply be sent to the person who arranged the interview. Business cards can be collected as records of the interviewers' names; otherwise, they should be written down. Another option is to write thank-you notes immediately after the interview and ask the receptionist to deliver them.

AFTER THE INTERVIEW

After the interview, one should take a few minutes to assess and determine what went well and what could use improvement. One should write down key questions with given answers, give thought to those answers, and consider whether, if one were given time, they would change. One should research questions where answers were unknown, keep the list of questions and answers for potential future interviews, and get feedback on one's performance. If the job was not offered, one should ask the interviewer what can be done to help increase the likelihood of getting a job there in the future. It may be presentation skills, and taking a speech or communication class can improve them. Or it might be that one does not meet the employer's job-specific criteria. One should find out what that criterion is and seek to meet it to be more marketable in the future.

PROFESSIONAL IMAGE

Professional speakers and trainers have long asserted that people make up their minds about people they meet for the first time within two minutes. Others assert that these first impressions about people take only 30 seconds to make. As it turns out, both may be underestimates. According to Malcolm Gladwell (2007), in *Blink: The Power of Thinking without Thinking,* the decisions may occur much faster—instantaneously, or in two seconds. According to Gladwell's research, we think without thinking; we "thin-slice" whenever we "meet a new person or have to make sense of something quickly or encounter a novel situation." Gladwell says, "Snap judgments are, first of all, enormously quick: they rely on the thinnest slices of experience . . . they are also unconscious." How keepers present themselves in public can have a huge impact on the institution in which they work and on their professional careers. This concept of thin-slicing also occurs daily in a zookeeper's job.

INSTITUTIONAL POLICIES: UNIFORMS, GROOMING, BODY DECORATIONS

In the zoo and aquarium setting, employee comfort and clothing that enables efficient work are key. Clothing must convey professionalism that is respectful to coworkers and visitors. There is great variance in what each institution provides for uniforms. Some provide everything from hats right down to the socks, while others may only provide a top garment, such as a T-shirt with the institution's logo. Some provide rain gear and steel-toed boots; others do not have the resources to provide any garments that offer protection from the elements.

Employees should wear clothing that is unwrinkled, clean, and maintained. Clothing that is torn, dirty, or frayed is generally not acceptable. Many keepers will find these standards difficult to achieve. Often, the only time a keeper is unwrinkled and clean is at the beginning of the day before they start their shift. The ability to keep uniforms neat may be dependent on one's job description; consider the difference between an aquarist and elephant keeper. Facilities that require their keeper staff to perform horticulture or exhibit maintenance may find that there is more wear and tear on uniforms. The climate in which people work may also have an impact. Staff who do formal or scripted presentations or shows may be required to maintain a "clean" uniform, as opposed to those who give impromptu presentations in the field. A suggestion is to keep a separate uniform for such presentations, or to wear outer gear such as coveralls when performing tasks that are particularly messy. If getting dirty is unavoidable, apologize to the audience and explain that it is "all in a day's work."

Individual institutional policies will also vary considerably in regards to grooming, piercings, and body decorations. Facial hair on men may not be appropriate. Long hair may be suitable for women but not men, and for safety purposes it may be required to be pulled back or secured away from the face. Piercings may be limited to one or two sets in the ear for women and none for men. Other body piercings may not be allowed by the institution. Tattoos may be limited by size, number, and location, and those that are visible may have to be covered.

THE INTERNET: MYSPACE, FACEBOOK, AND TWITTER

One's professional image can be instantly affected electronically. While self-expression on the internet is considered a personal right, it can have damaging or permanent effects on one's reputation. Prospective employers may visit an applicant's site to gain insight on how the individual conducts themselves when not at work. What is posted may cost the individual a potential job offering or the loss of a current job.

PROFESSIONAL MINDSET

"TRASH IS NOT MY JOB"

The keeper's job is extremely diverse and unlike any other. Depending on the size of the institution, the keeper's job description may vary considerably. In a typical day a keeper performs direct animal duties such as cleaning, feeding, and observations. At a smaller institution he or she may be expected to perform job duties not typical of animal care, such as exhibit maintenance. At larger facilities, keepers might focus only on animal care, as other staff is hired to perform non–animal-related tasks.

Many keepers inherently believe their only job is to "take care" of the animals. While it is not likely that a mechanic would be expected to do foot work on an elephant or that a keeper would change the oil in a truck, it is common that keepers perform tasks that are not animal-related. Keepers are thought to be behaviorists, conservationists, environmentalists, and researchers, but that is not all they do. In fact, it is standard for the keeper to also be gardener, handyman, architect, teacher, landscaper, records manager, and of course the "trash man." As mentioned above, the impression about an individual can be made in the blink of an eye; the same can be said of the impression about an institution. All employees should be concerned with the institution's image, as ultimately it can reflect on them as well. Keeping animal areas neat and tidy shows pride on the part of individuals who work in the area, just as keeping the grounds neat and tidy shows their pride in the institution. If that is not enough to convince a keeper that trash is their job, then the safety factor should also be considered. Trash can get into animal exhibits and cause harm, and ultimately the care of the animals is the keeper's job.

Being involved or exposed to many facets of the institution also helps build a well-rounded keeper. It allows for different perspectives and viewpoints. Everything at a zoo or aquarium is in some way connected to the animals, since without them it wouldn't be a zoo or aquarium. Because the keeper is the direct link to the animals, it is his or her responsibility to be familiar with how all of these components come together.

"AM I IN IT FOR THE RIGHT REASONS, AND ULTIMATELY, WHY AM I HERE?"

Individuals get into zookeeping for a variety of reasons. For some it is simply a job that pays the bills. While this reasoning is understandable, there is a good chance that people who subscribe to it will find fault with the job. Many individuals try to get into the zookeeping field because they think it would be "cool" to work with animals. Zookeeping work is not for the faint of heart, as it can be physically demanding, and potentially dangerous when one is working with dangerous animals such as elephants, carnivores, and venomous reptiles. It can also be uncomfortable work in extreme weather conditions. It is not a glamorous job to clean up animal waste, and a keeper's salary is not reflective of the level of education and experience that most keepers have.

Bunderson and Thompson (2009) state, "At the heart of the calling notion for these zookeepers, then, is a sense that they were born with gifts and talents that predisposed them to work in an animal-related occupation." This statement implies that many keepers believe that they were always destined to work with animals, and that becoming a keeper is a natural decision. Most keepers believe they are the voice of the animals, and that they have an obligation to create optimal living conditions for them.

CAREER TRACKS FOR THE ANIMAL CARE PROVIDER

In the workplace survey conducted by AAZK (Hansen 2000), it was determined that after an average of approximately four years of employment in the profession the keeper makes a life choice to either (1) become a lifetime keeper, (2) move up into management, or (3) leave the profession.

THE LIFETIMER

The "lifetimer" is the individual who chooses to make animal keeping the entire or principal work of their lifetime. In the United States this term has been historically used for keepers who work for a government entity. These individuals have often been covered by labor or trade unions which have helped to promote the security and longevity of keepers' employment. Because of the benefits often accompanying a union-supported position, these keepers have tended to stay within the workforce and become "lifers." Today, as many zoos and aquariums in the United States have become private entities, the unions have begun to diminish in number. Only a few countries, including Germany and Australia, recognize unionizing for keepers.

Without the benefit of labor or trade unions, keepers now remain lifetimers for different reasons. In the United States, the average keeper earns just under $25,000 per year. (Hansen 2000; Thompson and Bunderson 2008). In Australia the average salary is AUD$20,000–40,000 (Australian Zookeeping 2010). It would thus appear that the motivation to remain a keeper is not income, but rather the passion an individual feels for the profession. Maintaining the title or position of keeper does not mean, however, that an individual is not developing professionally or advancing in their career. The next section, "Professional Development Opportunities," will discuss further options for advancement for the lifetime keeper.

CLIMBING THE LADDER

The decision to further one's career in a different division of the institution—as a supervisor, administrator, or educator—may be a difficult one. Many keepers find it hard to separate themselves from the daily work of animal care. For the individual who wants to become a supervisor yet remain involved in animal care at a certain level, positions such as lead or assistant animal supervisor, head keeper, or animal care supervisor are good choices. Higher positions, such as curators and director, relate to management and business. If a keeper would choose to pursue higher positions such as curator or director, it might be expected that they would acquire additional degrees (Masters or PhD) or supplemental education in programs like business, management, or marketing.

WHEN TO CHANGE INSTITUTIONS

Keepers, in general, are a bit nomadic. As wild animals move with the changes in season and resources, so do keepers. Because opportunities for employment in a particular region or location are limited (not every city has a zoo or aquarium), keepers need to be willing to relocate. One keeper may choose to remain at an institution for the same reason that another may choose to leave. Family and geographic location are strong considerations when considering a facility. The following is a list of other priorities that keepers may consider when looking for a job:

- salary
- management philosophies
- opportunities to work with a particular species
- an institution's conservation and research reputation
- opportunities for professional development
- opportunities for supervisor/management advancement
- social climate
- opportunities outside of the institution (since life doesn't end when the workday does, having activities outside of work should be researched when considering a move)

There are two primary reasons why a keeper might consider a change in venue. One is a change in management and/or philosophy that might differ from that of the individual. The field of zoos and aquariums is constantly evolving and, as with any profession, a change in upper management can create the potential for a change in philosophy. The keeper must then decide how those changes may or may not conflict with his or her individual philosophy. It is important to remember that no institution is perfect and that each comes with its own challenges. A second reason for moving is that opportunities for advancement and development are either limited or non-existent. Moving to another facility might open new doors.

LEAVING THE FIELD, PURSUING OTHER OPTIONS

Historically, zookeeping positions were filled by men. It wasn't until the late 1960s that women started entering the profession. In surveys (Hansen 2000; Thompson and Bunderson 2008) it has been determined that nearly three-quarters of keepers today are female. One of the primary reasons a female keeper leaves the field (after approximately four years) is to start a family. The zoo and aquarium industry is a seven-day operation. Keepers may be required to work any day of the week, including weekends, thus making it more difficult for them to acquire child care. The keeper job can also be very demanding physically, making it difficult for one to work during pregnancy and early childhood development. The fact that a keeper job is physically demanding whether one is male or female is another reason why individuals leave the field. Back and joint injuries are a leading cause of absenteeism at work. (These concerns will be covered more thoroughly in chapter 7.)

Salary is another motivation for exiting the field. Only in the last 15 to 20 years have keepers been encouraged to pursue advanced degrees with emphasis in animal-related study. or to have attended programs that specialize in exotic animal care. In the United States, however, having the advanced degree does not seem to correlate to the average keeper salary of $25,000 per year. Hansen (2000) states that "it is difficult to analyze zoo keeper's salaries and draw a comparison to a similar profession. Most skilled labor/trades classifications do not have an 82% college degree rate. Most college graduates do not perform physical labor in the field, 40 hours per week. The profession is the proverbial square peg. In most regions though, the field is about 5–7% behind starting pay for college graduates."

The final reason why zookeepers leave the field is that the job is not what they expected. Many people come into the profession expecting a lot of hands-on work and "play" with the animals. The reality is that most of a keeper's day is spent cleaning and servicing animal holding areas and exhibits.

PROFESSIONAL DEVELOPMENT OPPORTUNITIES

There is a variety of opportunities for the keeper's professional development,. The following is a list of different categories for development and the opportunities within each.

EDUCATION

Many zoo, aquarium, and zookeeping associations play a leading role in promoting professional development for keepers. For example, the Association of Zoos and Aquariums (AZA) collaborates with the AAZK in offering the course "Advances in Animal Keeping," a component of AZA's professional development program. This course focuses on the very highest standards in animal husbandry in combination with problem-solving, team-building, and communication skills. More information on this course is available at the websites of the AZA (www.aza.org) and the AAZK (www.aazk.org). Many regional associations provide educational programs and advanced learning opportunities; information is available at their websites.

CERTIFICATION

Few regions have actual zookeeper certification programs that are above and beyond the fundamental education programs that are available.

MEMBERSHIP

Professional membership can be found in many animal-related organizations in addition to zoo and aquarium associations and zookeeper associations. Many of these organizations are specialty associations with a dedicated focus (e.g., training, specific taxa, conservation). Becoming a member of such an association allows for more networking connections as well as potential job opportunities.

CONFERENCES

Most if not all organizations and associations in the zoo industry host conferences and/or workshops with a variety of presentation formats. Papers are a format in which one or more individuals speak on a specific subject, sometimes using a Power Point program, and then take questions from the audience. Posters that detail an individual or group's presentation in writing, and which often include illustrations, may be publicly displayed during the conference. A poster is also specific to a subject, but does not allow for dialogue with the audience unless time has been allocated for them. A workshop can be organized in the form of a lecture or an open dialogue. Many lectures are based on themes and allow for much interaction and round-table discussion. A workshop may be led or moderated by one person while offering the opportunity for several others to participate in dialogue.

A keeper may simply attend a conference without expecting to make a presentation or interact with other participants. As a keeper develops within the profession, however, he or she may gain expertise that can develop into a presentation or poster to be shared with colleagues. Some institutions encourage keepers to participate in conferences by offering them financial assistance or professional leave to attend.

A poster is a good starting point for an individual who is nervous when talking in front of a group. A workshop is a great opportunity for an individual to participate a conference without having to make a formal presentation. But papers, posters, and workshops all have value in expanding a keeper's professional mindset, and all are generally recognized through some type of formal publication.

HUSBANDRY WORKSHOPS

Many Taxon Advisory Groups (TAGs) and Species Survival Programs (SSPs) provide workshops dedicated to specific taxa and species. These workshops might focus on husbandry, diet, training, enrichment, and research for a specific animal group. Check each regional association for information on classes and workshops.

ASSOCIATION OPPORTUNITIES AND POSITIONS

For keepers there are boundless opportunities to get involved in organizations and associations. For example, AAZK has about 85 local and regional zookeeper chapters, which generally are each affiliated with a particular zoo or aquarium. As a member of a chapter a keeper can hold an executive office such as president, vice president, secretary, or treasurer. The chapter may appoint coordinators for conservation, education, or fundraising. A chapter member may also participate in various committees such as recycling, or the American Association of Zoo Keepers' "Bowling for Rhinos" fundraiser. Each international keeper association has an executive board and committees that welcome member involvement. More information is available at the International Congress of Zookeepers website (www.iczoo.org).

ZOO ASSOCIATION PROGRAMS

Many zoo association programs such as TAGs, SSPs and studbooks provide opportunities for keepers to get involved as keeper liaisons, committee members, or program managers. Keepers may also participate with in professional development programs by becoming instructors and course administrators, or in committees dedicated to animal welfare or green practices.

EXPERIENCES, ACTIVITIES, HOBBIES

Keepers can also join or financially support animal-related interest groups and organizations such as the Snow Leopard Trust, the International Rhino Foundation, the Cheetah Conservation Fund, the Nature Conservancy, Save Nature.org, the World Wildlife Fund, or Polar Bears International. AAZK's international keeper exchange program lists individuals all over the world whom one can contact when traveling. These people may help provide lodging or simply be contacts at local zoos and aquariums.

SUMMARY

The zookeeping field is highly competitive, and it is essential that an individual who seeks a professional career consider many paths and options. Getting a foot in the door is just the beginning. Keepers no longer just "take care" of animals; they are professionals who build lifelong careers in a field dedicated to all aspects of animal husbandry and management.

REFERENCES

1. Australasian Zoo Keeping. March 2010. *How to Get That Job! Before You Start.* Accessed May 5, 2010. http://www.australasian zookeeping.org.
2. Bunderson, J. Stuart and Jeffery A. Thompson. 2009. "The Call of the Wild: Zookeepers, Callings, and the Double-edged Sword of Deeply Meaningful Work." *Administrative Science Quarterly:* March, 32–57.
3. Gladwell, Malcolm. 2007. *Blink, The Power of Thinking Without Thinking.* Back Bay Books/Little Brown and Company.
4. Hansen, Ed. 2000. *AAZK Workplace Survey Results (unpublished).*
5. *Merriam-Webster's Dictionary and Thesaurus.* 2006. Springfield, MA: Merriam-Webster, Inc.
6. Thompson, Jeffery A. and J. Stuart Bunderson. 2008. *AAZK Survey of Animal Care Professionals (unpublished).* Brigham Young University and Washington University in St. Louis.

3

Communication and Interpersonal Skills for Keepers

Judie Steenberg and Mark D. Irwin

INTRODUCTION

In a zoo or aquarium, "animal people" must work with people. Keepers will need to share the workplace with coworkers, supervisors, managers, the staff of other departments, and of course zoo visitors. They are the primary spokespeople for the animals and must effectively communicate information about them to others. This chapter is based upon a series of articles that appeared in *Animal Keepers' Forum* in 2010; the articles were updated from material originally published in 1976 (Steenberg 2010, 12–13, 68–69, 99–101, 261–62, 406–8). Just as with the basics of zookeeping, personal communication skills are timeless. Communication skills are important tools that enable keepers to do their job. Technology, which moves along quickly, has become a tool box for aiding in communications. Also, conflict may occur in any workplace and its resolution requires effective communication and skills that can be developed.

This chapter's purpose is to help keepers become better communicators in the workplace and provide a strong, clear voice for the animals. After studying this chapter, the reader will

- understand the concept of communication
- appreciate the importance and value of good communication skills
- identify when and where keepers will be involved in communication
- be provided with suggestions for improving their verbal and written communication skills
- be encouraged to seek out opportunities to improve their public speaking skills
- be provided with tips for effective listening and conflict resolution.

COMMUNICATION

It is important that keepers consider communication to be part of their job. They are the link between the care of the animals and the rest of the zoo or aquarium's operation, either directly or indirectly. They are the animals' voice. There are several types of organizational structures in zoos, as well as multiple levels of animal care staff, but the key to good zoo operations is the ability of all of its personnel to interact and communicate well.

Communication is key to human actions. To communicate is to impart, make known, tell, or transmit. People communicate with spoken and written words, through actions and attitudes, dress and gestures. The most basic and direct form of communication is talk. People talk at, to, over, across, down, up, past, and—the most difficult yet effective form—*with* other people.

It is ironic that something which is such a large part of one's life can be a most difficult problem, often at work. Years ago, newspaper columnist Sydney J. Harris said, "It is far easier to run your occupation or profession in a technical sense than to deal with all the 'people problems.'" It is not an easy task to really communicate, and for some people it may require concerted effort.

Communication is often confused with information, which is also important but different. Information can be given with no exchange or sharing of ideas and experiences. Information is merely a one-sided message, while real and effective communication involves the exchange of ideas to achieve understanding or meet an objective (e.g., success in the workplace).

Effective communication for keepers is both a personal and principal issue and cannot be ignored. It is a part of everything a keeper does. Ideas and experiences gain meaning when they are transmitted to others. Since communication is inevitable (to say or do nothing is still communicating), it is important that keepers develop their skills and become effective communicators.

Experts in the field of communication have developed guidelines for good communication. They are designed to aid in improving communication skills. The following basic guidelines can be found in many articles and publications on developing these skills.

1. Clarify ideas before communicating. Analyze the idea or problem, consider the goals and attitudes of all parties; don't fail because of inadequate planning.
2. Determine the true purpose of the communication. What objective is to be accomplished? Identify goals and adapt language, tone, and approach to meeting a specific objective; don't try to accomplish too much.
3. Consider the entire physical and human setting whenever you communicate. The impact of communication is affected by many factors: timing, setting (private or social), past encounters, and procedures.
4. Consult with others, where appropriate, in planning communications. Seek participation and consultation during preparation to lend insight and objectivity.
5. Be mindful of the overtones as well as the basic content of the message. Tone of voice, expressions, and receptiveness to the responses of others can all have great impact.
6. Take the opportunity to convey something of help or value to the listener. Consider the other person's interests and needs; see their point of view.
7. Follow up on communications. Find out how well the message has come across; get feedback.
8. Communicate for tomorrow as well as for today. Plan with the past in mind, but be consistent with long-range interests and goals.
9. Be sure that actions support communications. Often the most persuasive communication is not what is said, but what is done.
10. Be a good listener! It is as important to understand as it is to be understood; tune in to others.

IN-HOUSE COMMUNICATION

Consider communication in a zoo situation. Two major factors affect keepers and communication. One is the policy of the zoo and the administrative staff in providing the means for communication within the zoo's operation. Does the keeper have the opportunity to exchange ideas and information with coworkers and the administrative staff, to record important observations and data for the zoo's permanent record, and to be involved with the zoo's volunteers and visitors? Is the keeper encouraged to develop as a communicator within his or her own zoo and with other zoos? All of these things have an important impact on the success of a keeper's communicative efforts. But even in an ideal zoo, with all of these avenues open, a second and very important key factor is the keeper's attitude. Unless that individual makes an effort, none of this means much.

It is a fact of life that nothing is accomplished without effort. Good relationships among keepers and the administrative staff don't just happen; they're developed through communication. There are zoos where this is not possible, but

often, where effort is made, success follows. To communicate successfully, one must use the right approach and combine it with sincere effort and a constructive attitude. Persistence pays off . . . but be cautious about overcommunicating, which can cause others to tune out. Planning how and to whom the communication will be directed is essential to being successful.

For those who realize the value of keepers having strong communication skills, the possibilities of communicating are many. Zoo records are the first area of input. It is hard to imagine that something hasn't happened in an animal area during the day that requires notation. Perhaps it does not require an entry on the zoo's records, but often it does. Keepers must develop skills to perceive, interpret, and apply information about the animals in their care. Perception means noticing, observing, listening to and realizing what's taking place. A new keeper must develop this ability, while veteran keepers must take care not to become complacent or oblivious to what's happening.

In terms of workplace communication, interpretation refers to the analyzing of information according to importance and relevance. The application of information can be by oral or written communication, or through performance (how a keeper transfers or relates information to the daily care of animals). This could take the form of training a new keeper, or of working well in a team of coworkers. Probably the single most effective means of communication is making entries in the zoo's records. "Many species of animals owe their existence to facts learnt about them in zoological institutions" (Hediger 1964). Note the year on that reference; much is still being learned and the need to communicate information about the care of animals continues to be of major importance.

Communications between coworkers and administrative staff may not always be optimal. It is not reasonable to expect that everybody will always agree with or understand everything that's being communicated. It is important that keepers be thoughtful about what is communicated. Written or electronic communications in particular, can be widely transmitted and may be "permanent." Messages that are emotional, political, and otherwise "sensitive" should be relayed carefully and professionally.

If a keeper's attitude is constructive and the effort is rational, much can be accomplished. Keepers have a unique position in a zoo's operation; they have the opportunity to communicate with members of the administrative staff, auxiliary staff (maintenance, volunteers, etc.), and zoo visitors about their firsthand experiences with the animals. Although some zoo visitors can be difficult to deal with at times, keepers can do much to educate them and make their visit enjoyable. Because keepers can relay firsthand information about the animals in their care, they can generate sincere interest in and appreciation of them.

For a keeper, there is much to share and accomplish at all levels of zoo or aquarium communication; how much depends on the keeper and the facility. Some staff may only be interested in communication on a scientific level with colleagues, but there is opportunity for them to do much more for the zoo. At minimum, keepers have a professional obligation to communicate information about the animals in their care to those within the organization who can aid in their care and management, both directly (e.g., by coordinating browse collection with the horticulture department, or by planning medical treatments with the veterinary team) and indirectly (e.g., through fundraising or marketing).

A zoo's strongest asset is a good communication system. Keepers should be kept informed of everything that is happening, and in turn be expected to communicate pertinent information to the administrative staff about the animals in their care. Morning unit meetings and weekly crew meetings are just two ways to ensure everyone is up to date with what is going on in a given unit, section, or area of the organization. A daily keeper report provides the means to note data pertaining to the animals, facilities, and special activities. Each keeper should be expected to make entries regarding the animals in his or her area. Each routine or unit should also have a daily notebook that serves as an ongoing keepers' communication tool. It might contain some of the same information about the animals as the daily keeper report, but it should also note changes or variations in procedures in the area, things to watch for, and upcoming issues or events. It serves as a less formal briefing book for relief keepers who might not be up to date on what's happening in the unit. A keeper's notebook can also be a reminder or handy reference for recalling information. It can be kept on a computer, or simply as notes in a spiral notebook hanging from a hook in the work area. E-mailing or texting may be convenient, but information conveyed in that way is also temporary and not always available to all concerned.

Many zoos have a scheduled monthly meeting, the format of which varies according to the institution. Monthly meetings may cover topics from admissions to veterinary care and any other topic relating to the zoo's other departments. These meetings can be helpful for knowing what's going on throughout the whole organization. In a small zoo, a comprehensive monthly meeting can serve this purpose very well. However, this type of meeting at a large zoo can be disconcerting for the keeper when it deals with everything but the animals. There should always be a monthly meeting for the animal management staff, even if it has to be a separate meeting. For keepers to care about what's going on in other animal areas, they have to know what's happening in those areas. Ignorance breeds rumors and misinformation.

One way for a keeper to share information and generate interest in their area is to submit information to the zoo or aquarium's newsletter. Weekly or monthly in-house newsletters should be made available to all staff. A library with computer access should be available for keepers to review information from other facilities and use reference materials. Some facilities have unit libraries for use by keepers. Keepers should make use of these libraries to review periodicals from other institutions and current literature about the animals under their care.

Communication at some facilities extends outside the perimeter fence. Not only do some keepers relate well to zoo visitors and give brownbag talks ("lunch and learns") for staff, talks throughout the city to all types of groups, and interviews to radio and television, but they also give presentations at workshops and conferences for the professional zoo and aquarium community.

"GETTING THE WORD OUT"

Ideas are valueless if they are not conveyed. The information acquired by working as a keeper can be of great importance to colleagues, volunteers, zoo visitors, and others in the zoo community. Keepers are at the heart of the zoo or aquarium's activity and they often participate in many aspects of its operation beyond direct animal care.

Keepers should determine whether they are communicating their ideas, information, and knowledge to others effectively. If they are not doing so, why not? If a keeper has something to say to a group or in a paper, they shouldn't let discomfort or lack of confidence stop them from carrying it through. Keepers with a desire to communicate in these ways can learn to do it, and most likely will become very good at it. The key to a keeper's oral and written communication ability is held by the keeper. A keeper who wants to communicate, and works at it, can do it. Others already have learned to communicate effectively, simply because they had something they wanted to communicate.

Public speaking is probably the most difficult means of communication, yet it can also be the most effective. At a speaker's bureau workshop, the director of a local theater talked about how to speak to an audience. After a few short hours of listening to the theater director, several members of the audience who were quite inexperienced in public speaking were able to go out and deal with any situation and audience using the guidelines he had provided. The following are some of those ideas and suggestions.

In public speaking, most people experience three general problems:

1. stage fright
2. how to organize their thoughts
3. how to speak clearly and convincingly.

Stage fright is almost universal among public speakers. For some it is minor, and for others it is monumental. But in any case it is real; it cannot and should not be ignored. It is a reality that can be controlled but probably not completely eliminated. In fact, it can be *used* to enhance a presentation. Even experienced speakers can have stage fright, and the key to controlling it is to accept and confront the problem. It is analogous to an athlete being "up" for an important game. It is a natural reaction to a situation that is not dealt with on a daily basis. Often one's metabolic rate will increase to provide added alertness and strength. It is possible to take advantage of this and use the added energy to "get into the game." The anxiety will begin to leave as the presentation proceeds. It is possible to learn to control the physical and mental manifestations of stage fright.

Physically, a keeper can use the increased energy to take a firm grip on a chair, podium, or table, push their feet into the floor, clasp their hands tightly behind their back, and use it without being afraid. To alleviate a tight throat, one can hum a bit. Upon feeling unable to breathe, one can hold the breath for 30 seconds, thus guaranteeing that it will then be much easier to take a breath. These are simple techniques to manage anxious energy. To deal with the mental aspect of stage fright, one should remember that there is something important to be said, and a keeper who has been asked to speak and has something to say should focus on the message. Often the keeper has chosen to give the presentation because it represents an opportunity to share an idea, experience, or information about animals. It is a great opportunity to be infectiously enthusiastic! One should establish eye contact with the audience before starting to speak, as it conveys sincerity and a sense of something important to be shared. Public speaking is essentially a conversation with the questions deferred until after the presentation, although some speakers prefer to answer questions as they speak.

Organizing one's thoughts is an important first step in preparing for a presentation. Keepers must know the material, study it, and be sure of themselves. It is good to ask oneself these seven questions, and to answer them *in writing*.

1. What is the viewpoint (the general purpose for speaking)?
2. What should be accomplished? What are the objectives?
3. Who is the audience? This must be asked every time the presentation is given; no two audiences are alike.
4. What brings the audience members together? What is their common interest?
5. What knowledge do they have of the subject? What might they have read, heard, or experienced?
6. What are their attitudes and viewpoints?
7. What is to be expected from the audience? There will be different expectations of different audiences.

One should analyze each audience by its number of members, age, education, and experience. Its interest can be used as the base upon which to build the presentation and make it interesting within the first few sentences by appealing to that interest, revealing the subject's importance, or creating suspense. One should avoid telling jokes unless they relate to the subject. The human mind remembers best what it perceives at the beginning and at the end of a message rather than what's in the middle. The middle should be used to reinforce the message. The closing lines should tie into the opening sentence if possible.

A clear and convincing delivery will add to the impact of the speech. Good speakers are not those with golden voices, but those who have an interesting delivery, sound natural, and are easy to understand. Public speaking is enlarged conversation; one should pause before and after important ideas and change tempo, increasing the speed of delivery when excited or covering easily understood information, and decreasing the tempo when making a definite point. One should change the volume to emphasize ideas, but also use contrast: a whisper can have a tremendous effect, and it prevents monotony. Important words can be underscored with a pause, volume, or pitch. If the result sounds funny, the wrong words may be underscored, probably adjectives, adverbs and articles; nouns and verbs are usually the words to underscore.

Speakers come up against several common problems, but they all have solutions.

- When using a visual aid, such as a PowerPoint program, one should always present from the front of

the room. Even without direct eye contact it is possible to relate to the audience better and ensure that one's voice will carry more effectively. One should lead into the slides, explaining what they intend to cover before beginning.
- PowerPoint slides shouldn't be overloaded with all of the information to be read; one should instead expand on the talking points they contain.
- In the event of a lapse of memory, one should recognize it, deal with it, and move on to the next point.
- It is better to cover fewer points well than to force too much content.
- It is common for keepers to worry about what their peers will think. Being prepared and knowing the material well can get rid of the feeling. One should acknowledge that there are others present who also know about the subject, but also remember that they are there to hear new ideas and gain new information. The opportunity to discuss professional experiences, techniques, and philosophies, while sometimes unnerving, can have tremendous benefit for the individual, an institution, and the profession as a whole.
- Children's attention spans can be short and fragile. One should talk to children as people, not be phony, carefully choose vocabulary, and not talk down to them.
- With combined audiences of children and adults, one should focus on the primary purpose of the presentation. If necessary, one can acknowledge the children's presence or impatience and go on.
- Preparing a written speech can provide confidence, but it should be left at home. If the answer to an audience question is not known, one should admit it and offer to find it later. This will give added authority to one's answers to other audience questions.
- In the event of a voice quaver, one should take a deep breath and carry on; the quaver isn't as important as what is to be said. One should focus on the content and be enthusiastic.
- Signals that an audience is losing interest include noise, restlessness, and sleeping. These are good indicators that it is time to finish.
- The inquiry method (asking questions of the audience) is an acceptable approach, and may in fact be more effective for some audiences, but it will require more time to cover the material. One should be sure to have the answers and be able to redirect questioning.

One should be careful to avoid the "don'ts" when giving a presentation.

- Don't try to give all data to everyone and thus overcommunicate.
- Don't shift contexts, mingle ideas, and confuse the audience.
- Don't mumble or gesture, to the distraction of the audience.
- Don't laugh to conceal a lack of information, or wander through the dissertation in a disorganized manner.

One should be prepared, determine the purpose of the presentation, analyze the audience, evaluate the situation, organize the material, and then present it with a clear, convincing delivery and enthusiasm. Focus on the message.

THE WRITTEN WORD

There is a lot that keepers should consider writing about. This may include personal notes for future reference or communication to others to share ideas or experiences. Writing may be intended to inquire, to inform, or to persuade. These are three basic forms of written communication that apply to keepers.

Writing is really not all that difficult, especially in this day and age, with computers and spell checking. With a few basic guidelines, keepers can effectively use this medium to express themselves. To be effective, written communication requires organization and planning. To begin with, there must be an explicit meaning, a justification or reason for the communication. With that as the starting point, one can develop the message. One should think about the readers and how they will react to what is being said. One should speak their language, use clear and simple English, and keep in mind the basic principles of writing: unity, coherence, and emphasis.

Who is the reader? What is the purpose for writing? To inquire, inform, persuade, or perhaps entertain? One should think before starting to write: select a topic, list its possibilities, and select the most interesting. Organize around a single subject and in a logical manner. Write in a way that will be enjoyable to read.

For unity, one should be careful of sentence fragmentation, loose connections, word omission, and comma splices of main ideas. Coherence is the tying together of main ideas and topics. Linking words, phrases, and sentences will help to make their intended meaning clear. Emphasis gives power to the statement of key ideas. Also, one should strive for accuracy. Check facts, spelling, and statistics, verify the validity of references (especially information from the Internet), and look at the overall appearance of the copy. Writing is personal and should reflect oneself.

An important part of writing is rewriting. Few people can sit down and write an article, report, or technical paper without editing and rewriting it, often several times. The more important the message, the more care should be taken to get it across in the best way. Some of the great dangers of writing are wordiness (empty words and expletives), meaningless repetition, and clumsy overloading.

The development of writing skills takes time and practice; research skills will help but also take time to develop. Remember to review the document several times. Was it enjoyable? Why or why not? These are important questions to ask. The written word is a permanent, long-term method of communication. It is a most valuable way to convey ideas and share knowledge.

Of special note, keepers should consider writing articles or reports for professional journals. For example, the American Association of Zoo Keepers (AAZK) Animal Keepers' Forum (AKF) is a monthly periodical that publishes articles written by its members. This chapter is based upon a series of articles that had been published in AKF. There are a host of keeper, zoo, and specialty associations that may be interested in publishing papers by keepers. If a keeper is interested in developing an article, it is recommended that they investigate the journal and become familiar with it to be sure that their topic is a good fit. Also, it is important to follow the journal's submission guidelines precisely to improve one's chances of having the article published. Guidelines can often be found towards the front of an issue.

COMPUTERS AND CELL PHONES

Computers in the zoo have been described as "more dangerous than a loose gorilla." A keeper's primary responsibility is to care for the animals. Computer and cell phone use must not detract from proper animal care, safety, and observation. E-mail can easily be misused and waste time. The author has heard from supervisors whose keepers seem to spend more time at the computer than caring for and, especially, observing the animals. Yet another supervisor reported that she received e-mails from coworkers just down the hall, within earshot; she receives many, many questionable e-mails each day from zoo staff. Keepers who spend too much time at the computer, instead of caring for and especially observing the animals, are not doing their job.

Without a doubt, computers have been a great help in recordkeeping and preparing daily reports. Yet keepers have expressed concern that they are required to spend too much time at computers due to receiving too many e-mails. Information overload can become a serious problem in zoos, with every department sending everything to everybody.

E-mail can be an effective tool for communication, but keepers must remember to use it properly. It shouldn't be used in a "machine gun" manner, with frivolous messages sent in rapid succession, or to forward non–work-related messages. It should be used selectively. Due to its high potential for misinterpretation and sometimes impersonal feel, it should not be used as a means of communicating sensitive or complicated messages. Sometimes it is best to meet another person face-to-face. As a written document, an e-mail message is potentially permanent and could be seen by anyone. One should not make statements in e-mail that would be inappropriate to share with coworkers, supervisors, the public, or even the media. When used for work-related communication, e-mail should be written professionally and kept formal unless a more casual relationship with the recipient has already been established. It should be written as a formal letter, with no spelling errors or text message language. Written communication such as e-mail can establish or undermine a keeper's professional image.

Perhaps the most difficult problem with computers and getting information from the internet is that there is no control over the accuracy, or validity of the information out there. Keepers need to assess a source of information; is the source identifiable and reliable? Once incorrect data is on the "information highway," it's hardly ever corrected. When someone says, "I found it on the internet," it immediately becomes suspicious and can be a matter of opinion rather than fact.

While computers have unquestionably been a major means

of facilitating recordkeeping, maintaining animal inventories, documenting behavior and health information, and sharing information during the shipment of animals, to name a few uses, they are easily misused. Surfing the internet and game playing are the two most reported abuses of computers during the workday.

Some zoos enforce policies regarding the use of computers and cell phones for personal communication during working hours. These include complete prohibition of computer use for personal reasons at any time, including off the clock—the rationale being that the computers are owned by the zoo and their use ties up the broadband. One zoo reported that the staff had to accept a user agreement when they become employed but were allowed the use of personal phones and computers during breaks, but were not allowed to access restricted sites such as Facebook and chat rooms. At some zoos, hourly staff members carry zoo-issued cell phones for emergency purposes only. Another zoo reported that it permitted incidental and infrequent personal use of cell phones, but preferred that they be kept in the mute or off position while the keeper was on duty. One zoo reported that its keepers used personal cell phones instead of having lengthy conversations over the radio. Several zoos reported that its staff were not to use cell phones while operating vehicles or hazardous equipment.

It seems unfortunate that such policies even need to exist, and in some cases they are in place but not policed; cases of misuse instead are dealt with on a case-by-case basis. Do keepers really have time to spend on the computer for personal use or on cell phones instead of on observing, researching, or otherwise caring for the animals? Common sense should prevail. Also, it should be remembered that while new technology allows people to "connect" with others, this is not always "communicating." In fact, one should be careful, because misinterpretations occur quite easily.

CONFLICT RESOLUTION

Two strategies seem to stand out when considering conflict resolution training: (1.) listening and values, and (2.) steps for creating a safe environment for giving feedback and resolving differences. This is based upon a number of documents (authors unknown). It would be naive to say that there is one sure way to resolve conflict and issues within a zoo setting. There are too many factors involved, but keepers should make a conscious effort to make their workplace better for themselves and, first and foremost, for the animals.

A few key principles (lessons learned by the authors) should be noted:

1. It is not necessary to *like* everyone in the workplace, but a professional keeper must be able to *work with* everyone in the workplace for the benefit of the animals. It's not always easy.
2. It is important to discern whether a problem is "personal" or really about the animals. Keepers must work on being objective, not subjective, while still representing the best interests of the animals. It's difficult to be sure that personal and/or personnel issues don't cloud the problem at hand.

3. Keepers will not always succeed in every situation and therefore should choose their "battles."
4. The time will come for the keeper to "get over it, let go, and move on." Otherwise the keeper will be focusing on the past, not the future.
5. One should learn from past mistakes and not repeat them. Everyone makes mistakes, but the real "error" is not learning the lesson.
6. If something cannot be made better, one should not make it worse.

LISTENING AND VALUES (STEENBERG 2010, 407)

Surprisingly, the art of listening doesn't come easily to some people, and it may require a lot of work for them to really listen. Listening is a skill that can be developed, and it comes from a belief that

1. it is important to listen to people and worth the time and effort
2. insight can be gained from focusing on the other person
3. feelings and facts are equally important
4. listening is a powerful tool and can be empowering
5. it is possible for people to solve their own problems
6. empathy is important
7. everyone's thoughts and words hold value.

STEPS FOR CREATING A SAFE ENVIRONMENT FOR GIVING FEEDBACK AND RESOLVING DIFFERENCES

The ultimate goal in any workplace is to have a safe, enjoyable, and effective work environment. It is important to foster an environment that is safe for giving feedback and resolving differences.

- Each person should be encouraged to speak for themselves, focusing on facts and not hearsay.
- One should not take oneself too seriously.
- Feelings should be considered as important and as real as facts, but one should try to be objective, not subjective.
- One should sincerely listen to what others are saying. This can be very difficult to do. It is impossible to truly listen if one's mind is preoccupied with texting, or with planning what will be said next.
- One should respond not with accusations, but rather to what has been said.
- One should start statements with "I" rather than "you," and avoid hearsay.
- One should seek clarification of what was heard to ensure that the true meaning is clear and understood.
- Each party is equally important and valuable. Someone who is part of the problem will also be part of the solution.
- One should confirm information and investigate assumptions. It is unwise to rely on "war stories" or gossip as fact.
- Respect for others' opinions is essential.
- Ambiguity must sometimes be tolerated in others.

Being uncertain or undecided is often part of the process and should be expected; time might be needed for clarification and understanding.

• One should speak and act with compassion. Indifference and coldness are problems in themselves, while concern and kindness are often critical to resolving issues. For example, a supervisor should not be approached with a minor problem or request if they have urgent or stressful matters to deal with, or even if they are having a particularly bad day. In this case, considerate choice of timing can make for a better experience and outcome for all parties.

• One should be sensitive to issues of confidentiality. Confidences of the "who," "what," "why" of what is said stays between the parties directly concerned. Gossip is an occupational hazard, and it can worsen the situation when a workplace conflict needs to be resolved.

• Each person should be encouraged to participate and negotiate, but it should be their choice to determine whether and how much they will become involved.

• Agreements that are negotiated should be placed in writing and be quid pro quo (such that each party gives something to receive something else in return). This doesn't mean that they are "win/win" agreements.

• Rather than simply reporting problems, one should also try to propose possible solutions.

• Each person's self-esteem should be respected.

Some conflicts require professional mediation, and most large zoos have employee assistance programs (EAP) or human relations departments. EAP programs usually deal with subjects such as constructive conflict, stress management, managing workplace change, anger management, dealing with difficult people, and workplace mediation.

Conflicts with a boss are truly risky and can have long-term repercussions, no matter how well-prepared or well-meaning one is. They can be not only career limiting, but also career-ending. However, serious issues require serious action. One should try to keep communication focused on that which is constructive (likely to lead to a good outcome); unproductive venting or complaining doesn't make the situation better.

There is a real difference between today's zoo workforce and that of just 10 to 15 years ago, at all levels from keepers to managers and directors. Practical experience with zoo animals has steadily been replaced with academic knowledge about exotic animals, as well as business experience. It is not unusual for today's directors, curators, and collection managers to have never worked as keepers, or to have limited experience with animals. This can put them at a disadvantage in fully understanding animal management problems. It may require extra effort for everyone to get "on the same page" when dealing with an animal issue.

For a keeper, it is often best to resolve an issue at the lowest level before moving up the ladder. Sometimes miscommunication or a person's perception of a situation can be dealt with before a larger problem develops. Again, anyone who is part of the problem will also be part of the solution. Doing nothing is seldom the best option.

SUMMARY

Interpersonal skills are important in the workplace. The abilities to effectively communicate ideas and resolve conflicts are among the most important. Keepers should continually work to develop their communication skills; they should try new techniques, seek out more information, and consider taking classes on communication or public speaking. Basic personal communication skills are both timeless and invaluable. Communication is a complex and vital part of everything people do, and there is a vast amount of information on the subject. Conflict resolution skills are also important since interpersonal conflicts invariably arise in any workplace. It is important that keepers take a professional approach to conflict resolution to ensure a safe, productive, and enjoyable work environment. Professional animal keepers must work with people, and interpersonal skills are essential if they are to work safely as part of a team, support the mission of their facility, and effectively communicate the needs of their animals.

REFERENCES

Hediger, H. 1964. *Wild Animals in Captivity.* New York: Dover.
Steenberg, Judie. 2010. "Keepers and Communication, Part I." *Animal Keepers' Forum* 37(1): 12–13.
———. 2010. "Keepers and Communication, Part II." *Animal Keepers' Forum* 37(2): 68–69.
———. 2010. "Keepers and Communication, Part III." *Animal Keepers' Forum* 37(3): 99–101.
———. 2010. "Keepers and Communication, Part IV." *Animal Keepers' Forum* 37(6): 261–62.
———. 2010. "Conflict Resolution . . . Consider This." *Animal Keepers' Forum* 37(9): 406–8.

Part Two

Evolution of Zoos

4

Zoo History

Josef Lindholm III

INTRODUCTION

All zoos and zoo animals have their own stories, or at least they ought to. It is expected that keepers should know about the individuals they care for: how old they are, where they came from, whether they've bred or had a history of illness or behavioral problems. Beyond that essential information, however, is a much bigger story, which when understood may put these animals, the places they live in, and the work that is done with them in a completely different light. Zoo history is complex, and only a superficial overview can be presented in a single chapter. That said, a brief review of the history of animals in captivity follows. Rather than follow a strictly linear timeline, this will focus on various factors that lead people to keep animals in progressively complex ways.

- The evolution of the modern zoo from various forms of early collections is examined.
- Some factors affecting the species composition of modern zoos are presented.
- The changing role of zoo professionals is discussed.
- The idea is conveyed that zoo history is to be enjoyed.

THE BEGINNINGS OF CAPTIVITY

The naturalist and zoo historian James Fisher asserted that when people started keeping animals, "the initial reason . . . was plain curiosity" (Fisher 1967). Curiosity builds and maintains civilizations, but other less intellectual elements factor in also. The enjoyment of pets releases dopamine and endorphins into the human bloodstream, resulting in pleasure. Years before this process was understood, the eminent behaviorist and animal-keeping enthusiast Konrad Lorenz (1903–89) identified *innate releasing mechanisms*, responsible for the automatic responses humans display towards objects or situations, and not immediately governed by deliberate thought. These mechanisms are due to the chemistry and physiology of brains and nervous systems, part of the genetic

code—the sort of thing that kept ancestral humans alive long enough to raise the next generation. The recognition of a "child schema" is basic. Humans are cued for protective (rather than predatory or aggressive) impulses when they see animals that are small, round, and "soft-looking," with large, close-together eyes and no immediately visible teeth—in other words, the attributes of human children. (Hediger 1969, 119–22; Lorenz 1954, 184–86, 1971). In the same way, people are innately inclined to perceive "beauty," "strength," or "nobility" in certain animals such as eagles, lions, or horses due to their physical appearance. The preferences of zoo visitors are definitely influenced by innate releasing mechanisms.

People living as hunter-gatherers cannot necessarily afford the luxury of refusing to eat animals because they are cute or beautiful, and cultures with no agriculture keep few animals. An exception are dogs, whose domestication appears to go back possibly 12,000 years, to a time when the ice age was still in progress and all humans were hunter-gatherers (Clutton-Brock 1987). The domestication of plants totally changed the relationship between humans and animals. With a constant source of food at hand, humans, prompted by curiosity and innate releasing mechanisms, were free to observe and enjoy animals. They also had food to keep them alive.

Agriculture drew animals to people. As they killed "pests" to protect their crops, early farmers might find themselves in possession of orphaned young animals, the probable ancestors of today's barnyard breeds (Zuener 1963, 43–45). (Hoofed animals have their own innate behavioral mechanisms, resulting in their becoming imprinted upon humans and habituated to them.) As keepers know all too well, storing dry foodstuffs attracts rodents, which in turn draw small carnivores, the ancestors of domestic cats and ferrets (Zuener 1963). The history of domestic animals has been exhaustively (and engagingly) examined elsewhere (Clutton-Brock 1987; Diamond 2005; Zuener 1963). As human societies grew increasingly complex, their relationship to the animals they kept became increasingly varied. There were the animals that provided food, fiber, protection of persons and property, labor, and transportation, all essential to the existence of a society. Then there were animals kept for entirely other reasons.

THE EVOLUTION OF ANIMAL EXHIBITION

HOLY ANIMALS

Religion may be comprehended as a method by which humans comprehend their environment. Since no human culture arose in an absence of animals, it is natural that animals should figure profoundly in many religions. Raising holy animals in the midst of a community is an ancient tradition. It reached its zenith in the Egyptian culture, which existed for more than 3,000 years (2900 BC–AD 356; Houlihan 1996). Sacred animals were pampered in respective shrines all over Egypt. In 112 BC Horus, an Egyptian official, received instructions from his superior, Hermias, preparing for the visit of the Roman senator Lucius Mummius to the city of Crocodilopolis. Someone named Asklepiades was to see to it that the special food was ready for "Petesuchos," the holy crocodile (*Crocodilus niloticus*), so that the senator, who was sightseeing on the Nile, might enjoy seeing this famous animal being fed by his attendant priests (Toynbee 1973). Today, keepers expecting the visits of benefactors or celebrities may find this sort of thing familiar.

DISPLAYS OF WEALTH AND POWER

In the musical *Fiddler on the Roof*, Tevye the milkman envisions a life of wealth, including, among other things, a yard full of poultry "for the town to see and hear" (Harnick et al 1964). Their noise "would land like a trumpet on the ear, as if to say 'here lives a wealthy man.'" The same message is conveyed today by the presence of cassowaries (*Casuarius* sp.) in communities in various parts of New Guinea, or by mithun (*Bos gaurus* var. *frontalis*, domestic gaur) in hill tribe villages in northeastern India. Both animals are forms of currency and measures of wealth, and are killed and eaten only ceremonially (Clutton-Brock 1987; Fulch 1992).

Today in industrial nations, "status symbol" animals, such as purebred dogs or horses, mutations of various reptiles, or the inhabitants of saltwater aquariums or koi ponds demonstrate affluence. The Roman politician Cicero (106–43 BC) was exasperated by people who kept elaborate saltwater ponds at their houses and "[thought] themselves in heaven" when their pet goatfish (*Mullus barbatus*) ate out of their hands (Toynbee 1973, 210). Crassus held a funeral and went into mourning when his pet moray eel (*Muraena helena*) died. Quentius Hortensius valued his goatfish more than his carriage mules, and was as concerned about his sick fish as he was about sick slaves (Toynbee 1973; Varro 1934, 525–27). Gaius Hirrius, who devoted a pond to morays, put 2,000 of them on show at a triumphal celebration honoring Julius Caesar (100–44 BC) on the condition that they be returned to him alive (Toynbee 1973; Varro 1934, 523–25).

Zoos in modern cities have been traditional sources of civic pride. This has ancient roots. In 1100 BC, Tiglath-Pilaser I, king of Assyria, boasted of "live young of the wild oxen he caught, whole herds of them he brought together.... Live

elephants he caught and brought to his city of Assur" (Ley 1968). This same king received a crocodile and a hippopotamus (*Hippotamus amphibious*), from Egypt (Ley 1968), the same country from which his son, Assur-bel-kala, received a crocodile and a "large monkey" (Houlihan 1996). Assurnasir-pal, who ruled Assyria from 884 to 859 BC, recorded: "Fifty young lions [*Panthera leo*] I took, to the city of Kalach and to my palace, put them into my house and showed them. I let them bear young in large numbers." He continues with a varied list of "animals of the desert and of the mountains" which he "brought to my city and let the peoples of my lands view them" (Ley 1968).

Most published histories of zoos mention the enormous procession to honor the god Dionysius staged at Alexandria in 275 BC by Ptolemy II Philadelphos, the second Greek king of Egypt, who reigned from 283 to 246 BC (Belozerskaya 2006; Fisher 1967; Toynbee 1973). Another great collection of animals was displayed to celebrate the thousandth anniversary of the city of Rome in AD 248. The fact that the collection was assembled four years before the anniversary suggests that it was purely for show (Toynbee 1973). If so, it was an exception to the infamous Roman tradition of slaughtering rare and valuable animals in gladiatorial shows. Much has been written about the animals that died in the Roman arenas over at least seven centuries, commencing in 186 BC with a "canned hunt" of lions and leopards. The Roman Colosseum was inaugurated in AD 80 with the deaths of 9,000 animals (Toynbee 1973, 22).

More than a thousand years later, in another hemisphere, imperial power was demonstrated with captive animals in a different way. As with Ptolemy II's procession at Alexandria, any discussion of zoo history is likely to include Montezuma's collection of animals, which, in the first decades of the 16th century, was the finest in the world (Belozerskaya 2006; Hancocks 2001). Enormous series of birds, carnivorous mammals, and snakes were kept in the city of Tenochtitlan, where Mexico City now stands (Diaz del Castillo 1996). Hernan Cortes, who arrived at Tenochtitlan in 1519 and destroyed it, along with the zoo, in 1521, wrote that the birds alone were taken care of by 300 people, not counting veterinary services. The animals had been collected throughout the Aztec empire. Fray Diego Duran (1537–88), recorded firsthand accounts: "From each city and each province, every eighty days, a million Indians arrived with a third of the yearly tribute, laden with everything the land produces, even tiny creatures. This is not an exaggeration since we read that Motecuhzoma the Second . . . even had lice and fleas brought as tribute" (Duran 1994, 358–59). "Tribute was even demanded in centipedes, scorpions, and spiders. The Aztecs felt they were Lords of All Created Things; everything belonged to them, everything was theirs!" (Duran 1994, 203–204).

In contrast, the eighth-century Chinese emperor Zhongzong, and Zhu Jianshen, who reigned 750 years later, refused presents of lions from Arabia and Samarkand respectively (Belozerskaya 2006, xii; Schaefer 1963, 85). In the Chinese tradition of royalty setting good examples, neither emperor could justify the expense of keeping large carnivores. On the other hand, when the second of the two giraffes that arrived from Melinda (in what is now Kenya) in 1414 and 1415 respec-

Figure 4.1. The garden pool of Pharaoh Hatshepsut (1479–1458 B.C.), depicted in her mortuary temple in the ancient Egyptian city of Thebes. Hatshepsut is famous for her African expedition, which returned with a giraffe and other animals (Houlihan, 1996). From Loisel (1912a).

tively was received by the emperor Zhu Di, he explained that this blessing upon China was due to "the abundant virtue of the late emperor, his father, and to the assistance rendered to him by his ministers" (Hahn 1967).

THE PLEASURE OF COLLECTING

The Chinese ideal of royal virtue is expressed in a poem by the Confucian philosopher Mencius (Meng Zi, 372–289 BC) praising Wen, who died in 1056 BC:

> The king was in the marvellous park,
> Where the does were lying down—
> The does, so sleek and fat:
> With the white birds glistening.
> The king was by the marvellous pond—
> How full it was of fishes leaping about!
> (Bostok 1993)

The fact that the does were lying down signifies this was not a hunting park (Bostock 1993, 19; Hahn 1967, 41). Mencius commented that the common people "rejoiced in [their king's] possession of deer, fish and turtles. It was by sharing their enjoyments with the people that men of antiquity were able to enjoy themselves" (Bostock, 1993). This "marvellous park," also translated as "garden of intelligence," is often cited as the first true zoo.

Though the first century BC, Romans whom Cicero made fun of may have been the first to keep saltwater fishes (Toynbee 1973), the Egyptians kept tilapia (*Oreochromis* sp.) and other freshwater fishes in rectangular ponds in formal gardens more than twelve centuries earlier, at least as far back as the 18th dynasty (1550–1307 BC). Cichlid keeping has ancient roots. Tomb art documents them sharing their ponds with ducks. (Foster 1999; Loisel 1912a, plate II). Egyptians loved keeping ducks, geese, and swans, and at least 13 species are depicted in their art (Houlihan 1986), most famously a pair of red-breasted geese (*Branta ruficollis*) painted with other species of geese on a tomb wall sometime before 2465 BC (Houlihan 1986). This may reflect the way the birds were kept: a more than 4,000-year-old record of closely related species kept together or in close proximity—the sort of traditional zoo display that has recently lost popularity and is sometimes derided as "stamp collecting" (figure 4.1).

ANIMALS TO LEARN FROM

There is no real evidence that Wen's "garden of intelligence" was, as has been suggested, (Zuckerman 1980, 4) meant "for scientific and educational purposes" 3,000 years ago. Neither has much data survived from the zoo connected to the Museum and Library of Alexandria (Belozerskaya 2006; Houlihan 1996; Toynbee 1973), the presumed home of the animals in Ptolemy II's procession of 275 BC. Pheasants (*Phasianus colchicus*) appear to have been kept there (Toynbee 1973), along with six rock pythons (*Python sebae*), sometime between 283 and 116 BC (Belozerskaya 2006; Toynbee 1973). However, animals were studied in other royal collections. Huizong, emperor of China from AD 1100 to 1125, described the behavior of his ornate lorikeet (*Trichoglossus ornatus*), brought more than 2,500 miles from the Indonesian island of Sulawesi (Lindholm 1995). Roughly a century later, in Italy, the Holy Roman emperor Frederick II, who reigned from AD 1211 to 1250, recorded observations of an umbrella cockatoo (*Cacatua alba*), a gift of the sultan of Babylon (Wood and Fyfe 1943, 38, 59, 77). Frederick is famous for his book *The Art of Falconry* (Wood and Fyfe 1943). Great gray owls (*Strix nebulosa lapponica*) and arctic gyrfalcons (*Falco rusticolus*) traveled with the camels, big cats, and elephant that accompanied Frederick on his state visits through Italy and Germany (Ley 1968). The elephant was a gift from the sultan of Damascus, as was the first giraffe seen in Europe since Roman times. In return, Frederick sent the sultan a white peacock (*Pavo cristatus*) and a white bear (which could well have been a polar bear [*Ursus maritimus*] from the Greenland settlement).

By the 17th century, the Holy Roman Emperor resided in Prague, in what is now the Czech Republic. Rudolf II, who reigned from 1576 to 1612, relocated his court there from Vienna in 1583. He created a research center, including a natural history museum, a zoo, and the observatory where Tycho

Brahe and Johannes Kepler changed the way the universe was perceived. It is not clear that the dodo (*Raphus cucullatus*), in the museum by 1609, had arrived alive in Prague (Belozerskaya 2006; Fuller 2002, 81). Discovered in 1598, dodos were extinct before 1700. Living ones do appear to have reached England and India (Fuller 2002, 66, 69, 94). On the other hand, Rudolf's acquisition of the first cassowary (*Casuarius casuarius*) in Europe is well documented (Belozerskaya 2006). He commissioned an illustrated catalog of his museum and zoo. It records living parrots from around the world, including a salmon-crested cockatoo (*Cacatua moluccensis*), which died in 1608 after seven years in Prague (Belozerskaya 2006).

After Rudolf's death, the seat of the Holy Roman Emperor returned to Vienna, where royal menageries of one kind or another had long existed (Fiedler and Wessely 1976). In 1751 a new menagerie was created there, existing today as Tiergarten Schoenbrunn, recognized as the world's oldest continuously existing zoo. It is appropriate that this happened in the 1750s, as that decade witnessed a pivotal event in the history of zoology: the publication in 1758 of the 10th edition of Linneaus's *Systema naturae*, the book that laid the foundation for the classification of animals as recognized today. Carl Linneaus (1707–78) maintained a private zoo while he was professor of botany at Uppsala University in Sweden (Svan-

berg 2007). Among his animals were two species of lemurs, a cotton-topped tamarin (*Saguinus oedipus*), Barbary macaques (*Macaca sylvanus*, which bred there), a Diana guenon (*Cercopithecus diana*), a raccoon (*Procyon lotor*), an agouti (*Dasyprocta leporina*), and goldfish (*Carassius auratus*).

In France, Linneaus's rival, Georges-Louis LeClerc, Comte de Buffon (1707–88), achieved great, if not as pervasive, fame by publishing 36 volumes of his *Histoire naturelle,* commencing in 1749. Having attained early prominence in mathematics, which was followed by his pioneering studies of the structure of wood, Buffon was made a member of the French Academy of Sciences in 1734, and appointed director of the Jardin du Roi (the Royal Botanical Gardens) in 1739. His interests broadened to zoology, and he developed a research zoo at his family estate at Montbard, which included laboratories. Among the animals he kept there were a Knysna turaco (*Tauraco corythaix*) and a pied hornbill (*Anthracoeros malabaricus*; Loisel 1912b, 313; Buffon 1783). As director of the Jardin du Roi, Buffon was also expected to conduct studies of the animals at Versailles, but he found conditions there less conducive to research (Robbins 2002, 65). Created in 1665 as one of the grand projects of Louis XIV, the menagerie was designed to showcase large groups of ornamental animals, especially birds (figure 4.2; Loisel 1912b, 115; Robbins 2002). However, the Academy of Sciences, organized in 1666, was

Figure 4.2. The "Court of the Demoiselle Cranes" at the French Royal Menagerie at Versailles, created in 1665. A pair of great bustards (*Otis tarda*) are at the far right. From Loisel (1912b).

encouraged by the king to study the animals, and many scientific papers were published in the 1670s (Robbins 2002). Dissections often took place in the royal library. The king himself attended an elephant's dissection in 1681 (Robbins 2002, 45).

ZOOS FOR THE PEOPLE

Buffon hoped to create a more scientific live animal collection in Paris at the Jardin du Roi (Loisel 1912b, 316–17; Robbins 2002, 67), but he failed. In 1793, five years after his death, an animal collection was established there, but by that time the location had been renamed the Jardin des Plantes because the monarchy had been abolished and France was now a republic. The National Museum of Natural History was established to serve the people, and the new menagerie became one of its departments in 1794. Today the Menagerie du Jardin des Plantes is widely recognized as the first truly public zoo and the second oldest continuously existing one. While intended from the beginning to serve science, in 1793 the menagerie served the immediate purpose of clearing the streets of privately owned animal shows. Several animals from these shows were confiscated and brought to the museum and, by way of compensation, some of their owners and keepers were employed as the new zoo's first staff, reluctantly supervised by the museum's scientists (Robbins 2002). The last remnants of the menagerie at Versailles arrived in 1794 (Robbins 2002, 59, 222). By 1800 the collection had become fairly substantial, due in part to the looting of the Dutch royal menagerie in 1796 and 1798 (Loiselle 1912b, 38–49; Robbins 2002).

Other now defunct royal zoos formed parts of zoos that exist today. In 1831 the London Zoo, opened in 1828 by the Zoological Society of London (chartered in 1826), was given most of the animals in the Tower of London's menagerie, established in 1235 when Henry III was given three leopards by the aforementioned Holy Roman Emperor Frederick II (Keeling 2001). King William IV had already given the Zoological Society "nearly everything of note" in Windsor Great Park in 1830, though that collection was later revived by Queen Victoria (Keeling 2001). When Zoologischer Garten Berlin was opened by a nonprofit stock-holding company in 1844, its animals included the Prussian royal menageries from Peacock Island, dissolved by the new king, Friedrich Wilhelm IV, who also had provided land for the new zoo in 1842. (Strehlow 1996, 2001).

Though it would become the world's largest collection of animal species, for its first several decades Zoologischer Garten Berlin existed primarily as a forested retreat for the growing city of Berlin, with emphasis more on its natural setting than on its collection (Strehlow 1996, 2001). This ideal was expressed in a 1857 mission statement of the Frankfurt Zoological Society, which opened its zoo in 1858: "One turns away from the political and social strife of life to the contemplation of nature with a feeling of contentment, and one is revived and strengthened by the recognition and perception of nature's creativity and her eternal laws" (Strehlow 2001). This is very much in keeping with the German Romanticist movement, which had pervasive effects though the 19th century. The Frankfurt mission statement continues in the same spirit, but then concludes on a different note: "What a pleasant way to relax, walking in a lovely garden, looking at living creature from all areas and regions—each with its individual characteristics—and admiring their diversity."

LIVING MUSEUMS

In 1825, the year before the founding of the Zoological Society of London, its founders declared, ". . . It has long been a matter of deep regret . . . that we possess no great scientific establishments either for teaching or elucidating Zoology; and no public menageries or collections of living animals where their nature, properties and habits may be studied. . . . It would well become Britain to offer . . . to the population of her metropolis . . . animals from every part of the globe . . . as objects of scientific research, not of vulgar admiration . . ." (Vevers 1976, 15).

In the 19th century zoos became living museums to educate the general public and encourage research. Living and dead specimens often existed side by side. The Zoological Society of London maintained a zoological museum until 1855, when its specimens went to the British Museum of Natural History and other English museums (Blunt 1976). The Jardin des Plantes continues as a department of the French National Museum of Natural History. The zoos at Amsterdam and Antwerp, opened in 1838 and 1843 respectively, still maintain museums created by their first directors (Strehlow 2001).

An early exhibit of living animals in the United States was the Peale Museum, in Philadelphia, which existed from 1784 to 1849. Along with nearly 2,000 species of stuffed birds and a mastodon skeleton excavated in 1801, there were, over the years, living bears, monkeys, a bald eagle (*Haliaeetus l. leucocephalus*), African finches, and the black-billed magpie (*Pica pica hudsonia*) and black-tailed prairie-dog (*Cynomys ludovicianus*) sent to Thomas Jefferson by the Lewis and Clark expedition in 1805 (Kastner 1977; Sellers 1979). In 1859, ten years after the Peale Museum closed due to financial difficulties, the Zoological Society of Philadelphia was founded. The Philadelphia Zoo, opened to the public in 1874, is thus widely known as "America's First Zoo," though New York's Central Park Menagerie (1861), Chicago's Lincoln Park Zoo (1868), and the Roger Williams Park Zoo in Providence, Rhode Island (1872), had already come into physical existence (Kisling 2001). The National Zoological Park in Washington, DC, officially established in 1889, began as the Department of Living Animals of the National Museum of Natural History (itself a bureau of the Smithsonian Institution) in 1887, directed by chief taxidermist William Temple Hornaday (Kisling 2001). The animals were initially displayed a very short distance from the museum, in the National Mall.

In an effort to display as wide a variety of living animals as possible, zoos employed the latest technologies. London Zoo opened the first reptile house in 1849, the first aquarium in 1853, and the first insect house in 1881 (Keeling 2001). Along with fishes, London Zoo's Aquarium exhibited 162 species of marine invertebrates, in five phyla, during its opening year (Keeling 2001). Aquariums quickly became popular. Commencing in 1856, P. T. Barnum exhibited increasingly elaborate aquariums at his American Museum in New York, opened in 1841. By the time this building burned to the ground in 1865, these collections included exceptionally large brook trout (*Salvelinus fontinalis*), for which Barnum paid

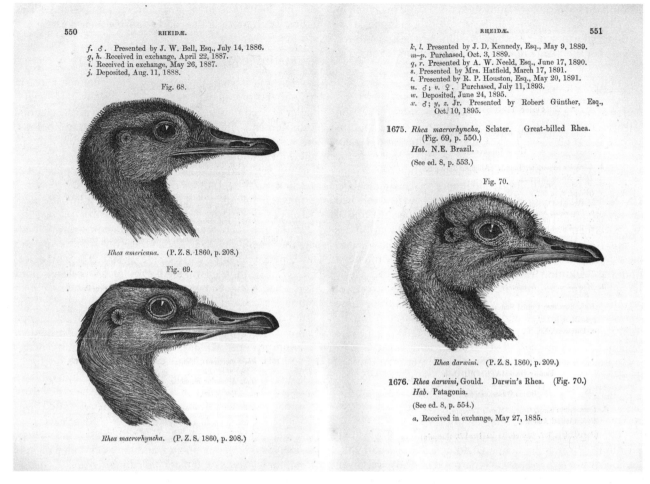

Figure 4.3. Two of the 681 pages (excluding the index) of the final edition of Philip Lutely Sclater's *List of the Vertebrated Animals Now or Lately Living in the Gardens of the Zoological Society of London,* published in 1896. Dr. Sclater described "*Rhea macrorhyncha*" from a living bird at the London Zoo. It was subsequently determined to be a specimen of the eastern Brazilian subspecies of the common rhea (*R. a. americana*). The subspecies of common rhea familiar to zookeepers and visitors is from Argentina (*R. american albescens*).

$100 each, fishes from the coral reefs of Bermuda and the Gulf of Mexico, and beluga whales (*Delphinapterus leucas*) from Quebec (Barnum 1871, 563–67). Barnum also exhibited an entire zoo indoors, including a hippopotamus.

As technology and prosperity advanced through the 19th century, enormous collections of animal species and subspecies, larger than any before, were created. The secretary of the Zoological Society of London, Philip Lutley Sclater, prepared nine editions of the *List of the Vertebrated Animals Now or Lately Living in the Gardens of the Zoological Society of London*, from 1862 to 1896 (figure 4.3). The first was 100 pages long; the last had 681 pages without the index. Over these 34 years Sclater determined that 770 species and subspecies of mammals, 1,676 taxa of birds, 420 taxa of reptiles, 80 taxa of amphibians, and 98 taxa of fishes had been kept at London Zoo at one time or another—a total of 3,044 vertebrate taxa (Sclater 1896). With the exception of fishes, he provided the date of acquisition or propagation of every specimen as well.

In its first two decades, Zoologischer Garten Berlin existed as a romantic nature park. In 1864, 20 years after its opening,

there were 166 species of birds and mammals (Strehlow 1996, 2001). In contrast, Sclater's 1863 *List* includes 229 sorts of mammals and 468 of birds "now or lately living" at London Zoo. By 1884, Berlin's bird and mammal taxa totaled 504. By 1901 there were 894 forms of birds and 402 forms of mammals (Kloes 1969). This was a time of exploration, discovery, and expansion of European influence, as a flood of new species were described for science. For instance, in 1876 three Père David's deer (*Elaphurus davidianus*) came to Berlin from China's Imperial Hunting Park, near Beijing, where this otherwise extinct species persisted until 1900. It only become known to Europe in 1865, when the missionary after whom it is named made observations from the Imperial Hunting Park's wall. The first such deer to leave China came to London in 1869, and others arrived in Paris (Crandall 1964). Although all Pere David's deer now living are descended from animals that the Eleventh Duke of Bedford obtained from Paris and Antwerp in the 1890s for his private collection at Woburn Abbey (Bedford 1949), it appears, from recent research by Marvin Jones and others, that all are actually descended from the Berlin importation (Strehlow 2001).

LIVING TEXTBOOKS

Carl Hagenbeck was a leading supplier of animals for zoos and circuses. From 1866 to 1886 alone he sold at least 1,000 lions, 1,000 bears, around 800 hyenas, more than 600 leopards, more than 300 tigers, and around 300 elephants, 150 giraffes, and 150 reindeer (*Rangifer tarandus*; Rothfels 2002, 58). Along with his 1901 importation of Przewalski's horses (*Equus ferus przewalskii*), among the handful of ancestors of all now living (Bouman and Bouman 1994), he was famous for such rarities as pygmy hippopotamuses (*Hexaprotodon liberiensis*; Reichenbach 1996). Today, however, he is most remembered for his innovation in animal exhibits.

In 1907, after having maintained his animals in a two-acre facility in the middle of the busy port city of Hamburg for more than 30 years, Hagenbeck opened his Tierpark on a 25-acre site in the Hamburg suburb of Stellingen. This zoo was an immediate international sensation as it featured Hagenbeck's "panoramas," which he patented in 1896 (Reichenbach 1996; Rothfels 2002) and premiered as temporary exhibits in various German cities shortly thereafter. These consisted of several enclosures separated by hidden moats so that all the inhabitants not only appeared to have access to a single area, but often appeared to not be enclosed at all. The term "Hagenbeck heart," meaning a medical condition brought on by encountering animals in places where one did not expect to see them, was soon put to humorous use (Munro 1914). Thomas Edison was a famous victim (Hahn 1967).

Enthusiasm for Hagenbeck's innovations resulted in the proliferation of moated enclosures in such places as London in 1914, Denver in 1918, Saint Louis in 1921 (Hancocks 2001), and San Diego in 1923 (Wegeforth and Morgan 1953). In 1934 the Cincinnati Zoo, in Ohio, engaged one of Hagenbeck's designers, Joseph Pallenberg, to design an African veldt. Featuring a massive kopje of artificial rock, it received international attention (Ehrlinger 1993). Similar open enclosures for hoofed animals featuring artificial rockwork were also constructed around this time at Paris, Detroit, the Chicago Zoological Park (Brookfield Zoo), and Saint Louis, all of which were by then engaged in sustained propagation of various African antelope.

The New York Zoological Park (Bronx Zoo), however, took the lead in naturalistic exhibits in 1941 when it opened its African Plains, an exhibit which today remains magnificent after seven decades. In marked contrast to the aforementioned displays, in which artificial rockwork was prominent, only preexisting glacial granite outcroppings were incorporated, showcasing the lions, whose hidden moat creates an excellent illusion of their sharing an enclosure in which lowland nyala (*Tragelaphus angasii*) have bred without interruption since 1941 (Crandall 1964). What set this enclosure apart from any previous attempt to create an African habitat was the unprecedented use of trees. Several North American species were pruned to resemble African ones (Hancock 2001).

While the 1940s were a decade of destruction and hardship for many European and Asian zoos (Kawata 2001; Solski 2001; Strehlow 2001), remarkable innovations were achieved at the Bronx Zoo. The New York Zoological Society (now the Wildlife Conservation Society) had been founded in 1895, and from its beginning had enjoyed considerable financial and political support (Bridges 1974). Thus, in contrast to many other American zoos, the New York Zoological Park, opened in 1899, experienced a consistent and orderly growth, and its master plan was fulfilled by 1914 (Bridges 1974, 271, 327, 474; Lindholm 1988). For more than a quarter century, no further major enclosures for living animals were constructed. The opening of the building containing the (preserved) "National Collection of Heads and Horns" in 1922 completed the vision of its first director, William Temple Hornaday (formerly chief taxidermist at the Smithsonian). Shortly before his retirement Hornaday, who served from 1896 to 1926, proclaimed his institution a "masterpiece" (Hornaday 1925). He died in 1937.

Along with the African Plains, 1941 saw the opening of a children's zoo, a three-acre moated exhibit for North American deer, and a moated display with trees and a natural glacial granite outcropping for giant pandas (*Ailuropoda melanoleuca*: Crandall 1964; Lindholm 1988). That same year, the office of director was discontinued for eleven years and the zoo was run by a planning committee. One of the committee's members was the French ornithologist Jean Delacour, who had arrived in New York as a war refugee in 1940. He returned to France in 1947 to oversee the reopening of his own Parc Zoologique de Cleres. Among the innovative exhibits he designed during his seven years in New York (Lindholm 1988) was a series of indoor aviaries representing different habitats around the world and featuring carefully tended living plants displayed with appropriate furnishings and selections of birds (Delacour 1943, 1945).

With the creation of Delacour's habitat aviaries in 1942 and 1943, the Bronx Zoo again set a new standard for animal exhibition, and it has continued to do so ever since. One after another, the Aquatic Bird House (opened in 1964), the World of Darkness (1969), the World of Birds (1972), Jungle World (1985), Congo Gorilla Forest (1999), and Madagascar (2008) presented animals in ways not seen before (Conway 1966; George 1975; Hancocks 2001; House and Doherty 1975; Ricciuti 1972).

The Aquatic Bird House was remodeled from a building present at the zoo's opening in 1899. The 1964 renovation had a definite goal: "The theme . . . is a simple one. We have attempted to tell the zoo visitor about birds dependent for their livelihood on wetland habitats, to show the *type of bird in the place it lives* and, where possible, to show some of what these birds *do* and *don't do,* and to tell why, clearly. Most of all, we hope simply to interest the zoo-goer in waterbirds . . . While the new building accents diversity and adaptation, it leaves for the most part the illustration of elementary biology to other zoo exhibits. It develops along an ecological theme towards a conservation conclusion" (Conway 1966, 131–32). It is well to keep in mind that Rachel Carson's *Silent Spring* had only been published in 1962, and the first Earth Day would not be celebrated until 1970. This early effort towards instilling environmental awareness in the zoogoing public was thus truly groundbreaking.

William G. Conway joined the staff at the Bronx Zoo, as curator of birds in 1956 at the age of 26. He was appointed zoo director in 1961, and retired as president of the Wildlife Conservation Society in 1999. His influence on the way things are done in zoos has been profound (Hancocks 2001). Of his many publications, perhaps the most well known is "How to Exhibit a Bullfrog" (Conway 1973), in which he presented an enormous and expensive building with surrounding outdoor exhibits, entirely devoted to displaying bullfrogs (*Rana catesbeiana*), exploring their life history and their place within the environment and zoological classification. While the building itself was never constructed, its principles of engagingly exhibiting animals within the context of their ecology and evolution were carried out in the above-noted Bronx Zoo exhibits of the last 50 years, and were widely adopted around the world.

Intentionally displaying the natural behavior of animals became an increasingly important feature of such exhibits. The acknowledged pioneer of this approach was Heini Hediger (1908–92), otherwise known as the "father of zoo biology." Prior to his 35-year career as director, successively, of three Swiss zoos, he conducted field research in Morocco and New Britain. He applied these observations to zoo animals, concluding that their well-being was profoundly influenced by the application of principles derived from knowledge of a given species' behavior and other aspects of its biology (Hediger 1950, 1955, 1969). With such concepts in mind, relatively simple things could be done through cage design and keepers' procedures to reduce animal stress and increase comfort, resulting in a decrease in abnormal behaviors and improvement in reproduction, longevity, and overall health, as well as a more positive experience for the zoo visitor (Gaedemann 1975).

While the concepts of enrichment and incorporation of training into daily husbandry are standard procedures today, it took years after the publication of Hediger's principles for them to become widely practiced, and exhibits continued to be designed which did not encourage natural behavior. As late as 1965 the Philadelphia Zoo opened the Rare Mammal House, a one-million-dollar "ultra-modern building" in which apes and other species were exhibited entirely indoors, on tile (of "various pastel shades. . . . to make the cages bright and cheerful"), with stainless steel furnishings and viewing windows electrified from the inside (Ulmer 1966). Every 90 minutes the floor of each exhibit was automatically flushed "to cleanse the cage floor of any debris." Sensory deprivation was imposed on zoo visitors as well. Emily Hahn (1967, 244) could see the siamang (*Symphalangus syndactylus*) hooting, but could not hear the amazing sound at all.

The Philadelphia Zoo was already controversial for its development and promotion of standardized diets, dating back to the 1930s. In his last and most elaborate book on zoo biology, Heini Hediger (1969) spent a major portion of a 27-page chapter titled "Food" vigorously criticizing Philadelphia's feeding techniques and the philosophy behind it. While it cannot be argued that the husbandry and propagation of New World primates, soft-billed birds, and other animals has been enhanced by the use of processed foods, the sole use of those foods without supplementation is now recognized as negative for both zoo animals and their visitors.

In 1978, 13 years after the opening of Philadelphia's Rare Mammals House, the Woodland Park Zoo in Seattle released gorillas into an exhibit converted a year before from old bear dens. This enclosure had stood empty for a year to allow the trees and other plants to establish themselves (Hancocks 2001). This was the first time a gorilla family had been displayed with an extensive collection of living plants, in an environment where they could behave as they might in the wild. Not only was the exhibit novel, but keepers' procedures had been altered as well, expanding to such practices as scattering seeds, nuts, and other enrichment around the enclosure several times a day.

Woodland Park's gorilla exhibit became a flagship for "landscape immersion" displays, designed to provide the illusion of seeing animals as one would in the wild. In the last three decades this has became an expected element in the zoo visitor's experience. While such exhibits have generally been greeted with enthusiasm, there has been some negative public response. As the conversion of a major portion of the Woodland Park Zoo proceeded, complaints were received about excessive vegetation and difficulty in viewing animals, and such criticism continues today. If one peruses the reviews in various user-generated-content websites such as Tripadvisor, one finds, together with praise for immersion-oriented zoos, comments that animals are too far away or not visible at all. There are occasional laments about the conversion of traditional enclosures to "habitats," and suggestions that zoos should make it easier to see the animals. In this age of health consciousness, it is amusing to note that commenters also express resentment over being made to walk "too much."

On the whole, however, immersion exhibits are well received. This reflects changing perceptions on the part of zoo visitors. Following the shock to the American psyche when the Soviet Union launched the satellite *Sputnik* in 1957, a much greater emphasis on science education was promoted by both the educational system and the media. Through the 1960s and 1970s, such corporations as Time-Life, the Golden Press, the National Geographic Society, and *Reader's Digest* published a flood of books and articles on natural history. Network television, through both special programs and such series as *Wild Kingdom* (hosted by Marlin Perkins, director of the Saint Louis Zoo), did much to educate people about animals during that same time, as did the growing medium of public television. Jane Goodall, Diane Fossey, and Jacques Cousteau became household names. Apes were no longer perceived as savage monsters or buffoons. The exploitation of cetaceans and pinnipeds became repugnant. Awareness of endangered species exploded, leading to widespread legislation in the late 1960s and 1970's. All of this led to changing expectations of the zoo-visiting experience. Displaying bored, stressed animals in unnatural conditions became generally unacceptable. Instead, there was a growing desire to see the animals as they appeared on television and in magazines. Zoo visits have increasingly become educational experiences, and zoos have become living textbooks, presenting lessons in ecology, ethology, evolutionary biology, and other elements of the life sciences.

David Hancocks, former director of the Woodland Park Zoo and the Arizona Sonora Desert Museum, is a key figure in the development of landscape-immersion exhibit techniques and believes they should be the primary consideration in zoo design so that any zoo animal should ideally be "seen in the context of its habitat" (Hancocks 2001, 146–48). Hancocks does not admire the tradition of exhibiting "stamp collections" of closely related species, and questions the value of such displays. As an example he cites New York's Staten Island Zoo as it existed in 1968, concluding that its rows of exhibits in hallways "were a perfect formula for monotony," resulting in the "inevitable boredom of the viewer" (Hancocks 2001, 78). In fact, a great many visitors to Staten Island were not bored but delighted to see 39 species and subspecies of rattlesnakes (*Crotalus* sp. and *Sistrurus* sp.; Kawata 2003, 59–62), including all 32 US taxa. Countless more young people who had never visited this zoo were inspired by its collection. Carl Kauffeld, who created it, arrived as curator of reptiles in 1936 (the year the zoo opened), and retired, as director, in 1973. Kauffeld authored two books and many articles on the captive husbandry of snakes and other reptiles (Murphy 2007, 189–93), which "motivated scores of his readers to follow in his footsteps" (Murphy 2007, 78–79). The zoo philosopher Stephen Bostock (1993, 115–17) suggests a balance between Hancocks's "story-driven" approach to zoo design and the more traditional "object-driven" exhibit philosophy, since, after all, most zoo visitors continue to be people who come primarily to enjoy animals.

THE EVOLUTION OF ZOO COLLECTION POLICIES

KEEPING ANIMALS ALIVE IN THE FIRST PLACE

In 1915 William Temple Hornaday wrote: "There is not the slightest reason to hope that an adult Gorilla, ether male or female, ever will be seen living in a zoological park or garden . . . It is unfortunate. . . . but we may as well accept that fact—because we cannot do otherwise" (Crandall 1964).

On 22 December 1956 a Western lowland gorilla (*Gorilla gorilla gorilla*) was born at the Columbus Zoo in Ohio, the first gorilla born in captivity. In 2010 Colo, the oldest captive gorilla on record, celebrated her 54th birthday. From 1968 through 1971 she gave birth to three offspring. Sixteen grandchildren were born from 1979 through 1993, four great-grandchildren since 1995, and two great-great-grandchildren since 2003 (anon. 2011). At the end of 2010 there were more than 750 captive Western lowland gorillas around the world (including those at 47 US zoos), most of them captive-bred.

When Hornaday wrote his famous words, he was contending with a dying gorilla. Although Dinah had arrived in New York on 23 August 1914 with a healthy appetite and appeared to be in "excellent condition" (Crandall 1964), her health began to deteriorate in November, and her death on 31 July 1915 was attributed to "malnutrition" and "rickets." Hornaday (1915) attributed this decline to the nature of the species, declaring the gorilla is "not an animal of philosophic mind, nor is it given to reasoning from cause to effect. What can we do with a wild animal that is not amenable to the pangs of hunger, and would rather die than yield?"

This gorilla's diet consisted of "fruits, cooked beef, chicken, or lamb, milk, and raw eggs" (Crandall 1964). No vegetables or leaves of any sort are mentioned.

Many animals familiar in today's zoos, like the gorilla, were once considered impossible to keep or highly demanding, but have likewise become established due to improvements in diet, medicine, and transportation and an understanding of their natural history that translates to the way they are accommodated and cared for. In the early 1960s, the best public zoo record for a squirrel monkey (*Saimiri* sp.) was six years, no captive howler monkey (*Alouatta* sp.) had lived more than five years, and no titi (*Callicebus* sp.) for more than four years (Crandall 1964). By 2005, five titis had lived at least 24 years, four howler monkeys had reached at least 22 years,; and four squirrel monkeys had lived for at least thirty years (Weigl 2005).

THE "ARK" IDENTITY

William Temple Hornaday is properly revered as a pioneer in promoting conservation through zoos (Bridges 1974). The year the New York Zoological Park opened, he proclaimed: "*Every large mammal species on earth is being killed faster than it breeds!*" (Hornaday 1899). He played a key part in the New York Zoological Park's unarguably successful role in rescuing the plains bison (*Bison bison bison*) from extinction (Bridges 1974;, Crandall 1964; Hornaday 1925). The establishment of the Wichita and Wind Cave herds from Bronx-bred stock is recognized as an early example of reintroduction of captive animals to their former range. Before the implementation of bird preservation laws (which he and the Zoological Society were instrumental in passing), he also oversaw programs to breed and release herring gulls (*Larus argentatus*), which were subjected to the unsustainable harvest of eggs for bakeries, and mourning doves (*Zenaida macroura*), which were threatened by unregulated hunting (Crandall 1915, 1917; Lindholm 2007).

It is thus startling to read Hornaday's 1920 annual report of the New York Zoological Society: "The first, the last and the greatest business of every zoological park is to collect and exhibit fine and rare animals. Next comes the duty of enabling the greatest possible number of people to see them with comfort and satisfaction. In comparison with these objects, all others are of secondary or tertiary or quaternary importance. The breeding of wild animals is extremely interesting, and the systematic study of them is fascinating, but both these ends must be subordinated to the main objects" (Bridges 1974, 414).

One must keep in mind that Hornaday began his career as a taxidermist, creating groundbreaking habitat displays (Hornaday 1925, 209–305). Furthermore, if one bears in mind that Martha, the last living passenger pigeon (*Ectopistes migratorius*), had died at the Cincinnati Zoo in 1914, ending a 40-year-old program to propagate that species there (Cokinos 2000; Ehrlanger 1993), and Incas, the last Carolina parakeet, had died there in 1918 (Ehrlanger 1993), one can see why Hornday might not have complete confidence in captive breeding as a conservation tool, as opposed to educating zoo visitors about the importance of preserving wildlife. On the other hand, being a person of his time, Hornaday probably could not imagine the scale and scope of captive propagation in zoos in the 21st century.

For that matter, much of what is routine in zoos today was unimaginable only a few decades ago. In 1970, for example, there were seven mountain bongo (*Tragelaphus euryceros isaaci*) in the United States. Four had arrived from Kenya only that year, as had all six in Europe. No captive-conceived birth had yet taken place anywhere (Bosley 2010). In July 2010, according to the International Species Information System (ISIS), mountain bongo were the most widely distributed African antelope in the United States, with 222 distributed among 43 institutions. A further 257 were present among 64 zoos around the world. It is estimated that less than 200 now exist in the wild (Bosley 2010). Another example is the red-ruffed lemur (*Varecia rubra*). Volume 12 of the *International Zoo Yearbook* cites one at London and four at San Diego Zoo as the entire captive population in 1971. In early 2011, ISIS listed 615 distributed among 171 zoos around the world.

Of course not all of the self-sustaining zoo populations

of conservation-dependent animals are of recent origins. Today's captive populations of Siamese fireback pheasants (*Lophura diardi*), Swinhoe's pheasants (*Lophura swinhoei*), and brown-eared pheasants (*Crossoptilon mantchuricum*) are partially descended from birds imported in the 1860s by the Jardin d'Acclimation in Paris, opened in 1860, expressly to establish captive populations (Delacour 1977; Osborne 1996). The aforementioned importations of Père David's deer in 1876 and Przewalski's horses in 1901 are classic examples of old zoo bloodlines. On the other hand, the last several decades have witnessed a phenomenal expansion in the range of animals maintained as conservation projects by zoos. Many of these resulted from last-minute efforts in the face of a range of ecological disasters. Programs for *Partula* snails; burying beetles (*Nicrophorus americanus*); cichlids of Madagascar, Lake Victoria, and Cameroon's crater lakes; *Atelopus* toads; the Louisiana pine snake (*Pituophis ruthveni*); birds of the Marianas Islands; Indian Ocean fruit bats (*Pteropus* sp.); the black-footed ferret (*Mustelus nigripes*); and the Chacoan peccary (*Catagonus wagneri*) just begin to represent this diversity.

Conservation significance has not been the only factor in the creation of managed populations in zoos. Today's zoo programs for such birds as touracos, hornbills, and toucans would have been unthinkable before a US outbreak of exotic Newcastle's disease caused the prices of such birds to increase exponentially due to import restrictions in 1972. Until then, the idea of investing much keeper time and finances in creating self-sustaining populations of fruit- and insect-eating birds that could be replaced for less than the daily wage of their keepers was inconceivable to zoo administrations. Only after the availability of such birds was restricted did programs to propagate them become established (Lindholm 2005a, b). From the 1970s onwards, legislation of all sorts around the world further affected the availability of animals. In the United States that decade saw the passage of the Marine Mammal Protection Act and an expanded Endangered Species Act, while on a global level the Convention on International Trade in Endangered Species (CITES) was created. Several major exporting countries ended or severely curtailed their animal trade. As the general public became far more environmentally aware, disapproval of large-scale removal of animals from the wild increased, and it was expected that zoos would "breed endangered species" and otherwise engage in conservation activities. Professional zoo organizations such as the AZA and EAZA encouraged and eventually compelled their members to create self-sustaining captive populations, to the point that Taxon Advisory Groups (TAGs) and corresponding programs cover most vertebrates and an increasing number of other animals.

As a result, there has been reduced institutional independence in animal acquisition, propagation, and other management in the last several decades. Communication and cooperation among collections is expected. This has resulted in decreased diversity in species holdings. Taxa targeted for programs have displaced those not recommended. The range of taxa within a genus or species is often restricted. On the other hand, TAGs and corresponding programs have resulted in the exhibition of animals in zoos where one would have not imagined seeing them before. Before such programs, the exhibition of "rare and endangered" animals was largely the prerogative of large, well-funded, and often famous zoos. Today, all AZA-accredited zoos are expected to participate in conservation programs, so that the presence of such animals as blue-eyed black lemurs (*Eulemur macaco flavifrons*) or Visayan warty pigs (*Sus cebifrons cebifrons*) in the zoos of smaller cities is not surprising.

THE EVOLUTION OF ZOO PROFESSIONALS

The changes in zoo animal collections are reflected in changes in the sort of people who are responsible for them. Animal keeping is now considered a skilled profession. Although there were exceptions among a few large zoos, this was not generally the case before the 1960s. Education was not considered a priority for zookeepers, and there was a degree of stigma attached to such a career choice. Today, zookeepers in the United States and Europe often find themselves the center of envy, and are frequently asked by members of the public how they attained so privileged a position. A keeper's relatives and educators are far less likely to react with disapproval or dismay than they might have done in earlier decades. As with changes in zoos themselves, this is in part due to increasing public environmental awareness, as well as other societal changes occurring from the 1960s onwards. As late as 1979, only 12.5% of American zookeepers held four-year degrees (Steenberg 1979). This figure rose to 82% by 2005 (Good 2010; Thompson and Bunderson 2005). This corresponded with a great demographic shift in the profession from a predominately male workforce to one in which 72.6% of keepers were women.

Keepers in the 21st century are expected to be far more knowledgeable than in previous times regarding the natural history of the animals they work with. It is understood that their responsibilities include the ability to interpret what they observe and record and to translate that knowledge to the things they do. In a time when a significant portion of the keeper's day often consists of devising and implementing enrichment, it is amusing to note that the proponents of standardized animal diets promoted them partially on the merits of relieving "the keeper of a lot of work" (Hediger 1969, 140)—with the result, as Hediger observed, that "the contact between man and animal, which is regarded as so important from the standpoint of zoo biology is thus reduced to a minimum." While in some cases animal management techniques have reduced the amount of physical contact between keepers and animals from previous levels, a keeper's awareness and comprehension of the animals under his or her care is often at a higher level than previously expected or thought possible.

A result of the aforementioned change in zoo collection policies is the management of animals inter-institutionally. In general, what might be done with the managed species in a given institution's collection is no longer primarily up to that facility's administration to decide, but is instead subject to policies agreed upon by committees organized from many such facilities. As zoo animal management has become an increasingly cooperative effort, opportunities for keeper involvement have increased dramatically. It is now commonplace for keepers to manage studbooks and participate in Taxon Advisory Groups, regional collection plans, and similar activities.

While keepers everywhere are increasingly expected to be

skilled professionals, there are still some regional differences in how this translates to management. In Europe, curators and directors are expected to hold advanced degrees and to have conducted research, and they often enter management positions directly from universities. While animal keepers in Europe often receive higher salaries than their American counterparts, they traditionally have been less likely to attain curatorial or directorial positions. While there was a trend in the 1970s to appoint US curatorial staff from academic backgrounds, this did not become standard practice; as of the first decade of the 21st century, most American curators have had experience as keepers, and this is expected for applicants to many advertised positions.

Although many American zoo directors have advanced from curatorial levels, there has not been as much of an emphasis on their possessing an authoritative knowledge of zoology as there has been in Europe. American directors may not have as much to do with direct animal management as their European counterparts, especially if the specimens in question are not "charismatic megavertebrates." European directors are more likely to take an interest in small freshwater fishes, rodents, and soft-billed birds whose management is often left to departmental specialists in American collections. They may be actively involved in programs for such animals, especially if, in the course of the daily rounds, they observe and appraise the daily routines in each area of their zoo.

Director's daily rounds, traditional especially in German zoos, are not standard practice in American collections. This distinction was noted long ago (Jones 1968). Visitors to Europe may also be startled to find zoo directors there writing guidebooks and annual reports, and authoring or coauthoring scientific papers. In America this is more likely to be an activity of curators. Furthermore, in America, articles and papers written or coauthored by keepers have become commonplace, to the point that the American Association of Zoo Keepers' (AAZK) journal *Animal Keepers' Forum* is a monthly publication. Publications by keepers appear in a wide array of other journals as well. Although keepers in Europe might be less likely to write papers, high standards are set for these keepers, and their training is often more formal and organized, resembling apprenticeship.

International differences in management style aside, the care of zoo animals has reached a level of sophistication and professionalism far beyond the way things were done several decades ago. The expectations of people who work in zoos, as well as of the people who visit them, are entirely different as well. One result is the routine collection of organized data. As it is collected and preserved it becomes zoo history to be studied, judged, applied, and hopefully appreciated.

REFERENCES

Anonymous. 2011. Colo (gorilla). Accessed at http://en.wikipedia .org/wiki/Colo_(gorilla).

Barnum, Phineas T. 1871. *Struggles and Triumphs; or, Forty Year's Recollections.* New York: American News Company.

Bedford, Hastings, Duke of. 1949. *The Years of Transition.* London: Andrew Dakars, Ltd.

Belozerskaya, Marina. 2006. *The Medici Giraffe, and Other Tales of Exotic Animals and Power.* New York: Little, Brown and Company.

Bosley, Lydia F. 2011. *International Studbook for Eastern/Mountain Bongo (Tragelaphus euryceros isaaci), Year 2009 Edition,* Vol. 24.

Bostock, Stephen. 1993. *Zoos and Animal Rights: The Ethics of Keeping Animals.* London: Routledge.

Bouman, D. T., and J. G. Bouman. 1994. "The History of Przewalski's Horse." In L. Boyd and D. A. Houpt, eds., *Przewalski's Horse: The History and Biology of an Endangered Species.* Albany: State University of New York Press, 5–38.

Bridges, William. 1974. *Gathering of Animals: An Unconventional History of the New York Zoological Society.* New York: Harper and Row.

Buffon, Comte de. 1783. *Histoire naturelle des oiseaux, tome 7.* Paris: Imprimerie Royale.

Clutton-Brock, Juliet. 1987. *A Natural History of Domesticated Mammals.* Cambridge: Cambridge University Press.

Cokinos, Christopher. 2000. *Hope is the Thing with Feathers: A Personal Chronicle of Vanished Birds.* New York: Jeremy P. Tarcher / Putnam.

Conway, William, G. 1966. "A New Exhibit for Water-Birds at the New York Zoological Park." *International Zoo Yearbook* 6:131–34.

———. 1973. "How to Exhibit a Bullfrog: A Bed-time Story for Zoo Men." *International Zoo Yearbook* 13:221–26.

Crandall, Lee S. 1915. "Breeding Results in the Bird Department, 1915." *Bulletin of the New York Zoological Society* 18:1270–71.

———. 1917. "Breeding Birds." *Bulletin of the New York Zoological Society* 20:1442–43.

———. 1964. *The Management of Wild Mammals in Captivity.* Chicago: University of Chicago Press.

Delacour, Jean T. 1943. "A Collection of Birds from Costa Rica." *Avicultural Magazine* Series 5, 8:29–32.

———. 1945. "Decorative Aviaries in the New York Zoo." *Avicultural Magazine* Series 5, 10:57–58.

———. 1977. *The Pheasants of the World, Second Edition.* Surrey: Spur Publications.

Diamond, Jared. 2005. *Guns, Germs, and Steel.* New York: W. W. Norton.

Duran, Fray Diego. 1994. *The History of the Indies of New Spain.* Doris Heyden, trans. Norman: University of Oklahoma Press.

Ehrlinger, David. 1993. *The Cincinnati Zoo and Botanical Garden: From Past to Present.* Cincinnati: Cincinnati Zoo and Botanical Gardens.

Fiedler, Walter, and Christine Wessley. 1976. *Schoenbrunn Zoo: History and Problems.* Vienna: Oesterreichs Wissenschaft.

Fisher, James. 1967. *Zoos of the World.* Garden City, NY: Natural History Press.

Foster, Karen Polinger. 1999. "The Earliest Zoos and Gardens." *Scientific American* 281 (no.1): 64–71.

Fulch, Anna. 1992. "Family Casuariidae (Cassowaries)." In Josep del Hoyo, Andrew Elliott, and Jordi Sargatal, eds., *Handbook of the Birds of the World, Volume 1.* Barcelona: Lynx Edicions, 90–97.

Fuller, Errol. 2002. *Dodo.* New York: Universe.

Gaedemann, Claus. 1975. "Heini Hediger of the Zurich Zoo." In *Our Magnificent Wildlife: How to Enjoy and Preserve It.* Pleasantville, NY: Reader's Digest, 210–11.

George, Jean. 1975. "Wonder World of Birds." In *Our Magnificent Wildlife: How to Enjoy and Preserve It.* Pleasantville, NY: Reader's Digest, 218–21.

Good, Shane. 2010. "The Future of Zookeeping and the Challenges Ahead." *Animal Keeper's Forum* 37:14–21.

Hahn, Emily. 1967. *Animal Gardens.* Garden City, NY: Doubleday.

Hancocks, David 2001. *A Different Nature: The Paradoxical World of Zoos and Their Uncertain Future.* Berkeley: University of California Press.

Harnick, Sheldon, Jerry Bock, and Joseph Stein. 1964. *Fiddler on the Roof.* New York: Crown Publishing Group.

Hediger, Heini. 1950. *Wild Animal in Captivity: An Outline of the Biology of Zoological Gardens.* London: Butterworth.

———. 1955. *Studies of the Psychology and Behavior of Captive Animals in Zoos and Circuses.* London: Butterworth.

———. 1969. *Man and Animal in the Zoo: Zoo Biology.* New York: Delacorte Press.

Hornaday, William T. 1899. Preface to *Taxidermy and Zoological Collecting, Seventh Edition.* New York: Charles Scribner's Sons, vii–viii.

———. 1915. "Gorillas, Past and Present." *Bulletin of the New York Zoological Society* 18:1181–85.

———. 1925. *A Wild-Animal Round-Up.* New York: Charles Scribner's Sons.

Houlihan, Patrick F. 1986. *The Birds of Ancient Egypt.* Warminster: Aris and Phillips.

———. 1996. *The Animal World of the Pharaohs.* London: Thames and Hussdon.

House, H. Bradford, and James G. Doherty. 1975. "The World of Darkness at the New York Zoological Park." *International Zoo Yearbook* 15:31–34.

Jones, Marvin L. 1968. "North American Zoological Gardens and Parks." In Rosl Kirchshofer, ed., *The World of Zoos.* New York: Viking Press. 225–30.

Kastner, Joseph. 1977. *A Species of Eternity.* New York: Alfred A. Knopf.

Kawata, Ken. 2001. "Zoological Gardens of Japan." In Vernon Kisling, ed., *Zoo and Aquarium History.* Boca Raton: CRC, 295–329.

———. 2003. *New York's Biggest Little Zoo: A History of the Staten Island Zoo.* Dubuque, IA: Kendall/Hunt.

Keeling, Clinton H. 2001. "Zoological Gardens of Great Britain." In Vernon Kisling, ed., *Zoo and Aquarium History.* Boca Raton: CRC, 49–74.

Kisling, Vernon. 2001. "Zoological Gardens of the United States." In Vernon Kisling, ed., *Zoo and Aquarium History.* Boca Raton: CRC, 147–80.

Kloes, Heinz-Georg. 1969. *Von der Menagerie zum Tierparadies: 125 Jahre Zoo Berlin.* Berlin: Haude & Spener.

Ley, Willy. 1968. *Dawn of Zoology.* Englewood Cliffs, NJ: Prentice-Hall.

Lindholm, Josef. 1988. "Captain Delacour at the Bronx (1941–1947)." *Avicultural Magazine* 94:31–56.

———. 1995. "Lories May Be Hazardous . . ." *AFA Watchbird* 22 (no.5): 22–27.

———. 2005. "Softbill Propagation in U.S. Zoos: A Thirty Year Perspective. Part I—The '70s." *AFA Watchbird* 32 (no.3): 27–38.

———. 2005. "Softbill Propagation in U.S. Zoos: A Thirty Year Perspective. Part II—The '80s." *AFA Watchbird* 32 (no.4): 27–41.

———. 2007. "Gulls." In Glen Holland, ed., *Encyclopedia of Aviculture.* Blaine, WA: Hancock House, 503–9.

Loisel, Gustave. 1912. *Histoire des menageries de l'Antiquite a nos jours I (Antiquite, Moyen Age, Renaissance).* Paris: Octave Doin et Fils.

———. 1912. *Histoire des menageries de l'Antiquite a nos jours II (Temps Modernes—XVIIe et XVIIIe siecles).* Paris: Octave Doin et Fils.

Lorenz, Konrad. 1954. *Man Meets Dog.* London: Methuen.

———. 1971. "Part and Parcel in Animal and Human Societies." In *Studies in Animal and Human Behavior,* Vol.1I. Cambridge, MA: Harvard University Press, 115–95.

Munro, Hector H. 1914. "A Defensive Diamond." In Hector H. Munro, *Beasts and Superbeasts.* London: Bodley Head.

Murphy, James B. 2007. *Herpetological History of the Zoo and Aquarium World.* Malabar, FL: Krieger Publishing Company.

Osborne, Michael A. 1996. "Zoos in the Family: The Geffroy Saint-Hilaire Clan and the Three Zoos of Paris." In R. J. Hoage and William A. Deiss, eds., *New Worlds, New Animals: From Menagerie to Zoological Park in the Nineteenth Century.* Baltimore: Johns Hopkins University Press, 33–42.

Reichenbach, Herman. 1996. "A Tale of Two Zoos: The Hamburg Zoological Garden and Carl Hagenbeck's Tierpark." In R. J. Hoage and William A. Deiss, eds., *New Worlds, New Animals: From Menagerie to Zoological Park in the Nineteenth Century.* Baltimore: Johns Hopkins University Press, 51–62.

Ricciuti, Edward R. 1972. "The World of Birds." *Animal Kingdom* 75 (no.3): 1–32.

Robbins, Louise E. 2002. *Elephant Slaves and Pampered Parrots: Exotic Animals in Eighteenth-Century Paris.* Baltimore: Johns Hopkins University Press.

Rothfels, Nigel. 2002. *Savages and Beasts The Birth of the Modern Zoo.* Baltimore: Johns Hopkins University Press.

Schafer, Edward H. 1963. *The Golden Peaches of Samarkand: A Study of T'ang Exotics.* Berkeley: University of California Press.

Sclater, Philip L. 1896. *List of the Vertebrated Animals Now or Lately Living in the Gardens of the Zoological Society of London, Eighth Edition.* London: Longman, Green and Co.

Sellers, Charles C. 1979. *Mr. Peale's Museum.* New York: W. W. Norton.

Solski, Leszek. 2001. "Zoological Gardens of Central-Eastern Europe and Russia." In Vernon Kisling, ed., *Zoo and Aquarium History.* Boca Raton: CRC, 117–46.

Steenberg, Judie. 1979. Animal Keeper Survey. Report to the American Association of Zookeepers, Inc.

Strehlow, Harro. 1996. "Zoos and Aquariums of Berlin." In R. J. Hoage and William A. Deiss, eds., *New Worlds, New Animals: From Menagerie to Zoological Park in the Nineteenth Century.* Baltimore: Johns Hopkins University Press, 63–72.

———. 2001. "Zoological Gardens of Western Europe." In Vernon Kisling, ed., *Zoo and Aquarium History.* Boca Raton: CRC, 75–116.

Svanberg, Ingvar. 2007. *"Dera Mistande Ror Mig sa Hierteligen": Linne och hans Sallskapsdjur.* Lund, Sweden: Grahns Tryckeri AB.

Thompson, J., and S. Bunderson. 2005. AAZK Survey of Animal Care Professionals. Report to the American Association of Zookeepers, Inc.

Toynbee, J. M. C. 1973. *Animals in Roman Life and Art.* Ithaca, NY: Cornell University Press.

Ulmer, Frederick A. 1966. "Philadelphia Zoo's Rare Mammal House." *International Zoo Yearbook* 6:119–20.

Varro, Marcus Terentius. 1934. "On Agriculture." In William D. Hooper and Harrison B. Ash, *Cato and Varro on Agriculture.* Cambridge, MA: Harvard University Press, 159–529.

Vevers, Gwynne. 1976. *London's Zoo.* London: Bodley Head.

Wegeforth, Harry M., and Neil Morgan. 1953. *It Began with a Roar!* San Diego: Zoological Society of San Diego.

Weigl, Richard. 2005. "Longevity of Mammals in Captivity: From the Living Collections of the World." *Kleine Senckenberg–Reihe* 48.

Wood, Casey A., and F. Marjorie Fyfe. 1943. *The Art of Falconry by Frederick II of Hohenstaufen.* Stanford, CA: Stanford University Press.

Zeuner, Frederick E. 1963. *A History of Domesticated Animals.* New York: Harper & Row.

Zuckerman, Lord. 1980. "The Rise of Zoos and Zoological Societies." In Lord Zuckerman, ed., *Great Zoos of the World.* Boulder, CO: Westview Press, 3–26.

5

Today's Zoos

Gordon McGregor Reid

INTRODUCTION

This chapter provides basic information about contemporary zoological gardens ("today's zoos") in relation to animal keeping. It covers the nature, purpose, standards, and general direction of zoos (the "vision," "mission," and "values") and the opportunities for external collaboration, especially in support of conservation and science. After studying this chapter the reader will understand, in the above context:

- vision, mission, and values
- conservation
- international partnerships
- science

Zookeeping is being professionalized across the globe. As employees (or sometimes as volunteers), it is important that keepers clearly understand the general framework within which they operate, the overall staff terms and conditions of service, and their own personal role, contract of engagement, and opportunities. In order to do a worthwhile professional job and build a good career, keepers need to understand the history, purpose, aims, structure, and organization of the zoo or aquarium where they work (both termed "zoos" from now on). Most zoos major in animal welfare, public recreation (or "entertainment"), education, conservation, and science. This chapter focuses on the last two topics, which closely relate to the mission, vision, and values of the great majority of modern zoos. It is crucial that the keeper generally understands why these aspects are important, what is involved in such areas, and how to play their part. All of the foregoing needs to be set in a strict ethical context (Hutchins, Dresser, and Wemmer 1995).

Typically there will be quite a lot of change in any five-year zoo planning period, so it is necessary to keep pace with the changes. Conservation and science are particularly fast-moving fields of endeavor. For continuous professional development (CPD), the zookeeper needs to keep up to date in their area of specialization and also with general developments. Much can be gained by talking to more experienced zoo colleagues, using the zoo library (if there is one), attending training courses run by the zoo (or another zoo or educational provider), and joining in external workshops and conferences organized by zoo and other organizations nationally and internationally. Wherever possible, keeper study is recommended to gain formal professional qualifications which involve conservation and science. This maximizes general benefit and personal fulfilment.

Zoos and aquariums may combine several different purposes or roles encompassing animal welfare, education, conservation, and science; they are organized in many different ways to achieve these ends. Zoos may be small or large, simple or complex. They may be publicly owned city or state bodies, independent private enterprises run commercially for profit, or not-for-profit charities. In reality, all zoos that depend on income from visitors have to be run in a businesslike way. The main difference between commercial and charitable organizations is in the application of the profits or financial surplus that is generated. In any event, both commercial and charitable zoos will typically raise money for and donate funds to worthwhile animal projects in the habitats and countries where the animals are indigenous.

Career opportunities in conservation, welfare, education, and science will be determined in large part by the nature of the animal collection. Today's zoos may be general or specialized in operation, and so too the keepers. The zoos are also of different sizes but are usually open to the visiting public, which is often charged an entry fee to help with the running costs. The term "zoo" covers a host of different facilities, with different animal collections and animal management requirements (Hosey et al. 2009). There will consequently be different zoo aims and objectives depending on the organization's vision, mission, and values. Zoo aims are also defined by the available budget, the priorities and scale, and the nature of the organization. Zoos range from traditional general collections to safari parks, specialized rain

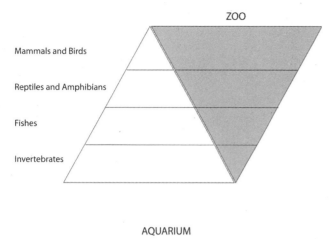

ZOO

Mammals and Birds

Reptiles and Amphibians

Fishes

Invertebrates

AQUARIUM

Figure 5.1. Diagram of key traditional differences between a zoo and an aquarium. Illustration by Gordon McGregor Reid.

forest displays (tropicariums), and even facilities dealing with microscopic life (micrariums). There are deer, bear, and bird parks; sanctuaries for elephants, monkeys, chimpanzees, orangutans, otters, and seals; serpentariums for snakes and lizards; bug houses, insectariums, and butterfly houses; and aquariums and dolphinariums.

The largest zoos and aquariums (with 500,000-plus visitors each year) may incorporate many of the above animal collection elements, but in other cases they are free-standing facilities. The biggest distinction in the "industry" is between traditional zoos (which mainly house terrestrial animals tended by keepers) and aquariums (which mainly house aquatic life tended by aquarists, who will for convenience also be referred to here as "keepers,"). There can be much overlap, but the key distinguishing features between a zoo and aquarium are the relative proportions of the different major classes of animals (taxa) that they hold (figure 5.1).

Some zoos have associated but separate foundations or institutes which specialize in conservation and scientific research and which have their own specialized strategy: for example, the Institute of Zoology, Zoological Society of London (ZSL, UK); the Institute for Zoo and Wildlife Medicine, Berlin (IZW, Germany); and the Center for Research on Conservation (CRC, Royal Zoological Society of Antwerp, Belgium). All of this tremendous variety means that there are many exciting and different keeper career opportunities.

Whatever their animal content and organizational structure, zoos usually do not work well in isolation and often need to collaborate with other zoos, zoo associations, and wildlife agencies to maximize progress in conservation, education, and science. This chapter deals with the collaborative partnership aspect. Much work in the foregoing areas may occur inside the zoo perimeter fence. When the work is outside of the fence, this is called outreach. It can take place locally, nationally, or overseas in the habitat countries from which many of the zoos' animals (and often plants) come—although this origin may be many breeding generations ago. Professional keeping often requires national, regional, and international communications and engagement. Being fully aware of the

external context for the zoo (local, national, international) is important. With the aid of core zoo resources, modern electronic communications, and information technology, much understanding and progress can be achieved—especially through professional zoo and wildlife networks and through scanning the World Wide Web.

VISION, MISSION, AND VALUES

Most modern zoos publish a "vision" and "mission" statement. In addition to having a vision and mission, zoos sometimes also use a brief phrase or "strap-line" to capture the essence of what they are about and to help market the organization to the public and the outside world. Examples include "Animals always" (Saint Louis Zoo, USA) and "Living conservation" (London Zoo [ZSL], UK). The vision statement briefly describes the higher institutional aims or aspirations—the good and worthwhile things that the organization is seeking to pursue in the long term. This is ultimately for the wider benefit of the planet, animals, and zoo stakeholders, including the visiting public. For example, the vision of Chester Zoo (the North of England Zoological Society, UK) is "a diverse, thriving and sustainable natural world." This is an inspiring idea which sets the tone and direction of the organization, but it will not be an absolute overall reality anytime soon. That is to say, a vision need not be "SMART": specific, measurable, achievable, realistic, and time-restricted (i.e., deliverable within a set period).

By contrast, the "mission" statement of a zoo describes the day-to-day function or aim of the organization, what it is established for, and what it sets out to do. It does need to be SMART if it is to effectively serve its purpose. For example the mission of Chester Zoo is "to be a major force in conserving biodiversity, world-wide." This grand and worthwhile intention is translated into a strategy and concrete actions (the strategy implementation plan, or SIP). In short, any zoo (and indeed any organization) needs to know what it is set up to do, and this must be spelled out in a strategy (the long-term goals and the means of achieving them). The SIP needs to be specific on practical plans to be implemented in any one year. This means determining what resources are needed, and who is going to do what by when. To be businesslike and purposeful, all of this needs to be realistic and achievable within a set time frame. Understanding the nature of the zoo collection, management purpose and structure, strategy development, and strategy implementation (involving individuals, teams, and tasks) is crucial to both business and mission.

Most zoos (either as public organizations or as companies) publish an annual report. In this, the zoo openly declares its vision, mission, organizational structure, governance and management, staffing, and business plan. The report will highlight performance (challenges, successes, and key indicators) in mission and business. This includes reports on conservation, education, science and veterinary health, and the animal stock lists (inventory). The lists usually include a multicolumn animal accounting system with the number of species and individuals, births and deaths, and conservation and welfare status. Not all keepers will typically read such a report, but they will find it very helpful if they do. The reports are designed to provide a full, all-round picture of the entire zoo's operation in any one year, as well as its plans going forward. Annual reports are often available to staff online, as personal hard copies, or via the zoo library where one exists. Comparison of performance across the years can also be very instructive in giving a sense of the history and development of the zoo, its potential, and its greatest achievements in conservation and science.

Today, many zoos work with staff to mentor individuals and build teams; and they develop a statement of shared "values,"—that is to say, the core operating principles for everybody. The values may be ethical (such as striving for "the best animal welfare"); scientific ("sound science"); related to education, marketing, and teamwork ("excellent communication"); or welfare- and business-oriented ("safe working environment," "effective use of resources"). Individual staff "behaviors" may be associated with these values. These are the normal personal working standards and attitudes expected from everyone such as: "hard work," "openness," "accountability," and "respect for the views and sensitivities of colleagues."

The mission is guided by the overall organizational values and supported by the individual roles of staff, as well as by the standards of behavior expected of them, whatever their level of seniority in the zoo. When things go wrong in a zoo (as they sometimes do!), it is often because the vision, mission, values, ethics, behaviors, and strategy are not clear from the start and are not aligned. When "things are not going according to plan," it is often because there *is* no plan, strategy, mission, or vision, or at least not an understood, agreed upon, and well-integrated one. Here, the board, management, and staff need to work hard together to resolve such issues. When things go right in a zoo it is usually because the vision, mission, values, and behaviors are shared and well understood, and work in concert. Everybody believes that what they are doing is worthwhile, and all have a sense of common purpose. To achieve that purpose they work well together as a team to complete important tasks. This feeling of the work being worthwhile and emotionally engaging often also communicates itself to zoo visitors and to potential financial donors or sponsors. Ultimately, animal food, staff wages, and exhibit and project funds depend on external sources of income. Public engagement creates a "virtuous economic cycle" of investment and conservation achievement. This in turn generates more visitors, donors, and sponsors and more jobs and valuable tasks for keepers and other zoo staff.

In summary, all of the above ideals can be conveniently

Figure 5.2. Vision, mission, and values. Illustration by Gordon McGregor Reid.

expressed in a triangle (figure 5.2). The vision directs the zoo (it's the "nose cone of the rocket ship"). The mission is what the zoo exists to do in line with this vision. The mission usually concerns animals, the natural environment, conservation, education, and science in some combination, but with differing emphases. However there may be essentially recreational objectives in profit-generating zoos. Having fun in a zoo can be a good thing, especially if it is educational. Discovery and learning through fun helps the paying visitor and children to be more sympathetic towards animals, science, conservation, and sustainable environments. It helps encourage repeat visits of the key audiences to be influenced, especially young families.

CONSERVATION

Today, conservation is usually the central mission of any one zoo or aquarium (Fa et al. 2010). Definitions vary, but "conservation" is probably best summed up as *actions that substantially enhance the survival of species and habitats, whether conducted in nature or outside the natural habitat* (WAZA 2005, 2010). There are other traditional dictionary definitions of conservation which focus on the more restricted *act of preserving, guarding, or protecting.* Today it is widely realized that the act of creating a nature reserve or national park with a perimeter fence is not in itself sufficient to guarantee the protection of animals and plants. Conservation essentially concerns the survival of wild species and habitats in the long term. Education and marketing studies indicate that there can be confusion (at least in the minds of zoo visitors and the general public) between animal welfare and conservation (see also "sustainability," below). While excellent animal welfare is an essential condition or precursor for conservation, it is not a synonym for conservation. Welfare mainly covers individual animals or small populations and their health, husbandry, nutrition, and veterinary care in the short to medium term (Hosey et al. 2009). An often-used alternative Latin term for conservation "in nature" is "in situ." Conversely, "ex situ" means "outside the natural habitat." The term "ex situ" can be validly applied to conservation work in zoos and aquariums, such as managed breeding programs to ensure species survival (see below). Nonetheless, many zoos now work both in situ and ex situ. Today, the distinction between the two categories is becoming blurred. Many would argue that what zoo and other conservationists should really be focusing on is the level of intensity of animal management and integrated activity, rather than the traditional concepts of in situ and ex situ.

The idea of zoos becoming engaged in the conservation of natural animal and plant diversity (i.e., biological diversity or biodiversity) developed from small beginnings in the early 1960s at a meeting of the International Union of Directors of Zoological Gardens (IUDZG). The IUDZG later became the World Association of Zoos and Aquariums (WAZA; see below). A true general awareness of what was required first came with the publication of the *World Zoo Conservation Strategy* (IUDZG/CBSG 1993). This landmark document still contains much of relevance, but is now superseded by two other publications: *Building a Future for Wildlife: The World Zoo and Aquarium Conservation Strategy* (WAZA 2005) and *Turning the Tide: The World Aquarium Conservation Strategy*

(WAZA 2010). These central documents focus on conservation issues such as impending species extinctions and what zoos can do about them. The documents have a far stronger focus on field conservation than did the original strategy. They also properly cover the threatening processes such as habitat destruction, overharvesting, poaching, pollution, the introduction of alien species, and human-induced climate change.

Encouraged by the global WAZA strategy, keepers today, as a vital part of their role, work to conserve biodiversity. The term "biodiversity" is actually very broad and is used to cover the genetic variation (chromosomes, genes, and DNA) that characterizes the uniqueness of species, populations, and individual animals. Biodiversity also includes the immense variety of microorganisms, ranging to local habitats and global ecosystems. Biodiversity conservation involving keepers is typically done in the zoo but can also be done in the field (see below). Zoo work takes place through the animal collections, by helping to communicate key conservation education messages to visitors, and also through science projects (see below). Keepers tend to be "species- or taxon-oriented" in their husbandry, conservation, and education work. That is to say that each keeper may have a preferred animal group from among mammals, birds, reptiles, amphibians, fishes, and invertebrates. Among these there will be favorite smaller groups (families or genera) that they like working with, such as chimpanzees, parrots, pythons, poison dart frogs, cichlid fishes, or corals. Deeper within these categories, they may focus on one particular species from a restricted geographical area, such as the Asian elephant (*Elephas maximus*), black rhino (*Diceros bicornis*), Grevy's zebra (*Equus grevyi*), Cuban Amazon parrot (*Amazona leucocephala*), or Cape seahorse (*Hippocampus capensis*)—all of which have formal binomial or two-word Latinized scientific names.

Species concepts vary and today are the subject of much debate in zoo and other circles. This is because of the importance of determining exactly which species and contained populations are being managed in zoos (Reid 2010). Conservation breeding programs are designed to establish representative genetic lines of populations and species. There is the long-term view that these populations serve as an "insurance policy" against extinction in the wild. Properly managed and in appropriate circumstances, they offer the option of reintroduction back to the wild. Hence it is vital to establish the species identity, genetic makeup, and geographical origin of the founding population of any one species. This is in order to maintain genetic integrity in the long term, avoiding hybridization.

According to the "biospecies" concept, species comprise animal populations that actually or potentially interbreed in nature. Nevertheless, for practically all formally classified animal groups (i.e., taxa) that exist in zoos, there are unresolved questions over the identity and precise geographical origin (provenance) of species and populations (Reid 2010). There is an abundance of zoo hybrids between apparently separate species. Hence we have some giraffes, tigers, jaguars, and chimpanzees that are genetically mixed and do not closely represent the wild species in nature. This is especially true in the case of taxonomically distinct geographical populations of species—sometimes called subspecies. These are formally recognized in trinomial or three-word scientific names such as Sumatran tiger (*Panthera tigris sumatrae*),

Rothschild's giraffe (*Giraffa camelopardalis rothschildi*), West African chimpanzee (*Pan troglodytes verus*) and Panama Amazon parrot (*Amazona ochrocephala panamensis*). Clearly the conservation value of such animals is minimized if they are not "purebred." They could not reasonably be used for reintroduction programs; nor, indeed, could they tell us much of conservation value about species biology in nature.

Keepers will often be engaged in collaborative breeding efforts, organized ex situ to ensure species survival. These take place within higher-level endangered species programs organized by national or regional zoo associations. The names of programs across the regions vary, but the basic operation is the same. In the European (EAZA) region they are called European Endangered Species Programmes or EEPs, while in the North American (AZA) region they are called Species Survival Plans or SSPs. These programs represent the most intensive kind of national and international population management. The supervising committee (which is selected from several zoos) coordinates breeding activities across the region. This includes recommending which animals should breed or not breed (to maintain genetic vigor or prevent inbreeding problems, and which animals should move to other collections. The management of the conservation breeding of individual taxa (e.g., elephants, rhinos, parrots, seahorses, and partula snails) is conducted within specialized Taxon Advisory Groups (TAGs). The TAGs report back, via their committee chairs, to the higher-level programs committee. Studbooks (detailed breeding records) are organized for particular species and studbook keepers keep track of these and report back to the TAG. Quite often, TAG chairs and studbook keepers are keepers whose zoo has kindly released them for this voluntary task. There have been major successes in conservation breeding. However, a contemporary global issue for collaborative breeding programs is that the majority of species they manage are not genetically viable in the long term. There needs to be a greatly increased zoo and wildlife community engagement to resolve this huge issue (see WAZA, below).

At home or abroad, keepers may work in the field on conservation programs such as reintroduction of animals back to the wild, as in the program for Californian condor (*Gymnogyps californianus*). There may be an ex situ component to this—for instance, the development of rearing and feeding techniques. This work may have an application in reintroduction, with "soft" releases in which the animals receive supplementary feeding in the wild until they become established. Biodiversity "hotspots" are often the focus of outreach activity. These hotspots are geographical areas of species richness, with high numbers that are endemic (i.e., which occur nowhere else in nature). For example, Madagascar is a hotspot for lemurs, Papua New Guinea for birds of paradise, Australia for marsupial mammals, and the Caribbean Sea for certain species of coral reef fishes and invertebrates. While the tropics contain the most remarkable hotspots, there is still much biodiversity in the temperate zone. There is important conservation work to be done in supposedly impoverished areas of biodiversity such as deserts, polar zones, and in the ocean depths, and some zoo animal collections reflect this, such as the Arizona-Sonora Desert Zoo and the Palm Desert Zoo, both in the United States.

Case Study 5.1. Assam Hatti Project: Reducing Conflict between Humans and Elephants in the Wild

The number of Asian elephants (*Elephas maximus*) in the wild has been declining dramatically throughout their natural range from India to Borneo. There are about 50% fewer elephants now than there were in 1986. Current IUCN estimates place the entire population as under 50,000 and it could be fewer than 35,000. The destruction of forest habitat means that elephant populations are becoming genetically isolated and also exposed to more frequent contact with people. This causes human-elephant conflict (HEC), resulting in negative outcomes for people, animals, or both. With less natural food available, hungry elephants will raid crops and damage houses, sometimes injuring or killing people in the process. The people who remain are deprived of accommodation and food to eat or take to market. Hence livelihoods and the local economy suffer, driving people into poverty. Understandably, people respond by shooting at elephants, injuring or killing them and sometimes leaving their calves as orphans, an additional welfare problem. One of the few remaining substantial single populations of elephants is in Assam, northeastern India, where there may be about 2,000 animals. Here Chester Zoo, with the assistance of a Darwin Initiative Grant, works with a local conservation organization called Ecosystems India on HEC mitigation. *Haathi* means elephant in the Hindi language, and so the partnership project was called Assam Haathi. It has involved tracking elephant movements using geographical information systems (GIS) and village-level surveys of elephant impact, all with keeper involvement. One very interesting finding was that people who brew beer in their houses are at greater risk of an elephant attack. Evidently the smell of alcohol resembles that of fermenting fruit in the forest, and can attract elephants from a great distance. The elephants break down the houses and drink the beer, and a drunken elephant is far more difficult to manage than a sober one! Simple advice has thus been given to villagers to brew beer somewhere other than in their own houses. There are many other low-tech means of reducing HEC, meaning that both humans and animals are placed in a better position in terms of welfare and conservation. Alternative crops, such as elephant-repelling peppers, and less rice-dependent lifestyles are among the practical ideas being implemented with the aid of training via local farmers cooperatives.

Figure 5.3. Reducing human-elephant conflict in Assam (Haathi). Courtesy of North England Zoological Society.

healthy environment and natural biodiversity" (WAZA 2010). Sometimes the term is used to refer to the economy and operation of a zoo—a financially stable organization that demonstrates an ethical concern for the environment. This might involve the promotion of economically sound green practices. Examples here are reducing pollution, recycling waste, and minimizing production of greenhouse gases that overheat the planet. Certainly, if humans live truly sustainable "carbon-neutral" lives, they can do much to help the environment and associated conservation efforts. Today's zoos will often act as a vehicle to demonstrate and communicate excellent practice in sustainability. Some zoos have "green teams" that include keepers who work together closely on issues of sustainability (see chapter 56). They suggest and implement better ways of operating, such as car sharing or cycling to work, in order to reduce both fuel costs and the release of damaging emissions into the environment. Around the world, leading zoos with green teams have been able to obtain certification to show that they meet the rigorous international environmental management standard ISO 14001. Increasingly, zoo conferences are organized to be certified as carbon-neutral.

INTERNATIONAL PARTNERSHIPS

Today's zoos are engaged globally in the pursuit of animal welfare and conservation. Zoos and zoo organizations respond to policy and program direction advocated by large global organizations such as the United Nations Environment Program (UNEP) and the International Union for the Conservation of Nature (IUCN). They often work in partnership with them (see case study 5.2). The United Nations, an intergovernmental organization, operates UNEP (UNEP 1995, 2011 a, b, c), established in 1972 with a mission "to provide leadership and encourage partnership in caring for the environment by inspiring, informing, and enabling nations and peoples to improve their quality of life without compromising that of future generations." The strap-line (tagline) is "Environment for development," indicating six priority areas: climate change, disasters and conflicts, ecosystem manage-

Reducing human-animal conflict (HAC; e.g., between jaguars and cattle ranchers, or between elephants and farmers) is a new and important area for outreach. This often involves partnership with several different agencies (see below). Increasingly, zoo programs will help tackle wider conservation-associated issues in poverty alleviation and human development (see case study 5.1).

"Sustainability" in today's zoos is a term closely associated with conservation but not synonymous with it (Dickie 2009; Tlusty et al. 2013). Environmental sustainability in general is about "meeting present human resource needs without compromising the needs of future generations or sacrificing a

ment, environmental governance, harmful substances, and resource efficiency. Hence, UNEP is more about sustainability and human development than about conservation (UNEP 2011a, b, c).

While engaged with UNEP, zoos typically work more closely with the IUCN. Founded in 1948, this intergovernmental organization has a vision of "a just world that values and conserves nature." The IUCN mission is "to influence, encourage and assist societies throughout the world to conserve the integrity and diversity of nature and to ensure that any use of natural resources is equitable and ecologically sustainable." The IUCN promotes and supports priority activities in biodiversity conservation, human well-being, climate change mitigation, sustainable energy, and the development of a green global economy. All of these topics relate closely to the vision, mission, and strategic interests of zoos and zoo associations (see above and below). Many of them are IUCN institutional members. Altogether there are more than 1,000 member organizations in 140 countries including more than 200 government, and more than 800 nongovernment organizations.

The IUCN has a seat on the United Nations Council, and is itself governed by a council elected by member organizations every four years at the IUCN World Conservation Congress. The Congress is the world's largest organized gathering of conservationists, environmentalists, and politicians, typically with 10,000 delegates including representatives from WAZA (see below). The headquarters are in Gland, Switzerland, alongside the headquarters of several other global conservation organizations, recently including WAZA. There are six major IUCN "departments" or commissions managed by paid executive staff and supported by nearly 11,000 volunteers. The volunteers may be experts on particular animal taxa or in specialized themes such as law, education, protected areas, environmental economics, and ecosystem management.

The main IUCN departments that zoos work with are the Commission on Education and Communication (Whitehead 2011) and the Species Survival Commission (SSC), the latter of which has 7,500 volunteer members. Many SSC volunteers are scientists, including people from zoos. The SSC advises the IUCN on technical aspects of species conservation, and mobilizes action for those species that are threatened with extinction. It also organizes and publishes the Red List. This is the official world list of threatened species whose conservation status has been rigorously assessed through internationally accepted processes and criteria (Mace-Lande criteria; IUCN 2011). Much of the work is conducted through IUCN-SSC specialist groups. These may be thematic or concern particular animal and plant taxa, such as elephants, rhinos, crocodiles, amphibians, turtles, freshwater fishes (Reid et al. 2013), and cacti. Indeed, the taxonomic IUCN-SSC specialist groups broadly correspond with the taxon advisory groups in zoos. Linkages between TAGs and specialist groups often exist.

IUCN-SSC thematic groups that are especially zoo-relevant are the veterinary group, reintroduction group, and conservation breeding specialist group (CBSG). The CBSG in particular has a major involvement with zoos, many of which are paying members. Among other aspects, the CBSG advises on the genetic management of organized breeding programs and facilitates conservation and extinction risk as-

sessments, covering both ex situ and in situ aspects. Vortex is the computer software used in making some of these assessments. Useful published products are various Conservation Assessment Management Plans (CAMPs) and Population and Habitat Viability Analyses (PHVAs).

The World Association of Zoos and Aquariums (WAZA) is the unifying global organization for the world zoo and aquarium community (WAZA 2011). Its strap-line is "United for conservation." WAZA's mission is "to guide, encourage and support the zoos, aquariums, and like-minded organizations of the world in animal care and welfare, environmental education and global conservation." WAZA was originally established in 1935 as the International Union of Directors of Zoological Gardens (IUDZG) in Basel, Switzerland. In 1949, the IUDZG became a founding member of the IUCN (see above), and after 1995 it was transformed into the World Zoo Organization. In 2000 the WZO was renamed WAZA to clearly embrace aquariums and to work at the highest international level. Its current organizational structure is given in figure 5.4.

WAZA's institutional membership includes the world's lead zoos, aquariums, associations, and affiliate organizations. Collectively these institutions attract more than 700 million visitors annually. The regional zoo associations are members of WAZA, including North America (AZA), Australasia (ARAZPA), Latin America (ALPZA), Europe (EAZA), Africa (PAAZAB), South Asia (SAZARC), and Southeast Asia (SEAZA). These associations represent individual zoos directly or through the national associations. The WAZA membership also includes the International Zoo Educators (IZE) and the International Congress of Zookeepers (ICZ 2011), which itself draws on regional and national animal-keeping organizations for its members. The vision of the ICZ is "a global network of zookeepers with the highest standards of professional animal care contributing to a diverse and sustainable natural world where neither wild animals nor their habitats are in danger." The mission is that "the ICZ will build a worldwide network among zookeepers and other professionals in the field of wildlife care and conservation. This exchange of experience and knowledge will improve the

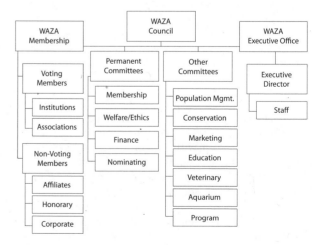

Figure 5.4. WAZA organizational structure. Courtesy of the World Association of Zoos and Aquariums.

professionalism of zookeepers for the benefit of the animals under their care and promote awareness and actions that will contribute to the preservation of wildlife everywhere." The ICZ works to a strategic plan (ICZ 2011).

All WAZA members are obliged to sign up to the organization's Code of Ethics and Animal Welfare and Ethical Guidelines for the Conduct of Research on Animals by Zoos and Aquariums (freely downloadable at http://www.waza.org). WAZA advocates in situ conservation, including through its strategy publications and "branded field projects." WAZA has a particular focus on the survival of animal species in the wild and would like to see zoos move into that area ever more effectively. It hosts a conservation database (originally developed by EAZA) which contains details on well over 400 field conservation projects involving more than 90 countries. WAZA also represents the zoo community in international conservation and environmental bodies such as UNEP and IUCN (see above). There is representation at conferences of the parties (CoPs) to global conservation and welfare conventions. These are legally binding international commitments covering, for example, the Convention on Biodiversity (CBD 2011), the Convention on Migratory Species (CMS), the Convention on Wetlands (RAMSAR), and the Convention on the Control of International Trade in Endangered Wild Animals (CITES). WAZA also helps organize and support collaborative international conservation campaigns such as "2008 Year of the Frog" (see case study 5.2).

Finally, WAZA helps build cooperative approaches to common needs and problematic or controversial issues. It shares information and advocates positions, including through its publications and official statements. For example, it issued a supportive "statement on behalf of the World Association of Zoos and Aquariums in reference to the recent conviction of staff of Zoo Magdeburg for the management euthanasia of three hybrid tigers" (freely downloadable at http://www.waza.org).

The regional associations between them manage some 1,000 other conservation breeding programs for species and subspecies (see Species Survival Plans and Taxon Advisory Groups, above). WAZA supports this ex situ process at a high level through their Committee for Population Management (CPM), which holds some 121 active international studbooks for 159 species and subspecies. It does so in collaboration with the regional and national associations, the Zoological Society of London, and the International Species Information System (ISIS). ISIS works particularly closely with WAZA and its institutional members. Established in 1973, ISIS is now a charitable (not-for-profit) membership network of more than 600 zoos on six continents. Members keep and share standardized and detailed electronic information on about two million zoo animal specimens in 10,000 or more taxa. They do this through the Zoological Information Management System (ZIMS), which is discussed in chapter 17.

There are many other global nongovernmental conservation organizations (NGOs), often with strong links to WAZA and the international zoo community. These include the World Conservation Monitoring Centre (WCMS), the Wildlife Conservation Society (WCS), the World Wildlife Fund (WWF), Conservation International (CI), Fauna and Flora International (FFI), and Wetlands International (WI).

In addition to international organizations, there is local and national involvement in animal care, welfare, conservation, and science. There are local and national regulatory bodies covering zoo legislation and animal welfare. National zoos and national zoo associations will likely assume the lead for provincial zoos in strategy development. There will be

Figure 5.5. Critically endangered Panamanian golden frog (*Atelopus zeteki*). Courtesy of the Amphibian Ark.

local and national conservation groups and government agencies. Working on conservation projects with cities and counties and with universities, veterinary schools, museums, and botanical gardens can be very productive.

SCIENCE

Science is *a body of knowledge based essentially on observation and experiment* (Reid et al. 2008). In today's zoos, its most important roles are in underpinning welfare and conservation work. It is a way of making practical, well-informed decisions in animal management based on established facts. Important science has been conducted in zoos worldwide since at least the 19th century. Many zoological societies have associated scientific institutions, such as those in Antwerp, Berlin, Delhi, London, Moscow, and New York. Scientific and cultural institutions such as natural history museums and botanical gardens will often incorporate live animal exhibits, and these facilities may be licensed as zoos, such as the Liverpool Museum in the United Kingdom, which houses a substantial aquarium and vivarium, and the Missouri Botanical Garden in the United States, which has a butterfly house and holds threatened species of frogs. Some aquariums are part of the public face of scientific research institutions, as in Genoa, Monaco, Naples, Plymouth, and Washington. Naturally, keepers are involved in managing such facilities.

Keepers are often in direct contact with animals, and are thus on the front lines in animal management. They are typically very observant and should know, from their own knowledge and experience and from simply looking, whether an animal in their care is prospering. They should be able to spot any small signs of parasites or other health problems, such as ragged fur, feathers, or fins or abnormal feces. They should know the likely maximum life span of a particular aquarium fish, what clutch size to expect from any one bird, or when aggression is running too high and how to reduce it in a herd of antelope or a monkey colony. Finally, they should know or find out which diets produce the best results in their animals, and how to vary them experimentally, perhaps with advice from a zoo nutritionist. Many keepers will write their observations in a daily record book or register them electronically in the zoological information management system (ISIS-ZIMS, a scientific database; see above). Others will write articles for publication in their areas of specialization. Indeed, authors with zookeeping experience have also written standard textbooks on modern zoo animal management (see, e.g., Hosey et al. 2009). In short, keepers (including those without formal qualifications) are in many respects scientific managers, with all of the necessary basic attributes, skills, and conditions to make worthwhile contributions to science. Many academically ambitious keepers have worked hard to obtain formal science-based zoo animal management qualifications, or bachelor's degrees in science (BSc). Some have acquired master's degrees (MSc) or doctorates (PhD).

"Research" is *the process to assemble and test knowledge using the scientific method* (Reid et al. 2008). This is the means by which science is created. "Zoo research" is *any scientific investigation involving a zoo or aquarium, their facilities, resources, programs and partnerships* (Reid et al. 2008). Research in zoos or zoo-associated institutions in the 19th to mid-20th century mainly involved traditional ex situ studies in anatomy and physiology, comparing different animals and species. While whole animal biology remains an important focus, today's zoos have far broader research activities, some conducted in situ at home and abroad. Many keepers are involved in research projects (Dibble 2010) and may do these as part of an animal management course or other formal study. Broad topics often concern animal welfare, veterinary medicine, animal behavior, and the conservation of biodiversity. Typical subjects include animal care, aging, assisted reproduction, behavior, gene banking, contraception, animal records management, diet, disease, parasites, hormonal monitoring, ecology, environmental enrichment, husbandry, radio tracking, and life histories (see case study 5.3).

Zoo research aims to be ethical and benign—that is, noninvasive and nonintrusive. For example, it is possible to monitor the level of stress hormones using samples of feces collected in the enclosure, or to use small ticks to collect tiny blood samples with a minimum of discomfort. There are various guidelines on the appropriate high ethical and welfare standards for animals used in research, teaching and for zoo education (Hutchins et al. 1995; Hare 2005; Reid et al. 2008; Hosey et al. 2009). Research of good quality creates a body of well-tested knowledge that is based on internationally accepted standards and suitable for publication and sharing. Many keepers have produced scientific papers on their own

---

Case Study 5.3. CORALZOO Project: Aquarium Research on Coral Husbandry and Conservation

Many species of reef-dwelling coral (plantlike colonial animals, sometimes surrounded by a stony or chalky skeleton) are kept in marine aquariums around the world. Maintaining them in good health in an aquarium is challenging, with high technical requirements for their life support (heating, lighting, water quality, water flow, and nutrition). Over the last two decades, technical skills in maintaining and propagating coral in aquariums have been built to a high level. For further improvement these skills are now being researched, developed, and applied to support species of coral that are under threat in the wild. Threats include marine pollution, a rise in CO_2 concentrations, and a rise in average sea surface temperature. The last two effects are associated with climate change and cause coral in water above 30 °C to "bleach," decline, and die. Bleaching is caused by the loss of microscopic life-supporting symbiotic algae normally present in the tissue of healthy corals. CORALZOO is a collaborative project involving the European Union (the funding partner), the European Association of Zoos and Aquaria, and several partner aquariums, universities, and research institutions. Keepers assist in developing methods, technologies, and husbandry manuals to better maintain, grow, and propagate coral ex situ, and in particular to obtain sexual reproduction (Jones 2006). With research helping to drive standards upward, aquariums are now in a position to rescue samples of threatened species of coral, and also to engage in reintroduction (coral gardening) programs where conditions are appropriate.

or in partnership with others. There are many publication outlets, including newsletters and journals such as *Zoo Biology*, *Bongo* (Journal of the Berlin Zoo), *Conservation Biology*, *Animal Behaviour*, *Applied Animal Behaviour Science*, *Animal Welfare*, *Animal Keepers' Forum*, *International Zoo Yearbook*, *International Zoo News*, *Zoo and Wildlife Medicine*, and *Zoo Vet News*. The kind of research that any one keeper conducts will depend on his or her personal training and capabilities; the nature of the animal collection and outreach projects; the size, resources, and policy of the zoo; and any ethical, legal, social, and environmental (ELSE) requirements.

The *World Zoo and Aquarium Conservation Strategy* (WZACS; WAZA 2005) defines conservation research as "any research that benefits conservation of species and habitats, directly or indirectly." An evidence-based and research-focused discipline of conservation biology involving zoos was first forged in the 1980s. This was in response to various local and global crises in species survival and the natural environment—everything from poaching and the introduction of alien species to pollution and pathogens. It involves a multidisciplinary, holistic approach, used to address various poorly understood conservation issues in situ and ex situ. From a zoo perspective it is about devising practical methods to halt or reduce rates of extinction and habitat loss. This includes studies of human impact on wild species and what can be done to minimize it. It can work, for example, to improve collections management and husbandry in zoos and aquariums; and it draws on academic studies in, for example, taxonomy, zoogeography, physiology, ecology, and, most importantly, small population biology and genetics. The huge scope of conservation biology creates many opportunities for keepers to be involved in big or small ways.

Two important subsets of conservation biology are conservation medicine and conservation psychology. Both of these topics originated in the mid 1990s and have grown rapidly since the beginning of this millennium. The key to tackling biological, medical, and psychological issues in conservation involves teams of professionals, who previously might not have come together, pooling specialized information and sharing constructive ideas. Here again, keepers can play a vital part.

Many universities now run courses of study in conservation biology, environmental science, wildlife medicine, and conservation medicine. Conservation medicine concerns "the relationship between human and animal health and the environment" (Alonso et al. 2002). This may place particular populations or species under threat, such as the masked palm civet (*Paguma larvata*) in southern China, which is thought to transmit severe acute respiratory syndrome (SARS). Emerging infectious diseases are a key focus and as many as 70% of those diseases that involve humans are believed to have originated in or have counterparts in wild animals; these include avian influenza and swine influenza (both of which threaten animals in many zoo collections and sometimes humans). Destruction of habitat can force wild animals into closer contact with domestic animal and human populations, thus increasing the risk of their exposure to wild diseases. Climate change may create warmer conditions for disease vectors such as the flies (*Culicoides* spp.) that spread bluetongue cattle disease virus (BTV).

Conservation psychology has been defined as "the scientific study of the reciprocal relationships between humans and the rest of nature, with a particular focus on how to encourage conservation of the natural world" (Conservation Psychology 2011). In short, it seeks to understand the negative behavioral processes involved in humans acting inappropriately towards nature, and the positive processes involved in caring for and valuing it (Winter and Koger 2004). For zoos, it is particularly important to know what it takes to get visitors emotionally engaged with animal care and conservation and, further, what encourages visitors to provide financial or other practical support. In this sense, conservation psychology is closely allied with zoo education and marketing.

Some scientific work for conservation involves the support of biotechnology (Reid 2001). This means *the application of biological processes to produce and improve materials in zoo biology and medicine*, such as hormones used as a therapy to assist reproduction in threatened species or as contraception to limit it. Sometimes standard studbook-based managed breeding programs are regarded as "traditional biotechnology." Gene banking is an area in zoos that involves biotechnology. Here, living materials (cells, tissues, sperm, eggs) are stored in support of conservation and other science. Cryobiology (ultra-low-temperature biology) is an effective tool in gene banking that allows the establishment of "frozen zoos" (Rawson et al. 2010). The Frozen Ark (Frozen Ark 2011) is an international collaborative initiative to collect, preserve, and store DNA and viable cells from animals in danger of extinction (see Red List, above). These materials are often supplied by zoos in the course of routine veterinary investigations.

SUMMARY

Keepers have a substantial role to play in advancing the aspirations, purposes, and standards of today's zoos. They can fulfill this role by understanding the ideas behind the vision, mission, and values of zoos. They can also fulfill it through engagement in practical animal husbandry and the management of breeding programs. All of this ultimately relates to conservation in the field, and to the pursuit of science in support of it. Where appropriate, keepers are encouraged to engage in conservation, education, and science initiatives both in zoos and in the wild.

REFERENCES

Alonso, Aguirre, Richard Ostfeld, Gary Tabor, Carol House, and Mary Pearl, eds. 2002. *Conservation Medicine: Ecological Health in Practice*. New York: Oxford University Press.

CBD. 2011. "Strategic Plan for Biodiversity 2011–2020 and the Aichi Targets: Living in Harmony with Nature." Montreal: Secretariat of the Convention on Biological Diversity. Accessed at http://www.cbd.int/doc/strategic-plan/2011-2020/Aichi-Targets-EN.pdf.

Conservation Psychology. 2011. Accessed on 25 March at http://www.conservationpsychology.org/about/definition/.

Dibble, Ivan. 2010. "Aquarists and Scientists Working Together to Save Species from Extinction." In Marie Carmen Uribe and Harry Grier, eds., *Viviparous Fishes*, Vol. II, 415–16. Mexico: New Life Publications/New Life Exotic Fish Inc.

Dickie, Lesley. 2009. "The Sustainable Zoo: An Introduction." *International Zoo Yearbook* 43:1–5.

Fa, John, Stephan Funk, and Donnamarie O'Connell. 2011. *Zoo Conservation Biology.* Cambridge: Cambridge University Press.

Frozen Ark. 2011. Accessed on 25 March at http://www.frozenark.org.

Hare, James. 2005. "Guidelines for the Treatment of Animals in Behavioural Research and Teaching." *Animal Behaviour* 69:i–vi.

Hosey, Geoff, Vicky Melfi, and Sheila Pankhurst. 2009. *Zoo Animals: Behaviour Management and Welfare.* Oxford, UK: Oxford University Press.

Hutchins, Michael, Betsy Dresser, and Chris Wemmer. 1995. "Ethical Considerations in Zoo and Aquarium Research." In Bryan Norton, Michael Hutchins, Elizabeth Stevens, and Terry Maple, eds., *Ethics on the Ark: Zoos, Animal Welfare and Wildlife Conservation,* 253–76. Washington: Smithsonian Institution Press.

ICZ. 2011. Accessed 25 March at http://www.iczoo.org.

IUCN. 2011. Accessed 25 March at http://www.iucn.org.

Jones, Rachel. 2006. "Aquariums and Research: The CORALZOO Project." In Bart Hiddinga, ed., *Proceedings of the EAZA Conference, Bristol, UK, 2005,* 142–44. Amsterdam: EAZA Executive Office.

Rawson, David, Gordon McGregor Reid, and Rhiannon Lloyd. 2011. "Conservation Rationale, Research Applications and Techniques in the Cryopreservation of Lower Vertebrate Biodiversity from Marine and Freshwater Environments." *International Zoo Yearbook* 45:108–23.

Reid, Gordon McGregor. 2001. "Biotechnology." In Catherine Bell, ed., *Encyclopedia of the World's Zoos,* 132–35. Fitzroy Dearborn: Chicago and London.

———. 2010. "Taxonomy and the Survival of Threatened Animal Species: A Matter of Life and Death." In Andrew Polasek, ed., *Systema Naturae 250: The Linnaean Ark,* 29–52. London: CRC Press, Taylor & Francis.

Reid, Gordon McGregor, Topiltzin Contreras MacBeath, and Katalin Csatádi. 2013. "Global Challenges in Freshwater-Fish Conservation Related to Public Aquariums and the Aquarium Industry." *International Zoo Yearbook* 47:6–45.

Reid, Gordon McGregor, Alastair McDonald, Andrea Fidgett, Bart Hiddinga, and Kristin Leus, eds. 2008. *Developing the Research Potential of Zoos and Aquaria: The EAZA Research Strategy.* Amsterdam: EAZA Executive Office.

Tlusty, Michael, Andrew Rhyne, Les Kaufman, Michael Hutchins, Gordon McGregor Reid, Chris Andrews, Paul Boyle, Jay Hemdal, Frazer McGilvray, and Scott Dowd. 2013. "Opportunities for Public Aquariums to Increase the Sustainability of the Aquatic Animal Trade." *Zoo Biology* 32:1–12.

UNEP. 1995. *Global Biodiversity Assessment: Summary for Policy Makers.* Cambridge, UK: Cambridge University Press.

———. 2011a. *Organisation Profile.* Cambridge, UK: Cambridge University Press.

———. 2011b. *UNEP Yearbook 2011: Emerging Issues in Our Global Environment.* Cambridge, UK: Cambridge University Press.

———. 2011c. *Towards a Green Economy: Pathways towards Sustainability and Poverty Eradication (a Synthesis for Policy Makers).* Cambridge, UK: Cambridge University Press.

WAZA. 2005. *Building a Future for Wildlife: The World Zoo and Aquarium Conservation Strategy.* Gland, Switzerland: World Association of Zoos and Aquariums. Freely downloadable in English, German, Russian, Swedish, Spanish, Portuguese, Chinese, Czech, and Hungarian at http://www.waza.org.

———. 2009. *Turning the Tide: A Global Aquarium Strategy for Conservation and Sustainability. Implementation of the World Zoo and Aquarium Conservation Strategy by the WAZA Aquarium Community and Partners.* Gland, Switzerland: World Association of Zoos and Aquariums. Freely downloadable in English, Spanish, Japanese, and Chinese at http://www.waza.org.

WAZA. 2011. Accessed 25 March at http://www.waza.org.

Whitehead, Malcolm. 2011. "The IUCN Commission on Education and Communication (IUCN CEC) Is a Network That You Need and That Zoos Need. It Also Needs You." *Zooquaria* 73: 28–30.

Winter, Deborah, and Susan Koger 2004. *The Psychology of Environmental Problems,* 2nd ed. Mahwah, NJ: Lawrence Erlbaum Associates.

6

Animal Ethics and Welfare

Joseph C. E. Barber and Jill D. Mellen

INTRODUCTION

Wild animals are kept in captivity for many reasons. Private owners of wild animals may talk about the companionship they gain from or can offer to primates, big cats, or even bears. Traveling menageries of animals, roadside zoos and circuses may exist because they can be profitable, offering entertainment to visitors eager to see wild animals or have their picture taken with them. Some research labs house colonies of primates as models for human disease studies, or for other types of research that increase our knowledge. Many modern zoos and aquariums see animals under their care as ambassadors that can connect people and nature, and seek support from the visitors to take some form of conservation action. As it relates to captive animals in these types of settings, ethics explores the reasons and justifications that people have for keeping wild animals in captivity, and investigates the issue of whether or not humans should have the right to use animals in any of these ways. This is a complex subject, with no black-and-white answers for the many ethical issues that arise.

There are many layers to explore when thinking about these ethical concerns, including the legal precedents that are created regarding the rights that animals have and what can be legally done to them; the moral issues of whether or not humans should behave in a certain way towards animals even if it is legal; and the philosophical issue of whether other species experience the world like humans, with a conscious sense of their own selves. The animals' perspective is important to consider, and it can be explored by combining scientific approaches, to understand what an animal may actually experience or feel, with some commonsense anthropomorphism, putting ourselves in the animals' place to determine how we might perceive the situation. Based on the broad spectrum of perspectives that people bring to a discussion of ethics, the issues of wild animals in captivity can potentially be reduced to two inter-related questions: should animals be kept in captivity, and how should they be treated if kept in captivity?

While captive wild animals can be found in many venues, this chapter will focus on zoos and aquariums, and specifically on those institutions that actively participate in regional zoo associations (e.g., AZA, BIAZA, WAZA, EAZA, SEAZA), such that the quality of care offered to the animals in these institutions follows clearly articulated standards of care (Barber et al. 2010). Most established zoos and aquariums that participate in these regional associations of professional organizations identify four main goals (or justifications) for housing wild animals: education, entertainment, research, and conservation. These topics will be discussed below in broad terms as they relate to the ethics of keeping wild animals in captivity. This section will address the following topics:

- why animals are housed in zoos and aquariums
- animal rights, animal welfare, and conservation
- common ethical and welfare issues faced by keepers in zoos and aquariums.

WHY ARE ANIMALS HOUSED IN ZOOS AND AQUARIUMS?

Humans are an intensely curious species, and the sheer diversity of animal species that differ so greatly in form, function, and behavior provides plenty to astonish and astound observers. Part of what has always driven humans to house wild species in captivity is the desire to experience something out of the ordinary. Before international travel was commonplace, those with the means to do so created their own menageries of animals, collected in foreign lands and shipped back to the home country for display. Many of these animals died quickly from chronic stress, or simply from lack of essential resources (e.g., heat, food, shelter, companionship, space; Hoage and Deiss 1996). They provided eye-opening experiences for people who otherwise would have seen only the limited number of species in their local environments. These experiences were a form of education—not formal or goal-driven, but certainly leading to a change in the way that spectators viewed the world around them. The exoticism and mystery of some of these species also led to their use as forms of entertainment. For example, bear pits were common in London in the 1600s, and staged fights between bears and dogs provided a ready source of entertainment. It was likely the size, power, and behavior of the bear that probably made this spectacle more entertaining than dog or cock fighting, which were also common. Even Shakespeare's *The Winter's Tale*, first published in 1623, included as the stage direction for one of the actors, "Exit, pursued by a bear" thus revealing some of the ways in which wild animals were influencing culture.

One reason why animals are still housed in zoos and aquariums, albeit in conditions very different from those in the old menageries, is that humans still benefit from this experience. Entertainment is still important for most modern-day visitors, whether it happens through fun family experiences, animal rides, immersion exhibits that mimic an animal's natural habitat, animal demonstrations, or human-animal interactions. In modern times, however, education has become much more critical to the mission of zoos and aquariums, even if it is not the main reason why visitors attend. Learning theory suggests that people learn more when they are actively engaged in the subject (Lindemann-Matthies and Kamer 2006), and when the process of learning is in itself enjoyable; many zoos and aquariums focus on the idea of "edutainment," in which educational messages or opportunities are integrated into entertainment experiences. These approaches are considered to be more effective in leading visitors to be more conservation-minded in their actions once they leave the zoo or aquarium (e.g., supporting conservation organizations, avoiding nonsustainable goods and resources, deciding against having wild animals as pets; Falk et al. 2007), although some researchers remain skeptical (Marino et al. 2010).

Collections of captive animals have always attracted scientists and other academically-minded individuals. Their curiosity and eagerness to learn has led to more formalized research approaches for captive wild animals, and this has been a significant driving force in the evolution of the modern zoo and aquarium. In the early days, much research could be conducted on either living or dead animals, since so much was unknown about the vast majority of exotic species that any information was good information. As interest in living specimens increased, probably in connection with the cost or difficulty of acquiring new specimens from the wild, research into how animals lived and what was needed to keep them alive also increased (Hediger, 1955). Much early research focused on the essential physical needs of animals, and only in more recent times have the behavioral and environmental needs of species in captivity received a greater focus—especially as meeting behavioral needs has been shown to directly influence their physical health (Kleiman et al. 2010). Research in zoos and aquariums has been given a much higher priority in modern times, and it focuses on a range of different aspects, some of which are listed below.

HUMAN RESEARCH

- Visitors: Studies look at how visitors perceive animals during their visits, and what educational messages they take away (e.g., Falk et al. 2007).

ANIMAL RESEARCH

- Veterinary care: The health of animals is always given a high priority, and research into diseases, treatment, and preventative care is regularly performed (Fowler and Miller 2007).
- Reproduction: Research to identify approaches to maximize breeding, or to prevent it through contraception, is common; in addition to natural breeding, scientists continue to focus on assisted reproduction techniques like artificial insemination and embryo transfer (Brown et al. 2004).
- Behavior: The behavior of animals can provide insight into how they perceive their environments, and whether those environments are meeting their needs (Mason and Rushen, 2006). Behavior also plays a role in reproduction, human-animal interaction, and visitor stay time (Kuhar et al. 2010), so it will usually be a component of other forms of research.
- Nutrition: Nutrition research in zoos and aquariums focuses primarily on providing a complete and balanced diet. Most diets are built from extensive knowledge of domestic animals' dietary requirements. Those animals in zoo or aquarium collections that do not have a domestic counterpart are often more challenging to feed (Fowler and Miller 2007).
- Enrichment: As a fairly recent scientific approach, research on enrichment seeks to identify initiatives, resources, and management approaches provided by keepers that best encourage species-appropriate behaviors and give animals choices in and control over their environment (Barber 2003).
- Animal training: Most research into animal training is less formal and more subjective than the other types of research listed here. Research into training looks at the effectiveness of different approaches used to train animals for shows or husbandry behaviors (Pomerantz and Terkel 2009).

Conservation is one of the more recent goals of zoos and aquariums, and it focuses on preserving and protecting wild animals and wild places through a variety of approaches (Wheater 1995). For example, many zoos and aquariums are involved in captive breeding efforts to maintain genetically viable populations of certain species within captive animal collections, although only rarely are captive-bred animals reintroduced back into the wild. In certain cases, very little of "the wild" still exists, and the individual animals reared in captivity are unlikely to possess the knowledge and experience they need to survive under natural conditions. Instead, money raised by zoos and aquariums through entrance fees or donations is often used to fund more direct conservation projects in natural environments, such as population and habitat viability assessments.

Like many of the original menageries, modern zoos and aquariums typically exhibit a cross-section of species from around the world that still satisfy our curiosity about their diverse forms and behaviors, but this is where the similarity usually ends. Over time there has been a significant and ongoing transformation of zoos and aquariums toward displaying native animals found locally. From a welfare and ethics perspective, these institutions have adapted to position themselves as fun, interesting, and informative venues while public attitudes concerning our relationship with wild animals have changed (Mullan and Marvin 1987; Bell 2001; Hanson 2002).

ANIMAL WELFARE, ANIMAL RIGHTS, AND CONSERVATION

The ethics of keeping wild animals in captivity, and more specifically in zoos and aquariums, can be discussed from three interrelated perspectives: animal welfare, animal rights, and conservation. These perspectives are described below, and keepers will find areas of each that will resonate with their own personal beliefs and approaches.

ANIMAL WELFARE

The subject of animal welfare involves a range of considerations, from political and economic to ethical and scientific, and is relevant to any situation in which animals are under the direct care of humans (e.g., as pets, farm animals, or laboratory animals or in zoos and aquariums). For example, the rearing of farm animals in many countries is economically important in terms of money brought in from exports and consumer spending. The larger the farming industry, the more political influence it can have. Related to the economic benefits of mass-producing animals for their meat, eggs, milk, or other byproducts are scientific approaches that investigate whether selective breeding or genetic modification can make animals better producers or breeders, and thus more productively efficient. Research is also performed under the umbrella of animal welfare to explore how captive environmental conditions or the consequences of selective breeding affect the animals themselves, regardless of any economic benefits. People who perform this type of research, and others who are aware of the conditions in which some farm, laboratory, or zoo and aquarium animals are housed and reared, often have concerns about the quality and effectiveness of certain housing and husbandry treatments. These ethical concerns, often loosely formed, stem from a feeling that the "needs" of the animals are not being met—but are not necessarily accompanied by a knowledge of exactly what needs those are, and how important they are to the animals, if at all. Animal welfare research attempts to answer some of these questions so that people can make informed ethical choices about the treatment of animals—about whether it is right or wrong to house them in certain ways. Three of the most important questions are:

- What needs do animals have?
- How can we assess how well these needs are being met?
- What new approaches can be developed to meet these needs?

It is not possible to understand the broad concept of animal welfare without exploring the dimensions mentioned above. From a keeper's perspective, the political and economic issues are probably not as relevant as the ethical and scientific ones. Since animal keepers work primarily with captive wildlife, it is likely that they have no problems with the overall idea of housing animals in captivity; they do not believe that animals should only exist in the wild. This means that keepers are likely not supporters of the concept of animal rights (discussed below). However, most of them do want captive animals to have their needs met, as research and decades of hands-on experience can identify what some of those needs are. The animal welfare standpoint is that humans can use animals for a variety of reasons but the treatment of those animals should meet their needs. The logical extension of this idea is that if the needs of animals cannot be met in a particular situation, then it could be unethical to house them in that situation.

Keepers in zoos and aquariums need to understand the scientific application of animal welfare, as this is something that they will implement every day. Animal welfare decisions are enhanced and supported by a scientific approach. Information about the behavioral, health, and physiological attributes of animals in captivity is key to their welfare. Thus, welfare is a scientific concept because it involves the measurement of biological variables, and one possible definition would be:

Animal welfare is the degree to which an animal can cope with challenges in its environment as determined by a combination of measures of health and of psychological well-being.

- Good health represents the absence of diseases or physical/physiological conditions that result (directly or indirectly) from inadequate nutrition, exercise, social groupings, or other environmental conditions to which an animal fails to cope successfully.
- Psychological well-being is dependent on there being the opportunity for animals to perform strongly motivated, species appropriate behaviors, especially those that arise in response to aversive stimuli.
- Enhanced psychological well-being is conditional on the choices animals have to respond appropriately to variable environmental conditions, physiological states, developmental stages and social situations, and the extent to which they can develop and use their cognitive abilities through these responses (Barber and Mellen 2008, 41)

Physiological responses, assessment of disease and injury, motivation, and cognitive abilities are some of the biological variables that can be included in welfare considerations and specifically measured. Some of these variables can be interpreted on the basis of extensive hands-on experience with animals in addition to more formal assessments of biologically relevant response (e.g., hormone levels, behavioral preference tests). By answering questions about whether an animal is sick or healthy or is experiencing well-being or discomfort, its welfare state can be determined. An animal that is healthy (physically and psychologically) can be said to have good welfare. If that animal becomes sick or cannot perform its strongly motivated behaviors, then it has poorer welfare. Comparing welfare assessments as they change over time in the same animal is relatively straightforward. However, these assessments will often change to differing degrees in different animals (even within the same species) in response to the same types of environments. For example, certain animals within the same species will respond to crowds of visitors in negative ways, whereas others seem to have no obvious behavioral or physiological responses (Wielebnowski et al. 2002; Carlstead and Brown 2005). Because of individual animal differences, welfare should always be assessed at the individual animal level, rather than at the group or species level. In general terms, however, what causes one animal to experience poor welfare can cause similar animals to experience poor welfare as well.

Ongoing research shows that keeper assessments of welfare, based on more subjective or qualitative approaches, can contribute useful information in determining the welfare state of animals (Wemelsfelder and Lawrence 2001). Keepers routinely observe the feeding habits, activity, and social interaction of animals along with their general appearance, and integrating this information may allow for a fairly complete picture of how an animal is coping with its captive environment on a daily basis (Meagher 2009; Whitham and Wielebnowski 2009). Researchers are using preliminary data to refine protocols for improving these types of subjective welfare assessments (Whitham and Wielebnowski 2009).

Scientific research and hands-on experience provide some insight into how individual animals respond from a behavioral, physiological, or health perspective to the captive conditions they face. At this point, however, we cannot say whether it is ethical to treat animals in certain ways on the basis of the presence or absence of a certain behavior, or higher levels of a hormone in their blood. If welfare depends on the degree to which an animal can cope with challenges in its environment, then what exactly is the difference between coping and not coping, and when does good welfare become poor welfare? The science reveals that some variable has changed, but it does not show whether that is good or bad from the animals' perspective, or how much of a change makes something that was good become bad, or change from acceptable to unacceptable.

Ethical questions can include "How much of a change in a physiological or behavioral variable is good?," "How much discomfort must an individual animal endure to be said to be suffering?," or "What are the indicators of positive welfare (e.g., play, exploration)?" Some questions can even be asked about whether certain animals can feel pain or experience suffering, although finding an answer requires a more scientific approach. If the scientific component of welfare involves measuring the actual experiences of the animals, then the ethical part is our (human) attempt to integrate the animals' experiences into our own moral perspectives. Ethics are never straightforward. Our ethical thinking is affected by many factors, such as culture, personal experience, religion, and nationality. Ethical concerns vary both from culture to culture and within a single geographic region. There are even differences in the ethical considerations the same people will give to different animals from the same species, depending on how and why they are used. Keepers who work in a petting

> ### Box 6.1. Five Key Points to Know about Animal Welfare
>
> 1. Animal welfare can be seen as a continuum that ranges from poor to very good, based on changes in scientific measures of health, physiological state, or behavior.
>
> 2. The welfare of animals should be assessed at the level of the individual animal, not of the group or species.
>
> 3. Welfare is a quality of the animal, not something that can be provided directly to it.
>
> 4. The overall welfare of an animal is dependent on its ability to cope with environmental, physical, and psychological challenges, and on the ability of keepers to provide the most appropriate environment and humane care for it to do so.
>
> 5. There is no magical cutoff point where "good welfare" becomes "bad welfare." Absolute levels of good and bad welfare are determined by human-centric and subjective ethical considerations.

zoo may be very concerned about the welfare of the chickens they care for, yet may at the same time be entirely comfortable eating chicken reared in intensive farming conditions where their welfare is often significantly poorer.

ANIMAL RIGHTS

If animal welfare can be seen as a combination of knowledge of and experience with specific animals, science, and ethics, then animal rights can be seen as a belief system that is based on providing other species with the same types of moral considerations as are given to most humans (or as would be given in an ideal world). The basic provision of this belief system or philosophy is that animals should have the right not to be harmed, and should not be used by humans for food, medicine, or any purpose at all that impinges on their freedom. As a purely philosophical standpoint, the mainstream animal rights perspective is that wild animals should not be housed in captivity because this impinges upon each individual animal's right to be free and often results in some measurable degree of harm to the animals. These animal rights–based freedoms are fundamentally different from those freedoms, often cited by supporters of animal welfare as the "five freedoms," that are used to guide the treatment and care of animals in captive settings.

The five freedoms commonly used as a basis for defining animal welfare are

1. freedom from hunger and thirst, provided by ready access to fresh water and a diet to maintain full health and vigor
2. freedom from discomfort, provided by an appropriate environment including shelter and a comfortable resting area
3. freedom from pain, injury, or disease, provided by prevention or rapid diagnosis and treatment

4. freedom to express normal behavior, provided by sufficient space, proper facilities, and company of the animal's own kind

5. freedom from fear and distress, provided by ensuring conditions and treatment that avoid mental suffering (Brambell 1965).

If humans and other species are moral equals, then humans should not assume that they have the right to use animals for their own benefit (or for the benefit of other animals), no matter how great that benefit may be (e.g., curing diseases, educating people, saving the species as a whole from local or global extinction). Although the general animal rights philosophy stands against the notion of speciesism (that certain species can be treated differently because they are different from humans), it is generally the more complex animals (e.g., mammal and bird species) that are given a greater priority because of their considered greater sentience (i.e., emotional understanding of the world around them, and ability to feel positive and negative subjective emotional states in response to their experiences).

The animal rights philosophy extends rights that have been developed by humans to protect human interests. Humans are considered to be "moral agents" because of their ability to develop these types of moral and ethical frameworks (Regan 1983). When these rights are provided to other species, those species can be considered to be "moral patients," as they play no role in developing these rights themselves (Regan 1983). Other species do not have the same types of moral frameworks within their natural environments; they have no specific natural rights. Depending on the species, individual animals within their natural environment can be preyed upon, attacked, eaten, chased off territories, or denied access to food and water, and these things can be done by conspecifics or members of another species on a routine basis. The only infallible right an animal has in its natural environment is to be able to use the behavioral, physical, and physiological adaptations developed over the course of evolution to survive and/or be reproductively successful—but there are no guarantees for this right either. Applying human ethical thinking to animals cannot work in natural environments, as enforcement of these principles is impossible (Moore 2002). The actions of one animal upon another (where both are considered to be "moral patients") cannot be classed as ethical or unethical. However, when animals are housed in captivity under the care and attention of humans, then the issue of rights and freedoms becomes more pertinent. In this case, human actions directly affect the lives of animals.

There are two lines of thinking when it comes to captive animals. If the question "Do you believe that animals should be housed in captivity?" is asked, a true animal rights supporter might answer "No." It would not matter to the animal rights supporter whether the animals were housed in captivity for farming, research, entertainment, education, or captive breeding for conservation; whether they were endangered animals on the brink of extinction; or whether the environment in which they were to be housed was designed specifically to meet their needs. However, there are different degrees of supporting the idea of rights, and some people see a distinction between having a dog at home (often as a

---

Box 6.2. Five Key Points to Know about Animal Rights

1. People will always have different beliefs about right and wrong. Beliefs are very personal.

2. Personal beliefs are affected by many factors, including nationality, culture, and religion.

3. There is a continuum of ethical concerns. At one extreme is the idea that animals have no rights and can be treated or used in any way without moral concern. Animal rights falls at the other end of this continuum.

4. There is nothing inherently right or wrong with the idea that animals should have the same moral rights as humans, but the feasibility of implementing this philosophical idea on a global scale may be very challenging.

5. Where people are found on a continuum of ethical concerns about whether animals should or should not be housed in zoos and aquariums is influenced in many cases by the type of animals housed and their perceived welfare in the particular captive situation.

---

companion or extended member of a family unit) and keeping walruses, elephants, penguins, and killer whales in zoos and aquariums. There is also the distinction between sentient animals and nonsentient species, although it is often hard to draw such a line between those categories (Broom 2010). The concept of sentience relates to the subjective emotional states of individuals. Sentient animals, like humans, have emotional feelings that can be negative and positive, and these feelings matter to us. Negative emotional states that result in suffering are not merely sensations but can have tangible effects on an individual's psychological well-being, health, and overall view of the world. Many would consider apes to be sentient beings. Few would consider the fleas or lice that live on the apes to be sentient.

The critical difference between animal welfare and animal rights is that while animal welfare supporters aim to provide animals in captivity with the best possible lives because they are under human care for various reasons, animal rights supporters generally believe that animals should not be in captivity under any circumstances. The concept of animal welfare looks at the way animals are treated in captivity; the concept of animal rights focuses on the issue of whether animals should be in captivity in the first place. The idea of animal rights has been much criticized because it has motivated the actions of individuals and organizations that are destructive or illegal in the eyes of the law (e.g., the release of mink from research facilities and fur farms by members of the Animal Liberation Front). Certain high-profile and aggressive advertising campaigns and activities sponsored by more mainstream animal rights organizations, such as People for the Ethical Treatment of Animals (PETA), have led some to ridicule the idea. There is also a lack of clarity about which organizations are welfare or rights organizations. For example, the International Fund for Animal Welfare (IFAW) is considered by some to be a rights group, and by others a welfare or even a conservation organization. Certain organizations (e.g., the Humane Society of the United States, Compassion in World Farming) may include elements of both approaches, and are often termed "animal protection" groups. Keepers looking to support these types of groups should educate themselves about the goals and mission of each organization to avoid confusion.

There is also confusion about the meaning of the term "animal rights" by members of the public. In a 2003 Gallup poll (Gallup 2003), some members of the public who supported the idea of granting animals the same rights as humans ironically went on to state that they opposed bans on medical testing of animals, and did not want stricter laws passed for improving the care of farm animals. The idea of granting animals the same moral considerations as humans only seemed to apply to certain animals (e.g., cats and dogs, but not chickens or rats). Those people who contradicted themselves obviously have some concerns for animals, but are not true "animal rights" supporters. These results also suggest that many people who support animal "welfare" also identify themselves as supporting "rights" for animals. The results of the survey support the idea that the general public often confuses the terms "animal rights" and "animal welfare." This is important for keepers, because a lack of clarity in the use of terminology can lead to miscommunication with visitors who have concerns about the treatment of animals at zoos and aquariums. There may be passionate supporters of zoos and aquariums who think they are animal rights supporters, yet who are more interested in appropriate care of the animals.

Keepers may also be faced with people who clearly oppose housing wild animals in captivity, especially for certain charismatic species such as whales or elephants. Attempts to justify the captive care of these species are often unsuccessful because of the purely philosophical stance often adopted by opponents of zoos and aquariums. It does not matter that the animals are being cared for appropriately, because what is at issue is the animals' right to be free and to not be used by humans: something that even optimal husbandry can never provide. However, there is a wide continuum of ethical concerns for animals. While one extreme is the animal rights philosophy, and the other is the idea that animals should have no rights at all, most people will find themselves somewhere in the middle. That is, they are more likely to see the housing of wild animals in captivity as unethical if the needs of those animals are not or cannot be met. Most zoo visitors who express concerns simply want to hear that the keepers are taking good care of the animals and that they care deeply about their well-being.

Attitudes toward the appropriateness of housing wild animals in zoos and aquariums and toward their care in captivity continually evolve. Changes in attitude often occur in conjunction with new scientific knowledge (e.g., about the sentience of animal species), shared knowledge, and the experience of keepers. More is certainly known now about the natural history and biology of many species than was known 50 years ago, thanks in part to research performed in captive environments. With this greater knowledge, however, comes a greater moral responsibility to ensure that the conditions in which those animals are kept in the future can meet their various newly identified needs.

CONSERVATION

The goal of conservation can add an interesting element to the discussion of welfare and rights. The focus of conservation initiatives is usually on species or populations, rather than just on individuals, which are the focus of both animal welfare and animal rights. Conservation projects typically involve only certain types of species (e.g., threatened or endangered ones) rather than all species. From a conservation perspective, an endangered species (e.g., the Guam rail) is considered more valuable than a common or invasive one (e.g., the brown tree snake or the rat). In managed natural environments, the eradication of common species that pose a threat to endangered ones is done with little consideration of any "rights" of the targeted species (Cruz et al. 2009), and often with less priority given to the welfare of that threat/pest species in the choice of eradication measures used (e.g., poison; Howald et al., 2007; Littin 2010). Endangered species are often given greater rights than other species, especially from a legal standpoint. Within zoos and aquariums, the implementation of conservation approaches is typically aligned with the idea of animal welfare (Fraser 2010). For certain species, individual animals currently need to be housed in captivity because the wild populations and their habitats are threatened and extinction is a real possibility. At a population level, it is in the best interest of the species to ensure a viable captive population in captivity. From an individual perspective, meeting the specific needs of each animal will most likely lead to healthier animals that can breed effectively and rear their young in a species-appropriate manner. Animals experiencing poor welfare are likely to decrease the effectiveness of conservation approaches within zoos and aquariums. While the value of good breeders may be greater than old or non-reproductive animals, the welfare of all animals is important.

COMMON ETHICAL AND WELFARE ISSUES FACED BY KEEPERS

The information above provides a brief summary of some of the key elements associated with animal welfare and ethics from a broad, often theoretical perspective. Keepers regularly must face issues at their own institutions that are relevant to this discussion. They may encounter or even cause many ethical or welfare concerns, including the examples that follow.

- Some traditional methods of rodent control (e.g., poison or sticky traps) are often considered inhumane (Mason and Littin 2003). Since high rodent populations may bring disease to animals in the collection, is it ethical to use methods such as sticky traps for pests when they would never be acceptable for non-pest species?
- Feeding live food (e.g., fish) as prey items may be highly enriching for the predators in the collection. What responsibility do keepers have for the welfare of the prey species being provided as food?
- Abnormal behaviors (e.g., pacing, self-stimulation, excessive grooming, self-harm, and regurgitation and reingestion) can sometimes already be engrained in an

animal before keepers are given the responsibility to care for it. What are the keepers' choices and responsibilities when an animal shows excessive abnormal behavior? What criteria can be used to decide what is abnormal or excessive? Whose job is it to make those decisions?

- What if animals at an institution are housed in spaces that appear not to meet their needs for prolonged periods of time? What role do keepers play in investigating and addressing these potential issues?
- When is it appropriate to pinion birds (or use other flight restriction methods) to prevent their flight? Who makes those decisions and how? What is a keeper's role in that decision-making process?
- Touch tanks that allow visitors to touch and feel living specimens (e.g., rays or sea urchins) are very popular with zoo and aquarium visitors, but they may be associated with increased welfare concerns about how the animals respond to these interactions. Is there an acceptable level of mortality for these program animals?
- Is it possible to determine what is "enriching" as opposed to "stressful" for contact animals (e.g., in swim-with-the-dolphin programs or petting zoos)? Who makes those determinations?
- If keepers witness another keeper engaging in questionable animal management approaches (e.g., hosing an animal that is not shifting into holding), what should they do?
- Euthanasia is used as a medical tool when seriously injured, ill, or aged animals are determined to be suffering. Who decides when it's time to euthanize an animal? What criteria are used?
- Euthanasia can also sometimes be used as a management tool (e.g., to cull surplus male antelope or tadpoles). Who decides the circumstances under which management euthanasia is used? What criteria are used?

Each item in this list represents a possible ethical or welfare issue. Whether or not these examples are actual issues depends on the individual circumstances of each situation. The most effective way to address ethical and welfare concerns within zoos and aquariums is through institutional processes designed to identify and deal with them. Professional zoos and aquariums have begun to develop these processes to enable staff at any level (e.g., keepers, veterinarians, or docents) to report issues, get feedback from managers, and work to implement solutions as part of a broader, formalized welfare approach. Curators, managers, veterinarians, and keepers routinely make multiple decisions about animal welfare every day. Since animal welfare is an ongoing, institution-wide responsibility, the development of some form of institutional process is an important part of being able to document and assess the positive impact that animal care decisions can have. Such a process provides a means for staff members within an institution to follow a well-defined protocol for bringing up questions and concerns to their supervisors, or to an independent body within the institution (e.g., an institutional animal

welfare committee) if necessary. It also helps in identifying welfare issues noticed by staff members, coordinating an appropriate response, educating staff members about welfare issues, and drawing attention to animal welfare successes throughout the institution. At least for institutions in North America's Association of Zoos and Aquariums (AZA), there is no specific protocol that each institution must adopt. Instead, each zoo and aquarium is required by the AZA accreditation standards to develop a protocol that works most effectively for it—based on the size of the institution and the complexities of issues that might arise. The 2013 AZA accreditation standards reflect this by stating:

The institution must develop a clear process for identifying, communicating, and addressing animal welfare concerns within the institution in a timely manner, and without retribution. Explanation: A committee or some other process must be identified to address staff concerns for animal welfare within the institution. The committee or process should include staff with the experience and authority necessary to evaluate and implement any necessary changes (Association of Zoos and Aquariums 2013).

Some issues that keepers face are complex, and it may not always be feasible for staff and visitors to agree on a desired long-term outcome. But the opportunity for people to become educated about the complexity of various issues and about the reasons for certain actions or inaction, can be very valuable. Each institutional process is likely to include some of the following key elements:

- a mechanism for keepers to report, anonymously or otherwise, such things as health concerns, social group concerns, housing or facility issues, behavior abnormalities, nutrition or diet concerns, or any questions related to the animals' welfare
- record-keeping of all issues reported, and the resultant responses, if any
- a process for understanding the issues identified, and for brainstorming approaches to address them
- a way to evaluate the long-term results of actions taken.

Keepers at zoos and aquariums that currently have no institutional animal welfare protocol can investigate what established processes other similarly-sized institutions have to address the types of issues listed above. Some animal protection and animal rights organizations have developed reporting processes for welfare or ethical issues that occur in zoos and aquariums, and encourage keepers with such concerns to anonymously provide information about them. For example, PETA has a page on its website that allows people to "report concerns about your zoo here." This opportunity is for zoo visitors and staff alike, and some people who report issues may do so because they believe it will help the animals. Since animal rights or protection groups often lack applied knowledge of the care and management of wild animals in captive settings, they will not be able to address the issue directly and the welfare of the animals involved will not be improved in the short term. Staff at the institution will be the only people with the power to investigate and address the identified issues, and will have access to the knowledge and skills needed to do so through the network of experts within their zoo and aquarium association. Developing a process that allows keepers to work through these issues in-house should be a goal of every zoo and aquarium. This institutional approach will be far more effective than reaching out to outside groups.

CONCLUSION

The animal rights position clearly states that animals should never be housed in captive conditions—whatever the rationale for doing so. This differs very significantly from the animal welfare position, which looks at how animals are treated in captivity with the goal of optimizing their treatment and care. Whether or not animals should be housed in zoos or aquariums is an issue that each person must consider for themselves. While animals remain in zoos and aquariums, however, keepers are directly responsible for meeting their needs. As keepers enhance their role as part of the larger animal care team (along with veterinarians, managers, curators, trainers, nutritionists, and scientists), the animals will more likely experience good welfare. It is important that keepers understand how much the welfare of the animals they care for is affected by what they do for those animals on a daily basis, whether that is cleaning, training, providing enrichment, observing their behavior, or reporting possible concerns to other zoo staff members. The more knowledgeable keepers are about their animals, the better the care they can provide, and the less likely it will be that ethical issues can arise. This publication is an important contribution to the knowledge base that keepers can use to be proactive in their care of animals. However, no single publication can provide all necessary information, or answer each person's questions about animal husbandry, welfare, and ethics. It was human curiosity that brought visitors to the first menageries of captive animals centuries ago. It is this same curiosity that keepers must cultivate to help themselves determine best practices of care, and clear ethical guidelines for how animals should be treated. If in doubt, ask. The hands-on knowledge of zoo and aquarium professionals represents hundreds of years of experience. If no information is available, then that can be seen as an opportunity for preliminary research to acquire some information, to collaborate with other keepers at different institutions who may be experiencing similar issues, to find ways to assess the animals' responses to situations, and to challenge preconceptions about how animal care should be done. The keepers who read this publication will be in an ideal position to build upon the experiences of others who have gone before them, and to help zoos and aquariums in their continuing evolution. Education, entertainment, research, and conservation will remain the four pillars of the zoo and aquarium experience, and as keepers work hard to make animal welfare considerations an integral part of each of these, then there will be little room for argument or concern about ethics.

REFERENCES

Association of Zoos and Aquariums. 2013. *Guide to Accreditation of Zoological Parks and Aquariums.* Silver Spring: Association of Zoos and Aquariums.

Barber, J. 2003. Making sense of enrichment and Auntie Joy's choice of presents. *Animal Keepers' Forum* 30(3), 106–10.

Barber, J., and Mellen, J. 2008. Assessing animal welfare in zoos and aquaria: Is it possible? In T. Bettingerand J. Bielitzki, eds.), *The Well-Being of Animals in Zoo and Aquaria Sponsored Research: Putting Best Practices Forward*, pp. 39–52. Greenbelt, MD: Scientists Center for Animal Welfare.

Barber, J., Lewis, D., Agoramoorthy, G., and Stevenson, M. F. 2010. Setting standards for evaluation of captive facilities. In D. Kleiman, K. Thompson, and C. Kirk Baer, (eds., *Wild Mammals in Captivity: Principles and Techniques for Zoo Management*, 2nd ed., pp. 22–34. Chicago: University of Chicago Press.

Bell, C., ed. 2001. *Encyclopedia of the World's Zoos.* Chicago: Fitzroy Dearborn.

Brambell, F. 1965. Report of the Technical Committee to Enquire into the Welfare of Animals Kept under Intensive Livestock Husbandry Systems. London: Her Majesty's Stationery Office.

Broom, D. 2010. Cognitive ability and awareness in domestic animals and decisions about obligations to animals. *Applied Animal Behaviour Science* 126, 1–11.

Brown, J., Goritz, F., Pratt-Hawkes, N., Hermes, R., Galloway, M., Graham, L., et al. 2004. Successful artificial insemination of an Asian elephant at the National Zoological Park. *Zoo Biology* 23(1), 45–63.

Carlstead, K., and Brown, J. 2005. Relationships between patterns of fecal corticoid excretion and behavior, reproduction, and environmental factors in captive black (*Diceros bicornis*) and white (*Ceratotherium simum*) rhinoceros. *Zoo Biology* 24(3), 215–32.

Cruz, F., Carrion, V., Campbell, K., Lavoie, C., and Donlan, C. 2009. Bio-economics of large-scale eradication of feral goats from Santiago Island, Galapagos. *Journal of Wildlife Management* 73(2), 191–200.

Falk, J., Reinhard, E., Vernon, C., Bronnenkant, K., Deans, N., and Heimlich, J. 2007. *Why Zoos and Aquariums Matter: Assessing the Impact of a Visit.* Silver Spring, MD: Association of Zoos and Aquariums.

Fowler, M., and Miller, R. 2007. *Zoo and Wild Animal Medicine*, 6th ed. Saint Louis: Saunders.

Fraser, D. 2010. Toward a synthesis of conservation and animal welfare science. *Animal Welfare* 19(2), 121–24.

Gallup. 2003. Retrieved 15 May 2005 from http://www.gallup.com /poll/8461/Public-Lukewarm-Animal-Rights.aspx.

Hanson, E. 2002. *Animal Attractions: Nature on Display in American Zoos.* Princeton, NJ: Princeton University Press.

Hediger, H. 1955. *Studies of the Psychology and Behaviour of Captive Animals in Zoos and Circuses.* G. Sircom, trans. London: Butterworth.

Hoage, R., and Deiss, W., eds. 1996. *New Worlds, New Animals: From Menagerie to Zoological Park in the Nineteenth Century.* Baltimore: Johns Hopkins University Press .

Howald, G., Donlan, C., Galvan, J., Russell, J., Parkes, J., Samaniego, A., et al. 2007. Invasive rodent eradication on islands. *Conservation Biology* 21(5), 1258–68.

Kleiman, D., Thompson, K., and Kirk Baer, C., eds. 2010. *Wild Mammals in Captivity: Principles and Techniques for Zoo Management*, 2nd ed. Chicago: University of Chicago Press.

Kuhar, C., Miller, L., Lehnhardt, J., Christman, J., Mellen, J., and Bettinger, T. 2010. A system for monitoring and improving animal visibility and its implications for zoological parks. *Zoo Biology* 29(1), 68–79.

Lindemann-Matthies, P., and Kamer, T. 2006. The influence of an interactive educational approach on visitors' learning in a Swiss zoo. *Science Education* 90(2), 296–315.

Littin, K. 2010. Animal welfare and pest control: Meeting both conservation and animal welfare goals. *Animal Welfare* 19(2), 171–76.

Marino, L., Lilienfeld, S., Malamud, R., Nobis, N., and Brogliod, R. 2010. Do zoos and aquariums promote attitude change in visitors? A critical evaluation of the American Zoo and Aquarium study. *Society and Animals* 18, 126–38.

Mason, G., and Littin, K. 2003. The humaneness of rodent pest control. *Animal Welfare* 12(1), 1–37.

Mason, G., and Rushen, J., eds. 2006. *Stereotypic Animal Behaviour: Fundamentals and Applications to Welfare*, 2nd ed. Wallingford: CAB International.

Meagher, R. 2009. Observer ratings: Validity and value as a tool for animal welfare research. *Applied Animal Behaviour Science* 119, 1–14.

Moore, E. 2002. The case for unequal animal rights. *Environmental Ethics* 24(3), 295–312.

Mullan, B., and Marvin, G. 1987. *Zoo Culture*. Chicago: University of Illinois Press.

Pomerantz, O., and Terkel, J. 2009. Effects of positive reinforcement training techniques on the psychological welfare of zoo-housed chimpanzees (*Pan troglodytes*). *American Journal of Primatology* 71(8), 687–695.

Regan, T. 1983. *The Case for Animal Rights*. Berkeley: University of California Press.

Wemelsfelder, F., and Lawrence, A. 2001. Qualitative assessment of animal behaviour as an on-farm welfare-monitoring tool. *Acta Agriculturae Scandinavica Section A-Animal Science, Supplement* 30, 21–25.

Wheater, R. 1995. World zoo conservation strategy: A blueprint for zoo development. *Biodiversity and Conservation* 4(6), 544–52.

Whitham, J., and Wielebnowski, N. 2009. Animal-based welfare monitoring: Using keeper ratings as an assessment tool. *Zoo Biology* 28(6), 245–60.

Wielebnowski, N., Fletchall, N., Carlstead, K., Busso, J., and Brown, J. 2002. Noninvasive assessment of adrenal activity associated with husbandry and behavioral factors in the North American clouded leopard population. *Zoo Biology* 21(1), 77–98.

Part Three

Workplace Safety and Emergencies

7

Workplace Safety

Ed Hansen

This chapter will address the specialized elements of working with exotic animals and combine those elements with standards addressed by the United States Occupational Safety and Health Administration (OSHA), thereby creating a reference that addresses practical employee safety in the zoo and aquarium profession and ensuring that keeper safety receives the same attention as animal welfare in the workplace. While OSHA regulations affect facilities within the United States, other countries have adopted similar and specific regulations regarding safety and employee welfare that will apply to animal facilities worldwide.

Each zoo and aquarium should have a basic safety plan in place for its employees. This safety plan will normally address the basic hazards of working with dangerous exotic animals. It should take the form of written policies defining the actions to be carried out by the keepers posted in their various animal care work areas. To compliment a basic safety plan, each zoo or aquarium should supplement its plans with new employee safety orientation and should continue safety education and training on a frequent basis. A facility safety plan is a "living" document, so frequent review of policies and procedures should be an integral part of orientation, initial training, and continuing education.

THE OCCUPATIONAL SAFETY AND HEALTH ACT

Zoos and aquariums are frequently unaware of the full scope of OSHA or of the enforcement of the Occupational Safety and Health Act within their facility and its potential impact on their day-to-day operations. Enacted by Congress in 1970 and signed into law, the act stipulates that employers are required to provide a safe working environment for employees.

The Occupational Safety and Health (OSH) Act applies to all employers, including zoos and aquariums, with limited exceptions for federal agencies. While exempt from the OSH Act, those federal entities have safety standards and policies, enforced by other government agencies, that mirror the OSH Act. The OSH Act is a simple statement that covers the ex-

pectation both that the employer will protect the employee and that the employee will follow all rules that will affect his or her own safety in the workplace.

Section 5
(a) Each employer—
(1) shall furnish to each of his employees employment and a place of employment which are free from recognized hazards that are causing or are likely to cause death or serious physical harm to his employees;
(2) shall comply with occupational safety and health standards promulgated under this Act.
(b) Each employee shall comply with occupational safety and health standards and all rules, regulations, and orders issued pursuant to this Act which are applicable to his own actions and conduct.

Due to its statement regarding employee protection, the OSH Act is frequently applied to a zoo or aquarium because of the specialized nature of the employee's job. In simple words, where regulations and laws are not specific to the safety of an animal keeper working directly with exotic animals, the employer still retains the duty to provide a safe environment for the employee and, where the hazards of working with animals are recognized, to train employees in safe work practices. Other OSHA standards are applied directly to the employee's everyday task, as in the use of chemicals and disinfectants, personal protective equipment, and hygiene and sanitation, to name a few.

ZOO AND AQUARIUM SAFETY PROGRAMS

Animal facilities garner positive attention and headlines for animal births, new enclosures, conservation programs, and green practices. Also, when a keeper has a serious or fatal accident, the facility and the profession receive negative press attention. A comprehensive safety program addresses and acknowledges the reality that an incident or accident will occur.

The goal is to lessen the frequency and seriousness of such incidents by taking reasonable and prudent precautions, and to provide frequent and meaningful training on the recognized hazards in the profession. The benefit to the employee is a healthier and safer working environment. The benefit to the employer is a lowered risk of accidents or incidents.

Both employee and employer must understand that a safety program is a "living" document. Zoos and aquariums experience frequent collection and personnel changes and often require employees to shift or rotate through different areas. Keepers may not be performing the same task on a consistent basis. For these reasons, a safety program must be updated frequently to keep up with the collection changes, personnel changes, and complex realities of an operation that requires staffing every day of the year.

The program should be managed by a dedicated individual who will update critical information, policies, and procedures on a frequent basis, ensuring that the information is communicated to the animal keepers and staff both verbally and in writing. In larger zoos and aquariums, this task may fall to the human resources department or manager, the risk manager or the safety coordinator. In small facilities this task may be delegated to a curator, secretary, or safety committee. A method of communication must be devised to ensure that every employee receives important information. Such notice can take of the form of memos, hazard reports, notebooks, or bulletin (white) boards.

Safety committees made up of management and workers are a very effective tool in any zoo or aquarium facility. It is difficult for any committee to *draft* policy, but it is very effective to have a committee *shape* policy to the reality of the workplace. Safety committees may also provide needed expertise in the investigation of any accident and may help to determine the root cause. The only way to ensure a safety committee's effectiveness is to exclude it from any aspect of employee discipline resulting from the investigation. This keeps the committee focused on the cause and prevention of accidents, not on how their investigations will affect co-workers.

Working safely around exotic animals begins with the research and education process. All keepers—especially entry-level keepers, students, and interns—must acknowledge the inherent danger, incredible strength, primal intelligence, and in some cases unpredictable behavior of the animals that will be placed under their care. The bridge between textbook learning and application of husbandry, training, and safety techniques begins with employee orientation and the initial employee training.

During the orientation process, new keepers should receive a copy of the facility's safety program and specific training on safety procedures. They should also receive specific instruction on where to find postings of any changes to safety policy. Most of these basic procedures will be discussed in the following section on applied keeper safety. Expect that all the training keepers receive and all policies they review will be documented with the keeper's signature for the record. The documentation, signature, and records are valuable in ensuring that all parties know their responsibilities.

Keepers should expect to spend months or even years in their early careers learning about their animals from experienced keepers. New keepers will witness both good and bad safety habits in their mentors, fellow employees, and even supervisors. The key is to assimilate the training, recognize safety habits, judge which habits are good or bad, and to incorporate the good (correct) behavior into their everyday work practice. A reality in every profession is that workers take shortcuts which often lead to employee injury or worse. A new keeper should exercise caution when he or she hears a fellow keeper say, "This is how it's supposed to be done, but this is how I do it," or sees them performing a task that looks unsafe. Rule of thumb: If it *looks* unsafe, under normal circumstances it *is* unsafe. New keepers should also spend this time absorbing all the information offered by their mentors in the initial training. It will be difficult but incredibly important for them to learn about their animals' personality traits and nuances, escapes, or history within the collection, all of which will offer clues about their behavior.

When observing the animals, trainees should mentally catalog visual clues, remembering critical information and asking themselves the following questions:

- Does this animal "challenge" an exhibit barrier frequently (daily)? Primates, bears, tapirs, and some other species will check padlocks frequently to see if they are secure, while felines check their enclosures daily to see whether exhibit furniture has shifted or storm debris has fallen to offer an avenue for escape.
- What is the animal's comfort zone (flight distance) as I work in its vicinity? Hand-raised hoofstock have a shorter flight distance and are less apt to turn and bolt from a keeper if stressed, while parent-raised hoofstock have a very large spatial requirement for comfort and will turn and run into fences or walls as if the barriers do not exist.
- Why is this animal studying me as much as I am studying it? Elephants, primates, bears, and cats are infamous for observing a keeper's routine behavior and taking advantage of any lapse in concentration or abject mistake. The results can be devastating and even fatal.

A facility that focuses on employee safety and has an excellent safety record will spend what seems (to the new keeper) an inordinate amount of time on safety-related topics. A comprehensive written safety program combining classroom and field training will provide a knowledge foundation for the keeper's entire career. A progressive keeper will continue to learn and apply safety knowledge and techniques; concentration is to the key to safety, and a keeper must stay vigilant and never let his of her guard down, or else the result may be catastrophic.

ZOONOTIC DISEASE

A zoonotic disease (Hansen 2008) is any illness that passes from animal to human. The list of zoonotic diseases is long and scary. It helps to understand the chain of events that may lead to illness. For example, the common cold virus is transmitted from person to person through intimate and extended contact. Other diseases or illnesses require only casual contact or a disease vector. Most zoonotic disease transmission results from a combination of human host, pathogen, and environmental factors. Keepers must recognize that when it comes to zoonotic disease, they hold the keys to prevention. Those keys are good hygiene and safe work practices, including the wearing of appropriate personal protective equipment.

When keepers first enter into the profession, the possibility of acquiring an illness from an animal may be of great concern to them. New keepers will practice excellent hygiene, washing their hands constantly, changing their shoes before leaving work, and storing their lunch in the staff room refrigerator. This practice may turn to complacency, however, as their fear of disease transmission recedes. In a few months their work shoes are worn everywhere, their hand washing becomes infrequent, and soon they are sharing food with the animals off the same plate or from the same bucket— a practice that is both unsanitary and dangerous.

To avoid the primary mode of disease transmission, keepers should wash their hands frequently with soap and water and concentrate on cleaning under the fingernails and the back of the hands. Soap and water removes a large percentage of the nasty stuff. To complete the process, keepers are advised to follow hand washing with an application of hand sanitizer (disinfectant) to kill any remaining bacteria, virus, fungi, or parasites.

If the zoo or aquarium does not provide a uniform cleaning service and uniform cleaning instead is the employee's responsibility, uniforms should ideally be segregated from the rest of the keepers' family laundry to prevent transmission of zoonotic disease. It is always wise to monitor the health and well-being of the collection and to know which types of zoonotic diseases are most likely to be passed between keeper and animal, especially within the facility. A number of keepers keep exotic pets, and disease transference, especially in birds and reptiles, while remote, is possible. Work shoes should never be worn outside the workplace, and especially should not be worn home.

Per the United States Department of Agriculture (USDA) Animal Welfare Act, food for human consumption shall be marked and segregated from animal food by isolation in separate vessels. This includes the establishment of separate cold storage devices (refrigerators/freezers) to keep human food away from animal food. Keepers should always consume their food away from the work environment in an area designated specifically for eating and drinking.

HAZARD COMMUNICATION

The Hazard Communication (HazCom) Standard is the most frequently cited OSHA regulation. The foundation of the HazCom Standard is to provide employees with "right-to-know" safety and health information regarding chemical use in the workplace by requiring familiarization and employee training on Material Safety Data Sheets (MSDS). OSHA recently has indicated a move to standardize the information listed in an MSDS with international systems used now, such as the Work Place Hazardous Materials Inventory System (WHMIS). So keepers are probably asking, "What does this have to do with me"?

Keepers work in a complex environment surrounded by disinfecting chemicals: chlorine and bromine for pools,

quaternary ammoniums and phenols for enclosure surfaces, and the old standby disinfecting agent, bleach—lots of it. Include the daily exposure to veterinary compounds, antibiotics, steroids, dewormers (anthelmintics), and a host of topical wipes and ointments. Now mix with the occasional radiograph (X-ray), and animal keepers are exposed to a daunting array of chemical and environmental hazards.

Keepers must quickly learn the differences between cleaning, sanitizing, and disinfection.

- Cleaning is the act of physically removing unwanted material through means such as hosing or raking an area.
- Sanitizing is the application of a chemical to enhance the act of cleaning, and will result in a kill rate for simple bacteria of up to 99.999%.
- Disinfection is the application of a specialized chemical that will kill 100% of bacteria, virus, fungi, and other microorganisms. Disinfection in a zoo or aquarium, outside the veterinary surgical suite, is virtually impossible to achieve.

Both sanitizers and disinfectants are products regulated by the US Environmental Protection Agency (EPA), which establishes the rules that govern these products, including their testing, claims, and direction for use. A disinfectant must completely eliminate all the organisms listed on its label. These organisms are not limited to bacteria, but could include viruses and fungi. Sanitizers need not eliminate 100% of all organisms to be effective, nor are fungi or viruses ever included in a sanitizing claim. For food contact surfaces, a sanitizer must reduce the bacterial count by 99.999%.

Keepers clean every day, and it is the focus of existence for some. They will often apply liberal amounts of chlorine bleach (sodium hypochlorite), a readily available over-the-counter product sold in solution (3.2% and higher), in a belief that they are disinfecting holding area floors, platforms, and enclosure furniture. Keepers will also apply granular chlorine, which is sodium or calcium hypochlorite (bleach) in powdered form, to holding areas. Granular chlorine is designed by the manufacturer to be used as a broadcast agent in pools for chemical disinfection and control of algae growth. Both practices are dangerous, as any form of chlorine, including chlorine bleach, reacts with ammonia (in urine/feces) and may result in a dangerous and noxious chemical reaction. Chlorine bleach and granular chlorine by themselves have acute and chronic effects on the user's lungs, and damage to the eyes and skin is frequent due to splash and "dust" exposure.

To achieve effective sanitization, chlorine bleach or chlorine products must be used in proper dilution and must be left on the surface being cleaned according to the manufacturer's specification for dwell time (sometimes called "contact" or "kill" time) prior to rinsing. Chemical disinfection of porous surfaces, such as floors and platforms, is literally impossible without the use of heat (steam) or of powerful disinfecting agents that, if not properly diluted and completely rinsed away, would be harmful to the animals.

Keepers are normally required to dispense medication in pill or liquid form to the animals under their care. They usually accomplish this by "hiding" or disguising the pill or liquid in a favored food item of the animal. Keepers are also asked to apply ointment or similar substances to tractable (compliant or trained) animals. A pharmacist will never handle a pill, liquid, or ointment. The primary reason, of course, is hygiene (somebody is going to swallow that pill or liquid and apply that ointment to their body). Another primary reason is that pharmaceutical compounds can be absorbed through the skin and eventually, with constant handling of drugs, a toxic reaction with the body may result, causing illness or other side effects. Handling pharmaceuticals, including topical sprays, should require the use of personal protective equipment in the form of disposable gloves. After dispensing medications, keepers should wash their gloved hands before removing the gloves, and again after removing them. When a radiograph (X-ray) needs to be obtained, keepers may be required to assist in the restraint or positioning of the animal. Proper protective equipment such as lead gowns, gloves, glasses, face shields, and thyroid protection must be worn. Disease, such as various cancers, can be traced to cumulative exposure over the course of an employee's career to radiation from X-rays, so keepers should always wear protective clothing. Pregnant keepers and technicians are at elevated risk while working in or around the animal hospital during radiograph procedures and while anesthetics are being administered. When radiographs are to be obtained, pregnant keepers should not be allowed to be in the same room as the X-ray machine, and in the field, a pregnant keeper should maintain a significant distance from the X-ray equipment. Hospitals, doctor's offices, and radiography clinics are constructed to protect employees from "scatter" radiation. Some veterinary facilities in zoos are not constructed to the same standards. This may be due to the age of the facilities or because the use of x-ray equipment was not part of the hospital's original design. In the absence of solid, grout-filled walls, pregnant keepers should be separated by one complete room from the X-ray machine while the machine is in operation. For portable or mobile X-ray equipment in the field, the pregnant keeper should consult the manufacturer's recommendations on distance or not participate in the process.

During animal procedures requiring anesthesia, veterinarians will insert an endotracheal tube into the trachea of an animal. These tubes are normally connected to exhalation exhaust systems which are ventilated to the outdoors. It is important to have a scavenging system, or a one-way positive ventilation system, for the safety of the staff attending to the animal. Prior to the insertion of an endotracheal tube, the veterinarian will routinely mask the animal to administer the anesthetic agent and relax the animal for tube insertion. Because exotic animals have specialized facial features, the typical canine/feline masks rarely fit properly and will leak anesthetic agents back into the room. The fumes from even the newest anesthetic agents can be especially harmful to pregnant keepers. Veterinary technicians and keepers can avoid this hazard by wearing a medically approved half-faced respirator, with appropriate cartridges if necessary.

PERSONAL PROTECTIVE EQUIPMENT

When a potential work hazard is recognized, changes should be made to routines and practices that will protect keepers.

There is a recognized hierarchy of response to potential safety hazards:

- policy
- engineering
- personal protective equipment (PPE).

A zoo or aquarium could, for example, set a *policy* that keepers would not be allowed in the X-ray room to position or restrain a sedated animal for the X-ray procedure. This policy would not work, however, as the animal must be positioned correctly to obtain optimum X-ray results. An *engineered* solution would place the keeper behind a lead barrier similar to a device seen in human hospitals. But this won't work either, for the same reason: the animal needs to be manually held in position. This leaves the only practical solution, which is to protect the keeper with *personal protective equipment* (PPE). PPE is always the last resort, because of the greater risk of error. For example, PPE may be forgotten, damaged, worn incorrectly, or ineffective for a particular hazard.

A hazard assessment, normally conducted by safety personnel, supervisor, or a safety committee is conducted on each work process, and PPE is assigned as required. For keepers the hazard assessment is quite basic and includes that listed in table 7.1.

Under specific medical circumstances, a keeper may need to wear a respirator in order to perform the essential job functions. Female keepers who are pregnant should avoid exposure to inhaled anesthetics during surgical procedures and to dried feces from felines to avoid potential exposure to toxoplasmosis, which could harm the fetus. Other keepers may have medically diagnosed allergies that require the use of a respirator or dust mask to reduce nuisance dusts and fumes.

According to OSHA standards (OSHA 2009, 1910.134), the

TABLE 7.1. Personal protective equipment (PPE) listed by the US Occupational Safety and Health Administration (OSHA)

Body part	Hazards	PPE
Hands	Cuts, abrasions Disease, chemicals, drugs	Leather gloves Latex/rubber gloves
Eyes	Dusts, debris, chemicals Animal fluids	Safety glasses Face shield[1]
Feet	Crushing Water, chemicals, feces	Steel-toed shoes[2] Rubber boots
Ears	Animal/machinery noise	Hearing protection[3]
Skin[4]	Sun exposure	Sunscreen

[1]A face shield, which protects the skin and mucous membranes (nose and mouth) in addition to the eyes, may be necessary where fluid splash may be encountered, to minimize the risk of viral disease transmission (from macaques), or when working with venomous animals.
[2]Steel-toed shoes, boots, and rubber boots are designed to prevent crushing injuries to or amputation of the toes, and are normally recommended for keepers of hoofstock or elephants. This type of safety shoe is also available with steel shanks to prevent puncture injuries to the bottom of the foot.
[3]The wearing of hearing protection (earplugs or earmuffs) to reduce ambient noise must be balanced against the keeper's need for awareness of the surroundings and/or of radio transmissions.
[4]Under specific risks of viral or bacterial transmission of disease, skin may need to be protected by long sleeves, long pants, or other protective covering such as coveralls or disposable suits.

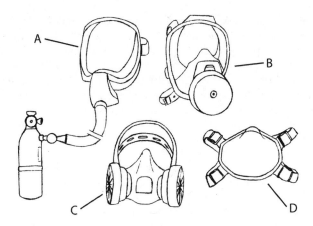

Figure 7.1. Four basic types of respirator are encountered in a zoo or aquarium: (a) air-supplied, (b) full-face air-purifying, (c) half-face air-purifying, and (d) disposable paper N95 (N, P, R95–100). Illustrations by Travis J. Pyland.

wearing of a respirator (figure 7.1), even a disposable paper respirator marked N95 or higher, requires the employer to maintain a written respirator program that defines and documents a medical professional's initial evaluation of the employee's respiratory health, an annual survey to evaluate any change in the wearer's respiratory health, and annual training on the use, care, and storage of the respirator. Keepers outside the United States should consult appropriate safety regulations for their country.

The keeper shall receive yearly training and an annual fit test by a safety professional or physician if he or she is to wear a respirator in the workplace. The respirator must then be either discarded (if disposable) or cleaned and disinfected after each use and stored in a clean, dry environment in a manner that will keep it free from contamination.

In the designation of a paper respirator, the letter—N (not resistant to oils), R (resistant to oils), or P (oil-proof)—indicates the type of material from which the employee will be protected, and the percentage of protection. For example, an N95 respirator will protect the wearer from 95% of particulate (0.3 micron penetrating size [MPPS]) exposure. Within the United States, OSHA (1910.134) does allow an employee to use an N95 (or higher) disposable paper respirator (a "dust mask," typically designed with one or two thin elastic straps and a conforming nose piece) on a voluntary basis without requiring the employer to have a formal respiratory program. Loose-fitting filtering facepieces such as surgical masks are not considered respirators by OSHA because they provide the wearer with very limited protection; they are actually designed to contain the wearer's germs, not to protect the wearer *from* germs. Hence, they are exempt from OSHA regulation.

HEALTH AND WELLNESS

New employees in any profession may tend take their health for granted, and keepers are no exception. Our well-being and ability to rebound from injury is something we may rarely if ever consider. However, if keepers take the time to consider how their difficult profession wears on the mind and body,

and thus plan for the future, their careers can be long and fulfilling.

Keepers are industrial athletes and should always warm up prior to any event. The start of the workday and the return from lunch break should be treated in a similar manner. Back injuries are debilitating and possibly career-ending. Stretching and movement prior to throwing of hay bales or restraint of an animal will help to keep a keeper healthy and fit. While keepers are subject to hand and arm injuries from the repetitive motions of raking, hosing, and using a shovel, the largest risk factor involves the back (OSHA 2009). Every aspect of a keeper's job has an impact on the muscles of the back. When a keeper is loading a wheelbarrow shoveling, crossing barriers, climbing ladders, or even just walking on wet, icy, or uneven surfaces, the risk of sustaining an injury to the back or neck is significant. According to a US Bureau of Labor Statistics (BLS 2010) analysis of worker's compensation cases, the average work-related back injury costs the employer an average of $7,600 (US) for evaluation, treatment, and therapy and from 6 to 13 days of lost time.

ERGONOMICS

Ergonomics derives from two Greek words: *ergon*, meaning work, and *nomoi*, meaning natural laws. Combined they create a word that means the science of work and a person's relationship to that work. There is considerable disagreement among physical therapists and ergonomists regarding proper lifting techniques. The Centers for Disease Control (CDC) published *Applications Manual for the Revised NIOSH Lifting Equation* (CDC 1994). This publication discusses proper lifting techniques, body positioning, repetitive motion injuries, and prevention.

For a number of years, people have been advised to approach an object to be lifted, squat down, place the strong foot slightly in front, keep the back straight and chin level, and lift with the legs. To visualize this posture, picture a toddler between the ages of 18 and 24 months attempting to master the nuances of walking and running. When the child approaches a toy or ball and picks it up, she never bends at the waist. If she did, she would fall over due to the weight of her head. A toddler squats, picks up the object, and continues on.

Adults forget that proper positioning of the head aligns the spine straight up and down. Keeping the chin level alleviates strain to the neck (cervical spine). When holding an object, bringing it as close to the body as possible before lifting it protects the chest muscles and thoracic spine from possible injury. Placement of the feet and lifting with the legs protects the lower back (lumbar spine) by using the largest, strongest muscle group, the thighs. Keepers should also use their brains, as well as their backs, by checking labels on boxes to assess their weight, test-lifting objects or animals to assess their weight, obtaining mechanical assistance, or simply asking for help when lifting.

A keeper's tools should be in good repair. Broken tools should be repaired or discarded. Hoses should be coiled and/or stored safely to keep them from becoming trip hazards. Tires on a wheelbarrow should be inflated and the wheelbarrow should never be overloaded. One should try not to "overextend" from the waist to reach for that last piece

Figure 7.2. Examples of ergonomically designed handles for common tools. Illustration by Travis J. Pyland.

of debris while raking and shoveling. Keepers who learn early in their careers to alternate their raking, shoveling, and hosing by using the "weak" or "off" hand will develop an additional skill set that will protect their bodies for years to come. Keepers should also recognize the value of tools with ergonomic handles. While uncomfortable and unfamiliar at first, the tools are designed to reduce injury (figure 7.2).

Finally, if the task is considered unhealthy or unsafe one should analyze the risks before proceeding. The task may require discussion with supervisors. Sometimes keepers get caught up in the mentality of "That's the way we have always done it," and are resistant to change or reluctant to consider it. For example: If your task frequently requires you to stack hay bales above waist level, the potential for injury will increase as the weight is lifted higher. Consider realigning the storage area to limit lifting, or possibly include a mechanical lift device to assist with positioning the hay.

COMMUNICATION

Communication in the zoo or aquarium profession takes two distinct forms. The primary form is the very important day-to-day conversation that takes place between keepers, and between keepers and their supervisor. The other form of communication takes place between keepers and the public. Communication regarding animal health and behavior is of primary importance for a safe working environment. It is essential to document the day's events in keeper reports, veterinary instruction, and maintenance requests. Communication about the movement of animals from enclosure to enclosure and from enclosure to hospital, as well as their restriction in animal holding, is needed to keep fellow keepers from harm. For example: before beginning work, it is imperative to know whether an animal was allowed access to an exhibit overnight or was secured on or off display.

Keepers must understand that conversations about animal movement that take place during lunch, during breaks, or over the facility radio do not replace the need for written documentation. Keepers who do not document animal movement in writing are assuming that spoken communication will be either remembered or passed on to another staff member. Keeper reports must be completed daily, and they should each contain a summary of the day's events in a specific work area. A keeper report that contains the word "routine" or "no change" when documenting everything that has taken place during an 8- to 10-hour workday is not, under most circumstances, engaged with their work and is usually prone to making mistakes. Safe keepers remain completely engaged with their work so that animal observations are frequent and are documented. Keepers should consider carrying notebooks (available for purchase with waterproof paper), noting their observations on whiteboards or logbooks located in animal holding areas, so that data can be transcribed at the end of the day.

As a component of a safety program, a zoo or aquarium will most likely have a form of electronic communication: radios, touch-to-talk cell phone, a paging system, or land-line telephones. All remote electronic devices need to be charged frequently, if not daily, and should be kept free from moisture, a difficult task for a keeper or aquarist. Communication devices should be kept on the keeper's person; If they are simply left in the vicinity while work is being performed, the keeper may not able to reach his or her communication tool in an emergency. The volume should be kept at a level that can be heard over the ambient sound of the workplace.

Zoos or aquariums with either radio or telephone communication may have code systems to alert their staff to emergencies. These codes, which would be defined in a facility's emergency response plan, generally come in three forms for an animal-related emergency:

- use of a specific number (e.g., "code 99" may mean an animal escape)
- use of a color (e.g., "code red" may mean an animal escape)
- use of a phrase (e.g., "Mr. Fox, please meet Mr. Polar Bear at the snack bar" means "Attention all staff: the polar bear has escaped and was last seen at the snack bar."

The important aspect of communication codes (number, color, or phrase) is that their use is limited to critical situations, practiced frequently, and applied consistently, so that all staff can recognize each code, match it to the appropriate situation, and execute the appropriate response. Communication codes may be limited to

- animal escape
- animal critical illness or death
- public injury or emergency
- fire/weapon/bomb threat
- lost child or missing person
- keeper emergency or injury.

During any emergency situation, keepers should maintain radio silence unless they have pertinent or critical information to contribute.

There is also a standardized numeric code system employed in North America called "ten-codes." Standard ten-codes that could be integrated into a numeric or color system applicable to zoos or aquariums include

10–4	understand (copy, OK
10–6	busy
10–7	out of service
10–9	repeat
10–18	urgent (quickly)
10–20	location
10–33	emergency
10–50	accident
10–70	fire.

APPLIED KEEPER SAFETY

Keeper safety starts with the basics, the job uniform. Be neat, be professional. Pull up the pants, tuck in the shirt, and either leave the jewelry at home or secure it in your locker for the day. Loose clothing and dangling jewelry may get snagged on wire, shrubbery, and machinery—or, more importantly, may be grabbed by animals. Wear work boots, long sleeves, and long pants, and apply sunscreen to lessen the cumulative effects of the sun for the duration of a career. For the same reason, cover the head with a quality hat approved for wear by the facility. Start each workday by grabbing the work gloves, taking a breath, taking a seat, and reading keeper reports from the previous workday. This is especially critical in facilities that have nighttime or second-shift keeper staff. While this routine may not necessarily have a direct impact on the specific work a keeper may perform during the day, it may have a related impact, and it serves to transition a keeper's mindset from home to work. Information is available from many sources: animal activity alerts, enrichment and training reports, summaries from veterinary rounds, and discussions with fellow keepers, supervisors, or other staff members. There are potentially three critical times during the workday when keepers are most prone to serious injury from interaction with animals: during exhibit entry and morning cleaning, during keeper talks or tours behind the scenes, and when securing animals in off-display holding for the night.

Prior to entering any off-display holding or service area, the keeper must enter a heightened state of mental awareness and preparedness. Before opening the service door or gate, look through it, or through a peep hole in it, and evaluate the surroundings. If the enclosure has an entry vestibule, a safe area protected by a second barrier, the process should be repeated. The lighting of the service area is critical; it has to be good enough so that any sign of an animal loose in the service area is readily evident. Tools in disarray, hoses uncoiled, and animal urine or feces in the service aisle are prime indicators that an animal may have compromised a barrier or that other doors were not secured the night before during lockup.

After safely evaluating and accessing an animal holding area, and prior to beginning keeper chores, another concentration technique is to safely interact with the first animal in holding by employing a training behavior, by providing an enrichment item, or by simple voice interaction. This technique again serves to clear the head of personal issues and

distractions from outside the workplace, and establishes the clear fact that work has now begun and keeper-animal safety is the first priority.

Depending upon facility policy or weather, animals may be routinely secured in their holdings (night quarters) overnight. This task might be performed by the same keeper for the duration of the work week, or by different people including night keepers or shift keepers. In facilities that are located in mild climates or which have seasonal policies for animal access to enclosures, the animals may have access to the exhibit enclosure overnight. In either case, it is the keeper's responsibility to ascertain that his or her animals are secure by verifying their location and making sure that the appropriate door(s) are secure before entering the enclosure. Failure to do so is one of the main factors in catastrophic incidents involving keepers and animals.

Verification of animal security means the keeper will

- count and visually verify the number of animals in the holding space
- physically place their hands on the locking device (e.g., padlock, shift door, hydraulic control lever) and test it to ensure that it is secure.

When an incident resulting in animal-keeper contact is investigated, the root cause is often traced to one specific action where a safety feature or policy was not followed. Examples include:

- Keeper encounters an unsecured animal due to having failed to verify that all holding doors were secured before entering a service area.
- Keeper failed to secure the enclosure's animal entry door, or failed to identify and count every animal secured from the previous day or night, before entering the enclosure to clean it.
- Keeper failed to secure the animal's exhibit entry door after releasing the animals on display, and then entered the holding area to clean it.
- Keeper failed to count and verify the location of every animal before entering the enclosure or holding area to clean it. Entering an enclosure is a particularly high risk when it requires the shifting of multiple animals through a chute or tunnel, or when multiple animals enter or exit through one exhibit entry door or shift door.

Where the design of a holding area makes it possible, keepers should always strive to have two doors between themselves and the animals. Under a *two-door policy* (or *double-door policy*), secure (empty) space is maintained between the keeper and the animals. This is especially critical when exhibits are cleaned and maintained while the animals are in holding. Sometimes, due to breeding situations and collection size, the ability to maintain two doors between a keeper and the animals may be compromised. Double-door security is achieved when an animal is shifted from the exhibit entry door into an adjacent holding area with a second secured and empty holding area between the keeper and the animal (figure 7.3).

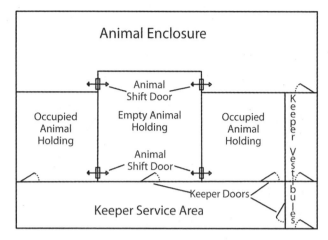

Figure 7.3. Optimum design: double–door holding and double–door safety vestibules. Illustration by Ed Hansen & Kate Woodle.

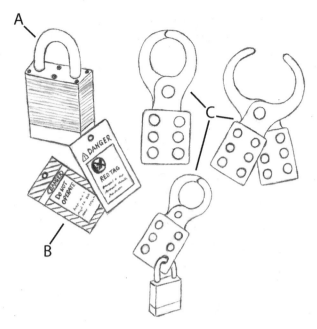

Figure 7.4. (a) Typical uniquely keyed padlock. (b) Example of danger tags. (c) Group lockout (LO) device. Illustration by Travis J. Pyland.

Under a double-door security policy, if a keeper fails to secure one door or the animal compromises one door, the second door will still provide a barrier and allow critical time for the keeper to escape. Exhibit design parameters are essential to consider when determining the need for a two-door policy, and in some facilities this design is only integrated into "dangerous" animal exhibits—for example, of carnivores and large primates.

While it is difficult and sometimes expensive to correct poor or unsatisfactory architectural design, keeper safety must be a primary consideration for zoo management and every attempt must be made to address or "engineer out" possible areas of concern. Contributing factors include the number of animals, their social hierarchy, and whether those animals will socialize while confined in a smaller holding space during cleaning or exhibit maintenance.

For enclosure maintenance performed by plumbers, electricians, horticulturists, and other crafts, and for facilities that have multiple keepers assigned to animal enclosures, OSHA's version of lockout/tagout (OSHA 2009, 1910.147), customized for the zoo and aquarium profession, should be applied. This policy would prevent animals from being inadvertently released into the display area while workers are still present.

The principle of lockout/tagout (LO/TO) places control of the "energy"—in this case the animals—under the direct control of the person(s) entering an enclosure to perform repairs or maintenance. While exhibit maintenance should always be performed with full knowledge of the keeper staff, the LO/TO program provides an additional and fail-safe method to further protect maintenance employees. The program can be also be enacted for keepers where multiple staff members may be involved in the cleaning of one enclosure prior to animals being placed on display.

A lockout device, normally a colored padlock keyed only to the person performing repairs or maintenance, is placed on the animal exhibit entry door (see figure 7.4). The padlock is keyed to only one individual or craft (backup keys are carried by the supervisor) and is used in place of the standard padlock to secure the door(s) from accidental opening by another individual. In every case, the unique padlock is ac-

companied by a "tag" that states the intent of the lockout; for example, "Danger—maintenance personnel on exhibit—do not open." For major exhibit projects, lockout devices can be deployed that have multiple openings for individual padlocks. Once the maintenance work is completed, the maintenance person(s) removes their unique padlock and tag, signifying that the work is complete, and replaces the standard padlock to secure the door. Any application of a LO/TO program in a zoo or aquarium setting requires a written policy outlining the program and extensive employee training.

Zoos and aquariums are also beginning to use interlock technology, which has been used by other industries for decades. A simple example of interlock technology is a light that comes on when any service door is left open (unsecured). A more complicated example would be a keeper door that does not open or unlock while an animal shift or exhibit door remains unsecure. The drawback with this technology in zoos and aquariums is that the technology normally requires electricity to power the interlock devices, and animal facilities use an abundance of water to clean holding areas. The abundance of moisture in the air has a negative effect on the electrical contact circuits.

Here are more tips for heightening awareness of safety and security in animal holding areas:

- Handles for shift doors and exhibit doors should be painted different or contrasting colors to emphasize the door type and whether the door is secured or open. The color pattern should be standardized throughout the facility. For example: red is the internationally recognized color for "danger," orange signifies "warning," and yellow denotes "caution." It should be noted that a significant number of people (especially males) are color-blind and may not readily identify this spectrum of colors. Many types of "safety tape" can be wrapped around the handles in color combinations (red/white, orange/black, or yellow/black) that will adequately convey the potential for danger.

- Plastic bubble-type mirrors should be placed in areas where blind corners, walkways, or other areas where animals may hide can be visually scanned before entry.
- Safety vestibules, an area of refuge that basically creates a double-door between the animal and freedom or between the keeper and the animal, should be employed in enclosures identified by the facility as containing "dangerous" species such as carnivores and large primates.
- Lights ("night lights") should be well maintained, and the switch should be located outside the keeper area so that it can be illuminated before entry.
- Barriers should extend completely from floor to ceiling to protect the keeper service area. This will prevent an animal that has escaped from a holding area from hiding on top of a holding cage in the service area and then attacking a keeper.

One of the things a keeper must consider is that for every exhibit entered, an action plan for escape must be developed in case of accidental entry into the enclosure by an animal, confrontation with an animal left on display, or even times when an animal may compromise a door that is otherwise secured. Keepers must plan for interactions with "dangerous" animals and even for accidental interactions with the most benign species, since a combination of surprise and flight distance may cause an unpredictable reaction in even the most tractable animal.

This is especially critical when multiple keepers, interns, students, or volunteers work together on exhibit cleaning and maintenance. Verbal communication between team members and physical accounting for every individual is extremely critical when working in groups. Accidents have occurred when a number of individuals, after cleaning a large display or pool, have gathered their equipment, exited the display, closed the keeper door, and released the animals while one person was still around a corner or down in the empty pool cleaning a last bit of debris. It is imperative to plan for every aspect of the profession, good or bad; so it is good practice to develop strategies to escape from an enclosure or identify areas within an exhibit that will provide refuge until rescue can be achieved: A professional keeper should develop the best plan in advance: scaling rockwork or an enclosure barrier, climbing the exhibit furniture, swimming the water feature, jumping into the moat, or—when all else fails—physically fighting for survival using work tools (rake, shovel, broom) as defensive weapons.

Keepers are wildlife educators and one of the most important resources that zoo or aquarium management can use to interact with the visiting public. Keepers volunteer or are tasked to do this in many ways, the most popular being "keeper talks" or tours behind the scenes. Behind-the-scenes tours give the public a rare opportunity to see how zoos manage and care for their collections. They are also an opportunity to provide visitors and donors with an up-close and personal glance into the zoo and aquarium world. These tours and the

animal behavioral enrichment they usually involve as part of their education message may also be dangerous for keepers due to distractions and/or the unpredictability of the visitor.

During tours that involve interaction with the animals, keepers are tasked with controlling the animals' behavior and monitoring that of the visitors. These behaviors are simultaneous and sometimes unpredictable. Since those tours are conducted in animal holding areas not normally designed to protect people unfamiliar with animal behavior, there is usually an increase in incidents until safe practices can be refined. A best-case scenario would involve two keepers, or a keeper paired with another zoo representative (e.g., an educator or docent), conducting tours and talks. The keeper would focus on the animal and the educational message while the second staff member would monitor the visitors' behavior, correcting aberrant behavior and removing noncompliant visitors from the tour. A maximum number of visitors on each tour should be established and enforced. Prior to talks and tours, keepers should immediately establish the rules and physical boundaries. When tours are conducted in areas where animals are present in off-display holding, lines indicating danger zones should be painted on the floor to reinforce the boundaries.

It is polite to face the audience while speaking. This can conflict with basic keeper safety training, which dictates that one should never turn one's back to the animal. Injuries have occurred when keepers have lost focus and leaned back into the cage mesh or turned to gesture toward the animal so that their fingers or clothing entered the animal's space. Awareness is critical when giving talks behind the primary safety barrier (the enclosure guardrail), or during behind-the-scenes events where animals are present in holding. It is good practice to select an area for the tour that allows one to face toward the animal at a 90-degree angle while addressing the group. This allows the keeper to keep all dangers in sight and to focus on the entire dynamic. It is also good practice to have lines on the floor to demarcate a visual barrier, and to fortify any animal barrier with smaller mesh to prevent accidental contact that could have serious consequences.

In some facilities, the workday is not complete until the animals are secured for the night. A number of factors come into play when exhibits are secured. The primary factor is that keepers may be tired after a long day and can be easily distracted. Once distracted, they can make mistakes. Keepers must develop a routine that includes positive visualization of the animals secured in the proper space. This is best accomplished by counting the animals physically and stating their names out loud. The verification is immediately followed by a physical check of all the security features (e.g., locks, hydraulics, electric switches). Keepers sometimes become "obsessive-compulsive" about security, but the keeper who remains attentive and practices excellent safety behavior that includes repetitive checking of animal security usually maintains an unblemished safety record.

PADLOCKS

Locking devices, normally padlocks, are the primary means of securing animals away from keepers and the public. An unsecured lock may result in an animal escape, injury to a keeper or visitor, or corrective action and retraining for the keeper. Evolving security techniques for a zoo or aquarium include a secondary safety latch that cannot be compromised by the animal, to compliment the primary padlock technique. A specially designed latch keeps a door or gate closed, and a padlock secures the door or gate. If a lock is inadvertently left unsecure, the door or gate is still held shut by the latch. It takes two independent actions to open any door with a primary lock and a secondary latch.

There are many different philosophies regarding animal keepers and padlocks. Some keepers believe that a lock is a lock, and regardless of what animals inhabit the enclosure, whether finches or polar bears, they will keep the lock secured at all times whether the enclosure is empty or not. Other zoos and aquariums have a more relaxed viewpoint and will allow a keeper to run the padlock through a door hasp (hinged strap), gate collar, or slide bolt without securing the lock if the animal is not present or if the animal is incapable of manipulating the lock and hasp. It is imperative for a keeper to understand the rules and procedures for lock security within the zoo or aquarium, and to know whether those procedures change depending upon the species.

Some facilities may use an alternate means of securing enclosures or cages, such as clips or carabineers. The simple fact is that padlocks come in a myriad of sizes that will fit any application. For simplicity, numerous padlocks can be all keyed to one single key for convenience or to limit access to qualified keepers who carry the key. Quality padlocks rarely fail when properly maintained, while clips and other devices will fail and can be easily compromised. A written policy on lock security is a foundation of safety policy for any zoo or aquarium.

Keepers new to the profession should know that some animals steal, hide, and eat padlocks. Some animals such as primates, elephants, and tapirs seem to know instinctively when a lock is left unsecured, and will immediately proceed to the unsecured door, slide the latch, and walk out. This is another reason why padlocks should always be secured. Keepers must also understand that padlocks need maintenance; they will break, and the lock mechanism will freeze in

cold climates. Because there are many ideas and suggestions on how to perform lock maintenance, keepers should check with their maintenance personnel or supervisors.

Keepers may find themselves working in smaller zoos or aquariums where exhibit maintenance, including padlock maintenance, will be a part of their regular duties. Lock maintenance should be performed at least quarterly and should include the application of lubricant to prevent padlock failure and, to a limited extent, prevent freezing during the winter. Lubricants that repel water, such as light oil or WD-40, work best on padlocks. The padlock keyhole contains a series of spring-loaded pins that conform to the key upon its insertion into the padlock. Lubrication of the shaft (hasp) of the padlock, not the keyhole, will lightly coat the lock pins. Applying lubricant directly into the keyhole will coat the key in oil. Dirt and debris from a keeper's work activity will coat the key, so that when the key is reinserted the dirt on the pins may jam the lock. Graphite, the substance found in pencils, is frequently used as a lock lubricant; though it may contain wax it also has a natural oily feel. Graphite powder is best used in "fixed" locks (like the locks on a door opened by a key) that have both drop pins and slides. Graphite powder is available from locksmiths or at any hardware store where keys are copied.

Keepers working in colder climates will probably have to thaw padlocks that are frozen by low temperature and ambient moisture. Depending on temperature, the options include holding the padlock between one's gloved hands to transfer warmth to the it, applying warm water to the exterior of the lock, or holding a cigarette lighter or small flame to the lock until it can be opened. The last example would require equipment knowledge, approval, and employee training and is not an option in wood frame enclosures or in the presence of flammable liquids or combustible gases. Keepers should also keep at least one extra padlock in each holding facility to quickly replace a damaged or malfunctioning lock. The author has observed keepers who carry extra padlocks on their belt loops for emergency use in the field, or who carry their personal lockout locks for daily use and for emergencies.

DEFENSE SPRAYS

Defense sprays have various names such as "pepper spray," "bear spray," or "OC" (*Oleoresin capsicum*) spray. These sprays have become common in the zoo community as a means of defense for animal keepers who find themselves in proximity to an escaped animal or in circumstances where an animal is accidently let on display while a keeper is still performing enclosure maintenance. Carried in a special pouch on the keeper's belt or in the pants pocket (the latter not recommended), the spray, which consists of aerosolized cayenne or red pepper, can be sprayed in the face of an attacking animal as a deterrent. The animal is effectively blinded by the burning spray and may break off the attack, allowing the keeper to escape.

As with padlocks, every zoo may have its own policy or philosophy regarding the use of defense sprays. Some zoos allow keepers to carry them; some allow only keepers working with specific animals to carry them, and some bar their use entirely. It is imperative that all keepers know and follow the policies of their employers prior to bringing and carrying defense sprays. When a zoo allows the use of such sprays, employees must be trained in the hazards associated with their deployment and in their proper use. Defense spray training programs should include an initial session in which keepers deploy the spray, judging the stream and its range, and experiencing its smell and the burning sensation it produces. Annual refresher training should includes inspection of the spray cans for damage and expiration dates. If the sprays are used, he manufacturer's recommendations regarding replacement or reuse should be followed.

Keepers should understand that when a defense spray is deployed, especially outdoors, either the wind or movement by the keeper or animal can cause the spray to go into the keeper's eyes, leading to incapacitation due to temporary blinding. It should also be understood that an animal temporary blinded by the spray may actually attack with increased fury.

VEHICLES

Keepers will usually operate a number of work vehicles in the course of their careers. Many of these vehicles, while having the same general characteristics as a car or truck, can be significantly different in how they steer and brake, and may be less stable due to their design and/or area of operation. It may be the keeper's responsibility to check such a vehicle for damage before operating it, and to perform basic vehicle maintenance including ensuring proper fluid levels, fuel type, and tire pressures.

Off-road work trucks come under varying names and in various types and styles. Most are similar to what is known as an all-terrain vehicle (ATV). An ATV may have three, four, or six wheels, and is built for off-road, all-weather capability. It may have an enclosed cab or a rollover protection device (ROP). Most work ATVs have seating space for a driver and passenger with a cargo box or dump bed behind the operator. Some models come with a trailer that can be attached to a hitch.

It is most important for a keeper to read and follow the manufacturer's operating instructions for any vehicle before operating it, and to become familiar with the zoo or aquarium's policy for vehicle operation. If the vehicle has a restraint device (all newer work vehicles will have at least

Figure 7.6. ATV vehicle, covered. Illustration by Travis J. Pyland.

a seat restraint), it should be worn by both passenger and driver. Surfaces in zoos are seldom level, and best practice is to drive the vehicle to the work, not haul the work to the vehicle. As a result, ATV-type vehicles are driven and parked on angles that may approach their design limits. If this happens they may tip over and possibly continue into one or more rolls. Vehicles should be driven up and down hills, not across the face of them, to decrease the likelihood of rollovers. Specialty vehicles such as snowmobiles would require additional employee training prior to operation.

As zoos and aquariums move towards "green" practices, electric vehicles are becoming preferred. Electric vehicles require battery charging from an electrical source. A keeper who is charging a vehicle or unplugging a charged vehicle should de-energize the cord by switching off or unplugging the source of electricity. One should not connect or disconnect an electrical device while standing in water, nor should one come into contact with the metal prongs within the plug of an energized charging cord when plugging it into the vehicle (the cord should be protected by a sheath that slides back during connection).

Keepers who are responsible for checking fluid levels on batteries for electric vehicles should do so while wearing the proper PPE. Safety glasses and a face shield should be worn to protect against damage to the eyes and skin from acid-filled batteries. A rubber or plastic apron to protect clothing, and rubber gloves to protect the hands against acid burns, are also recommended and may be required.

HAND AND POWER TOOLS

Hand tools are an integral part of a keeper's profession. The tools will include rakes, shovels, pliers, screwdrivers, wire cutters, knives, and hammers. The safe use of these tools is self-evident. However, it is imperative that the handles and working surfaces be kept in excellent shape. When a tool becomes damaged or broken it should be replaced. Ergonomic rakes and shovels that include shaped, padded, and curved handles are recommended. This type of ergonomic tool is fairly new to the zoo and aquarium market, but is worth investigating to combat repetitive motion injuries and musculoskeletal disorders.

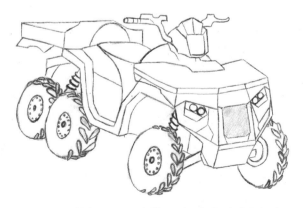

Figure 7.5. ATV vehicle, uncovered. Illustration by Travis J. Pyland.

Keepers will frequently be tasked with using power tools, especially when creating and installing enrichment devices or performing enclosure maintenance with lawn mowers, weed eaters, hedge trimmers, and chain saws. It is imperative that the operator follow the manufacturer's use and safety guidelines and wear all necessary PPE as recommended by the manufacturer or required by the facility. Many injuries occur because the operator is unfamiliar with a power tool and is reluctant to admit having little or no experience with it. Training in power tools varies according to the tool's complexity. A powered hand drill requires an overview and basic instruction, while the operation of a powered chainsaw requires intensive safety training, practical training, and, in some circumstances, certification.

SLIPS, TRIPS, AND FALLS

Other than sprains and strains, the most common injuries that will affect keepers are slips, trips, and falls. The nature of the profession is that the keeper will constantly be exposed to wet surfaces from cleaning or from the environment. The wet surfaces are then further complicated by the artificial substrates created in exhibits. These can be either smooth concrete surfaces or artificial rockwork. In either case, quality footwear is required to supply the needed traction for safety. Inattention and horseplay may lead to debilitating injury.

Keepers should strive to keep all work areas free from trip hazards such as hoses and tools. Hoses should be coiled and tools should be hung in their proper places when not in use. Holes in working surfaces should be filled or covered to prevent injury. Uneven surfaces (e.g., cracks in the floor or concrete heave) should be repaired as soon as possible. In the meantime the uneven surfaces should be marked with yellow, orange, or red paint.

In the United States, any employee performing a maintenance function greater than 1.2 m (4 ft.) from the ground or involved in a construction activity greater than 1.8 m (6 ft.) from the ground, unless working from a ladder, shall be protected from falling (OSHA 2009, 1910.23, 1926.500). This is normally accomplished with guardrails around elevated platforms for animal feeding or training, or with specialized equipment for lifting employees off the ground. Under special circumstances, specialized fall arrest systems may be required to protect keepers, maintenance, or construction personnel.

SUMMARY

In cataloguing media events and responding to media requests for comment over the last 10 years, the American Association of Zoo Keepers estimates that approximately 10 animal keepers lose their lives each year while working with exotic animals. Most of these fatalities happen in facilities outside North America, in circuses, or in what are usually described as "non-accredited facilities" such as zoos or animal sanctuaries in which staff normally consists of an owner and a team of employees or volunteers with limited training, if any.

Once or twice per year in the United States, a keeper in a recognized zoo or aquarium facility makes a mistake that results in grievous injury or fatality. With an estimated 10,000 keepers employed in US Department of Agriculture (USDA)-

licensed zoos and aquariums, the profession is a very safe one in which to work. Due to its nature, however, any serious incident is both scrutinized and sensationalized by the media and will receive nationwide attention.

Exotic animal keeping is and will always be a challenging profession. Under certain conditions it can be dangerous. Unlike in many professions where there may be no physical safety risks, mistakes can have serious consequences, and the animal will usually react and compound any keeper's error. The root cause of most animal-related incidents in the profession can be traced to two factors:

- the repetitive aspects of the work
- the unpredictable nature of the animals.

Keepers must strive to protect themselves against the former by being proactive and professional when it comes to safety. It is easy to become complacent, and it is sometimes difficult to embrace safety training, safety procedures, and policy. Keepers must embrace safety with the same attitude with which they embrace new enrichment and training concepts. Both keepers and animals embrace routine; both can find change stressful. It is good practice to vary routine, as subtle changes tend to bring tasks back into focus. But this variety must be practiced within limits, as too much change is a distraction and can also be unsafe.

Keepers can never forget that animals are unpredictable. They will develop a relationship with the animals under their care, but the animals' unpredictable nature will sometimes catch them with their guard down. Yet their bond with the animals is what makes the profession so unique. Keepers must foster that bond and sharpen all their skills, including safety skills, for the benefit and health of the animals under their care.

REFERENCES

Adams, Chris. 2010. "Ergonomic Definition." About.com guide to ergonomics. Accessed in February at http://ergonomics.about.com/od/glossary/g/defergonomics.htm.

Centers for Disease Control (CDC). 1994. Applications Manual for the Revised NIOSH Lifting Equation. NIOSH publication no. 94–110. Last modified January. http://www.cdc.gov/niosh/docs/94-110/.

Hansen, Ed. 2008. Zoonotic Diseases. In *A Summary of the More Common Maladies Affecting Animals and Humans*, 3rd Edition. Topeka, KS: American Association of Zoo Keepers.

Occupational Safety and Health Administration (OSHA). 2009. *OSHA Standards for General Industry*, 1st Edition. Washington: United States Department of Labor.

———. 2009. *OSHA Standards for General Industry*, 1st Edition. Promulgated by the United States Department of Labor. 29 CFR1910.134. Subpart I.

———. 2009. *Standards for General Industry*, 1st Edition. Promulgated by the United States Department of Labor. 29CFR 1910.147.

———. 2009. *Standards for General Industry*, 1st Edition. Promulgated by the United States Department of Labor. 29CFR 1910.23.

———. 2009. *Standards for General Industry*, 1st Edition. Promulgated by the United States Department of Labor. 29CFR 1926.500.

Official Ten-Code List. N.d. Association of Public Communication Officers.

US Bureau of Labor Statistics. 2010. Cases Involving Injury/Illness to the Back. Last modified January. http://www.bls.gov/iif/.

8

Emergency Readiness and Crisis Management

Donald E. Moore

INTRODUCTION

There is little zoo and aquarium professionals can do about the occurrence of some emergencies that have the potential to injure or kill humans or animals—extreme weather events, for instance, are forces of nature beyond their control. Keepers can take the appropriate precautions and try to wait out these crisis events. In other instances, keepers can try to reduce the probability of a crisis by double-checking doors, gates, and containment locks so that animals do not escape, or by providing plenty of water for animals, staff, and visitors during extremely hot weather to reduce the incidence of heat exhaustion in the animals and visiting humans. But in all cases, zoo and aquarium professionals can and should be prepared. Keepers can plan for extreme weather every year in their specific region of the world; we can know its effects and we can be aware of when it will be upon us. Hurricanes and other extreme storms in mid-latitudes and snowstorms at high latitudes are examples of such events. Keepers can prepare their facilities and animal housing to reduce risk of injury to animals and people during the extreme weather. Keepers can also plan their reactions to fire, to injuries to animals or people, and to animal escapes, even if they think those things will never happen to them. They can develop protocols for best practice reactions, learn from the effective reactions of others during emergency situations, and repeatedly drill reactions so that they have rapid response times and appropriate response behaviors.

Zoo professionals know from experience in school or other jobs that there is always a defined response to an emergency, and that individual responses need to be practiced. Schools often practice fire drills. The teacher acts as a classroom monitor and lines everyone up, then marches them out through a marked exit to a waiting area. Students might not know that their responsibility is simply to get in line, to be quiet to help ensure "order," and to get out as quickly as possible for their own safety in case of "the real thing." They might not know that their teacher's responsibility is to count heads to make sure everyone is accounted for, and to get them away from the possibly burning building in a clearly defined but quick fashion; or that the building manager's job is to count all groups at that meeting point to ensure that every teacher is accounted for and has actually counted heads; or that someone needs to report all of this information to a fire official, who also is going to use his or her staff to walk methodically through the building to ensure that everyone really is counted and outside. And this is all timed. In the case of a real fire, seconds count, so drills help ensure that everyone is out of a building within a certain amount of time, and that they all go to a designated meeting place where they can be counted. Hopefully, keepers understand all of these roles and responsibilities from regular fire drills at their zoo or aquarium.

Zoos and aquariums are sometimes large institutions with multiple buildings and many people, both staff and visitors, on any given day. To ensure success in emergency response, all emergency procedures must be provided in writing to staff and, where appropriate, to volunteers. Zoo personnel (professional staff, seasonal or temporary staff, and volunteers) must be trained to understand the zoo's emergency policies and procedures, and must train for their specific responses to an emergency. Emergency equipment and supplies, as well as written protocols, must be readily available in many appropriate locations (e.g., key keeper access points in a building or area) and at all times in the event of an actual emergency, which can happen anywhere and at any time of the day or night.

This chapter will provide basic information about emergency readiness as it affects and is affected by the keeper position. After studying this chapter, the reader will understand

- the basic types of emergencies that are important in the zoo or aquarium environment
- the importance of being prepared to respond to an emergency
- institutional versus individual responses to emergencies
- the keeper's role in emergency response.

CRISES TO CONSIDER

Written procedures for zoos and aquariums should deal with at least four basic types of emergency: fire, weather/environment emergency, injury to staff or a visitor, and animal escape. Other kinds of emergencies which may be just as important, depending on your zoo or aquarium include venomous animal bite (a very specific type of human injury which requires a very specific response), chemical spill, unauthorized human in animal exhibit, inappropriate human-animal interaction, bomb threat, missing child, armed robbery, and overaggressive visitor. In many cases, zoo personnel will call local police or fire personnel for some of these emergencies, but it is important that all members of the zoo staff understand their roles and responsibilities during the response to each and every type of event.

INSTITUTIONAL READINESS

There are multiple levels of institutional readiness. Every institution should have a risk management policy and an emergency management manual. These are used to help zoo personnel understand all of the risks zoo leadership has thought about, and how they intend for staff to respond to emergency situations. Although zoo and aquarium leaders may regularly review these policies—along with emergency training, drills, and actual event responses to ensure that the institution is ready for any challenge—keepers will need to read and understand them.

At the departmental or "zone," "section," or "line" level, keepers may also have some kind of annual training and drilling plan to ensure that they are ready to respond effectively in emergency situations. Staff roles and responsibilities are often defined at this level. Keepers can expect to go into more detail about individual facilities, animals and their behaviors, and so on during their training. They may also be asked for input on appropriate responses.

Preventing mistakes from becoming disasters may be accomplished through good planning for containment of a crisis. A minor hay fire in a barn might be contained through the judicious use of a fire extinguisher. An escaped animal might be contained within the zoo by effective perimeter fencing. For this reason, perimeter fencing should be separate from all exhibit fencing or other enclosures, and should be of good quality and construction. In USDA-licensed, AZA-accredited zoos in the United States, all facilities must be enclosed by a perimeter fence which is at least 2.4 m (8 ft.) in height or by some other acceptable barrier (e.g., physical geography); the perimeter fence should be constructed with no gaps below the fence or gates that would allow feral or wild animals to enter the grounds or allow a collection animal to exit (egress) in the event of its escape from a primary enclosure. A keeper's role is to help ensure that these secondary escape-preventative devices are in place and in good working order. Keepers may also be asked to help with regular checks of all fences in their work area, exhibit, and perimeter emergency containment systems, especially if they are asked to serve on the institution's safety committee.

CREATING A MODEL FRAMEWORK FOR AN INSTITUTIONAL CRISIS RESPONSE PLAN

A framework is any process used to create a self-sustaining program. One accepted "model framework" for zoo planning is the SPIDER model, first developed in the United States (Mellen and McPhee 2001), which is used for managing animal enrichment and training programs and is based on a good small business model. Using the SPIDER model as a framework or an institutional crisis response plan has also been beneficial.

The *S* in SPIDER stands for setting goals. In this phase, the safety committee or some other leadership working group has defined the goals for a particular program. If one goal is to prevent animal escapes, then there will be sub-goals of examining and implementing locking and containment systems, training keepers in basic safety and security, and so on. Another goal might be to have an escape response time of less than five minutes by appropriate staff, and to have other staff help visitors to "shelter in place" (i.e., take immediate secure shelter in the closest possible building) during that same five-minute period. Another goal might be to train and drill on animal escapes at least quarterly so that different kinds of animals can be the subjects of drills throughout each year, and also so that all staff are exposed to drills.

The *P* in SPIDER stands for planning. During the planning phase it is a good idea to identify all the resources—team members' abilities, tools, external help (police, fire), and so on—that are available for crisis response. It is also a good time to set a calendar for review of crisis response actions, training, drills, and such other necessary activities as safety committee meetings (a regular best practice for reviewing safety issues and emergency responses).

The *I* in SPIDER stands for implementation. The biggest need in the implementation phase is role clarification, also fondly known as "Who does what by when?" You need to completely understand your role and what is expected of you, by what date or time, so that you can be successful at crisis preparation or crisis intervention.

The *D* in SPIDER stands for documentation. Documentation gives safety teams proof of response times and other noteworthy events, gives the area team information for in-house review of crisis response, and may give the institution support in case the emergency is later reviewed by an external agency. Personal documentation, in keeper reports or other official notes, is very important to the self-sustaining program process; depending on the institution, sometimes it is better to document any concerns and offer them to the management team.

The *E* in SPIDER stands for evaluation. Evaluation of activities will help provide answers to questions about the zoo's crisis response program. It is best to meet and discuss response to an emergency situation immediately after the event. A good evaluation, which can be documented, can also be done through discussion in your area's regular safety meetings, or in the zoo's monthly or quarterly safety committee meetings. This is a time to openly and honestly discuss which activities went well and which need adjustment.

The *R* in SPIDER stands for readjustment. As noted above, you might discuss readjustment at your regular meetings.

If your team is not making steady progress toward crisis management goals, discuss what you need to adjust with the zoo's leadership so that the goals can be met.

Note that the SPIDER process can be used effectively in a proactive (annual program review and training) or reactive way (e.g., to plan for an extreme weather event). Other examples of proactive use of SPIDER would be to set time-of-response goals, or calendars for training and drilling for animal escapes or venomous animal contacts. Another example of a reactive use would be when a team trains for response to an oil spill or an event of increasing security concern.

POLICIES AND PROCEDURES

Development of an institution's risk management policy and associated crisis response procedures requires significant time and includes important basic safety goals that must be learned by all members of the team; this is a professional responsibility of all staff, including keepers. You can expect the policy to include an institutional philosophy and vision: the philosophy might be that all employees are wholly important and deserve to return home whole at the end of the day, and the vision might be that lost-time accidents will be kept to a minimum, and that risk to employees, visitors, and animals will continually be reduced.

All emergency procedures must be written and provided to staff and, where appropriate, to volunteers. Often this is done through an institutional safety or policy manual, and training is done via an accompanying employees' safety orientation. Information on appropriate emergency procedures must be readily available within the institution for reference in the event of an actual emergency. These emergency procedures should deal with at least four basic types of emergencies—fire, weather/environment, injury to staff or a visitor, and animal escape—and also for additional emergencies to which the institution may be particularly vulnerable (e.g., venomous snakebite).

It is best practice to have hard copies of institutional protocols available in each area and for each individual employee, so all policies and procedures can be accessed during emergencies, especially if the power goes out and computers are unavailable). Supervisors should require keepers to review their updated hard copies at least quarterly, and there should be an administrative documentation of this review.

TRAINING AND DRILLS

The training of staff in the above emergency procedures is undertaken regularly in every accredited zoo (and is encouraged in all other similar organizations), and records of such training are maintained. This training in emergency drills ensures that the institution's staff know their duties and responsibilities, and how to handle emergencies properly when they occur. Safety awareness training sessions should stress personal responsibilities for a safe and secure work environment, situational awareness during both standard operations and emergency situations, personal physical fitness, general communication techniques, and other elements required for a safe working atmosphere. Each training session should be tailored for an individual unit; curators, supervisors, keepers should work together to incorporate situational issues

---

> ### Case Study 8.1. Emergency Preparedness Training
>
> At Smithsonian's National Zoo, training and preparations for animal escapes now take place through classroom training and drills that are scheduled for several times per year in order to address the complexities of a large multidisciplinary staff working with and around a diverse living collection with multiple running, leaping, and flying taxa. Several species that are considered dangerous include large cats, great apes, venomous snakes, and large aggressive birds. When keepers and curators in the zoo's animal departments noticed that employees in nonanimal departments were not responding to animal escape drills, the zoo instituted department-specific training for everyone including nonanimal department personnel. This includes classroom training using an interactive PowerPoint presentation for maintenance staff, horticulture staff, and even seasonal or temporary guest services staff who work in cafes and parking lots. Every new employee receives animal escape training as part of his or her initial orientation, and this training is documented. Live-action on-site drills for animal escapes also now involve all departments of the zoo, going beyond the animal department, and include guest services and other staff. The responses and response times of employees are assessed immediately after each drill; the point of this evaluation is to readjust response procedures as necessary so that response during a real event can be most effective. This is not necessarily standard practice in all modern zoos, although it should be.

---

and specific challenges. Supervisors should be responsible for follow-up inspections to assure compliance on safety protocols, but keepers are sometimes asked to help in this, particularly if they are members of the zoo's safety committee.

Emergency drills should be conducted at least annually for each basic type of emergency (fire, weather/environment, injury to staff or a visitor, animal escape) to allow the zoo's leadership to determine whether all staff are aware of emergency procedures, as well as to identify potential problem areas. These drills must be documented and evaluated to ensure that procedures are being followed, that staff training is effective, and that what is learned is then used to correct and/or improve the zoo's emergency procedures. Records of these drills must be maintained and improvements in the procedures duly noted; these documents may be assessed by regulatory agencies. Note that professional associations around the world (e.g., AZA in the United States) often require institutions to annually train staff and drill on these four basic types of emergencies.

PREPAREDNESS FOR SITUATIONAL EMERGENCY RESPONSES

The institution must have a communication system that can be quickly accessed in case of an emergency. That is, there should be immediate access to designated persons via handheld radio, pager, mobile telephone, intercom, land telephone, alarm, or other electronic devices.

FIRE RESPONSE PLAN

Each zoo will have a plan to minimize the occurrence of an accidental fire; keepers should know this plan and act accordingly. Alarms for fire, security, and other safety alerts must be in place and in working order, according to local laws and most zoo accreditation standards. Routine maintenance records should be kept, detailing safety checks of the equipment.

FIRE DRILLS

Fire drills should be performed at least annually. The questions for each local zoo or aquarium will be: Should the fire drill be restricted to the biggest building in the facility? What about separated buildings? What about a fire drill for an external barn or storage shed? In any event, keepers must know their roles and responsibilities for each and every one of these situations. In some cases a keeper may be an active "building manager" responsible for counting the people who exit and meet in a designated area. In other cases, the keeper's responsibility may be to help direct fire trucks coming in through a perimeter gate that happens to be nearby. Remember that keepers' responses during drills will be very close to their responses during a real emergency. Take each drill as seriously as if it were a real emergency. A local zoo may choose to cede responsibility for animal evacuation (and perhaps even destruction) to keepers—and if this is the case, keepers should practice this action during each drill.

HUMAN INJURY

Accredited zoos and aquariums will have a written plan available to all staff for first aid and other various health emergencies involving staff or visitors. This plan should include a list of all qualified first responders working at the zoo, and emergency numbers for local ambulance, police, and fire departments. The institution should have an automated emergency defibrillator (AED) and provide cardiopulmonary resuscitation (CPR)/AED training to appropriate staff.

HUMAN INJURY DRILLS

Keepers should know what their roles and responsibilities are in stabilizing and getting help for an injured person. Human injury drills should be performed annually in each facility, and keepers' roles must be practiced during these drills. If a zoo or aquarium supports staff training in first aid and CPR, these valuable skills can be useful at home as well as at work, and it is rewarding for keepers and other staff to be able to help someone appropriately when they are injured.

Sometimes human injury drills in a zoo or aquarium environment are very interesting, because they can incorporate many different scenarios or emergency protocols. If the injury drills take place in conjunction with an animal escape drill, keepers may expect to see multiple injuries as the "escapee" attacks and injures more and more people. These kinds of realistic drills help keepers refine their roles as critical members of their facility's response team.

WEATHER PREPAREDNESS/RESPONSE PLAN

As a general rule (and, in the United States, as a regulation) the animal collection must be protected from any weather that may be detrimental to its health. Animals not normally exposed to cold weather should be provided with heated enclosures. Likewise, protection from excessive heat should be provided to all animals, particularly those normally occurring in cold climates. All animals (and people) need protection from extreme weather like snowstorms, extreme heat, hurricanes, tornadoes, ice and wind storms, and flooding.

In AZA-accredited zoos and aquariums, critical life support systems for the animal collection—including but not limited to plumbing, heating, cooling, aeration, and filtration—must be equipped with a warning mechanism in case of failure, and emergency backup systems must be available. All mechanical equipment should be under a preventative maintenance program evidenced through a record-keeping system. Special equipment should be maintained under a maintenance agreement, or a training record should show that staff members are trained for specified maintenance of special equipment. Therefore, facilities such as aquariums, tropical rainforest buildings, or other exhibits which rely on climate control for life-sustaining conditions must have emergency backup systems and a mechanism for warning if any of these systems are malfunctioning.

What is a keeper's role? First, it is to check the zoo or aquarium's emergency manual for his or her responsibilities. Second, it is to ensure the best possible animal welfare at all times. This means that keepers need to monitor incoming weather as much as leadership does, and to ensure that the zoo's facilities and life support systems are sound under normal working conditions. Keepers need to notify supervisors and fellow keepers in advance of incoming weather if they are concerned about facilities or system failure that could affect the lives of the animals in their care.

Keepers may be deemed "critical staff" members in times of adverse weather and other emergencies. "Critical staff" are expected to remain on duty providing care and protection for the animals during a weather crisis. To prepare for this, keepers should maintain a "go bag" with appropriate clothing and food; work colleagues and personnel from other agencies do not need an extra victim because keepers did not prepare adequately! Local emergency agencies may have a list of the items zoo personnel should have or carry during such an emergency situation. Keepers should always take emergency responsibilities personally and be prepared, and pass critical information on to appropriate supervisors or safety committee members so all can benefit.

WEATHER "TABLETOPS" AND DRILLS

A zoo or aquarium's leadership may do "tabletop exercises" for weather emergencies because it is so difficult to actually drill for these emergencies effectively. A "tabletop" is simply a detailed planning exercise in which all appropriate personnel walk through the crisis response plan (often with a large map of the facility on the table in front of them), and sequentially ensure roles and responsibilities for the plan, for personnel

> **Good Practice Tip:** A "go bag" is a good thing to have ready at home or work, in case you need to respond to a multi-day emergency. The type of emergency will dictate what is in a "go bag"; a veterinarian's "go bag" for an animal escape might be different from a keeper's "go bag" for staying multiple days in the zoo during extreme weather. And both will differ from a first responder's "go bag" for wilderness rescue. Check the bag regularly (monthly) for readiness.

> **Good Practice Tip:** Keepers can keep a copy of the zoo's emergency manual, including area protocols, and an up-to-date list of the zoo's emergency phone numbers at home or in a cell phone. This information could be needed at any time.

scheduling and placement, and for reaction and action ("who does what by when").

A "windstorm" (e.g., hurricane, tornado, or typhoon) will be used as an example. Let's say that windstorms are a recognized annual threat in an area. That zoo's emergency plan addresses windstorms and establishes roles and responsibilities on a departmental and individual basis. The zoo's leadership decides that it would be a good idea to have a "tabletop (planning) exercise" each year several months before the windstorm season, and an evaluation/readjustment exercise for all staff a couple of months after the windstorm season. The keepers' role in this is to ensure the integrity of their exhibit areas (e.g., by securing animals indoors and checking the area's overall security before the storm), and to report deficiencies to supervisors at least two months before the windstorm season. As the tabletop exercise team moves through the calendar and gets closer to the windstorm season, supervisors have "fixed" most facilities and keepers are scheduled to be part of the "critical staff" in place. Keepers check the "critical food list" to ensure that they have at least a two-week supply of food on site, and they check their "must-have list" for their personal "go bag" (for instance, clothing and canned goods as well as toiletries and other life supplies) knowing that they might be living at the zoo for one to two weeks. Every individual in the room has their personal role and responsibility in the weather event, and the zoo feels prepared. As the windstorm season gets closer, the zoo reviews keeper responsibilities and readiness on a weekly basis to ensure that everyone is ready; this is an extension of the tabletop.

EVACUATION PLAN

Each zoo and aquarium will have an evacuation plan. Evacuation plans are used for immediate threats including rapid-onset weather, bomb threat or other threat to human life, animal escape, and fire. As in other plans, each person will have a role and responsibilities in this procedure, and it is a keeper's responsibility to know theirs. The plan will provide for "zoo site" accesses and egresses, building security, who locks or secures areas and notifies the "communications team," and so on.

EVACUATION DRILLS

Each zoo and aquarium should have an evacuation drill on a regular basis. The keepers' job will be to ensure that "their"

animals are safe and accounted for, the area is locked down, and all personnel are evacuating with them (advance conversation about access to animals by emergency personnel should occur). As zoo personnel, keepers may be required to help evacuate visitors, possibly including seasonal staff and contractors. As in fire drills, the zoo should have a "designated meeting place" for all people evacuating the area.

ANIMAL-HUMAN INCIDENTS

Note that animal-human incidents include "unauthorized person in exhibit" as well as "unanticipated animal-human contact." Like many international zoo bodies, the AZA requires that institutions maintaining potentially dangerous animals (sharks, whales, tigers, bears, etc.) must have appropriate safety procedures in place to prevent attacks and injuries. Appropriate procedures must also be in place to deal with an attack resulting in an injury. These procedures must be practiced routinely according to the zoo or aquarium's emergency drill requirements. Whenever injuries result, a written account outlining the cause of the incident, how the injury was handled, and a description of any resulting changes to the safety procedures or physical facility should be prepared and maintained by the institution for a specific amount of time after the incident, as defined in the jurisdictional legislation.

Regional zoo associations like the AZA, the Canadian Association of Zoos and Aquariums (CAZA), and the European Association of Zoos and Aquaria (EAZA) may also require that all areas housing venomous animals, or animals which pose a serious threat of catastrophic injury or death (e.g., venomous snakes, polar bears, killer whales, or large felines), must be equipped with appropriate alarm systems and/or have protocols and procedures in place which will notify staff in the event of a bite injury, attack, or escape from the enclosure. These systems and/or protocols and procedures must routinely be checked, and periodic drills must be conducted. It is the keeper's responsibility to report any potential malfunction of these safety systems.

Institutions maintaining venomous animals must have appropriate antivenin available, and its location must be known by all staff members working in those areas. An individual on staff should be responsible for inventory, disposal, replacement, and storage of antivenin. Venomous animal protocols should be established with local medical experts, and must be practiced with local police and ambulance personnel and the local responsible emergency room at least once each year. Drills for getting humans out of exhibits or out of encounters with animals must also be conducted. Each zoo will have specific areas and specific animal spe-

Box 8.1. Accredited Zoo Best Practice: Example of Animal Escape Protocol for a Zoo

This zoo protocol prescribes policy, responsibilities, and procedures in the event of an animal escape.

Animal escape code. The zoo's "escape code" is a radio code that signifies that an animal is out of its primary enclosure. A zoo's code may be based on a color ("code red"), an event ("code lemur" or "code 99"), or may just be stated in plain English, as in: "I am reporting an animal escape of [species, number of animals, or identity of individual] at [location]").

Person-in-exhibit code. This code signifies that a human (unauthorized visitor or intruder, or staff member) is in an animal enclosure and is at risk from the animals in the exhibit, or that the animal is at risk from the person, and may be some modification of the escape code.

Recapture supervisor. Most often this will be the curator of the unit from which the animal has escaped. If the unit curator is unavailable, the area's senior keeper (the person most familiar with the animal) on the scene acts as recapture supervisor until another animal supervisor arrives. That supervisor then works closely with the keeper(s) most familiar with the escaped animal. In the event of an escape of an animal in the animal hospital, the veterinarian or veterinary technician on duty shall serve as recapture supervisor.

Refer to the following documents as needed: Zoo emergency contact list, contact numbers for animal emergencies, staff emergency telephone list, and other contact numbers located in crisis notebooks of senior staff (e.g., director or supervisors).

The zoo's policy is that the appropriate radio code shall be called for any animal escape or person-in-exhibit incident.

RESPONSIBILITIES

Senior supervisor for animal care department

- establishes and implements the policies, procedures, and responsibilities relating to animal escapes
- ensures that this zoo directive is reviewed annually with each keeper and all supervisors of zoo departments, the zoo's support organization, and police. Each new staff member shall be given a copy of this directive during orientation
- ensures that the animal care staff emergency telephone list is current and available to the appropriate zoo staff and police.

Recapture supervisor

- is primarily responsible for requesting the appropriate level of response, depending upon the species and individual animal, its location (e.g., outside its primary holding area but secondarily confined or approaching the perimeter fence), time of day, and proximity to visitors
- maintains control of all radio transmissions and direct employees during the event
- may designate another individual to assist with event logistics, including perimeter control and liaison with staff not directly involved with the recapture.

Veterinarian

- ensures that the veterinary staff maintains an animal escape protocol for equipment readiness, which shall be posted for hospital staff and revised annually, and which shall include anesthetic dosages for all species considered potentially dangerous at the zoo
- ensures that drugs and drug delivery systems suitable for recapture are available to the veterinary staff, that the veterinarians are trained for their use in escapes, and that contact numbers for the veterinary emergencies telephone list is current and available to the appropriate animal staff, veterinary staff, and police
- assists the recapture supervisor as necessary
- works with the recapture supervisor when chemical immobilization is required
- in some cases (e.g., absence of appropriate curator, supervisor, or keeper, animal escape from the hospital, or quarantine), assumes the role of recapture supervisor.

Police department or *security team* (some zoos have their own security teams, others work directly with police)

- ensures that these procedures are reviewed annually with each member of the security staff
- ensures that each new officer shall be given a copy of this directive and that the contact lists cited above are available to the security supervisor in case of an escape at night
- ensures that suitable caliber weapons and ammunition are available to the security team, and that security officers or curatorial and animal care staff are trained and qualified to use them to kill escaped animals.

Animal care (keeper) staff

- discusses passive containment strategies for their unit(s), and makes sure that the staff is familiar with behavioral management techniques that allow an animal to return to its home enclosure after an escape (e.g., graded response via passive opening of gates to allow animal to return on its own, baiting of the animal with food, negative reinforcement or "herding" of non-dangerous animals, veterinary use of tranquilizer darts, or finally destruction of a dangerous animal to protect human life).

DISCOVERY AND NOTIFICATION

Broadcasting the discovery. The person who discovers an escaped animal shall attempt to keep it under observation from a safe distance. Announcements of an animal escape shall occur as follows:

The person who discovers an escaped animal should announce the appropriate code on the radio to alert all zoo staff of the situation. The initial announcement shall be made on the zoo's animal department radio channel, if the zoo has multiple channels.

Information to be conveyed. The first person to spot the escaped animal (whether a keeper or not) should state in the initial escape code radio announcement:

- the caller's name
- the escape phrase "code (__)"
- the best possible description of the animal including species, age, sex, and number of animals if there are more than one
- the animal's specific location and any direction in which it may be heading
- whether medical attention is needed
- The level of response requested, if known.

The first keeper or animal care supervisor responding to this call should confirm by repeating the transmission.

Actions by recapture supervisor. The recapture supervisor should then announce the animal's escape over the radio and request immediate perimeter gate closure if necessary. They should then request any necessary assistance from keeper staff, veterinarians, police, facilities staff, and health unit staff. The recapture supervisor should request that both the veterinarians and the police be contacted by telephone if immediate assistance is required. Recapture supervisors or their designee should contact the zoo director to apprise them of the situation.

Response team. The following employees shall respond to the code announcement as requested by the recapture supervisor (to keep the escape scene as clear and unencumbered as possible, employees should not report unless requested): keepers, animal care staff and supervisors, veterinary staff, police, health unit personnel, facilities staff.

"Escape code" contact tree. The director's office shall notify zoo personnel as necessary, using the telephone numbers cited in the contact lists above. It is also best practice to include public affairs staff.

IMMEDIATE ACTION BY ANIMAL STAFF

The discoverer of the escaped animal should keep it under observation and not attempt to recapture it. The on-scene senior animal keeper with knowledge of the animal is the recapture supervisor until the arrival of the unit curator or manager, who shall then assume the role of recapture supervisor. All keepers from the escaped animal's area should report to the scene but approach cautiously and listen for directions from the recapture supervisor. No attempt should be made to capture the animal until the recapture supervisor initiates a plan. Nets, gloves, and other capture equipment should remain hidden at this time because it otherwise might scare the animal into a fight-or-flight response.

Excess and unrelated radio traffic must be avoided. Once the recapture supervisor is in place, he or she shall control the radio traffic and may repeat the following: "A(n) [escape code] has been announced, so radio traffic is restricted. No persons may send or acknowledge radio traffic on the code channel except those involved in the code action." After the initial radio an-

nouncements, the zoo's senior officer on duty shall rebroadcast the restriction announcement as needed.

All other animal care supervisors should report to the scene but stay at a safe distance unless and until needed. All other keepers should await instructions and report to the scene only if requested by the recapture supervisor. They may be asked to form a perimeter or otherwise assist in the recapture. Nets, gloves, and other capture equipment must remain hidden until their use is requested by the recapture supervisor. The initial goal is to establish a wide perimeter around the escaped animal to try to contain it while a recovery plan is being formulated, and to keep people safely away. Others may be involved in supervising the evacuation of visitors.

Recapture Supervisor

- directs keepers by radio in establishing the perimeter and requesting police assistance as necessary
- advises other zoo supervisors to keep visitors and non-involved zoo staff at a safe distance, inside buildings if necessary
- closes the zoo's perimeter gates by instructing other supervisors to have them closed if necessary (this should be done for any escaped carnivore or large primate)
- chooses a keeper with a radio to assist the veterinarian in carrying equipment during the recapture effort
- formulates a recapture plan and communicates details of that plan as necessary, requesting additional staff and recapture equipment as needed, and delegates any of the above actions to another staff member in order to focus directly on the recapture effort calls off the "code" when the animal is recaptured and secured.

Keeper or Animal Supervisor

- The escaped animal's unit should immediately implement a passive containment plan to allow the animal to return on its own to a home enclosure, and/or to prepare an enclosure for the escaped animal after it is recaptured.
- If the escaped animal is outside the secondary containment, and poses a risk of entering another animal enclosure, appropriate keepers should immediately try to shift all bears, big cats, wolves, apes or other dangerous animals into their secure holdings.
- The primary enclosure from which the animal has escaped must be checked for breaches, and cage mates, if any, should be counted and secured.
- At the direction of the recapture supervisor, keepers on the scene should begin to establish a perimeter.
- There should be no premature attempt to recapture the animal, and no nets, gloves, or other capture equipment should be visible to the animal unless directed by the recapture supervisor.
- In the case of a dangerous escaped animal (e.g., a bear, big cat, ape, or large monkey), only keepers and supervisors with radios should initially help to form the perimeter. In some instances this response must be made in vehicles affording protection to the responders. The perimeter should be well beyond the animal's flight distance. Other zoo staff may be asked to help form the perimeter.

Box 8.1. continued

- Once the perimeter is established, the recapture supervisor may direct the keepers to adjust the perimeter to allow or encourage the animal to return to its home enclosure on its own. The animal should not be approached closely, and there should be no waving, shouting, or running that might alarm the animal.

Veterinary Staff

- Veterinarians shall respond as quickly as possible if they judge that they are needed, or if they are requested to do so by the recapture supervisor.
- If capture or immobilizing equipment is needed, the veterinary staff shall assemble the equipment and transport it to the scene.
- The veterinarian(s) shall work with the recapture supervisor to capture the animal.

In case of nighttime escapes, available security officers or building engineers should help to safely monitor the animal's location until appropriate curators, keepers, and veterinarians arrive.

AFTER RECAPTURE

All staff shall remain on standby in alert status until the stand-down order is given. When the escaped animal has been recaptured *and* secured, the recapture supervisor shall give a direct stand-down command (e.g., "[escape code] is now called off, all staff stand down"), upon which the gates may be opened and the public may be allowed to use the zoo grounds. The recapture supervisor shall make an immediate verbal report to the director, and assist in preparing a statement for the media if necessary.

The recapture supervisor will request a debriefing of all staff involved immediately after the event. The recapture supervisor shall provide a written report of the incident to the director within 48 hours.

cies for which response protocols should be detailed and practiced.

ANIMAL ESCAPE PLAN

In all zoos and aquariums, it is best practice that all animal exhibits and holding areas be secured to prevent animal escape. Particular attention must be given to shift doors, gates, and keeper access doors to provide for staff and public safety. Locking or latching mechanisms as well as safety vestibules are necessary to meet this standard for dangerous animals. Two-person safety buddy systems for shifting and lock checks are good safety practice. All exhibit service areas must be safely lighted, free of debris, and spacious enough to allow for safe servicing. Service exit doors must be clearly marked, and all locks and shift doors must be in good working order.

In case of animal escape, capture equipment must always be in good working order and available to authorized, trained personnel. Best practice is that capture equipment should be located in designated areas throughout the zoo, clearly marked with access kept free of storage items and debris, and that keepers should inspect this equipment regularly to ensure that it is in working order. Appropriate emergency capture equipment includes safety gloves (different types for different species), nets (different types and sizes for different species), transport boxes, and noose poles.

This capture equipment should be dedicated to emergency recapture only, but if this is not possible due to limited resources, there should be an equipment sign-out system in which borrowing of equipment is clearly communicated to local staff, and the equipment should be returned to its proper location immediately after use and appropriate cleaning so that it is consistently accessible to staff. Keepers should practice quickly accessing and transporting appropriate equipment during drills. There are species-specific firearms and firearms usage protocols that will be specific to each zoo, and keepers should be familiar with them.

ANIMAL ESCAPE DRILLS

Like other drills, animal escape drills should not occur with untrained staff. Staff should first be trained in the animal escape emergency response process, and then practice it during announced drills. Then and only then should the zoo's administration have unannounced drills to test the real response time and abilities of the staff. The question of whether to involve the zoogoing public is complex. Nervous administrators can have personnel respond to a papier-mâché animal using plastic equipment or equipment tags rather than the real thing, but drills need to happen when the public is present in the zoo so that zoo staff know how to work around visitors safely. Visitors also need to be "evacuated" or "sheltered in place" from the dangerous animal that has escaped. Tools, locations and local emergency response maps will vary among institutions, and therefore local protocols will vary.

COMMUNICATION

One of the most important communication strategies for a keeper is to be open, honest, and direct with supervisors or police response teams, especially during a crisis. Keepers need to use their entire life experience, assess the developing crisis situation, and react to a potential problem with a thoughtful, proposed solution that will add to the success of the team. A good example might be the very public story of a tiger escape that occurred in San Francisco at around 5 p.m. on Christmas Day, 25 December 2007. Most of the staff had already left the zoo, but one employee who was also a member of the zoo's dangerous animal emergency response team was still on site. Although the zoo's retail employees refused entry to a young man who was bleeding and who claimed to have been attacked by a tiger, because they apparently did not believe his story, the animal care employee took the situation seriously and reacted to attain appropriate equipment to protect visitors and contain or eliminate the threat. When the police arrived in response to a phone call, this employee accompanied them in a squad car and helped direct them to the escaped tiger. The police then dispatched the tiger and ended an escalating emergency situation. This employee assessed the situation and explained it to the police calmly and clearly.

A zoo or aquarium's communications protocol may direct a keeper to respond or not respond to an emergency, and to maintain radio silence so that the appropriate personnel can react to the emergency. For this reason, if a keeper has experience that makes him or her uniquely qualified to assist in a zoo emergency, they should communicate that ability to zoo or aquarium leadership in advance of an emergency situation. Openly communicating expertise during emergency team planning exercises—a background in rifle marksmanship, chemical spill response, firefighting, mountain rescue, or any number of unique skills—can help in an emergency situation, and should be welcomed by zoo management.

COMMUNICATIONS PLAN, PREPARATION, POLICIES, AND IMPLEMENTATION

"Crisis" is defined as a situation that could be of significant concern to the general public and has a high potential of resulting in extensive news coverage. Examples might include the accidental injury or death of an animal, a zoo employee, or visitor; or civil disobedience, like a major animal rights protest at the front gate of the zoo.

"Crisis communications" is communication with the news media (and, through them, to the general public) during a crisis. The objective of a communications plan is to prescribe policies and procedures for the coordination of communications within the zoo, to the news media, and to the public in a time of crisis. When a crisis communications team is alerted, it should work quickly to develop a response for the news media that is as accurate and complete as possible. When media are being communicated with and there is more than one zoo or aquarium spokesperson for the crisis, it is critical that these institutional representatives coordinate their responses to ensure consistency of message. The crisis communications team shall establish a staging area for media to ensure the safety of the media personnel, zoo staff, and animal collections, as well as coordination of information flow to media and public.

NOTES ON A SAFETY CULTURE

Animal care professionals at all levels need to make safety a high priority for all of the zoo's staff. To provide outside perspectives and learning opportunities about safety in the zoo workplace, it may be desirable to bring in external animal care professionals with specialized safety expertise for the training and development of a zoo or aquarium's animal care staff. Participation by keeper staff should be mandatory for at least four training sessions per person per year (for animal escape, human injury, fire, and weather); managers and safety representatives from other areas of the zoo's governing organization should also be invited to attend these sessions to maximize cross-training by all personnel.

IMPROVING THE INSTITUTIONAL SAFETY CULTURE

A cross-departmental group representing management, front-line staff (keepers and curators), and occupational safety representatives may form a volunteer team to improve the culture of safety within the zoo. The team should work towards the following goals:

1. increasing the organization's safety culture (philosophy and practice of safe workplace conduct)
2. taking meaningful steps to improve safety and security
3. reducing accidents using personal commitments to daily safe practices
4. mitigating risks and injuries
5. focusing on and rewarding "near-miss" (near-hit) reporting in order to increase awareness of developing safety issues
6. developing stronger operational preparations for crisis and disaster situations, including both natural and man-made events.

A zoo's safety committee can form a series of smaller subgroups to target reviews of current safety directives and propose steps for improved disaster preparation. These subgroups should plan to regularly review policies from other accredited zoos and in similar-functioning occupations. The subgroups should also be tasked with reviewing recent safety directives, accident investigations, zoo-wide training, disaster planning, communication improvements, and general ideas for improving the culture of safety during seasonal and daily operations as well as during emergencies. Based upon their findings, the subgroups should make formal recommendations or present options to their zoo's management for potential improvements. In addition, they should offer training recommendations for the animal department, as well as for other departments within the zoo; seasonal and temporary employees may need training just as much as the full-time professional staff.

SUMMARY

Constantly improving an established program of risk assessment and occupational safety, animal escape training, and other crisis response awareness and procedures must be a regular part of a keeper's life. In this way, keepers will consistently contribute to improved safety protocols and general safety oversight throughout the entire zoo or aquarium. These safety activities are good for the animals, good for the zoo's visitors and good for zoo workers and their families, because they will help keepers go home whole at the end of each day.

REFERENCES

Mellen, J. and M. Sevenich MacPhee. 2001. Philosophy of Environmental Enrichment: Past, Present, and Future. *Zoo Biology* 20(3): 211–26.
Smithsonian's National Zoological Park. 2013. Zoo Directive: Animal Escapes.

9

Basic First Aid

Andrew A. Birr

INTRODUCTION

First aid is the immediate care given to treat an injury or illness. It does not include surgery or other advanced procedures and is performed by a lay rescuer until professional medical treatment facilities can be provided. Many minor illnesses or injuries such as small scrapes, cuts, bruises, insect bites, sprains, and strains do not require any treatment beyond basic first aid and often only require minimal equipment and action. However, these basic first aid procedures can have lifesaving results.

There are many benefits to implementing a first aid program. As has already been stated, simple first aid procedures may be lifesaving. An example of this would be a bee sting. Although to a majority of the world's population a bee sting is nothing more than a painful annoyance, to a select few who are allergic it can be fatal. Anaphylactic shock may occur in those who are allergic to bee stings. Anaphylactic shock is a life-threatening allergic reaction characterized by bronchial spasm, shortness of breath, and dilation of blood vessels with a sharp drop in blood pressure. Emergency treatment, including epinephrine injections, must be administered to prevent death. A person trained in first aid may be able to help the patient administer the epinephrine injection. Another benefit of a first aid program is that the immediate response may eliminate the need for an employee to be seen at a medical treatment facility. This is a benefit to the organization, as the loss in productivity and the potential need to hire temporary employees can affect operational budgets and revenues. Simply put, a first aid program can save money. Other benefits of first aid include rapid response time in emergencies and disasters. These may include animal escapes, fires, tornadoes, hurricanes, and any other emergencies that have the potential to cause severe or life-threatening injuries.

This chapter will provide basic information on first aid. After completing it, the reader should be able to answer the following questions:

- What steps can be taken to protect keepers responding to first aid emergencies?
- Can a keeper be held legally liable for responding to first aid emergencies?
- What are the common techniques used in cardiopulmonary resuscitation?
- What is an automated external defibrillator?
- What types of medical emergencies may be encountered, and how does a keeper respond?
- What types of trauma emergencies may be encountered and how does a keeper respond?
- What are the components of a basic first aid program?

KEEPER RESPONSIBILITIES AND SAFEGUARDS

Personal Protective Equipment (PPE) is necessary and should be required of those responsible for responding to first aid emergencies. PPE specific to first aid includes but is not limited to safety glasses, safety goggles, gloves, gowns, rescue breathing barrier devices, disinfecting and cleaning supplies, brooms, dustpans, tongs, sharps containers, biohazard bags designed for medical waste, hand-washing facilities, and any other supplies that may be required per employment or government standards.

PPE, along with good hygiene and frequent hand washing, will offer protection against bloodborne pathogens or other communicable diseases. Bloodborne pathogens are microorganisms in blood and bodily fluids that can cause disease. Examples of bloodborne pathogens and communicable diseases include human immunodeficiency virus (HIV), hepatitis, meningitis, tuberculosis, colds, and flu. Exposure to these pathogens can happen in several ways. First exposure can happen any time there is blood-to-blood contact with infected blood or bodily fluids. Second, accidental punctures from contaminated needles, broken glass, or other "sharps" may be a risk for exposure. Third, contact between broken or damaged skin and infected bodily fluids is another avenue for exposure. Finally, exposure may occur if blood-borne pathogens or communicable diseases make contact between mucous membranes and infected bodily fluids.

Other than the proper PPE, the greatest safeguards a first aid provider can use are "universal precautions." The United States Centers for Disease Control define universal precautions as precautions designed to prevent transmission of HIV, hepatitis B virus (HBV), and other blood-borne pathogens during first aid or health care (Centers for Disease Control 2011) Under universal precautions, blood and certain bodily fluids of all patients are considered potentially infectious for HIV, HBV, and other blood-borne pathogens. Simply treat all bodily fluids as potentially infectious and wear the proper PPE when administering first aid.

Aside from using proper PPE and exercising universal precautions when responding to first aid emergencies, responders must consider incident scene safety. This is different from "securing the scene," and it must be noted that many institutions require that accident scenes be left intact until a complete investigation, including photos, can be completed. Scene safety in relation to first aid will mean assessing the situation and eliminating all hazards that could harm employees responding in any capacity. You do not want to become a victim yourself when responding to any emergency. Prior to entering a scene, employees responding should look for hazardous materials, electric wires, fire, smoke, dangerous people, dangerous animals, falling objects, weapons, automobiles, power to machinery, and so on.

GOOD SAMARITAN LAWS

A concern many lay rescuers have is: Can someone be held legally liable for responding and administering first aid? Many countries and states have what are referred to as good samaritan laws, which state that a responder cannot be held liable for rendering aid during an emergency, provided that they are not negligent. One example of negligence would be a responder doing something that they know is wrong and which contributes to or causes further injury. Another example would be a responder doing something they have not been trained to do. However, litigation for rendering first aid is rare if responders use techniques learned in certified or accredited courses. If there are concerns or questions regarding legal aspects of first aid, one should obtain expert legal advice in the appropriate jurisdiction.

Even though good samaritan laws are in place, some may ask whether trained first aid responders are required to respond. The answer is both yes and no. If an institution requires that an employee be trained to provide first aid in emergency situations, then the employee must respond. Outside of an official capacity it is up to the individual to decide whether to provide first aid. In all cases the institution or local government agency should be contacted for guidance if concerns arise.

In most cases involving first aid in the workplace, institutions have policies and procedures in place that would require an injured employee to adhere to the first aid treatment and possibly seek medical treatment. A first aid provider generally would like the person's consent before helping them. Informed consent is such that the injured person consents to be helped, either verbally or in some other clear way, such

as nodding up and down. Implied consent is such that the patient is not responsive or coherent enough to express that they need help. In this case it is assumed that consent for treatment is granted. Injured persons may also deny treatment, and this is acceptable unless institutional policies state otherwise.

CARDIOPULMONARY RESUSCITATION

We can break cardiopulmonary resuscitation (CPR) into three words to define it accurately. "Cardio" deals with the heart, "pulmonary" deals with breathing, and "resuscitation" deals with restarting. Most countries offer their own accreditation standards for certification in the use of first aid and CPR. To find out what is required, you should contact your local health care provider for more information. The standards may vary internationally, and multiple organizations may offer different techniques, all being deemed acceptable. This section will focus on the general steps of CPR such as clearing the airway, checking for circulation, and proper chest compression. The ratio of chest compressions to breaths will not be discussed, as it may also vary on an international scale.

CPR should be included in some variation of what will be referred to as the chain of survival. The first link in the chain is early access, defined as activating the local emergency response system by phone or other means of communication to alert professional medical responders that assistance is needed right away. CPR is the next link in the chain of survival, and will be discussed later in detail. The third link is early defibrillation, which entails attaching an automated external defibrillator (AED) or instant paddles to the patient. Automated external defibrillators will also be discussed later in detail. The last link in the chain is early advanced life support. This is the point at which professional medical responders such as emergency medical technicians or doctors arrive and assume care of the patient.

First aid responders should also be aware of and ready for some of the pitfalls of CPR when administering it. Those receiving CPR may experience some of or all of the following conditions: fractured ribs, lacerated organs, punctured lungs, bruising, vomiting, loss of bladder control, and loss of bowel control. These conditions are not life threatening and can easily be treated in a medical treatment facility. Do not let them deter or inhibit your response.

Before a first aid responder begins CPR, they must first look for signs of cardiac arrest. If the patient is unresponsive, has blue skin, is not breathing, or has no signs of life, then CPR is necessary. There are a few situations in which a first aid responder will not begin CPR. The first would be injuries that are incompatible with life, such as decapitation. Next, if rigor mortis is present it would indicate that the patient has been deceased for some time and that CPR will not be necessary. The last things that would deter a first aid responder would be living wills or orders to not resuscitate. These are legal documents that indicate that the patient does not wish to be given CPR. These documents must be physically present and cannot be given by word of mouth. If no such documents are present, then CPR should continue. Check with your local health care provider or local government agency for living wills or do-not-resuscitate guidelines specific to

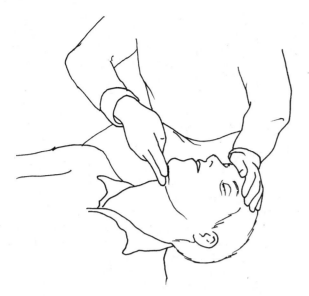

Figure 9.1. Head tilt and chin lift. Illustrations by Kate Woodle, www
.katewoodleillustration.com.

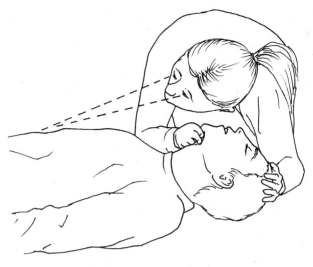

Figure 9.3. Look, listen, and feel.

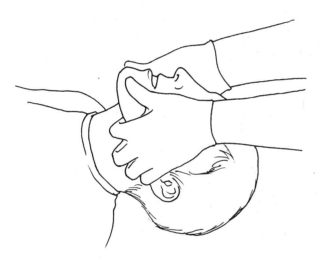

Figure 9.2. Jaw thrust maneuver.

your geographic area. Once a first aid responder has begun CPR they may stop if the person revives, if they are relieved by another qualified responder, if a doctor pronounces the patient deceased, or if the responder has become too physically exhausted to continue.

The steps of CPR are often referred to as the ABCs: airway, breathing, and circulation. There are two techniques for opening the airway. The first is called the "head tilt–chin lift"; responders must first place one hand on the patient's forehead and two fingers under the chin, and then push down on the forehead and up with the two fingers under the chin. The patient's mouth should open and expose the airway (figure 9.1).

The next technique is referred to as the "jaw thrust maneuver." The responder should place his or her thumbs at the base of the patient's chin and the other fingers at the top of the jaw on either side, just below the ears, and then push down with the thumbs (figure 9.2). Once the airway is

open, the responder should check for breathing by putting their ear close to the patient's mouth while looking toward the chest. The responder should look and listen for sounds coming from the mouth, for the rise and fall of the chest, or for any movement that would indicate breathing (figure 9.3).

If breathing is present, the responder should monitor the airway until professional medical help arrives. If no breathing is present, the responder should deliver rescue breaths. Rescue breaths are administered by applying the head tilt-chin lift technique, plugging the patient's nose, and blowing forcefully into the patient's mouth. The rescue breaths should be enough to cause rise and fall of the chest. The responder should next check for signs of blood circulation. One way to do this is to check for a pulse—most commonly the carotid pulse, in the neck, the brachial pulse inside the arm between the biceps and triceps, and the radial pulse in the wrist just below the thumb. Other signs of circulation include breathing, movement, and skin color. If the skin is pale or blue and the patient is not moving, one should begin compressions.

To deliver effective chest compressions, the first aid provider must use proper hand placement (figure 9.4). The point of reference will be at the center of the nipple line. This applies to infants, children, and adults. Infants will require that the responder place two fingers on the chest in between the nipple line, while children will require one hand, and adults will require two. A guide for depth of compressions is as follows; infants 1.25 to 2.5 cm (0.5 to 1 in.), children 2.5 to 3.8 cm (1 to 1.5 in.), adults 3.8 to 5.1 cm (1.5 to 2 in.). Compressions should be delivered at a rate of about 100 per minute. For adults, first aid providers should be sure to lock their elbows, keep their arms straight, and keep their shoulders directly over the patient's sternum. These guidelines, like those for the rate of compressions to rescue breaths, may vary internationally. A local health care provider should be contacted for assistance.

To summarize CPR, one can begin with the ABCs of cardiopulmonary resuscitation. A will signify airway, B will signify breathing, and C will signify circulation. The ABCs will always be followed in order: first open the airway, then

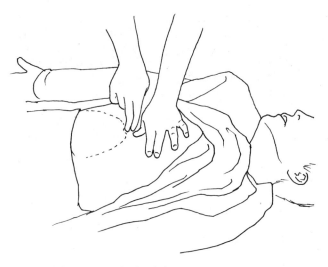

Figure 9.4. Chest compression hand placement.

check for breathing, and finally check for circulation. Begin CPR if necessary.

OBSTRUCTED AIRWAYS OR FOREIGN BODY AIRWAY OBSTRUCTION

It is possible that a provider of first aid or CPR may encounter an obstructed airway. This can have two results: either a good air exchange or a poor air exchange. In a good air exchange one will often hear sounds such as coughing coming from the patient's mouth and nose. In such an instance it is best to leave the patient alone and encourage them to cough. In a poor air exchange one will hear little sound or none at all. In many cases the patient may only be able to nod yes and no. If a first aid responder suspects that someone has a poor air exchange and may be choking, they should inform the patient that they are trained in first aid, and ask if they want assistance. We have referred to this earlier as informed consent.

If the patient acknowledges the responder's offer to help, then a technique may be applied to expel the foreign-body airway obstruction. Many health care providers refer to this as the Heimlich maneuver. To remove a foreign-body airway obstruction from an adult, the first aid provider must first stand behind the victim with his or own feet set at shoulder width apart. If the provider does not stand with a stable, wide base, the victim may fall back and cause the provider to fall as well. Next, the provider should make a fist and place it approximately two fingers' length above the victim's navel, then place the other hand over the fist and begin to pull up and in forcefully until the foreign object becomes dislodged from the victim's airway. For a child the same technique should be applied, except that the provider may need to kneel on one knee to apply the maneuver. For an infant, one technique involves holding the child head-down at a 45-degree angle and performing five back blows or slaps between the shoulder blades. This can be rotated with five chest compressions performed with the body at the same 45-degree angle until the foreign object is expelled. Never use fingers to attempt a blind sweep of the mouth and airway. If possible, try to use

tongs to remove the object. If the object is not visible, blindly searching with fingers or tools could worsen the obstruction.

There are a few advanced techniques that may be used for adults. The first is to stand the patient with their back to a wall and kneel down in front of them. The first aid provider should place one hand over the other at about two fingers' width above the patient's navel, and then push up and in until the object is expelled. The next advanced technique involves laying the patient on their back. In this case the first aid provider will straddle the patient and use the same hand placement as if they were against the wall. Again the provider should push up and in until the obstruction is expelled. Another technique is necessary if the patient is in a wheelchair. In this case the provider should position the chair against a wall or firm surface, lock the wheels, and use the same technique as if the patient were standing against a wall. If able, first aid providers may also remove the patient from the wheelchair, and lay them on the ground, use the same steps just described for use when straddling a patient. If an adult, child, or infant patient goes unconscious or unresponsive during the Heimlich maneuver, the provider should lower them to the ground and begin CPR.

AUTOMATED EXTERNAL DEFIBRILLATOR

In addition to the steps of CPR, an automated external defibrillator (AED) can offer additional lifesaving assistance for children and adults. There is insufficient data to recommend for or against the use of an AED in infants less than one year of age. (Kleinman 2008) An AED is a portable device that diagnoses life-threatening cardiac arrhythmias of ventricular fibrillation and ventricular tachycardia. It can be used to treat such a patient through defibrillation, which is the application of electrical therapy that stops the arrhythmia and allows the patient's heart to reestablish a normal sinus rhythm. AEDs are designed to be simple enough for use by a lay rescuer trained in first aid and CPR.

Before setting up an AED on a patient, the first aid responder should follow the same steps as in CPR. There are some special circumstances to consider when using an AED. The first is to make sure the patient is not lying in water. Water is good conductor of electricity, and first aid responders should remove patients from any water source before using the AED. The next special circumstance to consider is the patient's age. Many health care providers currently do not recommend that automated external defibrillators be used on children less than eight years of age. Again, one should contact a local health care provider for specific guidelines. Next, the first aid responder should look for transdermal skin medication patches or an implanted pacemaker or defibrillator the patient may have in or on their chest area. First aid responders should avoid placing the pads of the AED on these items.

Powering up an AED is done simply by turning the machine on. Once the AED is powered up, the electrode pads must be placed on the patient's bare chest. Many AEDs will have instructions specifying where to properly place the electrode pads. If not, use the following as a guideline. The right electrode pad should be placed to the right of the breastbone, below the collarbone, and above the right nipple; the left electrode pad should be placed outside the left nipple, with the upper edge of the pad several inches below the left armpit. To insure effective adhesion of the electrode pads, it is necessary to dry off a wet or sweaty chest with a towel and to shave a hairy chest. Most AEDs will then begin to analyze the patient's heart rhythm. It is important to avoid contact with the patient during the analyzing phase. If a shock is advised, the AED will ask the first aid responder to push the shock button. Again, one must not touch the patient. The AED will now advise additional shocks or ask the first aid responder to begin CPR.

AEDs can be lifesaving, and therefore should be maintained under some type of preventative maintenance program. Most are battery-powered. One should follow the manufacturer's recommendation for battery replacement or recharge. The unit itself should be inspected at least monthly or per manufacturer's recommendations, but more frequent inspections are better. Some units may have indicator windows that will show that a unit is in proper operating condition. Others may need to be powered up on a regular basis. The electrode pads are usually encased in a sealed bag, and will generally have an expiration date. The pads should be replaced once the seal is broken. Expired electrode pads can be kept for training purposes.

MEDICAL EMERGENCIES

For medical emergencies we will be discussing basic signs and symptoms, and the type of treatment a first aid responder can provide to a patient. As mentioned earlier in this chapter, it is critical to phone for help and request advanced life support from a medical treatment facility. First aid responders must also be cautious of scene safety. It is difficult to render help to someone in need if the responder becomes a victim as well. One should look for hazardous materials, electric wires, fire, smoke, dangerous people, dangerous animals, falling objects, weapons, automobiles, powered machinery, noxious gases, and so on.

SHOCK

Sudden illnesses, injuries, bleeding, or a combination of these may result in shock. Shock is the body's inability to supply the vital organs with sufficient blood. It is a life-threatening condition that can result even from minor injuries. Signs of shock include loss of consciousness, dizziness, lightheadedness, a feeling of passing out, trouble standing or keeping balance, and decreased alertness. A person in shock may seem confused when asked simple questions such as the day of the week or their birthdate. Treatment for shock includes dialing the local emergency number and requesting professional medical assistance, keeping the patient warm, calming and reassuring the patient, and elevating their legs 8 to 10 inches off the ground. This will help to increase blood flow back to the vital organs. One should monitor the patient's airway and breathing until help arrives.

HEART ATTACK AND STROKE

A heart attack is a blockage of a coronary artery that supplies blood to heart tissue. Signs and symptoms vary, but those commonly noted are chest pain, cool or clammy skin, shortness of breath, restlessness, fear, pain in the jaw or neck, and a feeling of indigestion. These symptoms may occur alone or in combination. First aid responders should immediately phone a medical treatment facility or emergency call line (in many parts of the United States this is referred to as calling 911). The first aid responder should then try to keep the patient calm and at rest. It may be necessary to assist the patient in administering medication. Medication for a heart attack can be as simple as an aspirin, or could require the use of nitroglycerin. Nitroglycerin comes in different forms, but is most commonly used in a pill that the patient places under their tongue, dissolving it quickly. Nitroglycerin dilates the arteries, making it easier for blood to flow to and from the heart. If helping a patient administer nitroglycerin, the first aid provider should wear rubber gloves, since moisture on the hands or fingers may cause them to absorb some of the nitroglycerin. This may cause dilation of blood vessels and arteries, as well as a migraine headache.

A stroke occurs when there is a lack of oxygen to the brain. The common signs and symptoms of a stroke are confusion, slurred speech, paralysis in one side of the body, or even the appearance of intoxication. First aid responders will phone for help, keep the patient calm, monitor the airway, and keep onlookers away.

SEIZURE

A seizure is a sudden alteration in behavior due to a temporary change in the electrical functioning of the brain. A seizure may be recognized by the following symptoms: convulsions, drooling, fluttering eyelids, eyes rolling back into head, falling down, incontinence, shaking, staring, stiffening of the body, clenching and grinding of teeth, and tremors. The first aid responder should not try to restrain the patient in seizure. They should remove objects from the area, such as furniture, that could cause harm to the patient. The patient's head should be protected with pillows, blankets, towels, clothing, or anything that can serve as a cushion to prevent further injury. Nothing should be placed in or near the patient's mouth. One should note the following information and provide it to the medical treatment facility upon their arrival: how long the seizure lasts, what parts of the body are involved, whether there is more than one seizure, and, if so, the amount of time in between seizures.

DIABETIC EMERGENCIES

First aid responders may encounter diabetic emergencies. Such a patient will have either a high or a low blood sugar level. High blood sugar has a slow onset, usually taking days to develop. Symptoms include warm, dry, flushed skin. The patient may also have a fruity odor on his or her breath. One should contact a medical treatment facility immediately and monitor the patient's airway and breathing. Low blood sugar has a rapid onset; its symptoms include cool, moist, pale skin and the patient will often appear confused and combative. If the patient is conscious and can swallow, they should be given candy, a high-sugar liquid, or oral glucose. If the patient is unconscious or cannot swallow, one should monitor the airway and breathing and wait for the arrival of medical personnel. Patients who have been diagnosed with diabetes will often wear or carry a medical identification device signifying that they are diabetic and indicating whether they have high or low blood sugar. If a patient shows symptoms of a diabetic emergency but one is unsure whether they have high or low blood sugar, one should treat them as if they have low blood sugar. Low blood sugar is much more serious and can end quickly with a seizure, coma, or even death. High blood sugar is generally less serious and can be treated easily at a medical treatment facility.

HEAT-RELATED EMERGENCIES

In warmer climatic conditions, heat-related medical emergencies are serious and can be life-threatening. However, they can be prevented with a few simple steps. Proper hydration and frequent breaks on warmer days are the best way to prevent them.

One should be aware of two types of heat-related medical emergencies. The first, heat exhaustion, is the less serious of the two. Signs of heat exhaustion include heavy sweating, pale skin, muscle cramps, tiredness, weakness, dizziness, headache, nausea, vomiting, and possibly fainting. One should seek medical attention if these signs are present. The patient should first be removed from the heat. A cool air-conditioned room is ideal, but if one is not available, a shaded area is sufficient. One should have the patient sip on cool water and place ice or cold packs on the neck, under the armpits, and in the groin area. Blood vessels are very close to surface of the skin in these areas. The application of ice or cold packs will help to cool the blood at a much faster rate than the body can do on its own. Fanning the patient is another technique that may be used. One should continue to monitor the airway and breathing until medical professionals arrive.

Heatstroke is the second heat-related medical emergency, and it is very serious and possibly life-threatening if not treated quickly. Signs of heatstroke include high body temperature, the absence of sweating, hot red dry skin, rapid pulse, difficulty in breathing, strange behavior, hallucinations, confusion, agitation, disorientation, seizure, and coma. One should seek medical attention immediately if any of these signs are present. Again, one should treat the patient first by removing them from the heat. A cool air-conditioned room or a shady area is sufficient. If able, the patient can sip on cool water. Ice or cold packs should be placed on the neck, under the armpits, and in the groin area. One should continue to monitor the patient's airway and breathing while waiting for medical professionals to arrive.

COLD-RELATED EMERGENCIES

Cooler climatic conditions will increase the likelihood of cold-related medical emergencies including hypothermia and frostbite. Hypothermia is an abnormally low body temperature. Shivering is the body's defense against cold, and is a key sign of hypothermia. Other signs include clumsiness, slurred speech, stumbling, confusion, drowsiness, lack of concern about one's well-being, loss of consciousness, weak pulse, shallow breathing, and poor decision making. Medical attention should be sought if these signs are present. First aid providers should then remove the patient from the cold environment, remove any wet clothing, and wrap the patient in blankets and towels. If available, heat packs should be placed on the neck, under the armpits, and the groin area of the patient. This will help to raise the patient's core body temperature. The patient's airway and breathing should be monitored until medical professionals arrive.

Frostbite is an injury to the skin and underlying tissue resulting from prolonged exposure to extremely low temperatures. It most commonly affects the nose, ears, fingers, and toes. The affected areas may have a waxy look and be hard to the touch. One should avoid rubbing the affected area and keep it warm; if medical professionals are not available for a prolonged period of time, the affected area can be soaked in warm tepid water at a temperature of 40 °C (104 °F).

ALLERGIC REACTIONS

First aid providers may encounter allergic reactions, which are fairly common and generally not life-threatening. An allergic reaction is sensitivity to a specific substance or allergen. Such substances include foods such as peanuts, foods containing sulfites (beer, wine, processed foods), and seafood. Other substances potentially causing allergic reactions are medications, pollens or plants, mold, animals with feathers or fur, metals, synthetic materials, tiny organisms such as bacteria, and insect stings. Signs and symptoms of a reaction to these allergens include rash or redness of skin; hives (raised swellings on the skin that itch); worsening of asthma or an asthma flare-up; swelling of the tongue, eyelids, or face; fever; joint or muscle pain; coughing; sneezing; nasal congestion; dizziness or lightheadedness; difficulty in swallowing; chest discomfort; and abdominal pain or cramping. These symptoms are generally easy to treat. In treating a bee sting, one should not use tweezers if a stinger is present; the use of tweezers can cause more venom to be injected into the patient and possibly worsen the reaction. One should instead use a hard flat object such as a credit card, name badge, or driver's license to scrape away a stinger. Most other skin reactions can be treated by simply washing the affected area with mild soap and water. Some antibiotic ointment along with a simple dressing can be applied if needed. For all other reactions one should continue to monitor the airway and breathing and call a medical professional if needed. Antihistamines (such as Benadryl) are drugs used to counteract the effects of allergic reactions; they may be necessary to prevent a simple reaction from turning into one that is life-threatening. A more serious allergic reaction is called anaphylactic shock. A person in anaphylactic shock may have itching, red raised blotchy skin, wheezing, confusion, weakness, or pale color, or be unconscious. They may often be unable to speak more than one or two words and may gasp for breath, purse their lips to breathe, or use their neck muscle to take breaths. Such a condition may require epinephrine, a controlled substance that must be prescribed by a physician. Epinephrine is administered through the use of what is commonly referred to as an epi-pen, an autoinjector that injects the epinephrine into the patient. To use an epi-pen on a patient that is incapacitated, a first aid provider must first be certain that its use has been prescribed for the patient. One should remove the cap, press the epi-pen firmly on the patient's thigh, and hold it there for at least 10 seconds. Anaphylaxis may be reversed within minutes if an epi-pen is readily available.

TRAUMA EMERGENCIES

BLEEDING CONTROL

A first aid provider may encounter cuts, scrapes, puncture wounds, or many other injuries that may require bleeding control. As mentioned earlier in this chapter, the use of personal protective equipment (PPE) is to be used by first aid responders to prevent the transmission of communicable diseases. It is especially important to use the proper PPE when dealing with bleeding, as the chance for fluid-to-fluid contact is greater in such situations. The first step in controlling bleeding is to apply pressure. This should be done by taking gauze or bandages and pressing them firmly against the wound. If the blood soaks through the gauze, it is important to not remove the gauze, since doing so may dismantle any blood clotting that has already started. Instead, one should apply more gauze or bandages. The next step in the control of bleeding is elevation. One should try to elevate the area that is bleeding above the level of the patient's heart. This should help to slow the bleeding. Next, cold applications may be used to restrict blood flow to the area. This is done by applying ice bags or cold packs over the gauze and bandages. If the bleeding is severe enough, pressure points may have to be used. Pressure points control bleeding by stopping blood flow from the artery to the point of the bleed. They should be used with caution, since indirect pressure can cause damage to extremities. One must not apply pressure to the carotid pressure points in the neck, as it can possibly cause the patient to go into cardiac arrest. Common pressure points include the brachial artery between the shoulder and elbow, the femoral artery in the groin area, and the popliteal artery behind the knee. When using pressure points, first aid providers should press on a point closer to the heart than to the wound. Pressing on a blood vessel further from the heart may have little or no affect on the bleeding.

SPRAINS, STRAINS, FRACTURES, AND DISLOCATIONS

Sprains and strains to areas of the body such as ankles, elbows, and shoulders are painful but generally easy to treat with no long-term effects. Sprains affect the ligaments that attach bone to bone. Strains affect the tendons that attach muscle to bone. Sprains and strains can be recognized by pain, swelling, bruising, limited mobility, muscle spasms, or even a pop at the time of the injury. To treat a sprain or strain, a first aid provider may use the RICE method (Elizabeth Quinn 2011):

- **R**est: the patient should be kept from applying any pressure to the affected body part.
- **I**ce should be used to control swelling and to alleviate some of the pain.
- **C**ompression: the affected area of the body should be wrapped in gauze or elastic medical bandage.
- **E**levation: the affected part of the body should be elevated to help keep the patient comfortable.

Sprains and strains are not life-threatening, but the patient may still need to be seen at a medical treatment facility and may not be able to do normal activities such as work or exercise. Fractures and dislocations are more severe than sprains and strains; they involve the bone breaking and the joint becoming dislodged or dislocated. A fracture is defined as a break, chip, or crack in the bone. Fractures can be caused by direct or indirect forces, by strong twisting, or even by muscle contractions. A dislocation is defined as a separation of the bone from its normal position at joint, such as in the shoulders or fingers. Dislocations are generally caused by strong forces. Many patients experiencing a fracture or dislocation will describe the following symptoms: pain, swelling, tenderness, inability to use the affected body part, skin discoloration, deformities, bleeding, snapping or popping noises, and a feeling that bones are grating together. First aid responders should keep the patient as comfortable as possible until professional medical help arrives. They should immobilize the affected body part in the position in which it has been found. The patient will generally have the affected body part in the most comfortable or least painful position. One should apply cold or ice packs and attempt to elevate the affected body part, if it does not cause the patient further pain or discomfort, One should also look for signs of shock, and treat for it if necessary.

SPINAL CORD AND NECK INJURIES

Spinal cord and neck injuries can be life-threatening and can cause permanent damage if the patient is not treated properly. For a first aid provider it is important to recognize that not all spinal cord and neck injuries are obvious. Numbness or paralysis can develop gradually as bleeding or swelling progresses in or around the spinal cord, so the fact that there is no initial numbness or paralysis does not rule out a spinal cord or neck injury. It is in a first aid provider's best interest to assume that a patient of any trauma injury has a spinal cord or neck injury, and to not move that patient unless directed to do so by a medical professional. If a first aid provider suspects a spinal cord or neck injury, it is important for them to call the local emergency medical assistance number. The provider should then try to keep the patient still and place towels or blankets on both sides of the head, or hold the head in place until professional medical assistance arrives. One should provide basic first aid, such as monitoring the airway and controlling any bleeding, without moving the head or neck.

ELECTRIC SHOCK

An electric shock occurs when a person comes into contact with a live electrical source. Electricity flows through part of the body, causing an electric shock. Exposure to electricity may result in nothing more than a painful reminder or it may result in electrocution, which is fatal. A patient who has experienced an electric shock may have little or no evidence of injury. Symptoms of electric shock may include burns, shortness of breath, abdominal pain, chest pain, and pain or deformities in the hands or feet. The first aid provider should first ensure that the source of the electric shock is no longer a hazard. They should then call the local emergency number requesting professional medical assistance. They should begin CPR, treat any burns, and treat for shock. Again, it is important that the source of the electric shock no longer be a hazard or be turned off completely. It may be necessary to contact a maintenance or utility worker to assure that the source of the electric shock is no longer an issue and that it is safe to treat the patient.

SPECIAL CONSIDERATIONS

Working in a zoo or aquarium puts first aid providers in situations that might not be discussed in a standard first aid and CPR course. Incidents involving venomous snakes, lionfish, stonefish, spiders, scorpions, jellyfish, and nonhuman primates, for example, require some additional first aid steps to assure the best possible care for the patient or injured person. For instance, when treating a patient bitten by an elapid species of snake (i.e., a coral snake or cobra), one should wrap the extremity with an elastic pressure bandage (Odum 2008), starting from the point closest to the heart and wrapping towards the fingers or toes. The bite should be kept lower than the heart. When treating a nonhuman primate bite, physicians may elect to leave a wound open, rather than suture it shut, until antibiotics have been administered to the patient. This is an attempt to prevent infection in the wound. With zoos and aquariums throughout the world having such diverse collections and species, it is important to work with one's own institution to develop first aid procedures specific to its animal collection. It is equally important that the local physicians and medical treatment facilities be familiar with any exotic species or special considerations that may require additional care or different first aid procedures.

CONTENTS OF A FIRST AID PROGRAM

A first aid program should be established, if not already in place, at a zoo or aquarium to aid employees in the steps that need to be taken when an injury occurs on the premises. A first aid program should consist of a written policy, incident reporting, incident analysis or investigation, first aid kits, and training. A written policy should be established and explained in an employee handbook that details the steps an employee should take if they are injured, and what to do if they come across someone that has been injured. An incident report form should be developed for conveying any information that will help to describe the incident in detail. It should, at a minimum, ask the following information: date, time, weather, location, nature of the illness or injury, steps or causes leading up to the illness or injury, whether the person was wearing personal protective equipment (if required), whether the person was trained, and recommended corrective action. It should also include a narrative section for all other information needed to properly describe the incident in detail. An incident analysis or investigation may be necessary to avoid similar incidents in the future. An incident investigation may include the incident report, photos of the scene, employee interviews, review of institutional policies, potential disciplinary action, and finally a recommendation on how to prevent similar incidents in the future.

First aid kits should be readily available to employees and those trained in first aid. They should be customized to each organization and should be maintained on a regular basis. Many first aid supplies have expiration dates. The kits should also be inspected to assure that all contents are present, even if no incidents requiring first aid have been reported. A basic first aid kit may include bandages, gauze, tape, scissors, tweezers, cold packs, heat packs, cotton swabs, sting swabs, burn spray, alcohol swabs, disinfectant swabs, and eyewash bottles, to name a few items. One should contact a local safety organization or medical treatment facility for recommendations or standards for what may be needed in a first aid kit specific to the organization. Finally, training should be conducted on annually to ensure that all employees and first aid providers are instructed on their roles in the first aid program.

RESOURCES

For more information on first aid, one should contact the local health care or safety organization, which will have information on certification classes and can help institutions with program development. In the United States there are several first aid organizations, including the American Heart Association (AHA), the American Red Cross (ARC), and the American Safety and Health Institute (ASHI). The AHA has global affiliates and many resources available.

SUMMARY

Again, first aid is defined as the immediate care given to person who is ill or injured. First aid can greatly benefit an institution as it speeds up response time in emergency situations, can potentially save lives, and can often save loss of productivity. First aid providers should be confident in their training, and assured that when they act as good samaritans there is little or no risk involved. First aid and CPR are relatively safe, but those who respond to first aid emergencies should be sure that the scene is safe and that they are wearing the appropriate personal protective equipment to avoid the transmission of communicable diseases. If a patient requires treatment beyond first aid, the first step should be to activate the local emergency response number and request assistance immediately. Each zoo or aquarium should have a first aid program in place. Keepers should study their institution's policy and know how to react if they encounter a first aid emergency; the policy should not be read for the first time during the emergency. Each institution should work with local safety organizations and medical treatment facilities to be sure that those outside organizations are familiar with the institution's first aid program.

REFERENCES

Centers for Disease Control. 2011. Accessed February 11. http://www.cdc.gov.
Kleinman, Monica. 2008. AEDs for Infants: Clarifying AHA/AAP Recommendations. *Currents in Emergency Cardiovascular Care,* 2008: 2.
Odum, Andrew. 2008. *Snakebite Protocol.* Toledo: Toledo Zoo.
Quinn, Elizabeth. 2011. About.com Sports Medicine. Accessed 11 January at http://sportsmedicine.about.com/cs/rehab/a/rice.html.

Part Four

Zoo Animal Management

10

Daily Routine and Basic Husbandry

John B. Stoner

INTRODUCTION

There are several reasons for developing a "daily routine" in a zoo or aquarium. First and foremost, a properly conceived and implemented routine will help maintain keeper staff relationships and ensure the best welfare possible for the animals. The more efficient a keeper becomes at completing the basic daily tasks, the more time and resources there will be for the other more interesting aspects of the job. Additional time may become available for implementing and monitoring enrichment programs, upgrading and changing exhibits, building nests and nest boxes, rearranging substrates to meet a fish's reproductive requirements, adjusting hot spots for reptiles, introducing rain showers for amphibians, or changing other containment furniture. Time may also become available to develop, implement, and complete research projects. Extra time could be used in developing and presenting poster presentations or papers at conferences, and even for being published in professional journals.

After reading this section, the keeper will have an insight into the daily husbandry procedures and day-to-day care of zoo and aquarium animals, especially as these things relate to the basic care of their enclosures and surrounding areas.

Topics will include

- the importance of understanding the zoo's geography
- the keeper's basic daily routine, including the need for observing animals, cleaning, feeding, watering, and record keeping
- the daily animal enclosure inspection
- the important components of a keeper's uniform
- the importance of shifting animals between containment areas (e.g., using remote gates and doors)
- commonly used tools and their use and care, including an overview of mechanical and motorized equipment
- ergonomics and the need to maintain one's own good physical condition.

KNOWLEDGE OF THE ZOO

It is imperative that when a new keeper joins an organization, however large, small, or sophisticated, he or she be given a thorough orientation to the facility. This should include introductions to as many of the zoo or aquarium's staff as possible. Many of these people may be required to help the keeper in work such as carrying out repairs, supplying plants and other husbandry materials, marketing the zoo to the public, or perhaps even acting as a backup person in an emergency. In many smaller zoos, and also in some larger ones, the keepers may have to assist with maintenance or groundskeeping (i.e., horticulture) or even do it themselves.

The new keeper should be made aware of the physical layout and structures of the zoo. This should include the location of the zoo's perimeter fencing, animal enclosures, access gates, landscape features, pathways, roads, buildings, and equipment storage locations. This knowledge is useful in emergencies such as animal escapes, intrusions by marauding wild or domestic animals or trespassers, and serious weather. For example, keepers will be better able to visualize the barriers between themselves, the public, and an escaped animal. Knowing that animals generally run along fence lines may allow the keeper to get a good idea of an animal's direction of travel. With this information, the keeper may be better equipped to inform colleagues of the animal's whereabouts and to move visitors out of harm's way. This knowledge will also make the keeper more comfortable when offering directions and other information to visitors.

In many zoos, the keepers will work within some form of daily work shift system (i.e., with some keepers working an early shift and others working a late shift) to allow the zoo to stay open longer to meet the animals' needs and the visitors' expectations; this maximizes the time that the animals are on exhibit. A new keeper might well find themselves working the late shift with fewer staff on duty. In a smaller zoo the keeper may be on duty with only one other person, and if an emergency develops they could become responsible for giving directions to the first responders and generally helping resolve the situation. If there is a fire in a building, all staff should be able to direct the fire truck by the quickest and safest route to the area of concern. The keeper may also find that they become a communication conduit between emergency services and other staff, so that the site knowledge becomes even more important.

When moving around the zoo, keepers must remain diligent. Monitoring the condition and state of repair of all the barriers and fences they pass will help ensure that any potential access and egress points, such as gaps under the fences or excessive gaps around gates (more than 5 cm [2 in.] wide) are remedied. These routine checks are particularly important after weather events, when wind may have blown down branches and trees that can damage fences. Heavy or persistent rainfall will cause erosion and washouts under fence lines. Heavy snowfall and the accompanying buildup may reduce the functional height of a fence. These situations must be reported and remedied before animals can be released into the enclosures.

DAILY ROUTINE

However many work areas ("strings," sections, or units) that a zoo may be divided up into, a daily routine for the animals' care will need to be developed. This routine must be established with the involvement of all the relevant keeping staff. Some zoos may only have a small number of keepers, and each person may be responsible for their own section of animals. In larger institutions, several keepers may be responsible for the same area. In either example, the daily routine will ensure that animal and staff safety, as well as the highest standard of animal care, is established. This routine will also help in maintaining good peer relations between keepers. Some staff may consider a routine to be uninteresting, unenriching for the animals, or even unfair to certain individual keepers' capabilities. This is not so, as the daily routine can be flexible, can be changed, and can recognize individual strengths and areas of special interest, as long as all parties who work in the area are involved in making the changes. This is especially important as more and more zoos and aquariums are developing mixed taxonomic and zoogeographic areas, thus requiring staff to have very diverse skills.

COMMUNICATIONS MEETINGS

A normal daily routine will usually start with a communications meeting. A review of the daily keeper report and topics such as animal births, deaths, and transfers, which animals have been left out on exhibit overnight, and whether an exhibit is closed for maintenance (with the animals off display) should be discussed. The daily and long-term weather forecasts and related precautionary activities, such as moving animals to their holdings to avoid lightning strikes or flooding, should also be discussed. Special educational and public relations events that may be scheduled, especially those involving the animals and the keepers' daily routine, would be useful agenda items. Staff attendance should also be taken at this time, as it is extremely important to monitor the number of keepers on duty and also to have a system established to account for their whereabouts and status before, during, and at the end of the day.

AREA INSPECTIONS

The keeper's first duty upon arriving at their work area is to check for notes in the keeper notebook or on whiteboards. The keeper will then inspect all their animals to ensure that they are accounted for, healthy, and secure. This will involve a sequential visit to each animal enclosure in their area. The enclosures should also be checked to ensure that nothing has been damaged overnight. Food and water consumption should be noted. It is also important to observe the amount and condition of feces and urine that has been excreted overnight. Temperatures and light timer settings should be checked. The animals may have been alone overnight; some will have been on exhibit and others in their sleeping quarters, depending on the weather or hours of operation. Notes should be made of any pertinent changes such as loose feces, rodent droppings in food bowls, or animals not shifting; should the keeper have any doubts, they should consult with more experienced staff or the area supervisor. It must be emphasized that daily tasks such as feeding and cleaning should not start until the head count and physical inspection of the animals has been completed. This will ensure that any newborn, injured, sick, or dead animals can be taken care of as soon as possible.

ANIMAL ACCESS AND CONTAINMENT SECURITY

The keeper will then usually travel through his or her section to sequentially begin the daily husbandry activities. Shifting the animals between holdings and exhibits to facilitate cleaning and feeding will be one of the first tasks. It is important to ensure that all enclosure access, shift gates, and doors are working properly, and that they run smoothly and are free from snow, ice, soil and any other detritus that may impede their proper operation. This will minimize the risk of trapping or injuring animals, as well as lessening the chances of keeper injury if the shift door jams when the keeper is moving it. Zoos will differ in their use of locking points and locking devices, but generally every animal area should be secured in such a way as to prevent escape, entry of intruders, theft, or any other mischief. Whatever locking system is selected by the zoo or aquarium, it must be adhered to by all staff. Proper securing devices must be in place for fish, reptile, and other small animal enclosures. It is not acceptable to place weighty objects on the tops of animal containers; these are unreliable and can damage the tops of the containers. National accreditation organizations and zoo licensing bodies expect "positive" locking systems to be in place. These positive systems are installed in such a way as to make it impossible for an animal or another keeper to open a door, shift, or slide when it is supposed to be locked. For example, in some older institutions some vertical slide doors were held in place by gravity. The controlling wire would run through a series of pulleys to a locking station to be secured. Not only was it possible (although considered unlikely) for an animal to lift the slide, but a person could also pull down on the wire without unlocking the lock. This could potentially have allowed an animal access to a person working on the other side of the slide. Today these pulley systems are still in place, but the slide has to be secured with a lockable bolt system.

ENCLOSURE INTEGRITY

The daily routine will be fundamentally similar whether the keeper is working in a large outdoor landscape immersion exhibit, an indoor aviary, a vivarium, or an aquarium. The enclosure will be checked for any sharp or injurious objects including wire, nail ends, and sharp edges of fiberglass in aquarium decor; all should be made safe, if possible, or be reported. It is possible that wild animals may have gained access to the enclosures; rodents (which frequently burrow in and around exhibits) and other pests such as ants, cockroaches, mites, or snails must also be reported. Areas between the public and exhibit barriers should also be inspected. Particular care must be taken following construction or repair of animal exhibits to ensure that all garbage, equipment, and other materials are removed. In some instances it may be necessary to scour the area with a magnet to find and remove metallic objects that have the potential to cause injury or death to an animal. "Hardware disease" occurs, particularly in ruminants, following ingestion of sharp metal objects such as pieces of wire, screws, and nails, which often settle in the reticulum compartment of the forestomach. The muscular contractions involved in digestion may push the object through the stomach wall and into the heart, killing the animal.

GENERAL CARE

As the keeper moves from one area to another, they should take the time to pick up any litter they see. It is every employee's responsibility to ensure that the zoo or aquarium is a clean and safe venue for the animals, staff, and visitors. Litter in its many forms, including metal cans, bottles, paper wrappers, plastic bags, and cutlery, presents a potential hazard to the animals and also sends a message of a lack of care to the guests.

DAILY HUSBANDRY

Once the area is secure and the animals are judged healthy, the daily cleaning can begin. This should be done as efficiently as possible with a minimal waste of time, materials, and resources. In large animal exhibits, using shift (shut-off) pens for the animals will ensure that the keeper can concentrate on their work without constantly looking over their shoulder to monitor an animal's whereabouts and behavior. If this is not possible, two keepers should work together so that one can act as a spotter and monitor the animal. In some aviaries, aquariums, vivariums, or small mammal displays, this separation may be impossible where the animals are permanently housed, so a skill sometimes referred to as "animal sense" has to be learned and used. This involves developing an understanding of the animals' behavior, comfort zones, flight distances, agonistic tendencies, and any other behavioral traits they display when the keeper is in their vicinity, and

then adjusting the husbandry to cause little or no discomfort to the animals.

FEED AND WATER

Servicing feed and water containers must be carried out daily. Food containers should be cleaned thoroughly, rinsed if detergents or disinfectants are used, and replenished. When the containers are returned they must not be placed under perches of birds or other arboreal animals, as they will quickly become contaminated with excreta. There should be enough feeders so that all animals feel comfortable when feeding and receive their rations. Food that is supplied ad libitum must be distributed in a way that prevents contamination. In mixed exhibits, or when weanling animals are present, the use of "creep" feeders may have to be considered. A "creep" can take many forms; it can be as simple as a box, or can be a set of horizontal or vertical barriers that prevent adult or larger animals from gaining access to the small or young animal's food. Nocturnal animals may not require feeding until the evening, although if they remain active during the day, foraging foods may be offered to them as appropriate. If animals are fed on the floor or other surface, or in large feeders or hay racks these locations and pieces of equipment will need to be checked for stale food, mold, pest species, general waste, and a buildup of other contaminants. They should be emptied, cleaned, and refreshed regularly. When cleaning these larger, fixed food stations, water use should be kept to a minimum to avoid moisture buildup and potential for contamination by bacteria and parasites in the surrounding area. If possible, these feed stations should occasionally be relocated to minimize wear and tear of surrounding areas.

Water containers come in many types, from a simple bowls or buckets to nonfreezing and automatic-filling containers. In cold climates, outdoor water supplies should be of the nonfreezing type to ensure that the animals have access to clean potable water all day; this is a requirement in many jurisdictions. If waterways or other water features are included in an enclosure, it is good practice to offer a separate supply of clean drinking water, as the water features may become contaminated.

INFORMATION FORMS

Current diet sheets including any food enrichment options should be posted in the holding area. Any changes to the diets should be discussed with other area keepers, the supervisor, and the person responsible for the diets. Cage cards—laminated index cards containing current information on the animal such as sex, age, parentage, identification, and International Species Inventory System (ISIS) number—should also be displayed.

CLEANING

There is a multitude of methods for daily cleaning and general day-to-day husbandry. Fundamentally, all waste food, urine, feces, broken or damaged furniture, litter, dust, cobwebs, and other unnecessary materials must be removed from the animals' living areas. Areas such as primate holdings may require cleaning with a prescribed disinfectant to kill micro-organisms. In some cases a pressure washer or steam cleaner may be employed. If one is required to use disinfectants and cleaning compounds, it is important to use them per the manufacturer's instructions. Over-usage does not improve a chemical's effectiveness; it just wastes resources and increases the zoo or aquarium's environmental impact. Other areas may have to be washed (sanitized) with a solution of detergent (soap or surfactant) and water. Some carnivore holdings, for example, have a significant amount of oily deposit on their floors, which may require detergent for removal.

When servicing large outdoor ungulate areas, it is always best to dry clean. Concrete hard standings will generally clean up quite well with dry brushing. It is important to inspect and clean drains before and after servicing. One should not remove the drain screens and allow waste to wash away into the sewage system, as this may overload or contaminate local waterways. It is becoming imperative to conserve water, as it is the planet's most important commodity, so one should dry clean wherever possible; if washing with water is absolutely necessary, the minimum amount should be used. If water is overused, it will cause water damage to buildings and other infrastructure. In some cases, water used at high temperatures or pressures can splatter or create steam that aerosolizes small particles and increases the risk of disease transmission.

The exhibit furniture and decoration should be inspected, cleaned, and refreshed as required. If fresh browse is being used, it should come from a known safe supplier, whether it is for feeding or for exhibit decoration. Furniture should be of natural materials whenever possible. One should scout the local area for tree stumps, rock sources, and substrate availability. Dead leaves should be collected in fall (autumn), bagged dry, and stored if possible so that habitats can be easily changed and enriched throughout the year. It is important to ensure that these resources are free from chemical and biological contamination such as herbicides and rodent and bird feces.

FOOD HANDLING

The daily routine includes the managing and monitoring of foodstuffs. If this is not carried out routinely, it can lead to poor-quality, perhaps contaminated food and a buildup of pests. New food or supplements should never be placed on top of old. Food storage containers must be cleaned inside and out, especially between uses. All containers must have labels describing the current contents and their use. Large storage bins should be on wheels, if possible, to allow easy movement for proper cleaning of floors. Spilled food must be picked up immediately; otherwise it will not only attract pests but possibly become a trip or slip hazard. Hay, straw, and similar baled materials must be kept off the floor, away from walls, and on racks if possible to minimize the opportunity for pest infestation and associated contamination. For convenience and safety, the bales should not be stacked too high and should also be delivered as needed, since "over-storing" can lead to spoilage, staleness, and pest infestation. (If pests do take up residence, the keeper may have to implement a control program, setting and maintaining humane traps and bait stations.) When opening bales, the keeper should make sure the wire or twine bindings are removed and disposed of properly. This will prevent entanglement, strangulation, ingestion, and other injuries to the animals. The wire or twine may be reused to secure perches or to tie up paper bundles, but it must be stored properly.

WASTE STORAGE

Keeping the manure bins and other waste disposal areas clean will prevent both noxious odors and the influx of flies, mosquitoes, wasps, rodents, and other pest species. Good hygiene practices can negate or minimize the use of chemical pesticide sprays.

DAY'S END

At the end of the workday the keeper will ensure that the light timers are functioning properly, and adjust them as required. They will also make sure that there is enough food, water, and other material available to the animals. Finally, they will check to ensure that all enclosures and holdings are properly locked. The keeper will then complete a daily keeper report. All animals left in their outdoor enclosures overnight should be noted and reported to the night keepers or security personnel so that they can move carefully around the facility and not be exposed to any surprises. All keepers should be accounted for before the zoo is considered secure.

USE AND CARE OF TOOLS AND EQUIPMENT

Ideally, one set of tools should be available for each work area or containment. The tools should not be distributed randomly or carried from one work area to another. The keeper should purchase quality tools and care for them properly. The tools should be cleaned after every use and stored correctly since they will have been exposed to water, chemicals, and animal waste, all of which are damaging to tools. Wall hangers in various styles are readily available for tool storage, or can be easily constructed. Long nails, driven in side by side, are suitable for hanging brooms, rakes, forks, and shovels. Alternatively, one can drill a hole through the handle of the tool, loop some binder twine through the hole, and knot the ends of the twine. The tool can now be suspended from one nail, allowing it to drain to the floor and dry more effectively. Smaller tools such as scouring pads ("doodle bugs"), window squeegees, nets, siphon tubes, and razors used for cleaning aquariums can all be handled in a similar manner. To identify tools and avoid their loss or removal from one area to another, it may be expedient to paint their handles in different colors. Storage areas, cupboards, and service aisles must always be kept clean and tidy in an effort to avoid cross-contamination. This will also help to keep the work area safe.

HANDHELD TOOLS

Smaller specialized and electrical hand tools should be available to the keeper in the work area or at a central location. A sign-out-and-in system may be implemented to help control the whereabouts of the tools.

Hammers of different sizes are useful for repairs, installation of perching, or the building of nest boxes. A large sledgehammer or maul may be useful for securing larger enrichment materials.

Crowbars and flat bars will be needed for pulling nails, prying apart boxes, and generally levering objects.

Pliers are useful for cutting and tying wires, removing stubborn nails, and undoing rusty bolts. There are regular pliers, needlenosed pliers for use in smaller spaces, and locking pliers, sometime called "mole grips," for really stubborn jobs. For fence repair side cutters, end cutters, fencing pliers, or hog ringers may also be necessary.

Screwdrivers come in a variety of types: cross-end (Phillips), flathead (straight), and square-ended (Robertson's) are the most common. They vary in size, and care should be taken when selecting the right screwdriver for the job. The wrong size will damage the screw head, making it difficult to remove.

Keepers frequently use knives, particularly when preparing food. Good quality chef's knives and sharpening steels are recommended. Learning to use them properly is most important. Many hand injuries result from their improper use. Some zoos may bring in local chefs to instruct the staff in their correct use.

A cordless electric drill with both drill and driving bits is an extremely useful tool. It can be used in the construction of nest boxes, food platforms, perching, and so on. Whenever such a drill is used, the operator must wear safety glasses.

A strong magnet suspended from a rope, on an extendable handle, or between a wheeled axle is useful to have in the zoo. It is used after any sort of maintenance work in an animal area to ensure that no metal parts, wire ends, or welding rods are left behind to endanger the animals' well-being through ingestion or external injury.

EXHIBIT MAINTENANCE TOOLS

There is a large variety of exhibit maintenance tools such as shovels, rakes, brooms, and wheelbarrows. It will be up to the keeping staff to select the tools they feel most comfortable with. The choice is further widened by the introduction of ergonomically modified equipment. In most cases the modifications involve the handles of the tools. Some have had various degrees of curve or bend added to allow easier or safer lifting techniques to be used. Keepers should be encouraged to try using this modified equipment.

The most commonly used tool is the push broom or brush. Some push brooms have hard plastic bristles and are best suited for use on concrete. Where possible, areas should be dry cleaned with the broom, because when fecal material and other dirt is wet it makes the broom very heavy and difficult to use. If working in the wet is unavoidable, banging the broom bristles vertically on the surface will dislodge the buildup. These brooms can also be flipped over and, with the use of a forward scrubbing movement, they will remove almost all adherent waste from the floor. In dry conditions this technique can help save water. Wider, softer-bristled versions of this broom are available, which cover more area in less time. They are particularly useful for drier indoor areas and on rubberized floors. Corn brooms, birch brooms, and other angled long-bristled lightweight brooms are also available, but are generally more suited to indoor areas. Smaller hand brushes may be used for dusting and removing cobwebs, or for scrubbing food and water bowls.

Shovels come in several shapes and sizes; a long-handled flat (spade- or square-mouthed) shovel is the most versatile. It can be used for digging, picking up piles of waste, and scraping floors to remove fecal material or ice. The long, straight handle offers more leverage and suits keepers of all heights. A round-nose (-ended) version is also available, but is more suited to digging. Larger aluminum, steel, or plastic scoop shovels are also part of the inventory. The aluminum or plastic versions are good for moving large amounts of lighter materials such as soft snow, wood chips, wood shavings, and sawdust. Because the plastic is a little softer than metal, it is preferred for use in areas with rubberized floors. The heavier steel shovel may be better for moving gravel and sand, but it should not be overloaded or it may result in strains. The snow shovel or scoop, generally reserved for more northern climes, is, as its name implies, used for removing snow. Snow shovels come in steel, aluminum, and plastic. They vary in shape and size. Some are shaped like conventional shovels, while others have larger, flatter push blades. They can be extremely useful as "barn scrapers," and they push large fecal boluses very well.

Different types of rakes are also available. Fan rakes, as their name implies, are fan-shaped and constructed of metal or plastic; they are sometimes known as leaf rakes. They are frequently used for raking feces of smaller ungulates as well as leaves and other light materials. Hard garden or landscape rakes also have their uses, particularly in drier areas, for leveling various stony or gravel substrates and for grading smaller exhibits.

Forks usually have long straight handles and metal or plastic tines or prongs. Mainly used for moving bedding materials, they come in three basic types. The metal-tined hay or pitchfork, which may have between three and five 30.5 cm (12 in.)-long tines spaced 5 cm (2 in.) apart, or the ensilage fork, with fifteen 38 cm (15 in.)-long metal tines spaced 2.5 cm (1 in.) apart, are more suitable for use with hay and straw bedding. The other group are the plastic-tined stall and basket forks, usually with eighteen 30.5cm (12 in.)-long tines spaced 1 cm (3/8 in.) apart, which are better suited for manipulating shavings and sawdust. They are particularly useful in hoofstock holdings, as they retain all but the smallest of fecal boluses while allowing the bedding material to fall back to the floor, thereby helping to minimize waste.

The wheelbarrow—or, as it is now sometimes called, the cart—has probably undergone the most change. It has evolved from being the heavy, one-wheeled, difficult to balance, easy to overload behemoth of the large-ungulate keeper to a well-balanced, lightweight two-wheeled vehicle used in food prep areas and small exhibits and as a general runabout. It is more ergonomically designed and easier to work with. It comes in a variety of sizes and styles, all based on the idea of a box with wheels. It generally has more carrying capacity than its predecessor, especially for lightweight materials. It must not be overloaded with heavy materials. When using wheelbarrows, keepers should be very conscious of their surroundings, especially when going forward. Keepers should judge passage widths and turning radii, and should avoid bumping into things. It is often better to pull wheelbarrows backward over humps and other obstacles.

A squeegee is constructed with a strip of rubber or similar material sandwiched in a metal holder, and its main purpose is to move water from a surface. Squeegees come in various sizes, including larger ones up to 60 cm (2 ft.) in width with long handles for floor surfaces. It should be noted that they often work far more effectively when pulled over the wet surface rather than pushed. The smaller, lighter short-handled version is most frequently used to clean glass exhibit windows and other finer surfaces.

Magnetic window cleaning devices are a unique and very useful tool for cleaning large aquarium glass. They take the form of two strong magnets, surfaced with cleaning cloth, which are placed opposite each other on either side of the window to be cleaned. The keeper can move the outer magnet, which in turn moves the inner one, thus cleaning the glass on both sides. This tool is particularly useful for cleaning algae off underwater viewing windows. The magnets can be left in place until the next use.

Hoses are constructed of various soft pliable materials in many lengths and diameters. The keeper must choose the diameter required to fit the taps, faucets, and hydrants in the area. Fittings are usually attached to a hose, and a keeper should know how to change them when they are damaged. Hose nozzles also come in various sizes, types, and degrees of quality, including metal twist nozzles as well as nozzles operated by a trigger mechanism. The twist nozzle is generally considered the most resilient as it usually has fewer working parts. Changing from jet to spray can easily be achieved by twisting the nozzle. Like an electrical cable or a rope, a hose has a "life" and can be difficult to coil or store on a reel. It may go its own way forming twists, kinks, and even knots. To resolve this problem, uncoil the hose and lay it out on a level surface, allowing it to straighten. After use, follow the same procedure; disconnect the hose from the tap and remove the nozzle. Allow the hose to empty, and then coil it. Larger, "looser" loops are easier to work with. Once this technique is mastered, it will save many sore wrists and much frustration. Better-quality hoses are easier to work with, and they last longer.

MECHANICAL EQUIPMENT

Keepers may have access to a large array of time- and labor-saving equipment, including vehicles ranging in size from the standard pickup truck to smaller, more versatile utility vehicles. The latter can move around easily on a variety of terrains and are well suited to zoo work. Some are fitted with a tipping "box" or bed, enabling them to carry and dump good-sized loads of hay, straw, wood wool, shavings, rocks, stones, logs and other enrichment materials easily. They can also dispose of large amounts of waste and manure. Some units may be fitted with equipment for snow removal and grading. The vehicles come in a variety of configurations including four- and six-wheeled, with covered or open driving positions, and with standard or tipping box. They can run on gasoline (petrol), diesel fuel, or electricity. Given the uneven nature of many zoo areas, they should be driven with great care by approved drivers.

Bobcats or "skid steers" are small bulldozer-like machines that can be fitted with a variety of equipment such as buckets for moving soil, forks for lifting boxes, digger units for trenching, and snow removal equipment—all very useful to the keeper. They require a skilled operator to be used effectively, so proper training is essential.

Snow blowers are available to keepers in many zoos that experience significant snowfall. Small electrical units are available, but the majority of heavier-duty units are powered by gasoline (petrol). They are reasonably easy to operate if staff members have experience with them or are given direction. Units should be self-propelled with directional controls, as they are heavy and may be difficult to maneuver. The blower unit and its screw blades are prone to damage if they hit or intake hard objects such as ice, stones, or broken tree limbs. Although the blower is fitted with sheer pins that will disable the machine should it ingest an object, extreme caution should be exercised when using the machine.

Pressure sprayers are usually either gasoline- or electric-powered pumps that force a stream of pressurized water through a variety of nozzles. Some units can heat water and produce steam under pressure. Protective clothing and equipment should always be worn when operating these devices, to prevent contamination due to "splash back" and inhalation of spray mist. This is particularly important when working with cleaning and disinfectant chemicals. The sprayers' steam capability allows for the thorough disinfection of holdings and service areas. The units are not generally used on a daily basis except perhaps in some large ungulate, elephant, and primate holdings. Frequently built as mobile units, they can be transported from one area of the zoo to another as required.

A boat may be required if a zoo has significant bodies of water to be managed. It should be lightweight and stored where required or in a central location, and may be fitted with an outboard motor. Oars and lifejackets should always be available and stored with the boat. Lifejackets should be worn at all times when the boat is in the water. The boat can be used for cleaning the waterways, managing water plants, and for rescuing people or animals that may inadvertently find themselves in the waterway.

OTHER KEEPING TOOLS

Keepers may use specialized tools for feeding animals, such as forceps and tongs of varying sizes, ranging from small surgical forceps to larger graspers (similar to those used in stores for removing objects from shelves). Restraint equipment might include a variety of nets, catchpoles, and snake hooks that will be discussed in other sections of this text. Most aquariums and some zoos employ keepers who either are or become qualified divers to service the aquatic animals and their environments. Countless other specialized tools may be required in different facilities. Keepers in and around aquatic environments will require a working knowledge of water testing instruments.

A hydrometer is a delicate instrument used for measuring the specific gravity of water. The resulting reading will give a good indication of water salinity in a marine environment. A refractometer can also be used to obtain similar information. Dissolved oxygen (DO) meters are sensitive hand-held electronic devices used for monitoring the concentration of dissolved oxygen in water. Many simpler test kits available for aquarium hobbyists may also be of use in smaller zoo or aquarium operations. These require the user to introduce the water sample into a simple tube containing chemicals, so that coloration changes can indicate the levels of ammonia, nitrate, and nitrite, for example, in the water. These measurements should always be confirmed if the results fall outside the expected levels (see chapter 33 for further information on aquarium science). As terrestrial exhibits become more complex, mixing animals and plants into rain forest habitats, keepers will need to become familiar with devices for measuring humidity (hygrometers), and temperature (thermometers). Electronic hydrometers are available which will give both relative humidity (RH) and temperature.

EMERGENCY EQUIPMENT

The daily routine will require that the keeper is cognizant of the facility's policies and procedures concerning the use of emergency equipment, as well as of the equipment's whereabouts. The most important emergency equipment is the communications system, which may consist of numerous handheld radios used to communicate with a central "control" person; or, in smaller zoos, a basic two-way communication system using handheld radios. Regular "landline" systems are often also available, with telephones situated in various locations throughout the facility. There should also be designated emergency systems even in smaller private zoos, with some telephones available to the public, usually in the vicinity of dangerous animals, and others kept specifically for keeper use. These telephones are usually colored red. The keeper emergency telephones and alarms should be situated in dangerous animal areas—for example, near venomous reptiles and large carnivores. Operable by the keepers, these systems should be set up in such a way that an alarm sounds in a central location when the system is triggered, and help is dispatched even if the injured keeper is incapacitated. Cell phones are also available and allowed for communications use in some institutions. A simple but effective system is for the keeper to carry a whistle or other noisemaking device that is small and loud, easily carried (perhaps on a belt loop), and readily available. Whatever system is in place, the new keeper must be aware of its associated codes and policies for use, and a consistent code of practice should be followed.

Many zoos will have emergency equipment situated in central locations. This may consist of ropes, ladders, and nets for helping people or animals out of wet or dry moats. A boat or life belts may also be included in this kit. Recapture equipment, large antelope nets, hand nets, and catchpoles, for example, may also be kept in a centralized location. Guns and anesthetic projectile weapons ("dart guns") will usually be stored under strict security. The new keeper should quickly become aware of the equipment's location and be trained in the attendant responsibilities. Zoos will often have a shooting team that is kept trained and up to date on firearms use and policy.

Fire can be one of the most dangerous events to affect an animal care facility, so many will have first aid equipment for firefighting in place. Frequently, building codes will dictate that smoke alarms and associated sprinkler systems be in place. In this case the keeper should ensure that these devices are in good working order and that their batteries are replaced as required. Fire extinguishers should also be in place, and it is important that the new keeper receives training in how and when to use them. Remember that the first action is always to notify others of the situation and ask for help.

UNIFORM AND PERSONAL EQUIPMENT

Many zoos require their staff to wear a uniform, and it will be up to the individual organization to decide and implement a dress code. What follows are some suggestions for that uniform. Long sleeves, long trousers, hats, and sunscreen are recommended to protect the keeper from the damaging rays of the sun. The weight and quality of the materials selected should reflect the climate in which the keeper is working. There are many new materials that suit different climatic conditions. For working in cold climates, it is suggested that the keeper dress in several layers of clothing rather than one really heavy warm coat. This will allow for the removal of layers as the day gets warmer, or the heat generated by work dictates. One should wear comfortable clothing.

The keeper should also keep gloves available: a strong, pliable pair for daily work routines and for restraining animals. Good-quality, lightweight, water-resistant, insulated gloves are suggested for keepers working outdoors in cold weather.

Protective footwear (safety shoes) may also be required, but may be expensive, and they can be damaged by repeated exposure to animal waste and water. It is in the keeper's best interest to make the effort to change between protective safety shoes and rubber boots as work conditions require. Insulated winter boots for cold climates are a must.

A keeper should also carry a small pocket-sized notebook and pen while going about their daily routine. Moisture-resistant notebooks are now available that are less vulnerable to the wet conditions a keeper frequently encounters. Notes taken during the day may be indispensable when one fills out the daily report at day's end.

A small, good-quality pocket knife or multi-tool is also indispensible for many unforeseen tasks such as cutting binder twine, repairing hoses, or recovering foreign bodies from those hard-to-reach corners.

SUGGESTIONS FOR BEST PRACTICE

What follows are a few suggestions to make the new keeper's workday safer.

Do not climb fences, as it is not a safe practice and will damage the integrity of many fences. There are many sharp ends to get caught on, which may snag the keeper or the keeper's clothing. This may in turn cause the keeper to lose their balance and fall. Larger mesh fences, such as the open wire types, will become stretched and distorted, increasing the chance of injury or opening escape routes for the animals. Even stepping over lower landscape ropes or chains can become dangerous. A misjudged trip may result in a fractured wrist. Use established access points, even if it requires a short walk.

In cold weather avoid handling metal locks with damp or sweaty hands, as skin may stick to the frozen lock and be painful to remove. Service the locks before the onset of cold weather. If possible, the locks should be protected with flaps of rubber (e.g., sections of old inner tubes) or acrylic sheets to stop rain or snow from entering them and freezing their moving parts.

If a keeper is working in a cold climate with frequent ice buildup, they should ensure that there is a good supply of sand, salt, or other ice-melting compounds available for their safety. Numerous attachments are also available for boots, to help minimize the risk of slipping.

A keeper should use hand wash stations in work areas and make sure that they are kept supplied with towels and soap as needed. Carrying a small bottle of hand sanitizer is also recommended for in-between cleanings. One should take advantage of any shower facilities and changing rooms that the zoo may offer. Work clothes should be left at work when possible, but if the keeper is responsible for cleaning them at home, they should be kept separate from other clothes.

The keeper should adjust parts of the daily routine to avoid repetitive physical movement and its associated consequences. Practice in changing hands when sweeping, shoveling, or raking, for example, will enable the keeper eventually to become competent with either hand. Keeping is a very physical occupation and whenever possible, the keeper should get into the habit of warming up and stretching before starting a physical task.

SUMMARY

This material only scratches the surface of the knowledge that a keeper will need to carry out the daily routine. These are just fundamentals of the job. Requirements will change and vary between facilities, between work areas within the facility, and between groups of animals, and even between individual animals. The wonderful thing about the zookeeping profession is that the keeper will never stop learning. They can continue to develop professionally and pass on techniques during their career and even after they retire. One should make sure that the animals' needs are addressed and that any changes are in their best interest.

11

Taxonomy

Gary L. Wilson

REASONS FOR UNDERSTANDING TAXONOMIC PRINCIPLES

Modern zoos and aquariums are partners in a worldwide effort to conserve the biodiversity of our planet. Each keeper has a role in this cause. For keepers, understanding taxonomic principles will help in caring for each species they are responsible for. Closely related species share characteristics and have similar needs. Understanding that a species new to the collection is in the same family as a species the keeper already has experience with will provide a starting point for learning how to care for this new species. Keepers play a critical role in educating the zoo visitor about the animals under their care. Knowing how the animal is classified helps the keeper teach the zoo visitor about the species. Understanding taxonomy will also aid the keeper in working with the zoo's curators and helping to manage the zoo's animal collection. Being familiar with how different species are classified and understanding the relationships between groups will make the scientific literature more accessible to the keeper, facilitating the application of discoveries from the field to better manage captive specimens.

After studying this chapter, the reader will understand

- what taxonomy is, and its value
- the difference between evolutionary taxonomy and phylogenetic systematics
- principles of systematics such as monophyly, polyphyly, and paraphyly
- the binomial system for scientific names
- the hierarchical nature of classification and the major taxa involved
- conventions followed for word endings of taxa and pronunciation of scientific names.

WHAT IS TAXONOMY?

Taxonomy is the science of identifying, naming, and classifying living things. More than 1.5 million animals have been identified to date. Thousands more are identified each year. Yet it is estimated that this represents less than 20% of the organisms that live on the Earth (Hickman et al. 2004, 190).

CLASSIFICATION

Humans love to classify things. Putting things into categories helps us to handle the onslaught of information faced each day. What is the first question the zoo visitor asks the keeper when looking for the first time at an unfamiliar species? Typically, it is "What family is it in?" or "What's it related to?" Faced with the unfamiliar, the human brain seeks to place the information into a framework of categories that already exist in the mind. Only in this way can the mind make sense of the new information.

According to Simpson, "taxonomy is the theoretical study of classification, including its bases, principles, procedures, and rules" (Simpson 1961, 11). "Zoological classification," he says, "is the ordering of animals into groups (or sets) on the basis of their relationships, that is, of associations by contiguity, similarity, or both" (Simpson 1961, 9).

SYSTEMATICS

One of the goals for any modern system of biological classification is for the scheme to illustrate evolutionary relationships. Different theories of taxonomy follow different rules in their attempts to reveal such relationships. The work of evolutionary taxonomy is to produce a phylogenetic tree, a diagram with a branching structure that indicates how taxa or organisms are related as well as the duration of lineages or the amount of evolutionary change that occurred in the lineages (Hickman et al. 2004, 196). Phylogenetic systematics or cladistics has the goal of producing a cladogram, a branching

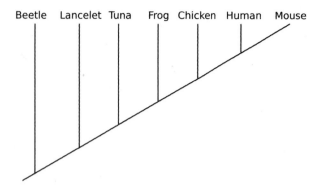

Figure 11.1. A cladogram. The fewer the number of branching points separating any two species, the greater the number of characteristics shared by those species. For example, mouse and human share more characteristics than mouse and tuna. Illustrations by Gary L. Wilson. Reprinted by permission.

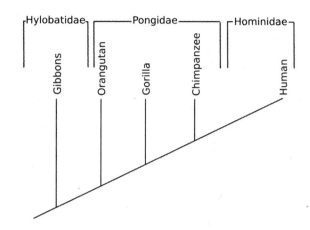

Figure 11.2. Phylogeny of apes. Rules of cladistics mandate the rejection of family Pongidae, since it is paraphyletic, and placement of the great apes into the family Hominidae along with humans (after Hickman et al. 2004, 201).

diagram similar to a phylogenetic tree wherein the emphasis is on shared and distinguishing characteristics (figure 11.1).

Evolutionary taxonomy and cladistics follow the same rules regarding monophyly and polyphyly. Monophyly describes a taxon that includes the most recent common ancestor of the group and all descendants of that ancestor. Polyphyly describes a taxon that does not include the most recent common ancestor of all members of a group—that is, a taxon that has at least two separate evolutionary origins (Hickman et al. 2004, 197). Marine mammals would be an example of a polyphyletic group, since it includes seals, sea lions, whales, and dolphins, but not the most recent common ancestor of all these animals. Under the rules of both evolutionary taxonomy and cladistics, monophyletic taxa are recognized while polyphyletic ones are rejected.

These two approaches differ in regards to paraphyly. Paraphyly describes a taxon that includes the most recent common ancestor of all members of a group and some but not all descendants of that ancestor (Hickman et al. 2004, 197). While cladistics rejects paraphyletic taxa, evolutionary taxonomy accepts such groups when they show distinct differences in characteristics indicating differences in adaptive zones. For example, the family Pongidae, containing the great apes, is recognized by evolutionary taxonomists because it occupies a different adaptive zone than the family Hominidae, containing humans. Under cladistics, the family Pongidae is rejected because it is paraphyletic, since it does not include a descendant (humans) of the common ancestor of the great apes (figure 11.2).

Throughout most of the history of taxonomy, the science has relied upon the ability of its practitioners to identify the key characteristics that should be used to distinguish different groups and to recognize relatedness. Since Darwin recognized that the similarities shared by related organisms were the result of descent from a common ancestor, taxonomists have endeavored to produce a tree of life that shows the true evolutionary relationships of its many branches. Yoon writes that in a "perfect world, the natural order is plainly marked by a nested series of pattern changes" and that any of the methods of taxonomy would "produce a tree that reflected"

evolutionary history. "Unfortunately," she observes, "evolution in real life is not nearly so orderly . . . traits can evolve multiple times in multiple lineages . . . creating confusing similarities" (Yoon 2009, 244).

In an attempt to cut through the confusion, cladistics requires that only real groups of evolutionary relatives (i.e., clades, from the Greek word for branch) be recognized. Such clades include all the descendants of an ancestor. As a result, many traditional groupings are being reviewed in light of newly available genetic data that provides the means to more accurately determine genetic closeness and distance of related organisms. For example, it has been proposed to replace the order Artiodactyla (the even-toed ungulates, such as pigs, hippopotamuses, giraffes, camels, antelope, cattle, sheep, and goats) and the order Cetacea (the whales and dolphins) with the order Cetartiodactyla. This is based on molecular evidence indicating that whales and dolphins are more closely related to hippopotamuses than they are to any other living mammals. If this is true, the order Artiodactyla is paraphyletic, since it does not contain all the descendants of the most recent common ancestor of the even-toed ungulates.

THE BINOMIAL SYSTEM

The system of classification that we use today is based on the system developed by Carolus Linneaus (1707–78) (Hickman et al 2004, 191). This system is hierarchical in nature, which is to say that each group or taxon is more inclusive than the taxon below it. The taxa (plural of taxon) that Linneaus included are kingdom, phylum, class, order, family, genus, and species.

An example illustrates this hierarchical arrangement. One can look at how the Toco Toucan is classified and identified (figure 11.3). One of the kingdoms is Animalia, the animals. The name Animalia calls our attention to one of the principal difficulties people encounter when learning scientific names—namely, that they are not in English. In Linnaeus's time Latin and ancient Greek were the languages of science, so the names are based on Latin and Greek. Some people

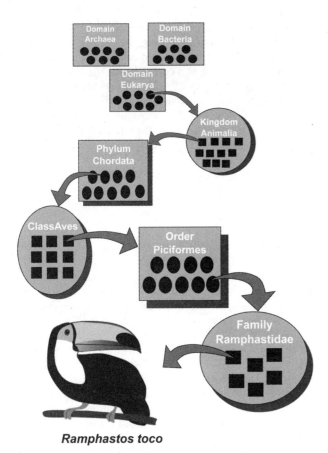

Ramphastos toco

Figure 11.3. Hierarchical arrangement of the binomial system, showing the relationship between the various taxa. Each taxon is more specific than the taxon above it. Illustration by Gary L. Wilson. Reprinted by permission.

find these names to be more understandable if they know the meaning of the Latin and Greek roots. If the keeper finds this to be the case, a dictionary of Latin and Greek roots will be helpful.

Within the kingdom Animalia, there are many phyla (plural of phylum) including Arthropoda (insects, spiders, crustaceans, etc.), Echinodermata (sea stars, sea urchins, sand dollars, etc.), and Annelida (earthworms, leaches, etc.). Phylum Chordata includes animals such as sea squirts and animals with backbones, the vertebrates. Class Aves is one of the classes within the phylum Chordata. Aves is from the Latin *avis*, meaning bird.

Class Aves contains all the birds, and there are many orders of birds. One of these is the order Piciformes (from the Latin *picus*, woodpecker, and *forma*, shape). So, Order Piciformes includes woodpeckers and relatives of the woodpeckers. Order Piciformes includes the family Picidae, the family of woodpeckers.

In this example, the focus is on another family within the order Piciformes, family Ramphastidae, the toucan family. The Toco Toucan is one member of this family. Toco Toucan is the common name of this brightly colored bird with the enormous bill. Scientific names consist of two parts, the genus name and the species name, and is therefore a binomial (or

binominal) system, or two-name system. The scientific name of the Toco Toucan is *Ramphastos toco*.

One can ask: Why not just use the common name Toco Toucan? As Philip Mortenson writes, ". . . 'common' does not necessarily mean 'universal'" (Mortenson 2004, 5). While the English common names of birds are recognized in many parts of the world, this is not the case with other animal groups. Besides varying from language to language, common names may vary even within the same country. Scientific names allow scientists, curators, keepers, and conservationists to communicate with confidence that they are referring to the same organism.

The name of the person who assigned the scientific name is often included when giving the name in a publication, as well as the year in which the person assigned the name. So, for the example of the Toco Toucan, the name would be written *Ramphastos toco* Statius Muller, 1776.

TAXA

The more closely related two species are, the lower the level of taxa they share. For example, the puma (*Puma concolor*) and the house cat (*Felis catus*) are both in the family Felidae (figure 11.4). The puma and the wolf (*Canis lupus*) are in different families, Felidae and Canidae respectively, but in the same order, Carnivora. The puma and moose (*Alces alces*) are in different orders, Carnivora and Artiodactyla respectively, but in the same class, Mammalia.

KINGDOM

Of the seven mandatory taxa, kingdom is the most inclusive. Until the late 1800s, every living thing was placed either into the plant kingdom or the animal kingdom (Hickman et al. 2004, 202). As scientists encountered more and more unicellular organisms, the need to place these organisms into either the plant or animal kingdom became increasingly difficult. For example, the single-celled protozoan *Euglena* has characteristics of both plants and animals. It can carry out photosynthesis as plants do, but it is mobile like most animals. Haeckel proposed the new kingdom Protista in 1866 for such single-celled organisms (Hickman et al. 2004, 203).

A five-kingdom system was proposed by Whittaker in 1969 (Hickman et al. 2004, 203). This scheme made a distinction

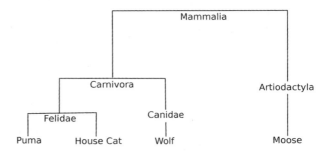

Figure 11.4. The level of taxa shared by different species indicates the degree of relatedness such that the lower the level shared, the more closely related the species are. Illustration by Gary L. Wilson. Reprinted by permission.

between prokaryotes and single-celled eukaryotes, kingdoms Monera and Protista respectively, and kingdom Fungi for molds, yeasts, and fungi along with kingdom Plantae for plants and kingdom Animalia for the animals. In 1990 Woese, Kandler, and Wheelis put the eukaryotes into the Eukarya, a domain—a level of classification higher than the kingdom. They also recognized two other domains, Bacteria and Archaea, containing the true bacteria and bacteria-like organisms with different membrane structure and ribosomal RNA sequences respectively (Hickman et al. 2004, 203).

PHYLUM

While there are almost 40 animal phyla, keepers typically deal with only a few. While most of the animals in a zoo belong to the phylum Chordata, most of the animals in the world are invertebrates and, therefore, belong to several other phyla. An increasing number of zoos are displaying invertebrates—mainly members of the phylum Arthropoda, which includes insects, spiders, scorpions, and millipedes. Aquariums frequently exhibit corals, sea anemones, and sea jellies, all members of the phylum Cnidaria. Members of the phylum Mollusca, including octopus, clams, and snails, are also frequently maintained in aquarium collections.

CLASS

The major classes dealt with in aquaria are Chondrichthyes (sharks, rays, and skates) and Actinopterygii (most of the bony fish). The vertebrate classes exhibited in zoos include Amphibia, Reptilia, Aves, and Mammalia.

While it is convenient to recognize birds and reptiles as distinct classes, cladistic taxonomists reject the separation because the class Reptilia, as traditionally defined, is paraphyletic. Birds are a sister group of the crocodilians (order Crocodilia), which is to say that both birds and crocodilians are more recently descended from a common ancestor than either is descended from any other extant (i.e., surviving) lineage (Hickman et al. 2004, 538). Evolutionary taxonomists keep the class Aves since it represents a different adaptive zone than that of the crocodilians and other reptiles.

ORDER

Taxonomic nomenclature is governed by a variety of codes that set out the rules for the particular group of organisms. For animals, the relevant rules are contained in the International Code of Zoological Nomenclature or ICZN (Groves 2001, 21).

Under the ICZN, the names of the orders of birds and fishes have the ending iformes. This makes it easy to recognize that the name refers to the order level of classification in these two groups. Unfortunately, this rule does not hold for other classes. For example, mammalia orders have names such as Primates, Carnivora, Artiodactyla, and so on. As new information is obtained about the genetic composition of organisms, taxonomists reexamine how the classification of these animals reflects their evolutionary relationships to other animals. For example, the order Insectivora was created for small, insect-eating mammals. Over the years, the order became a kind of "dumping ground" for new species. Lots of mammals have evolved to eat insects, a very abundant food source, and as a consequence they have evolved similar characteristics—a phenomenon called convergent evolution. The order Insectivora has now been abandoned in favor of several smaller orders which more accurately reflect the evolutionary relationships of their constituents. These orders include Afrosocicida (golden moles and tenrecs), Soricomorpha (shrews, moles, solenodons), Erinaceomorpha (hedgehogs and gymnures), Macroscelidea (elephant shrews), and Scandentia (tree shrews).

FAMILY

The taxon level of the family is perhaps one of the most useful. Just as one understands that the members of a person's family are individuals who are blood relations, the members of a taxonomic family are fairly closely related to each other. Because of this relatedness, it is expected that members of a family will exhibit many common characteristics, and this is generally true.

For example, members of the family Procyonidae typically have ringed tails and bold facial markings. This is true for raccoons (*Procyon*), coatis (*Nasua* and *Nasuella*), and the ringtail (*Bassariscus*). People who are familiar with raccoons quickly grasp the concept that coatis are in the same family because the shared features are obvious. Just as human families frequently have members who behave in distinctly different ways from most members of the family, taxonomic families also often have members that don't seem to fit in. The kinkajou (*Potos flavus*) is such a species in the case of the Procyonidae. Kinkajous lack the ringed tail and facial markings typical of the family, and even have prehensile tails which none of the other procyonids have. These differences are the result of the kinkajou's evolutionary divergence from the other members of the family and subsequent adaptation to a different niche (Koepfli et al. 2007, 1076).

Just as the ICZN specifies the endings for the order names of birds and fishes, it also specifies that the family names of birds, mammals, and fishes have the ending idae. This convention is frequently followed in the formation of the family names in other groups, but not in every case. For example, while most Reptile families end in idae, the family name of the worm lizards, Amphisbaenia, does not.

GENUS

The scientific name of an organism consists of the genus name and the species name (in zoology) or species epithet (in botany). The scientific name is written in italics or underlined. The first letter of the genus name is always capitalized. The root of the genus name often is used in forming the family name and order name. For example, the genus name of the bandicoots is *Perameles* while the family name is Peramelidae and the order name is Peramelemorphia. If a family includes only one species, the family name must be based on that genus name. For example, the red panda (*Ailurus fulgens*) is the only species in the family Ailuridae.

SPECIES

The second part of the scientific name is the species name (in zoology) or species epithet (in botany). It is never capitalized, even if it is based on a proper name. Under the rules of the ICZN, the scientific name is the only taxon that is a binomial name. That is, the names of all higher taxa must consist of a single word.

There are several different definitions of a species, often referred to as species concepts. The biological species concept says that species are "groups of actually or potentially interbreeding natural populations which are reproductively isolated from other such groups" (Mayr 1963, 19). The biological species concept does not help in determining species in organisms that reproduce asexually (Campbell et al. 2008, 492). Other species concepts include the morphological species concepts, the ecological species concept, the evolutionary species concept, and the phylogenetic species concept.

Since the scientific name of an animal consists of both the genus name and the species name, the species name is generally not written without the genus name. The first time a scientific name appears in a written document, the complete name is written out. If the name appears again in the same document, the genus name may be abbreviated as just the first letter. For example, since *Ramphastos toco* (toco toucan) appeared above, it could now be written as *R. toco* here.

Hybrids, the offspring resulting from the mating between members of different species, are problematic for the biological species concept. While a variety of reproductive isolating mechanisms typically prevent hybridization in nature, hybridization does sometimes occur in captive situations. When it does, the convention for the scientific name of the resulting hybrid is to use the scientific names of the parent species linked by an "×." For example, the scientific name of the offspring of the crossbreeding between a tiger (*Panthera tigris*) and a lion (*Panthera leo*) is *Panthera tigris* × *P. leo*.

SUBSPECIES

The seven taxa used by Linnaeus can be divided into taxa that are less inclusive than the taxon itself, but more inclusive than the next lower taxon. For example, animals are identified as subspecies because they show differences from other members of their species due to evolving adaptations to the locality in which they live. The subspecies "is a geographic segment of a species" (Groves 2001, 35). Groves describes four subspecies of the Geoffroy's (Black-handed) spider monkey (*Ateles geoffroyi*). The subspecies name is a trinomial, which is to say that it consists of three names. The subspecies of Geoffroy's spider monkey described in Groves's taxonomy of primates include *Ateles geoffroyi geoffroyi*, *Ateles geoffroyi yucatensis*, *Ateles geoffroyi vellerosus*, and *Ateles geoffroyi ornatus* (Groves 2001, 189).

Whether a group is identified as a subspecies or as a separate species is frequently influenced by the philosophy of the individual taxonomist. Some taxonomists are "lumpers," tending to lump subspecies together and minimizing the importance of geographic variation, while others are "splitters," tending to split groups apart into numerous subspecies based on small differences. While the debates between taxonomists can be viewed as a largely academic endeavor, conclusions about defining subspecies are likely to become increasingly important to conservationists as individual animals become more rare in the wild and decisions about preserving species or subspecies have to be made.

CONVENTIONS

As mentioned above, the naming of animals is governed by the International Commission on Zoological Nomenclature or ICZN. While most of the rules are of little relevance to the keeper, a few are useful to know as one may encounter various taxonomic terms in the process of searching for information about animals.

OTHER TAXA

Taxonomists frequently find it useful to divide a taxon into subgroups: subphylum, subclass, suborder, subfamily, and so on. A subclass or suborder may be divided into infraclasses or infraorders, respectively. For example, the class Mammalia is divided into subclass Prototheria, containing the egg-laying montremes, and subclass Theria, containing all other mammals. The subclass Theria is further divided into the infraclass Metatheria, the marsupials, and infraclass Eutheria, the placental mammals. In some cases, taxa are grouped together into supergroups. For example, the order Carnivora includes many families, and these families are grouped into the superfamily Canoidea (including the doglike carnivores such as dogs, bears, raccoons, and weasels), the superfamily Feloidea (including the catlike carnivores such as cats, civets, and hyenas), and the superfamily Pinnipedia (including the aquatic carnivores such as seals, sea lions, and walruses).

WORD ENDINGS

Some conventions exist for the formation of taxon names such that frequently the level of the taxon can be recognized even if the author does not explicitly identify it. The conventions for animal taxa differ slightly from the conventions for plant taxa, and for some of the animal taxa the conventions are not universally applied. The taxa for which the endings are consistently used include superorder (*-orpha*), superfamily (*-oidea*), family (*-idae*), subfamily (*-inae*), and tribe (*-ini*). The taxa for which the endings may vary include phylum (*-a*), subphylum (*-data*), class (*-lia*), subclass (*-ia*), order (*-iformes*), and suborder (*-morpha*); the most commonly used endings are in parentheses (Mortenson 2004, 9–10).

PRONUNCIATION

Most of the time, keepers will be dealing with scientific names and other taxon names in written form. Occasionally, the keeper may need to pronounce a name. Latin pronunciation of taxonomic nomenclature follows certain conventions which, unfortunately, do not match the conventions of Latin pronunciation in other realms. In the sciences, the Latin pronunciation follows that used in northern Europe (Covington

2010, 2). Therefore, *ae* is pronounced as long *e* (ē). For example, *-dae*, as in *Picidae*, is pronounced "dee," not "day" or "die." The stressed or accented syllable is the second or third to the last syllable (Covington 2010, 2).

REVISIONS

When one puts in the effort to learn the difficult scientific names of a number of species, one hopes that this information will be constant and unchanging for a long time. Unfortunately, this is not always the case. As the tools of science advance, scientists can probe more deeply or more accurately into the mysteries of nature. In the case of taxonomy, new tools have allowed taxonomists to reexamine how species and higher taxa have been classified in light of new information.

For example, Groves describes how different revolutions in the taxonomy of primates can be marked first by the ability to count chromosomes, then by the ability to compare proteins, and finally by the ability to sequence DNA. "As soon as the existence of substantial diversity in chromosome number and morphology in Primates became evident, new interpretations of interrelationships began to become widespread" (Groves 2001, 47). As new information is discovered and it is used to replace polyphyletic and paraphyletic groups with monophyletic ones, there is increasing pressure to replace the old Linnaean hierarchy. The PhyloCode is an example of an alternative system that is being developed (Hickman et al. 2004, 202).

STAYING CURRENT

Taxonomic revisions make it a challenge to stay up to date. The keeper's curator is an important resource for accomplishing this. Below is a list of resources on the World Wide Web that the keeper may find useful. The keeper should keep in mind that most zoos utilize computer database systems, such as ZIMS (as developed by ISIS), for maintaining records on the animals in their collections. There may be a delay between the time when a change in taxonomy is posted on one of the

websites (see appendix 2) and the time when the database service that the zoo is using adopts the change.

Remembering and pronouncing scientific names as well as staying current on changes in taxonomy can be difficult, but it is certainly worthwhile. Finding opportunities to use this information will help the keeper to become more proficient with it, just as using a foreign language becomes easier with practice. A wealth of valuable information about the many species keepers care for is available in the scientific literature. Familiarity with taxonomy makes this material much more accessible to the keeper. Keepers who continue to expand their knowledge in this way will continue to grow professionally, find new ways to further conservation efforts, and find fulfillment in their careers.

REFERENCES

Campbell, Neil A., Jane B. Reece, Lisa A. Urry, Michael L. Cain, Steven A. Wasserman, Peter V. Minorsky, and Robert B. Jackson. 2008. *Biology,* 8th ed. San Francisco: Pearson Education.

Covington, Michael A. 2010. Latin Pronunciation Demystified. Last modified 31 March. http://www.ai.uga.edu/mc/latinpro.pdf.

Groves, Colin P. 2001. *Primate Taxonomy.* Washington: Smithsonian Institution Press.

Hickman, Cleveland P., Larry S. Roberts, Allan Larson, and Helen l'Anson. 2004. *Integrated Principles of Zoology,* 12th ed. New York: McGraw-Hill.

Koepfli, Klaus-Peter, Matthew E. Gompper, Eduardo Eizirik, Cheuk-Chung Ho, Leif Linden, Jesus E. Maldonado, and Robert K. Wayne. 2007. Phylogeny of the Procyonidae (Mammalia: Carnivora): Molecules, Morphology and the Great American Interchange. *Molecular Phylogenetics and Evolution* 43(3): 1076–95.

Mayr, Ernst. 1963. *Animal Species and Evolution.* Cambridge, MA: Harvard University Press, Belknap Press.

Mortenson, Philip B. 2004. *This is Not a Weasel: A Close Look at Nature's Most Confusing Terms.* Hoboken, NJ: John Wiley and Sons.

Simpson, George Gaylord. 1961. *Principles of Animal Taxonomy.* New York: Columbia University Press.

Yoon, Carol Kaesuk. 2009. *Naming Nature: The Clash Between Instinct and Science.* New York: W. W. Norton.

12

Anatomy and Physiology (Part 1: Invertebrates)

Douglas P. Whiteside

INTRODUCTION

Increasingly, zoos and aquariums are displaying a greater number of invertebrate species as part of their collections. Keepers must provide care to these diverse species. Knowledge of comparative anatomy and physiology between the various taxa is useful for the keeper to develop an understanding of unique anatomical arrangements in less well studied species. This knowledge will better enable keepers to understand and meet the daily needs of these animals through husbandry and care procedures, and will serve as a valuable resource when making animal management decisions. This chapter will focus on terrestrial members of the phylum Arthropoda; the Mandibulata which includes insects (hexapods), the myriapods (millipedes [diplopods] and centipedes [chilopods]), and the Chelicerata which contains spiders and scorpions (arachnids).

At the conclusion of this chapter, the reader will appreciate and identify key features of the external and internal anatomy and physiology of terrestrial invertebrates.

EXTERNAL ANATOMY

Arthropods are characterized by a rigid, jointed exoskeleton that consists of a thick multilayered cuticle composed primarily of chitin. The cuticle consists of three layers: a thin waxy surface coat (epicuticle) that minimizes water loss, an intermediate thick middle layer (exocuticle), and a thick inner layer (endocuticle). Specialized cells in the epidermis include trichogen cells, which produce the bristle-like hairs (setae), and dermal cell glands, which produce secretions that are excreted through pores in the cuticle. At the joints (pleurites), the cuticle is thinner and more flexible, and internally, projections of the cuticle form points for muscle attachment (apodemes).

During growth, the exoskeleton is molted or shed periodically (ecydsis), with the number of molts varying with the species. The process is under endocrine control by the steroid hormone ecdysone. Once sexually mature, insects and many other arthropods cease growing and molting. While some arthropods undergo complete metamorphosis with one or more morphologically distinct larval stages (instar) between the egg and adult stage, others maintain the same form from hatching through to the adult stage. In most arthropods, if a limb is lost, it can be regenerated after several molts.

A tremendous range of external coloration exists in arthropods. Specialized color cells known as chromatophores contain five general classes of pigments and crystalline compounds: carotenoids (yellow and reddish orange), melanins (brown to black), ommochromes (brown to red and yellow), purines (yellow-white to silver) and pteridines (yellow-white, orange, and red). Color is imparted through expression of the pigments, or by the light refracting or scattering properties of the crystalline compounds (Anderson 2009a, 341, 349; Anderson 2009b, 528–29; Foelix 1996a, 13–21; Frye 1992a, 5–7; Frye 1992b, 31–36; Frye 1992c, 37–42; Frye 2006, 170–71; Gillespie and Spagna 2009, 941–42; Mitchell et al. 1988b, 179–81, 192–93; Pizzi 2006, 152–55; Williams 2009, 904–7).

Insects share many fundamental features despite their tremendous morphological diversity. The body of an insect is divided into three parts: the head, thorax, and abdomen. The head of an insect is the sensory and feeding center and usually consists of six segments, with a single pair of antennae, compound eyes, simple eyes (ocelli), and the mandibular and maxillary mouthparts. Three segments comprise the thorax, from which arises three pairs of legs and usually one to two pairs of wings. The abdomen consists of up to 11 body segments, and contains the majority of the internal organs. In many species a pair of mechanosensory structures (cerci) is found on the last abdominal segment, and these are highly modified terminal structures used in copulation and oviposition (egg laying; Cooper 2006, 206–7; Frye 1992c, 37–42; Headrick 2009, 13–20; Mitchell et al. 1988b, 179–81; Zachariah and Mitchell 2009, 17).

Millipedes are slow-moving myriapods characterized by their multisegmented cylindrical shape. The exoskeleton is calcificed, which provides for increased strength and durability. There are two pairs of jointed legs per diplosegment, with the exception of the head and anal segments, which do not bear legs, and the first few segments, which only have one pair of legs. The number of legs usually ranges between 150 and 200. Centipedes are fast-moving carnivores that are dorsoventrally flattened, lack a waxy epicuticle, and have one pair of jointed legs per segment (34–60 legs in total). Millipedes and centipedes have paired antennae located on their heads and jaws for sensory input. In addition, centipedes have a pair of venomous fangs (forcipules) on their first body segment, which are used for killing prey and for defense, and a pair of anal legs on the anal segment for tactile sensation, defense, and aggression.(Chitty 2009, 195–98; Frye 1992b, 31–36; Mitchell et al. 1988b, 179–81; Zachariah and Mitchell 2009, 16–17)

The bodies of arachnids have been reduced to two regions: the cephalothorax (prosoma), which is covered dorsally by flattened plates (tegites), which form a carapace, and ventrally by sternal plates (sternites) and the abdomen (opisthosoma). There are up to 12 eyes clustered on the craniodorsal aspect of the cephalothorax, and antennae are lacking. In spiders there is a clear distinction between the cephalothorax and abdomen, with the presence of a narrow waist (pedicel). In scorpions these two regions are fused, with a segmented opisthosoma that is divided into the anterior mesosoma and the posterior metasoma (tail). The tail is comprised of five to seven segments that terminate with a bulbous telson and its associated stinger (aculeus) (Foelix 1996a, 13–21; Frye 1992a, 5–7, 18–20; Frye 2006, 170–71; Gillespie and Spagna 2009, 941–42; Mitchell et al. 1988b, 179–81; Pizzi 2006, 152–55; Williams 2009, 904–7; Zachariah and Mitchell 2009, 15–16).

In many of the New World giant spiders, small barbed irritating or urticating hairs located on the dorsal aspect of the opisthosoma are used as a defense mechanism. The hairs are rapidly kicked off into the air by the caudal pair of legs, can cause severe irritation to a potential predator, and are replaced with each successive molt (Cooke et al. 1973, 130–33; Foelix 1996a, 13–21; Frye 1992a, 5–7, 18–20; Gillespie and Spagna 2009, 941–42; Pizzi 2006, 152–55; Zachariah and Mitchell 2009, 15–16).

All arachnids possess six pairs of appendages that arise from the cephalothorax. The first pair of appendages is the chelicerae, which can be modified as fangs with ducts that connect to the venom glands. In orthognath spiders (e.g., tarantulas), the chelicerae are parallel to one another and move in an up-and-down fashion, while in labidognath spiders (e.g., the black widow) the chelicerae oppose one another and move in a side-to-side fashion. The second paired appendages are the pedipalps. In spiders, the pedipalps have sensory and reproductive functions, while in scorpions they have been modified into large pincers for grasping prey. The last four paired appendages are the walking legs. Each leg has seven segments and ends with a tarsal claw. In many spiders there are dense tufts of hairlike structures (scopulae) associated with the tarsal claw that aid in climbing. Another important set of appendages in spiders is the three pairs of spinnerets, which are the external openings of the abdominally located silk glands and arise from the ventral aspect of the posterior abdomen (Foelix 1996a, 13–21; Frye 1992a, 5–7, 18–20; Frye

2006, 170–71; Gillespie and Spagna 2009, 941–42; Mitchell et al. 1988b, 179–81; Pizzi 2006, 152–55; Williams 2009, 904–7; Zachariah and Mitchell 2009, 15–16).

MUSCULAR SYSTEM

Although there is great variability in structure and performance among the musculatures of arthropods, there are also many basic features of structural organization, contractile performance, and composition that are shared between them and are even similar to the musculature of vertebrate species. Skeletal muscles attach internally to the cuticle of the exoskeleton through specialized projections of epidermal cells (apodemes). All of the musculature of arthropods is striated in nature, even though the visceral musculature behaves in a fashion similar to that of smooth (nonstriated) muscle in vertebrate species. In general, arthropods tend to have more morphologically distinct muscles than do vertebrate species, owing to their segmentation. Despite the number of distinct muscles, however, there are relatively few motor neurons that innervate them, with many muscles only receiving two to four motor neurons. This is considerably less than in vertebrate species. As invertebrates rely on the environmental temperature for body temperature regulation (i.e., are poikilothermic), their muscular activity is maximized in preferred optimal temperature zones (Anderson, 2009a 340; Foelix 1996c, 96–105; Josephson 2009, 675–77).

CIRCULATORY SYSTEM

All arthropods have open circulatory systems with noncontinuous vessels and no capillaries. The dorsally located hearts or circulatory pumps of arthropods are tubular in nature with muscular walls, and are surrounded by a pericardial sac with elastic connective tissue attachments to the body wall. Some insects also have pulsatile pumps at the limbs and other appendages to assist with circulation. The hemolymph is driven by peristaltic waves, in insects, or by nonperistaltic rhythmic contractions, in many other arthropods, as well as by pressure differentials and body movements. Hemolymph that is pumped by the heart and aorta is distributed by arteries to general body regions, where it flows directly into the hemocoel, which is composed of saclike sinuses and interstitial spaces. It then percolates through collecting sinuses or veins and enters back to the heart through specialized apertures (ostia) that have flaplike valves to ensure unidirectional flow. Diaphragms in the limbs divide the hemolymph into entering and exiting streams. The circulatory and respiratory systems are under the control of the autonomic nervous system.

In most insects and in the myriapods, the hemolymph contains no respiratory pigments, and gas exchange occurs by way of the respiratory system only, while in most arachnids the hemolymph is a light blue color due to hemocyanin, which contains copper instead of iron for binding oxygen. Hemolymph is important in the removal of waste from tissues as well as in thermoregulation (Foelix 1996b, 53–57; Miller 2009, 169–72; Mitchell et al. 1988a, 179–81; Mitchell et al. 1988g, 124–25; Pizzi 2006, 152–55; Williams 2009, 904–7; Zachariah and Mitchell 2009, 15–17).

RESPIRATORY SYSTEM

Owing to their impermeable, rigid exoskeletons, terrestrial arthropods have developed specialized respiratory structures for gas exchange. Insects and myriapods have evolved a unique tracheal system that may occupy up to half of the body volume. Air enters through valved apertures on the sides or ventrum of the body, known as spiracles, which are surrounded externally by hydrophobic hairs that prevent the influx of water and dust. Unlike insects, most myriapods cannot close their spiracles, which make them more sensitive to dessication. The spiracles open up into trachea, which then branch repeatedly to form minute terminal tracheoles that are intimately associated with the tissues. Fluid within these tracheoles is important for oxygenation during high demand. In smaller species the air diffuses in and out of the tracheal system, while in larger species differential gas pressures and muscular contraction of air sacs during body movement also aid in respiration (Cooper 2006, 207; Harrison 2009, 889–94; Mitchell et al. 1988a, 633–34; Mitchell et al. 1988h, 106–7; Zachariah and Mitchell 2009, 16–17).

Gas exchange in arachnids is accomplished by diffusion across book lungs located on the ventral surface of the abdomen. These lungs consist of fanned-out gas exchange surfaces with interdigitated air and hemolymph spaces, similar to the leaves of an open book. The spiracles open into four pairs of book lungs in scorpions, or one to two pairs in spiders. More advanced spiders also have developed a tracheal system similar in anatomy and function to that seen in insects (Foelix 1996b, 61–62; Frye 2006; 170–71; Gillespie and Spagna 2009, 942–43; Mitchell et al. 1988a, 633–34; Mitchell et al. 1988h, 106; Pizzi 2006, 152–55; Williams 2009, 904–7; Zachariah and Mitchell 2009, 15–16).

DIGESTIVE SYSTEM

Although the numerous variations in the digestive tracts of invertebrates reflect their various feeding strategies, all have a digestive system that is divided into three regions: the anterior or foregut, the midgut, and the posterior or hindgut. Similarly, mouthparts of the various taxa reflect dietary adaptations such as sucking, piercing, cutting, sponging, and chewing. The digestive system consists of the alimentary or gastrointestinal tract and the digestive glands. In all arthropods there is peristaltic and retroperistaltic activity to varying degrees.

In insects, the foregut is comprised of the mouth, buccal cavity, pharynx, esophagus, and, in some families, a crop and proventriculus. It is responsible for food storage and initiation of digestion. The midgut, which is the primary site for digestion and nutrient absorption, consists of the stomach and the majority of the intestines. Particularly in species that consume vegetation, two to six blind-ended sacs (ceca) branch from the anterior end of the stomach. The hindgut is composed of the distal intestinal tract (ileum and colon), rectum, and anus, and is the major site of water resorption and excretion of digestive and urinary wastes.

Depending on the species, the saliva of insects may contain no enzymes or a wide array of proteolytic enzymes. Digestive enzymes responsible for food digestion are found from the crop through the midgut (Bradley 2009, 334–39;

Cooper 2006, 207; Mitchell et al. 1988a, 630–32; Mitchell et al. 1988e, 89–90; Terra and Ferreira 2009; 273–76; Zachariah and Mitchell 2009, 17).

Centipedes and millipedes have a simple digestive system. Most of the alimentary tract length in centipedes is comprised of the pharynx and esophagus, while the midgut constitutes the major portion of the tract in millipedes (Mitchell et al. 1988a, 630–32; Mitchell et al. 1988e, 89–90; Zachariah and Mitchell 2009, 16–17).

In most spiders, the digestion of food starts externally through the regurgitation of digestive enzymes from the digestive glands onto the immobilized prey, which liquefies it. These digestive glands arise from the stomach. The liquefied food enters the foregut through the action of a sucking stomach in the prosoma. The midgut then traverses from the prosoma into the opisthosoma. In some species, the coxal (first) segment of the pedipalps is modified to form chewing mouth parts, or the pedipalps may possess a row of serrated "teeth" (serulla) for cutting into prey or bristles for straining food. In large mygalomorphs the fangs also aid in macerating the food, which allows the ingestion of some soft tissue. A large number of lobed diverticula which originate from the midgut increase the absorptive surface area of the digestive tract and possibly serve as digestive storage. Spiders have a short hindgut which terminates at the anus (Foelix 1996b, 45–50; Frye 1992a, 5–7; Gillespie and Spagna 2009, 942–43; Mitchell et al. 1988a, 630–32; Mitchell et al. 1988e, 89–90; Pizzi 2006, 154; Zachariah and Mitchell 2009, 15–16).

Scorpions also secrete enzyme-rich digestive fluids that initiate the digestive process, allowing them to imbibe, chew, and ingest their prey. Their chitinous chelicerae are serrated to facilitate the tearing of their food into smaller pieces that can be swallowed. The mouth opens into a pre-oral chamber that continues on to the pharynx, esophagus, stomach with ceca, intestines, and hindgut and terminates at the anus (Frye 1992a, 18–20; Frye 2006, 170–71; Mitchell et al. 1988a, 630–32; Mitchell et al. 1988e, 89–90; Williams 2009, 904–7; Zachariah and Mitchell 2009, 15–16).

HEMATOPOETIC AND IMMUNE SYSTEM

The hemolymph of arthropods serves as a storage pool for water, as a hydraulic fluid, and it has important roles in the transport of hormones, nutrients, and metabolites, and also in immune function, as it contains numerous protective proteins such as enzymes and antibodies. It differs substantially from the blood of vertebrate species as it lacks red blood cells (erythrocytes) or distinct white blood cells (leukocytes), and instead contains circulating cells known as hemocytes which play a role in wound healing, clotting, and infection control (Foelix 1996b, 58–60; Gillespie and Spagna 942–43; Kanost 2009, 446–48; Mitchell et al. 1988a, 633; Pizzi 2006, 154; Zachariah and Mitchell 2009, 15–17).

URINARY SYSTEM

In the vast majority of arthropods, the Malpighian tubules (ranging from two to hundreds in number, depending on species) are the site of urine formation and are the main excretory organs. The tubules arise as diverticula from the anterior aspect of the hindgut, and empty fluid into the hindgut for excretion via hydrostatic pressure. Urine production arises from the active transport of ions from the hemolymph across the epithelium and into the lumen of the Malpighian tubules. In addition, the tubules are responsible for elimination of nitrogenous waste and other by-products of metabolism. In most terrestrial insects and myriapods urea and uric acid are the predominant nitrogenous waste products, while in arachnids the predominant waste is guanine, along with lesser amounts of uric acid, adenine, and hypoxanthine. In arachnids, less conspicuous excretory organs include the coxal glands, which are associated with the coxae of the walking limbs, and nephrocytes, which are the largest cells in the arachnid body and store metabolites from the hemolymph (Bradley 2009, 334–39; Foelix 1996b, 50–52; Frye 1992a, 5–7, 18–20; Frye 1992c, 37–42; Frye 2006, 170–71; Gillespie and Spagna 2009, 942–43; Mitchell et al. 1988a, 634; Mitchell et al. 1988f, 159–60; Pizzi 2006, 152–55; Williams 2009, 904–7; Zachariah and Mitchell 2009, 15–17).

REPRODUCTIVE SYSTEM

Insects, myriapods, and arachnids have distinct sexes (i.e., are gonochoric) and practice internal fertilization, although various forms of parthenogenesis have been well documented in insects and a few spider species. In insects and myriapods, the female reproductive tract is comprised of a pair of ovaries, one or more saclike receptacles (spermathecae) for sperm storage, accessory glands, interconnecting ducts, and a genital aperture (gonopore). The ovaries are further divided into distinct egg tubes (ovarioles) that vary in number depending on the species. The spermathecae can store sperm for varying periods of time, and females can control or bias sperm usage. The accessory glands produce a variety of products that assist with egg packaging and laying as well as sperm maintenance, transport, and fertilization within the female reproductive tract. Females can either lay eggs (i.e., be oviparous) or gestate their eggs internally (i.e., be ovoviviparous). Many insects have modified abdominal appendages, known as ovipositors, for egg laying (ovoposition; Mitchell et al. 1988a, 637–39; Wheeler 2009, 880–82; Zachariah and Mitchell 2009, 17).

The female arachnid's reproductive tract is comprised of paired ovaries, oviducts, and spermathecae for sperm storage. The terminal oviduct (uterus externus) ends at the gonopore, located within the epigastric furrow just anterior to the book lung openings. Fertilization occurs in the terminal oviduct. Female spiders are oviparous, while female scorpions are ovoviviparous (Foelix 1996d, 177–89; Frye 2006, 171; Pizzi 2006, 147; Zachariah and Mitchell 2009, 15–16).

Internally, all male arthropods have either a single testis or paired testes. Sperm develops within membranous sacs and is transported through the ductus deferens to a common ejaculatory duct. The seminal vesicles are enlarged portions of the ductus deferens, which store sperm. Also present are sexual accessory glands, which produce seminal fluid to maintain the sperm; seminal plugs to prevent further breeding in females; protein-rich matrices, to packet the sperm (spermataphores); and compounds that exert behavioral and physiological effects on the female. In insects, the terminal end of the ejaculatory duct has been modified into a copula-tory organ (aedeagus). In male millipedes there are sperm-transfer organs (goniopods) on the seventh body instead of legs. In adult male spiders, the copulatory organ is present on the distal segment of the pedipalps (palpal organ). Sperm is collected from the gonopores and stored in the palpal organ until it is transferred to the female reproductive tract during mating. Male scorpions lack a palpal organ, and instead deposit a stalked spermataphore on the ground and then guide a receptive female over the spermataphore, which is taken up into her gonopore and the sperm released into the oviduct for fertilization (Foelix 1996d, 177–89; Frye 2006, 171; Klowden 2009, 885–87; Pizzi 2006, 147; Zachariah and Mitchell 2009, 15–17).

NERVOUS SYSTEM AND SPECIAL SENSES

The nervous system of arthropods is segmentally arranged with a marked tendency towards the fusion of the nerve cell bodies (ganglia) as compared with that of lower invertebrates. In more primitive insects and the arachnids, the central nervous system consists of a more anterior supraesophageal ganglion (preoral brain) that contains the optic lobes, and a larger ventral subesophageal ganglion (postoral brain). The two ganglia are linked by paired nervous tracts on either side of the esophagus, while in more advanced insects they are completely fused circumferentially around the esophagus. The ganglia of the thorax or cephalothorax and of the abdomen are distributed ventrally to the alimentary tract, and are connected to one another by the ventral nerve cord. Nerves radiate from the ganglia to form the peripheral nervous system.

The perception of sensory stimuli in insects is via hairlike receptors (sensilla) distributed all over the body, but concentrated in greater numbers on the appendages. Many insects have a tympanic organ for sound detection. The antennae and tarsi have numerous tactile, temperature, humidity, and chemical receptors. The simple eyes can detect changes in light intensity and aid in spatial orientation. The compound eyes are responsible for vision. Each visual unit (ommatidium) has its own lens and photoreceptor cells, and visual information is transmitted to the optic ganglia and then to the optic nerve for relaying to the optic lobes for processing. The combined action of the ommatidia produces a mosaic image (Cooper 2006, 206–7; Foelix 1996c, 96–105; Land 2009, 345–47; Mitchell et al 1988a, 634–35; Mitchell et al. 1988i, 270–91; Pizzi 2006; 152–55; Strausfeld 2009, 121–29).

The sensory structures in millipedes and centipedes consist mainly of the eyespots and antennae. Depending on the species, the eyespots consist of a varying number of ommatidia, but they are not clustered densely enough to form a true compound eye as is found in insects, and they likely cannot form images. Rather, millipedes and centipedes rely on photoreception and movement detection with their eyes. Many species can also detect movement through vibrations in their multiple limbs (Mitchell et al. 1988a, 635; Mitchell et al. 1988i, 270–91; Zachariah and Mitchell 2009, 16–17).

Spiders are conspicuously hairy, and different hair types function as mechanoreceptors allowing for the detection of tactile, seismic, and chemical stimuli. Numerous sensilla also cluster together to form lyriform organs which aid in sound detection. The simple eyes of spiders, up to 12 in number, aid in light and movement detection. With notable exceptions, such as the jumping spiders, most spiders are considered to have poor vision (Foelix 1996c, 68–92; Gillespie and Spargna 2009, 944–45; Mitchell et al. 1988a, 635; Mitchell et al. 1988i, 270–91; Pizzi 2006, 155).

In scorpions, located immediately behind the last pair of walking legs are paired comblike organs known as pectines, which are responsible for the detection of chemosensory stimuli, from prey or conspecifics, and seismic stimuli. Like spiders, scorpions have simple eyes, with a pair of median ocelli, and two groups of up to four lateral ocelli each (Frye 2006, 37–42; Mitchell et al 1988a, 635; Mitchell et al. 1988i, 270–91; Williams 2009, 904–7; Zachariah and Mitchell 2009, 15–16).

SUMMARY

A basic understanding of the normal anatomy and physiology of terrestrial invertebrates is essential for keepers. It aids them in husbandry decisions and allows them to promptly identify abnormalities that can be treated, thus improving the health and welfare of the species in their care.

REFERENCES

Anderson, Svend O. 2009a. "Exoskeleton." In *Encyclopedia of Insects*, 2nd edition, edited by Vincent H. Resh and Ring T. Cardé, 339–42. London: Elsevier.

———. 2009b. "Integument." In *Encyclopedia of Insects*, 2nd edition, edited by Vincent H. Resh and Ring T. Cardé, 528–29. London: Elsevier.

Bradley, Timothy J. 2009. "Excretion." In *Encyclopedia of Insects*, 2nd edition, edited by Vincent H. Resh and Ring T. Cardé, 334–39. London: Elsevier.

Chitty, John R. 2006 " Myriapods (Centipedes and Millipedes)." In *Invertebrate Medicine*, edited by Gregory A. Lewbart, 195–203. Ames: Blackwell.

Cooke, J. A. L., F. H. Miller, R.W. Grover, and J. L. Duffy. 1973. "Urticaria Caused by Tarantula Hairs." *American Journal of Tropical Medicine and Hygiene* 22(1):130–33.

Cooper, John E. 2006. "Insects." In *Invertebrate Medicine*, edited by Gregory A. Lewbart, 205–19. Ames: Blackwell.

Foelix, Rainer F., ed. 1996a. "Functional Anatomy." In *Biology of Spiders*, 2nd edition, 12–37. New York: Oxford University Press.

———. 1996b. "Metabolism." In *Biology of Spiders*, 2nd edition, 38–66. New York: Oxford University Press.

———. 1996c. "Neurobiology." In *Biology of Spiders*, 2nd edition, 68–109. New York: Oxford University Press.

———. 1996d. Reproduction." In *Biology of Spiders*, 2nd edition, 176–212. New York: Oxford University Press.

Frye, Fredric L. 1992a. "Arachnids." In *Captive Invertebrates: A Guide to Their Biology and Husbandry*, 5–29. Malabar: Krieger.

———. 1992b. "Chilopods and Diplopods." In *Captive Invertebrates: A Guide to Their Biology and Husbandry*, 31–36. Malabar: Krieger.

———. 1992c. "Insects." In *Captive Invertebrates: A Guide to Their Biology and Husbandry*, 37–63. Malabar: Krieger.

Frye, Fredric L. 2006. "Scorpions." In *Invertebrate Medicine*, edited by Gregory A. Lewbart, 169–77. Ames: Blackwell.

Gillespie, Rosemary G., and Joseph C. Spagna. 2009. "Spiders." In *Encyclopedia of Insects*, 2nd edition, edited by Vincent H. Resh and Ring T. Cardé, 941–51. London: Elsevier.

Harrison, Jon F. 2009. "Respiratory System." In *Encyclopedia of Insects*, 2nd edition, edited by Vincent H. Resh and Ring T. Cardé, 889–95. London: Elsevier.

Headrick, David H. 2009. "Head, Thorax, Abdomen, Genitalia." In *Encyclopedia of Insects*, 2nd edition, edited by Vincent H. Resh and Ring T. Cardé, 11–21. London: Elsevier.

Josephson, Robert. 2009. "Muscle System." In *Encyclopedia of Insects*, 2nd edition, edited by Vincent H. Resh and Ring T. Cardé, 675–80. London: Elsevier.

Kanost, Michael, R. 2009. "Hemolymph." In *Encyclopedia of Insects*, 2nd edition, edited by Vincent H. Resh and Ring T. Cardé, 46–448. London: Elsevier.

Klowden, Marc J. 2009. "Reproduction, Male." In *Encyclopedia of Insects*, 2nd edition, edited by Vincent H. Resh and Ring T. Cardé, 885–87. London: Elsevier.

Land, Michael F. 2009. "Eyes and Vision." In *Encyclopedia of Insects*, 2nd edition, edited by Vincent H. Resh and Ring T. Cardé, 345–55. London: Elsevier.

Miller, Thomas A. 2009. "Circulatory System." In *Encyclopedia of Insects*, 2nd edition, edited by Vincent H. Resh and Ring T. Cardé, 169–73. London: Elsevier.

Mitchell, Lawrence G., John A. Mutchmor, and Warren D. Dolphin, eds. 1988a. "Arthropods and Arthropod-like Phyla." In *Zoology*, 609–40. Menlo Park: Benjamin/Cummings.

———. 1988b. "Body Surface and Support Systems." In *Zoology*, 172–94. Menlo Park: Benjamin/Cummings.

———. 1988c. "Coordination: Hormones and Endocrine System." In *Zoology*, 244–69. Menlo Park: Benjamin/Cummings.

———. 1988d. "Coordination: Nervous System." In *Zoology*, 219–243. Menlo Park: Benjamin/Cummings.

———. 1988e. "Feeding and Nutrition." In *Zoology*, 75–96. Menlo Park: Benjamin/Cummings.

———. 1988f. "Internal Fluid Regulation and Excretion." In *Zoology*, 145–69. Menlo Park: Benjamin/Cummings.

———. 1988g. "Internal Transport and Defense Systems." In *Zoology*, 119–44. Menlo Park: Benjamin/Cummings.

———. 1988h. "Respiratory Gas Exchange." In *Zoology*, 97–118. Menlo Park: Benjamin/Cummings.

———. 1988i. "Sensory Systems." In *Zoology*, 270–96. Menlo Park: Benjamin/Cummings.

Pizzi, Romain. 2006. "Spiders." In *Invertebrate Medicine*, edited by Gregory A. Lewbart, 143–68. Ames: Blackwell.

Strausfeld, Nicholas J. 2009. "Brain and Optic Lobes." In *Encyclopedia of Insects*, 2nd edition, edited by Vincent H. Resh and Ring T. Cardé, 121–30. London: Elsevier.

Terra, Walter R., and Clélia Ferreira. 2009. "Digestive System." In *Encyclopedia of Insects*, 2nd edition, edited by Vincent H. Resh and Ring T. Cardé, 273–81. London: Elsevier.

Wheeler, Diana E. 2009. "Reproduction, Female." In *Encyclopedia of Insects*, 2nd edition, edited by Vincent H. Resh and Ring T. Cardé, 880–82. London: Elsevier.

Williams, Stanley C. 2009. "Scorpions." In *Encyclopedia of Insects*, 2nd edition, edited by Vincent H. Resh and Ring T. Cardé, 904–9. London: Elsevier.

Zachariah, Trevor, and Mark A. Mitchell. 2009. "Invertebrates." In *Manual of Exotic Pet Practice*, edited by Mark A. Mitchell and Thomas N. Tully, 11–38. Saint Louis: Saunders Elsevier.

13

Anatomy and Physiology (Part 2: Vertebrates)

Douglas P. Whiteside

INTRODUCTION

A wide array of vertebrate species is found in zoos and aquariums. Despite a tremendous range of morphological appearances, the general body plan of vertebrate species is fairly well conserved with fundamentally similar organ systems. Keepers must provide care to these diverse collections, and knowledge of the comparative anatomy and physiology of the various taxa is useful for developing an understanding of unique anatomical arrangements in less well studied species. This knowledge will better enable keepers to understand and meet the daily needs of these animals through husbandry and care procedures, and will serve as a valuable resource when making animal management decisions.

At the conclusion of this chapter, the reader will

- appreciate and identify key features of, and differences in, the external and internal anatomy and physiology of vertebrates
- understand directional terminology as it relates to the external anatomy of vertebrates
- be familiar with sites of blood collection in vertebrate species.

EXTERNAL ANATOMY

FISH

A wide variety of body shapes exist in the approximately 25,000 species of cartilaginous (chondrichthyans) or bony (osteichthyans) fish, although the most commonly recognized shapes are variations on fusiform or cigar-shaped fish (e.g., shark, salmon, goldfish), sagittiform or laterally compressed fish (e.g., angelfish), depressiform or dorsoventrally compressed fish (e.g., rays, skates), and anguilliform or elongated forms (e.g., moray eel). Despite the range of body types, the body is generally divided into three regions: head, trunk and tail. While fin morphology differs greatly across the taxa, fin arrangement is well conserved with dorsal, pectoral, pelvic, anal, and caudal (tail) fins found in all fish species, with an additional adipose fin in some species (e.g., salmonids, catfish).

Cartilaginous fish such as elasmobranchs (sharks, rays and skates) have five to seven gill slits, with the first slit modified as a small dorsolateral opening known as a spiracle. In bony fish the common gill opening is covered by an operculum.

The skin of a fish is thin and glandular, and it lacks a keratinized outer layer (*stratum corneum*). Most species produce a layer of mucous, from glands in the epidermis, to aid in protection. In addition, some species have specialized epidermal glands that produce toxins and venoms. There is a minimal space under the skin. In species with scales, the scales originate from the dermis, so that when they are lost the overlying epidermis is also removed. Four scale types are recognized: ganoid (e.g., gars), ctenoid (e.g., perch), cycloid (e.g., trout), and placoid (e.g., sharks; Bone and Moore 2008, 1–16; Harms 2003, 2–4; Miller and Mitchell 2009, 43; Mitchell, Mutchmor, and Dolphin 1988g, 705–11; Stoskopf 1993a, 2–30).

AMPHIBIANS

There are almost 6,800 species of recognized amphibians (Amphibiaweb 2011). The three orders of amphibians—frogs and toads (Anura), salamander, newts, and sirens (Caudata), and caecilians (Gymnophiona)—vary substantially in their external appearance. Frogs and toads are generally tailless as adults, lack external gills, and generally have hind limbs longer than their forelimbs, with webbed toes. The caudates are lizardlike in form, with four limbs (except sirens, which lack pelvic limbs) and a long tail. External featherlike gills may or may not be present. Caecilians are snake- or wormlike, with no limbs and sometimes a very short tail. True scales and claws are lacking, although a few species, such as the African clawed frog (*Xenopus laevis*), have modified epidermal clawlike structures.

The skin of amphibians is glandular, and it ranges from being smooth to having multiple rounded protrusions (i.e.,

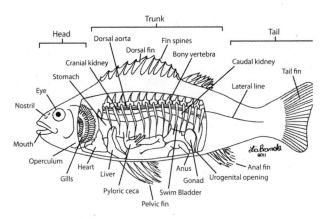

Figure 13.1. Fish: external and internal anatomy of a bony fish. Courtesy of Lia Brands.

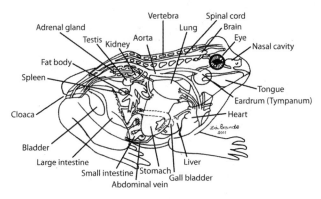

Figure 13.2. Amphibian: external and internal anatomy of a male frog. Courtesy of Lia Brands.

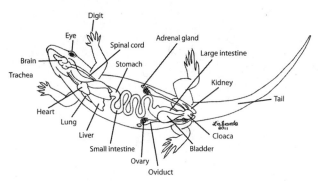

Figure 13.3. Lizard: external and internal anatomy of a female lizard. Courtesy of Lia Brands.

being bosselated), depending on the species. It is arguably one of the most important organ systems, as it functions in a protective capacity, acts as a sensory organ, and has vital roles in maintaining hydration and thermoregulation, respiratory gas exchange, gender recognition, and reproduction. In general the epidermis is thin and lightly keratinized, and thus it can be damaged easily if handled improperly or in contact with inappropriate substrates. A variety of specialized glands are present within the skin that may produce mucous or waxy substances to minimize water loss, or that may produce toxic or irritating protective substances. The ventrum serves as an important route for water uptake from the environment, with frogs and toads having specialized pelvic areas (drinking patches) that are responsible for up to 80% of water intake (Goin et al. 1978, 15–28; Helmer and Whiteside 2005, 7–13; Mitchell, Mutchmor, and Dolphin 1988a, 728–33; Pough, Janis, and Heiser 2009e, 222–32, 255–57; Wright 2001, 16–17).

REPTILES

The elongated body shape, presence of a tail, and the sprawling posture of the approximate 9300 species of reptiles (Reptile Database 2011) is well conserved across the various orders (figure 13.3). Compared to amphibians or mammals, the various species of reptiles—lizards and snakes (Squamata), turtles and tortoises (Testudines), crocodilians (Crocodylia), and tuatara (Rhynchocephalia)—have far fewer glands in their heavily keratinized skin, with the only glandular tissue being the pheromone-producing femoral and precloacal pores in some lizards. The epidermis is rich in lipid, which provides a water-permeable barrier. Formed from structured thickening and thinning of the epidermis, the scales are an integral part of the skin, and provide protection from abrasion and aid in locomotion in snakes (gastropeges). The thick skin of reptiles decreases cutaneous sensitivity, thus imparting a greater risk to thermal burns in captivity. In some lizards and in crocodilians, bony plates (osteoderms) that form within the dermis in turtles and tortoises (chelonians) have fused with the vertebrae to form the shell. Color is imparted by pigment-laden cells (chromatophores such as melanophores and iridophores) in between the dermis and epidermis and through iridescence (Kirchgessner and Mitchell 2009, 207; Mitchell 2009, 137–39; Mitchell, Mutchmor, and Dolphin 1988c, 190–93; Mitchell, Mutchmor, and Dolphin1988k, 750–60; Navarez, 2009a 114–15; Navarez, 2009b 165–66; O'Malley 2005b, 36–37; O'Malley 2005c, 73; O'Malley 2005d, 78; O'Malley, 2005e 42–44).

BIRDS

Unlike the tremendous morphological diversity that exists in mammals and reptiles, there is little variation in the basic anatomy of the approximately 9,700 species of birds, owing to the constraints of flight. The thin skin of birds is protected by the plumage or feathering, with feathers arising from specialized follicles that are arranged in organized tracts (pterylae) within the dermis. In many species there are also areas that lack these tracts (apteria), while most species have nonfeathered areas on their distal legs with keratinized scales (podotheca). The feathers are composed of keratin, with color imparted by pigments such as melanin, porphyrins, carotenoids, and the Tyndall effect by the feather barbules. Feather numbers may range from about 1,000 in hummingbirds to approximately 100,000 in penguins.

The avian skin has only three glands: the bilobed preen (uropygial) gland, located at the dorsal base of the tail, which is not present in all birds; the aural gland; and the vent gland. Since birds are homeothermic and lack sweat glands, heat is lost through the skin, in conjunction with a countercurrent

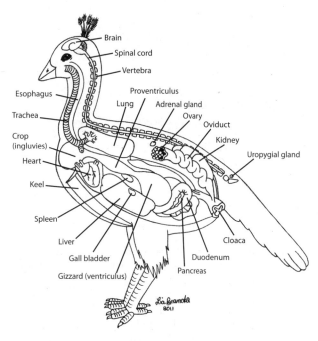

Figure 13.4. Bird: external and internal anatomy of a female peafowl. Courtesy of Lia Brands.

vascular system, and through the respiratory system (Mitchell, Mutchor, and Dolphin 1998, 778–83; O'Malley 2005, 149–58; Tully 2009, 257–58).

MAMMALS

Despite great variations in size and shape among the approximate 5,500 species of mammals (IUCN 2011), most external anatomical features are well conserved between taxa. An understanding of directional positioning on the body is important for the keeper to describe precisely where abnormalities may be present (figure 13.5).

There is a tremendous variety in the thickness and texture of the mammalian integument, with skin thickness ranging from that of only a few cells in small rodents to that of several hundred cells in "pachyderms" such as elephants, rhinoceroses, hippopotamuses, and tapirs, and ranging from very smooth to rough and wrinkled. Skin color is determined by pigments such as melanin, and by light absorption and scattering effects (Prum and Torres 2004, 2164–69). The various components of skin (epidermis, dermis, glands, sensory structures, and keratinized structures such as hair, nails, hooves, horns, and the outer covering of bony horns and antlers) are important in thermoregulation, water conservation, camouflage, communication, sensation, locomotion, display, and protection.

Several types of hair exist in mammals for body covering (undercoat, guard hairs), tactile sensation (whiskers or vibrissae), or defense (quills or spines). Hair color is imparted by pigment within the hair shaft, and the hair is lubricated and waterproofed by oily secretions from sebaceous glands. Three major types of skin glands are present in mammals, and they vary in location and number depending on the species: eccrine glands, which produce watery secretions; sebaceous glands, which produce oily secretions (sebum); and apocrine glands, which produce secretions that are important in chemical communication. Specialized scent glands, modified from sebaceous or apocrine glands for marking or spraying, are found in mammals. Unique to mammals are complex branching mammary glands for lactation (Mitchell, Mutchmor, and Dolphin 1988c, 190–93, and 1988j, 805–7; Pough, Janis, and Heiser 2009c, 533–35).

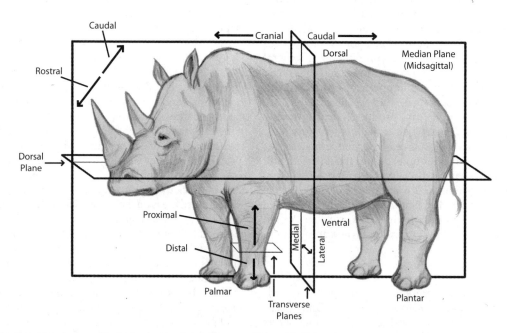

Figure 13.5. Directional terminology. Illustration by Lisa McLaughlin. www.McLaughlinWatercolor.com.

MUSCULOSKELETAL SYSTEM

FISH

Fish have a well developed axial skeleton (skull and vertebral column) with a reduced appendicular skeleton (pectoral and pelvic girdles). In cartilaginous fish the skeleton is composed of cartilage, although the axial skeleton is reinforced with calcium deposits. All other fish have a bony skeleton similar in composition to those of all other higher vertebrates.

All fish have two types of striated skeletal muscle fibers: red muscle, which is involved in sustained aerobic activity, and white muscle, which is used during bursts of anaerobic activity. Although the muscular arrangement in fish is in W-shaped blocks (myotomes) that run parallel to the long axis, the relative proportion and complex distribution of the red and white muscle fibers is species-dependent. Involuntary smooth muscle is similar to that of other vertebrate species, as is the striated cardiac muscle. As with other vertebrates, fibrous connective tissue structures support the musculoskeletal system; tendons are responsible for attaching muscle to bone, while ligaments hold the bones together and produce joints (Harms 2003, 2–4; Miller and Mitchell 2009, 48–49; Mitchell, Mutchmor, and Dolphin 1988g, 715–16; Stoskopf 1993a, 2–30).

AMPHIBIANS

As in their external anatomy, several variations exist in the skeletal elements among the amphibians. Salamanders typically have four limbs, with four digits on the forefeet and five on the hindfeet, although this can vary between species. They are capable of regenerating lost digits and limbs, and if threatened or injured, the tail can break free of the body and regenerate afterwards (autotomy). Anurans are well adapted to a lifestyle of jumping with elongated hind limbs. There are generally four digits on each forefoot, and five on each hindfoot. The vertebrae and pelvic girdle are fused into a structure called a urostyle, and a sacrum is lacking. The caudal vertebrae are absent. While tadpoles can regenerate limbs, adult anurans generally cannot. Caecilians lack a pectoral and pelvic girdle and sacrum; thus movement is achieved through wormlike contractions or lateral, eel-like undulations. As with other vertebrates, smooth muscle and two types of striated muscle (skeletal and cardiac) are present (Helmer and Whiteside 2005, 7–8; Mitchell, Mutchmor, and Dolphin 1988a, 733–34; Wright 2001, 17–19).

REPTILES

With the exception of snakes, worm lizards, and legless lizards, all other reptiles are quadrupeds with well-developed extremities and similar axial and appendicular skeletons. In some species of snakes (e.g., boids) there are remnants of the hind limbs, known as the "vestiges" or "spurs," which can be used during copulation. The spine is divided into presacral, sacral, and caudal regions, and the number of presacral vertebrae can range from 18 in chelonians to approximately 400 in snakes. The ribs are well-developed and are important in support, locomotion, and respiration. Unique to crocodilians are eight pairs of floating ribs (gastralia).

Most commonly, reptiles have been divided into two subclasses on the basis of their skull anatomy: the anapsids (chelonians), which have no temporal openings, and the diapsids (all other reptiles), with dorsal openings (superior temporal fossa) and lateral openings (inferior temporal fossa). As in birds, their skulls are cranially kinetic, which enables them to widely open their mouths; some species of lizards and the crocodilians have powerful snapping jaws, owing to their adductor jaw muscles.

It is important to avoid restraining some lizards, especially iguanids and geckos, by the tail, as it can undergo autotomy, an adaptation by which it will break off at predetermined fracture sites to facilitate escape from predators. The tail will then regenerate, but not to the original size, shape, or color. Other species do not undergo tail autotomy but may have prehensile tails (e.g., chameleons, prehensile-tailed skinks). (Kirchgessner and Mitchell 2009, 213–14; Mitchell 2009, 137–39; Mitchell, Mutchmor, and Dolphin 1988c, 182–90; Mitchell, Mutchmor, and Dolphin 1988k, 761; Navarez 2009a, 114–15; Navarez 2009b, 169; O'Malley 2005b, 21–23; O'Malley 2005c, 60–63; O'Malley 2005d, 81–83; O'Malley 2005e, 43–48).

BIRDS

The structure of the sternum divides birds into two subclasses: those with a flat sternum (ratites) and the rest of the avian species that have a large keeled sternum (carinates). The axial and appendicular skeleton is lightweight as a modification for flight, with fusion and reduction of bones to form a strong and rigid framework. Air sacs that are connected to the lungs extend into the medullary cavity of the long bones adjacent to the axial skeleton (pneumatic bone), and the skull is lightweight. The vertebral column has been regionalized into four segments: cervical (8–25 vertebrae), thoracic (3–10 vertebrae), synsacral (10–23 vertebrae), and coccygeal (5–8 vertebrae).

The pectoral muscle mass of flying birds represents the bulk of the muscular system, and it is concentrated ventrally to provide stability for flight. Dorsal musculature is sparse in birds. Flight is achieved through contraction of the pectoralis, which depresses the wing, and the antagonistic activity of the supracoracoideus muscle, which elevates the wing. Leg muscles are only well developed in flightless species (ratites), or in those that tend to spend more time ambulating than flying, such as large galliformes (e.g., turkeys). The main leg muscles are short with long, thin, lightweight tendons that extend to the feet (Macwhirter 2009, 38–39; Mitchell, Mutchmor, and Dolphin 1998b, 776–78; O'Malley 2005a, 100–109; Pough, Janis, and Heiser 2009a, 447–48; Tully 2009, 260).

MAMMALS

Most mammals are tetrapods, with a few notable exceptions in which the hindlimbs are vestigial (e.g., cetaceans, sirenians). More than 200 bones are found within the mammalian skeleton. The vertebral column has been regionalized into five segments: cervical (6–9 vertebrae), thoracic (12–19), lumbar (6–20), sacral (3–10), and coccygeal (4–49). In general, mammalians have a more upright posture than do lower terrestrial vertebrates, and a greater range of motion in their limbs. The axial and, to a greater degree, the appendicular skeleton are adapted to the lifestyle of the mammal (e.g., cursorial, saltatory, fossorial, arboreal, aquatic). As with other vertebrates, striated (skeletal and cardiac) and smooth muscle compose the muscular system, which is well developed (Mitchell, Mutchmor, and Dolphin 1988, 182–90, 806–8; Pough, Janis, and Heiser 2009c, 530–33).

CARDIOVASCULAR SYSTEM

FISH

Fish have a relatively simple circulatory system. The heart is considered to be two-chambered, comprising an atrium, with an associated sinus venosus, and a ventricle, with associated bulbous arteriosus. Oxygen-poor blood is pumped from the ventricle and bulbous arteriosus through the gill capillaries for counter current oxygen exchange before continuing to the dorsal aorta for systemic distribution via the arterial system. Blood returns to the sinus venosus and then into the atrium via the venous system. Depending on species, blood from the caudal body in the venous system is partially or completely shunted through the kidneys (via the renal portal system) before returning to the heart. Four distinct types of renal portal systems have been described. A hepatic (liver) portal system is present as well in some species, as it is in other vertebrate species. Blood is most often collected from fish through the caudal tail vein, though in sharks a superficial vessel cranial or caudal to the dorsal fin can also be used (Bone and Moore 2008c, 145–52; Harms 2003, 2–4; Miller and Mitchell 2009, 48–49; Mitchell, Mutchmor, and Dolphin 1988i, 127–33; Stoskopf 1993a, 2–30).

AMPHIBIANS

The circulatory system of amphibians is comprised of the arterial, venous, and well-developed lymphatic structures. A three-chambered heart (with two atria and one ventricle) represents the evolutionary step between the two-chambered fish heart and the more advanced three-chambered heart found in reptiles. The interatrial septum is complete in frogs and toads but has some openings (fenestrations) in caecilians and most salamanders, which allows for mixing of oxygenated and deoxygenated blood to varying degrees. Both a right and a left aortic arch exists.

In amphibians a renal portal system exists through which blood from the caudal half of the body passes before entering the posterior vena cava. The lymphatic system includes lymph hearts (lymph sacs) that beat independently of the heart and ensure unidirectional flow of lymph back to the heart. Peripheral lymph vessels run in close proximity to the blood vessels. Common blood collection sites in amphibians include the ventral abdominal vein, femoral vein, lingual vein, and heart, as well as the ventral tail vein which exists in salamanders (Helmer and Whiteside 2005, 8; Mitchell, Mutchmor, and Dolphin 1988a, 735–36; Pough, Janis, and Heiser 2009e, 250–51; Whitaker and Wright 2001, 102; Wright 2001, 27–28).

REPTILES

Reptilian species have a well-developed circulatory system. With the exception of crocodilians, a three-chambered heart (with two atria and one ventricle) is found within all reptiles. Although only one ventricle exists, it is more advanced than that in amphibians; it has an incomplete septum and muscular ridges arising from the ventricular wall that allow for shunting of oxygenated and deoxygenated blood to the appropriate arterial pathways to the body or lungs respectively. Crocodilians have a four-chambered heart (with two atria and two ventricles) similar to that seen in birds and mammals. All reptiles have right and left aortic arches, between which crocodilians can shunt blood via an opening (the foramen of Pizzi) during dives, to bypass the lungs. As in all vertebrates except mammals, a renal portal system in reptiles can shunt blood through the kidneys before reaching the systemic circulation. The common sites used for blood collection in reptiles include the tail vein (ventral in most species but dorsal in chelonians), abdominal vein, jugular vein, femoral vein, brachial veins, subcarapacial sinus in chelonians, heart, and occipital sinus (Kirchgessner and Mitchell 2009, 207–8; Mitchell 2009, 137–39; Mitchell, Mutchmor, and Dolphin1988i, 127–33; Mitchell, Mutchmor, and Dolphin1988k, 763; Navarez 2009a, 114–15; Navarez 2009b, 166–67; O'Malley 2005b, 23–25; O'Malley 2005c, 64; O'Malley 2005d, 83–84; O'Malley 2005e, 48–49).

BIRDS

To accommodate for their high metabolic rates, flight ability, and endothermic body temperature of approximately 40–42° C, birds have a highly efficient cardiovascular system and a heart-to-body ratio that is 50–100% larger than in a mammal of comparable size. The heart is four-chambered (with two atria and two ventricles), with a muscular flap for the right atrioventricular valve and a tricuspid valve for the left atrioventricular valve. Birds only have a right aortic arch. As with reptiles, the renal portal system is well developed. The cardiovascular system is important in thermoregulation, with development of large vascular plexuses in some species (e.g., Columbiformes) and countercurrent tibiotarsal arteriovenous "nets" composed of capillaries (*rete miribile*) for heat exchange (e.g., as in waterfowl). Common sites used for blood collection are the right jugular vein, which is larger than the left; the ulnar (basilic) vein; and the medial metatarsal vein (Macwhirter 2009, 37–38; O'Malley 2005a, 113–16; Smith, West, and Jones 2000, 141–46; Tully 2009, 260).

MAMMALS

Mammals have developed an efficient cardiovascular system to accommodate their endothermic nature. Specialized adaptations such as valves and complexes of distensible arteries accommodate changes in blood pressure. The heart is four-chambered with a tricuspid right atrioventricular valve and a bicuspid left atrioventricular (mitral) valve. A left aortic arch exists, and the renal portal system seen in other vertebrates is absent. In some species (e.g., cetaceans), there are artery and vein connections (arteriovenous anastamoses) for conservation and dissipation of heat. Sites for blood collection are species-dependent, but common sites for collection include the jugular vein, cephalic vein, saphenous vein, femoral vein, ear veins, and tail veins. (Mitchell, Mutchmor, and Dolphin 1998j, 808–11; Pough, Janis, and Heiser 2000, 537–38; Reidarson 2003, 442–43)

RESPIRATORY SYSTEM

FISH

The primary respiratory organ of fish is the gills. In elasmobranchs the gills originate within a common orobranchial cavity with five or more parabranchial cavities on each side, while in bony fish they are located within the two opercular cavities that are separated from the buccal cavity. Cartilaginous or bony gill arches (numbering four in most bony fish) support two rows of featherlike gill filaments per arch (primary lamellae). Arising at a perpendicular angle to the primary lamellae are the secondary lamellae, where respiratory exchange from the water flowing over the gills occurs in a fashion counter-current to that of the capillary blood flow. The gills also are important in monovalent ion regulation (e.g., chloride) and nitrogenous waste excretion (ammonia). Specialized cells produce mucus that coats and protects the gills.

Air-breathing fish have developed additional specialized respiratory organs to allow respiration above the water surface, such as the labyrinth organ in anabantid fish (e.g., gouramis, Siamese fighting fish) or lungs (e.g., as in lungfish, bichirs, and gar; Bone and Moore 2008c, 125–145; Harms 2003, 2–4; Miller and Mitchell 2009, 43–44; Mitchell, Mutchmor, and Dolphin 1988g, 127–33; Mitchell, Mutchmor, and Dolphin 1988l, 103–5; Pough, Janis, and Heiser 2009b, 78–79; Stoskopf 1993a, 2–30).

AMPHIBIANS

Gas exchange always occurs across a moist surface in amphibians. In general, larval amphibians use gill structures for respiration, while lungs are the main organ in adults. The glottis is the opening to the trachea, which leads to the lungs. In anurans and caecilians, gas exchange can occur via the lungs (pulmonically, the oropharynx (buccopharyngeally), and the skin (cutaneously). In caudates, an additional mode is branchial respiration from retained external gills in neotenic species such as axolotls or mud puppies. Some salamander species (e.g., plethodontids) lack lungs or have lungs that are reduced in size, and therefore rely upon cutaneous respiration (Helmer and Whiteside 2005, 9; Mitchell, Mutchmor, and Dolphin 1988a, 735; Pough, Janis, and Heiser 2009e, 251–53; Wright 2001, 24–26).

REPTILES

In most species of reptiles, gas exchange occurs across the epithelium of the lungs. The glottis is found at the base of the tongue and opens only for respiration to allow air to travel to and from the trachea. Chelonians are obligate nasal breathers, while all other reptiles can breathe interchangeably through the nasal or oral cavity. The coelom is not divided by a diaphragm into pleural and peritoneal cavities as in mammals, although more advanced species have a membranous postpulmonary septum that functionally divides the coelom. Despite the lack of a diaphragm, respiration is still accomplished by negative pressure breathing, with active inspiration and expiration accomplished by using intercostal muscles, in nonchelonian species, or strong trunk muscles, in chelonians.

The lungs of reptiles are important for respiration, buoyancy, defense, and vocalization. In general, the respiratory volume is large in reptiles but they only have about 1% of the lung surface area of mammals. The lung tissue is simple and saclike, with a honeycomb network of faveoli, which is analogous to alveoli in mammals. Primitive snakes (e.g., boids) generally have two lungs while most advanced species (e.g., colubrids) possess only a well-developed right lung. All other reptile species have paired lungs, which may be single-chambered (e.g., snakes) to multichambered (e.g. monitors, crocodiles, and chelonians).

Other important respiratory surfaces in reptiles include the skin (e.g., soft shelled turtles), cloacal bursae (e.g., freshwater turtles) and the mucosal surface of the oropharyngeal cavity in many lizards (Kirchgessner and Mitchell 2009, 209–10; Mitchell, 2009 137–39; Mitchell, Mutchmor, and Dolphin 1988k, 763; Navarez, 2009a, 114–15; Navarez, 2009b, 169; O'Malley 2005b, 26–27; O'Malley 2005c, 65–66; O'Malley 2005d, 84–87; O'Malley 2005e, 49–51).

BIRDS

The avian respiratory system is considered the most efficient among vertebrate species, and is important in gas exchange, in thermoregulation through panting and its extensive surface area, and in vocalization. Owing to the lack of a diaphragm, inspiration is achieved by the pulling the sternum away from the body via the costosternalis and external intercostal muscles, making it important to not compress the thoracic region during restraint. Air travels through the nares, located at the base of the beak, to the choana, located on the roof of the mouth. From here it passes through the slitlike laryngeal opening (rima glottidis) at the base of the tongue, down the trachea, and into the bilaterally paired airsac systems before traveling to the lungs through specialized openings (ostia). To move an air bolus through the respiratory system, it takes two breathing cycles rather than one as in other vertebrates. Most birds have simple to extensive air sinus development in their skulls, which communicates with the nasal cavity.

The trachea is relatively long relative to the size of the bird, is wider than in mammals of similar size, and ranges from curvilinear, in most birds, to highly coiled, in certain birds such as trumpeter swans and cranes. At the distal end of the trachea, where it bifurcates into the paired bronchi is the syrinx ("voicebox"), which is composed of modified tracheobronchial cartilage and two vibrating tympaniform membranes. Syringeal muscles are responsible for the vibrations involved in producing sound, from simple sounds heard in raptors, with one muscle pair, to the melodious song of many passerines, which have five to six pairs.

Unlike the lungs of other vertebrate species, which rely on a to-and-fro air exchange, the lungs of birds expand and contract minimally during respiration with gas exchange, rather relying on unidirectional airflow through parabronchi that is counter to the capillary blood flow. This does, however, make avian species more susceptible to infections and inhaled toxins (Macwhirter 2009, 33–34; Mitchell, Mutchmor, and Dolphin 1998l, 107–8; O'Malley 2005a, 118–29; Powell 2000, 233–39; Tully 2009, 258–59).

MAMMALS

Mammals have developed an efficient respiratory system that is important in gas exchange, thermoregulation, and vocalization as well as in nonrespiratory metabolic and endocrine functions. The chest (thoracic) cavity is separated from the abdomen (peritoneal cavity) by a muscular diaphragm, which allows for negative and positive pressure dynamics during respiration via contraction and relaxation of the diaphragm, intercostals, and abdominal muscles. Mammals breathe through their noses and/or mouths. In cetaceans, the single or paired nostrils (nares) are situated on the top of the head rather than on the anterior aspect of the face. The air bolus then travels to the trachea via the pharynx and larynx before continuing to the lungs. Just before the trachea reaches the right and left lungs, it bifurcates into right and left primary bronchi, which further bifurcate into secondary bronchi and bronchioles, with termination in the saclike clusters of alveoli where gas exchange occurs across the respiratory epithelium. Air flow in the lungs is bidirectional. In some species (e.g., bovids, pigs), there is an accessory bronchus that arises from the trachea just cranial to the bifurcation and connects to the apical lobe of the right lung. The lungs are variably divided into two to five lobes with more on the right side than the left. The pleural space between the lung and chest wall is variable depending on species, with elephants lacking a pleural space. In diving species of mammals, the diminished pleural space guards against pressure-induced collapse of the lungs (Mitchell, Mutchmor, and Dolphin 1988l, 108–10; Pough, Janis, and Heiser 2000, 537–38).

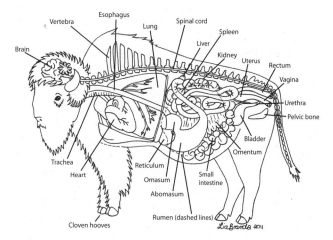

Figure 13.6. Bison: external and internal anatomy of a female bison. The rumen has been removed but its location is highlighted with dashed lines. Note the cloven or paired hooves typical of an artiodactylid. Courtesy of Lia Brands.

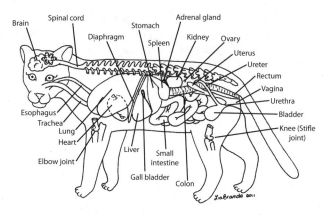

Figure 13.7. Cougar: external and internal anatomy of female cougar. Courtesy of Lia Brands.

DIGESTIVE SYSTEM

Readers are referred to chapter 16 for diagrams of the digestive systems.

FISH

Teeth are not present in all fish species, but when they are, they can be found on multiple structures such as the upper jaw, lower jaw, tongue, and fifth gill arch (as in the case of pharyngeal teeth). In elasmobranchs, the teeth originate from the skin and rest on the jaw rather than being embedded in the dermal bone of the jaws as in bony fish.

After food is ingested it passes down a short esophagus into the stomach. In some cyprinid species (e.g. carp, koi, true minnows, and goldfish) the stomach is not defined from the intestines, while in other species it may be straight, U-shaped, or J-shaped or may have a pouch (diverticulum) arising from it. The junction between the stomach and the beginning of the intestinal tract (pylorus) may not be clearly defined, or many diverticula may arise from the pyloric region. Depending on the dietary habits of the fish, the intestinal tract may be simple, be more complex with coiling, or have specialized modifications to increase surface absorptive area, such as the spiral valve present in elasmobranchs, sturgeons, gars, and lungfish. Emptying into the distal intestine in elasmobranchs is a rectal gland that is important in salt regulation. The intestinal tract concludes at the anus. As most fish species are poikilothermic, gastrointestinal transit time is significantly affected by environmental temperature.

With the exception of sharks, lungfish, and some catfish, most fish species do not have a discrete pancreas; rather, pancreatic tissue is found within the mesentery, liver, and spleen. The liver may or may not be divided into lobes, and a gall bladder is often present. Both organs contribute to the digestive process, along with the digestive enzymes from the gastrointestinal tract.

The swim bladder, which is responsible for adjusting buoyancy in many bony fish species, originates from the esophagus, and depending on the species, it may maintain a connection with the intestinal tract. It is located between the ventral aspect of the vertebral column and the peritoneal cavity. In some species it becomes compartmentalized. The gas gland regulates the contents of the swim bladder. In species without swim bladders, such as elasmobranchs or some deep sea fishes, lipid content in the liver or the ingestion of air helps to regulates buoyancy. The swim bladder also aids in the perception of sound and pressure fluctuations, and may help with respiratory gas exchange (Bone and Moore 2008c, 189–216; Harms 2003, 2–4; Miller and Mitchell 2009, 48–49; Mitchell, Mutchmor, and Dolphin 1988g, 716–18; Stoskopf 1993a, 2–30; Stoskopf 1993b, 53).

AMPHIBIANS

Depending on their diet (herbivorous through carnivorous), a variety of mouthparts can be found in tadpoles. In general, adult amphibians are carnivorous, with a wide variety of invertebrates and occasionally vertebrates composing their diet. Anurans and salamanders use sight as their primary sense for prey detection, while caecilians rely primarily on olfactory cues.

With the exception of bufonid toads, all other amphibians have teeth on one or both jaws of their mouth, which can be quite prominent in some species, such as ornate horned frogs (*Ceratophrys ornate*). Some species also have vomerine and palatine tooth patches on the roof of their mouths. The tongue of most salamanders and anurans flips and extends beyond their mouth for food capture. Once the prey is in the mouth, the eyelids are closed, forcing the globes ventrally to push food caudally.

The relatively short and simple gastrointestinal tract follows the common vertebrate plan, with food passing from the oral cavity into the esophagus, stomach, small intestines, and large intestines with feces being expelled into the cloaca, the common exit for the gastrointestinal, urinary, and reproductive systems. The pancreas is discrete and functions similarly to that in other vertebrates with respect to the production of digestive enzymes, including in the use of chitinase to digest invertebrate exoskeleton, and in glucose regulation. The liver is relatively large and usually bilobed, and it conforms to the body of the amphibian. It is variably pigmented due to specialized melanin containing phagocytic cells known as melanomacrophages. As in higher vertebrates, the amphibian liver is associated with numerous metabolic processes such as synthesis of nitrogenous compounds, glucose and lipid metabolism, and protein synthesis. The gall bladder is intimately associated with the liver (Crawshaw and Weinkle 2000, 165–67; Helmer and Whiteside 2005, 9–10; Mitchell, Mutchmor, and Dolphin 1988a, 734–35; Wright 2001, 21–23).

REPTILES

In general, the digestive tract is much shorter in reptiles than in birds or mammals, and it varies from a simple design in carnivores to a more complex tract in herbivorous species. Chelonians have no teeth and use their sharp horny beak (tomia) to grasp and tear food, while three types of teeth (acrodont, pleurodont, and thecodont) are found in the other reptiles. In most species, teeth are resorbed or lost and replaced throughout their lives.

The mouth and tongue are important in food apprehen-

sion, with varying degrees of mastication. In snakes and some lizards, such as monitors, the tongue is involved in odor differentiation by sampling the air and then inserting it into the Jacobson organ in the roof of the buccal cavity. Numerous oral secretory glands have developed to lubricate prey with mucous, with the most advanced development being the venom glands of snakes and some lizards (e.g. heloderms). Food passes from the oral cavity to the stomach, where it is digested before passing into the small intestines for further digestion and nutrient absorption. Herbivorous reptiles have a well developed cecum and large intestine with specialized valves to regulate the passage of digesta, which allows for microbial fermentation of plant material and breakdown of the plant cell wall to maximize digestive efficiency. Both peristalsis and retroperistalsis occur, especially in the distal colon to allow for water resorption. Feces then pass into the cloaca via the coprodeum before being expelled. Owing to their poikilothermic nature, digestion is maximized when reptiles are kept in their preferred optimal temperature zones.

The liver is relatively large and often simply lobated; it generally conforms to the shape of the reptile. A gall bladder is usually associated with the liver and stores biliverdin as the main bile pigment. In snakes the gall bladder is found caudal to the liver. The pancreas, located adjacent to the duodenum, secretes a variety of digestive enzymes and aids in glucose regulation. In snakes it is often intimately associated with the spleen (splenopancreas; Kirchgessner and Mitchell 2009, 210–11; Mitchell 2009, 137–39; Mitchell, Mutchmor, and Dolphin 1988f, 90–95; Mitchell, Mutchmor, and Dolphin 1988k, 761–63; Navarez 2009a, 114–15; Navarez 2009b, 166; O'Malley 2005b, 27–29; O'Malley 2005c, 67–69; O'Malley 2005d, 87–89; O'Malley 2005e, 51).

BIRDS

Among birds a wide variety of beak forms has evolved based on diet. In addition to food acquisition and prehension, depending on species, the beak also is important in courtship, nest building, feeding of young, and locomotion. In many species, the keratinized edges of the beak (tomia) are modified for cutting. Although in most species the tongue is not protrusible, its morphology reflects dietary needs and it is important in collection, manipulation, and swallowing of food. There are significantly fewer taste buds on the tongue in birds than in mammals.

The digestive tract is modified to accommodate dietary specializations, and owing to the reduction of weight for flight, it is relatively short compared to that in other vertebrates. The esophagus on the right side of the neck is highly distensible, and in most species there is an outpouching cranial to the thoracic inlet known as the crop (ingluvies), which is important in food storage, feeding of young (e.g., the crop milk of pigeons), and initiation of carbohydrate digestion. Ingesta continues to the stomach, which is divided into the cranial glandular portion (proventriculus) and the mechanical or storage portion (gizzard or ventriculus). The duodenum is looped around the pancreas, and the jejunum is demarcated by the remnant of the yolk sac (Meckel's diverticulum). Birds have paired ceca that range from well-developed structures in herbivorous species to minimally developed structure in others (e.g., pigeons). Peristaltic and retroperistaltic waves maximize digestive efficiency in the small intestines, and allow for water reabsorption in the distal colon. As in reptiles, the large intestine terminates at the coprodeum before feces are expelled through the cloacal vent.

The liver is bilobed, with the left lobe being smaller than the right in order to accommodate the stomach. A gall bladder is present in most species, though notably absent in others (e.g. Columbiformes). Biliverdin is the main bile pigment (Macwhirter, 2009, 30–33; Mitchell, Mutchmor, and Dolphin 1998b, 783–84; Mitchell, Mutchmor, and Dolphin 1998f, 90–95; O'Malley 2005a, 125–26; Pough, Janis, and Heiser 2009a, 458–63; Tully 2009, 259–60).

MAMMALS

Numerous adaptations to the digestive tract are reflective of the varied dietary specializations of mammals. The lips of mammals are more flexible than those in other vertebrate taxa, to aid with food prehension. Unique to mammals is heterodonty, with variable numbers and shapes of incisors, canines, premolars, and molars, depending on diet. Mammals also are the only vertebrate species that masticate their food and swallow a discrete bolus.

Food passes from the oral cavity down the esophagus on the left side of the neck to the stomach. The stomach can range from the simple or one-chambered (monogastric) stomach of many mammalian species, including carnivores, primates, and equids, to complex stomachs with ceca (e.g., peccaries) or sacculations (e.g., macropods), and to multiple-chambered stomachs, as seen in cetaceans (three compartments), pseudoruminants (e.g., the hippopotamus, with four chambers), and true ruminants (e.g., camelids, with three compartments [C_1, C_2, C_3]; and bovids, cervids, and giraffes, with four compartments [rumen, reticulum, omasum, abomasum]). In monogastrics, enzymatic digestion of food occurs in the stomach and the proximal small intestines. In herbivorous monogastrics, further digestion of plant material occurs in the cecum and large intestines via the action of resident symbiotic microbes (hindgut fermenters). In many herbivorous species with complex stomachs (foregut fermenters), fermentation of digesta occurs via microbes in the rumen or analogous chamber.

Absorption of nutrients can occur in the forestomachs of foregut fermenters in addition to the intestinal tract, as with monogastrics. Digestive enzymes are excreted by the pancreas, which is located in close proximity to the duodenum. Digesta continues from the stomach through the small intestines (duodenum, jejunum, and ileum), and then into the cecum (when present) and large intestines (colon with a spiral colon in some species) before feces are expelled via the rectum through the anus. Relative to their body size, mammals have the longest intestinal tracts among vertebrates, with the longest tracts found in herbivorous species.

The liver is lobated in mammals (two to eight lobes, depending on species) and its shape is determined by surrounding organs. Most species of mammals have a gall bladder, though there are notable exceptions (e.g., equids, rats, elephants, cetaceans, New World camelids, and deer). As in other vertebrates, the liver serves important roles in

protein, lipid and carbohydrate metabolism, detoxification, storage, and hormone production (Fowler 2003, 612; Miller 2003, 603; Mitchell, Mutchmor, and Dolphin 1988f, 90–95; Mitchell, Mutchmor, and Dolphin 1988j, 808–9; Morris and Shima 2003, 586; Pough, Janis, and Heiser 2000d, 562–69; Reidarson 2003, 442–43).

URINARY SYSTEM

FISH

In fish the kidneys are located along the ventral aspect of the vertebral column. Functionally they are divided into cranial (hematopoietic) and caudal (excretory) segments, and physically they are divided into distinct left and right in some species, such as elasmobranchs. The kidneys are responsible for divalent ion regulation and for the excretion of metabolic wastes other than ammonia.

The body fluids of freshwater bony fishes are more concentrated than their aquatic environment (i.e., they are hyperosmotic), and so they don't drink water, instead absorbing sodium and chloride through their gills and producing large volumes of dilute urine. Conversely, marine bony fish have body fluids that are less concentrated than their aquatic environment (i.e., they are hypoosmotic), and they absorb water from their digestive tract through linked passive diffusion of sodium and chloride. They actively excrete sodium and chloride through their gills, and produce small volumes of concentrated urine. Although the majority of elasmobranchs live in a marine environment, they have fluid homeostasis more similar to freshwater bony fish, with slightly more hyperosmotic body fluids than their aquatic environment, but they also actively secrete salt through their rectal glands (Bone and Moore 2008d, 161–83; Harms 2003, 2–4; Miller and Mitchell 2009, 48–49; Pough, Janis, and Heiser 2009b, 90–93; Stoskopf 1993b, 48–56).

AMPHIBIANS

Bilateral kidneys are present in amphibians. Their kidneys cannot concentrate urine above the solute concentration (osmolarity) of the plasma. Urine passes from the kidneys into collecting ducts, then into cloaca, and is finally stored in the urinary bladder, which is an evagination of the cloaca. In general, terrestrial amphibians excrete urea as their main nitrogenous waste, while larvae and aquatic adults excrete ammonia through their gills, skin, and kidneys (Helmer and Whiteside 2005, 10; Mitchell, Mutchmor, and Dolphin 1988a, 736–38; Pough, Janis, and Heiser 2009e, 253–55; Wright 2001, 23–24).

REPTILES

All reptiles have paired kidneys, which are located in the caudal aspect of the coelom. Ureters carry renal excretions to the urodeum, which is found within the cloaca. The kidney plays an important role in osmoregulation, fluid homeostasis, excretion of metabolic waste products, and the production of various hormones and vitamin D metabolites. Most reptiles excrete uric acid as their nitrogenous (uricotelic) waste, which precipitates to white pasty urates, although aquatic species excrete urea (i.e., are ureotelic) and ammonia. Like amphibians, reptiles cannot concentrate their urine above the osmolarity of their plasma, and thus rely on other adaptations to reduce water loss. Turtles and some lizards possess a bladder that is an evagination of the urodeum, and which allows for water and sodium reabsorption. Water and ions are reabsorbed from the distal colon of reptiles as well (Kirchgessner and Mitchell 2009, 211–12; Mitchell 2009, 137–39; Mitchell, Mutchmor, and Dolphin1988h, 153–54, 160–68; Mitchell, Mutchmor, and Dolphin1988k, 763; Navarez, 2009a, 114–15; Navarez 2009b, 167; O'Malley 2005b, 29–31; O'Malley 2005c, 69; O'Malley 2005d, 89; O'Malley 2005e, 52).

BIRDS

The paired kidneys of birds are found in the caudodorsal aspect of the coelomic cavity in the renal fossa of the synsacrum. Three distinct lobes (cranial, middle, and caudal) are present. As in reptiles, the kidneys have multiple roles. Owing to both reptilian-like and mammalian-like nephrons, the urine can be concentrated above the osmolarity of their plasma. Birds excrete their nitrogenous wastes as uric acid, which precipitates to pasty white urates, while water and ions can be reabsorbed from their distal colon. Urates and urine travel from the kidney to the urodeum of the cloaca via the ureters (Macwhirter 2009, 36; Mitchell, Mutchmor, and Dolphin 1998h, 160–68; O'Malley 2005a, 136–38; Tully 2009, 260).

MAMMALS

All mammals have well-developed paired kidneys located in the cranial to mid-dorsal abdomen. Urine travels from the kidneys via ureters to the urinary bladder, and then is excreted through the urethra. The kidneys are responsible for numerous functions including osmoregulation, excretion of metabolic wastes, and the production of various hormones and vitamin D metabolites. Mammals are able to concentrate their urine significantly above the osmolarity of their plasma. Nitrogenous waste is excreted as urea (Mitchell, Mutchmor, and Dolphin 1988h, 153–54, 160–68; Pough, Janis, and Heiser 2000c, 538).

REPRODUCTIVE SYSTEM

FISH

Distinct genders exist in most fish. The gonads (ovaries or testes) of fish are elongated structures within the caudal coelomic cavity that can vary in size on a seasonal basis. Fertilization is external in the majority of bony fish, the exception being live-bearing species such as guppies and mosquito fish, which have an elongate anal fin, known as the gonopodium, for sperm transfer to the female oviduct. In elasmobranchs, fertilization occurs internally with males transferring sperm to the female's oviduct during copulation via grooves in their external claspers, which are located near the anal fin. For oviparous species, eggs or sperm are passed out through the genital pore and develop externally after fertilization, while for ovoviviparous species the young develop internally before being born (Harms 2003, 2–4; Miller and Mitchell 2009, 43–44; Mitchell, Mutchmor, and Dolphin 1988g, 722; Stoskopf 1993a, 2–30).

AMPHIBIANS

Amphibians are gonochoric, with paired ovaries or testes that fluctuate in size and activity with their breeding seasons. Anurans are external fertilizers, while caecilians and the majority of salamanders are internal fertilizers. Caecilians copulate by the males everting a portion of their cloaca (phallodeum) and introducing it into the female's cloaca, while male salamanders lack an intrommitent organ and instead deposit sperm spermatophores on the ground, which the females uptake with their cloaca. Approximately 75% of caecilians are viviparous, in contrast to anurans and salamanders, which are oviparous (Helmer and Whiteside, 2005, 10–11; Mitchell, Mutchmor, and Dolphin 1988a, 741–44; Pough, Janis, and Heiser 2009e, 232–46; Wright 2001, 29).

REPTILES

Distinct sexes occur in reptile species, although sexual dimorphism is not always apparent, especially in immature animals. The gonads are paired, with ovarian and testicular size and activity being influenced by the breeding season. In addition, in male snakes and many male lizards the posterior segment of the kidney has become a sexual segment that produces seminal fluid. Two types of sexual determination can occur in reptiles: genotypic where the females are heterogametic (ZW) and the males are homogametic (ZZ), or temperature-dependent sexual determination.

A fleshy single phallus is present within the cloaca of male chelonians and crocodilians, while paired extracloacal hemipenes are found in snakes and lizards. The phallus is analogous to the penis in mammals, but is not involved in urination. The ovaries in females are found cranial to the kidneys. Ova are released into the oviducts, where fertilization occurs, while sperm travels to the phallus via the ductus deferens. Fertilization is always internal. Chelonians and crocodilians are oviparous, while squamates are mainly oviparous, though some are viviparous (e.g., vipers, boas, some chameleons, and skinks). In some species, females can store sperm for up to six years, and parthenogenesis has also been documented. Fat is stored in fat bodies that lie adjacent to the gonads and kidneys, and are used for yolk production in females. (Kirchgessner and Mitchell 2009, 212–13; Mitchell 2009, 137–39; Mitchell, Mutchmor, and Dolphin1988k, 766–68; Navarez 2009a, 114–15; Navarez 2009b, 167–68; O'Malley 2005b, 31–34; O'Malley 2005c, 70–72; O'Malley 2005d, 89–91; O'Malley 2005e, 52–54).

BIRDS

Although distinct sexes are present in birds, in many species there is no external sexual dimorphism (e.g., many psittacines), while in other species there are obvious differences (e.g., many waterfowl). Female birds are the heterogametic sex (ZW), whereas males are homogametic (ZZ). With rare exceptions, the female's reproductive tract develops only on the left side and consists of the ovary at the cranial pole of the kidney, and the oviduct with its five divisions (infundibulum, magnum, isthmus, uterus, and vagina). In males, the paired testes are found intracoelomically at the cranial pole of the kidneys, with ductus deferens delivering semen to the proctodeum in the cloaca. The testicular size can increase as much as 200-fold during the breeding season. There are no accessory sex glands. The breeding cycle in both sexes is influenced by photoperiod, food availability, and temperature, with the pineal gland and hypothalamus responsible for gametogenesis and breeding. Most birds accomplish insemination through cloacal contact, however in some species a nonprotusible or protusible phallus may be present for sperm channeling into the female's cloaca (Macwhirter 2009, 30–33; Mitchell, Mutchmor, and Dolphin 1998b, 788–90; O'Malley, 2005a 138–42; Pough, Janis, and Heiser 2009a, 467–75; Tully 2009, 260).

MAMMALS

Sexual dimorphism is common between the distinct sexes in mammals, with males often being larger than females. Males are the heterogametic sex (XY) with females being homogametic (XX). In most mammals the testes are external to the body and often are located within a scrotal sac to keep the testes slightly cooler than body temperature for spermatogenesis. A few species have internal testes located caudal to the kidneys (e.g., elephants, hyrax, manatees, cetaceans, hippopotamuses). Accessory sex glands in males are species-dependent and may include the prostate gland, seminal vesicles, ampullae, and bulbourethral glands. Intromission is achieved by introduction of the penis into the female's vagina. Female mammals have a vagina, a cervix, paired ovaries, and variable fusion of the paired uterine horns to form a common uterine body. Marsupial females have two uteri, cervices, and vaginas, while males often have a forked penis. Monotremes (e.g., echidnas, platypus) are the only mammals which lay eggs instead of bearing live young (Mitchell, Mutchmor, and Dolphin 1988j, 813–14; Pough, Janis, and Heiser 2000c, 533–57).

HEMATOPOIETIC AND IMMUNE SYSTEM

FISH

Bone marrow and lymph nodes are lacking in fish. Hematopoiesis primarily occurs in the cranial kidney with additional production in the spleen, and lymphomyeloid tissue in the coelomic mesentery. A distinct thymus for the production of T-lymphocytes is found dorsal to the gill arches. The immune system consists of a well-developed reticuloendothelial system of phagocytic cells, cell-mediated immunity, and humoral immunity (antibodies). Immune function, as in other poikilotherms, is linked to environmental temperature.

The cells that compose the cellular component of fish blood are nucleated red blood cells (erythrocytes), white blood cells (leukocytes-neutrophil, lymphocyte, eosinophil, basophil, monocytes), and thrombocytes (Bone and Moore 2008e, 383–99; Harms 2003, 2–4; Miller and Mitchell 2009,

43–44; Mitchell, Mutchmor, and Dolphin 1988i, 127–33; Stoskopf 1993a, 2–30).

AMPHIBIANS

The blood cells in amphibians consist of nucleated erythrocytes, thrombocytes, neutrophils, lymphocytes, monocytes, and poorly described granulocytic cells that stain similar to eosinophils and basophils. While bone marrow can be found in a number of terrestrial species, the liver, spleen, and kidneys are also important sites for the production of erythrocytes, leukocytes and thrombocytes. The thymus persists throughout life and is responsible for the production of T-lymphocytes. Lymph nodes are lacking, but aggregates of lymphoid tissue are present in the intestinal tract. As in fish and reptiles, the immune function of amphibians is optimized when they are within their preferred optimal temperature zone (Helmer and Whiteside 2005, 8–9; Plyzycz, Bigaj, and Midonski 1995, 15–120; Wright 2001, 28).

REPTILES

A highly developed lymphatic system found in close association with the blood vessels is present in reptiles, although lymph nodes are lacking. Smooth muscle lymph hearts pump the lymph throughout the body where it reconnects with the venous system at the base of the neck. The bone marrow, spleen, and thymus are all important in hematopoiesis, and the thymus does not involute with age. The reptilian blood cells consist of nucleated erythrocytes, thrombocytes, heterophils, lymphocytes, monocytes, azurophils, eosinophils and basophils (O'Malley 2005b, 25–26).

BIRDS

The lymph vessels are less numerous in birds than in mammals, and are found in close proximity to blood vessels. Lymph reconnects to the venous system through paired thoracic ducts. The blood cells of birds consist of nucleated erythrocytes, thrombocytes, heterophils, lymphocytes, monocytes, eosinophils, and basophils. The thymus is responsible for the production of T-lymphocytes, while B-lymphocytes are produced in the bursa of Fabricius, located as a diverticulum from the dorsal aspect of the cloaca. Discrete lymph nodes are not present, but aggregates of lymphocytic centers are distributed throughout the gastrointestinal tract. The bone marrow is the primary site of hematopoeisis (Macwhirter 2009, 25–29; O'Malley 2005a, 116–18).

MAMMALS

Mammals have a well developed lymphatic system, with discrete lymph nodes distributed around the body in addition to lymphoid aggregates in the gastrointestinal tract. Hematopoiesis occurs primarily in the bone marrow. The thymus is located in the mediastinum cranial to the heart and produces T-lymphocytes. The cellular components of mammalian blood consist of non-nucleated erythrocytes, platelets, neutrophils, macrophages, lymphocytes, eosinophils, and basophils (Mitchell, Mutchmor, and Dolphin 1988i, 127–33).

ENDOCRINE SYSTEM

FISH

Similar endocrine organs are found in fish as seen in higher vertebrate species, with numerous neurohormones and glandular hormones that are important for physiological function. The pituitary gland is located ventral to the midbrain. The thyroid gland is not a discrete organ; rather, it is distributed diffusely around the ventral aorta, branchial arteries, and retrobulbar tissues. Similarly, the adrenal gland is not a discrete organ as seen in higher vertebrates, with the cortical adrenal tissue that produces the glucocorticoids and mineralocorticoids found instead within the cranial kidney (thus making it an interrenal organ) and along larger blood vessels such as the dorsal aorta. The adrenal medullary elements that produce epinephrine and norepinephrine (i.e., the suprarenal organs) develop from the sympathetic ganglia and are found in association with the cranial kidney or dorsal aorta.

Two other important endocrine organs are the corpuscles of Stannius, found only within bony fish, and ultimobranchial bodies. The corpuscles are found within the caudal kidney and are responsible for electrolyte balance, while the ultimobranchial bodies produce calcitonin for regulation of calcium, and are located near the sinus venosus. Fish do not possess parathyroid glands (Bone and Moore 2008a, 255–88; Harms 2003, 2–4; Miller and Mitchell 2009, 43–44; Stoskopf 1993b, 48–56).

AMPHIBIANS

The endocrine organs in amphibians are similar to those found in reptiles, birds, and mammals. As in the higher vertebrates, the adrenal glands are discrete, are usually found in close association with the kidneys, and produce corticosteroids, epinephrine, and norepinephrine. The thyroid gland is responsible for the control of metamorphosis and ecdysis, and its function is controlled by the pituitary gland and hypothalamus. Obligate neotenic species such as the Mexican axolotl (*Ambystoma mexicanum*), which never undergo complete metamorphosis in nature, will complete the process when administered thyroid hormone.

The other endocrine organs (and their secretory products) are the pancreas (insulin), thymus (thymosin), pineal body (melatonin), parathyroid glands (parathyroid hormone), ultimobranchial bodies (calcitonin), and the gonads (estrogen, progesterone, testosterone; Goin, Goin, and Zug 1978, 15–28; Helmer and Whiteside 2005, 11; Mitchell, Mutchmor, and Dolphin 1988d, 250–67; Wright 2001, 28).

REPTILES

The endocrine system is well developed in reptiles, and the organs and their secretory products are similar to those in amphibians, birds, and mammals. The thyroid gland is found cranial to the heart and is unpaired and spherical in snakes and chelonians; in lizards it is most commonly bilobed but can be paired. Metabolism, shedding, and growth are all functions of thyroid hormones which are secreted under pituitary gland control. Parathyroid glands are paired and are responsible for control of calcium and phosphorus

levels. The adrenal glands lie dorsal to the gonads. The pineal gland of squamates and chelonians is closely associated with a parietal eye on the dorsal aspect of the head, which is well developed in some lizards and senses changes in light and aids in thermoregulatory shuttling. Crocodilians lack pineal and parietal glands (Mitchell, Mutchmor, and Dolphin1988d, 250–67; O'Malley 2005b, 34–35).

BIRDS

A well-developed endocrine system is present in birds; its organs and function are similar to those in other higher vertebrates. The pituitary gland has two lobes, with the anterior pituitary gland producing hormones responsible for regulating the actions of the gonads, thyroid glands, and adrenal glands, in addition to other hormones such as prolactin, which is involved with reproductive and nurturing activities (e.g., crop milk production in pigeons and doves), carbohydrate metabolism, and premigratory behavior. The posterior pituitary gland produces vasotocin and oxytocin, which regulate water conservation by the kidney and stimulate oviposition. The pineal body or gland, as in other vertebrates, plays a role in circadian rhythms, reproductive control, and photoreception.

The paired thyroid glands are found just caudal to the thoracic inlet, in between the trachea and jugular veins. Parathyroid glands are also paired, and are located just caudal to the thyroids; the ultimobranchial bodies are just caudal to these. The yellow-tan adrenal glands are located cranially adjacent to the kidneys and dorsal to the gonads. The pancreas lies within the mesentery of the looped duodenum and produces hormones responsible for carbohydrate metabolism, with glucagon being the most important rather than insulin (Macwhirter 2009, 36–37; Mitchell, Mutchmor, and Dolphin 1998d, 260–67; O'Malley 2005a, 143–45).

MAMMALS

The endocrine system of mammals is highly developed, with organs and respective functions similar to those seen in birds. The bilobed pituitary gland is under neurohormonal control from the hypothalamus, and produces hormones that govern the activity of other organs in the body. The pineal body is similar in function to that of birds. Lying on either side of the trachea in the cranial neck just caudal to the larynx are the paired thyroid and parathyroid glands. The adrenal glands produce glucocorticoids, mineralocorticoids, sex hormomes (androgens and progestagens), and catecholamines (epinephrine and norepinephrine). The gonads are the primary site for progesterone, testosterone, and estrogen production, although these are also produced in the adrenal cortex. Glucagon and insulin are produced by the pancreas, with the latter being more responsible for regulation of blood glucose levels (Mitchell, Mutchmor, and Dolphin 1988d, 250–67).

NERVOUS SYSTEM AND SPECIAL SENSES

FISH

The nervous system of a fish is similar in organization to those of other vertebrates, with a brain and spinal cord comprising the central nervous system and autonomic nervous system. The brain is divided into three sections: fore- mid-, and hindbrain, with well-developed optic and olfactory lobes. Ten cranial nerves are present.

Most fish have well-developed vision, with a spherical lens that is moved towards or away from the retina in order to accommodate. Rods (for monochromatic) and cones (for color) vision are both present in the retina.

Olfaction is well developed in fish, and taste-bud organs are found in the mouth, around the head, and the anterior fins. Magnetite particles have also been found in special sensory cells of the olfactory lamellae in some fish species, which allows for magnetoreception.

Fish possess a lateral line (acoustico-lateralis) system that enables them to respond to waterborne vibrations, gravity, and angular accelerations. This system is composed of specialized cells that form neuromast organs which are dispersed bilaterally over the surface of the head and within one or more canals that traverse along the sides of the body to the level of the tail approximately halfway between the dorsal and ventral surfaces. Like all vertebrates, fish have an internal ear that consists of semicircular canals and ampullae.

Six groups of fish (electric catfishes, electric eels, skates, elephant fishes, and stargazers) possess electric (electroplax) organs that are used for defense, communication, or stunning prey. In addition, many species of fish have sensitive electroreceptor organs (referred to as the ampullae of Lorenzini in elasmobranchs) for prey detection and magnetoreception (Bone and Moore 2008e, 348–74; Harms 2003, 2–4; Miller and Mitchell 2009, 43–44; Mitchell, Mutchmor, and Dolphin 1988g, 718; Pough, Janis, and Heiser 2009b, 83–89; Stoskopf 1993a, 2–30).

AMPHIBIANS

The amphibian brain is slightly more evolved than that of fish. The medulla oblongata controls most of the bodily activities, while the cerebellum, unlike in higher developed tetrapod classes, is responsible for controlling equilibrium rather than fine motor control. Ten cranial nerves are present. In anurans, the spinal cord ends in the lumbar region, as opposed to in caecilians and salamanders, where it extends to the tip of the tail. Aquatic stages possess a lateral line system similar to that in fish.

Most amphibians have good vision, with a well-developed retina that compensates for their relatively simple brain. Approximately 90% of the visual information is processed in the retina. The pupil is composed of striated muscle, and the lens is moved forward or backward to accommodate. Hearing ability varies amongst amphibian species, with anurans possessing a well developed ear structure.

Olfaction, touch and taste are well developed senses in amphibians. Taste buds are found not only on the tongue but also on the roof of the mouth, and the maxillary and mandibular mucosa. Like reptiles, amphibians have a specialized vomeronasal sense organ (Jacobson's organ), connected by ducts to the nasal cavity, which detects airborne chemical signals such as pheromones (Helmer and Whiteside 2005, 11–12; Mitchell, Mutchmor, and Dolphin 1988e, 234–42; Wright 2001, 19).

REPTILES

The reptilian brain is larger than that of amphibians and fishes, with more developed optic lobes for visual processing, and a neocortex similar to that of birds and mammals. Twelve pairs of cranial nerves are present, and the spinal cord extends the entire length of the body.

Senses are well developed in reptiles. Taste buds are present on the tongue and oral epithelium, and tactile papillae are numerous on the head and oral cavity. Only crocodilians have an external ear canal, while other reptiles may possess or lack bilateral tympanic scales. Only a single middle ear bone (columella) is present, rather than the three found in mammals, and it is more efficient at detecting low frequency than high frequency sounds. Olfaction is important in hunting, courtship, and mating. The Jacobson's organ is most highly developed in snakes, while it is only present in the early embryonic stages in crocodilians. In addition, some snake species such as vipers and boids have specialized infrared pit receptors (Kirchgessner and Mitchell 2009, 213–14; Mitchell, 2009 137–39; Mitchell, Mutchmor, and Dolphin1988k, 763–65; Navarez 2009a, 114–15; Navarez 2009b, 168–70; O'Malley 2005b, 35–36; O'Malley 2005c, 72–73; O'Malley 2005d, 91–92; O'Malley 2005e, 54; Schaeffer and Waters 1996, 165–67).

BIRDS

While the avian brain is more developed than that of most reptiles, it is still simpler and less developed that the mammalian brain. There is no folding of the cerebral cortex, as seen in mammals. Owing to an aerial existence, olfaction is not well developed in most birds, with some exceptions (e.g., vultures, kiwi), while vision is the most highly developed sense and is reflected by large optic lobes. The cerebellum is well developed for control of locomotion. There are 12 pairs of cranial nerves, and the spinal cord is the same length as the spinal canal, with no cauda equina as in mammals.

The bird eye is the most highly developed in the vertebrate world, with most birds having excellent color and binocular vision, and rapid accommodation. In some species the size and weight of the eyes is greater than that of the brain. The retina lacks visible blood vessels and, with the exception of a few species (e.g., nightjars, boat-billed heron), also lacks a reflective tapetum lucidum. A specialized feature of the retina is a black comblike structure known as the pectin, which provides nutrients to the retina and facilitates fluid movement. Hearing is the second-best sense in birds, with nocturnal species such as owls being able to pinpoint prey by hearing alone. External ear openings are covered by covert feathers. As in reptiles, only the columella is present in the middle ear. Compared to the same structures in the mammalian inner ear, the cochlea (for hearing) is shorter and uncoiled, and the semicircular canals (for balance) are larger. Touch also is a well-developed sense in birds, with widely distributed mechanoreceptors (Herbst corpuscles) in the beaks (beak tip organ), legs, and feather follicles. Taste buds are significantly fewer than those in mammals (Güntürkün 2000, 1–4; Macwhirter 2009, 29–30; Mason and Clark 2000, 39–46; Mitchell, Mutchmor, and Dolphin 1998b, 785–86; Necker, 2000, 21–24; O'Malley, 2005a 145–49; Pough, Janis, and Heiser 2009a, 463–67; Tully 2009, 260).

MAMMALS

The mammalian nervous system is highly developed and well preserved across taxa. The brain is large relative to body size compared with those in other vertebrate taxa, with a cerebral cortex that is variably convoluted (gyri and sulci). The cerebellum also is well developed for fine motor control. In placental mammals, a nerve tract (corpus callosum) links the two cerebral hemispheres. Twelve pairs of cranial nerves are present, and the spinal cord is highly segmented into cervical, thoracic, lumbar, and sacral regions. Unlike in other vertebrates, the spinal cord is shorter than the spinal canal and it terminates in the lumbar region. As a result, the paired spinal nerves are bundled together, forming the cauda equina as they continue down the spinal canal before exiting at appropriate segments.

All five senses are highly developed in mammals; however, there is variation that is taxa-dependent owing to lifestyle. Nocturnal or crepuscular mammals have retinas with a higher concentration of light-sensing cells (rods), for forming images in dim light, than cells for color and visual acuity (cones). The reverse is true in diurnal mammals. External ears (pinnae) are found in most mammals, and the middle and inner ear are more complex than those of other vertebrates. The middle ear contains three bones (stapes, malleus, and incus), while the cochlea of the inner ear is longer and more coiled. Insectivorous bats (*Microchiroptera*) and toothed whales (Odontocetes) use reflected high- and low-frequency sound to locate and identify objects (echolocation). Mammals have a keen sense of smell, and touch and taste also are highly developed (Mitchell, Mutchmor, and Dolphin 1988e, 234–42; Mitchell, Mutchmor, and Dolphin 1988j, 811–12; Mitchell, Mutchmor, and Dolphin 1988m, 276–94; Pough, Janis, and Heiser 2000c, 539–40).

SUMMARY

Developing a solid understanding of the normal anatomy and physiology of vertebrates is fundamental for keepers working in zoos and aquariums to maximize the health and welfare of the species in their care. This knowledge guides husbandry decisions, allows keepers to promptly identify abnormalities that can be treated, and is invaluable when keepers are involved with carrying out treatment.

REFERENCES

Amphibiaweb. 2011. Accessed January 7. http://www.amphibiaweb.org/amphibian/speciesnums.html.

Bone, Quentin and Richard H. Moore, eds. 2008a. "Endocrine System," in *Biology of Fishes*, 3rd edition, 255–88. New York: Taylor and Francis Group.

———. 2008b. "Food and Feeding," in *Biology of Fishes*, 3rd edition, 189–216. New York: Taylor and Francis Group.

———. 2008c. "Gas Exchange, Blood, and the Circulatory System," in *Biology of Fishes*, 3rd edition, 125–59. New York: Taylor and Francis Group.

———. 2008d. "Osmoregulation and Ion Balance," in *Biology of Fishes*, 3rd edition, 161–87. New York: Taylor and Francis Group.

———. 2008e. "The Diversity of Fishes," in *Biology of Fishes*, 3rd edition, 1–34. New York: Taylor and Francis Group.

———. 2008e. "The Immune System," in *Biology of Fishes*, 3rd edition, 383–408. New York: Taylor and Francis Group.

———. 2008f. "The Nervous System," in *Biology of Fishes*, 3rd edition, 347–82. New York: Taylor and Francis Group.

Crawshaw, Graham J., and Tristan K. Weinkle. 2000. "Clinical and Pathological Aspects of the Amphibian Liver," *Seminars in Avian and Exotic Pet Medicine*, 9 (3): 165–73.

Fowler, Murray E. 2003. "Camelidae." In *Zoo and Wildlife Medicine*, 5th edition, edited by Murray E. Fowler and R. Eric Miller, 612–25. St. Louis: Saunders.

Goin, Coleman J., Olive B. Goin, and George R. Zug, eds. 1978. "Structure of Amphibians." In *Introduction to Herpetology*, 3rd edition, 15–38. San Francisco: W. H. Freeman.

Güntürkün, Onar. 2000. " Sensory Physiology: Vision." In *Sturkie's Avian Physiology*, 5th edition, edited by G. Causey Whittow, 1–19 (1–4). San Diego: Academic Press.

Harms, Craig A. 2003. "Fish," In *Zoo and Wildlife Medicine*, 5tf edition, edited by Murray E. Fowler and R. Eric Miller, 2–20 (2–4). St. Louis: Saunders.

Helmer, Peter J., and Douglas P. Whiteside. 2005. "Amphibian Anatomy and Physiology." In *Clinical Anatomy and Physiology of Exotic Species*, edited by Bairbre O'Malley, 3–14. London: Elsevier Saunders.

International Union for Conservation of Nature. 2011. Accessed February 11. http://cmsdata.iucn.org/downloads/more_facts_on_mammals.pdf .

Kirchgessner, Megan, and Mark A. Mitchell. 2009. "Chelonians." In *Manual of Exotic Animal Practice*, edited by Mark A. Mitchell and Thomas N. Tully, 207–49. Saint Louis: Saunders Elsevier.

Macwhirter, Patricia. 2009. "Basic Anatomy, Physiology and Nutrition." In *Handbook of Avian Medicine*, 2nd edition, edited by Thomas N. Tully, Gerry M. Dorrestein, and Alan K. Jones, 25–55. Woburn: Saunders Elsevier.

Mason, J. Russell, and Larry Clark. 2000. The Chemical Senses in Birds." In *Sturkie's Avian Physiology*, 5th edition, edited by G. Causey Whittow, 39–56 (39–46). San Diego: Academic Press.

Miller, Michelle A. 2003. "Hippopotamidae (Hippopotamus)." In *Zoo and Wildlife Medicine*, 5th edition, edited by Murray E. Fowler and R. Eric Miller, 602–12. Saint Louis: Saunders.

Miller, Stephen A., and Mark A. Mitchell. 2009. "Ornamental Fish." In *Manual of Exotic Animal Practice*, edited by Mark A. Mitchell and Thomas N. Tully, 39–72. Saint Louis: Saunders Elsevier.

Mitchell, Lawrence G., John A. Mutchmor, and Warren D. Dolphin, eds. 1988a. "Amphibians." In *Zoology*, 727–48. Menlo Park: Benjamin/Cummings.

———. 1988b. "Birds." In *Zoology*, 772–93. Menlo Park: Benjamin/Cummings.

———. 1988c. "Body Surface and Support Systems." In *Zoology*, 172–94. Menlo Park: Benjamin/Cummings.

———. 1988d. "Coordination: Hormones and Endocrine System." In *Zoology*, 244–69. Menlo Park: Benjamin/Cummings. F 245–68.

———. 1988e. "Coordination: Nervous System." In *Zoology*, 219–43. Menlo Park: Benjamin/Cummings.

———. 1988f. "Feeding and Nutrition." In *Zoology*, 75–96. Menlo Park: Benjamin/Cummings.

———. 1988g. "Fishes." In *Zoology*, 704–26. Menlo Park: Benjamin/Cummings.

———. 1988h. "Internal Fluid Regulation and Excretion." In *Zoology*, 145–69. Menlo Park: Benjamin/Cummings.

———. 1988i. "Internal Transport and Defense Systems." In *Zoology*, 119–44. Menlo Park: Benjamin/Cummings.

———. 1988j. "Mammals." In *Zoology*, 794–820. Menlo Park: Benjamin/Cummings.

———. 1988k. "Reptiles." In *Zoology*, 749–71. Menlo Park: Benjamin/Cummings.

———. 1988l. "Respiratory Gas Exchange." In *Zoology*, 97–118. Menlo Park: Benjamin/Cummings.

———. 1988m. "Sensory Systems." In *Zoology*, 270–96. Menlo Park: Benjamin/Cummings.

Mitchell, Mark A. 2009. "Snakes." In *Manual of Exotic Animal Practice*, edited by Mark A. Mitchell and Thomas N. Tully, 136–63. St. Louis: Saunders Elsevier.

Morris, Patrick J., and Amy L. Shima. 2003. "Suidae and Tayassuidae (Wild Pigs, Peccaries)." In *Zoo and Wildlife Medicine*, 5th edition, edited by Murray E. Fowler and R. Eric Miller, 586–602. Saint Louis: Saunders Elsevier.

Navarez, Javier. 2009a. "Crocodilians." In *Manual of Exotic Animal Practice*, edited by Mark A. Mitchell and Thomas N. Tully, 112–35. Saint Louis: Saunders Elsevier.

———. 2009b. "Lizards." In *Manual of Exotic Animal Practice*, edited by Mark A. Mitchell and Thomas N. Tully, 164–206. Saint Louis: Saunders Elsevier.

Necker, Reinhold. 2000b. "The Avian Ear and Hearing." In *Sturkie's Avian Physiology*, 5th edition, edited by G. Causey Whittow, 21–38. San Diego: Academic Press.

O'Malley, Bairbre. 2005a. "Avian Anatomy and Physiology." In *Clinical Anatomy and Physiology of Exotic Species*, edited by Bairbre O'Malley, 97–161. London: Elsevier Saunders.

———. 2005b. "General Anatomy and Physiology of Reptiles." In *Clinical Anatomy and Physiology of Exotic Species*, edited by Bairbre O'Malley, 17–39. London: Elsevier Saunders.

———. 2005c. "Lizards." In *Clinical Anatomy and Physiology of Exotic Species*, edited by Bairbre O'Malley, 57–75. London: Elsevier Saunders.

———. 2005d. "Snakes." In *Clinical Anatomy and Physiology of Exotic Species*, edited by Bairbre O'Malley, 77–93. London: Elsevier Saunders.

———. 2005e. "Tortoises and Turtles." In *Clinical Anatomy and Physiology of Exotic Species*, edited by Bairbre O'Malley, 41–56. London: Elsevier Saunders.

Plyzycz, B., J. Bigaj, J., and A. Midonski. 1995. "Amphibian Lymphoid Organs and Immunocompetent Cells." In *Herpetopathologica*. Proceedings of the fifth international colloquium on the pathology of reptiles and amphibians, 115–27.

Pough, F. Harvey, Christine M. Janis, and John B. Heiser, eds. 2009a. "Avian Specializations." In *Vertebrate Life*, 8th edition, 439–84. San Francisco: Benjamin Cummings.

———. 2009b. "Living in Water." In *Vertebrate Life*, 8th edition, 77–103. San Francisco: Benjamin Cummings.

———. 2009c. "Mammalian Diversity and Characteristics." In *Vertebrate Life*, 8th edition, 519–52. San Francisco: Benjamin Cummings.

———. 2009d. "Mammalian Specializations." In *Vertebrate Life*, 8th edition, 553–79. San Francisco: Benjamin Cummings.

———. 2009e. "Salamanders, Anurans, and Caecilians," In *Vertebrate Life*, 8th edition, 220–65. San Francisco: Benjamin Cummings.

Powell, F. L. 2000. "Respiration." In *Sturkie's Avian Physiology*, 5th edition, edited by G. Causey Whittow, 233–64. San Diego: Academic Press.

Prum, Richard O., and Rodolfo O. Torres. 2004. Structural Colouration of Mammalian Skin: Convergent Evolution of Coherently Scattering Dermal Collagen Arrays. *Journal of Experimental Biology* 207 (Pt 12): 2157–72.

Reidarson, Thomas H. 2003. "Cetacea (Whales, Dolphins, Porpoises)." In *Zoo and Wildlife Medicine*, 5th edition, edited by Murray E. Fowler and R. Eric Miller, 442–59. Saint Louis: Saunders.

Schaeffer, Dorcas O., and R. Mark Waters. 1996."Neuroanatomy and Neurological Diseases of Reptiles," *Seminars in Avian and Exotic Pet Medicine* 5(3): 165–71.

Smith, Frank M, Nigel H. West, David R. Jones. 2000. "The Cardiovascular System." In *Sturkie's Avian Physiology*, 5th edition, edited by G. Causey Whittow, 141–231 (141–46). San Diego: Academic Press.

Stoskopf, Michael K. 1993a. "Anatomy." In *Fish Medicine*, edited by Michael K. Stoskopf, 2–30. Philadelphia: W. B. Saunders.

———. 1993b. "Clinical Physiology." In *Fish Medicine*, edited by Michael K. Stoskopf, 48–57. Philadelphia: W. B. Saunders.

The Reptile Database. 2011. Last modified March 3. http://www.reptile-database.org/db-info/SpeciesStat.html.

Tully, Thomas N. 2009. "Bird." In *Manual of Exotic Animal Practice*, edited by Mark A. Mitchell and Thomas N. Tully, 250–98. Saint Louis: Saunders Elsevier.

Whitaker, Brent R., and Kevin M. Wright. 2001. "Clinical Techniques," In *Amphibian Medicine and Captive Husbandry*, edited by Kevin M. Wright and Brent M. Whitaker, 89–110. Malabar: Krieger Publishing.

Wright, Kevin M. 2001. "Anatomy for the Clinician." In *Amphibian Medicine and Captive Husbandry*, edited by Kevin M. Wright and Brent M. Whitaker, 15–30. Malabar: Krieger Publishing.

14

Stress and Distress

Murray E. Fowler

INTRODUCTION

Stress is often spoken of in everyday conversation, frequently by persons with little or no understanding of what is actually happening in themselves or their animals. In fact, stress is necessary to make it possible for both domestic and wild animals to cope with an ever-changing environment. Each reaction to a stressor has adaptive significance.

Stress is the cumulative response of an animal to interaction with its environment via receptors (Fowler 2008; Moberg 1987). Intense or prolonged stimulation induces detrimental responses (distress) that may be fatal (Breazile 1987, 1988). Keepers should be ever-vigilant to conditions that may contribute to the development of distress in their charges.

The objectives of this chapter are to

- enable zookeepers to understand the principles and concepts of stress and distress in animals
- provide practical examples to illustrate principles
- encourage zookeepers to consider animal stress in all animal management activities.

An animal has many nerve receptors located in strategic regions of the body that alert the animal to changes in its environment. These are the physical senses (sight, sound, odor, taste, and touch). Stimulation of any of these senses produces a specific effect, such as becoming alert when a strange sound is heard. That same stimulus may cause a general response that may not be apparent outwardly, but may have a profound long-term effect on the welfare of the animal. A general response is mediated through neurohormonal pathways, which are the primary focus of this chapter. It is important to understand that stress effects are cumulative, so that all factors in an animal's life must be considered when assessing its welfare.

A stress-producing factor (stressor) is any stimulus that elicits a response when perceived by an animal (Arnemo

and Caulkett 2007). Stimulation of physical senses include temperature changes; strange sights, sounds, touches, or odors; traumatizing of muscles during restraint procedures; close confinement; thirst; and hunger. Psychological stressors include anxiety, fright, terror, anger, rage, and frustration. Closely allied are behavioral stressors, including overcrowding, lack of social contact, pecking order (hierarchical) upsets, unfamiliar surroundings, transporting, and lack of habituated foods. It is becoming more and more important to recognize that stimulation of visual and auditory senses has a marked bearing on accumulative stress. Miscellaneous stressors include malnutrition, toxins, parasites, infectious agents, burns, surgery, drugs, chemical and physical immobilization, confinement, excessive physical exertion, and injuries.

There is marked species variation in how organisms process and act upon stimuli. There may even be varying responses within an individual, depending upon which stimuli are acting upon it at a given time (experience, training, adaptation, hierarchical status, nutrition). These influences may reduce or increase a response to the stressor.

Animals respond in appropriate ways to stimulation of specific receptors. For instance, when cold receptors are stimulated in mammals, the body experiences a sensation of coolness. Various general and behavioral changes occur that conserve heat and stimulate increased heat production. The animal is adjusting to a new situation. That adjustment is called homeostasis or homeostatic accommodation (figure 14.1).

PHYSIOLOGY OF STRESS

Response to the stimulation of a receptor may follow one of three pathways: (1.) muscular reflex pathway (voluntary motor); (2.) hormone pathway (autonomic nervous system, epinephrine); or (3.) stress pathway (neuroendocrine). See figure 14.1. The most common stress pathway is complex, with nerve reflexes that ultimately stimulate the adrenal gland cortex and produce cortisol, a hormone. This pathway is il-

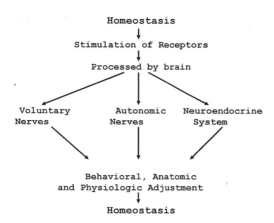

Figure 14.1. Pathways for regulation of homeostasis. Illustration by Murray Fowler.

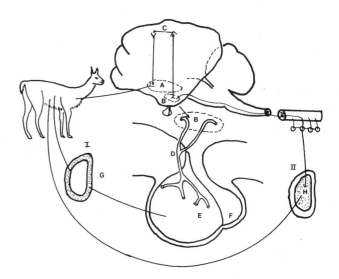

Figure 14.2. A diagram illustrating the neurohormonal pathways of (I) the cortisol-stress response and (II) the alarm response. (A) thalamus, (B) hypothalamus, (C) cerebral cortex, (D) hypothalamic pituitary vascular portal, (E) anterior pituitary gland, (F) posterior pituitary gland, (G) adrenal cortex, and (H) adrenal medulla. Illustration by Murray Fowler.

lustrated diagrammatically in figure 14.2 and described as follows.

A stimulus is received from a receptor in the animal. A nerve impulse reaches the thalamus (A) and is relayed to the cerebral cortex (C), and then on to the hypothalamus (B). The hypothalamus is a "switchboard" located at the base of the brain, just above the pituitary gland (E). Neural stimulation of the hypothalamus produces releasing factors that are carried to the anterior pituitary gland by a connecting blood vessel (D). The releasing factors cause the production of compounds that in turn reach the adrenal cortex (G) via the general blood circulation, which stimulates production of appropriate hormones. The example illustrated and described is for stimulation of the adrenal cortex to produce cortisol. Similar neurohormonal action may involve any or all of the endocrine glands (thyroid, parathyroid, ovary, testicle, pancreas), which likewise may have a profound effect on their target organs.

Disruption of appropriate hormonal action on the ovary or testicle caused by stress may initiate reproduction problems, such as failure to come into heat (estrus; Moberg 1985b). Other metabolic actions may likewise be initiated from intense or prolonged stress episodes.

The alarm reaction is an adaptive response to a perceived threat. The pathway is illustrated in figure 14.2. It is also called the "flight-or-fight" response. The alarm response may be initiated by strange sights, sounds, fear, odors, or other animals, particularly predators. The initial pathway is similar to the stress response up to the hypothalamus. From that switchboard a neural stimulus is transmitted through the spinal cord to the sympathetic trunk of the autonomic nervous system, and on to the adrenal medulla (central region of the adrenal gland). That stimulus produces the hormone adrenaline (epinephrine) and other hormones responsible for the alarm reaction. Specific effects include an increased pain threshold, elevated blood glucose, heightened alertness, opening (dilatation) of the pupil, opening of the airways (bronchiolar dilatation), increased blood pressure, diminished secretion of saliva, and increased blood flow to skeletal and heart muscle with a decreased blood flow to the skin and intestines. An important visible effect is the raising of the hair on the back and neck (piloerection). Not all species

are capable of piloerection, but if present, it is a sure sign of an alarm response.

For many years it was thought that only the effects of excess cortisol were responsible for stress and distress, but modern researchers of the stress phenomenon have discarded that concept for the broader view that stressor stimulation may affect a number of endocrine pathways (Selye 1973, 1976). This is not to say that cortisol is not involved, but the picture is more complex than once supposed. Blood cortisol level is still the generally used measure of stress in animals. More detailed information may be found in the references cited (Arnemo 2007; Fowler 2008).

Continuous adrenal cortex stimulation and excessive production of cortisol elicits many adverse metabolic responses. Behavioral as well as physical changes occur. The clinical syndromes of adrenocortical stimulation have been identified in some species (human, dog, horse, laboratory animals). There is much still to learn about the effects of hypercorticism in wild animals. However, the basic biologic effects of cortisol should be understood (Breazile 1987; Moberg 1985a).

Cortisol contributes to the breakdown of protein and fat, making the products available for the manufacture of glucose. Slight to moderate elevations in blood glucose (hyperglycemia) has a diuretic effect, producing urination and thirst. Prolonged hyperglycemia stimulates the pancreas to produce more insulin.

Cortisol reduces the heat, pain, and swelling associated with the inflammatory response, an effect useful in the treatment of many diseases. Cortisol also decreases the ability of individual cells to combat infectious and parasitic agents. Within a few hours of cortisol stress response, reduction in the number of circulating lymphocytes in the blood of mammals is 50% or greater. Lymphocytes are one line of defense against infectious agents. Other changes in the blood cells include an increase in neutrophils and a decrease of eosinophils. These changes in the numbers of white blood cells are an indication of stress, but they must be differentiated from similar effects that may occur and that are associated with infectious diseases (Fowler 2008).

Cortisol interferes with DNA (deoxyribonuclic acid) synthesis by causing a loss (atrophy) of lymphoid tissue throughout the body. Some immune responses (cell-mediated responses) are diminished, an effect that may interfere with tuberculin testing programs and the animal's ability to mount an immune response to vaccines.

Gastric and intestinal ulcers are common in some animals. Stress ulceration of the gastrointestinal system is a well-known syndrome in humans, rats, and marine mammals. Whether or not stress is a factor in other wild animal ulcers is unknown, but one should be mindful of the basic effect of cortisol on the digestive system. The development of gastric stress ulcers in humans and marine mammals is influenced by many factors, including hormonal and bacterial components. Elevated cortisol levels cause overproduction of acid and digestive enzymes, which have a detrimental effect on the lining of the stomach (mucous membrane). Adrenaline (epinephrine) may also contribute to the production of gastric secretions, so stimuli that increase the levels of this hormone (fear, anxiety, frustration, anger) may have a potential effect on the development of ulcers.

SIGNS OF DISTRESS

Signs are highly variable in different species, and may be vague and nonspecific but still suggestive. Collectively, animals with distress may have muscle weakness, loss of hair (bilateral alopecia), increased susceptibility to infection, poor antibody response, vaccination failure, and poor wound healing. An important consideration that may be unrelated to excessive cortisol is the stress effects on the ovaries and testicles that may contribute to poor reproduction (Moberg, 1985b). Behavioral changes that may be noted include increased aggression, antisocial tendencies, depression, refusal to eat or overeating, and increased or decreased sexual activity. A troop of baboons at a major zoo was studied by an eminent behavioral biologist and reported to be hypersexual. When that same species was studied by biologists in the field, no such hypersexuality was noted. For the zoo baboons, overcrowding and inability to establish troops created a stressful situation.

Damage to tissue (lesions) produced by harmful stress are difficult to document. Clinicians and pathologists often negate a diagnosis of death caused by stress. Many of the effects of stress are functional, leaving no definitive lesion to mark their presence. Nonetheless, it is known that tissues and organs are weakened by prolonged insult, lowering resistance to disease. Classic lesions are lymphoid tissue atrophy, adrenal cortical hyperplasia, and gastrointestinal ulceration. Though the actual cause of death may be pneumonia, parasitism, or starvation, stress may have paved the way for development of these terminal ailments.

Husbandry practices should be evaluated, and those that may be harmful should be corrected. Social animals may be intolerant of isolation for therapy or recuperation. Malnutrition is a stressor, as are repeated restraint episodes. More detailed information about stress may be obtained in the references (Arnemo 2007; Fowler 2008; Moberg 1985b, 1987).

EXAMPLES OF THE EFFECTS OF DISTRESS IN WILD ANIMALS

Marsupial macropods (kangaroos and wallabies) are commonly maintained in groups by zoos. These are social animals, but when their population numbers and group dynamics exceed the parameters of what is needed for a healthy interaction between individuals, stress plays a part in returning balance to the group. Keepers and veterinarians have noted that when overcrowding becomes an issue, the mortality rate climbs. Necropsy findings do not follow a pattern, but seem to be nonspecific with reports such as mild parasitism, malnutrition, or infection from generally nonpathogenic bacteria. When population levels decrease, so does the mortality.

Neither should it be assumed that distress is limited to wild animals maintained in captivity. The population cycle of arctic rodents is known to be governed by stress factors. When a population explosion of lemmings (*Lemmus* spp.) occurs in conjunction with a limited food resource, a population crash is precipitated. Studies have shown that with overcrowding and starvation, cortisol levels in the lemmings are inordinately high. A variety of disease conditions are found to cause the death of the animals. The snowshoe hare (*Lepus* sp.) cycle is likewise governed by similar stress-related responses.

A classic example is that of the marsupial mouse (*Anticinus* sp.). Biologists have known for years that this short-lived species has a unique mortality pattern wherein nearly all of the males die shortly after the breeding season. Various theories were given for the high mortality, including parasitism (toxoplasmosis, hemoparasites), specific infectious diseases, and even toxicities. Ultimately a team composed of biologists, microbiologists, parasitologists, pathologists, endocrinologists, and veterinary clinicians approached the challenge from an ecological perspective.

The conclusion they reached had a logical explanation. This species has a unique biological cycle. The males are extremely aggressive during the breeding season. Fights are continuous between males competing for breeding privileges. Stress level in the males is high. Necropsies revealed as much as a fivefold increase in the size of the adrenal gland with corresponding high levels of cortisol present in the blood. At the conclusion of the breeding season, the males were exhausted and in a state of distress. Thus, they were susceptible to disease conditions that they normally would be able to resist. But this population survives because the pregnant females produce male offspring that reach sexual maturity quickly enough to impregnate females in the following season.

It must be emphasized here that both free-ranging and wild animals maintained in captivity have similar stressors acting upon them. However, wild animals born and reared in captivity may have a diminished reaction to stress stimulation.

CAPTURE MYOPATHY

Exertional muscle damage (capture myopathy, capture stress) is another stress-induced condition that may occur while wild free-ranging animals are captured, but also while zoo animals are restrained or transported.

PREDISPOSING FACTORS

Excessive muscular activity, usually accompanied with the alarm response, sets off a complex cascade of biochemical reactions that lead to muscle cell death (necrosis) and changes in the blood chemistry, such as an increase in acidity, that may be incompatible with life. For example, a small gazelle was hand-captured and placed in a crate for transport to another zoo several thousand miles away. The crate was small, but adequate for the animal; however, the gazelle was agitated and kept trying to escape.

When the gazelle was unloaded at the new destination, keepers noted that it refused to stand and had severely traumatized hocks. They refused to accept the animal in this condition and returned it to the sender. The gazelle was exhibiting classic signs of myopathy, including inability to stand, because of the loss of muscle function. The muscle actually ruptured. While the gazelle struggled in the recumbent position, the hocks were traumatized.

If an animal with exertional myopathy is able to walk, it does so stiffly and painfully. Urine from severely affected animals may appear reddish because the ruptured muscle cells release myoglobin into the blood, which is excreted in the urine.

Birds may also suffer from capture myopathy. Long-legged birds (storks, cranes, flamingos) may be placed in a small canvas sack with their heads protruding while they recover from anesthesia. If they are left too long in this confined space, their struggling may result in capture myopathy. A flock of ducks or geese captured under a cannon net may struggle against the net's weight. Signs of capture myopathy in birds include incoordination and paralysis.

PREVENTION

Great strides have been made in training animals to accept confinement in crates for necessary transport. Feeding the animals inside a crate for a period of time before the transport helps to minimize their apprehension and fear. Wild animals become habituated to certain feeds and sources for drinking water. When changes occur, such as when animals are moved to a new enclosure or to a new zoo, it may be a highly stressful and they may need time to adapt.

MALNUTRITION

Zoos have advanced the science and practice of feeding wild animals, but simply supplying the correct amount of protein, fat, fiber, vitamins, and minerals may not be all that is necessary. The correct amino acid composition of protein or the correct form of a vitamin may be crucial. The lack of those essentials may be stressful for any animal. An example of malnutrition as a stressor is sudden death (peracute mortality) in giraffe. This can happen because giraffe are specialized ruminants with special nutrient requirements that may be overlooked. Other predisposing factors include transporting, introduction to a new herd, and concurrent diseases in the herd. Signs include sudden (overnight) death of a giraffe that has shown no prior evidence of illness. The malnutrition and other accumulated stresses can lead to distress and death.

WATER DEPRIVATION

The deaths of several feral horses in a desert area in the western United States illustrates how animals may be unwilling to drink from a novel water container. A cluster of dead and dying horses was noticed near a newly installed water container. Initial evaluations mistakenly ascribed the deaths to noxious vapors from a government installation that had been established nearby to work with biological and chemical warfare agents.

A more critical evaluation of the situation revealed that the weather had been extremely hot—above 45 °C (113 °F)—for several days. Furthermore, a stream that had been their only source of water for many miles was dry. The spring that had fed the stream had been capped and the water piped to a large plastic container, supposedly as a better source of water for the horses. When the drain of the tank was opened and water allowed to flow into the old streambed, the horses that were still alive struggled to their old water source and attempted to satiate their thirst by eating mud. The horses had died from dehydration and hyperthermia because they either refused to drink from an artificial container or did not recognize that the water was there. This same situation may occur in a zoo when an animal is introduced to a new enclosure or transported from another facility.

HEAT STRESS

Keeping wild animals in an appropriate environment during the hot and humid season is frequently a challenge in certain regions of the world. Excessive heat and humidity are significant stressors during those months. Zoo managers and keepers must understand the consequences of failure to provide optimal living conditions. Factors include enclosure and housing design for management, and facilities for shade, water and other cooling opportunities provided by keepers.

HYPERTHERMIA

ETIOLOGY

The predisposing factors for hyperthermia (heat exhaustion, heatstroke, sunstroke, and heat stress) include prolonged high environmental temperatures, especially with high humidity, muscular exertion, fever, dehydration, mycotoxins that inhibit thermoregulation, and drugs that depress thermoregulation. Activities that may contribute to the production of body heat include breeding, fighting, being transported, prolonged restraint, being chased by dogs, or (in breeding males) pacing a fence (Fowler 1994).

The normal core body temperature of adult mammals is highly variable, ranging from 33 to 36 °C (91.4 to 96.8 °F) in marsupials and edentates, and 37.5 to 38.6 °C (99.5 to 101.5 °F) in most other mammals. Normal temperature of young animals may be a degree Celsius higher than that of the adults. Marine mammals are particularly susceptible to hyperthermia when being transported between facilities; steps must be taken to ensure cooling. Birds may have core body temperature one to three degrees Celsius higher than that of mammals. Reptiles are generally close to ambient temperature, but some have the ability to partially control their body temperature. Fish may not be able to properly respire in water that is too warm, as it may not hold enough oxygen for their metabolic needs, although this varies greatly with species.

HEAT EFFECTS ON ORGAN SYSTEMS

The degree and effects of hyperthermia on organs and tissues may vary according to the duration of exposure to excessive heat and humidity and the presence of other conditions, such as medical conditions that cause elevated blood acidity (metabolic acidosis), cardiovascular dysfunction, chronic disease, excessive hair covering or other insulating factors (Fowler 1994, 2008).

Brain. The central nervous system (CNS) is highly sensitive to hyperthermia. Effects of heat on the brain may be direct, causing death of nerve cells (neurons); by secondary factors, such as low blood pressure (hypotension), causing lack of oxygen flow to the brain (cerebral hypoxia); or by electrolyte alterations, causing neurotransmission dysfunction. Lesions in the CNS may also be caused by hyperthermic damage to the cardiovascular and blood (hemic) systems (hemorrhage, disseminated intravascular clotting [DIC]). Signs exhibited are determined by the area of the CNS damaged, but generally there is decreased mental function and possibly convulsions. Damage to the brain centers concerned with temperature regulation may predispose animals to relapses or subsequent increased sensitivity to heat.

A frequently overlooked but serious consequence of heat stress in a pregnant female is fetal brain damage, resulting in various congenital anomalies or even death of the fetus. Congenital brain defects associated with prenatal prolonged hyperthermia in humans and other animals are the result of excessive heat acting on the embryonic cells of the brain at a crucial time.

Reproductive system. Heat stress may have a marked effect on the adult female, including diminished intensity of receptivity and anestrous. During pregnancy, the more profound effects are seen as fetal damage, including inhibition of embryonic cleavage and implantation, initiation of congenital defects, and abortion.

General effects on the fetus may result in reduced birth weight, which may be caused by placental retardation. Fetal effects have been noted in llamas and alpacas when the core body temperature of the dam rises above 40.1 °C (104.2 °F) for prolonged periods. These are temperatures routinely recorded in clinically unaffected llamas and alpacas; during hot weather the body temperature can elevate during the warmer time of the day and return to lower temperature at night. Hyperthermic effects on an embryo are dependent on the degree of hyperthermia, duration of hyperthermia, and the stage of development of the embryo. Abortion may be the result of placental death, direct effects on the fetus causing death (vascular leakage, edema, hemorrhage), or, in near-term fetuses, a stress response causing elevated cortisol levels.

In all males, excessive heat is spermicidal. In some species, such as camelids, this may be a serious problem. In hot climates it is common for camelid males to become infertile during the hot weather of summer. The scrotum of camelids is not pendulous as it is in most domestic ruminants, so core body temperature changes have a rapid and profound effect on developing sperm cells. At least 35 to 60 days are required for new spermatogenesis to produce mature viable sperm once the heat stress has decreased.

Respiratory system. A 1 °C (1.8 °F) rise in the body temperature increases by 10% the requirement for oxygen to maintain normal function of the body's energy systems. If the body temperature rises to 41 °C (105.8 °F), the respiratory system can no longer supply sufficient oxygen by normal respiration. Heat stress causes increased breathing, respiratory acidosis, and open-mouth breathing.

Digestive system. Signs of colic are commonly seen in heat-stressed animals. Elevation of the core body temperature initiates a shift in the blood supply from the internal organs to the skin. Decreased blood flow to the stomach and intestine causes decreased digestive function. Decreased oxygen (hypoxia) supply to liver cells (hepatocytes) and decreased liver function results from decreased blood flow to the liver. In severe cases there is a failure of the production of elements necessary for blood coagulation (coagulation cascade). Persistent low-intensity hyperthermia may cause decreased digestive function, which in turn may cause poor growth rates in juveniles, and poor appetite and less efficient feed utilization in adults. Hyperthermia also has detrimental effects on other body systems including the heart and circulatory system, blood forming-organs, and kidneys.

It is often assumed that when a hyperthermic animal has been cooled, all organ systems begin functioning again at their normal capacity. That may be true if the heat stress has been of short duration and moderate intensity. But if the heat stress has been severe or prolonged, many residual effects may alter organ function and even kill the animal long after the core body temperature has returned to normal.

SEQUENCE OF EVENTS DURING HYPERTHERMIA

1. elevation of the core body temperature
2. accelerated heart rate
3. increased respiratory rate (panting)
4. redness of skin surface
5. sweating (only in some mammals)
6. increased concentration of blood cells
7. body fluid shift from intestine, lungs, and muscle to skin
8. decreased urination
9. dehydration
10. decreased blood pressure
11. effects on the brain
12. effects on the embryo and fetus
13. coagulation defects (disseminated intravascular clotting)
14. other organ system damage

Some animals are at increased risk of hyperthermia because of low heat tolerance. Older animals may have decreased sweat gland activity, be in poor physical condition, or have deterioration of cardiovascular function. Obese animals are generally less physically fit than those of normal weight. Evaporative cooling in mammals is determined by the number of sweat glands per square meter of skin surface. Metabolic heat is produced in proportion to body mass, but is dispersed in proportion to skin surface area. Thus, obese animals are at a distinct disadvantage in hot weather. These animals must be observed frequently and special steps must be taken to ensure their ability to remain cool. Some zoos provide fans and even air-conditioned stalls in particularly oppressive conditions.

HYPOTHERMIA

Hypothermia (low body temperature) is also a stressor to animals.

PREDISPOSING FACTORS

Neonates are particularly susceptible to hypothermia because they have poorly developed mechanisms of thermoregulation and a higher metabolic rate. Their relatively greater proportion of skin surface allows for rapid dissipation of heat. Neonates may lack a shivering reflex. Even adults under anesthesia or in shock are prime candidates for hypothermia if in a cold ambient environment. Insufficient food intake reduces metabolic heat production. Restricted muscular activity prevents heat generation. A poor fiber coat provides less insulation capacity and contributes to heat loss. A coat that is soiled with feces or dirt provides little or no insulation. California sea otters (*Enhydra lutris*) are particularly susceptible to hypothermia when being transported if not kept warm and given the opportunity to clean and groom their fur.

SIGNS

Clinical thermometers record body temperatures only as low as 33.3 °C (92 °F). With more sensitive thermometers, temperatures as low as 29.4 °C (85 °F) have been recorded in living mammals. Other signs of hypothermia include depression progressing to coma. In contrast to hyperthermia, the hypothermic animal may live for hours. A decrease in body temperature is accompanied by a decrease in cardiac output, heart rate, blood pressure, and urine production.

THERAPY

Total body immersion in warm water 40.5–45.5 °C (105–14 °F) is the fastest way to warm a mammal. Total body immersion is impossible with most large adult animals, but possible with a neonate. Running the fingers through the fiber coat will keep warm water flowing over the skin. If the whole body cannot be immersed, warm water should be applied to the legs, and the legs should be massaged.

A warm-water enema is highly effective, though the ability to monitor the body temperature with a rectal thermometer is temporarily lost. A hair dryer may be helpful in warming a neonate. Covering the animal with blankets will help conserve heat, but if the temperature is below 32.3 °C (90 °F), metabolic heat production is proportionately reduced, and endogenous rewarming is slowed.

Intravenous infusions of warm saline are effective. Surgical exposure of a suitable vein may be necessary to effect intravenous administration because of vasoconstriction. Circulating-water-type heating pads are effective in preventing hypothermia in neonates during surgery and in treating accidental hypothermia, but electric heating pads have caused skin burns and sloughs. Hypothermic and shock patients normally suffer from skin vasoconstriction and exhibit a reduced ability to carry heat away from the skin. One should be cautious when applying heat directly to the skin. Measure the temperature between the skin and the pad, and keep it below 42 °C (107.6 °F). Hot-water bottles may be used to raise the ambient air temperature in a small enclosed area. Plastic milk cartons or plastic bags may be substituted for hot-water bottles. The air surrounding the patient may be warmed with infrared heat lamps, forced-air driers, or electric floor heaters.

PREVENTION

Shelter from wind and rain should be provided. Deep straw bedding may minimize heat loss from the thermal window in extremely cold climates. A shelter should be small enough to be warmed by a group of animals huddling together. A box stall can be made smaller by blocking it with bales of hay or straw. Insulated and heated barns may be required in particularly harsh, cold climates. High-quality feed should be provided, including concentrates if the animals are used to eating them. Water must be available. For birthing in cold weather, a maternity stall with provisions for supplemental heat should be available.

REFERENCES

Arnemo, J. M., and N. Caulkett. 2007. Stress. In G. West, D. Heard, and N. Caulkett,. eds., *Zoo and Wildlife Immobilization and Anesthesia*, 103–9. Ames, IA: Blackwell.

Breazile, J. E. 1988. The Physiology of Stress and its Relationship to Mechanisms of Disease and Therapeutics. *Vet. Clin. North Am., Food Anim. Pract.* 4(3): 441.

———. 1987. Physiologic Basis and Consequences of Distress in Animals. *J. Am. Vet. Med. Assoc.* 10:1212.

Fowler, M. E. 1994. Hyperthermia in Llamas and Alpacas. *Vet. Clin. North Am., Food Anim. Pract.* 10(2): 309–18.

———. 2008. Stress. In M. E. Fowler, *Restraint and Handling of Wild and Domestic Animals*, 3rd ed. Ames, IA 65–69. Wiley-Blackwell.

Moberg, G. P. 1985a. Biological Response to Stress: Key to Assessment of Animal Well-Being. In G. P. Moberg, ed., *Animal Stress*. Bethesda, MD: American Physiological Society.

———. 1985b. Influences of Stress on Reproduction: Measure of Well-Being. In G. P. Moberg, ed., 27–50. *Animal Stress*. Bethesda, MD: American Physiological Society.

———. 1987. Problems of Defining Stress and Distress in Animals. *J. Am. Vet. Med. Assoc.* 191:1207–11.

Selye, H. 1973. The Evolution of the Stess Concept. *Am. Sci.* 61:692–99.

———. 1976. *Stress in Health and Disease*. London: Butterworth.

15

Physical Restraint and Handling

Murray E. Fowler

INTRODUCTION

Animal restraint and handling practices coevolved with the domestication of animals for food, fiber, work, sport, and companionship. Domestication occurred before recorded history for most animals. Close association with the animals being domesticated necessitated some degree of control. Trial and error resulted in the development of satisfactory methods for handling animals. When people began bringing wild animals into captivity, a different set of methods was necessary. Even today, those who handle animals are experimenting with new nets, snares (also called catchpoles or dog nooses), and restraint cages to provide safer and more efficient restraint.

Whether the animals were domesticated or were wild animals maintained in captivity, the keeper needed to understand the biology and behavior of each species under their care. Early in the history of the relationship between humans and animals, the derived animal handling methods were based on the personal experiences of the keeper. Later, the art and even the science of animal handling could be committed to the written word and taught to students who wished to become more skilled in the handling of animals.

The objectives of this chapter are to

- enable zookeepers to understand the principles and concepts of animal handling
- provide practical examples to illustrate principles
- encourage zookeepers to consider animal welfare in all restraint procedures.

RESTRAINT OF ANIMALS

Any time an animal is restricted in its movements, its behavior and biological processes are affected. Everyone should be concerned about animal welfare (Anonymous 2005). It must be understood that the minimum amount of restriction should be applied that is consistent with the procedure to be carried out.

Each time an animal is to be restrained, the following questions should be asked:

1. Why must this animal be restrained?
2. What procedure will produce the greatest benefit with the least hazard to the animal?
3. Who is the most qualified person to accomplish the task in the least amount of time and with the least amount of stress for the animal?
4. Is the contemplated restraint in the animal's best interests (Fowler 2010)?

Keepers are responsible for the welfare of their animals. They will be required to either apply the necessary restraint or assist in the preparation and execution of procedures for transporting, housing, feeding, examining, treating, and monitoring their charges. Keepers should accept that responsibility, acquire knowledge, and practice to gain the experience necessary to become proficient and successful in performing a task. The author uses a formula for success in dealing with all kinds of situations with animals, including restraint:

$$\text{Plan} + \text{prepare} + \text{practice} + \text{produce (do it)} \\ + \text{persist in doing it} = \text{Success.}$$

BASIC CONCEPTS

PLANNING

Assuming that the previous questions have been answered favorably, the next step is to select a method of restraint based on the following criteria.

1. Is the procedure *safe* for the person doing the restraint and their assistants?
2. Can the method be applied *safely* to the animal to be restrained?

3. Will the method allow the task to be accomplished effectively and efficiently?
4. Will the method allow adequate exposure of the head, body, or limbs that require examination or treatment?
5. Will the method allow sufficient time to complete the examination or treatment without jeopardizing the animal's welfare?

PREPARATION FOR A PROCEDURE

When dealing with a new species, one should communicate with someone who has experience in dealing with that species. The literature is voluminous now on handling and restraint, and it should be used before a procedure. This is part of the planning and preparation phase of restraint.

TOOLS

All those responsible for animal restraint should understand the tools needed for restraint and their availability within the zoo or aquarium. It may be necessary to acquire specific tools to successfully perform a particular procedure. Tools include voice communication with an animal, the keeper's hands and arms, ropes, snares, nets, squeeze cages, chutes, and shields.

VOICE

Some docile animals may respond to a keeper's soothing voice sufficiently to allow limited physical examination. Veterinarians are usually not included in an animal's acceptance of a soothing voice because their voices are usually associated with unpleasant experiences. Loud noises, casual conversation, and shouting should be avoided during restraint procedures. Such sounds are stressors. One may plug the animal's ears with cotton, but it should become a habit to keep sound to a minimum during restraint. If earplugs are used, remember to remove them before releasing the animal.

HANDS

The keeper's hands are an important tool and are easily injured by unwise actions near mouths, feet, claws, beaks, and talons. Grasping with the hands may be the only tool needed in handling some small mammals, birds, reptiles, amphibians, and invertebrates. Trained birds may allow general handling, but if more restrictive restraint is necessary they may anticipate being grasped by the hands, so some degree of hiding the hands, such as with a towel, may be necessary. Birds often recognize gloved hands as well as bare ones.

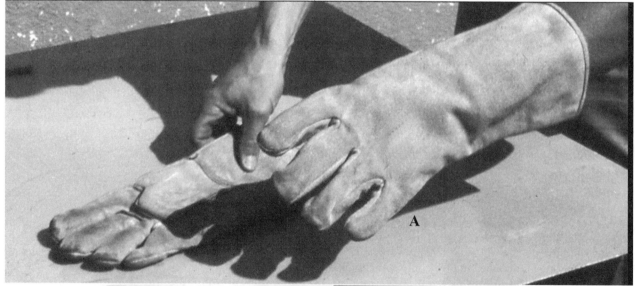

Figure 15.1. Gloves. (A) leather welder's gloves, (B) gauntleted double-thickness gloves, (C) chain mail glove. From *Restraint and Handling of Wild and Domestic Animals, 3rd Edition*, M. E. Fowler, © 2008. Reprinted with permission of John Wiley and Sons, Inc.

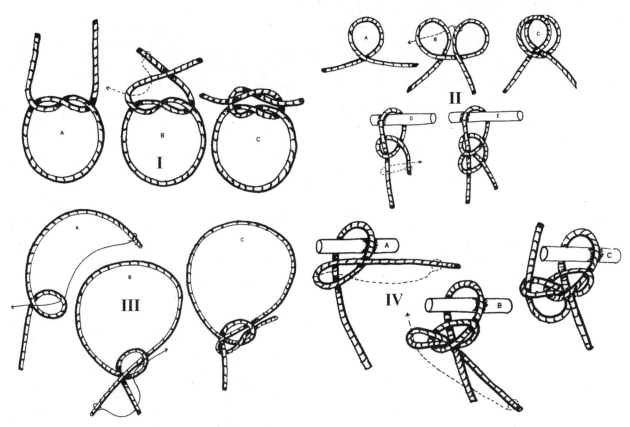

Figure 15.2. Knots. (I) Square knot, (II) Clove hitch, (III) Bowline, (IV) Halter tie. A, B, C = sequence for tying the knot; D, E = sequence for tying the clove hitch around a rope. From *Restraint and Handling of Wild and Domestic Animals, 3rd Edition*, M. E. Fowler, © 2008. Reprinted with permission of John Wiley and Sons, Inc.

The hands may be protected with gloves (figure 15.1), but this may decrease one's tactile discrimination, and may cause unnecessary or excessive pressure to be exerted while holding an animal. Gloves provide limited protection, but most carnivores and some primates are capable of biting through even heavy leather gloves. Gloves also don't prevent crushing injuries from animals with strong jaws, such as hyenas, large carnivores, macaws, and eagles. Immediately after initial capture the gloves should be removed, and the animal held with bare hands.

ROPES AND KNOTS

Ropes in various diameters, lengths, and types of fiber are tools commonly used by keepers for restraint and handling of animals. Natural fibers used for the construction of rope may be soft (e.g., linen, cotton, hemp) or hard (e.g., horsehair, sisal, manila). The degree of hardness is determined by the strength of the twist applied while constructing the rope. Synthetic fibers include nylon, polypropylene, terylene, and other plastic bases. Cotton fibers are the softest and least likely to cause rope burns, and are easier to tie knots in. Hard fibers and synthetic fibers are stronger than cotton, but may be more abrasive. Nylon fiber ropes will stretch, so tension has to be adjusted. Synthetic fibers are not used with pulleys (in a block and tackle), but they do have the advantage of being

waterproof and easier to clean. Ropes should be kept clean, in good repair, and coiled and stored properly.

A few basic knots will serve the needs of keepers: the square knot, clove hitch, bowline, halter tie, and temporary rope halters (Anderson and Edney 1991; Fowler 2008, Sheldon, Sonsthagen and Topel 2004; Sonsthagen 1991). Keepers should practice these knots until they can tie them while blindfolded (figure 15.2).

Square (reef) knot. This is the basic knot used for securing cages and crates and in general restraint procedures. When the knot is completed, both strands are parallel with each other. A square knot should not be used when it will be subjected to linear tension, as it may then change into a slip knot. If the strands are not parallel it is called a granny knot, which is much more difficult to untie if the knot has been under tension.

Clove hitch. Frequently used to begin other procedures, such as tying around a leg or a post. A clove hitch will not stay secure unless tension is constantly applied.

Bowline. This is a universal knot used in animal restraint as a starting point for other procedures such as temporary rope halters, casting ropes, and slings. The bowline is a secure knot, but it may easily be untied despite excessive tightening. This knot should be practiced until tying it becomes automatic. One should forget the shortcuts and verses that are supposed to help a person remember how to tie this knot, and instead learn the principle of tying it.

Halter tie. The term implies securing an animal on a lead rope to a post, ring, or fence. This is not usually appropriate for zoo animals. However, there are numerous instances where it is necessary to secure an animal's leg or arm to an object. This knot is important because it can be secure but is easily untied. It must be tied close to the post or ring to avoid losing tension when the knot is tightened. Bringing the end through the formed loop is only of value when securing an animal that may pull the end of the rope and untie itself.

Temporary rope halters. These have numerous applications in both physical and chemical restraint. If tension is slackened, the nose loop may slide off the nose or become low on the muzzle; if tension is tightened, air flow may be cut off. A safer method is to form the neck loop by tying a bowline around the neck. If the nose loop then comes off, the head is still secured without danger of the animal being strangled.

ADDITIONAL PLANNING

Time and place for restraint. One does not always have a choice as to where a restraint procedure will be carried out. When possible, however, the site should allow sufficient space for the animal and personnel to perform the restraint. It should be free of exposure to sunshine, wind, and inclement weather. The animal's comfort should always be the highest priority.

The timing of the procedure is important. One should avoid restraint procedures in hot or humid weather. If necessary, the procedure should be conducted early in the morning. One should be aware of projected ambient temperature and humidity. If necessary, methods of cooling should be kept available (fans, cold water to appropriate areas of the body, ice packs, and shade). Some procedures may be best done at night to diminish vision in diurnal animals. The author handled wallaroos (*Macropus robustus*) at night by restricting them to a night house and then moving in quietly among them to grasp one by the tail and move it out of the building. Deer may also be managed and moved within darkened enclosures. Diurnal birds such as parrots may be approached more easily with a net in subdued lighting.

Prolonged restraint procedures should not be begun late in the afternoon. When an animal is released, it may be somewhat disoriented. Such a release should not be made at night. Social animals maintained in a group establish a hierarchical relationship. When an animal is removed from that pattern for restraint and then returned, some adjustment, including fighting that needs to be sorted out during daylight hours, may ensue. Releases should be monitored by keepers until social interaction among the animals is back to normal.

Confinement. Domestic animals may tolerate confinement to a household, pasture, barn, stall, or stanchion, but to a wild animal confinement is a definite challenge. Wild animals may adjust to enclosures with no outward evidence of stress, but physical restraint will add another layer of stress. Physically catching and holding an animal may be extremely stressful to some animals, but if it is carried out quickly, the animal will compensate through homeostatic processes. A keeper should take steps to minimize stretching, exertion of pressure, loud noises, and extension of procedures longer than absolutely necessary. Squeeze cages (figure 15.4), holding cages, and chutes are essential to proper management of wild animals, but prolonged confinement may be detrimental.

Nets are a form of confinement and are important tools for handling a variety of zoo animals (figure 15.3). The material and mesh size used for nets are dependent upon the species to be caught. It is important that the animal cannot push its head through the mesh. The depth of a hoop net should be sufficient to completely encompass the animal with enough additional room so that the hoop can be flipped over to prevent the animal from climbing out. An animal may bite and/or claw through the net. Primates with dexterous hands are difficult to net. Nets are frequently damaged by animals biting and clawing the fabric. One should inspect a net for

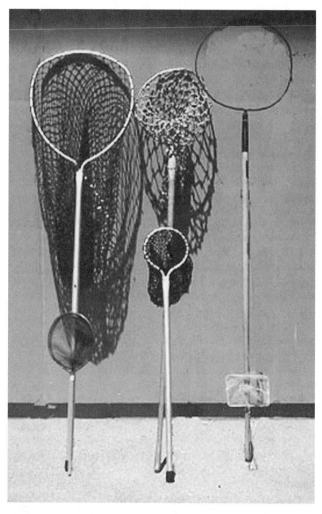

Figure 15.3. A variety of nets used for capture and restraint of mammals, birds, reptiles, amphibians, and fish. The net is attached to a hoop like the one on the far right. From *Restraint and Handling of Wild and Domestic Animals, 3rd Edition*, M. E. Fowler, © 2008. Reprinted with permission of John Wiley and Sons, Inc.

holes both before and after it is used to avoid the escape of an animal that has been thought to be secure.

BEHAVIOR RELATED TO RESTRAINT

An understanding of animal behavior is crucial to the successful application of restraint procedures that minimize stress (Fowler 2008; Hediger 1955, 1964, 1969; Houpt 2005). Each species or animal group has a repertoire of actions that astute observers are capable of evaluating. This author defines behavior as all aspects of an animal's total activity, especially that which may be outwardly observed (Fowler 2008). Behavior may be controlled by genetics, in which case the action is innate; but behavior may also be learned or modified by an individual's experiences.

All wild animals possess a special social distance called flight distance ("fight-or-flight" distance), which is that distance at which, when an enemy or predator approaches the animal, it will either flee or attack. When attempting to restrain an animal, knowledge of its flight distance is vital. The flight distance may be reduced by taming, training, or habituation to people, but the instinct remains present.

An example of the use of flight distance follows. As a keeper approaches a capuchin monkey (*Cebus apella*) in a group cage to capture it with a net, the monkey may begin charging around the cage. If, however, the keeper backs up slightly and extends a long-handled net toward the monkey, it may not consider the net as a violation of its flight distance. This may allow capture without causing as much stress.

The key to learning about behavior is observing normal behavior. An understanding of a species' behavior may enable one to select an appropriate restraint procedure or detect incipient illness or distress, and most of all, it may assist in maintaining the animal's welfare. Behaviors important to understand for restraint purposes include the methods animals use for offense and defense, communication (vocalization, body language, facial expression), hierarchical status, locomotion, recumbency, and the sequence of getting up and down. More detailed information is provided in the references (Fowler 2008; Hediger 1955, 1964, 1969; Houpt 2005).

OFFENSE AND DEFENSE

A person restraining any species of animal should know how it defends itself or may respond to a perceived threat. A llama may regurgitate its stomach contents, while an elephant may strike a person with its trunk or step or kneel on them. In many mammal species, ear and tail position is a clear indicator of impending action.

The method an animal uses to lie down and arise may provide vital information for providing it sufficient space and opportunity to rise after a restraint procedure, or for caring for it during hospitalization. The animal's position of normal recumbency should be known in order to assess abnormal positions associated with various illnesses or injuries. Llamas and camels have a pronounced callosity on the sternum that allows them to remain recumbent for days or even weeks without impairing their health, while most animals would be harmed by prolonged recumbency. It is important to understand that the abnormal behaviors of animals may

> **Good Practice Tip:** Select one member of the restraint team to be responsible for monitoring the animal.

be an extension of their normal behaviors, so the degree to which a behavior is expressed may be vital knowledge.

STRESS

All restraint procedures impose some degree of stress on an animal (Fowler 2008a; Moberg 1984). It is critical that those who plan and implement a procedure are aware of the possible consequence of their actions, and take steps to minimize stress. This section is provided to help zookeepers understand the factors that act on an animal when it is restrained, so that restraint procedures do not impose distress. Keepers should understand the basic physiology of stress and distress. It is recommended that chapter 14 be reviewed when studying restraint.

Every animal is subject to stress whether domestic or wild (free-ranging or in captivity). The biological responses brought about by stress are adaptive, and are directed at coping with environmental change. Stresses are accumulative. Intense or prolonged stimulation may induce detrimental responses (distress; Moberg 1987).

All animals have the capacity to cope, to varying degrees, with undesirable actions on the body or adverse psychological stimulation. This is called homeostatic accommodation (see figure 14.1). This is crucial for coping with actions caused by restraint. Restraint procedures should be carried out as quickly as possible to avoid prolonged stress action, which may become distress. Necessary but repeated restraint episodes are particularly stressful. Potential negative influences on a restrained wild animal include strange sights, sounds, and odors, lack of oxygen, being touched, position changes, stretching muscles and joints, medication, pain, heat, cold, pressure, and trauma. In addition to the physical stressors, psychological stressors such as anxiety, fright, terror, anger, rage, and frustration may have a profound effect on a restrained animal. The behavior of the restraint team may be carried over into the animal being restrained. Keepers should remain calm and avoid loud conversation.

TRAINING FOR RESTRAINT PROCEDURES

Keepers may train and establish a strong bond with these animals, but casual visitors, such as administrative personnel and veterinarians, may not be given the same acceptance. Training animals to voluntarily cooperate in restraint and veterinary procedures is an important cornerstone of a modern zoo's animal care program, and it provides numerous benefits to the animal and those responsible for caring for them (Ramirez 1999; Reichard 2007). The keeper may not be the person conducting the training, but surely he or she must be aware of the degree of training that animals within the institution have and how this may impact the degree of restraint necessary to carry out a procedure successfully.

Primates, elephants, and marine mammals have been

trained to allow repeated collection of blood samples. Such training is not restricted to mammals. Komodo dragons (*Varanus komodoensis*) and even crocodilians (*Crocodylus* spp.) have been trained to tolerate minimally invasive procedures. Female nonhuman primates may be trained to allow a keeper to handle their newborn infants. Training animals to accept medications is extremely important to reduce stress and provide safety for the keeper. Training protocols should be standardized within a zoo. One keeper should carry out the initial training, with others brought into the process as required. Training is labor-intensive and it requires support from the zoo administration to enable dedicated and trained keepers to spend the time and effort necessary for training.

MEDICAL PROBLEMS

A zookeeper will often be assigned to monitor a restrained animal and be the first to communicate with other members of the medical team about potential or real medical problems. It is important to learn to recognize the signs of the following conditions and be prepared to assist in correcting them (Fowler 2008b).

HYPERTHERMIA

Refer to chapter 14 for details on hyperthermia (overheating, sunstroke, heat stress).

HEMORRHAGE

Predisposing factors are lacerations, contusions, and fractures. Signs include visible blood on the surface. Spurting blood indicates arterial bleeding; seeping or flowing blood indicates venous or capillary bleeding. Internal bleeding may not be visible. With sufficient blood loss, the mucous membranes become pale and the heart beats faster in order to compensate for the lowered oxygen capacity of the blood.

HYPOXEMIA AND ANOXIA

Predisposing factors for hypoxemia and anoxia (diminished oxygen and lack of oxygen) include abnormal positions that restrict air flow to the lungs, obstruction of the nostrils, excessive pressure on the chest, or a tight rope around the neck. Birds breathe by sternal movement, so pressure on the sternum should be avoided. Signs include labored breathing, bluish mucous membranes, and accelerated pulse. When lack of oxygen becomes critical, an animal will struggle, which may be confused with struggling against the restraint. Ultimately the hypoxic animal becomes unconscious.

TRAUMA

Trauma may occur in any physical restraint procedure. Constant attention is necessary to prevent lacerations, contusions, abrasions, and fractures. The clinical signs are variable, depending on the location and severity of the trauma, but they include lacerations, bruises, hemorrhage, and abnormal positioning (angulations) of the limbs.

CARDIAC IRREGULARITIES

Predisposing factors of cardiac irregularities include the alarm reaction, bradycardia, and hyperthermia. Signs include depressed or rapid heart rate, irregular heart rhythm, and a weak pulse. Bradycardia (slowing of the heartbeat, fainting reflex, diving reflex) is the mechanism used by marine mammals when diving. It also happens if a person faints when confronted with a startling or fearful situation. The alarm reaction has been explained in chapter 14 as a stimulation of the sympathetic component of the autonomic nervous system. Under certain circumstances, instead of sympathetic stimulation with an increase in heart rate, a parasympathetic stimulus is initiated which slows the heart rate. Taken to the extreme, the heart stops beating and oxygen deprivation of the brain (hypoxemia) causes unconsciousness. This is seen in the North American opossum (*Didelphis virginiana*), which enters a torpor state (plays dead) when mauled by a predator or restrained by a keeper. This may also be what happens when a wildebeest (*Connochaetes* spp.) is finally overpowered by a lion (*Panthera leo*) and gives up fighting. Predisposing factors include severe fright and restricted movement with inability to cope with the situation or being subjected to an extremely painful situation. Clinical signs include struggling as hypoxemia worsens, unconsciousness, and death. Chemical restraint may minimize stress, but the choice of selecting chemical over physical restraint rests with the veterinarian in charge, though the keeper may be asked to express an opinion (Fowler 2008b).

HANDLING SPECIFIC ANIMAL GROUPS
HOOFSTOCK

This group includes a diverse collection of species, ranging in size from a small antelope weighing less than 50 kg (100 lbs.) to the Asian rhinoceros (*Rhinoceros unicornis*), weighing up to 4,000 kg (8,800 lbs.). The larger hoofed animals include rhinoceroses, horses, and tapirs in the order Perissodactyla, and hippopotamuses and giraffe in the order Artiodactyla (Certartiodactyla; Kock, Meltzer, and Burroughs 2006).

All hoofed animals act defensively and offensively by kicking, striking, or charging. A few species (camels, pigs, horses, tapirs) are capable of biting. Giraffe (*Giraffa camelopardalis*) swing the head and use it as a battering ram with considerable force, killing large carnivores in the wild and breaking oak beams in captivity. Caution is advisable when working around the limbs of hoofed animals. For example, even under chemical sedation (with the drug xylazine), zebras, eland, and giraffe may kick reflexively when their limbs are touched. It is important to know the radius of these animals' limbs to avoid being kicked. One should not underestimate the strength and agility of hoofed animals. Even a large antelope such as the eland (*Taurotragus oryx*) may jump over a 2.5 m (8 ft.) fence or wall.

Sedation or immobilization may be necessary for involved procedures, but many of the larger species may be handled for some minor procedures within restraint devices (chutes, squeezes, alleyways). Portable squeeze chutes have been used

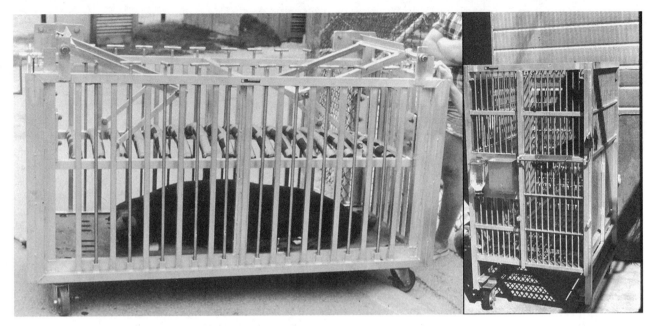

Figure 15.4. Squeeze cages for a pinniped, on the left, and small primates or carnivores, on the right. From *Restraint and Handling of Wild and Domestic Animals, 3rd Edition*, M. E. Fowler, © 2008. Reprinted with permission of John Wiley and Sons, Inc.

for decades when working with zoo animals. Several companies in the United States specialize in making cages designed for specific animals, such as sea lions (figure 15.4). Construction of new facilities for elephants and giraffe usually have built-in squeeze facilities that include scales. Such facilities should be incorporated into the alleyways normally used by the animals for moving to an outside yard or coming in. The animals may have to be trained to enter a squeeze chute first, without being confined. Commercial livestock chutes may be modified to prevent injury to horns, antlers, and limbs. A special chute for handling bison (*Bison bison*), with a front-end crash gate (a moveable metal frame placed 0.3 m [1 ft.] in front of the chute), was constructed to prevent the bison from escaping before their heads could be contained. In this chute the crash gate is moved aside to allow work on the head, or to release the animal (Haigh 1999). Another livestock chute was modified to allow passage of Watusi cattle (*Bos taurus*) with huge spreading horns. It, too, had a crash gate to prevent premature escape. Some facilities may have excellent handling facilities that incorporate holding pens, curved alleyways, sorting gates, and squeeze chutes. The squeeze chutes may be quite sophisticated, with hydraulically operated front and rear gates, lateral padded squeeze panels, and floors that drop from beneath an animal once it is secured in the squeeze. These chutes are usually custom-designed and unique to a given facility. The design possibilities are endless. Many chutes and squeeze cages are illustrated in Fowler 2008.

Special crates are used to transport hoofed animals. Crates must be designed and made for specific taxonomic groups and sizes of animals; they must provide adequate ventilation, yet not have openings through which a foot or horn may be thrust. A non-slip floor is a must. If a crate is meant for an animal to face in a single direction, it should be loaded into a vehicle so that the animal's hind end faces in the direction

of travel. This will avoid head and neck injuries in the event of a sudden stop.

Many facilities now train larger species to allow manipulation of their limbs for examinations and radiographs while they are contained in a chute or squeeze. This is particularly helpful to keepers when handling giraffe as well as deer and elk (cervids). Opaque plastic sheeting or burlap may be used to direct individuals or groups of deer, antelope, and sheep into smaller enclosures, trucks, or handling facilities. Nets and snares are seldom used on hoofstock except when free-ranging species are captured with drop nets or collapsible nets placed in their path while they are being driven by a helicopter. Sheep, goats, and deer may be captured by projecting a net over them from a hand-held projector or from a helicopter. Peccaries may be captured by using a cone net and then grasping the head and hind limbs.

One should be particularly attentive to horned or antlered artiodactylids during physical restraint. In addition to being used in offense and defense actions, the horns may serve as thermoregulatory organs. Antlers in velvet should not be grasped as handles, because the soft bony tissues may be easily fractured or the velvet damaged, thus resulting in hemorrhage, fly infestation, and deformation.

NONHUMAN PRIMATES

Primates may range in size from a pygmy marmoset (*Cebuella pygmaea*), weighing less than 60 g (0.13 lb.), to a gorilla (*Gorilla gorilla*) male weighing more than 275 kg (600 lbs.). All primates are capable of inflicting serious bite wounds, including amputation of a finger. A few species are adept at scratching and pinching. The great apes may grasp any part of the human anatomy within their reach and inflict contusions or pull it toward the mouth to bite it. All primates may be transported within a zoo or between zoos in crates commensurate with their size and strength.

Gloves are often used when handling small primates (10 kg [22 lb] in size), but many species are capable of biting through light leather gloves. Heavy welder's gloves with long gauntlets are necessary with some species (figure 15.1). Larger primates are capable of crushing fingers with bites that don't necessarily penetrate a glove. Gloves should be removed as soon as possible after capture so that one can grasp the arms with more dexterity, without using excessive force.

Nets of varying sizes and meshes are frequently used to capture primates weighing less than 15 kg (33 lbs.). Snares (e.g., catch poles, dog nooses) are not useful for primates because they can grasp the snares with their hands. Free-ranging primates can be enticed into cages with food. Small primates may be removed from nets and squeeze cages, and can be manually held by keepers who grasp both arms behind their back. The larger species must often be chemically restrained.

Blood may be collected from accessible vessels (cephalic, jugular, saphenous, and femoral). The femoral vein is accessible in all species and has the same landmarks for location (Fowler 2008). Gastric intubation is possible in manually restrained small primates by holding the mouth open with a dowel speculum (a round wooden rod with a hole bored across at the center).

CARNIVORES

Carnivores include cats, wolves, foxes, bears, raccoons, weasels, mongoose, hyenas, and pandas. They range in size from a skunk (*Mephitis mephitis*) weighing 0.75 kg (less than 2 lbs.) to a polar bear (*Ursus maritimus*) at 800 kg (1,760 lbs.). Biting and clawing are the defensive and offensive weapons. Blood may be collected by methods similar to those used for domestic dogs and cats. Again, the femoral vein is always a source for blood collection.

Small carnivores may be captured with nets and snares and then manually restrained, perhaps with gloves. Some species may be grasped by the skin over the neck and withers. The skin over the neck may be so flexible that a carnivore—for example, an otter (*Lutra* spp.) —might still turn its head around and bite.

Squeeze cages are used to physically restrain larger carnivores, but chemical restraint has become the standard protocol for handling most large carnivores during significant procedures. Caution is necessary when squeezing any animal, because the power (mechanical advantage) of the squeeze may crush the animal. Cats also may be able to reach out through the bars of a squeeze cage to claw a keeper.

MARSUPIALS AND MONOTREMES

Monotremes (echidnas [*Tachyglossus* spp.] and platypuses [*Ornithorhynchus anatinus*]) are native only to Australia and New Guinea and are rarely exhibited in other countries. Marsupials (kangaroos, wallabies, possums, etc.) are found primarily in Australia and nearby countries but also in the Americas, and many are exhibited worldwide. The North American opossum (*Didelphis virginiana*) is native to the southeastern United States but has become naturalized in most states and southern Canada. Marsupials range in size from small marsupial mice (*Antechinus* sp.) weighing a few grams to large red kangaroos (*Macropus rufus*) weighing up to 70 kg (154 lbs.). Omnivorous and carnivorous marsupials defend themselves by biting. The herbivorous kangaroos and wallabies may punch with their powerful hind limbs. The koala may bite or scratch.

Marsupials are transported in crates. One can manually restrain a kangaroo or wallaby by grasping its tail or, in the case of smaller species, by lifting it off the ground. One should keep the animal's legs directed away from one's body to avoid being kicked. In some cases gloves may be used. Nets and snares may be used for initial capture. Koalas (*Phascolarctos cinereus*) may be grasped by the arms while they are perching.

SMALL MAMMALS

Small mammals are included in nine different mammalian orders. They vary in size from shrews weighing a few grams to capybaras (*Hydrochaeris hydrochaeris*) weighing 50 kg (110 lbs.). Their offensive and defensive methods are variable depending on the order, but include dermal spines (in hedgehogs and porcupines), clawing (in anteaters), dermal plates that allow animals to curl up into a ball (in armadillos and pangolins) and biting (in bats and other carnivorous species).

When handling bats, one must always be aware that the rabies virus has been isolated from at least 23 species of insectivorous bats in the United States, as well as from the vampire bat (*Desmodus rotundus*) in South America. It is recommended that keepers working with bats be vaccinated against rabies. Protection against biting is paramount.

Some species of rodents may be captured and handled manually. Others may be handled using gloved hands. Initial capture may be with nets, snares, or specially designed squeeze cages. Giant anteaters must be handled in special squeeze cages or sedated to prevent serious injury from clawing. North American porcupines (*Erethizon dorsatum*) may be handled manually, since they do not project their quills, but must slap them into a victim with the tail. By restricting the porcupine's forward motion and approaching its tail hairs from underneath, one can prevent the movement of the tail and discharge of the quills. African crested porcupines (*Hystrix cristata*) have a different behavior. They raise the massive quills and rapidly propel themselves backwards to impale a victim with the quills. They must be handled in a special metal or wooden box, or sedated. Sedation is commonly used in handling many small mammals, usually after initial manual restraint.

BIRDS

More than 8,600 species of birds are grouped in 27 orders worldwide. They vary in size from the Anna's hummingbird (*Calypte anna*), weighing less than 5 g, to the ostrich (*Struthio camelus*), weighing up to 160 kg (350 lbs). Avian respiration is different from the in-and-out flow of air characteristic of mammals. The lungs are intimately associated with the chest wall, and are assisted in respiration by a series of air sacs. Furthermore, birds lack a complete diaphragm. Inspiration and expiration is accomplished by a bellows-like

Figure 15.5. Proper method for holding a small bird: without encircling the sternum, so as not to inhibit its breathing. From *Restraint and Handling of Wild and Domestic Animals, 3rd Edition*, M. E. Fowler, © 2008. Reprinted with permission of John Wiley and Sons, Inc.

action brought about by movement of the sternum. Bird ribs are jointed in the center so that movement may occur. Any restriction of the free movement of the sternum inhibits breathing and must be avoided during any restraint procedure. Do not completely encircle the chest when grasping a small bird. The proper method of holding a small bird is illustrated in figure 15.5. The location of birds' nostrils is highly variable, and keepers must avoid covering the nostrils when restraining birds. Offense and defense is by pecking with the beak (in most species), scratching with toenails, impaling with talons (in raptors), and kicking (in ratites). A net is a common tool used for initial capture for most species of small to medium-sized birds. Bird nets should have fine mesh to prevent feathers from poking through and being damaged. After the initial capture, the bird's head and body should be grasped through the net; this is followed by careful insertion of the other hand beneath the net to grasp the bird again. The legs and feet must be controlled if the bird has talons or sharp toenails. The net is then removed.

Parrots and macaws housed in small cages may be approached from behind with an outstretched towel to obscure the hand. The heavy, strong beak should be avoided by initially grasping the back of the head and neck quickly.

Penguins. Small penguins may be captured with a net. The wings, modified as flippers for swimming, should be controlled so that one is not slapped in the face. Penguins have sharp serrated bills to aid in capturing fish. Controlling the head and legs is important. Large penguins may be captured with an overturned plastic garbage container with the bottom cut out.

Ratites. The large ratites (ostriches and cassowaries) require special alleyways and squeeze panels to avoid the powerful kick that may cause significant injury. Keepers should avoid entering ostrich enclosures during the breeding season, as otherwise docile males may become aggressive. Cassowaries can be particularly aggressive, and they have a sharp medial claw they use to slash an enemy. Ratites are now being farmed for meat, oil, and feathers; this has spawned many methods of handling these potentially dangerous species (Fowler 2008, 386–87). Adult ostriches may be safer to handle if they can be hooded. Hoods may be constructed of any material the bird cannot see through. Adult ratites should always be approached from the rear. Species other than ostriches may be herded into a corner, then grasped by the wings while straddling the rump.

Raptors. Raptors (hawks, falcons, owls, eagles) are capable of inflicting serious tearing bites and impaling fingers and arms with their talons. Nets are usually not employed for capturing raptors, because the talons may become entangled in the mesh. Once a raptor grasps something in its talons, it may not release its hold until the hock (tarsus) is straightened out. This is the normal perching mechanism that allows a bird to grasp a branch and hold tightly to it when the hock is flexed (perching position). A special tendon pulley mechanism flexes the digits at the same time that the hock is flexed. Should a keeper or an assistant be grasped by a large eagle, the eagle's leg must be straightened out to effect a rescue.

Gloves are usually used when grasping a raptor, approaching from behind the bird. A towel or small blanket may be tossed over the bird to disorient it and allow it to be grasped. Large owls may lie on their back with outstretched talons, defying capture. A free glove or a towel may be dangled above the talons. While the owl grasps the object, a hand may reach in and grasp both legs. The owl may be quickly lifted and the head grasped from behind.

Medium-sized raptors may be effectively restrained after initial capture by placing them in a stockinette or nylon hose. The stockinette may serve as a hood to prevent visual stimulation and control the wings. Birds restrained for a long period in a nylon hose may overheat, as the nylon retains heat better than does a cotton stockinette.

Large eagles require two or more persons for safe restraint. The legs and talons must be controlled. The author was examining a raptor being held by a student who lost her grasp on a leg, which was being directed at my hand. The hand was quickly retracted, but unfortunately into the face of another student who was observing the procedure over my shoulder. This resulted in a bloody nose.

Long-legged, long-billed birds. The primary defense of long-beaked birds (herons, storks, cranes, hornbills, and flamingos) is pecking with their long, sharp-pointed bills directed at the face and eyes of handlers. Flamingos have large recurved beaks, but the margins of the beak are serrated and may lacerate the skin. Long-legged birds may be herded into a holding enclosure using outstretched opaque plastic sheeting. If pecking is a probability, one can use safety goggles or visors or even a fencer's face mask to protect one's eyes. The neck should be grasped as close to the head as possible, then the bird is scooped up by the body and the upper legs grasped with the other hand. The legs may be fragile, so the lower legs should never be grasped. One should direct them away from one's body to avoid scratching or entangling them in clothing. Long-legged birds may also be restrained while standing by grasping the neck with one hand and bringing the wings together over the back with the other hand. One should never peek into a crate or enclosure containing one of these birds, lest the bird focus on the eye to peck it and cause serious injury. Once these birds are captured, a cork, rubber stopper, or tape may be applied to the tip of the beak to avoid accidental pecking of personnel.

Hummingbirds. It is extremely difficult to capture a flying hummingbird with a net, because it can reverse its direction instantaneously. Mist nets may be used to capture hummingbirds and other small swift-flying birds. Once captured, these tiny birds may be held gently in a cupped hand or placed in an appropriately sized stockinette. One must not restrict its sternal movement. None are aggressive or capable of injuring a keeper.

It is recommended that the heads of all birds be held during manual restraint. The sternum should always be free to move for respiration. Darkening the room during initial capture may aid in the capture and reduce the bird's stress. Blood may be collected from the right jugular vein, the brachial vein, or medial tarsal vein. The clipping of a toenail to obtain blood is discouraged.

This has been a basic overview of bird restraint; more specific details can be found in the cited references (Anderson and Edney 1991; Fowler 2008; Sheldon and Sonsthagen 2007; Sonsthagen 1991).

REPTILES

Reptiles comprise several different groups with distinct morphology and physiology, requiring different methods of restraint: crocodilians (alligators, crocodiles), chelonians (turtles, tortoises), and squamates (lizards and snakes). Some species are aquatic, some are terrestrial, and others are amphibious. Most are poikilothermic (unable to regulate the body temperature internally, also known as ectothermic), so restraint practices should be conducted in an optimal temperature environment.

Crocodilians. Crocodilians (alligators, crocodiles, gavials, caimans) are all carnivorous and capable of inflicting serious bites even as new hatchlings. Small crocodilians less than 0.6 m (2 ft.) in length may be grasped manually from above. Animals up to 2 m (6 ft.) in length may be captured with a snare. The tail should be quickly grasped and held to the snare to prevent the spinning that crocodilians are prone to do when grasped around the neck. If such spinning does occur, the person holding the snare must spin it with the animal to avoid tightening the snare and strangling the animal. Commercial snares usually have a built-in swivel to avoid this problem. Larger specimens may be handled manually by experienced personnel or in a squeeze cage that is lowered into the water. The animal is enticed into the squeeze cage

with food. Blood may be collected from the supravertebral plexus or from the tail vein.

Chelonians. Chelonians (turtles, tortoises, terrapins) don't have teeth to bite with, but some species have hard dental plates and strong jaws that can administer crushing injuries. Snapping turtles (*Chelydra serpentina* and *Macroclemys temminckii*) are large carnivores (up to 90 kg [198 lbs.]) that may be particularly aggressive when handled. Snapping turtles also have a strong hooked upper beak that they use to catch fish. A bite from one of these turtles may inflict a disabling injury. Small snapping turtles may be approached from behind and lifted by the tail. A larger individual may be removed from a water tank by approaching it from behind, moving a hand over its upper shell (carapace), and then grasping the shell directly above the head with one hand and at the rear with the other. Large species may require two persons, one on either side, to lift and move the animal together while avoiding its flailing legs and head.

Tortoises are usually handled manually. Generally they are not prone to bite. Small tortoises may be grasped by placing the fingers under the lower shell (plastron) between the front and hind limbs and placing the thumb on top. Medium-sized tortoises may require a hand on each side. Giant tortoises are usually not picked up unless they are being moved. Placing a tortoise on its back will render it immobile for a few minutes. Rapid movement should be avoided when turning it over to prevent twisting (torsions) of the intestines.

Adults of large species of sea turtles may weigh as much as 800 kg (1,900 lbs.). They are generally herbivorous and not aggressive. Small individuals may be netted or grasped manually from a tank. The tank must be drained for access to large individuals. The flippers are strong and may strike a keeper. If a sea turtle is tipped upside down, it tends to relax and quietly allow examination and even minor surgery. The turtle may be made more comfortable by placing it on top of the inflatable inner tube of a tire or by supporting it with rolled towels. Blood may be collected from the jugular vein or the tail vein.

Lizards. Lizards vary in size from a few grams to the Komodo dragon (*Varanus komodoensis*) that weighs up to 135 kg (300 lbs.). Small individuals may be grasped with a bare hand, or with a light glove if the individual is prone to bite. The dermal scales of some species are rough and require the use of a gloved hand. A lizard should not be caught by the tail, as some species are able to discard the tail as a device to distract predators. Some moderate to large species (iguanas, monitors) may also have formidable claws and teeth that require the use of heavy, gauntleted gloves. Care should also be taken with the long tail in larger species, as they may lash out at a perceived threat and cause severe lacerations. Common iguanas (*Iguana iguana*) should be grasped at the back of the head while one quickly grasps the tail to avoid the lashing that may ensue and pins the legs against the body to prevent scratching. Large monitor lizards may be initially captured with a snare, but with caution to avoid injury to the neck.

The Gila monster (*Heloderma suspectum*) and Mexican beaded lizard (*Heloderma horridum*) are venomous but

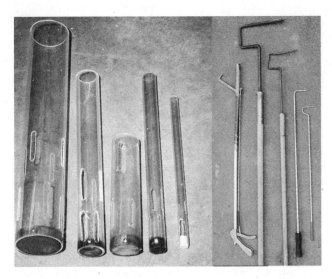

Figure 15.6. Snake handling tools. Left to right: snake tubes, a Pillstrom tong, and snake hooks. From *Restraint and Handling of Wild and Domestic Animals, 3rd Edition*, M. E. Fowler, © 2008. Reprinted with permission of John Wiley and Sons, Inc.

rather slow-moving (phlegmatic), and may be restrained by grasping at the base of the head. Some Komodo dragons (*Varanus komodoensis*) have been trained to tolerate some manipulative procedures. Untrained Komodo dragons are dangerous and should be handled in a squeeze cage, in a specially designed crate, or under sedation.

Snakes. All snakes are capable of biting but some are more aggressive than others. Many will not bite unless agitated. One should control the head of a large snake, such as a python, particularly when manipulating it for examination. It is important to support a snake's body while holding it so that it feels secure. Snake hooks are the principal tools for working with these reptiles (figure 15.6). They come in various sizes and shapes and are used singly or in pairs to move a snake from one cage to another or to conduct an examination. Keepers should be specially trained before working with snake hooks, and snake tongs (Pillstrom tongs) should be used only by expert personnel to avoid crush injuries to the animal.

Vipers (Viperidae), cobras (Elapidae), and some colubrids (Colubridae) are venomous and should be handled only by experienced keepers. Two qualified keepers should be present when venomous snakes are handled. Venomous snakes may be directed into an appropriately sized plastic tube with a diameter approximately equal to the thickest area of the snake's body, to prevent the snake from turning in the tube and crawling out (figure 15.6). The snake may feel that the tube is a place of refuge. A Pillstrom tong is used to hold the tube in front of the snake, and a hook used to direct the snake. When the snake has progressed into the tube to approximately one-third of its body length, the tube and the snake are grasped with one hand at the spot where the snake enters the tube. Do not hold the tube with one hand and the snake with the other, because then the snake may back out of the tube. This procedure should be practiced with nonvenom-

ous snakes until the keeper is comfortable before using it on a venomous species. The tube method should not be attempted with extremely agile or swift snakes such as the black mamba (*Dendroaspis polylepis*). Special squeeze cages have been designed that attach to an exhibit enclosure to allow for examination, medication, or sedation (Fowler 2008).

AMPHIBIANS

The class Amphibia is comprised of over 5,000 species including frogs, toads, salamanders, and caecilians. Frogs are currently experiencing a worldwide catastrophic mortality associated with a fungal infection and other environmental pressures. Most amphibians are inoffensive, but larger species may bite. Some species of salamanders (e.g., giant salamanders and amphiumas) are rather pugnacious. No amphibian has a venomous bite.

The skin of amphibians protects the body from dehydration, as it does in mammals, but the aquatic environment is particularly challenging. Skin glands secrete substances (i.e., mucus) that coat and protect the skin. It is important during restraint to avoid harsh handling that may remove that protective coating. The skin gland secretions of toads are particularly unpalatable and even toxic to would-be predators. For this reason it is unlikely that a puppy will mouth a toad a second time.

The skin of certain South American frogs (*Dendrobates* spp.) contains lethal substances that were projected by natives using blowpipes. Several species of large toads— the Colorado river toad (*Bufo alvarius*), cane toad (*Bufo marinus*), and Columbian giant toad (*Bufo blombergi*)—have a parotid gland behind the eye that produces a poisonous substance which may be lethal when ingested by a dog or other carnivore. These species also may squeeze the gland to project the secretion towards a keeper's eye. The secretion causes inflammation and temporary blindness unless it is washed out quickly.

Physical restraint. Amphibians should always be handled with moistened hands or moistened light gloves. They can be initially captured manually or with a net. When fully aquatic species are grasped in the water, they may feel slippery. Avoid the natural tendency to squeeze them too tightly. A moistened 4 × 4" gauze sponge may allow more secure grasping. It is also possible to use an unpowdered rubber glove. Salamanders and newts are grasped over the back and shoulders. Toads and frogs may be grasped by the body or by the rear limbs close to the body; this prevents them from jumping.

FISH

Over 25,000 species of fish have been identified worldwide. Aquariums are popular in homes and educational venues, with fish typically being moved between tanks by nets. Fish have been harvested from the wild and farmed as food for millennia. Commercial fish farms are sophisticated operations, in which the basic handling is carried out with well-designed raceways and intricate netting systems. Sport fish are raised in government and private commercial hatcheries, and are placed in streams by specialized tank trucks. Fixed-wing aircraft and helicopters are used to place fish in remote lakes.

Ornamental freshwater and tropical fish are the most popular pets in the United States. Ornamental fish in a pet shop tank are caught in a net and transferred to a plastic bag containing water from the tank for transport to the home aquarium. Several species of such fish have venomous spines projecting from the fins, such as the lionfish (*Pterois volitans*). The author consulted with a physician who cared for a pet shop owner who had used a dip net to transfer an Indo-Pacific catfish (*Plotosus lineatus*) from a stock tank to a container for a client. The fish flopped out of the net and the shopkeeper instinctively caught it in his hand. The immediate sharp pain caused him to fall to the floor (Fowler 1992). Even the non-venomous spines found in many species can be quite painful.

Physical restraint. Fish do not lend themselves to physical restraint. Nets are the primary tool for handling fish, although clear acrylic containers may also work for smaller wary species as they are almost invisible underwater. Prolonged restraint for examination or surgery requires anesthesia and bathing of the gills with water. Care must be taken with larger marine species such as sharks, skates, and rays, as their skin is quite abrasive, requiring the use of neoprene gloves for manual restraint. Large fish (not just sharks) may also have formidable teeth, and they can easily bite when in nets.

INVERTEBRATES

The animal kingdom is represented by approximately 50,000 species of vertebrates, but there are millions of species of invertebrates (Frye 1992; Gunkel and Lewbart 2007; Lewbart 2006). Zoos and aquariums are catering to patrons interested in insects, arachnids, and aquatic invertebrates by constructing special facilities for housing and exhibiting many of these species. Butterfly houses are common in the United States and worldwide. One zoo in Japan constructed a butterfly house with a roof shaped like a butterfly with outstretched wings.

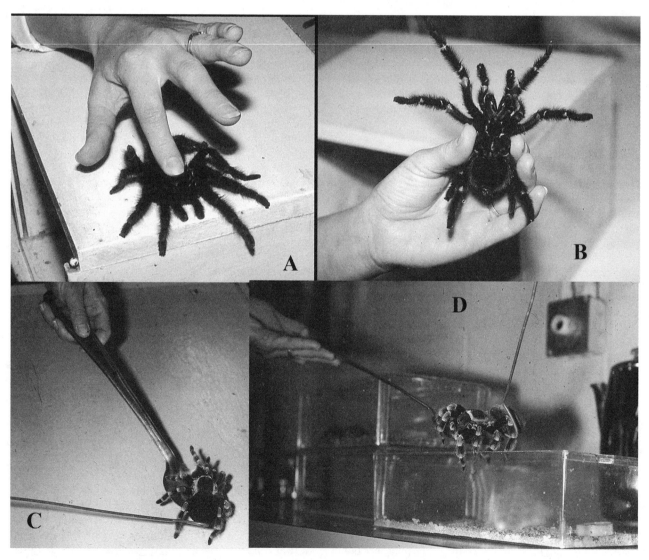

Figure 15.7. Physical restraint of a spider: (A) Pressing with a forefinger, (B) grasping with the thumb and middle finger, (C) scooping with a spoon, and (D) lifting with a spoon. Courtesy of Murray Fowler.

Marine invertebrates are also often exhibited in aquariums and zoos, many of which provide opportunities for patrons to touch some species.

Arachnids and insects. The life spans of arachnids (spiders) and insects are generally shorter than those of most mammals. Invertebrates are generally fragile and require gentle handling when they must be moved to different enclosures or restrained for examination. Generally only the larger species of spiders and scorpions are exhibited. All spiders are predators and possess a pair of biting mouth parts called chelicerae, with associated venom glands ducted to the hollow, cornified fangs at the tip. While the venom of most spiders is effective only against their normal prey species, a few are highly venomous to people and other animals, and should be handled with extreme caution. All species can bite, but a few docile species of tarantulas can be handled by trained keepers.

Spiders undergo a periodic molt (ecdysis) of the exoskeleton. When seen in a terrarium, the molted exoskeleton looks at first like a dead spider. The live spider is hiding somewhere in the terrarium to allow its new exoskeleton to harden and become more protective. A spider should not be handled during or for several hours after molting.

Before a large spider is handled, its aggressiveness should be assessed by gently stroking its foreleg with a small artist's brush. If one decides to proceed, one should gently press an index finger to the top of the cephalothorax and pin the spider to the surface (figure 15.7). The thumb and middle fingers are used to grasp the spider's cephalothorax. Alternatively, a large mixing spoon may be used to scoop up the spider. A small snake hook or another spoon can be used to guide the spider into the spoon and then to cover it to prevent it from jumping out (figure 15.7). If a large spider accidentally falls to the floor, its abdomen may burst open. One can also restrain a spider by placing a transparent plastic container over it, then slipping a sturdy piece of flat plastic sheeting beneath the container and the spider, being careful not to traumatize the spider's legs. Some species of spiders have urticarial hairs on the abdomen, which they can rub with their hind limbs and fling at an enemy. Keepers handling these species should wear protective goggles to prevent hairs from entering their eyes.

Physical restraint is limited to moving a spider to a new enclosure or to a cursory examination. More invasive examinations or treatments should be conducted after the spider has been immobilized with an inhalant anesthetic agent. Various methods may be used to administer the inhalant agent, such as placing a cotton ball soaked with the agent into a closed container with the spider. A safer method for both the spider and the restrainer is to use a calibrated vaporizer for the specific inhalant and to flow the mixture through a clear plastic chamber containing the spider.

Scorpions are eight-legged arachnids with an elongated abdomen ending in a bulbous telson containing paired venom glands and a single, recurved, hollow stinger. A small scorpion can be directed into a small glass or plastic vial (test tube) for close inspection. Some keepers pick up the larger species by grasping the tail close to the telson with a gloved hand. Preferably, a pair of long thumb forceps is used to gently grasp the tail. Scorpions also molt.

Insects. Some insects (ants and large beetles) may bite, and a few species (bees, wasps, and ants) possess stingers on the caudal abdomen that are connected to venom glands. Many people are allergic to bee venoms and several hundred people die each year from bee stings in the United States. Most restraint of insects is related to husbandry of the various species. Physical restraint of ants, bees, butterflies, praying mantises, beetles, and cockroaches is limited. Species capable of jumping or flight may be captured with a fine-meshed net. Crawlers are usually grasped gently with a thumb and forefinger, or carefully with forceps. Bees may be rendered less aggressive with the use of artificial smoke. Small insects may be scooped up with a spoon and transferred to a container with a cap. Ventilation ports must be provided of appropriate size to prevent escape. Some caterpillars (larvae of flies, moths, or butterflies) have urticarial hairs that may irritate human skin.

Aquatic invertebrates. Numerous phyla, dozens of orders, families, and genera, and thousands of species of invertebrates inhabit marine and freshwater environments. At least five phyla contain venomous species, including jellyfish, shellfish (conesnails, octopuses), starfish, and urchins. Each has its unique method of administering the venom for capture of prey species. Keepers must know the biology of the species they care for in order to work safely around the animals. In addition to the venomous species, some marine invertebrates possess appendages capable of inflicting painful pinches (as in crabs and lobsters) or grabbing human body parts (as in octopus). Safe species of starfish are popular in tide pool exhibits, and are often picked up and handled.

Some species (jellyfish, octopus) are transferred by scooping them into small perforated containers held under the water. Nets may be used on the more robust specimens.

REFERENCES

Anderson, R. S., and A. T. B. Edney, eds. 1991. *Practical Animal Handling*, 1st ed. Oxford: Pergamon Press.

Anonymous. 2005. *Animal Welfare Act and Animal Welfare Regulations*. Washington: US Department of Agriculture, Animal and Plant Health Inspection Service.

Fowler, M. E. 1992. *Veterinary Zootoxicology*. Boca Raton, FL: CRC Press.

———. 2008. *Restraint and Handling of Wild and Domestic Animals*, 3rd ed. Ames, IA: Wiley Blackwell.

Frye, F. L. 1992. *Captive Invertebrates*. Malabar, FL:, Krieger.

Gunkel, C. and G. A. Lewbart. 2007. "Invertebrates." In G. West, D. Heard, and N. Caulkett, eds., *Zoo and Wildlife Immobilization and Anesthesia*,147–58. Ames, IA: Blackwell.

Haigh, J. C. 1999. The Use of Chutes for Ungulate Restraint. In M. E. Fowler and R. E. Miller, eds., *Zoo and Wild Animal Medicine, Current Therapy*, 5th ed., 657–62. Philadelphia: W. B. Saunders.

Hediger, H. 1955. *The Psychology and Behaviour of Animals in Zoos and Circuses*. New York: Dover.

———. 1964. *Wild Animals in Captivity*. New York: Dover.

———. 1969. *Man and Animal in the Zoo*. New York: Seymour Lawrence/Delacorte Press.

Houpt, K. A. 2005. *Domestic Animal Behavior for Veterinarians and Animal Scientists*, 4th ed. Ames, IA: Blackwell.

Kock, M. D., D. Meltzer, and R. Burroughs. 2006. *Chemical and*

Physical Restraint: A Training Field Manual. Greyton, South Africa: International Wildlife Veterinary Services.

Lewbart, G. A., ed. 2006. *Invertebrate Medicine.* Ames, IA: Blackwell.

Moberg, G. P. 1985. "Biological Response to Stress: Key to Assessment of Animals. In G. P. Moberg, ed., *Animal Stress.* Bethesda, MD: American Physiological Society.

Ramirez, K. 1999. *Animal Training: Successful Animal Management through Positive Reinforcement.* Chicago: Shedd Aquarium.

Reichard, T. A. 2007. "Training Animals for Veterinary Procedures. In M. E. Fowler and R. E. Miller, *Zoo and Wild Animal Medicine,* 6th ed., 66–67. Saint Louis: Elsevier.

Sheldon, C. C., T. Sonsthagen, and J. A. Topel. 2006. *Animal Restraint for the Veterinary Professional.* Saint Louis: Mosby/Elsevier.

Sonsthagen, T. F. 1991. *Restraint of Domestic Animals.* Goleta, CA: American Veterinary Publications.

16

Nutrition

Eduardo V. Valdes

INTRODUCTION

Meeting the nutritional needs of wildlife outside their natural environment is a complex process if they are to be maintained in a healthy condition and to reproduce. The conservation and husbandry objectives of North American zoos and aquariums require the design of sound nutrition and feeding programs to provide adequate nutritional support for all animal species while meeting their physiological and psychological needs throughout development.

This chapter will provide basic information about animal nutrition as it applies to the husbandry and care of zoo and aquarium species. After reading this chapter, the reader will understand

- factors to consider when formulating diets for zoos and aquariums
- basic concepts in animal nutrition
- the role of nutrition in preventive medicine
- basic concepts in food preparation
- organization of an ideal zoo food production area (e.g., commissary)
- importance of assessing body condition in zoo and aquarium animals
- some important nutritional diseases
- the role of the animal keeper in feeding zoo and aquarium animals
- design of practical diets.

CHALLENGES

There are many factors to consider when formulating diets for zoo animals.

1. Ideally it would be important to know what the animals are eating in their natural habitats: especially how much of each of the type of food is consumed

and its chemical composition. Zoos and aquariums will not be able to mimic the exact natural diet.
2. Knowledge of the animals' gastrointestinal tract anatomy and physiology is also important.
3. Although the nutritional requirements for domestic and companion animals are well understood, the nutrient requirements of most wild animals still remain a mystery, despite the progress made in recent decades in the area of comparative nutrition. The use of domestic or companion animals as nutritional models for their wild counterparts can be useful, but information acquired in that way should be used with caution. Nutrient requirements are not fixed; they will be affected by physiological stages (e.g., growth, reproduction, lactation, maintenance, and disease).
4. How are the animals housed and managed in their zoo enclosures? Are they exposed to browse, forage, soil, or other environmental stimulus that they might be able to consume? Are they kept alone or held in multispecies displays? Are they exposed to natural sunlight or extreme weather conditions?
5. Knowledge of the availability and chemical composition of feeds that can be used in zoo diets is important so that the composition of diets can be matched to the animals' nutrient requirements.

Animal nutrition is the science that studies the nutrients that are needed by animals, both domestic and wild. It involves determining how much of each nutrient is required in an animal's healthful diet, how the nutrients can be supplied to the animal, and how the animal's body uses the nutrients for different physiological processes (e.g., growth, lactation, egg laying, maintenance, and disease). Nutrients are defined as chemical substances that the animal's body uses as sources of energy or as part of its metabolic machinery, and which are essential for carrying on its various life processes. Metabolism can be defined as the series of chemical changes that take place in an animal, through which nutrients are used and processed and waste materials are eliminated. Feeds are sources and carriers of nutrients. Nutrients can be classified as organic and inorganic. Organic nutrients contain carbon molecules, while inorganic nutrients have none. The organic nutrients include protein, carbohydrates, fats, and the fat- and water-soluble vitamins. The inorganic nutrients include macro- and micro-minerals. Essential nutrients are nutrients that animals need but cannot synthesize at all in their bodies or cannot produce in the amount needed for body function. Therefore, essential nutrients (e.g., the amino acid taurine for felines) must be present in the feeds offered to animals.

WATER

Water is critical to life and good health. It is the most important nutrient, though it is often called the "forgotten nutrient." Animals will suffer quickly and more severely from a lack of water than from a lack of any other nutrient. Water is the universal solvent of life; it comprises about 70% of a lean adult animal's body. Fat tissue contains very little water, so as the animal matures and body fat reserves increase, water as a proportion of total body weight decreases. Some of its functions include transporting nutrients and hormones around the body via the blood, lubricating joints, and helping the body to maintain its proper temperature. Some animals need a continuous supply of drinking water; others can obtain all their water requirements from their food. However, all zoo animals should have a constant supply of safe, clean, fresh water. Animals living in arid places have adapted to using water from their metabolic processes, and can store water for long periods of time. Water quality is a problem in many places around the world, including North America, and there are concerns about contamination with harmful bacteria and abnormal mineral levels. Water should be available to animals on a free-choice basis. A restriction in water supply to the animals will result in a drop in their feed intake.

PROTEIN

Protein, an organic nutrient, is essential to the body because it provides the building blocks, called amino acids, with which the body performs vital functions such as initiating chemical reactions and rebuilding its tissues. In nature there are about 22 amino acids present in proteins. Animals can produce some of these amino acids in their bodies, using the metabolic breakdown products of other nutrients. However, about 10 of them, the essential amino acids, cannot be synthesized in sufficient quantities by animals, and therefore must be provided in the diet or, in the case of ruminants, must be synthesized by microbes in the gut. The essential amino acids have been well defined for domestic animals, and it is assumed that these will be similar for most of the wild animal species. For example, pigs, rats, chickens, and fish require arginine, histidine, leucine, lysine, isoleucine, methionine + cystine, phenylalanine + tyrosine, threonine, tryptophan, and valine. Chicks will also require glycine + serine and proline. Cats require the amino acid taurine, while dogs do not. Different feeds with the same overall protein content can differ in protein quality. The term "protein quality" refers to the concentration of essential amino acids in feed in relation to an animal's requirements. A low-quality protein will indicate an absence or low concentration of specific essential amino acids needed for proper animal metabolism. For example, the protein in corn, a common feed ingredient, is very low in lysine, one of the essential amino acids. Providing a balanced diet with levels of essential amino acids that meet the nutrient requirements of animals is essential for their proper body function.

LIPIDS

Fats and oils (lipids) are organic nutrients, concentrated sources of energy that contain 2.25 times the energy per unit weight of carbohydrates. Fats are composed mainly of two structural components: glycerol and fatty acids. Dietary fat is also the source of essential fatty acids, required by the animal, including but not limited to linoleic acid and—in some animals, such as cats—arachidonic acid. Complete removal of fat from the diet may result in poor growth and the onset of skin inflammation (dermatitis) in animals; these symptoms can be reversed by feeding small amounts of linoleic and arachidonic acids. Thus, an animal's requirement for dietary fat relates to meeting its energy needs and its requirements for essential fatty acids. Dietary fats also serve as carriers of fat-soluble vitamins, and are needed for the absorption of these vitamins. Foods such as fish, red meat, and avocados or supplements like salmon oil have a higher risk of rancidity due to their high lipid content, and should be stored at proper temperatures with minimal exposure to air.

CARBOHYDRATES

Carbohydrates are organic compounds that form the largest part of most animals' food supply. They make up 75% of the dry weight of the plant world, upon which animal life primarily depends. The carbohydrates in plants are produced by means of photosynthesis—the most important chemical reaction, as the initial source of all food is solar energy. Sugars, starches, and fiber are all carbohydrates. The basic units of carbohydrate structure are sugars such as glucose, which are called monosaccharides (simple sugars). More complex carbohydrates such as cellulose and starch are composed of large numbers of simple sugar molecules joined together. Starch is a carbohydrate found in many grains that are used in animal feeds; it is stored in the seeds or fruits of plants as energy, and can be digested down to its sugar molecules in most animals far more easily than the fibrous components of plants. Cellulose and hemicellulose are major components of plant fiber. They are important components of fibrous feeds such as roughages, and of agricultural products such as milling by-products, soy hulls, and oat hulls. High-fiber feeds are important for proper rumen function in ruminant animals. There is, in general, no definitive requirement for fiber in the diets of animals, but fiber facilitates proper functioning of the gastrointestinal tract and should be included in the diet. In animals with simple digestive systems, the physical role of fibers like cellulose and hemicellulose may often be of greater importance than the contribution they make to meeting the animal's energy demands. High-fiber diets are "bulky" diets, having lower energy concentration per unit weight. The bulkiness of higher-fiber diets in animals with simple digestive tracts encourages peristaltic movement (contraction of smooth muscles) in the small and large intestines, by which food residues are propelled through the intestinal tract.

ENERGY

Energy is a primary requirement of animals. A major function of most feeds is to serve as a source of energy for vital processes in the body. This energy represents an animal's ability to perform work and other productive processes. Energy is not a nutrient, but is a property of some nutrients, with carbohydrates and fats being the most important sources of energy. Protein, when it is fed in excess, can also become a source of energy for an animal. All forms of energy are converted into heat. Thus, energy as related to body processes is expressed as heat, measured in calories or joules. Energy density in an animal's diet is defined as the amount of energy per unit of weight (e.g., calories [or joules] per gram). Thus, a pelleted feed or grain may be classified as an energy-dense feed, unlike grass hay or a head of lettuce, which is considered a feed with low energy density. Obese or overconditioned animals can lose weight on low-density feeds. Both types of feeds are important in the practical feeding of animals, for the supply of nutrients and for body mass control.

MINERALS

Classification of the dietary minerals is given in tables 16.1 and 16.2. Minerals are classified into two categories: macro-

> **Good Practice Tip:** It is important to consider the shelf life of each feed material, as well as its labeling and the conditions under which it is stored. For example, one should be aware of vitamin loss in frozen fish; one should rotate products such as produce by their date of arrival; and one should be aware that pelleted feeds degrade in hot environments. It is important to label feeds with "best use before" dates, and to mark on feed bags when they were opened, if their contents are not used immediately.

minerals, which are required in relatively large amounts measured in percentages or grams per kilogram, and microminerals, which are present in much smaller quantities and are measured in terms of parts per million or milligrams per kilogram. Approximately 26 of these minerals have been identified as essential for an animal. In a general way, the quantity of each mineral present in the body of an animal reflects the amount needed in that animal's diet. In most animals, calcium and phosphorus are the minerals present in the highest concentrations. For example, approximate mineral composition in pigs expressed in percentages of the total macrominerals will be 48.1% calcium, 32.1% phosphorus, 6.4% potassium, 5.1% sodium, 3.5% chlorine, and 4.8% sulfur. The microminerals found in highest concentrations would be iron at 56.8%, zinc at 34.1%, and copper at 3.4%. Macrominerals generally function as constituents of skeletal structure, whereas microminerals function in many of the

TABLE 16.1. Macrominerals: Summary of individual functions, interrelationships, and toxicities

Major macrominerals	Major functions	Sources	Interrelationships, deficiencies, toxicities
Calcium (Ca)	a) Bone and teeth formation b) Nerve function c) Muscle contraction	a) Oyster shells b) Limestone c) Dicalcium phosphate d) Milk e) Bone meal	a) Vitamin D is important in absorption and bone deposition b) Excess P and Mg result in decreased Ca absorption Ca:P ratio should not be below 1:1 and over 7:1 (calves) and 1:1 to 2:1 (monogastrics)
Phosphorus (P)	a) Bone and tooth formation b) Phosphorylation c) High-energy phosphate bonds (ATP) d) Component of RNA, DNA and several enzyme systems	a) Dicalcium phosphate b) Monosodium phosphate c) Bone meal d) Most cereal grains and their by-products	a) Vitamin D in renal reabsorption and bone deposition b) Excess Ca and Mg decreases P absorption c) Ca:P ratio as specified above
Magnesium (Mg)	a) Bone and tooth formation b) Enzyme activation in energy metabolism c) Decrease of tissue irritability	a) Magnesium oxide b) Magnesium sulphate c) Magnesium carbonate d) Magnesium chloride	a) Excess Mg interferes with Ca and P metabolism b) Excess; likely not toxic c) Deficiency affects Mg absorption
Sodium (Na)	a) Major element of extracellular fluid b) Involvement in osmotic pressure and acid-base balance c) Cell permeability	a) Salt, added to diet	a) Salt toxicity that can be accentuated with water restriction
Potassium (K)	a) Major element in extracellular fluid. b) Involvement in osmotic pressure and acid-base balance c) Carbohydrate metabolism	a) Potassium chloride b) Potassium sulfate c) Forages	a) Excess K affects Mg absorption
Sulfur (S)	a) Sulfur containing amino acids b) SH groups in tissue respiration c) Found in insulin, biotin, etc.	a) Sulfate	a) Related to Cu and Mo metabolism

TABLE 16.2. Microminerals: Summary of individual minerals' functions, interrelationships, and toxicities

Major Microminerals	Major functions	Sources	Interrelationships, deficiencies, and toxicities
Copper (Cu)	a) Component of blood proteins b) Component of enzyme systems c) Synthesis of hemoglobin	a) Cupric form of sulfates b) Cupric form of carbonate c) Other cupric forms	a) Excess of Cu is toxic and accumulates in the liver b) Interrelation with Mo and S
Cobalt (Co)	a) Component of vitamin B12 b) Needed by microorganisms in rumen for synthesis of B12 c) Needed for the growth of rumen microbes	a) Cobalt salts	a) Related to B12
Chromium (Cr)	a) Decrease of serum cholesterol in animals b) Promotion of glucose uptake and metabolism	a) Magnesium oxide b) Magnesium sulphate c) Magnesium carbonate d) Magnesium chloride e) Small doses in many feeds	a) Cr-deficient diets in rats: impaired glucose tolerance b) Cr-deficiency might affect insulin sensitivity
Flourine (F)	a) Role in metabolism and fortification of bone and teeth	a) No supplementation needed in animal diets	a) Ca and Al salts protect against toxicity b) Harmful cumulative effects (e.g., ingestion of waters high in fluorides)
Iodine (I)	a) Component of thyroid hormones T3 and T4	a) Stabilized iodized salts (K-iodide, Ca-iodate)	a) Intake of large amounts might affect thyroid uptake of I
Iron (Fe)	a) Cellular respiration (hemoglobin, cytochromes, myoglobin) b) Oxygen transport c) Peroxide scavenging d) More than half the iron present in the body is in the form of hemoglobin.	a) Ferrous sulfate b) Ferrous carbonate c) Leafy plants d) Meats e) Legume seeds and cereal grains f) Milk is not a good source, as it is low in Fe	a) Cu required for Fe metabolism b) Too much Fe interferes with P, Cu, and Se c) Iron deficiency can cause anemia similar to that produced by low protein intake
Manganese (Mn)	a) Bone formation b) Energy metabolism c) Fatty acid synthesis	a) Manganous sulfate b) Manganous oxide	a) Excess Ca and P decreases absorption
Molybdenum (Mo)	a) Component of many enzymes b) Production of uric acid	No supplementation necessary	a) Toxic levels of Mo interfere with Cu absorption
Selenium (Se)	a) Linked to vitamin E to maintain tissue integrity (antioxidant properties) b) Component of enzyme glutathione peroxidase	a) Sodium selenate b) Sodium selenite c) Barium selenate	a) Deficiencies can cause white muscle disease in cattle and exudative diathesis in chicks b) Toxic levels can cause blind staggers in cattle c) Deficiency in pigs results in mulberry heart disease and liver necrosis
Zinc (Zn)	a) Component or cofactor of many enzymes (e.g., peptidases) b) Also a part of the hormone insulin		a) Parakeratosis (dermatitis) in pigs accentuated by excess Ca intake, causing skin lesions

body's chemical reactions and are often integral to activating those reactions. Different species have different levels of tolerance to minerals, especially some of the microminerals such as iron and copper.

VITAMINS

These are distinct dietary essentials that differ chemically and on the basis of their physiological functions. Vitamins are classified as either fat-soluble (e.g., vitamins A, D, E, and K) or water-soluble (e.g., vitamin C and all B vitamins). Their classification and functions are given in table 16.3. An excess of water-soluble vitamins such as vitamin C is unlikely to be of concern, since most can be excreted in the urine. There is however, a greater potential for toxicity with fat-soluble vitamins. For example, there would be concern about feeding

an excess of organ meats such as liver, which can be very high in vitamin A. It is important to note that the presence of vitamins in feeds is affected by the environment in which the feed is kept. Many vitamins can be destroyed by exposure to light, by high storage temperatures, or just through time. This is why it is a good practice to label feeds with "best use before" dates, and to mark on feed bags when they were opened, if their contents are not used immediately. This promotes awareness of using feed that is of the highest quality. Another good practice tip involves consideration of vitamin loss in frozen prey species used as food. Fish especially will lose two important vitamins, thiamin and vitamin E, when frozen. These nutrients should be added as supplements when feeding frozen fish to primarily fish-eating (piscivorous) species. It is also important to thaw frozen prey items properly in a refrigerator, to minimize the loss of nutrients.

TABLE 16.3. Summary of individual vitamins' functions, interrelationships, and toxicities

Vitamin	Major function(s)	Sources	Interrelationships, deficiencies, toxicities
Vitamin A (fat-soluble) *Units* (Retinol) 1 IU = 0.3µg vit. A alcohol 1 IU = 0.344 µg vit. A acetate 1 IU = 0.55 µg vit A palmitate *Note* Vitamin A is readily destroyed when exposed to light, heat, or moisture. The destruction can be accelerated if the vitamin is in close contact with trace minerals or unstable fats.	a) Development, protection, and regeneration of skin and mucosa b) Health and fertility c) Increased resistance to infections d) Regulation of metabolism of carbohydrates, protein, and fat e) Presence in visual pigment f) Bone development	a) Whole milk b) Fish liver oil c) Pro-vitamin A, beta-carotene, in plants (e.g., alfalfa meal)	*Deficiencies* a) Depression of growth b) Pathological changes of skin and mucosa c) Stillbirth d) Sterility e) Low fertility f) Watery eyes (in cattle) g) Night blindness *Toxicities* a) Internal hemorrhaging (in primates, birds, and pigs) b) Deformed embryos c) Neonatal bone fractures d) Poor growth, loss of weight e) Bone deformation (e.g., bone resorption)
Beta-carotene *Units* mg per kg feed or mg per animal per day	a) Precursor of vitamin A b) Essential role in reproductive performance of male and female cattle, horses, goats, and sows c) Function is independent of vitamin A effect	a) Commercial stable forms b) Green plants (but concentration depends on harvesting time, degree of wilting, and preservation method; normally concentration reduces with passage of time after harvest)	*Deficiencies* Symptoms similar to those of vitamin A deficiency a) Silent oestrus b) Delayed ovulation c) Low fertility d) Embryonic death, early abortion e) Increased susceptibility to juvenile and infectious diseases
Vitamin D (D$_3$) (cholecalciferol; fat-soluble) *Units:* 1 IU = 0.025 µg *Active form* 1α,25(OH)2D3 (important in Ca-binding protein) *Note* a) Vitamin D$_3$ is formed by *ultraviolet* irradiation of 7-dehydrocholesterol (animal origin) b) Also called antirachitic factor c) Vitamin D$_2$ (ergocalciferol) is another form of vitamin D and originates from plants via the precursor ergosterol.	a) Normal calcification of bone b) Amount needed depends on amounts of Ca and P present in the diet and animal species c) Essential role in proper absorption of of Ca and P in gut	a) Forms of the pro-vitamin D are widely distributed in nature. b) Active form has limited distribution in natural foods c) Egg yolk d) Some fish meals f) Ergosterol (pro-vitamin D) occurs commonly in plants but the active form (ergocalciferol) is not present in plants. g) Animal feed supplements (D$_3$)	*Deficiencies* a) Rickets (in young animals) (relating to impaired (Ca and P metabolism) b) Congenital malformations c) Decrease in egg production and hatchability d) Thin eggshell *Toxicities* a) Weight loss (in ruminants) b) Stiffness of forelimbs c) Arching of back d) Calcification of kidney and lungs
Vitamin E (tocopherol; fat-soluble) *Note:* Most biologically effective: *Units* mg per kg feed or mg per animal per day (Vitamin E acetate) *a)* 1 mg dl-α-tocopheryl acetate = 1.00 IU vitamin E *b)1 mg dl-α-tocopherol = 1.10 IU vitamin E* *c) 1 mg d-α-tocopherol = 1.49 IU vitamin E*	a) Essential role in normal muscle activity and reproduction b) Preservation of membrane cells from oxidation (anti-oxidant) c) Protection against hepatic necrosis and muscle degeneration	a) Animal and plant feeds (*dl-α-tocopherol*; most important in animal nutrition) c) Cereals (mostly less effective forms such as B and Y) d) Plants e) Good-quality hay (alfalfa) and browse f) Oils	*Deficiencies* a) Muscle damage to myocardium and skeletal muscles (dystrophy, myopathy) b) Yellow fat disease (steatitis) c) Muscle weakness d) Liver damage e) Testicular degeneration (in dogs, hamsters, pigs, rabbits, roosters, fish) f) Sudden death *Toxicities* Unlikely
Vitamin K (Menadione, anti-haemorrhagic vitamin; fat-soluble) *Units* mg menadione per kg feed or mg per animal per day 1 mg vitamin K$_3$ = 2.0 mg menadione sodium bisulphite MSB. *Note* Vitamin K$_3$ can be made water-soluble with about 50% menadione.	a) Coagulation of blood	a) Green forage (present in the unstable form K$_1$ [phylloquinone]) b) K$_3$ synthesized by most animals in the digestive system (not chickens)	*Deficiencies* a) Increased clotting time b) Increased bleeding and hemorrhage c) Above deficiencies reported in birds (chicks, geese, pigeons, ducks)

continued

TABLE 16.3. continued

Vitamin	Major function(s)	Sources	Interrelationships, deficiencies, toxicities
Thiamin (vitamin B$_1$; water-soluble) *Units* mg per kg feed or mg per animal per day Official unit: (I.U and U.S.P) Biological activity of 3 mcg of pure thiamin hydrochloride *Note* The enzyme thiaminase has an antagonistic effect on thiamin.	a) Regulation of carbohydrate metabolism b) Important role in energy metabolism	a) Cereals, milling by-products, extracted oilseed cakes, milk products, and dry yeast b) Meat and bone meal (contain small amounts) c) Best sources are brewer's yeast and wheat germ d) Content decreases in hay as it matures e) Synthesized by rumen microbes	*Deficiencies* a) Depression of growth, stunting b) Disorder of nervous system c) Cerebrocortical necrosis (CCN) d) Polyneuritis (in birds) e) Impairment of digestion (in pigs, chickens); Chastek disease (in foxes) f) fern or bracken poisoning (in horses) g) Destruction of thiamin by thiaminase enzyme
Riboflavin (vitamin B$_2$; water-soluble) *Units* mg per kg feed or mg per animal per day Commercial forms: B$_2$ pure compound	a) Involvement as co-enzyme in the metabolism of protein, fat, and nucleic acids b) Co-factor in energy metabolism c) Development of fetus	a) Synthesized by plants, yeasts, fungi, and some bacteria b) Animal tissue is a poor source) c) Synthesized by rumen microbes d) Browse (leaves) e) Yeast is a good source	*Deficiencies:* a) Curled-toe paralysis (in chicks) b) Low egg production c) Poor hatchability d) Crooked and stiff legs (pigs) e) Thickened skin, skin eruptions, cataracts (in pigs) f) Anestrus (in young pigs)
Vitamin B$_6$ (pyridoxine, pyridoxal, pyridoxamine; water-soluble) *Units* mg per kg feed or mg per animal per day	a) Central role in protein metabolism b) Involvement in metabolism of fats and carbohydrates c) Breakdown of tryptophan and synthesis of niacin d) Incorporation of iron in hemoglobin synthesis e) Role in more than 50 enzymes	Widely distributed in feeds a) Yeast, muscle meat, milk, cereal grains, and their by-products b) Synthesized by rumen microbes	*Deficiencies* These symptoms are not considered specific to B$_6$ deficiency: a) Dermatitis (in rats) b) Growth retardation (in pigs, chicks) c) Stunting d) Skin inflammation e) Hepatic and cardiac damage f) Poor egg production and hatchability g) Hyperirritability, erratic swimming, anorexia, greenish-blue color h) Inhibiting factor in linseed *Toxicity* Very low
Vitamin B$_{12}$ (cyanocobalamin; water-soluble) *Units* mcg per kg feed or mcg per animal per day *Note:* Metabolically related to choline, methionine, and folacin	a) Protein metabolism (formation of individual amino acids)	a) Synthesized by micro-organisms, but apparently not by yeast b) Foods of animal origin (meat, liver, kidney, milk, eggs, fish) c) Plants are devoid of B$_{12}$	*Deficiencies* a) Cobalt is an integral molecule and must be available to avoid signs of deficiency. c) Growth depression d) Anemia, rough hair coat, poor coordination of hind legs (in pigs).
Biotin (water-soluble) *Units* mcg per kg of feed or mcg per animal per day	a) Energy and carbohydrate metabolism b) Fat metabolism c) Protein synthesis d) Nitrogen excretion e) Maintenance of hair, skin, nerves, and sex gland	a) Cereal grains b) Maize, soybean meal, animal protein feeds c) Substantial intestinal synthesis in most animals	*Deficiencies* a) Avidin, a naturally occurring compound in egg whites, binds with biotin and makes it unavailable. b) Hoof lesions, lameness (in swine) c) Growth retardation d) Fertility disorders e) Poor feather condition f) Hair loss (alopecia) g) Inflammation of the skin (dermatitis) of the beak (e.g., spectacle eye), shanks, and toes (in birds) h) Fatty liver and kidney syndrome (in birds) i) Dermatitis j) Perosis (but also can result from deficiency of Mn, choline, and folic acid) Mn, choline and folic acid *Toxicities:* No obvious symptoms

Vitamin	Major function(s)	Sources	Interrelationships, deficiencies, toxicities
Niacin			
(nicotinic acid, nicotinamide; water-soluble) *Units* mg per kg of feed or mg per animal per day	a) Presence in two important coenzymes, NAD and NADP, associated with carbohydrate, lipid, and protein metabolism	a) Widely distributed in many feed items b) Animal and fish by-products. c) Distiller's grain d) Oil meals e) Cereals contain low amounts) f) Maize, rye, and milk are poor sources g) Niacin can be synthesized from tryptophan.	*Deficiencies* a) Black tongue disease(in dogs; similar to pellagra in humans) b) Diarrhea, vomiting, dermatitis, anemia (in pigs) c) Poor feathering, mouth symptoms d) Scaly dermatitis (in chicks) e) Poor hatchability and egg laying
Folic acid (folacin; water-soluble) *Units* mg per kg of feed or mg per animal per day	a) Metabolism of protein and nucleic acids b) Role in preventing anemia c) Involvement, with vitamins C and B_{12} in production of red cells and hemoglobin	a) Dried yeast b) Soybean and fish meals	*Deficiencies* a) Reduced number of white cells b) Poor growth, poor feathers, reduced hatchability c) Perosis and crossbeak d) Anemia and alopecia (in pigs)
Pantothenic acid (vitamin B5; water-soluble; antidermatitis factor) *Units* mg per kg of feed or mg per animal per day	a) Component of coenzyme A b) Role in metabolism of fat, protein, and carbohydrates c) Synthesis and degradation of fats d) Synthesis of steroids	a) Milk products b) Fish solubles (by-product) c) Dried yeast d) Milling by-products	*Deficiencies* a) Changes in skin and mucosa b) Loss of pigmentation, rough coat, deficient plumage, loss of hair or feathers c) Goose stepping (in pigs) d) Growth and reproductive failure e) Lesions to nervous system
Choline (water-soluble) *Units* mg per kg of feed or mg per animal per day 1 mg choline = 1.15 mg choline chloride *Note* Choline does not really qualify as a vitamin; it is an structural component of fat and nerve tissue.	a) Synthesis of phospholipids (e.g., lecithin) b) Transport and metabolism of fats c) Role in the transmission of neural impulses (acetylcholine) d) Synthesis of acetylcholine e) Prevention of abnormal accumulation of fat in liver (lipotropic factor) f) Maintenance of cell membranes	a) Synthesized in the body from methionine and serine b) Animal protein c) Fats	*Deficiencies* a) Fatty liver b) Perosis (in chicks) c) Growth depression and fatty liver (in trout) d) Methionine, cystine, folic acid, and vitamin B_{12} (interrelationships with choline)
Vitamin C (antiscorbutic factor; water-soluble)	a) Transport of hydrogen in cellular respiration b) Tissue respiration c) Enhancement of iron absorption in gut d) Normal development of odontoblast e) Metabolism of tyrosine f) Conversion of tryptophan to serotsonin	a) Citrus foods c) Leafy vegetables c) Green peppers d) Strawberries	*Deficiencies* a) Scurvy (in humans, guinea pigs, monkeys, red-vented bulbul birds, Indian fruit bats) b) Defect in dentin during tooth formation

FEED FORMULATION

To formulate a diet for any zoo and aquarium animal, a nutritionist will need to know the following basic information:

1. the ecological aspect of the diet, or the animal's interaction with its natural environment
2. the anatomy and physiology of the animals' digestive tract
3. feed availability and its chemical analysis (for quality control)
4. the animal's nutrient requirements
5. how the species is managed in its enclosure
6. the animal's physiological parameters (serum values, etc).

1. Knowing what the animal eats in the wild is important for a better understanding of how to select dietary ingredients for it in captive conditions. We can seldom mimic animal's natural diet, but knowing the chemical composition of "natural feeds" can help us in formulating a nutritionally appropriate captive diet.

However, only a few natural diets have been defined "chemically." This is because we cannot always know what the animal eats in the wild, as it is more difficult to measure in the wild how much of each food item is consumed.

2. Knowing the morphology of an animal's gastrointestinal tract helps one to understand how the animal obtains and processes its food, and to know what type of food it should be offered in captivity. Digestion is associated with the gastrointestinal tract's morphology. For example, herbivorous animals have more complicated and capacious digestive tracts than do carnivorous or omnivorous species. Felines and canines have simple digestive systems with reduced microbial fermentation, unlike herbivores (e.g., ruminants like cows, giraffes, and wildebeests, or nonruminant herbivores like horses, rabbits, rhinoceroses, and elephants). Herbivores have a great capacity to ferment plant carbohydrates, thanks to the presence of large microbial populations in their foreguts or hindguts. A ruminant has microbes in its foregut—mainly in the rumen, which is part of its four-chambered stomach. Besides obtaining nutrients from plants, herbivorous ruminants will use their gut microbes and microbial by-products (e.g., volatile fatty acids produced by the microbes) as food as well. Nonruminant herbivores, such as the horse and its wild relatives the zebra and the Somali wild ass, have microbes in their hindguts (large intestine and cecum) and tend to have less microbial fermentation than ruminant animals. These two examples represent two different strategies for digesting and extracting nutrients from plants. In the animal kingdom we can find a variety of gut morphologies and strategies to deal with natural food.

3. In zoos and aquariums, animals are commonly fed diets that are very different in chemical composition from foods available in the wild. These will include agricultural products produced for humans (e.g., seeds, fruits, vegetables, and meats), as well as legumes and grass hay produced for the domestic animal industry. It is important to analyze all diets and feeds used for captive animals whenever possible. This is part of a quality control program that would aim to monitor the nutritional value of the diets and ensure that it is maintained and understood. Zoos and aquariums should establish regular sampling schedules; some will follow HACCP (Hazard Analysis and Critical Control Points) procedures for each food item used in the production of the diets. Feed samples should be taken regularly and submitted for chemical analysis to a dedicated laboratory; the results would be used by the technical staff (hopefully a nutritionist) to develop or change animal diets. Some institutions will rely on the commercial companies that manufacture the food (mainly pellets) for chemical information based on the company's analysis. When nutrition is part of a zoo or aquarium's preventive medicine program, then a good quality control program that includes chemical analysis of feeds becomes essential, and the institution should ideally have dedicated funds for this purpose.

4. Nutrient requirements are known for most of the domestic and commercial animal species, including fish. The National Research Council (NRC) of the National Academies of Sciences periodically revises these requirements and publishes the results. However, we do not know very much about the nutrient requirements of wild animal species. For large herbivores, we can extrapolate with caution from the information given for cattle and horses. For birds, domestic species are not generally appropriate for comparison, as the available NRC information relates to domestic poultry (chickens, quail, ducks, and turkeys), which might not be the best model for wild birds or for common pet birds. No officially recognized and scientifically justified nutrient requirements for reptiles or amphibians have been published. Information on feeding zoo animals and on nutrition research with wild animals that has been published in the last few decades includes proceedings from the Dr. Scholl Conference on the Nutrition of Captive Wild Animals, the Nutrition Advisory Group (NAG) of the AZA, and the Comparative Nutrition Society (CNS). Please see further readings for access to these references. The latter two are good references for the zoo and aquarium nutritionist and other technical staff. Nutrient requirements of animals are not fixed; they depend on species, sex, activity, environmental conditions, disease, and the animal's physiological state (growth, lactation, maintenance, etc.).

5. Management of the species in the enclosure is an important factor when feeding animals. Frequently, animals are brought into their holdings at night when most of the food (e.g., pellets and hay) is offered. It is important to know if the animals will share the enclosure with conspecifics, or with other animals as part of a multispecies exhibit. Furthermore, it is important to know if the animals are able to consume plants and soil from the exhibit. All these factors might have an effect on the actual nutrients the animals will consume, and should be taken into consideration when preparing the diets.

6. Physiological parameters that track animal health could be important for correct diet formulation and diet assessment. For example, blood and tissue (such as liver) levels of macrominerals such as calcium, phosphorus, magnesium, micro-minerals (e.g., copper, zinc, and iron), and vitamins (e.g., vitamins E and A) can be important for determining biochemical status before the animal shows clinical signs of deficiencies or toxicities. However, blood values are not universal in their usefulness for assessing biochemical status; they should be used with caution and in conjunction with other parameters. The nutritionist will work with the veterinarians in assessing these physiological values.

Box 16.1. Hay Quality

Hay is an important dietary ingredient for herbivores kept in zoos. When designing herbivore diets, one must consider the quality of the hay. Normally, hays are evaluated by visual inspection based on plant type (legume, grass, mixtures); maturity (e.g., prehead, early head, head, posthead or pre-bloom, early bloom, midbloom, full bloom); leafiness; amount of weathering; dust; mold; heat exposure; and contamination from foreign objects, weeds, and toxins (e.g., pesticides). The color of the hay might indicate how it was handled and stored. Faded color of hay can often indicate bleaching in the sun, or excessive plant maturity before baling. However, visual evaluation is not precise and does not provide a true analysis of the hay's nutrient content. Thus, it is important to add quality control testing to all hay used in feeding programs. Quality parameters such as the amount of fiber, expressed as ADF (acid detergent fiber) or NDF (neutral detergent fiber), or the amount of protein are important nutrient parameters to check in hay samples. These parameters can be used in the prediction of digestibility (via ADF and NDF) and dry matter intake. Hays will vary in their NDF and ADF concentrations. Legumes (e.g., alfalfa or lucerne) have lower amounts of NDF than grass hay, but often have higher levels of protein. Grasses such as timothy, bromegrass, and orchard grass (temperate and cool season grasses) show less hemicellulose than tropical and subtropical grasses (e.g., bermudagrass, bahiagrass). Animal intake will be higher on legume hays and lower with subtropical grasses, as legume hays are generally more palatable. Many species of grass hay are fed in zoos, depending on the zoos' geographical location. Northern US zoos will use mainly timothy, orchard, or bromegrass, while zoos in southern states might use bermudagrass hay, and some western states will use sudangrass. The composition of selected hays is given below.

Species	Stage of maturity	Composition (% dry matter basis)[1]					
		Protein	NDF	ADF	Lignin	Ca	P
Alfalfa	Early vegetative	23	38	28	5	1.80	0.35
Alfalfa	Full bloom	17	49	39	10	1.19	0.24
Bermudagrass	Early vegetative	16	66	30	4	0.49[2]	0.27
Bermudagrass	43–56 d regrowth	8	78	43	7	0.26	0.18
Timothy	Early bloom	10.8	61	35	4	0.51	0.29
Timothy	Late bloom	7.8	70	40	7	0.38	0.48

[1]Adapted from the *Nutrition Advisory Group Handbook*, fact sheet 001, 1997.
[2]*Nutrient Requirements of Dairy Cattle*. Washington: National Research Council, National Academy Press, 2001.

Good Practice Tip: The best investment that a zoo or aquarium can make is to purchase scales and balances to weigh animals, the food offered to them, and the food they refuse. Without knowing how much food is actually being offered to and eaten by animals, there is no way to know whether the animals are receiving the proper nutrition. A centralized kitchen operation can be important for preventative medicine purposes so that the diets are consistent and well tracked.

DIET SHEET

A diet sheet is a very important document generated by those responsible for tracking nutrition for each animal or species in a zoo or aquarium (e.g., the animal nutrition center or commissary). It will detail the amount and types of feed ingredients that the animal is offered, based on the animal's daily consumption. It will also indicate the feeding schedule and the purpose of each dietary ingredient (e.g., maintenance, lactation, growth, enrichment, training, medical, or supplement). The diet sheet will give the animal's latest body weight and, if possible, a target body weight and body condition score. Notes regarding diet changes, transaction information, medical status, and behavior can also be added. The diet sheet is an important document used by all the zoo staff, including nutritionists, veterinarians, animal managers, registrars, curators, and keepers, when they need to know about the feeding or health of the animal. It is also important during the transfer of an animal to or from another institution. Although the nutritionist (when one is present) should have the final say on the diet, it is important to have input from all staff levels in the development of the diet sheet.

FOOD PREPARATION

In order to have a scientific and effective feeding program, it is important to have a well-organized and meticulously maintained control system for diet preparation. Many zoos will have centralized facilities, normally called commissaries, animal nutrition centers, or diet kitchens, in which food and other nutritional components are stored and most of the handling takes place. This is also where the highest standards of hygiene should be maintained. All diets are prepared and food items weighed out in this area, with just minor preparation in the animal-keeping areas of the zoo. In some zoos these diet preparation facilities will deliver prepackaged diets to the animal areas, while in other institutions the keepers will arrive to pick up the prepared diets. Ideally, these commissary or kitchen facilities will have separate unloading and shipping docks; the capacity to hold frozen meat, fish, other frozen products; and proper space for handling and preparing diets. Institutions can follow HACCP guidelines regarding food handling safety. From a budgeting perspective, centralized facilities work better at avoiding waste of resources. The commissary or nutrition center becomes the place where all nutrition knowledge is applied to the production of animal diets. It is important to establish good communication between

All the information described above will be necessary to facilitate the correct formulation and assessment of zoo and aquarium diets. Nutritionists and keepers must try to be proactive in reviewing and changing diets based on this relevant information.

the nutrition center staff and the animal keepers and zoo managers to assess food consumption and minimize food wastage. The best investment that a zoo or aquarium can make is to purchase scales and balances to weigh animals, the food offered to them, and the food they refuse. Without knowing how much food is actually being offered to and eaten by animals, there is no way to know whether the animals are receiving the proper nutrition.

BODY CONDITION SCORING

Body condition scoring (BCS) is an important management practice used by domestic animal producers and some veterinarians that is now being applied to zoo animals as well. It is a useful tool that helps in evaluating the health and nutritional status of animals. The BCS will help evaluate the amount of body reserves (including body fat and protein stores in muscle) that an animal has at a particular time. In zoos, the BCS should be done routinely, perhaps twice a year, as well as at particular physiological stages such as during pre-breeding, mid-gestation, parturition, lactation, weaning or the growth period for young animals. The BCS assessment is best performed by a team of people including keepers, managers, curators, vets, and the nutritionist. It is performed by a thorough visual examination and by palpation of musculature and body fat if possible—for example, during a routine medical examination. To assign a body condition score, people performing the evaluations will look at different anatomical points. For mammals these will be the neck, shoulder, withers, loin, back, tailhead with hips, and ribs. The BCS assessment will be based on the amount of fat and muscle cover on the animal. Certainly, we can expect variations in fat cover in specific body locations for animals that have similar body condition scores. Also, different species will tend to deposit fat and muscle in different anatomical locations. That is why it is important to develop specific BCS systems for each species. An example of a BCS system for yellow-backed duikers is given in table 16.4. Zoos and aquariums are encouraged to develop their own BCS systems taking into account seasonal variations as well as physiological states (growth, lactation, pregnancy, old age, etc.) as part of the management strategies for a species.

TABLE 16.4. Summary body condition scores for yellow-backed duikers in Disney's Animal Programs (Disney's Animal Kingdom™)

Score	1: Emaciated	2: Thin	3: Good	4: Fat	5: Obese
Outline depictions					
Neck and shoulders	☐ Emaciated ☐ Bone structure easily visible ☐ No fat	☐ Thin neck ☐ Decreased girth	☐ Thick neck ☐ Flat shoulders	☐ Thick neck ☐ Fat deposits evident ☐ Shoulders slightly rounded	☐ Fat evident along neck ☐ Bulging fat ☐ Thick neck ☐ Neck blends into shoulder ☐ Rounded shoulders
Withers	☐ Emaciated ☐ Bone structure easily visible ☐ No fat	☐ Thin ☐ Evident bone structure	☐ Fat deposits on withers ☐ Decreased visibility of bone structure	☐ Fat deposits evident	☐ Fat deposits which make withers appear flatter and less discernible
Loin and back	☐ Emaciated ☐ Spinous processes easily identifiable	☐ Spinous processes not individually identifiable, but spine still prominent ☐ Transverse processes faintly discernable	☐ Back sloped to withers	☐ Fat deposits present ☐ Back appears flatter	☐ Wide back ☐ Patchy fat ☐ Flat back
Tailhead and Hips	☐ Pelvic bones very prominent	☐ Pelvis bones at the point of hip rounded, but still evident ☐ Pelvic bones at rump may be slightly discernible	☐ Fat present around tailhead ☐ Pelvic bones flat	☐ Hips rounded	☐ Hips and thighs very round
Ribs	☐ Emaciated ☐ Rib spacing appears wide and depressed	☐ Ribs still discernible, but fat discernible by touch	☐ Ribs not visible, but discernible by touch	☐ Ribs not visible ☐ Fat deposits perhaps evident	☐ Fat deposits may be present, easily evident

NUTRITIONAL DISEASES IN ZOOS AND AQUARIUMS

A good nutrition program is part of any preventive medicine program. Sound nutrition is the key to disease prevention, and prevention of disease is essential if we are to be successful in keeping wild animals in zoos and aquariums. A description of some selected nutritional diseases will follow. Inappropriate diets generally result in disease, because maintaining the proper nutrients in foods are fundamental for basic metabolism and good physical health. Nutrient deficiencies or excesses are often a main cause of diseases in captive animals. One of the most prevalent dietary issues in zoos is the relationship between calcium, phosphorus, and vitamin D. Imbalances in these nutrients result in a syndrome called metabolic bone disease in many animals including amphibians, reptiles, birds, and mammals. Metabolic bone disease encompasses a number of conditions that develop as a result of prolonged deficiencies of calcium or vitamin D, or because of an improper ratio of calcium to phosphorus in the diet. Deficiencies in vitamin A (hypovitaminosis A) have been reported in amphibians, causing "short tongue syndrome" or squamous metaplasia of the tongue. Spindly leg syndrome has also been seen in amphibians, especially metamorphosing tadpoles and young frogs, and has been widely reported in the literature. Spindly leg syndrome is thought to be possibly caused by multiple nutritional deficiencies and poor water quality. Serum vitamin E deficiencies have been reported in many zoo animals including elephants, black rhinoceroses, and giraffes in comparison to their wild counterparts. Deficiency in vitamin E (hypovitaminosis E) has also been reported in reptiles; it results from deficient diet and can result in muscular dystrophy when combined with selenium deficiency. Vitamin D and A toxicities have also been reported in many species, including tarantulas, because of inappropriate diet or excessive use of mineral and vitamin supplements. Alternatively, the underuse of proper supplementation can commonly occur in some institutions, causing other problems such as metabolic bone disease and other nutritional issues in herbivorous reptiles. Vitamin A deficiency (hypovitaminosis A) is also commonly observed in juvenile semiaquatic turtles, with symptoms such as the sloughing off of skin. Goiter, a swelling of the thyroid gland (neck region), has been reported in terrestrial turtles. This is caused by a lack of iodine in the diet and/or by the use or overuse of foods known to suppress function of the thyroid gland (i.e., foods that are goitrogenic), such as cabbage, kale, broccoli, or cauliflower. Iron storage disease, in which an excess of iron is stored in the body and can lead to organ damage, has been reported in birds (e.g., toucans) and mammals (e.g., black rhinos, lemurs, and dolphins). This syndrome may have multiple causes besides diet;, it can be related to stress, genetics, infections, intoxications, anemia, or other problems. Nutritional imbalances have been reported to be related to sudden death (peracute mortality syndrome) in giraffes; the specific signs are not clear, but they often include the deterioration of fat stores and most likely have multiple causes. Acidosis and ruminitis—conditions that result from diets with high levels of soluble carbohydrates, mainly starch and sugars—have been reported in many wild ruminants kept in zoos, particularly browsers (e.g., kudus, giant elands, bongos, nyalas, and giraffes). For example, feeding an animal a high-starch pellet or grain in large quantities will stimulate reduction of the rumen pH, thus creating an acidic environment (acidosis) and rumen upset. It is important when shifting diets to change pellets and grain gradually due to these concerns, but most importantly to avoid grains or pellets with high starch for browsing ruminants. A sudden switch to a high-starch diet from a mainly pasture- or browse-based diet commonly will lead to acidosis. Urolithiasis (stones) or mineral deposit in the bladder and urethra, causing obstruction leading to bladder rupture and death, has been reported particularly in male ruminants in zoos (e.g., giraffes and bongos), and may be caused by dietary mineral imbalances.

OBESITY

One of the most common problems in zoos is obesity in animals. Obesity is caused by an intake of energy (calories) which exceeds expenditure, in conjunction with a lack of proper exercise. It is very important to monitor feeding, and especially the items fed, as overfeeding contributes greatly to this issue. Making sure that high-calorie items are limited for sedentary animals that do not need the excessive calories can help, as can monitoring body weights of the animals. As previously discussed, the development of a body condition scoring system has become an important husbandry tool in the monitoring and management of obesity in zoo animals. Obesity is a serious concern due to the chain reaction of problems it creates in many species. In primates it is associated with heart disease. Many species are more likely to experience body trauma and develop stereotypies when obese, often because of trouble walking in their exhibit or because of muscle degradation. Excess protein/energy intake can cause rapid weight gain in young birds, producing a condition called "slipped" or "airplane" wing. Excess protein in birds and reptiles can also cause gout, a condition in which uric acid accumulates most often in the joints and tendons. Obesity is a common problem that requires the animal care team to better address and monitor an animal's feeding and exercise routine.

Anorexia (loss of appetite) in snakes, caused by inadequate diet or incorrect presentation of the food, has been reported in many institutions. It can also be related to environmental conditions (e.g., temperature or humidity) that play a major role in reptiles' feeding response. Marine mammals in aquariums are fed a more limited diet than what they would be exposed to in the wild. Thiamin and vitamin E deficiencies in marine mammals can be related to improper supplementation of a fish-based diet with thiamin and vitamin E. These two nutrients are degraded in fish that have been frozen and stored. Vitamin C deficiency in fish—which causes deformed gills, scoliosis (crooked spine), lordosis (inward curvature of vertebral column), and broken back syndrome—has been reported in aquariums. Vitamin B deficiencies in fish can cause erratic swimming, hyperexcitability, anorexia, and other abnormalities. Vitamin A deficiency in fish can cause "pop eye" (exopthalmia), cataracts, and displaced lenses.

THE ROLE OF THE KEEPER IN FEEDING CAPTIVE ANIMALS

Keepers play an important role in feeding animals in zoos and aquariums. They are the main contact between the animal and the rest of the animal care team (i.e., veterinarians, managers, nutritionists, etc). They will know how the animal behaves on a daily basis, and they are responsible for reporting any abnormal behavior and changes in the animals. They should know how well the animals are consuming and accepting the diet, and they are important in assessing the animal's body condition. Their work can include weighing the food offered and tracking the weight of food left uneaten to measure consumption and minimize waste.

Public feeding is a common practice in many zoos and aquariums. The keeper's most important role in public feeding is to select the best feeds: those which will not negatively affect the animal's overall health. Examples of feeds to avoid would be high-starch products given to giraffe or other ruminants, which can upset their rumen and digestive tract, or fish with high fat content, such as herring to piscavores. A suggestion would be to feed lettuce or browse, instead of high-energy feeds such as grain, to such ruminant animals as giraffes. Public feeding plays an important role in attracting guests, but keepers should aim to attract the public without harming the animals' health.

PRACTICAL DIETS

GENERAL GUIDELINES

The following are examples of practical diets that can be fed in zoos and aquariums. The ingredients do not represent commercial feeds and names are just given as examples. For large herbivore species it is important to note that the balanced part of the diet should be a "safe" pellet with no more than 5 % starch and other soluble carbohydrates (e.g., sugars). Pellets can be deemed "safe" for large herbivores when they are low in these soluble carbohydrates, because eating large quantities of them should not result in death or extreme

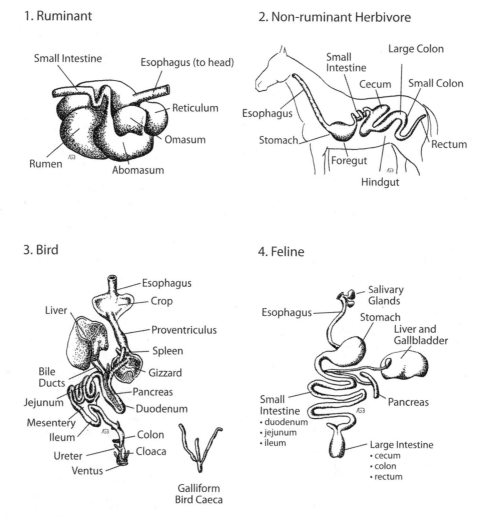

Figure 16.1. Digestive tracts of selected animals: (1.) ruminant; (2.) nonruminant herbivore; (3.) bird; (4.) feline. Courtesy of Anthony Galván III (c).

digestive tract upset. The ratio of hay (plus browse) to pellet will depend on the pellet's "safety" features. A safe pellet, containing less than 5% soluble carbohydrates, can be used in higher amounts (more than 50% of the diet), unlike pellets containing more than 10% starch, which must be limited in the total diet. For most species the amount of high-starch pellet in the large herbivore's diet should not exceed 25% of the total, depending on the animal species. Hay should be free of mold and of good quality (see appendix 1). Both pellets and hay should routinely undergo chemical analysis. Browse should be free of herbicides. Analysis of all pellets and chows should be routinely performed for quality control and nutritional balancing of the diet. The selection of the pellet or chow should be based on sound nutritional knowledge of the animals to be fed. Diets also should be practical and economically sound, allowing for the best quality at the least cost. This implies that the institution makes its decisions about animal diets on the basis of investment in preventive medicine, rather than just on cost.

LARGE HERBIVORES

AFRICAN ELEPHANT (*LOXODONTA AFRICANA*)

The African elephant is a nonruminant herbivore (figure 16.1). It needs a large volume of food, mainly forage, which should include browse to mimic its natural diet. Total amounts will range from 120 to 170 kg of fresh food per day for adult animals, depending on body weight.

Ingredients	Amount (kg, as fed)
High-fiber pellet	8–10
Grass hay[1]	50–80
Browse[2]	50–100
Enrichment[3]	0.5
Training[4]	0.5
Supplements[5]	If needed
Water	Ad libitum

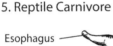
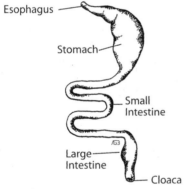

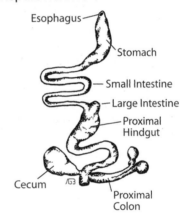
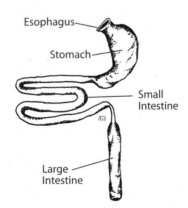
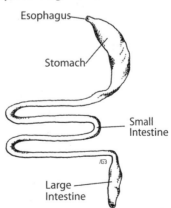

Figure 16.2. Digestive tracts of selected animals (continued): (5.) reptile carnivore; (6.) reptile herbivore; (7.) amphibian, toad (*Bufo americanus*); (8.) amphibian, tiger salamander (*Ambystoma tigrinum*). Courtesy of Anthony Galván III (c).

Approximate composition (dry matter basis, DMB)

Protein	10–12%
Fiber[6]	35–40%
Starch	3 (less than 5 %)
Calcium (Ca)	0.8%
Phosphorus	0.4%
Ca : P ratio	2 : 1
Vitamin E	150 IU/kg
Digestible energy[7]	1,800.0 kcal/kg

Notes: [1]Grass hay: North American institutions can use bermudagrass, timothy, or orchard grass hay. [2] Browse can vary depending on the zoo or season. Some zoos do not have access to browse, but are encouraged to use it and develop sources. Elephant grass (*Penissetum purpureum*) is a good source of browse during the summer months in some institutions. [3,4]Enrichment and training food items could be small amounts of produce and vegetables. Sometimes very small amounts of commercial diets can be used for this purpose. Alfalfa hay should be used in limited amounts only as an enrichment or training item, or should be mixed with grass hay to maintain body condition, as alfalfa hay can be very high in protein and calcium. Large amounts can unbalance a diet and lead to overconditioning. [5]Supplements such as vitamin E can be added to the diet, depending on animal serum values as evaluated by the veterinarian or nutritionist. Psyllium or bran can be added a few days per month to avoid sand colic. [6]Fiber refers to acid detergent fiber (ADF). [7]Digestibility coefficient (dry matter) = approximately 40 %. Total daily food consumption should be around 1.2–1.5 % of body weight (on an as-fed basis) Increase or decrease in amount of food should be based on BCS and physiological stage. With a pregnant elephant, it is important to adjust diet based on body weight (if possible) and BCS.

GIRAFFE (*GIRAFFA CAMELOPARDALIS*)

The giraffe is a ruminant browser, rather than a grazer (fig. 16.1). Its diet should include browse species to meet its nutritional and behavioral requirements. Browse can include the leafy branches of trees such as willow, acacia, mulberry, and hundreds more examples, depending on geographic availability. The pellet should be adequate in terms of the amount of soluble fiber; it should have reduced (< 5%) soluble carbohydrates (e.g., starch and sugars) and adequate protein (12–14%). Dietary fat levels can be increased in winter months or to improve body condition.

Ingredients	Amount (kg, as fed)
High-fiber pellet	6–12
Grass/legume hay[1]	4–6
Browse[2]	5–10
Enrichment[3]	0.5
Training[4]	0.5
Supplements[5]	If needed
Water	Ad libitum

Approximate composition (DMB)

Protein	12–14%
Fiber[6]	32%
Starch	3 (< 5)%
Calcium	1.1%
Phosphorus	0.6%
Ca : P	1.8 : 1
Vitamin E	250 IU/kg
Digestible energy[7]	2,640 kcal/kg

Notes: [1]Grass / legume hay: North American Institutions can use bermudagrass hay, timothy hay, orchardgrass hay and alfalfa (a type of legume) mixes. Most giraffes will consume alfalfa hay only, but their hay consumption is not as high as that of most browsers. The use of hay mixes, rather than just alfalfa, is encouraged. [2]Browse availability can vary depending on the zoo and the season. Some zoos do not have access to a supply of browse, but they are encouraged to develop a source. Willow, apple, and different acacia species are good sources of browse when available during summer months in some institutions. Some zoos are now harvesting and cold storing browse for the winter, when adequate space is available. [3,4]Enrichment and training food items may consist of limited amounts of produce and vegetables. Sometimes very small amounts of low starch herbivore pellets can be used for this purpose. [5]Supplements such as vitamin E can be added to the diet, depending on animal serum values. [6] Fiber refers to acid detergent fiber (ADF) in this context. [7] Digestibility coefficient (dry matter) = approximately 65 %. Total daily food consumption should be around 1–2% of body weight (on an as-fed basis). The amount of food may be increased or decreased based on body weight and BCS.

CARNIVORES: MAMMALS

Felines, such as domestic cats and their wild relatives the tigers and cheetahs, are the most specialized meat eaters. They are defined as "strict or obligatory" carnivores, unlike the canines, such as dogs and their wild counterparts (wolves, foxes, and other similar carnivores such as bears and hyenas), which are "facultative" carnivores or can also be classified as omnivores. Canines have broader feeding habits, and their diets may include fruits and other plant materials. Felines have lost certain metabolic pathways and have no tendency to consume anything but meat (prey items). Cats cannot use plant material; they require higher protein and certain amino acid levels in their diets. The digestive systems of all carnivores are universally simple (figure 16.1). The feline digestive system consists of a short small intestine. Thus, the ideal feline diet must be concentrated, highly digestible, and low in residues, as its digestive tract is primarily designed to digest fat and protein. In contrast, dogs have a relatively large small intestine. Cats require high levels of the amino acids methionine and cystine, possibly required for the synthesis of taurine, an essential amino acid for cats. Unlike dogs, they also require an external (dietary) source of taurine. Cats lack glucokinase, a liver enzyme needed for glucose metabolism, unlike dogs, in which that enzyme is present. Cats need preformed vitamin A in their diet (they cannot convert the plant pro-vitamin A) and niacin, a B vitamin (since they cannot convert tryptophan, an amino acid, into niacin). Cats also require arachidonic acid, a long-chain fatty acid. Cats obtain all these nutrients from their prey. There is no evidence that dogs need this fatty acid. Thus, cats are specialized carnivores, unlike dogs, which have a greater degree of dietary adaptations.

TIGER (*PANTHERA TIGRIS*)

The tiger is a strict carnivore (body weight: ~ 130 kg).

Ingredients	Amount (kg, as fed)	Frequency
Complete feline diet (wet)[1]	2	Monday, Wednesday, Saturday, Sunday
Shank beef bone	1	Thursday only
Horse tail	0.6	Tuesday only
Rabbit	1	Friday only

Approximate composition (DMB)
Protein	65%
Fat	18%
Calcium	1.9%
Phosphorus	1.6%
Ca : P	1.2 : 1
Vitamin E	120 IU/kg
Metabolizable energy	5,100 kcal/kg

Notes: [1]Commercial feline diets for zoo carnivores are available in some markets, though not in all countries. The regimen outlined above is a typical zoo diet for felines. Some institutions can use commercial pellets, but the value of these for zoo animals is unproven and would be of concern, especially due to high carbohydrate levels, which do not mimic natural diets. Feline meat diets can consist of horse, beef, or other sources (e.g., deer), with bones of any kind, though beef and horse are the most common.

GRIZZLY BEAR (*URSUS ARCTOS*)

The grizzly bear is a facultative carnivore-omnivore (body weight: ~300 kg).

Ingredients	Amount (kg, as fed)	Frequency
Complete canine diet (wet)[1]	2	Daily
Herring	0.7	Daily
Romaine lettuce	4	Daily
Parsnips	2	Daily
Celery	0.9	Daily
Apples	0.2	Wednesday, Friday (training and enrichment)
Yams	0.2	Monday, Saturday (training and enrichment)
Dog chow (dry)	0.2	Tuesday, Thursday (enrichment)

Horse necks	0.8	2× month (enrichment)
Vitamin E	100 IU/kg	1 g (1 capsule in fish)
Thiamine	50 mg/kg	0.6 g (half tablet in fish)

Approximate composition (DMB)
Protein	43%
Fat	25%
Calcium	0.94%
Phosphorus	0.80%
Ca : P	1.2 : 1
Vitamin E	110 IU/kg
Metabolizable energy	3,800 kcal/kg

Note: [1]Commercial canine diets for zoo carnivores are available in some markets.

SPECIALIZED CARNIVORES (INSECTIVORE-MAMMALS)

Insectivore animals such as giant anteaters (*Myrmecophaga tridactyla*) and tamanduas (*Tamandua tetradactyla*) present a greater challenge in zoos with regard to mimicking their natural diets. These animals have very low metabolic rates. In the past they have been fed captive diets that had little or no resemblance to their natural diets, with negative consequences to their health (e.g., hypervitaminosis A and D). Mixes of commercial high-fiber primate chows and domestic dry cat food have been used by some zoos. However, the high levels of starch in these diets might be a concern (e.g., for diabetes). Recently, commercial insectivore pellets have been formulated on the basis of information from wild diets. The diet outlined below, which includes beef heart and chicken breast as main ingredients, has been used by some zoos with positive results.

TAMANDUA (*TAMANDUA TETRADACTYLA*) (BODY WEIGHT: 5.8 KG)

Approximately 400 g of the following liquid diet can be used on a daily basis. Amount offered will depend on body condition and weight.

Ingredients	Amount (g, as fed)	% of total liquid diet mix
Lean ground beef heart[1]	220	52.5
Peeled banana	20	4.9
Hard-boiled egg (no shell)	17	4.0
Honey[2]	10	2.4
Safflower oil or flax oil	4.0	1.8
Vionate[3]	17	4.0
Vitamin E (40,000 IU/kg)	1.5	0.3
Calcium carbonate	0.9	0.2
B-complex (1 tablet)	0.5	0.1
Ascorbic acid mg/g	0.6	0.1
Chitin	0.3	0.07
Acetic acid (vinegar)[2]	0.5 (1 drop)	0.1
Water	127	30.3

Approximate composition (DMB)

Protein	41%
Fat	15%
Calcium	1.2%
Phosphorus	0.8%
Ca : P	1.5 : 1
Vitamin A	12,000 IU/kg
Vitamin E	100 IU/kg
Metabolizable energy	4200 kcal/kg

Notes: [1]Chicken breast can be used as well, but it has to be well ground to avoid a fibrous texture. The amount of water can vary and will depend on the animal's preference. The amount of water added to the mix should be recorded, and ingredients adjusted to maintain consistency, avoiding too much dilution. [2]Honey is used for flavor, as well as vinegar. [3]Vionate (a commercial vitamin/mineral supplement) can be replaced with other supplements, but adjustments to ingredients might then be necessary. The diet can also include a variety of invertebrates (local termites, mealworms, superworms, wax worms, etc) for maintenance, enrichment, or training purposes. Small amounts of fruits (e.g. mango, papaya, and avocado) can be also added. This diet can be used in combination with commercial insectivore pellets. [4]Commercial feline diets for zoo carnivores are available in various markets.

BIRDS

Choice of food for birds varies from species to species. Birds occupy every nutritional niche. Energy requirements are substantial for birds despite their small size, and are higher for smaller birds and birds in flight. Birds will consume between 3% and 200% of their body weight daily. Most are general feeders or omnivores (e.g., quail, pheasant, and tinamous), consuming both animal and plant material, and in the wild their food choices will depend on many factors such as seasonal availability and changing nutrient requirements (e.g., for breeding and raising chicks). Other bird groups include specialized feeders such as carnivores or faunivores (e.g., penguins, herons, and albatrosses), insectivores (e.g., cuckoos, woodpeckers, swallows, and thrushes), crustacivores (e.g., crab plovers and some rails), piscivores (e.g., pelicans, loons, storks, and mergansers), terrestrial vertebrate consumers (e.g., hawks, owls, and eagles), herbivore-browsers (e.g., ostriches and some ducks), herbivore-grass eaters (e.g., geese and swans); herbivore-folivores (e.g., hoatzins), concentrate selectors and grain eaters (e.g., finches, parrots, and pigeons), concentrate selectors and frugivores (toucans, birds of paradise, tanagers, and bulbuls), and concentrate selectors and nectarivores (e.g., hummingbirds, sunbirds, and lorikeets), among others. Regardless of their adaptations, all birds appear to need the same nutrients, although in different proportions. Captive diets should be practical and take into consideration the birds' dietary preferences. As noted above, it is difficult to mimic animals' natural diets in captivity, and commercial balanced (pelleted) diets should be used to meet nutritional requirements, together with fruits, vegetables, seeds, and other agriculturally produced foods. The pellet is the easiest way to provide all essential nutrients and it is generally easier to feed, store, and clean. Other ingredients (e.g., fruits, vegetables, and seeds) can be added in small amounts to stimulate intake and provide occupational enrichment therapy. A disadvantage of the pelleted diet may be the cost, as well as the production of excessive waste if not measured and presented properly. Birds have preferences for certain food items. These preferences appear to be influenced by habit, shape, size, texture, color (birds generally favor yellow, followed by blue and then red), and taste. Practical zoo diets for omnivore and fruit-eater birds might include approximately (on an as-fed basis) the following:

1. Dark green and yellow vegetables   25–30%
2. Fruits   15–20%
3. Pelleted diets   50–65%

However, these diets, which contain considerable amounts of fruits and vegetables, should be managed carefully to avoid spoilage. It is also very important to ensure that the birds avoid extreme selectivity, which can render them deficient in certain nutrients. In some aviary exhibits the balanced-diet pellet is offered before the fruits and vegetables, to promote a balanced nutrient intake. The exclusive use of seed mixes can cause obesity in birds because of the high fat content of seeds and might cause benign fatty tumors (lipomas) and fatty liver disease. Seed diets have an inverse calcium to phosphorus (Ca: P) ratio, which can cause metabolic bone disease, and they may also produce other vitamin deficiencies. Seed diets can also be addictive for the birds, due to the good taste of the high fat content. The digestive system of a bird is shown in figure 16.1.

BLUE AND YELLOW MACAW (*ARA ARARAUNA*) (BODY WEIGHT: 1,010 G)

Ingredients	Amount (g as fed)
Parrot maintenance	30–40
Bird produce mix[1]	45–55
Peanut with shell	4
Enrichment[2]	10–20
Training[3]	5–10
Supplements[4]	If needed
Water	Ad libitum

Approximate composition (DMB)

Protein	15–18%
Fat	4–8%
Calcium	0.8%
Phosphorus	0.7%
Ca : P	1.1 : 1
Vitamin E	150 IU/kg
Metabolizable energy	3,300 kcal/kg

Notes: [1]Bird produce mix could consist of mixed vegetables such as corn, sweet potatoes, green beans, broccoli, yellow squash, pear, honeydew, and apple. [2,3]Could consist of any item indicated in the bird produce mix, or small amounts of nuts. [4]If needed or recommended by nutritionist or veterinarian. Also, the pellet given is just an example. Institutions must choose on the basis of the animal's physiological state and the availability of pellets in the local market.

AFRICAN PYGMY FALCON (*POLIHIERAX SEMITORQUATUS*) (BODY WEIGHT: 72 G)

The African pygmy falcon is a carnivorous bird.

Ingredients	Amount (g, as fed)	Frequency
Juvenile mice (3–4 g) [1]	10–15	Monday, Wednesday, Saturday
Commercial carnivore diet [2]	12–15	Tuesday, Thursday, Friday, Sunday
Mealworms [3]	1–2	Friday (training)
Crickets [3]	2–4	Monday, Wednesday (enrichment)
Supplements	If needed	
Water	Ad libitum	

Approximate composition (DMB)

Protein	50–57%
Fat	17–20%
Calcium	0.8%
Phosphorus	0.7%
Ca : P	1.1 : 1
Vitamin E	250 IU/kg
Metabolizable energy	3,800.0 kcal/kg

Notes: [1]A diet of terrestrial vertebrates (e.g., mice, rats, and chicks) is generally similar in major nutrient composition. [2]Total balanced carnivore diet. [3] Insects and invertebrates (e.g., crickets, mealworm larvae, wax moth larvae, and earthworms) used in mammal diets show extremely unbalanced calcium-phosphorus ratios. Supplemental calcium should be provided. Calcium carbonate can be dusted on invertebrates.

PISCIVORE (FISH-EATING) BIRDS

If possible, a variety of species of fish should be fed to compensate for nutrient variability. Fish will undergo deterioration of nutrients (oxidation) once they have died. This oxidation of nutrients such as fat is combated by vitamin E, which is used up and destroyed in protecting the other nutrients and counteracting the rancidity of fat. Supplemental vitamin E is recommended in these diets (50–150 IU per kg of fresh fish), since the vitamin E that was previously in the food has been used for oxidative protection. Thiaminase, an enzyme present in some fish (e.g., herring, smelt, and mackerel) and mollusks (e.g., clams) destroys the B-vitamin thiamin, so diets should be supplemented (50–100 mg thiamin per kg of fresh fish) to avoid deficiencies. Whole fish are good sources of most nutrients, and are similar in many ways to terrestrial whole prey. The shipping, handling, and storage of fish in zoos and aquariums is extremely important because the nutritional value and composition of fish depends on how these operations are carried out. It is important for institutions to follow HACCP guidelines.

SADDLE-BILLED STORK (*EPHIPPIORHYNCHUS SENEGALENSIS*) (BODY WEIGHT: 6.6 KG)

Ingredients	Amount (g, as fed)	Frequency
Trout	590	Daily
Lake smelt	380	Daily
"Fuzzy" mouse	20	Daily
Vitamin E [1]	100 IU	Daily
Thiamine	50 mg/kg	Daily
Water	Ad libitum	daily
Superworm [2]	10	Sunday, Tuesday, Thursday (training)
Crickets	size 1.9 cm (3/4 in.) × 10	1× / week (enrichment)
Waxworm	4	1× / week (enrichment)

Approximate composition (DMB)

Protein	58%
Fat	26%
Calcium	2.4%
Phosphorus	2.0%
Ca : P	1.2 : 1
Vitamin E	120 IU/kg
Metabolizable energy	5,100 kcal/kg

Notes: [1]Supplements commonly added to fish (put into the mouth or gill so that it is ingested). [2]Insects and invertebrates (e.g., crickets, mealworm larvae, wax moth larvae, and earthworms) show severely unbalanced calcium-phosphorus ratios. Supplemental calcium should be provided, although this is less important if other foods are consumed. Calcium carbonate dusted on invertebrates.

REPTILE DIETS

Reptiles can be classified as carnivores, herbivores, or omnivores, based on their feeding strategies and differences in their gastrointestinal tract morphology. Besides diet and management of the diet, the environmental conditions, including husbandry techniques and group dynamics, will influence the efficiency with which they use their diets. Most nutritional problems will be caused by inappropriate diet selection and poor management of the species in the exhibit. Reptiles will need appropriate ambient temperatures and basking spots. Reptiles kept at below-optimum temperatures cannot maintain active metabolism, and this in turn negatively affects their digestive efficiency. On the other hand, an underfed reptile in a warm environment will not maintain its body weight because of an increase in its metabolic demands. Poor environmental management conditions may cause stress for the animal, resulting in reduced feed intake. It is important to maintain reptiles in an optimum environment to ensure that they thrive.

REPTILES: CARNIVORES

Carnivorous reptiles have short digestive systems similar to those of carnivorous mammals, and are adapted to a high-fat, high-protein, low-carbohydrate diet (figure 16.2). Included in this group are snakes, crocodilians (i.e., caimans, crocodiles, alligators, and gavials), many lizards (e.g., monitors, tegus, and many skinks), anoles, and juvenile marine turtles. Diet consists of whole prey: fish, amphibians, reptiles, birds, and mammals. Many tropical geckos and other small lizards mainly consume invertebrates.

ROYAL/BALL PYTHON (*PYTHON REGIUS*) (BODY WEIGHT: 1,600 G)

Ingredients	Amount (g, as fed)	Frequency
Adult mice[1]	52	Friday, first and third week of the month
Vitamin E	1 g	2 capsules in mice

Approximate composition (DMB)	
Protein	46%
Fat	26%
Calcium	2.9%
Phosphorus	2.0%
Ca/P	1.4
Vitamin E	60 IU/kg
Metabolizable energy	4,400 kcal/kg

Notes: [1]Offered as dead prey. Other whole dead animals can be used as well, including chicks and immature mice and rats ("fuzzies" and "hoppers"). Some animals can be used live, but this may create problems because live prey can attack a predator in a captive situation. Live prey feeding must be monitored by keepers. Prey size is related to the size of animal being fed, and is based on animal needs and body weight. It is important to routinely analyze for parasites (e.g., every six months) in the prey (food) supply. It is also important to get commercial suppliers who abide to HACCP guidelines.

REPTILES: HERBIVORE

Herbivorous reptiles are represented by terrestrial tortoises and lizards (green iguanas), chuckwallas, spiny-tailed lizards (genus: *Uromastyx*) (figure 16.2). In zoos they will consume a variety of vegetables, greens, and fruits. Some will accept chopped hay, cacti, and flowers. Increasing the amount of vegetables, fruits, and chopped hay in the diet can help manage body weight. High-fiber herbivore pellets can also be used in the feeding program to supply fiber and other nutrients. A "green or reptile salad" plus supplements (e.g., calcium carbonate and vitamin-mineral supplements) may also be fed. The use of pellets mixed within the salad will reduce the need for such supplements, though it is always good to monitor for selective feeding. Normally these reptiles can be fed three times weekly, but if they are very active, daily feeding will be more appropriate. The most common nutritional concerns with herbivorous reptiles are related to calcium deficiency, which causes metabolic bone disease when animals are not properly supplemented. Other nutritional concerns would be related to overall energy and/or protein deficiency.

EGYPTIAN TORTOISE (*TESTUDO KLEINMANNI*) (BODY WEIGHT: 165 G)

Ingredients	Amount (g, as fed)	Frequency
Reptile salad[1]	5 g	Monday, Wednesday, Friday
Alfalfa hay (leaves)	3 g	Monday, Wednesday, Friday
Herbivore pellet[2]	2 g	Monday, Wednesday, Friday

Approximate composition (DMB)	
Protein	12%
Fat	3%
Calcium	0.6%
Phosphorus	0.4%
Ca : P	1.5 : 1
Vitamin E	90 IU/kg
Metabolizable energy	2,000 kcal/kg

Notes: [1]Reptile salad might consist of a mix of vegetables and other produce that can include kale, romaine, cabbage, nappa, spinach, banana, apple, sweet potato, carrot, tomato, hardboiled egg, and avocado. Vitamin-mineral supplements plus calcium carbonate can be mixed with the salad. [2]Commercial high-fiber pellets.

AMPHIBIAN DIETS

Adult amphibians are carnivorous and consume diets very different from those consumed by the larval forms (fig. 16.2). Adult amphibians, including salamanders, caecilians, and frogs, mainly consume invertebrates. The most common and commercially available foods used in captive diets are earthworms, bloodworms, white worms, glassworms, brine shrimp, springtails, mealworm larvae, common crickets, waxworms, fly larvae, and feeder fish (e.g., guppies, goldfish, minnows, and whole smelt), among others. Occasional use of neonatal mice (pinkies) may be incorporated into the diet of adult amphibians. Fat-soluble vitamins (e.g., vitamins A and D_3) must be carefully assessed to avoid metabolic bone disease. Tadpoles require drastically different diets, as they are generally herbivorous or omnivorous. Tadpoles can be fed commercially available fish food flakes or pellets, as well as commercial spirulina tablets, aquatic vegetation, and algae. Supplementation of these diets with B-complex vitamins has been suggested in order to reduce or avoid the spindly leg syndrome previously described. Also, water quality must constantly be monitored to ensure that the tadpoles are in a healthy environment.

DIETS FOR PUERTO RICAN CRESTED TOAD (*BUFO LEMUR*) TADPOLES[1]

Aquarium herbivore diet	24.84%
Tetra FD-Menu 4-in-1 blend	24.84%
Sera San Color Enhancing Flakes	24.84%
Murex Spirulina Flakes	24.84%
Ascorbic acid (coated)	0.063 g

Approximate Composition (DMB)

Protein	51.1%
Fat	5.8%
Calcium	2.5%
Phosphorus	1.45%
Ca/P	1.70
Vitamin A	5,800 IU/kg
Vitamin D	3,300 IU/kg
Vitamin E	400 IU/kg

Note: [1]Brands listed may not be available as described originally in Lentini 2000. B-vitamin complex (including thiamin) can be added to the above mix (e.g., one tablet).

ADULT PUERTO RICAN CRESTED TOAD (BODY WEIGHT: 60 G)

Ingredients	Amount (g, as fed)	Frequency
Cricket, 1.9 cm (3/4 in.)	3.0	Monday, Wednesday, Friday
Red worms (wigglers)	0.6	Saturday only
Fruit flies[2]	0.02	2 × / week (enrichment)
Amphibian supplement[3]	Dusted on crickets	

Approximate composition (DMB)

Protein	61%
Fat	19%
Calcium	3%
Phosphorus	0.8%
Ca : P	3.5 : 1
Vitamin E	60 IU/kg
Metabolizable energy	3,100 kcl/kg (calculated)

Notes: [1]Crickets (*Acheta domestica*: approximately 0.32 g each (3/4" size). [2]Enrichment item. Other invertebrates can be also used for this purpose. [3]Amphibian supplement might be similar to the following composition on a dry matter basis: crude protein 8.6 %; fat 1.4 %; calcium 23 %; phosphorus 0.4 %; vitamin A 280 IU/g; vitamin E 300 mg/kg; vitamin C 7,000 mg/kg.

AQUARIUM DIETS

There is a lack of information about fish nutrition and diets for wild fish maintained in aquariums. Most of the information available is related to commercially farmed species. Very little is known about the nutrient requirements of wild fish. Keepers are learning on a daily basis, and they play an important role in feeding fish, particularly in mixed-species exhibits. An example of a mixed-species fish-exhibit diet fed to 60 different fish species in the aquarium at Epcot (Walt Disney World, Orlando, FL) is given below. The composition of the whole diet is also given. The keeper's role is important in target-feeding different species of fish due to their different physiological and digestive tract morphologies (differences between carnivores, omnivores, and herbivores). A centralized kitchen operation can be important in keeping the diets consistent and well-tracked.

MIXED-SPECIES FISH EXHIBIT DIET (GIVEN DAILY)

Broadcast diet	% of diet (by weight)
Krill	4.7%
Clam meat	5.0%
Spirulina flakes	0.94%
Aquatic gel[1]	46.83%
White shrimp	2.41%
Peas	0.95%

Mixed fish food	
Squid	4.95%
Capelin	9.74%
Glass minnow	2.73%
Lake smelt	2.73%
Silverside	2.85%

Other food	
Romaine lettuce	6.70%
Commercial pellet	9.53%

Approximate composition (DMB)

Protein	58.9%
Fat	9.5%
Calcium	1.9%
Phosphorus	1.5%
Ca : P	1.3 : 1
Vitamin E	52 IU/kg
Gross energy	4,870 kcal/kg

Notes: [1]Commercial aquatic gel might include fish meal, as well as mineral and vitamin supplements.

SPOTTED EAGLE RAY (*AETOBATUS NARINARI*)

Ingredients	Amount (g, as fed)	Frequency
Squid	480 g	Daily
Shrimp	250 g	Daily
Clam meat	300 g	Daily
Supplement	1.5 g	Daily
Calcium carbonate	1.0 g	Daily

Approximate composition (DMB)

Protein	81%
Fat	4%
Calcium	1.25%
Phosphorus	1.05%
Ca : P	1.2 : 1
Metabolizable energy	4,470 kcal/kg

ACKNOWLEDGMENTS

The author would like to acknowledge Kathleen Sullivan for her editing assistance and for her contributions to the chapter.

REFERENCES

AZA. 1999. "NAG, Nutrition Advisory Group, Fact Sheet 010, Quality Control of Feedstuffs: Nutrient Analyses." Nutrition Advisory Group (AZA) proceedings.

Comparative Nutrition Society (CNS) proceedings: 1996, 1998, 2000, 2002, 2004, 2006, 2008, 2010.

Crissey, S . 2005. "The Complexity of Formulating Diets for Zoo Animals: A Matrix." *International Zoo Yearbook* 39:36–43.

Fact Sheet 001. Accessed online at http://www.nagonline.net/Technical%20Papers/NAGFS00697Hay_Pellets-JONIFEB24,2002MODIFIED.pdf.

Fowler, Murray, and Rebecca E. Miller. 1999. "Zoo and Wildlife Medicine" Saunders, 1978, 1986, 1993, 1999, 2003 Veterinary Clinics of North America. *Exotic Animal Practice* Vol 2. 1999.

Lentini, Andrew. 2000. *Puerto Rican Crested Toad (*Peltophryne lemur*) SSP Husbandry Manual.* Keeper and Curator Edition. Scarborough, ON: Toronto Zoo.

Maynard, L., J. Loosli, H. Hintz, and R. Warner. 1979. *Animal Nutrition.* McGraw-Hill.

McDowell, L. R., and J. Arthington. 2005. *Minerals for Grazing Ruminants in Tropical Regions.* Gainsville: Institute of Food and Agricultural Sciences, University of Florida.

National Research Council. 2001. *Nutrient Requirements of Dairy Cattle.* Washington: National Academy Press.

Oftedal, O., and M. Allen. 2001. "Nutrition and Evaluation in Zoos." In *Wild Mammals in Captivity: Principles and Techniques.*1996. Edited by D. Kleiman, M. Allen, K. Thompson, S. Lumpin, and E. Stevens. The Digestive System of Vertebrates. CD. North Carolina State University, College of Veterinary Medicine, Biomedical Communications.

Valdes, E V. 2010. "Nutritional Resources to Meet the Challenges of Feeding Captive Wildlife." Proceedings of North American Veterinary Conference, Orlando, Florida.

Wright, Kevin, and B. Whitaker. 2001. "Diets for Captive Amphibians." *Amphibian Medicine and Captive Husbandry.* Krieger.

17

Recordkeeping

Jean D. Miller

INTRODUCTION

KEEPERS HAVE THE BEST KNOWLEDGE OF THE ANIMALS IN THEIR CHARGE

Of all zoo employees, keepers have the closest association with the animals in an institution's collection. As such, they are the ones who know the details of daily feeding, the normal behaviors and activities, and the physical condition of the animals in their charge. This information is of no overall value unless it is shared by colleagues, both at the facility and at other facilities around the globe. The best mechanism for sharing this information is the written record.

This chapter will provide basic information about animal record keeping as it affects and is affected by the keeper position. After studying this chapter, the reader will understand

- the history of recordkeeping, its growth, and its importance
- the keeper's role in the creation of animal records
- the basic principles that make good records
- the importance of keeper reports and what they should include
- how the information a keeper records is used at the keeper's institution and by others outside the institution
- the role of regulatory permits and their importance to the institution
- recordkeeping as it relates to animal shipping and animal medical records.

WHY KEEP RECORDS?

Records gathered over time can provide an archive on individuals, and a larger number of records gives a more accurate picture of a species. The records of births and parentage provide the genetic information on which studbooks and management decisions are based. Including the identification numbers at previous and/or subsequent institutions creates

a link to information at those facilities and helps provide an animal's entire history, thus adding value to a record. These written records, by themselves or when combined with other information, provide justification for actions and are the foundation for annual reports, both institutional and governmental.

Written records also provide an efficient communication tool, both within the facility and between institutions. Remember the game of "telephone," in which each child whispers to the next a sentence he or she has just heard? Very rarely does the sentence heard by the last child exactly match the sentence said by the first. Zoo and aquarium keepers must have accurate information, and that information is more reliable when written.

RECORDS HAVE CHANGED OVER TIME

In the late 19th and early 20th centuries, zoo acquisitions and dispositions were logged in large journals, usually one year per journal. Keepers often kept information they needed or felt was important on scraps of paper in their pockets; some may have kept journals of daily information, which they took with them when they left the institution. Regardless of what was recorded, the fragility of the paper containing these records was also an issue: it had to be safeguarded from deterioration or loss. So by the end of the last century, many facilities had begun using the computer to maintain records as text documents or in simple databases. Gradually, too, animal managers began realizing that valuable information that could help them solve current problems was disappearing as staff retired from or left the facility. They also realized the advantage of being able to obtain information from other zoos. So these forward-thinking individuals began keeping uniform records on the animals and activities in their facilities.

In the early 1970s, Drs. Ulysses S. Seal and Dale Makey took the lead in standardizing zoo animal records, and proposed an organization called the International Species Inventory System (ISIS)—now named the International Species Information System (www.isis.org). ISIS in turn developed a computer program to gather animal information from zoos and aquariums around the world into a global, centralized database which could be used for computerized animal management. The program, called the Animal Record Keeping System (ARKS), did accomplish its goal of gathering information in a standardized format, but especially in the early years, program limitations required the information to be recorded in such a way that it did not always reflect the actual facts. For example, the program allowed only one sire (male parent) to be recorded for an animal. So even if a keeper knew that the sire was one of the three males with the female, the sire had to be recorded as "unknown" and a note would have to be entered elsewhere in the record about the three males.

From its inception, ARKS has required that each institution create a new record when it acquires an animal, so that each facility at which an animal has lived has its own records for that animal. Thus, a single animal may have two, three, or more records, each representing a portion of its life. These independent portions of a specimen's history are linked together in the global database through the use of ISIS numbers assigned by an institution to each specimen and reported to ISIS, to thus provide the animal's complete history. This also meant that an animal would have as many ISIS numbers as institutions at which it had lived; an animal that had been kept in three institutions would have three different ISIS numbers. These ISIS numbers also make available basic information about each ISIS member's specimens to all other ISIS members via the ISIS website's global database.

Through the intervening years, however, succeeding upgrades to ARKS have allowed ever more accurate recording of information, and greater numbers of institutions have joined the ranks of ARKS users. Some larger institutions, while remaining ISIS members, have created records programs that meet their own particular needs; but most nevertheless transmit their information to ISIS in a format compatible with that database. In early 2013, ARKS was used in almost 850 zoos and aquariums in 84 countries, and the ISIS database contained more than 2.6 million specimens of 10,000 species and subspecies (www.isis.org). The current version of ARKS will soon be replaced by the comprehensive online Zoological Information Management System (ZIMS), being developed by the global zoological community. ZIMS is an extremely complex program that will store all information about an animal in one easily accessible place and will eventually include a medical module that will replace the existing medical program, MedARKS. A facility acquiring a specimen will simply add its information to the existing record rather than creating its own new record; an animal will have only one primary ID number in its lifetime. ZIMS will be a single, global, real-time, web-based database. However, because of the program's complexity, ISIS can only convert a few institutions at a time from ARKS to ZIMS. This conversion, called "deployment," began in March 2010, and by November 2012, 420 institutions around the world were using ZIMS.

KEEPER DAILY REPORTS

WHAT KEEPER DAILY REPORTS ARE AND HOW THEY ARE USED

All institutions have, or should have, a system of keeper daily reports, also called "keeper dailies" or "keeper reports": a form, completed each day, into which keepers from each area write detailed information about the lives of the animals in their charge. There is no standard format, and keeper daily reports can be handwritten or electronic. Ideally they should follow the same format within an institution, and if electronic, they should be in a searchable format at an easily accessible location. An institution usually designs it own keeper daily report form, but designs are often willingly shared among facilities. All daily report forms should include sections for changes to the area's inventory, medical information, and enrichment and training activities. Also included could be information about supply requisitions and deliveries, "housekeeping" activities, maintenance requests and accomplishments, and any other information the keeper feels is important to document. Two examples of keeper reports are shown in tables 17.1 and 17.2.

The keeper daily report has come to be an essential in the zoo field: it is the means by which keepers transmit essential information to those who need it to fulfill other duties. A keeper returning from a weekend off can read the reports for those days to get caught up on what has transpired during the time away. Senior animal management staff is apprised of births, deaths, illnesses, breeding, and other information necessary for the management of the collection. The registrar enters this same information into ARKS (or the institution's inventory system, if it does not use ARKS), and depending on staffing resources at a facility, may also enter other detailed information. Information about the health and condition of collection animals, including whether they are taking medication as prescribed and the effect of those medications, is gleaned by the veterinary staff. The commissary ("diet kitchen") staff learns of diet changes that may require more or fewer food items or larger or smaller quantities of food. The zoo director can keep current on all animal department activities, and the public relations staff can learn about newly acquired animals or those that have died or been shipped out.

Please note, however, that not all information should wait to be put on a keeper daily report. Critical information—information that might affect the well-being of an animal or staff if not passed along quickly—should be reported immediately to a supervisor or the veterinary staff. It should later be placed on the keeper daily report as well.

WHY KEEPER DAILY REPORTS ARE SO IMPORTANT

Keepers often transmit information verbally—they discuss it among themselves, update a person who has just returned from days off, or discuss issues with senior staff, curators, or veterinarians. However, the keeper daily report is the formal record by which information about the collection is documented. Past keeper daily reports can be referred to when there is a question about when an activity started, what an animal did at the same time of year in the past, or the

TABLE 17.1. Sample keeper report. Courtesy of Buffalo Zoo.

Tuesday, April 20, 2013

VANISHING SOUTH DAILY REPORT

KEEPER: Keeper A

CHANGES IN STOCK (births, deaths, arrivals, departures, transfers):

ACTION	ISIS / ID	NO. & SEX	SPECIES	SIRE / DAM	COMMENTS

ANIMAL HEALTH   (diet, illness, injury, treatments):

Joan Celebes	Prenatal vitamin 1x, Retrovir 1x

SUPPLIES NEEDED:

HOUSEKEEPING:

MAINTENANCE REQUIRED:

REMARKS (breeding activity, behavioral changes, observations, other information):

ENRICHMENT: (using enrichment codes)

All –	Lemur 2 – 27 (shelled) (2x)
Celebes – 3w/27	Snow1 – 1 (pine tree), 27 (shelled)
DeBrazzas – 1 (pine), 27 (shelled)	Snow 2 – 1 (pine tree from S1 exhibit), 27
Lemur1 – 89/96, 16	

Celebes	DeBrazzas	Lemur 1	Lemur 2	Snow monkey 1	Snow monkey 2
Joan - 86M265	Colonel – 88M57	Kid – 84M41	Zolton – M03050	Billie – 97M131	Ki – M01081
Shanta - 88M181	Suzette – 91M130	Roy – M0039	Zelda – M03051	Bobbie – 97M132	Betty – 97M133
Suze - M0086	Houdini – M0034		Hermione – M03052	Debbi – 98M125	Cathy – 97M134
Sandy - M02035	Keely – M01046		Alexander – M03053	Eric – 99M54	
Tango - M02055	Ashley – M02042		Jamie – M03054		
TJ - M03056					
Marie – M03071					

details of a past action. These past reports serve as primary sources of information in case of discrepancies in other records, provide proof of compliance with various regulations, and indicate who has worked in a given area. They can also provide historical information for use in retrospective scientific studies. Keeper daily reports should therefore be kept as part of the institution's permanent records in environmental conditions that protect their integrity, are secure from natural or man-made disasters, and are accessible when required. Backup copies of electronic keeper daily reports should also be archived. For more on records retention policies, contact a local certified records manager (CRM) or a chapter of the Association of Records Managers and Administrators (ARMA) International.

TABLE 17.2. Sample keeper report.

Keeper Daily Report	Date: 2011-04-10

Section/Area *Gorilla South*

Keepers Jane Doe, John Doe

Daily Events

Health

1.0 Snow monkey m 08072 "Yakou" Acting very "off" this am. Not eating, looking very uncomfortable. Was immobilized, X-rayed and ultrasounded. Was found to be retaining a large amount of urine. Bladder emptied and flushed. Waking and held in restraint cage for follow up injections and treatments. Kidney biopsy scheduled for Wednesday.

0.0.1 Gorilla baby (no i/d assigned) looks good and strong, only observed nursing once today. Not much time on observations. Dam and baby separated in holding. Gorillas in exhibit with no creep.

Maintenance Require creep to be constructed in gorilla exhibit to facilitate introduction of dam and baby to the group

Supplies

Housekeeping

Horticulture Browse required for primates, some large leaves for gorillas.

Remarks

Enrichment
Gorillas: whole produce & treat tubes. Lemurs: trail mix. Celebes: corn husks.
Snow monkeys: Trail mix. Meerkats: scattered live crickets.
Responses have been recorded in the enrichment log.

WRITING KEEPER DAILY REPORTS

Keeper daily reports should be considered a type of scientific paper: after all, they record events, activities, and observations that will be used in the science of animal management. As such, *reports should be accurate, include solid facts and precise measurements, and provide complete and unbiased descriptions of events and observations.*

Above all, *accuracy* is expected in daily reports. A good rule is to write down notes about events and observations *as they occur.* Memory is notorious for losing information—if data is not written within a short time, there is risk of the details being inaccurate when recalled later, or lost completely (Miller and Block 2004, 6). For this reason, the keeper should carry a small notebook and pencil and write down as soon as possible the details of treatments, change in an animal's routine, an incident, or anything else that is noteworthy. (Some larger institutions provide palm-sized computers, which can automatically download their contents into the keeper daily report.) Refer to these notes when completing the keeper daily report.

Similarly, don't guess. Record only the known facts. For instance, suppose an exhibit contains two male basilisk lizards and four females. Male #1 is seen breeding female #1, and then several days later a clutch of eggs is found. While it is a reasonable conclusion that male #1 and female #1 are the parents of these eggs, the truth is that *any* of the animals in the exhibit could be the parents. The parents cannot be recorded as "unknown," because the animals in the exhibit are known. Unless there is some proof that eliminates any of these other animals, both males and all four females must be recorded as the parents. Note: parentage for the purposes of

Good Practice Tip: When returning to work after time off, always review the area's keeper reports to see what has happened on missed days.

Good Practice Tip: Carry a notebook and pencil to record events and observations as they occur.

record keeping is determined at the time of conception, not at the time offspring hatch or are born.

There should be *consistency* across all keeper daily reports at an institution: use the same terms and references. The reader should not have to know that keeper Jane always refers to *Podocnemis unifilis* as the "yellow-spotted river turtle" but keeper John refers to it as the "Amazon turtle"; the use of the scientific name leaves no doubt. When using comparative scales, always use the same end of the scale as "the best" and the other as "the worst." In other words, one keeper shouldn't use 1 as the best and 10 as the worst, while another keeper uses 10 for the best and 1 as the worst. Use the same measuring system: don't report an animal's weight in kilograms at one time and in pounds at a different time, or report length in inches for one lizard and in centimeters for another.

Writing should be efficient, allowing the reader to grasp and understand the facts using the fewest words possible. Reports should convey to the reader the exact situation, without interpretation.

Information should be specific and provide details, but only those details that help explain the event or information. A report that says "Showing courtship behavior" is insufficient. Instead, provide the details, as in: "Following female closely and displaying flehmen." Don't use unnecessary words, and write in the active voice rather than in the passive voice; write "Offered Metacam twice with food; ate it all." Be careful and mindful to use pronouns correctly: a pronoun refers to the last noun used. The following is unclear: "The keeper went into the sheep's stall to clean and she slipped." Depending on the intended meaning, write instead: "The keeper slipped and fell when she went into the sheep's stall to clean" or "The keeper went into the sheep's stall to clean and the sheep slipped and fell."

Know the normal behaviors of the animals in your charge. For example, a lip-curl in a dog species may be either a snarl or an indication of submission. If an unusual behavior or change in behavior is recorded, there should also be a report of when behavior returns to normal. It is acceptable to write "Treatment will be continued daily until further notice"; but once the treatment is discontinued, remember to record that fact. Don't forget to also record changes in condition or behavior during and after any medical treatment, diet change, or vet visit.

Reread entries to make sure they make sense and say what is intended. The facts should be stated clearly and without ambiguity, and with the assumption that the reader has no knowledge of the immediate circumstances or of what has happened recently. Some examples:

- "Laid 5 eggs on exhibit; eggs not being incubated." Does this mean that the parent has abandoned the eggs, or are the eggs being discarded as a management tool? Reporting "Eggs were discarded since parents are siblings" leaves no doubt.
- "The monkey was scared and ran to the bottom of the cage and the armadillo attacked him." Report instead: "The monkey ran to the bottom of the cage, where the armadillo bit the monkey's tail."
- "Gave Metacam and Baytril twice." Does this mean that both medications were given twice, or that Metacam was given once and Baytril twice? Say instead: "Received Metacam once and Baytril twice today."
- "Ate full diet; weight = 539 g." Is this the weight of the diet or of the animal being fed? Was the weight taken before or after the feeding? "Was weighed prior to feeding: 539 g; then ate all 50 g of diet" is much clearer.
- "The one treated yesterday was given medicine again today." Specify the animal's identity and the medication each day. Be clear if the medication was completely consumed (received) by the animal or only partially consumed.
- "Veterinarians prescribed .10 ml Baytril to be given . . . ." Write instead "Veterinarians prescribed 0.10 ml of Baytril to be given . . . ." Always include a zero before the decimal place to be sure that the reader doesn't miss the decimal and inadvertently give a tenfold (or greater) excess of a drug.

Identify a subject animal by *multiple means of identification*: use a combination of species, sex, ISIS number, other identifier, and even age. In other words, referring to "the female Gouldian finch" is not sufficient when there are two females in the collection. Instead, use a second (or third) identifier: "the female black-faced Gouldian finch" is better, but "female Gouldian finch ID #B234" or "the female black-faced Gouldian finch, band NZ 3 right leg" leaves no doubt as to the bird's identity. When only one or two identifiers are used and one is incorrect, confusion on the part of the reader is almost certain; using a third identifier, however, will verify one of the other two. So in the case of specifying an individual, more identification is better.

ABBREVIATIONS AND SLANG

Though abbreviations and slang terms are very tempting and are commonly used verbally, most have no place in keeper daily reports. There are several reasons for this: terms may be misunderstood by those readers not familiar with slang terms used in animal husbandry, similar abbreviations may exist in other disciplines and have different meanings, and slang or abbreviations may change over time. *Keeper daily reports are meant to be lasting records and need to be abso-*

lutely clear in meaning, regardless of who reads them or when they are read.

Similarly, shorthand and abbreviations (including those used in texting, Twitter, etc.) generally have no place in keeper daily reports. There are, however, a few exceptions. One of the few "abbreviations" that can be used is the symbol for an animal's sex: the universally recognized symbol for a male is ♂ and the symbol for a female is ♀. In addition, the zoological community has adopted a shorthand to refer to the number and gender of animals: two or three numbers separated by periods or slashes, where the first number refers to the number of males, the second refers to the number of females, and the third refers to the number of animals of undetermined gender. For example, a group composed of two males, four females, and their seven offspring of undetermined sex would be represented as 2.4.7 or 2/4/7. A single animal of undetermined gender is written 0.0.1 or 0/0/1. If there are animals of only one gender, the other two sections are filled with zeros, as in 0/0/4 (four undetermined) or 2.0.0 (two males). In the event of zeros in the third place, the number may be shortened, as in 0/4 or 2.0. Other acceptable abbreviations are those for the universal units of measure (American, Imperial, metric, etc.) and some of the more common medical terms, such as qid (four times per day). Medical abbreviations, though, should be used very sparingly.

In summary, when writing keeper daily reports,

- be accurate
- report complete information but provide only the relevant facts, without interpretation or embellishment
- use multiple identifications when referring to an animal
- avoid using slang and abbreviations.

ZIMS AND KEEPER DAILY REPORTS

ZIMS will in essence reverse the reporting role of the keeper daily reports. Currently, keepers enter information into their keeper daily reports and the registrar or records keeper copies relevant information into the appropriate ARKS specimen record. Once an institution converts to ZIMS, however, the expectation is that keepers will do all data entry directly into ZIMS and the registrar, curator, director or other zoo staff member can then create a report of activity for a selected time period, such as the previous day. In effect, ZIMS will create the keeper daily report based upon the data entered by the keeper, rather than the keeper entering information into the keeper daily report.

INDIVIDUAL AND GROUP RECORDS

Most animal records are individual records; they contain information only on a single animal (Miller and Block 2004, 14). The animal is identified from all others of the same species by means of some identifier: a natural marking or physical characteristic or an artificial device that makes the animal unique. Natural identifiers include coat or skin markings, and defects such as broken horns or kinked tails. Artificial identifiers include ear tags, leg bands, transponders (microchips, PIT [passive integrated transponder] tags), tattoos, and ear notches. Keepers and staff often refer to an animal by a combination of identifier types, as in "the green-tagged brown-spotted male goat."

In some situations, a group record is more practical. A group record contains information on several or many individuals of the same species or subspecies. This type of record is used when

- it is not possible to tell one individual from another because body color is uniform and there are no distinguishing markings (e.g., with vampire bats);
- when the number in the group is very large (e.g., with a colony of honeybees); or
- turnover in the group is rapid (e.g., with insects that have a life span of a matter of weeks).

Note: this record type is not to be confused with that used for animals physically housed together—for example, a herd of addax. Because the animals in that herd each have identifiers, each will have an individual record.

Regardless of the type of record, it is associated with an accession number: a string of numbers and/or letters "assigned by the recording institution, unique to one specimen (or group), and used to identify that specimen (or group) in the files of the recording institution. The accession number is tied to the physical characteristics (tag, tattoo, identifiable color patterns) and transaction information about that specimen (or group). In essence, it is a key or code to a specimen and its history" (Miller and Block 2004, 15). If the facility is an ISIS member, the accession number is almost always the same as the identification number forwarded to ISIS through ARKS. (Note: in ZIMS, information is not forwarded to ISIS as it is in ARKS; rather, data is entered directly into the central database.) All official reference to a specimen or group, whether in-house or external to the institution, should be by accession number.

INSTITUTIONAL ANIMAL RECORDS

PERMANENT ANIMAL RECORDS

The institution maintains permanent animal records: documents pertaining to the individuals and groups at the facility. These include, but are not limited to, pedigree information, transaction documents, permits and (where applicable) import/export documents. Historic information may be sparse, since it has been only in the last decade or so that animal records have been given attention. But recent information may include breeding summaries, rearing information, or enrichment and diet logs; and if an animal has been the subject of behavioral studies, the results of the study are often included, sometimes as copies of scientific papers. Medical records are also considered to be "animal records," though they are usually maintained separately by the veterinary staff. As the designation indicates, these records are not discarded even after the animal dies or leaves the institution.

The method by which these permanent records are filed at the institution will vary. Some facilities maintain "specimen folders," which contain copies of all papers relevant to a particular individual or group. Other facilities may keep separate files for separate types of documents, such as current permits, outdated permits, transaction documents, and breeding loan agreements. Sometimes both systems are used, with originals in one location and copies in other locations.

OTHER INSTITUTIONAL RECORDS

Keeper-provided information also forms the basis of other types of records and reports maintained by a facility.

- **Annual inventories.** The registrar prepares a census of animals at the beginning and end of a time period, usually corresponding to the institution's fiscal year (e.g., 1 January to 31 December). Also included in the annual report are numbers of births and other acquisitions, and deaths and other dispositions during the time period. Most of this information comes directly from the keeper staff via keeper daily reports.
- **Breeding summaries.** These are complete descriptions of methods and procedures undertaken in efforts to breed a species. Specific details and observations by keepers are essential, and information about parameters in which reproduction fails is just as important as those that succeed in producing offspring: both add to the body of knowledge about captive species. Keepers usually prepare these reports.
- **Diet information and feed logbooks** ("logs"). These are records about the intake or refusal of food, specific diet items and amounts, presentation of diets, special diets, and so on.
- **Enrichment logs.** These detail items presented to animals to enhance their captive activities and behavior, the acceptance of the items by the individuals, and the reactions of the animals to the items. Depending on institution policy, these records might be kept in the keeper's area or in a central location.

INFORMATION SUPPLIED TO OTHER ORGANIZATIONS

Keeper-reported information is also used outside the institution.

- The protection of native and exotic wildlife is overseen by national, state, territory, and local government agencies, all of which may require zoos and aquariums to be licensed to carry out their normal activities and operations, such as holding endangered species.
- All permit processes involve an application justifying an activity, fact-finding by the permitting agency, and the issuance (or denial) of a permit. Information about the animals involved is an essential part of the application.
- Reports may be required as a condition of permits or licenses. These reports are a summary of how the permit or license was used, and most often contain inventory information and/or transaction information.

In order to manage species on a wider basis than at the institutional level (e.g., at the regional level), information on the individuals in that region must be gathered from each facility that holds or has held the species. A studbook keeper maintains records on all animals of a single species in a defined region. (Many zoos and zoo associations encourage keepers to become studbook keepers.) This genealogical database, the studbook, primarily includes information on the parentage of the specimens and the facilities at which they have been or can be found, and is the database on the basis of which breeding (or nonbreeding) recommendations are made. The studbook keeper periodically requests updates from the institutions that hold the species—information that ultimately arises from the keeper daily reports. Each animal listed in a studbook is assigned a studbook number, by which it is identified when discussing management of the species; this number will be unique to specimens of that species within the scope of the studbook. Each institution holding a specimen in a studbook should include that studbook number in the specimen's permanent record, and it is the responsibility of participating institutions to verify the accuracy of their specimens' information in the studbook. If a studbook animal is transferred to another facility, the studbook number should always be forwarded to the receiving institution. (Note: when ZIMS is fully implemented, studbook numbers will not be necessary since each animal will already have a unique global number.)

RELATED TOPICS

ARKS AND MEDARKS

Depending on the institution, keepers may or may not enter information into the institution's ARKS records, for which special training is required. The Medical Animal Records Keeping System (MedARKS) is the companion program to ARKS, and it uses the ARKS database as a framework upon which medical information for each animal is recorded. Because of the complexity of the program and the highly technical data entry required, MedARKS will most likely not be part of the keeper's normal knowledge base. It is mentioned here so that the keeper-in-training will have at least some general information about it. (As mentioned above, MedARKS will eventually be phased out as medical records become part of the ZIMS program.)

GOVERNMENTAL REGULATION

The zoo and aquarium field has become highly regulated by many agencies at various levels of government. National agencies have authority over certain activities throughout a country as well as over imports into and exports from that country. Individual states, territories, or provinces may also have agencies that generally mirror the national level, but which regulate imports into, or activities that take place within, a single state, territory, or province. Many of these regulations have direct or indirect bearing on the keeper position.

- The protection of native and exotic wildlife. As mentioned previously, one of the main charges of these agencies is the regulation of the possession and import/export of species determined to be endangered or threatened. They also regulate possession of nonendangered native wildlife. If a country is a party to the Convention on International Trade in Endangered Species of Wild Fauna and Flora (CITES), these agencies also enforce restrictions placed by that treaty.
- A safe work environment. Governments often regulate the physical workplace and the chemicals and other substances a worker uses in the performance of daily duties. Keepers need to be cognizant of safety procedures when working with their animals, and they should be constantly on the lookout for items that need repair and tasks that could be modified for their own increased safety and that of other employees and animals. Dangerous conditions should be reported immediately to a supervisor, but should later be placed on the daily report as well.
- Diseases and public health. An important disease and public health concern in zoos and aquariums is zoonosis, the transmission of disease from animals to people. Regulatory agencies may restrict the import or transport of certain species known to carry diseases such as plague, rabies, and hepatitis. Certain animals, especially primates, are susceptible to human diseases, so keepers who are not feeling well need to consider their animal charges when deciding whether or not to come to work.
- Animal welfare and the protection of commercial agriculture. Many governments regulate animal welfare: the physical, psychological, nutritional, and social aspects of captive animals' lives. These same agencies may also be charged with safeguarding the country's agriculture industries and natural resources from possible diseases or plant pests that could be spread by the transport of animals.

LOAN AGREEMENTS

A zoo or aquarium (or its operating body) usually owns the animals exhibited at its facility, but there are two exceptions.

- When a facility exhibits or holds specimens that are owned by another entity, those specimens are said to be "on loan in" to the facility.
- Conversely, when an institution's specimens are being held elsewhere, they are "on loan out" at the holding facility.

In each case, legal title to the specimen is retained by the owner while the animal is physically at another facility. A loan agreement is the document detailing the rights and responsibilities of each party, and may be for exhibit or breeding purposes. A holding institution must inform the owning institution of all changes to the loaned animal's status and/or condition, and must defer to the owner's directives in all decisions. It is therefore important for keepers to be aware of the ownership of the animals in their areas. Breeding loans also contain specific directions for the allocation of ownership of any offspring born to or sired by the animal on loan. Offspring born at the facility but owned by another institution are termed "born on loan in"; an institution's offspring born elsewhere are termed "born on loan out." Loans usually contain a requirement that the holding institution provide to the owner a yearly update on the animal's condition together with information about any offspring the loaned animal may have given birth to or sired during the year. The information for these annual loan update reports originates in the information that keepers supply.

Occasionally, the government of a country will claim ownership of all individuals of species found in the wild in that country, even when those specimens are held in captivity in other countries; these are referred to as government-owned species. Usually a specified institution is authorized by the foreign government to manage the individuals (i.e., to act as that government's agent). Examples are golden-lion tamarins, owned by the Brazilian Institute of Environment and Renewable Natural Resources (IBAMA) and managed by the regional associations; tuataras, owned by the minister of conservation of the government of New Zealand, also managed by regional associations; and radiated tortoises, owned by the government of Madagascar and managed by the Durrell Wildlife Conservation Trust. In these cases, a holding institution has the same responsibilities to the managing facility as are normal with other loans.

ANIMAL SHIPPING

The physical transfer of animals from one institution to another is a necessary and routine part of a keeper's life. While the registrar or shipping coordinator and veterinarian are responsible for the legal documentation that accompanies an animal, the keeper responsible for the animal provides the document that can help ease an animal's transition to the routine at a new facility: a summary about the animal and its diet, enrichment, training, and general preferences. (In the United States and Canada, this is an American Association of Zoo Keepers [AAZK] animal data transfer form [ADTF]). This is the written vehicle by which the current keeper communicates personal observations and knowledge to the future keeper; it represents probably the only instance in which the keeper can be less than scientific in the presentation of information. The current keeper's contact information is also included, and the keeper at the new facility is encouraged to communicate directly with the previous keeper as necessary.

COLLEAGUES LOCAL AND DISTANT

The animal staff at an institution no longer works in isolation from the rest of the zoological community. Many links exist between zoological institutions at all levels, between zoological facilities and the commercial industries that supply them, and between zoological facilities and academic institutions. Regional zoo and aquarium associations are now found in all parts of the world, and, thanks to the internet, individuals and institutions can develop relationships with other institutions and conservation organizations around the globe. For example, associations of records keepers in several regions are dedicated to promoting standardized records practices in their regions, and provide information to each other. They also serve as primary resources should a keeper want to learn more about recordkeeping.

SUMMARY

Unlike the animal keepers of long ago who often functioned in isolation from the other keepers at their facilities, the keeper today must know much more than simply the husbandry of the animals in their own care. The modern keeper must also understand that knowledge needs to be shared for the good of each animal, the species, and the institution. Therefore the keeper must know how to keep reliable, concise, pertinent records about daily activities and the interactions between the animals in his or her care, always keeping in mind the types of information that must be gathered and the various ways in which that information is shared and used.

REFERENCES

Alley, Michael. 1987. *The Craft of Scientific Writing*. Englewood Cliffs, NJ: Prentice-Hall.

Miller, Jean, and Judith Block. 1992 (revised 2004). Animal Records Keeping. http://www.buffalozoo.org/E-manual_short_form _Word_format.pdf.

18

Identification

Erika K. (Travis) Crook

INTRODUCTION

Animal identification (ID) is an important part of successful animal management in zoo and aquarium collections. If possible, each individual animal should be identified so that it can be tracked throughout its lifetime for husbandry, medical, reproductive, and management-related activities. Each zoo and aquarium will determine how they want to identify their animals, as the International Species Information System (ISIS) has no standardized identification methods. Identification is crucial when animals are being transferred between zoos and aquariums, or scheduled for import or export. Some regulatory agencies require the use of permanent identification for the species under their jurisdiction. Identification is also necessary for animals that are in species survival plans (SSP), are important for breeding, or are traveling away from the zoo. Another benefit of individual identification is to recognize a collection animal in case of theft. The goal is to provide a permanent method of identification, and it is acceptable to use multiple methods for one individual. Primary identification is the method that is used most frequently to definitively identify the animal (e.g., microchip) or is easiest to see or read (e.g., ear tag). Secondary identification will provide additional information about the individual and help identify it if the primary method fails. Confirming both primary and secondary identification at frequent intervals is strongly recommended. Sometimes it is not feasible to identify each animal, such as with a large aquarium population or insect group. In these cases, the exhibit name or number becomes a method of group identification.

The ideal identification method (Loomis 1993, 21; Anon. 2009, 38–39; Elasmobranch Husbandry Manual 2010)

- uniquely identifies each animal
- remains unaltered on an individual throughout its lifetime
- has no effect on growth, behavior, breeding, or movement

- is easy and fast to apply, with minimal stress or pain to the animal
- is unambiguous and easier for zoo staff to read, while not being readily apparent to the public
- is cost-effective and easily obtained
- is adaptable to animals of different sizes
- cannot be tampered with or removed by the animal
- does not cause the animal injury
- does not influence the animal's behavior
- can be read in only one way (i.e., cannot be read incorrectly)
- may be temporary on some occasions (e.g., on an individual that is growing, that will be shipped, or that is under medical treatment).

This chapter will discuss methods of identification for zoos and aquariums to use to identify their animal collection. After reading this chapter, the reader should be familiar with

- components of an ideal identification method
- methods for passive identification
- methods for external active identification
- methods for internal active identification
- recommended sites for microchips or transponders
- at least one identification method per major taxon.

PASSIVE IDENTIFICATION

Animal identification can be passive or active. Methods of identification, as well as their advantages and disadvantages, are listed in table 18.1. Passive identification is a noninvasive way to identify an individual animal based on external details or distinctive information. It primarily relies on unique features of the animal. Some of these features are external, such as overall appearance, facial characteristics, body or fin conformation, movement, coloration, distinct markings, or pattern. They may also include behavior and habits that differentiate the individual from others of the same species.

TABLE 18.1. Identification methods and their advantages and disadvantages. Regardless of which ID method is chosen, the identification and records must be confirmed regularly.

Passive identification	Advantages	Disadvantages	Species / miscellaneous
Photograph (black-and-white or color)	• Noninvasive	• May become inaccurate over time (because of changes in the animal due to season, maturation, trauma, or surgery)	• All species
Drawing and diagram	• Noninvasive	• May become inaccurate over time (because of changes in the animal due to season, maturation, trauma, or surgery)	• All species
Written description	• Noninvasive	• May become inaccurate over time	• All species

External active identification	Advantages	Disadvantages	Species / miscellaneous
Ear tag	• Comes in different colors, shapes, sizes and numbers • Can be seen at a distance	• Can fall out of animal's ear or get caught on an object • Is visible to the public • Number may rub off	• Hoofed species (most often) • Carnivores and marsupials
Ear notch	• Can be seen at a distance • Allows tissue to be saved as a genetic sample	• Lacks standardized zoo numbering system • Takes practice to learn to read • May change shape as animal grows • Ear trauma can render number illegible	• Hoofed species (most often) • Rodents and carnivores • A numbering system is suggested in this chapter, but is not officially recognized by AZA or ISIS
Leg band or ring	• Comes in different sizes, colors, letters, numbers, and materials to accommodate multiple species	• Number can wear off or plastic color can fade • Inappropriate size or placement can cause compression or injury • Can get caught on an object • Can be crimped by bird • Keratin can accumulate under the band • If metal, predisposes animal to frostbite • Requires skill for removal	• Avian species (most often) • Small mammals
Wing or flipper band	• Comes in different colors and patterns that can be seen at a distance	• May require replacement of the colored tape or zip ties / cable ties • Same as for leg bands (see above)	• Avian species in general, but most often penguins • Extra vigilance necessary during molting, due to wing swelling
Wing tag or neck band	• Large colorful plastic tags can be seen at a distance	• May be distracting to zoo visitor • Same as for leg bands (see above)	• Free-ranging birds more often than captive birds
Flipper tag	• Can be seen at a distance	• Is visible to the public • Requires care to avoid blood vessels • May fall out or get pulled out • Numbers and color may fade	• Free-ranging pinnipeds and sea turtles
Thumb band	• Can accommodate different color and number combinations • Can be seen at a distance	• Subject to removal or compression by animal • Same as for leg bands (see above)	• Bats
Shell notching (by drilling or filing)	• Can be seen at a distance	• Lacks standardized numbering system • Can cause scute regrowth necessitating renotching • Can cause shell injuries in notched area	• Turtles and tortoises • A numbering system is suggested in this chapter, but is not officially recognized by AZA or ISIS
Living tag	• Entails unique placement of grafted shell pieces	• Surgical procedure • Causes discomfort • Pieces may not adhere	• Released sea turtles
Fin tag • anchor tag • dart tags • trans-body tags • tail loop tags (for temporary use only)	• Can be seen at a distance	• Visible to the public • Can fall out or get caught • Numbers and color may fade • Can cause infection or inflammation	• Fish, including elasmobranchs
Tattoo	• Can incorporate unique international studbook number	• May fade or become distorted • Is painful upon application • Is readable only when animal is in hand or in close proximity	• Many mammal species • Not appropriate for stingrays
Heat branding or freeze branding	• Can be seen across large pastures or exhibits	• Is painful during application and healing • May develop an infection • Is not aesthetically pleasing	• Hoofstock • Generally not used in zoos

TABLE 18.1. continued

External active identification	Advantages	Disadvantages	Species / miscellaneous
Toe clipping	• Provides tissue usable for genetics, histopathology, infectious disease exam, or age study	• Can cause pain, disfigurement, or infection • May impair normal behaviors • May be followed by regrowth of toes • May affect survival	• Wild adult amphibians, mostly frogs and toads • Should be avoided when possible

Internal Active identification	Advantages	Disadvantages	Species / Miscellaneous
Microchip, PIT tag, transponder	• Is permanent and unalterable • Includes unique alphanumeric code • Is small enough to be suitable for most species	• Can migrate or fall out • Requires close proximity to scanner for reading of the microchip • May result in swelling or infection	• Many species • Some reports of cancer

This type of ID, based on physical characteristics, requires familiarity with the animal's external appearance as well as its daily activity and demeanor. Individual variations in color, pattern, body condition, or anatomy can be documented via black-and-white photographs, color photographs, or drawings. If the animal undergoes seasonal changes, or changes as it matures, additional photographs may be necessary. It is recommended that the photographs be printed and laminated or protected and that they be made available near the animal's enclosure, within the keeper record system (folder, binder, etc.), and probably in the medical records. Drawings are used to capture distinctive scale or skin patterns on reptiles and amphibians, but they can also be used for mammals with unique facial or body markings. A keeper familiar with the features used to distinguish conspecifics may be able to draw those important traits for easy identification. Some zoos and aquariums may prefer photographs to drawings because they capture the individual's exact likeness, but others may prefer drawings in order to concentrate on particular details. Some of the unique features, such as behavior or movement, can also be recorded in written descriptions. It is best to combine passive ID methods with other forms of identification.

EXTERNAL ACTIVE IDENTIFICATION

There are many methods for active identification of zoo and aquarium animals in which staff apply an identification device or perform a procedure to ID the animal. Active identification can be external or internal; each method is outlined in table 18.1. External active identification methods currently used in zoos and aquariums include ear tags, ear notches, leg bands, wing or flipper bands, flipper tags, thumb bands, shell or scute notches, fin tags, tattoos, heat branding, freeze branding, and toe clipping. Each of these methods will be discussed along with its advantages and disadvantages. Some zoos and aquariums will place external identification on a particular side of the animal's body to indicate gender. Many zoos and aquariums identify males on the right side of the body and females on the left. Other forms of external active identification are also available, such as wing tags, neck bands, shell tags, living tags, necklaces, fin clipping, toe tags, tail notches, and dye marking; but since they are used less commonly in zoos and aquariums, they will not be discussed here in detail.

EAR TAGS

Ear tags are most often used in hooved animals (ungulates), but are sometimes also placed on carnivores or marsupials (figure 18.1). The tag is placed on the pinna of the ear, which is the flesh-covered cartilage protruding from the head. It is inserted through the cartilage, and visible blood vessels are avoided to minimize bleeding. Plastic or flexible polyurethane tags are available in multiple colors, shapes, and sizes with different numbers. Some of the most common ones in zoos are circular buttons or tags that hang below the ear. Metal tags are usually aluminum, steel, or nickel alloy and come in various colors, shapes, and numbering systems. A specialized tool is necessary for placement of most ear tags. The advantage of an ear tag is that it can be seen at a distance; the disadvantages are that it could fall out, get caught on a fence or exhibit furniture, be visible to the public, or have the number eventually fade or rub off. Debris can accumulate between the ear tag and the ear, so that cleaning may sometimes be necessary. Ear tags should be inspected often so that concerns can be addressed before they become a problem. Some zoos with large hoofstock collections prefer ear tags to ear notches (described below) because they may be easier to see at a distance.

EAR NOTCHING

Ear notches are often used to identify hoofstock. They can also be used for rodents and occasionally for carnivores. A specialized tool called an ear notcher is used to cut a piece of tissue from the ear pinna. This will leave a triangular or round-shaped defect on the ear, depending on the type of ear notcher used. The tool is available in different sizes, enabling different-sized notches to be cut, depending on the size of the animal. The same shape of notch can be cut using a scalpel blade or scissors, but using an ear notching tool is easier and quicker. Ear notching is frequently done in young animals, using manual restraint. Gauze and cauterizing powder may be needed to stop the bleeding. There is no standard method for ear notching, but the method described below is used by zoos with large hoofstock collections. A specific number is assigned to each of four different locations on each ear (figure 18.2). The four positions on the left ear correspond to the numbers 1, 2, 4, and 7, while the four positions on the right correspond to the numbers 10, 20, 40, and 70. The left and right ears are on the

> **Good Practice Tip:** For certain forms of identification (ear tags and leg bands), the right side is often used for males and the left side for females.

animal's left and right sides, not the keeper's. When a notch is created, the number corresponding to that area is added to the values of all the other notches on that animal's ears. The sum of the values on both ears provides a unique identification number. Some zoos may prefer to use the commercial swine industry method of notching the right ear with a herd or litter number, and the left ear with the individual's number. The swine method assigns different numbers to different locations on the ear. If an ear-notched animal is transferred between zoos, it is important that the numbering system information be communicated to the receiving institution.

Some institutions save the ear notch tissue pieces as a way to bank genetic information. These tissues can be frozen, preferably in an ultracold freezer at minus 70°C. The advantage of ear notching is that the unique number can be seen at a distance. The disadvantages are that there is no standardized zoo numbering system, it takes practice to learn the numbering system, the ear notch may change shape or not be as obvious as the animal grows, and ear trauma can affect the notch, possibly causing it to be read as indicating a different number.

OTHER BANDS AND TAGS

Leg bands are most commonly used for avian identification, but they can also be used on small mammals. The

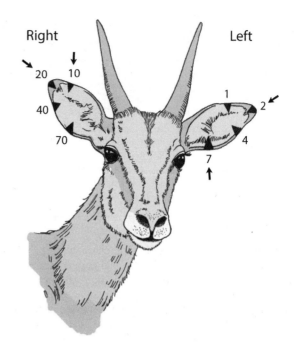

Figure 18.2. The locations and corresponding numbers on the right and left ear for ear notches. An example of notching the number 39 is shown by the arrows. The notched numbers on both ears are added to equal the number 39. Illustration by Kim Lovich, San Diego Zoo Global.

bands may also be called rings; they come in many different sizes to fit a variety of species. Leg bands are often placed at the bottom part of the tibiotarsus bone, just above the ankle (figure 18.3). Long-legged birds may have bands placed at the bottom part of the tarsometatarsal bone just above the toes (figure 18.4). With wading birds the ankle location is preferable, as the band can then be seen while the bird is in the water. The band will usually be numbered, but may also bear a letter code for the name of the zoo or aquarium. Bands can either be plastic or metal; the most common metal is aluminum, but stainless steel or nickel alloy is available. The numbers and letters stamped on steel bands are more durable than those on aluminum. Stainless steel is harder for a bird to crush and disfigure; this should be considered for parrots (psittacines), because their beaks are very strong. Color choices are available regardless of the band material. As in the methods described above, some zoos and aquariums prefer to identify males on the right side of the body and females on the left. The most important part of placing any band is to achieve the proper fit. The band should be loose enough on the leg to move freely, but not so loose that stray wire, branches, or enrichment devices can get caught underneath. The band should be checked frequently to ensure that it is not constricting the leg tissues. If temporary identification is needed, a colored zip tie (cable tie) can be placed on the leg.

Metal bands are of two main types. A closed band is a complete circle with no gap. It is usually placed on a hatchling bird, between several days to a few weeks of age, by being slid over the foot and onto the lower leg. The band will stay in place as the foot grows too large for the band to be removed. A band of this type indicates that the bird was captive-bred, and it is con-

Figure 18.1. An ear tag. Illustration by Kim Lovich, San Diego Zoo Global.

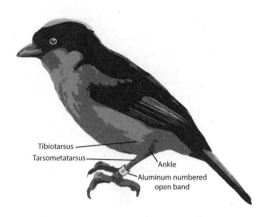

Figure 18.3. Leg band on a short-legged bird at the bottom part of the tibiotarsus bone just above the ankle. Illustration by Kim Lovich, San Diego Zoo Global.

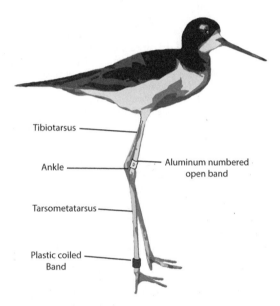

Figure 18.4. Long-legged birds may also have a leg band at the bottom of the tarsometatarsal bone just above the toes. However, the ankle location is preferable for wading birds. Illustration by Kim Lovich, San Diego Zoo Global.

on different parts of the leg. As mentioned above, the most important part of banding is a proper fit.

Disadvantages of leg bands include the possibilities that the number may wear off, that inappropriate size or placement may lead to compression injuries or restricted blood flow to the foot or leg, the band may get caught on cage wires or exhibit materials, that accumulation of keratin under the band can cause constriction, that a large bird may crimp or collapse the band with their beak, and that a metal band may predispose the leg to frostbite in a cold climate. The color of a plastic band can fade. When band removal is necessary, sedation may be used to lessen the chance of the animal moving suddenly, which could cause a soft tissue wound, bone dislocation, or fracture. Great care must be taken to apply a controlled force to the band itself and not to the leg. An open band can be removed using two pairs of pliers or similar medical instruments and applying opposing forces to the opening. Once the opening is large enough, the leg can slip out. Small closed bands of aluminum or plastic may be cut with scissors. One cut may be enough to then allow the use of pliers as described above. Another method is to cut on opposite sides of the band (two cuts total), which will allow the band to fall off in two halves.

A wing or flipper band around the mid-humerus is another way to identify an avian species, but this method is usually reserved for penguins (figure 18.5). Penguin colonies are sometimes identified via numbered aluminum bands, but are most often identified using color. A colored cable tie or zip tie may be all that is necessary if the penguin colony is small. If additional color options are needed for a larger colony, colored tape can be wrapped around part of the cable tie or aluminum band to create a unique color pattern. Alternatively, with large penguin colonies of multiple species, the same color of cable tie is placed around the wing to identify the species. Then different colored small cable ties are placed

sidered permanent identification as it can only be removed by being cut off. An open band is a piece of metal shaped to form a circle. The ends of the band do not meet, and are opened so that the band can be placed around a mature bird's leg. After placement, the ends are carefully squeezed back together until they touch without the band being too tight around the leg. The closure of the band can be done by hand, but a special tool made to fit the band is helpful. An open band can be easily removed or substituted, so it is considered a less permanent form of identification. Some plastic bands are coils that wrap around the leg. They can expand as the leg grows and require no special equipment to apply, as the plastic can be uncoiled by hand. Some birds may be able to unwrap these types of bands, which may cause them to fall off or loosen enough to get caught on exhibit objects. For additional identification combinations, colored plastic and aluminum bands can be used on different legs, and long-legged birds can have bands

Figure 18.5. Penguin wing bands can be aluminum bands with color tape or cable ties of different colors. Illustration by Kim Lovich, San Diego Zoo Global.

around the primary cable tie to identify each individual. The cable tie fastener should face outward so as not to irritate the wing. All extra cable tie material should be cut off, and water-resistant adhesive (e.g., superglue) should be applied to secure the connection, so that it doesn't inadvertently tighten. Most zoos and aquariums only band one wing, but placing identical bands on both flippers allows identification from each side and is helpful if one band is lost. Zoos and aquariums with large colonies may need to band each wing uniquely to create more color combinations for identification. The disadvantages of wing bands are the same as those listed for leg bands. In addition, the colored tape or zip ties may need to be replaced if they fall off or loosen. Vigilance is necessary, especially during molting, as the penguin's flipper will swell and the band may become too tight.

Wing tags are not often used in zoo and aquarium collections, but they are a proven method to ID free-ranging birds. Large colorful plastic tags are attached to the feathers (in which case they will fall off when the bird molts), or are anchored more permanently through the wing membrane. Sometimes plastic or metal ear tags (described above) are used as bird wing tags; the metal tags are smaller and less conspicuous. Neck bands are large colorful plastic bands that encircle a bird's neck. These are usually reserved for large-bodied birds or those with longer necks. The large colorful plastic wing tags and neck bands are quite prominent; most zoos and aquariums prefer less obvious forms of ID. As with any band-type identification system, the right fit is essential so as not to injure the wing or neck. Many of the disadvantages of leg bands also apply to wing tags and neck bands.

Flipper tags are similar to wing tags in that they are plastic tags of various colors and numbers used to identify wild seals and sea lions (pinnipeds) as well as wild sea turtles. Seals (phocids) are tagged in the webbing of their rear flippers, while sea lions (otariids) are tagged on their front flippers. Seals and sea lions in captivity are identified more often by external characteristics and behavior than by flipper tags. The tags may fall out over time, or get caught on objects and be pulled out. Furthermore, fading may affect the numbers and colors. Care needs to be taken to identify and avoid blood vessels in the flipper when placing the tag. The vessels may be easier to visualize in a pinniped than in a sea turtle.

Colorful plastic thumb bands are one method used to identify bats. Multicolored bands can be used and numbers can be imprinted on the plastic for additional identification information. The band should be loose enough to turn easily, and should be monitored regularly. The bat may try to remove the thumb band and, in the process, cause crimping or constriction. The band can also fall off on its own. Many disadvantages listed for legs bands apply to thumb bands as well.

SHELL NOTCHING

Shell notching (drilling or filing) can be used on turtles and tortoises (chelonians) for identification (figure 18.6). The top shell (carapace) is usually the one that is "notched" with a square or triangular cut into one or more marginal scutes. The notches are made by a handheld file or a rotary electric tool (Dremel is one brand). There is no standard zoo/aquarium numbering method recognized by North America's Associa-

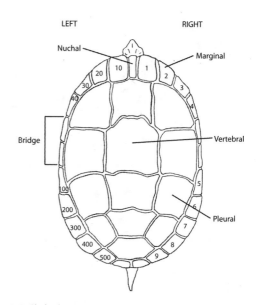

Figure 18.6. Chelonian carapace notches, each corresponding to an assigned number, can be created with a drill or file. The marginal scutes along the bridge (midshell) are not used. Since not all chelonians have the same number of marginal scutes, the caudal-most scutes near the tail may or may not be used. The numbers of the notched areas are added to produce the animal identification number. Illustration by Kim Lovich, San Diego Zoo Global.

tion of Zoos and Aquariums (AZA) or ISIS for shell notching. A numbering system is suggested in this chapter on the basis of chelonian work at the Georgia Sea Turtle Center through Dr. Terry Norton. The numbers 1, 2, 3, and 4 are on the right cranial marginal scutes, the numbers 5, 6, 7, 8, and 9 are on the right caudal marginal scutes, the numbers 10, 20, 30, and 40 are on the left cranial marginal scutes and the numbers 100, 200, 300, 400, and 500 are on the left caudal marginal scutes. The marginal scutes along the bridge (at mid-shell) are not used for numbering, because notches in this area may cause shell instability. Not all chelonians have the same number of marginal scutes, so the most caudal scutes near the tail may or may not be used for notching. As with ear notching, the numbers notched on all parts of the shell are added together to create a unique number for the individual turtle or tortoise. For instance, if the scutes representing the numbers 4, 300, and 10 are notched, then the individual is #314.

Whatever numbering scheme is used, the information must be communicated when a turtle or tortoise is being transferred to a different zoo or aquarium. A photograph may be helpful to illustrate the locations which have been notched. The scutes have nerves and blood vessels, so the chelonian may feel discomfort during the notching procedure. Scutes grow continually, so notches may need to be recut multiple times over the chelonian's life. If the animal has a shell injury, its notches may be affected; this individual would then be identifiable on the basis of how the shell appears after the injury, especially if the original notches are not readable.

SHELL AND FIN TAGS

Shell tags are not commonly used in captivity. When wildlife biologists place "tags" on sea turtles, they often use epoxy to secure a radio transmitter or other tracking device to the top shell. It is possible to create a living tag for identification in sea turtle rehabilitation and release. A living tag is created by harvesting a small piece of tissue from the carapace and bottom shell (plastron). The piece is then grafted with surgical glue to the opposite shell. The final result is that the carapace has a light-colored area and the plastron has a dark-colored area, and both will increase in size as the animal grows. Shell pieces can be grafted to unique locations for each individual. Disadvantages of living tags include the need for surgical skill to perform the technique, the animal's discomfort from the procedure, and the possibility that the shell pieces may not adhere to their new location. Written records and pictures should illustrate which parts of the top and bottom shell have been used.

Fin tags can be used for fish, including sharks and rays (elasmobranchs), but many large aquariums do not perform external active identification on their display specimens. Animals in large aquarium exhibits are often managed as a group, but some fish have unique features that make them easy to passively identify. A variety of external fin tags are available. Some tags are small and unobtrusive, especially when placed on a bigger aquatic species. The most widely used are anchor tags (with T-shaped ends) and dart tags (with V-shaped ends), which are anchored into muscle or bone at the base of the dorsal fin (figure 18.7). The trailing end of the tag has identification information. Transbody tags pass through both sides of the body in the muscle at the base of the dorsal fin. They can be circular plastic discs separated by a metal pin, or colorful vinyl flagging tape threaded through a needle and passed through the muscle. Tail-loop tags are soft-material (ribbon or string) or cable ties loosely tied around the caudal part of the body just in front of the tail fin (caudal peduncle). Tail-loop tags are used for short-term identification only. Jaw tags are metal rings placed around the lower jaw of a fish, and are similar to the open bands used for bird legs. The fish needs to be watched closely to make sure it can eat with the jaw tag in place. Anesthesia may be required if the fish is large or the procedure can't be performed quickly. As with many ID methods, fin tags of all varieties can fall out over time, get

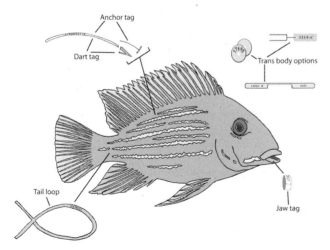

Figure 18.7. Fin tags used in fish including elasmobranchs. Illustration by Kim Lovich, San Diego Zoo Global.

caught in the exhibit, or fade so that numbers and colors are no longer obvious. Keepers need to monitor these changes and also watch for signs of infection or inflammation near the site of identification.

TATTOOS AND BRANDING

Tattoos can be a primary identification method, but are best when combined with other forms of ID (e.g., microchips). The carnivore and tiger AZA husbandry manuals recommend tattoos, but many zoos prefer other forms of identification. Accepted tattoo locations are the underside of the wing in bats and birds, the inner ear or inner thigh in carnivores, the inner thigh or arm in primates, the areas around the eyes or face of some primates in large groups, the inner ear in hoofstock or marsupials, and the chest, nose, or inner lip in other suitable species. The most appropriate information to tattoo on an individual is the international studbook number. The numerals 6 and 9 should be avoided if they are the only digit in the tattoo, as they are easily confused when seen from different angles. Some zoos will tattoo males on the right side of the body and females on the left. Disadvantages of tattoos are the pain they cause during application, their visibility only when the animal is in hand or in close proximity, and their tendency to fade or distort over time. Furthermore, tattoos are not appropriate for all species; ink leakage in a stingray species was believed to have caused death in three individuals (Raymond et al. 2003, 210).

Branding techniques developed for domestic animals such as goats, sheep, cows, and horses can also be used to externally identify zoo hoofstock. Branding is not used very often in zoo collections because it will mark the skin and be visible to the public. A branding iron (metal rod) has on its end a number, letter, design or combination of symbols that will uniquely identify the branded animal. The brand is applied on the neck or rear part of the body. Zoo veterinarians may wish to anesthetize or sedate a zoo animal for branding. Heat branding requires the iron to be heated in a fire or stove; electric branding irons are also available. The heated iron is applied to the skin until the hair is removed and a permanent mark is created. The thickness of the skin will determine how long the brand has to be applied. The final product is a visible scar as identification. Freeze branding requires the branding iron to be cooled with dry ice or liquid nitrogen. For freeze branding, the hair must be shaved to expose the skin and the site prepared with alcohol. The hair will grow white in the area that was freeze branded. Branding is a good method if animals need to be identified at a distance in large pastures. Its disadvantages are pain upon application and during healing, infection if the brand has not been properly applied, and its lack of aesthetic appeal to zoo visitors.

TOE CLIPPING

Toe clipping, the removal of part or all of one or more digits, has been used in field studies of small rodents. The information presented here, however, pertains to the use of toe clipping in adult frogs and toads (anurans). Toe clipping is an inexpensive and quick form of ID, but it is considered one of the least desirable methods for amphibians (Wright

2001, 275–76). The toes can be removed using a small pair of scissors, a medical clamp, a stainless steel clip, or a suture. Detailed instructions on the toe clipping procedure for amphibians are available through the US Geological Survey National Wildlife Health Center, and they recommend the use of scissors (USGS 2010). Ideally the amputation should occur at the joint between the bones of the toe. If the procedure is performed in a zoo collection, disinfection of the amputation site, anesthesia, and/or pain medications are suggested. Bleeding is usually minimal, without need for tissue glue or coagulant powders. The removed tissue can be preserved and used for genetic analysis, histopathology, infectious disease surveys, or age studies (Anon. 2009, 38–41). The disadvantages of pain, disfigurement, infection, and the regenerative ability of amphibians to regrow toes make toe clipping an infrequently used method in captive populations. Some species depend on their toes for grooming, breeding, feeding, climbing, and communication, so toe clipping could interfere with their necessary behaviors. Specifically, the first digit (thumb) of the front feet in males may be needed during breeding behaviors. Toe clipping may affect the survival and return rates of wild amphibians, and it is a controversial topic among amphibian scientists. Amphibians may lose toes due to predation or trauma, which can complicate the interpretation of missing toes. The Association of Reptile and Amphibian Veterinarians recommends that toe clipping be avoided when possible (Anon. 2009, 38–41). Passive identification methods or PIT tags may be better than toe clipping for zoo and aquarium amphibian collections.

TEMPORARY IDENTIFICATION

Occasionally, temporary identification may be necessary. It can be used for a growing animal until permanent ID is possible, to identify an animal scheduled for shipment, or to recognize an animal that is under treatment while housed with conspecifics. Temporary ID methods include shaving or cutting an area of hair or fur, using a colored spray or marker on an area of hair or body part, placing a colored zip tie/cable tie on an appendage, or trimming feathers in birds. If the situation requiring temporary ID becomes prolonged, then permanent identification is recommended.

INTERNAL ACTIVE IDENTIFICATION

MICROCHIPS

Internal active identification is achieved by placing a material within the body or underneath the skin (subcutaneously). The best-recognized form of internal active identification is a microchip. Microchips are also known as passive integrated transponder tags (PIT tags) or radio frequency identification devices. Microchips are considered a permanent, nonprogrammable, unalterable method of identification that provides a unique code of 9 to 15 characters. The code is alphanumeric (mostly numbers, though it can also contain letters). The microchip is a small electronic chip enclosed in a biocompatible glass cylinder about the size of a rice grain. Since a microchip is only 12 mm long, it can be used in very small species; amphibians as small as 30 mm in length and weighing

> **Good Practice Tip:** A microchip should be placed on the left side of the animal, in accordance with CBSG and AZA guidelines.

as little as 2 g have been implanted with microchips (Wright 2001, 276). The microchip transmits a unique identification code to a scanner/reader which displays the information on a screen. Most microchips placed in the United States are 125 kHz, but 134.2 kHz is the international standard. For ease of insertion, microchips are usually supplied with a single-use specialized syringe or within a needle that can be screwed onto a multiple-use applicator.

There are three primary microchip manufacturers: Trovan, AVID (American Veterinary Identification Devices), and Destron Fearing. Trovan may be marketed as InfoPET, and Destron may be marketed as Biomark or Home Again. A 2001 document produced by the Conservation Breeding Specialist Group (CBSG) of the International Union for Conservation of Nature (IUCN), recommended the Trovan system as the global standard (CBSG 2001, 23–24). However, a 2010 guideline for transponder placement and recording from AZA does not state a preference for any microchip company (AZA 2010). The manufacturers mentioned above offer products and services abroad, as well as "universal" scanners that should detect all microchip frequencies. Aquariums or facilities with aquatic animals can purchase a scanner/reader that is waterproof and submersible.

Regarding microchip placement, table 18.2 is a compilation of the information presented in the 2010 Guidelines for Transponder Placement and Recording from the AZA and the 2001 CBSG Transponder Working Group Report. The AZA document does not include as much descriptive information on specific locations as the CBSG report, so table 18.2 includes the more detailed information. All transponders should be placed on the left side of the animal whenever possible. In order to reduce the chance of migration, the CBSG recommends shallow intramuscular (IM) placement of microchips. Despite this recommendation, many microchips are placed underneath the skin (subcutaneous or SQ). Some microchips are placed in the abdominal/coelomic cavity. Fish, reptiles, amphibians, and birds have a coelomic cavity which combines the chest and abdominal cavity without the presence of a diaphragm. If another type of microchip implant is also necessary (e.g., a contraceptive implant), it should be placed on the right side of the animal to avoid confusion with the identification microchip. Once a microchip has been used and the number has been assigned, it should not be reused on another recipient.

The technique for SQ implantation is as follows:

1. Review the medical record and scan the patient to check for other microchips.
2. Scan the microchip within the needle or application device, to make sure it is emitting a signal and showing the unique ID number.
3. If thick fur or hair is present, part it to better see the skin, or clip away some fur or hair to create a small window.

4. Pinch and pull up the skin at the desired placement location (table 18.2).
5. Slide the needle, with its open tip facing up, under the raised skin at a 45° angle until the tip of the needle is no longer visible.
6. Depress the plunger on the insertion device, and keep it held down while applying pressure to the injection site as the needle is withdrawn. This will help keep the microchip from coming back up out of the small hole.
7. Scan the insertion site to check for the microchip.
8. Check to make sure the microchip is not hidden under the animal's hair, feathers, fur, or scales in the insertion area.
9. Optional step: A drop of tissue glue or a suture can be used to reduce the chance that the microchip will come out through the injection site during healing.
10. Record the following information: date of placement, location of placement, manufacturer, and unique alphanumeric code.

11. Confirm the microchip at future procedures and opportunities. Note: Reptiles and amphibians may also have microchips placed SQ. The insertion needle is inserted under the skin or scales at a 10° angle, and this often can be accomplished without any need to pinch or pull up the skin. The correct angle is one that will allow the needle to go under the skin but not penetrate deeper muscle or other structures.

Some of the general steps outlined above are followed when the microchip is placed IM or in a body cavity. Intramuscular placement is preferred in birds, because they have very thin skin and SQ tissues. The recommended avian microchip location is the left pectoral (chest) muscle or the left thigh muscle. A few other avian locations are listed in table 18.2. The feathers can be moistened with a small amount of alcohol to make the insertion location more visible. The microchip is placed in the thickest part of the selected muscle, and the needle is inserted at a 10° to 20° angle, depending on the muscle mass. It is inserted only until its bevel (open-

TABLE 18.2. Guidelines for transponder placement per AZA and CBSG. The AZA document (2010) does not include as much information on specific placement location as the CBSG report (2001). This table includes the more detailed information. According to the CBSG, the microchip should be placed in a shallow intramuscular position to reduce the chance of migration whenever possible. Transponders should be placed in the left side of the animal whenever possible.

Species	AZA and CBSG guidelines
Fishes	• Small (< 30 cm): coelomic cavity • Large (> 30 cm): left base dorsal fin
Amphibians	• Lymphatic cavity • Coelomic cavity ❖ Boreal toad: subcutaneously (SQ) on either side of the backbone ❖ Puerto Rican crested toad: SQ between the left shoulder and parotid gland
Lizards	• Small (< 12.5 cm, snout to vent): coelomic cavity • Large (> 12.5 cm, snout to vent): lateral left body anterior to inguinal region
Turtles and tortoises (chelonians)	• Left hind limb socket • Leg • Alternative methods may need to be considered for animals less than 10 cm in length
Crocodilians	• Left side anterior to nuchal cluster • Left hind leg
Snakes	• Left side dorsal to vent (cloaca) • (The AZA document is less specific, and only states "body")
Birds (aside from exceptions below)	• Left pectoral muscle • Left thigh
Ratites	• Pipping muscle in chicks • Left lateral neck in adults
Vultures (old and new world)	• Left base of neck
Penguins	❖ SQ in loose skin on back of neck
Micronesian kingfisher	❖ SQ in the left pectoral muscle
Mammals (aside from exceptions below)	• Behind left ear • Left of the spine between scapula
Elephants	• Left tail fold
Hyrax	• Left lumbar area
Loris	• Left lumbar area
Carnivores	• In some species (e.g., cheetahs in Africa and Mexican wolves in North America), microchips have been placed at the left tail base
Bats	❖ SQ in the dorsal posterior part of body, to avoid interference with flight muscles between the shoulder blades
Otters	❖ SQ above bridge of nose over forehead

❖ Information written with this symbol is present in specific AZA husbandry manuals.

ing) is covered by the muscle. The plunger is depressed and gentle pressure is applied over the site after the needle has been removed, to prevent bleeding. If the bird is small or if the muscle does not look thick enough, a microchip is not recommended.

A separate IM technique is available for domestic horses, in which the microchip is placed in the left side of the neck. This will not be described here, as the CBSG and AZA documents do not describe it for exotic animals. Table 18.2 should be consulted for species-specific recommendations on microchip location.

The coelomic cavity is the recommended microchip site for some fish, amphibians, and lizards. The animal is held with its underside (ventrum) up and the microchip needle is inserted at a 20° to 30° angle. A layer of muscle must be penetrated before the needle will enter the abdominal or coelomic cavity, but a shallow angle is recommended so as not to penetrate vital organs. The placement should not be along the ventral midline of the animal, as large blood vessels can be encountered there. There are not many situations in which microchips would be placed abdominally in mammals; however, the right ventral side may be best for such placement, as the spleen is located on the left side.

Microchips are considered permanent forms of ID, but there are situations in which they can fail. One disadvantage of microchips is that they can move or migrate away from their insertion sites. Such migrations are most common when they are placed SQ or in the abdominal/coelomic cavity, and are not as common with IM placement. Occasionally microchips can migrate outside of the body; that is, they can fall out. Another disadvantage is that the scanner must be close to the microchip to read it. Animals can be trained to present the body part that contains that microchip, so that the keeper can safely scan that area and confirm the ID. On occasion the microchip itself may fail to transmit the number, but manufacturers market them with an indefinite life span. Sometimes microchips can cause infection or swelling, and a few cases have been reported of microchip-associated cancer in wildlife (Siegal-Willott et al 2007, 352; Pessier et al 1999, 139).

Microchips are visible as bright white on radiographs (X-rays), and should not be confused for something abnormal. Even though a microchip is seen on a radiograph, its number should be confirmed by scanning with a reader. Newer microchip technology is becoming available that can detect an animal's body temperature and transmit that information to the reader (www.destronfearing.com). Such chips are currently marketed for horses; placed in the neck, they will read temperatures between 77 and 122 °F (25–50 °C). Perhaps zoos may use this type of microchip in the future.

Other techniques for internal active identification that have been developed for the fish industry may not be appropriate for most zoos and aquariums. They are decimal coded wire tags, visible implant elastomers, and visible implant alpha tags, available through Northwest Marine Technology (www.nmt.us/index.htm). Decimal coded wire tags won't be discussed here because that equipment is specialized, expensive, and usually used to ID thousands of animals. The visible implant elastomer uses silicone material available in 10 different colors to provide an externally visible form of ID. The colored liquid is injected SQ and cures into a rubberlike visible mate-

Good Practice Tip: Make sure to check the identification at least once a year and confirm the number and proper fit so as to not cause harm. Examples include, but are not limited to, checking that leg bands are neither too tight nor too open, inspecting for microchip migration or loss, and seeing whether a tattoo is readable.

rial. This technique can be used in reptiles and amphibians, and is mentioned in the *Boreal Toad Husbandry Manual* (www .wildlife.state.co.us). Visible implant alpha tags are placed SQ in fish, reptiles, and amphibians; they have alphanumeric codes written in black with a fluorescent background color. This method of ID is preferable to toe clipping for amphibians.

RECORD KEEPING

Record keeping is a very important part of animal identification. The identification should be checked regularly to make sure that it is accurate, still in place, and is not causing a problem for the animal. The type of identification used will dictate what information is recorded. A general rule is to record as much information as possible, such as date of placement, location of placement, product manufacturer, identification material (e.g., plastic vs. metal), unique band or tag number, color, alphanumeric code, photos, and drawings. All pertinent identification information should be listed in multiple locations. Each zoo should have a system of entering identification information into ZIMS (or, in the past, ARKS) specimen reports, health certificates, medical records, and animal data transfer forms. ZIMS will have a thorough list of body locations suitable for identification placement in all species. Keepers should log the information in their reports and notebooks, including all passive and active forms of identification. If there is a change in identification information, all records should be updated to reflect the change. Proper animal identification and recording is important during animal shipments, especially if the shipment includes multiple individuals of a species.

SUMMARY

Accurate animal identification and record keeping is essential for effective animal management. It is imperative to know the identity of the animal for which exceptional husbandry and health care is being provided. There are many systems for identification, and the best approach is to combine several methods so that if one fails, the individual can still be recognized in other ways. One goal of this chapter is to promote some consistency in animal identification methods to benefit the zoo and aquarium community.

REFERENCES

Anonymous. 2009. Toe-Clipping in Amphibians. *Journal of Herpetological Medicine and Surgery* 19:38–41.
AZA. 2010. Guidelines for Transponder Placement and Recording. Accessed 23 December. http://www.aza.org/uploadedFiles

/Animal_Care_and_Management/Animal_Management/Animal
_Data_and_Recordkeeping/IDMAG_Documents/Transponder
Statement2010.pdf.

Boreal Toad Husbandry Manual. 2010. Accessed 23 December.
http://wildlife.state.co.us/NR/rdonlyres/23912565–3F5F-4026
-A4C3-BD34C29BB564/0/FinalHatcheryManual122402.pdf.

CBSG. 2001. Transponders Working Group Report. *CBSG News*
12, no. 1: 23–24.

Elasmobranch Husbandry Manual. 2010. Accessed 21 December.
http://www.elasmobranchhusbandry.org.

Loomis, Michael. 1993. Identification of Animals in Zoos. In *Zoo
and Wild Animal Medicine: Current Therapy 3*, ed. by Murray E.
Fowler, 21–23. Philadelphia: W. B. Saunders.

Pessier, Allan P., Ilse H. Stalis, Meg Sutherland-Smith, Lucy H.
Spelman, and Richard J. Montali. 1999. Soft Tissue Sarcomas
Associated with Identification Microchip Implants in Two Small
Zoo Mammals. Paper presented at the annual American As-
sociation of Zoo Veterinarians meeting, Columbus, OH, 9–14
October.

Raymond, James T., Freeland Dunker, and Michael M. Garner. 2003.
Ink Embolism in Freshwater Orange Spot Stingrays (*Potamortry-
gon motoro*) Following Tattooing Procedure. Paper presented at
the annual American Association of Zoo Veterinarians meeting,
Minneapolis, 4–10 October.

Siegal-Willott, Jessica, Darryl Heard, Naime Sliess, Diane Naydan,
and John Roberts. 2007. Microchip-Associated Leiomyosarcoma
in an Egyptian Fruit Bat (*Rousettus aegyptiacus*). *Journal of Zoo
and Wildlife Medicine* 38:352–56.

USGS. 2010. Toe Clipping for Amphibian Research Procedures. Accessed 23 December. http://www.nwhc.usgs.gov/publications/
amphibian_research_procedures/toe_clipping.jsp.

Wright, Kevin M. 2001. Surgical Techniques. In *Amphibian Medicine
and Surgery*, ed. by Kevin M. Wright and Brent R. Whitaker,
273–83. Malabar: Krieger.

19

Reproduction

Linda M. Penfold

INTRODUCTION

In nature, the goal of animal reproduction is to ensure that an individual's genes are perpetuated, preferably in multiple offspring. Competition with other animals, combined with an array of environmental and social pressures, generally prevents the likelihood of any single animal's genes dominating a population, and thus results in healthy, genetically diverse populations. Reproduction not only is an important part of species propagation but is also considered to be part of the normal repertoire of animal behaviors; the keeper's role is to be vigilant in observing those behaviors to time breeding introductions correctly and anticipate births. Similarly, it is good practice that animals in captivity are carefully managed to ensure that their genes are sustained in the population while not becoming over- or underrepresented, for the sake of maximal gene diversity and a genetically healthy population.

However, with a limited number of animals in captivity it is critical that every animal reproduce for maximum genetic diversity of the population, and an understanding of the species' basic reproduction can greatly aid managers in accomplishing this. For example, colonial species such as the naked mole rat need to be maintained in colonies, and solitary species such as the okapi and the cheetah generally need to be managed singly. Because animal holding spaces in zoos are limited, the challenge is to continue producing offspring to maintain a genetically diverse sustainable population while minimizing the production of surplus animals. This is accomplished through diverse management strategies including recommended breeding, contraception, animal moves between institutions, and animal separations.

This chapter will provide basic information about animal reproduction as it affects and is affected by the keeper position. After studying this chapter, the reader will understand

- the diversity of reproduction strategies in the animal kingdom
- the keeper's role in noting reproductive behaviors

- the importance of recording relevant mating and breeding behavior data to predict birth, egg laying, and so on
- the similarities and differences between pseudopregnant and pregnant animals
- the usefulness of contraception to manage populations and individuals
- the relevance of appropriate environment and social housing for reproduction
- the scope of assisted reproductive techniques, and how they might complement natural breeding programs.

ASEXUAL AND SEXUAL REPRODUCTION

Asexual reproduction produces a new individual, derived from a single parent animal, that is genetically identical to that parent. Forms of asexual reproduction include budding (as in sponges and many corals), where a new individual begins by developing from a small part of the parent organism that may or may not remain attached to the parent, and regeneration from fragments of the parent animal (as in corals, anemones, and hydroids, collectively known as cnidarians). Parthenogenesis, literally meaning "virgin birth," is the development of an embryo without fertilization from the male. In brief, the oocyte contains two sets of chromosomes (2n) rather than one set (n) of chromosomes being eliminated as the polar body to yield a haploid germ cell. For species dependent on the XY sex determination (e.g., whiptail lizards [*Cnemidophorus* sp.]), parthenogenetic populations are usually entirely female as a result. For those that use the ZW sex determination (butterflies, moths, and komodo dragons), offspring are all male. Parthenogenesis has been seen in many insect species and some amphibians and lizards. It was also recently discovered that the hammerhead shark (*Sphyrna mokarran*) and the komodo dragon (*Varanus komodoensis*) are able to reproduce by parthenogenesis. Most vertebrate species practice sexual reproduction, which requires both a female and a male, each of which produce different sets of gametes (oocytes and spermatozoa, respectively). Fertilization may be external, as in many amphibians, such as frogs, and in bony fish (teleosts), whose eggs are released into the water to be fertilized by sperm released at the same time; or internal, wherein sperm is deposited inside or picked up by the female, as is the case with salamanders and caecilians, reptiles, birds, monotremes, marsupials, and placental (or eutherian) mammals.

MALE REPRODUCTIVE ANATOMY

Testes produce the male gamete, or spermatozoon (plural spermatozoa). The testes are located inside the abdomen of most vertebrate males, and the presence of paired testes and a glans penis distinguishes male mammals (Kunz et al. 1996). In most mammalian males the testes descend into a scrotal sac, allowing them to stay a couple of degrees cooler than body temperature for sperm production. Exceptions to this are the monotremes, some insectivores, marine mammals, sloths, armadillos, and elephants, all of which have internal testes. A few species have testes that descend temporarily during the breeding season (bats, squirrels, and some primates). Lateral or bilateral cryptorchidism ("hidden testes") is a reproductive disorder whereby one or both testes do not descend into the scrotal sac. This has been associated with inbreeding and was documented in four maned wolves (*Chrysocyon brachyurus*) following father-daughter matings (Burton and Ramsey 1986).

In most fish, frogs, and amphibians, fertilization is external and sperm and eggs are shed into the water simultaneously. For species that lay shell-covered eggs or carry the embryo internally, internal fertilization is necessary and the sperm must reach the ovum inside the female for fertilization to occur. In many salamanders, newts, and invertebrates, including most arachnids, a spermatophore or sperm packet is deposited by the male onto a surface, such as a leaf, or the ground, where it is then picked up by the female. Other species use a male copulatory organ to deposit the sperm inside the female. This may be as simple as a brief touching of the male and female cloaca. From the Latin word meaning "sewer," the cloaca is an external opening for the elimination of excretory products such as urine and feces, but also serves as the opening to reproductive tract. Other specialized appendages developed to deliver sperm include claspers in male sharks and rays, formed from modified pelvic fins that are inserted into the female cloaca to transport the sperm. These claspers may contain cartilaginous hooks that "clasp" the inside of the female's oviduct. Fertilization in nearly all frog species is external and is characterized by a behavior known as amplexus, whereby the male grasps the female from above and sheds sperm from his cloaca while the female is shedding eggs from hers. An exception to amplexus is found in the tailed frog (*Ascaphus truei*), which uses a tail-like extension of the male cloaca to deposit sperm directly into the female's cloaca. Male lizards and snakes have specialized reproductive organs known as hemipenes: paired copulatory organs that are grooved to facilitate sperm transport. A retractor muscle pulls the hemipene back into a pocket behind the cloacal vent after reproduction. Bird species generally rely on the transient touching of their cloacas—termed a "cloacal

kiss," such as is found in turkeys (galliformes), parrots (psitticines), and songbirds (passerines)—but a few bird species, such as waterfowl (e.g., ducks), and ratites (e.g., ostriches), have a penislike phallus. The phallus carries the sperm into the female along the surface in multiple grooves and ridges. In some duck species the phallus can be quite elaborate and surprisingly long, almost the length of the bird. This can be problematic, as trauma to the phallus is not uncommon, resulting in swelling that in turn prevents its retraction into the body. Fortunately for waterfowl species, the removal of part or all of phallus can occur with no apparent ill effects to the duck, other than limiting reproduction. This procedure is not uncommon in captivity.

All eutherian mammals have a penis that, with the appropriate stimulus, becomes erect either by spongy sinuses becoming enlarged with blood or by muscular control, as in killer whales (*Orcinus orca*). The penis of bats, rodents, carnivores, and many nonhuman primates also contains a bone, known as a baculum or os penis, that helps to stiffen it. In several felid and flying fox species, the penis surface is covered with small spines thought to be important to help stimulate the females and induce ovulation (see below) during the mating process. In some marsupial species the penis is forked to fit inside two lateral vaginae of the female. In kangaroos and wallabies (macropod species), the penis retracts neatly away inside the urogenital sinus when not in use, and also is not as closely associated with the more ventrally located testes as in placental mammals.

FEMALE REPRODUCTIVE ANATOMY

The female reproductive organ is called the ovary, and is always located internally in a pair. It contains the female gametes, ova (singular: ovum). Unlike the testes that produce spermatozoa continually, by the time a female mammal is born, the ovary contains a finite number of ova that will be steadily depleted during her lifetime. In most vertebrates the eggs are released from one or both of the paired ovaries. But in birds, most bats, and the platypus (*Ornithorhynchus anitinus*), although two ovaries are present, only one is functional. In egg-laying (oviparous) animals such as amphibians, fish, reptiles, birds, and monotremes, the ovum develops protective layers as it moves through the female tract (e.g., leathery or calcified shells are laid down in reptiles and birds). In birds, the basic structure of the egg is composed of the yolk, the albumen (or "white"), the shell, and the germinal disc. The yolk and albumen provide nutrients and water, and the germinal disc develops into the embryo if fertilized. The shell encloses and protects the developing embryo while allowing water to move across the shell through pores. The ovum is fertilized internally in monotremes, birds, and reptiles, and externally in fish and most amphibians. In all mammals it is fertilized internally, where it develops into an embryo.

In egg-laying birds and reptiles the ova make their way out of the body from the ovary through the oviduct, where the shell is laid down. In bird species, the left ovary and oviduct continue to grow while the right ovary and oviduct regress after hatching. Occasionally two ovaries may be noted in birds of prey, but the right ovary is usually inactive.

Marsupial females have paired lateral vaginae that are

equally patent, and give birth after a short gestation period to offspring that are underdeveloped and must find their way into the marsupial pouch, where they attach themselves to a nipple and continue to develop. Only the females have pouches, and pouches are absent in some marsupials such as the Tasmanian devils (dasyurids) and some opossums.

In eutherian mammals, the embryo implants into the uterus and continues development while nourished through the placenta, a specialized fetal-derived structure that connects to the fetus through the umbilical cord. The placenta is usually detached from the mother's uterus soon after birth and eliminated via the vagina. Some species may eat the placenta—presumably to "hide" evidence of a newborn, but also to regain some of the energy invested in the pregnancy. Where possible, a keeper should visually confirm that the placenta has been passed, and alert veterinary staff in case of a "retained placenta," which can result in uterine infection.

In many vertebrates, males and females are visibly different. This is termed sexual dimorphism. For those species that do not display sexual dimorphism, it may be difficult to tell the sex of the animal. At least in mammals, external genitalia generally allow for sex identification, though one exception is the spotted hyena (*Crocuta crocuta erxleben*). In this species, females display masculinization of the external genitalia with a pseudopenis and pseudoscrotum that, along with their similar size and ability to display erections like the males when greeting other hyenas, render them difficult to differentiate from the males. Females can urinate, mate, and give birth through the pseudopenis, though first-time mothers often have trouble passing the offspring through the reproductive tract (dystocia) during the birth of their first litter. All other internal sexual organs are similar to those of other carnivore species.

In spider monkeys (*Ateles* sp.), males and females again are not sexually dimorphic, and the female has a pendulous clitoris (thought to be used for scent deposition [Klein 1971]) that can be confused with a male penis. In contrast, sexual differentiation is easy in adult female chimpanzees (*Pan troglodytes*), macaques (*Macaca mulatta*), and baboons (*Papio hamadryas*), all of which have bare, pink, or red skin around the anus and genital (anogenital) region that swells during estrus, effectively advertising their sexual status.

ESTRUS AND OVULATION

When female animals are receptive to breeding by the male, they are described as being in "heat" or estrus. During this period an ovum matures in the ovary until ovulation, which is defined as the release of the ovum from a follicle triggered by a surge of luteinizing hormone (LH), usually around peak estrus. A single ovum or multiple ova will be ovulated, depending on the species. The ovum travels from the ovary through the oviduct, where fertilization takes place, and on to the uterus where it will eventually implant, if fertilized, before developing into a fetus. In many antelope and primates, estrus occurs repeatedly every three to four weeks, depending on the species, until the animal becomes pregnant. The species that ovulate irrespective of mating are referred to as spontaneous ovulators. Once a follicle has released an ovum, it develops into a yellowish mass on the ovary called a corpus

luteum, or CL, which produces the hormone progesterone. If the ovum is fertilized, the new embryo produces a specific hormone that sends a signal to "rescue" the CL and continue progesterone production to support pregnancy and prepare the uterus for implantation. If fertilization does not occur, the CL has a finite life span and diminishes over a few days, allowing another follicle to be recruited and the cycle to start again. In species like canids and felids, the LH surge that causes the release of the ovum from the follicle is actively induced by the physical act of mating. If these animals are not physically stimulated by the breeding process, there will be no LH surge and ovulation will not occur. Similarly, a lack of sufficient physical stimulation during the breeding process will result in no ovulation. For example, removing a breeding cheetah male (*Acinonyx jubatus*) after only one mating can sometimes hinder the LH surge, preventing ovulation in the female even though a good breeding, including penetration, might have been observed. These species are referred to as induced ovulators. If the ovum is fertilized, the female will become pregnant.

In spontaneous ovulators the life span of the CL is only a couple of weeks, but for many induced ovulators it is much longer, up to a couple of months. In these cases, if the ovum is fertilized, the female will become pregnant, but if it isn't fertilized the female may still appear "pregnant" (i.e., with a weight gain and abdominal swelling), a condition is referred to as pseudopregnancy. It is usually impossible to tell the difference between pseudopregnancy and early-stage pregnancy by visual assessment alone. Little can be done until the CL comes to the end of its life span and the female comes into estrus again. Carnivore species such as the wolves (canids), cats (felids; Brown et al. 1994), weasels (mustelids), many rat and mice (rodent) species, and also rabbits (lagomorphs), may undergo pseudopregnancy. Little is known of the estrus cycles of monotremes, but echidnas (*Zaglossus* sp.) and the platypus are probably monoestrus animals, having only a single estrus cycle per year. Most marsupials are polyestrus animals, having multiple cycles per year, though Tasmanian devils and banded anteaters (*Myrmecobius fasciatus*; dasyurids) and marsupial mice are monoestrus. In most marsupials the estrus cycle is actually longer than the gestation, though a long lactation period is also usually associated with marsupial species.

Species may be seasonally monoestrus, as are most canids and bears (ursids), including the giant pandas (*Ailuropoda melanoleuca*) that display estrus only once annually at approximately the same time each year. Seasonally polyestrus animals, such as mustelids, most felids, and some antelope, come into estrus several times annually, but only for part of the year, as in the case of the Pallas cat (*Otocolobus manul*), the clouded leopard (*Neofelis nebulosa*), Jackson's hartebeest (*Alcelaphus buselaphus jacksoni*), and the scimitar-horned oryx (*Oryx dammah*). Onset of estrus in seasonal species may be dictated by day length or sometimes resource availability (i.e., it is timed around the rainy season and/or an increase in available food). It is important to know the reproductive strategy of a species, as one study revealed that winter lights put up at one institution had the effect of changing the "day length" for Pallas cats, preventing their reproduction. Once the cats were moved to an area not affected by the lights, their breeding began again at the appropriate season (Brown et al.

2002). Species may also be aseasonally polyestrus and cycle all year round, such as the sun bear (*Helarctus malayanus*), the cheetah (*Acinonyx jubatus*), the lion (*Panthera leo*), the Bengal tiger (*Panthera tigris tigris*), and the gerenuk (*Litocranius walleri walleri*).

Primate species are polyestrus and may be seasonal, like many lemur species, or nonseasonal, like gorillas and chimps. Generally speaking, the reproductive cycle lasts 28 to 30 days, like that of the human, and is termed a menstrual cycle. The endometrium that has formed throughout the cycle is shed as a bloody discharge, rather than reabsorbed as in those placental mammals that display estrous cycles. The menstrual cycle, or menarche, starts at about four years of age in many Old World primates and around six to seven years in captive great apes. Estrus duration varies from two to three days in the gorilla, four to six days in the orangutan, and up to two weeks in chimpanzees.

The estrus cycle of the elephant is 14 to 16 weeks long, and is different from that of most other mammals in that it consists of two LH surges two to three weeks apart. The first LH surge is thought to develop accessory CLs, and the second stimulates ovulation. This unique feature has been put to good use by endocrinologists, who use the first LH surge to predict when ovulation will occur for timing artificial insemination. This has greatly aided the development of a successful artificial insemination (AI) program for this species.

The keeper's role in detecting behavioral or physical changes associated with estrus is important for facilitating introductions, watching for breeding activity, or ascertaining that the female is cycling regularly. Breeding introductions of males to females can be somewhat aggressive interactions for several species, such as Indian rhinos (*Rhinoceros unicornis*) and some okapi (*Okapia johnstoni*). This can be exacerbated if the female is not receptive to the male; so knowing when the female is in estrus can aid the breeding process. Also, for females that come into heat only for a short period each year, knowing when to introduce the male is important so that a breeding year is not missed. Zoo and aquarium breeding programs rely on keeper staff to judge when to introduce males and females.

EGG LAYING

In egg-laying or oviparous species, the ovum is fertilized by the sperm (in the case of internal fertilization) and transported through the oviduct. All birds and most reptiles are oviparous, though some reptiles do give birth to live young hatched from eggs that develop internally, and are thus referred to as ovoviviparous. In birds and reptiles, the fertilized egg moves through the oviduct and receives a coating deposited by shell glands, which creates a barrier to sperm. (Thus, it can be fertilized only before the shell is laid down, and not afterward.) The egg is then held in the oviduct until it is time for it to be laid (ovipositioned). In many birds and reptiles, sperm can be stored in the female oviduct for several days or even months, and a single insemination may fertilize multiple eggs over several days. Many reptiles are oviparous, but most do not incubate the eggs, instead burying or concealing them in a nest and relying on the ambient temperature to incubate them. Although most vertebrates have genetically fixed sex determination, for some species of fish and reptiles the incubation temperature of the eggs determines the sex of the offspring. Temperature-dependent sex determination (TSD) has been demonstrated for several species including loggerhead (*Caretta caretta*), green (*Chelonia mydas*), leatherback (*Dermochelys coriacea*), and olive ridley (*Lepidochelys olivacea*) sea turtles, as well as geckos and crocodiles. For example, in fresh water turtles a temperature of 31 °C or above will produce females, whereas temperatures of 24 to 27 °C will produce males (Standora and Spotila 1985). It has been shown in green sea turtle nests at pivotal temperatures that eggs in the center of the nest will develop as females while those at the edges will develop as males. Thus the temperature gradient within the nest may dictate whether the offspring develops into a male or a female.

In bird species, with a few exceptions that bury their eggs in heat-producing rotting vegetation or geothermally heated soil, such as the maleo (*Megacephalon maleo*), eggs are usually incubated or brooded by the female. This may involve sitting on the eggs to keep them warm or, in the case of the Emperor penguin (*Aptenodytes forsteri*), keeping the egg on top of the feet and under a "brood pouch," a featherless flap of skin, to insulate them and keep them warm. In a few cases (e.g., black swans, whistling ducks, king penguins, emus, and cassowaries) the male will brood the eggs. Signs that the eggs are close to hatching include the female staying on the nest, moving only to get food or not at all, or relying on the partner to bring her food. In some arctic nesting duck species the female may consume nothing during the entire incubation period, and male Emperor penguins will go without food for two to three months while brooding an egg.

PREGNANCY AND PARTURITION

With the exception of monotremes, all mammals are viviparous in that they give birth to live young. Placental (eutherian) mammals are characterized by the formation of a placenta, a structure formed of both fetal and maternal tissues that supports the nutritional and developmental needs of the fetus. The fetus completes its growth and development inside the uterus until birth or parturition. In some eutherian mammals, implantation of the early embryo may be postponed for weeks or months, extending the length of gestation, while the mother completes nursing of a previous litter or if resources are scarce. This process is known as delayed implantation and is found in bears (ursids) and some weasels (mustelids). The duration of gestation, defined as the development of the fetus in the uterus, varies dramatically among species (table 19.1) and can range from 13 days in the American opossum (*Didelphis marsupialis*) to 22 months in the African elephant (*Loxodonta africana*).

In most macropods fetal diapause, similar to delayed implantation in eutherian mammals, occurs, whereby the embryo does not develop while the previous pouch young is suckling. So even though a female kangaroo or wallaby may be separated from a breeding male, it may still have an embryo waiting in the uterus that will resume its development once the pouched young has left the pouch.

Marsupials are most clearly different from other (eutherian) mammals in their reproduction. In marsupials, the

TABLE 19.1. Reproduction parameters for diverse nondomestic species

Species	Sexual maturity	Gestation length	Number of offspring	Estrus type	Ovulation
Platypus	2 years	28–30 days	1–3 eggs	Seasonally monoestrus	Induced
Echidna	5 years	21–28 days	1 egg	Seasonally monoestrus	Induced
Rednecked wallaby	Female: 15 months Male: 24 months	30 days	1	Aseasonally polyestrus	Induced
Gray kangaroo	Female:17 months Male: 20 months	30.5 days	1	Seasonally polyestrus	Spontaneous
Coatimundi	Female: 2 years Male: 3 years	75 days	3–5	Seasonally polyestrus	Induced
Spring hare	2–3 years (when reaches 2.5 kg)	77 days	1	Aseasonally polyestrus	Induced
Pallas cat	12 months	75 days	3–5	Seasonally polyestrus	Induced
Cheetah	2 years	90–98 days	1–3	Aseasonally polyestrus	Induced
Lion	2 years	110–20 days	2–4	Aseasonally polyestrus	Spontaneous
Maned wolf	1 year	63 days	2–3	Seasonally monoestrus	Induced
African wild dog	Female: 2 years Male: 1.5 years	69–73 days	6–16 (avg. 12)	Aseasonally polyestrus	Spontaneous
Polar bear	5–6 years	195–265 days	1–3	Seasonally monoestrus	Induced
Gerenuk	Female: 1 year Male: 1.5 years	196–210 days	1	Aseasonally polyestrus	Spontaneous
Scimitar-horned oryx	1.5–2 years	224–55 days	1	Seasonally polyestrus	Spontaneous
Grevy's zebra	Female: 2 years Male: 6 years	380–90 days	1	Seasonally polyestrus	Induced
Somali wild ass	2 years	330–65 days	1	Seasonally monoestrus	Induced
Gorilla	Female: 8 years Male: 12 years	240–55 days	1	Aseasonally polyestrus	Spontaneous
Bonobo	Female: 10 years Male: 9 years	240–55 days	1	Aseasonally polyestrus	Spontaneous
African elephant	10–11 years	660–720 days	1–2	Aseasonally polyestrus	Spontaneous
Indian elephant	Female: 10 years Male: 15 years	660 days	1–2	Aseasonally polyestrus	Spontaneous
White rhinoceros	Female: 6–7 years Male: 7–10 years	480 days	1	Aseasonally polyestrus	Spontaneous
Indian rhinoceros	Female: 4 years Male: 9 years	470–531 days	1	Aseasonally polyestrus	Spontaneous
Bottlenose dolphin	Female: 5–10 years Male: 10 years	365 days	1	Aseasonally polyestrus	Spontaneous
Killer whale	Female: 6–10 years Male: 10–13 years	510 days	1	Aseasonally polyestrus	Spontaneous
Manatee	Female: 5 years Male: 9 years	390 days	1	Seasonally polyestrus	Spontaneous

Sexual maturity indicates the age at which some animals are potentially capable of reproducing, and not the average age of first reproduction.

young are born extremely immature and underdeveloped, requiring further nourishment for continued development, and must find their way in most cases to a pouch, where they attach to a nipple and continue their development. Gestation is generally short in marsupials; the longest is 40 days, in the red-necked wallaby (*Macropus rufogriseus*).

Conception and pregnancy are difficult to determine visually, except perhaps for late pregnancy in many mammal species, when the female becomes obviously larger and heavier. In hoofstock species the development of the udder several weeks before parturition might be noted (and referred to as "bagging up"); many females may act "differently" immediately before parturition. Also, the fetus may "drop," moving lower into the birth canal nearer to the time of birth. An experienced keeper will recognize these behavioral changes and know that parturition is near. For example, a white rhino (*Ceratotherium simum*) female or herd antelope female may distance herself from the herd. Felid and canid species may start to investigate areas in the enclosure for denning, and squirrels (rodents) and rabbits (lagomorphs) may pluck their fur to build a "nest." It is also anecdotally recognized that a drop in air pressure, usually associated with a change in the weather (e.g., stormy weather or a cold front moving in), can precipitate the birth process, though no published evidence in any species has been found to support this hypothesis. Similarly, it has been noted that many animals tend to give birth at night or early in the morning, though this may be somewhat coincidental, because if births are evenly distributed throughout a 24-hour period, it would follow that 66% of them would occur outside of normal working hours. Again, no scientific evidence supports either claim.

Ultrasound may be a useful tool for pregnancy diagnosis or for examining potentially gravid reptile females. Hormone analysis of serum, urine, and feces for pregnancy diagnosis has been used with much success for many mammalian species, especially in felids, canids, and hoofstock. However, a simple and effective method of confirming pregnancy in most mammalian species can be to observe breeding and count the days to see whether the female cycles back to estrus. For species that exhibit pseudopregnancy, it may be several weeks before a female cycles again. This should not be confused with a lack of estrus cyclicity, as in some sensitive females a stressful event may interrupt reproductive cycling. Generally speaking, if there are no known fertility problems, and especially if the female has previously had offspring, this can be as reliable a method of pregnancy diagnosis as any, and an important tool for the keeper.

PUBERTY AND AGING

Anecdotally it is assumed that first-time mothers may be less successful at raising their offspring than multiparous females. Age effects are well described in many bird species; first-time breeders and young birds tend to be less able to produce and rear offspring than are older, more experienced birds. In ungulate species it has been documented that the condition of red deer calves (*Cervus elephus*) improves with the age of the mother, and that older bison females (*Bison* sp.) are apparently more tolerant of offspring and spend more time nursing their calves than younger females (Green 1990).

Male ungulates often start sperm production at about one year of age, though they may be unable to compete with an older and more experienced male to breed females. Studies have shown that gerenuk antelope (*Litocranius walleri walleri*) as young as one year old produce spermatozoa (Penfold et al. 2005), and studbook records show that an animal as young as 11 months old has sired offspring. Similarly, cheetah males as young as one year old may have started to produce sperm but are usually still too immature to breed a female. Nonetheless, if reproduction by young males is undesired, most antelope, felid, and canid males should be removed from breeding groups no later than 10 to 12 months of age. Most mammalian species continue to reproduce throughout their lives, but with increased knowledge in animal management, many animals are living longer than ever in captivity, and more signs of reproductive aging are being noticed. Reproductive senescence or aging has been documented in several species, including gorillas (*Gorilla gorilla gorilla*), olive baboons (*Papio anubis*) and African lions (*Panthera leo* sp.). As mammalian females age, they may cease to become pregnant or the uterus may become less able to sustain a pregnancy. This may manifest itself as an increase in the incidence of abortions, pregnancy resorptions (whereby the pregnancy is not sustained), or pseudopregnancies in carnivores, and/or a reduced ability to raise offspring. As most mammalian offspring depend on

the female for survival, reproductive cessation may be more apparent in females than males.

> **Good Practice Tip:** Age of puberty onset is an estimate. An animal may become sexually mature sooner or later than average, depending on its social setting. For example, a young hoofstock male may mature quickly if housed with females but no adult male. According to studbook data, the youngest gerenuk male to sire offspring was only 11 months old.

SEASONAL EFFECTS ON REPRODUCTION

Many species are seasonal breeders, and breeding may be dictated by day length or available resources. The breeding season may be relatively short and discrete, as is found in many bird species including cranes and waterfowl, and in mammalian species such as ursids, canids, and cervids. Or seasonal effects may be more subtle, as in some antelope species (scimitar-horned oryx, Morrow et al. 1999; Jackson's hartebeest, Metrione et al. 2008) that stop cycling for a few months in the spring even though males continue to produce sperm.

Waterfowl generally breed once annually—especially during the spring, in species from the Northern Hemisphere, or when resources such as water are abundant, in tropical and subtropical species. In regions where rainfall is unpredictable, such as Australia, nesting may be triggered by rainfall and flooding. Seasonal effects such as rainfall and availability of resources (e.g., food) also can be reproductive triggers for amphibian, reptile and fish species.

ENVIRONMENT AND SOCIAL EFFECTS

Understanding the basic reproductive biology of a species can greatly aid our ability to manage reproduction. For example, understanding that for many bird species removal of the eggs will promote a second clutch enables us to "double-clutch" a species to maximize its offspring production. As amphibians rely heavily on environmental cues for reproduction, understanding these cues allows us to provide the appropriate conditions to promote its reproduction. Many institutions have used audio recordings of thunderstorms, artificial rainfall, and other large sounds to induce amphibian breeding. A landmark study by Wielebnowski et al. (2002) showed that the usually solitary cheetah may cease reproductive cycling when housed with another female, as chronic stress suppresses reproduction, thus demonstrating the importance of housing animals according to their social needs if they are to breed. Providing nest boxes for palm cockatoos (*Probosciger aterrimus*) or greater hornbills (*Bucero bicornis*) helps to promote their breeding and nesting behavior. Secretive species, like the clouded leopard (*Neofelis nebulosa*), benefit from adequate places to hide; and many crocodilians mate in the water, so that providing an adequate body of water is important. Appropriate management of a species is the best way to promote its reproduction. The importance of social cues on reproduction is best described in the naked mole rat

(*Heterocephalus glaber*): one female can monopolize a colony of up to 100 animals, breeding with only one to three breeding males and reproductively suppressing the other females (Faulkes and Abbott 1997). The suppression is thought to be mediated by aggressive or agonistic interactions between the queen and the nonbreeders, rather than by pheromonal effects. Similarly, in African wild dog groups (*Lycaon pictus*), more than 75% of litters are produced by the alpha female (Creel et al. 1997), and subordinate males mate less often than dominant males. The subordinates are thought to be suppressed by social stress, although interestingly, in wild dogs dominant animals have higher stress hormone levels than do subordinates. Larger groups of white rhinos (*Ceratotherium simum*), a herd species, tend to be more successful at producing offspring than groups of two or three (Metrione 2010). Lastly, there is evidence that animals raised in stimulating environments may reproduce better than animals raised in more barren environments, and that some level of "upset" may actually promote reproduction (Carlstead and Shepherdson 1994). The concept of introducing a new male into an area to "stimulate" a breeding male to breed is widely practiced in zoological institutions, and testosterone spikes in males have been measured in rhinos and cheetahs (Penfold, unpublished data), thus supporting this theory. Replicating the animal's environment as far as is practical, together with managing the animal in a way that mimics the social structure of its wild counterparts, will greatly aid successful captive reproduction.

FEMALE REPRODUCTIVE BEHAVIOR

When a female comes into heat or estrus, a signal is generally given out to the male that she is receptive to breeding. Signs that a female is ready to reproduce vary among species. In birds it may include gathering of nest material and nest building, or it may entail elaborate courtship displays. In Old World primates it may include an overt sign, such as sexual swelling and reddening of the genital region, or may be more subtle, as in hoofstock females that stand to be bred when checked by the male by foreleg kicking or chest pressing. In these cases, a hoofstock female is said to be in "standing estrus," which may be accompanied by tail twitching or holding the tail to one side. Hoofstock species, especially antelope and cattle species (bovids), may also isolate themselves from the herd or attempt to mount other females. A small amount of clear mucus may sometimes be noticed at the vulva. In some small cat species and rodents, downward arching of the spine (lordosis) may be observed. In species like the cheetah, estrus behavior may be more subtle, but by careful observation it may be determined that rolling, rubbing, calling, or urinating behavior may be associated. In canid and felid species estrus can last days, but in hoofstock species it generally only lasts a few hours. Although there are many similarities between related species, each species may have unique behaviors associated with courtship, copulation, and parenting, and keepers should watch carefully for these differences.

MALE REPRODUCTIVE BEHAVIOR

Elaborate courtship displays in waterfowl species ensure mating with conspecifics, reducing the chances of hybridization with other species. For some duck species the drive to mate in the spring can sometimes result in excessive pursuit by the males, which can be a problem for females in captivity with no room to escape aggressive amorous advances. Waterfowl are generally monogamous, though duration of pair-bonding varies, from seasonal bonding in many duck species to long-term and even lifetime bonding in swans and geese.

Males of several bird species, including hornbills and birds of paradise, may solicit breeding behavior by offering food items.

Male ungulates may taste the urine of a female, curling their lip back in a process known as "flehmen." Further male advances, especially in hoofstock, may include repeated following of the female, foreleg kicking (lafschlag), chest pressing, and mounting.

Mounting of the female by the male is an obvious sexual behavior, but it is important to remember that some sexual behaviors may not be linked with reproduction. Mounting of males unrelated to breeding (sociosexual mounting) has been observed in primate species like the bonobo (*Pan paniscus*), in the bottlenose dolphin (*Tursiops truncatus*), and in antelope species like the gerenuk (*Litocranius walleri walleri*). It is thought to act as an alternative to aggression.

Musth in elephants is characterized by the discharge of temporin, a fluid high in testosterone, from swollen temporal glands on the sides of the head. During this period the male is usually very aggressive, and the behavior is linked to sexual arousal and dominance behaviors. However, elephants are able to mate and fertilize females outside of musth. Another seasonal change in some male hoofstock species is rut. Typical in many deer species but also in the camel, rut is noted by increased aggression and antler fighting and by an increased musky odor that also may be noticed. Less is known about the reproductive behaviors of cetacean species, though it is thought that the song of the humpbacked whale (*Megaptera novaeangliae*) is part of its mating behavior.

CONTRACEPTION

Despite careful management, surplus animals are occasionally produced. For example, only one breeding male is required at a time for a herd of hoofstock, though the ratio of calves produced is usually 50:50. Since spaces for holding individuals in zoos are limited, the production of offspring must be carefully controlled, either through separation of breeding animals or by the use of planned contraception. If space is available, the easiest option is to separate the animals. But when space is an issue or disruption of a breeding pair's social bonds must be prevented, or with larger social groups such as in primates, contraception may be considered. Contraception can be permanent if the individual is not required to breed. Available techniques include vasectomy, tubal ligation, and ovariohysterectomy. If the animal may be required to breed in the future, or if surgical contraception techniques are too risky (as in many marine mammals), then a reversible contraception is warranted, which is usually a hormonal pill given orally every day (as is commonly used for primates) or an implant, inserted subcutaneously and lasting for approximately 6 to 12 months (in the case of Suprelorin) or for two years (such as melengesterol acetate, MGA) or immunological treatment (such as the use of porcine zona pellucida, PZP). Hormonal contraceptives usually work by shutting down production of key reproductive hormones at the level of the pituitary (in the case of Suprelorin) or gonads (in the case of progestin-based pills or implants). The most extensive contraceptive research has been conducted in mammals, and the largest array of potential contraceptives is available for these taxa (Asa and Porton 2005; table 19.2). Broadly speaking, progestin implants and human birth-control pills work well for primate species. The use of progestin implants and progestin milled in feed has met with success in hoofstock females. The hormone agonist Suprelorin, a gonadotropin- releasing hormone agonist, has been used in males and females of many mammalian taxa. Unfortunately, it is ineffective in male hoofstock and has quite variable effects in male lion-tailed macaques (*Macaca silenus*), and therefore is not recommended for those animals.

It is important for keepers to recognize potential changes in an animal treated with a contraceptive. For example, male animals treated with Suprelorin may display increased aggressive behavior for several days after treatment before testosterone decreases. This aggressive behavior is likely testosterone-mediated, so no increase in aggression should be expected in females, although it is likely they will come into estrus and thus should be housed separately or treated with another contraceptive for about the first two weeks (note that anecdotal reports suggest Depo Provera can interfere with Suprelorin, so their combined use is contraindicated). Castrated males or males treated with contraceptives that reduce sex-steroid hormones will likely lose secondary sex characteristics, such as a lion's mane, or the dark-brown coloration of Nile lechwe (*Kobus megaceros*) or Javan banteng (*Bos j. javanicus*). If this loss is undesirable, permanent contraception by vasectomy is recommended, as it will not affect secondary sex characteristics. It is important to note that vasectomy as a method of contraception should never be used in species that are induced ovulators, as that would promote repeated pseudopregnancies and likely lead to abnormal uterine pathology.

> **Good Practice Tip:** Watch for transient increased aggression in males, and use a secondary contraceptive for seasonal breeding females while down-regulation of the reproductive system occurs (about two weeks in duration) in animals treated with Suprelorin as a contraceptive.

Most females treated with PZP will continue to display estrous behavior, and some treated with progestin implants may even continue to produce enough estrogen-secreting follicles to stimulate estrus, though they should not become pregnant. It is important to document these behaviors in case of a contraception failure. It is critical that booster injections required for contraceptives like PZP are given on time, as there are countless examples of pregnancies that occurred when booster injections were administered late. Conversely, although the average effective life span of contraceptive implants is estimated at two years, individual differences may be such that the implant's efficacy is longer than stated. For this reason it is imperative to remove progestin implants after the expiration date, if reversal is desired.

Contraception is usually reversible when the individual has not received it for more than one or two subsequent years. Prolonged periods of female contraception may be equated with similar periods of keeping an animal "open" or not pregnant, and may make pregnancy less likely for the individual later in life.

Less research has been conducted on bird and reptile species, so the availability of contraceptives is extremely limited outside the mammalian taxa. Until more is known, simple egg removal prior to incubation works well, and dummy eggs are usually placed in the nest if a second clutch is undesirable.

Finally, many contraceptives that reduce sex steroid hormones (testosterone) in males have been used for aggression control rather than as a contraceptive. The same physiological changes occur, and the male will be unable to breed, but con-

TABLE 19.2. Proven and recommended contraception methods for zoo animals

Taxa	Progestins	PZP	GnRH analogs	Reversible vasectomy
Felids	Contraindicated long-term	Contraindicated	Proven/recommended[1]	Unknown
Canids	Contraindicated long-term	Unknown	Proven/recommended	Proven
Bats	Proven	Unknown	Proven/recommended	Unknown
Antelope	Proven/recommended	Proven	Ineffective in males	Unknown
Equids	Proven	Proven/recommended	Ineffective in males	Proven
Elephants	Unknown	Proven	Unknown	Unknown[2]
Rhinos	Proven	Proven	Unknown	Unknown
Primates	Proven[3]	Unknown	Proven/recommended	Unknown
Great apes	Proven/recommended	Unknown	Proven	Unknown
Birds	Contraindicated	Unknown	Proven/species-dependent	Unknown
Pinnipeds	Proven	Proven	Proven/recommended	Unknown[2]
Cetaceans	Proven	Proven	Proven/recommended	Unknown[2]

[1] Associated with significant weight gain in black-footed cats and clouded leopards.
[2] Surgical procedures are particularly challenging in large mammals with internal testes.
[3] Contraindicated in *Callimicos*.

traceptives also have been used to manage bachelor groups or control undesirable testosterone-mediated behavior. For example, fallow deer (*Dama dama*) fed MGA showed decreased aggression (Penfold et al. 2005). In contrast, gerenuk (*Litocranius walleri walleri*) treated with MGA implants failed to show any decrease in aggressive behavior even though their testosterone concentrations decreased, suggesting that in some species, aggressive behavior may be "hardwired" (Penfold et al. 2005).

THE POTENTIAL FOR ASSISTED REPRODUCTION

Artificial insemination (AI) has only recently been used as a tool to assist reproduction in nondomestic animals, and even then it has only been repeatable as a management tool in a handful of species, namely black-footed ferrets, elephants, giant pandas, and dolphins. In brief, artificial insemination may involve depositing semen collected from the male at the entrance to the female's cervix (e.g., with the gerenuk), may include passing a catheter through the cervix to deposit the semen into the uterus (e.g. with the rhino or elephant), or may require surgical laparoscopy in which semen is deposited high into the uterine horns via a catheter through the oviduct, or via a needle through the uterine wall (e.g., with the cheetah). It is most useful to overcome mate incompatibilities, and in the case of the elephant AI has proved to be a logistically viable alternative to shipping animals between institutions. More recently it has been used for sex selection, and to date 13 dolphins have been produced using sex-selected sperm for offspring of the desired sex. A long-term goal for AI is to move frozen sperm globally instead of translocating animals. This is now theoretically feasible for carnivore species, but there are more challenges to be overcome for hoofstock species because of disease risk issues with domestic hoofstock in the agricultural industry. The keeper's role is critical to the success of these efforts, as keepers can identify the best candidates for these types of techniques and may be required to work with the selected animals to condition them for certain procedures. For example, a keeper may run hoofstock through a chute system for AI, or condition rhinos or elephants for rectal palpation and AI.

Other reproductive technologies include in vitro fertilization, in which ova are recovered from the female and fertilized in a Petri dish with semen from the male. The resulting embryo is then transferred to the female or surrogate female's uterine horn. This technique has been used successfully once each in the Siberian tiger (*Panthera tigris altaica*), the Indian desert cat (*Felis silvestris ornatus*), and the serval (*Leptailurus serval*). Embryo transfer has been used with more success, and endangered ocelot kittens (*Leopardus pardalis*) have been produced in the United States and Brazil by transporting frozen embryos internationally and transferring them to recipient females. Similar success has been seen in Pallas cats.

Cloning has been covered extensively in the media as a technique for saving species. In fact, of all the reproductive techniques, cloning requires the most resources—namely, hundreds of ova to achieve a single embryo for transfer, because the technique is currently so inefficient. It is important to note that a cloned animal is a genetic copy of an individual and thus does not necessarily expand genetic diversity. It also

is a fruitless exercise to clone animals, such as thylacines (*Thylacinus cynocephalus*) and woolly mammoths (*Mammuthus primagenius*), for which there are no other animals with which to breed. Nevertheless, cloning may have some potential application in the propagation of certain species that produce thousands of ova, such as frog species, so it should not be dismissed completely.

SUMMARY

The role of zoos and aquariums in species conservation continues to expand, and so must the role of the animal keeper, to keep pace with these changes. For species about which little of their reproductive biology is known, careful observation by keepers can increase our basic knowledge and provide insights into the most appropriate ways to manage them. Keepers are encouraged to communicate and report what they observe, and to participate in research that can aid reproduction. When natural breeding fails, tools such as artificial insemination may then be available to assist in ensuring that individual animals reproduce.

ACKNOWLEDGMENTS

The author is grateful to Holly Huffnes, Deb Morse, and Josh Watson for their input. Special thanks to Lauren Stamatis and Michael Clifford for their enthusiasm and input for this chapter. Thanks also to Dr. Cheryl Asa and Ana Maria Cepeda for their contributions.

REFERENCES

Asa, Cheryl, and Ingrid Porton, eds. 2005. *Wildlife Contraception: Issues, Methods and Applications*. Baltimore: John Hopkins University Press.

Brown, Janine, Sam Wasser, David Wildt, and Laura Graham. 1994. Comparative Aspects of Steroid Hormone Metabolism and Ovarian Activity in Felids, Measured Non-invasively in Felids. *Biology of Reproduction* 51:776–86.

Brown, Janine, Laura Graham, Julie Wu, Darin Collins, and William Swanson. 2002. Reproductive Endocrine Responses to Photoperiod and Exogenous Gonadotropins in the Pallas' Cat (*Otocolobus manul*). *Zoo Biology* 21:347–64.

Carlstead, Kathy, and David Shepherdson. 1994. Effects of Environmental Enrichment on Reproduction. *Zoo Biology* 13:447–58.

Creel, Scott, Nancy Creel, Michael Mills, and Steven Monfort. 1997. Rank and Reproduction in Cooperatively Breeding African Wild Dogs: Behavioral and Endocrine Correlates. *Behavioral Ecology* 8:298–306.

Green, Wendy C. H. 1990. Reproductive Effort and Associated Costs in Bison (*Bison bison*): Do Older Mothers Try Harder? *Behavioral Ecology* 1:148–60.

Klein, L. L. 1971. Observations on Copulation and Seasonal Reproduction of Two Species of Spider Monkeys, *Ateles belzebuth* and *A. geoffroyi*. *Folia Primatologica* 15:233–48.

Kunz, Thomas, Chris Wemmer, and Virginia Hayssen. 1996. Sex, Age and Reproductive Condition of Mammals. In *Measuring and Monitoring Biological Diversity: Standard Methods for Mammals*. Eds. Don Wilson, F. Russell Cole, James D. Nichils, Rasanayagam Rudran, and Mercedes S. Foster. Washington: Smithsonian Press.

Metrione, Lara, and Harder, John. 2011. Fecal Corticosterone Concentrations and Reproductive Success in Captive Female Southern White Rhinoceros. *General and Comparative Endocrinology* 171:283–92.

Penfold, Linda M., Steve L. Monfort, Barb Wolfe, Scott B. Citino, and and Divid E. Wildt. 2005. Reproductive Physiology and Artificial Insemination in Wild and Captive Gerenuk. *Reproduction, Fertility and Development* 17:707–14.

Penfold, Linda, Marilyn Patton, and Wolfgang Jöchle. 2005. Contraceptive Agents in Aggression Control. In *Wildlife Contraception: Issues, Methods and Applications*, Asa and Porton, eds. Baltimore: John Hopkins University Press.

Standora, Edward, and James Spotila. 1985. Temperature-Dependent Sex Determination in Sea Turtles. *Copeia* 3:711–22.

Wielebnowski, Nadja, Karen Ziegler, David Wildt, John Lukas, and Janine Brown. 2002. Impact of Social Management on Reproductive, Adrenal, and Behavioral Activity in the Cheetah (*Acinonyx jubatus*). *Animal Conservation* 5:291–301.

20

Population Management

Linda M. Penfold

INTRODUCTION

Populations are dynamic, responding to influences such as environment, disease, and demographic fluctuations. So in zoos and aquariums, why is it important to manage populations rather than leave them to their fate? The answer is that generally speaking, zoos and aquariums manage a limited number of individuals for every species, and population problems become magnified in small populations. For a species to survive, it is important that it contains the genetic diversity to be "evolutionarily flexible." The most famous example of evolutionary flexibility is the rise of the mammals in terrestrial communities following the mass extinction in the late Cretaceous period 65 million years ago. It is likely that a catastrophic event occurred, such as an asteroid colliding with the planet, and that the mammalian species were able to adapt to the changed environment whereas the dinosaurs were not. Even today, the word "dinosaur" is synonymous with a failure to adapt to a changing environment.

The primary focus of this chapter is to give a very general overview on the concepts of population management. After studying this chapter, readers should understand

- the hazards associated with small populations
- the idea that a critical number of animals is required if a population is to be sustainable
- the definition of a founder animal
- the role of demographics in population management
- different tools that managers may use to manage populations

HAZARDS OF SMALL POPULATIONS

LOSS OF GENETIC VARIABILITY

A critical number of individuals are required for species of animals or plants to maintain genetic diversity. Without this number, negative effects on the population become magnified. This is because heritable differences (traits passed down from parents to offspring) between individuals will influence how they interact with the environment and within the ecosystem. These differences allow a species to cope with environmental changes, reduce the chances of harmful (deleterious) genes becoming expressed in the population, and confer a flexibility to adapt to future environmental uncertainty. For example, the highly contagious disease rinderpest, known to have killed up to 90% of ruminants in Africa in 1885, is described as the "Great Pandemic." However, a handful of animals were resistant to the disease and survived it. If all the species had had similar genes, they all would have responded to the disease in exactly the same way, and thus 100% would have succumbed. A lack of genetic diversity in a population can result in a downward cycle of its reduced ability to withstand environmental pressures and novel conditions, which in turn can result in further reductions in population size and further loss of genetic diversity. This is described as an "extinction vortex" (see below).

A landmark study (Shaffer 1981) showed that a critical (absolute minimum) number of animals were required for a population to be self-sustaining for a long time into the future (beyond 100 years). The Minimum Viable Population (MVP) size is the smallest number of individuals required to withstand the various pressures that affect individual persistence over time. If the population number dips below the MVP, the chances of extinction increase. The tendency of small populations to decline towards extinction has been compared to a vortex (like a spiraling tornado or whirlpool), whereby faster-spiraling events occur closer to the center of the vortex. At the center of the "extinction vortex" is extinction (Gilpin and Soulé 1986). Captive populations in zoos and aquariums are derived from initial groups of animals, many of which may have been imported from their native range countries decades ago. Those individuals are referred to as "founders." Any new animals brought into the population that are not related to animals produced from the original group are similarly referred to as founders. If the population is large enough and is managed correctly (i.e., in accordance with studbook recommendations), a population can be self-sustaining with little need for the introduction of new founders. This type of population is referred to as "closed" because animals do not flow in and out of it as they might do in a wild population. Because of the finite number of animals and plants maintained in zoos and aquariums, it is quickly apparent why each individual needs to be genetically represented in the population: to maintain maximum genetic diversity. With an insufficient number of animals reproducing, there is a risk that deleterious genes will be expressed, resulting in health and reproduction problems—or that heterozygosity (genetic variation) will be insufficient to allow some animals to survive a disease epidemic or other catastrophe. This situation is referred to as an "inbreeding depression." In contrast, outbreeding effects occasionally may be seen, such as when two subspecies are bred but are not genetically compatible, and the offspring have reduced fertility or are even sterile. This has been described in the Kirk's dik-dik (*Madoqua kirkii*; Howard et al. 1989).

DEMOGRAPHIC FLUCTUATIONS

Demography is the scientific study of populations, especially with regard to their size, sex ratios, births, deaths, and so on. A balanced demographic population would include all age groups, and the average birth rate would equal the average death rate. In reality this seldom occurs; and for certain species, demographic fluctuations may greatly influence the population, making it more susceptible to extinction than it would be by chance alone. For example, in species that reproduce slowly like the elephant, early infant loss or a failure to produce offspring may have dramatic effects on the demographic age structure. Similarly, the overproduction of male or female offspring in a given year may result in a sex skew that over time can have detrimental population effects. In captive settings, although the percentage of production of offspring (at least in mammalian species) is usually 50:50, there are times when chance can result in a different ratio of anywhere from 100:0 to 0:100. When considering the sex ratio of an entire population, it is important to remember that these ratios may occur by chance and may not necessarily represent a sex skew. It is prudent to review the birth ratio over an extended period of time and for a minimum of, say, 10 years to check whether a sex skew is actually occurring. For example, production of a significant majority of male versus female antelope in a single year is alarming for animal managers who only require one breeding male for a herd of females, and may seem like a "sex skew," but production of a majority of female antelope in a given year doesn't usually generate comment.

Recently, facultative parthenogenesis (asexual reproduction, possible under certain conditions) has been reported in Komodo dragons (Watts et al. 2006) and hammerhead sharks (Chapman et al. 2007). Since the reproduction was parthenogenetic, however, these offspring would be genetic copies of their mothers. If not managed appropriately, parthenogenesis in captive populations could result in unexpectedly high inbreeding, loss of genetic variation, and changes in founder contribution (Hedrick et al. 2007), and thus would be important to consider in the overall population.

ENVIRONMENTAL FLUCTUATIONS

In most zoos, many environmental conditions such as temperature, food availability, water, and light can be controlled. However, disease epidemics may be more challenging to contain. The chytrid fungus (*Batrachochytrium dendrobatidis*) in amphibians is a good example of where stringent quarantine methods and animal handling will be required to maintain the health of amphibian populations in captivity. Similarly, the H5N1 influenza (bird flu) is highly contagious to birds and remains a serious concern to bird collections and human health worldwide.

MANAGING POPULATIONS
RECORD KEEPING

The International Species Information System (ISIS) is an international database of zoological information, containing information from more than two million animals. Institutions that use ISIS contribute data on age, sex, parentage, place of birth, circumstances of death, and so on to generate a database that is used to manage genetic and demographic populations in animal collections around the world. ISIS software is recognized as the world standard in best practice for zoological record keeping. Accurate record keeping is critical to the success of this system, and is facilitated by the animal keeper's contribution of regular informational updates on animals.

ACQUISITION AND DISPOSITION

To appropriately manage animal collections, it is necessary to periodically introduce new animals into a zoo or aquarium's collection. This may be accomplished by buying, trading, donating, loaning, capturing, breeding, or rescuing an animal. The term for bringing in new animals is "acquisition." Similarly, animals may need to be permanently removed from a population for reasons such as population management, reintroduction, behavioral incompatibilities, sexual maturation, ending of a breeding loan, or death. This process is referred to as "disposition." It is important to remember that animals should be moved, traded, loaned, and donated only to institutions that have the expertise and appropriate facilities to adequately care for the animal.

CONTRACEPTION, EUTHANASIA, AND SEX SELECTION

In spite of careful population management, animals surplus to the population are still produced. This is especially true for many herd ungulates, where a single breeding male may be needed for a herd of multiple females. It may also result when an individual has reproduced sufficiently in a population and its genes are well represented, but continues to breed. For example, if males are not separated from females, due to space limitations or other reasons, offspring will continue to be produced. This could have the effect of flooding the population with genes from the overrepresented individual, and it is counterproductive for genetic diversity. Surplus animals can be managed in several ways; they can be treated with a contraceptive that may be permanent or reversible (see chapter 19), sold or donated to institutions outside the managed collection, maintained as a separate population if space allows for zoological research programs, or euthanized. Euthanasia as a management tool is commonly practiced in many countries, depending on social mores, and it has the advantage of keeping a female in more of an active breeding condition. It is widely known, though undocumented to date, that keeping females "open" (unbred) for several years can hamper efforts to get them to reproduce when it is finally desired.

Cutting-edge research is investigating the use of artificial insemination using sex-selected sperm to produce offspring of known sex. This has been particularly effective in the bottlenose dolphin (*Tursiops truncatus*), where 13 female offspring of predetermined sex have been produced (O'Brien et al. 2009; O'Brien and Robeck 2010).

The nucleated blood cells of birds allow sex determination to take place, and some institutions have refined the technique of carefully drilling a hole into a fertilized egg, collecting a tiny amount of blood from a blood vessel, sealing the hole, and continuing incubation. The blood can be used to determine the sex of the embryo, and then a decision can then be made as to whether to hatch the egg.

TOOLS

Collection planning determines, sometimes on the basis of staff expertise or geographical region, which assortment of species the institution will focus on. In zoological institutions belonging to organized associations such as the Association for Zoos and Aquariums (AZA), the European Association for Zoos and Aquariums (EAZA), the British and Irish Association of Zoos and Aquariums (BIAZA), and the Zoo and Aquarium Association (ZAA) in Australia, captive populations are managed in a more formal and cooperative way. A collection of experts, called a taxon advisory group (TAG), determines which species require formal breeding programs to be self-sustaining. These programs, such as AZA's Species Survival Plans® (SSP) or EAZA's European Endangered Species Program (EEP), will make formal recommendations to breed individual animals in order to ensure the viability of the population. They often use specialized software such as the Single Population Analysis and Records Keeping System (SPARKS) and Population Management 2000 (PM2000) to provide a detailed analysis. Population biologists and studbook analysts work with the SSPs and EEPs to confirm population sizes, meet conservation objectives, and ensure that captive populations do not grow beyond the ability of zoos and aquariums to manage them. Decision-making is facilitated by the use of studbook data including births, deaths and parentage: information essential to determining which animals should breed. Also, husbandry manuals, often written by animal care staff, outline proper care and breeding of a species and can be instrumental in the success of a breeding program. For details on these tools, see chapter 55.

SUMMARY

The zookeeper can play an important role in managing animals by recognizing that individuals may need to be transported to other institutions in accordance with breeding recommendations. Compliance with such recommendations supports the long-term sustainability of the species by ensuring that it remains genetically diverse. Similarly, prompt removal of offspring once they get close to puberty prevents father-daughter or mother-son matings which can result in genetically useless animals that take up space in the institution. For example, a bias of male hoofstock offspring several years in a row should be scrutinized within the context of a 10-year period to prevent a coincidental occurrence being confused with a genuine sex skew. Lastly, keepers wishing to learn more about population management or to become more involved in it are encouraged to take on studbooks and/or attend various professional training courses supplied by AZA and EAZA.

REFERENCES

Chapman, D. D., M. S. Shivji, E. Louis, J. Somner, H. Fletcher, and P. A. Prodöhl. 2007. Virgin Birth in a Hammerhead Shark. *Biol Lett.* 3: 425–27.

M. E. Gilpin and M. E. Soulé. 1986. Minimum Viable Populations: Processes of Species Extinction. In *Conservation Biology: The Science of Scarcity and Diversity*, M. E. Soule, ed., 19–34. Sunderland, MA: Sinauer Associates. Excellent summary of extinction vortices facing small populations.

Howard, J. G., B. L. Raphael, J. L. Brown, S. B. Citino, M. C. Schiewe, and M. Bush. 1989. Male Sterility Associated with Karyotypic Hybridization in the Kirk's Dik Dik (*Madoqua kirkii*). Proceedings of the American Association of Zoo Veterinarians Annual meeting, 58–60.1997.

Meffe, Gary K., and C. Ronald Carroll. 1997. *Principles of Conservation Biology.* New York: Sinauer Associates.

O'Brien, J. K., K. J. Steinman, and T. R. Robeck. 2009. Application of Sperm Sorting and Associated Reproductive Technology for Wildlife Management and Conservation. *Theriogenology* 71:98–107.

O'Brien, J. K., and T. R. Robeck. 2010) The Value of Ex Situ Cetacean Populations in Understanding Reproductive Physiology and Developing Assisted Reproductive Technology (ART) for Ex Situ and In Situ Species Management and Conservation Efforts. *International Journal of Comparative Psychology* 23:277–48.

Shaffer, M. L. 1981. Minimum Population Sizes for Species Conservation. *Bioscience* 31:131–34.

Watts, P. C., K. R. Buley, S. Sanderson, W. Boardman, C. Ciofi, and R. Gibson. 2006. Parthenogenesis in Komodo Dragons. *Nature* 444:1021–22.

Kleiman, D., K. V. Thompson, S. Lumpkin, M. E. Allen, and H. Harris, eds. *Wild Mammals in Captivity.* Chicago: University of Chicago Press.

21

Management of Neonatal Mammals

Harmony B. Frazier, Janet Hawes, and Karla J. Michelson

INTRODUCTION AND HISTORY

Techniques and philosophies regarding neonatal care of zoo animals continue to evolve. The cornerstone of neonatal care should be reproduction in species-appropriate family groups that meets the physical and behavioral needs of zoo animals and supports healthy and sustainable captive populations. Historically, zoo infant care practices have included removing infants or neonates from their dams for hand-rearing to "tame" them for public programs, to increase reproduction, or to decrease infant mortality caused by parental neglect. In many cases the parental neglect might have been attributed to inadequate animal husbandry. Infant care nurseries were once common and provided us with information on improved hand-rearing formulas and early reintroductions of hand-reared infants to their natal groups. Public display nurseries began to fade from view in the 1990s due to improved husbandry techniques and species population management resulting in a less frequent need to provide hand-rearing. The Zoo Infant Development Project (ZIDP), which focused on parent-reared zoo mammals, was created in 1994 by the American Association of Zoo Keepers (AAZK) (Frazier and Hunt 1994). The profession began striving to provide infants requiring hand-rearing with care that better met the needs of normal physical and behavioral development. Healthy family groups and empty "nurseries" were viewed as successes, and that philosophy is still being built upon today. Keepers need to think about infant care as sustaining the normal maternal and social group dynamics (i.e., being proactive), rather than from a neonatal critical care (i.e., being reactive) perspective. The goal should be to provide support to the family group so that neonates stay with their mothers as a best-care practice. Creating the optimum situation for a positive outcome is an active process, and it requires action and planning. "Hoping for the best," "leaving the dam and infant alone," or "letting nature take its course," while appropriate in some circumstances, do not in themselves constitute supportive care management. Team pre-planning with an action outline can be critical.

This chapter will address

- birth planning
- neonatal assessment and considerations
- hand-rearing alternatives
- hand-rearing guidelines
- ethical considerations.

PLANNING AHEAD FOR NATURAL FAMILY GROUPS

A zoo or aquarium's institutional collection plan is a great place to initiate breeding plans for the upcoming year. From there, it's best to work in teams that include animal health and animal management staff to create a Birth Management Plan (BMP) for each of the priority species (Frazier et al. 2009). The goal of a BMP is to create a working document that provides guidelines for the majority of contingencies and preparations in a future birth. A BMP should serve as a guide to urgent intervention and informed animal care. Consider all staff who might be involved before, during, and after the birth and what they need to know. Have the team review and comment on the plan. Make the information easily accessible and clear. Planning ahead means potentially avoiding hand-rearing, illness, injury, or death.

The following criteria should be considered when creating a BMP (see table 21.1):

1. the natural history of the species, including normal gestation, birthing, postpartum behaviors, and weaning information
2. Species Survival Plan (SSP) recommendations on breeding and care
3. the individual reproductive histories of the animals
4. what new technologies and resources are available for pregnancy detection, and how they might enhance preparations (e.g., fecal hormones, ultrasound)
5. the intended birth environment (does it need any

upgrades or changes such as heat source, video monitoring, or den preparation?),with clear definition of which animals should or should not be present during the birth

6. the time of year the birth will take place (are there potential weather issues?)
7. expected normal birth weight and appearance of the neonate, helpful even for "visual-only" exams
8. medical evaluation, wellness checks, and vaccines (an early medical evaluation can provide valuable information that may go unnoticed by nonmedical staff)
9. supplemental care and support for dam and infant, and how it might take place if needed
10. hand-rearing information (formula, frequency, feeding apparatus, housing)
11. reintroduction plan or alternative, with strategies

for how reintroduction might take place at 24 to 48 hours, at two to four weeks, or at weaning.

Create a flow chart for assessment of birth progress and possible need for additional care. This type of quick reference flow chart provides a checklist that informs the team about what to look for during labor and soon after (Ensley 1995; Ensley and Meier 1978). The early days are critical; 80% of all mortality in lambs between birth and weaning occurs in the first week of life (Rings 1995). A good assessment chart will suggest what steps to take based on the stage of labor, birth, or following care; it should include information on lines of communication (figures 21.1 and 21.2).

One of the best ways to ensure a positive outcome is to gather all information, supplies, equipment, and staff well ahead of the birth. Emotions may run high during and after a birth, even for the most seasoned keeper. A solid plan pro-

TABLE 21.1 Example of a Birth Management Plan (BMP) form used by Woodland Park Zoo. Courtesy of Woodland Park Zoo, Seattle.

BIRTH MANAGEMENT PLAN for _____
Date created: _____

Species:		EDITOR:
Scientific name:		TEAM:
Plan for	DAM # 　　　　　 SIRE #	MANAGER:
Expected birth date:	Actual birth date:	VETERINARIAN:

Seattle, WA, USA

BIRTH NOTIFICATION: TWO people notified directly (no phone messages)

	NAME	PHONE TREE DETAILS OR COMMENTS	BEST WAY TO CONTACT
1			
2			

If you are unable to talk with them directly, please call the following back-ups:

SPECIES REPRODUCTIVE CHARACTERISTICS:

Gestation period:		Estrus cycle:	
Interbirth interval:		Estrus duration:	
Number in litter:		Weight at birth:	
Weaned:		Independent:	

TABLE 21.1 continued

BIRTH MANAGEMENT PLAN for _____

PRE-PARTURITION

HOUSING	
Begin separating dam and sire:	
Holding space:	
Bedding:	
Drinkers:	
Disinfection:	
Building must be kept warm & quiet:	
No visitors:	
Additional considerations:	

ENVIRONMENTAL PARAMETERS	
Birth area temperatures:	
Specific environmental needs:	

PARTURITION

Signs of pending birth:	
Parturition:	
In-house observations:	
Level of intervention:	

(continued)

vides focus and structure for the team and helps ensure a positive start for everyone.

Some resources that are helpful when creating BMPs include

1. AAZK biological values (Danzig 1992)
2. resources for species-specific natural history
3. AZA animal care manuals (Sodaro 2006)
4. the species veterinary advisor through American Association of Zoo Veterinarians (AAZV)
5. ISIS information, noting which other institutions have the same species
6. zoo infant development project notebooks (ZIDP, AAZK)
7. hand-rearing manuals and books (Gage 2002; AAZPA 1985).

CONSIDERATIONS FOR SUCCESS

In zoo infant care, lack of a successful birth and/or insufficient postnatal care generally falls into one of the following categories.

MATURITY, EXPERIENCE, AND GROUP COMPOSITION: APPROPRIATE AND INAPPROPRIATE BEHAVIORS

An immature or first-time (primiparous) dam that is naive to parturition and neonatal care may require human intervention or assistance. Bonding (care and protection of the neonate) is needed to provide physical and emotional support to the neonate so that it can thrive. In cases where it is appropriate to leave the dam, sire, siblings, and/or group members together during a birth (e.g., with gorillas or tama-

TABLE 21.1 continued

BIRTH MANAGEMENT PLAN for _____

POSTBIRTH

Husbandry access:	
Monitoring:	
Housing:	
Environmental parameters:	
Postnatal monitoring and care:	

NEONATAL EXAM

Initial Neonatal Exam:	

DEVELOPMENTAL NORMS FOR THIS SPECIES
PREVIOUS WEIGHTS FOR FULL SIBS OR OTHERS OF SAME SPECIES:

Average weight at birth:		Average weight at one week:	

Developmental "Milestones" with past infants

Age in Days	Developmental Changes
Data From:	
Day	
Day	
Day	
Day	
Day	
Day	
Day	
Day	
Day	
Day	
Day	
Day	
Day	
Day	

Age in Days	Example Weight ranges in grams
1	
4	
6	
12	
23	
40	
55	
70	
98	
109	
120	

Hand rearing outline on file at AHD:	DATE:	BY:
Special considerations:		

TABLE 21.1 continued

BIRTH MANAGEMENT PLAN for _____

CALL LIST:

Team member	Office: Ext, DID	Work Cell	Home Cell	Home landline

rins) or place them together soon after, the animals' previous experience and individual disposition will play a part in the success or failure of bonding.

Dominance and rank can play a role. Is the dam ranked high enough in the group so that she will be supported during and after the birth, or will she or the infant be harassed by the other group members? An inexperienced individual may act inappropriately by becoming overly excited or aggressive or generally interfering with the dam and neonate. There is the potential for either well-intentioned or aggressive "kidnapping" by any of the group members, so they should be monitored. In some species it may be appropriate for experienced individuals to exhibit "aunting" behavior or to assist with care. Some species, such as fruit bats, do better with more conspecifics; and some, such as felids, do better with fewer animals present during the actual birth. Separating a highly social animal during birth and labor may cause anxiety. Knowing the species' natural history as well as the individuals' personalities will help in the decision-making process.

Equally important, inexperienced animal care staff may miss subtle but vital signs from the birth family, causing them to either react too soon or too late. When the species or group giving birth is new for keepers, they should keep their supervisors informed of the progress and behaviors they are seeing in a timely manner.

Examples:

1. "The infant is nursing." Nursing is not the same as taking in nutrition. An infant can suckle without receiving milk. Watching for signs that the infant is receiving nutrition (appearing bright, alert, responsive, and neither fussy nor lethargic) is vital.
2. "The infant is vocal and following dam." In some breeds of ungulates the infant should initially remain hidden. If the infant is following and vocalizing, it could be hungry. Has the dam been seen allowing it to nurse?

ENVIRONMENTAL HAZARDS: STRUCTURE, TEMPERATURE, SUBSTRATE, FURNISHINGS, SANITATION

Hazards within the exhibit, lack of appropriate birth areas, inappropriate temperatures (which can be too hot, as with red pandas, or too cold, as with carnivores), inappropriate substrate that moves leaving an exposed and cold floor, or use of materials that may be ingested all need to be taken into consideration. Eliminating drafts in the birth location

> **Good Practice Tip:** An infant can suckle without receiving milk. Watch for signs that the infant is receiving nutrition (appearing bright, alert, responsive, and neither fussy nor lethargic).

is essential. Consider that wind speed of 19.3 km (12 mph) at a temperature of 12.7 °C (55 °F) on a wet, newly born lamb has the same effect as a –31.6 °C (–25 °F) temperature (Rings and Martin 1995).

An infant with a reduced body temperature (hypothermia) is at risk of pneumonia or sepsis. Conversely, caution must also be taken not to overheat the infant (hyperthermia) which can create dehydration or heat exhaustion.

Exhibit furnishings all need to be examined with the infant in mind. There has been reported evidence of infants or juveniles becoming tangled in ropes, sometimes fatally. With the trend toward mixed species exhibits in zoos today, it is also important to consider what other animals may be present in the exhibit that might interfere or attack an infant should it become separated from the dam.

When cleaning, it is important to avoid getting the bedding wet, which could possibly lead to hypothermia, bacterial growth, fungal growth, or irritations due to ammonia buildup from urine. It is equally important to avoid introducing harsh chemicals into the environment.

DISEASE, CONGENITAL DEFECTS, AND DIFFICULT OR PREMATURE BIRTH

Disease or defects in the dam's reproductive organs or limbs can make giving birth difficult. A neonate with a cleft palate or immature lungs may have difficulty nursing or breathing. An infant that is malpositioned for delivery, or even the dam's health or emotional state, may cause the birth process to slow or the infant to be born prematurely.

Birth Management Plan
Pre-Labor and Labor Assessment Checklist – Mammals

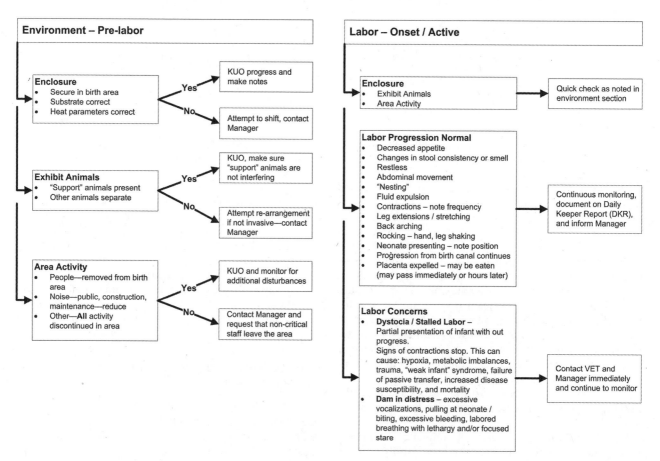

Figure 21.1. Example of an assessment chart for reference during labor. Assessment charts are a useful reference for monitoring critical stages during the birth process. Courtesy of Woodland Park Zoo, Seattle.

Birth Management Plan
Birth to One-Week Assessment Checklist – Mammals

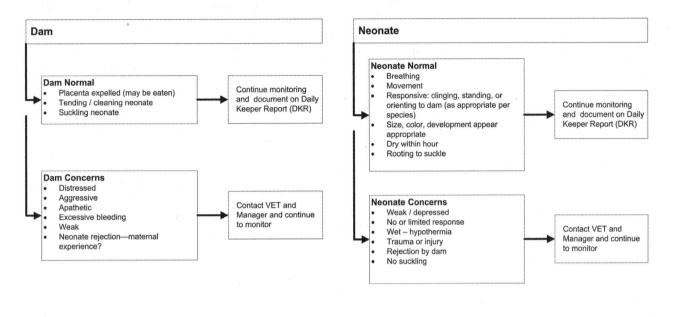

Figure 21.2. Example of an assessment chart for use at birth through the first week of life. Creating an assessment flow chart for a particular species and keeper experience level will assist observers in monitoring the critical steps during prelabor through the first few weeks of life, the most critical time. Courtesy of Woodland Park Zoo, Seattle.

NUTRITION

Proper nutrition for the dam to accommodate gestation and growth of the neonate is essential. Although additional foods may be appropriate, monitoring the dam's weight to avoid obesity, which may lead to a difficult labor and birth, is just as important. The neonate requires appropriate volume and concentration of milk or formula through the weaning process.

KEEPER OBSERVATION AND COMMUNICATION

Ongoing monitoring of the dam and neonates during the critical first days cannot be overemphasized. An infant that looks fine at birth can be found dead three days later. Observations can often be accomplished through the use of remote cameras, and this is often the least invasive way to monitor progress. Cameras should have a clear image and include an auditory component. Some species or individuals (e.g., many felines, or nocturnal mammals) will usually accept a quick neonatal health exam performed by the veterinarian. Monitoring the infant's weight consistently can help the assessment process.

Maintain effective verbal and written communication between all keeper staff, veterinary staff, nursery staff, and management while following the BMP. Lack of communication can be hazardous to the infant; for example, if formula is not labeled properly. It is also important to note the young animal's growth and health trends, and to know who is scheduled to care for the infant next.

If the above considerations and potential hazards have been examined and are openly discussed, most birth failures can be avoided and infant death or injury may be prevented.

GENERAL HAND-REARING GUIDELINES

When presented with a neonate for hand-rearing or supplemental care, consider the following steps.

ASSESSMENT

Why is the infant separated from the dam? Has it been interfered with by exhibit mates, staff, or visitors? Is the dam injured, ill, or dead? Has the dam rejected the infant? Is the infant weak due to lack of nutrition? If the infant is removed from the dam, the infant should undergo a veterinary assessment as soon as possible to rule out injury, illness, or prematurity. Can the infant be stabilized and returned to the dam for rearing?

ADMISSION PROCEDURES

A veterinarian should conduct thorough examinations on all neonates as soon as possible. A schedule of ongoing veterinary exams for any infant being hand-reared is highly recommended.

The umbilical cord should be ligated (tied) as necessary, and a mild tincture of iodine (2%) or a chlorhexidine solution (4%) should then be applied to the umbilical cord stump. Application of the solution is continued three times daily for three days or until the entire stump has dried or fallen off. Exercise caution when using the iodine solution to ensure that it does not contact the skin of the infant, as it can cause dryness and irritation. Continue to monitor the umbilicus for any signs of infection or swelling, which could indicate a hernia.

HOUSING AND SECURITY

Housing for hand-reared infant animals must provide a nurturing environment to promote health, security, and growth. The nursery area must be considered a quarantine area and not a gathering place for interested employees. Only essential personnel should be permitted into the nursery area. A footbath should be installed outside all entrance doors. Make sure the hand-rearing environment is dry and draft-free, and that it meets the infant's need for body temperature maintenance (thermoregulation).

Consider how the infant may be cared for by the dam, and attempt to replicate that scenario: carrying the infant and allowing it to grasp (in the case of a primate), keeping it in a quiet and darkened space (in the case of a macropod), and providing soft toys, blankets, or animal contact (surrogates) for comfort. Initially having one keeper care for the infant until it is successfully nursing and stabilized may help the infant to adjust more quickly.

When an infant is moved from the incubator to a larger space, a surrogate "parent" should be provided. There are excellent commercially available pet beds made of fake fur and stuffed with batting material. Routine regular checks of all bedding and cage furniture items (blankets, towels, surrogates, etc.) are recommended. Strings or small holes in the bedding can contribute to serious injury. Infants can entangle themselves in small holes or on frayed edges of blankets while they root and search for a nipple. Strangulation and death may result. Any bedding with holes, frayed edges, or strings should be repaired or replaced.

Providing access to natural sunlight is necessary for conversion of vitamin D. The recommendation is a minimum of 70 minutes of direct sunlight each week. Outdoor natural sunlight is best. If that is not practical, "sun time" using full-spectrum lighting may be substituted. Review safety precautions and discuss them with a veterinarian.

As the infant becomes more capable of moving about, it begins to explore its environment. A selection of suitable enrichment and toys should be provided. Toys need care-ful inspection for safety, and should be rotated to provide variety and novelty. Browse, wood, and natural items also make good enrichment items, provided that they are sanitary and cannot be ingested by or cause injury to the neonate. Toys should be removed if they are being chewed apart or ingested. A curious infant that does not have normal parental supervision might become trapped or injured by items that would not be hazardous for a parent-reared infant. Adequate safe space for exercise is important. During the first visits to a larger enclosure, familiar cage items are transferred to the new location along with a surrogate and a covered heat source for warmth and shelter.

Occasionally when neonates are hand-reared together, there are problems with the neonates suckling from each other in an attempt to find a nipple. This can be problematic as the suckling causes chafing, dry skin, and irritation. Occasionally when the suckling attempts are concentrated on the genitals, the trauma to the area is more severe because the tissues are fragile and delicate. Application of "bitter orange" may be effective to temporarily discourage non-nutritive suckling on nongenital areas. In some cases it may become necessary to separate infants temporarily.

WARMTH

Normal body temperatures range from 37.22 to 39.44 °C (99–103 °F) for most mammals, depending on the species. Rectal body temperatures should be recorded twice daily for the first three days, then once daily until the temperature is consistently within normal range. Taking the temperature at about the same time each day provides consistency and accuracy. Exceptions are made for animals that are very wiggly or have extremely small anal openings, to avoid causing injury. Elevated body temperatures can indicate infection or simply an excited or overactive animal. Wait until the animal is calm to ensure the most accurate body temperature reading. If low body temperature (hypothermia) is detected, placing a neonate next to a covered hot water bottle or inside an incubator set at 26.67–29.44 °C (80–85 °F) will safely restore normal body temperature. For larger neonates, hanging a heat lamp or using a ceramic brooder lamp in a small enclosure may suffice. Monitor all heat sources closely to avoid dehydration or burns.

If a neonate cannot maneuver on its own or cannot maintain its body temperature, it may need special housing upon arrival in the nursery setting. Supplemental heat may be required for the first few weeks, and can be supplied by infant incubators, animal intensive care unit (AICU) incubators, or by using hot water bottles or reheatable heat discs, which are available at pet supply stores. Generally the temperature of the environment is lowered incrementally as the animal is able to maintain its body temperature. Health and low birth weight can affect the length of time an animal needs to be housed inside an incubator. Sparsely furred neonates should be fed in a warm room free of drafts and returned to the incubator promptly to avoid chilling.

HYDRATION

Before starting formula, one should make sure that the infant is properly hydrated. Oral fluids of any kind should not be offered to the neonate until normal or near-normal body temperature has been restored, as the digestive system is not fully functional during a hypothermic event. The first few feedings usually consist of an electrolyte solution (e.g., Pedialyte), followed by several feedings of diluted formula before the full strength formula is offered. This is also a safer way of assessing an infant's suckling response before graduating to formula. Aspiration of formula can lead to pneumonia.

Low blood sugar due to lack of nursing (hypoglycemia) is a common concern when a neonate first arrives for hand-rearing, or when it does not take in enough nutrition. The infant may be cool to the touch, lethargic, and unable to nurse properly.

If an infant is unable to suckle, fluids can be provided in several ways; oral with a nasogastric (gavage) tube, under the skin (subcutaneous or SQ), in the vein (intravenous or IV), into the abdominal cavity (intraperitoneal or IP), or even into the bone marrow (intraosseous or IO, a technique used in very dehydrated and debilitated infants).

Poor early nutrition and hypothermia can lead to sepsis (bacterial infection of the blood). Onset of infection frequently begins in the first few days after birth.

GROWTH

Collecting an accurate body weight each day is essential in determining nutritional needs, but it is also perhaps the single most important piece of empirical information used to determine general condition. Each animal should be weighed in the morning, before the first feeding. An accurate, good quality gram scale is an essential piece of equipment. With low weight gain, an infant may have difficulty combating environmental or biological stresses. Conversely, a rapid rate of gain may be detrimental if the immature bones cannot accommodate the body weight.

Monitoring weight gain and condition is typically a subjective evaluation. For the novice keeper, for species not previously reared, or for evaluating formulas, it can be advantageous to have additional methods to analyze the rate of gain in individuals. Weight gain in the neonate can be more objectively evaluated with the use of a theoretical growth curve by using a regression equation ($Y = 0.0766W^{0.71}$, where Y is the average daily gain [ADG, g/d] and W is the mature, or adult, body weight [g]; Robbins 1983). The growth rate for ungulates during this initial phase is linear, and it can confidently be estimated by using the regression equation. In any particular species, using both published adult weights and mean weights of individuals in the collection, one can use the theoretical growth curve to effectively evaluate hand-rearing protocols and milk replacers.

NUTRITION AND GENERAL FEEDING

Proper feeding formula is essential, as incorrect composition can result in diarrhea, gastric upset, constipation, or lack of proper bone development, and poor general growth and health. Zoo and aquarium nutritionists are a great resource for formula composition information. Contacts made through the zoological associations (i.e., AAZK, AZA Species Survival Plans [SSPs], EAZA European Endangered Species

> **Good Practice Tip:** It is important not to overwork or wear down the neonate while attempting nursing sessions. Ten- to fifteen-minute sessions are sufficient. When the neonate loses interest or becomes agitated, stop. Give both parties time to recover. This will help to alleviate frustration and decrease negative feedback to both the neonate and the keeper.

Programmes, ICZ, and Zoo Biology groups, or companies like Pet Ag, Wombaroo, Biolac, or Bio-Serv; see references and Gage 2002) are an excellent resource. Monitor growth closely, offering the appropriate percentage of body weight in formula per day, and reassess daily to adjust needed volume (Hoch 1972).

Work with the infant to find the most acceptable nipple (figure 21.3). Consider the infant's mouth shape and the mother's nipple shape. It can often take a few different presentations before one finds a nipple that is accepted. Be patient.

Neonatal animals can be selective eaters during the bottle adaptation process. Once nursing reliably, some are vigorous when nursing from the bottle. If the hole in the nipple is too large, they can easily suckle formula too quickly and inhale milk into their lungs, causing aspiration and difficulty breathing (dyspnea). If this occurs, an experienced person must stand and "swoop" the infant in a half-arc using centrifugal force to help remove milk from the airways. One must be sure to cradle the neck and body of the neonate carefully, to stabilize the body position and prevent injury.

A strong emphasis should be placed on not overenlarging nipple holes for "speed" feeding. Select the smallest possible hole in the nipple: one that allows the animal to feed without becoming frustrated. Feeding position should mimic the natural situation. If the chosen nipple does not have a pre-stamped hole in it, one must be created. The procedure for making a hole in a latex or rubber nipple is as follows. Wearing heatproof gloves, heat a fine (26 gauge) needle or wire over a hot flame from a gas stove or Bunsen burner until it is red-hot. Push the needle through the latex rubber in the center of the nipple to form a small hole. If a hole of slightly larger diameter is desired, a larger needle can be used. Careful trial and error using several nipples may be necessary. As the animal grows, the need may arise to enlarge the nipple hole. This should be done slowly and incrementally. Each enlargement should be only slightly larger than the hole previously used.

When bottle feeding, warm the milk to the proper temperature (the typical body temperature of the species being fed) just prior to feeding time. To heat the formula to the proper temperature, use warm water—never a microwave. Microwaves heat foods unevenly and can cause serious thermal burns. The formula is safely warmed by placing the filled feeding bottle in a receptacle (such as a measuring cup) filled with hot water. Agitate the feeding bottle well to mix and equalize the temperature of the formula before offering the bottle.

Any formula not consumed is measured and then discarded after that feeding. Formula should never be returned to the refrigerator or offered later; once it has been warmed,

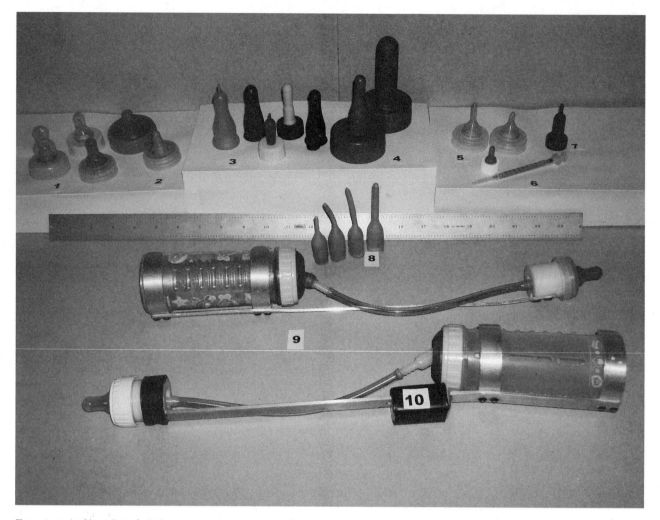

Figure 21.3. A wide variety of nipples are commercially available: (1) human nipples, which work well for primates and felines; (2) human preemie nipples; (3) lamb and goat nipples; (4) calf nipples; (5) miscellaneous silicone nipples, good for small hoofstock such as waterdeer; (6) puppy/kitten nipples; (7) rodent nipples; (8) marsupial nipples; (9) remote feeding bottles used for great apes at Woodland Park Zoo, Seattle (Customanimalcaresolutions@comcast.net); (10) a special locking mechanism designed to release if an ape tries to pull the bottle into the exhibit, thus detaching so as not to deliver the milk bottle reward. Photo by Harmony Frazier, Woodland Park Zoo, Seattle, WA.

milk formula has an increased risk of bacterial growth. When filling feeding bottles, measure the correct amount and immediately return the stock bottle to the refrigerator. Do not allow milk to sit at room temperature. Formula that clings to the fur around the mouth or forelegs of an infant following a bottle feeding should be cleaned off using a soft damp cloth after each feeding to avoid irritation, hair loss, and bacterial growth.

Infants generally progress from frequent feedings (every two to three hours) to greater intervals between bottle feedings. When adjusting the feeding schedule, recalculate gastric capacity.

WEANING

Adjust weaning according to the animal's acceptance of solid food. Some individuals make the transition to solids easily, and may choose to wean themselves by preferring solids and refusing bottles. Conversely, other infants may require lon-

ger bottle-feeding periods and dedicated encouragement to consume solids while they tenaciously cling to their bottles. Offering formula mixed with the intended solid food may be necessary as a first step towards weaning. If it is practical, keep an older, weaned animal that will provide the stimulus for the neonate to eat solids. Sometime neonates will need to observe this behavior to learn it.

FORMULA CALCULATIONS

Body weight is used to calculate the amount of formula offered to the neonate. Use an accurate gram scale. Determine the target percentage of body weight in formula that is to be offered per day. Generally this is 20–25% for carnivores, 15–18% for primates, and 15–20% for hoofstock. Multiply the infants' gram weight by the target percentage to yield the total volume of formula that is to be offered each day. The total volume is then divided equally into the number of feedings offered.

For example, to feed 22% to a 324-gram infant in six equal feedings per day: 324 (g) × 0.22 (% to feed) = 71.28 (ml to feed). The animal should be fed 71 ml of formula per day.

In turn, 71 (ml) divided by 6 (feedings) = 11.8 (ml/feeding). The animal should be fed 12 ml per feeding, six times daily.

An extra ml of formula may have to be added to each bottle to compensate for the milk that coats the inside of the bottle and collects in the nipple ring.

Each feeding should not exceed the animal's gastric capacity. The guideline for determination of gastric capacity is 50 ml of formula per each kilogram (kg) of body weight: 50 ml × 0.324 kg (body weight) = 16.2 ml (stomach capacity). Therefore, a 324 g animal should not be fed more than 16 ml at any one feeding.

Overfilling the stomach causes gastric upset and increases the risk of aspiration. Feeding amounts are to be adjusted as the animal grows.

EMPHASIS ON MONITORING AND OBSERVING

Neonates exhibit dramatic changes quickly when ill. By the time the untrained eye has noticed a problem, it may be too late for successful medical intervention. This is especially true for hoofstock, as they are genetically wired to hide injury. Make note of even subtle changes in appetite, stool, posture, vitality, and stamina. Communicate concerns immediately, and make thorough notations in the animal records. Report any suspicious changes to a veterinarian immediately so that appropriate investigations can begin promptly. High respiration can indicate stress, hyperthermia, or a pulmonary abnormality.

VACCINATIONS

Begin vaccinations at the appropriate times to insure disease protection. Contact a veterinarian to set up the appropriate vaccination schedule. In addition to vaccinations, regular fecal checks are important to detect the presence of enteric infections and parasites. Adhere to flea treatment recommendations made by a veterinarian once the animal is exposed to areas where flea infestation is possible.

STOOL CONSISTENCY AND ELIMINATION

The neonate's stool is an excellent indicator of its health. Within the first 24 hours of birth, the initial stool that is passed is called meconium. Unlike later feces, meconium is composed of materials ingested during the time the infant spends in the uterus. It will likely be a dark green to black-brown color of tar consistency. As the neonate nurses, the milk will form a curd in the stomach and intestine and push the meconium through, also stimulating normal peristalsis of the intestine. The milk stool will be a dark orange to light brown in color, with pasty consistency. As the neonate starts to eat solid foodstuffs, the stool will darken to brown or black with a pelleted, clumpy, or formed consistency. By the time the infant is weaned, its stool should resemble that of the adult of the species. Any variance from the above can be a result of a medical condition or nutritional inadequacies. Blood or mucus in the stool, or liquid or watery diarrhea should be reported and investigated as causes for concern.

With young carnivores and hoofstock, manual stimulation for urination and defecation must be provided at each feeding. There are two simple and effective methods for manual relief. The first is to prepare a damp, warm rag (or a cotton ball, for small neonates) and provide direct stimulation around the genital area. Be sure to select a soft cloth and use a gentle touch, since anal and genital tissues are sensitive and can become abraded. Use only mildly warm water, to prevent thermal burns. The second method is to provide manual stimulation under warm running water. Hold the neonate's hind end under the faucet, allowing the water to flow over the anal-genital region while massaging with one or more fingers. When using either method, apply stimulation until the neonate is no longer straining. Be sure to clean and dry the area around the anus and genital region to prevent irritation, but be gentle when doing so, as the cleaning itself may also cause irritation. Manual stimulation is routinely continued until the neonate is reliably urinating and defecating on its own.

SANITATION

A neonate does not have the reserves or the immune protection of the adult. Until fully vaccinated, it is not protected against certain infectious diseases. Some of these diseases are transmitted on items (fomites) such as clothing, skin, or other objects that come into contact with the neonate. Therefore, it is important that all personnel working in the nursery change clothing before starting work. Wear fresh uniforms and footwear dedicated only for nursery use, and wash hands before working with the neonates. Proper sanitation in the nursery setting is extremely important to hand-rearing success.

Milk and meat are excellent culture media for bacteria. When feeding or handling these foods, keepers must use caution to prepare the food to avoid bacterial contamination. Use caution when making, measuring, or handling formulas. Proper sanitation is particularly important when handling milk products: wash bottles, rings, and measuring equipment in warm water with soap and 2% bleach or other appropriate disinfectant. Soak the items for three to four minutes before using a soft bottle brush to scrub any residue from the surface. Sterilize over, but not in, boiling water using a double boiler. Nipples should not be heat sterilized, as this causes them to break down prematurely. Promptly refrigerate formulas when not in use, and check refrigerator temperature often to insure safety.

IDENTIFICATION

When two or more infants of the same species are being hand-reared, they should be marked for easy identification. This eliminates confusion and makes feeding and record keeping reliable and accurate. Marking can be safely and easily accomplished by using a livestock marker or shaving a small patch of hair. Several brands of markers (e.g., Twist-Stik, All-Weather Pianistic) are available through livestock suppliers. These products are available in a variety of bright colors and are made of nontoxic beeswax. Caution should be exercised in application, as excessive or haphazard use may cause some hair loss. A small dot of color is applied to a visible part of the body such as a tail tip, ear tip, or top of the head. Reapplication of the marker or shaving is periodically necessary, as the product wears off or fades and the hair regrows. Ultimately, a more permanent identification system should be applied: a microchip, ear tag, or tattoo.

BEHAVIOR

If possible, raise the infant near the family (natal) group. Information that the infant would normally receive through visual, olfactory, and auditory stimulation in the natal group is important to its normal development. If rearing next to the natal group is not practical, consider rearing the infant with a conspecific, or cross-foster it to a different adult of the same species. Other zoos that may also be hand-rearing similar species might serve as a resource of conspecifics that could be raised with the infant. Domestic animals (such as dogs or hoofstock) have also been used with success as live surrogates for developing infants. Being exposed to more than a human caregiver will help the infant to be better socialized and prepared for eventual reintroduction.

REINTRODUCTION

Actively plan for reintroduction as soon as the neonate enters nursery care. If the reintroduction is done too late, the infant may not be accepted by the family unit. Reasons for delaying reintroduction might include first training the infant to come to a keeper for its nutritional needs, or training another animal from the group to bring the infant to the keeper for the same reason. A sequential plan for who is introduced to the infant first and then thereafter is important to strategize before starting.

Once a way to provide warmth and nutrition to the infant is established, consider how this supplemental support might be continued with the infant reintroduced to its own species. Bottles with extensions (Custom Animal Care Solutions 2010), as shown in the nipple photos in figure 21.3, have been used with great success to feed an infant while it remains with its family group. More discussion on training an infant to come for the bottle is noted in hand-rearing examples below. Be sure to use available informational resources to create a safe plan for reintroduction.

Survey any environment the infant will be in without supervision. Water sources and any escape or entry locations can be extremely hazardous.

Hand-reared animals can generally begin reintroduction to their intended exhibits once they are vaccinated. Check with the zoo or aquarium's veterinarian or animal manager for recommendations. This transition from the nursery to the exhibit is accomplished gradually, and may start with a brief supervised visit to the adult sleeping quarters after they have been cleaned. This may continue until the youngster is comfortably exploring the new space, at which time the nursery keeper should begin encouraging independence by leaving the youngster alone or with a conspecific, but continuing to monitor it to ensure its safety. The next step may include safe visual and olfactory introduction to some of the adult animals. Monitor the infant for signs of distress, and check its body weight often during this period. Where appropriate, a gentle female or older animal might be selected for the first tactile introduction to the youngster, so that the animals can gradually acclimate to new cage mates.

ANIMAL AMBASSADORS

When rearing a young animal that is slated to become an "animal ambassador" for an education department's outreach program, it is important to identify it early in the hand-rearing process. Such animal ambassadors should be trained and handled differently from other infants. These animals will need extensive training and acclimation to their eventual trainers early in life. Development of a plan to incorporate the training of staff into the nursery routine should begin early, one week after the first vaccination. All staff should exercise the same care to avoid cross-contamination and disease transfer. If the trainers are allowed access early on, they must adhere to the same sanitary and isolation procedures as the nursery keepers do.

Trained behaviors that are beneficial for animal ambassadors to begin early include harness/collar adaptation, leash training, and crate training.

RECORD KEEPING

Clear, concise, and complete record keeping is essential to the continuity of successful infant care. Make sure that what is recorded is clear and does not need further explanation. Records are the only way to monitor progress and collect information to assist with next-step decision-making. Creating charts for recording food intake, elimination, medications, behaviors, and vital signs (weight, heart rate, respiration, and temperature) help to make the recording of information more consistent between caregivers. The recording of behavior frequencies (using behavioral ethograms and behavioral budgets) can indicate problems or success during evaluation of care and reintroductions.

HAND-REARING STAFF

Plan who will care for the neonate until weaning, and consider the number of keepers to be involved. Multiple keepers may create infant stress and insecurity and create a lack of continuity. If there are too few keepers, the infant may be so bonded with them that it does not transition well to changes in the environment, new keepers, or new animals.

Being an excellent keeper does not automatically translate to being an excellent "surrogate mother." Some of the positive attributes of an infant caregiver include being

- patient
- calm and quiet
- flexible and focused at the same time (able to think ahead to next steps while remaining open to changes that might be necessary on a daily basis)
- consistent, since all infants do better and feel more secure when they know what to expect
- an open communicator and a good team player
- an innovator, because sometimes meeting a unique need requires thinking "outside the box" and creating a solution

- skilled at different nursing techniques
- professional and realistic
- able and willing to manage their ego, keeping the infant's needs at the forefront and stepping aside as needed to ensure the infant's success.

Tremendous variability exists between species. Care techniques vary with regard to nearly every aspect of neonatal care: temperature requirements, bottle-feeding schedules, solid food introduction, weaning periods, housing, introduction, and enrichment. For example, some infants are weaned in as little time as six weeks, while others may be bottle-fed for as long as 18 months. When developing a hand-rearing plan, it is advisable to collect as much information as possible from a variety of sources. Begin by contacting other institutions that have enjoyed success with hand-rearing. Remember that care plans are best offered as guidelines, with the clinical information (animal health and weight) guiding each step.

Providing all the details for hand-rearing of the many species represented within zoos and aquariums is beyond the scope of this chapter. The above information is applicable to all infants, and following are a few examples of hand-rearing of specific animals (carnivores, primates, and ungulates), along with some of the specific needs for each group. See further reading and references for additional resources.

GUIDELINES FOR HAND-REARING CARNIVORES

SANITATION AND ISOLATION

Sanitation and isolation are particularly important when hand-rearing carnivores (Ensley 1987). Before a young carnivore has been vaccinated, it is vital for keepers who have dogs and/or cats at home, or those who volunteer in animal shelters or have contact with other carnivores, to be mindful of the potential for infectious disease transmission.

CLIPPING CLAWS

Upon arrival and periodically thereafter, claws should be clipped and filed smooth. Long, curved, or ragged claws can become entangled in bedding, resulting in injury. A small human or dog nail clipper works best for this purpose. When shortening claws, caution must be used to avoid the blood supply. After clipping, any rough edges can be filed smooth using an emery board.

WARMTH

If hypothermia is detected, placing the carnivore neonate next to a covered hot water bottle in an incubator set at 31.1–32.2 °C (88–90 °F) will safely restore normal body temperature. Rewarming must be done slowly, to protect tissues and the nervous system, so one must never plunge a hypothermic neonate into warm water to correct hypothermia.

HOUSING

During the first few weeks of life, many hand-reared carnivores need supplemental heat to maintain normal body tem-

> **Good Practice Tip:** Keepers must be careful not to overfeed neonates, as overfeeding is perhaps the most significant health risk when hand-rearing carnivores. Many neonatal carnivores will nurse vigorously and continue to suckle even after ingesting enough milk to fill their stomachs. Amount of formula to be fed should be determined prior to feeding.

perature. Larger carnivores generally require supplemental heat for a shorter period of time than smaller ones (a healthy tiger cub may need only a day of supplemental heat, while a tiny Pallas cat [*Otocolobus manul*] may require two weeks inside an incubator).

Once the incubator has been turned down to near room temperature, the neonate can begin to transition (in short supervised periods) into a new larger space. The transition should be gradual, allowing a few short (one-hour) supervised visits each day.

OVERFEEDING

Overfeeding is perhaps the most significant health risk when hand-rearing carnivores. Many neonatal carnivores will nurse vigorously and continue to suckle even after ingesting enough milk to fill their stomachs. Since the neonate may not show signs of being full (satiation), it is easy to overfeed a young carnivore if one gauges the amount of food offered only by the infant's appetite or nursing enthusiasm. It is, therefore, prudent to calculate an appropriate amount of formula to be offered during an entire day, divide that amount equally among the feedings, and not exceed that amount. Serious medical problems can arise when carnivores are overfed. These include but may not be limited to the formation of excessive gas, diarrhea, vomiting, bloody stool, gut stasis, lactobezoars, and painful bloat. The effects of overfeeding should be discussed with staff members so that everyone clearly understands the importance of adhering to recommended nutritional guidelines.

FEEDING POSITION

An infant carnivore must be fed in sternal recumbency with its head elevated. Nursing a carnivore lying on its back can be harmful to digestion. A heat lamp or space heater placed in the immediate area will keep the air warm during feedings.

MEAT PRODUCTS AND WEANING

When thawing meat products or carnivore diets, never leave the meat to thaw at room temperature. Instead, place the frozen product in a refrigerator to thaw slowly. When offering meat to a young animal, take care to prevent it from choking. Tear or cut the meat apart into small, suitably sized pieces, removing lumps, fat, and gristle. Offer the meat one piece at a time. Solid foods can also be mixed with milk formula or water to form slurry, and offered in a dish or shallow pan. After meat feedings, all uneaten product should be collected and discarded. Food pans and the floor should be sanitized.

> **Good Practice Tip:** If using a lure toy to encourage appropriate play in carnivores, do not store it in the animal enclosure, as it may present a strangulation or choking hazard.

PROVIDING MANUAL STIMULATION

Neonatal carnivores are unable to urinate or defecate on their own (see above). A litter box can be offered to help keep the enclosure clean once the neonate is eliminating on its own. Paper strips, old receiving blankets, or towels are good options for lining the litter box, as bed-o-cobs or cat litter are frequently ingested.

BEHAVIOR AND DISCIPLINE

Guidelines for consistent handling should be established prior to or early in the hand-rearing process. It is not advisable to allow a carnivore to suckle on any part of the keepers' hands, skin, or arms. Occasionally a finger can be used to elicit the nursing reflex just prior to feeding, but allowing prolonged nonnutritive suckling bouts on one's skin should be discouraged. Suckling interactions become increasingly dangerous and unpredictable as the carnivore grows. The neonate associates the comfort behavior with the keeper. Should it become necessary to end the suckling bout before the animal is willing, the young carnivore can become frustrated and aggressive.

It is not advisable to allow a young hand-reared carnivore to bite or swat the keeper, even in play. Instead, select toys, surrogates, or other items for play sessions. One method that works well is to attach a favorite toy with a short length of rope to the end of a pole, such as a broom handle. The keeper can encourage the animal to exercise by chasing this toy while maintaining an appropriate distance from the handler's body.

Some hand-reared carnivores will become aggressive or experience aggressive moods before weaning age. When this occurs, it will be necessary to formulate a standard method of discipline. Whatever method is agreed upon, all keepers should apply the rules consistently. If one keeper encourages rough play, the behavior will be repeated. If all personnel involved do not follow a consistent discipline policy, the animal will be confused about which behaviors are acceptable.

REINTRODUCTION

After the animals have been given the required vaccinations, they may begin gradual introduction to the intended exhibits (see above).

GENERAL GUIDELINES FOR HAND-REARING PRIMATES

The decision to hand-rear a primate is never an easy one. It is obvious to the modern-day animal care manager that hand-rearing should never be considered for primates unless there is no reasonable chance for survival with the natural mother or other conspecific. The goal is to produce a well-adjusted, socially competent individual. Assist the natal group in rearing the youngster, and provide only as much support as necessary. Human intervention is minimized while social introductions with conspecifics are elevated to the highest priority. Flexibility, innovation, and empathy are essential. Each event is unique and must be addressed separately using the motif of early socialization. The event must be planned carefully, with steady progress toward full integration into a family group.

PROCEDURES FOR EVALUATING THE NEONATE

After birth, close monitoring of the infant will ensure that any problems are caught early. There are several circumstances that might warrant removing an infant from its mother. If the female does not clean or hold the infant, if the infant shows signs of weakness, or if no nursing is seen in 48 to 72 hours, an examination is indicated. The neonate may require correction for sepsis, dehydration, poor early nutrition, hypothermia, or injury. If the neonate is healthy, however, reintroduction should be considered immediately. In some cases, an "auntie" or other gentle group member may be selected as a surrogate. If no such surrogate exists, the infant should be removed for assessment. It is often possible to provide formula to an infant while it remains with its mother.

HOUSING AND VISUAL PROXIMITY

Most newborn primates (except great apes) will need to be housed in an incubator or be supplied with an external heat source for two weeks or longer. Whenever possible, the incubator should be placed within visual proximity of the natal group, even while the animal is undergoing critical or supportive care. Maintaining continuous visual, auditory, and olfactory contact with the group starts immediately and continues until a full integration is achieved.

It is usually necessary to start with a brief introduction. If the interactions are positive and the infant tolerates them well, time spent with the family should be carefully but steadily increased. Make all changes to the neonate's daily routine one at a time. Body weight is one important piece of empirical evidence that can be used to assess the acceptance of changes in daily routines. If interactions are negative, or if aggression is suspected, a "howdy" cage, or creep, can be used to permit only limited contact.

After the infant has adjusted to being outside the incubator, construct a small introduction cage that affords a safe environment with easy visual access to the group. The cage can be used whenever the infant is separated from the family. Provide safe climbing structures so that the neonate can properly develop motor skills. When the youngster has mastered perching that is low and "easy," move the perches to higher locations to create new challenges. Enrichment is important for all animals, but perhaps especially important for young primates. Novel items encourage activity and stimulate mind and body.

FEEDING STRATEGY

Keepers must use caution when handling and feeding young primates. Keeping a protective barrier between the keeper's regular work uniform and the neonate and wearing a mask (especially during the first month of life) will ensure that the infant is not exposed to infectious agents. Primate protocols must be strictly adhered to, as the immune system of the neonate is not fully developed. If a keeper suspects that he or she is ill, exchange of duties with another keeper is best.

Primates are fed using various artificial milk substitutes intended for human infants. The general rule is to offer 15 to 18% of total body weight per day for growth, and 10% for maintenance. Newborns are usually fed every three hours around the clock.

GUIDELINES FOR HAND-REARING HOOFSTOCK (UNGULATES)

THE UNGULATE DIGESTIVE SYSTEM AND IMMUNITY

Unlike most placental mammals, ungulate neonates receive no transplacental immunity. At birth, the neonate is totally without protection from infectious disease. The neonate receives all immune protection from nursing the mother's first milk or colostrum. There are specific sites in the gut that absorb immunoglobulins (IgGs) through passive transfer which provide systemic protection against disease. However, these absorptive sites in the gut shut down 48 to 72 hours after birth. After that, the dam's colostrum continues to provide immune protection, but only locally in the gut. A hoofstock neonate deprived of colostrum during this window of time will subsequently be at risk of disease or even death. Some institutions elect to substitute maternal colostrum with artificial preparations, fresh bovine colostrum, or administration of a serum transfer. Following its natural physiology, the neonate should be bottle-fed 100% prepared colostrum or replacer for the first 24 hours during the gut closure period when IgGs are absorbed. After 24 hours, adding a colostrum product at a rate of 10% to the daily amount of milk formula for an additional 21 days can provide localized gut protection.

In the ruminant neonate, it is important that the infant suckle the formula. Chewing and swallowing should be discouraged. Nonnursing behavior can inhibit the esophageal groove mechanics that are required to send the milk to the abomasum. Milk diverted into the rumen accidentally can cause life-threatening rumenitis and ulcers. Reflexive closure of the esophageal groove, also called the reticular groove, channels milk from the esophagus to the abomasum via the omasal canal in the young ruminant. Activation of the reflex depends upon the chemical stimulation of milk and the mechanical stimulation of nursing to activate the receptors situated in the oral cavity, the pharynx, and the most cranial part of the esophagus. Water consumed does not activate this reflex, and therefore goes directly to the rumen. The sensitivity of the reflex to milk remains intact until about two years of age (Allen 1992).

Note: because of the physiology of the ruminant stomach, tube feeding of ungulate neonates is not recommended.

ASSESSMENT OF BLOOD SAMPLES

1. A blood serum gluteraldehyde agglutination test is a good indicator of early nursing status. If the test result is negative, additional medical intervention may be required, such as a plasma transfusion to assist the infant in passive transfer of immunoglobulins (IgGs).
2. A simple blood glucose test can also be used as a tool to determine the nursing and medical status of the neonate. The animal should have enough energy to nurse (blood glucose above 25), yet it should not have a blood glucose level above 150 (via intravenous fluids, for example), which would cause it to feel full.

EMPHASIS ON MONITORING AND OBSERVING

The survival of hooved animals in the wild depends in part upon their ability to hide injury. Animals that are sick or injured are predated; therefore, hoofed animals do their best to hide symptoms of illness. Keepers need to be ever-vigilant. Clinical signs that may indicate problems are teeth grinding and high respiration, depressed behavior, anorexia, and a "fluffed" hair coat. Teeth grinding can indicate that the animal is experiencing pain.

One should continue to periodically palpate the umbilicus. Some umbilical hernias may not be obvious until the animal becomes more active. Watch for a wet umbilicus, which may indicate that the neonate has a patent urachus. Failure for the lumen of the urachus to be filled in after birth leaves a patent (open) urachus. The telltale sign is leakage of urine through the umbilicus. A patent urachus must be surgically repaired.

HOUSING, WARMTH, SECURITY

If it is not feasible to raise the neonate within the natal group, the young or newborn neonate should be housed in a warm area at a temperature of 21.1–26.6 °C (70–80 °F), free of drafts. In the case of neonates that are weak or have trouble standing, nonskid mats should be provided to prevent splaying of the legs in attempts to stand. Neonates that are nervous or easily frightened should be provided with padded floors and walls to prevent self-trauma. An area approximately four by four feet in size should suffice for most small to moderately-sized ungulates.

Daily outdoor exposure is important for the neonate's health and development. If the neonate is metabolically stable, it should be moved outside during the day into a small, enclosed yard with natural lighting, natural substrate, and bedded areas under shelter. This will encourage exercise and interaction with other conspecifics or hand-reared neonates. After the neonate has matured (at three to four weeks of age), it can be moved full-time to an outdoor enclosure with a supplemental heat source.

If the neonate has no immediate health concerns, or once it is deemed stable, it can be housed in a warm enclosure adjacent to the natal herd. Visual, auditory, and olfactory access should precede full tactile introductions. If available, an adjacent barn with supplemental heat and no drafts can be used. This would be essential for neonates born in cold climates. A visitation screen can be erected across an open barn door, and the adult feeding stations can be relocated nearby to promote interaction between the herd and the neonate.

Another way of providing protective contact is to set up a "howdy pen" inside the adult exhibit. This pen is a safe, secure, fenced-in space for the neonate, situated near the keeper service entrance. The pen should be equipped with a shelter, such as a lean-to or an appropriately sized shed.

BEDDING

Clean, dry, absorbent bedding should be supplied for the neonate. Natural sources such as grass hays (e.g., Sudan or Bermuda) are preferable. Soiled bedding should be removed daily. Sometimes the neonate will nibble on whatever substrate it happens to be lying on; therefore, it is important to provide a natural source of food for its bedding that will be easily digested. This will also encourage the random picking up of solid foodstuffs.

FREQUENCY IN FEEDINGS

Two major categories are observed in the behavior of ungulate neonate that may be determined by their natural history. These two categories are the "hider" or "tucker" and the "follower" or "grazer."

THE "HIDER" OR "TUCKER": A LESS FREQUENT NURSER

A neonate that is from the "hider" and "tucker" group is hidden or tucked in a "safe" place by the dam while she goes away to feed. The dam returns every few hours to prompt the neonate to nurse by giving it a vocalization cue and physical stimulation, usually nudging or licking its anogenital regions. The neonate will then nurse till satiated and find a place to tuck in again. The natural milk compositions of species that fall into this category tend to have a higher percentage of fats and solids that will sustain the neonate over a longer duration. Typically, these neonates are started on a schedule with three-hour intervals between nursing. The frequency of nursing decreases with age, to encourage solids consumption.

Some of the species that fall within this description are deer and bovids such as cattle, gazelles, duikers, and antelopes (Oftedal and Jennes 1988; Jennes and Sloan 1988).

THE "FOLLOWER" OR "GRAZER": A MORE FREQUENT NURSER

A neonate from the "follower" or "grazer" category is usually on the move with the dam shortly after birth. Since the dam may constantly be migrating from one grazing area to another, the neonate is conditioned to follow. Short, frequent bouts of nursing are common in this group, which contains the equids, elephants, goats, sheep, and some plains or steppe species of antelope such as the Russian saiga (Michelson and Rubin 1994). Typically, neonates within this group are fed every one to two hours. Milk composition data indicates that species in this category have a lower percentage of total solids and a higher percentage of carbohydrates (Oftedal and Jennes 1988).

UNGULATE FEEDING TECHNIQUES

If a neonate is initially reluctant to nurse, a variety of techniques can be used to help it feel more natural and comfortable.

1. Whenever possible, keep the animal in a standing position.
2. Nursing often coincides with stimulation by the dam. Lightly rub the neonate's tail and anogenital regions. Lean over its head to encourage natural nipple searching. Slip the nipple into its mouth as the animal searches.
3. The keeper can drape a towel over his or her head and the animal's head, thus mimicking the dam's shadowing belly.
4. Try to mimic species-specific vocalizations.
5. Some neonates prefer to face the keeper; others eat well when they are given extra standing support by backing up to the keeper.
6. Many neonates require support around the muzzle. The hand that holds the bottle is folded like a taco shell underneath the jaw and against the sides of the mouth. This also helps to keep the tongue in place.
7. Sometimes, when all the above techniques have been unsuccessful, a surrogate animal is used to stimulate the neonate's sense of smell and texture. Use a larger, calm animal that will tolerate this type of interaction. Stand the neonate in a position facing the surrogate, and gently press it into the surrogate's side. The natural feel and smell of fur may stimulate the neonate to start nibbling and looking for a nipple. Once stimulated and nibbling on the surrogate's fur, drip a small amount of milk in the area to encourage a stronger searching response. Gently slip the nipple into the neonate's mouth while it is actively searching.
8. Another less frequently used practice involves offering milk in a pan, dish, or bucket. This is usually saved for neonates that will absolutely not suckle from a nipple.
   a) Fill a low sided pan with formula and present it to the neonate.
   b) Dip a finger into the formula and try to get the neonate to suckle off of it by lightly inserting it into the neonate's mouth.
   c) Bring the pan up to the level of the neonate's muzzle and draw the finger closer, down into the formula. Sometimes the neonate will continue to sip the formula.

WEANING

Weaning is typically a learned behavior cued by the dam. By five days of age, hoofstock neonates should be offered small amounts of the adult diet. To promote interest in solids, chopped apple, carrot, and yam can be sprinkled on top of the grain and herbivore pellets. Solids (such as fortified pellets or hay) and a shallow water tub should be offered fresh daily. Eructation in ruminants may be observed as early as two weeks of age. This signifies the development of a functioning rumen.

The weaning process should start before the animal is actually weaned from the bottle. The process should include an intermediate housing area with older, weaned neonates, where human contact is minimized. This reduces the animals' psychological dependence on the keeper while increasing natural interactions with other ungulates. The animal should be allowed sufficient time to gain maturity and independence

> **Good Practice Tip:** If it is practical, keep an older group of weaned animals that will provide the stimulus for the neonate to eat solids. Sometime neonates will need to observe this behavior to learn it. It is typically a learned behavior cued by the dam. This method often works well for ungulate species.

from keepers in order to better handle stressors when they are eventually integrated into their natal groups or exhibits (Michelson 1986).

WEANING AGES

Generally, larger species require a longer weaning time, and smaller species require a shorter weaning time. For example, dam-reared giraffe calves may nurse up to nine months of age. Dam-reared dik-diks (Genus *Madoqua*) or duikers (Genus *Cephalophus*) may only nurse until six to eight weeks of age.

PARAMETERS FOR EARLY WEANING

Early weaning is loosely defined as that which takes place one to four weeks sooner than a protocol indicates. Generally, when the daily total amount of formula offered has dwindled to 3% of body weight, an animal is approaching the appropriate time to wean. There are four situations that may warrant early weaning once the 3% point is reached:

1. if the bottle-feeding is interfering with the social interaction between the neonate and its herd or enclosure mates
2. if the animal does not readily come to the bottle, or has to be caught up or restrained for bottle feedings (this is more common with animals orphaned after three weeks of age)
3. if the animal is consuming solid food well, and appears to prefer solids to the bottle
4. if the animal is still nursing periodically from the dam, and retains a sufficient growth curve.

There are two reasons to continue bottle feedings past weaning age:

1. if a prescribed oral medication can be administered easily by bottle
2. if a bottle is used as a training tool (e.g., in crate training or shifting), in which case water or a dilute formula can be used.

IMPRINTING

The bond between keeper and neonate should only be reinforced for the feeding of bottles. Interactions other than feedings should be discouraged. Neonates should have conspecifics or other ungulate neonates to interact with, to learn species-appropriate behaviors. Hand-raised animals, in particular, can become dangerous as adults if not raised with this practice in place.

HAND-REARING IN THE NATAL GROUP

Hand-rearing an animal within the natal group can improve hand-rearing success rates. The neonate "grows up" as part of the herd, eliminating the perils of reintroduction. To accomplish a hand-rearing event within the natal herd, some elements within the facility must be present:

1. There must be an adjacent pen or barn in which the neonate can be safely bottle-adapted and monitored. This area should allow visual access to the herd, but should also safely contain the neonate.
2. There must be adequate heat. Overnight lows should not be less than 18 °C (approximately 65 °F).
3. The exhibit should be large enough to provide the herd with ample flight distance while bottle feedings take place.
4. The neonate must be solidly imprinted to the bottle-feeding process prior to release within the group.

CLICKER TRAINING

If a neonate is clicker-trained to come to a keeper for bottle feedings, there will be minimal disturbance to the natal herd. Training begins once the neonate is consistently nursing. Using a clicker when the animal takes its first sips from each bottle reinforces the connection between the "click" sound and the act of nursing. The next step is for keepers to "click" as they stand on the opposite side of the small enclosure or howdy pen from the neonate. The goal at this stage is to train the animal to get up and move across the small area to eat.

For "tuckers/hiders," use the bottle-walk method: Nudge the animal to its feet. Once it latches on to the nipple, click the clicker. Walk backward, causing the animal to walk forward as it nurses. Periodically remove the nipple from the animal's mouth and allow it to take a few steps forward to retake the nipple. Click the clicker each time the nipple is retaken. At each feeding, increase the distance the animal must walk before it can retake the nipple. When an animal is readily coming to the keeper for bottles, it can be released into the herd. For consistency, a bottle-feeding station near the exhibit service entrance should be designated. One "click" should attract the animal's attention, and a reinforcement "click" should still be used at each feeding as the animal takes its first sips from the bottle.

REINTRODUCTION TO THE HERD

Timing of a reintroduction attempt depends on many factors, including herd behavior toward the neonate, general health of the neonate, exhibit design, and weather. Introductions can begin with supervised limited access to the exhibit and herd. A small section of the "howdy pen" is opened into a "creep" design (which selectively discriminates access to animals of a certain size) to allow the neonate to enter the exhibit on its own. The pen opening should be large enough for the neonate to pass through, yet small enough to keep adult animals out. This provides a safe place for the neonate to return to if it feels threatened. The duration of exhibit access is increased each day, as long as the herd is not abusive. Eventually the animal's need for the comfort of the "howdy pen" will diminish. The pen can then be removed.

In species that show aggression through an introduction screen or during supervised visits, the neonate's release may have to be delayed. If space permits, a few nonaggressive individuals may be separated from the herd to form a second herd. The two herds can be reunited when the youngster is older. Releasing conspecifics together into a group can also prove successful.

GUIDELINES FOR HAND-REARING MARSUPIALS

By their very nature, marsupial neonates require very specialized hand-rearing techniques. An infant marsupial, called a joey, is born in an immature state after a short gestation in a fetal state. Blind, furless, and tiny, these neonates require a pouch, specialized nipples, and specialized formulas which change in concentration as the infant matures.

The authors recommend going to the source of greatest experience, Australia. A number of excellent zoos and wildlife care centers in Australia specialize in hand-rearing marsupials and can provide many valuable resources for marsupial hand-rearing information (see the list of further readings and Staker 2006).

ETHICAL CONSIDERATIONS CONCERNING HAND-REARING

Before being faced with the decision to hand-rear an animal, it is important to discuss the ethical considerations that surround such a decision. Keeping in mind that the goal is "a well-adjusted socially competent individual and the maintenance of normal social dynamics of the family group," it is important to assess whether it is possible to provide the care needed to reach that goal.

Some considerations:

- Are the necessary resources available either at the current institution or at another? These will include the availability of experienced staff, time, money, and continuity of care over the long term.
- What are the SSP recommendations?
- Will the animal be able to live and breed with conspecifics? If not, does this individual have value as an educational animal or for exhibit purposes?
- Would the animal be releasable, and would its release be appropriate and desirable from a conservation standpoint?
- For this species (e.g., an elephant), is there an obligation to intervene regardless of circumstance? How will public communication and opinion be handled?
- Is hand-rearing the best choice for this particular animal? Consider its health, any injuries, its immune status, and its particular species. Hand-rearing a langur to reach the stated goal will be much more challenging than hand-rearing an armadillo. Can these challenges be addressed through supplemental care or early reintroduction?
- What is the zoo or aquarium's philosophy of care? Is it focused on the individual animal, on the species' population dynamics, or both?
- What is the zoo or aquarium's euthanasia policy?

These are not easy questions to consider, nor should they be. A frank and open discussion should be held by the whole team (senior staff, colleagues, veterinarians, population managers, and keepers) before or during collection planning to help alleviate painful misunderstandings later. Even if the zoo or aquarium will support hand-rearing under any circumstance, it is important to always evaluate each case on the basis of what is best for the individual animal.

SUMMARY

Hand-rearing has changed significantly over the years. The modern keeper must focus on the total rearing picture: medical, dietary, behavioral, and social success. Attention to detail, careful record keeping, effective and clear communication, and a focus on team versus individual contributions lead to successful outcomes. A good team of hand-rearing professionals is both prepared and flexible, allowing the details of each unique hand-rearing event to dictate appropriate action. Hand-rearing does make an important contribution in attaining a self-sustaining captive population, and is a professional responsibility to animals in the care of zoos and aquariums when no other alternatives are available. The hand-rearing process no longer must hinder an individual animal's success in leading a full and productive life.

REFERENCES

Allen, J. L. 1989. "Fluid Therapy at the Wild Animal Park." *Zoological Society of San Diego Neonatal Symposium Paper Presentations.*

Allen, Jack L. 1992. "Development and Diseases of the Pre-ruminant Stomach." Neonatal Symposium, Zoological Society of San Diego.

AZA Animal Health Committee. N.d. 1980s *Infant Diet Notebook.*

Baker, A. 1994. "Variation in Parental Care Systems of Mammals and the Impact on Zoo Breeding Programs." *Zoo Biology* 13:413–21.

Bauman, J. E. 2009. "Pregnancy Diagnosis in Exotic Animals Using Hormone Analysis." *ICZ/AAZK 2009 Proceedings*, Seattle, WA.

Custom Animal Care Solutions. 2010. customanimalcaresolutions@comcast.net.

Danzig, T. 1992. *Biological Values for Selected Mammals*, Third Edition. Topeka: American Association of Zoo Keepers.

Durham, K. and J. Hawes. 1998. "Unformed to Unmanageable: Hand-raising a Malayan Sun Bear." *Zoological Society of San Diego Neonatal Symposium Paper Presentations.*

Edwards, M. S., and J. Hawes. 1997. "An Overview of Small Felid Hand-Rearing Techniques and a Case Study for Mexican Margay (*Leopardus wiedii glaucula*) at the Zoological Society of San Diego." *International Zoo Yearbook* 35:90–94.

Ensley, P. 1987. "Recommended Guidelines for Animal Movement into and out of Children's Zoo Nursery and CZ." San Diego Zoo.

———. 1995. "Hoofstock Nursery Protocol Seminar No. 4." Wild Animal Park Nursery Personnel Emergency Care and Assessment. *Animal Keeper's Forum* 22 (10).

Ensley, P, and J. Meier. 1978. "Neonatology and the Monitoring of Neonates." Zoological Society of San Diego, *Ann. Proc. Am. Assoc. of Zoo Vet.*

Frazier, H. 1981. "The Importance of Data Collection in the Nursery Setting." *Association of Zoo Veterinary Technician Annual Proceedings*, 1981.

Frazier, H. et al. 2009. "The Evolution of Infant Care at Woodland Park Zoo, Utilizing Birth Management Plans for More Positive Outcomes." *Conference Proceedings ICZ/AAZK*, Seattle, WA.

Frazier, H. Hunt, K. eds. 1994. *Zoo Infant Development Notebooks.* Topeka: American Association of Zoo Keepers.

Frazier, H., and W. Karesh. 1989. "Supplemental Care in a Maternally Reared Snow Leopard (*Panthera uncia*)." *International Snow Leopard Yearbook.*

Gage, L. 2002. *Hand-Rearing Wild and Domestic Mammals.* Ames: Iowa State Press.

Hoch, J. 1972. "Management Principles of Nursing Exotics," San Diego Zoo, AAZV Annual Convention, Houston TX.

Janssen, D. 1992. "Scientific Guidelines for Diet Formulation of Newborn Mammals." Neonatal Symposium, Zoological Society of San Diego.

Jennes, R., and R. E. Sloan. "The Composition of Milks of Various Species: A Review." *Dairy Science Abstract* 32(10): 599–612.

Liker, J. 2004. *The Toyota Way.* Columbus, OH: McGraw-Hill.

Liker, J., and D. Meier. 2006. *The Toyota Way Fieldbook.* Columbus, OH: McGraw-Hill.

Mehren, K., W. Rapley, and M. Cranfield. 1986. "Supportive Care, Evaluation and Reintroduction of Neonatal Zoo Mammals." Metro Toronto Zoo.

Michelson, K. 1987. "Survival of Hand-raised Hoofstock." *Animal Keeper's Forum* (February): 60–61.

———. 1989. "Case Scenario of a Hand-raised Domestic Water Buffalo (*Bubalis arnee f. bubalis*)." *AAZK Animal Keeper's Forum* (January): 27–28.

Michelson, Karla. 1992. "The Hand-Rearing of Ungulate Neonates." AAZK National Conference Workshop.

Michelson, K. and E. Rubin. 1994. "Nursing Behavior in Dam-reared Russian Saiga (*Saiga tatarica tatarica*) at the San Diego Wild Animal Park." *Zoo Biology* 13:309–14.

Oftadel, Olav T. and R. Jennes. 1988. "Interspecies Variation in Mil Composition among Horses, Zebra and Asses (Perissodactyla: Equidae). *Journal of Dairy Research* 55:57–66.

Ogden, J., K. Killmar, and J. Hawes. 1997. "Hand-Rearing Primates in Zoos: The Times They are a-Changing." *American Journal of Primatology* 42 (2).

Olow, P. M. 1992. "Milk Proteins and the Neonate." *Zoological Society of San Diego Neonate Symposium.*

Oosterhuis, J. 1989. "Non-Domestic Neonatal Husbandry and Medical Care Conference: Summary of Proceedings." *Animal Keepers Forum* (July): 235–38.

PATH. 2009. "Chlorhexidine Reference." Seattle: Technology Solutions for Global Health.

Rings. 1995. "Neonatal Management Small Ruminants: Sheep." 77th Western Veterinary Conference, Las Vegas.

Siebert, J. R., B. Williams, D. Collins, H. Frazier, L. A. Winkler, and D. R. Swindler. 1997. "Sporadic Appearance of Cleft Palate in a Full-Term Gorilla (*Gorilla gorilla gorilla*)." Poster presentation at the Gorilla Workshop, Pittsburgh Zoo.

Sodaro, C. 2006. *Orangutan Development, Reproduction and Birth Management.* Orangutan SSP, AZA.

Zuba, J. 1991. "Factors Influencing Neonatal Infections." Neonatal Symposium, Zoological Society of San Diego.

22

Management of Geriatric Animals

Cynthia E. Stringfield

INTRODUCTION

With the increasing knowledge and implementation of proper husbandry and medical care for zoo and aquarium species, many animals in captivity now live well past their average life expectancy in the wild. Many facilities may have a number of animals that are considered geriatric (aged). Animal keepers should know the longevity data for the species they care for, so that they know when that species becomes geriatric. They should also be aware of the accurate age of the individual animals they are responsible for. Old age itself is not a disease or diagnosis; however, geriatric animals have special care requirements due to changes that occur in the body due to aging and medical conditions that are commonly seen, and so they often require more time and attention than young healthy animals. Gerontology is the study of aging and geriatrics; human physicians have this formal specialty, but veterinarians do not. Zoo veterinarians, however, are usually very skilled and experienced in geriatric medicine, due to the high numbers of geriatric animals they care for. Proper husbandry throughout an animal's life can not only extend its life but prevent or delay some common geriatric diseases. Additionally, the most difficult of decisions, euthanasia, must often be made when assessing quality of life for an animal that has a terminal disease. Animal keepers that are emotionally and professionally bonded to an animal need to manage the grief, loss, and sense of responsibility they feel when an animal dies.

This chapter will provide basic information about geriatric animals and the care they need from their keepers. After studying this chapter, the reader will understand

- the importance of the keeper knowing the typical longevity of the species and age of the individual animal
- common health concerns of geriatric animals
- the effect of proper husbandry in preventing or managing common geriatric diseases
- the current care requirements of geriatric animals in zoo settings

- nutrition and feeding changes of geriatric animals
- the importance of diagnosis in determining treatment, palliative care, or euthanasia for a geriatric animal
- the what, when, who, how, and ethics of euthanasia of geriatric animals
- grief management and help resources for the animal keeper.

AGING AND LIFE SPAN

When evaluating an animal's life span, resources should include species-specific books and articles, and studbook or husbandry managers. Species-specific longevity data is available in many cases (Weigl 2005). Information exists on the internet, but it must be scrutinized carefully and its origin must be determined. Often differing information or age ranges will be found. For some species, information about life span in the wild and/or in captivity may not be well known. It is important to realize the differences between life span in the wild, life span in captivity, and longevity records. Depending on the species, animals may typically live either longer or shorter lives in captivity; however, in modern facilities with proper animal management (husbandry), nutrition, and veterinary care, the majority of species live longer (and may breed longer) in captivity. For example, camels used to be considered old in their late teens, but now they frequently live past the age of thirty. A longevity record refers to the age of the oldest individual of a species ever documented. For example, the Guinness world record (Guinness World Records 2009) for the oldest human is a French woman who died at 122 years of age; however, the United Nations reports that the average life expectancy for a woman in France is 84. In the United States it is 81 for women and 76 for men. In Swaziland it is 32. Animal managers and keepers may unrealistically expect their animals to reach the same age as the "record holder," but it is important to be realistic about an individual's life span. It will vary between individuals and can be heavily influenced by environmental factors.

COMMON HEALTH CONCERNS: THE BIG FIVE

A keeper may have a geriatric animal that is "higher-maintenance," or requires more time and care, than a younger animal. Diet modifications, husbandry changes, time for administration of medication, and extra coordination with veterinary staff may be required. Additionally, individuals of social species may have challenges relating to social dynamics, and may require separation or special management at feeding times. Aged animals may move slower, eat slower, or be reluctant to leave a warm area on a cold day. They may require a special diet that is different from that of other individuals they live with. All these things may require additional time on the part of the keeper. While by no means all-inclusive, the following problems are the most common ones geriatric animals experience as they age.

ARTHRITIS

Age-related joint deterioration and inflammation (arthritis) is a very common problem in geriatric zoo mammals, birds, and reptiles. In addition to the "wear and tear" that comes with age, this problem may be accelerated or exacer-

bated by improper substrate, especially for heavy animals and hoofstock. Animals housed on hard surfaces and who receive minimal exercise and/or have had improper trimming and care of hooves, toes, and claws commonly develop arthritis prematurely, before other organ systems present with abnormalities. The limbs and vertebral column are the most common locations for arthritis, and the animal may show a variety of signs, ranging from dramatic (severe lameness and reluctance to move) to subtle (decreased activity, appetite, and/or energy level). Often these signs will worsen with colder weather. It may be difficult to know whether the changes are due to behavior or other medical conditions, and often they require a thorough veterinary examination, radiographs, and other diagnostic testing to determine their cause and severity. In addition to medical and surgical treatment, husbandry changes can improve the arthritic animal's quality of life dramatically. The routine tilling of dirt or decomposed granite substrate and the provision of softer substrate, rubber mats, ramps to eliminate jumping and falls (figure 22.1), extra wood shavings, straw, blankets for bedding, heat pads, and increased ambient temperatures are important husbandry techniques that may help these animals remain comfortable.

Figure 22.1. This geriatric baboon, "Rosie," has arthritis in her back. The ramps allow her to move safely around her enclosure. Photo by Cynthia Stringfield.

Pain management is an important part of arthritis treatment, and different treatments are commonly used by zoo veterinarians. Nutraceutical chondroprotectives (nutritional supplements to support joint health) like glucosamine and chondroitin are added as a dietary supplement to provide building blocks for cartilage repair and joint (synovial) fluid production. Other nutritional supplements may also be used to provide some relief. As such, these supplements often take several weeks to provide an effect. Immediate effects for relief of pain may be accomplished by nonsteroidal anti-inflammatory drugs (NSAIDS), steroids (powerful anti-inflammatory hormones), and narcotic pain relievers. While there are no known negative side effects of the nutraceutical chondroprotectives, other medications may have side effects and may need to be managed carefully to give the lowest effective dose. Careful monitoring, husbandry changes, and nutraceuticals may decrease the amount of these drugs that are necessary.

NEOPLASIA

Neoplasia literally means "new growth," and refers to individual cells that have multiplied at an abnormally high rate. These growths may occur only locally (as in the case of benign tumors), or may spread to other parts of the body (as in the case of malignant tumors, also called cancer). Growth rates vary from slow to fast (also termed "aggressive"), and tumors can occur anywhere in the body. Neoplasia is a common problem in geriatric animals and, when malignant, is a common cause of death. Oncology is a subspecialty in veterinary medicine wherein a veterinarian who is board-certified in internal medicine specializes in the diagnosis, prognosis, and treatment of neoplastic disease. Zoo veterinarians often consult with these veterinary oncologists, and also with the clinical pathologists who examine and interpret tissue samples to give a diagnosis. Surgery, radiation, and chemotherapy (drug therapy) are successful options for some types of neoplastic disease, with the most successful outcomes occurring when the disease is caught early and is in only one location (as is the case of benign tumors). More extensive treatment for more aggressive tumors is often possible, depending on the general health of the animal and the resources and logistics involved. Keepers can help identify external lesions early by being aware of any changes in the animal's external appearance, and routinely checking for swelling and growths.

AGE-RELATED ORGAN FAILURE

Chronic renal failure (CRF) is a decrease in the kidneys' ability to function, and it occurs over a long period of time (i.e., it results in a slow decline). Kidneys filter the blood and eliminate circulatory waste products in the form of urine or urea, in mammals, and uric acid, in reptiles and birds. Kidneys are also important in the regulation of blood minerals and blood pH, and in mammals they make a hormone, erythropoietin, that stimulates production of red blood cells in the bone marrow. In mammals urine is normally concentrated in the kidneys to avoid excess water loss, and this ability is lost over time in chronic renal failure. Consequently the animal will need to drink more water and will produce a larger quantity

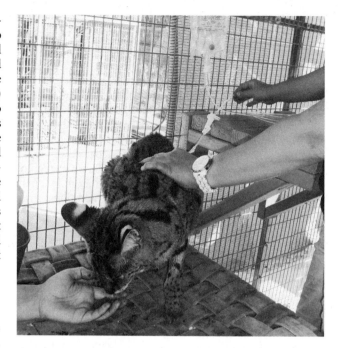

Figure 22.2. This geriatric serval, "Mazoe," had been previously trained to accept a subcutaneous injection. When she developed chronic renal failure, this behavior was used to routinely administer subcutaneous fluids. Courtesy of Tia Orbin.

of more dilute urine, thus "compensating" for the kidneys' inability to concentrate urine the way it used to. When a change like this is noticed by the keeper, the veterinarian should be notified. The animal can be checked with blood and urine testing to diagnose kidney disease early or to diagnose other disease that may cause these same signs. Stomach upset, poor appetite, and weight loss may be other signs of failing kidneys. While there is no cure for chronic renal failure, the decline of the animal can be slowed with appropriate dietary, fluid, and medical management (figure 22.2).

Animals in CRF should be allowed and encouraged to drink plenty of fluids, and if they suddenly decrease their drinking and/or appear lethargic or dehydrated, a veterinarian should be notified immediately. The ability to obtain urine easily from mammals is an important husbandry goal and will permit routine monitoring of the animal's kidney function. Of all mammals, felids most often suffer chronic renal failure as they become geriatric. Chronic renal failure related to improper protein levels and other dietary problems may be prevented by a properly balanced diet during the animal's earlier phases of life.

The heart is a muscular organ responsible for pumping blood to the lungs and throughout the body. As it ages it can lose the ability to pump efficiently, causing decreased blood circulation through the lungs, the body, or both. This causes fluid to build up in the lungs, abdomen, or both; this is referred to as congestion. In the lungs this congestion is specifically called pulmonary edema and in the abdomen, it is called ascites. This age-related disease is generally called congestive heart failure (CHF). Animals may show decreased energy, cough, exercise intolerance (becoming "winded" with minimal exertion), noisy breathing, pale mucous membranes,

Figure 22.3. This geriatric barn owl, "Sweetpea," has a visible cataract in the left eye. Courtesy of Tia Orbin.

and abdominal distension. Renal failure may also result secondarily due to decreased blood flow to the kidneys, and consequently blood pressure problems can result from both CHF and CRF. While there is no cure for the failing heart, there are medical and dietary techniques to help it function better. An animal in CHF should not be exercised or stressed. Any lethargy, inappetance, abnormal breathing, or change in mucous membrane color should be immediately reported to the veterinarian. Difficulty in breathing or poor mucous membrane color (i.e., if it is very pale) is an immediate emergency.

LOSS OF SPECIAL SENSES

Special senses are the senses to which specialized organs are devoted. While many senses may dull with age, the two most common geriatric losses of special senses are of hearing and vision. The most common vision problem is cataract development (figure 22.3). A cataract occurs when the normally transparent lens of the eye becomes cloudy, thus impeding vision. An observant caretaker can observe changes in the lens when looking at the pupil (black center) of the eye, because the lens sits behind it (in contrast, cloudiness in the cornea, or the clear front part of the eye, will obscure both the iris—the colored part of the eye—and the pupil). Bright unfiltered sunlight may lead to lens (and corneal) opacity changes in species not adapted to a sunny environment if they are not provided adequate shade (this has been proven in humans, who need to be allowed to wear sunglasses when they work in the sun routinely). It is important to provide the appropriate environment for these species. One common occurrence is in pinnipeds that have historically lived in sunny exhibits while in captivity. These animals commonly develop cataracts as they age, and excess sunlight is thought to be one possible cause.

Medical and surgical management may be possible, and treatment will depend on the severity of the cataract and the type of animal. Training an animal to accept eye medication and examination is very important for cataract treatment. It is also important for the keeper of an animal with hearing and/ or vision loss to change their expectations of that animal and accommodate their loss. An animal with impaired sight, for example, may need enclosure modifications; consistent placement of food, water, and enclosure furnishings; and auditory cues instead of visual ones. Conversely, a deaf animal may need visual cues instead of auditory ones. An animal formerly in education programs that now "spooks" easily because it cannot see as well as it used to may need to be retired, or handled differently. Before assuming the animal has a behavioral problem, one must evaluate its special senses carefully. It is of paramount importance that a keeper knows how animals normally perceive the world around them in order to be able to detect subtle changes in their behavior that can serve as important clues in evaluation of their special senses.

DENTAL DISEASE

Tooth pain or loss of teeth in mammals and some reptiles may need to be addressed via dental care and dietary modifications. Proper dental care during an animal's life will decrease the problems it will have as it ages, especially in the prevention of pain, tooth loss, and bacterial infections. These conditions are completely treatable, and veterinary dentistry is a subspecialty in veterinary medicine, just as it is in human medicine. Mammals should have their teeth cleaned when necessary; they can have broken teeth repaired, nerves or teeth removed to eliminate pain, and infections treated before they become severe and infect the surrounding bone. Zoo veterinarians may use veterinary dentists and/or human dentists and equipment to treat these patients. Animal keepers can train animals to allow oral exams, routine tooth brushing, and oral topical medication administration. Routine tooth brushing will decrease the need to anesthetize the animal for teeth cleaning. Keepers should be alert to any difficulty in eating, reluctance to eat, chewing only on one side, dropping of food or increased chewing time, reluctance to drink cold water, swelling of gums or of areas where a tooth root lies under the gums, or foul odor from the teeth and gums, as these are all common signs of tooth-related disease. Catching these problems early will allow them to be treated sooner and limit further problems (e.g., infection of kidneys, heart, or jaw bones, or deterioration of body condition). Unfortunately, ungulates may simply wear down their teeth from a lifetime of chewing fibrous feed. These animals can be very difficult to manage, because such fiber is needed for their digestive tracts to function properly.

NUTRITION AND FEEDING CONSIDERATIONS

As animals age, they may need modifications to their daily rations due to decreased metabolism and/or decreased activity. Depending on the species, they may benefit from changes in the composition of their diet based on recommendations by veterinarians and zoo nutritionists. For an animal with a certain disease, prescription diets may be used. Prescription diets made for dogs and cats may be used in similar nondomestic patients. These diets must be prescribed by a veterinarian, just as medication is. Also, the diets can be specially formulated by animal nutritionists to provide the proper nutrients that geriatric animals require. Dietary supplements (or nutraceuticals) can also assist in the treatment of many chronic diseases. One example is the previously mentioned use of nutraceutical chondroprotectives for arthritis (Stringfield 1999). Finally, diets may need to be presented differently for animals with chronic disease. Some adjustments might include the softening of hard food for an animal with missing teeth, addition of extra water for an animal in CRF, improvement of flavor to encourage an animal to eat, and so on. For example, a geriatric fox in chronic renal failure that has lost molars due to previous tooth disease may benefit from having its prescription dog chow soaked or ground up, or may be switched to the canned version of the diet. Sometimes, as animals in decline become picky eaters, keepers may be challenged to provide diet items they will eat while still maintaining a balanced diet for them.

OLD AGE IS NOT A DISEASE

It is extremely important to remember that just because an animal is geriatric and becomes ill, this does not mean it has a serious terminal disease. Many diseases or problems may be minor, or very manageable, but there is no way to know what the actual problem is unless the animal is evaluated by a veterinarian and a diagnosis (even a tentative diagnosis) is made. Many animals considered to be at the end of their life spans may actually have significant time, even many years, left.

Even when a terminal disease process is found, many diseases may be managed with palliative care. Palliative care (from the Latin *palliare,* meaning "to cloak") is any form of medical care or treatment that concentrates on reducing the severity of disease signs rather than striving to halt, delay, or reverse progression of the disease itself or provide a cure. The goal is to prevent and relieve suffering and to improve the animal's quality of life. Palliative care is not dependent on a prognosis, and is offered in conjunction with curative and all other appropriate forms of medical treatment.

EUTHANASIA

Euthanasia translates to "the good death." When an animal's quality of life is no longer good and the prognosis for recovery or improvement is poor, the decision may be made to humanely end the animal's life. This can be a straightforward decision or an extremely difficult one. Personal opinions and feelings about quality of life, euthanasia, death, and when to euthanize an animal can conflict with those of other people in some cases. Human feelings about euthanasia in general, or desire to keep an animal alive for a person's own reasons should never be placed above the animal's needs and its quality of life. Quality of life can be a very subjective thing, different for each animal; but such problems as struggling for breath and constant, unmanageable pain should always be of the utmost immediate concern. When an animal has more "bad days" than "good days," that may also raise concerns for its quality of life. When time allows, in most facilities, the decision to euthanize is made by a team that includes the veterinarian, keepers, and management staff. In this author's experience, the best approach involves a meeting away from the animal (a "quality-of-life meeting") where all present share their objective information about the animal. It is the author's experience that when such a meeting is conducted appropriately, consensus is almost always reached. If not, new management ideas and euthanasia criteria involving the animal's condition can be discussed. Depending on the facility's chain of command, someone may have more authority than the others to make a final decision. Some keepers will not want to be the decision-maker, due to the responsibility and guilt they may feel. Making a group decision, or allowing a veterinarian or manager to make the final decision, can allow some people to voice their information more objectively without feeling as though they are giving a beloved animal a death sentence. Conversely, keepers may feel frustration if they feel it is time for euthanasia but a decision is made to continue the animal's life. Less common but better is the practice of being proactive instead of reactive. Deciding on criteria for euthanasia ahead of time removes the tendency to backslide on the decision, and removes some of the natural emotion from the actual decision-making process. For example, the group decision for a geriatric wolf with severe hip arthritis, where no further medical or husbandry options exist, could be that when the animal cannot rise and stand on its own, it will be euthanized. In situations that require an immediate decision (e.g., in an emergency, after a drastic decline, or when information is found during a workup for which the animal has been anesthetized), that decision is usually made jointly by the animal manager and the veterinarian. In almost 30 years in the zoo field, this author has never seen an animal euthanized too soon. However, it is not unusual to feel, after seeing the pathology (abnormalities) at necropsy (the postmortem examination), that the euthanasia could have been done sooner. It is important to remember that since wild animals are programmed to hide signs of illness in order to survive at all cost, when they do show signs of pain or disease it is often further advanced than might be expected. Recently veterinarians have been working to develop objective scoring systems to evaluate pain, quality of life, and physical condition (Gaynor 2009; Villalobos 2006; Follmi 2007). Veterinarians should share necropsy information with keepers to reassure them that the proper decision has been made, and so that they and the keepers can critique and learn from each other to make better decisions in the future.

The American Veterinary Medical Association (AVMA) and American Association of Zoo Veterinarians (AAZV) publish methods of humane euthanasia, which are available on their websites. In a zoo setting, euthanasia is usually accomplished by chemical means, most often by using a drug referred to as "euthanasia solution," which is a very concentrated pentobarbital (anesthetic) solution. This drug is administered intravenously and is commonly colored pink, red, or purple so that it is not confused with any other drug. The term "putting an animal to sleep" refers to this method, which is truly an anesthetic overdose that stops the lungs and heart and causes the animal no pain or distress. Experienced keepers and veterinarians can often devise a plan to preanesthetize the animal quickly and without discomfort

Figure 22.4. This Abyssinian ground hornbill, "Wilhemina," survived being run over by a rhino as a young bird. She is now 38 years old, years beyond being considered very geriatric. Photo by Cynthia Stringfield.

before this final solution is given, thus creating a peaceful and nontraumatic experience for the animal.

Ethical considerations abound when euthanasia is considered for a geriatric animal. Is the cost of the needed palliative or curative care too large for the facility? Is the space required to house the animal now needed for a reproducing animal (which will contribute to conservation), or an animal needed for education purposes? Is there enough staff time to properly care for an ailing geriatric animal? These questions are of real significance and concern in addition to the animal's quality of life, and different facilities and individuals will arrive at different answers.

GRIEF MANAGEMENT

Animal keepers become keepers because of their love, care, and concern for animals. Many people feel this way, but few decide to make animals their career. Because of this, loss of an animal often causes grief for keepers. The loss may be compounded by other things: a long relationship with the animal, a close bond due to the intense care the animal has needed in the final part of its life, emotions related to

euthanasia, personal emotions involving other losses of people or animals that have not been processed, or too many losses in a short period of time. Grief management is a very important aspect of preventing professional "burnout" and dealing with the stress inherent in a keeper's job. Psychology professionals, veterinarians, and pet loss support groups are recommended resources. Keepers should also remember that managers and veterinarians may also have had special relationships with the animal, and that they too may be struggling with their own emotions. Veterinary personnel often have additional stress due to the responsibility of actually ending the animal's life, and must answer to their own consciences when agreeing to perform euthanasia. It is only human to lose control of one's emotions during times like this. Utmost care must be taken to care for oneself, and to be empathetic or sympathetic with coworkers who are also experiencing stress and grief. People manifest these emotions differently, and they may not show them or handle them in the same way. Maintaining a professional mindset and still meeting the responsibilities of the job safely can be extremely challenging, but they are necessary parts of the position of animal keeper. It may be a further challenge to have to communicate information about the euthanasia to the public, docents, and other concerned people. A geriatric animal can often have quite a "following," and a keeper may have the role of explaining the end-of-life situation to these other people who have cared about the animal but have not been as aware of its situation.

SUMMARY

Knowledge of longevity data, prevention of disease, and proper husbandry, nutrition, and medical care of geriatric animals are very important for keepers. Geriatric animals have specialized requirements for husbandry and nutrition, and may have medical problems that require intensive involvement by their keepers. A caring and knowledgeable keeper can make a tremendous difference in the quality of an animal's life in its final stages. However, euthanasia may be necessary to prevent needless suffering when the animal's quality of life is no longer good. Keepers are often emotionally attached to these animals, and need to be aware of the importance of caring for their own needs during these times as well.

REFERENCES

Follmi, J., A. Steiger, C. Walzer, N. Robert, U. Geissbuhler, M. G. Doherr, and C. Wenker. 2007. A Scoring System to Evaluate Physical Condition and Quality of Life in Geriatric Zoo Mammals. *Animal Welfare* 16, no. 3: 309–18 (10). Wheathampstead, UK: Universities Federation for Animal Welfare.

Gaynor, J. S., and M. W. Muir. 2009. *Veterinary Pain Management.* Saint Louis: Mosby Elsevier.

Stringfield, Cynthia E., and Wynne Janna. 1999. Nutraceutical Chondroprotectives and Their Use in Osteoarthritis in Zoo Animals. *Proceedings American Association of Zoo Veterinarians,* 63–68.

Villalobos, Alice, and Laurie Kaplan. 2007. *Canine and Feline Geriatric Oncology: Honoring the Human-Animal Bond.* Ames, IA: Blackwell Publishing Professional.

Weigl, Richard. 2005. *Longevity of Mammals in Captivity: From the Living Collections of the World.* Stuttgart: E. Schweizerbart Science Publishers.

23

Transportation and Shipping

Andrea Drost

INTRODUCTION

Live animal shipping can be challenging. Whether animals are being moved domestically or internationally, the shipments must occur quickly and efficiently without jeopardizing their well-being. The primary goal is to limit stress on the animal, as well as to ensure an efficient shipment so that all goes well in a timely manner. No two shipments are ever the same. Instead of providing a step-by-step account of a particular shipment, a more general sequence of steps will be described to demonstrate what is involved in live animal shipping. The steps are as follows:

- the relevance of obtaining background information on the proposed animal to be transported
- permitting and health requirements
- crates and crate training
- modes of transportation
- shipment date selection
- notification of proposed animal transport to all relevant parties
- documentation package preparation
- "day of" arrangements
- tracking of the shipment
- notification of successful animal transport.

After studying this chapter, the reader should understand

- methods of and approaches to the transportation of common zoo animals
- the importance of planning in animal transportation, and the types of preparation required
- specific considerations relating to each step of the live-animal shipping process
- general approaches to transportation of common zoo taxa.

OBTAINING ALL RELEVANT INFORMATION ON THE ANIMAL

Once it's been decided that a specific animal will be transferred out of a particular zoo's collection, internationally or domestically, the first step is always the same. A copy of the animal's record should be provided from the sending zoo, the consignor, to the receiving zoo, the consignee. These records will outline the animal's history, which includes its date of birth, sex, parentage, and any other pertinent information. Most accredited zoos in North America compile this information in a document referred to as the "specimen report." One of the key pieces of information in the animal records and/or the specimen report is the country of origin. This information is pertinent when crossing international borders, as it could influence permitting applications. (This will be covered in further detail below.) In addition to the specimen report, a diet sheet and medical records should be supplied to the consignee. The medical records will also provide information needed when addressing what pre-shipment health testing is required.

PERMITTING AND HEALTH TESTING REQUIREMENTS

As a general rule, every country will have a minimum of two federal bodies that regulate the movement of live animals across its borders. One of these federal bodies will usually be a wildlife office. In Canada, this is Environment Canada's (EC) Canadian Wildlife Service (CWS). In the United States, all zoos are familiar with the US Fish and Wildlife Service (USFWS). It is through offices like these that Convention on International Trade of Endangered Species (CITES) permits are issued. Animals will be listed as either CITES I, CITES II, CITES III, or non-CITES. A non-CITES animal requires no CITES permitting. A CITES II and III animal requires that an export permit be issued by the exporting country before transport of the animal can occur. A CITES I animal requires that an export permit be issued by the exporting country and that an import permit be issued by the importing country before transport of the animal occurs. These permits must travel with the animal while it is in transit from one country to the other. CITES permits are only required when crossing international borders. At the time of shipment, the original CITES permit must be handed over to the exporting countries' wildlife authority to be validated. For example, a CITES I listed animal being moved from the United States to Canada will have its US CITES I export permit validated by USFWS during its exit inspection. Validation simply means that the USFWS officer will fill in information in a box located at the bottom of the permit, and then stamp the permit with the authorizing USFWS stamp. This occurs while the animal is still in the United States. When the animal reaches the first Canadian port of entry, the validated original US CITES I export permit and the original Canadian CITES I import permit must now be surrendered to the Canadian Border Service Agency (CBSA), Canada's customs officials. Now the CITES-listed animal has legally left the United States and has legally entered Canada.

In addition to the CITES permits, some federal bodies also require further import or export permits, depending on the species of animal being proposed for transport. The USFWS, for example, issues endangered species/threatened species permits if the species is listed as endangered or threatened in the US Endangered Species Act (ESA). The USFWS will also issue import/export permits if the transaction is commercial—that is to say, if money is being exchanged as a condition of the animal's movement. Another potentially required USFWS-issued document is the designated port exception permit, if a USFWS-designated port of entry/exit is not being used. All zoo and aquarium animals imported into or exported out of the United States must be declared with USFWS. As such, USFWS has designated specific cities for the purpose of processing these declarations. Ports (cities) other than the designated ports can be used, but only with the designated port exception permit, on which will be listed the specific nondesignated ports allowed for use.

The second federal body that must be contacted is the agricultural agency. For example, Canada's agricultural agency is the Canadian Food Inspection Agency (CFIA) and the United States' counterpart is the US Department of Agriculture (USDA). Agencies such as these issue their own import permits (normally issued at the agency's national headquarters) and export certificates (normally issued at the local office in the city where the zoo or aquarium arranging the animal's movement is located). These permits and/or certificates are required for some animals, but not all. Agricultural agencies tend to be focused on the protection of their country's livestock industry and are therefore most interested in hoofstock, swine, poultry, and similar domestic species. They are less interested in regulating animals such as amphibian and reptile species, with exceptions (e.g., in Canada, where the import of turtles is regulated by the CFIA to ensure humane transport practices). These agriculture agencies are responsible for deciding what health testing an animal is required to undergo before and after shipment has occurred.

For example, the CFIA requires that a white rhinoceros (*Ceratotherium simum*) being imported from South Africa undergo testing to certify that it is free of diseases such as bluetongue, trypanosomes, and bovine tuberculosis. The animal must then be retested for these diseases at least 21 days after the initial test date, but within 30 days of its intended importation date into Canada. If these requirements are not met, the rhinoceros will be refused entry into Canada. The same species being imported from the United States to Canada will require no preshipment testing, but only a CFIA inspection at the first port of entry. This demonstrates how testing requirements for a specific species of animal can change depending on the country from which the animal is being imported. This is a direct result of the health status of countries (i.e., the types and prevalence of diseases present) being different from one another.

Different zoos have different quarantine capabilities. In Canada, the CFIA has three categories of quarantine that an animal is regulated to undergo: nonquarantine, minimum quarantine, and medium quarantine. Any zoo can import an animal directly to its facility if there is no federal CFIA quarantine requirement. Both the minimum and the medium quarantine requirements dictate that the quarantine area must be approved by a federal veterinarian from CFIA before the animal's arrival. Most zoos can do a minimum quarantine as it simply requires that the importation testing requirements be met and that the related information be provided to CFIA. Very few zoos have medium capability for quarantine, as this requires staff to shower when entering and exiting the quarantine area. In addition to this, other requirements of the building must be met, such as a separate air exchange for the quarantine area. When importing nonhuman primates into the United States, the animal must first be quarantined in a Centers for Disease Control (CDC)–approved facility for a minimum of 30 days, and until all import testing requirements have been met. This is because of these species' potential to carry serious zoonotic pathogens and diseases such as Ebola Reston, herpes B virus (Cercopithecine herpesvirus 1), monkeypox, yellow fever, simian immunodeficiency virus, and tuberculosis. It is only after meeting these quarantine requirements that the primate can then be transferred to the zoo of import, where it will then undergo that zoo's non-federally regulated quarantine, normally for an additional 30 days.

It should be noted that once an animal is transported into the importing country, there are specific quarantine requirements to be met for different species. In Canada, for example, an imported toad species requires no federal quarantine even though the importing zoo will have its own import quarantine requirements. The CFIA does not have any federal quarantine requirements for the Puerto Rican crested toad (*Peltophryne lemur*), but the Toronto Zoo has a 60-day in-house quarantine requirement that must be met before the animal is moved into the collection. An imported red panda (*Ailurus fulgens*) from Japan will have a federal CFIA minimum quarantine requirement. This means that the CFIA must inspect the quarantine holding facility before the animal can be imported into the country. Once in Canada, the red panda must be held in quarantine for a minimum of 30 days, and must undergo further testing before being released from quarantine. A swine species like the warthog (*Phacochoerus africanus*) must undergo a federal CFIA medium quarantine. Again, the CFIA must inspect and approve the quarantine facility prior to import, but in this case the requirements are more restrictive. The airflow must be regulated within the quarantine space, and keeper staff will be required to "shower in" prior to working with the quarantined warthog and "shower out" after that work is complete. Separate work clothes will be used and kept within the quarantine area. In this way, staff will avoid carrying pathogens out of the quarantine area on themselvColumbus, OH: es or their work clothes.

In addition to federal regulations for permitting, jurisdictional requirements must also be met. Some provinces or states require their own separate import and export permits, and also require that a certificate be issued, while other jurisdictions require only that an import number be issued and placed on shipment documentation. In-transit permits might be required if the animal has a connecting flight in a country that is neither the importing nor the exporting country. For example, if cheetahs (*Acinonyx jubatus*) are being moved from South Africa to Canada with a flight change in Germany, the German officials may require an in-transit permit indicating the purpose and length of the animal's stay in Germany. It should be noted that this in-transit permit does not cover care and location for any unforeseen delays. It simply informs the in-transit country that the animal is going through that country on a specified date and time. If any unforeseen delay happens to occur in the in-transit country, that country will want the importer or the exporter to hire a local freight forwarder to feed, water, and care for the animal during the delay.

CRATES AND CRATE TRAINING

The International Air Transport Association (IATA) publishes the Live Animal Regulations (LAR). The IATA-LAR stipulates the requirements that must be met for shipping container construction and design for a specific species of animal. This reference material is updated every two years and must be adhered to strictly if the animal is being moved by air, or if the animal is CITES-listed and is being moved internationally. USFWS now requires that all CITES-listed animals being transported internationally must be moved in compliance with the IATA-LAR for that species, even if they are being moved by land. The IATA-LAR regulations are such a useful resource that it is strongly recommended they be used for all non-CITES domestic land transports as well. General considerations to remember when selecting a crate for an animal are that (1) the crate has external access for food and water, (2) the crate is leakproof, (3) the animal can be monitored while it is in the crate, (4) the crate is constructed of appropriate and sturdy material (e.g. wood or metal) which will safely contain the species, and (5) the crate is comfortable (e.g., contains the appropriate amount of bedding) for the animal. One of the most common mistakes when selecting an appropriate crate for a specimen is to provide too much space, the assumption being that a larger crate will provide the animal with increased comfort. Unfortunately, increased space also increases the risk of self-inflicted injury to the animal, which might try to run at the front or side of the crate in panic or in an attempt to escape.

As mentioned before, it is essential that a crate used to transport an animal be well constructed, clean, and leakproof. It must be able to contain the animal at all times, must prevent unauthorized access so that accidental opening of the crates during transit cannot occur, and must be appropriately ventilated. Generally, the container must be ventilated on at least three sides, with most of the ventilation being provided on the upper portion of the sides. It should be noted, however, that different species have different ventilation requirements. IATA regulation #34 for the gorilla, for example, requires that all four sides and the top of a crate must be ventilated. Ventilation holes must always be small enough to prevent the protrusion of the animal in any way, and in some cases they must be covered with fine mesh, wire, or muslin (a woven cotton fabric that allows for ventilation but restricts visual access). In general, the crate must be made so that the animal inside can stand, turn, and lie down in a natural manner. There are exceptions to this; for instance, some bird crates must incorporate a perch, and the bird must be able to stand and turn in a natural manner.

Crate training for the animal begins the process of providing it with physical and psychological comfort during transport. Key to successful crate training is the development of a plan before starting the process. Much information on training is readily available, but the following are very basic steps that may be followed. Generally, the crate is introduced to the animal and secured safely so that the animal can become used to seeing and/or smelling this novel item. The next step is the introduction of food at the entrance of the crate. This food, which can simply be a portion of the animal's diet, is then placed further and further into the crate over a period of time. The final goal is to ensure that the animal is eating its food with its body fully inside the crate. At this point the animal will usually display behavior indicating that it is comfortable eating in the crate. Generally the animal is not locked into the crate unless there is additional time for training prior to shipment, as once the animal is locked in, it is usually startled and will not go back to the crate again for some time afterwards. One of the main benefits to crate training is that it relieves much of the stress the animal might otherwise experience should it have to be forced into the crate. At the end of crate training the crate will be very familiar to the animal, with the animal's own smell present, and can be essentially the animal's "home away from home." Crate training can also eliminate the need to use sedatives for transport.

It should be noted that for some species using a trailer is a better choice than using a crate. General rules for crates still apply to the trailer; the animal must still be able to stand, lie down, and turn around naturally. Depending on the species, more than one animal can be transported in a trailer. Moving five female bison in a 6 m (20 ft.) trailer is completely acceptable, and can be less stressful on the small group than moving them separately.

TAXON-SPECIFIC TRANSPORTATION

While specific methods of shipping animals will vary depending upon the taxa, some basic principles apply in most situations:

- Shipping containers must be secured to prevent the animal from escaping and unauthorized people from gaining access to the animal. This applies to the primary container, which immediately holds the animal, as well as to the secondary and tertiary containers, when applicable.

- No part of the animal should be permitted to extend outside the crate (e.g., through a ventilation hole).
- The shipping containers must be ventilated adequately for the animal being shipped. Specifics may be mandated by IATA.
- Temperature must stay within a range that is healthy for the animal.
- With some exceptions (noted below), animals should have space to move around within the shipping container.

REPTILES

Crates used to transport reptiles need to be adequately ventilated, although it should be noted that reptiles may require less oxygen than other species. A standard reptile crate is made of wood and is usually lined with polystyrene foam. A thin wire mesh or muslin should be fixed between the wood and polystyrene foam, covering the ventilation holes. In some cases a polystyrene foam box within a cardboard box will suffice. Transport of most reptiles will require that a primary container is placed within a crate. It is important to note that reptiles are generally packed dry, with no moisture making direct contact with the animal, although there are exceptions to this general rule. In the case of snakes the primary container can be a cotton bag; for small turtles it can be a cotton bag or a clear plastic ventilated box. Turtles and tortoises must be transported in a natural position, with the plastron on the bottom. No stacking of turtles and tortoises is permitted. Heat packs can be used if there is a concern that the animals might be too cool during transport. If heat packs are used, they cannot come into contact with the primary container or with the animals themselves. Larger reptiles such as crocodilian species must be packed singly, and the direction of the head should be indicated on the outside of each crate. This is in direct contrast to the general rule that an animal must be able to turn around in a crate. The crate containing some of the larger reptile species (e.g., crocodiles or large monitor lizards) must ensure that the animal cannot turn around so that the direction of the head is known at all times. This becomes important when the animal is released from the crate, since not knowing which end of the animal will exit the crate first will pose a greater risk to the keepers' safety.

In the shipment of a venomous snake, a translucent fabric bag should be used as the primary container, which then can be placed within a ventilated clear plastic box as the secondary container. The plastic box, in turn, will be placed in the polystyrene foam–lined wooden crate. Thus the venomous animal is said to be "triple contained," which is an important safety measure. It is essential that each of these containers be sealed securely; the bag must be tied tightly and can be secured with an electrical zip tie for added security. Always remember that it is possible for the venomous animal to see motion through a translucent bag, and therefore can strike and bite through the bag; hence the importance of the secondary container of transparent solid material. The translucent bag should be labeled with the animal's common name, scientific name, and current body weight. It is always a good idea to provide information on the outside of the wooden crate explaining exactly how the venomous animal is contained within.

AMPHIBIANS

The transport of amphibian species is quite similar to the transport of reptile species. The major difference is that amphibians are generally transported on a moist substrate, such as moistened moss, sphagnum, or even moistened paper towels inside their primary container. The primary container can be as simple as a polystyrene foam cup or a clear plastic container. Again, the primary container can be placed inside a polystyrene foam–lined cardboard box or wooden crate. It should be noted that the primary container must be large enough to allow the entire ventral surface of every animal to make contact with the bottom of the container. As with reptiles, heat packs can be used as long as they do not come into direct contact with the primary container or the animal.

BIRDS

Bird species should be crated with great care, as they have the ability to injure themselves greatly within a crate or rigid plastic pet container (kennel). Because of this, the interior must be safe for the bird, all edges must be smooth, and there must be no sharp projections of any kind. Wooden perches must be provided for the majority of bird species and must be placed in a position to ensure that excreta will not fall into the food or water dishes. Ground-dwelling birds normally do not require wooden perches. A nonperching bird should have non-slip flooring and the bird should be able to stand in a natural position. It is recommended for most birds that padding is added to the interior roof to prevent injury should the bird attempt flight within the container. It is particularly important that crates are securely closed, since once a bird escapes during transit, it may be very difficult to catch it again. Larger birds such as ostriches (*Struthio camelus*), emus (*Dromiaus novaehollandiae*), and tall cranes can be transported overland via trailer instead of being crated.

MAMMALS

Hoofstock. Most larger hoofstock species can best be moved via trailers overland. If space and time allow, the animal can easily be acclimated to the trailer before its shipment. Simply feeding the animal inside the trailer can help with the acclimation process and make the trailer a safe and secure surrounding for the animal. If the animal is being transported by air, then using an IATA-specified crate will be mandatory. This may require the use of a forklift or crane to lift larger, heavily constructed crates. Crates for hoofstock are another exception to the recommendation that a crate be large enough for an animal to turn around in. For hoofstock the crate should be narrow enough to prevent the animal from turning around, so that when it is released from the crate its direction can be predicted. For this reason, the direction of the head should be indicated on the outside of the crate.

Carnivores. The larger carnivores such as tigers, lions, and bears should be transported in sturdy, solidly constructed metal crates. As well as ensuring that no part of the animal can fit through a ventilation hole, one must ensure that no person can inadvertently get too close to the crate and come into direct contact with the animal. Padlocks should always be used to secure these crates.

Primates. Smaller primates can be transported in modified pet carriers. Some primate species will require either branching to hang from or shelves to perch upon. Most medium-sized primate crates must have very good ventilation on three sides of the crate. For larger primate species, the roof must be adequately ventilated as well. Padlocks are recommended to secure these crates.

FISH AND AQUATIC INVERTEBRATES

Fish and aquatic invertebrates are generally packed in a strong plastic bag containing two-thirds air (to provide oxygen) and one-third water. Most zoos and aquariums will then place this bag inside another plastic bag of the same size, thereby "double-bagging" the fish or invertebrate. The bag is the primary container and is placed inside a polystyrene foam–lined cardboard box or wooden crate that must have adequate strength to contain the weight of water and resist crushing. Fish are usually fasted for a 24-hour period prior to shipment, so as to reduce excreta and ammonia formation that will foul the water. Fish are directly affected by ambient temperatures, so the shipper must ensure that a suitable temperature is maintained during the entirety of the transport.

TERRESTRIAL INVERTEBRATES

Invertebrate species are often placed in a primary container which can be clear plastic or cardboard. They can be packed with some food, which will provide nourishment as well as moisture. The primary container is then placed inside a polystyrene foam–lined cardboard box. Due to the greatly reduced amount of oxygen required by these species, the air in the primary containers at the time of packing is normally sufficient for transport. If additional ventilation holes are required, a thin material should cover them so as to allow the exchange of air but still prevent the escape of animals.

MODE OF TRANSPORTATION

When considering what mode of transportation to use, consider what will get the animal to its destination safely and as fast as possible. Always use the most direct route available. If possible, avoid inter-airline transfers. The risk of something going wrong increases significantly if two different airlines are used. Usually a freight forwarder is required, to physically collect the animal from the first airline and deliver it to the next. A freight forwarder is a company that acts as an agent on the zoo's behalf to arrange an airline booking or to switch an animal between different airlines if necessary. Most airlines now require up to five hours to remove the animal from one plane and move it to the connecting plane, even if the actual physical transfer takes only an hour of real time with the remaining four hours of time spent waiting. It is always best to have minimal layover times, but making that happen has become more and more of a challenge. Also, most airlines now require the animal to be dropped off at least two hours in advance of the departing flight if the shipment's destination is domestic, and four hours in advance if the destination is international. So what appears to be a quick trip for the animal by air could actually involve as much time as moving it by land, if not more. For example, to move a capybara (*Hydrochoerus hydrochaeris*) from the Toronto Zoo to the Cleveland Metropark Zoo in Ohio by air could require a route of Toronto-Chicago-Cleveland, with four hours of drop-off time, one hour of flight time to Chicago, five hours of downtime in Chicago to switch planes, one hour of flight time to Cleveland, and one hour of recovery time at the Cleveland airport. All told, the animal would be in transit for up to 12 hours. It would be better for the animal to make the trip via a five-and-one-half-hour drive with a one-hour stop at the border for customs clearance. It is also important to make sure that all live animal restrictions for a particular airline are known. Some airlines will not carry venomous animals. Some airlines that will not fly animals internationally will fly them domestically. Some airlines will fly animals only on direct flights and not on connecting flights. The list of restrictions is extensive.

Moving an animal by air or by land (road) is preferable to moving it by rail or water, simply because of the time involved. It can be argued that moving an animal by road can take just as long as moving it by rail, but during a road trip a transporter is directly responsible for the care of the animal. An animal being moved by rail is considered perishable cargo and will not necessarily receive the same level of care while in transit. Moving an animal by sea may take days, which means that it is crated for a very long time. Airlines now have the capability to move even the largest of zoo animals, and can do so in a fraction of the time it takes to move by water. Most airlines do not have animal-care specialists for layovers, but a freight forwarder or a zookeeper in the layover city can be recruited to care for the animal.

SHIPMENT DATE SELECTION

Before a shipment date can be set, several things should be considered. One of the foremost considerations for both the sending and receiving zoos is the weather. Winters can be too cold for live animal shipments, and summers can be too hot. The zoo or aquarium responsible for transporting the animal has to be ready to send the animal, and it will require time to complete crate training if that is deemed necessary. A gorilla (*Gorilla gorilla gorilla*) might be crate trained in just seven days, while a moose (*Alces alces americana*) might require up to four weeks. Another consideration is whether the animal is old enough for transport. The IATA-LAR has very strict guidelines about not moving primate species while they are still nursing from the mother. AZA Species Survival Plans (SSP) also have guidelines for various animals. Both the white rhinoceros SSP and the giraffe SSP recommend that these species not be removed from their mothers before the age of two years. The receiving zoo has to be ready to receive the animal as well. Perhaps it is receiving a species new to their collection, and a new exhibit must be completed first. Or perhaps it needs to move a specimen out to another zoo before it can receive the new animal. Once all these considerations and any others have been addressed and resolved, it's time to select a tentative shipment date.

NOTIFICATION OF ALL INVOLVED

The tentative shipment date must be selected by the zoo sending the animal and agreed upon by the zoo receiving the animal. Once this date is selected, an airline booking is made or land transporter is arranged. Once the mode of transportation is decided upon and booked, the wildlife and agricultural agencies responsible for monitoring animals in transit need to be notified and/or booked for inspections. Inspections are often required in both the exporting and the importing countries. Most agencies require a minimum of 48 hours notice, but some can require as much as 72 hours. Which specific agencies are required for inspection will depend on the species being transported and the regulations of the countries involved. Increasingly, customs brokers have to be used to complete the customs clearances at the first port of entry of the importing country. A customs broker is an employee of a brokerage firm who acts as the zoo's representative during import or export. This involves the preparation of electronic or nonelectronic submissions to customs for clearance. Some ports still allow zoo personnel to present shipment paperwork for customs clearance, but more ports are insisting that brokerage firms be used, and this will therefore add a fee to the overall cost of shipment. Therefore, it is advisable to contact the customs offices of both the importing and exporting countries for verification. In some instances a freight forwarder will have to be used. Most airlines will only allow bookings from "known shippers," which most zoos are not, as they simply do not use airlines often enough to maintain "known shipper status." In this case a freight forwarder would be used to make the booking on the zoo's behalf. This is unfortunate, as the freight forwarder is now a third party, which causes the cost of transportation to rise and could cause a delay in shipment if the freight forwarder is not readily available. Of course, if the animal is being moved domestically, with no international borders being crossed, then the notification process will be simpler, involving the sending zoo, the receiving zoo, and the selected mode of transportation. The state or province from which the animal is being moved will also need to be contacted, in case there are any state or provincial requirements that must be met.

PREPARATION AND FORWARDING OF THE DOCUMENT PACKAGE

Documentation packages should be prepared and forwarded to all parties when they are notified of the shipment. It is at this point that any necessary corrections to the shipment paperwork should be made. Federal veterinarians of every country are very particular about the wording on health certificates and will supply the correct language to be used on documentation. The following is a list of some documents that might be included in a shipment package for an animal

being moved from Canada (shipper/consignee) to the United States (receiver/consignor).

1. **permits and licenses:**
   - CITES I import permit
   - CITES I, II, or III export permit
   - USFWS endangered/threatened species permit
   - USFWS designated port exception permit
   - USFWS migratory import/export permit
   - marine mammal transport permit
2. **health certificates:**
   - standard zoo health certificate issued by the zoo veterinarian for non–federally regulated species
   - federally issued health certificates for federally regulated species
3. **air waybill (for animals transported by air):** a document issued by the airline which serves as a means for identifying and tracking the shipment until shipment has been turned over to the consignor
4. **manifest (for animals transported by land):** a document issued by the land transporter that serves as a record of employment by the zoo to transport the animal from the consignee to the consignor
5. **handbill of transporters (for animals transported by land):** a document issued by the transporter that serves as a record of employment by the zoo to transport the animal from the consignee to the consignor; also the invoice the transporter will issue to the consignor for payment of services
6. **certificate of origin:** a document issued by the shipper certifying the country in which the animal was born
7. **specimen report:** a document issued by the shipper which provides pertinent information about the animal being transported, including the animal's place of origin, date of birth, sex, identification information, parentage, microchip numbers, tags, bands, tattoos, etc.
8. **diet sheet:** a document issued by the shipper outlining the animal's diet at the consignee zoo
9. **medical records:** a document issued by the shipper outlining the animal's medical history
10. **declaration of import/export:** a document filed by a zoo in the United States to the USFWS outlining that a particular species of animal is scheduled for transport on a particular date, and including the US port of entry or exit, the consignee's address and contact information, the consignor's address and contact information, the animal species (both scientific and common names), the permit numbers required to move the species, the animal's origin (wild-born, captive-born, or unknown), the number of species being moved, the animal's monetary value, and whether or not the species is considered venomous
11. **IATA shippers certificate:** a document issued by the shipper certifying that the shipment has been packed in accordance with the IATA live animal regulations, and specifying whether the animal(s) have been properly acclimatized if taken from the wild,

if the animal(s) is a CITES-listed animal; and in the case of reptiles and amphibians, if the animal(s) is healthy and free of any apparent injury and external parasitic infestation; this document also states the air waybill number, airport of departure, and airport of destination

12. **commercial invoice:** a document issued by the shipper certifying the animal's monetary value, which must be declared for customs purposes even if the animal is a donation or on loan

13. **AAZK animal data transfer sheet:** a document issued by the shipper which provides a plethora of information on the animal's identification, diet, medical history, enclosure/holding history, training history, and behavioral history.

"DAY OF" ARRANGEMENTS

One of the most important arrangements to make for the day of shipment is to ensure that the most experienced keepers are on hand for the crating. These might include the keeper who has been crate-training the animal, or those who have been involved with past shipments of the same species. One must arrange for the appropriate number of staff to assist with moving the crate on the day of shipment, and ensure that all appropriate equipment will be on hand. Moving rhinoceros, elephants, and other large species requires front-end loaders, forklifts, and/or cranes, which may have to be supplied by a company offsite and will require arrangements made in advance. Though an inspection appointment will have already been set upon notification of the regulatory officials, some inspectors also require a phone call on the morning of the shipment day to narrow down an exact time for the inspection. If the animal is being moved internationally, at least one inspection on the day of shipment will always be required. If the animal is being moved domestically, inspections are usually not required by regulatory bodies.

TRACKING OF THE SHIPMENT

Most if not all airlines have made the tracking of shipments much easier by adding tracking systems to their websites. The tracking number, which often is simply the air waybill number, can be entered on the airline's website to produce an update of where the animal is at any moment in the shipment process. Airlines can also be contacted directly by phone to obtain the same information. If the animal is being moved by road, the driver should carry a cell phone. When the animal is crossing international borders by road, communication between the driver and either the shipper or the receiver is of the utmost importance and should be required for all shipments. If the driver will be late for previously booked inspections, the inspectors need to be notified that the shipment is behind schedule. Once the animal has cleared the border, the driver should notify either the exporting or the importing zoo or aquarium of the delay and provide an updated estimate of arrival time. There can be times when the transporter is moving several animals to multiple zoos. In this case, the routing should have already been established and agreed upon by the transporter and the facilities involved. Generally, the facility closest to the border crossing will unload its animal(s) first, the next closest facility will unload its animal(s) next, and so on.

NOTIFICATION OF ARRIVAL

A step sometimes overlooked in an animal shipment is for the receiving zoo to notify the sending zoo of the animal's safe arrival. The welfare of every creature undergoing transfer from one facility to another is very important to the team at the facility of origin. A quick e-mail from the receiving zoo or aquarium indicating the animal's safe arrival will be much appreciated!

SUMMARY

As indicated in the introduction to this chapter, this information on animal shipping is by no means exact or comprehensive; rather, it provides a formula with which to begin. Obtaining background information on the animal proposed for transport will allow for the receipt of appropriate permits and follow-up on the required health testing. Next, selecting an appropriate crate and proceeding with crate training is imperative to ensuring a successful shipment. Being knowledgeable in the different modes of transportation is also important. Once a choice is made, a shipment date can be selected. At this point, notification of the proposed animal transport must be sent to all relevant parties, followed by the delivery of documentation packages to each. "Day of" arrangements can then be made. Once the animal is in transit, the sending and/or receiving zoo or aquarium should actively track the shipment. Finally, notification of the animal's arrival at its destination should be sent by the receiving institution. Live animal shipping, particularly across international borders, can be quite overwhelming. But following the shipping steps outlined above can ensure that the animal's transfer need not be a stressful experience.

24

Exhibit Design

Patrick R. Thomas

INTRODUCTION

The keeping of animals in living collections has been practiced in one form or another by a number of different cultures
for at least 4,000 years, but exhibiting animals to the public
only arose in the 18th and 19th centuries (Polakowski 1987,
18–20). These early menageries put little emphasis on exhibit
design, and it wasn't until Carl Hagenbeck opened Tierpark
Hagenbeck in Hamburg, Germany, in 1907 that zoos began
featuring species in large, more natural habitats with no
obvious form of containment (Reichenbach 1996, 59–61).
Hagenbeck was also a proponent of keeping species from
the same region in mixed-species exhibits, and separating
predators and prey using concealed dry moats (Hancocks
2010, 124). His vision was not instantly universally accepted.
It wasn't until 1941, for example, that the Bronx Zoo opened
its African Plains exhibit—the first naturalistic predator-prey
exhibit in North America (Bridges 1974, 452).

Once naturalism gradually became the norm in the zoo
community, pioneers like Heini Hediger (1955) and William
Conway (1968) argued that exhibits should also address the
biological and behavioral needs of animals. The concept of
landscape immersion arose in the 1980s and took naturalistic
exhibits one step further by providing visitors with the sense
that they were in the same habitat as the animals (Hancocks
2010, 126). The Zurich Zoo's Masoala Rainforest exhibit,
which opened in 2003, advanced the immersion principle
by actually placing zoo visitors in the same exhibit space as
the animals, enabling them to experience intimate views of
a diverse array of species from Madagascar (Bauert et al.
2007, 204).

In the past 20 years, zoos increasingly have begun incorporating enrichment into exhibits to encourage animal
activity (Shepherdson 2003, 120–22). The Bronx Zoo's Tiger
Mountain, for example, was developed with enrichment in
the forefront of the design, and includes innovative features
such as training walls, where visitors can observe keepers

conducting training sessions with the tigers (Thomas and
Chin 2004). More recently, elements are being designed into
exhibits that provide animals with choice and control to enhance their well-being (Coe and Dykstra 2010, 207–8). The
Lincoln Park Zoo's Regenstein Center for African Apes provides gorillas (*Gorilla gorilla gorilla*) and chimpanzees (*Pan
troglodytes*) with a variety of devices such as water spritzers,
air fans, and heaters that the animals can turn on and off
(Perlman et al. 2010, 315).

Polakowski (2001, 446–47) believes that the direction
of exhibits in the 21st century will lead to a more holistic
approach to exhibit design, with a focus on biological diversity, environmental interactions, habitat conservation, and
the opportunity for visitors to connect with live animals.
The Bronx Zoo's Congo Gorilla Forest is illustrative of an
exhibit that highlights the diversity, threats, and conservation of the species and habitats of the Congo Basin, while
also effectively providing visitors with powerful face-to-face
encounters with western lowland gorillas and many other
species (Ehmke 1999). It is these intimate experiences with
wildlife that inspires zoo visitors to want to help conserve it.
But a word of caution about exhibit designs: as they become
more specialized and intricate, they become somewhat more
restrictive with respect to the species that can be housed in
them, and they may limit which species can be added to or
removed from a zoo's collection plan.

Designing an effective exhibit requires a team approach
that draws upon the diverse expertise of a variety of zoo
staff. The planning team should consist of curators, field
biologists or others with detailed knowledge of the species'
natural history, architects, construction personnel, educators,
horticulturalists, veterinarians, and keepers who work with
the species on a regular basis. According to zoo design specialist Jon Coe (1999, 2), keepers should be integral members
of the design team because they know the animal "clients,"
they often have knowledge of what has worked before, and
they and the animals will be the primary users of the exhibit.

The goal of this chapter is to identify the elements that go into the creation of successful animal exhibits. After studying this chapter the reader will comprehend

- the design components that address the physical, behavioral, and social needs of the animals, encourage the expression of species-typical behaviors, and enhance well-being
- the essential details that relate to the keeper's work routine and safety
- which aspects of the design inspire visitors and positively influence their perception of the animals and zoos.

> **Good Practice Tip:** When planning a new exhibit, for each design feature under consideration ask: How will this feature impact the animals? How will it impact a keeper's ability to service the exhibit? Will visitors benefit from it?

PLANNING AND DEVELOPMENT OF ZOO ANIMAL EXHIBITS

An exhibit can be a stand-alone display with no obvious links to other exhibits in the immediate vicinity, a series of large exhibits that relate to each other thematically, or an area of the zoo or building that either places a group of taxonomically-related animals in a particular context or displays diverse species that share similar behavioral or physical attributes. One example of a stand-alone display is Zoo Atlanta's Giant Panda Conservation Center, whose goals are to facilitate research, inspire and educate visitors about giant pandas (*Ailuropoda melanoleuca*), and provide a high-quality home for the animals (Wilson et al. 2003). Singapore's Night Safari (Rodriguez-Herrejon 2001, 874) is illustrative of a series of habitats that have a common thread running through them—in this case, a diverse array of large and small animals in a nocturnal setting. Examples of exhibits that place taxonomically related animals in a specific context or display species with similar attributes include the Cincinnati Zoo's cat exhibit (Brady, Huelsman, and Maruska 1990, 169–72) and the Frankfurt Zoo's Grzimek House for small mammals (Scherpner 1982).

Regardless of its style, an effective exhibit must successfully meet the needs of three "clients"—the animals in the exhibit, the staff members (e.g., keepers, horticulturists, maintainers) who have to work in and maintain it, and the zoo guests who see it. Successful exhibits provide a suitable habitat for the animals and allow them to engage in a myriad of interesting and appropriate behaviors. They are created with the best interests and safety of the animals in the forefront of the design, and they take the species' biology and physiology into account in order to contain it effectively. They afford the animals adequate space, the ability to interact with conspecifics or other species (in the case of mixed-species exhibits), refugia from other animals, protection from the elements, and an adequate water source.

Exhibits also must be safe for staff to work in and service in a reasonable amount of time. They should provide good visibility so that the locations of animals can be ascertained without great difficulty, and should be easy for staff to access. Good exhibits also convey interesting stories and educate visitors about animal behavior and the threats the species face in the wild. The importance of having effective graphics at visitor viewing areas should not be underestimated as a critical component of the exhibit design. Exhibits should give visitors the

sense that they are good for the animals, and inspire care and compassion for wildlife while positively influencing visitor attitudes about animals (and zoos). Hancocks (2010, 126–32) believes excellent exhibits should create the look, feel, sound, and smell of nature.

DESIGNING FOR THE ANIMAL

The first consideration when designing a new exhibit should be the species that will occupy that space. A detailed knowledge of the species' biology and behavior is essential. A successful exhibit will consider the species' habitat, its tolerance of temperature and humidity, the water quality (for aquatic species), whether it is diurnal or nocturnal, its behavioral ecology and social needs, its feeding behavior, its activity patterns, and its space requirements. First-rate exhibits should not only be aesthetically pleasing to visitors but also replicate as many aspects of the natural environment as a species would regularly interact with (Hutchins and Smith 2003, 134–37). This will encourage the animals to be engaged and active and to display natural behaviors, while minimizing unwanted ones (figure 24.1). For example, Mellen, Hayes, and Shepherdson (1998, 191) determined that complex exhibits reduced stereotypic behaviors, such as pacing in small felids. Exhibit "furniture," however, must be safe for the animals to use and not difficult for keepers to maintain.

Exhibits should provide as much space as possible for the animals (Hancocks 2010, 135), but unfortunately this simple rule is often overlooked. Making exhibits as large as is possible or practical enables increased environmental complexity to be designed into the habitat, which in turn encourages the expression of a wider range of behaviors and an improvement in animal welfare (Veasey, Waran, and Young 1996, 143–51). Exhibits should be big enough so that the animals do not feel threatened by visitors or by keepers and other staff who may need to be in the exhibits at the same time as the animals. The design should also consider how many individual animals will be in the exhibit, so that it doesn't appear empty (so that it is difficult for visitors to locate the animals) or, worse, overcrowded. As much as possible, design features in the exhibit (e.g., artificial rockwork, water features, moats) should not take away usable animal space.

Exhibits should also provide animals with refugia away from other animals in the exhibit, and/or give them the sense that they are safe from visitors whenever they want. These refugia can be in the form of rock outcroppings, dens, heavily planted areas, or elevated pathways or perches (in the case of arboreal or semiarboreal species). Providing sufficient refugia can minimize agonism in zoo animals. Offering African wild dogs (*Lycaon pictus*) multiple denning sites, for example, helped mitigate female aggression in the Bronx Zoo's pack

Figure 24.1. The Bronx Zoo's Himalayan Highlands exhibit. Well-designed exhibits encourage the expression of a variety of species-typical behaviors, enabling visitors to see wonderful animals doing amazing things. Photography by Julie Larsen Maher, Wildlife Conservation Society. Reprinted with permission.

during the birthing season (Thomas et al. 2006, 475). Refugia can also encourage natural behavior. The chimpanzee exhibit at the Chester Zoo uses plantings to create visual barriers that facilitate the expression of the chimpanzee's "fission-fusion" society, wherein members of a troop spend some time together as one unit and some time apart in temporary subgroups (Wehnelt, Bird, and Lenihan 2006, 313). Natural plantings can not only serve as forage and enrichment but also may provide some medicinal benefit. Apenhuel Zoo Primate Park and Howletts Wild Animal Park use herb gardens and medicinal plants in primate enclosures to enable the animals to "self-medicate" for certain ailments (Cousins 2006, 341–50).

Exhibits must also be safe for their inhabitants. In the creation of a new exhibit care must be taken to ensure that any plants, shrubs, or trees added to it are not toxic to the animals. Careful consideration should also be given as to whether to include vegetation with long and sharp spines that could injure animals, keepers, or other staff who must work in the exhibit. Keepers should regularly inspect outdoor exhibits to make sure that poisonous plants do not establish themselves, and that vegetation does not become overgrown and provide an escape route for species that can climb.

Outdoor exhibits must also have features that provide animals with potable water and shade throughout the day. If water features (e.g., ponds or streams) are part of the design, care must be taken to design them so that their depth, slope, and texture do not pose a hazard to the animals or staff. Terracing the bottom and sides of a water feature may

help minimize "slip or fall" accidents. Outdoor exhibits in northern climate zoos should also be oriented so that they are protected against prevailing winds and receive sunlight during as much of the day as possible (e.g., see Bogsch 1990, 149), which is especially important during winter months. The inclusion of exhibit furniture such as rocky overhangs, caves, or "hot rocks" (artificial rocks with electric heating coils inside them) can help shelter animals during rain or snow, while thickly planted stands of vegetation (e.g., bamboo) can serve as effective wind blocks.

A good exhibit design will allow visitors to see active, engaged animals because it will stimulate the animals' physical activity and enhance their psychological well-being by providing them with greater choice and control. All new exhibits should incorporate animal enrichment into their design. Enrichment can either be designed into exhibits, and serve as nonvarying components of the habitat (e.g., pools, elevated resting sites) to be used by the animals, or can take the form of items placed in exhibits every day to stimulate activity and the expression of species-typical behavior patterns (Cipreste, Schetini de Azevedo, and Young 2010, 175–78). It is important to strategically site attachment points for at least some of the enrichment items so that visitors can see the animals interacting with them. Animals should receive as much variety as possible to allow them to choose items they want to interact with.

Exhibits can also be designed to be versatile enough to allow different species to be rotated through the exhibits at different times (Polakowski 2001, 445; Coe and Dykstra 2010, 204–5). The Denver Zoo's Predator Ridge, for example, has two exhibits in which African lions (*Panthera leo*), spotted hyenas (*Crocuta crocuta*), or African wild dogs can be exhibited. This design enables the three species to be rotated through the exhibits and allows each species an opportunity to investigate the scents left behind the previous day by one of the other species. It also creates the opportunity for unique behaviors like overmarking, whereby one individual marks over the scent marks of a competitor (Ferkin and Pierce 2007, 107). The design also provides the staff with some flexibility if one of the species cannot be exhibited for a management reason.

MIXED-SPECIES EXHIBITS

There are several advantages to designing exhibits to hold a variety of different species simultaneously. For one, they can provide an efficient use of resources (both space and manpower) and enable a zoo to exhibit a greater number of species than if all exhibits contain only a single species. Mixed-species exhibits can also increase the behavioral repertoire of the species in the habitat and enable visitors to observe not only intraspecific interactions but interspecific encounters as well. These interactions can provide enhanced enrichment and welfare to the animals (Veasey and Hammer 2010, 151). Different species can also use different parts of the exhibit (e.g., including both arboreal and terrestrial species in a habitat), making for a more dynamic visitor experience.

There are some risks associated with maintaining more than one species in an exhibit. Closely related species may occasionally hybridize (e.g., see Tenaza 1985), and one species may be able to transmit infectious diseases or parasites to other species (McAloose 2004; Deleu, Veenhuizen, and Nelissen 2003, 10–11). There is also a risk of interspecific aggression (e.g., see Moreno 1990; McAloose 2004), but the reality is that aggression is almost always going to be greater among conspecifics because the competition for resources (e.g., access to potential mates) is likely to be greater between individuals of the same species. According to Popp (1984, 217) interspecific aggression increases when species are taxonomically more distantly related to each other, possibly because these species have more difficulty interpreting each others' warning or threat displays and therefore do not behave in a manner that might mitigate aggression. Two or more incompatible species (e.g., predator and prey) can be made to appear as if they are sharing an exhibit by the use of carefully concealed barriers to keep them apart (figure 24.2).

Two general rules of thumb when selecting species for mixed-species assemblages are that the species should come from the same general geographical region (which is more for the visitor's educational benefit than an animal requirement), and that one species should not have a significant negative impact on the well-being of another (Thomas 1992, 70). Additionally, certain species (e.g., topi and hartebeest) should not be included in mixed-species exhibits unless the exhibits are very spacious, because of the inherent agonism they direct towards other species (Thomas and Maruska 1996, 205).

Figure 24.2. The Bronx Zoo's African Plains exhibit. The exhibit, opened in 1941, was the first predator/prey exhibit in North America. The lion cub and her pride are separated from the nyala herd by a carefully concealed dry moat. Photography by Julie Larsen Maher, Wildlife Conservation Society. Reprinted with permission.

CONTAINMENT

A number of different structures can be used to contain animals. The choice of what type of barrier to use for an exhibit will largely depend upon the species to be displayed in that habitat. One of the most commonly used animal containment features is fencing. It comes in a wide array of styles and materials (e.g., chain link, park and paddock, woven mesh, welded metal) and is inexpensive compared to other barriers. Fences used along the back and sides of an exhibit can blend well with surrounding vegetation, making it very difficult for visitors to discern the exhibit's boundaries. This is especially true if the fencing material is painted black. If it is used at the front of an exhibit, it may detract from the overall aesthetics of the habitat unless a clear viewing window is incorporated into the design. Care must be taken when using fences with species that can climb, and the use of overhangs, clear acrylic sheets along the top of the fence, or electric fence may keep species from climbing out of an exhibit. While all forms of barriers are vulnerable to vandalism or storm damage, fencing may be more susceptible than most because it is less sturdy than many of the other forms of containment.

Another common means of containing animals is the use of solid walls, which can also be constructed from a number of materials (e.g., wood, block, cement, gunite). They are normally sturdier than fences with respect to withstanding environmental conditions, but may be more difficult to hide from visitors unless they are covered by rock facades. If solid rock is used, care must be taken to ensure that the rockwork does not provide escape routes for species that can climb. Thick stands of vegetation can sometimes be used to hide or soften the effects of a wall barrier.

Dry moats (sometimes called "ha-has") are another frequently used form of containment. If designed properly, they can provide an unobstructed view of the animals and the exhibit, and they require relatively little upkeep. They do reduce the amount of usable exhibit space, and therefore are not a good choice when space is limited. Also, animals that are resting or hiding in the moat may be partially or completely out of view for zoo visitors. One way to discourage animal use of these areas is to place large, uneven rocks in the moat, although care must be taken to ensure that there are access and egress routes for both staff and animals.

Water features can also be used to contain animals. They can be incorporated into exhibit design to offer an unobstructed and extremely naturalistic view of the animals while simultaneously serving as a form of enrichment and a potential source of potable water. If possible, the water should be filtered and/or recirculated to minimize waste. Water features are not a good option for species that are adept swimmers, because they may provide escape routes, or for species that are poor swimmers, because they may drown. They also may not be effective in northern climate zoos unless they are equipped with efficient heaters and/or bubblers to keep them ice-free. Iced-over areas at the water's edge are especially dangerous for ungulates.

Tension wire can allow visitors to have intimate views of animals while also being able to hear and smell them. It also can be an extremely effective barrier between two abutting exhibits if the species are not aggressive towards each other.

The spacing between the wires should not allow animals to become entangled in them, or be so wide that animals (especially carnivores) can stick their forelegs through it. Tension wires must be regularly checked to ensure that they are not damaged and proper tension is maintained.

Glass is widely used in animal exhibits. In smaller exhibits, glass panes may make up the entire front of the exhibit; in larger ones they are often incorporated with other types of barriers to provide unobstructed views of the animals and exhibit. Glass allows visitors to safely get extremely close to animals. However, it can be very expensive (especially when laminated); it may be scratched by animals or visitors, and must be regularly cleaned. Glass can break or crack. Condensation can form on it under particular environmental conditions, and light can be reflected off it. This is especially true in indoor exhibits. Most importantly, animals may not recognize glass as a barrier and thus may be injured or killed if they hit it with force. This last factor is especially a concern when animals are new to an exhibit and are not yet fully aware of or comfortable with their surroundings. In these instances, covering the glass or soaping it to make it more visible will enable animals to gradually become familiar with it.

Bars are a sturdy form of containment, but they do not provide unobstructed views of animals or their exhibits. They perpetuate the stereotype of older, sterile zoo exhibits, and zoo guests typically view them negatively. Therefore, their use should be carefully considered.

Electric fences ("hot wire") should never be used as a primary means of animal containment, but they are very effective as a secondary means of containment or to exclude animals from certain areas of the exhibit. Electric fences require regular maintenance and checking, and, depending how they are used, may require backup power systems for use if electrical power goes out. Certain individual animals may be less deterred than others by hot wires, so care must be taken when deciding where and when to use them.

OFF-EXHIBIT HOLDING AREAS

Off-exhibit holding areas are essential for proper animal management (Rosenthal and Xanten 2010, 164). They should be designed to hold animals overnight or for extended periods of time if necessary. While off-exhibit areas may "eat into" usable exhibit space, the trade-off is worth it if the facilities are properly designed. For one, tropical or subtropical species living in northern climate zoos may spend a significant amount of their time in indoor areas during winter months, so providing spacious off-exhibit areas will greatly enhance their quality of life in cold weather. Powell (2010, 55) also noted that these areas allow staff far more control during animal introductions, largely because they tend to be smaller and environmentally less complex than exhibits, and have shift doors that can be used to separate animals when necessary. Indoor spaces should be equipped with ultraviolet (UV) light sources, or skylight panels that allow UV light to pass through them (Hosey, Melfi, and Pankhurst 2009, 188).

Off-exhibit areas also provide keepers with ideal locations in which to conduct husbandry training sessions without visitor interference, and relatively private locations for potentially delicate animal management events (e.g., delivery areas for offspring) or for conducting sensitive procedures (e.g., veterinary examinations). They also provide space for individuals that may need to be separated from conspecifics for social or medical reasons (Hosey, Melfi, and Pankhurst 2009, 199).

Where appropriate, the inclusion of restraint devices in off-exhibit areas should be carefully considered, because they can greatly enhance animal management and limit the need for chemical immobilization. Restraints can counter the size and strength of larger animals and restrict their movement while providing staff with safe access to various body parts (Christman 2010, 40). Restraint devices work most reliably when they are sited so that animals must pass through them on a routine basis without any negative stimuli, and when the animals have become habituated and desensitized to human activity around the restraint while they are in it (e.g., see Calle and Bornmann 1988, 250).

All indoor exhibit and off-exhibit holding facilities require adequate heating, ventilation, and air conditioning (HVAC) to keep the animals' environment at the proper temperature, to minimize odors, and to allow the facility to properly dry in order to retard the buildup of mildew, algae, and bacterial growth (Rosenthal and Xanten 2010, 168). Some ranges of certain variables (e.g., temperature ranges and the number of air exchanges) are mandated by regulatory agencies such as the US Department of Agriculture (USDA) Animal and Plant Health and Inspection Service (APHIS), and the Department for Environment Food and Rural Affairs (DEFRA) for the United States and the United Kingdom, respectively.

DESIGNING FOR THE KEEPER

While the animals are the principal users of exhibits, it is essential that exhibits also be designed with keepers in mind (Simmons 2005); the ease with which exhibits are maintained and managed is important to their overall success and effectiveness. Therefore the layout of the exhibit, the animal holding and keeper areas, and the way in which animals and keepers will move through all these areas during the course of the day should be carefully considered early in the design phase. The location and size(s) of keeper entrance doors and of other access doors that will be used for the addition or removal of exhibit furniture is important, as are the locations of animal shift doors. Vestibules incorporated into the design of keeper entrances can greatly reduce the chance that an animal in the exhibit will escape past a keeper when the keeper door is opened.

Keepers should have reasonable access to all areas of the exhibit for daily perimeter checks, putting out enrichment, and cleaning. The exhibit's daily upkeep should not be too complex to fit into the keeper's normal routine. Less complex upkeep will prevent keepers from rushing their work, and will minimize the chance of accidents or mistakes.

Exhibits should be designed so that they are safe and secure not only for the animals but for the keepers and any other zoo employees who need to work in them. Whether an exhibit is indoors or outdoors, it should be well lit and provide the keeper with good visibility throughout. It should not be difficult for keepers to check on the locations of their animals. Naturalistic exhibits typically provide places for animals to spend some time out of view of keepers and visitors, but these blind spots should be kept to a minimum (or, ideally, eliminated). The use of video cameras, convex "traffic" mirrors, or viewing portals at different locations on a solid barrier are all means by which keepers can monitor the location of their animals. This will help keepers avoid startling animals inadvertently or getting too close to them unexpectedly.

All potential trip, fall, or slip hazards should be considered, along with the equipment (e.g., wheelbarrows, hoses, and tools) a keeper will require to service an exhibit. Exhibits with complex or uneven terrain should have pathways that keepers can negotiate reasonably and safely, and should be equipped with handholds if necessary. In indoor exhibits lacking an organic substrate, consideration should be given to textured or nonskid floors, especially if the exhibits are likely to be wet. In all indoor exhibits, floors should be sufficiently pitched towards the drain(s) to eliminate standing water and improve hygiene.

Exhibits containing dangerous animals should have signs on the keeper doors that let staff know whether (1) animals are in the exhibit or (2) it is safe to enter the exhibit. All doors to exhibits with dangerous animals should also be equipped with double locks for safety (Rosenthal and Xanten 2010, 170), and the locking device design should enable keepers to easily see that the locking mechanism is engaged. Careful consideration should be given to the location of emergency equipment, including telephones, alarms, fire extinguishers, and animal capture equipment.

A well-thought out exhibit will also take into account the location of controls for essential exhibit or holding area components (e.g., light switches, hose bibs, HVAC equipment), the easy accessibility of all areas of the exhibit for routine maintenance (e.g., replacing light bulbs or servicing enrichment items), and the proper labeling of all mechanical equipment. The keeper's section in the off-exhibit area should have wide alleys for safety and ease of working, and suitably-sized storage areas for supplies.

Pests can be a problem in animal exhibits and off-exhibit holding areas. Not only can they consume food meant for collection animals, but they can damage facilities, spread diseases to the animal collection and/or staff, and, in exhibit areas, give zoo visitors negative impressions about the cleanliness of the facility. Control of pests can increase the amount of time staff spends in exhibits. Because it is difficult to keep food and water sources away from pests, eliminating harborage (the locations where pests can safely hide and still get to the resources they require) is one of the most effective ways of controlling them. Identifying potential pest havens as much as possible during design and construction can help reduce the chance that pests will become established in an exhibit. For example, filling in the hollow areas of structures such as artificial rocks and trees, or carefully sealing seams that prevent pests from gaining access to air spaces between walls, will serve to limit infestation.

DESIGNING FOR THE VISITOR

Zoo exhibits, whether intentionally or not, imbue visitors with distinct feelings, beliefs, and views about how well the zoo cares for its animals. Exhibits also influence visitors' perceptions about the zoos themselves. Good exhibit designs, therefore, can positively shape people's attitudes and behavior about wildlife and zoos, and can inspire visitors to care about wild animals and help conserve them. In at least one study (Hosey, Melfi, and Pankhurst 2009, 186) visitors believed that highly naturalistic exhibits were more appealing and better for the animals' welfare than more sterile environments.

Most zoo visitors will never have the opportunity to see the wildlife they view in zoos in the animals' native environment. Properly designed zoo exhibits replicate the essence of nature and place animals in appropriate habitats that encourage species-typical behavior patterns. In effect, high-quality exhibits provide windows in which visitors can see wild animals doing real and wonderfully interesting things in an environment that simulates their native habitat (Seidensticker and Doherty 1996, 180–88). In doing so, well-done exhibits educate visitors not only about the biology of a species, but also about its behavior and ecology, because it enables them to see animals interacting with the biotic and abiotic components of the exhibit.

Local species of plants and trees can be used to simulate the vegetation found in the animal's native range and give visitors the sense that they are looking at an environment different from the one found at the zoo's location (Moore and Peterkin 2010, 196–97; Jackson 1996, 180–81), while simultaneously providing enrichment, refugia, and/or shade for the animals. If local trees or shrubs do not resemble the vegetation found in the animal's habitat, they may be artfully pruned to give visitors the sense they are looking at nonnative flora. Various artifacts (e.g., termite mounds, rock formations, lianas) can also be created to further enhance the illusion of being in an environment where the animal is found in nature.

Containment features can be incorporated into the exhibit design, further adding to the design elements that enable visitors to feel as though they are looking at part of the species' environment. They must be designed to safely keep animals in the exhibit, ideally without detracting from its aesthetics. Whenever possible, off-exhibit animal facilities and support systems (e.g., filtration systems for aquatic features) should be situated behind containment features so they are not in the visitors' sightline or within hearing distance, or otherwise detract from the exhibit design.

Visitor viewing areas should be situated in discrete areas so that there are overlapping lines of sight which encourage a sense of intimacy with the exhibit and prevent the feeling of viewing animals while part of a large crowd (Hancocks 2010, 134), but which are not so small that people feel they must compete with other visitors to get a view. This design will also minimize the chances of visitors viewing other visitors. As much as possible, the viewing areas should include the same elements as are used in the exhibit (e.g., the same species of trees and shrubs, rock formations, vines) to foster a sense that the visitors are sharing the same space as the animals. These visitor areas should also be situated slightly below the grade of the exhibit so that the animals are at or above the visitor's eye level, which will make them appear more impressive. Coe (1985, 202–4) believes that visitors who look down at zoo animals as a result of the exhibit design are more likely to subconsciously perceive them as subordinate beings. Finally, the viewing areas should be designed to pre-vent unwanted interactions or direct contact between visitors and the animals.

CONCLUSION

The best-designed exhibits successfully meet the needs of the animals in the exhibits, the keepers and staff who have to maintain them, and the visitors who view them. They provide stimulating habitat for the animals and are created with their best interests in mind. They also are safe for the animals, and for the staff to work in and service within a reasonable time. These exhibits convey interesting stories to visitors and give them the sense that they are good for the animals. They also inspire visitors to care about conserving wildlife while positively influencing visitors' attitudes about zoos.

REFERENCES

Bauert, M. R., S. C. Furer, R. Zingg, and H. W. Steinmetz. 2007. "Three years of experience running the Masoala Rainforest ecosystem at the Zurich Zoo, Switzerland." *International Zoo Yearbook* 41: 203–16.

Bogsch, Ilma. 1990. "A new house for apes at the Budapest Zoo." *International Zoo Yearbook* 29: 148–53.

Brady, Barbara, Jack Huelsman, and Edward Maruska. 1990. "Cats in contrast: Cincinnati Zoo cat exhibit." *International Zoo Yearbook* 29:169–74.

Bridges, William. 1974. *Gathering of Animals: An Unconventional History of the New York Zoological Society*. New York: Harper and Row.

Calle, Paul P., and John C. Bornmann. 1988. "Giraffe restraint, habituation, and desensitization at the Cheyenne Mountain Zoo." *Zoo Biology* 7:243–52.

Christman, Joe. 2010. "Physical methods of capture, handling, and restraint of mammals." In *Wild Mammals in Captivity: Principles and Techniques for Zoo Management, 2nd Edition*, edited by Devra G. Kleiman, Katerina V. Thompson and Charlotte K. Baer, 171–80. Chicago: University of Chicago Press.

Cipreste, Cynthia F., Cristiano Schetini de Azevedo, and Robert J. Young. 2010. "How to develop a zoo-based environmental enrichment program: Incorporating environmental enrichment into exhibits." In *Wild Mammals in Captivity: Principles and Techniques for Zoo Management, 2nd Edition*, edited by Devra G. Kleiman, Katerina V. Thompson, and Charlotte K. Baer, 171–80. Chicago: University of Chicago Press.

Coe, Jon C. 1999. "An integrated approach to design: How zoo staff can get the best results from new facilities." Paper presented at the First Annual Rhino Keeper Workshop. Orlando, Florida, May 7–8.

Coe, Jon. 1985. "Design and perception: Making the zoo experience real." *Zoo Biology* 4:197–208.

Coe, Jon, and Greg Dykstra. 2010. "New and sustainable directions in zoo exhibit design." In *Wild Mammals in Captivity: Principles and Techniques for Zoo Management, 2nd Edition*, edited by D. G. Kleiman, K. V. Thompson, and C. K. Baer. 202–15. Chicago: University of Chicago Press.

Cousins, D. 2006. "Review of the use of herb gardens and medicinal plants in primate exhibits in zoos." *International Zoo Yearbook* 40:341–50.

Conway, William G. 1968. "How to exhibit a bullfrog: A bed-time story for zoo men." *Curator* 4:310–18.

Deleu, R., R. Veenhuizen, and M. Nelissen. 2003. "Evaluation of the mixed-species exhibit of African elephants and hamadryas baboons in Safari Beekse Bergen, the Netherlands." *Primate Report* 65:5–19.

Ehmke, Lee. 1999. "Congo Gorilla Forest: It's all in the details." Paper presented at the annual meeting of the Association of Zoos and Aquariums, Minneapolis, Minnesota, September 24–28.

Ferkin, Michael H., and Andrew A. Pierce. 2007. "Perspectives on over-marking: Is it good to be on top?" *Journal of Ethology* 25:107–16.

Hancocks, David. 2010. The history and principles of zoo exhibition. In *Wild Mammals in Captivity: Principles and Techniques for Zoo Management, 2nd Edition*, edited by Devra G. Kleiman, Katerina V. Thompson, and Charlotte K. Baer, 121–36. Chicago: University of Chicago Press.

Hediger, H. 1955. *Studies of the Psychology and Behaviour of Captive Animals in Zoos and Circuses.* London: Butterworths.

Hosey, Geoff, Vicki Melfi, and Sheila Pankhurst. 2009. *Zoo Animals: Behaviour, Management, and Welfare.* New York: Oxford University Press.

Hutchins, M., and B. Smith. 2003. "Characteristics of a world-class zoo or aquarium in the 21st century." *International Zoo Yearbook* 38:130–41.

Jackson, Donald W. 1996. "Horticultural philosophies in zoo exhibit design." In *Wild Mammals in Captivity: Principles and Techniques*, edited by Devra G. Kleiman, Mary E. Allen, Katerina V. Thompson, and Susan Lumpkin, 175–79. Chicago: University of Chicago Press.

McAloose, D. 2004. "Health issues in naturalistic mixed species environments: A day in the life of a zoo pathologist." Paper presented at the 55th annual meeting of the American College of Veterinary Pathologists and 39th Meeting of the American Society for Veterinary Clinical Pathologists. Middleton, Wisconsin, November 13–17.

Mellen, Jill D., Marc P. Hayes, and David J. Shepherdson. 1998. "Captive environments for small felids." In *Second Nature: Environmental Enrichment for Captive Animals*, edited by David J. Shepherdson, Jill D. Mellen, and Michael Hutchins, 184–201. Washington: Smithsonian Institution Press.

Moore, Merle M., and Don Peterkin. 2010. "Zoological horticulture." In *Wild Mammals in Captivity: Principles and Techniques for Zoo Management, 2nd Edition*, edited by Devra G. Kleiman, Katerina V. Thompson, and Charlotte K. Baer, 192–201. Chicago: University of Chicago Press.

Moreno, Abelardo. 1990. "The African veld exhibit at the Havana National Zoological Park." *International Zoo Yearbook* 29:206–11.

Perlman, Jaine E., Victoria Horner, Mollie A. Bloomsmith, Susan P. Lambeth, and Stephen J. Schapiro. 2010. "Positive reinforcement training, social learning, and chimpanzee welfare." In *The Mind of the Chimpanzee: Ecological and Experimental Perspectives*, edited by Elizabeth V. Lonsdorf, Stephen R. Ross, and Tetsuro Matsuzawa. Chicago: University of Chicago Press.

Polakowski, Kenneth. 2001. "Zoo design." In *Encyclopedia of the World's Zoos*, edited by Catharine E. Bell, 441–47. Chicago: Fitzroy Dearborn Publishers.

Polakowski, Kenneth J. 1987. *Zoo Design: The Reality of Wild Illusions.* Ann Arbor: University of Michigan School of Natural Resources.

Popp, James W. 1984. "Interspecific aggression in mixed ungulate species exhibits." *Zoo Biology* 3:211–19.

Powell, David M. 2010. "A framework for introduction and socialization processes for mammals." In *Wild Mammals in Captivity: Principles and Techniques for Zoo Management, 2nd Edition*, edited by Devra G. Kleiman, Katerina V. Thompson, and Charlotte K. Baer, 49–61. Chicago: University of Chicago Press.

Reichenbach, Herman. 1996. "A tale of two zoos: The Hamburg Zoological Garden and Carl Hagenbeck's Tierpark." In *New Worlds, New Animals: From Menagerie to Zoological Park in the Nineteenth Century*, edited by Robert J. Hoage and William A. Deiss, 51–62. Baltimore: Johns Hopkins University Press.

Rodriguez-Herrejon, Francisco. 2001. "Nocturnal animal exhibits." In *Encyclopedia of the World's Zoos*, edited by Catharine E. Bell, 872–74. Chicago: Fitzroy Dearborn Publishers.

Rosenthal, Mark, and William A. Xanten. 2010. "Structural and keeper considerations in exhibit design." In *Wild Mammals in Captivity: Principles and Techniques for Zoo Management, 2nd Edition*, edited by Devra G. Kleiman, Katerina V. Thompson, and Charlotte K. Baer, 162–70. Chicago: University of Chicago Press.

Scherpner, Christoph. 1982. "The Grzimek house for small mammals." *International Zoo Yearbook* 22:276–87.

Seidensticker, John., and James G. Doherty. 1996. "Integrating animal behavior and exhibit design." In *Wild Mammals in Captivity: Principles and Techniques*, edited by Devra G. Kleiman, Mary E. Allen, Katerina V. Thompson, and Susan Lumpkin, 180–90. Chicago: University of Chicago Press.

Shepherdson, David J. 2003. "Environmental enrichment: Past, present and future." *International Zoo Yearbook* 38:118–24.

Simmons, Lee. 2005. "Zoo and aquarium design: Playing 'the what if game.'" Paper presented at the 6th International Symposium of Zoo Design, Paignton, UK.

Tenaza, Richard. 1985. "Songs of hybrid gibbons (*Hylobates lar* × *H. muelleri*)." *American Journal of Primatology* 8:249–53.

Thomas, Patrick R., David M. Powell, Glen Fergason, Brenda Kramer, Keri Nugent, Catherine Vitale, Anne Marie Stehn, and Tina Wey. 2006. "The birth and simultaneous rearing of two litters in a pack of captive African wild dogs (*Lycaon pictus*)." *Zoo Biology* 25:461–77.

Thomas, Patrick, and Susan Chin. 2004. "The Bronx Zoo's Tiger Mountain: Enrichment at the forefront of exhibitry." Paper presented at the Association of Zoos and Aquariums Annual Meeting. New Orleans, Louisiana, September 18–23.

Thomas, Patrick R., and James G. Doherty. 1992. "Considerations for exhibiting large ungulates in mixed species habitats at the New York Zoological Park." Paper presented at the regional meeting of the American Association of Zoological Parks and Aquariums, Baltimore, Maryland, March 15–17.

Thomas, Warren D., and Edward J. Maruska. 1996. "Mixed-species exhibits with mammals." In *Wild Mammals in Captivity: Principles and Techniques*, edited by Devra G. Kleiman, Mary E. Allen, Katerina V. Thompson, and Susan Lumpkin, 204–11. Chicago: University of Chicago Press.

Veasey, Jake, and Gabriele Hammer. 2010. "Managing captive mammals in mixed-species communities." In *Wild Mammals in Captivity: Principles and Techniques for Zoo Management, 2nd Edition*, edited by Devra G. Kleiman, Katerina V. Thompson, and Charlotte K. Baer, 151–61. Chicago: University of Chicago Press.

Veasey, J. S., N. K. Waran, and R. J. Young. 1996. "On comparing the behavior of zoo housed animals with wild conspecifics as a welfare indicator, using the giraffe (*Giraffa camelopardalis*) as a model." *Animal Welfare* 5:139–53.

Wehnelt, S., S. Bird, and A. Lenihan. 2006. "Chimpanzee Forest exhibit at Chester Zoo." *International Zoo Yearbook* 40:313–22.

Wilson, Megan, Angela Kelling, Laura Poline, Mollie Bloomsmith, and Terry Maple. 2003. "Post-occupancy evaluation of Zoo Atlanta's Giant Panda Conservation Center: Staff and visitor reactions." *Zoo Biology* 22:365–82.

25

Zoo Horticulture

Jay H. Ross

INTRODUCTION

In zoos and aquariums, the objective is to provide the best possible environment for the animals displayed. Creating and maintaining such a habitat will involve an understanding of the role plants play.

After studying this chapter, readers will

- understand the benefits of horticulture at a zoo or aquarium
- understand the aspects involving horticulture from exhibit design to regular day-to-day plant maintenance
- have guidance on how to operate in a manner that benefits both the animals and the plants associated with the animal's exhibit.

Many zoos and aquariums try to focus on providing a good representation or interpretation of an animal's native habitat, both for the education of their visitors and for the well-being of the animals. These institutions are also an excellent place for the conservation of plants that humans and other animals rely on. The botanical aspect of an exhibit provides, in the words of the late American journalist Paul Harvey, "the rest of the story." The creation and maintenance of this botanical aspect is the work of the zoo's horticulturists if the zoo is fortunate enough to have such professionals on staff. Otherwise, these responsibilities may fall on the keepers. Understanding the elements associated with these endeavors will make a keeper more effective in managing the animal's environment. Zoo horticulture also includes the development and maintenance of the habitats and environments in the zoo's public areas. Plants are a very noticeable and extremely important asset in a zoo or aquarium. Keepers can use their knowledge of plant life to help interpret the conservation and biology of the animals in their care, and to expand on the dynamic of animals and plants' codependency in the natural world.

Zoo horticulture will include exhibit and landscape design and maintenance, management of a plant collection, and the supplying of browse for animal enrichment programs. It may also include lawn maintenance, forest management (arboriculture), and the propagation of new plants for use in or around an animal exhibit. A better appreciation of the role that horticulture plays in zoo and aquarium operations will aid a keeper in providing the highest quality of animal care.

EXHIBIT DESIGN FROM A HORTICULTURAL PERSPECTIVE

From the very outset of design and development of a new zoo exhibit or renovation of an existing exhibit, either the zoo's horticulture department or an individual with good horticultural knowledge should be involved. Even when developing the concept and message that the exhibit intends to send to its audience, horticulturists will be able to give valuable input. They can help determine, from a biogeographical standpoint, which plant species can be used that will represent the desired habitat or region of the world. A given plant species for a zoo animal exhibit may also be selected for the following reasons:

- habitat simulation
- animal viewing enhancement
- animal behaviors and safety
- plant maintenance needs
- potential plant hazards.

When developing the message desired for an exhibit representing a geographic region of the world, a horticulturist will evaluate and advise on which plants can contribute most effectively. Frequently, an animal's natural habitat can be quite different from the habitat of the zoo's geographic location. This makes it necessary to select plant species that can best simulate the natural habitat but can also survive the climatic conditions of the zoo or aquarium's location. The plants selected should emulate growth habit and other plant characteristics (e.g., leaf size and shape, bark type) that are found in the area to be depicted in the exhibit. Does the plant retain its foliage all year round (evergreen) or drop it at some point in the year (deciduous)? Does it have soft fleshy tissue that dies back at the end of the growing season (herbaceous) or a persistent woody tissue that supports new growth from year to year (woody plant)? An exhibit can gain authenticity when indigenous plants (plants from the depicted geographic area) are available and will survive climatic conditions at the location of the zoo or aquarium. A keeper must be cognizant of introducing a nonnative plant species that could be dispersed outside the zoo or aquarium and could create a problem or harm the native ecosystem. This dispersion can occur due to movement of seed by wind or animals (e.g., birds) or an aggressive growth habit. Invasive plant species, a major concern today, are exotic species that, when introduced to an area, disperse and become a problem. These invasive species can dominate and sometimes cause the elimination of native plant species, and this can in turn affect native animal species. Kudzu (*Pueraria lobata*), a woody vine introduced into the United States from Asia in the late 1800s to help with erosion control and as a possible forage crop, has become a major problem in the southeastern United States. Its aggressive growth can smother surrounding vegetation, including trees. Treatment to eradicate kudzu vine can be expensive and labor-intensive, especially if it has become well established over a large area. Herbicides are available, but they must be used judiciously and with an understanding of what they can affect other than the targeted plant. Small areas can be controlled by the cutting back, digging up, and removal of rootstock.

The plant selection process also includes the areas adjacent to an exhibit that serve to set its environment or atmosphere and/or create a transition from other exhibit areas of the zoo. This concept, termed landscape immersion, has become very popular. It is best described as immersing the zoo visitor in a total environment. This may include the seasonal placement of true indigenous plants that can be interpretive tools for the exhibit, either in containers or as in-bed plantings. These seasonal plants can be set out during the warmer months and then removed for the winter and kept in a greenhouse, if one is available. For example, this practice is appropriate in parts of the world where palms can't grow outdoors all year. Sometimes a large portion of a zoo's botanical collection may fall into this category. Moving and storing plants in a greenhouse is quite an undertaking, especially if the plant specimens are large. Often these larger specimens are planted in containers to allow for easier moving and to avoid the shock a plant can experience from being dug up and containerized before being moved to the greenhouse. If a zoo or aquarium is large enough to have these types of facilities, it more than likely also has horticulture staff to manage the greenhouse. A keeper should be aware that some plants in an exhibit may need periodic relocation, and that he or she may be involved in the procedure. These seasonal plantings can provide a great opportunity to introduce specific animal/plant dynamics to zoo visitors. For instance, if a plant in the exhibit is a vital part of an animal's diet in its native environment (i.e., by providing fruit or foliage), that information can be conveyed in signage or by keepers or horticulturists when they interact with zoo visitors. Keepers can play a vital role in presenting this aspect of the plants in an exhibit when the opportunity arises.

Plants in and around an exhibit can be positioned to create viewing opportunities, screen off undesirable views, and encourage preferred traffic flow. A zoo exhibit is like a stage production in which the animals are the actors and the plants and other exhibit elements serve as the staging. The zoo visitors expect to see the animals. Using plants to frame their views of the animals is a common zoo practice that also takes animal behaviors into account. For example, suppose that an exhibit for a bear species is being designed with two primary viewing areas. The exhibit has a feeding area, a water feature, and some open ground for the bears to relax in the sun. It is surrounded by a deep concrete dry moat to prevent the bears' escape, but this detracts from the natural habitat effect desired. To enhance the visitors' experience of discovery of the exhibit, the viewing areas can be flanked with dense plantings which dictate that a visitor views the exhibit from the designed viewing area. The plantings should be of a height that does not obstruct the view even for children, but that pushes the viewer's sight line up and over the moat so that the moat seems less present. The resting area can be planted with grasses, sedges, and wildflowers that can withstand the bears' treatment and give the exhibit the appearance of a meadow. Periodic pruning of plant materials may be necessary to maintain views and prevent obstructions. Inclusion of plants in these ways can benefit the animals while offering visitors the best experience. Shade trees can be added to cool the exhibit viewing area or the animal. Banks of vegetation can also be used to obstruct the public's view of service buildings or barns. Such an exhibit is an attractive location for the animal to present itself and be in good view of the visitors.

The size a plant can attain at maturity should be considered when determining its placement. This is especially important when considering an animal's potential escape routes from the exhibit. Plants in the exhibit that could grow over the containment barrier must be considered during the exhibit's design, as must plants that may grow into the exhibit from outside. The selection of plants for an exhibit should include consideration of their growth rate. This will include both the plants' expected size at maturity and how long they will take to reach maturity. A fast-growing tree, such as a poplar or silver maple, will have weaker wood and thus be more prone to breaking due to wind or ice and snow loads. Such species should be avoided in favor of a stronger species that grows more slowly. The plant's mature size should direct where it is placed in an exhibit. Regular monitoring of its growth by both horticulture and animal care staff will dictate whether it must be pruned to prevent an animal's escape. It is important to remember that plants (particularly fruits or flowers) on the outside of the exhibit can be attractive to the animals and can provide motivation to escape. Pruning time can vary from one plant or tree to another. Generally, the pruning can be carried out before growth resumes in the spring; at this time one can terminate the plant's growth in one direction and redirect it in another. This is done by pruning just above the growing point, angled in the desired direction of growth.

For a better understanding of correct pruning methods and scheduling, one should consult the zoo horticulturist or a local arborist. If a plant becomes a persistent challenge, sometimes its removal is the best option. However, good planning and plant selection can prevent this from becoming necessary.

Animal behaviors are very important to consider when determining the placement of plants in an exhibit, to avoid injury to either the animal or the plant. A good example would be the placement of plants in a roosting bird's exhibit. The plants should not be positioned directly under the bird's favored perches, because the high nutrient and salt content of the bird's feces will have a detrimental burning effect on the plant roots. Also, regular cleaning of the plant's foliage can result in waterlogged soil and the suffocation of roots, causing plant decline or death over time. Establishing perching locations for the bird away from sensitive plants will help minimize these negative effects and reduce the daily maintenance required for both the plants and the bird.

Plant placement can also help elicit animal behavior, such as breeding, while still providing effective botanical interest. Some animals breed better in the secure situations that plant cover can provide. Plant placement can create a retreat or secure area away from the intrusion of constant viewing by visitors. However, it has to be managed or created so that the secure area is not a hiding place where the animals stay out of view a great deal of the time. A balance between visitor experience and animal comfort must be reached. Knowing the typical behaviors of a specific animal species can be very beneficial in designing or creating an exhibit, as well as in maintaining its performance over time. Some captive animals just cannot be displayed with live plant material. A good example would be parrots or other psittacines, which have a tendency to defoliate live plants and chew on branches to the point of killing them. Parrots would be an excellent group of animals for which to consider the use of plants outside the exhibit in containers or planters. The plants can still help build the feeling of habitat without being directly subjected to the pressures of destructive animal behavior.

Exhibit accessibility is important, both for daily animal maintenance and for occasional plant or utility maintenance. This includes access for the maintenance staff to repair a fence or dig up a utility line, and for a keeper to get around the exhibit for routine tasks of feeding and cleaning. Keepers will also want access for moving the animals in and out of the exhibit and performing routine husbandry duties. Horticulture personnel will need reasonable access for pruning, pest or disease control, and general maintenance (watering, weeding, mowing, etc.). Trees or other plant material should be placed where they don't obstruct anyone's access or hinder the use of any equipment necessary for moving animals or otherwise servicing the exhibit. Plantings should be placed where they will not interfere with access to utilities (water, gas, electric, and sewer) for repairs. If this is impossible, plants should be chosen that can easily be moved and replaced. The accessibility and availability needs of all staff should be addressed during the design process to prevent problems or difficulties. Coordination is extremely important for achieving the best-managed exhibit. For example, the keeper should ensure that animals have been moved off the exhibit before mowing the grass.

Finally, the process of selecting plants needs to include consideration of any attributes a plant may possess that could be detrimental to the animals and/or visitors. These may be toxic chemicals or physical features, such as thorns, that could cause injury. Plants have evolved many of these features over time to protect themselves from predators. Chemical poisons or toxins that plants possess are a means of protection developed to dissuade consumption. These toxins can sometimes cause skin reactions. Such plants should be avoided or placed where the visitor or animal will have no direct interaction with them.

Phytotoxicity refers to the poisonous reaction that can result from an encounter with a plant. Knowledge of the toxicity of plants being considered for use in an animal exhibit is very important. A captive exotic animal collection is a very valuable asset, and its well-being cannot be left to chance. Toxicity can occur at different stages in a plant's life or in different parts of the plant, and can affect animals in a way totally different from how it affects humans. A coordinated process for screening plants for phytotoxic properties should be part of all zoos' management practices. This screening should include personnel from the zoo's veterinary, horticulture, and animal care staff. Each of these people brings vital information and perspective to the selection process. Potentially toxic fruit can be an additional concern in the zoo if it is planted where visitors can gain access to it. There is always the threat of someone, especially a child, "tasting" an inappropriate fruit. But a fruit picked on zoo grounds and thrown into an animal enclosure is an added threat. "Better safe than sorry" is a good rule of thumb if information validating a plant's safety cannot be obtained. Each zoo or aquarium should establish its own lists of allowed and prohibited plants. Phytotoxicity has been an issue in zoos for years, and it is better understood and documented every day. Some good references on the subject include *A Manual of Poisonous Plants* by L. H. Pammel, *Common Poisonous Plants and Mushrooms of North America* by Turner and Szczawinski, and the Cornell University Animal Science web site, http://www.ansci.cornell.edu/plants/. In Murray E. Fowler's *Zoo and Animal Medicine*, chapter 14, "Plant Poisoning in Zoos in North America," gives a very good overview on this issue, explaining the different elements of a plant that can affect the degree of its toxicity.

Animals sometimes use the physical features of plants (e.g., thorns) to provide protection from predators. Predators won't be an issue in a captive animal exhibit, but selecting plants that provide such protection is a good interpretive opportunity. Use and placement of such plants must be considered carefully, as they could present a hazard to keepers.

After establishing a list of plants for the exhibit, horticulturists, animal care staff, and other members of the design team need to consider whether the plants need to be protected. Feeding, rubbing, and trampling are the most common animal behaviors that threaten plants in an exhibit. These behaviors can be related to the amount of space allowed per animal on exhibit, the animal species itself, and sometimes the individual animal. Such organizations as the Association of Zoos and Aquariums (AZA) provide recommendations of the space a particular species will need. These recommendations can often challenge an institution's desire to give visitors as much exposure to the animals as possible. And a captive animal exhibit will almost always provide only a fraction of the area that would be available to that animal in the wild. The concept of carrying capacity for animal/plant relationships must always be kept in mind. Carrying capacity is the maximum population (as of an animal species) that an area can support without experiencing damage or deterioration. Good management practices—such as rotating animals on and off exhibit areas, keeping them off an exhibit when adverse conditions such as heavy rains occur, and routine replanting to keep vegetation present—are all beneficial. However, it is almost inevitable, despite the best design and understanding of animal behavior, that plants will need physical protection.

Zoos and aquariums have come a long way in softening their approaches to plant protection. Many times, ideally, these barriers are incorporated into exhibitry so that they are undetectable to the viewer, and that is the best type of protection. The use of electric wire (hot wire) or fencing to create plant protection has been developed to the point where it can be done in a much less conspicuous manner. The development of grass and vinelike electrical barriers to protect trees, planting beds, and other types of vegetation has greatly advanced plant protection (figure 25.1). There are some instances when more substantial barriers are the only option. These can sometimes be temporary situations, while plants become established or mature to a point where they can withstand the pressure animals may place on them. The first thing one must consider is the type of damage one is trying to prevent. Is it the eating of foliage off a plant or group of plants? Is it the chewing or rubbing of a trunk? Is it the compaction of the ground around a plant from animal traffic? These can all be looked at similarly, in that the common objective is prevention of access to the plant(s) or area around the plant(s).

One must consider what animal(s) the protection is designed for. A rhinoceros or other animal of great size is more difficult to keep back from an area than a deer. A giraffe and its long reach pose unique challenges in protecting plants. Regardless of the animal, knowledge of its behavior, both strengths and weaknesses, is very helpful. A giraffe is not

Figure 25.2. Tree protected with wood wrap. Photo by Allyson Whalley, Virginia Zoo.

confident in negotiating uneven ground, and it has difficulties with depth perception. With this knowledge, a barrier can be created using materials, such as uneven stone, that it won't try to cross. A change in elevation can also make a giraffe uncomfortable and serve as a barrier.

Trees are important elements to protect in an exhibit because of their overall presence and the time vested in their establishment and growth. Simply put, they are not easily or cheaply replaced if lost. Electric vines have been mentioned, and they can be very effective for protecting trees from smaller, meeker animals. When a protective barrier around a tree's trunk is needed, there are numerous methods to employ. These can include simply wrapping the tree with wire fencing, using snow fencing material, or wrapping tree bark slabs from a sawmill around the tree (figures 25.2 and 25.3). Sometimes a tree wrap may also utilize electric wire to more strongly persuade an animal not to bother or damage a tree. All these applications will keep an animal from having direct access to a tree trunk. None of these types of protection should be attached directly to the tree. Nailing or screwing them directly into a tree opens the tree to entry by diseases or insects. These types of protection need to be monitored over time. As a tree grows, the wraps can become too tight and damage the tree. Regular adjustments are necessary. Also, inspecting for insects underneath the protection is advised. Sometimes individual protection is not as practical as protecting a group of trees or plants. If actual fencing is the method employed, it should have as low a visual impact on the exhibit as possible. Another idea is the placement of obstacles to keep animals away from plants. The use of deadfall (tree limbs and trunks) may be enough to keep plants protected and looking very natural in an exhibit. In some situations, large boulders may be necessary to hold back an animal (figure 25.4). Planting within these boulders can help soften this approach and disguise the appearance of an intended barrier.

Even with the best design and planting plan in place, damage may still occur. This can more routinely be caused by general animal traffic, overgrazing, or animals establishing

Figure 25.1. Tree protected with hot vine and grass. Photography by Tony Range, Saint Louis Zoo.

Figure 25.3. Plastic mesh around a tree in a zebra exhibit. Photo by Tom Frothingham, Little Rock Zoo.

paths or trails within an exhibit. Routine replanting may be necessary, both as a manner of repair and as a management tool to support a desired appearance. This is commonly used for grass maintenance in outdoor exhibits to alleviate compaction and overgrazing. Compaction is a situational problem resulting from the confinement of captive animals to a restricted area. Exhibits for hoofstock and large animals, especially pachyderms (elephants and rhinos), are prone to problems of substrate compaction.

Compacted soils are less capable of absorbing water and oxygen, necessary elements for healthy root growth. The actual plants are crushed as well, causing damage or death. A good management program will include periodic aeration of an exhibit, especially in the high-traffic areas. Aeration is a process of physically breaking up the soil to loosen it and to regain a structure that will allow good water infiltration and gas exchange to support healthy plant growth. Aeration can be achieved simply by hand, with a fork or tiller, or with a walk-behind aerator. Larger areas may require the use of a tractor-pulled aerator. There are core aerators that penetrate the soil and remove a core from the ground, and slit aerators that cut the ground and open it up. The core aerator is a more effective type. Regardless of the equipment used, the objective is to break the soil up to a depth of 5 to 10 cm (2 to 4 in.). Caution should be exercised when aerating around established trees and shrubs, to minimize damage

to the roots of these plants. Following the aeration process, top-dressing with compost and then overseeding with grass seed is a good practice. Depending on the animal species being dealt with, the practice of aeration and rehabilitation (replanting) may be simple or more complicated. A routine annual replanting is usually sufficient for most situations. Sometimes more frequent aeration may be necessary (and beneficial), and some unique situations may require multiple plantings a year. When replanting, removing the pressure of animal(s) will be necessary for a period of time. This may mean that the animal remains on exhibit but is restricting from accessing the replanted areas.

Managing the time animals are kept in an area can greatly relieve or reduce the problems of compaction and overgrazing. It can be achieved by moving animals off an exhibit for a period of time to allow vegetation to regenerate. If this is not an option, an alternative solution is to rearrange the exhibit to disrupt traffic patterns. This can be done by erecting temporary electric fencing or moving logs in the exhibit, or by placing boulders that will change the exhibit's layout. Changing the type or species of plant material may also be an option. For example, converting a problem turf grass area to ornamental grasses or a ground cover may be a solution. These types of changes will require going back though the review process for approval by all involved, with the well-being of the exhibit and its inhabitants as objectives.

Figure 25.4. Plant protection with boulders in elephant exhibit. Photography by Mark Sparrow, Chester Zoo.

To review, horticulture plays an important role in the design or redesign of a zoo or aquarium exhibit. The selection and placement of plant material can help in creating the best simulation of a given region of the world. The location of plants in an exhibit can help enhance visitors' views of the animals and screen undesirable views, while making it easier for keepers to manage the animals and maintain the exhibit. The selection of plants must take into account the chemical or physical traits that could make them injurious to the animals and the viewing public. It should also take into account the mature size of plant species selected and the protection they may need from animals. All these considerations will help create a better exhibit.

PLANT AND EXHIBIT MAINTENANCE AND CARE

Some elementary points about plants need to be understood before continuing. Plants are living organisms that for the most part cannot get out of harm's way. They have needs similar to those of animals: water, nutrition, oxygen, and treatment of injuries and diseases. And though plants broadly share the same basic needs, their specific needs can differ from one species to another. Therefore, basic plant identification is a must for a keeper, with or without a horticulture department's assistance. A keeper should be able to easily recognize most of the plants in an exhibit and should have some knowledge of their problems, such as toxins, diseases, and pest infestations. Good plant identification skills will aid a keeper in the maintenance of these plants and in identifying unwanted plants or "weeds" before they become established. These invading plants could compete with the desired plants in the exhibit or could be a danger to the animal(s). If keepers can identify them, these unwanted plants can be controlled before a problem develops. There are many ways of learning how to identify plants. The first would be to make use of the horticulture staff and its knowledge. If a zoo is too small to have an in-house horticulture staff, keepers can learn plant identification from books, the internet, or local authorities on the subject. Many times an exhibit's original design and landscape plan can be a resource, if an actual design was done that included horticultural plans. Often, however, this is not the case. A good plant reference book that includes photos or line drawings of a wide range of plants and the environments they grow in, like Michael Dirr's *Manual of Woody Landscape*

Plants, can be a start. There are also many plant identification web sites, such as the USDA's plant database (http://plants .usda.gov/java/), which enables a keeper to search for a plant on the basis of many different criteria. One can either try to identify an unknown plant, learn more about one that is already known, or learn about other plants that may grow well in a particular environment or exhibit. It is good for a keeper to understand what is present in an exhibit, how it will grow, and its potential for causing problems. A keeper should also be able to recognize different pests or diseases associated with plant species. At the least, he or she should be capable of determining whether a plant is suffering from an abnormality. This is easily done by monitoring the plants in an exhibit and noting any appearance outside the norm. Notable observations would be discoloration, missing plant parts, abnormal growth, or the presence of insects. Sometimes the injury to the plant may have been caused by the animal on exhibit. Such an injury can make the plant more susceptible to pests or diseases.

A keeper who knows how to identify plants can provide them with the essentials of life (water, food, oxygen, and injury treatment). Plants receive water either from precipitation or from irrigation. Irrigation can be totally automated, with timers programmed to supply a set amount of water at a predetermined time. Or it can be supplied manually by opening a valve or running a hose from a source to a sprinkler. Automated irrigation offers a number of advantages. An application can easily be scheduled either to avoid conflict with the animals or to let them enjoy or benefit from exposure to the water. Keepers should remember that since much damage to an exhibit can occur when the substrate is wet, managing animal access is very important. When an irrigation system is used in an animal exhibit, keepers should know where its water supply lines, valves, timers, and sprinkler heads are located. For reasons of animal safety and exhibit maintenance it may be expedient, where possible, to locate the irrigation system outside the exhibit. Knowledge of the irrigation system's location will be needed to avoid it when digging or erecting anything in the exhibit that could damage the system. It should be included in the zoo's "dig policy," which spells out procedures to be followed before digging on zoo grounds. If a leak or line break occurs, the keeper should be able to shut off the system before extensive water damage occurs. Curious animals can sometimes become interested in the sprinkler heads, and this may lead to damage. This possibility should be recognized in the exhibit's design. For example, sprinkler heads can be mounted on moat walls or exhibit fences, where they will be out of the animals' reach. This can be a matter of very specific concern with some animals and of no concern with others.

Knowing how to provide water for an exhibit's plant life is just the beginning. As noted earlier, knowing the needs of plants will help a keeper know how much and how often to water them, whether manually or by automation. Some plants—succulents and cacti, for example—are like camels in that they preserve water well and can survive periods of drought. Other plants—sedges, willows, and cypress—can withstand great amounts of water, even standing water. Water stress is a basic concept applied to the water needs of plants. It can occur at either end of the spectrum, as a result of too much water or too little. Every plant species has its own specific range of water intake at which it grows best and does not experience water stress. Sometimes an overflow of water can't be avoided when rains come often and in large quantities. A lack of sufficient water can be remedied by irrigation.

Whether water is too much or too little, drainage is the next water dynamic that must be understood, and building and/or managing it is important. Water can be very powerful and destructive. It can erode the soil from under the foundations of structures or plants. It can wash animal excrement out of an exhibit into areas where it is problematic. It can collect or pool in an exhibit, creating a potential breeding site for mosquitoes and other undesirables, or just rendering part of the exhibit unusable until it dries up. These problems should be prevented in the exhibit's planning and design, but often they are not. In case of excess water runoff and erosion, steps should be taken to divert or slow the water's flow or collect it and pipe it away from the area. The keeper should consider how to change the exhibit's drainage patterns. If a steep grade carries water into the exhibit area during either a natural or human-made flood, the keeper should look to divert, slow, or break up the water flow. If a grade is too steep, terracing with the use of berms and swales can slow down, collect, and control the path of water. A berm is a small hill of any length that acts as a dam to block the movement of water in a particular direction. A swale is just the opposite: a depression of any length that collects water and moves it in a desired direction. Berms and swales are often used together in the management of water drainage. Water can be either directed to a certain point for collection or allowed to disperse across the ground where it will do no damage and possibly be beneficial. Impediments in its path can either spread it out or dam it and let it move on more slowly after its accumulation has reached a certain level. Vegetation slows water flow very effectively and acts as a stabilizer for soil, so it can reduce the amount of water that enters the exhibit.

Soil compaction resulting from animal traffic can affect water flow patterns and cause erosion. It can also reduce the amount of oxygen available for plant roots, and thus inhibit a plant's ability to grow. Many larger animals tend to establish routes or paths along which they repeatedly move within their exhibits. The resulting compacted ground can develop into channels for water runoff and thus cause erosion.

Periodic aeration of areas in an exhibit where compaction is a problem will help loosen the soil and improve its composition, which in turn will better support plant growth. Aeration is any method that can be used to break up the soil. It can be done on a small scale with a shovel or a fork. On a larger scale, one will want to employ either a walk-behind-type aerator or a pull-behind-type used with a tractor. If compaction has led to the elimination of vegetation, replanting will be necessary. Grasses can be replaced by reseeding or resodding. Sod consists of living "rolls" of grass that can be placed where desired to get instant vegetative cover. In either case a keeper must restrict animal access to the newly planted area until it becomes established and is capable of withstanding the animal traffic without damage. Many animals develop regular paths or trails to get around their living space. When this occurs, a keeper can break up their routine with the placement of obstacles that will cause them to choose different routes. These obstacles could be boulders, deadfall (limbs or trunks from dead trees), or other deterrents. It must always be kept in mind to try to create obstacles that work well with the exhibit and don't create hazards for the animals. For some animals it may be necessary to move the obstacles regularly to prevent the establishment of paths and compaction. Whenever an exhibit can be cleaned without water, its physical attributes are disturbed less and water is conserved.

Plants have nutritional requirements just as animals do, and healthy plants have the best defenses against pests and diseases. Plant fertilization in a zoo exhibit must be coordinated with animal management, especially if a synthetic or manufactured fertilizer is going to be applied in a dry granule formulation. Depending on the animal species, concern may center on the possibility of an animal ingesting the fertilizer granules. Dry-formulation fertilizers should be thoroughly "watered in" after application to reduce loss from vaporization and to make the material available to the plants' root systems for uptake. This also reduces or eliminates any potential contact with animals. Liquid or sprayed fertilizers should be thoroughly researched for any potential dangers from ingestion following application. Keepers can assist in this process by keeping animals off-exhibit during the application and "watering-in" period. A good method of feeding turf (grass) and the majority of an exhibit is to aerate it thoroughly and then top-dress it with compost. This will help with compaction issues and add organic material and a slow-release fertilizer source to the soil. Top-dressing means applying a thin (less than 2.5 cm) layer of material (e.g., compost or sand) on the surface of the ground.

Many zoos generate their own compost using the green waste created in their day-to-day operations. This includes animal manures, animal bedding, and landscape waste. By using compost, a zoo shows a commitment to operating in a sustainable manner by reducing landfill waste and reusing waste materials to create a nutrient-rich soil amendment or fertilizer. Some zoos are capable of creating more compost than they can use, and can create a revenue stream from this product. The basics of creating compost are quite simple, but considerations must be made for containing a composting operation to prevent possible contamination of surrounding areas. Many times there are state or local regulations that must be met to legally operate a composting site. One should check with local authorities on these requirements before making a significant commitment to such an endeavor. The rewards are great, but public safety is a necessary concern.

Animal feces can be left in an exhibit, at least periodically, to break down and add nutrients to the soil. This should be done carefully to control problems that could arise with pathogens, pests, and runoff contamination. As previously mentioned in regard to compost production, contaminated runoff is an issue that also needs to be addressed in animal exhibits. The potential for the transmission of infectious or zoonotic disease organisms is the primary concern. The organisms that can be carried from an exhibit (or from a compost operation) via water runoff to an adjoining exhibit or to a natural body of water can potentially be problematic for other zoo animals, native animals, or the public. Containing this runoff in its original location with the use of barriers, either natural or artificial, is one option. This can sometimes be accomplished with the construction of a natural bio-filtering collection pond, either as part of an exhibit or just beyond its perimeter. Collection and removal of runoff by way of underground piping connected to the sanitary sewer system is another mode of control. Regardless of the method employed, awareness of the potential impact of contaminated water runoff and its management is essential.

As with fertilizer use, keepers need to be well aware of methods used for disease, pest, and weed or vegetation control in an exhibit. The ability to read and interpret a Material Safety Data Sheet (MSDS) is a must for a keeper. An MSDS gives information about potential hazards, precautions, personal protective equipment (PPE) needed during application, and actions to take in case of an accident (exposure or spill). A zoo facility should maintain definitive records of all fertilizer materials used and make them accessible to all staff. These records should include the material used or action taken, when the application was made, the quantities applied, and whom the applicator was. Many diseases and pests can be prevented or controlled by good sanitation and physical maintenance. This can include corrective pruning or thinning of plant material to allow good air circulation, and timing of irrigation to permit foliage to dry. Foliar diseases are more common in damp situations that can be either cool or warm.

Another key consideration is plant selection. Many plants can be specifically selected for resistance to disease, pests, or other maladies. Especially in a zoo environment, the practice of Integrated Pest Management (IPM) should be used. IPM involves identifying pests that could be present, monitoring their population levels against an established threshold of allowable damage, and then implementing different degrees of control. Control measures progress from less aggressive to more aggressive methods. Less aggressive methods can include the introduction of predatory insects, applications of oils or soaps, or the physical removal of infected plant parts. For example, spider mites are small pest insects that prefer hot, dry conditions and, given those conditions, can reproduce at an extreme rate. Periodic misting or spraying of foliage can help physically remove spider mites and create an environment that will reduce their outbreak. Due to their soft bodies, they are vulnerable to the desiccating action of insecticidal soaps. The introduction of a predatory mite species like *Phytosieulus persimilis* can be used to both reduce and control a spider mite infestation. Use of predatory insects can be more challenging outdoors than indoors. The predators are vulnerable to environmental conditions (temperatures, precipitation, wind, etc.) that can reduce their numbers. However, once their populations become established, they can effectively control the pest insects. There may be situations in which the exhibited animal is an insect eater and could contribute to the control of pest insects. If the animal eats the plant to get to the insects, one should find a different solution. Finally, if a pest population becomes too widespread and/or is a persistent problem, the physical removal of part or all of the plant may be necessary.

All these different actions are based on being aware of the potential plant pest species, being able to identify the pest and its effect on plants, and then monitoring the population level to know when to implement the necessary control. Sometimes a problem may reach a level that can't be controlled with the softer methods. If the plant species is considered of high enough value, a stronger material, such as a synthetic chemical insecticide, may be necessary to reduce the pest population to an accepted level. A softer method can then be reinstituted to maintain those levels. However, good monitoring can prevent the need for using the less selective and more toxic materials. Different animal taxa have different levels of sensitivity to control methods or products, specifically chemical insecticides. Fish, bird, and amphibian species exhibit the highest sensitivity to exposure. Amphibian and reptile species cannot break down or eliminate some toxins, and are thus more vulnerable to chronic buildup of those materials. If a material or practice cannot be employed that will work with the resident animal, substituting the plant species in its enclosure for another that doesn't have the problem is necessary. Regardless of the method used in IPM, keepers should make themselves knowledgeable of what is being used and its potential impact on the animals, both immediate and residual. Some materials are just too toxic for use around animals. The weather should be conducive to the type of application being made; otherwise, the application's effectiveness can be compromised. Rain, for example, can dilute or remove the application, thus resulting in a waste of time and material. Such aspects as wind drift, long-term residual presence, and potential secondary poisoning should also be known in advance. Wind can carry a material outside the target area and have undesirable effects. A material that persists for a long time or has a long-term residual presence can result in exposure to untargeted species or require that an area be kept unavailable until the material is considered safe. A material with secondary poisoning attributes can be ingested by the original target pest, after which the pest is ingested and potentially poisons the second consumer. These situations accentuate the need for good communications between keepers and pest control personnel.

A practice that is gaining in popularity and shows very good potential for use in zoos is the use of compost tea, an aerobically brewed extract of compost, water, and microbial foods such as kelp, fish hydrolysate, and humic acid. The result of this brewing process is exponential growth of a diverse population of active beneficial soil organisms. Finished compost tea can be applied as a soil drench or as a spray application to plant foliage. It presents little risk to either humans or other animals, and has some very beneficial attributes. The microbes present in it are important members of a community of organisms known as the soil food web. The term "soil food web" refers to all organisms that live all or part of their lives in the soil, and the transfer of energy between species within the soil ecosystem. These organisms, (fungi, bacteria, protozoa, etc.) support plant health as they interact with each other to decompose organic matter, cycle nutrients, enhance soil structure, and control plant pests by attacking disease causing organisms in soil and on foliage (Ingram 2005).

As mentioned earlier, pruning is a big part of plant maintenance in many ways. It is done to control a plant's growth for better structure, to change the direction of a plant's growth so that it does not serve as an escape route for an animal or obstruct the visitors' view, or to control a plant's overall size. It is also done to remove infected areas or damage, or to allow better air circulation, which can prevent disease and pest problems.

All the different aspects of plant maintenance and a good understanding of their practice are important to a good zoo exhibit. They can mean the difference between a drab animal holding area and an effective habitat that simulates an animal's natural environment.

BROWSE AND ANIMAL ENRICHMENT

Growing plant material specifically as food for the animal collection is very important and takes both understanding and buy-in from horticulture and animal care staff. Browse (plant parts from woody plants) is becoming widely used as a form of captive animal enrichment and, in some cases, as an important nutritional supplement to an animal's regular diet (Moore 2005). Some of the more commonly grown species are willow (*Salix nigra* or *Salix babylonica*), hackberry (*Celtis occidentalis* or *Celtis laevigata*), and hibiscus (*Hibiscus* spp.). Many other species can also be used, but careful research should be done before feeding out anything. A browse program can be as simple or as intricate as a zoo has the interest and resources to develop. Browse can be grown as part of an exhibit, in a specific area of the zoo, in multiple locations around the zoo, or in any combination of ways as determined by the specific zoo.

The first step in creating a browse program is to choose the plant species that will most benefit the animal collection as a whole. Keepers, animal curators, and veterinary staff can determine which parts of the chosen plants should be used, and in what quantities. This will help determine how much space will be needed for growing, and how many different species will be grown. This list of desired plants should be cross-referenced with lists of plants that are the most economical and efficient to grow. The intent would be to select species that grow fast and will regenerate quickly, though certain species may also be chosen that have a unique value to specific animals.

Once the browse species are chosen, procedures for harvesting, storage, and distribution will need to be established. Horticulture and/or commissary staff may harvest and process the browse, or keepers may do so. Keepers must be taught the best ways to prune the plants while harvesting them, to help promote new growth and provide the desired amount of browse without injuring the plants or reducing their ability to regenerate. Browse can often be collected on zoo grounds during regular maintenance, but this practice requires the creation of a list of plant species that can be fed to the zoo's animals. Every zoo will have its own unique collection of plants on grounds and will therefore need to establish its own feed/no-feed list. As mentioned above, keepers need to learn to identify plants, at least the ones that occur in and around their exhibits on the zoo grounds, and the ones on the feed/no-feed list. The list can be created with the assistance of horticulture staff, animal care staff, nutrition staff, and veterinary staff. A number of publications contain information on what others have found to be safe for browse material. A good source is the internet website entitled The Foragers Source (www.foragerssource.org), which explains the development of a browse database and provides nutritional and identification information on many plants commonly found in zoo environments. This site is the result of a coordinated effort between Colorado State University, the Denver Zoological Gardens, and a number of other zoos.

SUMMARY

A good working relationship between keepers and horticulture staff has multiple benefits including better exhibits, optimum animal and plant health, and improved daily work conditions. Learning plant identification, the basics of plant care, and the plant support systems in an animal exhibit will strengthen a keeper's overall exhibit management skills and make him or her a better keeper. Having these skills along with open communication with the zoo's horticulture personnel will result in better animal care, better plant care, and a better zoo overall. Horticulturists are an important and valuable resource at any zoo. Many zoos have professional horticulturists on staff, but some don't have this luxury. In the latter case, the Association of Zoological Horticulture (AZH), an organization dedicated to the advancement of horticulture in zoological parks, gardens, and aquariums, is a great resource for zoological horticulture practices and procedures. They can provide contacts with AZH members who can be consulted on specific horticultural situations or issues. This group also provides training in many areas of zoological horticulture that can further strengthen the knowledge base of a keeper or horticultural staff. No zoo fortunate enough to have horticulture staff should overlook this resource.

REFERENCES

Ingram, Elaine. 2005. *The Compost Tea Brewing Manual, 5th Edition.* Corvallis, OR: Soil Foodweb.

Moore, Merle. 2005. *Development of a Browse Database.* Accessed 18 October at http://www.foragerssource.org/content/view/13/6/.

Part Five

Zoo Animal Husbandry and Care

26

Husbandry and Care of Small Mammals

Donald E. Moore and Michelle R. Farmerie

INTRODUCTION

There are over 4,500 species of mammals, and they are found in the oceans, high in the mountains, in the hottest deserts, and in the coldest places on the planet. Some are adapted for burrowing, some for swimming, some for running at high speeds or for long distances, and some for climbing, leaping, gliding, or flying—but most are not specialists at moving around the environment, and can move quickly in many of the above-mentioned ways to find food or for flight/fight purposes. Most are secretive, cryptically colored, and nocturnal in nature and therefore are difficult to observe or even find in their natural habitats. They range in size from the tiny bumblebee bat (*Craseonycteris thonglongyai*), weighing in at less than two grams, to the massive blue whale (*Balaenoptera musculus*), which can weigh more than 120,000 kg (264,600 lbs.). Most are less than one-half meter long—and these and other mammals zoo professionals call "small mammals" are sometimes identified as being "smaller than a bread box."

Larger mammals such as hoofstock and primates will be covered in their own chapters, but basic concepts for management of all mammals run throughout small mammal husbandry, including guidelines for species-appropriate exhibit size, bedding, and nutrition. In modern zoos, for instance, small mammals have been managed in taxonomically specialized areas with names like "Small Mammal House," "World of Darkness," or, more recently, as part of the diversity in habitat immersion exhibits, "Amazonia," "Australia," or "American Prairie." The best of these exhibits recreate natural habitat and diet so that the animals are "up close and personal," so that zoo guests can observe the nuances of mammalian adaptations and species-appropriate behaviors, and so that guests are engaged and inspired to conserve these unique and charismatic animals. Although only small collections of small mammals continue to exist in zoos and aquariums, much research has been done on small mammal relatives from different taxonomic groups due to their use in laboratories, as nutrition models, and as companion animals (or "pocket pets"). For some mammals, research is routinely done to aid agriculture or commercial activities, and information about them is found in a variety of specialty publications and websites maintained by a variety of small mammal interest groups (see Resources, below). This chapter will emphasize management of small mammals in human care.

After studying this chapter, the reader will understand

- the interrelationship of mammal behavioral needs and keeper behavior, and important safety precautions keepers should take to protect themselves and their animals, specifically small mammals as mammalian examples
- best practices for furnishing and cleaning small mammal enclosures
- key physical and behavioral traits of small mammals.

NATURAL HISTORY

Mammals are defined by their mammary glands and their obvious hair, whether in the subclass Prototheria (monotremes), the infraclass Metatheria (marsupials) or the infraclass Eutheria (placental mammals). Mammary glands that provide milk nutrition to newborns may be in slits (as in cetaceans), in multiple smallish teats (as in multimammate mice and others), or in relatively large breasts (as in sloths). Their hair is often an adaptation for temperature control, and naked mole rats (*Heterocephalus glaber*), which live underground in sandy soils of East Africa, have few hairs scattered over their wrinkled skin; they can thermoregulate by finding appropriate temperatures underground. Tropical mammals have less hair, and different types of hair, than do thick-coated mammals from arctic or montane regions. Many mammals actively thermoregulate; they can bunch up their bodies or their hair coats can fluff up (piloerect) to create a warm blanket of air around the body. Specialized hairs include the whiskers that help nocturnal mammals find their way in dark undergrowth,

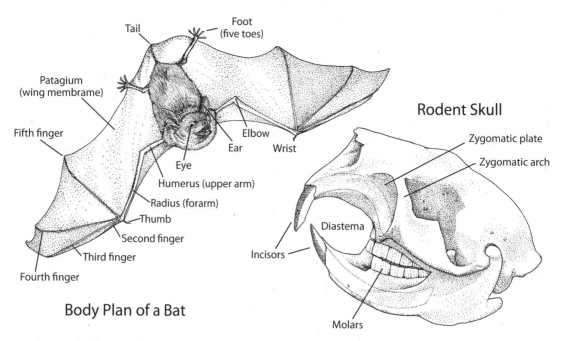

Figure 26.1. Basic anatomical features of small mammals: (1) Body plan of a bat; (2) a rodent skull. Most members of this order (1700-plus species) have incisors with sharp enamel ridges and a gap (diastema) left by the absence of canine and premolar teeth. Illustrations by Kate Woodle, www.katewoodleillustration.com.

odor/chemical-holding hairs that aid mammals in marking territories, and quill-like defensive hairs in mammals like porcupines.

Some small mammals have become domesticated by humans (e.g., dogs, cats, ferrets of the order Carnivora), are commensal with humans (e.g., rats, mice of the order Rodentia), or are in the process of being domesticated (e.g., foxes [*Vulpes* sp.] on fur farms, and "pocket pets" like hedgehogs [*Erinaceus europaeus*]). Whether domestic or wild, the diversity of mammalian natural history and habitat has resulted in behavioral specialties, and diets must be carefully considered in the development of exhibitry and husbandry techniques for each species. Canids are generally omnivores, felids are obligate carnivores, mustelids are specialized feeders, and so on. Just within the rodent group of 30 families and more than 1,700 species, there are animals with a wide variety of feeding strategies, from generalists to specialists. Similarly, a diversity of feeding strategies is found across the entire class of mammals, as well as within smaller taxonomic units. Of course there is also a diversity of adaptations that need to be considered when developing husbandry. Small mammals adapted to the Arctic are not capable of thermoregulating in what for them is extreme warmth, and conversely, small mammals adapted to the tropics may be hypersensitive to drafts or cold.

Natural histories of mammals provide a framework for working with these species. Although vertebrate species represent only about 5% of all animal life on earth, and mammals are only 12% of all vertebrate species, mammals are very diverse and can be found in just about every ecological niche that will support life. Some of their defining hairs are modified into sensory whiskers that help them survive a nocturnal lifestyle in these varied environments. Their hair

and skin is often modified for attack or defense; examples are the horns of rhinoceros, the scales of armadillos, and the quills of porcupines. Some mammals have almost eliminated hair to help themselves keep cool, as in the naked mole rats that live underground in Africa, or the hippopotamus that remain underwater to stay cool during the day. Arctic mammals like the arctic fox (*Alopex lagopus*) have the most hairs per skin area of any mammal, to help themselves stay warm in their cold environment. Mammals are much more often nocturnal than birds, but exceptions are such primates as golden lion tamarins (*Leontopithecus rosalia*), such squirrels as Prevost's squirrels (*Callosciurus prevosti*), several viverrids such as meerkats (*Suricata suricatta*), and larger mammals including hoofstock. Even mammals typically thought of as being mostly nocturnal, such as cats, foxes, and weasels, may be active during the daytime.

Some mammals, like raccoons (*Procyon lotor*) and bobcats (*Felis rufus*), have very broad diets and the ability to live in a variety of habitats, and therefore have very extended ranges. Others, like Oriental water shrews (*Chimarrogale* sp.), which live in water and eat a relatively narrow range of aquatic invertebrates, and koalas (*Phascolarctos cinereus*), which live in trees and eat only eucalyptus, have very narrow ranges. Herbivores eat food that cannot escape (plants), while omnivores and carnivores are more adapted to eat prey that can and does escape. Even so, herbivores are often equipped (adapted via the teeth and gut) to most efficiently use only certain types of vegetation, such as roots, seeds, leaves of bushes, leaves of grasses, or leaves of trees. Some bats (order Chiroptera) are fruit eaters and can feed from a roost, while other bats and insectivores are insectivorous and need to feed on moving insects. Carnivores may show hunting specializations like the crouch-wait-rush-and-bite behavior of cats or

> **Good Practice Tip:** Small mammals mark their territories in nature and in human care. When cleaning their enclosure, a keeper should leave at least one-fourth of its area unscrubbed and undisinfected, so that the residents' odors remain in the exhibit. This should minimize stress in the animals.

> **Good Practice Tip:** Understand the animals. Knowledge of an animal's natural history, individual history, and captive care needs can make a keeper very successful. Careful study of published research on the species, or on similar species, to fully understand an animal's environmental, behavioral, and nutritional needs, will help keepers enhance the lifelong welfare of the animals in their care.

the front foot dexterity of otters (*Lutra* sp.) and raccoons, or they may be distance runners like wolves (*Canis lupus*). Some herbivores may show hoarding behavior, and may need to store large piles of food to feel "comfortable"; some carnivores may show caching behavior, in which they eat part of their prey and bury the rest. Each species has its own tooth, fang, and claw architecture and strength adaptations based on its diet and habitat.

Many mammals have evolved shelter-building behavior and construct dens, tunnels, tree nests, and other elaborate structures. Some of these behaviors are so innate and so specific that keepers will need to provide specific materials to allow appropriate shelter construction and associated activities (e.g., reproduction) to occur. For instance, many small mammals make nest sites within their shelters, and often prefer different types of bedding for their nests (or sometimes prefer a "clean" nest with no bedding at all). As an example, North American red tree mice (*Phenacomys longicaudus*) build their nests only in Douglas firs (*Pseudotsuga menziesii*), and their preferred food is the needles of that tree; a keeper of this species could potentially provide an artificial nest, but might need to consider the provision—and method of presentation—of fresh fir branches for their needles (as food for the mice) and branches (to serve as trail systems and runways).

Mammals communicate in a variety of ways including visually, auditorily (by sound), olfactorily (by scent), and with tactile (touch) stimuli. A mammal's highly developed facial musculature, control over erection of hair, and ability to assume a variety of positions enables it to communicate in a visual way that is not observed in many other animals; primates, canids, and hooved animals are especially well-studied in this respect. Think of a cat that lays its ears back, opens its mouth, and hisses; a dog laying back its ears and lifting its lip to show its entire length of upper teeth while growling; or any animal that simply shows an open mouth—a clear threat without any noise. Elephants make sounds below the level of human hearing ("infrasound"). Small primates and medium-sized carnivores generally communicate within the range of human hearing. Bats and shrews use sound well above the level of human hearing ("ultrasound"), and bats' use of echolocation ("radar") is well studied. Certainly any person would be well aware of the intent of a small cat or dog that growls; other small mammals challenge a keeper's awareness, but also growl or even grind or chatter their teeth to signal that the keeper is too close. Rodents use ultrasonic vocalizations in a variety of social ways: for example, as calls by infants to adults, as inhibitions of adult-infant and adult-adult aggression, and even as territorial announcements. Mammals communicate olfactorily via specific scents called pheromones, and we know these scents communicate in-

formation including precise species, sex, age, reproductive status, and territory boundaries among a variety of mammals. For the keeper it is important to remember that full cleaning with detergent or disinfectant of an environment marked by an olfactorily-oriented mammal such the tree shrew (*Tupaia* spp.) may create some stress for the creature; it is better to spot-clean inside the enclosure and then fully clean about 25% of the area at a time, to limit the stress. As for tactile communication, the precopulatory behavior of many mammals may include a male laying its chin on a female's rump, nuzzling the female's genitalia, or touching various body parts of a conspecific. A dog (*Canis lupus familiaris*) may greet its "friends" with a mouth-greeting. Many aspects of mammalian communication have not been studied in detail, and keepers who are interested in it may find it rewarding to spend part of their days helping communication researchers.

Keepers need to remember all of this great diversity and address it effectively to be successful when working with mammals.

SMALL MAMMALS AND THE ANIMAL KEEPER

Keeper attitude and behavior around mammals, including delicate small mammals, is important. Keepers need to thoroughly understand the small mammals with which they work, so that they can work around them safely and effectively. Some such mammals can be "delicate" and flee at breakneck speed at the slightest unexpected movement, some can be aggressive toward keepers at relatively unpredictable times, and some can be very situationally hypersensitive to keeper activity (during feeding time, in breeding conditions, or when nesting with young). One must take great care to not injure these animals or to injure oneself when working around them.

Keepers need to be concerned about safety around small mammals, whose social behaviors range along the continuum from asocial to social. Relatively asocial small mammals include carnivorous and highly territorial ones (marsupials such as kowari [*Dasyuroides byrnei*], placentals such as weasels, and some rodents), which are really only social during reproductive periods of male-female courtship and mating and female infant-rearing; some of these are frequently intolerant of conspecifics outside breeding season, so keepers need to take great care with introductions. Other species' natural histories include living in bonded pairs or family groups, and these animals may be tolerant of one another and even of their keeper when they are kept in small groups; examples include otters and North American beavers (*Castor canadensis*). Still other animals are highly social, live in large groups, and are highly tolerant of group members but perhaps not tolerant of "outsiders," keepers, or group members which have been separated from them for a certain amount of time; examples include bats, meerkats (*Suricata suricatta*), and spiny mice (*Acomys* sp.). The most extremely social of all small mammals are naked mole rats, in which there are even breeding "queens," as there are in social insects.

Small mammals, like other mammals, may seem very quiet. This is a good adaptation to nocturnal life, and to life in different habitats in general. Small mammals are different from birds and insects in this way. But they communicate with scents, and with body posturing. So keepers need to be aware of the mammals' body odors, and if they change, it may indicate fear, aggression, illness, or a change in their reproductive status. An animal's body posture, ear movements, or lip shape may send a message to the observant keeper that it is ready to flee or attack (the "flight/fight" response to a stimulus).

Small mammals' flight distances vary from species to species, on the basis of their ability to flee. Naked mole rats have not evolved an ability to flee; rather, they can dig quickly. In nature they live in stable ground, so in zoos, keepers need to ensure that they have a stable habitat. Squirrels, on the other hand, have evolved an arboreal ability to run and jump; so keepers need to ensure that squirrels in zoos have appropriate high places to move away to, and they also need to establish good relationships with the squirrels, to help decrease their flight distance. Blue duikers (*Cephalophus monticola*) are some of the world's smallest hooved animals, and are often considered "small mammals"; like maras (Patagonian cavies [*Dolichotus patagonum*]), hares, and other running and jumping animals, they have relatively large flight distances. They can break their very thin legs if keepers do not work hard modifying their behavior with the goal of calming and habituating them to human care.

Small mammals can be surprisingly aggressive. This behavior may be regulated primarily by limited resources, including food or space for flight away from a keeper or conspecific perceived as a threat, as well as by reproductive season. These factors are all modulated by internal motivations such as hunger or hormonal readiness. New techniques in endocrinology may give modern keepers a huge advantage over keepers of the past who were solely reliant on external physical cues from animals; we now have the ability to gauge changes in hormonal status and to ascertain whether animals are undergoing seasonal changes, are stressed, have diseases, and so on.

RESTRAINT AND HANDLING

Animals must be handled appropriately. Keepers should understand as much as they possibly can about the species they care for and every procedure used in handling or transporting them. One should *plan* all handling by reviewing the animal's natural history and individual history, as well as the best handling practices for the species. The appropriate equipment should be available and in working order. Is a handling box, crate, or vehicle needed? The best possible handling technique for mammals is to box-train or injection-train the animals well in advance of the need to transport them in boxes or give them injections. Sometimes a keeper will need to net a small mammal or other animal, but if the

keeper is the animal's primary trainer or keeper they should attempt to maintain their positive relationship with the animal by having someone else do the netting. One good idea is to practice netting techniques with a lacrosse stick and lacrosse ball, then with the working net and the same fast-moving ball; if the keeper is not a quick netter but the team really needs to use nets, the keeper should allow the best netters to do the job for the sake of the animal's welfare. One should know where all emergency equipment is located in the work space before starting a procedure, and know the level of the veterinarian's experience. A keeper should discuss normal respiration for the species with the veterinarian, and then provide backup monitoring of breathing and heart rate to help the veterinarian know when the animal's breathing changes from normal. If the animal is being transported to a different location for a procedure, the keeper should be familiar with that facility and know where to find emergency supplies. Building doors should be closed and locked prior to the procedure and a radio should be nearby in case keepers need to call for additional help. Keepers should use the appropriately sized net, use good and accurate netting techniques, know how to spin (twist) the net to control the animal, and know where the animal is located in a net at all times. If the animal is in a social group, the keeper should try to move a short distance out of the animals' line of vision, and hearing. One should try to use a net-to-box transfer, which is a transfer directly from the safety of the net to the safety of the box. The animal's abilities should not be underestimated; a keeper should remain calm and communicate effectively with the animal care team during handling procedures. An animal's behavior can change in a non-routine situation: an animal that is normally "tame" toward keepers may become aggressive if members of its social group are being captured. If keepers lose control of a handling situation, there can be multiple bite wounds to people. If a keeper loses his or her grip or is bitten, he or she needs to communicate this calmly so that another team member can help (meanwhile, he or she should try to not let go of the animal until a colleague can help with the restraint). Safety of the animals and people must be the primary consideration; when actually handling the animal, one should get control of the head near the back of the neck and get control of the limbs as appropriate. Keepers should stay focused and constantly be aware of their surroundings when handling or transporting an animal; they should pay attention to their gut instincts and use common sense during these times. If keepers have pre-crated the animal to save time for the veterinary team, it is their responsibility to constantly monitor the animal until the veterinarian takes over and gives monitoring instructions. A good keeper should also clean the crate when the procedure is done, and not leave the dirty crate for others. An appropriate number of trained people should be stationed outside the enclosure to manage the door.

OBSERVATIONAL SKILLS

Small mammal keepers need excellent observational skills to identify individual animals and determine changes in their condition and daily behavior, because these parameters are scaled down to the sizes of the small mammal species under their care. Animals might be identified through passive in-

Case Study 26.1. Small Mammals at Smithsonian's National Zoo

Despite the nocturnal nature of most small mammals, National Zoo's Small Mammal House is operated on a "local" (not reversed) light cycle because of the presence of numerous diurnal small mammals in the collection. In consideration of the known Vitamin D needs of diurnal tamarins (Callitrichids) and an iguana (mixed with diurnal agoutis, nocturnal sloths [*Choloepus* sp.], and other animals) in the exhibits, staff members requested the installation of roof window glass that would allow the "fullest" spectrum of light into the animal enclosures. Other diurnal small mammals in the house included a variety of diurnal rodents from around the world, as well as meerkats and other small carnivores. The rationale for the request for UV-transmissible glass roofing was from the *Callitrichid Husbandry Manual* (AZA 1999): "In captivity, callitrichids should be provided with 12 to 14 hours of daylight. The distribution of callitrichids in the neotropics where little variation in day length occurs has led several authors to recommend a 12 light / 12 dark photoperiod (Snowdon et al. 1985; Stevenson 1975; Savage 1995; DuMond 1972). Beck et al. (1982) preferred a photoperiod of 14 light / 10 dark for *Callimico*; at least five hours of light were provided after the afternoon feed. Rettberg-Beck (1990) recommends 12 to 14 hours of daylight for *Leontopithecus rosalia*, as well as 30 to 60 minutes of exposure to ultraviolet light to provide Vitamin D3."

Some curators have noticed dramatic changes in the coat color of golden lion tamarins (*Leontopithecus rosalia*) when they are exposed to direct outside light versus fuller-spectrum interior light versus filtered light. The mechanics behind the change is not entirely understood; the most plausible explanation is that some part of the UV spectrum is affecting the melanin in the tamarins' skin, creating a much deeper and richer golden color in those receiving a fuller spectrum of light. The question is: "What do we not know about the positive effect of this lighting on other diurnal small mammals, or even on nocturnal animals that sleep exposed to these light rays?" We have long known that it is part of the B-range of UV light that is most important for reptiles, but there are still debates among reptile professionals about which part of UV is important and why. Less is known about the small mammal species kept in zoos. We do know from glass company technical representatives that the UVB-levels are often the first levels to drop off, over time, in both glass skylights and the UV-emitting bulbs.

The product called Starfire glass seemed to be among the best on the market for roof windows that would allow the fullest spectrum of light into animal enclosures, and last the longest. National Zoo curatorial experience was that all glass tended to lose UV transmitting properties over time, and that some became opaque after a few years. Thus, that zoo's engineers chose glass that would retain its clarity and ability to transmit UV in the Small Mammal House. There was some concern that the brighter light might increase the heat in the building, but since the installation of new glass, the building has not experienced any thermal problems. Given the still-debated discussion of specific parts of the light spectrum and the known drop-off in transmission levels, staff opted for the broadest spectrum of lighting possible. It is best practice to measure indoor exhibits' light levels and wavelengths with good-quality light meters on a regular basis to ensure that glass and lightbulbs are transmitting wavelengths for the appropriate distances and for the species within the space.

REFERENCE

Sodaro, V., and N, Saunders, eds. *Callitrichid Husbandry Manual*. 1999. Chicago: AZA Neotropical Primate Taxon Advisory Group, Chicago Zoological Park.

tegrated transponder (PIT) tags, individual characteristics, tattoos, or body part clipping. Even the least visible identification method (PIT tags) requires the keeper to have the time and expertise to get close to the animal with a digital reader. Condition changes are relatively easy to observe in large animals like deer and antelope, but difficult to observe in small mammals, unless the keeper takes the time to observe each animal as it moves toward food and around in the environment; these observations need to take place at the appropriate time every day. It is difficult, for example, to determine the condition of a sloth or mouse housed in a diurnal environment if that nocturnal animal is asleep in a nest box every time the keeper tries to observe it. So it might be better if occasionally the keeper were scheduled on an evening shift (e.g., during an evening event at the zoo) to observe the animal while it is active.

THE ZOO ENVIRONMENT

ABIOTIC ENVIRONMENTAL FACTORS

Light. Effects of light and public activity on the behavior of small mammals need to be considered. In small mammal exhibits in a diurnal house, it may be best practice to have glass that allows full-spectrum light penetration, and for keepers to keep this glass clean. Small mammal exhibits may be reverse-lighted to be dark in the daytime. In "nocturnal small mammal houses" with light-dark cycles reversed from the outdoors, keepers will need to ensure that full-spectrum light bulbs are used and make sure they are cleaned and replaced as necessary.

Shelter. Naturalistic shelters and other environmental necessities must be carefully researched and provided to small mammals in captivity. Most small mammals, diurnal or nocturnal, have evolved to use shelter of some kind during different life stages, seasons of the year, or times of day. Squirrels (sciurids) make tree nests or use branch hollows, carnivores like raccoon relatives (procyonids), members of the mongoose family (viverrids) and weasel family (mustelids) use natural caves or tree hollows, and some rodents and insectivores use tunnel systems. The unifying qualities of all of these refuge areas may be darkness and quiet. While a keeper might turn on a radio to create comforting ambient sound for hooved animals, the unique sensory abilities of small mammals suggests that keepers in these areas should provide quiet surroundings, as well as species-appropriate types and amounts of nest boxes. For example, sloths may need one or two open-faced nest boxes per animal, because in nature they sleep in branch forks high in trees. Tree shrews are known to need at least two nest boxes per reproductive female.

Humidity and temperature. Small mammals are found in hot dry areas, cold dry areas, and hot humid areas. In winter they can even be active in swampy areas: American star-nosed moles (*Condylura cristata*) can be observed swimming and foraging under ice during winter and early spring in their native wetlands habitat. The point is that provision of supplemental heat or cooling and humidity or low humidity needs to be carefully planned on the basis of the species' natural history, and with consideration of its life stages. Supplemental humidifying or dehumidifying may also be needed for small mammals in captivity. Adult sloths, for instance, may remain in excellent condition in a normal building humidity of 65%, but reproductive females and their infants may need more than 80% humidity for adequate milk production and weight gain (author's personal observation).

Barriers, gates, and their maintenance. Barriers and gates should be constructed of stainless steel or some other nonchewable, impervious material, and can be installed to slide vertically or from side to side. In some cases (e.g., for sloths or prehensile-tailed porcupines that move relatively slowly), swing gates should be used. Small mammals will be very active when keepers are not present, and latching mechanisms should be routinely checked for security.

Substrates and their maintenance. Wood shavings, bark chips, sand, and other materials are often used as substrates; cleaning practices vary by species. Keepers need to be observant with meerkats and other species that maintain "latrine" areas in nature, because the amount of urine in the latrine area in captivity may compact sand or other substrate to a significant depth (almost one meter, in the author's experience) while being subject to "cave-in" on top of an exhibit specimen that makes a burrow underneath it. (The senior author's case resulted in an animal surviving for two days buried in a pocket under the compacted sand and being found only when keepers carefully dug up the entire exhibit one shovelful at a time). Some small mammals, mostly from montane or arctic regions where they have not evolved with our temperate parasite and bacterial biota, may need to be kept on stainless steel mesh of appropriate size; this is true for snowshoe hares (*Lepus americanus*) and some other delicate animals that need to be continuously monitored and treated for parasites. Stainless steel mesh is strong enough to be scratched and chewed without damage, and resists rust and corrosion better than other mesh types. Animals like the delicate arctic species can urinate and defecate through the mesh, so that their urine and feces are removed from contact with them; this reduces infestation or reinfestation of the animals by parasites and other harmful biota that might build up or arrive in more organic bedding.

Bedding and its maintenance. Bedding is different from substrate. However, preferred bedding can be made to mimic the bedding and substrates that animals prefer in nature. Gerbils and other small rodents from desert areas, for instance, can be kept on sand, but their burrow bedding may be wood shavings or some other organic compound they can use to make "nests." Generally, small mammals should not be kept on cedar or pine bedding, which can release volatile compounds and cause contact dermatitis, respiratory disease, and even death. However, fir and aspen shavings have been found to be safe for a variety of small mammals. Fir shavings also have good absorbency (a urine consideration) and low dust levels (a respiratory consideration). Oak leaves have been used for giant elephant shrews (family Macroscelidae) to create nests for their offspring, and seem to be one key to good reproduction of natural habitat for these relatively difficult-to-care-for small mammals from African forest areas (personal observation). Keepers need to watch their animals' hair coat condition closely, because each species and each individual may react differently to different types of bedding. Cleaning practices for bedding must be based on species' natural history needs; some species have evolved to have deep litter in their bedding sites, others have evolved to keep their bedding areas immaculately clean.

Pools, furniture, and other enclosure amenities. All enclosure furniture, pools, and other amenities should be species-appropriate. If the mammal is an otter, platypus, or water-dwelling shrew, its enclosure should have a pool with adequate poolside drying areas for absorbent bedding or substrate, appropriate denning areas that are monitored (or even on public view), and a life support system that keeps the water clean and at the appropriate temperature and pH for the mammal and for the live prey for which it naturally forages. Note that the term "life support systems" often means aquarium systems, but keepers can use it in a more general way for delicate small mammals. The term can and should be expanded to include all temperature, humidity, lighting, and water quality systems that help to keep delicate small mammals within their minimum and maximum critical values, and therefore physically and behaviorally healthy. Keepers should be adequately trained in the maintenance of these systems, or should at least be able to work with facilities staff to understand when a life support system is challenging environmental quality (e.g., water quality) and beginning to affect an animal's well-being. Burrowing animals need noncollapsible substrates in which to create burrow systems, or clear species-sized burrows that are built into the exhibit so that visitors can see their fascinating tunneling behavior. They also need solid bottoms on these enclosures; the anecdotes about prairie dogs digging out of their exhibits are legion! A concrete or metal bottom under the diggers' substrate is a good idea.

NUTRITION

It is important to consider natural diet versus captive diets and their supplements, with a focus on small mammal diversity. Calltrichids in captivity seem especially prone to vitamin D deficiencies (calcium deficiency, brittle bones, etc.). Most zoos give them a canned marmoset diet that is very similar to the canned primate diet—with some exceptions. The canned marmoset diet is much higher in vitamin D3 than is the regular primate diet (this can cause a problem in mixed species exhibits, where another animal might be exposed to the marmoset diet; there have been cases of vitamin D toxicity in those other animals that ate from the same pan). Marmosets and tamarins also need to be provided higher amounts of vitamin C than exists in most commercially prepared diets. The companies that manufacture prepared diets for zoo animals generally warn us of known potential nutrition issues like these; we need to understand and heed their warnings.

CONSIDERATIONS FOR FEEDING CAPTIVE DIETS

Keepers need to understand the impact of sociality on food consumption, solo and group feeding, and pest control. Dominant individuals may selectively eat favored foods, or more food, than do their conspecifics. It is important that keepers observe this carefully and manage it if necessary. Small mammals are often messy feeders. They sometimes spill small mammal diet out of dishes and into the substrate where it can become a highly nutritious food for cockroaches, rodents, and other pests. Integrated pest management (IPM) is important to reduce shelter and access to loose food and water for pests in a zoo's small mammal house; keepers should collaborate with pest control professionals to design the best-possible IPM plan for a zoo environment.

Husbandry staff must be concerned about standard diets and "diet drift." Scientifically based diets are created through careful research and trials for acceptance by diverse species. It is important that keepers achieve consistency of presentation so that the animals in their care achieve consistency of consumption. "Drift" away from the standard diet, even though it might be well meant (e.g., the provision of "treats" to an animal during training, or the provision of the animal's favorite foods), should be minimized, and any changes to the diet should be approved by nutritionists, veterinarians, and/ or the area's curator.

FEEDING BEHAVIOR AND ENRICHMENT FEEDING FOR SMALL MAMMALS

Different species feed in different ways. Ground-dwelling mammals may feed easily out of ground-based bowls or platforms. Tunneling mammals like mole rats may eat and behave normally only in tight burrows. Arboreal mammals like ruffed lemurs (*Varecia* sp.) and sloths may prefer to eat while hanging down in some way. Many herbivorous mammals forage during a large proportion of their daily activity budget in nature, and a keeper may help stimulate long foraging activity in such an animal by hiding its food or placing it in some kind of artificial or naturalistic feeder that requires manipulation by the animal, like a puzzle feeder. Note that these manipulable feeders will need to be placed in accordance with an individual's preferred feeding heights and locations. Some animals are so sensitive to novel objects that keepers will need to take great care in presenting their diet items in food containers known to (and perhaps marked by) them.

WATER PROVISION

Mammals are very diverse, so they may drink in different ways just as they eat in different ways. Though some arid land mammals may not drink much at all, note that water should still be offered to them all the time (ad lib). Some small mammals may only lick water off leaves and may not drink out of bowls. One study of arboreal lorises and "enrichment waterers" showed that some arboreal mammals may prefer to drink from elevated drinking bowls and may not drink from bowls on the ground (Chepko-Sade and Miller, pers. comm.).

SEASONAL CHANGES IN NUTRITIONAL REQUIREMENTS AND DIETS

Many species from temperate areas, or from ranges where food availability changes seasonally, have evolved metabolic rates that lower during winter or other times of seasonal change. Keepers, curators, and nutritionists may determine that these animals can receive reduced amounts of food at these times. In some cases it is best practice to at least offer minimal dry food to ensure that the animal has the option of eating if it feels hungry (e.g., in the case of a hibernating bear, *Ursus* sp.). It is also best practice to offer water ad lib even if it is apparent that the animal is hibernating.

CAPTIVE BEHAVIORAL MANAGEMENT PRACTICES

SEPARATION VERSUS GROUP HOUSING, AND DOMINANT ANIMAL MANAGEMENT

Keepers may witness social group breakdown within an animal group, and may not always understand the underlying reasons for the social behavior change. The animals may be experiencing a change in dominance hierarchy, or the keepers may be causing a problem by feeding or training a subordinate animal before feeding the group's dominant individual. An animal's motivation to avoid a terrible altercation with a dominant animal is often stronger than its motivation to obtain favored food through training. Keepers may find little or nothing in the scientific literature about social interactions in "nonsocial" or "solitary" species, but keepers' observational skills may reveal that individual animals are being social—behavior does not lie! Keepers should carefully observe and report changes in animal behavior, because such changes often suggest an underlying issue. For example, when subordinate animals take food from a dominant animal, hindsight through review of thorough animal records may show that the dominant animal was sick at the time.

TECHNIQUES FOR SHIFTING AND SEPARATING ANIMALS

The easiest way to shift a mammal is to prepare a shift cage that contains a portable nest box already known to the animal, and to use this box as a shift area by regularly securing the animal in it while cleaning and checking its home enclosure. Modern care methods often suggest that animals should be transferred to new enclosures or zoos inside their own nest boxes; this means that some nest boxes should be designed to double as transport boxes. Cottontop tamarin induction boxes for anesthesia have been designed by staff members of the Pittsburgh zoo (Farmerie, pers. comm.) and other zoos; the animals can then be trained to enter these boxes and remain calm in them. These husbandry training techniques for mammals and other animals with good learning capabilities have led to a major improvement in humane care of captive mammals during routine and nonroutine care, and keepers should learn them thoroughly.

MIXED-SPECIES EXHIBITS

Benefits and complicating factors. Mixed-species exhibits of small mammals have benefits for the animals, because there may be social relationships between species and individuals that include olfactory, tactile, auditory, and other behavioral interactions. This is then more engaging and interesting for the staff and visitors, as well as for the animals, than single-species exhibits, because the animals are often more active at all times of day and night than are animals in more traditional exhibits. This kind of management may allow keepers to maintain more conservation-managed species and individuals in a given area. The complications inherent in this kind of management may include agonistic behaviors between species or between individuals. Individual candidates for multispecies enclosures need to be evaluated carefully by keepers who know the different species very well and who can interpret individual behavioral characteristics (e.g., whether the animals are individually aggressive, shy, extroverted, or introverted). At worst, seemingly benign individuals of a given species may be aggressive at night when keepers are not around; or adults of one species (e.g., neotropical rodents) may be aggressive toward infants of another species (e.g., sloths).

Guidelines for intra- and interspecific animal introductions. One-on-one introductions via "howdy" cages within enclosures are important to make after considering which species is least dominant, which species should be introduced into the area first to establish "territory," and which individual might be most likely to be calm and not aggressive with the other. Note that introductions should take as long as the animals' responses dictate, and should not be rushed. It is good practice to communicate about needs and intentions with as many experienced keepers from other institutions as possible, in order to most effectively introduce species and individuals. It is also important to be able to recognize stress in individuals of each species prior to an attempted introduction, and to be observant for these signs during the introduction itself.

REPRODUCTIVE MANAGEMENT

For best possible reproductive management, keepers should be familiar with small mammal behaviors related to reproductive seasonality, breeding, pregnancy, birth, rearing, and the introduction of infants to a group or exhibit. Small mammals, like all mammals, are diverse in their reproductive strategies and may be born naked and helpless (altricial) or fully furred with their eyes open and the ability to move like adults (precocial). Keepers must take these strategies into account, and to minimize stress on the animals they should schedule their neonatal exams only when altricial young are fully bonded with their mother (dam) or when precocial young are calm before being handled. Keepers should take great care to avoid being injured by protective mothers; even the smallest mammalian mother is willing to attack with tooth and claw to protect her offspring, and due to maternal hormonal and behavioral changes she may even attack familiar keepers at this time.

VETERINARY CARE

DISEASES OF IMPORTANCE IN MANAGEMENT

It is known that some diseases (zoonoses) can be transferred between small mammals and humans. Small mammal keepers should consider using rubber gloves and dust/surgical masks as best practice to prevent disease transfer between themselves and their small mammal charges. This is especially true for keepers who are prone to cold sores (viral herpes simplex lesions); this particular human disease can cause death in New World tamarins and marmosets, just as macaque herpes transmitted in fresh body fluids can cause death in human caretakers.

CHALLENGES IN VETERINARY TREATMENT AND HANDLING

Small animals, especially the "delicate" small mammals that have thin legs for leaping and long-distance running, may be prone to long-bone fractures. Inferior handling of domestic rabbits may cause broken backs or other physical problems. The message for keepers is that knowledge of the species' anatomy and potential stressors is always necessary, and that one must always take care to handle each individual appropriately for its species, age class, muscular power, and individual health condition.

PREVENTIVE CARE FOR CLAWS, HOOVES, AND TEETH

Small mammals with good nutrition in zoos may grow long claws and even long teeth, so any overgrowth should be managed by keepers and veterinary staff. Preventative care is easiest: hard items, such as nontoxic hardwood branches or blocks for chewing, allow the ever-growing teeth of rodents and rabbits to wear down naturally; appropriate substrate and digging areas allow their toenails to wear down naturally. If there is overgrowth of teeth or claws, it can be trimmed by appropriately experienced and careful personnel using professional nail trimmers, or even a small electric filing tool.

ENRICHMENT AND TRAINING

Each institution needs to have a formal enrichment approval process. Small mammal enrichment items should be species-appropriate and naturalistic, and all must be evaluated as safe. There should be no latex rubber for small carnivores, and no cotton rope toys, which may cause gut torsion or blockage in dogs and small cats.

OPERANT CONDITIONING WITH SMALL MAMMALS

Prey animals have a lower fear threshold, and therefore need more time to build trust in keepers. Animals that have not been husbandry-trained and for which keepers do not have

Good Practice Tip: Care should be taken to assess risk and safety of enrichment items. Raisins used for "preferred enrichment food," for instance, may get caught in mammal molars and cause tooth decay. Size of vines, ropes, and other habitat enrichment items may cause entanglement of hands, hooves, toes, or toenails. This caveat is meant not to discourage enrichment, but to encourage keepers to use their institutional risk management assessment through the enrichment approval process to create the best possible conditions for their animals.

accurate individual histories, may be difficult to train due to lack of trust-based relationships. Keepers need to move slowly, deliberately, and quietly in order to build these relationships. When offering treats to a prey mammal, a keeper should have enough of them so that if the animal drops one while taking it, the keepers does not need to bend over and risk looking like a predator to the animal. The keeper should watch where the animal is looking and be aware of its potential next movement; the keeper may be standing at some boundary where the animal may be ready to either target forward or flee backwards. Keepers should also note that many small mammals have cheek pouches in which they may store training treats without swallowing them. Such an animal may suddenly end an operant conditioning session, especially if it is in a social group. Smaller treats or more time between treats may need to become part of the keeper's training plan.

CRATE AND INDUCTION BOX TRAINING

Mammals are different from other wild animals due to their level of intelligence. While other animals often are driven by innate behavior, the duration of maternal care and sociality in many mammals creates a solid basis for learning as a behavioral adaptation. Even the brief duration of mother-offspring relations in asocial mammals is enough to result in learning. Consequently, training and learning should not be ruled out for most mammals, including those normally considered "flighty." Mara and cotton-topped tamarins are recent, well-known examples of animals that have been trained to calmly enter transport or induction boxes.

PROGRAM ANIMALS

Small mammals are a particularly popular group of mammals frequently and appropriately used as "program animals." Program animals deserve enclosure conditions as good as those of similar animals kept in "traditional" zoo exhibits, and it is appropriate that keepers caring for them should have extra training in nature interpretation, safe handling of animals, and public speaking.

SUMMARY

Small mammals are challenging to work with. Many are rare in the wild due to threats to their diverse environments, and many are rare in zoos and aquariums because they are difficult to care for appropriately. Keepers who choose to work with these animals will have rewarding experiences, but will need to constantly be attentive to safety, and to all the diverse factors that affect the animals' physical and behavioral health. These factors include the challenge of keeping animals from deserts, dry forests, tropical rain forests, wetlands and other areas, often in the same "small mammal house." They also include the provision of nutritious and palatable diets for animals whose diet has not yet been thoroughly evaluated in nature. But the reward of keeping these animals physically and mentally healthy, in breeding condition, and in species survival programs for their benefit and for that of future human generations is huge and incredibly satisfying.

27

Husbandry and Care of Hoofstock

Brent A. Huffman

INTRODUCTION AND NATURAL HISTORY

Hoofed mammals, also known as ungulates, display an incredible diversity of forms, adaptations, and lifestyles—features which make them an enduring part of zoo collections around the world. More than 250 living ungulate species are currently recognized (although recent revisions suggest that there may actually be more than 450 distinct species), with representatives found in nearly every zoogeographic region and biome on earth. They range in size from rabbit-sized chevrotains (family Tragulidae) to the six-meter-tall giraffe (*Giraffa camelopardalis*) and 3,600-kilogram white rhinoceros (*Ceratotherium simum*), with social groupings ranging from solitary species like tapirs (*Tapirus* spp.) to immense herds of more than a million Serengeti wildebeest (*Connochaetes taurinus*).

Although united by their common possession of enlarged, weight-bearing toenails (hooves), ungulates do not form a taxonomic group: the hooves have evolved several times independently. Modern hoofed mammals are classified either as "odd-toed ungulates" (order Perissodactyla) like horses, rhinoceroses, and tapirs, or as "even-toed ungulates" (artiodactyls) like pigs, peccaries, hippopotamuses, camels, and the diverse ruminants (deer, cattle, antelopes, and giraffes, among others). Genetic evidence also includes whales and dolphins (formerly Cetacea) within the even-toed ungulate family tree (formerly Artiodactyla); combined together, they form the new order Cetartiodactyla. Despite the disparate origins of hoofed animals, two common traits warrant their treatment as a group for husbandry purposes. First, all ungulates feed primarily on plants, using specialized strategies to deal with fibrous foods. Second, all ungulates have similar physical and behavioral adaptations for avoidance of predators. Although these common traits have been ecologically successful, they present challenges in the care of hoofstock (as ungulates are called in captivity).

Hoofed mammals have a long history of human care: evidence of captive sheep exists in the remains of 9,000-year-old settlements (Herre and Röhrs 1990, 585). More than a dozen ungulates have since been domesticated, including the horse (*Equus caballus*), the pig (*Sus domesticus*), the goat (*Capra hircus*), four cattle species (*Bos* spp.), the water buffalo (*Bubalus bubalis*), camels (family Camelidae), and the reindeer (*Rangifer tarandus*). With 4.5 billion domestic ungulates (livestock) worldwide (FAO Database 2009, 2007 figures), the experience with their husbandry is extensive. This knowledge base is an important resource for those caring for exotic ungulates.

This chapter will elaborate on the challenges and techniques of working with ungulates in captivity. After studying this chapter, the reader will understand

- anatomical terms specific to ungulates
- impacts of species-specific biology on housing, nutrition, and social management
- effects of ungulate behavior and keeper demeanor on animal and keeper safety
- best practices for encouraging species-appropriate natural behaviors
- principal issues involved in the reproductive and medical management of ungulates.

BASIC EXTERNAL ANATOMY

The basic four-legged (quadrupedal) mammalian body plan has evolved for a running (cursorial) existence in ungulates: elongated legs provide speed when fleeing from predators. Because the limb joints of ungulates and humans are in different relative positions, the joints have specific names, detailed in figure 27.1 along with other important ungulate anatomical features. The threat of predation has also molded keen threat-detecting senses; sensory emphasis varies between species, but all ungulates have eyes on the sides of their heads, providing an arc of vision approaching 360 degrees.

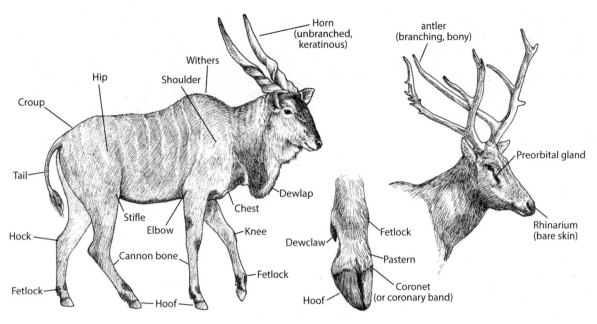

Figure 27.1. Basic anatomical features of hoofstock. Illustrations by Kate Woodle, www.katewoodleillustration.com.

THE ZOO ENVIRONMENT

As a result of their cursorial tendencies and relatively large body size, zoos often exhibit hoofstock in large outdoor enclosures (paddocks). Smaller yards and indoor housing are commonly provided to facilitate animal management. Enclosure requirements vary greatly between ungulate species and zoo locations, but several key considerations are universal.

SUBSTRATE

Local soil is the most common substrate in paddocks, since replacing large areas of ground is prohibitively expensive. However, rates of hoof growth are substrate-adapted: the coarser a species' native substrate, the faster its hoof growth must be to compensate for wear. Ungulates adapted to rough terrain, like wild goats, are therefore prone to overgrown hooves when kept on softer surfaces. Hoof wear can be increased by adding abrasive substrates like decomposed granite, limestone screenings, or roughened concrete to high-traffic exhibit areas, such as around feeders and along pathways. Holding yards are often covered exclusively with rough substrates, in part because of their relative stability in varying weather conditions. In all cases, the extent of this "hard-standing" should be determined on the basis of species' biology, as ungulates with sensitive feet (e.g., hippos and rhinos) may develop foot abrasions and injuries if confined on rough ground.

Whatever the enclosure substrate, it should provide good traction and even footing. Proper drainage is necessary to prevent erosion ruts and areas of deep mud: these uneven surfaces can cause injuries to the long, slender legs of hoofstock. In cold climates, ice may form in poorly drained areas; similar slippery areas may form with compacted snow, and clearing fresh snow from frequently-used areas should be a priority. Traction over slip hazards can be provided by spreading sand, but rock salt and other potentially caustic or poisonous ice melters should be avoided. During spring thaws or after prolonged rain, even well-drained substrates may become waterlogged. If this occurs, hoofstock should be held in barns or yards until the ground is dry and firm. Soft terrain is readily mired by hooves, creating potentially dangerous uneven surfaces when dry; uneven ground can be smoothed with rakes or harrows, but prevention is preferable.

CONTAINMENT

Ungulates may jump over, crawl under, climb through, or dig beneath obstacles (Fitzwater 1972, 52), so it is vital to research physical adaptations and behavioral repertoires when choosing barriers. Visually solid barriers like walls and stockades are "understood" by hoofstock; even fabric walls can be effective short-term barriers. Enclosure boundaries that can be seen through carry the risk of collision-related trauma, so injury-reducing features should be employed whenever possible. Dry moats should slope gently downwards to solid walls; steep drops should be avoided, lest an ungulate run over the edge. A water moat requires an additional barrier on the public side to prevent escape should an animal choose to swim or panic and jump into the water, and a sloped area is essential to provide an easy route back to the enclosure. By far the most popular hoofstock barrier is wire fencing due to its availability, low cost, and ease of installation. Fences tend to "disappear" into landscaping: a benefit for visitors, but a potential hazard for hoofstock. Flexible fencing, like chain link, can stretch to absorb impacts and thereby reduce the potential for injury; it is generally preferable to rigid or ungiving fence materials.

Keepers should check for containment weaknesses daily, as these may cause injury, permit escape, or allow free-roaming predators and native species (which may bear disease) to enter the enclosure. Barriers may be the target of behaviors like sparring and ramming, so secondary barriers like bumper rails can be employed to minimize damage by keeping large, strong, or belligerent ungulates back from the primary containment. Electric fencing ("hotwire"), which delivers a shock on contact, is a common secondary barrier choice due to its effectiveness and unobtrusive appearance. The ease with which slender hotwires are broken makes them generally unsuitable for primary containment (especially for large ungulates), while the low visibility increases the chances of entanglement during panicked flight.

The risk of collision with barriers is highest when hoofstock are introduced to unfamiliar enclosures. Burlap, shade cloth, and other materials are commonly attached to fences during initial introductions to give them a solid appearance and reduce visual distractions from beyond. Electric fences are usually "flagged" by tying strips of cloth or plastic (e.g., caution tape) at regular intervals; for naive animals, the stress of initially encountering hotwire in a new exhibit can be reduced by exposing them to sections of flagged fencing in a familiar environment. Further introductory precautions, such as reducing water depth in wet moats and pools, are covered in detail by Kranz (1996). As the new animals become established, hazard warnings can be gradually removed until the exhibit returns to its normal appearance.

REFUGE

Flight responses, as well as stress and conflict among group members, can be reduced by providing hoofstock with options for privacy. Refuges for smaller individuals (called creeps) can be created using secondary barriers that exclude larger animals. Visual barriers like plantings, exhibit furniture, and rolling terrain allow animals to retreat from their exhibit mates, keepers, and the public, thereby imparting a sense of security. Man-made visual barriers include wooden lean-tos, stacked straw bales, and sections of wall, but these may be inappropriate in naturalistic exhibits. Fallen trees (deadfall) provide natural, multipurpose furniture, which—unlike live vegetation—does not need to contend with soil compaction and frequent browsing. Live plants often need to be protected by secondary barriers such as electric fence ("hot wire") or rings of boulders if they are to survive in hoofstock exhibits; trunks of established trees are commonly wrapped with chain-link fencing to prevent bark abrasion and browsing. Thorny or unpalatable (but nontoxic) plantings (e.g., Barberry [*Berberis* spp.] and Hawthorn [*Crataegus* spp.]; Hohn 1986, 10) are used with mixed success to create a natural look without protective barriers. However, grasses (including bamboos) tend to be the most resilient to damage from ungulates, even without protection: some grasses grow tall enough to provide cover, and even short species can enhance enclosure aesthetics. Wherever vegetation is present, regular checks for and removal of toxic plants should be performed; these may include nightshade (*Solanum* spp.) and alsike clover (*Trifolium hybridum*). Extensive toxic plant lists are available online (see web links).

SHELTER

Every enclosure should include protection from wind, precipitation, and sun, whether in the form of shade structures, open sheds, or enclosed barns. Group-housed ungulates should have numerous shelter locations; multiple entrances and visual barriers further ensure that dominant individuals cannot exclude others. If needed, supplemental heating (heaters, bedding) or cooling (misters, fans) can keep animals comfortable, while significant differences between local and native climates usually require temperature-controlled indoor holdings. Delicate species, such as gerenuk antelope (*Litocranius walleri*), may need indoor facilities regardless of the region.

A typical hoofstock barn has a series of box stalls with sliding doors that connect pens and lead to outside yards. Floors must be slightly roughened to provide hooves with traction, and good drainage is needed to prevent water and urine from becoming slip hazards. Hard floors can be made more comfortable (cushioned and insulated) through the use of bedded areas. These are best located along walls or in pen corners—locations that impart a sense of security—and should be visible from keeper areas to permit observation. Wood shavings, straw, and (in some regions) grass hay are common bedding materials. Roughage-feeding species like camels and horses may consume forage-based bedding like hay or even coarse straw, and as a result can experience dietary upset, gain excess weight, or contract parasites or diseases. Where these issues occur, other bedding options exist, such as peat moss and mulch. Rubber mats, spray-on cushioned surfaces, and other novel livestock flooring products are also increasingly popular in zoo settings. Specifically developed for hoofstock, these synthetic materials increase animal comfort and reduce the need for disposable bedding.

When extra cushioning and traction are needed—for a sore or unsteady animal, or in preparation for a birth or immobilization—a "deep bed" can be made. A dense, soft "mattress" is formed using a layer of cohesive materials like soil, mulch, or moistened shavings over an entire stall; this base is usually covered with regular bedding to facilitate cleaning. Similarly, "hot beds" have a base of bedding and manure which produces heat as it decomposes. Helpful for providing warmth in buildings without power, hot beds need proper ventilation to eliminate fumes and regular cleaning of the top bedding to maintain hygienic living conditions.

HUSBANDRY

Ungulate management practices vary depending on the local climate, facilities present, institutional goals, and species' natural histories. Regardless of whether hoofstock are loosely or tightly managed, the knowledge and skills of their keepers are universally important. The diversity of ungulates prevents a one-size-fits-all approach to their management, and one of the best investments a keeper can make is to research the species under their care.

OBSERVATIONS

Throughout the daily routine, keepers should be on the lookout for slight changes in their charges, remembering that ungulates hide signs of weakness that might attract predators. Slower reaction times, glassy eyes, or ears that fail to prick up all indicate that an individual may be feeling "off." Subtle changes in locomotion, such as a barely perceptible limp or a slight reluctance to rise, suggest a potential issue with hooves, feet, or joints. Minor gastrointestinal upsets may be evidenced by changes in fecal consistency, fecal volume, or appetite; visible bloating and awkward positioning may be seen in more advanced cases. In herd situations, aggression may be directed towards sickly individuals, causing them to withdraw and isolate themselves. In all cases, keepers must recognize *changes*—to do so, they must be familiar with what is "normal" with their animals. Documenting observations for coworkers is important, regardless of whether or not a problem is obvious; effective communication, such as identifying the individuals involved and using proper anatomical terms (figure 27.1), is essential. Early recognition of issues greatly improves the prognosis of ill individuals.

Undertaking detailed visual checks and separating animals of concern is easily accomplished if hoofstock are brought into holdings on a routine basis; this practice also helps exhibits to recuperate from grazing and trampling. Daily separation of individuals permits keepers to monitor their food consumption and is also useful for delivering medications to specific animals. These shifting routines are best established if new animals are given time to associate holdings with food and shelter before being introduced to larger exhibits. In more loosely managed species that do not shift off-exhibit, keepers must be able to approach their animals close enough to identify and monitor individuals. Doing so safely and effectively requires an understanding of ungulate behavior.

FLIGHT RESPONSES

Ungulates avoid predation by maintaining space between themselves and predators; their flight distances dictate how close a potential threat can approach before they retreat. Daily nonthreatening exposure to people habituates many zoo ungulates to humans, reducing their flight distances and allowing closer approaches than would normally be possible. Building (or rebuilding) this trust may take weeks upon the introduction of new keepers, enclosures, or routines, but such tolerance needs only a split second to be overridden by wild instincts in novel or stressful situations.

Ungulates inherently recognize being cornered, being approached rapidly, and being separated from their herd as predatory scenarios, and they react with instinctive fight-or-flight responses. When there is space to flee, hoofstock may run desperately or unpredictably to reach security or rejoin herdmates. In captivity, barriers are a common source of trauma for panicked ungulates, and collisions may result in facial trauma, broken limbs, or even death. With no option for escape, they may also show extreme aggression towards the threat as a last-chance survival strategy. Keepers should avoid creating these predatory scenarios for the safety of humans and animals alike.

Predictability in environments, routines, and keeper movements allows hoofstock to anticipate future events and respond calmly. Wary hoofstock often allow closer approaches if keepers walk a zigzagging path perpendicular to the direction of actual movement. Surprise advances can be prevented by creating deliberate noises like jingling keys, pronounced footsteps, or quiet talking. A softly-playing radio can be used to provide constant background noise in holdings, thereby reducing the impact of startling sounds. Desensitization to keepers is advisable in order to diminish flight responses during daily care; animals in smaller enclosures will generally habituate more readily than those that have extensive space to flee. However, there is a fine balance between desensitization and becoming overly friendly, and fearless ungulates can draw keepers into dangerous situations.

KEEPER SAFETY

Ungulates have the potential to harm humans in a variety of ways. Horns and antlers are efficient and effective weapons. Strong, hoofed limbs can deliver crushing blows: horses kick backwards, giraffes kick forward, deer may rear up and "punch" with their front feet, and camels can kick their legs in all directions. Several ungulates, including pigs, tapirs, camels, and tiny chevrotains, have sharp tusklike teeth. Body weight alone may pose a safety hazard, should part of a keeper's body be caught between the animal and a solid object. It is vital for keepers to understand the potential dangers involved with their animals, and to work wisely to avoid injury.

Moving an animal to another enclosure before servicing is the best way to ensure keeper safety. Gates should be firmly latched and locked, since charging and butting may force unsecured doors open. When working close to barriers, keepers must remain aware that charging animals may cause fences to bulge outwards. Horns, antlers, mouths, and hooves can also be extended into keeper spaces; as an extreme example, the long horns of oryx antelope (*Oryx* spp.) can spear an object a meter away through a chain-link fence.

If keepers must enter an enclosure with hoofstock, they should pay constant attention to animal behavior—starting before entering the enclosure and continuing until after servicing is complete and the enclosure is secured. The safety of both keepers and animals depends on recognizing changes in behavior, responding proactively, and knowing escape routes: keepers should immediately remove themselves from any potentially unsafe situation. If necessary, servicing can be delayed to allow the animals to calm down, or other keepers can be sought for backup. The presence of two or more keepers provides increased opportunities for observation and an immediate source of assistance in case of emergency; in many cases, ungulates will also keep a greater distance from multiple people than from one. A keeper should never hesitate to ask a colleague for assistance if uncomfortable in the presence of hoofstock.

Ungulates normally tolerant of keeper presence may be more flight-prone during and after a traumatic event such as a severe storm, construction, or veterinary work. Their temperament may also be affected by changes in season, social structure, and hormonal activity; male ungulates, particularly deer, can be very aggressive during the breeding season (rut). Space and social considerations also play a role, and animals that are isolated or in confined spaces may behave differently than their counterparts in larger enclosures or more social settings. Habituated hoofstock are more likely to approach keepers when seeking food or attention, or when defending mates, infants, or territory. Although keepers may desire this close connection, maintaining a healthy respect for—and distance from—all ungulates is prudent, since all animals have the potential to be unpredictable. The best practice is to reduce the motivation for approaching closely, such as by cleaning around feeders before bringing food into the enclosure. Bold keeper movements, which usually cause hoofstock to flee, can invite combat in unafraid ungulates. Brooms, rakes, and other tools can be used as passive barriers to keep animals back, but physical contact (striking an animal) may incite aggression and should be avoided as a method of defusing a potentially dangerous situation.

NUTRITION

Most hoofstock are exclusively herbivorous, although a few groups, like pigs (Suidae) and duiker antelope (Cephalophinae), will consume animal matter opportunistically. Since mammals do not produce fiber-digesting enzymes, ungulates rely on symbiotic bacteria and protozoa (microbes) to digest (ferment) plant fibers, gaining the added benefit of microbe-produced vitamins and energy-rich fatty acids. These fermentation by-products are best absorbed in the upper intestines, so that species which accommodate microbes in their enlarged multichambered stomachs before the intestines (foreguts)—such as camels, hippos, and ruminants—are more efficient at extracting nutrients than those with expanded lower intestines (hindguts), like pigs, peccaries, and all odd-toed ungulates. The result is that foregut fermenters typically require much less food than similarly-sized hindgut fermenters (monogastrics).

Plant-produced fiber comes in many different forms, which has led to a range of dietary specializations in ungulates. Roughage feeders like wild horses (*Equus* spp.) and wildebeests (*Connochaetes* spp.) have high-crowned teeth and muscular stomachs to cope with fibrous plants like grasses. A reasonable approximation of a grazer's diet can be achieved with fiber-rich grass hay like timothy or Bermuda grass. Conversely, browsers such as dik-dik antelope (*Madoqua* spp.) and musk deer (*Moschus* spp.) selectively forage on leaves, buds, and other high-protein, low-fiber plant parts. These concentrate selectors have low-crowned teeth and smaller stomachs, and therefore avoid fibrous stems and usually refuse grass hay. Legume hay like alfalfa is a better browser choice, but it is far from optimal due to the hay's large stem fraction and small, sparse leaves. Ungulates with intermediate feeding strategies, such as gazelles (*Gazella* spp.) and Père David's deer (*Elaphurus davidianus*), feed on a wide variety of vegetation in the wild and benefit from a mixture of the two hay types in captivity. When both hays are fed together, the amount of protein-rich (and therefore palatable) legume hay should be restricted to encourage consumption of grass hay; for grazing species, legume hay should be fed only in limited quantities to prevent digestive distress.

Captive browsers frequently develop oral stereotypies and digestive issues because of the poor match with their natural diet: browse is their optimal fiber source. Formerly used strictly for enrichment, browse is slowly being integrated into nutritional plans thanks to production and preservation innovations. Keepers should consult a horticulturalist or browse reference (see further reading) before feeding any browse: toxic species may cause vomiting, inappetence, diar-

rhea, excessive or foamy salivation, paralysis, or even death (Rietschel 2002, 110–12).

Hay is the cornerstone of hoofstock diets, providing fiber and occupation in the form of chewing. However, local growing conditions often result in nutritional deficiencies. Supplementation with pelleted concentrates (which contain all major nutrients) helps to ensure that captive diets are nutritionally balanced. A variety of pelleted formulas are available commercially to meet species-specific requirements. For instance, Dall's sheep (*Ovis dalli*) do well on standard ruminant pellets, but European mouflon (*Ovis aries musimon*) require a low-copper version. New browser formulas are also being developed to address the common nutritional issues of the group. Balancing nutritional and occupational needs of hoofstock requires an appropriate ratio of hay and concentrate, typically 25% to 40% pellets by weight (Lintzenich and Ward 1997). Drinking water must be provided for proper digestion of these dry feeds: automatically filling reservoirs and manually filled water bowls are two popular options. Where freezing is a concern, heated water sources are needed to ensure constant availability; ungulates are also more likely to drink water that is not extremely cold.

Nutritional demands may fluctuate between seasons and (for females) may increase dramatically during pregnancy and lactation, thus requiring keepers to adjust diets on the basis of their observations of food consumption and body condition. The amount of food offered can be easily increased or decreased, but changes to the diet's composition or to the relative proportion of feeds must be made gradually (over several weeks) to allow the gut microbes to adapt. Adding or removing components too quickly—including giving unaccustomed animals free access onto rich pasture—can lead to diarrhea or constipation. Some nutritional issues can be resolved with supplements like salt (sodium chloride) and coat conditioners; foregut fermenters can also benefit from trace minerals like cobalt, thanks to their symbiotic microbes (monogastrics cannot absorb mineral-based by-products). Fruits and vegetables (produce) are rarely used for nutritional purposes, due to their high sugar levels and low fiber content. However, their palatability can be used to encourage consumption of supplements or medication and to provide motivation during training and enrichment. Since produce ferments rapidly once ingested, only very limited amounts should be fed.

FEEDING

Feeding captive diets in an appropriate manner is important for encouraging consumption. Exposed locations may discourage feeding in inclement weather; sheltered locations have the added benefit of minimizing nutrient leeching and spoilage (pellets in particular disintegrate when wet). Feeding methods should cater to a species' foraging ecology: browsers like giraffes (*Giraffa camelopardalis*) may be reluctant to feed at low feeding stations, and grazers like white rhinos (*Ceratotherium simum*) may be unwilling or unable to feed from elevated feeders. Ground-level feeding can result in substrate ingestion, a risk that can be minimized by feeding pellets from troughs or bowls. Such feeders can also discourage freeloading pests (e.g., rodents and waterfowl) that might otherwise consume a large proportion of the rations. In group situations, multiple feeding sites are often needed to ensure that dominant animals do not exclude subordinates from food or water. Creeps or exclusion feeders are another solution, using body size or the presence of horns or antlers to provide less competitive animals with access to food. By confining feeds, feeders allow good estimations of food intake. Leftover food should be removed daily and the amounts offered should be adjusted to prevent excessive waste. Feeders are usually dry-cleaned (swept out) and sporadically washed, but wet cleaning may be needed daily with animals that salivate copiously, like giraffes. Feeders should be dry before being filled, since moisture accelerates spoilage of pelleted feeds.

The variety of mouth sizes, feeding positions, and horn and antler morphologies in exotic hoofstock creates a number of species-specific challenges when selecting feeders and waterers. Horns, antlers, and tusks are prone to entanglement. Hay nets are unsuitable for hoofstock with these features, and hayrack bars should be spaced to accommodate only the muzzle (since horns and antlers can catch on wider bars, entrapping the head). Conversely, water bowls often need larger openings and should be located away from obstructions to ensure that horned individuals can freely access water. Pigs, which tend to root, should be fed and watered from weighted troughs to prevent overturning; a nipple-style waterer is another option successfully used with several wild pig species.

Keepers should always pay close attention to food items. Pellets should be dry and firm; dustiness, clumpiness, and the presence of mold indicate spoilage. Flakes of hay should be pulled apart and examined using sight, smell, and touch: it should optimally be soft, green, leafy, and sweet-smelling, while inferior hay may be brittle or brown. Quality checks will also highlight hazards like wire, broken baling equipment, and baling twine, which may cause trauma to the mouth or gut, or carcasses, bird droppings, mold, and weeds, which may be sources of diseases or toxins. Problematic feeds should be discarded and reported; ingestion of inappropriate material can have severe health consequences.

CLEANING

The feeding ecology of ungulates makes cleaning an essential duty of hoofstock keepers. Parasites and diseases sustained by fecal-oral transmission are easily passed among confined ungulates feeding from fecally contaminated areas. Feces should always be removed daily from around feeders; regular removal from the entire enclosure is also important, particularly for grazing species. Most hoofstock produce hard, pelleted feces which can be raked or swept up easily, leaving minimal residue. Dry cleaning is often acceptable, conserving water, reducing potential slip hazards, and prolonging the life of bedding materials. Species with soft feces or messy toilet habits, such as hippos, are an exception: hosing and scrubbing may be needed daily.

HANDLING

Hoofstock instinctively avoid potential threats by maintaining their flight distance, and will typically move away if approached by a keeper. Used carefully, this innate avoidance behavior can be harnessed as a low-stress handling tool to move ungulates between enclosures and to recapture escaped individuals; Grandin (2005) provides an excellent perspective that all hoofstock keepers should read. Patience is necessary when working with ungulates. Walking at a slow pace behind hoofstock usually incites calm, directional movements, so long as keepers approach only as close as is needed to get the animals moving forward. Trying to hurry ungulates by making loud noises or by approaching too closely or too quickly is counterproductive: fear increases stress, and results in erratic fight-or-flight behavior.

RESTRAINT

Although hands-off handling is largely sufficient when managing hoofstock, physical handling is sometimes required, such as for medical purposes. To reduce the inherent stress of restraint, blindfolds and earplugs can be used on the animals to minimize sensory input and thereby induce calm and minimize struggling. Keepers should also work efficiently to prevent prolonged handling. Before any restraint procedure, the staff involved should review their individual roles, the intended movement of animals, and potential hazards (including horns, tusks, and hooves) to ensure everyone's safety.

Understanding the risks to the animals is also vital. Capture myopathy, in which extreme stress and exertion cause muscle cells to die, is a serious and sometimes fatal risk in hoofstock restraint. Treatment is difficult once the condition begins, making preventative measures essential: high ambient temperatures, extended duration of chases and restraint, and excessive restraining force should all be avoided. Supplemental selenium and vitamin E may be administered during restraint procedures to prevent deficiencies that can exacerbate muscle issues (CAZWV 2009, 9.20–9.23). Another principal concern is regurgitation and aspiration of stomach contents, which may cause fatal pneumonia in restrained hoofstock; ruminants are particularly at risk, due to their large foreguts. Veterinarians may suggest that ungulates be fasted to reduce their regurgitation, particularly before being chemically immobilized. A 12- to 24-hour fast is typical for monogastric ungulates, but recommendations vary for foregut fermenters because their stomach chambers always retain fluids. During restraint procedures, keepers should watch carefully for signs of regurgitation such as green froth in the nose or mouth, or heavy, wet breathing (CAZWV 2009, 9.15). Whenever possible, hoofstock should be restrained belly down (sternally); the head should be elevated, with the nose pointing downward to allow any fluids to drain out of the mouth. Lateral positioning carries greater regurgitation risks for ruminants: when the animal's left side is on the ground, downward pressure from the body's weight can force stomach contents up into the esophagus, but when the animal is restrained on the right side, the force of gravity can draw fluids out.

PHYSICAL RESTRAINT

The capturing and handling of small ungulate species is often done by hand; this method is faster and poses fewer risks to delicate limbs than the use of nets. Bush (1996, 33) suggests a 15 kg maximum body size for manual restraint, although hoofstock weighing 45 kg can be successfully restrained by a coordinated team of keepers. Captures are quickest in small enclosures, which limit mobility; the natural tendencies of hoofstock to move as a group and run around the perimeter can be exploited to quickly catch individuals. Once the animals are in hand, their struggling can often be reduced by lifting at the groin, so that the rear hooves do not make contact with the ground; species or individuals that continually struggle are better suited to other restraint methods. Horns are convenient handles to hold during catch-ups, but forceful twisting should be avoided lest the outer sheath detach from the bony core beneath, particularly in young animals. Deer antlers, which are grown and shed annually, are unsuitable handles: while growing they are sensitive, vascularized, and easily injured, and when mature the gradual weakening of the connection to the skull may permit them to be broken off with minimal force.

Mechanical restraint devices are commonly used for larger and tough-to-handle ungulates. Standing chutes, essentially narrow hallways, are easily created in existing corridors and work well with cooperative training programs. Physical restraint can be performed using drop-floor chutes and hydraulic tamers, which gently squeeze an animal with moveable floors and walls, and have access panels to reach various body parts. Restraint devices are typically associated with a series of pens and alleyways to sort and separate animals. Walls in such facilities should be solid (with peepholes for monitoring) to eliminate external stimuli that may cause ungulates to balk; similarly, curved runways promote better forward movement than those with visible dead ends.

CHEMICAL RESTRAINT

For situations in which physical restraint is impossible or invasive procedures are involved, immobilizing drugs may be used. Anesthesia should be performed by a veterinarian, and keepers should be prepared to follow their directions throughout the procedure. By disrupting muscular function, immobilizations can hamper normal processes, leading to regurgitation or stalled breathing. On occasion, fermentation gases may build up in the stomachs of foregut fermenters and may further hamper breathing and stimulate regurgitation. This condition, observed as abdominal bloating, can usually be resolved by shifting the animal's body position to permit the gases to be burped up (eructed). After the immobilizing drugs are reversed, the ungulate should be kept in a dark, quiet stall as its poor post-procedure coordination increases the risk of self-injury. Depending on the drug, the recovery period may last for up to 72 hours, as signs of sedation may reappear after an animal has apparently recovered (this is called renarcotization or resedation).

BEHAVIOR

TRAINING

While shifting can be accomplished using herding techniques, formal training programs can be used to reduce flight responses and facilitate cooperative medical treatment. Ungulates are adept learners and can be taught numerous behaviors. A sampling of published articles is cited in the resource section, highlighting the training process for behaviors such as voluntary blood draws, ultrasounds, semen collection, hoof care, and tusk trimming.

A stumbling block in using positive reinforcement to train hoofstock is in finding effective reinforcers. Food may not be a strong motivator, especially for ungulates with constant access to forage. Common food rewards for ungulates include concentrates, browse, and produce. Other novel dietary items may also increase motivation: primate leaf-eater biscuits, for instance, tend to be popular with many browsers. Veterinarians and nutritionists can help adjust diets to accommodate training additions, and can also highlight concerns (onions, for instance, are toxic to most hoofstock). Other reinforcers can be found in training articles and husbandry manuals (see additional readings); tactile reinforcement has proven effective with several ungulates, including pigs, tapirs, and rhinos.

The close proximity to animals during training can pose a danger to keepers. This risk can be reduced by minimizing opportunities for undesired contact (e.g., biting or crushing force), such as by using long-handled brushes to touch animals. When training without a protective barrier, a two-keeper policy is highly recommended. Training through a barrier is sometimes seen as a hindrance, but it can be a lifesaving precaution when keepers work with large or aggressive hoofstock. The training of timid species may even be accelerated with a barrier, as physical separation from keepers can give the animals a sense of security, reducing flight distance and nervousness.

BEHAVIORAL ENRICHMENT

Encouraging natural behavior is a principal goal of enrichment, and Burgess (2004) provides a wealth of enrichment ideas for hoofstock. Occupied animals are less likely to exhibit stereotypical behaviors as a result of their captivity. Severe problems such as self-mutilation are rarely seen in hoofstock, but captive ungulates spend far less time feeding than their wild counterparts, and may therefore develop oral stereotypies like object-licking and tongue-rolling. Multiple feedings per day and the use of feeders (from simple hay racks to more complex puzzle toys) help eliminate these issues by promoting foraging. Providing additional hay or browse also helps by increasing chewing opportunities.

Many natural behaviors can be stimulated environmentally. A variety of substrates provides options for dust bathing, mud wallowing, and grazing; topographical diversity increases exercise while creating lookout points and sheltered refuges; and exhibit furniture like rubbing posts and deadfall encourages grooming and play. In addition to these choices, movement and exploration can be encouraged by thoughtfully positioning food, water, and shelter around the enclosure. Safety concerns must be addressed before enrichment is offered. Hanging items should be used with caution; lengths of rope, chain, and cable should be sheathed with pipe to prevent strangulation and entanglement. For the safety of the public, heavy enrichment and furniture for powerful species should be anchored to prevent them from being tossed wildly around.

The alertness and suspicion of ungulates can make providing them with novel stimuli a challenge. New objects may be viewed as a threat, and should not be placed near gates through which animals are expected to move. Nervous animals may derive enrichment from observing items placed outside of their enclosure; this is a good initial step before the objects are brought into the enclosure. As ungulates gain experience with novelty, their suspicion towards new things becomes less severe: enrichment makes change a part of their routine and helps reduce stress in unplanned unusual situations.

Group housing is arguably the best enrichment for herd-living ungulates, as it encourages social behavior like herding, hierarchical establishment, and breeding. Social interactions can also enrich typically solitary species, and need not be limited to the same species: mixed-species exhibits, in which hoofstock are housed with other animals, are common in modern zoos. Several databases (e.g., on the AZA Antelope TAG website) document experiences with mixed-species combinations. Mixing of ungulates should be attempted with an understanding of risk, as some groupings may be ill-advised due to behavioral incompatibility, enclosure setup, or even possibilities of hybridization.

HOOFSTOCK INTRODUCTIONS

Understanding a species' natural social tendencies and the demeanor of the animals involved is key to planning a successful introduction. For many species, females tend to be easier to maintain in groups; males, which must compete for access to mates in the wild, are usually more aggressive and may be intolerant of other males. Young ungulates are easier to integrate into established herds, on account of their sexual immaturity and the minimal threat they pose to existing hierarchies. In contrast, new adults are often harassed, especially by individuals of the same sex, in order to establish dominance.

Aggression in the initial introduction stages can be buffered by allowing restricted contact through a barrier. Even after ungulates have become accustomed to each other's presence, sparring matches and chases are common when they are introduced to the same space for the first time. Introduction locations should have sufficient space for the animals to get away from each other, with circular routes to prevent individuals from being cornered. Agonistic encounters help establish social order and should be allowed to occur; separating animals during a fight or chase can increase their aggression in later introduction attempts. However, excessive aggression between incompatible individuals may lead to injury or death. Keepers should therefore closely monitor introductions and keep records of behavior. Dominant or aggressive individuals can be preemptively impeded by blunting their tusks, sawing off their antlers, or sheathing their horn tips with rubber hose, tennis balls, or resin spheres. Tranquilizing drugs can be used to disrupt social patterns and permit new animals to integrate into a group, but timing introductions with the animals' reproductive cycles—when their sexual activity overrides other social factors—is often the most successful method.

REPRODUCTION

Captive environments do not always suit innate reproductive cycles: hoofstock infants born during freezing winters or scorching summers may experience high mortality. To maximize infant survival, it is common practice to time breeding introductions so that, based on the species' gestation period, births occur during the optimal birthing season (often in spring). Males are usually added to a group when females begin to enter their period of reproductive receptiveness (called estrus or "heat"). A variety of physical and behavioral signs can be used to detect estrus, including swelling and mucus discharge from the vulva, mounting by other females, and often increased vocalization. The behavior of males can also provide important clues: they are adapted to detect subtle reproductive signals, and often show increased agitation and competitiveness when near a receptive female.

PREGNANCY AND BIRTHING

After breeding introductions occur, recorded observations of mating are useful for determining when males can be separated, which females may be pregnant, and when births should be expected. Among pregnant ungulates, physical signs of impending birth (parturition) include a prominent udder, a swollen vulva, and a shift in how the fetus is carried. Hoofstock births usually occur at night or in the early morning, and as labor begins, expectant mothers tend to seclude themselves and become restless. Females near parturition are sometimes kept in maternity stalls, which should be well bedded to provide cushioning to the newborn and to absorb fluids discharged during birth; these might otherwise cause infants to slip or splay, potentially causing life-threatening injuries.

Single infants are typical of many ungulates, but twins, triplets, and (in the case of wild pigs) litters up to twelve may occur. Hoofstock infants are universally precocious, able to stand and nurse soon after birth. This is obvious in "follower" ungulates like wildebeests (*Connochaetes* spp.), in which infants closely accompany their mothers (dams) from the moment they gain their footing. In contrast, the monitoring of mother-infant interactions is more challenging with "hider" species like white-tailed deer (*Odocoileus virginianus*), in which the dam leaves the infant in a concealed spot and visits it two to four times per day for nursing; only after a period of days, weeks, or sometimes months is the infant consistently seen in its mother's presence. Knowing the species' biology is therefore essential in interpreting whether an isolated infant has been abandoned or is acting normally.

Identifying the dam (and the sire, if known) is important for management programs; in herd situations, new mothers can be identified by physical evidence, like fluid stains on the hind legs and afterbirth hanging from the vulva, and behaviorally, using cues such as nursing and defensiveness. Observations of nursing are important for assessing the infant's health, as the neonatal immune system depends on milk-borne antibodies during the first few days after birth (the antibody-rich milk is called colostrum). Proper positioning at the udder does not itself indicate nursing; milk acquisition is better inferred from an enthusiastically wagging tail during nursing or a milky muzzle afterward. Maternal behaviors to watch for include grooming, tolerance of nursing, and licking of the neonate's anogenital region to stimulate defecation. Keeper observations should be made from a respectful distance, since perceived threats can discourage normal mother-infant interactions. Remote video may be the best option for monitoring highly sensitive species.

INFANT CARE

Hoofstock neonates usually receive a veterinary exam 24 to 48 hours after birth. Performing the exam any sooner risks disrupting the mother-infant bonding process, and performing it any later can allow the cursorial skills of infants to outmatch those of keepers. Infants are almost always caught and restrained by hand; the capture should be done quickly, since neonates tire easily and can severely injure themselves while running on unsteady legs. Defensive mothers pose a risk to keepers during infant capture and examination, and even typically shy animals may be very bold in response to an infant's distress cries. For the safety of everyone involved, the dam and any other group members should first be separated into another enclosure. If this is not possible, the infant should be taken to a protected area for the examination.

Young infants usually require minimal restraint. Smaller hoofstock will often rest quietly when held in a keeper's arms, while larger individuals are usually held in a prone position. During the examination, veterinary staff will confirm the infant's gender, apply permanent identification (such as ear tags, ear notches, or microchips), and check for congenital problems like cleft palate or imperforate anus. A small blood sample is frequently drawn to test for glucose and antibody levels (to confirm nursing), and injections of antibiotics, vaccines, and other supplements may be given. The way an infant is returned to its mother after the examination depends on the species. Followers set down in view of their mothers will usually run directly back to them, while hiders should be returned to their caching spot. Keepers should then watch to ensure that the infants are successfully reunited with their mothers and that maternal care has not been disrupted.

Difficult decisions must occasionally be made during neonatal checkups. Untreatable conditions may warrant euthanasia; culling is also used by some institutions for population management, especially with surplus males. Treatable medical concerns or maternal neglect may require infants to be hand-raised. Some zoos also purposely hand-rear skittish species like duikers (Cephalophus spp.) and gazelles (Gazella spp.) to facilitate their habituation and reduce trauma-related mortality. Hand-raised infants can become imprinted on humans if reared in isolation, preventing successful integration with conspecifics. Socialization with other ungulates is important in promoting species-typical behaviors, and it reduces the likelihood that aggression or courtship will be directed towards humans in adulthood. To this end, keepers should avoid "roughhousing" with young ungulates, since it can encourage habits that are dangerous in adults. If hand-raising is needed for reasons besides desensitization to humans, it may be possible to maintain an infant in its natal group while providing it with supplementary feedings or treatment (Read and Meier 1996, 43); this strategy minimizes the potential for imprinting.

Several resources are available for hand-rearing hoofstock (e.g., Greene and Stringfield 2002), providing guidance on milk formulas and amounts, feeding schedules, nipple sizes, and weaning times. As juveniles transition onto solid foods, they must acquire fiber-digesting microbes. Many of these are ingested from the environment, but if body condition and fecal consistency suggest that an infant is not digesting fiber properly, its gut can be inoculated by adding a sample of screened feces or stomach contents (for ruminants) from the natal group to the infant's food.

CONTRACEPTION

The limited space in zoos requires planning to avoid the production of surplus animals, but preventing reproduction can cause physical and behavioral issues. Indeed, female ungulates that do not breed for several years may become effectively sterile with the onset of physical and hormonal changes that prevent conception; this phenomenon nearly destroyed the North American population of Przewalski's horses (Equus ferus przewalskii). Because each contraceptive option has different costs and benefits, a combination of methods is the best choice for sustainable population management.

Breeding is most readily controlled by separating males and females. However, disrupting natural mixed-sex herd structures can lead to unstable hierarchies and increased aggression among females. Similarly, isolated males may lose their normal social behaviors; if separated from the herd at a young age, they may become socially incompetent and unable to successfully court or breed females when eventually placed into a breeding situation. Bachelor groups comprised solely of males are one way to provide socialization, and short-term successes have been achieved with several ungulates, including Speke's gazelle (Gazella spekei) and Grevy's zebra (Equus grevyi). To circumvent natural aggressive tendencies among males, these groups are best created with similarly-aged, sexually immature animals, and should optimally be kept away from visual, auditory, and olfactory contact with females. Monitoring the behavior of these groups is important as males age and mature: bachelor herds are rarely stable over the long term.

Surgical contraception of males (castration or vasectomy) eliminates the need to separate the sexes to prevent reproduction. Castrated males do not develop testosterone-induced characteristics; this reduces aggression but also reduces sexual markings, manes, and musculature. Vasectomized males, on the other hand, retain these physical and behavioral traits; their aggression can even be increased, since they often become competitive and aggressive to other individuals whenever a female comes into estrus (a frequent occurrence when pregnancy is prevented). Surgical contraception of females is significantly more invasive, and is usually done only for medical reasons.

Chemical contraception is currently only effective for female hoofstock, where most options work by interrupting estrus cycles. This not only allows the sexes to remain together, but eliminates many of the behavioral consequences described above. However, while usually reversible in the short term, some contraceptives can cause sterility when used for extended periods, thus tempering their social benefits. The AZA's Contraception Center (listed in suggested websites) is a primary resource for chemical contraception.

TRANSPORTATION

Transportation options are limited for most exotic hoofstock. Crates are the most practical method for transporting powerful ungulates, and are usually necessary for moving animals by airplane. A crate should be sized appropriately to allow the animal to stand up and lie down, but too much space can allow a stressed individual to injure itself. Horns and antlers also need to be considered, as they may require significant additional height or width. When groups need to be relocated, livestock trailers provide an efficient way to move compatible individuals together. Bedding should always be provided to provide cushioning and traction during transportation.

Most ungulates will not willingly enter a strange transportation container, but can be gradually desensitized to do so using dietary rations. Desensitization to the closing of crate doors is beneficial, but must be done slowly and well in advance of shipment; following a negative experience, hoofstock often balk at re-entering an enclosed environment. When training is not feasible, manual handling can be used to crate smaller species. Larger species may require mild sedation or closer approaches within their flight zone to get them to enter a confined transportation space.

During transportation and after arrival at the destination, darkness and quiet can help calm hoofstock. Mild sedation of nervous individuals can further reduce stress. Excited individuals should be given time to calm down before being offloaded, to prevent them from rushing blindly into the new environment. Crated hoofstock should be released from the rear of the crate: backing them out minimizes the chances that a traumatic collision will occur. As further insurance, some arrival stalls have padded walls to reduce the risks of trauma; unpadded walls can be lined with straw bales if it is deemed necessary.

VETERINARY CARE

The veterinary care of exotic hoofstock draws heavily on the techniques developed for domestic livestock, although the similarities between these groups must be considered alongside significant and sometimes unexpected differences. Injuries are particularly challenging to treat, making the elimination of potential sources of trauma (e.g., uneven substrates and unsafe barriers) the better option. Proactive training is highly recommended to facilitate medical management.

GASTROINTESTINAL ISSUES

The most frequently encountered hoofstock health issues occur along the digestive tract. Colic is an umbrella term for symptoms such as bloating, abdominal discomfort, and diarrhea or constipation. Colic has many causes, including impactions of substrate, hair, and other foreign materials; insufficient water consumption; rapid dietary changes; excessive fermentation; or twisting of the intestines. In mild cases, keepers can relieve the symptoms by withholding highly fermentable foods (e.g., produce and concentrated feeds), providing warm water to stimulate drinking, and encouraging exercise. Severe cases may require emergency surgery.

Colic tends to be acute, while intestinal parasites are often a chronic problem. Many ungulates harbor parasites without showing ill effects, but high parasite numbers may cause loose stool, poor body condition, weight loss, or even sudden death.

Suspected infestations should be confirmed by examining feces for eggs and parasites prior to treatment with medications like fenbendazole or ivermectin. Prophylactic treatment in temperate climates is often performed in spring and late summer when parasite egg counts typically rise. After medication, follow-up tests will confirm whether the treatment was effective and whether additional treatment is needed. Parasites are best controlled with good hygiene around feeding areas. In severe cases, parasite transmission can be hindered by restricting the animals' grazing opportunities, regularly mowing grass, providing rocky or sandy terrain, and bringing the ungulates off paddocks at night. Moving animals to different enclosures can also break the cycle.

If an ungulate has trouble eating, salivates excessively, or chews more frequently than usual, there may be issues with its teeth. Older animals are more prone to overgrown teeth, which are corrected by rasping (or "floating") excess enamel ridges to realign the chewing surfaces. Other oral issues, like gum abscesses or trauma to the mouth, may develop into extensive swelling, known as "lumpy jaw." Because swelling is seen only in chronic cases, early diagnosis is challenging; treatment usually requires that teeth from the affected area be removed.

HOOF CARE

Activity levels, genetics, and injuries may make some animals prone to hoof problems even if they are housed on appropriate substrates. If left untreated, an overgrown hoof may crack or separate from the sole of the foot, thereby straining the underlying bone structures. Vascular issues, including high blood acidity levels from the rapid fermentation of rich feeds, may cause the hoof structures to become inflamed and to separate from each other. In severe cases, this laminitis can develop into a painful condition known as founder, in which the terminal bone of the foot rotates away from the hoof.

Hoof problems can affect an animal's quality of life and its ability to breed, and may even result in death. Early detection and treatment is far easier than dealing with progressed hoof disease. Exotic ungulates can be trained to allow voluntary footwork, but usually this work is performed under anesthesia. The resulting unusual positions and species-specific morphologies can make trimming overgrown hooves a challenge even for experienced personnel (in some regions, all hoof work must be performed by licensed farriers). When a hoof is trimmed, its keratinous wall should be shaved down in numerous thin passes to prevent the sensitive living tissues within the hoof from being exposed.

DISEASES

Exotic ungulates are susceptible to many of the same diseases that affect domestic livestock, and the risk of transfer between these two groups (and to humans) has resulted in tight medical regulations for hoofstock. Zoos that house multiple ungulate species must also contend with the transfer of disease (e.g., malignant catarrhal fever) from asymptomatic carrier species to neighboring susceptible species. Important hoofstock diseases (Rovid-Spickler and Roth 2006, 113–245; Junge 2007, 1–2) include

- transmissible spongiform encephalopathies (prion diseases), like bovine spongiform encephalopathy (BSE or "mad cow disease"), scrapie, and chronic wasting disease (CWD)
- bacterial diseases, including anthrax, brucellosis, leptospirosis, bovine tuberculosis (TB), and Johne's disease (paratuberculosis)
- viral diseases, including foot and mouth disease (FMD), bluetongue, malignant catarrhal fever (MCF), rinderpest, equine encephalitis, and West Nile virus.

These diseases are rarely seen in healthy, well-managed zoo collections due to government-enforced quarantines, which maintain new animals in isolation for a period of testing lasting at least 30 days. Permanent quarantines, like Permanent Post-Entry Quarantine (PPEQ) in the United States, may be mandated for ungulates arriving from the wild or other high-risk areas. Quarantine regulations are in place to ensure the health and safety of human and animal populations, and they must be closely followed.

MEDICATING

When an illness or disease requires treatment, a principal challenge is delivering medications to hoofstock. Noninvasive oral medications are usually offered on favored foods like pellets or produce; consumption can be further encouraged by holding back other rations until the medication is consumed. Oral medications tend not to be used with foregut fermenters, as their stomach volume hampers timely absorption and the foregut microbes may neutralize drugs. Long-term oral antibiotics also risk destroying the microbial population, requiring gut inoculation after treatment. In contrast, injectable drugs ensure that therapeutic levels are achieved without harming the digestive microbe balance, but regular delivery can be nearly impossible unless animals have been trained for voluntary injections. Where nervous ungulates require extended treatment, long-lasting tranquilizing (antipsychotic) medications can make the procedures safer and less stressful.

CONSERVATION AND RESEARCH

Many wild ungulates are of conservation concern, as their relatively large size makes them vulnerable to habitat loss and hunting. Unfortunately, several recent ungulate species are already extinct: in 1883 the last quagga (*Equus quagga quagga*, a relative of the zebra) died at the Artis Zoo in Amsterdam, and in 1938 the last Schomburgk's deer (*Rucervus schomburgkii*) was killed in a temple zoo in Thailand. Despite these losses, several ungulates owe their continued existence to captive breeding, including the Przewalski's horse (*Equus ferus przewalskii*), the Père David's deer (*Elaphurus davidianus*), the Arabian oryx (*Oryx leucoryx*), and the European bison (*Bison bonasus*). These species were once extinct in the wild, but zoos have preserved them all and reintroduced them to their native ranges.

Zoos are developing more partnerships with in situ projects, providing funding and expertise to help conserve ungulates in the wild. In captivity, research is being conducted on assisted reproductive technologies like artificial insemination.

These technologies, common in the livestock industry, have been developed for several exotic ungulates including the bongo (*Tragelaphus eurycerus*), the banteng (*Bos javanicus*), and the gerenuk (*Litocranius walleri*). Their widespread use is limited by the species-specific nature of hormones, anatomy, and physiology, as well as by the expense involved. Once developed, however, these assisted reproductive techniques allow zoos to transfer gametes instead of animals and, through gamete preservation, involve deceased individuals in breeding programs. In the future, gamete transfer may permit gene flow between zoos and the wild, although tight quarantine regulations on biological samples (including semen) remain a major hurdle.

SUMMARY

The diversity of ungulates makes generalizing many aspects of their husbandry a challenge. Providing appropriate care in regard to housing, group size, and diet requires hoofstock keepers to research and understand the natural history of the species they care for. Keepers must work in a calm, predictable manner and develop keen observational skills to overcome the survival adaptations of ungulates. Although hoofstock is sometimes challenging to work with, the benefits of maintaining these species in zoos are immense, and captive conservation programs continue to directly enhance populations of ungulates in the wild.

REFERENCES

Burgess, Amy. 2004. "The Giraffe in Captivity: Enrichment." In *The Giraffe Husbandry Resource Manual*, ed. Amy Burgess, 139–52. Silver Spring, MD: Association of Zoos and Aquariums Antelope/Giraffe Taxon Advisory Group.

Bush, Mitchell. 1996. "Methods of Capture, Handling, and Anesthesia." In *Wild Mammals in Captivity: Principles and Techniques*, ed. Devra G. Kleiman, Mary E. Allen, Katerina V. Thompson, and Susan Lumpkin, 25–40. Chicago: University of Chicago Press.

CAZWV (Canadian Association of Zoo and Wildlife Veterinarians). 2009. *The Chemical Immobilization of Wildlife, 3rd Edition*. Winnipeg: Canadian Association of Zoo and Wildlife Veterinarians.

FAO (Food and Agriculture Organization of the United Nations) Database. 2009. Global Stocks of Domestic Hoofed Mammals for the Year 2007. Available online at http://faostat.fao.org/site/573/default.aspx.

Fitzwater, William D. 1972. "Barrier Fencing in Wildlife Management." Proceedings of the 5th Vertebrate Pest Conference, 49–55. Available online at http://digitalcommons.unl.edu/vpc5/.

Grandin, Temple. 2005. "Principles for Low Stress Cattle Handling." Utah State University Cooperative Extension Paper AG/beef/06. Available online at http://extension.usu.edu/files/publications/beef6stress.pdf.

Greene, Kelley, and Cynthia Stringfield. 2002. "Exotic Ungulates." In *Hand-Rearing Wild and Domestic Mammals*, ed. Laurie J. Gage. 256–61. Ames: Iowa State University Press.

Herre, Wolf, and Manfred Röhrs. 1990. "Domestic Mammals." In *Grzimek's Encyclopedia of Mammals*, Volume 5, ed. Sybil P. Parker, 529–35. New York: McGraw-Hill.

Hohn, Timothy C. 1986. "A Brief Survey of Plants Used in Hoofed Stock Exhibits." Longwood Graduate Fellowship Report. Available online at http://dspace.udel.edu:8080/dspace/bitstream/19716/3140/1/hohn_1986.pdf.

Kranz, Karl R. 1996. "Introduction, Socialization, and Crate Training Techniques." In *Wild Mammals in Captivity: Principles and Techniques,* ed. Devra G. Kleiman, Mary E. Allen, Katerina V. Thompson, and Susan Lumpkin, 78–87. Chicago: University of Chicago Press.

Lintzenich, Barbara A., and Ann M. Ward. 1997. "Hay and Pellet Ratios: Considerations in Feeding Ungulates." Nutrition Advisory Group Fact Sheet 006. Available online at http://www.nagonline.net/Technical%20Papers/technical_papers.htm.

Read, B. W., and J. E. Meier. 1996. "Neonatal Care Protocols." In *Wild Mammals in Captivity: Principles and Techniques,* ed. Devra G. Kleiman, Mary E. Allen, Katerina V. Thompson, and Susan Lumpkin, 41–55. Chicago: University of Chicago Press.

Rovid-Spickler, Anna, and James A. Roth, ed. 2006. *Emerging and Exotic Diseases of Animals, 3rd Edition.* Ames, IA: Institute for International Cooperation in Animal Biologies, Iowa State University College of Veterinary Medicine.

28

Husbandry and Care of Carnivores

Adrienne E. Crosier and Michael T. Maslanka

INTRODUCTION AND NATURAL HISTORY

The mammals classed in the order Carnivora are extremely diverse. There is a total of 274 species in this order comprising mammals of diverse sizes, structures, and behaviors. Additionally, animals in the Carnivora group reside in a wide variety of ecosystems and occur naturally on every continent except Antarctica and Australia (Carnivora species have been introduced to Australia by humans over approximately the last 5,000 years). Members of the order live in a variety of ecosystems, from tropical rain forests (e.g., jaguars [*Panthera onca*]), to arid savannah (e.g., cheetahs and wild dogs), to mountains (e.g., snow leopards) and the polar tundra (e.g., polar bears). The pinnipedia are aquatic members of this order (living primarily or solely in water), but they will be discussed in chapter 31. Although the name of the order Carnivora literally means "eaters of flesh," this group includes a wide variety of foraging and digestive strategies, well beyond carnivory. The term "carnivore" will be used throughout this chapter, but in this usage the term does not refer to specific taxonomy or diet, but will be used as a general grouping of animals based on husbandry and care needs (and thus will not include the pinnipeds; see table 28.1).

This taxon maintains a variety of social structures. Some carnivore species (e.g., tigers, leopards, polar bears, pandas, ferrets) live singly in the wild with the exception of females with their young. In these species, the males and females are together only during mating. After the young are born, the female raises them with no assistance from the male. Other carnivores live in groups such as packs (wild dogs), clans (hyenas), mate pairs (maned wolves), or prides (lions). In these species, the males and females remain together after the young are born, and both (or, in some species, all animals in the pack) contribute to rearing the offspring.

All members of this order are well equipped for hunting and eating other animals, and have teeth, claws, and binocular vision adapted for the task (this does not necessarily dictate their foraging strategy, however). Carnivores have comparatively large brains, and the structure of the skull and dentition (teeth) make this taxon different from others (Christiansen and Adolfssen 2005). The type of uterus and placentation and the position of the nipples also assist in classifying species into this taxon. Interestingly, successful "taming" or behavioral modification of an individual (Driscoll, Macdonald, and O'Brien 2009) is possible for certain carnivore species, including but not limited to (hand-raised) cheetahs, raccoons, and some foxes. In comparison, other species such as spotted hyenas, large bears (polar and grizzly) and the great cats (e.g., lions) are not regarded as being easily tamed by humans (Driscoll, Macdonald, and O'Brien 2009). Carnivores "mark" or demarcate their ranges by urinating and defecating in specific areas. Demarcation also is used by many species during mating, as the sense of smell is heightened in most carnivores and chemical cues in urine and feces are a critical mode of communication.

Carnivores are primarily either nocturnal (active mostly at night) or crepuscular (active during the dawn and dusk hours), although a few species are considered diurnal (active during daylight). Carnivores in general hunt a variety of prey including small mammals, birds, antelope, and deer. However, members of this order can be omnivorous (canids, ursids, viverrids) or even primarily herbivorous (giant and lesser [red] pandas). In general, a primary role of the carnivorous members of the order Carnivora within their ecosystems is to hunt prey species, thereby managing the numbers of these populations. Although most carnivores are opportunistic and will eat whatever they can hunt, they often target sick, weak, or elderly prey animals, thereby maintaining the overall health of these populations. Interestingly, members of the Hyenidae and Procyonidae families can act as scavengers within their ecosystems, a unique role not shared by species from many other orders. There are also members of this taxon, such as maned wolves (*Chrysocyon brachyurus*) and some Viverridae species, that supplement their diet with

TABLE 28.1. Eight families in the order Carnivora are the focus of this chapter.

- Canidae (e.g., dogs, foxes, coyotes)
- Ursidae (e.g., bears)
- Mustelidae (e.g., otters, mink, ferrets)
- Procyonidae (e.g., raccoons)
- Felidae (e.g., cats)
- Viverridae (e.g., genets, civets, linsangs)
- Herpestidae (e.g., mongooses, fossas)
- Hyenidae (e.g., hyenas, aardwolves)
- Unique members of this order for which taxonomic orientation has recently been under contention include the well-known giant panda (*Ailuropoda melanoleuca*) and the lesser panda (*Ailurus fulgens*, also known as the red panda). These two panda species have been classed in both the "bear" (ursidae) and "raccoon" (procyonid) families. Currently, most experts have them classified in the ursidae group (Olaf and Bininda-Emonds 2004).

fruits and other plant matter, often with seasonal regularity. Many species within the carnivores have specific and even limited dietary preferences or requirements. For example, the giant panda consumes primarily bamboo, although it will also eat insects, birds, and small mammals. The black-footed ferret dines almost exclusively on prairie dogs (*Cynomys ludovicianus*) while the fishing cat (*Prionailurus viverrinus*) has unique behavioral and skull modifications for hunting fish, shellfish, and other aquatic animals (Macdonald, Loveridge, and Nowell 2010, 54–55).

This chapter will describe basic principles for working successfully with a variety of carnivore species in a zoo environment. After reviewing this chapter, the reader will understand

- the basic anatomy of carnivore species
- guidelines for housing and caring for carnivores in zoos
- effects of species biology on enrichment and training programs
- specific reproductive and veterinary issues for mammals of this order
- key conservation initiatives for carnivore species.

BASIC EXTERNAL ANATOMY

The external anatomy of each family of carnivore is unique. Most carnivores are sexually dimorphic, meaning that the male is larger than the female. Felidae are native to every continent except Antarctica and Australia, requiring a variety of adaptations for each species to thrive in its individual ecosystem. Virtually every species of the felid family has distinct markings, such as spots or rosettes that provide camouflage, except for the puma (*Puma concolor*), jaguarundi (*Herpailurus yaguarondi*), and lion (*Panthera leo*), which are uniform in color (Werdelin et al. 2010, 78–80). The 37 total species of felids are divided generally into large (e.g., lion, jaguar, leopard, tiger) and small species (e.g., ocelot, black-footed cat, fishing cat). The claws of each felid species except the cheetah, are completely retractable. In contrast, the claws of canidae are not retractable. Canids are known for long legs, pointed

ears, and a long muzzle, which improves their sense of smell. Species in this family live and hunt in packs and occupy almost every major type of ecosystem worldwide. Ursidae are short-tailed, large bodied carnivores with shaggy coats and nonretractable claws. Most bear species are omnivores and have an excellent sense of smell.

The hyenidae family contains only four species, but they likely have the most unique anatomical and social structures of all carnivores. In a clan (group living together) of hyenas, the females are dominant over the males (Holekamp 2006). Hyenidae are found in Africa and Asia and are primarily scavengers, although they are also very skilled hunters (Holekamp 2006). They are known for extremely strong jaws and heavy musculature of the skull, which enables them to crush and consume the entire carcass, including bones and hooves, of their prey (Tanner et al. 2010). The bones and other calcified parts of the prey are digested in the hyena's large (capacity of up to 14.5 kg) stomach.

While most Mustelids are strictly carnivorous, the range of items they consume is quite broad, and it occasionally includes plant material. Mustelids as well as herpestids have scent glands and completely nonretractable claws. Herpestids, being more arboreal than other members of the order, are regarded for their ability to attack and kill poisonous snakes, but more often consume other small mammals, birds, eggs, and occasionally fruit. The claws of the procyonid species are short, curved, and either nonretractile or semiretractile. These animals usually have a single coat color with facial and/or tail markings. Some procyonids have prehensile tails, such as the kinkajou (*Potos flavus*), or semiprehensile tails, as in the coatis (genera *Nasua* and *Nasuella*), which help with balance and climbing. Viverrids are likely the least understood of the carnivores; species of this family live in small areas, usually in heavy forest and dense vegetation. Some viverrids that are found in zoos include binturongs (also known as "bear cats") and the fossa (*Cryptoprocta ferox*). The secretions from the musk glands of the civet species (members of the viverridae family) are used in the perfume industry, making these carnivores economically important in some regions.

THE ZOO ENVIRONMENT

Because this taxon contains mammals of diverse size and behavior, it is logical that there is great diversity of adequate enclosure size and structure. When considering the ideal temperature, lighting, substrate, and fencing for any carnivore, the natural biology as well as the ecosystem where the species lives in the wild should be taken into careful account. For example, the clouded leopard (*Neofelis nebulosa*) lives secluded, primarily in treetops of evergreen tropical rainforests of Southeast Asia (Macdonald, Loveridge, and Nowell 2010, 16–17). In contrast, the polar bear (*Ursus maritimus*) resides strictly in the arctic where it hunts in the sea ice and survives primarily on seals (Wiig, Aars, and Born 2008).

In the way that each species maintains unique structure and behavior, the ideal substrate and enclosure type also varies. Some examples of substrates that may be used alone or in combination include concrete, gravel, grass, dirt, mulch, and sand. Safety of animal care staff is of utmost importance

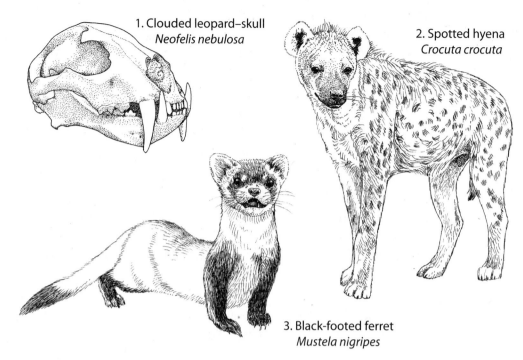

1. Clouded leopard–skull
Neofelis nebulosa

2. Spotted hyena
Crocuta crocuta

3. Black-footed ferret
Mustela nigripes

Figure 28.1. Examples of the diverse morphology of species within the order Carnivora: (1) the skull of the clouded leopard (*Neofelis nebulosa*) depicts that this species has the largest canines of any living Felid, relative to body size; (2) it is very difficult to differentiate a male from a female spotted hyena (*Crocuta crocuta*), due to the masculinization of the female genitalia; (3) the black-footed ferret (*Mustela nigripes*) of the North American plains dines almost exclusively on prairie dogs (*Cynomys ludovicianus*) and this species was once reduced to only 18 total individuals due to indiscriminate removal of their preferred prey. Illustrations by Kate Woodle, www.katewoodleillustration.com.

when considering housing large carnivores in zoos. The larger species of this order can be housed successfully in a combination of indoor and outdoor enclosures. The strength of the fencing, doors, buildings, and dens must be adequate for each individual species. Also, the size of the holes in the fencing (if mesh is used) must be carefully considered, so as to prevent limbs or heads (especially of offspring) from becoming tangled or stuck in the mesh. Some species are prone to chewing, and breakage of teeth on metal fencing can occur. A covered shelter should be provided at all times for all species. Indoor holdings with adequate heating sources are necessary for any warm climate animals living in cold environments. In general, every carnivore should have access to an "off-exhibit" or secondary holding area in which it can be secured by the keeper. This allows the keeper to enter and clean the primary holding area safely. Also, use of a secondary holding area allows the keeper to feed carnivore species safely (discussed below) and can provide a safe, secure, and secluded area for the animals away from the public.

For certain species, such as lions or large bears, the ability to bring individuals off of an exhibit and into a secure structure each night is a requirement for safety, especially for zoos in urban areas. Many felids (e.g., jaguars, leopards) climb extremely well, and for them completely covered enclosures are imperative. Other large carnivores (e.g., cheetahs, wild dogs) are contained adequately with fencing systems that have overhang of fencing angled back 45 degrees inward over the enclosure (Ziegler-Meeks 2009, 15). Electric or barbed wire is not indicated for primary containment, but is used simply to keep animals away from specific parts of the enclosures

(Tilson 1995, 25). Buried fencing that extends one meter (three feet) vertically parallel to the bottom of the fence (also known as a dig-out barrier) is essential in exhibits for species prone to digging. More modern-appearing zoo exhibits have creatively and safely used thick glass, moat systems, or other similar setups. Staff should carefully consider the most current recommendations, while always taking into account species biology, for facility design and animal containment for each individual carnivore species (e.g., depth and width of moats for large felids and certain bears) (Tilson 1995, 25; Polar Bear Care Manual 2009, 15).

A system of doors and chutes is extremely helpful in moving large carnivores safely from one enclosure to another (often referred to as "shifting"). Small animal entry systems either between enclosures or between interconnecting buildings with outdoor enclosures are beneficial. Secondary containment (e.g., two fences around an enclosure) is recommended for any large felid, due to risk of injury to humans if direct contact occurs. Specific species are primarily terrestrial (e.g., canids and hyenids), while others are primarily arboreal (e.g., clouded leopards, jaguars, some viverridae and procyonidae), and enclosures should be constructed to reflect the natural tendencies of each species. This can also help with reducing stress or boredom.

Live plants are recommended for the enclosures of most carnivores. Even though most members of the order Carnivora are considered carnivores, they will consume some plant material. For this reason, all plant material included in exhibits should be evaluated for toxic properties associated with ingestion. Shade in the form of trees, bushes, and

> **Good Practice Tip:** For safety, keepers should wear steel-toed boots and always carry a communications radio when working with carnivores. Keepers should not wear dangling jewelry, loose clothing, long unsecured hair, or any equipment, such as name badges or sunglasses hanging on strings or chains, that could get caught in fencing or doors or grabbed by an animal.

live plants (such as pampas grass) are recommended for all species housed outdoors or with outdoor access. Individuals benefit from natural materials such as logs for sharpening claws, mounds or elevated spots for surveying territory, climbing trees for enrichment and privacy, and tall grasses and bushes for concealment. A combination of substrates is recommended for carnivores, but more than 50% natural substrate is ideal. A floor made of hard substrate (e.g., concrete) is useful for helping to wear down nails, and can also be well disinfected. But limiting carnivores to only a concrete (or similar manmade substrate) floor can cause damage to foot pads, especially in species prone to pacing. All substrate used in facilities housing carnivores should provide adequate drainage so as to prevent any standing water. All natural substrates should be checked regularly for signs of digging near fence lines. For some species, especially bears and some canids, digging allows for denning and other natural behaviors.

Smaller species of felidae are frequently housed in indoor enclosures very successfully. Habitats where each species naturally occurs in the wild should be considered in the care of any zoo animals, and staff should be aware of the most current housing and facility recommendations for each species or taxon. For example, the sand cat (*Felis margarita*) is native to extremely dry and hot areas of Africa (the Sahara) and Central Asia (the Arabian Desert). These environments have a wide range of temperatures, as the daytime temperatures can easily reach 45°C (113°F) in summer while the nighttime temperatures can plunge to 0°C (32°F). An enclosure for this species would need to be carefully maintained for low humidity, natural substrates (sand), and higher temperatures than for most other species of this order.

Other small carnivore species are also housed completely indoors with great success. In general, the indoor enclosure temperature should be maintained at 13 to 18 °C (55 to 65 °F) in winter and 18 to 21 °C (65 to 70 °F) in summer. However, the specific temperature requirements may vary with specific species' needs on the basis of the environment they occur in naturally. Tropical and subtropical forms should be housed indoors if the temperature is expected to fall below 60 °F. A general recommendation is for relative humidity to be between 30 and 70% indoors for mustelid, viverrid, and procyonid species. The higher end of this range (55–70%) is needed for the tropical forest species and the lower end (approximately 30%) is required for the desert species (Procyonid Care Manual 2010; Viverrid Care Manual 2010). The families herpestidae and mustelids are comprised of animals that primarily reside in burrows (Mustelid Care Manual 2010). Species of these families depend on intricate burrow systems for shelter, food, and protection from larger carnivores. Some species thrive on manmade burrow systems;

however, other species seem to need to dig and produce their own burrows (John Stoner, pers. comm.). All indoor facilities for any carnivore should be well ventilated.

Nest boxes are widely used for many smaller carnivores. These are typically wooden or plastic structures that provide safe places for animals to hide and are ideal for the birth of young. For smaller species, these boxes should be elevated above the ground to help keep out pests and provide additional security. Many zoos will place a variety of structures, called furniture, in an enclosure. For carnivores these structures include logs, tree limbs, pools, rocks, walking and climbing ropes, and platforms, among other items. This furniture should be cleaned and rearranged regularly, and use of specific items is dependent on the species. For example, introduction of a climbing rope would not be beneficial for a lion and could even be potentially dangerous, whereas a red panda would benefit greatly from the exercise and enrichment a rope offers.

BASIC HUSBANDRY

Special precautions inherent in caring for carnivore species are associated with the safety of the keeper staff and the individual animals. The integrity and security of each enclosure should be examined very carefully each day. Each individual animal should be located and accounted for, and each enclosure should be visually examined carefully by the keeper first thing each morning. This is especially important before a keeper enters any area that could contain a large carnivore species, to ensure that there has been no breach in security. The keeper should look for foreign or sharp objects that could cause injury, as well as any potential breaches to fencing. All doors and shifts within the enclosure of any large carnivore must be lockable. The enclosure should be carefully checked for any signs of vomiting or diarrhea, which indicate possible illness. Healthy animals will be alert and responsive to noise and visual stimuli. Disinterest in food, lethargy, poor skin or hair coat, and a depressed attitude are all signs of possible illness. All animals should be checked carefully by the keeper at the end of each day. All gates and doors should be checked to ensure that they are closed and locked. All the animals should have adequate access to shelter, food, and water, and each individual should be comfortable and secure.

Both metal and rubber food and water bowls can pose risks to carnivores. For certain species, especially those prone to chewing, metal poses a risk of breaking or damaging teeth. Rubber receptacles are very enticing for animals to chew on, and the risk of ingestion of rubber pieces is high. Hard plastic bowls are easy to clean and disinfect, and they are too sturdy for most carnivores to break or chew. Many zoos use stainless steel bowls, which are very unlikely to be broken by a large carnivore and can be sterilized easily. For lions or tigers, feeding on bare substrate such as concrete may be appropriate. Bears naturally forage for their food in the wild, and mimicking this in zoos can help reduce boredom and provide enrichment. Food for ursid species is often scattered or hidden throughout an animal's enclosure to stimulate this foraging behavior.

A basic rule for feeding a carnivore is to move it to a holding area where it can be well secured away from the keeper. Food is then placed into the enclosure where the keeper wants the animal to eat, and the individual animal is allowed back into the enclosure after the keeper has safely departed. Some groups of carnivores will need to be separated and fed individually to prevent fighting over food. This feeding strategy also allows the keeper to make sure each animal is consuming its prescribed diet, observe any change in its eating habits (e.g., not finishing the offered diet, eating more slowly than usual, regurgitation), and measure the amount of diet consumed. Other species will eat successfully in groups; the decision whether to feed an animal individually or in a group should be based upon its natural history as well as its individual temperament.

The goal of feeding carnivores in a zoo setting is to offer them a diet that meets their nutrient needs in a palatable and readily digested form. Forms of carnivore diets include commercial raw meat mixes, canned and dry pet foods, and whole prey (carcasses). Many carnivores eat only once every few days in the wild, and gorge themselves on portions of a comparatively large carcass after a hunt. However, in most zoo settings where such large whole carcass feedings are neither practical nor accepted by staff or visitors, each individual may be fed many times per week, if not every single day. Many facilities have adopted "fast days" for carnivores, either to loosely mimic the fact that they do not eat every day in the wild or to encourage them to focus their efforts on bones or similar hard "food" items for their improved dental health. Numerous published guidelines are available through species and taxa specific interest groups (AZA, SSP, and TAG programs) and scientific advisory groups associated with zoo and wildlife nutrition.

Obtaining regular body weights is an objective and extremely useful way to monitor adequate food intake as well as overall health. Smaller carnivore species can often be weighed easily and readily by using a small transport carrier to place the individual on a scale. A free-standing platform scale can be used for some larger species that can be trained to stand on the device for food or a similar reward (discussed below). Other species that are larger or difficult to train may require the use of squeeze or crush cages fitted with scales. In situations when regular measurements of body weight are not readily available, assessing body condition is one way to determine whether each individual is receiving an adequate diet (thus maintaining its appropriate body condition). Body condition scoring is the practice of visually assessing the amount of tissue covering specific bony structures on the body as a measurement of the body's condition. This method does require some degree of training and experience to be done accurately and consistently. Carnivores have simple stomachs designed to process readily digestible food items (meat, fish, whole prey), as compared to the more complex digestive systems of herbivores. Recent work has suggested that some species of felids may actually have the ability to ferment some of the less easily digested components of their diet (hair, cartilage, digestive tract contents) in a fashion similar to that of herbivores. Interesting to note within the carnivores is the giant panda: a simple-stomached carnivore in form and function, but an obligate herbivore (nearly 100% bamboo) in practice.

Carnivores require access to clean, fresh water in clean receptacles. Each species will consume a different amount of water, in accordance with its natural dietary intake and metabolic requirements. Most cat species get the bulk of their moisture from the food they eat. Some carnivores tend to urinate or defecate in standing water, so cleanliness of their water supply must be checked daily. In cold temperatures it is necessary to check that water sources are not frozen. If necessary, heated, plug-in water bowls can be used in animal buildings during periods of cold temperatures for certain species. Care should be taken with species that are prone to chewing, and any electric cables should be protected to prevent damage.

EQUIPMENT AND HANDLING

Knowledge of the species, and especially of each individual's behavior and temperament, is critical when discussing the handling of a carnivore. Three types of restraint are generally used for carnivores: physical, chemical, and behavioral. Physical restraint involves using force alone to prohibit an individual's movement (Christman 2010, 39). Large carnivore species are not usually handled directly, except when they are very young and can be controlled safely. For restraint of large carnivores such as bears, the great cats, and hyenas, use of a remote mechanical squeeze cage or chute is recommended. These are usually permanent structures that are integrated into the animal's enclosure. This integration enables its regular use and training by keepers, and enhances each animal's comfort with the restraint device. Certain carnivore species may be tractable enough for keepers to work with directly in the enclosure throughout adulthood (some examples include cheetahs, smaller felids, maned wolves, ferrets). However, some specific individuals of a species may never be considered safe enough for a keeper to have "free contact" (no barrier between keeper and animal) with them, and push boards, proximity sticks, and/or shields may be required for working with them in a free-contact situation. If the offspring of larger felids are to be handled for weighing, vaccines, or other routine exams, thick leather or mesh gloves are required. Nets, nooses, kennels and ropes can be used for physical restraint of small to midsize carnivores (Christman 2010, 43–46). Smaller members of the order Carnivora tend to fight back and become aggressive when being handled or restrained, and there is a high risk of animal care staff being bitten or scratched. Wire cages, thick gloves, and often face shields or visors are recommended for restraint of small carnivores such as mink or ferrets (Christman 2010, 43–45).

Chemical restraint is the use of sedatives or anesthetics by a veterinarian (Christman 2010, 40). These agents should be administered and used only by licensed veterinarians, so they will not be discussed further in this section. Behavioral restraint involves the use of training and/or behavioral modification to enable the keeper to handle, manipulate the activity of e.g., in weighing), or administer medications to an individual animal. Behavioral restraint will be discussed in the following section.

BEHAVIOR

Stressors are stimuli an animal experiences that cause a biological response. The first reaction of most animals to a stressor is a state of heightened awareness. If exposure to the stressor (either positive or negative) continues or increases, there is often an associated physiological response of increased glucocorticoid (cortisol) concentrations. For example, moving to a new enclosure can cause a felid to have a spike in glucocorticoids. This occurs when the new environment is novel and stimulating, not necessarily because it is frightening or intimidating, and it could be considered a positive stressor or stimulus. In contrast, transportation to the veterinary hospital for a routine exam can also cause an increase in glucocorticoids, because the animal is nervous and unsure. This would be considered a negative stressor or stimulus for that individual. All animals react to sudden changes in their environments and are generally categorized into having either a "fight" or a "flight" response. This refers to *how* animals naturally respond to changes in their environment. Carnivores generally respond by fighting, and rarely fleeing, from danger or novel stimuli.

Some species (small and mid-size felids) display a combination of these behaviors, in that they will initially move away from a perceived threat (flight) but respond with aggression (fight) if cornered. Large felids, canids, bears, and hyenas can be very aggressive and can pose a high risk to keeper safety. These natural behavioral tendencies are important to understand, and such an understanding will improve the keeper's ability to manage and work safely with carnivores.

TRAINING

Traditionally, when managers needed to handle, weigh, vaccinate, or medicate zoo animals, the animals were restrained either physically or chemically. In recent years, however, the zoo community has moved towards behavioral restraint through training programs and behavioral modification. Carnivores, like most mammals, are successfully trained by positive reinforcement. They are naturally motivated by food, so preferred dietary items are often ideal rewards during training exercises. Human safety is a primary concern when training any carnivore, especially when rewarding animals with food. Keepers should always be prepared to quickly end any training session if an emergency arises; a quick release or escape from the area, for both animals and keepers, should always be available if necessary. Animals can become agitated during training sessions, increasing the risk of injury to themselves or keepers. As a general rule, veterinary staff should be notified before training sessions begin.

ENRICHMENT

Enrichment for carnivores is varied depending on species, but is deemed a critical part of managing these animals in captivity. Allowing for physical and psychological stimuli in an otherwise static environment can improve their overall health, well-being, and interactions with both conspecifics and keepers. Knowledge of species biology as well as individual animal temperament should inform enrichment programs. Felids are naturally curious and will usually investigate a new addition to their enclosure. Some species are primarily motivated by action (e.g., cheetahs are stimulated by watching or chasing) while others (e.g., canids) are more enticed by smell. Large "boomer balls," Kong® toys, boxes, bags, specialty food items (only with approval of the institutional nutritionist and/or veterinarian), spices, perfume, feathers, melons, and coconuts are all good sources of enrichment for carnivores. Not all enrichment is appropriate for each species. Individual species biology as well as animal temperament should be carefully considered before any enrichment item is introduced. An item that provides positive enrichment for one bear may be damaged and then pose a risk to another. When an enrichment item is first introduced to an individual animal, the keeper should monitor it closely to be sure it uses the item as intended and does not consume any parts of it. A plan should also be in place for retrieval of the enrichment item if it becomes absolutely necessary.

For smaller carnivores, large soil-filled enclosures can provide extensive enrichment (e.g., for mustelids and herpestidae). The soil is rearranged daily by the animals, providing enrichment, tunnels, and extensive exercise. Tubes also provide enrichment and a place to hide and feel secure. Additional enrichment items such as golf balls, boomer balls, or boxes can be provided if they can be disinfected. Herpestidae are particularly known for opening hard food items (nuts, eggs, crustaceans) with tools such as rocks, and allowing for this natural feeding behavior can provide them with enrichment.

The diet should always be considered enrichment, in possibly its purest form. Even though a commercial meat mix may not appear as outwardly stimulating as a whole carcass, the diet can still be presented in a way that allows the animal to exhibit natural foraging and search behaviors. Whole carcass feeding, whether small or large) is finding increasing acceptance throughout North America as a means to meet not only the nutritional but also the behavioral needs of carnivores.

TRANSPORTATION

When necessary, carnivores may need to be moved out of their enclosure, thus requiring a transport vessel in which to move them. Reasons for transport can include transfer to a new enclosure not accessible through a door or chute, a veterinary exam, or even relocation to a new institution. Each individual should be made comfortable with the carrier or crate prior to transport whenever possible. Desensitization to the shipping compartment will reduce the animal's stress and make travel easier for it. Smaller carnivore species (mustelids, herpestidae, and smaller felid and canid species) are often transported in carriers used for domestic cats and dogs (e.g., sky kennels). Large carnivores need special cages, with specific requirements for each species. Slatted flooring (with slats that are spaced adequately to prevent toes from getting caught), mesh for proper ventilation, secure locks, and easy visual access to the animal are all required in shipping containers. For lengthy trips, food, and water receptacles must be present. All possible relevant regulatory agencies (e.g., states, the US Fish and Wildlife Service, and CITES) should always be checked for shipping, health, and permit requirements before transporting animals. For general, taxa-specific comments, see the appropriate AZA care manuals. The International Air Transport Association (IATA) also publishes rules for individual species, including specific guidelines for animal transport containers.

REPRODUCTION

Extensive research has been conducted in the field of carnivore reproductive physiology (see Kleiman et al. 2010; Macdonald and Loveridge 2010). In general, a carnivore produces a single litter annually, but some species can produce multiple litters each year. Larger species such as bears and the great cats will have gaps of two to three years between litters, as the females are caring for their young. Many carnivores, such as mustelidae, canidae, and some felidae, are highly seasonal. In species that breed seasonally, the females come into estrus or heat only during a specific time of the year, usually for about four months. Males of these seasonal breeders also produce sperm only when females are receptive to mating. Some felid and mustelid species are induced ovulators, meaning that an oocyte is released from the ovary only after mating. Canidae, mustelidae, ursidae, and procyonidae species have a well-developed baculum or penile bone. The baculum, used for and during mating, allows the male to extend the length of the copulatory period. The average gestation or pregnancy period ranges from 50 to 115 days, after which time a female gives birth to 1 to 13 young, depending on the species. Species of the ursidae and mustelidae families have delayed implantation, whereby an egg (oocyte) can become dormant for a period of time after its fertilization. This phenomenon makes accurate determination of gestation length very difficult, as it can extend pregnancy six to nine months beyond the normal period.

Carnivores are born underdeveloped, with eyes and ears closed. Cub mortality of carnivores is deemed the biggest contributing factor to poor sustainability in captive populations. Maternal neglect or aggressive behavior towards the newborn can be observed (Ziegler-Meeks 2009, 47–55). General husbandry guidelines for most carnivores suggest a hands-off approach to the management of parturition and neonate care. Specifically, remote monitoring with cameras or other equipment is ideal. Limited keeper presence and involvement is highly recommended so as to reduce stress of the mother and encourage maternal neonate bonding. Carnivore females nurse their young for several weeks or months and often will care for them for several months (e.g., in mustelids and viverrids) or years (e.g., in large cats and bears).

The hyenidae family has the most unique reproduction and social structure of any carnivore (see Glickman et al. 2006; Holekamp 2006). The female spotted hyena (*Crocuta crocuta*) is the only mammal that lacks an external vaginal opening (Glickman et al. 2006). The genitalia of the male and female of this species are remarkably similar, making sex determination by simple observation difficult. The female spotted hyena has an enlarged clitoris, also referred to as a pseudopenis, through which she urinates, copulates, and also gives birth (Holekamp 2006). The sexes have a linear dominance hierarchy, the lowest female outranking the highest male. The dominant male in a clan has access to the most females for breeding. The dominant female monopolizes carcasses, which results in better nutrition for her cubs (Hole-

kamp 2006). A female cub inherits the dominance status of her mother (Holekamp 2006).

VETERINARY CARE

In general, carnivores are treated for internal and external parasites and also receive vaccinations on a regular schedule. Zoo carnivores are susceptible to the same external parasites as domestic species (ear mites, fleas, ticks), and treatment is achieved using the same agents as would be used for domestic dogs or cats. Animals should be treated for internal parasites on a monthly or quarterly schedule. Fecal samples should be routinely screened to determine the presence of parasites. Vaccinations routinely given to members of this taxa include but are not limited to rabies, distemper, parvovirus, corona, leptospirosis, and feline panleukopenia. Mustelids have varying species- and exposure-dependent sensitivities to feline panleukopenia, canine distemper, rabies, and leptospirosis. Most resources recommend vaccination of mustelids for rabies and canine distemper, as well as for canine hepatitis if prevalent (Mustelid Care Manual 2010, 46). Vaccination administered to any group, species, or individual is the final decision of the veterinarian in charge of that specific collection. Species with outdoor housing or access should routinely be administered heartworm preventative in areas where that parasite is endemic.

Felids are susceptible to many feline viral diseases, such as feline immunodeficiency virus (FIV), feline leukemia virus (FeLV), panleukopenia (parvovirus), feline infectious peritonitis (FIP), feline herpes virus, and parasitic diseases such as toxoplasmosis. Wild felids carry toxoplasmosis (*Toxoplasma gondii*), and although that parasite is often carried harmlessly by most mammals, including people, it can have serious adverse effects on humans with compromised immune systems. Pregnant women are advised to not handle cat feces at all, and they should strive to avoid felid enclosures in general (Rosenthal and Xanten 2010, 77). Cats also suffer from feline respiratory disease, a complex of viral contagions including rhinotracheitis, pneumonitis, and influenza, marked by fever, sneezing, and running eyes and nose. Mortality is low, but recovery from severe cases may be difficult and prolonged, with relapses. Antibiotics are used to prevent secondary bacterial infections.

Cheetahs are highly susceptible to herpes virus (Ziegler-Meeks 2009, 51–52). Young cubs normally do not show any symptoms until after they are four days old. Early detection offers the best chance for treatment. Minimal signs of infection may clear up on their own in young cheetah cubs. If the dam is actively shedding the virus, injury to the cubs can become quite severe, including lesions of the eyes, nose, and/or mouth. Most symptoms appear by about one month of age. Severe lesions can lead to permanent scarring, including that of the cornea and prolapsed third eyelid (nictitating membrane).

Cystinuria has been found in a significant number of both free-ranging and captive maned wolves. The disease also occurs in humans and canids, and is transmitted genetically. Cystinuria is characterized by excretion of amino acids, especially cystine, in the urine (Bush and Bovee 1978), and can result in difficulty with urination. If the condition becomes severe enough, the animal may become "blocked" (i.e., the urinary tract clogged with insoluble and impassable solids such as crystals) and lose the ability to urinate completely, which can eventually be fatal. A screening program for early detection is recommended.

Epizootic plague kills both prairie dogs and ferrets, and is a major factor limiting recovery of the highly endangered black-footed ferret (Matchett et al. 2010). For captive-reared individuals, this has been combated through a combination of vaccinations and treatment for fleas (Matchett et al. 2010).

CONSERVATION PROGRAMS

Species of the order Carnivora are normally thought of as highly charismatic and usually are public favorites when housed on exhibit in zoos. Unfortunately, most carnivores are viewed as pests and threats to human safety in their wild habitats, and therefore most species are threatened or vulnerable to extinction. Because of their unique place in their respective ecosystems and their vulnerability to human persecution, there are multiple conservation and research programs for carnivores worldwide. Two specific conservation programs will be discussed in this section as examples of successful initiatives for species of this order. There is a plethora of very active and successful programs for carnivore species around the globe, and we encourage anyone working with such species to investigate these programs.

The Global Tiger Initiative aims to protect wild tigers and their ecosystems. The initiative proposes to accomplish its goals through increased community education, reduction of illegal offtake or poaching, and direct protection of tiger habitat. Much of the this program's success will depend on changing people's attitudes. People living with tigers will hopefully learn to value them and desire to maintain them in their ecosystem. In addition, people must see a financial benefit to themselves and their communities from helping to protect tigers. To be successful, the initiative must increase knowledge and understanding of tigers so that people will fear them less and appreciate them more.

The black-footed ferret is an excellent example of a carnivore species that has recovered in the wild due to intense conservation and research efforts. Once thought extinct in the wild, a small population of this species was found in Wyoming in the 1980s. In 1985, a total of 18 individuals were taken from the wild and into captivity. These few individuals became the foundation for the entire future of the species. After years of research and assisted reproduction programs, the species has been reintroduced into the wild. Additional animals are released every year, bolstering the free-ranging population. The success of the black-footed ferret program is due largely to management in zoos and captive breeding facilities, public education, and intense biological research.

SUMMARY

Species of the order Carnivora have diverse anatomy, physiology, and behavior, presenting unique challenges and considerations for managing these animals in zoos. Keepers should research the natural history of any species under their care. A better understanding of a species' biology will improve its daily care, training, and enrichment through the use of best practice methods. Keepers and others involved with the care and management of these species should also investigate and be familiar with current conservation and research programs. The preservation of endangered species benefits from heightened public awareness and appreciation, and the zoo community plays an important role in bringing it about.

REFERENCES

AZA Bear Taxon Advisory Group (TAG). 2009. *Polar Bear (Ursus maritimus) Care Manual.* Silver Spring, MD: Association of Zoos and Aquariums.

AZA Small Carnivore Taxon Advisory Group (TAG). 2010. *Mustelid (Mustelidae) Care Manual.* Silver Spring, MD: Association of Zoos and Aquariums.

———. 2010. *Procyonid (Procyonidae) Care Manual.* Silver Spring, MD: Association of Zoos and Aquariums.

———. 2010. *Viverrids (Viverridae) Care Manual.* Silver Spring, MD: Association of Zoos and Aquariums.

Bush, M., and K. C. Bovee. 1978. "Cystinuria in a Maned Wolf." *Journal of the American Veterinary Medical Association* 173:1159–62.

Christiansen, P., and J. S. Adolfssen. 2005. "Bite Forces, Canine Strength and Skull Allometry in Carnivores (Mammalia, Carnivora)." *Journal of Zoology* 266:133–51.

Christman, Joseph. 2010. "Physical Methods of Capture, Handling, and Restraint of Mammals." In *Mammals in Captivity, Principals and Techniques for Zoo Management,* 2nd edition, edited by Devra Kleiman, Katerina Thompson, and Charlotte Baer, 39–48. Chicago: University of Chicago Press.

Driscoll, Carlos, David Macdonald, and Steven O'Brien. 2009. "From Wild Animals to Domestic Pets, an Evolutionary View of Domestication." *Proceedings of the National Academy of Science* 106:9971–78.

Glickman, S. E., G. R. Cunha, C. M. Drea, A. J. Conley, and N. J. Place. 2006. "Mammalina Sexual Differentiation: Lessons from the Spotted Hyena." *Trends in Endocrinology and Metabolism* 17:349–56.

Holekamp, K. 2006. "Spotted Hyenas." *Current Biology* 16:944–45.

Macdonald, David, Andrew Loveridge, and Kristin Nowell. 2010. "*Dramatis Personae*: An Introduction to the Wild Felids." In *Biology and Conservation of Wild Felids,* edited by David Macdonald and Andrew Loveridge, 3–58. Oxford: Oxford University Press.

Matchett M. R., D. E. Biggins, V. Carlson, B. Powell, and T. Rocke. 2010. "Enzootic Plague Reduces Black-Footed Ferret (*Mustela nigripes*) survival in Montana." *Vector Borne Zoonotic Disease.* 10:27–35.

Olaf, R., and P. Bininda-Emonds. 2004. "Phylogenetic Position of the Giant Panda." In *Giant Pandas: Biology and Conservation,* edited by Don Lindburg and Kara Baragona, 11–33. Berkeley: University of California Press.

Rosenthal, Mark, and William Xanten. 2010. "Safety Considerations in a Zoological Park." In *Mammals in Captivity: Principals and Techniques for Zoo Management,* 2nd edition, edited by Devra Kleiman, Katerina Thompson, and Charlotte Baer, 76–80. Chicago: University of Chicago Press.

Tanner, J. B., M. L. Zelditch, B. L. Lundrigan, and K. E. Holekamp. 2010. "Ontogenetic Change in Skull Morphology and Mechanical Advantage in the Spotted Hyena (*Crocuta crocuta*)." *Journal of Morphology* 271:353–65.

Tilson, Ron, Gerald Brady, Kathy Traylor-Holzer, and Doug Armstrong. 1995. *Management and Conservation of Captive Tigers.* 3rd edition,edited by Doug Armstrong, 1–136. Apple Valley: Minnesota Zoo.

Werdelin, Lars, Nobuyuki Yamaguchi, Warren Johnson, and Steven O'Brien. 2010. "Phylogeny and Evolution of Cats (Felidae)." In *Biology and Conservation of Wild Felids*, edited by David Macdonald and Andrew Loveridge, 3–58. Oxford: Oxford University Press.

Wiig, O., J. Aars, and E. W. Born. 2008. "Effects of Climate Change on Polar Bears." *Science Progress* 91:151–73.

Ziegler-Meeks, Karen. 2009. *Husbandry Manual for the Cheetah* (Acinonyx jubatus). Silver Spring, MD: Association of Zoos and Aquariums.

29

Husbandry and Care of Primates

Colleen McCann

INTRODUCTION

Humans belong to the order primates, and the many similarities they share with their closest living relatives make primates extremely popular species among zoo visitors. The primate order is diverse in its species and characteristics. The order includes the prosimians (e.g. lemurs, lorises, and tarsiers), New World monkeys (e.g. marmosets, tamarins, titis, sakis, capuchins, and squirrel, howler and spider monkeys), Old World monkeys (e.g. baboons, colobus, guenons, langurs, macaques, and mangabeys), apes (e.g. gibbons, siamangs, bonobos, chimpanzees, gorillas, and orangutans), and humans. Primates range in size from 30 grams for the smallest of the mouse lemurs, Berthe's mouse lemur (*Microcebus berthae*), up to 200 kilograms for the largest of the great apes, the mountain gorilla (*Gorilla beringei beringei*), and they have an equivalent range across the species in locomotor patterns, foraging strategies, mating systems, social structure, and behavior (Strier 2011, 30–58). However, primates as an order share a suite of characteristics that unite them as a taxonomic group and distinguish them from other mammals.

The two major taxonomic groupings of primates are the strepsirhines (wet nose connected to upper lip) and haplorhines (dry nose not connected to upper lip). It is important to note that throughout the literature, the terms prosimian and anthropoid are often used interchangeably with the terms stepshirine and haplorhine, although they are not synonymous (Fleagle 1999). The difference lies in the classification of the tarsier, which is a prosimian but shares many derived molecular and morphological features with monkeys and apes, although its ecological niche is akin to that of other nocturnal prosimians.

Although there is much debate on the number of species and their taxonomic assessment, table 29.1 contains a generally accepted classification of the extant species of primates (adapted from Groves 2001).

This chapter will cover the primary techniques and best practices in husbandry for keepers caring for primates in zoos. After reading this chapter, the reader should have a comprehensive understanding of

- the natural history of primates and the importance of incorporating key features into primary husbandry practices
- the basic anatomical features used in describing primates
- a general knowledge of the complexity of primate behavior and social systems, and the importance of understanding species-typical behaviors
- a general understanding of the key features of primate enclosures, important for maximizing the expression of primate behavioral repertoires
- the technical skills important for keepers working with primates and for developing a rapport with individual primates in order to provide optimal care
- basic observational skills needed for managing primate social and clinical health
- the importance of incorporating operant training techniques into husbandry practices
- principles and techniques for managing primate reproduction and infant care
- knowledge of the conservation status of primates and important in situ and ex situ efforts being taken by various organizations to further our understanding of their needs to ensure their survival.

NATURAL HISTORY

Primates evolved in tropical habitats; they retain many characteristics that are adaptations to this environment and to an arboreal lifestyle. Morphologically, primates have a generalized skeleton and dentition; however, the primary characteristics that define the order are the opposability of the digits, a large brain relative to body size, and a long developmental period with an increased learning and socialization period. Napier and Napier (1967) provide a complete list of

TABLE 29.1. Classification of living primates

Order Primates

Suborder Strepsirhini: prosimians (excluding tarsiers)
    Family Cheirogaleidae: dwarf lemurs and mouse lemurs
    Family Daubentoniidae: aye-ayes
    Family Lemuridae: lemurs
    Family Lepilemuridae: sportive lemurs
    Family Indriidae: woolly lemurs
    Family Lorisidae: lorises and pottos
    Family Galagidae: galagos

Suborder Haplorhini: tarsiers, monkeys, and apes
    Family Tarsiidae: tarsiers
    Family Callitrichidae: marmosets and tamarins
    Family Cebidae: capuchins and squirrel monkeys
    Family Aotidae: owl monkeys (douroucoulis)
    Family Pitheciidae: titis, sakis, and uakaris
    Family Atelidae: howler, spider, and woolly monkeys
    Family Cercopithecidae: colobines (colobus and langurs) and
       cercopithecines (baboons, guenons, macaques, and mangabeys)
    Family Hylobatidae: gibbons and siamangs (lesser apes)
    Family Hominidae: great apes and humans

characteristics that define primates; however, the following are the most notable features that are key to understanding their needs:

- shoulder joints that allow high degrees of movement in all directions (brachiation)
- five digits on the fore and hind limbs with an opposable first digit (hallux) and grasping hands and feet
- the replacement of claws with nails, a flat nail on the hallux, and sensitive tactile pads
- a trend towards a reduced snout and flattened face, attributed to a reliance on vision at the expense of olfaction (most notably in haplorhines, and less so in strepsirhines)
- a complex visual system, including stereoscopic and color vision
- a large brain in comparison to body size, with a well-developed cerebellum and enlarged cerebral cortex
- two pectoral mammary glands
- typically one young per pregnancy
- a trend towards holding the torso upright, leading to bipedalism
- a long gestation, developmental, and learning period with concomitant socialization.

It is important to note that not all primates exhibit these traits, not every trait is unique to primates, and no one trait defines the order. Rather, primates are generalist mammals, and as an order they are characterized by a suite of features which are well adapted to living in trees (arboreality).

Primates inhabit the equatorial regions of three continents: (1) from the rain forests of southern Mexico to northern Argentina, (2) from the archipelago of Indonesia to the mountains of southwest China, and (3) from the sub-Saharan bushlands and savannas to the equatorial rain forests of the Congo Basin and south to the scrub forests of South Africa and the spiny forests of Madagascar (Lehman and Fleagle 2010, 1–6; Wilson and Reeder 2005, 111–84). Primates play an import role in seed dispersal, and as a result they occupy an important ecological niche in tropical forests and woodland savannas. Most nonhuman primates spend some portion of their day in trees either foraging, sleeping, or taking refuge from predators. The majority of primate species are diurnal and arboreal; however, some species have evolved a nocturnal niche (lorises, galagos, owl monkeys) while some are primarily terrestrial (baboons and macaques). Primates exhibit great diversity in foraging strategies, and a wide range in food item preferences across species. For instance, lorises, nocturnal lemurs, and tarsiers primarily consume insects (i.e., are insectivorous); marmosets, tamarins, and galagos exploit tree sap and gum (i.e., are gumnivorous); geladas are unique in being grass and seed eaters (i.e., are graminivorous); langurs and colobus and howler monkeys feed primarily on leaves (i.e., are folivorous); mountain gorillas have a restricted diet of leaves, roots, and shoots from herbaceous plants (i.e., are herbivorous); spider monkeys, guenons, gibbons, and orangutans feed predominantly on fruit (e.g., are frugivorous); and capuchins, baboons, and chimpanzees include both plants and animals in their diets (i.e., are omnivorous; Fleagle 1999; Strier 2011).

Primates are highly social species, with most individuals living their entire lives in groups. As with other primate adaptations, the nature of those groups differs among species with regard to group size, ratio of adult males to females, mating system, and social structure (Strier 2011). The structure of primate social groupings fall into the following catgories:

- **solitary**: characterized by solitary adults and females with dependent offspring. Adult males have large home ranges that overlap that of several adult females. This type of social grouping is found in mouse lemurs, galagos, and orangutans.
- **monogamous**: a mated pair and their immature offspring. Species that exhibit this social structure are highly arboreal and territorial. Adults have a low tolerance for adults of the same sex and there is little sexual dimorphism. Monogamous primate species include gibbons, siamangs, indris, tarsiers, titis, and owl monkeys.
- **polyandrous**: one adult female, more than one adult male, and their immature offspring. This structure is rare among primates and is most notably observed in marmosets and tamarins, although they exhibit facultative polyandry.
- **polygynous**: characterized by a single adult breeding male, several adult females, and their dependent offspring. Females form the nucleus of the group; a high degree of sexual dimorphism is exhibited. This is the most common type of social structure found in primates. Species exhibiting polygynous social groups include capuchins, guenons, langurs, and gorillas.
- **multimale/multifemale**: several adult males, several adult females, and their immature offspring. Dominance hierarchies and marked competition over access to mates and limited food resources is evident in both males and females. This type of structure is found in some lemur, squirrel monkey, colobus, macaque, and baboon species.

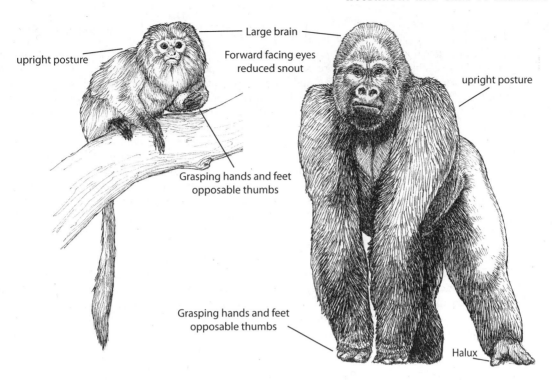

Figure 29.1. Basic anatomical features of primates. Illustrations by Kate Woodle, www.katewoodleillustration.com.

• **fission-fusion**: group size and composition change throughout the year with varying environmental fluctuations (fruiting seasons) and reproductive cycles (females in estrus). Individuals enter and leave communities from time to time, as is the case in spider monkeys and chimpanzees.

Many primate social behaviors have evolved to facilitate group living. Affiliative behaviors, such as allogrooming—when one individual grooms another of the same species—serve a hygienic purpose but primarily function as social bonding mechanisms. And grooming frequencies among individuals can be used as indicators of an individual's dominance status within the group. Agonistic behaviors, which include patterns of aggressive-submissive interactions and displays, are also commonly exhibited in primate groups. Dominance hierarchies (a hierarchy of ranked status among individuals, maintained by agonistic behaviors) are established among group members and often result in greater access to resources, such as sleeping sites, preferred food items, and mates for those ranked highest. In captive conditions it is important to acknowledge dominance hierarchies within primate groups; the challenge is to respect the hierarchy while ensuring that each individual group member's basic needs are being met.

PRIMATES IN CAPTIVITY

Primates have a long history of being kept in captivity in some form or another. Illustrations and references to primates appear in literature dating back to the era of exploration and colonization of the African and Asian continents, when primates and other exotic wildlife were often given as gifts to royalty for private menageries. In the late 1800s and early 1900s, zoological parks began acquiring primates for exhibition purposes. However, a lack of understanding of primate dietary and social requirements resulted in a high rate of mortality. Over the ensuing decades, as information from field scientists on primate behavioral ecology emerged, the survival rate of primates in captivity gradually increased, and in 1956 zoos achieved their first captive-born gorilla (Crandall 1964, 167). In the 1960s the fields of psychology and anthropology flourished, and primates featured prominently in studies on language acquisition, beginning with the seminal ape sign language study with the chimpanzee Washoe and subsequently extending to studies with gorillas and bonobos (Gardener et al. 1989). Additionally, primates also became the focal species for use in biomedical research, resulting in eight US regional primate research centers being established by the National Institute of Health for studies on human health. Primate collections in zoos also flourished between 1960 and 1970, with many different primate species being acquired from the wild for the establishment of zoo collections. Three important legislative regulations affected US captive primate collections: (1) the 1985 Amendment to the US Animal Welfare Act (1966), regulating the treatment of animals in research, exhibition, transport, and by dealers, and requiring a physical environment that "promotes the psychological well-being of primates"; (2) the US Fish and Wildlife Service Endangered Species Act of 1973, which put in place restrictions for the importation of endangered species "to prevent the extinction of endangered species and their habitats," and prohibiting "activities with these protected species unless authorized by a permit from the US Fish and Wildlife Service"; and (3) the Convention on International Trade of

Endangered Species of Wild Flora and Fauna (CITES). As a result of increasing regulations on the acquisitions of primates from the wild, regional zoo associations initiated cooperative captive breeding programs to establish sustainable breeding populations of species for zoo collections (1980: AZA Species Survival Plan [SSP]; 1985: EAZA European Endangered Species Programme [EEP]).

THE ZOO ENVIRONMENT

Primates are adapted to a natural habitat in which their survival depends upon a complex behavioral repertoire and high levels of vigilance, problem solving, and awareness of social group dynamics. Primate enclosures should be designed to meet the physical, psychological, and behavioral needs of the species. The unique arboreal adaptations of primates make them adept at climbing, leaping, and in some cases swimming; thus adequate containment measures for each species' physical propensities must be considered in any primate enclosure and exhibit design, to prevent an animal's injury or escape.

The main factors that are of primary importance in primate enclosures are the size, in terms of usable space; the containment barrier (glass-fronted, mesh, or moated); the basic design in relation to capture and husbandry methods; and the provision of a complex environment to meet the animal's physical, cognitive, and social needs (Kleiman et al. 1996, 2010). Primates will use three-dimensional space; therefore, enclosures should make use of all available space, with its complexity and quality taking precedence over its total size (Wolfensohn and Honess 2005, Hosey et al. 2009).

Enclosures should be large enough and contain ample furnishings to provide for the group housing of conspecifics as well as providing the option for temporarily maintaining individuals singly or in subgroups due to medical and/or social situations (McCann et al. 2007). Individuals should be able to exhibit normal postural and locomotor patterns and express species-typical behaviors (e.g., exploration, brachiation, foraging, resting, playing, and allogrooming). They should have a secure and suitably complex environment, with food and water easily accessible and adequate opportunities to negotiate diverse social interactions with group members. Attention should be made to the species' size, use of vertical space, and other important species-specific characteristics. Other species-specific characteristics may include perching preferences (arboreal versus terrestrial), sleeping sites (nest-boxes versus shelving), foraging devices (group feeders versus timed dispensers), and refuge areas (areas within enclosures for safe distances from the public, keepers, and/or other group members). All primates have a vertical flight response when alarmed by an unfamiliar or threatening stimulus. Thus, the vertical dimension of the enclosure is of great importance. The enclosure should be equipped so that individuals are able to retreat above human eye level (Reinhardt and Reinhardt 2000). Furthermore, many primates—particularly marmosets, tamarins and lorises—prefer the highest available areas and rarely use the lower half of their enclosures. Thus, key resources such as food, nestboxes, and heating sources should be placed high in the enclosure so that individuals can access food and shelter at preferred enclosure heights (Buchanan-Smith et al. 2002).

HUMIDITY, VENTILATION, TEMPERATURE, AND LIGHTING

The majority of primate species are found in tropical habitats that experience high humidity, and this aspect of their environment should be approximated as closely as possible in captive conditions. Humidity levels can affect the condition of the skin and coat and should be kept at above 50%, although many primate species can tolerate lower levels. Primate enclosures in the United Kingdom and the United States require humidity levels that range between 30% and 70% (Wolfensohn and Honess 2005, 26). Special considerations should be made for young, geriatric, or clinically compromised individuals. Shade structures, shelters, air coolers, and fans can provide relief from high humidity, and misters can be effective for increasing humidity levels where needed.

Indoor primate enclosures should be sufficiently ventilated at all times when animals are present such that odors, drafts, ammonia levels, and moisture condensation are minimized (USDA/APHIS 3.76). Air changes inside housing should be conducted at a rate of 10 to 15 per hour, with at least 10% fresh air in recycled air when outside ambient temperatures are below 13 °C (55 °F) and up to 50% when outside ambient temperatures will not cause a significant decrease in indoor temperatures (AWR 2005; Wolfensohn and Honess 2005, 25).

The majority of primate species should be maintained within temperature ranges of 18 to 28 °C (65 to 82 °F; Wolfensohn and Honess 2005, 24). In some notable exceptions (e.g., *Colobus guereza*, *Macaca fuscata*, *Theropithecus gelada*, *Gorilla beringei beringei*) their geographic range includes temperate forests and/or high altitudes, and therefore they can tolerate lower ambient temperatures. Most species will acclimate to lower temperatures, but should be exposed to them gradually. This could include allowing individuals access to outdoor enclosures only during the warmest times of the day and then gradually increasing the amount of time they are allowed outdoors, or keeping indoor areas at slightly lower temperatures to allow animals to acclimate to colder ambient temperatures outdoors. Primate extremities (tails and digits) are particularly susceptible to cold temperatures. In ambient temperatures that fall below the average range for primates, individuals should have access to heat sources and shelters for protection against wind and rain. There must be enough heat sources so that all the animals have access to them and are not prevented from using them by more dominant individuals. Extreme care should be taken in the placement of heating devices (heat lamps and radiant heaters) to ensure that an individual cannot be burned by touching them. In temperature ranges above 32 °C (90 °F), cooling mechanisms such as fans and misters should be incorporated; shaded areas should always be available to prevent prolonged exposure to the direct sun, and enough sheltered areas so that dominant animals cannot monopolize them.

Light intensity, duration, and spectral requirements for captive primates should approximate those in the wild (Wolfensohn and Honess 2005, 26–27). Most zoos are in regions of the world that differ from the near-equatorial light cycles of the tropics, and thus natural daylight hours are reduced. For diurnal primates, access to light for 12 hours out of each 24-hour period is recommended. For nocturnal species, a reverse light system should be implemented. Timers can be used to mimic natural light cycles. For indoor enclosures, windows or skylights transparent to full-spectrum ultraviolet (UV) light are recommended in lieu of artificial light. It is strongly recommended that primates have access to natural lighting to prevent the developmental problems associated with vitamin D deficiency, such as metabolic bone disease. Although exposure time to direct sunlight need only be for short periods (10 to 15 minutes per day), the benefits for normal developmental growth are paramount.

SUBSTRATE AND VEGETATION

Natural substrates are recommended for primate exhibit enclosures, as they provide a myriad of behavioral opportunities, and more closely approximate substrates the species would encounter in their natural habitat (McCann et al. 2007). They also provide a soft surface to protect infants that may fall to the ground as they develop their locomotor skills and learn to navigate through arboreal pathways. For indoor holding enclosures, cemented or sealed floors are the norm for practical sanitation reasons; however, supplemental materials are recommended to provide enhanced foraging opportunities and nesting materials (e.g., hay, wood wool, woodchips, and leaf litter). Traditional tiled and concrete floor enclosures provide a sterile environment for ensuring the clinical health of primates; however, they lack the features needed to meet an individual's psychological health. With appropriate cleaning and husbandry routines, enclosures that are stimulating and complex can meet both important aspects of primate well-being (Wolfensohn and Honess 2005, Hosey et al. 2009).

While providing natural vegetation in primate enclosures is recommended, care should be taken in species selection to ensure that the plant species and parts (leaves, fruit, seeds, or flowers) used are not toxic. Most primates will actively forage on live plant material; however, it is a fundamental husbandry practice to provide additional plants as browse in diets for leaf-eating species. Though browse plants provide an important component of a folivorous primates' diet, secondary plant compounds such as tannins and alkaloids may be present within these plants, and depending on the plant and primate species, they may cause digestive problems. Thus, browse species selection should be carefully decided among the veterinary and animal staff. Plants provided in enclosures as browse should be offered in an appropriate and safe manner. Bark and stems from fibrous plant species can cause physical obstructions in the gastrointestinal tract (Janssen 1994; Calle et al. 1995). Methods for avoiding ingestion of the stem or bark have included their removal prior to distribution, or the placing of the stems in a protective PVC tube where they cannot be stripped by the animal. Also of concern are ropes of indigestible acacia fiber, which have resulted in impaction and subsequent death in langurs offered a diet including acacia leaves (Ensley et al. 1982).

HUSBANDRY

Primates should be provided with fresh, uncontaminated water continuously throughout the day, and it must be accessible to all individuals in the enclosure. It can be presented in a variety of ways, depending on the species and features of the enclosure, but the watering devices should be easy to disinfect to minimize the spread of disease. Watering devices can include rubber pans, bottles, or automatic drinkers ("lixits"), and should be checked daily to ensure that they are operational. Automatic waterers should be positioned to run continuously for individuals that have not been exposed to this method previously, until they have learned their operation. In addition to a clean potable water supply, pools or waterfalls in enclosures can be supplementary sources for water when they can be cleaned or filtered daily.

General animal care knowledge and experience is a basic foundation for keepers working with any species, including primates. It is paramount that keepers have a general understanding of primate natural history as well as the unique characteristics of the species under their care (e.g., husbandry requirements for a family group of tamarins will differ from those for nocturnal lorises or a troop of baboons). Additionally, keepers should educate themselves about the special care requirements of primate species that are readily available from studbooks, husbandry manuals, and primate husbandry care guidelines (Wolfensohn and Honess 2005; McCann et al. 2007). Primate husbandry care differs from those for other taxa in that special attention must be given to individual needs and the role complex social dynamics play in the management of the species. Thus, primate keepers should be highly observant of individual differences in physical and social needs, and should possess a keen understanding of group dynamics to be able to act appropriately and implement acceptable management techniques to provide the best care for individuals while respecting social complexities.

Primates are highly social; the cohesiveness of groups continually changes with intrinsic events, such as sexual maturation, and extrinsic events, such as the removal of a dominant animal. Thus it is most important for keepers to attain an astute awareness of the compatibility of individuals within groups, such as when a change in social dominance results in heightened aggression among group members. They should also be flexible and adapt husbandry routines to accommodate the changes in social dynamics that occur when change in dominance relationships result in a change in a certain individual's access to food. Finally, they should have the patience to work with individual animals that may have differences in responsiveness to management routines. Often this may result in an individual animal being reluctant to routinely shift between enclosures due to a change in the social relationships among the other group members. As the primary caregiver of a primate group, a keeper must display appropriate demeanor when working with primates; developing a strong rapport with individual primates will form a foundation for their successful management. It is counterproductive and in some cases harmful for a keeper to become overtly agitated or show frustration in response to a breakdown in a management routine as a result of changes occurring in the primate social group. It is more productive to understand what motivates the incompatibility and find ways to mitigate any negative effects on the individuals' daily care.

It is important to emphasize that managing aggression is one of the most challenging aspects of primate captive care. While it is a natural part of primate social life, aggression within captive groups requires consistent monitoring of group dynamics, an understanding of the root of the aggression and whether it is acute or chronic, and ongoing communication among keepers, animal managers, veterinary staff, and in some cases the species' population manager, to arrive at a plan to manage the social incompatibility. Ultimately, management decisions must balance individual welfare needs, the long-term stability of the group, and the population goals for the species program; and the keeper features prominently in this process as the individuals' primary caregiver.

One of the primary responsibilities of a keeper is to be aware of the health status of the individuals in their care and to be able to detect the external signs of an individual's condition. Assessing the health condition of primates can be determined by the following indicators:

- changes in the quantity and quality of an individual's stool
- rapid weight gain or loss, or chronic deviations from the optimum weight
- coat condition and overall appearance (well groomed versus unkempt, with attention to any signs of hair loss)
- changes in food/water intake rates and appetite, excessive salivation, or difficulty in the ability to process food items
- activity level, lethargy, separation from group members, nonresponsiveness, tendency to seek heated areas
- changes in locomotion (gait) or posture, inability to negotiate all accessible areas of the enclosure
- changes in spatial distance from group members and/or lack of participation in daily activities
- self-directed behaviors such as excessive autogrooming or self-mutilation
- stereotypies or repetitive behaviors.

Keepers should know the normal range of physical attributes and social behavior of a species, and of the individuals in their care, so that they can notice deviations that may indicate other underlying processes.

HANDLING AND RESTRAINT

The handling of primates can be greatly facilitated by operant conditioning training. The use of positive reinforcement techniques are preferred as they can decrease stress for individuals. Primates can be reliably trained using positive reinforcement to cooperate with capture and immobilization procedures such as shifting into nests, transfer and/or induction boxes, crates, or squeeze cage systems; stationing on scales; presenting body parts for hand injections or blood sampling; and positioning for veterinary procedures such as ultrasound examinations.

While operant conditioning methods benefit an individual's well-being, traditional capture and restraint methods must also be employed in emergency situations, or when behavioral training is not successful. For small and medium-sized primates, netting is the preferred method of capture. Handling gloves should be used when removing the animal from the net, in addition to a face mask and eye protection. Immobilization drugs can be administered by hand while the

individual is in the net, and then the individual can be removed once the drug takes effect. For larger primates, netting is not recommended; shifting the animal into a smaller confined area (e.g., a chute, squeeze, or isolation cage) for hand injection or darting with an immobilization drug is recommended instead. The method of manual restraint to use for a primate varies with the size of the species. Particular caution must be given to the head and arms, to prevent bites and scratches. For basic restraint and handling methods for a variety of species, see Fowler (1995), Bush (1996), and Christman (2010).

ENRICHMENT

Optimal care for primates requires a program that maintains their clinical health as well as their psychological health and well-being (Cipreste et al. 2010), and in some regions this is mandated by regulatory agencies. For instance, in 1991 the US Department of Agriculture/Animal and Plant Health Inspection Service (USDA/APHIS) regulations require primate facilities to develop, document, and carry out a species-specific plan for environmental enhancement to promote the psychological well-being of nonhuman primates. Many enrichment opportunities can be used with primates. Primates are highly social beings, and providing them with compatible conspecifics is the most important social enrichment to consider, as it can stimulate the expression of their behavioral repertoire (Young 2003, Hosey et al. 2009).

Providing primates with a complex, stimulating environment is key to a successful husbandry program (Young 2003). Enrichment programs should be well planned to provide the appropriate items and frequency of use to make them most effective. Enrichment items should be chosen with care to prevent exposure to hazardous materials or injury, and to stimulate natural behaviors continually. If enrichment items become permanent fixtures in a primate's enclosure, then they rapidly lose their desired effect of stimulating new behaviors. Enrichment activities fall into different categories, based on the behavior patterns they elicit:

- activities that elicit locomotor behaviors (e.g., using ropes, swings, vines, logs, branches, hammocks, or climbing structures), important for exercise and facilitates the full repertoire of locomotor agility
- activities that elicit increased foraging behaviors (e.g., variation in feeding schedules, food delivery systems, and foraging times) to encourage individuals to spend more time searching, locating, and processing food items
- activities that provide novelty and a variety of objects to stimulate cognitive abilities (e.g., barrels, balls, baskets, boxes, puzzles) and enable primates to explore, be challenged by tasks, and solve problems
- activities that enable primates to exercise preferences and some degree of control in a suitably variable environment.

In the wild, primates are faced with a multitude of choices every day. Providing choice-making opportunities in captive environments can enhance this important aspect of their adaptation (Hosey et al. 2009).

TRAINING

All primate keepers should be familiar with the basic tenets of operant conditioning training and should use positive reinforcement techniques wherever possible. Routine husbandry often involves moving and positioning animals for cleaning, exhibit maintenance, and veterinary examinations. It has been demonstrated that these procedures can be greatly facilitated by positive reinforcement training (Laule et al. 2003). Primates should be trained to reliably shift between enclosures and exhibits, to station on scales for weighing, to enter crates for transfer between areas, to station for physical inspection and the administering of oral medications, to separate from group members for medical procedures, and to present body parts for hand injections. Housing social groups of primates poses additional challenges for behavioral management, and training techniques can also be applied to facilitate the socialization of conspecifics, reduce aggression, and eliminate abnormal or unwanted behaviors (Prescott and Buchanan-Smith 2007). Cooperative feeding is an effective technique for facilitating the compatibility of conspecifics, the introduction of unfamiliar individuals, and managing the dominance effects in a group by training the dominant animal(s) in a group to allow subordinates to safely approach. Training programs require keepers to work in close proximity with animals, so extra safety protocols should be in place to prevent risks to keepers or animals. It is good practice to avoid hand feeding during training sessions, as the risk for injury to hands and fingers is much greater at that time. Many primate trainers drop food items directly into the primates' hands, use a utensil, or offer only food items greater than six inches in length in order to keep their fingers away from the animal's mouths.

REPRODUCTION

Nonhuman primate females are cyclically or seasonally receptive; and this period of receptivity in many species is associated with visual changes of the anogenital region and conspicuous behavioral changes (e.g., presenting, staring, tongue flicking, head shaking), making it evident to a male when females are in estrus (Strier 2011). Outside of this general pattern, there exists a wide range in reproductive parameters across primate taxa. Most lemurs and lorises are characterized as seasonal breeders, with very short mating seasons which typically last from one to three weeks, with females being receptive for only a few days within this period (Campbell et al. 2011). The breeding season is correlated with highly seasonal climatic changes and availability of food resources. In captive environments in the Northern Hemisphere, the breeding season occurs at opposite times of the calendar year than those in the Southern Hemisphere due to the influence of environmental factors such as temperature, rainfall, and day length (Wilson and Reeder 2005). In general, reproductive cycles in monkeys and apes can occur continuously throughout the year with few exceptions (e.g., squirrel monkeys, macaques), although birth peaks are characteristic of species that inhabit environments with marked seasonality of rainfall (Campbell et al. 2011). Zoo breeding programs (SSP, EEP) make recommendations for pairings of animals

and the resulting transfer of individuals among zoos with knowledge of a species' breeding season, in order to facilitate the introduction of individuals before the next breeding season. Additionally, a variety of reversible contraception methods have been successfully used to prevent conception in managed primate populations while maintaining intact groups. Breeding in primates can be associated with heightened agonistic behavior among group members, and keepers must be vigilant to these changes in group dynamics and individual demeanor, as they could result in overt aggression if not carefully managed. Heightened awareness should be applied during daily husbandry routines to facilitate breeding while mitigating any potential aggressive bouts (e.g., disputes over preferred spaces and food items, or difficulties in shifting individuals).

GESTATION AND PARTURITION

Relative to their size, gestation lasts considerably longer in primates than in other mammals. It varies from 54 to 68 days in mouse lemurs, from 110 to 133 days in galagos, from 126 to 168 days in diurnal lemurs, from 130 to 155 days in callitrichids, from 135 to 230 days in cebid monkeys, from 166 to 190 days in lorises, from 150 to 210 days in cercopithecine monkeys, and from 210 days in lesser apes to between 230 and 258 days in great apes (Campbell et al. 2011). Pregnancy in most primates can be detected by the absence of the visual signs of estrus and a cessation of breeding cycles, or through more direct means such as vaginal cytology, ultrasound examinations, or hormonal assays of reproductive hormones. Pregnant primates should be closely monitored in preparation for parturition. Primates typically give birth among group members, and females should not be separated from their groups unless the management of a particular species requires alternate management strategies to increase infant survival rates in captive conditions. Upon parturition, privacy for the mother and newborn is paramount to allow the appropriate mother-offspring bond to occur. Keepers must display the appropriate demeanor and maintain a safe distance while observing the mother-infant pair closely enough to see any problems that require intervention.

INFANT CARE

A newborn primate is born fully furred and with eyes open, and with grasping hands and feet that enable them to cling to their mothers (exceptions to this are some species of lemur and loris; Fleagle 1999; Strier 2011). The first three days of an infant's life are critical to its ultimate survival, the most important factor being the mother's ability to appropriately care for it. Keepers must closely monitor the mother's handling of the infant, noting the mother's positioning of the infant and the mother's reaction when the infant roots to locate the nipple, and they must be able to confirm the frequency and duration of nursing bouts. Nursing does not always occur on day one but should be seen by day two, as the infant will otherwise progressively loose strength and be unable to cling to the mother. There is a significant learning component to maternal behavior in primates. Primiparous females should have the opportunity to watch other females in their group care for their infants to obtain this critical knowledge. The importance of the appropriate social environment for the acquisition of maternal skills, as well as the experience of various breeding programs, necessitates that primates should not be hand-reared as a general rule. In extraordinary circumstances (e.g., death of the dam, needs of the population, genetic value of the individual), hand-rearing may be recommended. When an infant must be hand-reared, a plan should be in place to swiftly integrate it into an appropriate social environment. From the comprehensive experience of captive breeding programs in zoos, hand-rearing protocols are well established for the majority of primate species.

TRANSPORTATION

The transport of primates must follow the International Air Transport Association (IATA) regulations, which should be referred to before any shipping arrangements are made. In some cases, there may be additional requirements imposed by regional regulatory agencies (e.g., the US Fish and Wildlife Service). A primate must be carried in a closed container that is well constructed to secure the animal safely and protect it from unauthorized access through the entire transit process. Crates used for primates can be constructed of plastic, wood, or metal, depending on the size and strength of the species, and should only include nontoxic materials. Small primates can adequately be shipped in wooden crates or appropriate-sized kennel crates with added tamper-proof locks; large primates like great apes should be shipped in metal crates. Meshed ventilation openings must exist along the sides of the crate, and the front door must include a solid panel with ventilation openings in the top third. Muslin or burlap material should cover all ventilation openings. The crate's floor must be solid and leakproof to prevent excreta from leaking from the crate during transit. Primates must be provided with food and water during transit, and depending on the type of crate and length of transit, their food and water containers will vary.

Preparing an individual for transport is critical for its acclimation to a new environment. Primates can be readily trained to voluntarily enter crates and accept hand-injections for immobilization for preshipment examinations, thus reducing the stress involved in these events. Additionally, the transfer of all pertinent information to the receiving institution is a critical role of keepers in any transfer of collection animals between zoos. Primates being moved to a new facility are faced with new surroundings and keeper staff. In some cases, particularly with ape species, keepers can accompany the animals through the transfer to ensure a more successful acclimation to the new environment.

VETERINARY CARE

The medical management of primates involves clinical practices to protect the animal collection from common zoonoses affecting primate species, to protect the keeper staff that provides their daily care, and to prevent disease transmission. Primates are susceptible to a variety of bacterial, viral, parasitic, and fungal zoonoses that are infectious or contagious, and different diseases have varying degrees of

prevalence among certain primate taxa (ILAR 2003; Fowler and Miller 2003). Common primate gastrointestinal bacterial pathogens can include: *Salmonella*, *Shigella*, *Campylobacter*, *Klebsiella*, *Yersinia*, and *Clostridium*. Clinical signs include diarrhea, weight loss, vomiting, and lethargy; treatment is often an antibiotic therapy. Successful administration of oral antibiotics requires individually medicating the affected animal, and this need underscores the importance of operant conditioning training in primate medical management. The best methods for the prevention of bacterial diseases are the implementation of biosecurity measures (footbaths, gloves, face masks) and general cleanliness in food preparation and enclosure disinfection. Perhaps the most commonly known bacterial pathogen of concern in primates is *Mycobacterium* (i.e., tuberculosis or TB). There are many different species of mycobacteria; however, clinical signs may not be detectable and TB testing, radiographs, blood sampling, and cultures may be required for diagnostic confirmation (Fowler and Miller 2003). Preventative measures require all primates entering a facility to be tested for TB, to prevent introducing individuals infected with *Mycobacterium* into the collection.

Different primate species are known to carry different viruses; and it is important to note that not all viruses cause disease in the host species. Herpes virus, similar to that found in humans, is common in prosimians and callitrichids; and great apes are susceptible to the same viruses as humans. Many viruses are carried by cercopithecine monkeys, including Foamy Virus, Simian Immunodeficiency Virus (SIV), Herpes B, and Herpes B-like virus; these have varied etiologies and clinical signs (e.g., lethargy, respiratory or gastrointestinal infection, and ulcers on the mouth or face; Murphy et al. 2006). Most viral diseases are not treatable; preventative measures are the focus instead, and they include knowing the health status of individual primates entering a collection and ensuring best practices in biosecurity.

The potential for the transfer of diseases between humans and nonhuman primates should always be considered a risk factor, and protocols should be in place to prevent zoonotic disease transmission both to and from nonhuman primates (McCann et al. 2007). The transmission of disease can be substantially reduced by ensuring good personal hygiene (e.g., frequent sanitization of hands and the use of footbaths between areas housing primates) and management protocols that include personal protective equipment (e.g., face shields, goggles, masks, and gloves). Information on known primate zoonotic diseases and their prevention should be made available to all keepers who work with primates and they should be familiar with preventive measures that are in place to reduce risk factors. Additionally, primate care staff should be screened annually for tuberculosis. It is also recommended that they be vaccinated against tetanus, polio, rabies, measles, and hepatitis. Facilities housing macaque species should have preventive measures in place to protect keepers from the risk of infection by Herpes B virus which may be carried by macaques. The Centers for Disease Control and Prevention (CDC 1987, 1999), the Institute for Laboratory Research (ILAR 1996, 2003) and regional zoological associations can be used as resources for best practices in primate management guidelines.

Specific diseases will manifest in primates at different stages in their lives. Metabolic bone disease often prevails early in the growth process due to insufficient calcium and vitamin D, and presents as poor growth rate and inability to locomote appropriately (commonly described as "bunny hopping"). Treatment can include oral or injectable calcium, appropriate vitamin D levels in the diet, and exposure to full-spectrum light (Wolfensohn and Honess 2005, IPS 2007). Providing young primates the opportunity for exposure to full spectrum light for vitamin D synthesis, for as little as 10 minutes per day, is of critical importance for normal growth processes. At the opposite end of the spectrum in the developmental process are diseases that commonly manifest in aged primates. Of particular concern are degenerative diseases such as arthritis, spondylosis (spinal osteoarthritis), and heart and renal disease.

The last particular concern is sepsis as a result of trauma. The social nature of primates and the management of their group dynamics will often necessitate the clinical management of wounds due to aggression from conspecifics. These wounds can be in the form of lacerations, punctures, or fractures, and clinical signs of sepsis can include lethargy, depression, lameness, anorexia, and drowsiness (somnolence). Keen observation and inspection of individuals for wound sites is basic to prevention, as treatment of sepsis requires early medical intervention with antibiotics.

Keepers play a vital role in the medical management of primates. Their responsibilities include providing a clean, enriched, and safe environment; closely observing the behavior and appetite of individuals; knowing the animal's preference for medical compliance (e.g., form of medications, frequency of dosing, individual administration); and training individuals for voluntary participation in clinical health procedures (e.g., regular weighing, entering of anesthesia induction boxes, and accepting hand injections). In summary, sound primate management strives to prevent disease transmission both to and from nonhuman primates. Cleanliness and biosecurity are the two most important factors in the prevention of infectious and contagious diseases in primate collections. A keeper's observation and communication skills are imperative for identifying illness in an animal and preventing the spread of disease.

CONSERVATION AND RESEARCH

Primates are humankind's closest biological relatives; humans share 98.4% of their DNA coding with chimpanzees. The majority of primates live in tropical forests, where they play an integral role in ecosystem function as plant pollinators, seed predators, and seed dispersers. Increasing forest fragmentation, human encroachment, and illegal capture of primates for the bush meat and pet trades have resulted in more isolated populations. According to the International Union for the Conservation of Nature (IUCN) Primate Specialist Group (PSG), there are currently 612 recognized species and subspecies of primates (www.iucn.org). Since 2000, 46 new species have been discovered or redescribed, and many more isolated populations have yet to be discovered by scientists (Lehman and Fleagle 2010). Due to their reliance on tropical forests and due to human extraction activities, primates are considered to be amongst the most endangered taxa worldwide, and in most urgent need of conservation measures.

In situ conservation efforts are aimed at mitigating threats

TABLE 29.2. A sample of primate societies across various regions of the world

American Society of Primatologists
Associação Portuguesa de Primatologia
Associacion Primatologica Espanola
Associazione Primatologica Italiana
Sociedade Brasileira de Primatologia
Congolese Society of Primatologists
Deutsche Primatologische Gesellschaft
European Federation for Primatology
Groupe d'étude et de recherche sur les primates de Madagascar
International Primatological Society
Mexican Association of Primatology
Pan African Sanctuary Alliance
Primate Conservation Inc.
Primate Ecology and Genetics Group
Primate Society of Great Britain
Primate Society of Japan
Societe francophone de primatologie
South East Asian Primatological Association

to fragile populations and protecting the integrity of the landscapes they inhabit. As the majority of primate species have evolved specialized dietary and habitat preferences, the destruction of their forests and the depletion of key plant species make them exceptionally vulnerable to these unprecedented environmental perturbations. The advancement of our understanding of a species' ecological requirements from primate behavioral ecology studies provides critical information necessary to inform conservation management plans (see Rowe and Meyers 2013 for a comprehensive online database of primates).

Ex situ primate research ranges in topic from a species' basic biological description to assisted reproduction and management techniques. Captive breeding programs at zoos provide an opportunity for describing the life history traits of species that may not be known from studies in the wild; and basic biological parameters provide the foundation for understanding species' unique adaptations (e.g., Beehner and McCann 2008; Villers et al. 2008; Shelmidine et al. 2009). Managing captive primates requires continued enhancement of husbandry practices, and many research efforts are focused on ways to evaluate and improve current standards (e.g., Stoinski et al. 2004; Ballou et al. 2010). Research initiatives are often collaborative efforts administered through regional zoological association species programs and local universities. There are many primate organizations whose mission includes scientific research, educational outreach, and conservation of primate species (table 29.2). Keepers are encouraged to become members of primate organizations and their regional zoological associations to become involved in the broader aims of primate research and conservation.

SUMMARY

The primates are a diverse order, with a wide range of ecological and behavioral adaptations that make the techniques used in their husbandry care equally diverse. Best practices for their management and care can be achieved when key elements of primate natural history are incorporated into daily husbandry routines. The complexity of primate behavior and social structure makes them popular species for zoo exhibition; and it is this aspect of their behavior that presents the most challenges to keepers who are charged with their daily care. The key to becoming a proficient primate keeper is acquiring the observational and technical skills needed to manage the social and clinical health of primates on a daily basis. These include a general knowledge of primate natural history and of each species' ecological adaptations, an understanding of their species-typical behavior and the complexities of their social interactions, and acquisition of the technical skills needed to develop a rapport with individual primates and facilitate operant conditioning training in order to provide optimal care.

Keepers should be encouraged to become members of their regional zoological associations. They should become educated about population management of the species in their care, ongoing research projects relating to primate taxa, educational outreach programs, and ways to contribute to in situ conservation efforts. There are also many primate associations working on various aspects of primate research, education, and conservation in which membership would provide keepers with a means for professional development in the field of primatology.

REFERENCES

Animal Welfare Regulations (AWR). 2005. Animal Welfare Act, 7, USC Animal Welfare Regulations, 9 CFR chapter 1, subchapter A, parts 1–4.

Ballou, J. C. Lees, L. Faust, S. Long, C. Lynch, L. Bingaman-Lackey, and T. Foose. 2010. Demographic and genetic management of captive populations for conservation. In D. Kleiman, K. V. Thompson, and C. Kirk Baer (eds.), *Wild Mammals in Captivity: Principles and Techniques for Zoo Management, Second Edition.* Chicago: University of Chicago Press, 2010.

Beehner, Jacinta, and Colleen McCann. 2008. Cortisol levels in captive and wild geladas (*Theropithecus gelada*). *Physiology and Behavior* 95:508–14.

Buchanan-Smith, Hannah M., Carole Shand, and Keith Morris. 2002. Cage use and feeding height preferences of captive common marmosets (*Callithrix j. jacchus*) in two-tier cages. *Journal of Applied Animal Welfare Science* 5:139–49.

Bush, Mitchell. 1996. Methods of capture, handling and anesthesia. In D. G. Kleiman, M. E. Allen, K. V. Thompson, and S. Lumpkin (eds.), *Wild Mammals in Captivity: Principles and Techniques.* Chicago: University of Chicago Press, pp. 25–40.

Campbell, Christina J., Augustin Fuentes, Katherine C. MacKinnon, Simon K. Bearder, and Rebecca M. Stumpf, eds. 2011. *Primates in Perspective.* Oxford: Oxford University Press.

Crandall, Lee S. 1964. *The Management of Wild Mammals in Captivity.* Chicago: University of Chicago Press.

Christman, Joe. 2010. Methods of capture, handling, and restraint. In D. G. Kleiman, K. V. Thompson, and Charlotte Kirk Baer (eds.), *Wild Mammals in Captivity: Principles and Techniques for Zoo Management, Second Edition.* Chicago: University of Chicago Press, pp. 39–48.

Cipreste, Cynthia Fernandes, Cristiano Schetini de Azevedo, and Robert J. Young. 2010. How to develop a zoo-based environ-mental enrichment program: incorporating environmental enrichment into exhibits. In D. G. Kleiman, K. V. Thompson, and Charlotte Kirk Baer (eds.), *Wild Mammals in Captivity: Principles and Techniques for Zoo Management, Second Edition.* Chicago: University of Chicago Press, pp. 171–80.

Ensley, Philip K., Thomas L. Rost, Marilyn Anderson, Kurt Benirschke, Diane Brockman, and Duane Ulrey. 1982. Intestinal obstruction and perforation caused by undigested *Acacia* sp. leaves in langur monkeys. *Journal of American Veterinary Medicine Association* 181:1351–54.

Fleagle, John G. 1999. *Primate Adaptation and Evolution, Second Edition.* San Diego: Academic Press.

Fowler, Murray E. 1995. *Restraint and Handling of Wild and Domestic Animals, 2nd Edition.* Ames: Iowa State University Press.

Fowler, Murray E., and R. Eric Miller. 2003. *Zoo and Wild Animal Medicine, 5th Edition.* Philadelphia: W. B. Saunders.

Gardner, R. Allen, Beatrice T. Gardner, and Thomas E. Van Cantfort. 1989. *Teaching Sign Language to a Chimpanzee.* Albany, NY: SUNY Press.

Groves, Colin. 2001. *Primate Taxonomy.* Washington, DC: Smithsonian Institution Press.

Hosey, Geoff, Vicky Melfi, and Sheila Pankhurst. 2009. *Zoo Animals: Behaviour, Management and Welfare.* Oxford: Oxford University Press.

Institute for Laboratory Animal Research (ILAR). 1996. *Guide for the Care and Use of Laboratory Animals.* Bethesda, MD: National Research Council.

———. 2003. *Occupational Health and Safety in the Care and Use of Nonhuman Primates.* Washington: National Academies Press.

Janssen, Don L. 1994. Morbidity and mortality of douc langurs (*Pygathrix nemaeus*) at the San Diego Zoo. American Association of Zoological Veterinarians (AAZV) Conference, St. Louis, MO.

Kleiman, Devra G., Mary E. Allen, Katerina V. Thompson, and Susan Lumpkin. 1996. *Wild Mammals in Captivity: Principles and Techniques.* Chicago: University of Chicago Press.

Kleiman, Devra G., Katerina V. Thompson, and Charlotte Kirk Baer, eds. 2010. *Wild Mammals in Captivity: Principles and Techniques for Zoo Management, Second Edition.* Chicago: University of Chicago Press.

Laule, Gail E., Mollie A. Bloomsmith, and Steven J. Schapiro. 2003. The use of positive reinforcement training techniques to enhance the care, management and welfare of laboratory primates. *Journal of Applied Animal Welfare Science.* 6:163–73.

Lehman, Shawn M., and John G. Fleagle, eds. 2010. *Primate Biogeography: Progress and Perspectives.* Chicago: Springer.

McCann, Colleen, Hannah M. Buchanan-Smith, Mark Prescott, Lisa Jones-Engel, Kay Farmer, Helena Fitch-Snyder, and Sylvia Taylor. 2007. IPS International Guidelines for the Acquisition, Care, and Breeding of Non-Human Primates, 2nd Edition. Available online at http://www.internationalprimatologicalsociety.org.

Murphy, Hayley W., Michele Miller, Jan Ramer, Dominic Travis, Robyn Barbiers, Nathan D. Wolfe, and William M. Switzer. 2006. Implications of simian retroviruses for captive primate population management and the occupational safety of primate handlers. *Journal of Zoo and Wildlife Medicine* 37(3): 219–33.

Napier, John R., and Prudence H. Napier. 1967. *A Handbook of Living Primates.* New York: Academic Press.

Nowak, Ronald M. 1999. *Walker's Mammals of the World, Vol. 1, 6th Edition.* Baltimore: John Hopkins University Press, 490–631.

Prescott, Mark J. and Hannah M. Buchanan-Smith. 2007. Training laboratory-housed non-human primates, part 1: A UK survey. *Animal Welfare* 16:21–36.

Reinhardt, Viktor and Annie Reinhardt. 2000. The lower row monkey cage: An overlooked variable in biomedical research. *Journal of Applied Animal Welfare Science* 3:141–49.

Rowe, Noel, and Mark Myers, eds. 2013. *All the World's Primates.* Charlestown, RI: Primate Conservation Inc. Accessed on 11 January 2013 at www.alltheworldsprimates.org.

Shelmidine, Nichole S., Carola, Borries, and Colleen McCann. 2009. Patterns of reproduction in Malayan silvered leaf monkeys at the Bronx Zoo. *American Journal of Primatology* 71(10): 852–59.

Stoinski, Tara, Kristen E. Lucas, Chris W. Kuhar, and Terry L. Maple. 2004. Factors influencing the formation and maintenance of all-male gorilla groups in captivity. *Zoo Biology* 23(3): 189–203.

Strier, Karen B. 2011. *Primate Behavioral Ecology, Fourth Edition.* Upper Saddle River, NJ: Prentice Hall.

Suzuki, Shuji. 1999. Selection of forced- and free-choice by monkeys (*Macaca fascicularis*). *Perceptual and Motor Skills* 88:242, 250.

Villers, Lynne M., S. S. Jang, C. L. Lent, S.-C. Lewin-Koh, J. A. Norosoarinaivo. 2008. Survey and comparison of intestinal gut flora in captive and wild ring-tailed lemur (*Lemur catta*) populations. *American Journal of Primatology* 70(2): 175–84.

Wilson, Don E. and DeeAnn M. Reeder, eds. 2005. *Mammal Species of the World: A Taxonomic and Geographic Reference, 3rd Edition.* Baltimore: John Hopkins University Press.

Young, Robert J. 2003. *Environmental Enrichment for Captive Animals.* Oxford: Blackwell Science.

30

Husbandry and Care of Elephants

Chuck Doyle and Daryl Hoffman

INTRODUCTION

On the internet there is an abundance of information on elephants. In fact, if you search for the word "elephants" you will get more than 11 million results. Granted, the results cover everything from natural history to elephants in human care to elephant collectibles. As with any other subject, some of the information is excellent and some is less than credible. The purpose of this chapter is to provide an entry-level zookeeper with an overall view of elephant management. Please note that the complexities and diversity of elephant management cannot be covered in one chapter, or even in a complete book on the subject. Sometimes it seems that there are as many ways to care for elephants as there are internet results. Most novice keepers spend two years in an apprentice program before they become proficient as elephant keepers. Each institution's elephant management program varies and must meet the institution's individual goals and needs. There are many ways to care for elephants, but it is critical that each institution has the behavior components necessary for a successful program, such as those presented in the Association of Zoos and Aquariums (AZA) course Principles of Elephant Management.

After studying this chapter, readers will understand

- anatomical adaptations of elephants
- effects of elephant biology on housing, nutrition, and social management
- basic components of elephant behavior as they relate to animal and keeper safety
- practices for promoting species-appropriate behaviors
- management methods for elephants.

NATURAL HISTORY

The first thing any prospective keeper needs to learn about elephants is their natural history. African and Asian elephants are two distinct species which belong to separate genera. They are generally similar to each other in size, appearance, physiology, and social behavior. The Asian elephant is considered to be a single species, *Elephas maximus*, with four extant subspecies: *E.m. hirsutus* (Malayan elephant), *E.m. indicus* (Indian elephant), *E.m. maximus* (Sri Lankan elephant), and *E.m. sumatranus* (Sumatran elephant). The African elephant is considered to be a single species, *Loxodonta africana*, with two subspecies: *L.a. cyclotis* (forest elephant) and *L.a. africana* (savanna elephant). Viewed in depth, elephant herd structure is complicated, but put simply it is a matriarchal herd structure. Mother, daughters, and their offspring make up the basic family group, usually led by the oldest female. Males leave the group when they approach sexual maturity and are solitary or form loosely structured male herds.

The African elephant is the largest living land mammal, with the Asian elephant coming in as a close second. The males are larger than the females, and both sexes continue to grow throughout their lives. The African elephant has larger ears than the Asian. In both species, the ears are used for communication, behavioral and auditory, and in regulating body temperature. The tusks are upper incisors that grow throughout the elephant's life. Both male and female African elephants can have tusks, while it is usually only the male Asian elephant that carries large tusks. The female Asian elephant's tusks, called "tushes," seldom extend beyond the upper lip. In both species, "tuskless" elephants have been observed.

The trunk is an elongated nose combined with the upper lip. The elephant uses it to breathe, explore its environment, communicate with conspecifics, pick up, push, carry, drink water, and give itself a shower of water, mud, or dirt. The trunk is important to the elephant's survival, although some elephants are able to successfully adapt their feeding and drinking behavior after severe trunk injuries. The tip of the trunk of the African elephant has two finger-like projections, while the Asian elephant's trunk has only one.

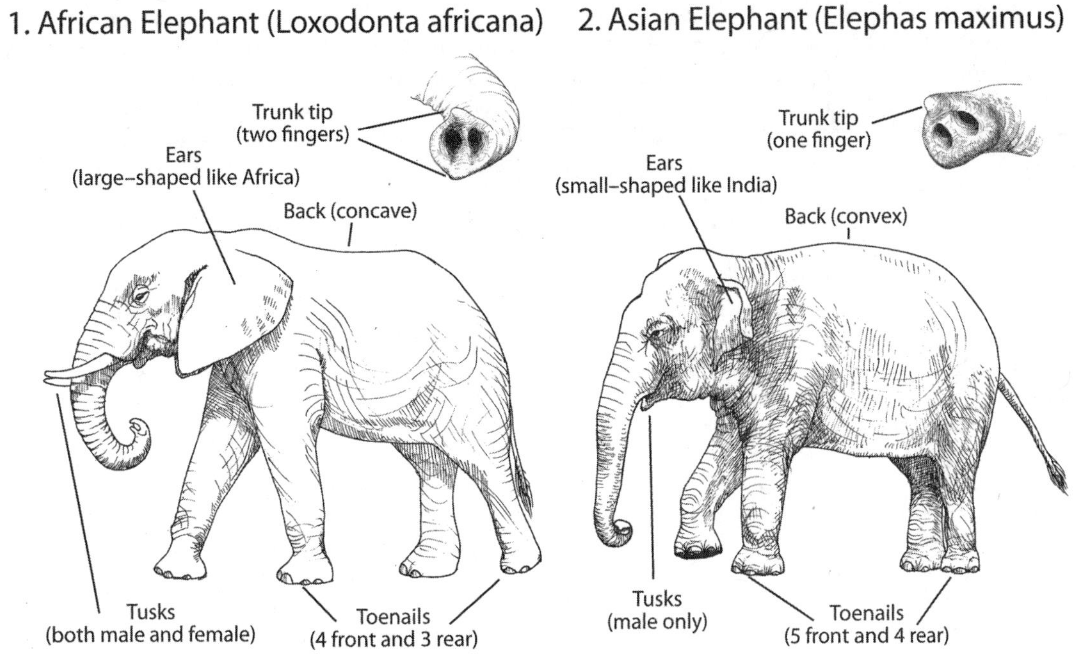

1. African Elephant (Loxodonta africana)   2. Asian Elephant (Elephas maximus)

Figure 30.1. Basic anatomical features of elephants: (1) African elephant; (2) Asian elephant. Illustrations by Kate Woodle, www .katewoodleillustration.com.

The feet of both species of elephants are round, with a circumference larger than that of the legs. The elephant's weight rests on a pad, which cushions the toes. This pad grows continuously and is worn down by the elephant's natural movement. The number of toenails on both species of elephants may vary by individual. Typically, Asian elephants have five toenails on each forefoot and four on each hindfoot. The African elephant has four toenails on each forefoot and three or four on each hindfoot.

THE ZOO ENVIRONMENT

Elephants need contact with other elephants to develop species-appropriate social behavior. They need sufficient space to interact with conspecifics, as well as the means to exercise. The zoo enclosure must be appropriate for the climate and provide protection from the elements. It must provide the means to secure the elephants, safely meet the goals and needs of the elephant training program, and provide for the safety of the elephants and staff.

It is recommended by North America's AZA that all facilities maintaining and breeding elephants provide separate enclosures for the males. The North American population of both the Asian and African elephants is not self-sustaining, so increased breeding efforts are needed. Increased efforts require many more facilities to house adult male elephants for breeding, as well as to provide space to house the male calves that will result from increased reproduction. Adult bull elephants must be housed in facilities that have elephant restraint devices and staff with experience in caring for male elephants. This is especially important when bull elephants are in "musth." Musth is an intermittent period of elevated

testosterone hormone levels, which usually produces erratic and aggressive behavior. Elephants often become unresponsive to commands while in musth.

INDOOR HOUSING

Every elephant-holding facility in North America is required to have indoor housing of some type. Facilities should be able to maintain an indoor temperature of at least 12.8 degrees Celsius (55 degrees Fahrenheit) during cold weather. For very young, sick, and/or debilitated animals, at least one room of the indoor facility should be able to be maintained at 21.1 degrees Celsius (70 degrees Fahrenheit).

Indoor space can be designed as individual or group stalls. Each stall should provide adequate room for the elephant(s) to move about freely and lie down without restrictions. Each facility must take into consideration its climate, herd composition, and compatibility; the needs of its elephant management program; and that program's future objectives when determining the stall sizes most appropriate for their elephants. The AZA Standards for Elephant Management and Care (2011) require a space not less than 56 sq m (600 sq ft.) for males or females with calves, and not less than 37 sq m (400 sq. ft.) for females.

Floors should be impervious to water; they should drain rapidly and dry adequately to prevent chronic damp conditions or standing water. A recommended slope is 6 mm per 30 cm (about 1/4 in. per ft.), which is equal to 13 cm of slope per 6 m (about 5 in. of slope per 20 ft.) of floor. A minimum recommended drain circumference is 20 cm (about 8 in.). Rubber flooring, sand stalls, and alternate bedding are being used and investigated for the cushioning of floor surfaces.

Ambient light cycles are generally appropriate for elephants. Indoor areas should be well-illuminated during daylight hours, followed by a dark period with a night light to simulate moonlight so that animals are not in complete darkness. Good lighting in the elephant stalls and staff workspace is required for both elephant comfort and keeper safety. Natural lighting provided by an ample number of skylights and/or windows, in conjunction with artificial lighting that provides a broad spectrum of illumination, is recommended for use in indoor holding areas. Timers are recommended in areas of short winter days to extend the day length.

Proper ventilation is very important for the elephant's health. Ventilation should exchange the heavier humidified air quickly, maintain a constant temperature, eliminate some of the odor, and not promote drafts. Indoor facility ventilation should be provided to accommodate at least four air exchanges per hour.

Other considerations include a closed-circuit video monitoring system that is capable of recording several days' events at a time. It is favorable for such camera systems to be monitored from remote locations such as the homes and offices of keepers and veterinarians. Built-in scales for weighing elephants should be in every barn, in a location where the elephants have to walk on them daily. These scales should have a weight capacity of up to 6,800 kg (about 15,000 lbs.). Barns should contain equipment sufficient to lift and support an elephant in the event of an emergency. The hoist should be mobile and should access as much of the barn as possible. It must have a lifting capacity of at least 6,800 kg (about 15,000 lbs.). An elephant restraint device (ERD) is a required for all facilities housing adult male elephants, and is also strongly recommended for facilities with intractable females. The ERD should be in a location where the elephants have to walk through it daily, and every elephant should be comfortable being restrained in it. It should have multiple access doors, and a moving wall is a definite benefit if further restraint is necessary. Every elephant barn should allow access to a stall area with an elephant transport trailer, as well as vehicle access into the indoor facility and outside exhibit yard for exhibit maintenance.

OUTDOOR HOUSING

Each facility should have a large communal yard, to encourage species-appropriate behaviors and stimulate exercise, and a separate enclosure for mature male elephants. An additional holding yard or yards should be available, preferably adjacent to the large communal yard, in case there is a need to separate elephants due to aggression, illness, introductions, or a female with a new calf.

Outside yards should be as large as possible, so that the elephants can move about freely and individuals can separate themselves from the herd if they so desire. AZA Standards for Elephant Care (2011) require a minimum of 500 sq m (5,400 sq. ft.) per elephant in the habitat. Yard elevation should provide drainage but not be so steep that the elephants have difficulty walking or using the entire area. The yard surfaces must be of a natural substrate such as grass, soil, clay, or a coarse sand-type material that provides good drainage and adequate footing with a cleanable, dry area for feeding.

Additionally, the following enclosure features are required: shade (trees or artificial cover, sufficient for each animal), furnishings (e.g., logs secured to allow some movement), scratching posts, a water source (such as a pool, misters, or shower), visual barriers to allow animals to get away from one another if needed, and the availability of escape routes from conspecifics. Finally, activities that provide behavioral enrichment should be considered in the design of the outside yard, such as hanging hay feeders, large brushes, and tires. Designing enrichment for these very strong and sometimes destructive creatures can be very challenging.

PERSONNEL

Each elephant-holding facility should have established goals for its elephant management program, including a written elephant management protocol approved by the elephant management team as well as the director, CEO, or owner. This document should have at least an annual review by the entire elephant team. A protocol is an effective training tool, as well as a means to verify that the staff understands the facility's elephant management plan, including its husbandry standards, handling techniques, and safety protocols. New elephant staff should be given a copy of the protocol prior to their first day of working with the elephants, and all elephant staff must understand it and should have full access to the most recently updated version of the document at all times.

All elephant-holding facilities must make safety their highest priority. It is recommended by the AZA that a formal safety assessment program is established, and that a safety inspection occurs at least twice a year. A written record should be kept detailing each meeting, the inspection proceedings, the actions to be taken and their priority, and the date by which the changes are to be completed.

Each facility should have an elephant manager, who is directly in charge of the elephants' daily care and training and who supervises related personnel. The elephant manager should be responsible for instructing the staff in the standard methodology of elephant training and husbandry as described by each facility's elephant management protocol. The elephant management team should meet regularly to discuss relevant issues such as training, husbandry, enrichment, facility maintenance, veterinary concerns, and safety issues. Minutes of the meetings should be recorded and maintained to document any training and behavior issues or management changes. The supervisor of the elephant manager should review the meeting minutes.

Elephant keepers must be institutionally qualified, and two keepers must always be present to work with an elephant. A qualified elephant keeper is a person the facility acknowledges as a trained, responsible individual who is capable of, and experienced in, the maintenance of elephants. Each facility must determine the level of training and experience it requires of those who work with elephants, and must update those requirements as necessary. Each facility should develop a qualification program that quantifiably assesses each keeper's knowledge of the institution's training practices, of all elephant cues and associated responses, of each elephant's behavior toward conspecifics and keepers, and of husbandry requirements.

Elephants in human care should be afforded the highest standard of treatment, and elephant holding facilities should maintain strict policies regarding their standards of elephant care.

MANAGEMENT SYSTEMS

The management of elephants in North America has evolved in recent years, as elephant keepers have developed new ways—or have modified old techniques—to improve the care they provide to each individual elephant. Given the wide range of facilities that house elephants, each must develop its own elephant management program based on its own specific set of circumstances. As a facility develops its program, it should consider its goals in regard to the elephants; the design of its enclosure; the experience and ability of its keepers; the number, age, gender, and demeanor of its elephants; and all its administrative directives and education, conservation, and research activities. Protocols and action plans must be developed to reflect the elephant management styles that are adopted.

Elephant management and behavior training have developed into a continuum of management techniques. These range from the keeper working immediately next to the elephant (free contact), to the keeper working with the elephant only through or from behind a barrier (protected contact), to a combination of the preceding two techniques, with a varying amount of direct physical contact between the elephant and keeper. Within a single facility, different combinations of elephant management may be used, on the basis of each elephant's disposition or the keeper's level of training. The management technique or techniques a facility uses should allow the keepers to safely meet or exceed the established minimum standards of elephant care.

When a keeper works with an elephant through or from behind a barrier (by protected contact), their physical contact is restricted to specific locations through which the elephant can extend a foot, ear, or trunk at the keeper's request. The elephant is trained to respond and change location or position through the use of targets, cues, guides, and reinforcements. This is the technique recommended for use with adult male elephants, and also with female elephants that are less tractable. It may also be established as an organization's sole management system, regardless of the elephant's disposition.

Another alternative is a program in which the keeper and the elephant share direct space (free contact). As with other approaches within the management continuum, targets and reinforcements are primary training tools; but in free contact, the keeper also carries a guide, a tool used to teach, guide, and direct the elephant. Working with elephants in this management style therefore requires additional training and skills, and will facilitate certain activities such as the movement of highly tractable elephants from one location to another.

Many facilities operate in the middle of the management continuum. The keepers may not share the same space with the elephants, but they may have physical contact with them to differing degrees during training and husbandry care. Facilities will develop their management systems to maintain strong relationships between elephants and keepers in consideration of the temperament and the tractability of the individual elephants and the techniques and tools used for behavior modification. Ultimately, a facility should develop an approach that enables keepers to provide the best care for each individual elephant and permits a wide variety of exercises and behaviors, the ability to conduct thorough physical examinations, and the provision of medical treatment when needed.

Regardless of which type(s) of elephant management program a facility employs, it is critical that the facility management team and the elephant manager understand the need to develop a qualified, well-trained staff and a consistent elephant-training program. The elephant must be trained to be responsive to all cues given, and the keeper must be able to obtain a reliable response from the elephant. As a general rule, a keeper should not ask an animal for a behavior that they do not think it will execute. It is important to create an opportunity for success for both the keeper and the elephant. Establishing defined cues and criteria for each behavior and maintaining consistency between the different keepers who work with the elephant will help to accomplish this. Though there is a diversity of approaches, all of the methods within the management spectrum share many of the same techniques and qualities.

- All training processes use both classical and operant conditioning.
- In all approaches, the elephant's behaviors are determined by their consequences. This may be the definition of operant conditioning, and it is true of all interactions with an elephant. Desired behavior is reinforced, and undesired behavior is not.
- The training techniques and tools are interchangeable throughout the management continuum. Guides, targets, and restraints can be used by any qualified elephant keeper.
- In all approaches the success of the elephant's management depends upon a comprehensive program of staff training, an understanding of elephant behavior, and the use of proper elephant husbandry techniques, including a solid understanding of training theory and the appropriate use of training techniques.
- In all three approaches, all aspects of the management system must be evaluated constantly, including both the keeper and the elephant.

TRAINING

Working with elephants is inherently dangerous. The elephant's sheer size and ability to use its trunk as well as the rest of its body aggressively makes it especially dangerous. People have worked with elephants for centuries and have often perceived them as "gentle giants," but the animals must be treated with respect for their unique abilities. The combination of the elephant's strength, intelligence, dexterity, and massive size makes it unique from most other animals in a zoo. The principles of training an elephant are the same as that of training any other animal. The trainer must have a strong understanding of training theory and terminology, and also of the animal's natural and individual history. Elephants are extremely intelligent, and when communicated with properly by the trainer, they can learn new behaviors very quickly. Elephants are trained for a variety of reasons including husbandry, medical procedures, education, research, exercise, and enrichment. A good elephant trainer must be well versed in the proper use of all the training tools, and must know the advantages and disadvantages of each tool.

Every elephant should respond to a minimum number of cues and stimuli in order for the trainer to properly care for it. Some of the most basic behaviors an elephant should be trained to do are as follows.

1. "Steady." Probably one of the harder behaviors to teach any animal is to stand still and do nothing. Elephants must learn to stand steady in various positions and locations so that staff can perform routine husbandry behaviors such as blood collections or baths. Most cues presented to an elephant are followed by a verbal cue of "steady," which instructs the elephant to remain in that position until released.

2. "Come" (when called). Animals need to be moved and must be responsive to trainers, especially when being trained in a group setting. The verbal cue for this behavior usually involves the elephant's name followed by the verbal cue "come here."

3. "Open" (the mouth). This is one of the first behaviors a young elephant is trained to execute. Oral inspections on elephants of all ages are a very important part of their daily care and health checks. The verbal cue for this behavior is "open."

4. "Foot." Elephants require regular foot care, and in order to receive it, they must learn to lift each foot on cue. The verbal cue for this behavior is "foot." Most facilities train the elephants to respond to the verbal cue "foot," coinciding with a visual cue of the hand gesturing towards the desired foot or the use of body positioning to indicate the desired foot.

5. "Lean in." Many elephants are trained to present the sides of their bodies, perpendicular to the trainer. This behavior allows trainers working in protected contact to inspect, scrub, and manipulate every inch of the elephant's body. The cue for this behavior includes the verbal cue "lean in" in conjunction with the trainer holding an arm outstretched in the direction where the elephant is to move its hindquarter.

These behaviors provide the foundation for more advanced ones, and enable staff to care for the elephants appropriately. Each elephant should be trained to perform all of the standard behavioral components as outlined by the AZA's Principles of Elephant Management (PEM) class, which are listed here.

• The elephant must be able to receive complete daily body exams and proper skin and foot care, including trimming and foot radiographs when needed. Elephants in a captive environment often require foot care. The nails and pads of the foot need to be trimmed on occasion. Proper skin care includes daily scrub baths, providing appropriate substrate for dusting and large objects for scratching. Staff must be able to perform daily eye, ear, mouth, and tusk exams.

• The elephant must be trained to allow for the collection of biological samples such as blood, feces, urine, saliva, trunk washes, skin biopsies, and temporal gland secretions.

• The elephant must be trained to be comfortable receiving injections, accepting oral medications, and allowing staff to treat any wounds.

• The elephant must be trained to enter and stand comfortably in a restraint device. If a restraint device is not available, the zoo staff must demonstrate the ability to restrain the elephant on tethers.

• The elephant must be weighed regularly.

• The elephants must be trained to accept rectal and transabdominal ultrasound exams, urogenital exams, and (if male) semen collection.

The staff must demonstrate the ability to manage the elephant herd's social structure. They must be able to manage social compatibility issues, introductions, calves, and the separation of animals. They must also be able to address an elephant's psychological and physiological welfare by providing it with sufficient mental stimulation, through environmental and behavioral enrichment, to promote activity and proper social behaviors, while also providing it with sufficient physical exercise to develop and maintain muscle tone, flexibility, agility, stamina, and a healthy weight.

TOOLS AND EQUIPMENT

The management of elephants encompasses various methods and means of training. The following are the primary tools in the trainer's "toolbox" of elephant management. This list is not all-inclusive, as the different tools are simply too numerous to list. As components of elephant training and management, all tools and equipment should always be well cared for and used only for their intended purposes. All new keepers should be instructed and knowledgeable in the use of each tool prior to working with an elephant.

An elephant restraint device (ERD) restricts the elephant's movements while allowing keepers access for routine husbandry and medical care. An ERD restricts most but not all of the elephant's mobility. Although the access for husbandry and medical care is safer when an elephant is restrained in an ERC, it is not risk-free; keepers do come into contact with the elephant and vice versa. Despite the many variations in design, certain basic elements are shared by all ERDs. A properly designed ERD should allow the keeper access to all four feet, both tusks, the trunk, the face, both ears, both sides, both hindquarters, and the back. The ERDs should be designed with multiple access doors and a moving wall to maximize access and minimize risk to both the keeper and the elephant. The ERD must open easily and quickly to free an elephant in the event of an emergency. Some ERDs are even equipped with walls that swing open to free an elephant easily if needed. The ERD should be fitted with winches and associated slings to lift and support an elephant if needed in the event of prolonged medical treatments where the elephant's health is compromised. It should also be able to comfortably contain an elephant for extended periods of time should the need arise for an ongoing or lengthy medical or husbandry procedure. Most importantly, it must be able to contain the facility's largest elephant safely. It should be located in an area of the holding facility where it is routinely accessible to the elephant and where it can be used 365 days a year regardless of weather conditions. Preferably the ERD should be placed in an aisle, so that each elephant must go through it every day to access its outside yard or another space within the facility.

A guide (ankus) is a tool used to teach, guide, and direct the elephant into the proper position or to reinforce a verbal cue. The tip of the guide consists of a point and hook mounted on one end of a shaft made of fiberglass, plastic, or wood. The design of the hook allows for the elephant to be physically cued with either a pushing or pulling motion. The point and hook are tapered to elicit the proper responses from the elephant efficiently, with the keeper exerting very little pressure. The end of the hook should not tear or penetrate the skin. The guide provides the elephant with additional information and is intended to complement verbal cues. It is an aversive stimulus from which the animal is conditioned to move away. When training the behavior "move up," a keeper would give the verbal cue "move up" and touch the elephant behind the front leg to initiate movement. Once the elephant starts to move, the pressure from the guide ceases and the elephant is given positive reinforcement. The ultimate goal of the elephant keeper is to have the elephant respond to verbal cues alone, using the guide as little as possible. When paired with verbal cues, the guide can be used to teach an elephant a variety of behaviors including lifting a leg, moving forward, and moving backward.

A target is another tool used in the behavior modification of elephants. It differs from the guide in that the animal is conditioned to move toward it rather than away from it. Once the animal is trained to initiate contact with it, a target can be a very effective tool for training new behaviors. A target is typically a pole with a consistent and defining feature on the end that is presented to the elephant. The end that is presented as the "target" to the elephant may have a differently shaped object attached to it, or a painted tip. Targeting is an effective way to manage and manipulate less tractable elephants. There is potential for the target to be grabbed and eaten by the elephant. Therefore, consideration should be given to the material selected as the target, to minimize harm if it is ingested.

Leg restraints or tethers are an acceptable, necessary, and useful tool in the management of captive elephants. They provide a means to limit an elephant's movements and permit safer handling. This can facilitate footwork, feeding, veterinary procedures, transportation, introductions, parturition, scientific investigation, training of new keepers, and training of new behaviors. Tethering is just one component of an elephant management program. The decision to tether should be made taking into consideration the best interest of the elephant. Elephants should not be tethered for most of a 24-hour day. Elephants under medical care, in transit, or with other special circumstances may need to be tethered for longer periods. Elephants that require prolonged tethering with chains should have their leg chains wrapped in fire hose or leather to provide protection from abrasion. Elephant keepers should be well versed in the various and proper uses of ropes, chains, and hardware such as shackles (clevises) and

cold shuts. All facilities should develop tethering protocols and all elephant keepers should be familiar with them. This will ensure that the tethers are used correctly, efficiently, and humanely.

An up-to-date elephant profile should be kept on file for each elephant maintained by a facility. The purpose of an elephant profile is to track a specific elephant's history, record training data, provide reproductive information, and, most importantly, identify behavior trends in relation to the elephant's keepers and male and female conspecifics.

CONSERVATION STATUS

The conservation status of each elephant species is published in the current International Union for Conservation of Nature (IUCN) Red Data List, in US Fish and Wildlife Service (USFWS) listings and by the Convention on the International Trade of Endangered Species of Wild Fauna and Flora (CITES). Wild population status is estimated by the IUCN/ Species Survival Commission (SSC) African Elephant Specialist Group and the Asian Elephant Specialist Group.

Populations of both elephant species continue to decline in the wild (IUCN 2010). Human encroachment, habitat loss, and poaching pose major threats to the extant populations. Human-elephant conflicts are frequent, as the population of humans increases and suitable habitat for elephants decreases. Human or elephant fatalities are often the result.

Asian elephants have disappeared entirely from western Asia, Iran, and most of China. Remaining populations are usually restricted to hilly and mountainous areas. It is estimated that at the turn of this century there were more than 100,000 elephants in Asia. The number of Asian elephants surviving today is estimated at between 41,000 and 52,000. Loss of habitat is the primary reason for the decline of the Asian elephant, as migration routes have been disrupted and new settlements and agriculture confront herds; resultant conflicts with humans are inevitable (IUCN 2010). *Elephas maximus* is listed as an endangered species with the USFWS, and is classified under Appendix I with CITES.

African elephants once ranged throughout Africa. By the 16th century the species had become extinct in northern Africa, primarily due to the ivory trade. Much of the population that survives today is fragmented by human activities that disturb the elephants' traditional migratory routes. The population of this elephant, estimated to be at about 1.3 million in the early 1970s, dropped by more than half by 1995 (IUCN 2010). In 1989 the African elephant was listed as Appendix I (endangered) by CITES due to uncontrolled poaching, and consequently an international trade ban on elephants and elephant products was enacted. In 1997 the African elephant was downlisted by CITES to Appendix II in some southern African countries, as a result of rebounding populations and protection programs. This remains the case today. Appendix II classifies these populations as threatened and allows some limited trade in elephant products with certain restrictions, quotas, and permits. Today the African elephant population is optimistically placed at near 500,000 by some, but census analysis by the IUCN African Elephant Specialist Group suggests that the true numbers may be lower.

Box 30.1. Elephant Endotheliotropic Herpesvirus (EEHV)

Martina Stevens

Elephant endotheliotropic herpesvirus (EEHV) is an infectious and often fatal disease that attacks the cells that line blood vessels (endothelial cells), causing widespread hemorrhaging. EEHV primarily targets juvenile elephants and has been detected in both Asian and African elephants. There have been approximately 40 confirmed clinical cases of EEHV in North America, including 37 in Asian elephants and 3 in African elephants. Twenty-nine of the 40 cases resulted in death. Of these 29 cases, only 2 were African elephant deaths. Confirmed cases resulting in death have also been discovered in wild populations in Thailand, India, and Cambodia. The total impact of EEHV on wild populations at this time is unknown, but these findings prove that EEHV is not just a disease found in captivity.

EEHV is a betaherpesvirus. There are eight known types of EEHV: EEHV1A, EEHV1B, EEHV2, EEHV3, EEHV4, EEHV5, EEHV6, and EEHV7. Of the eight types, five have been confirmed as causing death, with EEHV1A and EEHV1B as the most common types associated with the fatal disease in North America. The transmission route of the virus is still unknown, but current research has found EEHV in trunk wash secretions from Asian elephants in zoos that have not demonstrated symptoms (i.e., that are asymptomatic carriers).

EEHV infection manifests very rapidly, and death can occur within hours of observation of the first clinical symptoms. Those symptoms include, but are not limited to, lethargy; colic; low numbers of circulating red blood cells (anemia); blue-purple discoloration (cyanosis) of the tongue; swelling of the head, neck, and limbs (edema); lameness; and not eating (anorexia). Elephant care professionals should carefully monitor any change in a juvenile elephant's behavior or appearance. Infection is confirmed with a polymerase chain reaction (PCR) test that detects herpesvirus DNA in an elephant's whole blood. Early detection of EEHV is imperative. Elephant care professionals who suspect an EEHV infection should not wait for a diagnostic confirmation to start treatment. Aggressive supportive care and antiviral drug therapy are recommended. Historically, there has been some success using the antiviral drug Famciclovir. More recently, the drug Ganciclovir has been used successfully to treat two elephants.

Researchers, elephant care professionals, and veterinarians are trying to find answers to the many unknown questions surrounding EEHV. EEHV research has become a worldwide effort involving zoos, private elephant owners, and in situ and ex situ research facilities. The development of a vaccine for EEHV is a long-term research goal. A vaccine would have a major impact on both captive and wild elephant populations.

SUMMARY

The purpose of this chapter is to provide the entry-level keeper with a basic overview of the care and management of elephants in human care at zoological facilities. In addition to this information, there are some basic skills that everyone must possess in order to be a successful elephant keeper:

- *Ability to communicate with others.* A team approach is necessary for a successful elephant program. The ability to communicate is more than being able to express one's thoughts in a concise, logical manner. A keeper must also be able to listen to others, both peers and supervisors, with respect and consideration even when their views differ.
- *A working knowledge of operant conditioning.* Operant conditioning is a type of learning in which the probability of a behavior recurring is increased or decreased by the consequences that follow.
- *Excellent problem-solving skills.* The ability to use the scientific method and critical thinking must be developed to solve training issues.
- *Ability to educate.* One goal of every zoo is to educate its guests on conservation issues. Every keeper needs to be able to speak to the public each day to meet this mission.
- *Ability and willingness to work hard.* An elephant can eat and defecate 200 pounds per day. Its living area needs to be cleaned, disinfected, raked, and scrubbed

on daily basis. Because of the animal's size, every maintenance task is larger than life: bathing, foot care, and so on. Every task is important, from cleaning the elephant to cleaning the shovels and rakes used every day.

These are just a few of the skills needed by a successful elephant keeper. If someone is seriously considering a career as an elephant keeper (or as any kind of animal keeper), the authors would encourage them to try to find an institution that accepts animal care volunteers. This would give the person an opportunity to see and experience the profession.

REFERENCES

AZA. 2003. AZA Standards for Elephant Management and Care. Modified May 2003. Accessed at http://www.elephanttag.org /Professional/AZA%20Standards.pdf.

IUCN. 2008. Choudhury, A., D. K. Lahiri Choudhury, A. Desai, J. W. Duckworth, P. S. Easa, A. J. T. Johnsingh, P. Fernando, S. Hedges, M. Gunawardena, F. Kurt, U. Karanth, A. Lister, V. Menon, H. Riddle, A. Rübel, and E. Wikramanayake. "*Elephas maximus.*" Accessed April 2010 at http://www.iucnredlist.org/apps/redlist /details/7140/0.

————. 2010. Blanc, J. 2008. *Loxodonta africana.* IUCN Red List of Threatened Species. Version 2010.1. Accessed April 2010 at http://www.iucnredlist.org.

IUCN Red List of Threatened Species. Version 2010.1. Accessed April 2010 at http://www.iucnredlist.org.

31

Husbandry and Care of Marine Mammals

Gerard H. Meijer

INTRODUCTION AND NATURAL HISTORY

This chapter does not limit itself to a specific order or taxon; animals from several taxa are lumped together here because of the environment in which they live. When referring to "zoos and aquariums," specialized zoos such as marine mammal parks or dolphinariums are included. Marine mammals all live at least part of their lives in salt water. However, in each of the groups we find some species that live in fresh water.

This chapter will provide basic information about the husbandry and care of marine and semiaquatic mammals. After studying this chapter the reader will understand

- the enclosure and environmental needs of marine mammals
- food and food preparation for marine mammals
- their propagation and maternal care
- their environmental enrichment and training needs
- their handling and transportation.

PINNIPEDIA

The first group to be discussed is the superfamily Pinnipedia ("fin footed" mammals) from the suborder Caniformia ("dog-like"), which is divided into the following families.

PHOCIDAE (SEALS)

The most well-known members of this family are kept in zoos and aquariums.

- The harbor seal (*Phoca vitulina*) ranges from temperate waters to arctic areas, and is found in both the North Atlantic and North Pacific Oceans.
- The gray seal (*Halichoerus grypus*) lives in the temperate and subarctic coasts of the North Atlantic Ocean.

OTARIIDAE (SEA LIONS AND FUR SEALS)

- The Californian sea lion (*Zalophus californianus*) lives along the Californian coast to Mexico, including Baja California. Separate populations live on the Galapagos Islands and in the southern Sea of Japan, although the latter is thought to be extinct.
- The Patagonian or South American sea lion (*Otaria flavescens*) lives on the South American coasts from Rio de Janeiro, on the Atlantic Ocean, and Peru, on the Pacific Ocean, down to the most southern part of South America.
- The Steller's sea lion (*Eumetopias jubatus*) lives along the North Pacific coasts of Japan, Russia, Canada, and the United States.
- The Australian sea lion (*Neophoca cinerea*) lives on offshore islands of western to southern Australia, and is seldom seen in zoos or aquariums.
- The South American fur seal (*Arctocephalus australis*) lives off the neotropical Pacific and Atlantic seacoasts from southern Peru down to Cape Horn and up to southern Brazil.
- The South African fur seal (*Arctocephalus pusillus*) lives in and around the southern and southwestern coasts of Africa as well as the southern and southeastern coasts of Australia, and is the fur seal more commonly found in zoos and aquaria.

ODOBENIDAE (WALRUSES)

- The walrus (*Odobenus rosmarus*), which is found throughout the entire arctic region, is not often seen in zoos or aquariums.

ODONTOCETI

The second group is the suborder Odontoceti or "toothed whales": the dolphins, porpoises, and orcas.

Figure 31.1. Basic anatomical features of marine mammals. Clockwise, from upper left: Californian sea lion (*Zalophus californianus*), spinner dolphin (*Stenella longirostris*), West Indian manatee (*Trichechus manatus*), harbor seal (*Phoca vitulina*), walrus (*Odobenus rosmarus*). Illustrations by Kate Woodle, www.katewoodleillustration.com.

DELPHINIDAE (MARINE DOLPHINS)

- The bottlenose dolphin (*Tursiops truncatus*), which lives everywhere except in polar waters, is the most well-known dolphin species in zoos and aquariums.
- The Pacific white-sided dolphin (*Lagenorhynchus obliquidens*), which lives in cool to temperate waters of the North Pacific Ocean, is also commonly found in zoos and aquariums.
- Representatives of the larger delphinid species are the orcas or killer whales (*Orcinus orca*). These are found living in all oceans, but they prefer colder waters.
- The beluga whale (*Delphinapterus leucas*) is found in arctic and subarctic waters along the Canadian, Alaskan, Russian, Greenland, and Norwegian coasts. A group of approximately 500 also live in the Gulf of Saint Lawrence.
- One of the smaller species maintained in zoos and aquariums is the Commerson's dolphin (*Cephalorhynchus commersonii*), which lives along the Patagonian coast to Cape Horn and through the Strait of Magellan to the Falkland Islands. A second population is found around Kerguelen Island in the southern Indian Ocean; these dolphins are different in appearance and are not generally found in zoos and aquariums.

PHOCOENIDAE (PORPOISES)

- Harbor porpoises (*Phocoena phocoena*) are found in coastal regions of the North Atlantic, Arctic, and North Pacific Oceans, and also in the Baltic, Mediterranean, and Black Seas. They are seldom seen in captivity, but there have been recent breeding successes in Europe.

SIRENIA

TRICHECHIDAE (MANATEES AND THE DUGONG)

- The West Indian manatee or sea cow (*Trichechus manatus*) is found off the coast of Florida, along the southeastern coast of the United States, south along the Gulf of Mexico to Texas, along the coast of Central and South America, and in the Greater Antilles. Of sirenia, this is the species most commonly found in captivity.

This is in no way a complete list of all marine mammal species kept in zoos and aquariums; it is just an overview of the most familiar species a keeper can expect to care for (Klinowska 1991; King 1983; Animal Diversity Web 2010; Grzimek 1973; Coffey 1977; Bateman 1986).

BASIC EXTERNAL ANATOMY

PINNIPEDIA

The pinniped species are very similar anatomically, but there are differences between them. They all are well adapted to living and hunting in the world's seas and oceans, a hostile and dangerous environment. The main differences between seals, sea lions, and walruses are as follows.

- Seals use their front flippers to move on land; they "crawl" on their bellies using their front flippers. They do not have external ears (pinnae). When swimming, they use their hind flippers and their lower backs. Unlike sea lions, seals quite often have spots, rings, or patches in their fur that vary in color from gray to brown. The males are slightly larger than the females.
- Sea lions and fur seals can walk on land. They turn their hind flippers to the front and use them in conjunction with their front flippers for walking on land, and their front flippers are generally larger than those of seals. All sea lions have external ears, even though they are sometimes difficult to see. Sea lions are brown to tan in color. Spots are not found in sea lion fur.
- The walrus is in between the seals and sea lions in the use of its flippers, on land as well as in the water. Of course, its tusks are a very clear difference. Walruses with tusks are seldom seen in zoos and aquariums, because they wear the tusks down too rapidly, which can lead to their infection and removal. External pinnae are not present. The fur seen on the seals and sea lions is almost absent in the walrus. Instead, it has a very thick skin, which in mature males can be 2.5 to 4 cm (1 to 1.5 in.) thick around the neck area. All species of pinniped have whiskers (vibrissae); in the walrus these are more like bristles (heavier and thicker), but they are still called vibrissae. They are used for locating food on the seabed. Seals and sea lions also use their whiskers for locating food in murky waters. They sense the vibrations from moving fish and hunt and catch them in this way.

DOLPHINS AND PORPOISES

These mammals have adapted to living their whole lives in water, and they give birth in the water. Dolphins use their tails and horizontal flukes for swimming. In dolphins, the dorsal fin is very distinctive. The beluga whale lacks a dorsal fin but has a slight ridge along its back. The dolphin's blowhole or "nose" is on the top of its head and is a special adaptation to life in water. The melon between the blowhole and the tip of the snout is an outstanding feature; it is part of the nasal system, and is seen in all dolphin and porpoise species. The melon is the bioacoustic part of the echolocation system in dolphins and porpoises, and an environmental adaptation which is still not completely understood. When above water, the dolphin's tail is very distinctive; it differs from that of a fish in that the fluke is horizontal instead of vertical. This adaptation means that the dolphin swims by moving its tail up and down, whereas fish move theirs from side to side.

MANATEES

These are the only vegetarians in the group of marine mammals. The manatee's big rounded tail fluke is different from those of the other species previously discussed. The shape of the mouth is also characteristic: the upper lip is very thick and protrudes upward, ending in a disc-like structure. The nostrils are on the upper part of the snout and can be opened and closed at will. The tips of the flippers can be touched together at the chest, and this adaptation allows the manatee to use its flippers in feeding. The manatee has a very thick skin, with only a sparse covering of hair.

CONTAINMENT, ENCLOSURE, AND ENVIRONMENTAL CONSIDERATIONS

Each species included in this chapter needs to be kept in a minimum of at least one pool. Many of the species' needs are similar, although there are also some differences. For example, land areas, which are important for seals and sea lions, are not needed for dolphins and manatees. Some institutions , however, have installed shallow concrete shelves with imbedded smooth stones into the exhibits to allow the cetaceans to beach themselves for a scratch. Exhibits have to be built and furnished not only to accommodate the basic needs of the species kept in them, but also to allow the animals to exhibit and exercise their species-specific behaviors.

Unlike seals, sea lions, walruses, and manatees, marine dolphins cannot be kept in fresh water. They will become exhausted swimming, because of the difference in specific gravity between salt and fresh water, which affects their buoyancy. Saltwater is more dense than fresh water. Secondly, their skin will start to slough, much like the skin of a person who stays in water too long. Dolphins should be housed in a saltwater pool with a salinity between 20 to 30 parts per million. Freshwater dolphins are kept in fresh water. Legislation imposes the use of salt water for all marine mammal species in some countries.

Seals, sea lions, walruses, and manatees can be kept in fresh water. It can be discussed what is the best for the animals; while they all (except the manatee) originate from salt water, in practice they do not seem to have problems living in fresh water. However, the author strongly recommends the use of saltwater in all marine mammal exhibits.

Water and air temperatures should be kept within a range that approximates the animals' natural environment (table 31.1), although most sea lions do not have problems living for short periods of time in water at 0 °C, with outside air temperatures well under 0 °C. Seals and fur seals also live and even give birth in very cold conditions. For example, the ringed seal (*Phoca hispida*) gives birth in a snow and ice lair in which water and air temperature can vary to a great extent. One should keep in mind that pinnipeds can, in general, be more endangered by heat than by cold. Sea lions and fur seals exposed to direct summer sun can suffer from heat prostration within an hour. Shade should be provided in all cases.

TABLE 31.1. Acceptable temperature ranges for some marine mammal species

Species	Temperature range
Seal	0–25 °C (32–77 °F)
Sea lion	5–25 °C (41–77 °F)
Bottlenose dolphin	10–32 °C (50–89 °F)
Manatee	24–29 °C (75–84 °F)

Dolphins are similar to the pinnipeds in their sensitivity to variation in temperature. They can stand a great variety of temperatures as long as they have time to acclimate to them. For example, a marine mammal should never be moved from an exhibit with a temperature of 20 °C to a pool with a temperature of 10 °C. Manatees, which live only in tropical areas, need heated water. In countries with cool temperatures they must have indoor exhibits, although they are to a certain extent capable of coping with a wide variety of temperatures.

POOL SIZES

In several countries, such as the United States, Germany, and Belgium, there are rules and regulations describing minimum sizes for pools and land areas for many marine mammal species kept in captivity. The European Association of Zoos and Aquaria (EAZA) husbandry guidelines also set minimum standards for these marine animals. It has to be noted that these standards are only minimums, and that keepers should strive to improve upon them where possible.

All animals should be provided with as much usable space as possible. Both the size and the shape of a pool and land area are important considerations. "Kidney-" or irregularly-shaped pools are preferred. Irregularly-shaped pools need to have rounded contours, with corners of not less than an angle of 90 degrees, to prevent animals from being cornered during fights or chases. For seals, sea lions, and walruses there should be a shallow sloped side for easy access and egress for all animals, including young and very old individuals. In many pools the land areas are made of concrete and are painted or coated, but in some cases the adjacent land areas for seals and sea lions are of sand, soil, and rocks. Planted areas should be considered, as they can make the environment even more interesting. Dolphins can also be enclosed in areas of natural bays or harbors, although this is not a common practice for zoological facilities.

A maternity pool and associated holdings should be provided for all species. In a dolphin area, a maternity pool fitted with a lifting platform will be beneficial for the animals and keepers when mother or baby needs to be handled. The maternity area should in all circumstances offer the female a sense of security and allow her, when necessary, to be able to easily defend her young from conspecifics.

A separate area for sick or injured animals is also a necessity. Keep in mind that separating a social animal from its group can affect the group dynamics and could perhaps cause negative ramifications within it. A close companion animal can help to alleviate behavioral concerns, and may even help speed up recovery.

Natural daylight is preferred for all species, although quite a few zoos and aquariums have only indoor exhibits and pools, in which case natural light cycles should be provided. These, of course, should supply enough light to imitate the animals' natural daylight cycle by means of windows and artificial lighting. When considering outdoor exhibits, especially in areas with a lot of sunshine and high temperatures, shade should be provided over a part of the exhibit pool. Dolphins especially can suffer from severe sunburn when left without shade. In colder climates some form of shelter should be provided. Some facilities have land areas raised on pillars so that animals can shelter underneath as needed. This makes the most efficient use of space and exhibit volume.

In indoor exhibits and holdings, one should keep in mind that a good exchange of fresh air is essential. Most marine mammals breathe just above the water surface or, when on land, close to the substrate. It is therefore very important that the exchange of air or ventilation has appropriate movement at floor and surface levels. This is even more important when chlorine is used for the treatment of water or cleaning. Chlorine and other unwanted derivatives that originate in a pool tend to appear just above the water surface.

FOOD AND WATER

What kind of food can a keeper be expected to use to feed marine mammals? The wide variety of places and diversity of habitats that the animals come from, and the fact that zoos and aquariums are spread widely over the world, mean that a massive variety of fish species are available to choose from. Manatees have their own completely different dietary needs, which will be discussed later, but other marine mammals will eat most of the fish species we can offer them. In general, it is possible to feed them species that are locally available. It can take some time to get them used to food species other than those they normally eat, however, so the keeper should make the changes slowly.

In general, the most common species of fish fed to captive marine mammals are herring (*Clupea harengus*), mackerel (*Scomber scombrus*), whiting (*Merlangius merlangus*), blue whiting (*Micromesistius* poutassou), capelin (*Mallotus villosus*), and sprats (*Sprattus sprattus*). Although not fish, squid (*Loligo* sp.) are also commonly fed. Some institutions feed their animals a mix of these species ad libitum (at will). Other establishments will weigh the daily food intake. Still another method is to feed on nutritive and caloric value. This last method means, however, that the feeder fish from each shipment must be analyzed before they can be fed to the mammals, and this can be rather expensive. Weighing the feeder fish and monitoring the mammals' daily intake is a well-balanced method. In most cases, the use of common sense and the monitoring of feeding practices by experienced keepers is a very good way to ensure the feeding of good quality fish. On average, the daily food intake of a normal healthy animal is between 5 and 8% of its body weight.

In some species there are seasonal differences in food intake. For instance, a bull in mating season will commonly lose its appetite; also, females will not eat during the first period of lactation. If the aquarium is not analyzing the feeder fish, the keeper should pay more attention to their quality, and should ensure that they come from dependable suppliers. Frozen fish should be as fresh as possible; the dates of catch are known by the suppliers. One should open boxes of feeder fish and check the contents. There should be no damage, or any excessive blood between the fish. The fish should not look dull or off color; they should not smell sour or abnormal. When thawed, they should have bright red gills, clear eyes, and firm, elastic skin. Old or refrozen fish are dull and soft, and their eyes will be red-bordered and cloudy; after they thaw, one can poke them with a finger and the indentation will remain. Fish should be ordered and bought as fresh as

Good Practice Tip: After feeding, temporarily store leftover fish in a cooler or immediately dispose of it appropriately. Fish can spoil very fast and quickly become a breeding ground for bacteria. Feeding old or improperly stored fish can harm the health of the animals.

possible and stored in a freezer at the correct temperature of −30 to −18 °C (−22 to −0.4 °F). The storage of fish should be organized in a system of "first in/first out" to ensure that it is as fresh as possible when fed, and no old fish stays in the freezer. Once removed for thawing, fish must be used within 24 hours, and kept refrigerated between 2 to 6 °C (35.6 to 42.8 °F). Hygiene in the fish preparation area and during all handling is very important (Chrissey 1998).

There are different methods of thawing fish. Thawing it by submerging it in slow flowing water is one method, but it is one in which the loss of vitamins and nutrients can be significant. Thawing fish by leaving it packed or unpacked in open air can cause it to dry out rather easily. Thawing in a cooler unit with forced air can minimize the loss of nutrients and vitamins and keep the drying out to a minimum. Variations are possible in all methods; the last method is preferred by the author.

As all fish eaters are hunters; they should be fed several times a day. Feeding them as much as they can eat on four or five occasions per year is no problem, and might even be desirable. In the wild, the animals will sometimes find a jackpot of food and will eat until they are satiated. Many marine mammals are used in presentations and shows; this provides a good opportunity to spread their feeding times throughout the day. But since a day is 24 hours, one should feed not only during presentations, but also at the beginning and end of the working day.

Keepers will use fish as a reward for aquatic mammals during presentations, and this must be considered when calculating the mammals' daily feed intake. As most of the fish will be deep frozen and stored for months before being fed, they will lose vitamins. They should be supplemented in the daily diet as prescribed by a nutritionist or veterinarian. Keepers should know, however, that over time new knowledge about animal diets and supplements is obtained; the composition, quantity, and types of vitamins prescribed may change. The subject requires more research and discussion.

Manatees are not carnivores (hunters), but herbivores (plant eaters). Depending on their geographic location, they may eat seagrasses, algae, roots, fallen fruits, or leaves. In captivity they feed on a variety of food items, with lettuce and endive as major food sources. Spinach, grass, aubergine (eggplant), carrots, bananas, cauliflower, beets, apples, and sweet potatoes are also fed. The diet fed will sometimes depend on the individual animal's preference. The keeper must remember that the floating foods can offer a challenge for the exhibit's filtration system. Food should be fed frequently throughout the day, and preferably in such a way that it will not float directly into the filter system. Feeding in racks and putting grates in front of the outlets will help to prevent this. As these animals are "grazers," a keeper will have to feed a large amount of greens over the course of the day: approximately 18 to 25 kg (39.5 to 55 lbs.) for an adult animal. Feeding this amount over the day will not only benefit the animals but also prevent floating parts of the food from getting directly into the filtration system.

Marine mammals do not drink, as they obtain the water they need from the food they eat, partly by extracting the moisture from their food and partly by metabolizing their body fat. This is why it is so important to ensure the best quality and freshest food for them. Sometimes dolphins may need extra hydration. It can be given to them by injecting fresh water into their feeder fish, or by giving them water through a stomach tube. The latter should only be done by experienced keepers in consultation with or on the direction of a veterinarian. Seals and sea lions do not drink, but fasting bull sea lions during mating season may sometimes appear to drink seawater.

Observing marine mammals can be challenging compared to monitoring the behaviors of the terrestrial animals. Once marine mammals are in the water, they may be out of sight; light refraction and wave movement will further hinder observations. First the keeper must learn to recognize their normal swimming patterns. Experienced keepers are a valuable resource for advice about what is seen or not seen. Most animals will show the first signs of ill health or other problems by moving about in a manner different from what is normal (e.g., hanging vertically at the water's surface, showing a curve in the back, tilting sideways, swimming very slowly or rapidly). Together with other signs, these behaviors can indicate problems. One should be careful with conclusions—for example, either slow or rapid swimming can indicate pain. Alternatively, movement in a healthy bottlenose dolphin may be completely different from movement in a healthy manatee. Always be aware that some behaviors are species-specific, and also that sometimes there are differences between individual animals. New keepers and visitors are always surprised if a keeper walking alongside a pool, seeing nothing more than part of a head and back, greets an animal by name. The author remembers being surprised that even an experienced relief keeper didn't see the difference between the light and dark female sea lions that he was regularly responsible for. Observation is a very important skill that has to be learned. One must keep watching all the time, not only for little things but also for the overall appearance of the animals and their habitat. One should not hesitate to ask or inform the more experienced keepers about anything one may observe.

Animal training is an essential duty for a keeper working with marine mammals. It allows the animal to be in close proximity to the keeper for observation of all external body parts such as the mouth, teeth, ears, and eyes. Changes in skin and fur condition, bumps, or lumps can be better examined when the animals are close.

Swimming with closed eyes is often a sign of eye problems, and is particularly common in pinnipeds. It is a problem seen not only in captive marine mammals, but also in the wild. Keeping pinnipeds in chlorinated and/or fresh water can contribute to eye problems. Also, reflection of sunlight on the water and walls might contribute to eye problems; more research is necessary. When an animal withdraws itself from the group it can be a sign of discomfort. Pregnant females due to deliver their young will also often withdraw from the group. A keeper will generally know which animals are pregnant, but some animals may not show signs of pregnancy.

TOOLS AND EQUIPMENT

Being involved with the upkeep and maintenance of water treatment systems (life support systems or water filtration) is one of the major differences between keeping marine mammals and caring for other mammals. A keeper working with marine mammals needs a basic understanding of water treatment, even in institutions where water treatment systems are cared for by maintenance specialists.

In some seal and sea lion exhibits, a system of "drop and fill" is still in use. This system requires that once or twice a week, the pool is drained, cleaned, disinfected when necessary, and then refilled. In the case of dolphins and manatees this simple procedure is not feasible, although during a short period at the end of the 1960s and the beginning of the 1970s in traveling dolphin shows this was the procedure of choice. This would have caused stress and discomfort in these animals. All marine mammal species should have good water treatment systems.

What needs to be in a water treatment system and what are the standards for controlling water quality? First of all, water should be cleaned mechanically. This process entails a system whereby water is pumped out of the pool, via a prefilter (for large particles), towards a filter. The filter is usually filled with a sand-gravel mixture (a "filter medium" or "filter bed"). Prefilters are for easy removal of hair, leaves, and fish parts. The water flows through the filter medium, leaves behind most of the unwanted material, and then flows back into the pool. In general, this is not enough to keep pool water completely clean. Very small particles will flow through the filters. Maintenance of clean water requires the use of ferric sulfate or aluminum sulfate mechanically injected in small doses into the water stream towards the filter, which binds the particles that normally would go through the filter bed, causing them to clump together and be trapped by the filter media, and in this way removing them from the water. This process is called flocculation.

To prevent the filter from becoming overloaded with detritus from the pool, the filter is cleaned by reversing the flow ("backwashing") in the filter. This process does not flush the accumulated dirt back into the pool; instead it flushes the dirt out of the system to a sewer, or into a recycling system. When this process is completed, the system is set to the normal direction of flow and it starts filtering again.

In conjunction with the mechanical filtration, which has filtered most of the solids out of the system, an ozone installation can be used to not only remove residues in the water flow but also kill bacteria. As the ozone gas is dangerous to both man and animal, these systems should be operated by experts only, and all precautions to prevent exposure must be adhered to. A fail-safe safety system should be installed, which would include an alarm and an automatic shutoff system in the event of a leak or malfunction. Ultraviolet light is also used for killing or inactivating bacteria and protists in filtration systems.

In general, two systems are used to filter pool water, more or less on the basis of the outline just described. One is a me-

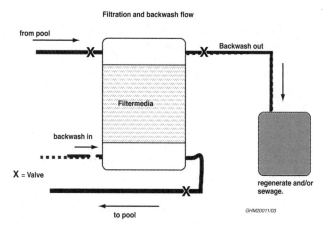

Figure 31.2. Filter backwash to flush the accumulated dirt and debris from the sand filters. By closing and opening the valves at the filter, the normal flow is stopped and a reverse flow is started. This flow loosens the sand bed and takes the dirt from the sand bed to a recycling system or to the sewage. In some systems an airflow is used to loosen the sand bed. Courtesy of G. H. Meijer.

chanical/chemical system in which chlorine in some form is used to disinfect. This is the system commonly used in public swimming pools. The other system is a mechanical/biological system in which bacteria is used to take care of the residues that are left in the water after the mechanical filtration process is complete. Experts disagree about which of these systems is best to use with marine mammals. In spite of all precautions, safeguards and low-level usage in chlorinated systems, chlorine and its derivatives should not be in the environment for the animals in our care, if it can be avoided and other options are available. Chlorine disinfection of the system, however, is found to be important for maintaining an acceptable bacterial level in the system. Bromine is used as well, but is considered not as effective as chlorine. In a biological system the chlorine disinfection is missing; ozone and/or ultraviolet disinfection can fill the niche, but it is often used with protein skimming (foam fractionation) in salt water. A keeper and all others responsible for maintaining the animals' environment should, however, keep looking for the best methods of securing a healthy and safe environment for the animals.

POOL CLEANING

To keep pools clean and in most cases free from algae, a water vacuum cleaning system should be in place, and can be connected to the filtration system. Independent vacuum systems are also available. One should be aware of the potential dangers when using electrical equipment in the vicinity of water. Water and electricity are not a good combination and can lead to electric shock. Also, animals and electrical cables can lead to potentially dangerous situations in which animals may bite or become entangled in the wires. Animals should not be in a pool that is being cleaned with electrical devices. With algae growth in a pool, it is often necessary to scrub the pool with a brush. Brushes on long poles, such as those made for cleaning swimming pools, are often used. Brushing by hand while using diving equipment is also a common method of cleaning, especially in larger tanks.

LAND AREA CLEANING

Old-fashioned scrubbing with soap and water on land areas for seals and sea lions is still a good practice. This process would be appropriate for demonstration stages and land-slide areas in dolphin pools. Food preparation areas, especially fish preparation areas, should be kept clean and disinfected at all times. The chance of fish being contaminated with all kinds of bacteria is always there, especially in summer. High-pressure cleaning devices should be used with care, especially indoors. Quite often they spread dirt particles and dangerous bacteria instead of removing them.

DIVING

In jobs that require working with marine mammals, being an experienced swimmer and diver is often a necessity. Diving is a valuable skill when working in animal shows, as well as when handling animals in water. There can be rules and regulations for diving that do not allow one to work under water without proper training (e.g., in the Netherlands any work underwater with the use of diving equipment may only be done by qualified professional divers). Governments may also have extensive regulations embedded in the health and safety legislation that is in place for work areas. The keeper diver must be aware of these rules and also of the potential dangers of working in water and underwater with live animals. Keepers should only work within the established safety protocol and take all necessary precautions, which probably involves adhering to a two-person policy when working in the water.

HANDLING

Sometimes it is necessary to move an animal from one area of the facility to another, to another zoo or aquarium, or perhaps to the veterinary clinic. Sometimes it is necessary to restrain an animal. For example, seals, sea lions, and walruses can be restrained using a suitably sized net or a "squeeze" or "crush" cage, depending on the size of the animal. To drive or push the animal, one can use plywood boards that may be fitted with handles to allow for more control. It is important to realize that many of the animals being manipulated weigh much more than the keepers. A male walrus, for example, weighing more than 1200 kg (2640 lbs.), is not an animal that can be pushed into a corner with a plywood board. It is always preferable to train animals to tolerate being handled or moved, instead of forcing them. However well trained or "friendly" an animal is perceived to be, and however well-behaved it is, if in distress it can become totally unapproachable and possibly very dangerous.

With sea lions or seals, an inside area or separation area is necessary. When the animal is confined to that area, the keeper may, with the help of a net, be able to catch the animal or use sheets of plywood to corner it or drive it into a squeeze cage. A squeeze cage can be incorporated as part of the exhibit so that animals become used to it and can be trained to move through it as they enter and leave the holding. Some restraint cages may incorporate weigh scales.

When using a net to catch an animal in water, the risk of drowning it is always there. One should keep in mind that seals and sea lions are well equipped for defending themselves. It would be extremely dangerous to be cornered or in a pool with an angry 300 kg (660 lb.)-plus sea lion male. Even a 50 kg (110 lb) youngster can become more than a handful, especially on a wet and slippery surface.

A lift platform or a lifting floor in one of the pools is a very good tool when a keeper needs to handle a dolphin. Here, training is also most important. Training and voluntary cooperation in difficult situations is by far the best way to manage these totally aquatic animals. Using nets to catch a dolphin is possible, but there is always a possibility of entanglement and of drowning the animal. Draining a pool when possible will allow for an easier approach.

Manatee handling requires many of the same precautions as outlined previously. Catching the animal with a net and lifting it out of the pool may be possible, and a lifting platform may also be considered. The use of a stretcher when the animal needs to be lifted out or away from the pool is recommended.

Before handling and moving an animal, one should first make a plan:

- Develop and follow a protocol.
- Have all the required equipment ready and complete.
- Ensure that enough experienced staff is available.
- Do not be afraid to stop a procedure if there are too many unexpected complications.
  (Osinga and de Wit 2002; Joustra 2003; Heukels and von Leeuwen 2008; Elk 2009; Holland 1999; Dierauf 2001; Geraci and Lounsbury 1993.)

BEHAVIOR TRAINING

Training of animals is a tool that has been refined and expanded upon to become an essential component of marine mammal husbandry and care. Some keepers responsible for maintaining manatees say that they need to be particularly inventive, since manatees are very independent and are not as interested in interacting with the keeper as most marine mammal species will be.

Basic target training is an essential starting point. If an animal can be further trained to come to the side of the pool and is willing to turn around for the keeper or, in the case of seals and sea lions, to come out of the pool, and if it allows close examination and touching of body parts, then the keeper can make a good assessment of the individual animal's health and condition. This training is not a substitute for honing of the keeper's observational skills, as the animal's behavior in the pool is the first thing to be assessed and is very important. All wild animals will hide their health problems as a part of their natural survival technique. By observing from a distance, the keeper can see changes that cannot be seen when an animal is aware that it is being observed.

Training has been an essential part of marine animals' care for as long as they have been kept in captivity. The sight of a sea lion balancing a ball on its nose is etched in almost every aquarium visitor's memory. Much background information about training will be found in chapters 42 and 43.

Some of the more important behaviors to be taught

to marine mammals include training for simple medical procedures, like removal of foreign bodies such as small balls or plastic cups from a cetacean's blowhole or mouth, or fecal and urine sampling, especially in dolphins whose feces quickly dissipate before they can be collected from the pool. Veterinarians usually collect blood, so it is important to include them regularly in an animal's training program. No matter how well-trained or behaved an animal is when a "stranger" is nearby or making contact, it can become anxious. A veterinarian whom the animal associates with discomfort or pain can further aggravate the situation. Time spent in training will be more than paid back in the future. Training is based on trust between the animal and the keeper; it also requires a thorough knowledge of the animal's behavior. A bull in breeding condition can behave in a manner completely different from normal. This reinforces the need for objective daily observations. If the keeper is unsure of an animal or does not trust it, he or she must follow those feelings and stop working with that animal. A keeper who feels uncomfortable may also act in a manner different from normal. The animal may sense this and it may add more tension to the situation. An accident or loss of trust will be the detrimental result.

Other procedures requiring more advanced training, such as hydration techniques, sperm collection, and artificial insemination, may be possible as the animals' level of conditioning progresses. An enthusiastic and resourceful keeper can achieve a great deal and improve the welfare and care of the animals in his or her charge.

ENRICHMENT

Daily training for shows and presentations, as well as husbandry training, is a great form of enrichment. A keeper should also be looking for ways to make the animals' environment more interesting. Water jets introduced randomly both above and below the water can be enriching, especially if the animals cannot predict their presence. The incorporation of waves into the pool is another valuable enrichment, and can be left running day and night. Addition of a high-speed stream of water into the pool with the help of a well-screened propeller will be very enriching and fun for the animals, and will also build up their swimming muscles and stamina. Introducing fish into the pool without keepers in sight can also be enriching.

A keeper should change her work program every day, as a repetitive routine can become boring both for the keeper and for the animals. Some simple changes may include starting work at a different time or changing the times for cleaning. A keeper who does presentations for the public should vary their starting times as much as possible. Predictable shows or presentations are boring for all involved. One should watch the animals, who sometimes invent their own games. The author has seen sea lions bouncing and catching stones against an exhibit wall, even bouncing them from one wall to another and then catching them again. One should try to think "outside the box." It can be challenging to find new enrichment that will keep the attention of marine mammals in the long run (Pryor 1999; Ramirez 1999).

REPRODUCTION

The gestation period (pregnancy) of most pinnipeds is approximately 11 to 12 months, with the Australian sea lion (*Neophoca cinerea*) having a breeding cycle of 17 to 18 months (it is not clear how long the delayed implantation stage is as opposed to the growing stage). Within days to weeks after birth, mating again takes place, on the same beaches where the previous young have been born. After mating, the fertilized ovum develops into a blastocyst (a hollow ball of cells) and instead of continuously developing, as in most mammals, the pinniped blastocyst does not establish contact with the wall of the uterus. There is then a period of inactivity for approximately three months before the blastocyst imbeds itself in the uterus and continues developing into the embryo. The pinnipeds normally give birth to one pup at a time; twins are occasionally seen, and under some circumstances they survive. In most pinnipeds, one male mates with more than one female (i.e., it is polygynous), with bull sea lions often having harems and defending significant territories. In contrast, a gray seal male may have just one female (i.e., it may be monogamous).

There are great differences in the lactation periods of seals and sea lions. In seals the lactating (suckling) period may be quite short, and it can vary from between three and five days in the hooded seal (*Cystophora cristata*) up to two years in the walrus (*Odobenus rosmarus*). The harbor seal (*Phoca vitulina*) suckles for about four to six weeks. Gray seals (*Halichoerus grypus*) are weaned at around 16 to 21 days. Sea lions and fur seals have more uniform lactation periods than the other seal species, suckling their young for between 6 and 12 months. During the first period of parental care, the mother stays with the pup and suckling occurs several times a day; later she will leave to feed and suckling becomes less frequent. This period lasts through the summer, and in some cases until the next pup is born. For a keeper, weaning a pup is in most cases not that difficult. Most youngsters will play with fish when they are available, and start eating them as a continuation of play. Some facilities offer additional live fish to make the pups develop greater interest in the fish and start eating them. Some pups do not so easily understand the concept of eating fish. Sometimes the previous year's offspring disturb the relationship between the new offspring and the mother. When the offspring is separated from the mother, it may still sometimes refuse to eat and will need to be forced to eat fish.

Gestation in the bottlenose dolphin lasts approximately 12 months; usually just one offspring is born. Lactation lasts between 18 and 20 months. The reproductive interval is between three and six years, with females becoming pregnant soon after their calves are weaned. Females care for their young, but other females in the group will normally assist with teaching, playing, and defending them. The orca has a gestation period of between 16 and 18 months and its lactation lasts approximately 12 months, with a reproductive interval of between 6 and 10 years. The female raises her young alone within the pod (group) that she lives in. These pods consist of related females and males.

The beluga whale has a gestation period of 14 to 15 months

and the lactation period lasts from 1.5 to 2 years. Females reproduce every 2 to 3 years. Little is known about the reproduction of Commerson's dolphin, but the gestation period is thought to be a maximum of 12 months. The total nursing period is not known, but it may last for at least 4 months. The reproductive interval is not known. There is also much to learn about the harbor porpoise, but its gestation period is thought to be around 11 months, and its lactation period between 7 and 8 months. Their reproductive interval is not known. It is possible that the recent births of harbor porpoises in Europe will offer more knowledge about this species' reproductive cycle (Klinowska 1991; King 1983; Animal Diversity Web 2010; Dierauf 2001).

TRANSPORTATION

For transportation of marine mammals it is very important to have a good plan, including procedures for unforeseen contingencies. Keepers with experience in transporting marine mammals should attend all transportation situations. Good preparation contributes significantly to the success of any animal transport.

PINNIPEDS

All pinnipeds can be transported in the same way. Crates should be well ventilated. This can be achieved by using welded wire mesh. For young animals, "sky kennels," commonly used for dogs and cats, are a good means of transportation. Transport crates should be of dimensions that allow the animal the ability to move around (IATA-LAR container requirement 76 should be referred to).

It is strongly advised to start crate training an animal as early as possible, since an animal that is used to the crate is much less stressed during travel. The animal should not be fed for 24 hours preceding the transport. This will decrease discomfort from pressure on its stomach and intestines. During road transportation, staff must be aware that inside a vehicle that is standing still (e.g., in a traffic jam or travel break) and is exposed to direct sunlight, the temperature can easily rise above 50° C (122 °F); so the vehicle should be well ventilated or air conditioned. Also, drafts on wet skin can overchill the animal to cause stress and predispose it to pneumonia. Pinnipeds cope much better with low temperatures than with high ones. Maximum transport temperatures should be maintained between 20 and 25 °C (between 68 and 77 °F) and, if possible, travel should take place at night as temperatures are cooler. Water and ice should be available as part of the equipment, although water to cool the animals may not be needed when temperatures are at normal levels. Even if a transport operation is well prepared, overheating (hyperthermia) may still be caused by stress, so the accompanying keeper must be watchful and aware of the signs. These may include lethargy, an increased respiratory rate, open-mouthed breathing, or hind flippers that are warm to the touch. The keeper should avoid unnecessary handling and attempt to lower the ambient temperature, increase ventilation, and cool the hind flippers. One should be well prepared and plan ahead to ensure a successful outcome.

DOLPHINS

Both dolphins and manatees have the same crate requirements, which are outlined in the IATA guidelines (IATA-LAR container requirement 55). A watertight crate is required, with enough space in which to hang the dolphin or manatee in a stretcher. There must be sufficient space on all sides to prevent the animal from hitting the sides, front, or back of the crate. The crate should be partly filled with salt water; this will not only lower the body weight pressure on the dolphin or manatee while it is in the stretcher, but will also help to prevent hyper- or hypothermia. The crate should not be overfilled with water, as it may splash out of the crate with any sudden movements of the transportation vehicle. This is especially pertinent in aircraft, as salt water can be very damaging. For a dolphin, one should keep the water cool by adding ice cubes or cooled replacement water. A manatee should be moved on a closed celled foam mattress and water. In Europe, manatees have been moved in water alone, with good success (B. Klausen, pers. comm.). Manatees are more tolerant of warmer temperatures than of cooler, but it is important that the water level and water temperature be monitored and stay within normal ranges within the vehicle, which should be air-conditioned with good ventilation and no drafts.

A mechanical hand sprayer should be available during all transport. Water must not be sprayed over the head nor near the blowhole or nostrils. A sufficient number of experienced attendants should accompany the transport, including an experienced veterinarian and preferably at least one attendant per animal.

VETERINARY CARE

Every zoos and aquarium should have at least a part-time veterinarian, or one hired by contract for consultations. In the daily routine, keepers should check their animals' overall appearance, their food intake, and, if possible, their feces. Any changes noted in these daily observations may be the first signs of illness and should be noted in a daily report.

Marine mammals can carry intestinal parasites such as tapeworms and roundworms; therefore, a regular program of control has to be in place. Additional laboratory tests of blood and fecal samples or Röntgen photos (radiographs or X-rays) can offer more information in cases of potential health problems.

All marine mammals can also carry a wide variety of diseases from viruses, bacteria, fungi, and other pathogens (causes of illness). If a keeper practices simple, proper hygiene techniques like washing hands after contact with food, animals, feces, or blood and wearing clean clothes and footwear, the risk of spreading or catching a disease is much reduced. Most institutions have hygiene and health protocols, which a keeper should always follow.

Still, there are some infections that a marine mammal keeper should be aware of. For example, *Mycoplasma* in seals can cause "sealer's finger" in humans; this is a very painful bacterial infection that can occur after a seal bite. If properly treated, it will usually heal well. Other diseases seen in marine mammals can also be transferred to humans, including leptospirosis, brucellosis, and tuberculosis. A keeper should inform their physician that they work with zoo animals. In case of a bite wound, a keeper should be aware that the bacteria in the animal's mouth may carry a high risk of infection.

CONSERVATION AND RESEARCH

Conservation in marine mammals is frequently associated with conservation programs in coastal areas. Most marine mammals are not on the IUCN Red List, but destruction of their habitat could change this very rapidly. For instance, just one accident in the oil industry can change the habitat of a species of marine mammal. Climate change and pollution are also major threats to the marine environment.

As a conservation strategy, protection of these areas makes the most sense, since all species are part of a bigger system and are connected to other species all over the world. Some protected areas include the US national marine sanctuaries; parts of the Wadden Sea along the coast of the Netherlands, Germany, and Denmark; and areas of the Mediterranean Sea. The IUCN website can provide more information for keepers who are interested in marine mammal conservation and habitat protection.

A great amount of research is being carried out on marine mammals both in the field (in situ) and in zoos and aquariums (ex situ). In the field, the research is often focused on population trends, outbreaks of diseases, climate change, and the effects of pollutants. In captivity, research is often on individual animals, and it focuses to physiology (e.g., hearing in seals and sea lions, the use of sonar in dolphins, or the use of vibrissae in walruses), food intake, and behavior (e.g., comparative psychology), to name a few areas. Publications included in the list of further readings in this volume are valuable resources of information.

SUMMARY

Marine mammals are a wide variety of animals that live partly or completely in water (usually seawater). Seals and sea lions give birth to their offspring on land and stay ashore for longer or shorter breeding periods, while dolphins and manatees are adapted to breeding in water. Enclosures should be built to give marine mammals places where they can show their natural behavior, with water and land areas built to their needs.

Good water quality and shelter from the sun and from high or low temperatures should be provided at all times. Great care should be given to providing a good quality of feeder fish or, in the case of manatees, vegetables, including storage and preparation. Good hygiene is also important to prevent transfer of diseases between animals and from animals to keepers.

Observing and understanding marine mammals can be difficult, because most of the time they stay in water. Also, their restraint and transport has its own unique challenges and therefore requires experienced keepers. Training and enrichment are of particular importance in the day-to-day care of marine mammals, and are important to their well-being. A knowledge of filtration technique and water chemistry is very important. Marine mammals are very interesting and uniquely challenging animals for a keeper to care for properly.

REFERENCES

Baxi, S. 1999. "Cephalorhynchus commersonii." Animal Diversity Web. Accessed 13 January 2011 at http://animaldiversity.ummz.umich.edu/site/accounts/information/Cephalorhynchus_commersonii.html.

Bonner, W. Nigel. 1994. *Seals and Sea Lions of the World.* New York: Facts on File.

Burnett, E. and K. Francl. 2009. "*Orcinus orca.*" Animal Diversity Web. Accessed 15 January 2011 at http://animaldiversity.ummz.umich.edu/site/accounts/information/Orcinus_orca.html.

Carling, M. 1999. "*Odobenus rosmarus.*" Animal Diversity Web. Accessed 15 January 2011 at http://animaldiversity.ummz.umich.edu/site/accounts/information/Odobenus_rosmarus.html.

Coffey, D. J. 1977. *The Encyclopedia of Sea Mammals.* London: Hart-Davis MacGibbon.

Crissey, S. D. 1998. *Handling and Fish Fed to Fish-Eating Animals: A Manual of Standard Operating Procedures.* Washington: US Department of Agriculture.

Dierauf, L. A, and F. M. D. Gulland. 2001. *CRC Handbook of Marine Mammal Medicine, Second Edition.* Boca Raton, FL: CRC Press,

Edwards, H. 2000. "*Trichechus manatus.*" Animal Diversity Web. Accessed 15 January 2011 at http://animaldiversity.ummz.umich.edu/site/accounts/information/Trichechus_manatus.html.

Elk, N. 2009. *Standards and Guidelines for the Management of Bottlenose Dolphins under Human Care.* European Association of Aquatic Mammals. Accessed at http://www.eaam.org/jdownloads/Documents%20and%20Guidelines/eaam_standards_and_guidelines_for_the_management_of_bottlenose_dolphins_under_human_care_sept_2009.pdf.

Geraci J. R., and V. J. Lounsbury. 1993. *Marine Mammal Ashore: A Field Guide for Strandings.* Galveston: Texas A&M University Sea Grant College Program.

Gonder, M. 2000. "*Eumetopias jubatus.*" Animal Diversity Web. Accessed 15 January 2011 at http://animaldiversity.ummz.umich.edu/site/accounts/information/Eumetopias_jubatus.html.

Grzimek, B., ed. 1990. *Grzimek's Encyclopedia of Mammals,* Vols. I–IV. Series. New York: McGraw-Hill.

Hammond, G., and A. Masi. 2000. "*Phocoena phocoena.*" Animal Diversity Web. Accessed 15 January 2011 at http://animaldiversity.ummz.umich.edu/site/accounts/information/Phocoena_phocoena.html.

Heukels, M., and L. van Leeuwen. 2008. *Draft-Husbandry Guideline of the West-Indian Manatee.* Amsterdam: European Association of Zoos and Aquaria.

Hiller, C. 2002. "*Arctocephalus pusillus.*" Animal Diversity Web. Accessed 15 January 2011 at http://animaldiversity.ummz.umich.edu/site/accounts/information/Arctocephalus_pusillus.html.

Holland, C. 1999. *Australian Sea Lion: Husbandry Manual for Marine Mammal Department.* Zoological Parks Board of New South Wales, New South Wales, Australia.

Jenkins, J., and P. Myers. 2009. "*Tursiops truncatus.*" Animal Diversity Web. Accessed 15 January 2011 at http://animaldiversity.ummz.umich.edu/site/accounts/information/Tursiops_truncatus.html.

Joustra, T. 2008. *Husbandry Guidelines for True Seals.* Amsterdam: European Association of Zoos and Aquaria.

Kiehl, K. 2001. "*Lagenorhynchus obliquidens.*" Animal Diversity Web. Accessed 15 January 2011 at http://animaldiversity.ummz.umich.edu/site/accounts/information/Lagenorhynchus_obliquidens.html.

King, J. E. 1983. *Seals of the World.* London: British Museum (Natural History).

Klausen. B. 2010. Odense Zoo, Denmark (pers. comm.).

Klinowska, M. 1991. *Dolphins, Porpoises and Whales of the World.,* Gland, Switzerland, and Cambridge: IUCN.

Liu, S. 2000. "*Otaria flavescens.*" Animal Diversity Web. Accessed 15 January 2011 at http://animaldiversity.ummz.umich.edu/site /accounts/information/Otaria_flavescens.html.

Osinga, N., and N. de Wit. 2008. *Husbandry Guidelines for Eared Seals.* Amsterdam: European Association of Zoos and Aquaria.

Price, R. 2002. "*Zalophus californianus.*" Animal Diversity Web. Accessed 15 January 2011 at http://animaldiversity.ummz.umich .edu/site/accounts/information/Zalophus_californianus.html.

Pryor, K. 1999. *Don't Shoot the Dog: A New Art of Teaching and Training.* Lydney, UK: Ringpress Books.

Ramirez, K. 1999. *Animal Training: Successful Animal Management through Positive Reinforcement,* Chicago: Shedd Aquarium Society.

Smith, J. 2008. "*Halichoerus grypus.*" Animal Diversity Web. Accessed 15 January 2011 at http://animaldiversity.ummz.umich .edu/site/accounts/information/Halichoerus_grypus.html.

Swolgaard, C. 2002. "*Arctocephalus australis.*" Animal Diversity Web. Accessed 15 January 2011, at . http://animaldiversity.ummz.umich .edu/accounts/Arctocephalus_australis/.

Steinway, M. 2003. "*Phoca vitulina.*" Animal Diversity Web. Accessed 15 January 2011 at http://animaldiversity.ummz.umich.edu/site /accounts/information/Phoca_vitulina.html.

Williams, S. 2002. "*Delphinapterus leucas.*" (On-line), Animal Diversity Web. Accessed 15 January 2011 at http://animaldiversity .ummz.umich.edu/site/accounts/information/Delphinapterus _leucas.html.

32

Husbandry and Care of Birds

Ted Fox and Adrienne Whiteley

INTRODUCTION

Members of the class Aves are the most widespread of vertebrate animals on the planet. Birds can be found in every habitat on the continents, and some even spend the majority of their lives at sea, returning to land only to reproduce. All birds share certain characteristics: feathers; beaks; sturdy, lightweight skeletons; and egg-laying. After completing this chapter, the reader will have a basic understanding of:

- avian families
- avian physiology
- exhibit design and maintenance
- the formulation and presentation of diets
- reproduction
- the form and function of bird eggs
- major disease concerns
- management strategies, including mixed-species exhibits
- avian-specific terminology.

TAXONOMY

There is some dispute regarding the number of orders that exist in the class. The classification standard for decades has been that of James Clements (2007). Recently, Sibley and Monroe (1997) published a revised taxonomy based on DNA sampling. The DNA taxonomy is gaining slow acceptance among aviculturists, although there is growing evidence in its favor. Depending on the source used, the number of orders is either 27 or 22. As the Clements standard is still more widely used, however, it is the classification described in table 32.1.

GENERAL LIFE CYCLE AND ANATOMY

All birds hatch from eggs. An egg is self-contained and includes all the nutrients necessary to allow the embryo to grow to hatching. Eggs are porous and lose weight throughout incubation ("drying down"). At the end of the incubation period, the chick breaks into the air cell and begins to breathe prior to hatching. Using a specialized egg-tooth, it cracks the shell around the perimeter of the egg, usually at the widest point, on the end with the air cell. The hatch muscle provides the force for the beak to penetrate through the shell. The chick then begins stretching its body until it can push free of the surrounding egg.

Most chicks are raised by one or both parents, although a few species practice parasitism: laying their eggs in another species' nest, then leaving them to be raised by the nest-owners. The chicks of these species are typically larger than the nest's resident chicks and outcompete them. For example, cowbirds (*Molothrus ater*) may lay their eggs in a chipping sparrow (*Spizella passerina*) nest. The chipping sparrow nestlings are much smaller than the newly hatched cowbirds, and soon starve as the cowbirds reach the food brought to the nest first. Interestingly, cowbird chicks will make sparrow vocalizations to beg from their "parents." After weaning, they acquire vocalizations appropriate for cowbirds.

Other species are colonial and allow helpers at the nest. White-crested laughing thrushes (*Garrulax leucolophus*) are one example of this. Closely related birds may help with nest building and chick feeding, occasionally even taking turns with incubation. Altricial chicks must be fed by their parents; precocial chicks are able to feed themselves. Altricial chicks are usually nidiculous (staying in the nest while they grow). Precocial chicks are usually nidifugous, (leaving the nest soon after hatching, following the parents to find food). Altricial chicks are typically bare, or nearly so, on hatching. As the chicks grow, their feathers develop. Time in the nest varies by species. Generally, smaller birds spend less time in the nest; a cardinal (*Cardinalis cardinalis*) fledges 9 to 13 days after hatching. Bald eagle (*Haliaeetus leucocephalus*) chicks spend nearly three months in the nest, building their flight muscles while perched on the nest edge for the last few weeks before taking off.

Most chicks learn their vocalizations from their parents.

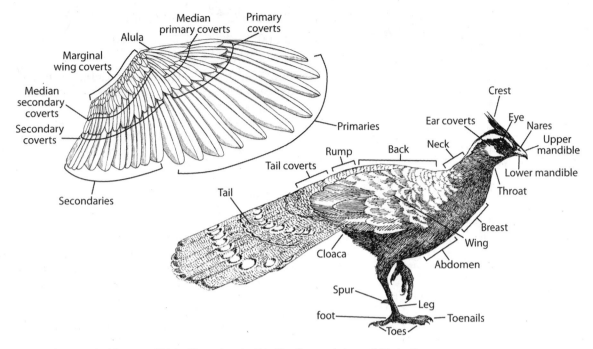

Figure 32.1. Basic anatomical features of birds. Illustrations by Kate Woodle, www.katewoodleillustration.com.

Some, like mockingbirds (*Mimus polyglottos*) increase their repertoire throughout their lives. Song vigor and plumage coloration directly correlate to mate selection in many species. Female mate choice is prevalent among birds, the dominant females often choosing the best singers with the brightest colors. In species that exhibit sexual dimorphism, male birds are frequently much "showier" than females, many going to great lengths to impress a prospective mate, literally singing and dancing their way into her favor. Courtship rituals vary widely among species; some are the most dramatic displays in the animal world. For species in which song and plumage plays a lesser role in courtship, the mechanism behind mate selection is more mysterious. Much of a bird's life is spent in courtship, bonding, chick-rearing, and territorial defense. What time is left over is devoted to feeding and preening. Keeping feathers in good condition is critical to a bird's survival. Some birds, of course, do find time to enjoy themselves. The astounding aeronautical maneuvers of crows (*Corvus brachyrhynchos*) on a windy day can be considered a display of play behavior.

GUIDELINES FOR THE AVIAN KEEPER

INTERACTION BETWEEN KEEPERS AND BIRDS

It is paramount for keepers to have an understanding of normal species behavior. Birds can be habituated to human presence, and food is a great motivator to accomplish this. Feeding live food or presenting food in a way that allows birds to come close to the keeper is extremely beneficial in allowing close observation. It is obviously important to move slowly and carefully when working in proximity to birds, since they are sometimes easily alarmed. One should always know where the birds are when moving among them, and allow them an

escape route. Birds that are frightened out of hiding spaces or off perches may fly blindly into enclosure barriers, risking injury or even death.

In addition to taking care to avoid injuring a bird, it is important to remember that some birds are very capable of inflicting damage on a careless keeper. Large birds, like cranes, use their beaks as weapons and will frequently aim for a person's face. Birds of prey have formidable beaks and talons, with which they are capable of inflicting serious harm. Psittacines use their beaks to open very hard objects like nuts, and have no difficulty with a relatively soft object like a human finger. Large parrots like hyacinth macaws (*Anodorhynchus hyacinthinus*) can easily break the bones in a person's hand if given the opportunity. Even smaller birds can cause injury with the right motivation; waterfowl may peck or hit keepers with their wings if their nest or offspring are approached. Even small passerines may attack by flying at a person who ventures too close to their nest. Keep in mind that a bird's behavior can change dramatically with changing circumstances (e.g., restraint, breeding, chick-rearing), resulting in a bird that no longer acts in a familiar way.

SPECIAL CONSIDERATIONS FOR WILD-CAUGHT SPECIMENS

Wild-caught specimens must be closely monitored for health and behavioral problems. Their food consumption should be documented. It is important to feed newly acquired birds as close to a natural diet as possible. If changing over to a different diet is necessary, it should be done gradually. Holding areas should have adequate perching and visual barriers to allow birds to become acclimated to their new surroundings. It is paramount for keepers to move calmly and quietly. Night lights should be provided when room lights are not in use.

TABLE 32.1. Classification of birds. Each order name is followed by some or all of the common names of families that are included in it; partial family listings are followed by an asterisk (Clements 2007).

Struthioniformes: ostriches, rheas, emus, cassowaries, and kiwis

Tinamiformes: tinamous

Sphenisciformes: penguins

Gaviiformes: divers

Podicipediformes: grebes

Procellariiformes: albatrosses, petrels, shearwaters

Pelecaniformes: tropicbirds, pelicans, gannets, boobies, cormorants, darters, frigatebirds

Ciconiiformes: herons, hamerkops, storks, shoebill, ibises, spoonbills

Phoenicopteriformes: flamingos

Anseriformes: screamers, ducks, geese, swans

Falconiformes: New World vultures, ospreys, hawks, eagles, secretary birds, caracaras, falcons

Galliformes: megapodes, guans, chachalacas, curassows, pheasants, grouse, hoatzins

Gruiformes: buttonquails, cranes, limpkins, trumpeters, rails, coots, sunbittern, seriemas, bustards*

Charadriiformes: oystercatchers, avocets, stilts, plovers, sandpipers, gulls, terns, skimmers, auks*

Pteroclidiformes: sandgrouse

Columbiformes: pigeons, doves

Psittaciformes: lories, cockatoos, parrots, macaws

Cuculiformes: turacos, cuckoos

Strigiformes: owls

Caprimulgiformes: oilbirds, frogmouths, potoos, nightjars

Apodiformes: swifts, tree swifts, hummingbirds

Coliiformes: mousebirds

Trogoniformes: trogons

Coraciiformes: kingfishers, motmots, bee-eaters, rollers, hoopoes, hornbills

Galbuliformes: jacamars, puffbirds

Piciformes: barbets, toucans, woodpeckers

Passeriformes (perching birds): broadbills, cotingas, pittas, larks, swallows, leafbirds, wrens, chickadees, cardinals, tanagers, finches, sparrows, starlings, orioles, bowerbirds, birds of paradise, crows, jays*

Many people advocate the use of softly playing music to help keep the birds calm and to help mask unfamiliar ambient sounds. As soon as the bird is stable in the new environment, medical tests (physical exam, hematology, radiographs, and cloacal and choanal cultures) should be performed to ascertain baseline health.

OBSERVATIONS: VISUAL AND VOCAL

A healthy bird will appear to be bright and alert with clean, unbroken plumage. The feathers will lie close against the body, and the bird will preen to keep the feathers in good condition. Most birds are relatively lively throughout their respective activity periods (diurnal, nocturnal, etc.).

ENCLOSURE DESIGN AND MAINTENANCE

INDOOR VERSUS OUTDOOR ENCLOSURES, WITH CLIMATE CONSIDERATIONS

Birds can be exhibited in indoor enclosures with controlled light and climate, or in outdoor areas with exposure to the elements. The best choice depends on the species being exhibited and the ability to keep the birds safe and secure. Some choices are obvious: for example, one would not keep an Antarctic penguin species in an outdoor enclosure in Arizona. Bird species typically found in climate parameters similar to those in which the zoo is located may usually be kept outdoors with access to appropriate shelter. Birds capable of flight should be kept in exhibits that are netted or meshed in a suitable size to prevent escape or injury. It is possible to render a bird incapable of flight either permanently by surgical means (pinioning or tendonectomy) or temporarily by feather clipping to keep it contained in an area that is not completely enclosed. This option is most successful with birds that are typically found on the ground or in water when not flying (waterfowl, wading birds, cranes, etc.). Birds that do not fly can be restrained by walls or fences. In temperate areas, tropical birds can be displayed outdoors seasonally.

WALK-THROUGH AVIARIES

Mixed-species aviaries are very popular with the public and can be used to illustrate not only individual species, but the ways in which multiple species can interact with a similar environment. These aviaries provide a great deal of interest, both visual and auditory. Care must be taken to provide adequate space to meet the needs of all the species chosen to inhabit the enclosure. Knowledge of the behavioral parameters of each individual species is the key to selecting birds that will successfully share the area allotted to them. It is useful to have a variety of perching opportunities at multiple levels, plentiful nesting material, and multiple feeding areas. Live plants and water features will enhance the effect of the space and provide enrichment for the birds.

CONTAINMENT MATERIALS

Birds housed indoors will also need containment of some type: examples include free-standing cages, glass-fronted rooms, aviaries that employ double door entries, and exhibits with piano wire or "invisible" netting. It is sometimes possible to house diurnal birds in a lighted exhibit that zoo visitors can view from a darkened hallway, as the birds will be reluctant to enter an unlit space.

SHELTERS AND CLIMATE CONTROL

All birds exhibited outdoors must have some shelter from inclement weather. This may take many forms, such as a simple windbreak, a covered roof, or a nest box. Climate control in enclosed habitats is used to help create an environment that replicates the natural habitat as closely as possible. The previously mentioned Antarctic penguins kept in Arizona would need a refrigerated habitat to keep them comfortable and healthy.

WATER FEATURES

Water is an essential environmental component for most bird species. It is necessary for drinking and bathing. Some birds swim and/or find their food in water. Adding a water feature to a habitat for many birds enhances the environment and the birds' well-being, while also increasing interest for the visitor by allowing the birds to display a broader range of natural behavior. Moving water (waterfalls, misters, etc.) may attract the birds' attention, motivating them to make use of the feature. The water should be monitored for quality and cleaned frequently enough to prevent bacterial growth. Fresh, clean drinking water should be provided daily.

SUBSTRATES

A variety of substrates can be used, depending on the species and type of enclosure. A freestanding cage should allow the fecal material to drop through the cage bottom and collect on a surface that can be cleaned (cement, newspaper, wood shavings, etc). Larger spaces may have grass, dirt, sand mulch, or other natural substances. Fecal material should not be allowed to build up, regardless of the substrate used. Some consideration must also be given to pest control. Mice and rats in particular have an amazing aptitude for finding their way into animal exhibits with their unending food supply. It is helpful to bury wire mesh under the substrate where feasible, to prevent rodents from tunneling their way in. Integrated pest management is essential to keep all pest populations under control without causing harm to the collection.

TEMPORARY HOLDING, CAPTURE, AND INTRODUCTION UNITS

It is sometimes necessary, especially in large enclosures, to use a temporary cage. Such a cage may take one of many forms, but it should have enough space to comfortably house a specimen for the period of time required. When introducing a bird to a new space or to new exhibit mates, an introduction cage can be used to allow the birds to see and hear each other without making physical contact. A temporary holding may be used to protect a nesting female and her chicks prior to fledging in a mixed-species environment. A capture unit is very useful in a large aviary, where it can be very difficult to net an individual bird that may need attention. The unit can be baited with food and rigged with a trapdoor to isolate a specimen that may need to be examined. If the capture unit is left in the environment, the birds will become desensitized to it, thus simplifying the process. The netting should have a small enough mesh to contain the smallest species in the aviary, while being large enough inside to accommodate the largest species. Plant material (natural or artificial) can be used to camouflage the unit and make it less obtrusive in the environment.

AVIAN DIETS: PRESENTATION AND FORMULATION

COMPLETE DIETS

Many commercial diets are formulated by reputable companies to meet the needs of specific types of birds. These diets often come in the form of an extruded pellet available in a variety of shapes and size that allow easy consumption by the bird for which it is designed. A nutritionist or veterinarian can assist in the selection of appropriate diets for the species in the collection. Knowledge of natural diets will also be helpful in choosing an adequate diet. One should keep in mind that many birds in the wild have access to a large assortment of food choices. Feeding one single complete feed can lead to boredom and subsequent behavioral problems. In addition, birds can be quite picky about what food they will consume. Obviously, a perfectly designed diet will be of no benefit to a bird that refuses to eat it. It may be necessary to supplement the food with something more interesting (e.g., chopped fruit, diced hard-boiled egg, or insects) to entice the bird to eat it all. Some birds that may lay eggs more frequently (e.g., poultry and pigeons) have higher calcium requirements. A dish of mineral-rich grit in the enclosure will allow them to consume minerals when they are depleted.

LIVE FOOD

Numerous birds consume live food, insects in particular, every day. When possible, insects should be part of the food offered to them. Crickets, mealworms, and moth larvae are all readily available through commercial sources. Pet stores frequently stock these, albeit at fairly high prices. Some large vendors sell insects in bulk at more reasonable prices. Live crickets can be stored in bins with screened lids for extended periods of time if they are provided with air circulation, food, and water. A high-quality cricket food will increase the nutritional content of the crickets as food for the birds. Mealworms can likewise be kept in bins with food (sliced yam and pelleted poultry food work well) for a fairly long time. The insect bins should be cleaned weekly to prevent the buildup of waste. It is possible to store live mealworms and moth larvae in the refrigerator for weeks without feeding them (they will not eat when cooled in this way). When removed from refrigeration and warmed, these insects will become active again. The advantage of refrigeration is longer storage without metamorphosis to the next life stage (to beetles for the mealworms and to moths for the moth larvae). Live food makes an excellent enrichment for many bird species. Insects are also fed exclusively by some species to their hatchlings, and thus are crucial to the survival of these young.

FROZEN FOOD

Special consideration must be given to proper thawing techniques for frozen food items. In almost all cases, frozen foods should be brought to room temperature before they are fed to birds. Ideally, frozen food should begin thawing in a refrigerator. Once it is thawed, it can be brought to room temperature for a short period of time in an enclosed container prior to feeding. Refrigerated thawing reduces the risk of bacteria growing in the food item. Previously frozen food should be fed to birds as soon as possible after reaching room temperature, or should be kept in a refrigerator for no more than two to three days.

SUPPLEMENTS

Many nutrients in food break down during storage. It is sometimes necessary to add supplements to it to maintain its nutritional quality. For example, fish loses some of its essential vitamins during storage in a freezer. Specially formulated vitamins for piscivores can help replace these lost nutrients. Certain bird species, like flamingos and scarlet ibis, need carotenoids in their diet to maintain their plumage coloration. These carotenoids can be given as a supplement if the birds' diet does not contain sufficient amounts.

BROWSE

Supplying browse to birds may serve a variety of purposes: enrichment, food, camouflage, or nesting material. Psittacines will benefit from a steady supply of branches to tear apart, satisfying their urge for destruction. The provision of fruiting branches (e.g., hawthorn [*Crataegus* spp.] or autumn olive [*Elaeagnus umbellata*]) will be greatly appreciated by fruit-eating birds, as it will give them the opportunity to forage for their food. Twigs and branches of varying sizes can be scattered about the enclosure to allow the birds to select nesting material. Any plant species added to an aviary should first be checked against a reliable plant list to determine its level of toxicity.

SPECIES MANAGEMENT

SINGLE VERSUS MIXED

Unquestionably it is simpler to care for a single species within a given exhibit than to manage multiple species. Some species, such as large birds of prey, are generally incompatible with other bird species and should therefore be housed by themselves. However, many bird species can be held in mixed-species exhibits with other birds, or even with reptiles or mammals. Although such exhibits present a greater management challenge, they also offer greater opportunity for educating the visitor and create a richer and more natural environment for the animals being displayed. Single-species exhibits focus all the attention on the individual bird species and their requirements. In the absence of competition for space and resources, the single species housed can have their biological needs more easily met. Mixed-species displays must have resources (food, space, nesting opportunities) available that meet a wide variety of needs, so that all the species can function without undue stress or aggression. Once again, knowledge of the biological requirements and natural behavior of all the species being considered is critical in making wise decisions about which species will be compatible with each other. Communication with other zoo professionals is invaluable in avoiding poor selections that may have disastrous consequences. For example, small primate species may make a lovely display with a rainforest passerine, But if the bird is not given nesting areas that the primates can't reach, its eggs or young chicks may become a tasty monkey treat. Discussions with colleagues can help avoid fatal errors.

SPECIAL CONSIDERATIONS FOR SINGLE SPECIMENS

Most bird species are social creatures, but circumstances may dictate that a particular bird should be housed individually. This could happen for a host of reasons. One likely example would be a parrot that is donated to the zoo after having been a family pet. Psittacines that have been raised in isolation are often difficult to introduce to other parrots, as they are generally imprinted on their human family. These birds may have to be housed alone. In such a situation, enrichment is extremely important. Birds that are bored and socially segregated may become psychologically impaired and act with increased aggression towards their keepers. Feather-plucking, stereotypical movements, and screaming are just some behaviors that may be seen in a parrot that is not adequately stimulated.

CAPTURE, RESTRAINT, AND HANDLING

Medical attention, transport, and beak or nail trims are just a few reasons why a bird may need to be captured, restrained, and physically manipulated. Capturing a bird improperly can cause injury to the bird or the keeper. A bird's lightweight bones are easily fractured, and blood feathers can be broken or air sacs ruptured by careless handling. It is essential to restrain a captured bird securely without placing undue pressure on it to avoid such injuries. Restraint attempts may cause many species to aggressively defend themselves from the perceived attack. Potential captives may peck, bite, scratch, kick, or talon the person attempting to control them.

A skilled keeper may be able to hand-catch a bird. This is easiest with a bird that is habituated to approaching people (hand-feeding is one way to accomplish this), or one that has been maneuvered into a confined space. Trap units like those described above can also be used to confine a bird for capture. Nets are indispensible tools that extend the keeper's reach and help to reduce injury; they are available in many types and sizes, and it is important to match the net type to the job. Small, lightweight mesh nets for small passerines and large woven nylon nets for waterfowl are two examples. Some nets are made with extendable handles, which may be useful for catching birds in large aviaries.

If possible, it can help to extinguish lights just before making contact with the bird. Generally a bird will remain motionless when it is suddenly subjected to darkness. Obviously one must know exactly where the bird is located, and must be able to reach it in the dark. Once the bird is captured, it is important to keep its wings folded against its body to keep it from flapping and potentially injuring itself or the handler. Some birds, like penguins or swans, can deliver painful blows with their wings. Particular attention should be paid to keeping the legs of long-legged birds from hitting against each other and causing damage to the skin or underlying tendons. This is done by keeping a finger or hand between the two legs. Some birds may stay calmer when being held if their eyes are covered by a hand, towel, or hood.

A handler often must restrain a bird's head to keep it from biting. The most common way to do this is to hold the bird by the back of the head with one hand, keeping one's thumb and middle finger on each side of the bird's mandible and one's index finger over the top of the bird's head and controlling its movement, while taking care that the throat is not constricted and allowing access for a physical exam. Birds, like all animals, may overheat if held too long, especially if they are struggling.

INTRODUCTIONS

When introducing birds to a new environment or conspecifics, one should try to anticipate their negative reactions. It may be necessary to add visual barriers, extra enrichment, or hiding spaces while they become acclimated. An anxious bird may fly headlong into a glass window, resulting in injury or death; it thus may be necessary to cover the glass on a glass-fronted exhibit with paper or soap until the bird becomes aware of the barrier. Introduction cages can be used to enable new birds to assess one another before making any physical contact.

STRESS

Birds are subject to many types of social, environmental, or medical stress, which can lead to illness or death. A keeper must guard a bird against severe stress by observing it every day. A stressed bird may show poor appetite; listless behavior; dull, broken, or piloerect feathers ("fluffed" appearance); open-mouthed breathing; or pacing. It is paramount to ascertain the cause of the stress and alleviate it as soon as possible. If reducing the stressor is not effective, medical intervention may be necessary. Birds are masters of camouflaging illness, so by the time symptoms are observed, there may not be much time for a cure.

BEAK AND NAIL MAINTENANCE, FEATHER CLIPPING, AND PINIONING

Captive birds sometimes need help in keeping their beaks and nails in good condition. A varied environment is the best way to help a bird keep itself in an optimal state. Perches of varying thickness and texture will allow a bird to naturally maintain its beak and nails. Occasionally the beak and nails may need to be trimmed if they are growing abnormally. Overgrown portions can often be clipped with human or dog nail trimmers; one should be careful not to cut too close to the blood supply. Thick beaks or nails can be coped with a high-speed rotary grinder, such as a Dremel® tool. This electric tool works quickly, and can be stressful for the bird; it may take more than one session to restore a beak to its natural condition. Knowledge of the natural shape and length of the beak and nails will guide a keeper in knowing how much trimming is necessary.

In some circumstances it may be desirable to render a full-winged bird flightless. Birds housed in outdoor exhibits that are not fully enclosed can have their wings clipped to prevent them from flying off. This is done by clipping the primary feathers on either the left or the right wing (not both) just under the tip of the wing coverts. The lack of primaries on only one side will prevent the bird from gaining lift and keep it off balance so that it cannot take off. The cut is usually made with strong scissors at a 90-degree angle to the feather shaft. This solution is only temporary, as new feathers will grow in each year as a normal part of molting and must be clipped again as soon as the blood supply of the new feathers dries up. New or developing feathers (called blood or pin feathers) should never be trimmed, as the resulting blood loss can compromise the bird's health. If a blood feather is cut inadvertently, it can be pulled out with a pair of pliers or forceps to prevent excessive bleeding. Pinioning is a method of permanently rendering a bird flightless as a hatchling, by removing the end of its wing at the wrist joint when it is three to five days old. Done properly, this amputation is quick and relatively painless with little blood loss. The young bird heals quickly and will not need to be caught and have its wings trimmed annually. Pinioning is most commonly used on waterfowl, cranes, flamingos, and other birds that are often housed in open outdoor exhibits.

CRATING AND TRANSPORT

Birds can be trained to enter a crate on their own volition, or they can be captured and placed in a transport crate. A transport crate should have a perch or a nonslip substrate, depending on the needs of the species. IATA regulations specify the types of containers required for air transport of birds. Food and water should be included in the crate for long-distance travel. It is wise to cover the doors and windows with an opaque material that will allow air flow to decrease stress to the bird. Padding can also be added, especially on the ceiling, for birds that are likely to jump up and injure their heads (cranes, ratites, pigeons, and pheasants, for example).

AVIAN REPRODUCTION

COMMON BREEDING STRATEGIES

Bird breeding strategies vary widely from species to species. Some species practice seasonal or long-term monogamy, in which two birds in a pair reproduce only with each other, or polygamy, in which one bird has multiple partners. Monogamous birds, like rock pigeons (*Columba livia*) and Humboldt penguins (*Spheniscus humboldti*), generally share incubation and chick-rearing. In polygamous species, the female often rears the chicks on her own. Megapodes (brush and scrub turkeys) have a unique breeding strategy: the males construct large nest mounds in which the females lay their eggs. The heat produced by the decomposition of the vegetation in the mounds incubates the eggs. The male monitors and adjusts the nest and the chicks hatch in a superprecocial state, able to survive without parental care. Ostriches (*Struthio camelus*) gather in groups and all the hens lay eggs in one nest, where they are incubated by the dominant male and female in the group.

SEASONALITY AND PHOTO PERIODS

Most birds are seasonal breeders. They lay eggs and hatch and rear their young when light and food are abundant. Increased day length is often the trigger for the start of reproduction; it can be replicated in an indoor zoo exhibit by increasing the amount of time during which artificial lights are on, or by providing the birds access to increasing ambient sunlight.

Emperor penguins (*Aptenodytes forsteri*) are obvious exceptions to this, as the males incubate the eggs during the darkest, coldest times of the year in Antarctica. However, this seasonal timing allows the young penguins to make their way to the ocean when conditions are more favorable for them. It is important to understand which breeding strategy is favored by the birds being considered for a zoo's collection, if it is a goal to ensure that the breeding requirements will be met.

NEST SITES AND NESTING MATERIAL

Bird nest types range from the simple "scrape" on hard-packed soil made by an ostrich to an elaborate platform constructed of sticks and grasses which can be reused from year to year to support a family of osprey (*Pandion haliaetus*). Many birds use nests made of grass woven together to support the eggs. Others may lay their eggs in tree hollows. Barn swallows (*Hirundo rustica*) build their nests out of mud and plant fibers which they attach to vertical surfaces just under ceilings. Captive birds may use an artificial nest if provided with one of the appropriate type, size, and shape. Making nesting material available, again of the proper type and size, may allow the birds to construct a suitable nest. Researching the type of nest used in a natural setting will guide the selection of materials to offer.

EGGS

All birds lay eggs. An egg comprises an ovum and yolk sac, albumin, an allantois, amniotic fluid, chalazae cords which hold the yolk in place, a blood supply, an air space, and a shell. The yolk provides nourishment for the developing embryo and is eventually pulled into the chick through the umbilicus. The outer shell contains pores through which respiration takes place.

INCUBATION: ARTIFICIAL AND NATURAL

Whenever feasible, natural incubation is preferable to artificial. Incubation and hatching is a complicated process that requires detailed knowledge for replication in an artificial setting. Birds instinctually know how tightly to brood, when to turn their eggs, and how to vocalize to the chicks while they are hatching. They create the ideal temperature for embryos to develop within the eggs. A newly hatched chick imprints on its parents and begins learning the behaviors appropriate for its species soon after leaving the egg. There may be circumstances which lead a keeper to select an artificial means of incubation, such as a desire to imprint a bird on humans, a nest that has been abandoned after eggs have been laid in it, or a high risk of predation. If it is necessary to artificially incubate and hatch bird eggs, one can choose from a number of incubators. Standard models work well for poultry or waterfowl eggs with a moderate amount of monitoring. Specialized models with more precise temperature and humidity controls and adjustable rollers that can be set to the desired turning time are also available, and are well worth acquiring if eggs from rare, endangered, or delicate species need to be hatched artificially. Artificial incubation will require more staff time, expertise, and resources for monitoring the equipment and the eggs, and for successfully raising the chicks.

PARENT-REARING VERSUS HAND-REARING

Parent-reared chicks will naturally acquire the species-specific knowledge that is essential for normal behavior. Hand-reared chicks always become imprinted on humans to some degree. In some birds this may not prevent them from interacting with conspecifics. However, hand–rearing may impede the chick from connecting with others of its own species, reducing the possibility of it socializing or breeding when it reaches maturity. For birds that are planned to be used exclusively in outreach or education programs, hand-rearing may reduce stress and abnormal behavior by acclimating the bird to its keepers from the beginning.

VETERINARY CARE

COMMON MEDICAL CONCERNS

Birds are susceptible to a number of diseases to which they may quickly succumb if the symptoms are not recognized and treated quickly. Certain diseases affect species that are naturally found in cold or very dry climates when they are housed in temperate regions. Fungal infections or insect-borne diseases like malaria (*Plasmodium* spp.) are not common in cold or dry habitats, so species from those habitats may have no immunity to them. Snowy owls (*Nyctea scandiaca*) are particularly susceptible to malaria. Humboldt penguins may readily contract malaria as well as aspergillosis, a fungal infection that is extremely difficult to treat successfully. West Nile virus has become a concern in North American zoos, especially those housing swans, corvids, raptors, and penguins, which seem to experience very high mortality when exposed to this recently introduced virus. *Mycobacterium avium* (sometimes known as avian tuberculosis) gives cause for concern when diagnosed in a zoo bird. However, it is being better managed as its prevalence decreases and more is known about its transmission. Parasitic infections may also be a problem in birds. Roundworms, tapeworms, coccidia, giardia, cryptosporidium, mites, lice, and ticks are commonly detected in birds.

IDENTIFICATION OF ILLNESS

Each of the medical conditions mentioned above has symptoms that can be detected by a knowledgeable keeper. Once specific symptoms are observed or even suspected, medical assistance should be sought as soon as possible before the disease is out of control.

PREVENTIVE MEDICINE

Many medical problems can be prevented or controlled with a proactive approach. Vaccines are available to build immunity to West Nile virus, prophylactic medicines to guard against malaria can be given during mosquito season, and regular deworming will help control parasites. It is very important to develop a comprehensive preventive medicine program, and every zoo should have one in place.

MANAGEMENT OF INFIRM BIRDS

Aggressive treatment may be necessary when a bird has been diagnosed with an illness. The sick bird may require supportive care, which might include a secluded treatment area, quiet surroundings, and most importantly a constant source of heat. Birds have difficulty thermoregulating when fighting disease, so a warm environment may prevent deterioration or even death.

ZOONOTIC CONCERNS

Some avian diseases are transmissible to humans and can cause illness or death. Strains of avian influenza periodically transfer to human populations, often from infected poultry, resulting in flulike symptoms, which may be severe. Bacterial infections like salmonellosis and psittacosis can also be contracted if contact is made with an infected bird or its secretions or feces. Good hygiene and sanitation is the best way to ensure that disease transmission is unlikely to occur.

CONSERVATION

The International Union for the Conservation of Nature (IUCN) recognizes 10,027 bird species. Of these, 190 species are listed as critically endangered, 372 as endangered, and 678 as vulnerable, for a total of 1240 (12%) species at risk out of the total population (IUCN 2011). The greatest threats facing birds are global climate change, habitat loss, and ecosystem degradation and exploitation by human populations. Zoos assist in conservation in many different ways: an excellent example is the California Condor Recovery Program. In cooperation with the US Fish and Wildlife Service, California condors have been brought back from near-extinction. In 1985 the last few remaining wild California condors were brought into captivity at the San Diego Wild Animal Park and the Los Angeles Zoo. These condors have a very low reproductive rate, usually producing only one chick every other year. But through the use of foster parents and artificial incubation, their reproductive rate was dramatically increased. By 1992

some of the captive birds were being reintroduced to the wild. The total population of California condors now stands at about 350 birds, with nearly all of these living back in the wild (Zoological Society of San Diego 2011). Other zoos and agencies have joined the effort, ensuring that the California condor species has a much greater chance of surviving into the future. Providing optimal care for bird collections and educating the visiting public can aid in the global conservation effort by increasing awareness and maintaining genetic diversity.

SUMMARY

Birds are everywhere, and they have adapted to thrive in a wide variety of habitats. They vary greatly across species, ranging from the bee hummingbird (*Mellisuga helenae*), which weighs less than two grams, to the ostrich, which can weigh up to 160 kilograms. Birds come in numerous sizes, shapes, and colors, but they all have a few traits in common: all have feathers, lay eggs, are endothermic, and have wings (although not all birds fly). The care and management of exotic birds requires patience, dedication, and an understanding of what birds need to thrive. Keepers must be willing to seek information from any available source, such as mentors, colleagues, books, or journals, to build their proficiency in avian husbandry. There is still much to be learned about comprehensive management practices. The desire to observe birds and disseminate what they have learned can lead keepers to contribute to the body of avian knowledge and improve standards for zoo bird collections.

REFERENCES

Clements, James. 2007. *The Clements Checklist of Birds of the World.* Ithaca, NY: Cornell University Press.
Monroe, Burt L., Jr., and Charles G. Sibley. 1997. *The World Checklist of Birds.* New Haven: Yale University Press.
Zoological Society of San Diego. 2011. Accessed 3 April 2011 at http://cacondorconservation.org.

33

Husbandry and Care of Reptiles

Aaron M. Cobaugh

INTRODUCTION

Reptiles comprise a highly diverse class of approximately 9,300 species (Reptile Database 2011) in four distinct taxonomic orders: Crocodylia (crocodiles), Testudines (turtles), Rhynchocephalia (tuataras), and Squamata. The order Squamata is comprised of three distinct suborders: Sauria (lizards), Serpentes (snakes), and Amphisbaenia (worm lizards). Traits shared by reptiles include skin covered in keratinous scales, lungs for respiration, and having an ectothermic ("cold-blooded") metabolism. Most also have four limbs (are "tetrapods") and lay shelled eggs. Reptiles are an ancient lineage with ancestors dating back more than 300 million years, but 30% of all reptile species are now estimated to be "threatened" or worse (IUCN 2010), due to factors that include habitat loss, exploitation, and ecological interference (e.g., introduced species or disease).

Reptiles have been historically linked to amphibians due to their ectothermic metabolism, with the field of herpetology devoted to their shared study, and the term "herptiles" (or "herpetofauna") often used to describe species in zoological collections. This is somewhat unfortunate, as amphibians and reptiles are distinct biologically and taxonomically, and they are not the only ectothermic vertebrates; fish are ectothermic as well, possibly confusing the uninformed. However, one of the most prestigious scientific societies devoted to the study of reptiles and amphibians, the American Society of Ichthyologists and Herpetologists (ASIH), actually includes the study of fishes as well, and there are numerous other societies devoted solely to the study of herpetology.

Reptiles also have been historically exhibited in zoos in sparse and minimal conditions (as were many other species) due to their ability to survive on fewer resources, as well as due to the "biological ignorance" of early facilities and keepers who simply did not know what is now well known by professional and well-educated zoo and aquarium staff. The last few decades have seen continued improvement in the exhibitry of many species, with elaborate naturalistic settings and even mixed-species exhibits providing visitors a more well-rounded and educational experience. This chapter includes the basic general care guidelines for most reptile species, with provisions to account for their ectothermic metabolism, as it is probably the single most important "limiting factor" (as it limits success if not provided for) to be considered by new keepers. The unique care considerations necessitated by reptile thermoregulatory behavior (thermoregulation) cannot be overlooked. This chapter will also briefly discuss the skills and tools needed for a keeper to properly and safely maintain venomous species in captivity.

After studying this chapter the reader will understand

- that the unique behavior and physiology of reptiles affects the housing, feeding, and reproduction of these animals in the zoo and aquarium environment
- the best practices for daily care, handling, housing, and transport of reptiles
- the need for specific tools, enclosures, and training to work with certain dangerous reptile species
- the key habitat and environmental requirements of reptiles— particularly temperature and lighting, but also the availability of water and proper substrate
- the principal issues involved in medical management of reptiles
- the need for conservation and continued research into threatened species.

GENERAL CHARACTERISTICS OF THE TAXA

The order Crocodylia is comprised of alligators, caimans, crocodiles, and gharials, and includes the largest living reptile, the saltwater crocodile (*Crocodylus porosus*) at 7 m (23 ft.) and 1500 kg (3300 lbs.) or more. Most of the 24 species in this order are tropical in distribution, with both the American and Chinese alligators (*Alligator mississippiensis* and *Alligator sinensis*) extending their ranges into temperate regions. Crocodilians are well-adapted for a semiaquatic existence, with

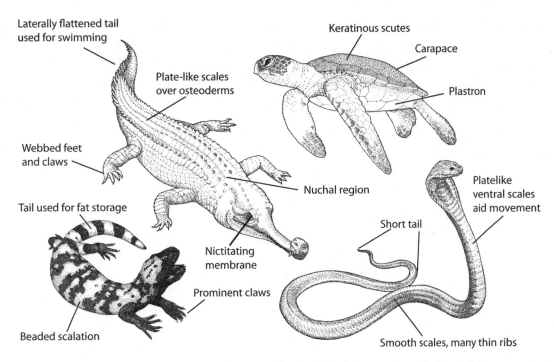

Figure 33.1. Basic anatomical features of reptiles. Clockwise from upper left: gharial (*Gavialis gangeticus*, order Crocodylia), green sea turtle (*Chelonia mydas*, order Testudines), king cobra (*Ophiophagus hannah*, order Squamata), gila monster (*Heloderma suspectum*, order Squamata). Illustrations by Kate Woodle, www.katewoodleillustration.com.

elongate heads, throat valves (so they can open their mouths underwater to grab prey), valved nostrils, eyes (with clear nictitating membranes) dorsally placed so that the animal can remain submerged during hunting, webbed digits, and literally flattened tails for efficient swimming. They are also the only reptiles with four-chambered hearts, which allow them to bypass the lungs while submerged (as they are not being used) so that blood can be sent elsewhere. The crocodilian's heart is also more efficient than that of other reptiles, allowing crocodilians to reach greater sizes and, along with higher levels of hemoglobin in the blood, allowing them to remain submerged for relatively long periods. Crocodilians tend to be apex predators in their environments, with some species considered "generalists" (alligators and caimans have broad snouts and blunt teeth) and other "specialists" (gharials have very thin snouts and thin sharp teeth to catch fish). They have very powerful jaws with muscles biased for closing pressure (but relatively weaker opening strength), and some species can feed on relatively large prey items, which they dismember before swallowing. Their teeth can be replaced constantly throughout their life span by teeth growing in the sockets below, and all species are carnivorous. Most species will bask to aid thermoregulation, and all species lay eggs in covered nest sites. Although widely distributed, crocodilians face significant conservation pressure due to humans, with half of all species probably suffering population losses in the wild.

The order Testudines is comprised of turtles and tortoises, and is noteworthy because of the unique bony shell, incorporating the skeleton, that allows most species to withdraw entirely within it and offers protection from predation. The shell comprises two parts, the dorsal carapace and the ventral plastron. The bone of the shell in most species is covered by keratinous plates known as scutes. The carapace and plastron are fused together, with an anterior opening for the head and forelimbs, and a posterior opening for the hindlimbs and tail. Shell morphology differs among the 13 families and 300-plus species, with some species having a heavy, thick shell and others having a thin, flexible shell. Some aquatic species will exhibit a greatly reduced plastron coupled with a heavily fortified carapace, and others will exhibit one or two plastral hinges that allow them to completely close the plastron against the carapace after withdrawing. A few species even exhibit hinged carapaces. Another trait unique to testudines is the absence of teeth. Instead, they have keratinous sheaths that cover the jaws. These horny (hardened) sheaths can be quite sharp in carnivorous species, while in herbivorous species they may be serrated. Limb morphology also varies among testudines, with larger terrestrial forms (tortoises) exhibiting "elephantine" clubbed feet to better support their weight, while semiaquatic terrestrial species (terrapins, softshell turtles) exhibit significant webbing of the digits and streamlined shells for swimming. Sea turtles have evolved paddle-like limbs for efficient underwater propulsion, as they spend their entire lives at sea (except for the females). All species lay eggs, usually in secluded nests. Testudines are widely distributed in many different habitats including deserts, grasslands, forests, remote islands, fresh water, and the ocean. However, they also face significant conservation pressure, with more than 50% of the species considered threatened due to human behavior. The term "tortoise" usually refers to a terrestrial, herbivorous turtle with a heavy shell, although the use of this term varies with locality. The term "terrapin" usually means a semiaquatic omnivorous turtle with webbed feet and a streamlined shell from freshwater or brackish environments in North America. Technically, they are all simply turtles.

The order Rynchocephalia is comprised of only two species of tuataras (*Sphenodon* sp.), odd lizardlike "living fossils" endemic only to New Zealand and rarely exhibited outside of that region. They are actually found only on 32 offshore islands, with a small population only recently reintroduced onto the mainland in 2005 (Zealandia 2008). Although they superficially resemble lizards, they are part of a distinct evolutionary lineage that flourished some 200 million years ago, of which they are the only living remnants. Their unique characteristics include two rows of teeth in the upper jaw, functional hearing without external ears, primitive spine and rib bones, and the ability to remain active at temperatures usually too low for other terrestrial reptiles (they are actually intolerant of the higher temperatures used by many other reptile species for proper thermoregulation). They also have a very slow reproductive rate, probably due to the harsh cooler climate they inhabit; they lay eggs and are the only reptiles without a copulatory organ. They have been considered endangered for over a century due to relatively small populations on the various islands, where they seem to thrive only in the absence of mammalian predators. Introduced rats have posed a continued threat to some populations, but efforts at eliminating rats from these islands have met with some success.

The order Squamata comprises the lizards, snakes, and amphisbaenians in three suborders and accounts for almost 96% of all reptile species, with more than 5,400 species of lizards, 3,300 species of snakes, and 181 species of amphisbaenians (Reptile Database 2011). With such a large number of species distributed in so many diverse habitats, this order's morphological diversity is also extensive, probably because most lizards are under 30 cm (1ft.) in length, which allows them to take advantage of the many ecological niches available to smaller animals. This fact is coupled with the environment being able to support many smaller animals rather than fewer large ones, and smaller animals not being able to move as far, thus encouraging speciation in smaller geographic areas (Bauer 1998, 126). Snake diversity is probably due to similar abilities, coupled with the success of limbless locomotion and the evolution of venom to subdue prey. Most lizards have four limbs with five clawed digits on each, eyelids, external ears, and a relatively long tail, although exceptions to each trait exist. Snakes lack each of those traits, although a few primitive species do have remnants of the pelvic girdle and limbs. Snakes do have unique adaptations to a limbless lifestyle, such as a higher number of ribs (more than 400 in some species) due to a lengthening of the body (with a relatively short tail) and a highly flexible skull that enables many species to consume relatively large prey items whole and consequently feed less often than similarly-sized lizards. All snakes are carnivorous, whereas there are numerous herbivorous and omnivorous lizards. Further differences include a single functional lung in most snakes, an extendable trachea (protrusible windpipe) to aid respiration when swallowing prey in most snakes, autotomy (tail loss and regrowth as a predator defense) in many lizards, greater thermoregulatory activity in many lizards, and specialized ventral scales in snakes to aid locomotion. Lizards and snakes both exhibit hemipenes, paired eversible structures within the cloaca

(a shared terminal pouch for the digestive, urinary, and reproductive tracts) in males used for internal fertilization of the female; oviparity (egg laying) is their primary mode of reproduction, although viviparity (giving birth to live young) is not uncommon in both suborders. Both lizards and snakes exhibit a well-developed vomeronasal organ (also known as Jacobson's organ), an auxiliary olfactory sense organ which is probably well known by the tongue flicking in many snakes and some lizards, especially varanids (monitor lizards). Lizards and snakes also both exhibit a cyclic pattern of skin shedding (ecdysis), with snakes usually shedding their skin in one piece (rare in lizards). Also, while a quarter of all snake species are known to be venomous, only two lizard species are venomous: the beaded lizard (*Heloderma horridum*) and the Gila monster (*Heloderma suspectum*), with the venomous apparatus in each species located in the lower jaw, quite differently from snakes.

Finally, amphisbaenians, or worm lizards, are specialized for a fossorial (underground) lifestyle, with heavy skulls to aid in burrowing, recessed mouths, reduced or absent eyes, shorter tails, loose skin, and ringlike skin scalation to aid in traction underground. Most species appear to be oviparous, although little is known of many species due to their fossorial lifestyles, with some known from only a single specimen. The family Bipedidae contains the only three species that possess forelimbs, with all other species limbless. Amphibaenians are efficient predators with muscular jaws that can deliver a powerful bite, with some species able to tear flesh from prey too large to be swallowed whole. They tend to be restricted to tropical regions, and are rarely exhibited in zoos.

ORIENTATION TO THE KEEPING AND HUSBANDRY OF REPTILES

Reptiles occupy a tremendous variety of habitats: tropical to temperate (some even arctic, as in the European viper [*Vipera berus*]), arboreal/terrestrial/fossorial/aquatic (many freshwater and some marine), desert/savanna/forest/montane (many in isolated island populations), and some even troglodytic (living deep in subterranean caves). Reptiles share ectothermic metabolism with amphibians and fish, and with invertebrates for that matter, but they tend to exhibit more active thermoregulatory behavior, actively and quickly seeking proper temperatures for their metabolic needs, driven by digestion or reproduction. Some species are "passive conformers," living in environments or microclimates that vary little in temperature (e.g., caves), but the majority are either heliothermic (basking in the sun to absorb heat), thigmothermic (lying on warm surfaces to absorb heat), or both, depending on environmental conditions. Whether a reptile is diurnal or nocturnal often determines its mode of thermoregulation. Keepers should make use of microenvironments and the opportunity to offer variable temperatures (and humidity) to captive reptiles so that they can choose the environment that best suits their needs.

Keepers must know their species' origins in order to design proper care regimens and habitats. Proper species identification is important, as an animal's natural history and behavior should be used to shape its captive conditions. Especially to the untrained or inexperienced eye, many species may look similar to others that are closely related but come from distinctly different habitats or environments. Basing husbandry on incorrect environmental parameters can greatly stress certain species and can even lead to death. For example, many people think of chameleons, especially the "true chameleons" of the subfamily Chamaeleonidae, as creatures living in dense tropical rain forests subject to high humidity and rainfall. However, many species within this taxa live not in rainforests but in quite arid climates, and continuous moisture at consistently warm temperatures would be harmful to them. Similarly, keeping cave geckos of the genus *Goniurosaurus* in the same manner as leopard geckos of the genus *Eublepharis* (very common in the pet trade) would be a mistake, even though they appear very similar in morphology and belong to the same subfamily (Eublepharinae). *Goniurosaurus* sp. originate from Asian subtropical forests with high humidity, and they are usually stressed by higher temperatures and lack of consistent moisture. *Eublepharis* sp. originate from Middle Eastern deserts with greater temperature swings and, of course, drier habitats. Leopard geckos are known to be less "delicate," but this is probably due to the fact that they have adapted to greater environmental extremes and can tolerate less precise conditions in captivity.

Species common to the pet trade (sometimes referred to as "domesticated") often make interesting exhibit animals, especially if kept in naturalistic enclosures meant to mimic their original environment. These species can also provide clues to husbandry that can be extrapolated towards husbandry methods for less common species. In the example above, leopard geckos can be kept in an enclosure setup meant to mimic a desert escarpment, where they would shelter under rocks and in crevices underground during extremes of heat or cold. These underground or subsurface areas will usually retain moisture, which the geckos need to prevent dessication. Keepers and experienced hobbyists have known for quite some time that if these animals are kept completely dry, they will have difficulty shedding and may possibly suffer a lack of appetite, probably due to stress from dehydration; so they provide them with areas of moisture (dampened sand beneath a large flat rock, for example). One could use this example to judge the needs of other species from arid climates, such as other lizard species and snakes that shelter in similar microclimates but other distinct geographic areas. In fact, many reptile species from arid climates will dig their own tunnels to access such subsurface moisture. It also typically follows that "generalists" tend to fare better in captivity than "specialists," as the generalists can adapt to change more easily, exhibiting less stress. However, captive zoo and aquarium conservation programs may require facilities to keep species with challenging needs that can be met with appropriate research and attention to detail. Such research must include learning about the environmental parameters of microclimates or ecological niches the animals inhabit, as well as about available food items they consume throughout the year. Also, seeking previous successful husbandry methods or new ideas from other experienced keepers can greatly aid a new keeper when he or she is faced with caring for a potentially challenging species.

Not only must keepers pursue such background information to be successful, they must also learn and acquire the attitude and skills necessary to work with certain dangerous species such as large crocodilians and venomous snakes. Proper mentoring and training is a must, and dangerous species should never be underestimated or taken for granted, no matter the level of experience and training accrued by the keeper. Training for working with venomous snakes is usually a graduated process, with inexperienced keepers learning from observation of more experienced personnel. This is followed by work with nonlethal species, so that the keeper can learn how to use the tools required as well as how the animals may respond to being manipulated by them. This continues with progressively more "challenging" or "difficult" species, and the training may go on for some time, especially as the various venomous snake species exhibit diverse behavior and capabilities. Approximately 25% of snakes (more than 700 species) are considered venomous, with more than 30% of those (approximately 250 species) considered to possess venom of medical significance to humans.

THE ZOO AND AQUARIUM ENVIRONMENT

Reptile keepers must consider a variety of environmental factors to ensure proper captive conditions, but three are usually considered most important: temperature, lighting, and humidity. These factors may vary significantly throughout an exhibit, especially if it is large enough, but each species will have "preferred target levels" of each factor. Keepers will need to compare natural conditions to the artificial or captive conditions the exhibit can provide and ask: "Can the proper conditions required of this species be accomplished in captivity, or at least closely mimicked?" Although many species can tolerate less than optimal conditions, keepers need their charges to thrive, not just survive.

TEMPERATURE

Temperature is of course important for ectotherms, as proper environmental temperature will allow these species to carry out essential metabolic processes such as digestion, reproduction, immune response, and basic muscle activity to allow movement. They cannot generate metabolic heat themselves, so they rely on the environmental heat around them, which can vary significantly even over short distances (e.g., from sun to shade) and is naturally provided by the sun. In captivity, keepers will usually use artificial light sources or substrate heat emitters for indoor exhibits. Lighting fixtures work well for heliothermic species, as they would normally bask in the sun to warm themselves; however, the type of light fixture and bulb used will be important. Fixtures and bulbs will become very hot, so the animals should not be able to come into contact with them as serious burns could result. Fixtures should be constructed of ceramic, as plastic will degrade with heat over time. Fixtures and bulbs should also not be able to be contacted by any water (including water sprayed by keepers), because of the risk of electrocution and also because hot glass bulbs will shatter when quickly cooled by water. Standard incandescent PAR bulbs can be used to transmit heat, as their shape provides a focus to transmitted heat and light. However, keepers should be aware there are two basic types: flood and spot. Floodlight bulbs have a wider light spread and spread heat over a wider area; this may be necessary in smaller exhibits or with species that do not bask in higher temperatures. Spotlight bulbs have a narrower light spread, concentrating heat and light into a narrow beam; they can be useful in larger exhibits where the basking areas are some distance from the light fixture, but they have very little temperature range in smaller exhibits and can burn animals quickly as they try to bask. Some light bulbs, such as some self-ballasted mercury vapor bulbs, emit not only heat but also useful light spectrums (such as UV-B), to be discussed in the next paragraph. Thigmothermic species rely on heat absorbed from the substrate they are on, and this can be provided by overhead lighting sources or substrate heating sources. Various heat pads, tapes, and cables can be placed to warm the substrate, but must be closely monitored to ensure that they do not cause the substrate to become too warm as burns can result. Keepers should strive to measure heat regularly with thermometers (digital infrared models can be quite useful) at designed basking sites both before and after their installation, so that animals are not harmed by incorrect temperatures. They should also strive to provide a gradient of heat throughout the exhibit, so that the animals can choose the thermal environment they need. An exhibit with a proper temperature gradient may have not just a warm and a cool zone but several warm zones of various temperatures, as well as multiple cool zones. This is admittedly difficult in smaller enclosures, and that is why keepers must be aware of the animals' natural thermal range and seek to provide as much of it as possible. Keepers must also remember that temperature will vary throughout the day in the natural environment, typically dropping at night; mimicking this variance may benefit their animals or even be required.

LIGHT

In Earth's biosphere light and heat are linked, and in captive conditions they can be linked as well, but the quality of artificial light can be quite different from one source to another. Natural sunlight is considered to be "full-spectrum" light, and includes visible light, infrared (IR), and ultraviolet (UV) wavelengths. Artificial light that contains all the wavelengths of visible light is also considered full-spectrum light even though it may not include much IR, which has a longer wavelength than visible light and registers as heat, or UV, which is important to many reptiles. Many vertebrates use UV light as a catalyst for vitamin D formation in the skin, but there are three different wavelengths of UV to consider. UV-A, the longest of the UV wavelengths, is beneficial to many diurnal reptiles as it may drive activity levels and social behavior, and is often emitted by many full-spectrum bulbs. It also passes through normal glass, such as windows or skylights. UV-B is of even greater importance, as it is needed to synthesize vitamin D in the skin, and vitamin D is needed to metabolize calcium for proper bone growth. Without proper exposure to UV-B, many species may suffer from metabolic bone disease (MBD), in which the affected animal's skeleton is weak due to a lack of metabolized calcium, even if calcium is provided in the diet, as without vitamin D it cannot be metabolized properly. UV-B is usually not found in common full-spectrum lamps or typical PAR spot and floodlamps, and it also cannot pass through regular window glass. It can, however, be provided by specialized lamps, both incandescent and florescent. Keepers should determine the appropriate exhibit lighting needs for each species, which may consist of a combination of lamp types. UV-C is the shortest of the UV wavelengths, and the most dangerous, as it quickly damages living cells. It is usually filtered out by the atmosphere's ozone layer and is not found in typical lightbulbs. Keepers should note that both UV-A and UV-B can cause damage to human skin (and the DNA within the skin cells) with sufficient exposure, and they should therefore limit their exposure to these light sources. The animals should also be given the opportunity to remove themselves from exposure. Not all species, including most snakes or nocturnal species, require such high-quality full-spectrum light. Photoperiods should ideally follow those the animals would experience in the wild.

HUMIDITY AND WATER

Atmospheric moisture can be provided through the use of moving water features and intermittent spraying or misting, combined with a reduction in ventilation. However, keepers should take care to allow sufficient ventilation and periodic drying to prevent mold growth in the exhibit. Exhibits should be equipped with sufficient drains so that any substrate does not become constantly saturated if exposed to continued water inputs. Ventilation can also be controlled to maintain humidity at desired levels, and the use of a hygrometer (which measures humidity), either analog or digital, is recommended. A variety of substrates may be used (soil, sand, gravel, etc.) as long as they are not harmful to the animals, especially if ingested, and as long as they do not present a drainage problem. Live plants can be used for many reptile species, except for certain herbivores which may eat them, and the plants can increase humidity in the exhibit as well through transpiration (plant respiration). Water sources must remain clean, with smaller nonmoving pools periodically drained and flushed when fouled, and moving water features or larger pools treated appropriately with mechanical, chemical, and biological filtration, which must be regularly maintained so that water quality does not suffer or degrade. Keepers should also be aware that not all species will drink from water dishes or pools. Many lizards (including chameleons) will only drink from falling water as they normally drink during rains, or from water droplets as they would when covered with morning dew. Some species also drink very slowly, so the keeper should ensure water is provided for them over a period of time, and leaving misters running for several minutes or more might be necessary.

ENCLOSURES

Reptiles are generally less active than similarly-sized endotherms, and therefore can often be housed in smaller enclosures. Enclosure materials resistant to water are preferred because they will not degrade when exposed to water used to clean or to maintain humidity. Most exhibit or enclosure shells are constructed of some combination of concrete, plastic, fiberglass and glass, with glass preferred over clear plastics for exhibit windows as it resists scratches. Glass aquariums can be used to house smaller species away from exhibits. Ventilation openings should of course be screened to prevent escape, and keepers should periodically inspect them, as their integrity may be compromised with age. Snakes and some lizards are excellent at using relatively small openings to escape, so enclosures should be designed or implemented with this in mind. Holding areas and vestibules should be designed with possible escape in mind, with secure secondary barriers, and with clutter kept to a minimum to facilitate recapture. Weapons may also be considered in case recapture is not a possibility and the animal has to be destroyed; however, further discussion about the specifics of such usage is beyond the scope of this chapter.

Many reptile species are cryptic or more comfortable when shelter is present, but providing shelter and materials that allow cryptic species to "blend in" may make it difficult for the public to view the animals. Many keepers would prefer an exhibit that is easy to work in, but such an exhibit may not look appealing, whereas a lushly planted exhibit may look very nice to the public, but be difficult for the keepers to work in. Keepers must weigh the needs of the animals against the needs of the visitors who would like to observe the species on display. Often a compromise can be struck in which the shelter in which the animals hide is presented in such a way as to allow visual observation. For example, a small hollow log can be cut lengthwise and placed against the exhibit glass so that the sheltering species can be seen. Materials can also be used in such a way as to allow the public to view cryptic species, especially if exhibit signage explains the animal's capabilities and morphology. For example, lizards that perch on vertical small trees can be highlighted in an exhibit if the trees are near the viewing window and signage explains where visitors should look. However, exhibit material should never present a way for the reptile to escape, or compromise the keeper's ability to work safely with it in the enclosure. For example, a sectioned slab of a log used by the reptile on exhibit to hide should be positioned carefully. If it is too close to the door of the exhibit, the keeper might have to move it every time the exhibit is serviced or entered. Conversely, if it is too distant from the door, the keeper might leverage his or her position to check on the species sheltering under it, and might be unable to react properly if the animal attempts to escape through the open exhibit access door (which may happen in smaller exhibits) during inspection of the shelter. In larger exhibits, keepers can close the access doors after they enter. Arboreal species will require branching and even large tree limbs, but they should be kept away from the keeper access doors so that keeper can safely open the doors without startling the animal and can maintain an appropriate distance from it. Also, one must evaluate whether the exhibit design is flexible enough to accommodate changing needs such as seasonal reproduction: whether eggs can be laid on exhibit in removable or easily accessible nesting areas that enable the keepers to take them out of the exhibit easily. Not having to remove certain animals from exhibits for reproductive purposes frees up resources and space for other species that may require it, and certain species may even continue to allow juveniles to inhabit the same space.

DANGEROUS SPECIES AND TOOL USE

Many species present a risk to keepers when kept in captivity, including crocodilians, large lizards (e.g., monitor lizards) and nonvenomous snakes (e.g., large boas and pythons), and venomous snakes. Crocodilians and large lizards can inflict significant bites and can also strike with their tails, causing wounds or knocking keepers off-balance. Large lizards may also have very sharp claws used to inflict deep wounds. Large boas and pythons are famous for their ability to constrict, and they can indeed kill humans in that way, but they also have multiple rows of large teeth that can inflict serious wounds. Large species should never be worked alone, and keepers who work with them should be well trained. Dangerous species can and should be shifted off exhibit if the exhibit needs to be serviced for cleaning or maintenance. Some species can be trained for this purpose, but others may need to be tempted with food to move from an exhibit to a holding cage, and keepers may have to be patient.

Certain species can be moved with tools such as snake hooks or tongs, and crocodilians can be pushed with long blunt wooden poles if necessary. Keepers should always use caution with tools and should be well-versed and experienced in their usage. While snake hooks rarely pose a danger to the animal, they require practice to use properly, as the keeper must balance the animal on one or two of them, and the animal may not want to remain balanced. Certain snakes, such as arboreal boas, pythons, and vipers tend to remain on hooks (though they may try to climb the hooks), while other species, like some colubrids (racers, boomslangs, etc.), may never feel comfortable while on them, thus requiring the keeper to work quickly. Heavy-bodied species also present a challenge, as the hooks tend to be narrow, causing the animal discomfort. Keepers might also find these heavier species uncomfortable to work with, since holding their weight away from one's body requires more effort and endurance, particularly in the arms. Tongs can be used with lizards as well as snakes, but can cause injury to the animal if not used properly. They consist of two opposing metal segments at the end of a pole that collapse together when the handle at the other end is squeezed, but because the two metal lengths are joined at a pivot joint, they can pinch, and they can also present a crush problem, especially to moderately sized snakes and their many thin ribs. Keepers may be tempted to "tail" certain species—grasping the animal's tail while keeping the rest of the body, especially the head, away from them—but the author considers this risky, as the tail may be injured or may slip from the keeper's grasp.

Venomous snakes present a unique problem, because their bite can initially cause little physical damage, but the venom can cause significant tissue damage, and in many species may be enough to cause death. Venomous snakes should be remotely shifted between enclosures using cable-actuated or levered shift doors whenever possible. Even the most experienced keeper can be bitten, and it only takes one "bad" bite to end a life. Shifting is not always possible, so keepers may need to move the animals using tools and should therefore be well trained and experienced before working with venomous species. They should also never work these animals alone. Clear acrylic tubes can be used to restrain snakes that cannot be restrained "freehand" without risk. The keeper guides the snake partly into the tube and then restrains its posterior half against the tube, preventing it from crawling through or backing out. Sometimes a snake will resist entering even a clear tube, but it can be placed into a large open container, such as a large plastic trash can, with the tube above it and slid over it as it tries to crawl out of the can. With the snake trapped in the tube, the keeper or veterinarian can draw blood, inject medicines, or do a close visual inspection. The tube may also have small openings or slits cut into it to allow wound treatment through the tube and near the dangerous head region. If the facility has a large collection of variously sized snakes, multiple tubes of various sizes may be needed, but tubing should only be performed by highly trained keepers.

If venomous snakes are to be removed from a secure enclosure or exhibit and placed into temporary quarters, such temporary housing must still be appropriately secured. Keepers will commonly use large plastic trash cans for temporary placement, and these cans usually come with snap-on lids. However, these lids are not secure enough on their own and should be taped to the can body with strongly adhesive duct tape. Colored identification tags or cards (usually red) should always be used with venomous species, both on the exhibit and on temporary enclosures, so that all keepers know which animals are present in a given enclosure. Many zoos will assign a tag to a particular animal, and that tag will follow the animal wherever it goes within the zoo.

Finally, provisions should exist for the possibility of a venomous bite sustained by a keeper. Appropriate species-specific antivenin should be on hand so that it can travel with the keeper to the hospital if a bite is sustained. Not all bites result in envenomation or will warrant antivenin treatment, but if they do, the antivenin must be administered in a hospital setting as it may itself cause a serious reaction. Local hospitals should be trained in venomous bite treatment and be prepared to accept a bite victim, with a medical liaison trusted by the zoological facility to treat the victim accordingly. As one might imagine, this requires significant advance planning. Antivenin must be stored properly in a dedicated refrigeration unit and must be kept current, as it loses effectiveness over time. Zoo staff should be trained to respond to a venomous bite accordingly, and snakebite alarms should be placed throughout the zoo and tested regularly, so that their activation alerts other staff members. Keeper response to a possible venomous bite should be practiced, with regular drills used to ensure that all staff members know their particular roles, which will vary by institution. Trained staff members should make every effort to secure the animal if possible after a bite, so as not to compound the situation with an escape. Holding areas and vestibules should also be designed and constructed with escape in mind. Floors should be kept clear of clutter, doors and windows should seal tightly, walls should be solid, and drains should be securely screened.

FEEDING AND NUTRITION

As reptiles are ectothermic, they require only about 10% of the dietary energy intake of similarly-sized endotherms. Because of this, they require less food, and this is what allows many species to occur in habitats of lower productivity in nature, or to occur in higher numbers in areas of greater productivity. Being an ectotherm is not necessarily a handicap when compared to an endotherm in different environments. However, keepers may overlook the importance of this difference, and

as a consequence, overfeeding and obesity occur. Weight and body condition should be monitored and recorded regularly, especially for juveniles but also for erratic or periodic feeders like some snakes, and adjustments made as necessary. Species fed ad libitum, especially in group settings, should be closely monitored, as should dominant individuals that may consume a disproportionate amount of food, as gut distention can result in death. Temperature should also be monitored regularly, as digestion is compromised when temperature drops below a certain threshold (usually a species-dependent one). Below this threshold, food cannot be digested, and if allowed to remain in the gut it can cause illness and death.

Certain species can be shifted into holding quarters to feed and to allow keepers to service their exhibits; this is especially useful and warranted for dangerous species. If species are fed while on exhibit, care must be taken to ensure that the substrate does not present a risk of accidental ingestion. Feeding tongs or forceps can be used with many species, especially carnivores, but care must be taken so that the animal does not injure its mouth on the forceps or tongs when lunging for food items. The tongs or forceps should also be long enough so that the keeper's hand and fingers are not at risk of being accidentally bitten, but not so long as to be unwieldy; the keeper should be able to move them quickly. Certain carnivores and insectivores may exhibit feeding responses only for live or moving prey items; a keeper can use forceps or tongs to mimic movement in prekilled prey items, but he or she should take care that the animal does not injure its mouth when attempting to feed on the grasped food. Juvenile snakes and sick or injured species may also need to be "teased" with food to elicit a feeding response, and this entails slowly moving the food items near the animal's head or along the jaws, causing the animal to reflexively bite at the food. Once it has bitten the food, the animal will often continue reflexively to swallow it.

Live food items commonly used for carnivorous reptiles in zoos and aquariums include various species of crickets, fruit flies, beetle larvae (e.g., mealworms, superworms), cockroaches, earthworms, and moth larvae (silkworms, waxworms). Prekilled food items, usually for larger species, include various species of rodents (usually commercially-raised mice and rats), rabbits, birds (usually domestic chickens), fish, and even domestic pigs. The size of the prey item, of course, is determined by the size of the animal feeding on it, usually corresponding to the size of its head. Snakes can and will proportionally feed on relatively larger single prey items, due to their skull and jaw morphology. Certain prey species can be cultured on site, but feasibility will depend on the zoo or aquarium's needs, and it may be more convenient and efficient to procure food animals from commercial vendors. Keepers should still familiarize themselves with how to culture various species, in case the need arises.

Herbivorous and omnivorous species will feed on many commercially available fresh green vegetables, including various lettuces, endive, kale, collard greens, mustard greens, dandelion, and spinach. They will also consume various fruits and vegetables, including apples, grapes, various berries, pears, papaya, mangos, melons, carrots, squash, sweet potatoes, corn, peas, and beans. A number of commercially produced dry pelleted foods produced by various companies, usually for the pet trade, can be used by zoos to supplement a diet of fresh produce.

While keepers should research the nutritional profiles of a diet before deciding on supplementation, certain diets will ultimately require that additional substances be added. Vitamins and minerals are common supplements, especially to carnivorous diets, as prey items may be deficient in one or another. Commercially produced insects tend to have a poor calcium-to-phosphorus ratio, basically being very low in calcium. Keepers can "gut load" the prey species before offering them to the reptile, by feeding them a diet rich in vitamins and minerals, and they can also "dust" them with a vitamin and mineral powder. Insects prepared in this manner should be fed to reptiles immediately, as the dusted material will wear off and ingested material will be passed with subsequent digestion. Feeding reptiles with tongs will ensure that they receive the proper amount of supplemented prey items. Snakes rarely need supplementation as they are usually fed whole prey, and omnivorous reptiles should be given a varied diet so that deficiencies are minimized. Herbivores can also be fed varied diets to minimize the need for supplementation, and keepers should remember that more is not always better, as oversupplementation can be just as harmful as undersupplementation. Keepers should also remember that water is important for proper digestion in many species, especially to those that are fed dry prepared diets, but that not all species will drink from standing water.

BEHAVIOR

Many reptile species do not exhibit easily identifiable behaviors such as those for submission, aggression, injury, and illness. Keepers must learn to look for subtle actions in many species, as they will not often act like cobras that spread their hoods when threatened. Many species will become habituated to their keeper's presence, in effect tolerating it, but that does not mean that they are without stress, and they may lash out unexpectedly as a result. This lack of identifiable clues is especially problematic in cases of illness or injury. If there are no identifiable behavioral changes, what does the keeper look for? Most often, keepers will look for changes in feeding behavior, but changes in thermoregulatory behavior (such as extensive or limited time spent basking by a heliothermic species) or a lack of response to stimuli (such as not moving when prodded lightly by the keeper) may mean that the animal's health is compromised. In fact, if it is determined that the animal is suffering from an illness, the keeper may be able to aid the animal by providing additional sources of variable heat, as many species will spend more time basking to increase their internal body temperature and system metabolism, thus boosting their immune response. In effect, this may also provide an inhospitable environment for pathogenic organisms, in much the same way as a fever would for a mammal.

Aggressive behavior by these animals, toward both conspecifics and keepers, is to be avoided, and keepers must learn and note its causes. It can be due to various factors including territoriality, reproduction, hunger, and defense. Reproductive changes in behavior are often seasonal in most species; they tend to coincide with egg laying or birthing in females and

the exclusion of competition in males. The author once cared for a large male green iguana (*Iguana iguana*) at the Buffalo Zoo that was usually a wonderful nonaggressive animal, often basking in the middle of the exhibit for the public to view, and presenting little difficulty when the enclosure needed servicing. In fact, the iguana would often slowly approach the service door and present its throat for a good scratch from the keeper staff. However, usually around April, it would become belligerent when approached, often bobbing its head and puffing itself up to become larger, and occasionally gaping and hissing and even attempting to bite. This behavior, of course, was driven by reproductive hormones and their subsequent effect on behavior in many males, and it lasted for about a month. Some species also become overly excited at feeding times, and in their rush to feed they become problematic. Keepers can alter feeding schedules and locations so that the reptiles do not become fixated on a particular time and location, although the need for this depends on the species. For example, a group of medium-sized tortoises do not pose much of a threat, but a group of larger monitor lizards might.

The potential for enrichment exists for many ectothermic species, although their response to it may be more tempered than that of endotherms. Food is still the most common form of enrichment, although it does not have to be the only one used. When it is used, food can simply be presented differently to the animals (e.g., given whole instead of cut up, or live instead of prekilled) or presented in such a manner as to challenge the animals and cause them to spend extra time trying to consume it (e.g., strawberries hung by strings for tortoises, or live fish placed into an exhibit for aquatic reptile species). Other reptiles can be trained, like mammals and birds, using operant conditioning, usually consisting of target training and desensitization. For example, a large tortoise may reach a weight of 135 kg (300 lbs.) or more and present a challenge to keepers if they want to move it within an exhibit or off the exhibit. Rather than have people lift the tortoise, the keepers could employ target training to have it follow a brightly colored foam ball on the end of a stick. The keepers could then use the ball to move the animal wherever they wanted, within reason and of course with patience. Keepers should remember that enrichment may not be warranted or practical for all species, including "sit and wait" predators (e.g., many boas, pythons, and vipers) that are not "bored" by sitting in one position for a period of time, since that is simply what they do.

REPRODUCTION

Sexual dimorphism between the sexes varies greatly in reptiles and is not always obvious (e.g., in many lizard species including monitor lizards, *Varanus* sp.). This is especially true with young animals. Keepers often use secondary sex characteristics such as behavior, relative size, color, and shape and size of the tail (if males have hemipenes, the tail may be larger) to identify the different sexes. A keeper can also use a reptile-specific ball-tipped probe to search the animal's cloaca for hemipenes in squamates (the probe can be inserted further into the cloaca if hemipenes are present), although it must be done carefully to avoid injury to the delicate tissue. X-rays and ultrasounds can also be used in some species to identify specific sex organs. Some species have specific morphological differences between the sexes, such as the longer forelimb claws some male terrapins (*Trachemys* [*Chrysemys*] sp.) use to stroke the female's neck during courtship, or the throat dewlaps many male iguanids (such as *Anolis* sp.) use in territorial visual displays. Young tend to resemble adults in morphology and diet, although behavior may change with maturation, and keepers should remember that young venomous snakes still possess venom.

Reproductive behavior is usually stimulated via external environmental cues such as light, temperature, humidity, rainfall, and food availability. Behavioral cues driven by the environment (e.g., calls or vocalizations in some geckos, male posturing or displays in many lizard species) can cause further stimulation in some species. Even many tropical regions experience climatic variability, in which animals have evolved to time their reproductive behavior so that their young stand a better chance of survival, or so that the females have adequate resources to produce viable eggs and young. By controlling the environmental parameters in zoo and aquarium exhibits, keepers can control reproduction in many species. For example, many temperate species require a period of cooling before their gametes will mature and become viable within the male and female. These reptiles experience this cooling period in the wild, but do not experience it in captivity unless keepers manipulate the environment. Keepers can subject their animals to a controlled period of cooling to induce successful breeding, but must first account for their health. Reptiles should not be cooled when they are stressed or ill or have recently been fed. Their cooled environment should not promote dessication; they may be offered water during cooling. Often, light schedules should be changed in concert with ambient temperatures, as the reptiles might respond best to multiple cues.

The laying of external eggs (oviparity) is the most common strategy in reptiles, although live birthing (viviparity or vivipary) is also widespread, probably due to the presence of egg predators or harsh environmental conditions (i.e., cold temperatures) inhibiting egg incubation. Temperature sex determination (TSD) is widespread in crocodilians, turtles, and many lizards. With TSD, the temperature in which the egg develops determines the embryo's sex, and this may enable keepers to predetermine the sex of a clutch (or of a certain percentage of it). This could be of importance to certain conservation efforts, as females may be considered more "valuable" because a single male can fertilize multiple females. Parthenogenesis is also found in a few species of lizards (e.g., whiptails, some geckos, and even Komodo dragons [*Varanus komodoensis*]) as well as snakes (blindsnakes and boas); it is the asexual reproduction process in females that results in female offspring in the majority of examples. Parthenogenesis, which is considered odd and rare for vertebrates, has the practical aspect of not needing a male for reproduction. Egg incubation conditions vary in the natural environment, mirroring local conditions usually at the microclimatic level, as eggs are usually buried or hidden, with temperature and moisture around the eggs the most important factors. Keepers should attempt to mimic the natural microclimates for temperature and moisture, and should take care when moving eggs so as not to rotate them as many birds do, since that can kill the developing reptile embryos. Keepers can mark eggs lightly on their dorsal surfaces for reference, and common incubation mediums include moistened vermiculite, perlite, sand, and paper towels. An incubator may be needed to ensure consistent temperature; some models have alarms for temperature problems. Keepers should also provide proper egg deposition sites within zoo enclosures to prevent females from "holding" their eggs too long, which can lead to egg binding or dystocia ("difficult birth") and a need for surgical intervention.

TRANSPORTATION

Reptiles are moved between facilities or institutions as needed for various management strategies, including North America's Association of Zoos and Aquariums (AZA) or similar breeding programs. The goals for proper and efficient transport are to limit stress, reduce time in transit, limit exposure to extreme temperatures, and prevent dessication (especially in some delicate species). Under International Air Transport Association (IATA) guidelines, containers should be insulated to protect from temperature extremes and should be ventilated, clearly labeled, and shock-absorbent (i.e., cushioned). They also need to be secure, especially for dangerous species, with multiple layers of confinement. Moisture requirements for reptiles are modest for most species, unlike in the case of amphibians, although juveniles of some delicate reptile species (small lizards and snakes) will dessicate quickly without a dampened substrate. Larger crocodilians usually require custom-built crates that are just large enough to hold an animal but do not allow it to turn within the crate; this ensures that the animal's head stays at the labeled "head end" of the crate, and keepers are not surprised or endangered when it is allowed to exit. Venomous snakes should be explicitly labeled, with at least three confinement layers, and keepers should never open such containers without proper backup and tools. The final exterior crate for venomous snakes must be completely secure; wooden crates are often screwed shut. Snakes are often bagged, but because many venomous species can bite through the bag material, the bag should be placed in a rigid (usually plastic) secondary container that is properly ventilated. Keepers should refer to the proper IATA guidelines for particular species, as well as to chapter 23 in this volume.

VETERINARY CARE

Signs of illness in reptiles include reduced or absent feeding response, regurgitation, odd fecal consistency, odd skin color or texture, odd posture, increase in basking behavior (or the opposite), localized swelling or bloating, gaping of the mouth, trembling, and reduction in normal reflexive behavior. Common diagnostic techniques include fecal/blood/tissue exams, radiography, ultrasound, and endoscopy. Certain diagnostics, specifically fecal exams, should be performed regularly. Diseases are often dietary, environmental, genetic, or infectious, with the infectious agents including bacteria, viruses, fungi, and parasites (with both internal and external possibilities). Keepers should be well versed in disease possibilities and specific symptoms, so that veterinary personnel can be notified properly. The importance of quarantine cannot be overemphasized as a means to prevent transmissible diseases from moving from new arrivals to infect the rest of the zoo or aquarium's collection. Not all health problems are the result of infectious disease, however, as difficulties with ecdysis are commonly encountered as a result of environmental factors such as humidity. Severity will depend on the species involved and the extent of the problem. Shed skin that adheres to the ocular scales of snakes ("eyecaps") can ultimately lead to blindness if allowed to continue over time, so keepers should make every effort to ensure that snakes do not retain the eyecaps after shedding. Stuck eyecaps can be loosened with gentle streams of water, but keepers should take care not to simply pull them off, especially if they are dry and tightly adhered, as it can damage the underlying delicate tissue of the eye. Adhered shed on digits can constrict over time, especially in growing juveniles, and impede proper blood circulation, so keepers should periodically inspect the digits for these tissue ringlets.

Specific commonly encountered diseases include infectious stomatitis ("mouth rot"), respiratory infections, metabolic bone disease (MBD), vitamin A deficiency, cryptosporidium, salmonellosis, ectoparasites (mites and ticks), endoparasites (pentastome "worms" and nematode roundworms), and a host of other bacterial and viral diseases. Infectious stomatitis and respiratory infections are usually the result of stress from inappropriate temperatures, overcrowding, and poor nutrition. MBD is usually the result of improper calcium-to-phosphorus ratios and lack of exposure to UV-B light. Cryptosporidium is a highly contagious disease passed via fecal contamination that infects the stomach and intestines, preventing digestion; it is quite problematic due to a lack of effective treatment options. Salmonellosis is usually only a problem for keepers; it appears that *Salmonella* sp. bacteria occur normally in the gastrointestinal tracts of many reptile species as possible symbionts, but can cause illness in humans. Parasitic mites feed on the blood of their hosts and can function as disease vectors; every effort should be made to prevent their introduction to the collection, because they can be difficult to eradicate safely, and they are quite mobile for their size. Ticks are larger parasitic arachnids related to mites, and should be quite visible to the trained eye. They do not move around as often as mites, but like them they can function as disease vectors. Both mites and ticks can complete their entire life cycles within zoo enclosures, posing a continued and growing threat to reptiles without proper treatment. Endoparasites such as pentastomes and nematodes can have complex life cycles with multiple hosts, but many rely on fecal contamination to reach their final hosts. Pentastomes are actually not worms but arthropods, related to arachnids, that infect the lungs and respiratory tract of reptiles (usually snakes) and can actually infect humans through fecal contamination. They infect reptiles after the reptiles consume an intermediate host (e.g., a bird or rodent) that has contracted the parasite through contact with contaminated reptile feces. Again, viable treatment options are limited once the reptile has the parasite, but feeding parasite-free prey is the surest way to prevent infestation. Nematodes are roundworms that can be found infesting various tissues of a reptile including the intestines, circulatory system, respiratory tract, and coelomic cavity. They are usually contracted through fecal contamina-

tion, although some may be transmitted by other animals or by a vector. Numerous publications exist that keepers should refer to for additional information, and treatments should always be performed in concert with animal care supervisors and veterinary staff after positive confirmation of the organism.

CONSERVATION AND RESEARCH

Three of the four reptile orders (Crocodylia, Testudines, and Rhynchocephalia) are at great risk of continued loss of numbers, species, and populations due to human influence. The reasons include habitat loss (e.g., through climate change), overcollection (overharvesting), introduction of nonnative species (which either consume natives or outcompete them), introduction and spread of foreign diseases, accidental bycatch in commercial fisheries (e.g., sea turtles), and pollution. Island populations have been and continue to be at great risk due to their naturally low population numbers, their inability to relocate on their own, and their lack of evolved responses to foreign predators. Zoological institutions and their larger management organizations play a vital role in fostering education and research within the zoos and aquariums (ex situ) and in the natural environments of the reptiles themselves (in situ). Zoos and aquariums also preserve populations of species and their unique genetic biology, acting as reservoirs for future breeding and reintroduction plans. Keepers should share the information they derive from their work, including dietary information and reproductive successes, so that collectively their efforts help the zoological community meet its goal of conservation. Zoological facilities and management organizations have partnered with various local, national, and international organizations to protect terrestrial and semi-aquatic turtles in Asia, sea turtles worldwide, and crocodilians in South America, Asia, and Australia. Only their continued efforts will prevent further losses as humans continue to negatively affect the global and local environments needed by reptiles.

SUMMARY

Reptiles share an ectothermic lifestyle with amphibians but do not rely on environmental water to the same degree because their scaled exteriors minimize dessication. With more than 9,000 species in four distinct orders, their morphological diversity is significant and is reflected by the diversity of their habitats. Reptiles tend to be active thermoregulators and not simple thermoconformers; providing appropriate and various microclimates in captivity can be vital to their success, with heat and light being vital to most species. Keepers must be familiar not only with the natural history of the species in their care, but also with the unique tools used in their husbandry, including snake hooks and tongs. Keepers of venomous and large dangerous species need to account for the unique challenges posed by their charges, and must learn how to properly care for them while remaining safe themselves. Keepers must also account for reptiles' ectothermic metabolism in management of the planning and implementation of their diet, reproduction, transport, and veterinary care. Finally, keepers should always remember why they are caring for the various reptile species in captivity, as these animals will serve as ambassadors of education and conservation, and in some cases as the founders of future generations both in captivity and in the wild.

REFERENCES

Bauer, Aaron M. 1998. Lizards. In *Encyclopedia of Reptiles & Amphibians,* 2nd edition, edited by Harold G. Cogger and Richard G. Zweifel, 126–173. San Francisco: Fog City Press.

IUCN. 2010. IUCN Red List of Threatened Species, Version 2010.4. Accessed 26 March 2011 at http://www.iucnredlist.org.

The Reptile Database. 2011. Species Numbers as of January 2011. Accessed 26 March at http://www.reptile-database.org.

Zealandia. 2008. New Zealand's "living fossil" confirmed as nesting on the mainland for the first time in 200 years! Accessed 26 March 2011 at http://www.visitzealandia.com/site/zealandia _home/inside/news/media_releases_2009/tuatar_eggs.aspx.

34

Husbandry and Care of Amphibians

Andrew M. Lentini

INTRODUCTION

The amphibians are a highly diverse class comprising more than 6,600 species in 45 families. There are three orders in the class amphibia: Gymnophiona (Apoda), the legless amphibians; Caudata (Urodela), the newts and salamanders; and Anura (Salientia), the frogs and toads. Although there are exceptions, most amphibians—as suggested by the Greek root of the name *amphibios*, which means "double life"—have a two-phase life cycle with an aquatic larval stage and a terrestrial adult phase. All families share a number of characteristics. The main shared characteristic is the permeable skin that serves as a major surface for water absorption and gas exchange. Some species, such as the Plethodontid salamanders, rely entirely on their skin for the transfer of oxygen and carbon dioxide and have reduced or eliminated the lungs or gills.

Gymnophiona: These are legless amphibians known as caecilians. Caecilians are adapted primarily for burrowing, and some also have adaptations for an aquatic lifestyle. There are approximately 181 species within the order Gymnophiona. Caecilians have reduced tails, reduced eyes, segmented skin with small dermal scales, strongly built and heavy boned (ossified) skulls for burrowing, and acute olfactory systems. These animals have an elongated body form and lack even the internal elements associated with limbs (a pelvic or pectoral girdle); many look more like giant earthworms than typical amphibians. Caecilians are found in tropical South America, Africa, and Asia. They possess a penis-like intromittent organ (phallodeum) for internal fertilization. Some species of caecilians exhibit a level of parental care in which the young feed inside the female before birth and on the female after birth (e.g., the Kenyan caecilian, *Boulengerula* sp.). Most caecilians have lungs and also use their skin for gas exchange. Caecilians have retractable tentacles, which are sensory organs, on either side of the head between the eye and the nostril.

Caudata: These are tailed amphibians, salamanders, and newts. They resemble the early generalized amphibian

body form. There are approximately 584 species within the order Caudata. The body is elongated and divided into head (cephalic), trunk (thoracic), and tail (caudal) regions. Some salamanders can lose their tails (tail autonomy) as a defense mechanism to distract potential predators. Once a tail has been lost, it can be regenerated. Larvae in this order resemble adults but have external gills. Most caudates have nonfunctional auditory and vocal structures, but also have a well-developed olfactory communication that uses pheromones from glands in the cloaca and the skin. Most rely on internal fertilization, in which the male deposits a spermatophore that the female picks up with her cloaca.

Anura: These are "tailless" amphibians, the frogs and toads. Anura is the largest order of amphibians, with approximately 5,834 species. Frogs and toads have reduced bone in the cranial (head) region and distal limbs, and as adults most have no ribs present. They are specialized for jumping, and have a short bodies and elongated hind legs. The sacral (pelvic) vertebrae are fused into a unique pelvic skeletal structure, known as the urostyle, that facilitates this mode of locomotion. Most of these animals rely on external fertilization. This is the most successful order of amphibians, with a geographic distribution that includes all the continents except Antarctica.

Nearly one third of the world's amphibian species (more than 1,800 species) are threatened, making amphibians a priority conservation group. Many zoos are expanding or modifying their amphibian collections in response to International Union for Conservation of Nature (IUCN) recommendations to address the worldwide decline in amphibian populations. Further, many zoos actively participate in "rescue" projects, where threatened populations of amphibians are brought into captivity in the face of catastrophic losses in the range country. The hope is that rescued species can be returned to the wild once the threats to their survival have been mitigated. This requires maintaining amphibians in biosecure facilities to ensure that they can be used in future reintroductions.

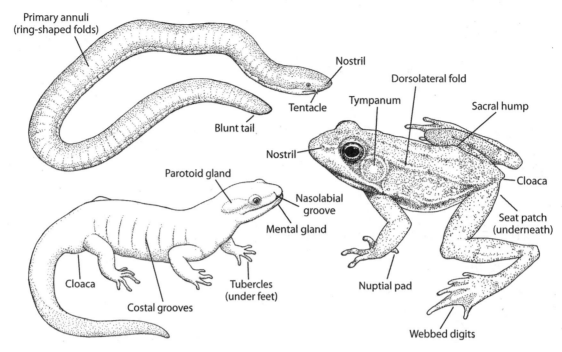

Figure 34.1. Basic anatomical features of amphibians. Clockwise from upper left: Caecilian (legless, order Gymnophiona), frog (tailless, order Anura), salamander (tailed, order Caudata). Illustrations by Kate Woodle, www.katewoodleillustration.com.

This chapter provides fundamental amphibian care guidelines and the basic keeper skills required to successfully manage amphibians in a captive setting. After studying this chapter, the reader will understand

- the anatomical and physiological terms used in describing the three different orders of amphibians
- that the unique physiology of amphibians affects the housing, nutrition, and reproduction of these animals in the zoo and aquarium environment
- best practices for daily care, handling, and transport of amphibians
- key habitat and environmental requirements, particularly water quality and temperature
- principal issues involved in medical management of amphibians.

ORIENTATION TO KEEPING AND HUSBANDRY OF AMPHIBIANS

In order to manage populations of amphibians for conservation outside of their natural surroundings (ex situ) and to monitor the well-being of individuals, keepers must be able to identify the individual animals in their care. Marking and identification techniques for individuals include tags, microchips, and elastomer (a polymer with the elastic properties of natural rubber) dyes that can be injected just below the translucent skin of many amphibians. The least invasive method of individual identification relies on differences in the appearance of individual animals. In many species, each individual has a unique pattern (dorsally, ventrally, or on the limbs) that can be photographed and used to identify it. In some species these patterns can change over time, so the

keeper should update identification photographs periodically. Individuals or groups can also be kept isolated from each other and identified by their enclosures.

Amphibians are ectothermic (often referred to as cold-blooded) and their body temperature is dependent on the environment. Ectothermic animals obtain the energy they need to raise their body temperature from the sun or from radiant heat, and they regulate their body temperature by exchanging heat with their surroundings. Ectothermic animals require 10 to 14% of the energy used by similar-sized endothermic animals (birds and mammals), a point to consider in keeper-controlled diets. This enables them to redirect the energy that endotherms would use to maintain a constant high body temperature to other activities instead, such as growth or reproduction. Ectothermic vertebrates, therefore, can successfully exploit environments of low biological productivity that could not support similarly sized endotherms. Most amphibians are thermoconformers; that is, they exist at the same temperature as their surroundings. Others engage in behavioral thermoregulation to maintain their body temperature within preferred limits. Ectothermy and thermoregulation affect the ecology, behavior, morphology, and physiology of amphibians.

Temperature tolerance: Species-specific temperature adaptations have been well documented. Some temperate species, such as the wood frog (*Rana sylvatica*), are adapted to low temperatures and are actually freeze-tolerant. Their embryonic temperature tolerance ranges from 2 °C to 20 °C (36 °F to 68 °F), and adults can freeze with their body temperature dropping to as low as −7 °C (20 °F). Some tropical amphibian species, such as the waxy monkey tree frogs (*Phyllomedusa sauvagei*), are active at temperatures ranging from 22 °C to 41 °C (72 °F to 106 °F). The salamanders are generally active at lower temperatures than frogs and toads, a fact which is reflected in their predominantly temperate distribution. The "web-toed" salamanders (genus *Bolitoglossa*) are an exception and are found in the New World tropics. Amphibians generally select a microhabitat to achieve a preferred (but variable) body temperature, since they have to balance increasing body temperature with increased evaporative water loss from their permeable skin.

Water Balance: Amphibians face several challenges in maintaining water balance, because of their highly permeable skin. In aquatic environments they have a greater amount of dissolved solutes in their tissues than the surrounding water (hyperosmotic), and therefore they face a loss of ions and an increase in water content (i.e., they absorb water). Amphibians therefore have adaptations that allow them to excrete excess water. In most amphibians the kidney produces large quantities of dilute urine to this end. In a terrestrial environment, amphibians constantly lose water and face the threat of lethal dehydration. Amphibians living in extremely dry (xeric) environments have water-conserving adaptations that allow them to avoid dehydration. Such adaptations include the ability to store and reabsorb water in their bladders, mechanisms for efficient uptake of water from the environment, specialized secretions that reduce skin permeability, the ability to produce insoluble uric acid for efficient nitrogenous waste excretion, and various behaviors to avoid desiccation. The waxy monkey tree frog (*Phyllomedusa sauvagei*) inhabits arid areas of South America and is constantly faced with the threat of desiccation. This species has evolved the ability to secrete a waxy (lipid) substance that it spreads over its entire body by grooming with its limbs. In this way, the frog covers its normally permeable skin with a waterproof coating that reduces evaporative water loss to 5 to 10% of that seen in most other anurans—a rate similar to rates seen in desert lizards.

Most amphibians do not drink. Most of their fresh water intake takes place through their skin. The "seat patch," an area of highly vascularized thin skin on the pelvic region of anurans, facilitates dermal water absorption from free water or damp substrates.

KEEPING AMPHIBIANS

Amphibians are a highly diverse group that includes both generalists and specialists that pose challenges to ex situ husbandry and conservation. To meet these challenges, keepers need to adopt approaches very different from those required for mammals and birds. The keeper is responsible for creating, monitoring, and maintaining an environment that provides thermal and water (hydrological) features which allow amphibians not only to maintain homeostasis (regulate their internal environment so as to maintain a stable state), but thrive in the captive environment.

DAILY MAINTENANCE

Daily exhibit and off-exhibit servicing includes checking lighting, temperature, water quality, and humidity levels, and correcting them if they are outside the desired limits for the species. All fecal material and uneaten food items should be removed daily to reduce the biological load on the system and prevent excessive bacterial contamination. Water features (pools, filters, recirculation pumps, etc.) should be checked daily and maintained per an established schedule. Pools should be free of excessive algae. Water bowls and unfiltered pools should be scrubbed daily and refilled with dechlorinated (aged, carbon filtered, or chemically dechlorinated) fresh water. Water should be at the appropriate temperature for the species. Its temperature can be adjusted by mixing hot and cold dechlorinated water, or by maintaining a reservoir of water at the correct temperature. Most amphibians will consume their own shed skins; however, remnants of shed skid will occasionally adhere to the glass of a tank, particularly for some arboreal species, and these should be removed daily. Dead and overgrown plant material should also be removed. The final step in daily maintenance should be a thorough cleaning of the viewing glass. Water droplets on glass are unsightly, and if not cleaned they can leave permanent mineral deposits that will require the replacement of the entire glass. Daily drying of the glass with a lint-free cloth or squeegee after servicing or misting will prevent unsightly mineral buildup. A mild white vinegar solution (one part household vinegar [3–5% acetic acid] and one part water) can be used. One should remove the animals from the enclosure, wipe down the spotted glass with the vinegar solution, and then rinse thoroughly and squeegee the area dry.

Since amphibians have delicate skin, which they use for gas exchange and water balance, care in the use of disinfectants is essential. Keepers must avoid transferring irritants such as oils, soaps, lotions, or acids from their hands to the animals. Disposable gloves (powder-free vinyl or latex) are recommended. Most amphibians secrete toxins through their skin or through the enlarged parotid glands behind their heads. Keepers should always wash and thoroughly rinse their hands before and after handling any amphibian species. Aquatic species can be handled using a soft fish net or simply by scooping them up in a plastic bag to avoid traumatic injury to their delicate skin. If bare hands are used, they should be wet to prevent loss of the protective mucus from the amphibian's skin.

THE ZOO AND AQUARIUM ENVIRONMENT

Amphibian keepers face the challenge of creating microhabitats that meet the needs of the animals and the visitors. Enclosure design must provide adequate humidity, access to clean water, refugia (shelters and plants), and lighting that meets behavioral and physiological needs. To meet the needs of the captive amphibian, a keeper must first research and understand the animals to be cared for. Researching the geography and microhabitats they exploit in the wild is a good place to start. Climate data from the regions where the animals are found is readily available from credible government or academic internet sites, from libraries, or from colleagues. Knowing the seasonality and temperature range and precipitation that a species experiences in the wild will guide the keeper in designing an ex situ environment suitable for that species. Knowing a species' natural history and habits (terrestrial/arboreal, aquatic/fossorial, nocturnal/diurnal, etc.) will also aid a keeper in designing a habitat structure that meets its behavioral and reproductive needs. In the following paragraphs the fundamental principles of amphibian habitat design, as well as more advanced and specialized multihabitat automated systems commonly used to house amphibians, will be discussed. Specialized multihabitat rack units with automated misting systems and false-bottomed flow-through enclosures are commonly used to house amphibians.

THE ENCLOSURE

Enclosure types available to house amphibians are varied. Due to the diversity of species involved, the most appropriate enclosure will depend on the particular species and life stage of the animals being housed. The most readily available and versatile enclosure is the glass aquarium. Aquariums are generally inexpensive and work very well as the foundation of a captive habitat for terrestrial and aquatic amphibians. Most amphibians are very good climbers and can escape through a regular aquarium lid or hood. Therefore, a secure escape-proof lid is essential. Such lids can be made of screen in a metal or plastic frame that fits snugly into the plastic trim of the aquarium. The lid should be secured by latches or hasps; one should not rely on weights to keep a lid in place.

The complexity of the habitat depends on the purpose of the enclosure. An exhibit enclosure with natural substrate and live plants will usually be more complex than a holding enclosure that may be set up in a more sterile manner for improved hygiene or treatment of sick animals. The fundamental aspects to include in any enclosure design are water, lighting, temperature control, refugia, and substrate, with appropriate variation or gradients in each (fig. 34.2). Glass aquariums may not be sufficient for many complex modern exhibits, which may instead require custom-designed and built enclosures.

A popular enclosure setup technique for exhibit and holding tanks is the false-bottom design. Using plastic (polystyrene) light diffusers (often called "egg crate diffusers") covered

with fiberglass (fine mesh) window screen, a permeable and snugly fitting elevated bottom is created to fit in an aquarium with a drilled drain hole in the tank's glass bottom. This allows for the liberal use of water in the system and also facilitates the use of advanced filtration. Tanks with false bottoms can also be decorated with naturalistic backdrops of expanding foam insulation sealed with silicone sealant (which is inert when cured) and natural long-lasting substrates such as coir (coconut fiber), peat moss, sphagnum, mulch, or gravel. Detailed instruction on enclosure setup (false bottoms, glass drilling, etc.) can be found on the internet (www.amphibiancare.com or the AZA and EAZA Amphibian Husbandry Resources at www.aza.org).

WATER SOURCES

Amphibians are so closely tied to water that water sources, availability, and quality are of utmost importance in the ex situ environment. Amphibian enclosures benefit from the addition of water features such as pools, waterfalls, streams, and drip walls that support live plants. These enhance the appearance of exhibits, help to increase humidity, and often encourage natural breeding behavior. Water features can be open systems that constantly introduce fresh water or closed systems in which water is retained. They may employ some type of recirculation with or without filtration. Closed systems with low biomass, where water circulation is more for aesthetic purposes than for the benefit of the animals, may not require filtration.

Substrates used in terrestrial amphibian enclosures must facilitate drainage that is adequate for the water features used. If a substrate lacks proper drainage it will become stagnant, resulting in accumulation of waste products (ammonia) and bacterial overgrowth which can lead to a toxic environment for amphibians. Stagnant substrates can also result in parasite infestation. Mixes of coco husk chips, haydite (heat expanded shale, slate, or clay) or other lightweight aggregate and activated charcoal provide a safe substrate that provides good drainage, breaks down slowly, provides for good aeration, and retains moisture. Other substrates include peat moss, sheet moss, pebbles, gravel, and potting soil. The use of perlite (a lightweight heat expanded volcanic glass) in a substrate mix is not recommended, since it is somewhat sharp and can also be high in fluoride. Amphibians occasionally ingest some substrate when eating, so keepers should observe the animals in their care to see whether the substrate poses a threat. If potting soil is used, it should not have perlite or added fertilizers. Long-lasting and well-drained substrates meet the needs of the animals and also support plant growth.

Water can be obtained from natural sources such as wells, springs, or rainfall; however, natural water sources can be contaminated both chemically and biologically. Deep wells are often oxygen-deprived, and cold water sources can be supersaturated with other gases. Well water is usually high in minerals and metals and can often have some unwanted salinity. Rainwater is naturally soft, but it can have variable pH. Rainwater is also prone to environmental contamination from pollutants and runoff (e.g., from a metal roof). Further, none of these natural water sources meet biosecurity requirements, because they can be contaminated with bacteria, viruses, fungal spores, or parasites.

Fundamental elements of the captive environment

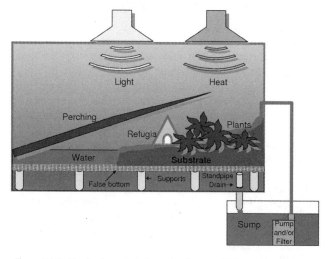

Figure 34.2. The fundamental elements of an amphibian enclosure: water, lighting, temperature control, refugia, and substrate, with appropriate variation or gradients for each. Courtesy of A. Lentini.

Bottled water has also been suggested for captive amphibians. The quality of bottled water is uncertain, since monitoring requirements are not as stringent as are those for municipal tap water. In many Western countries, bottled water is seldom of higher quality than tap water. Further, all plastic bottles leach synthetic chemicals and their effects on amphibians are uncertain.

Reverse osmosis (RO) filtration systems yield nearly pure water by forcing water through a semipermeable membrane. Clean water passes through the pores in the membrane and the impurities are left behind. However, RO removes not only harmful components but also the beneficial, naturally occurring minerals in water. Therefore, RO water must be reconstituted (important minerals must be added back) prior to use in amphibian husbandry. Products for reconstituting RO water are commercially available from aquarium supply companies. A recipe for reconstituting RO water is available in the Amphibian Husbandry Resource Guide from the AZA website (see AZA and EAZA amphibian husbandry resources). However, RO systems are relatively expensive and also waste a large quantity of water. Most systems generate approximately two to three liters of waste water for every liter of RO water they produce. In areas where the only water available is of very poor quality, RO may be the only option available.

Municipal water is generally highly regulated, monitored, and safe, but variable in composition. The disinfectants used—chlorine and/or chloramines (a combination of chlorine and ammonia), which are both toxic to amphibians—should be removed from municipal water before it is used in amphibian husbandry. Chlorine is volatile and will dissipate if water is vigorously aerated or "aged" (left to stand in an open container for 24 hours or more to allow the chlorine to dissipate). Aeration and aging also de-gas municipal water, which is often supersaturated with gases (N, CO_2, O_2), but they do not remove chloramines. Filtration with activated charcoal and chemical treatment with inorganic reducing agents, such as sodium thiosulfate, are both effective in removing chlorine and chloramines. However, when chloramines are removed with either of these methods, toxic ammonia is released. Thus, thiosulfate or carbon filtration alone is not adequate for eliminating toxicity from chloramines, and ammonia must also be removed using a chemical binding agent, a de-ionizing resin filter, or biological filtration. Treated municipal water appears to be the best option for use in ex situ amphibian care, given biosecurity concerns surrounding amphibians in ex situ conservation programs.

LIGHTING

The proper quality and quantity of light will help meet the physiological requirements of animals and will promote natural behavior, promote proper thermoregulation, facilitate plant growth, and improve the aesthetics of exhibits. Proper lighting also provides ultraviolet (UV) radiation that facilitates vitamin D synthesis. Many amphibians benefit from exposure to controlled safe levels of UV light—especially after metamorphosis, when their bone growth and development is increased. UV light can be provided by using commercially available fluorescent blacklight lamps. Blacklight lamps differ from standard fluorescent lamps only in the composition of the phosphor which radiates most of its energy in the near ultraviolet region, peaking at about 350 nanometers. These lamps are safe to use in conjunction with regular lighting over all amphibians. Black lights are effective if placed within 50 cm (20 in.) of the animals. At greater distances, using timed exposure of higher UV output from self-ballasted mercury vapor lamps, such as those manufactured by Westron of Canada Inc. (Dorval, QC, Canada H9P 1H2; Westron Lighting Corp., Oceanside, NY 11572–5829) or modified Eiko brand halogen bulbs (Eiko Ltd., Shawnee, KS 66227) is effective (see the AZA Amphibian Husbandry Resource Guide). The use of UV meters is recommended to verify UV levels and required exposure time. Keepers can also refer to Gehrmann et al. (2004) for details on the use of UV light in animal husbandry. See AZA and EAZA amphibian husbandry resources and follow manufacturers' guidelines for instructions and exposure precautions.

TOOLS AND EQUIPMENT

WATER QUALITY PARAMETERS

Creating and maintaining a healthy aquatic environment for amphibians requires the keeper to employ the same monitoring and maintenance procedures used in fish husbandry to ensure adequate filtration that provides a safe and stable environment. The equipment, filtration, and life support principles are the same for larval and adult aquatic amphibians as they are for freshwater fish.

A number of water quality parameters should be considered when developing a husbandry plan for amphibian species. Where data exists for water quality in situ, it should be used as a guideline for water quality parameters in captivity. These parameters can be measured regularly, using specialized meters or commercially available color-changing (colorimetric) aquarium test kits that are relatively simple, inexpensive, and effective.

HARDNESS

Water hardness measures the presence of dissolved minerals, specifically calcium (Ca) and magnesium (Mg), in water. Hardness can be measured with simple inexpensive aquarium testing kits. Adjustments can be made by adding Ca and Mg salts to increase hardness, and by adding RO, distilled or deionized water to lower hardness. Hardness can be quantified as soft (0 to less than 60 parts per million (ppm), medium hard (60 to 120 ppm), hard (120 to180 ppm), or very hard (more than 180 ppm).

PH

Alkalinity and acidity are measured as pH. Pure water has a neutral pH of 7. The acceptable pH range for most amphibians is between 6 and 8. Some amphibian species are found in ponds and in water with high organic content, which lowers pH, and these would likely benefit from slightly acidic water (pH 6–7). Adjusting pH can be accomplished by adding buffers or tannins such as almond leaves (*Terminalia catappa*), aquarium peat. or commercially available extract solutions.

AMMONIA

All living creatures give off nitrogenous waste (ammonia) produced by protein metabolism. Ammonia is toxic and must be removed from aquatic systems. Fortunately, a natural process in which biological conversion of ammonia into relatively harmless nitrogen compounds can be used to do this (fig. 34.3). Several species of bacteria convert ammonia (NH_3) to nitrite (NO_2), while others convert nitrite to nitrate (NO_3) which is relatively nontoxic. Since ammonia is highly toxic, concentrations of it should be kept below 0.01 milligrams per liter (mg/L). Ammonia above 0.01 mg/L requires immediate correction. Ammonia should be tested following any suspicious mortality. Nitrites are slightly less toxic; concentrations should be less than 0.1 mg/L. Nitrates are the least toxic, and concentrations below 10mg/L are safe. Nitrifying bacteria are present everywhere and once ammonia appears in a system, the desired bacteria will establish a colony in the filter media and substrate; this is called biofiltration.

It is important to be aware of the relationship between ammonia toxicity, pH and temperature (fig. 34.4). Ammonia (NH_3) is much more toxic than the ammonium ion (NH_4^+). The relationship between ammonia, ammonium, pH and temperature is described in figure 34.4. At a higher temperature and pH, more of the nitrogen is in the form of toxic ammonia than at lower temperature and pH. It is therefore preferable to maintain pH at or slightly below neutral (pH = 7) for most species.

A keeper's goal is to start with clean water and keep it clean by not overfeeding or overcrowding an enclosure. When animals are kept at high densities, the biomass in the enclosure is often too high for any filtration system to cope with the amount of ammonia present. Water quality should be monitored using appropriate test kits or meters, and regular water changes should be scheduled to maintain water quality by preventing nitrate and organic waste accumulation. Regular cleaning of filters (since they do not eliminate detritus but merely trap it) and use of live plants will also help maintain

Effect of Temperature and pH on Ammonia Toxicity

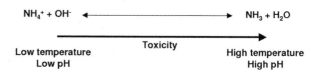

Figure 34.4. An illustration of the relationship between ammonia toxicity, pH, and temperature. Courtesy of A. Lentini.

good water quality; the plants use wastes as nutrients and contribute oxygen during photosynthesis.

FEEDING AND NUTRITION

NUTRITIONAL REQUIREMENTS

As ectotherms, amphibians obtain the energy required to raise body temperature from their environment and therefore require less food energy than similar-sized birds and mammals, which are endothermic. This allows amphibians to exploit environments of low biological productivity, and it means that they require less food in captivity. The daily energy requirements of an amphibian, calculated on the basis of a standardized metabolic rate at 25° C, indicate that a typical 40 g frog requires 1.25 kcal per day. Domestic crickets have 1.9 kcal of energy per gram; therefore, a 40 g frog needs approximately 4 g of crickets per week (only two large crickets per day). Overfeeding and obesity are common in captive animals, and weights of individuals should be monitored and recorded regularly (at least four times per year) so that intake can be adjusted as necessary. Metabolic rate is temperature-dependent. For every 10 °C drop in temperature, there is a corresponding 50% drop in the standard metabolic rate of an ectothermic animal (Wells 2007). Therefore, amphibians kept at lower temperatures need less food. Bear in mind that if an amphibian is kept too cold, it may not be able to digest properly, resulting in putrification of food in its gastrointestinal tract. The amount of food offered should be controlled, since many amphibians will overeat when food is abundant, resulting in gastric distention, which can be fatal.

Amphibians employ diverse feeding strategies, as both herbivores and carnivores, and their preferred diet varies according to life stage. Aquatic amphibians (adult and larval) will accept prepared diets (gels, pellets, tablets, etc.); however, most terrestrial species require live prey. Amphibians respond to moving prey because their neural processing and brains are "hardwired" to detect motion, which will elicit a feeding response. Some species (tree frogs, aquatic frogs, and salamanders) can be trained to tong feed; however, most will require live prey. Commercially available live prey includes mealworms (*Tenebrio molitor*), gray crickets (*Acheta domesticus*), wax worms (*Galleria mellonella*), cockroaches (usually *Blaberus craniiferus*), wingless fruit flies (*Drosophila melanogaster* and *D. hydeii*), and earthworms (*Lumbricus terrestris*). These and other invertebrate species can be cultured by the amphibian keeper. In order to maintain a reliable

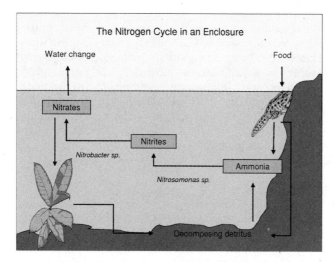

Figure 34.3. The nitrogen cycle in an amphibian enclosure is the bacterial processing of toxic metabolic waste (ammonia) produced by animals into relatively nontoxic compounds (nitrates). Courtesy of A. Lentini.

supply of appropriately sized (generally no longer than the width of the amphibian's head) live foods for the animals in their care, amphibian keepers should be familiar with invertebrate culturing techniques. Amphibians possess a digestive enzyme (chitinase) in the stomach and pancreas that assists in digestion of the chitin found in the exoskeletons of many insects (Oshima et al., 2002). However, care should be taken regarding impaction and perforation when feeding prey with heavy exoskeletons or highly chitinous parts. For example, the large ovipositor of adult female crickets or the spurs on the legs of certain species of cockroach can puncture the gastric wall, and mealworms can cause impaction in certain species that may not commonly feed on such chitinous prey. Also, wax worms are somewhat high in fat and can lead to obesity if fed in excess.

SUPPLEMENTATION

Many invertebrate prey species are generally deficient in vitamins and have a low calcium-to-phosphorous ratio. For this reason, vitamin and mineral supplementation is beneficial. Gut loading (feeding prey species vitamin- and mineral-rich foods) and dusting with powdered supplements are methods commonly used to improve the nutritional value of invertebrate prey. However, externally applied supplements wear off (i.e. crickets quickly groom themselves and shed any adhering supplement) and nutrients in the gut are transient. In order to prevent captive amphibians from feeding on nutrient-poor prey, any uneaten food items should be removed from the enclosure the next day. Extending the periodicity of feeding (once or twice a week) ensures that hungry amphibians will quickly consume the supplemented food. Ideally, the food should be consumed within one to two hours of being offered. For amphibians that can be trained to accept food from tongs, this method of feeding also ensures that food items are consumed with supplements. Tongs used for tong feeding should be blunt-ended to prevent injury to the animals, which often will lunge at the presented food item and grasp it and the tongs at the same time.

BEHAVIOR

Amphibians engage in a variety of interesting and often unique behaviors that enable them to go about the daily business of survival. They feed, find mates and reproduce, protect their young, defend territories, and avoid predators. Some of the behavior is geared towards maintaining water balance, thermoregulation, and communication.

Amphibians use behavioral adaptations to maintain their body temperature within acceptable limits (thermoregulate) and to maintain their water balance (hydroregulate). Amphibians are highly susceptible to desiccation because of their thin and highly permeable skin. These animals have been shown to alter their habitat selection and daily movements in response to changes in humidity. Hydroregulation involves moving from areas of different water availability, humidity, and air currents and altering body posture and activity levels, thereby selecting microhabitats in which conditions are more favorable. The keeper must therefore provide an ex situ environment that allows amphibians to display the full array of

behaviors associated with homeostasis. Such environments can employ sloped habitats that go from a cool aquatic to a warm terrestrial area within the same enclosure.

Amphibians use chemical, acoustic, tactile, and visual methods of communication. Many of the tailed amphibians possess mental glands (the glandular patches under the chins of some salamanders) and use pheromones and olfaction to communicate. Frogs and toads were the first vertebrates to possess vocal chords, and the anurans rely primarily on vocalization to attract mates, announce territory, or defend themselves when frightened. Visual signaling using the limbs or body (semaphoring behavior) is employed by several amphibians. These use their feet, legs, and bodies as "flags" to signal to other animals; this semaphoring behavior is most common in species found in environments where sound does not carry well. As defense, this behavior is often known as the "unken reflex" and is characterized by reverse flexing of the body to display warning colors of the ventral surface of the limbs and body.

Operant conditioning is not common; however, some tailed amphibians (salamanders and newts) and anurans have been target trained for food rewards. Varying the substrates, basking sites, prey species, feeding stations, and feeding schedules can serve to enrich the ex situ environment for amphibians.

Many amphibians will exhibit territoriality in defense of limited resources such as food, shelter, and reproductive (calling and egg laying) sites. Most aggression is associated with courtship, and vocalization by male anurans is a major component of this. Aggression associated with parental care (defense of eggs or larvae) is common in some frogs and salamanders. Male plethodontid salamanders may engage in biting, chasing, and aggressive displays such as open-mouth threats when placed in close quarters. Aggressive territorial behavior can be avoided by not overcrowding and limiting the number of males in a group, and by providing visual barriers and multiple retreat sites within an enclosure.

TRANSPORTATION

Cooperative management requires animals to move between institutions, and transportation methods for amphibians are designed to reduce stress, reduce time in transit, and reduce exposure to adverse conditions (temperature extremes and desiccation). When arranging a shipment of amphibians, the keeper should anticipate problems that might delay the shipment or expose the animals to unfavorable conditions. Potential problems include weather or traffic delays. When shipping by air, keepers should avoid the last flight of the day, multiple connecting flights, and weekend flights, since problems can result in lengthy delays.

Following IATA (International Air Transport Association) guidelines, clearly labeled, rigid-sided, insulated shipping boxes should be used. A polystyrene foam inner box placed in a cardboard outer box or insulated wooden crate can be used. Crumpled newspaper or polystyrene foam packing chips should be used around the containers housing the animals. This will prevent jarring and support the containers in the box during transit. Most species should not be shipped if they will be exposed to extremely hot or cold temperatures (above

30 °C or below 8 °C). To protect animals from extremes in temperature, heat or cold packs may be useful.

Containers used for housing amphibians during transit should provide adequate moisture and a secure shock-absorbing environment for the animals. Clear plastic containers (such as Glad-ware or Rubbermaid) work very well since they allow for easy inspection. New or thoroughly washed and rinsed delicatessen or margarine containers also work well (fig. 34.5). Ensure that lids are secure by tying them down with cable ties (encircling the entire container) or closing them with strong and durable adhesive tape (e.g., duct tape or electrical tape). All containers should be clearly labeled with species name and number of animals. Ventilation holes (1/8 to 1/4 in. in diameter) should be punched or drilled in the sides and tops of the plastic container. Be sure to drill from the inside out to prevent rough edges inside the container that could injure the animal.

Each animal container should be filled with an "airy" and damp substrate in which the animal can nestle comfortably. Acceptable packing substrates include slightly dampened sphagnum or sheet moss that has been teased or pulled apart to create air spaces, dampened paper towel, dampened sponge pieces, and chips (aquarium filter sponges work very well and are known to be safe and nontoxic). Whichever substrate is used, it should not be saturated with water, since the weight of saturated substrate can crush, trap, or drown small animals.

Aquatic species can be packed in the same manner as fish. Keepers can use standard plastic fish bags. To prevent small animals from getting trapped in the bottom corners, one can use square-bottomed bags with their corners taped down, or bags with rounded corners. Each bag should be filled one-third to one-half full with clean dechlorinated water. For trips longer than 18 hours, oxygen should be added to the bag before sealing. Products that absorb ammonia, like Poly-filter, (Poly-Bio-Marine, Reading, PA) can also be added to the water for long trips. A keeper should always double-bag an animal, since a leak can result in its death. Further, since airlines will refuse to carry any container that is wet, a shipping container should be lined with a plastic bag.

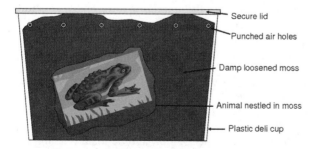

Sample Transport Container

Figure 34.5. An example of an amphibian transit container that provides adequate moisture and a secure shock-absorbing environment for the animal. Courtesy of A. Lentini.

REPRODUCTION

Some of the more fascinating aspects of amphibian biology are the variations on the typical life history model of egg-to-tadpole-to-metamorphosis. Some species have evolved life histories that tend towards parental care and direct development without an aquatic tadpole stage. These include terrestrial nesting species such as the eyelash leaf frog (*Ceratobatrachus guentheri*), in which the tadpole develops within the egg and the metamorphosed adult form emerges from it without ever having gone through a free-swimming larval stage. Other variations include species in which either the male or female carries the eggs for a period of time and either deposits them when they are ready to hatch (e.g., the midwife toad, *Alytes obstetricans*) or carries them until the larvae have fully metamorphosed and emerge as the adult form (Surinam toad, *Pipa pipa*).

Amphibians are sexually dimorphic to varying extents. Keepers can determine the sex of an individual by examining its size and color, and looking for the presence of secondary sex characteristics. A key secondary sexual characteristic in anurans and some newts and salamanders (caudates) is the nuptial pads. These are raised mucus glands located at the base of the thumb and forearm on males. Nuptial pads are used to grasp females during amplexus. Amplexus (from the Latin "embrace") is the copulatory embrace of amphibians in which a male grasps a female with his front legs as part of the mating process. In the anurans throat color can differ between the sexes, with males having a darker and looser skin around the throat. This reflects the fact that generally only males produce mating and release calls. The external ear drum (tympanum) is also larger in the males of some species of anurans. In the caudates, males will often have much larger crests and more glandular or enlarged cloacae. Female amphibians generally appear to be larger and rounder as a result of the eggs they carry, particularly when they are carrying eggs (gravid).

Prior to breeding, some form of environmental cycling may be essential for gamete (egg/ova) maturation. Seasonal changes for most amphibians involve either brumation or aestivation. Brumation, a state of torpor in response to cooler environmental conditions, is similar to hibernation, in which endothermic vertebrates lower their body temperature and pass the winter in a dormant or torpid state. Brumation in amphibians involves a period of reduced metabolism and food intake in response to temperatures that are too cold to allow for normal activity. Keepers can use incubators, refrigerators, or wine refrigerators to replicate the cold temperatures that temperate amphibians would experience in the wild. It is important to ensure that sufficient water or moisture is available to prevent desiccation in a cooler environment. Temperate species can be cooled to 6 °C to 8 °C (43 °F to 46 °F) for 4 to 12 weeks. Although tropical species do not normally brumate, a brief two- to three-week cooling period with temperatures ranging from 16 °C to 20 °C (61 °F to 68 °F) may be beneficial for successful breeding. Aestivation is a similar period of inactivity, seen in some species from xeric environments, that allows animals to conserve energy and moisture when their supply of food and water is low.

The danger of fatal desiccation can be mitigated by having a water dish always available while allowing the substrate to dry out. In order to reduce these animals' metabolic rates, temperatures should be kept 4 °C to 6 °C (39 °F to 43 °F) cooler than during the active season. In captivity, brumation and aestivation may be triggered by reduced lighting, temperature, humidity, and food availability. Prior to inducing brumation or aestivation, keepers should stop feeding two weeks in advance and perform a health and weight check to confirm that animals are of breeding age and in good condition. A breeding plan that specifies the desired temperature profile, photoperiod, humidity, and duration should also be prepared. Changes in temperature and humidity should be made slowly over a period of several days or weeks.

Following a period of inactivity, external environmental cues such as increased temperature and extended day length stimulate hormone secretion by the hypothalamus, pituitary, and gonads in amphibians. In females this triggers gonad maturation and spawning behavior. In males it results in activation of the testes and release of sperm into the urine, and stimulates both calling and amplexus. In captivity, the use of recorded calls, rain chambers, and increased humidity and temperature can provide the cues to initiate reproduction. However, cycling captive amphibians at lower temperatures does carry some risk. As mentioned earlier, desiccation can be fatal. Further, at low temperatures the immune system can be compromised, resulting in infections and parasite infestations. Artificial hormone techniques can also be used to stimulate reproductive behavior. Assisted reproduction using hormones is an effective method of inducing or coordinating reproduction without the need to cycle some animals. The use of hormones will not be effective if the animals are not in breeding condition. If animals are in poor condition—for example, if females do not have sufficient energy stores to produce eggs—hormone injections will not be effective.

Animals selected for breeding can be left in an existing enclosure if that environment is suitable for mating and egg laying. Often, however, it is more effective to move the selected animals to a special breeding enclosure. A commonly used breeding enclosure is the "rain chamber" (fig. 34.6). This setup involves the use of a misting or watering system that replicates a rainy environment. Misting and watering systems generally recycle water through filtration in order to ensure that the water is of a suitable quality and temperature. For terrestrial and semiaquatic species, the water level in a rain chamber can be 10 to 20 cm deep. Keepers should provide small "islands" of floating vegetation and perching that extends above the water line so that the animals can rest. Other cage furniture, such as plastic plants or PVC tubes that serve as egg deposition sites, should also be provided. To augment the environmental stimuli provided to anurans, keepers should also provide auditory cues. This can be accomplished by playing a recording of males calling in order to simulate the breeding choruses that are heard in the wild.

MATING

Amphibian species employ a variety of strategies to accomplish successful mating and fertilization. Most anurans

Breeding Chamber

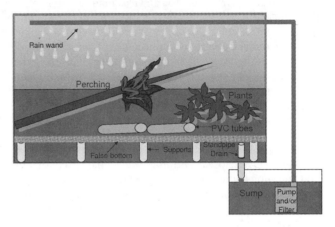

Figure 34.6. A commonly used breeding enclosure, the "rain chamber," uses a pump to recirculate water through an enclosure to simulate a rainy environment. Courtesy of A. Lentini.

typically use a method of external fertilization in which the male clasps the female with the male dorsal to the female, aligns his cloaca with hers, and simultaneously releases sperm as the female deposits her eggs. This clasping posture and behavior is known as amplexus. Referred to as "amplectant positions," a variety of amplexus postures are used by different species. The typical amplectant position is the axillary clasp (e.g., in the American toad, *Anaxyrus* [*Bufo*] *americanus*), in which the male places his forelimbs under the female's so that their cloacae are next to each other (in juxtaposition). Another amplectant position is the inguinal clasp (e.g., in the Surinam toad, *Pipa pipa*), in which the male places his forelimbs around the female's hind legs and groin. Yet another amplectant position is the cephalic clasp, in which the male grasps the female about the head. In the caudates, internal fertilization occurs when the male attaches a spermatophore (a packet of sperm) to substrate, twigs, branches, or stones in the water or on land, and the female then pulls them into her cloaca. Courtship leading up to fertilization can take the form of amplexus or "liebeespiel" (love play), in which both sexes engage in nudging and bumping, guided by olfactory and visual cues. Caecilian breeding has not been described in detail, but it likely includes olfactory and tactile communication. Caecilians are the only order of amphibians to rely entirely on internal fertilization. The male inserts the penis-like phallodeum into the cloaca of the female for several minutes or hours to inseminate the female.

Eggs are usually deposited singly or in masses, and can be laid in water, in ground nests, or attached to vegetation. Eggs laid in water are often free-floating or attached to submerged vegetation, and are protected by several mucus layers. The eggs vary in size and color with different species. They can be dark or pale. They can be laid in long stands (toads' eggs look like a bead necklace) or in a jellylike mass (some frogs lay eggs in a floating mass, and some salamander eggs form firm gelatinous balls). Structures such as plastic aquarium plants, live plants, branches, or stones provide egg deposition sites in the captive environment. When eggs are present, it is important not to disturb them. Some frog and caecilian species protect their eggs and will aggressively defend them. Filters and return water should be moved away from any eggs, and filter intakes should be covered with fine netting (panty hose work well) or sponges to avoid damaging the eggs. Covering the filters in this way also protects tadpoles from being trapped inside them. Once the tadpoles have hatched, keepers should delay feeding until the yolk has been consumed (usually takes 24 hours after hatching) which coincides with free swimming behavior; prior to this the tadpoles typically rest motionless, clinging to either the sides or the bottom of the tank. Larval rearing involves close attention to water quality parameters and correct feeding. Tadpoles engage in a variety of feeding strategies. They can be opportunistic omnivores, herbivores, carnivores (in the case of most of the caudates), cannibals, and even filter feeders.

For herbivorous tadpoles, tank algae that consist mostly of blue-green algae (*Leptolyngbya cf. boryana*) and diatoms (green-brown algae) have been found to provide good nutrition. Algae should be cultured for several weeks prior to a planned breeding to provide an adequate source of food for herbivorous tadpoles. Tank algae can also be harvested and stored frozen until needed. The use of algae from ponds or other non-biosecure sources should be avoided, in order to avoid possible pathogen introduction. Alternatively, fresh high-quality commercial fish foods such as Sera-micron,® Tetramin,® Sera-san Color and Growth Enhancing flakes,® Tetra 4 in 1 FD menu,® aquaria herbivore diet, and spirulina flakes can also be used. For grazing tadpoles, mixed flakes can be moistened with a few drops of water to form a firm paste that can be dropped into the water for the tadpoles to feed on. Alternatively, by adding more water to a flake mixture, keepers can create a slurry that can be "painted" onto clean glass microscope slides and allowed to dry. The coated slides can then be placed in the water for tadpoles to graze on. Tablet fish foods can also be used. Frozen or heated (either microwaved or blanched, to rupture the plant cells) spinach and lettuce can be fed to supplement the diet. For filter feeders, flake foods can be finely ground and sprinkled on the water for the tadpoles.

Some anuran tadpoles and all larval salamanders and caecilians are carnivorous and will benefit from live foods such as microworms, daphnia, brine shrimp (hatchlings and adults), whiteworms, and tubifex worms. "Survival of the fittest" applies to most carnivorous tadpoles and larval amphibians since they become cannibalistic, consuming any smaller siblings they can overpower. Dependent on program goals, in cases were only a few individual animals are to be raised from a large clutch, cannibalism can be allowed to proceed and the more robust animals will survive. Where the goal is to raise as many of the young as possible, larval amphibians should be reared individually to prevent cannibalism.

As tadpoles of terrestrial species approach metamorphosis (when one or both of the front limbs have appeared) they should be moved to a rearing tank to prevent drowning. Rearing tanks are set up so that they are sloped with shallow water at one end and easy access to a nonslip land area at the other. There should be plenty of cover at the water-land interface to allow metamorphs to comfortably rest at this point until they are ready to move to land. Metamorphs are quite susceptible to desiccation because of their small size. Even when water is available, they are driven by instinct to leave the natal pond. Therefore, tanks should be securely covered and ventilation should be controlled to maintain high humidity (80–90%).

VETERINARY CARE

New animal quarantine for amphibians is vital in light of new emerging diseases (e.g., chytridiomycosis caused by the fungus *Batrachochytrium dendrobatidis*) in this group of animals, which require high biosecurity standards. By adhering to sound quarantine procedures, keepers can decrease the risk of introducing disease to resident animals and also protect newly arrived animals from acquiring disease from resident animals. This is of particular importance when zoo animals may be used for reintroduction programs. The quarantine period also allows for adaptive management of individuals and diet adjustments, thus allowing the keepers and the animals to get used to each other. Keepers should maintain a quarantine area, with tools dedicated to each enclosure within that area. Proper disinfection procedures for tools and cages, hand washing, and the use of disposable gloves between quarantine enclosures are all essential and should be mandatory. Disinfection using a 1% sodium hypochlorite solution with a contact time of 10 minutes is sufficient to disinfect hard surfaces. Household chlorine bleach ranges from 3% to 5% sodium hypochlorite, and commercially available bulk solutions can be of much higher concentration (up to 12%). It is therefore vital to confirm the concentration on the container label or with the manufacturer or supplier before preparing a dilute working solution. After disinfection, rinse everything thoroughly with hot tap water and leave to air-dry. Porous materials such as bark or driftwood are best replaced if excessively soiled. Iodine toxicity has been seen in amphibians from iodine reversibly bound to plastic holding containers. Because of this potential binding with plastics, it is recommended that iodine not be used to disinfect any material that will come in contact with the amphibians. Other disinfectants are not recommended for use on amphibian tanks or furniture. Care should be taken not to expose animals to chlorine vapors present when concentrated solutions are used for cleaning and disinfection.

The quarantine period should last at least 30 days. However, based on the incubation period of certain diseases, a longer period (60 days) is preferable. Ideally, an "all in/all out" system, where no new animals are introduced until the quarantine period for the first group has ended, is best. During quarantine, temperatures in the range of 18 to 25 °C are safe for most species. Temperatures should be at the cooler end of this range for temperate or montane species, and at the higher end for tropical species. Appetite is one of best indicators of health, and keepers should weigh all animals upon arrival and periodically throughout the quarantine period in order to ascertain whether they are feeding and maintaining their weight. Keepers should also be observing the animals daily and taking note of any obvious changes in their physical appearance or in activity level or posture that may suggest illness. Fecal samples should be submitted to a veterinary laboratory for parasite checks. Animals should not be allowed to clear quarantine until three negative fecals or two negative post-treatment fecals have been produced.

Keepers should always look for the common signs of illness in amphibians under their care. These include loss of appetite, changes in skin texture or color, changes in behavior or posture, reduced righting reflexes, bloating, and change in eye clarity. Diseases of special concern include viral, bacterial, and fungal infections. New emerging diseases (ranaviruses, chytridiomycosis) have been associated with population declines and have been diagnosed in zoo collections. Some disease-causing parasites, such as nematodes, have life cycles that are completed in the soil or water. Regular substrate changes and air drying of the exhibit and props after regular disinfection is often all that is required to prevent reinfestation or transmission of parasites from one animal to another. A schedule for periodic substrate changes and enclosure disinfection should be established and followed. When animals are being treated, it is beneficial to have an identical clean and simple tank already set up, so that the animals can be transferred to a fresh tank once a week. This prevents the buildup of parasite loads and reduces the chance of reinfestation following treatment. Wild-caught amphibians often have heavy loads of parasites (e.g., nematodes such as *Rhabdias ranae*). Antiparasitic drug therapy (e.g., Fenbendazole and Ivermectin) accompanied by tank changes can take several weeks to successfully eliminate the parasite infestation. Due to the potential toxicity of some antiparasitic drugs, dosing should be prescribed by a veterinarian familiar with their use in amphibians.

CONSERVATION AND RESEARCH

Amphibian declines and extinctions are well documented. The International Union for Conservation of Nature (IUCN) Global Amphibian Assessment of 2004 has found that 42 % of all amphibian species are declining in population, and 32 % of the world's amphibian species are threatened with extinction. These declines are a result of emerging infectious diseases such as chytridiomycosis, toxins, climate change, land use change, unsustainable collection and trade in amphibians, and the introduction of exotic species. Professionally managed zoological institutions play a role in conservation through public education, scientific research, professional training and support of in situ conservation projects (Hutchins et al. 1995). Furthermore, zoos are in a position to preserve threatened species in captivity over long periods, thus providing a reservoir of genetic and demographic material that can be used periodically to reinforce, revitalize, or reestablish populations in the wild.

Keepers can participate in amphibian research utilizing the species they care for. Research involving ex situ populations can be used to guide and develop recovery actions and adaptive management strategies. Ex situ populations can also aid in the development of technologies and research techniques that will benefit wild populations. Moreover, databases from ex situ populations can complement research conducted on wild populations. We know little of the detailed biology of some of the many amphibians in zoos. Keepers can contribute to our knowledge by conducting research into the life history,

behavior, and biology of the animals they care for. By sharing their experiences and findings with zoo visitors, colleagues, and the scientific community, keepers are able to fulfill their roles as educators as well.

Many zoos actively participate in "rescue" projects in which threatened populations of amphibians are brought into captivity in the face of a catastrophic loss in the range country. Habitat loss and fragmentation remains the primary cause of amphibian declines. The hope is that these rescued species can be returned to the wild once the threats to their survival have been mitigated. This requires maintaining them in biosecure facilities to ensure that they can be used in future reintroductions. Ex situ conservation and management of threatened amphibian species in zoos is one component of a worldwide coordinated amphibian conservation response to the global decline of amphibians. The recommendation for an ex situ population of certain threatened amphibian species can come from a number of recognized sources such as the Global Amphibian Assessment (www.globalamphibians.org), the IUCN Red List (the IUCN Technical Guidelines for the Management of Ex Situ Populations recommend ex situ populations for all critically endangered species), local, regional or national government requests. A successful amphibian conservation program requires institutional support and an advocate to speak on behalf of the species. Keepers can fill this role and champion the species they work with.

In situ conservation activities, with the participation of in-country local partners, are also essential for an effective amphibian conservation program. With local involvement, zoos can contribute to in situ capacity building, so that rescued population can be maintained in country. Keepers can participate in in situ conservation by traveling to rescue centers and sharing their expertise with local conservationists. Keepers have also been instrumental in fund-raising to direct needed resources to in-country projects. This type of support is extremely beneficial to the long-term success of amphibian conservation.

SUMMARY

Amphibians are a group of animals whose life history is intricately associated with water. These fascinating animals come in a variety of forms, from completely aquatic to terrestrial. Some amphibians are among the most vibrantly colored of animals and make beautiful exhibit species. The largest can weigh over 40 kg, and the smallest just a few grams. Amphibians are of high conservation value, since almost one-third of species are threatened with extinction. The specialized needs of this diverse group of increasingly popular animals makes keeping them a challenge, requiring knowledgeable keepers who can create the often complex captive environments that will ensure their well-being.

REFERENCES

Clayton, Leigh Ann, and Stacey R. Gore. 2007. Amphibian emergency medicine. *Veterinary Clinics Exotic Animal Practice* 10:587–620.
Gehrmann, William H., J. D. Horner, G. W. Ferguson, T. C. Chen, and M. F. Holick. 2004. A comparison of responses by three

broadband radiometers to different ultraviolet. *Zoo Biology* 23:355–63.

Hutchins M. and William G. Conway. 1995. Beyond Noah's Ark: The evolving role of modern zoological parks and aquariums in field conservation. *International Zoo Yearbook* 34(0): 117–30.

Oshima, H. et al. 2002. Isolation and sequence of a novel amphibian pancreatic chitinase. *Comparative Biochemistry and Physiology B-Biochemistry and Molecular Biology* 132(2):381–88.

Wells, Kentwood David. 2007. *The Ecology and Behavior of Amphibians*. Chicago: University of Chicago Press.

Wiese, R. J., et al. 1994. Is genetic and demographic management conservation? *Zoo Biology* 13:297–99.

35

Aquarium Science: Husbandry and Care of Fishes and Aquatic Invertebrates

Bruce Koike

INTRODUCTION

Many parallels exist between terrestrial animal care and the husbandry of fishes and aquatic invertebrates. Minimizing stressors by implementing husbandry best practices should be the common goal of all animal keepers.

Because aquatic animals and the environment in which they live are inherently different from their terrestrial counterparts, understanding these differences and consistently applying industry standards of care are keys to being a successful keeper. Though the term "keepers" will be used throughout this chapter, individuals who care for fishes and aquatic invertebrates at a zoo or aquarium are often called aquarists.

This chapter serves as a starting point to understand

- important processes that occur in the aquarium, and ways of managing water quality
- life support system principles and operations
- core aquatic animal husbandry responsibilities
- husbandry requirements of selected species.

Refer to appendix 3 for a list of institutions that offer formalized training in aquatic animal husbandry or aquaculture.

THE AQUARIUM HABITAT

Most public aquariums and many zoos have both freshwater and marine exhibits. The terms "marine," "saltwater," "seawater," and "artificial seawater" refer to water that contains a variety of dissolved salts. Only the term "seawater" is appropriately applied to water taken from the natural environment. Artificial seawater is made either with a commercial mix or by mixing individual salts according to a customized recipe. Both methods should include trace elements such as strontium chloride, potassium iodide, and other compounds which are present in concentrations less than 100 parts per million (ppm). Trace elements are essential for maturation, growth, and physiologic functions.

Water conservation is a growing issue at zoos and aquariums from a cost control and resource sustainability perspective. Aquariums located along the coast might have the option to pump seawater in directly from the ocean or bay, then filter and temporarily store it before use. Though natural seawater varies in salinity, proportionally the composition of dissolved salts and trace elements is constant. Seawater can be stored in enclosures made of fully cured concrete, fiberglass, acrylic, high-density polypropylene, or glass. Material selection should take into account the volume of water to be held, its application, and financial resources.

WATER QUALITY

Considerable staff time and resources are devoted to monitoring water quality with the goal of maintaining all water quality parameters within the animal's range of tolerance. Animals kept in poor water quality conditions experience physiological stress. Chronic or acute exposure to free un-ionized ammonia can result in organ dysfunction, tissue damage, reduced growth rate, poor food conversion, a depressed immune system, abnormal blood chemistry, and death (Randall and Tsui 2002, 17–25).

Sensitivity to particular environmental conditions varies by species. Amazonian fishes are typically intolerant to saltwater, while different freshwater fishes in estuarine regions with brackish water regularly enter and stay in the marine environment. Likewise, many marine fishes have the ability to thrive in freshwater habitats (Hoese and Moore 1998, 274–75). The animal's overall health is another factor that influences its ability to overcome poor environmental conditions and stress. Compromised individuals are less able to adjust to environmental fluctuations than are healthy individuals.

The aquarium environment and its inhabitants are interwoven. The normal biological functions of aquatic animals and plants alter the aquatic environment by adding metabolic waste, reducing trace element concentrations, and altering the dissolved oxygen (DO) concentration and pH. Various water parameters need to be tested and monitored on a regular basis to ensure the water's suitability.

DISSOLVED GASES

The ability of water to dissolve oxygen is limited by the water's salinity, temperature, and altitude. Aquariums should be regularly checked with a DO meter. The concentration of DO can be reported as a percentage of saturation (%) or as milligrams per liter (mg/L). Because the DO probe is placed directly in the aquarium, one should prevent the introduction of parasites and pathogens (cross-contamination) by rinsing the probe with tap water before testing a different aquarium system. Fish behavior can also serve as an indicator of oxygen shortage. Swimming at the water's surface, rapid respiration, or gaping mouths can indicate a low oxygen condition. If these behaviors are observed, the DO needs to be checked and the water aerated if conditions require. Normal respiration by aquatic animals can reduce the DO to stressful and even lethal levels. Other processes that consume oxygen are plant respiration, and decomposition of organic material by bacteria.

A properly designed system maintains DO saturation by allowing water to contact the atmosphere. Within the aquarium, this air-water interface occurs with active aeration, with photosynthesis, at the water surface, and where wet/dry filters are used. In each of these instances oxygen reenters the water while carbon dioxide (CO_2) is released into the atmosphere.

The majority of fish species extract oxygen by pumping water over their gills. As blood and its associated red blood cells circulate through the gills, hemoglobin binds oxygen molecules for eventual delivery to the cells. In contrast, some fishes such as the walking catfish have a spongy (lunglike) swim bladder that can extract oxygen from the atmosphere. Many species of shark must constantly swim forward with their mouth open so that oxygenated water can contact their gill tissue. This technique of respiration is referred to as ram ventilation. Aquatic plants, including algae, can also lower the dissolved oxygen concentration during respiration at night. Heavily stocked ponds during hot, windless nights can experience low oxygen levels because both the animals and algae are consuming oxygen. What constitutes a dangerously low oxygen level depends upon species.

The uptake of oxygen by fishes is coupled with the release of carbon dioxide (CO_2) into the water. Carbon dioxide is a highly soluble gas that readily combines with water to form carbonic acid, which lowers pH. To lessen the pH decline, water can be directed through a stripping or bio-tower, which releases carbon dioxide into the air. Carbon dioxide will be discussed further in managing pH.

$$CO_2 + H_2O \rightarrow H_2CO_3$$
(carbon dioxide)   (water)   (carbonic acid)

Nitrogen (N_2) is another gas that enters the aquarium through the water-air interface and through the reduction of nitrate by anaerobic bacteria (Spotte 1991, 66). This gas is principally responsible for gas bubble disease, the result of an environmental condition in which water is supersaturated with dissolved gases. Unfortunately the dissolved nitrogen is readily absorbed into the fish's circulatory system, where it coalesces into bubbles and lodges in various tissues such as the skin, eyes, gills, brain, and kidneys. This condition is parallel to the "bends" in scuba diving and can be fatal. Supersaturation occurs when gases and water mix while under pressure. Air can gain entry into the circulating water if a pipe fitting is loose or the water level at the pump intake is too low. In this low water situation, a vortex is created as the pump actively draws in water and air. To prevent supersaturation, one should monitor water levels and check the snugness of threaded fittings regularly. When water is allowed to contact the atmosphere, excess gases diffuse back into the atmosphere, making the dissolved gas levels acceptable.

pH

The term pH expresses the relative concentration of hydrogen ions (H^+) and hydroxyl ions (OH^-) in a fluid at a specific moment. If a water sample contains more H^+ ions than OH^- ions, the sample is acidic and has a pH less than 7.0. A neutral pH environment (pH = 7.0) has an equal concentration of H^+ and OH^- ions, while a pH greater than 7.0 reflects a basic environment (more OH^- ions than H^+ ions).

Like other water parameters, pH must be maintained within a range that is suitable for each species. Amazonian fishes prefer acidic water (pH < 7.0) while marine fishes thrive in pH 8.23, the pH of natural seawater. The calcium carbonate ($CaCO_3$) buffering system is principally responsible for this pH level in marine systems (Moe Jr. 1992, 33; Spotte 1991, 9). Aquarium systems tend to decline in pH as buffers such as calcium carbonate or sodium bicarbonate are consumed while the influx of carbon dioxide molecules continues. Bicarbonates serve as buffers to resist abrupt pH changes by either giving up H^+ ions or accepting H^+ ions. This dynamic process is depicted in the following equation:

$$H_2O + CO_2 \rightarrow H_2CO_3 \rightleftarrows HCO_3^- + H^+ \rightleftarrows CO_3^{-2} + H^+$$
(carbonic acid) (bicarbonate)   (carbonate)

The buffering capacity of the water can be increased with the addition of crushed coral and sodium bicarbonate. Because pH is dynamic, daily monitoring is recommended. Handheld or bench-top meters are available. Before using a meter, one should rinse the pH electrode and shake off any excess water. One then places the probe in the water sample and gently moves it back and forth until a stabilized value is displayed. Keepers should adhere to the water quality parameters recommended by their supervisor. The pH is adjusted by changing the water or by adding calcium carbonate, sodium bicarbonate, crushed shell or coral, or commercial pH adjusting products.

SALINITY

Salinity and temperature are water quality parameters that are not altered by animals but do affect their well-being. In the case of marine fish, a considerable amount of energy is used to excrete excess salts. Elevated salinity induces stress by increasing the influx of salt into the body. The excess salt must then be excreted, resulting in an energy cost to the animal. Salinity is monitored with a refractometer or a hydrometer, and is recorded in parts per thousand (ppt) or as specific gravity. The specific gravity of pure freshwater is 1.0, while saltwater has a value greater than 1.0 because of its salt content. Specific gravity is the ratio between the density of salt water and that of pure fresh water, and therefore it has no unit of measure. A hydrometer uses the water's increasing density, as the salt content increases, to suspend a float to be read against a scale. In contrast, refractometers use the property of water to bend light. This refraction changes according to the amount of dissolved salts in the water (figure 35.1). One can ensure accuracy by calibrating the refractometer with distilled water before using it. A set screw on the refractometer adjusts the reading if the calibration sample does not read zero. To use it, one should hold the refractometer level and lift the daylight plate off the prism, then place three to four drops of water on the prism and close the plate. The water's salinity can be determined by looking into the viewing port and identifying where the blue and white interface lies on the scale. Some models contain scales for both salinity and specific gravity.

Cross-contamination of tanks can be prevented by rinsing the measuring instruments between samples with tap water. After use, the instruments should be air-dried or wiped with a soft cloth before storage. All aquariums undergo evaporation, and the rate of evaporation depends upon water and air temperature, humidity, and air and water circulation. Water levels need to be checked daily and water added when necessary. As water evaporates from a marine aquarium, the remaining salts are concentrated, thus increasing the water's salinity. When the salinity rises 3 to 4 ppt above the acceptable range, dechlorinated tap water or reverse osmosis water should be added gradually to allow adequate mixing between the water types. Adding water at the location furthest away from the animals is advised.

WATER TEMPERATURE

Water temperature is another parameter not altered by the animal population. Handheld thermometers, infrared thermometers, or temperature probes are used to determine water temperature. Any unexpected temperature values should be investigated and reported. Water temperature is influenced by room temperature, lighting, and equipment such as pumps, heaters, chillers, or heat exchangers. If heaters or chillers are

Figure 35.1. An aquarium science student at Oregon Coast Community College monitors an exhibit's salinity, and will add dechlorinated freshwater if the salinity is elevated. Note the two-liter microalgae cultures on the windowsill. Photo courtesy of B. Koike.

adjusted, water temperatures should be monitored throughout the day. A wide range of heaters and chillers are available. Titanium heaters can heat large volumes of water, while glass heaters are more appropriately used for small aquariums. Glass heaters must be handled carefully to avoid breaking the thin housing. One should physically separate animals from heaters to prevent potential burns. Identifying a heater's control dial, heating element, indicator light, and water line mark is essential to using it appropriately. The temperature is set with the control dial and the thermostat will automatically turn the heater on and off. The heating element occupies the submerged lower portion of the heater and the indicator light is illuminated when the heater is on. Heaters will overheat and burn out if not in water. If a hot, dry heater is found, it should be unplugged and allowed to cool to room temperature before being reinstalled. Glass heaters that are hot will shatter if placed in water. One should always check the water line mark on the heater to determine whether the unit can be fully or only partially submerged. On very large systems a heat exchanger can be used to maintain temperatures. Heat exchangers work on the principle of transferring heat from one medium to another, and can be used to heat or chill water.

In contrast to heaters, chillers extract heat from water (still considered heat transfer), thus lowering its temperature. The refrigerant used in chillers alternates between liquid and gaseous states as it is pumped through the chiller. As the liquid refrigerant quickly evaporates, the coils that are in direct contact with the water cool and absorbs heat from the water. The chiller's efficiency can be maintained by keeping the condenser free of dust and debris. The condenser allows the absorbed heat to be released into the surrounding atmosphere. As with heaters, the set temperature is determined and monitored by the keeper. Chiller repairs and recharging with environmentally safe refrigerants must be carried out by a licensed or certified technician.

ALKALINITY AND WATER HARDNESS

Alkalinity is the net charge of all negatively charged ions (anions) that could bind with H^+ ions. These anions aid in creating a stable water pH by binding with hydrogen ions. Once these negatively charged ions begin to decrease in number, the pH level will begin to lower due to the relative increase of free H^+ ions in the water. Alkalinity can be expressed by one of three ways; milligrams per liter of calcium carbonate (mg/L), milliequivalents per liter (meq/L), or degrees of carbonate hardness (dKH). The total alkalinity of natural seawater is between 2.1 and 2.5 meq/L (Moe 1989, 34; Spotte 1991, 10), though a higher range (2.5 to 3.6 meq/L) is recommended for marine systems (Brightwell 2007, 46). These units can be converted to others by using conversion factors (Brightwell 2007, 45). The term carbonate hardness (KH) refers to the concentration of several different anions in solution, including carbonate and bicarbonate molecules. As these specific molecules quickly combine with available hydrogen ions, a shift in pH is delayed or prevented, thus resulting in a pH-stable environment. As the alkalinity drops in concentration, the aquatic environment becomes more susceptible to pH changes.

In contrast, water hardness reflects the total amount of dissolved minerals in the water, and is particularly important in freshwater aquaria as species prefer hard or soft water. Providing a water hardness that mirrors the species' natural habitat is the best objective. Total hardness (denoted as GH) is the amount of divalent (+2) ions, principally calcium and magnesium, that are dissolved in both freshwater and marine systems (Brightwell, 2007, 50). Distilled water contains no minerals (i.e., is very soft) while very hard water contains more than 530 ppm of calcium carbonate. Total hardness is less of a concern in marine aquariums because of the high concentration of dissolved minerals with $^{+/-}1$ charge. Having an adequate level of calcium carbonate in reef aquariums is still an important consideration.

Because of the lack of buffering capacity, soft water is more susceptible to abrupt pH fluctuations. Ground and surface water vary in their mineral content, so it may be necessary to obtain a water analysis from the local public works department. Hardness can also be determined with aquarium test kits or more sophisticated instrumentation. The removal of minerals from hard water can be accomplished with water softeners, with ion exchange resins, or with reverse osmosis systems (selective membranes that remove specific dissolved compounds).

NITROGEN-CONTAINING COMPOUNDS

Understanding the relationship between ammonium, ammonia, nitrite, and nitrate and their effect on organisms is crucial. These nitrogen-containing compounds are produced by animal or bacterial metabolism and are measured in ppm or mg/liter. The sum of free un-ionized ammonia and ionized ammonium equals total ammonia-nitrogen (TAN). This relationship between these distinct compounds is illustrated by the following equation.

$$NH_3 \quad + \quad NH_4^+ \quad = \quad \text{total ammonia-nitrogen}$$
(un-ionized   (ionized   (ammonia)
ammonia)   ammonia)

The breakdown of proteins either within the body or in the environment results in toxic, free un-ionized ammonia (NH_3). In either of these aquatic environments, a portion of free un-ionized ammonia combines with water to form a relatively nontoxic, ionized ammonium (NH_4^+) ion. This equilibrium relationship is expressed by the following equation.

$$NH_3 \quad + \quad H_2O \rightleftarrows NH_4^+ \quad + \quad OH^-$$
(free un-ionized (water) (ionized   (hydroxyl ion)
ammonia)   ammonia)

The double arrows indicate that free un-ionized ammonia and ionized ammonium can change back and forth depending upon pH, salinity, and temperature of the water. Since free un-ionized ammonia only passively diffuses from the fish's body, any accumulation of ammonia in the environment slows the release of toxic ammonia. Total ammonia-nitrogen concentration can be determined with ammonia salicylate test kits or ammonia probes. Accuracy is necessary to achieve valid results, so one should use a syringe or graduated cylinder to measure a water sample and follow the test manufacturer's

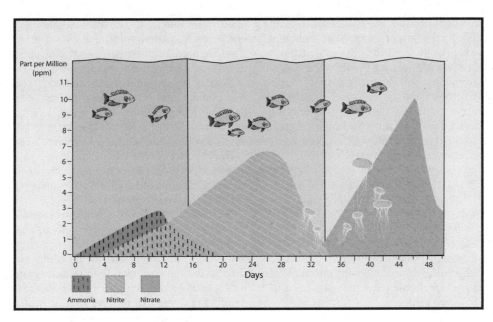

Figure 35.2. Represented is the nitrogen cycle for a hypothetical new aquarium system. In practice, the number of days required for the nitrogen containing compounds to cycle through depends upon many variables. Likewise, the actual concentration of ammonia-nitrite-nitrate will vary with each aquarium system. On approximately day 47, a water change is initiated which drastically reduces the concentration of nitrate. Illustration by Amy Burgess.

directions. The presence of green coloration in the sample indicates the presence of TAN. TAN at 0.24 ppm should initiate a water change or the addition of a commercial ammonia binder. The actual concentration of the toxic, free un-ionized ammonia is determined from a data table that contains the percentage of free un-ionized ammonia present at any combination of pH, salinity and temperature (Spotte 1991, 118–26).

Precise results are achieved by using instrumentation such as a spectrophotometer. This instrument can also test for many other compounds, including nitrite and nitrate. Each test requires test-specific reagents and an exact wavelength of light produced by the spectrophotometer. Based on the percentage of light absorption, the spectrophotometer determines the concentration of the compound being tested. If less precision is acceptable, color cubes or test strips are an option. This method requires the keeper to compare the resulting test color against the manufacturer's color standard produced by known concentrations.

The breakdown of nitrogen-containing compounds from the aquatic habitat is accomplished by nitrifying bacteria that are beneficial and which oxidize and derive energy from these molecules. Although this sequential breakdown results in less toxic forms, daily monitoring is recommended.

$$\text{free un-ionized ammonia} \rightarrow \text{nitrite} \rightarrow \text{nitrate}$$
$$(NH_3) \qquad\qquad (NO_2) \quad (NO_3)$$

This entire process is called the nitrogen cycle, ammonia cycle, or nitrification. Nitrifying bacteria colonize all surfaces that are in contact with water, including the substrate, inside of pipes, sides of the enclosure, and filter media. Different families of bacteria, such as *Nitrosomonas* and *Nitrococcus* species, convert ammonia to nitrite while *Nitrobacter, Nitro-*

spina, and other *Nitrococcus* species oxidize nitrite to nitrate. These aerobic bacteria must first attach to a substrate before converting these nitrogenous compounds. Having sufficient biological filtration and nitrifying bacteria is essential to achieving good water quality. Establishing a bacteria population large enough to reduce the concentration of nitrogenous compounds requires time (figure 35.2).

Seeding the aquarium by adding substrate or filter media from an established aquarium reduces the time required to cultivate the bacteria. The donor system should be pathogen-free so that accidental contamination of the new system does not occur. An alternate method is to inoculate the system with a commercially grown bacteria concentrate. Overstocking an aquarium before this beneficial bacterial population is established results in the accumulation of ammonia: a condition known as "new tank syndrome." To avoid this situation, initially stock a few hardy fish that have successfully passed through quarantine. Such species include mollies, guppies, and goldfish for freshwater aquariums, or damselfish and gobies for marine systems.

The best way to reduce the concentration of undesirable compounds is to exchange a portion of the old water with new water of better quality. Depending on the system's design, the water can be changed by pumping, draining, or siphoning. Clear flexible tubing is the material of choice when siphoning, because it allows the water to be seen filling the tube. Though there are several ways to start a siphon, the selected method depends upon several variables including personal preference. First, one should select a length of tubing long enough to have a downward loop, which aids in the process. One fills the tube by placing its entire length in the aquarium or by filling it manually. Once the tube is full, a thumb should be placed over each of its ends, one in the water and the other below the level of the tube's opposite end. Simultaneously opening both tube ends will begin the flow of water. When siphoning, animals should be kept away from the tube so that they're not accidently sucked into it. Directing the siphon at organic debris removes material that will otherwise decompose into ammonia. The amount of water to be removed depends upon the target objective. For example, if the concentration of ammonia is to be reduced from 0.24 ppm (starting concentration) to 0.10 ppm (desired concentration), the percentage of water volume to be removed is determined thus:

$$\left(\frac{(\text{starting concn}) - (\text{desired concn})}{\text{starting concn}} \right) \times 100$$

$$= \% \text{ of water to be removed}$$

The calculation indicates that 58% of the water volume needs to be changed to achieve the desired 10 ppm concentration. After the water change, one should retest the water to ensure that the objective has been achieved.

Depending on its design, an aquarium can also be "flushed" with water to reduce the concentration of unwanted compounds. In this case, no water is actually removed before the higher-quality water is added. Excess water leaves the aquarium through an overflow drain. Whenever filling a tank, the keepers must remember to return before the tank overflows. Overflowing a tank is an easy mistake to make.

Commercial products are also available to adjust parameters such as ammonia, nitrate, phosphate, calcium, pH, and trace elements. When using commercial additives, one should observe animals for signs of stress, such as increased respiration and color changes. The desired result can be confirmed by testing the water once the procedure is done. To avoid the accumulation of compounds, one guideline is to change 10% of the water volume in established aquariums every two weeks, though recommendations vary (Spotte 1991, 94; Fenner 2001, 127; Tullock 2001, 141); the best choice may depend on things like the amount of biomass in the system, the amount of food introduced at feedings, and the availability of water.

In freshwater aquariums, commercially available natural minerals such as zeolite can be used to remove ammonia. This type of removal is a form of chemical filtration known as ion exchange. The zeolite surrenders a sodium ion in exchange for an ammonia ion from the water. It should be contained in a mesh bag, so that it can be removed easily when it becomes saturated with ammonia ions. A saltwater soaking will then remove the ammonia ions so that the zeolite can be reused.

Plants (in freshwater aquariums) and algae (in marine aquariums) use nitrate and phosphate as an energy source, effectively reducing the concentration in the water. In marine systems, an algal "turf scrubber" (algae growth tray) can be integrated into the life support system's design. Scrubbers grow algae on removable grids or trays and require intense lighting and water movement for the algae to flourish. Maintaining this component will require periodic harvesting of a portion of the algae to remove nitrate and phosphate reserves. If plants and algae are allowed to decay, the nitrates and phosphates will be reintroduced into the environment.

The incorporation of refugia in marine systems promotes good water quality by providing a safe habitat for invertebrates, such as amphipods, that consume organic material. Another method to reduce nitrate concentration couples a deep sand bed with a plenum, an open space between the bottom of the aquarium and the substrate. This design, which was first advocated by Dr. Jean Jaubert, enables anaerobic bacteria to convert nitrate to nitrogen gas (denitrification) in marine systems. Brightwell (2007) recommends substrate depth based on particle sizes. Unmaintained beds can result in the production of hydrogen sulfide and a lower level of dissolved oxygen. Other denitrification processes use the bacterial metabolism of sulfur in a separate reaction chamber (Aikens 2009, pers. comm.).

Since the chemical composition (hardness, pH, minerals, etc.) varies depending upon region, one should obtain a water analysis from the local municipality. Potable water will require treatment due to the addition of disinfecting chemicals. One strategy is to use reverse osmosis water (RO water) as the major source of water. RO systems consist of membrane filters that remove specific minerals and metals based on their size. Chloramines, metals, and phosphates are examples of compounds removed by reverse osmosis. Some considerations in selecting a RO system are the initial and maintenance costs, the water output capacity, and the amount of waste water that is generated. In spite of these concerns, RO water is recommended for use in reef aquariums. Additional details about RO water and municipal water can be found in chapter 34.

LIFE SUPPORT SYSTEMS

The life support system (LSS) encompasses a wide array of equipment that maintains a healthy environment for the aquatic animals (figure 35.3). It is appropriately named, because without it the animals' well-being would be jeopardized. The organization of Aquatic Animal Life Support Operators (www.aalso.org) is an effective resource for developing a detailed understanding of the LSS components and their application. Identifying these components and understanding how they function are crucial to aquatic animal husbandry. This knowledge enables the keeper to recognize any problems and adjust or repair the components.

Another essential skill is tracing the flow of water throughout the system. Prominent LSS components include pipes, valves, fittings, pumps (circulation pumps and powerheads), filters, temperature control units (chillers or heaters), aerators, foam fractionators (protein skimmers), disinfection units (ozone generators and ultraviolet radiation units), reservoirs (sumps), and the main enclosure. Circulation pumps move water through each LSS component and are either air-cooled or water-cooled (submersible). Examples of submersible pumps include sump pumps, bilge pumps, and powerheads. Each type of pump has a slightly different application, and the pump outputs (measured in gallons or liters per unit of time) vary according to model. Sump pumps are often used when the water contains solid debris, and bilge pumps are typically operated by DC voltage. Powerheads or similar pumps are frequently used in smaller reef aquariums to create the water currents preferred by corals. These submersible pumps will overheat if not operated in water. Air-cooled pumps are the

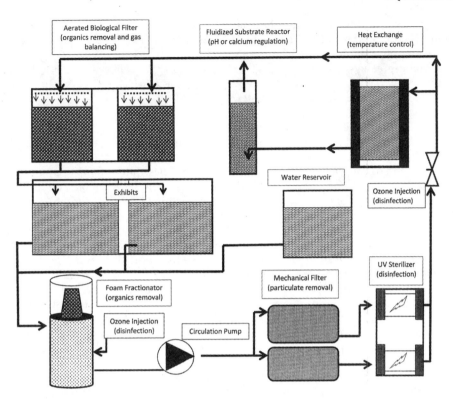

Figure 35.3. A theoretical marine life support system designed to support fishes and invertebrates. Courtesy of Mark Smith, Cosestudi.

heart of many public aquarium life support systems and are used in both large and small aquariums.

The manufacturers of pumps publish data about their output potential (performance curve), which is useful for determining the appropriate size pump to install. The performance curve of an individual pump depends on the height to which water must be pumped, the pipe diameter, and any friction loss inside the pipes. Pump output decreases as pumping height increases, pipe diameter decreases, and the number of pipe fittings and pumping distance increase.

Air bubbles can also move water, though the height to which water can be lifted is limited. This air lift principle can be observed in small aquariums, in quarantine tanks, and with sponge filters. Water movement is created by first pumping air into the bottom of the air lift tube. As these bubbles rise, water is pushed ahead of the bubbles and returned to the aquarium. A circular water pattern results as water leaving the air lift tube is replaced with water that is pulled through the substrate and back into the air lift tube.

PLUMBING SYSTEMS

If a pump is the heart of a water circulation system, then the plumbing is its veins, arteries, and capillaries. Pipes enable water to move throughout the facility and through an aquarium system's LSS component. Plumbing skills are essential, since facilities frequently design and build their own systems. Commonly used materials include PVC (polyvinyl chloride) pipe, flex tubing, polypropylene, acrylic and other nonmetal materials. PVC comes in different diameters and is either schedule 40 (white pipe) or 80 (gray pipe). Schedule 80 has a

thicker wall and is better suited for outdoor and high-volume or high-pressure applications. PVC pipe can be cut with any of a number of hand or power tools. Adjacent sections of PVC are joined together by fittings with specific functions such as redirecting water or reducing the pipe's diameter. Identifying these fittings by name, diameter, schedule type, and whether the fitting is a slip or threaded increases efficiency in the workplace. A slip fitting is permanently attached to a pipe after both elements have been first primed and then glued together. The primer and glue are applied to the pipe with an applicator, using a circular motion. The primer etches the PVC, increasing its surface area for the glue. Once the PVC is primed, the glue must be applied when the primer is still moist. Since the pigmented primer leaves a permanent stain, the work area should be covered for protection. One should follow the manufacturer's recommended sequence in gluing PVC pipes for use in a pressurized water system. Such glued bonds are permanent.

In contrast to slip fittings, threaded fittings can be reused because they are screwed together rather than glued. To prevent leaking, Teflon® tape is wrapped around the "male" fitting multiple times. These fittings are hand-tightened, then leak-tested. If a leak develops, the fitting should be further tightened. If the leak persists, the fitting will need to be disassembled and the fitting retaped.

Each aquarium system consists of multiple valves which determine the volume of water that passes through them. Aquatic systems consisting of pipe 5.08 cm (2 in.) in diameter or smaller commonly use ball valves. These valves possess an internal "ball" that can either stop the flow of water in the pipe or allow water to pass through the open port of the ball.

To allow maximum flow through a valve of this type, the long axis of the valve handle is positioned parallel with length of the pipe. Conversely, flow is stopped by positioning the handle perpendicular (i.e., at a 90-degree angle) to the pipe length.

Depending upon the system's design, changing the flow to one part of the system could affect the amount of water delivered to other parts of the system. Because of this, one should observe the water level and flow rate for several minutes after adjusting a valve to ensure that the aquarium will not overflow or be deprived of water. Butterfly valves are used with pipes that are larger than 7.62 cm (3 in.) in diameter and can be motorized. These automated valves can be actuated from a remote location via computer control.

FILTRATION

Another feature of a well-designed life support system is adequate filtration. Filtration is the process of removing, altering, or destroying undesired compounds or organisms. Filtration types include mechanical, biological, chemical, and disinfection (Moe Jr. 2009, 60–89). Such systems should be "overdesigned" with extra capacity to accommodate increased biomass in the future.

The goal of mechanical filtration is to remove solids by trapping them in filters. Mechanical filters include rotary drum filters, diatomaceous earth (DE) filters, filter bags, pleated cartridge filters, rapid rate sand filters, sponge filters, and pre-filter pads. Rapid rate sand filters, sponge filters, and undergravel filters can also function as biological filters. Both surface area and pore size are important when selecting filters. Surface area indicates the amount of filter medium that will contact water, while pore size dictates the maximum size of particle that the filter will trap. Largest surface area and smallest pore size represent the greatest filtering capacity. Flow rate and water pressure are two other factors that also influence filter efficiency. When these rates are too high, solid particles can be forced through the filter matrix, reducing its efficiency. All filters require regular monitoring, and cleaning or replacement when needed.

Bag filters and pleated filter cartridges are frequently used mechanical filters. A bag filter traps solids on the inside of the bag as water passes from the inside to the outside of the filter. It can be suspended inside a sump, and can easily be removed for servicing and replacement. In contrast, a cartridge filter traps particles on the outside of the filter surface. This type of filter is placed inside a housing which receives pressurized water that passes to the inside of the cartridge in order to exit the housing. Both bag and cartridge filters eventually plug with debris and require cleaning. Dirty filters should be hosed, then disinfected overnight with a 2% solution of household bleach (Spotte, 1991, 473). Before reuse, the filters are hosed off, and then the bleach is neutralized with a solution of sodium thiosulfate ($Na_2S_2O_3$). The active ingredient in household bleach, sodium hypochlorite, decomposes over time and when organic matter is broken down or oxidized. The bleach solution should be changed or recharged weekly to ensure that it remains effective. One should handle bleach carefully, following the manufacturer's recommendations and using appropriate personal protection equipment, and recognize that nylon nets and other materials weaken with repeated or extended exposure to bleach.

> **Good Practice Tip:** Check resources such as Material Safety Data Sheets (MSDS), and reduce exposure to chemicals in the workplace. Clean up any spills or debris left behind after a project or activity.

Rapid sand filters, the most commonly used mechanical filters are often used in multiple banks to support mega-size exhibits. These filters contain layers of different-sized particles including sand, pebbles, or plastic beads. Water flows from the top of the filter through the filter bed and out via "laterals," or perforated pipes, under the sand. These filters are cleaned when the flow of water is reversed, in a process called backwashing. The change in flow direction dislodges trapped particles, and the dirty water leaves the filter to be either discarded or recycled. Backwashing also helps prevent channelization, a condition that reduces the filter's effectiveness. Since water seeks the path of least resistance, channels within the filter bed can form, allowing the passage of particulates as well as undesired dissolved compounds.

There are two basic methods to determine when to backwash. The first method is based on operational history or experience. The second method requires monitoring the "Δ" (delta) or change in water pressure (in kilograms or pounds per square inch) or flow rate between that of a newly backwashed filter and the present pressure or flow rate. As a filter gets dirty, the flow rate decreases and the internal pressure increases. The pressure differential that triggers a backwash procedure will be determined by each aquarium facility, following the filter manufacturer's recommendations.

A good practice is to visually inspect and feel the filter bed annually. Hardened or foul-smelling media should be replaced, and the filter scheduled for more frequent backwashing. Filter sand can be lost if the slotted pipes (laterals) located inside the filter bottom have broken, or if the backwashing flow rate is too high. If the laterals are broken they must be repaired, as filtration capacity is reduced by sand loss.

Undergravel filters function both as mechanical and biological filters. Essential parts include slotted plastic plates, air lift tubes, and substrate. Undergravel filters can also be operated on a "reverse flow" basis by pumping water down the air riser tube using a small powerhead. In this operation, water is forced up through the substrate. Substrate choices include crushed coral, sand, and pebbles. Aerating these systems is recommended because the bacteria population can reduce the dissolved oxygen concentration to less than optimal levels. The substrate should also be gravel-vacuumed to remove organic debris. This vacuum device consists of a tube connected to a larger-diameter cylinder. After the siphon is started, the plastic cylinder is pushed into the substrate and slowly lifted up. During this process, the denser substrate falls out of the cylinder while the lighter suspended organic matter is removed by the siphon.

Sponge filters also serve as biological and mechanical filters and typically use the air lift system to move water through the filter matrix. Maintenance involves first removing the filter and then rinsing and squeezing the foam to remove particulate matter. The cleaned filter is then reinstalled into the same aquarium. A filter should be disinfected before being placed in a different aquarium.

A wide variety of bio-media can be used to increase filter capacity. Bio-balls, bio-rings, and bio-beads are made of plastic or ceramic and feature highly porous surfaces. In a similar manner, marine reef rock, also called live rock, promotes good water quality because of the biological activity that occurs in and on it. "Live rock" is the term used to describe rock, traditionally the skeletal remains of coral, that is colonized by a wide variety of invertebrates. A closed recirculation aquarium system with too little biological filter capacity will never achieve good water quality. In contrast, open or flow-through aquarium systems benefit from constant flushing of undesired compounds. These open systems also provide animals with constant access to trace elements. One drawback of an open system is that the source of water may fluctuate in salinity, temperature, and silt content. Man-made disturbances such as oil and chemical spills could also affect the water quality. A semi-closed system combines features of open and closed systems. This hybrid system has a life support system and is flushed by a steady flow of water.

Another form of filtration, disinfection, uses ultraviolet radiation (UV) and ozone (O_3) as disinfecting agents to reduce or eliminate suspended organisms such as algae, bacteria, viruses, and protozoa. Ozone has the added benefit of oxidizing undesired organic compounds. The effectiveness of these disinfectants depends upon system design, exposure time, concentration, and intensity. UV lamps produce only high-energy, short-wavelength light. UV lamps, which look like fluorescent tubes, are housed in waterproof sleeves around which water is circulated. Water clarity is important when using UV disinfection, because UV rays do not penetrate solid particles. These lamps also degrade over time and must be changed according to manufacturer recommendation (typically every six months to a year). One should never look directly at UV light, and one should always disconnect the electrical system before working on it.

Ozone gas is a highly reactive and unstable molecule that is generated by passing air or oxygen across high-voltage electrodes. The electrical energy breaks apart oxygen molecules and fuses three oxygen atoms together, creating an ozone molecule. The action of ozone has been likened to punching holes in the cell, resulting in cytoplasm leakage and death. Exposure to ozone can irritate and damage sensitive tissue such as eyes, nostrils, throat, and lungs. Keepers should recognize the smell of ozone and avoid entering areas where it is present. In large-scale application, ozone generator rooms are equipped with detection systems linked to visual and audio alarms.

Ozone generators are rated in terms of their production (in grams or pounds per day). In a large system, approximately 10% of the aquarium's volume is treated with ozone. This "side stream" is physically separated from the main exhibit, in order to safeguard the animals from ozone exposure. Like UV, ozone requires adequate contact time and concentration to be effective. It also clarifies water by breaking down the organic compounds that cause yellowing. To increase efficiency, ozone can be introduced into the contact chamber via a venturi. This specially designed tube increases water flow rate, and the corresponding reduction in pressure creates a suction by which gases can be introduced into the water. A properly designed life support system also allows the ozone-treated water to de-gas by coming into contact with the atmosphere before the water reaches the animals. The ozone-air mixture is passed across a heated element in the ozone destruction unit before being vented outside.

Ozone interacts with the bromine and chlorine ions present in salt water to produce hypobromous acid and hypochlorite respectively. In significant concentration these ozone produced oxidants (OPOs) are harmful to fish and invertebrates, but are effective disinfectants in marine mammal exhibits. How much ozone is too much? One recommendation is to monitor the oxidation-reduction potential (ORP) with an ORP meter and set a target range of 250 to 400 millivolts (mV; Moe 1992, 58). The presence of organic material is reflected in lower ORP levels. As these compounds break down, the water becomes more reactive and ORP increases. Water can be overtreated with ozone; it will appear sparkly, and may even have a bluish tint. Fishes in this environment swim erratically and have increased respiration rates. To correct this situation, one should shut off the ozone generator and increase aeration until the ORP levels return to an acceptable level. If an ozone system is being used for the first time, the generator should be operated only when multiple personnel can monitor the ORP and observe the animals. The amount of ozone used each day can be adjusted according to water clarity, animal behavior, and ORP reading.

While commercial products eliminate specific compounds (ammonia, nitrites, nitrates, chloramines, and phosphate) to create better-quality water through chemical filtration, foam fractionators (protein skimmers) and activated charcoal remove a broad array of organic and inorganic compounds. The use of foam fractionators is limited to marine systems or freshwater systems that have a high organic load. A feature of foam fractionators is the mixing of air or ozone with water to form bubbles. As these bubbles become coated with hydrophobic compounds and coalesce with adjacent bubbles, foam is formed. This foam rises up the contact chamber, eventually spilling over into a collection cup. Greater efficiency is achieved with smaller bubbles and increased contact time. While large foam fractionators have automated wash-down and on-demand drains, keepers are responsible for emptying the collection cups of small units, monitoring water flow, and insuring sufficient bubble generation. Ozone can increase the efficiency of foam fractionators, yet too much ozone destabilizes bubble formation. If the aquarium is being treated with medication, the foam fractionators should be turned off so that a therapeutic level can be maintained.

Activated carbon (AC) also serves the dual function of mechanical and chemical filtration. Several features contribute to the efficiency of activated carbon, including high porosity, varying pore size (microscopic) and shape, a high number of surface charges, and variable particle shape. AC will have different properties, depending upon the manufacturing process and base material (bone, coal, nutshell, or wood). It removes chloramines, heavy metals, organic molecules that cause odor and discoloration, and medications such as antibiotics from domestic water. Because of these abilities, an AC filter must be bypassed when therapeutic compounds are used. Failure to do so will likely prevent a therapeutic concentration from being achieved. Eventually, AC needs replacement because all the pores and surface charges are occupied. The best way to determine whether it needs replacing is to test the water for the compound of critical concern. Samples of water before and after the AC filter should be tested for the compound of concern, and the AC should be replaced if the values are the same or above an acceptable level in the samples. To avoid introducing fine black powder into the aquarium, keepers need to thoroughly rinse new activated charcoal before use.

AERATION

Aeration is an essential feature of the aquarium system, as most fishes and invertebrates require dissolved oxygen. All aquariums should be of a type that can be aerated during a power outage. Effective emergency aeration systems have been developed since experiences with hurricanes Katrina and Gustav (Arnold 2010; Whittaker 2010). Air blowers are used to produce high-volume, low-pressure air while air compressors produce high-pressure air. This pressurized air eventually enters the aquarium tank through air lines and diffusers ("air stones") which create fine bubbles. Smaller bubbles increase the dissolved oxygen levels more efficiently than larger bubbles, because of the greater surface-area-to-volume ratio. The flow of air is controlled by a valve and should be set so that a steady stream of bubbles gently breaks the water surface. Air stones need replacing when the flow of air becomes restricted. A small diffuser can be checked by gently blowing on the stem of the air stone. If restriction to the flow is felt, then the air stone should be replaced. Larger packed-glass diffusers can be soaked in a dilute solution of muriatic acid to oxidize organic compounds and dissolve mineral deposits, then rinsed thoroughly before reuse. The water surface of the tank is another important source of oxygen. Deep, narrow tanks have a lower surface-to-volume ratio than do shallow tanks that have large footprints. Keepers should be aware that electrical components need to be kept well away from water. An air pump should be placed above the water line, as this prevents water from possibly back-siphoning into the pump if air flow is lost.

LIGHTING

Effective lighting not only enhances viewing but is essential for freshwater planted aquariums and for marine organisms such as algae, corals, and giant clams. The keeper must provide a lighting scheme that fulfills the organisms' needs. Important lighting criteria are spectrum, intensity, and photoperiod (Fenner 2001, 52; Delbeek and Sprung 1994, 179).

Lights can be categorized in different ways including bulb type, spectrum, intensity (lumens per watt), and color temperature (Fosså and Nilsen 1996, 195–211). Popular types of lighting include high-output (HO) or very-high-output (VHO) fluorescent lights, metal halide lamps, and light-emitting diodes (LED). An appropriate spectrum for freshwater plants of 400 to 450 nm and 600 to 650 nm is recommended (Moe 1990, 155). In contrast, lamps that approximate the color temperature of daylight (10,000–30,000 Kelvin) and have peak outputs of 420 and 450 nm (actinic and daylight spectrum) are recommended for reef aquariums (Delbeek and Sprung 1994, 187). Giant clams and corals require lighting because they possess symbiotic algae, called zooxanthellae, within each cell. The zooxanthellae produce sugars and oxygen during photosynthesis, which is then used by the host organism. Multiple light fixtures can be used to provide light of the desired wavelength blend and intensity. By placing these lights on timers, a photoperiod can also be simulated. The amount of light reaching the intended organisms is maximized with clear water, little surface water disturbance, shallow depth, and use of light reflectors. Lighting that is too intense can result in tissue damage and zooxanthellae death. Lights also generate heat, which can increase water temperature and reduce the life of the lamps. An exhaust fan or water chiller can counteract the increase in water temperature. Replacing lamps every 6 to 12 months is recommended due to the reduced quality of light output. LED lights have a reported life span of 5 to 10 years.

CORE HUSBANDRY ACTIVITIES

Keepers should contact husbandry colleagues when technical questions arise. Good resources for information are the Regional Aquatics Workshop (www.rawconference.org), the annual gathering of public aquarium personnel, and the Aquaticinfo listserv (aquaticinfo-owner@mail.seaplace.org) hosted by the New England Aquarium, for those who are actively working in an aquatic animal husbandry capacity.

A clean exhibit contributes to an overall positive experience for aquarium guests. Though most tanks do not need daily cleaning, keepers are expected to have exhibits ready each day before the aquarium or zoo opens its doors to the public. Abrasive pads or brushes are suitable choices for cleaning tanks, except for acrylic tanks. One should avoid scratching the soft acrylic by using only acrylic-safe white cleaning pads, cotton diaper, or chamois skin. Scratches provide algae with footholds to grow in, making cleaning more difficult. If a cleaning pad contacts rocks or the substrate, one should wash any hard particles out of it before continuing, as it would scratch glass. Cleaning is best done in a methodical pattern to ensure that the entire surface is covered.

To clean the public side of an acrylic window, one should first mist it with water or an approved acrylic cleaning solution, then gently wipe it with a cotton diaper to remove fingerprints and smudges. Ammonia-based cleaners and exposure to sunlight damage acrylic and could lead to catastrophic failure. Scratches in acrylic can be removed through the tedious process of using progressively finer grit emery paper (grit: 1800-6000-8000), ending with buffing compound and a buffing pad.

Another core duty of aquarium keepers is exhibit diving to clean, perform maintenance, conduct in-water feedings, or give interpretive presentations. Scuba (self-contained underwater breathing apparatus) certification is often a job eligibility requirement at aquariums, and applicants without scuba certification are frequently not considered for interviews. Uneaten food and dead or sick organisms should be removed when found by divers . Any dead organism should be placed in a plastic bag, labeled (with date, species and exhibit), and placed in the a designated refrigerator until a necropsy can be conducted. The specimen should also be deaccessed from the

TABLE 35.1. Common diseases and their causative agents. The likelihood of encountering one of these diseases will depend on many factors, including the animal's origin, its post-capture care, the effectiveness of quarantine, and biosecurity measures. List adapted from Carpenter 2005.

Disease agent/"disease"	Environment	Appearance/notes	Method of diagnosis	Treatment
Virus: *Lymphocystis virus*/"lymphocystis"	Marine and freshwater	External cluster of greatly enlarged cells	Wet mount histology	Stress reduction; prophylactic antibiotic treatment
Bacterium: *Mycobacteria* sp./ "fish tuberculosis"	Marine and freshwater	Nodules in various internal organs; zoonotic potential	Cultures and staining	Euthanasia; reduction of chronic stressors
Bacterium: *Vibrio* sp./"vibriosis"	Marine	Open lesions	Cultures and staining	Naladixic acid, oxolinic acid
Bacterium: *Aeromonas, Flexibacter, Flavomonas*/"gram-negative infection"	Freshwater	Open lesions on skin, gills	Cultures and staining	Chloramphenicol, chloramine-T, ceftazadine, aztreonam
Ciliate: *Ichthyophonus multifilis*/ "ich"	Freshwater	Skin, gill, eye irritation; fish rubbing against objects	Wet mount of skin or gills	Formalin, copper sulfate, malachite green
Ciliate: *Uronema marinum*/"uronemiasis"	Marine	Initial external inflammation progressing to deep tissue lesion	Wet mount of skin or gills	Formalin bath; minimizing of organic debris
Ciliate: *Cryptocaryon irritans*/ "white spot disease"	Marine	Powdery appearance on body and eyes; fish rubbing against objects	Wet mount of skin or gills	Freshwater bath, formalin, copper sulfate
Dinoflagellate: *Oodinium*, or *Amyloodinium ocellatum*/ "marine velvet disease"	Marine	Powdery appearance over body, gills, and eyes; fish rubbing against objects	Wet mount of skin or gills	Freshwater bath, formalin, copper sulfate
Dinoflagellate: *Piscinoodinium* sp./"freshwater velvet disease"	Freshwater	Powdery appearance over body, gills, and eyes; fish rubbing against objects	Wet mount of skin or gills	Formalin
Flatworm: monogene or digene trematodes/"flatworm infestation"	Marine	Skin, gill, and eye damage; fish rubbing against objects	Wet mount of skin or gills	Trichlorfon, copper sulfate, formalin, mebendazole, praziquantel, freshwater bath
Crustacean: copepods/"sea louse, anchor worm infestation"	Marine and freshwater	Skin erosion	Direct observation of skin, fins, and eyes	Diflubenzuron, trichlorfon, luferuron

animal inventory. A number of aquariums also offer guided "dive experiences" to scuba-certified guests.

Animal health management is a core duty that spans many husbandry activities, including biosecurity. The goals of biosecurity are to limit pathogen entry to the facility and to contain any agents to prevent their spread within the facility so that pathogen-free areas remain free of disease. Keepers must adhere to standard operating procedures (SOPs) that keep the animal collection healthy. The SOP is a living document that should be revised as new information and understanding develops. Breaks in the protocol will expose the animal collection to potential disease causing agents such as viruses, bacteria, fungi, protozoa, and multicellular organisms. Commonly observed conditions are listed in table 35.1, though many other diseases and parasites are also encountered. Depending upon an institution's organizational structure, keepers may be responsible for carrying out treatments, which may include short term dips (e.g., fresh water, potassium permanganate, or formalin), medication (given orally or by injection), prolonged bath immersion (30-day treatment with copper sulfate and citric acid), or vaccination.

All new animals should be quarantined. Some cases will only require close observation while the animal acclimates to the new setting, while others should be given prophylactic or specific treatments to alleviate particular conditions. Another quarantine protocol might allow the animal to acclimate to the new environment; active feeding is considered to be a positive sign of adapting to the captive environment. If a treat-ment is initiated, it must be fully completed unless conditions contraindicate its continuance. When such a situation arises, an open discussion between keepers and veterinary service staff should occur. Keepers are a source of valuable husbandry information, just as the veterinary staff is a source of other health information. Some species are hypersensitive to particular medications and will require alternate treatments.

Disinfecting equipment and tanks with bleach or other disinfectants is another biosecurity measure. One should rinse and neutralize the treated water residue with sodium thiosulfate if bleach is used. Many different disinfectants are available, and one strategy is to alternate the use of disinfecting agents (Danner and Merrill 2006, 91–128).

Cleaning food containers with hot soapy water and using dedicated cleaning equipment for each system will help reduce cross-contamination of animal enclosures. Washing and drying of one's hands can also prevent the spread of disease agents.

HANDLING

The capturing and transporting of specimens should be planned in advance. Flawed transfers can result in stress, injury, or death of healthy specimens. Depending on the fish species, slow and stealthy use of a large net may be more effective than a fast netting attempt. In some cases, highly energetic species should be partially anesthetized prior to handling, because even soft netting will abrade their skin.

Removing structures from the tank and lowering the water level will provide easier access to the animals. The best practice is to move animals in water-filled containers rather than suspending them in nets. Such containers can include buckets, fish shipping bags, or other suitably-sized plastic containers. Gentle handling is required, because fish lack protective eyelids, and the loss of scales, or of the body's slime layer, can render fish susceptible to bacterial infection. Aerating the transfer container should be considered, depending upon the species and the length of time required to take the specimen to the new location. When animals are shipped, the plastic shipping bag should contain the animals, one-third water, and two-thirds oxygen (either pure oxygen or air with oxygen added). After being sealed, the bag should be placed inside an insulated Styrofoam® box and outer cardboard box, and shipped overnight to minimize transit time. The package may require either ice or heat packs to maintain an appropriate temperature. Having animals fast one to two days in advance of the trip will reduce their fecal material and their potential for regurgitating food. Various additives, such as ammonia neutralizer and buffers to minimize pH change, can be added to the shipping water as needed.

Fish can be anesthetized to facilitate their capture, transport, health examinations, or surgeries. The anesthesia is most commonly delivered by injection or bath. Tricaine methanesulfonate (MS-222), the most commonly used anesthetic compound, is delivered via a bath. The drug crosses the fish's gill membrane, so it is important to observe its gill movement. The specific amount of medication needed will depend upon the species, the volume of water required, the level of sedation required, and other factors. Roberts (2009, 166–71) provides a comprehensive list of anesthetic compounds and doses. To avoid a sudden drop in pH, one part MS-222 should be combined with two parts sodium bicarbonate. Both these compounds are dissolved directly into the treatment container. Besides the treatment container, an aerated recovery container with nonmedicated water should be in place before an animal undergoes anesthesia. The keeper must be aware of the level of sedation and closely observe the specimen during this process. Gill movement must always be observed, as this water movement enables nonmedicated water to pass over the animal's gills during recovery. A solution of MS-222 sodium bicarbonate can also be effectively sprayed directly on the gills of larger specimens.

NUTRITION

Providing for the animal collection's nutritional needs is an essential task that requires high-quality food, proper handling and storage, and appropriate feeding technique. New shipments of food should be inspected for freshness and rejected if unsuitable. One should always check the daily ration for texture (which should not be slimy), appearance (normal color), and odor (lack of foul smell). Daily rations should be thawed in the refrigerator or cooler. If food is thawed in water, then watertight bags should be used to reduce any loss of water-soluble vitamins. Nutritional supplements are used to increase the food's value. The addition of iodine to reduce the incidence of goiter in sharks is an example; when an iodine deficiency occurs, the thyroid gland, located in the soft tissue between the lower jaws, becomes swollen. This condition is reversible by supplementing the diet with potassium iodine.

A wide array of invertebrates such as clams, mussels, shrimp, krill, and squid may be part of the dietary offering. Likewise, many fishes including capelin, smelt, herring, trout, cod, mackerel, and bonito may be chopped to size and offered to the animal collection. Nutritional information about these species is available through the fish supply company or the scientific literature. Placing leafy greens such as romaine lettuce, broccoli, and nori (*Poryphra* sp.) in the aquarium allows herbivorous fishes to graze throughout the day.

Besides using fresh and frozen foods, some facilities make their own gelatin diet, or purchase a commercial "add-water" formulation. This offers a broad spectrum of nutrients in a single bite. In-house gelatin diets are labor-intensive, but they produce a sinking food that can be cut into various sizes. Once all the ingredients are homogenized with dissolved gelatin, the warm mixture is poured into plastic trays and refrigerated until the gelatin hardens. Once the gel diet is firm, it can be cut into daily portions, wrapped in plastic, and frozen.

Proper feeding is much more than just chopping food and tossing it into a tank. Other considerations include particle size, frequency of feeding and feeding technique. Typically, food needs to be cut into different sizes so that all the animals in the tank have a chance to consume it. In general, young animals need more frequent feeding than adults, and consume a greater percentage of their body weight per day. Large sharks can be fed 1% to 4% of their estimated body weight per week with meals coming two to three times a week (Janse, Firchau, and Mohan 2004, 183–200). This food allocation does not take into account any incidental predation that may occur. Overfeeding can occur, particularly in mixed species or community exhibits, as aggressive fish outcompete more passive individuals. Overeating can result in the development of degenerative lipidosis, a condition in which fat deposits displace functional liver tissue. To ensure that specific fish receive their share of food, target training has been incorporated into feeding protocol. Facilities have trained ocean sunfish (*Mola mola*), zebra sharks (*Stegosoma fasciatum*), California bat rays (*Myliobatis californica*), whale sharks (*Rhincodon typus*), and others to feed at specific stations.

Providing greater access by distributing food simultaneously throughout the exhibit is another way to feed passive species. Presentation of food should cease when the feeding response slows. Some exhibits will be given food on a daily basis, so these fishes do not need to be fed to satiation. Failure of individuals to feed may indicate an underlying problem; early detection through detailed observation improves the chances of correcting the condition. Bottom-dwelling and sessile animals may require food delivered via feeding poles, tongs, or tubes.

Cultured live organisms such as paramecium, brine

shrimp, and rotifer may be required as food. The nutritional value of these animals can be increased by soaking them in a highly unsaturated fatty acid (HUFA) supplement, or by gut loading them just before feeding them out. Using live food animals from the wild can introduce parasites or pathogens into the collection, so it should be minimized if possible.

WATER QUALITY TESTING

Another essential duty is testing of water quality, with recording and interpretation of the test results. One can make water quality testing as efficient as possible by collecting samples on a single pass through the facility, and by arranging the tests so that the results can be determined within minutes. Keepers are responsible for adjusting the water quality parameters, and they may need to calculate the amount of additive needed for a particular exhibit. To reduce the chances of an incorrect figure, a coworker should be asked to make the same calculation. Proceed with the additives only if both parties confidently reach the same conclusion.

The task of troubleshooting extends into all aspects of husbandry work. Knowing what to expect under normal conditions is key to diagnosing a problem. Deviation from the norm should raise the possibility that a problem exists. One should consider the following questions: What are the expected odors, temperature, and sounds in a particular work area? What behaviors are normal for a particular species? What is the normal operating water level for this aquarium? A keeper should use all senses and observational skills to assess the environment. Once a problem has been detected, one should determine the possible cause and execute a remedy.

THE ANIMAL COLLECTION

Simply stated, people come to zoos and aquariums to see the animals. Because of this, considerable resources are committed to assembling and maintaining the animal collections. Though some aquatic animals are easily obtained through capture fisheries or propagation, none should be considered expendable. The purpose of this section is to familiarize keepers with conspicuous aquatic animals found in zoos and aquariums. Significant differences exist among fishes, especially between bony fish and sharks (figure 35.4).

Though there are 488 species of sharks and 513 raylike species (McEachran and de Carvahlo 2002, 509; Campagno 2002, 360), only 31 species of shark and 25 species of rays were housed at zoos and aquariums worldwide as of 2008 (American Elasmobranch Society 2008). The term "elasmobranch," which translates to "elastic gills," indicates that the structural component of the gills and the entire skeletal system is flexible cartilage rather than calcified bone. Rays include stingrays, skates, sawfish, and other species. Sharks have a reputation for being tough, but many are more sensitive to environmental disturbances and handling than bony fishes. Some sharks are obligated to swim continuously with their mouth slightly agape, which allows oxygenated water to flow across their gills. Many species also exhibit a pattern of movement that alternates between active swimming and passive gliding. When these sharks are restrained, oxygenated water must irrigate their gills or else they suffocate.

Sharks also present several physical traits unseen in rays and bony fishes. They possess rigid pectoral and dorsal fins, which are susceptible to skin erosion due to repetitive rubbing

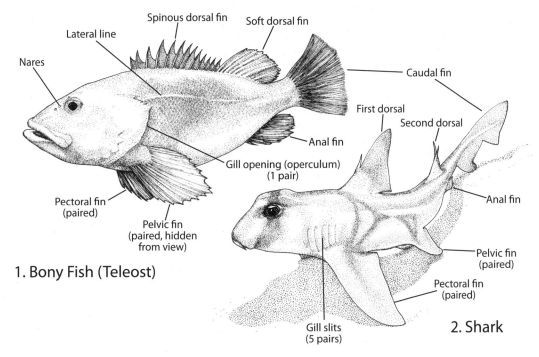

Figure 35.4. Basic anatomical features of fishes. Many external features differentiate sharks from bony fishes (teleosts). Sharks possess rigid fins, protective membranes over their eyes, replaceable teeth, and dermal denticles (modified scales). Typical fish, in contrast, have scales, fins that can fold against the body, fin spines, and eyes that lack a protective covering. Illustrations by Kate Woodle, www.katewoodleillustration.com.

against the insides of their enclosures. In contrast, the paired fins of a bony fish are flexible and can be laid flat against the fish's body. Elasmobranchs differ from bony fishes in that the teeth of sharks and rays are replaced periodically. This enables them to always have sharp, unbroken teeth. This trait is not typically found in bony fishes. The skin and scales illustrate another difference. While some fish lack scales, most have scales covered with epidermis, which protrude from pockets in the dermal layer of the skin. In contrast, the structure of the dermal denticles in sharks is similar to that of their teeth. A portion of each denticle protrudes above the epidermis, giving a sandpaper-like texture to the skin. The snouts of sharks are also susceptible to repeated bumping, which results in trauma to the area. Specimens with skin erosion should be relocated, and their wounds should be monitored and treated. Shark-friendly exhibits are spacious and incorporate chamfered (rounded off) corners with angles greater than 90°.

Courtship amongst sharks and rays (including freshwater stingrays) commonly results in the female being bitten. These wounds can be severe but tend to heal without complication. Care is required when handling stingrays because of their sharp, serrated barbs, and the associated poison gland located on the tail. These barbs can be surgically removed, but they will regrow. One should exercise care when handling sharks. Large sharks will be restrained in stretchers made of cargo strap netting or heavy tarp with drain holes. Smaller species, such as the whitespotted bamboo shark (*Chiloscyllium plagiosum*) and epaulette shark (*Hemiscyllium ocellatum*), can be netted and or held firmly with one hand placed between the gills and pectoral fin and the other near the pelvic fins. Never hold a shark by the tail, and realize that more people have been bitten by dead sharks than by live ones.

Popular petting pools can incorporate small bottom-dwelling sharks and stingrays. Keepers should inspect these exhibits for coins left behind by guests, which rapidly corrode in saltwater, leaching out toxic metals. For a comprehensive treatment of husbandry considerations for elasmobranches, consult the comprehensive *Elasmobranch Husbandry Manual* (Smith, Thoney, Warmolts, and Heuter 1994).

Another prominent group of animals are members of the seahorse family (Syngnathidae). The tiny mouth is a physical trait shared by all members of this family. The weedy seadragon (*Phyllopteryx taeniolatus*), leafy seadragon (*Phycodurus eques*), and ribbon pipefish (*Haliichthys taeniophorus*), all from the waters around New Zealand, are spectacular display animals. Zoos and aquariums are making concerted efforts to breed them. Water depth may play an important role in enabling a courting pair to complete its courtship, mate, and transfer the eggs (Branshaw 2009, pers. comm.). At this time, syngnathids are usually fed live mysid shrimp and brine shrimp. Removal of organic debris through frequent cleaning and gravel vacuuming appears to aid in reducing the population of *Uronema marinum*, a pathogenic ciliate.

An array of invertebrates including jellies (formerly known as jellyfish), corals, and cephalopods are found in zoos and aquariums. Invertebrates are generally sensitive to heavy metals, so one should avoid exposing metal to water, particularly salt water. Though they appear to be simpler than vertebrates, invertebrates require the same level of care and commitment by the keeper.

Since the early 1990s, jellie husbandry has developed to the point that many zoological collections include these invertebrates. These animals have a life cycle that involves an inconspicuous, sessile polyp stage (the asexual stage) and the better known "jellyfish" or free-swimming medusa stage (the sexual stage). Aquariums have succeeded in maintaining various species through their complete life cycles in captivity (Widmer 2008).

A specially designed enclosure called a kriesel (German for carousel), or variations of this design provide a continuous circular flow pattern that keeps medusa suspended in the water column. This flow pattern is essential for jellie medusa to thrive. Successful jellie culture requires a keeper who has a passion for these animals.

Tropical marine invertebrates encompass a considerable array of different species, many of which can be displayed in reef aquariums (Sprung and Delbeek 1994, 1997 and 2005; Fosså and Fosså 1998, 2000, and 2002). Possibly the most popular of these animals are corals, which have colonized substrate referred to as "live rock." Newly shipped live rock should be placed in a dedicated system so that the organisms decomposing in it will not cause an influx of ammonia into an exhibit of healthy animals. The "curing" process, during which injured organisms die off, can take two to three weeks. Corals can be divided into two main groups: hard corals (which possess calcified structures) and soft corals (which lack hard skeletons); both require appropriate lighting and exceptional water quality and water currents in order to thrive. The keeper must monitor and adjust each facet as needed. In addition to standard water quality testing, calcium and phosphate concentrations should be monitored regularly. Integrating a calcium reactor that includes a calcium source, a cylinder of carbon dioxide (to acidify and dissolve the calcium source), and pH monitoring into the system—or adding *kalkwasser* (limewater)—will increase the amount of calcium available for coral growth. *Kalkwasser* also has a high pH, so one should gradually add moderate amounts via a drip line or metered system (Delbeek and Sprung 1997, 212.). Maintenance of life support components is essential for corals to thrive. Keen observational skills will enable keepers to detect subtle changes in the coral's condition. Commercially formulated trace element complexes should be added according to manufacturer recommendation. Portions of a thriving colony can be cut off or fragmented, resulting in "frags" or cuttings. These new colonies are genetically identical to the mother colony, and can be transferred to other exhibits. Some corals use stinging tentacles or sweeper tentacles to keep adjacent colonies from encroaching. Biological controls in the form of peppermint shrimp (*Lysmata wurdemanni*), raccoon butterflyfish (*Chaetodon lunula*), the nudibranch (*Berghia verrucicornis*) and several members of the wrasse family are useful in controlling parasites and noxious animals such as the *Aiptasia* anemone.

Another high-profile group of animals are cephalopods (primarily octopuses in zoos and aquariums), which display complex behaviors and intelligence. Keepers play an active role in behavioral enrichment for octopuses. These solitary and short-lived animals interact with keepers and respond to environmental stimuli, hunting/investigative opportunities, and tactile and interactive exploration (Rehling 2010). One should always allow an octopus to initiate contact and keep away from the animal's hard parrotlike beak, which can inflict a deep bruise or laceration. These animals are escape artists, so enclosures must be equipped with inflexible, secured lids. Additional safeguards should include wide bands of artificial turf immediately above the water line and drain systems that cannot be dislodged. The bristle-like texture of the artificial turf will typically prevent a octopus from traveling past it. A rule of thumb is that an octopus can pass through a hole of the same diameter as its hard beak. Keepers must be diligent in keeping these tanks secure. An escaped octopus has a limited life span out of water because of desiccation, the lack of oxygen, and pressure on the internal organs.

SUMMARY

Fish and aquatic invertebrate keepers are responsible not only for providing direct care to a diverse group of animals with unique husbandry requirements, but also for managing the animals' environment. It is important that they understand how these animals alter their environment and how changes in water quality affect their health. Through observations and regular testing, changes in water quality can be detected before adverse effects are observed in the animal collection. Keepers must also be competent in their understanding of life support systems, because when these components fail to perform optimally, water quality can be compromised. In the event of malfunctions, troubleshooting must be timely. Less than optimal conditions could induce stress and even lead to the death of the entire population. A crucial facet of husbandry work is the flow of information between keepers, which can occur face-to-face or through record-keeping. Observations about behavior, feedings, water quality, health care, and other procedures are important to document. By committing to these duties, one can provide an exceptional level of animal care.

REFERENCES

Aikens, Andy. 2008. Personal communication.

American Elasmobranch Society. 2008. Captive Elasmobranch Survey. Accessed on 7 July 2010 at http://www.elasmo.org/census.php.

Arnold, James. 2010. "Emergency Schemes and Hurricane Preparations for the Audubon Aquarium of the Americas." Paper presented at the annual Life Support Operators Symposium, 2–5 May 2010 in Galveston, Texas.

Branshaw, Paula. 2009. Director of husbandry, Dallas World Aquarium, personal communication.

Brightwell, Chris. 2007. *Marine Chemistry: A Complete Guide to Water Chemistry in Marine Aquariums.* Neptune City, NJ: TFH Publications.

Campagno, L. V. 2002. *The Living Marine Resources of the Western Central Atlantic.* Vol 1, ed. K. E. Carpenter. Rome: Food and Agriculture Organization of the United Nations.

Carpenter, James W. 2005. *Exotic Animal Formulary*, 3rd edition. Saint Louis: Elsevier Saunders.

Danner, G. Russell, and Peter Merrill. 2006. "Disinfectants, Disinfection, and Biosecurity in Aquaculture." In *Aquaculture Biosecurity: Prevention, Control, and Eradication of Aquatic Animal Disease,* ed. A. David Scarfe, Cheng-Sheng Lee, and Patricia J. O'Bryen, 91–128. Ames, IA: Blackwell.

Delbeek, J. Charles and Julian Sprung. 1994. *The Reef Aquarium: A Comprehensive Guide to the Identification and Care of Tropical Marine Invertebrates.* Vol. 1. Coconut Grove, FL: Ricordea Publishing.

Fenner, Robert M. 2001. *The Conscientious Marine Aquarist: A Commonsense Handbook for the Successful Saltwater Hobbyist.* Neptune City, NJ: TFH Publications.

Fosså, Svein A., and Alf Jacob Nilsen. 1996. *The Modern Coral Reef Aquarium.* Vol. 1. Bornheim, Germany: Birgit Schmettkamp Verlag.

———. 1998. *The Modern Coral Reef Aquarium.* Vol, 2. Bornheim, Germany: Birgit Schmettkamp Verlag.

———. 2000. *The Modern Coral Reef Aquarium.* Vol. 3. Bornheim, Germany: Birgit Schmettkamp Verlag.

———. 2002. *The Modern Coral Reef Aquarium.* Vol. 4. Bornheim, Germany: Birgit Schmettkamp Verlag.

Hoese, H. D., and Richard H. Moore. 1998. *Fishes of the Gulf of Mexico, Texas, Louisiana, and Adjacent Waters.* Second edition. College Station: Texas A&M University Press.

Janse, Max, Beth Firchau, and Pete Mohan. 2004. "Elasmobranch Nutrition, Food Handling, and Feeding Techniques." In *The Elasmobranch Husbandry Manual: Captive Care of Sharks, Rays and Their Relatives,* ed. Mark Smith, Doug Warmolts, Dennis Thoney, and Robert Hueter. Columbus: Ohio Biological Survey, 183–200.

McEachran, J, D., and M. R. de Carvahlo. 2002. *The Living Marine Resources of the Western Central Atlantic.* Vol, 1, ed. K. E. Carpenter. Rome: Food and Agriculture Organization of the United Nations.

Moe, Martin A., Jr. 1992. *The Marine Aquarium Reference: Systems and Invertebrates.* Plantation, FL: Green Turtle Publications.

———. 2009. *Marine Aquarium Handbook: Beginner to Breeder.* Neptune City, NJ: TFH Publications.

Randall, D. J., and T. K. N. Tsui. 2002. "Ammonia Toxicity in Fishes." *Marine Pollution Bulletin* 45:17–25.

Rehling, Mark. 2010. *Octopus Enrichment Notebook.* Cleveland: Metroparks Zoo and Aquarium.

Spotte, Stephen. 1991. *Captive Seawater Fishes: Science and Technology.* New York: John Wiley and Sons.

Sprung, Julian, and J. Charles Delbeek.1997. *The Reef Aquarium: A Comprehensive Guide to the Identification and Care of Tropical Marine Invertebrates.* Vol. 2. Coconut Grove, FL: Ricordea Publishing.

———. 2005. *The Reef Aquarium: A Comprehensive Guide to the Identification and Care of Tropical Marine Invertebrates.* Vol. 3. Coconut Grove, FL: Ricordea Publishing.

Tullock, John H. 2001. *Natural Reef Aquariums: Simplified Approaches to Creating Living Saltwater Microcosms.* Neptune City, NJ: TFH Publications.

Widmer, Chad L. 2008. *How to Keep Jellyfish in Aquariums: An Introductory Guide for Maintaining Healthy Jellies.* Tucson: Wheatmark, 192.

Whittaker, Greg. 2010. "Hurricane Planning in Retrospect: Lessons Learned." Paper presented at the annual Life Support Operators Symposium, 2–5 May 2010 in Galveston, Texas.

Wikipedia. 2010. "Hard Water." Accessed 1 July 2010 at http://en.wikipedia.org/wiki/Water_hardness.

36

Husbandry and Care of Terrestrial Invertebrates

Tom Mason and Aaron M. Cobaugh

INTRODUCTION

The invertebrate world is one of diversity. There are literally millions of invertebrate species, many of them terrestrial, so this one chapter is somewhat insufficient for a thorough discussion of this fascinating group of animals. Invertebrates fill many niches in terrestrial ecosystems; they are essential to these ecosystems and to the vertebrate taxa that share them. To maintain specimens in captivity, the keeper must understand how and where the animals fit into their own ecosystem and what they feed on in the wild—or, if that is not known, what their closest relatives feed on. The keeper should know the animals' social behavior (or lack of it) and life cycle, and also the animals' ancestry or origin if possible. Another essential point to remember is that terrestrial invertebrates are very small creatures (the largest tarantula only weighs as much as a small rat). They are also all ectothermic and therefore can suffer quickly from adverse conditions. Temperatures, humidity, light levels, and food availability are all factors that must be assessed, just as with vertebrates. Ensuring that the animals are kept contained is essential, and sometimes it is a challenge due to their size. Daily maintenance cannot be ignored simply because they are considered to be less advanced than vertebrates; even though they may not show stress as vertebrates often do, their size may put them at a disadvantage, especially with regard to water balance (smaller animals often dessicate more quickly).

Maintaining a consistent routine is the first step to keeping invertebrates. At certain times of their development, species such as caterpillars (larval butterflies or moths) deprived of food for just two or three hours cannot recover and ultimately die without developing into the adult form. Conversely, certain tarantula species have survived 100 days without water, or a year without food. Although this would not be done in zoos, it is a testament of the hardiness of some species. Researching the animals one is caring for is essential to understanding their husbandry. This chapter will concentrate on the species regularly kept in zoos and institutes. Most species are

arthropods (possibly well over one million species, including insects, arachnids, and crustaceans), but some mollusks (e.g., snails) are also maintained. In many cases, groups will be generalized. Specific techniques will have to be obtained from the source of the animals if they are captive-bred through specific protocols. The husbandry of wild-caught specimens that have not been kept before must be learned through detailed data collection and trial and error. A few books on species' husbandry do exist, but they mostly specialize in the care of those species that are most often kept as pets (and which may be kept in zoos). With few exceptions (e.g., honeybees, silkworms, mealworms, and domestic crickets), invertebrates in captivity are maintained for public display. This is a relatively new field and thus information is very scattered and, for certain species, difficult to obtain. Maintaining good records and dispersing information to others in the field is essential if populations are going to be successfully kept for display, research, and education in the future.

After studying this chapter the reader will understand

- the general anatomical terminology, similarities, and differences among the varied taxa
- how the anatomy, physiology, and behavior of terrestrial invertebrates affects the housing and feeding of these animals in a zoological setting
- the best practices for the daily care, handling, housing, and transport of these animals
- the general environmental requirements of these varied taxonomic groups; specifically, the availability of water and proper substrate
- the unique requirements of certain species and the challenges they pose to keepers.

ORIENTATION, TAXONOMY, AND GENERAL HUSBANDRY

No single system or method can describe the husbandry of invertebrates. To cover as much of the subject as possible, the chapter will use various examples. The two phyla covered in this chapter are the Arthropoda and the Mollusca. There are four subphyla of importance in the phylum Arthropoda and in this chapter: Chelicerata (spiders and scorpions), Crustacea (crabs), Hexapoda (insects), and Myriapoda (centipedes and millipedes). The arthropods are noted for having a chitonous exoskeleton, jointed appendages, compound eyes (in most species), and a segmented body.

- The subphylum Chelicerata is noted for its chelicerae (appendages before the mouth) and two main body segments (cephalothorax and abdomen). Spiders (approximately 40,000 species) have eight legs and multiple eyes, lack antennae, inject venom through their modified chelicerae ("fangs"), and often produce silk from spinnerets on the abdomen, which is distinct from the cephalothorax. Scorpions (approximately 1,400 species) are noted for their large claws and distinct tail with a venomous barb (telson). They also have eight legs and multiple eyes, but their cephalothorax is not distinct from the abdomen.
- The subphylum Crustacea (approximately 40,000 species) is noted for having biramous appendages (branching into two parts), two pairs of antennae, and nauplius larvae (which use their antennae for swimming). They can have either two or three body segments. Most of the commonly exhibited terrestrial species (crabs) are within the order Decapoda, having five pairs of biramous appendages.
- The subphylum Hexapoda encompasses the insects (more than one million species) and is noted for its

three distinct body segments (head, thorax, abdomen), single pair of antennae, and three pairs of legs in adults. The adult forms also often have wings.
- The subphylum Myriapoda (approximately 13,000 species) is noted for having multiple body segments (more than three), a single pair of antennae, and numerous legs (most species have between 20 and 400). Centipedes (approximately 3,000 species) are noted for having one pair of legs per segment, flattened bodies, unique venomous forcipules (which are the modified first pair of limbs [maxillipeds]), and are fast-moving predators. They also lack the waxy cuticle layer of the exoskeleton, and therefore are restricted to moist environments. Millipedes (approximately 10,000 species) are noted for having two pairs of legs per segment, for being slow-moving detritivores, and for usually having cylindrical bodies. Many species in this subphylum also have odiferous glands that secrete poisonous chemicals to ward off predators.

The first step in husbandry of an animal is to identify it to species. If the animal has come from another zoo, it will most likely be accompanied by proper identification and a protocol for care. This is a good start, but a protocol designed for Florida, with a high ambient temperature and humidity levels of 70% or more, may be different than a protocol designed in North Dakota. If possible, keepers should contact others who maintain the species to learn of any existing protocols and adapt them to the specifics of their own facility to create the best microhabitat for the animal. Invertebrates maybe difficult to acquire from captive sources, so often a keeper will need to develop original care protocols. After identifying the species, or at least its closest relatives, it is important to determine how to house, feed, and maintain it.

When it is time to design an enclosure for a particular species, several basic questions must be addressed. The best way to begin is to find out where the species is from, including its natural climate, but especially its ecological

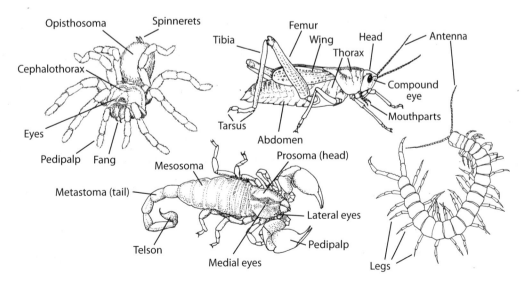

Figure 36.1. Basic anatomical features of terrestrial invertebrates. Clockwise from upper left: spider (subphylum Chelicerata), indect (subphylum Hexapoda), centipede (subphylum Myriapoda), scorpion (subphylum Chelicerata). Illustrations by Kate Woodle, www.katewoodleillustration.com.

A Guide to Invertebrate Husbandry Needs

Taxon	Food																									Water			Substrate						Décor			
	crickets	Mealworms	Fruitflies	beetles	other insects	moths	aquatic insects	caterpillars	shrimp	live fish	Earthworms	lettuce	Fruit	banana	mushrooms	Natural foods	Foodplants	Rodent pellets	Dog food	Mice	Leaflitter	Leaves	Honey	Dry Fish Food	Cereals	Aquatic	Water dish	Daily Mist	Sand	Peat	S/Pmix	Mulch	gravel	Hydrostone	Branches	grass	Rocks	Live plants
Crayfish							●		●	●	●						●				●			●		●			●				●				●	
Shrimp							●		●		●						●							●		●			●				●		●		●	●
Land Crab	●	●							●	●	●	●					●		●		●			●			●	●	●				●		●		●	
Land Hermit Crab									●	●	●						●				●			●		●	●	●	●			●	●		●		●	
Daphnia																●										●												
Centipedes	●	●		●	●	●		●			●									●							●	●		●	●	●					●	
Millipedes											●	●	●	●	●	●	●				●			●	●					●	●	●			●	●		
Scorpions	●	●	●	●	●	●		●			●																●	●		●	●						●	
Whip Scorpions	●	●	●	●	●	●		●			●																●	●				●	●				●	
Whip Spiders	●	●	●	●	●	●		●			●																●	●		●	●	●	●				●	
Solpugids	●	●	●	●	●	●		●																			●			●	●	●					●	
Tarantulas	●	●	●	●	●	●														●							●				●	●			●		●	
True spiders	●	●	●	●	●	●																					●	●	●	●	●	●	●		●	●	●	●
Dragonfly nymphs							●		●	●	●															●							●				●	●
Dragonfly			●	●																												●	●		●	●		●
Roaches												●	●	●		●	●	●	●		●			●	●		●			●		●			●			
Mantids	●	●	●		●	●		●																				●			●	●			●			●
Termites																●												●			●	●			●	●		
Crickets												●	●	●		●		●	●		●			●	●		●				●	●	●		●		●	●
Katydids *												●	●	●		●	●		●				●	●	●		●	●		●		●			●		●	●
Katydids	●	●	●	●	●	●						●	●			●			●			●	●				●	●			●	●			●		●	●
Grasshoppers												●	●	●		●	●					●		●	●		●	●		●	●	●			●		●	●
Sticks																	●					●					●	●			●	●			●	●		●
Water scorpions	●	●	●		●		●			●	●															●							●		●		●	●
Water striders	●		●		●		●																			●							●		●			●
Backswimmers	●		●		●		●																			●							●		●			●
Toe-biters	●	●	●		●		●			●	●															●							●		●		●	●
Assassin Bugs	●	●	●	●	●	●		●																			●	●		●	●	●			●			●
Plant bugs *													●				●					●						●				●	●					●
Lacewings		●	●																									●				●			●	●		●
Velvet Ants																			●								●					●			●	●		●
Ants	●	●	●	●	●	●		●				●											●							●	●			●	●		●	
Leaf cutter ants												●	●			●							●		●			●						●				
Tiger Beetles	●	●	●	●	●	●		●																			●	●		●	●				●		●	
Ground Beetles	●	●	●	●	●	●		●																			●	●		●	●				●			
Fungus Beetles															●													●		●	●				●			
Darkling Beetles												●	●	●		●		●	●		●			●	●					●	●				●			
Assorted Scarabs												●	●	●	●	●					●	●		●	●			●		●	●				●			
Diving beetles	●		●		●		●		●	●	●															●							●		●		●	●
Whilygigs	●		●		●		●																			●							●		●		●	●
Longhorn Beetles													●	●		●	●											●			●	●			●			●
Caterpillars													●	●			●												●	●	●	●						●
Land Snails & slugs												●	●	●		●	●	●			●	●			●			●		●		●						
Aquatic snails												●				●								●		●							●		●		●	
Earthworms												●				●					●	●		●	●			●			●	●						●
Flatworms										●																		●		●		●	●				●	●

Figure 36.2. Guide to invertebrate husbandry needs.

niche microclimate, and specifically its preferred humidity. It is also important to know what else the species requires in its microhabitat. There are many possible scenarios, so the information provided here can only be general. For basic information, please refer to figure 36.2 for information on humidity (misting), water supply, substrate, and food preferences.

CRUSTACEANS

Although most crustaceans are marine, they can be found in a broad array of habitats, including terrestrial environments. Almost all species have gills for respiration (some use highly modified structures, such as woodlice, order Isopoda, which use pleopods), but that doesn't mean that all are aquatic (for example, there are isopods adapted to desert habitats). However, the life histories of many terrestrial forms can be complicated, specifically with regards to reproduction, as many female crustaceans must return to the water to deposit their eggs, and the larvae may undergo various stages of development. Only terrestrial forms that are commonly used for display will be discussed here, and it is only the adults that are displayed.

The most commonly displayed terrestrial crustaceans are land hermit crabs, various land crabs, and (most recently) certain highland forms of crabs (*Geosesarma* sp. from Indonesia, and *Eudaniella* sp. from the New World) found in rain forest areas of the world. For husbandry purposes, these groups can be divided into two broad categories. Most land crabs and all land hermit crabs depend on the ocean for at least part of their life cycle. These species tend to require some "hard" water (i.e., water with high mineral content, usually calcium and magnesium) or even brackish water (a mix of fresh and marine water) in captivity. The highland crabs (variously known as manicou crabs, mountain crabs, spring water crabs, and carnival crabs) are found well away from salt water and do not require any contact with a marine habitat. They do well with fresh nonchlorinated water sources. For all crustacean species, water quality and the control of microbes in the exhibit is important. Regular water changes and/or proper water filtration are required.

Most crustacean species maintained in captivity are found in tropical or subtropical areas. Knowing exactly where a species is found will help determine its temperature preferences. A good search through meteorological data would provide the keeper with seasonal temperatures and rainfall. Relative humidity is also an important factor for some species. Knowing the chemistry of the substrate the animal lives on and whether the substrate drains, sheds, or holds water is also something to determine before building an environment for the animal. Knowing how the animal uses that substrate (whether it digs, sifts, etc.) is also important in deciding what depth of substrate to provide. Once these factors have been determined, the keeper can design the exhibit. When examining the environmental features, one should look at the moderate levels and not the extremes. The complete replication of an environment is impossible; therefore, all things need to be done in moderation.

To determine the size of the enclosure and its decor, knowing how the species uses its environment is essential. Species that live in leaf litter may spend their entire life moving from damp to dry sites within the litter. The opposite can be seen in the Caribbean land hermit crab (*Coenobita clypeatus*), whose adults can be found in trees, several meters off the ground. Knowing how and where the species use the habitat will help determine whether perching and live plants are possible or whether a sloped flat landscape is required. Also knowing whether the species is solitary (like the land hermit crab) or social (like the fiddler crab) will also help determine the size and shape of the exhibit.

Most terrestrial crustaceans are predominantly scavengers and detrivores. Dead leaves and wood are two important sources of nutrients for these animals. Some fruit, root vegetables, insects, and small amounts of animal protein (e.g., fish or shellfish) will be readily taken. However it is very easy to overfeed. These animals have to compete heavily for sufficient sources of protein and easily digestible carbohydrates in their natural environment. Dead leaves constitute a large portion of their food. For species such as fiddler crabs (*Uca* sp.) and the species found on mudflats, microbes sifted out of the wet silt are a major portion of their diet. It also must be remembered that the exoskeletons of crustaceans contain high amounts of calcium. It is therefore very important that calcium is available to these animals, usually in the form of cuttlebone.

ARACHNIDS

Members of the class Arachnida are among the most popular species used in public display. They are in general hardy, long-lived, and popular in displays. The most popular forms are the spiders (tarantulas, widows, and golden silk orb weavers), scorpions (emperor, desert hairy, and flat rock), whip spiders (Tanzanian, *Damon* sp.) and whip scorpions (American, *Mastigoproctus*; or Thai, *Typopeltes* sp.). Other species and orders are also displayed, but to a much lesser degree; therefore, this chapter will discuss the most commonly displayed animals.

TARANTULAS

There are approximately 900 species of tarantulas in the family Theraphosidae. All are predators, and all live in tropical to warm temperate areas. They differ, however, in what they need in the way of optimal temperatures, humidity requirements, and habitat. They also vary in the level of their aggression, the toxicity of their venom, and the presence or absence of urticating hair.

Tarantulas (also known as baboon spiders, earth tigers, or bird-eating spiders) can be found in habitats ranging from desert to rain forest. Desert species either burrow or find previously made shelters. Rain forest species are either terrestrial or arboreal. Forest-dwelling terrestrial species are often the same as desert species in their habits. Again, exhibit design should meet the needs of the species. Arboreal species need vertical hides, terrestrial species that hide require "caves" or ready-made burrows, and burrowers need substrate of a depth that allows them to dig. To ensure that a species can be seen, a tip would be to design its cage so that its prime hiding sites are at the front of the exhibit nearest the viewing area. Various substrates have been used over the years, including soils and mulches. Hatchling spiders can be kept in pill jars on vermiculite or perlite. As they mature and are moved into larger enclosures, they can be kept on peat moss or on a mixture of peat and sand, but sharp silica sands must not be used. Aged mulch can be a useful substrate for arboral species. The substrate and decor should be free of any pesticides, fungicides, preservatives, paints, or dyes, as chemicals of any form can be hazardous to the spiders' health.

Abiotic conditions vary. Tarantulas require variation in temperature and humidity. Desert species should be kept drier, but their enclosures should have areas slightly higher in humidity that the spider can retreat to. Forest species require higher humidity, and there should always be a level of moisture within the substrate, but keepers should also ensure that there is an area that remains dry. Desert species require warm days and cool nights. They also often prefer warmer, damper summers (26–29 °C [79–84 °F]) and cooler, drier winters (15–18 °C [59–64 °F]). For the reproductive success of temperate and high-altitude species, seasonal change is essential. Forest species also need seasonal variation, but in some cases the level of humidity combined with the temperature is more critical. As far as lighting is concerned, no requirement has been found for tarantulas. Lighting should be determined by what works best for the exhibit. Obviously, planted exhibits would need more light. In general, however, tarantulas tend to be somewhat photophobic. Therefore, the display sites one wants them to spend the most time in should be made the darkest. All tarantula species should have a source of water. Newly hatched specimens can get moisture from damp substrate, but as the spider gets larger a water source should be provided. A small shallow dish is best. It can be left open, but this often becomes a magnet for crickets that drown in it. To prevent this, keepers place a sponge or pea gravel in the water container to give the crickets a surface to walk on.

Crickets are probably the most common food item for tarantulas, but they are not a complete diet. In the wild a tarantula may eat large insects, scorpions, other tarantulas, small frogs, lizards, snakes, and small mammals. They will even take small birds. In captivity, cockroaches, caterpillars, stick insects, mice, and chicks can be used as food. Small specimens should be fed lightly twice a week. Older specimens can be fed once a week to once a month, depending on the amount of food given. A spider should always be observed before it is fed. If the spider has covered the entrance to its retreat with silk, one must not feed it. This action is often the precursor to moulting or egg laying; the spider will not feed and is at a weakened state at this time. Crickets finding a spider in this defenseless, freshly moulted condition can use it as food. Even a small wound on the leg while the spider is recovering can lead to excessive fluid loss and death.

Tarantulas are solitary to somewhat social in nature. Most specimens, especially females, never travel more than a meter away from their hide, as their food comes to them. A few New World species have developed what appears to be a symbiotic relationship with certain species of microhylid frogs. However, this relationship has never been maintained or displayed in captivity. Some tarantula species are capable of living in association with conspecifics. This is always risky, however, and there is no guarantee that the spiders will remain compatible over time.

Once the exhibit is set up, it should be disrupted as little as possible. Leftover and dead food or any rotting organic matter should be removed. Tarantulas make their shelters to last a lifetime; excessive cleaning and disruption creates unnecessary stress for them, and it will end up affecting their quality of life. With a New World spider, one should wear disposable gloves due to their urticating hairs, which can cause discomfort. Though some people handle tarantulas, it is not recommended. Some species will bite (with various reactions) and can be severely injured or killed in a fall. When a tarantula must be moved, the easiest method is to use a tube or a small container, such as a clear plastic cup. One should set the cup in front of the spider and then, using the lid, gently touch the abdomen and push the spider into the cup. PVC tubes are often used as hides for tarantulas. With both ends of the tube covered, it is easy to transport the spider for short periods with no danger to the spider or the handler.

"MODERN" SPIDERS

There are approximately 40,000 known spiders, most in the suborder Araneomorphae. Given such a large group, this section will concentrate on the species most often seen in displays: members of the genus *Latrodectus*, known as the widows, and the golden silk orb weavers of the genus *Nephila*. Although spiders in both genera are web spinners, their husbandry needs are quite different. One thing that they have in common is the strength of their silk, as the webs of both have great tensile strength.

The enclosures in which these species are kept require minimum cleaning, usually consisting of the removal of refused food. A keeper must not use any soaps, pesticides, fungicides, bleach, or other chemicals. A surface such as glass should be cleaned with water used sparingly. If a problem occurs, such as an invasion of mites, one should remove the spider and clean the enclosure thoroughly. Quaternary ammonia-based disinfectants are safest in emergencies, but the container must be thoroughly rinsed and allowed to dry before the spider is returned to the enclosure.

Lighting is not something that has been studied much in spiders. Many species seem to be photophobic. Widows are included in this group, so one must not put strong lighting over them. Again, this is a situation in which lighting or the lack thereof can be used to position spiders where they can be best observed. On the other hand, *Nephila* are often observed in the middle of the day sitting at the middle of their large orb webs, in sun or shade. They are also at the middle of their webs at midnight.

Both these spiders use their webs to catch prey. The black widow attaches its lines to the ground in what appears a random manner. Although flying insects are sometimes caught in the web, most of the prey are ground insects. For these spiders, two or three small crickets or a small cockroach or grasshopper twice a week or less is a sufficient diet. The golden silk orb weaver designs a large flat orb web to catch flying insects. Sometimes a few strands are haphazardly placed on one side of the orb. The silk is strong enough to catch dragonflies, butterflies, beetles, and even a small bird. In captivity, crickets, cockroaches, grasshoppers, or any insect of the appropriate size can be used as food. Although they can live on a small amount of food, these species do best if fed daily. Spiders are not a group that does well in mixed exhibits. If fed well enough, however, both of these types of spider can live with conspecifics if enough space is available. This coexistence is occasionally observed in the wild.

As for water requirements, widows and orbweavers are at opposite ends of the scale. The golden silk spider is found in areas of high humidity, where dew falls heavy and rainfall comes often. Without high humidity and regular spraying of the web, *Nephila* will dehydrate. Widows are adapted to drier regions, where dew fall is rare and rainfall is not guaranteed. An occasional spray to the web is sufficient for them, as they get enough moisture from the food they eat.

Spiders should always be handled carefully. Their legs are long and delicate. If pinched, a leg can be taken off or an abdomen crushed. Spiders are best moved in small containers. It is also possible to have them walk onto a stick or forceps. They will set silk onto the substrate and use it as an anchor to keep from falling. They can be moved short distances in this way. When capturing the spider in a container, it is easiest to simply place the container over the spider on one side of the web while holding the lid on the other side, and then bring the two together. If one wants to move the spider but save the web, the container must be placed beside or below the web. Then, using forceps or a long prod, one must force the spider into the container and then cover it with the lid. In this way the exhibit can be safely cleaned and the freshest web left untouched. The spider can be returned to the web as soon as the exhibit is clean.

Although *Nephila* is large, the genus is not known to be dangerous to humans. The widows, on the other hand, are known as dangerous spiders due to their venom, which holds medical significance for humans. Species vary in what type of bite they can inflict. The southern black widow (*Latrodectus mactans*) and the red back of Australia (*Latrodectus hasselti*) are considered two of the most dangerous, and can cause death. Both species are known for their highly toxic neurovenom, which attacks the nervous system. In extreme cases, the bites of these species and others in the genus require antivenom. Luckily, the species are not overly aggressive and bite only if extremely provoked. Female widows will also protect their egg sacs if they are threatened, and only female widows are large enough to cause a serious bite.

SCORPIONS

Most scorpions are tropical to subtropical, and are found from deserts to rain forests. They may live in trees, under rocks, or in burrows. Some are social, while others are solitary. Ecosystems in the Middle East have been known to contain 25,000 scorpions per hectare (2.471 acres) of land. In captivity, many species are offered by dealers. The species recommended by the North American's Association of Zoos and Aquariums (AZA) Terrestrial Invertebrates Taxon Advisory Group (TITAG) is the African emperor scorpion (*Pandinus imperator*). Other species often displayed are the African flat rock scorpion (*Hadogenes* sp.) and the desert hairy scorpion (*Hadrurus arizonensis*) from the American southwest and Mexico; however, the diversity of habitat use by scorpions is by no way covered by these three species, of which only the emperor scorpion is consistently bred in captivity.

Each of these three highlighted species needs a different kind of substrate. The emperor scorpion is a communal forest dweller that digs its own burrows; it needs a substrate that can hold the burrows without collapsing and also maintain a relatively high humidity. Pure peat moss, potting soil, aged mulch, or a mixture of peat and sand are often used. In the wild, emperors spend much of their time below ground, but replicating this in captivity would make them difficult for zoo visitors to see, so the depth of their substrate is often reduced. For best viewing, a sheet of clear glass or acrylic can be placed over a shallow depression in the substrate. The emperor will hide under this as a form of security, even though it can be clearly seen from above. This works as long as ambient light levels are minimal.

The desert hairy scorpion is also a burrowing species. The harsh habitat it inhabits forces it as much as a meter (39 inches) underground for many months of the year. In the desert scrub the substrate is normally coarse sand. In captivity, a small amount of peat moss mixed with coarse sand helps to hold the sand in position. This species may spend long periods of time buried underground. The lower levels of its substrate should show some signs of humidity, while the upper layers should be dry. Again, a glass cover placed over a shallow depression in the substrate can be useful for exhibiting this species. The rock scorpions (*Hadogenes* sp.) are not burrowers. This African genus specializes in rocky habitats. They squeeze into cracks between rocks, or find flat rocks under which to dwell. They do not dwell outside the rocky outcrops. When housing them, it is best to have a shallow absorbent substrate. Aside from that, layers of rock set up as shelves will keep this species happy.

Scorpions as a whole are photophobic. They may come close to the surface to heat up, but they will always stay under cover. The best way to view them is in subdued light. One point of interest is that all scorpions have a waxy substance on their exoskeleton that reflects ultraviolet light. Many exhibits are designed to go dark at the push of a button and then use ultraviolet light to make the scorpion glow. This is a very dramatic method of display, but some reports state that prolonged exposure to ultraviolet light is dangerous to scorpions. It is best to spot clean these containments, removing leftovers, with minimal disturbance to the animals. No chemical product should be considered safe with these animals. If the exhibit glass needs cleaning, try to dry clean it first. If this does not work, use a small amount of water.

All scorpions are predators, and they will eat anything they can overpower. In most cases, the prey is half the scorpion's size or less. Some species will take larger prey, using their venom to subdue it. In the wild, the prey can be ants, termites, beetles, cockroaches, spiders, or other scorpions; for larger species it can also be small lizards, snakes, or other small vertebrates. Larger species tend to overpower their prey, using their strong claws (pedipalps) to grab and subdue it. Smaller species grasp the prey with their pedipalps, but these claw-like organs are not powerful enough to subdue the prey. The prey is then quickly dispatched with a sting. For scorpions in captivity, crickets are the main staple, though they are not considered a complete diet. To ensure that specimens are kept healthy, one should feed them a variety of food. Mealworms, mealworm beetles, cockroaches, and other insects can be used. Larger species will also take small mice. One must not feed larger live mice to large scorpions, as the mice can fight back and cause damage. Feeding a scorpion once a week is sufficient. If it starts to look fat, it may be about to moult or produce young. It must only be fed what it will eat in a short time period. Many scorpion species feed on other scorpion species as a normal part of their diet. In fact, many species are solitary and are socially compatible with other scorpions only when a female is receptive. On the other hand, species such as the emperor scorpion are social and live in family groups. Both the male and female emperor scorpion are known to feed their young for up to 18 months, and the offspring may live with their parents for three years. Reproduction involves ritualistic courtship, and females give birth to live offspring, which will remain on the female's back until they molt.

Scorpions are quite diverse in their need for water. Forest species such as the emperor scorpion should have a source of moisture supplied regularly. Open water is not advised, as scorpions cannot swim, so pea gravel or a sponge is often placed in their open water containers. It is extremely important for a keeper to know the preferred habitat for a species. Too much exposure to water may be fatal to the scorpion. Desert species are very well adapted to living in areas with little water. For example, desert hairy scorpions allowed to walk over a water source can actually absorb too much water, and this can prove fatal. These scorpions are quite capable of procuring enough water from their food, and thus they should have very limited contact with water. It is therefore extremely important to research each species and know what range of humidity and moisture is normal for it.

Of the 1,500 species of scorpions, only a number of species in one family (Buthidae) are considered dangerous to humans (i.e., possessing venom of medical significance). The three species highlighted in this chapter are not considered highly venomous, and their stings range from a pinprick feeling to that of a bee sting. As some species can be deadly, one should always be careful when handling a species one does not know. The best way to move any unknown scorpion is to use a container that can be placed over it or in front of it, and then pushing the scorpion into the container using forceps or the container's lid. Scorpions can be picked up by grasping the last segment of the tail, preferably with forceps and not fingers. Large species can lift themselves up and use their claws in defense, and can draw blood with them.

WHIP SCORPIONS

The whip scorpions are a small order (Thelyphonida) of arachnid that is becoming more popular. Unfortunately, most still come from the wild. Two species are found in the zoo world. The most common is known as the vinegaroon (*Mastigoproctus* sp.), from the southern region of North America, and the other is from Thailand (*Typopeltes sp.*). The latter species has been bred in captivity, but captive-born numbers are low. Most species tend to come from forests ranging from upper scrub and pine to humid forest. Although they look dangerous, they are harmless. Their one defense is their ability to spray acetic acid into the faces of their enemies. If the acid contacts the eyes, it will sting and perhaps cause temporary blindness. Whip scorpions have the same basic anatomy as other arachnids. They have eight legs, eight eyes, and two major body parts. They differ from scorpions in having a flagellum and not a tail ending with a venomous sting. The first pair of legs are antenniform (elongate with sensory organs). They cannot be used for walking, but have the same use as antennae in insects.

Whip scorpions are mainly terrestrial. Some may climb, but they prefer flat horizontal surfaces. Most of their time is spent underground in burrows. They are photophobic, so subdued lighting is preferred in a zoo display. A shallow substrate of a sand/peat mixture or a sterile cactus soil is fine. For viewing, place their shelter against the viewing glass or use a piece of glass over a depression in the substrate. The substrate should be dampened regularly, but should never get too wet. The lower levels of substrate should always contain some moisture, with the top staying dry. This will enable the whip scorpion to find the humidity level it desires. The diet for this group is similar to that of scorpions, but the food should be smaller. Whip scorpions can only physically overpower their prey, not having a stinger, so they are unable to take larger meals. They have been known to eat small spiders, scorpions, a variety of insects, small lizards, and snakes. Of course, whip scorpions should also be given a source of clean water.

When cleaning, it is best to disturb the animal's habitat as little as possible. Dead and refused food should be removed and surface fecal material cleaned up if possible. If the animal is held off display, one should let it burrow and not disturb or feed it if it fills in the entrances to its burrow. Covering of the burrow means the specimen is going to moult, aestivate, or have young. Disturbing it will only stress the animal. No precautions are required when handling these species, except to keep them away from the eyes. If a specimen is placed into a small container and sprays acetic acid within the container, it can be fatal to the animal. So one should ensure that the animal is not disturbed when contained, or it should be allowed to spray beforehand. Reproduction in the whip scorpion is similar to that in scorpions, and the sexes should be kept separate except for breeding.

WHIP SPIDERS

The whip spiders (order Amblipygi) are the closest relatives of the true spiders. They were once known as tailless whip scorpions, but the common name was changed to help avoid confusion. Whip spiders differ from true spiders in not having the ability to produce silk and in not having secondary reproductive organs on the pedipalps. They are tropical to subtropical and are photophobic and, for the most part, antisocial. Pairs with young will live together for some time, but this compatibility will break down over time.

Whip spiders are extremely vertically compressed (flattened). Like spiders they have two body parts, eight legs, and eight eyes. Their vision is poor but their first pair of legs are very long and antenniform. Most often they are found on vertical surfaces, and during the day they can be found in caves or wells. At night they venture out and may be seen on tree trunks or flat rock faces. Smaller species sometimes use rodent burrows as their residence. The species most often available is the Tanzanian whip spider (*Damon variegates*).

The substrate can be shallow, as they do not dig; peat moss

or pea gravel can be used. Keeping the substrate damp will help maintain the high humidity they prefer. When properly set up, whip spiders will only visit the bottom of their displays when there is no comfortable vertical surface to sit on. They prefer vertical surfaces and are often positioned upside down. A single animal appears to do well on a surface that is 30 × 30 cm (one square foot) in size. Whip spiders prefer flat surfaces with a slight roughness, and tree bark (e.g., cork) or wood is fine. Rock can also be used, but its texture may cause the claws to wear down.

Whip spiders will keep the vertical surfaces they walk on clean by defecating at the bottom of their enclosures. One should clean the substrate periodically and wash any fecal matter from vertical surfaces with clean water. No chemicals of any sort should be used, unless it is planned to remove the animals and break the entire exhibit down. Once a whip scorpion is set up, it is best to not move it. Physical contact will increase stress and reduce longevity. As with scorpions and whip scorpions, subdued lighting is definitely preferred.

These animals are predators. Whereas a spider must eat its food as a liquid, whip spiders chew and masticate their prey. Only the hardest materials are left behind. The large raptor-like pedipalps are used to capture and subdue the prey. Whip spiders feed on any living thing of the appropriate size. This could include crickets, cockroaches, mealworms, moths, caterpillars, small lizards, or one another. In captivity they should be offered a variety of food, and fed twice per week.

These animals should also always have a dish of clean water. They are capable of going underwater when frightened, but one must ensure that they can maintain a grip so they can pull themselves out. It is best to put a sponge or gravel in the water container so they can drink without having to risk going in. Place the water container close to the vertical surface, as this will ensure that they can find it.

It is possible to place a pair of whip spiders together as long as they are given sufficient food. The males have longer pedipalps than the females. It is not advisable to try more than one pair together. The young will remain on the female's back after hatching until their first shed. Groups of these species living together have been noted in captivity. In the wild, it is unusual to see more than whip spider at a time, but sometimes pairs are observed together, or a small group can be seen in a well or a cave where conditions are optimal. Handling and transporting these arachnids and all the others mentioned previously is basically the same. They should be placed in a small container that restricts their movement. They should be able to sit in a comfortable position but have cushioning above, below, and all around them. The animal must be protected from any jarring or shaking, as abdominal injuries are often fatal.

OTHER ARACHNIDS

The orders Opiliones (harvestmen, daddy longlegs), and Solifugae (wind scorpions, camel spiders) have occasionally been used for displays. Opilionids can be maintained on dead insects for prolonged periods, and some are impressive in appearance; however, they are not available commercially. Solifugids are available through the pet trade, but there is little available information that explains how to maintain this order in captivity. Experimental trials on this group may help develop maintenance strategies for these animals, but none have yet been successful.

INSECTS

More species of insects are known than all other species of plants and animals put together. Insects make up 85% of all known animal species, and other invertebrates make up a further 12%. Described here are the insect species most commonly maintained in public displays.

ORTHOPTEROIDS

Orthopteroids are part of a large group of insects from various orders that include the katydids, grasshoppers, cockroaches, mantids, and stick insects. Most have two pairs of wings, although there are species that cannot fly or have entirely lost their wings. They also have chewing mouthparts, characteristic of more primitive insects. Members of four orders are often kept for exhibit. They cover a large variety of shapes and sizes, and are found on every continent except Antarctica. In all there are over 36,000 species. Their importance to humans has been noted since biblical times. They have competed with humans for crops in plague proportions; they use people's homes as shelters, and sometimes as food. Some have been recognized as symbols of luck, while others have been used in competitions to fight to the death.

The cockroaches (order Blattodea) are well known for sharing residences with humans throughout the world. They are also part of the detrivore (organic decomposers) cycle from temperate zones to the tropics. A few feed on flowers and vegetation, but most specialize in consuming dead organic matter. Most are nocturnal by nature and vertically compressed, with cryptically-coloured bodies, usually brown in color, although green, black, and multicolored species are known. Most are considered semisocial, and some species, such as the wood roaches of Canada or the burrowing roaches of Australia, are classified as monogamous, with pairs mating for life. Several species have been kept for display. The most commonly kept are the American cockroach (*Periplenata americana*), the hissing cockroach (*Gromphadorhina portentosa*), Brazilian giant cockroaches (*Blaberus giganteus*) and the Cuban roaches (*Panchlora* sp.). All of these species are basically kept in the same manner. The main differences are their humidity requirements and the enclosures required for them. Food, substrate, temperature, and furniture are all similar.

Cockroaches in general should be maintained on a dry absorbent substrate, such as peat moss. Slight dampening will help keep the dust down, but if the substrate is too wet, fungus will proliferate and this could jeopardize the health of the animals. In the species previously mentioned, less than 2 cm (4/5 in.) of substrate is required. Juveniles will burrow and hide under the substrate if it is loose enough. Furniture requirements can be a variety of materials, depending on the environment being portrayed. American roaches are often displayed in household settings, while Brazilian giants and the Madagascan hissers tend to be depicted on dead wood or forest floor settings. The way an animal is displayed can tell the public a lot about the animals' habits. If cockroaches are being kept off display, a good medium to hold them on is cardboard egg cartons. This medium is absorbent and easily replaced if necessary. Cockroaches will basically eat any organic matter. Food can include carrot, sweet potato, banana, apple, orange, lettuce, kale, rabbit pellets, dog chow, cereals, any human foods, dead wood, dried leaves, or dried animals. The food should not be allowed to become moldy, but otherwise if it is organic it can be eaten. A source of water is important, but the water should not be an open source as the cockroaches may drown in it.

Cockroaches should be kept shaded as they do not like light. If the exhibit contains too many dark crevices, the specimens will disappear. The temperature should stay between 21 and 25 °C (between 70 and 77 °F). Temperatures that are too high appear to cause premature ejection of egg masses. Species such as American roaches and hissing roaches are quite capable of climbing glass, while others can fly; so one should ensure that the enclosures are properly and securely covered, with metal screen over ventilation openings, as cockroaches are quite capable of chewing through nylon or fiberglass screen.

Praying mantises (order Mantodea), although taxonomically close to the other insects mentioned here, are very different in behavior and morphology. They are solitary hunters that only come together for a short time to mate. When mating, at least in captivity, the female mantis often kills and consumes the male after fertilization. Compared to that of other orders, husbandry of praying mantises is relatively simple. All are ambush predators that will feed on anything that they can overpower. The smallest species make a meal of a house fly, while large species such as the Chinese mantis (*Tenodera sinensis*) have been known to capture and consume hummingbirds.

For best results, a praying mantis should be reared singly. A specimen does not need a large space. A young specimen will do well in a small plastic box 15 × 10 × 10 cm (6 × 4 × 4 in.) in size. As the mantises grow, they can be moved to larger containers. A large praying mantis could be kept in an aquarium 60 × 30 × 30cm (24 × 12 × 12 in.) in size. Most species prefer being off the bottom, so perches should be placed both vertically and horizontally. If possible, do not place perches in a manner that will allow crickets to climb to the top of the container. The mantis can climb to the top by walking up the side, and this gives it a "safe" spot to allow it to moult without being disturbed. A freshly moulted mantis is defenseless and open to attack by feed insects such as crickets. Feeding the mantis is quite easy, and the larger the mantis the larger the prey should be. Flying insects are the preferred diet. Fruit flies, house flies, wax moths, and even honeybees can be fed to mantises. Crickets, grasshoppers, and beetles will also be taken if the mantis is hungry, but they are not what mantises prefer. The species with the heaviest raptorial forelegs are capable of taking the largest prey. For water, misting is the best method. All species prefer to drink droplets of water, and normally one spray of water a day is sufficient. Known rain forest species should also get an evening shower as well. On average, mantises are short-lived, with three to four months being the average, although occasionally they can reach six months or possibly a year. For breeding, one should use a young adult female and feed it very well before

adding the male, and use a larger space than they are normally kept in, as this will give the male a greater chance to survive.

Some stick and leaf insects (order Phasmatodea) are possibly the most displayed insect species in North American zoos. The Macleay's spectre's (*Extatosoma tiaratum*) only competition may be the Madagascan hissing cockroach, but the Macleay's spectre is much more impressive in appearance. Stick and leaf insects are superb exhibit animals and educational tools. Some are the longest insects in the world. Many are sexually dimorphic, but some are parthenogenic with female-only populations produced from one individual, and colonies that may last for years. Some are winged, others totally flightless, and their ability to "disappear" in the environment has given them their ghostly name, the phasmids; yet a few specimens display aposematic (warning) coloration. Although all stick insects are similar in their husbandry needs, there are also differences. Some are generalist feeders, while others require specific foods. Some stay in branches to hide, while others retire to cracks or under rocks for the day. Good references are available through groups such as the Phasmid Study Group in the UK (www.phasmid-study-group .org). One stick insect species that is recommended to keepers by the AZA Terrestrial Invertebrate Taxon Advisory Group (TITAG) is the Peruvian firestick (*Oreophoetes peruana*). This is the one available stick insect with warning colors of red and yellow on black. Although not a large insect (6–9 cm/ 2.5–3.5in), its striking colors always make an impression on the visitor. The species is also unique in that it requires ferns as a food source. Stick insects are often displayed in large numbers and make an impressive display. However, one should make sure that enough space is available; otherwise accidental cannibalism may occur. If limbs are lost, the number of animals in the enclosure should be reduced. Limb problems may also occur if the humidity is kept too low as well. If this is the case, pieces of old sheds will remain attached at the legs. If limbs are lost in the earlier stages of development, they can regrow, but adults that lose limbs will not grow replacements.

For all species the need for good browse is essential, especially during the early life stages. Favorite plant species for generalist feeders are raspberry, bramble, rose, white oak, leather leaf viburnum, and ficus. For many Australian species, eucalyptus can be added. Mango leaves are also taken by many stick and leaf insects, but that plant appears difficult to maintain in northern climates, so it is not recommended as a key food source. Generalists include some of the favorite species including Macleay's spectres, thorny devils (*Eurycantha calcarata*), Annam sticks (*Medauroidea extradentatum*), and Malayan wood nymphs (*Heteroptera dilatata*). When offering food, one should keep the leaves on branches, place the branches in a container of water, and block any access to the water; otherwise the insects can drown. Ensure that the branches are long enough to support them. Stick insects prefer to hang upside down from the branches, and enough space must be given to allow for this. It is sometimes difficult to get newly hatched juveniles to eat. Raising the humidity helps, but newly hatched "sticks" prefer to feed at sites that have already been chewed, so to get them started they should be placed on the same food plant with older specimens. If a young insect is the first specimen in the collection, one should tear (not cut) a piece out of a leaf and offer it to the insect. Changing the species of plant offered as food can sometimes cause problems in a collection. For example, losses may be as great as 50% when a species is moved from bramble or raspberry to any of the fig species. Once converted, the insects appear to be fine. Changing from a hard leaf to a soft leaf tends to lead to fewer problems, but problems still can occur. A gradual change is best. It is not usually advisable to feed stick insects on potted plants; they will use the pot as a place to lay their eggs. This leads to problems later when the plant is moved elsewhere to "rest" and regenerate, as babies may end up hatching in the wrong place. Potted plants should only be used if the soil is covered or if the plant is large enough and the stick insects are in low enough numbers not to cause major damage. Reproduction is relatively straightforward as long as males are present with the females, except in parthenogenic species. Finally, when working with stick insects, keepers should be extremely careful when exchanging food. Newly hatched stick insects tend to sit on branches and often get overlooked and discarded with the refuse. Also, keepers should always know how many insects are supposed to be in the container and count them before discarding any old and refused food.

Grasshoppers and katydids (order Orthoptera) have species with different requirements, some of which are very specific. Grasshoppers include the locusts, and their infamy dates back to biblical times, as swarms of locusts and grasshoppers can destroy large tracts of farmed crops. Despite that, a few species are maintained for display, whereas some species, such as the desert and migratory locusts, may only be kept in labs under strict quarantine. The species that appears to be most popular for displays is the eastern lubber grasshopper (*Romalea microptera*). This is a colorful (aposomatic) nonflying species that is hardy and reliable as a display and as an educational tool for children. The species can be maintained in front-opening screened insect cages at least 40 × 40 × 60 cm (16 × 16 × 24 in.) in size, at a temperature of 23–30 °C (73–86 °F). These grasshoppers are diurnal and enjoy basking under a hot spot as do many reptiles. In the wild they will eat a wide variety of foods including some that are quite noxious to poisonous, and they are capable of using these chemicals after they are ingested. In captivity they can be fed kale, sliced apple, carrot, and sweet potato. The food is best presented to them hanging from hooks, as grasshoppers tend to like vertical surfaces to sit on. The food should be replaced approximately three times a week, or more often if it gets consumed faster. Shallow dishes of dry rolled oats, bee pollen, and bran can also be used. It is a good idea to not use lettuce unless it is 100% organic, as the pesticides used on farms often target grasshoppers, and residue appears to be on lettuce quite often. As for watering, most species tend to get enough moisture from the food they eat. A light misting is recommended for forest species. Eggs are laid in the substrate (females have an ovipositor), and adults die shortly after mating.

Katydids have diverse life histories. Among them there are herbivores, frugivores, food specialists, carnivores, and omnivores. Most katydids live in trees and bushes, but some like grasslands. Their eggs can be laid in the ground or in trees, in dead or living plants, or sometimes on the edges of leaves. Some katydids can also inflict an extremely painful bite. They differ from grasshoppers by having longer antennae. In general katydids prefer an arboreal habitat, so tall enclosures are best

suited for them, similar to those used for grasshoppers. Most katydids are solitary by nature and nocturnal, but a few appear to be semisocial and will hide together in small groups. Green katydids tend to hide on leaves in the open while many of the large brown forms find cracks or hollow logs to hide in during the day. They tend to come from moister habitats than most grasshoppers, so twice daily misting is recommended. Food will depend on the species kept. Most, even carnivores, will chew on sliced apple or cucumber. Some can be fed the same as grasshoppers, while others may need crickets or pollen added to their diet, and keepers should always keep their food fresh. The reproduction of katydids is similar to that of grasshoppers. Some katydids can live more than a year, so they can make good exhibit specimens. However, keepers should note that the adults are capable of flight, and that all have extremely good jumping abilities.

TRUE BUGS (ORDER HEMIPTERA)

The true bugs most often displayed are the aquatic forms such as giant predacious water bugs, back swimmers, and water striders. Two terrestrial species predominate the captive displays. These are the large milkweed bug (*Oncopeltus fasciatus*) and the two-spotted assassin bug (*Platymeris biguttatus*). The first of these is an herbivore and the second a predator. As in all true bugs, the mouthparts in both these species have been modified to form a long, jointed tube capable of penetrating the protective skin layer of a plant or of prey. Both species require enclosed containers, as the adults are capable of flight. Increasing the surface area to walk on will allow for a greater concentration of specimens living together. The species are often found just above the substrate, so a thin cover of substrate is all that is required. For optimal activity, the temperature should remain between 23–26 °C (73–79 °F), although the large milkweed bug is capable of surviving in much cooler temperatures.

Both these species are quite capable of living without water, although they have been known to drink if the opportunity arises. They suck food and water through their proboscis. The large milkweed bug can be fed on milkweed plants; it especially favors the pods and seeds. Stored dried seed is fine. In laboratories a strain of this species has been developed that feeds on shelled sunflower seed. This form is nontoxic, unlike the milkweed-feeding wild types that use the toxins within milkweed to become toxic themselves. Two-spotted assassin bugs can be maintained on a diet of crickets. The more they are fed, the greater the number can be maintained together, as less cannibalism will occur. Both species also display aposematic coloration. The bright orange and black color of the large milkweed bug is a warning to predators of its noxious, poisonous nature. The two-spotted assassin warns of its venomous bite, which is extremely painful and can be lethal to people if an allergic reaction (anaphylaxis) occurs. The two-spotted assassin is also capable of "spitting" venom, which is very painful if it hits a person in the eye.

BEETLES (ORDER COLEOPTERA)

The beetles are the largest order of animals in the world. There are approximately 10 times more species of beetles than of vertebrates. The diversity in their life history is just as broad. There are carnivores, herbivores, carrion feeders, aquatic dwellers, desert dwellers, fossorial species, and arboreal species. Some are adaptable to living among humans while others suffer due to human activity. The methods of care for the beetles are just as diverse. Certain species kept in captivity require specific diets to survive, whereas others are generalists. The most important and popular species in North American institutions are the scarab beetles. Temperatures for all should be maintained between 20–25 °C (68–77 °F).

Several species of scarabs are regularly maintained in collections. This large family contains the dung beetles, the flower beetles, and the giant rhinoceros beetles. As larvae all feed on decomposing plant matter, while adults become sap, fruit, or leaf feeders. One popular species is the yellow-bellied chafer (*Pachnoda flaviventris*). Another species recommended as a species to be maintained by the AZA Terrestrial Invertebrate Taxon Advisory Group (TITAG) is the Atlas beetle (*Chalcosoma atlas*). Although quite different in size, these species have similar husbandry requirements. As adults, they both need to be able to move off of the substrate. Adults live on branches or any surface above ground at a temperature of 23–26 °C (73–79 °F). It is possible to keep an individual in a small space (30 × 15 × 20 cm, or 11.8 × 5.9 × 7.9 in.), but if there is more than a single animal a space of 60 × 40 × 20–40 cm (23.6 × 15.8 × 7.9–15.8 in.) is required. The base should be covered with a 10 cm (4 in.)-deep mixture of manure, dead leaves (oak, beech, or mixed deciduous), and soft (moist) rotting hardwood. In chafers the manure can be sterile commercial cattle or sheep manure, but for larger species such as Atlas beetles fresh cattle, zebra, or rhino manure results in higher production rates. Zoos are lucky in having such material readily available. Larval chafer beetles can be maintained in crowded conditions with 50 to 80 within an 80 × 40 × 15 cm (31.5 × 15.8 × 5.9 in.) space. Larger species, however, need to be given more space or cannibalism will occur, and usually only two to four larvae can be maintained within the same space. The medium should be changed when a high proportion is made up of larval frass (insect feces) or the medium looks very consistent in nature. Metamorphosis may take place after four months (in chafers), or after three years (in some of the larger rhinoceros beetles). Adults live from three to six months. All of the adults feed on soft fruit. The chafers and some other small species also feed on fresh leaves, and in captivity kale is often fed. The species contained within the scarabs are among the most unusual and colourful species of beetles.

The last species of beetle that should be considered is the American burying beetle, *Nicrophorus americana*. This endangered species is one of the SSP species of the AZA, and a complete protocol for their care is available. Adults are set up in pairs for captive breeding. They are given a deep substrate of loose damp soil, and they feed on insects or pieces of meat. For breeding, the pair is given a carcass of appropriate size. Working together, the pair buries the carcass, and once underground the carcass becomes a nursery for baby beetles. The female (and rarely the male) takes care of the brood until it disperses into the soil to pupate. This program has been underway for several years and self-perpetuating wild populations now exist from the captive born stock. This is one of the AZA's best examples of successful captive propagation and release of invertebrates.

BUTTERFLIES AND MOTHS (ORDER LEPIDOPTERA)

Complete books have been written on the husbandry and maintenance of this group, so only the basics are covered here. For displaying adults, large spaces are required. Butterfly displays range from 122 to more than 3,355 square meters (from 400 to more than 11,000 square feet), and most species tend to do better in the larger spaces. Feeding and watering requirements vary tremendously. Some moths do not feed as adults, while others have the longest "tongues" for their size of any animal in the world. Many flowers have coevolved with the moths and butterflies to maximize their chances of procreation. Those moths and butterflies that feed require a liquid diet. Besides nectar, species have been known to drink tree sap, urine, muddy water, tears, fecal material, and liquefied fruit. Offering a variety of feed types is always important.

The normal way of transporting and receiving lepidopterans is in the pupal stage. At this stage the animals are undergoing tremendous changes internally. They are relatively immobile for days, and thus can be moved without having to be fed and watered. They normally arrive in tightly well cushioned compartments, or individually wrapped in some cases. Each specimen must be unwrapped and hung by the pedicel (the tip of the pupa normally attached to the substrate). In most cases a small amount of silk or a piece of material remains. This allows a pin to be inserted through it and into the medium from which the pupa is being hung. If there is nothing to use the pin on, a small amount of carpenter's glue (wood glue or white glue) or a tiny amount of material from a hot glue gun can be used to attach the pedicel to a piece of paper, which can then be pinned like the others. It is important to keep the pupa warm (24–28 °C/ 75–82 °F) and humid (above 80%). Pupae can hatch within hours or, in certain species, only after months. Pupae should be maintained in a confined space, as they may be carrying parasites. These parasites hatch out, killing the pupae, and should be destroyed as soon as they are discovered. If any pupae become black or start weeping, they have probably been damaged or infected, and it is therefore best to separate them from healthy stock until they are destroyed. All surfaces should be sterilized before bringing in new pupae.

Larval care is more complicated. Caterpillars usually are quite specific as to which plant material they will eat, and those that will eat several food types usually can only change their food choices directly after a moult. Caterpillars do best on fresh leaves. If space is sufficient, a fresh batch of food can be placed beside the old batch thus allowing the caterpillars to move over themselves. Leaves are best presented on a branch that allows the caterpillars to climb. To maintain their freshness, branches can be placed into a container of water, but it is extremely important to plug the open spaces between the branch and the water; otherwise the caterpillars will follow the branches, continue into the water, and drown. If a caterpillar has to be moved, the best way is to allow it to walk from one branch to the other. It should not be picked up and moved; the exoskeleton of a caterpillar is extremely fragile, and what is common in handling other insects is deadly for lepidopteran larva. Their skin bruises easily, and the oils that are normally on the keeper's hands can suffocate them. During certain periods (usually just before a moult) caterpillars become dormant; they anchor themselves to branches and stay still until the moult is complete. If they are removed from their site, they are unable to reattach themselves and will then be unable to complete their moult. These specimens ultimately wither and die. If it is necessary to move such animals, one should leave them attached and move the caterpillar attached to its anchor. Scissors or clippers usually work.

MYRIAPODS

Although centipedes and millipedes are not as obvious in the environment as butterflies or grasshoppers, they can and do make interesting exhibit animals under the right conditions, and they can live several years. Keepers should be aware that many species are nocturnal or semifossorial, and that many of these animals like to hide, and thus present challenges to display similar to those presented by scorpions. Centipedes are carnivorous and can consume a wide variety of prey (including crickets, earthworms, mealworms, and even small mice), but keepers should refrain from overfeeding them as it will shorten their lifespan; usually once or twice a week is sufficient. They must also be kept singly, except for reproductive purposes, as they will feed on other centipedes. Water can be provided by misting, but larger species will also use a shallow water dish, and keepers should ensure that their substrate (usually a soil mix) does not become too dry (especially for forest dwellers), as dessication is a concern. Some species are found in arid climates, but they shelter in moist microclimates during the day, such as underground and under large rocks. Most species are nocturnal, but a reverse light schedule can be used to enable viewing during the daytime. Enclosures must be secure, with metal screening used for ventilation, as the larger species have been known to chew through weaker materials such as fiberglass and plastic. Species from the genus *Scolopendra* are most commonly exhibited, with the South American *S. gigantea* relatively popular in zoos and among private hobbyists due to its larger size (up to 30 cm [12 in. long]) and its red body with yellow legs. They are quite fast and aggressive, with powerful venom, and should not be handled without tools. The females are known to protect their eggs.

Millipedes are herbivorous, consuming decaying plant material in the wild, but they can feed on produce such as lettuce, vegetables, and fruit in captivity. They can be fed ad libitum, with the food presented in a dish for ease of presentation and removal, and uneaten food removed daily; they may also require an additional source of calcium. Most species require sufficiently moist substrate (usually a soil mix) and higher humidity, but keepers should provide sufficient ventilation to prevent mold growth. Millipedes can be kept in groups with both males and females present; they lay their eggs in the substrate. The adults are usually dark-colored (black or brown, although some species can be bright orange), and the juveniles are usually white with few segments and legs, although additional segments and legs are added with growth. Most species are slow-moving and nocturnal, but again they can be displayed with a reverse lighting schedule. They are not aggressive, often coiling up when threatened, but they do possess paired glands on each segment that emit a noxious fluid mix of chemicals as an antipredator defense. Because of this, keepers should take care to wash their hands after handling them and prevent the fluids from contacting their eyes and mouth. The giant African millipede (*Archispirostreptus gigas*) is the most commonly exhibited species, which can reach up to 28 cm (11 in.) in length. Keepers should note that this millipede will burrow, and should account for its behavior when designing its enclosure.

TERRESTRIAL SNAILS AND SLUGS (CLASS GASTROPODA)

The European striped slug (*Limax maximus*), the banana slug (*Ariolimax columbianus*), the European garden snail (*Cepaea nemoralis*), the Roman snails (*Helix* sp.) and various giant land snails are often displayed because of their impressive size or bright colors (the giant land snails are not displayed in the United States, due to various regulations). They represent a large important taxa not often displayed. The unique mouthparts (rasping radula) and mode of locomotion make these species great educational tools. They can be maintained in enclosed plastic or glass containers with dampened substrate and no need for extra light. Only the tropical giant snails require warmer temperatures (25 °C/ 77 °F). Most gastropods are nocturnal, but if the substrate is moist and the light is not overly strong, they can be stimulated to move during the day.

Moisture should be offered through the food or by regular misting. Water in an open dish is not to be used, as the animals may enter and drown. As for food, these species are detritivores and herbivores. Lettuce, sliced yams, carrots, and apples can be placed with dead deciduous tree leaves and rotting wood. For a calcium source, cuttlebone works well (slugs require calcium too). This ensures proper shell growth and is required for metabolic purposes. Occasionally a protein source should be offered. In the wild this would be provided by carrion or fecal material. In captivity, a small amount of mouse chow or dog kibble could be used.

Reproduction is possible in captivity with these species. Some are hermaphrodites (in which the individuals have both male and female reproductive organs), so that any two animals could breed, while in other species male and females are separate. In the giant snails, an appropriate diet and warm humid conditions are all that are required for successful reproduction. Temperate species may reproduce in their first year, but a seasonal cooling is required to be continually successful. Eggs are laid under cover or buried in the substrate. The eggs hatch within a few weeks to a few months (some eggs laid in the fall may not hatch until the next spring). Young can be treated as the adults, but they are extremely sensitive to desiccation and overwatering. Their growth is very quick as long as the appropriate food and water is available. The food must be plentiful, but rapid fungal growth must be avoided.

ENCLOSURES, TOOLS, AND TRANSPORTATION

Most of the species considered and discussed in this chapter will require relatively modest enclosures, typically constructed of glass and transparent acrylic (plastic) for optimal viewing and ease of cleaning. Glass may be preferable over acrylic, due to its greater resistance to scratching (a concern when cleaning), but it is also heavier than acrylic and can break if struck or dropped, whereas acrylic is impact-resistant. All enclosures must be ventilated, but openings must be screened (preferably with metal screening that cannot be chewed through) to prevent escape. Keeper access for husbandry efforts must be considered when choosing or designing an enclosure, and keepers should be able to access the enclosure without fear of escape or harmful interaction with the species within. There are many possible designs, including the use of commercial aquaria (especially for "holding" areas where aesthetics are not important) and even commercially available reptile/invertebrate enclosures with integrated screened openings. Exhibits may also be designed and constructed in-house by the facility's staff using materials such as fiberglass and gunnite and possibly incorporating intricate water features for certain species such as crabs. Other animals can be kept and displayed in quite large and unique exhibits such as the leafcutter ant exhibit at the Cincinnati Zoo, where lengths of acrylic tubes traveling throughout the World of the Insect exhibit building allow the ants to forage for leaves and return to their nest. As noted throughout this chapter, however, many species can be kept quite simply and without occupying much space.

Tools used in invertebrate husbandry usually consist of forceps used for the feeding of prey to certain species, small scoops to remove fecal material and uneaten food from the enclosure substrate, variously sized plastic cups used to capture and move species that cannot be "free-handed" due to venom or delicacy of the invertebrate, and spray bottles used to mist enclosures. As is done with tools used for other groups, tools used with invertebrates must be cleaned after use and stored so that they are accessible and free from damage.

Transportation needs are usually quite modest for most species discussed in this chapter, many of which can be moved in simple clear plastic cups ("deli cups") with snap-on lids. Moisture, perching, and cushioning needs should be addressed within these plastic containers, with moistened paper towels often used, and small twigs placed within for perching if necessary. These plastic containers are then placed within expanded polystyrene (Styrofoam®) boxes for insulation, and materials such as crumpled newspaper can be placed among the plastic containers for additional cushioning. Small ventilation holes in the cups can be created with a hot soldering iron tip or a punch, but the keeper should ensure that no sharp edges are present. If ventilation openings are needed in the larger shipping container they should be screened to prevent escape, and the lids on individual plastic containers can be taped to prevent their separation from the containers during transport.

VETERINARY CARE

Very little is known about the veterinary requirements of invertebrates, and invertebrate medicine is still in its infancy.

The small size of invertebrates makes pathology difficult, as does the fact that their tissue begins to autolyze (break down and liquefy) soon after they die, leaving little for the pathologist to work with. Also, with only their chitonous exoskeletons to support them, little can be done to treat their wounds. Research has been done on commercially important species such as the honeybee (*Apis* sp.), and on some pest species; indeed, much invertebrate veterinary care is based on treatments used for bees. Little has been done with spiders and insects. In 2009 work was begun with millipedes; nematodes once thought to be part of their normal gut fauna were eliminated with positive health effects.

CONSERVATION AND RESEARCH

As invertebrates are basically the workhorses of the environment, their presence in a zoo should serve as the perfect tool for education about the importance of conservation and the delicacy of many ecosystems. Terrestrial invertebrates are harmed by the same forces that are problematic for vertebrates: habitat loss, introduction of alien species, and pollution. Many populations, especially of island species, exist in limited areas, and modest environmental pressure can cause rapid population declines. Zoos in general do not participate in terrestrial invertebrate conservation very often. One genus of spider (*Brachypelma* sp.), a species of beetle (*Nicrophorus americana*), a genus of snails (*Partula* sp.), and a few species of butterflies are currently the subjects of conservation initiatives at zoos in North America, but with increased awareness of the needs of terrestrial "microfauna," further efforts will be required.

SUMMARY

Even with an enormous diversity of species in four different subphylums, terrestrial invertebrates share many husbandry needs and can serve as wonderful exhibit specimens. Many zoos incorporate at least a few invertebrate exhibits into their collections, and some devote significant resources to their display. Simple displays can have great impact on people unaccustomed to viewing "exotic" species, and even local species can be fascinating when viewed up close. Not all keepers will have entomological backgrounds or will view the task of invertebrate care as desirable as that of caring for some of the vertebrate taxa, but if they devote enough time to research and preparation, they will often find the fascinating behavior and lifestyles of their invertebrate charges as enjoyable as their public visitors do. Many invertebrate species may look "otherworldly," yet may respond as a vertebrate would to environmental cues such as food and shelter. Their sheer numbers and diversity suggest that more species may be exhibited at zoos in the near future, and one hopes that zookeepers and zoo visitors will continue to be fascinated by them.

Part Six

Animal Behavior, Enrichment, and Training

37

Introduction to Animal Behavior

Michael Noonan

INTRODUCTION

In its own way, the American Museum of Natural History in New York City is an example of a truly great zoological institution. Included among the many other things that it does well, the museum displays naturalistic dioramas that depict wildlife in faithful reproductions of their natural habitats. From darkened walkways, the visitor is afforded successive glimpses of nature that are truly spectacular. In the African Hall, for example, one sees a life-sized family of mountain gorillas on a thickly forested hillside. A few steps away, one is afforded a view of musk oxen in a flawless recreation of the high Arctic. The animals, plants, and environmental elements are displayed so perfectly that when one stands close to the viewing windows, it is possible to get the impression of viewing the scenes in nature.

Readers of this text will immediately recognize the similarity of goals shared by museums of this type and modern zoological parks and aquariums. The goal in both types of institutions is to give visitors educational experiences that inform them about the natural world. Like museums, modern zoos make great efforts to display animals in realistic recreations of their natural habitats. Moreover, zookeepers can be rightly jealous of museums because they have many advantages that zoos do not. For one thing, the delicate plants contained in their dioramas are not continuously picked to pieces by their animals. For another, the preserved museum animals do not have any of the special dietary or veterinary needs that living animals do. Nor does the museum have concerns about containment, backup holding facilities, and so on.

So why do we bother with live-animal displays in zoos and aquariums? Why do we not all convert to museum-type displays? The answer is quite simple: animal behavior.

ANIMAL BEHAVIOR IS CENTRAL TO ANY ZOO OR AQUARIUM'S MISSION

In viewing a gorilla at a zoo, one obtains a richness of experience that cannot be captured in a still diorama. The visitor gets to experience the gorilla's furtive catch of the eye, followed by the equally quick look away. There is the experience of watching gorillas pick through their browse in search of preferred food items. There is the experience of long periods of lazy tranquility punctuated by dramatic outbursts of mock aggression. In short, the visitors get to experience gorilla *behavior*. As a consequence, the visitor to a zoo or aquarium gets a different—and arguably better—experience than the one at the museum.

The museum dioramas are spectacular. They really are. But they are static. Even at their best they are merely the equivalents of really good photographs. Museums can never match the zoo and aquarium experience of being in the presence of a living, breathing animal that looks back at you. Thus, it is the breathing and looking—the foraging and socializing—that distinguishes zoos and aquariums from museums. In brief, it is animal behavior that makes the zoo or aquarium experience unique.

This is a very important point. The display of behavior—of exotic animals *behaving*—can be described as the single thing zoos and aquariums add to our culture that other institutions do not. Thus animal behavior should be uniquely at the center of any zoo's mission. In fact, it is argued here that the provision of animal behavior to the public is the very raison d'être of zoos and aquariums.

If one accepts that argument, then it follows that providing opportunities for animals to display their full behavioral repertoires should be uppermost on the minds of the exhibit designers and curators who set the zoo and aquarium stages. It should also be uppermost on the minds of every keeper who manages exhibits on a day-to-day basis. After all, in a modern zoo the visitor should experience much more than seeing a curled-up animal staring back from within a barren cage. The ideal should be that zoo and aquarium animals are provided with rich habitats that allow them to behave in very much the same way as they do in the wild.

ETHOGRAMS AND BEHAVIORAL BUDGETS

In addressing this goal, two terms are used by animal behaviorists that should be familiar to every keeper. One is ethogram; the other is behavioral budget. Both can be thought of as ways of assessing progress toward the goal of displaying the full repertoire of animal behavior.

An ethogram is a catalog of all the behaviors performed by any given species. Ideally, it is a lengthy document that describes every unique behavior seen in that species under every conceivable circumstance. In practice, ethograms usually take the form of compact lists of brief descriptions of behaviors, along with shorthand abbreviations that can be used when taking rapid observational notes. Usually such shorthand ethograms consist of discrete behavioral events: short acts with discrete starting and stopping points, such as the fluffing of feathers in courtship displays, the stripping of leaves from twigs, stereotyped territorial vocalizations, and so on. As a case in point, an ethogram used by the author to assess behavior in captive beluga whales is provided in table 37.1.

A behavioral budget is a tally of all of the different behavioral states a given animal enters over an average time period. Unlike the discrete events typical of an ethogram, behavioral states usually last longer and are primarily defined by the animal's inferred motivation. In practice, this involves watching the animal's behavior over a given time span and making inferences about the overall goals it is trying to achieve. Examples of behavioral states include foraging, play, socializing, and resting. It is useful to think of a behavioral budget as a pie chart that summarizes how a given species spends its day. Behavioral budgets are often published that describe the average behavior of an entire population. However, it is sometimes even more interesting to separately present a species' behavioral budget broken down by season, or separately for males and females, or for juveniles and adults.

The bottom line is, that for a keeper, both ethograms and behavioral budgets based upon wild animals are invaluable tools that allow us to know what our captive animals are *supposed to be* doing. They serve as benchmarks against which the suitability of zoo-based conditions can be judged. If animal behavior is indeed the unique thing that separates zoo and aquarium displays from museum displays, and if the preservation and presentation of animal behavior is indeed a central mission of a zoo or aquarium, then keepers can use wildlife reports to learn about what they should be preserving and presenting. The logic is simple. If wildlife researchers have reported that wild gorillas forage approximately 55% of their time, but the gorillas in a zoo are spending only 10% of their time picking food out of their pans, then the zoo's goal is not being met. Some facilities solve this problem by using a "scatter feed" technique in which food is spread out over an entire exhibit and then covered with straw. The gorillas are then obliged to pick through the straw in search of tasty items (figure 37.1). The goal should be to get them to the point where they are looking for food items for about the same 55% of their time as is observed in the wild. Similarly, if wild dolphins have

TABLE 37.1. Shorthand ethogram used by the author to investigate behavior in a captive population of beluga whales

BRE	Breath
URI	Urinate
DEF	Defecate
REG	Regurgitate
HEA	Heave
QUI	Body quiver
HBK	Humpback
BBN	Backbend
SSH	S-shape
CSS	Corkscrew swim
HRJ	Herky-jerky Swim
HNO	Head nod
VOC	Vocalize (type)
LEU	Left eye up
REU	Right eye up
LFT	Left turn
RTT	Right turn
LAT	Look (at target)
JAW	Jaw pop (at target)
OPM	Open mouth (at target)
CHS	Chase (target)
BUB	Bubble (from where) (bubble type)
SUC	Suck (thing)
SPI	Spit (thing) (at target)
PAR	Pec aerial (intensity)
TAR	Tail aerial (intensity)
HFA	Head first aerial (extent) (landing sky side) (landing speed)
TOU	Touch (body part of focal whale) on (target whale) (body part of target)
TOU	Touch (body part of focal whale) on (BIT) against (target)
TOU	Touch (body part of focal whale) on (BUB) (effect)
TOU	Touch (body part of focal whale) on (bottom, rock, wall, gate, object)
RUB	Rub (as any of the above)
PSH	Push (as any of the above)
ATT	Attempted touch (as any of the above)
RET	Receive touch on (body part focal whale) from (other whale) (body part of other whale)
REP	Receive push (as any of the above)
AVO	Avoid touch (as any of the above)

been observed echolocating as they explore their surroundings, yet the same behavior is largely absent in captivity, then the caretaker of the captive dolphins should try to increase the complexity of their exhibits so that they have the opportunity and need to exercise that important aspect of their natural behavior. If gibbons in the wild normally produce territorial calls every morning, then in captivity they should be provided with high branches spaced far enough apart so that they are stimulated to make the same morning displays.

If an animal in a zoo is observed to be in a resting state 80% of its time, would that be acceptable and how would one know? If it were a species of cat like a leopard, then the

Figure 37.1. Gorilla at Detroit Zoo foraging after a scatter feed. Photo by M. Noonan.

behavior probably would not cause concern. This is because naturalists report that wild cats spend about that same proportion of their time resting. If, on the other hand, the animal in question were a marmoset, then it would be reasonable to suspect that the conditions of captivity were limiting it in some way, since wildlife observers report that free-ranging marmosets ordinarily rest only 30% of their time. Consider, by contrast, an animal that is observed to be in motion 80% of its time. If it were a bald eagle, this would reasonably be judged as exhausting and as a probable sign of distress, because in the wild, members of that species remain perched and quite still for long periods. On the other hand, if the animal were a water shrew, the shorter periods of rest would be considered not only normal but desirable. Shrews are species in which nearly nonstop motion is the norm. In each instance, then, it is possible to evaluate the behavior of zoo and aquarium animals and the efficacy of displays by comparing them to the ethograms and behavioral budgets compiled for wild populations. Wherever the animal's behavior in captivity differs from that reported for conspecifics in the wild, the keeper should strive to eliminate the difference.

As an aside, this logic leads instantly to an interesting debate. Many aspects of nature are harsh, and are not often on display in zoos and aquariums. For example, it can be asked whether prey animals should be deliberately subjected to frightening circumstances so that they have an opportunity to exercise and demonstrate the watchful/avoidance sides of their natures. Similarly, it can be asked whether predatory animals in captivity should be allowed to attack and feed upon live prey. Another particularly difficult question pertains to whether animals should be allowed to breed, even if it would lead to surplus offspring.

In any event, every keeper should become an avid reader of wildlife publications pertaining to the species for which they are responsible. Every keeper should also allocate time for formalized data collection so that they can know the ethogram and behavioral budgets of their particular animals. (When keepers lack the luxury of time to spare for such an endeavor, students and/or volunteers should be recruited to carry out the data collection.) Every keeper should make it their mission to adjust exhibit furnishings, food presentation procedures, social groupings, and all other environmental factors so that the ethograms and behavioral budgets of their animals come as close as possible to matching those for wild animals of the same species.

BEHAVIOR CAN SERVE AS AN ANIMAL WELFARE WINDOW

There is another reason why keepers should be continuously attuned to the behavior of their animals. Animal behavior often serves as the principal channel by which keepers monitor the health and welfare of their charges. Since nonhuman animals cannot use words to inform us about their health or mental state, we must rely on their behavior to give us clues.

An apparent loss of appetite is the most obvious example. Every keeper knows that a decline in food intake is almost always a cause for concern. But many more subtle, but equally important, examples also exist.

Regulatory behavior can tell us a great deal about the suitability of the environmental variables in animal enclosures. For example, an excess of huddling behavior can indicate that temperatures are too cold, just as the observation of excessive sprawling often suggests animals that are too warm. When animals scratch themselves more than usual, it is often because the ambient humidity has fallen or risen inappropriately. Keepers of marine mammals are always on the watch for signs that their animals are rubbing themselves on pool walls, because that is often the first sign of an imbalance in water chemistry.

There are also many instances in which observations of behavior are critical in managing reproduction in captivity. This is particularly important in species that are characteristically solitary in lifestyle. For example, the females of most cat species will respond with aggression to the approach of a new male, except when they are in estrus. An observant caretaker can sometimes detect the optimal time for an introduction by watching for an increase in solicitous behaviors. In cats, a sudden increase in behaviors like rolling can indicate the onset of sexual receptivity, the critical time when a pairing can be safe and effective. Similarly, in the Indian rhinoceros

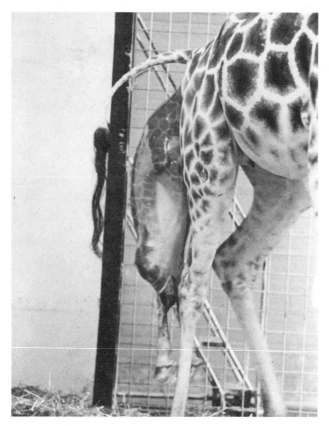

Figure 37.2. Female giraffes give birth while pacing, with the result that the neonate's first experience is a drop to the ground. Photo by M. Noonan.

male-female introductions at the wrong times can lead to serious fighting and injury. These are long-lived animals that have exceptionally slow reproductive cycles in nature; keepers must be particularly patient and observant of the females. It is an increase in whistling that is a sign of estrus in this species. Timing an introduction to a period of behavioral whistling can mean the difference between success and disaster.

Consider too the problem of managing births in captivity. Parturition is a critical time for any species. It is a particularly dangerous time for giraffes. Giraffe mothers characteristically give birth in a standing position while slowly pacing (figure 37.2). The neonate has to drop approximately two meters to the floor, and then it has to rapidly learn to stand and walk. Concrete floors can be a particular hazard, both because of the risk of injury upon impact and because they present very slick footing when made wet by birth fluids. Ideally, the keeper would provide a parturient giraffe with sufficient dry

bedding so that the new baby can have both a soft landing and a firm footing, but the challenge is in anticipating the time of parturition. Since it is neither practical nor sanitary to maintain thick bedding for long periods, the keeper must look to the behavior of the pregnant giraffe for signs that the birth process is imminent. Systematic observations made by the author at the Buffalo Zoo during seven pregnancies and deliveries indicated that drinking can be the indicator behavior in this species. During the 24 hours just before birth, the seven pregnant giraffes rarely, if ever, drank. A keeper who attended carefully to that behavior would know exactly when to prepare for the newborn's arrival.

Of course, these are only examples. The interpretation of each species behavior in captivity depends upon a careful understanding of that same animal's behavior in the wild. To appreciate this, consider just one more example.

Some years ago, there was a case in which a pair of zoo-based birds suddenly began to show what appeared to be problematic behavior. They had been living together for years and they had been thriving. But suddenly both birds significantly decreased their food intake. Instead of eating, they started mashing and smearing their food items all around their cage, particularly in the corners. The key to understanding this sudden change was in making a connection to the natural history of the species. The animals in this case were great hornbills (*Buceros bicornis*), a species that characteristically uses mud to modify tree holes for nesting. It was recognized that the smearing of food mash into the cage corners was reminiscent of smearing mud in nest cavities. It was therefore suspected that the pair had entered into a breeding cycle. It was recommended that a hollowed-out tree trunk be moved into the exhibit and that a pan of mud be provided. Since the zoo staff was already struggling with the mess the birds were making with their food, the idea of providing the birds with a pan of mud was not initially well received. Yet, once it was tried, it proved to be the ideal solution. The birds immediately went back to eating their food, and both of them used the mud to prepare a perfect nest. In short order, the female entered the cavity and successfully brooded a clutch of eggs.

This last example nicely illustrates three themes pertaining to animal behavior in zoos. One, cavity nesting in the hornbill is a rich behavioral sequence that visitors to this zoo were able to experience in a way that would never be possible in a museum. Two, provision of the correct environmental elements can promote the welfare of captive animals by allowing them to complete their life cycles. Three, the proper interpretation of behavior can lead to an effective management decision. In all three ways, this illustrates the importance of attending to animal behavior in the zoo setting.

38

Applied Animal Behavior

B. Diane Chepko-Sade

INTRODUCTION

An understanding of animal behavior is invaluable to a keeper in maintaining his or her safety around animals, obtaining the animal's cooperation in basic husbandry procedures, and in interpreting the animals' behavior to the public. In maintaining safety, keepers use their knowledge of animal behavior to "read" the animal's body language so as to correctly interpret, for example, whether it is likely to attack or interact in a positive manner, or whether it is showing an inclination toward reproductive behavior or territorial behavior. Similarly, a keeper applies his or her knowledge of animal behavior to train animals to cooperate in their daily care and medical examinations, thus transforming potentially stressful situations into routine and mutually satisfying activities. Finally, keepers are called upon to present information about different animals to the public, and a common question of zoo visitors is: "Why is the animal doing what it is doing now?" A detailed knowledge of animal behavior will help the keeper to answer such questions accurately and in a way that stimulates the visitor's fascination with animals.

The mission statements of many zoos include education of the public and quality care of the animals. Both of these goals can be well served by understanding the behavior of animals. We can best interpret an animal's behavior to the public if we understand it ourselves, and we can provide the highest quality of care for animals if we understand what their behavior is telling us about their level of satisfaction with their environment. After studying this chapter, the reader will understand

- the science of ethology, or animal behavior
- the difference between trained and domesticated animals
- classifications of behaviors based on their function
- classifications of behaviors based on how they are organized within the animal

- the importance of early experience in animal development, and how hand-rearing can impact development
- how to interpret species' specific behaviors to promote safety and animal care
- how to recognize and mitigate stress in the lives of captive animals
- the role of behavioral research in zoos, and how keepers can facilitate and learn from it.

ETHOLOGY

Biologists who study the natural behavior of animals are called ethologists. Ethologists travel to wherever a species lives naturally to study the habitat the animals live in, how they use their environment, what food they eat and how they find it, how they protect themselves from predators, how they find mates, and how they raise their young. Their accounts are recorded in scientific journals and are available either online or in biological libraries. Details of some of the more unusual behaviors and behavior of some of the more dramatic or exotic animals have been videotaped and aired on television in shows such as *Planet Earth* (BBC 2007) or *The Life of Mammals* (BBC 2003). While such videos are valuable resources, it is best to realize that the picture they portray of an animal's life represents only the tip of the iceberg, the "exciting parts" edited from hours and hours of more routine behaviors such as those keepers are likely to observe in the animals in their care. It is important to know both the dramatic behaviors and the more routine ones, as well as knowing how much of each day a healthy animal spends in each type of behavior. The original accounts of ethologists can provide a keeper with an accurate picture of how the wild counterparts of their captive animals live. For example, an ethologist will sometimes publish a "behavioral budget" (also called an activity budget) for a species. This is a breakdown, usually by percentages, of how much time a species spends resting, feeding, engaged in social interaction, and in other activities. A keeper can compare such a behavioral budget with the amount of time their captive animals spend in the same activities. Any serious discrepancies may provide valuable insights into changes that might be made in the management of the animals that could improve their health.

DOMESTICATED VERSUS TRAINED ANIMALS

Domesticated animals are animals that have been bred over many generations to be docile and cooperative with humans. Dogs, cats, cattle, horses, and camels are examples of animals that have been under domestication for thousands of years, during which breeders have selected only those individuals with the characteristics they wanted to sire the next generation. Animals bred for food (cows, sheep, goats, chickens) were selected for large size or high milk or egg production. Animals bred for transport (horses, camels, llamas) were bred for strength and stamina, but also for docility and willingness to follow commands. Companion animals (dogs, cats) were bred for their willingness to interact socially with humans as well as for attractiveness. Different goals have led to different breeds of animals. For example, dogs have been bred for many different types of work including guarding, herding, hunting, and racing, each giving rise to specific characteristics in different breeds of dogs, each of which continues to be "improved" with selective breeding. Although many domestic animals can still breed with their wild counterparts, they seldom do so, and most are completely dependent on man for their food and shelter. They have lost many of the genetic characteristics that allow their wild counterparts to survive in the wild, most notably the fear of humans.

Although many animals in zoos and aquariums today were born in captivity, most are within 3 to 10 generations of a wild ancestor. This would not be long enough for domestication to occur even if institutions were actively selecting for docility and cooperation. Although animals that do not tolerate captivity well often do not survive to reproduce in captivity, it is not the goal of zoos or aquariums to domesticate the animals they display. The goal is to let the animals retain their wild behaviors and characteristics, both for public display of those characteristics and for the maintainance of captive breeding populations from which we could ultimately replace wild populations of their species should they become endangered or extinct in part or all of their range.

Most zoos and aquariums do train their animals. Unlike domestication, however, which takes many generations of genetic selection, training occurs within the lifetime of a single animal and can be accomplished in an afternoon for some behaviors. Training is behavioral modification based on a schedule of rewards carefully timed so that specific behaviors will be reliably performed when the animal is presented with a specific cue (Pryor 1999).

TYPES OF BEHAVIOR

Behavior can be categorized in a number of different ways. One way of categorizing it is by the function of different behaviors. Functional categories include behaviors related to feeding, resting, socializing, reproduction, play, self-maintenance (such as grooming and preening), and self-

defense. Each of these larger categories can be broken down further into more specific categories (e.g., behaviors related to feeding include searching for food, obtaining food, extracting food from its coverings, and eating food). The sum of all of the behaviors expressed by any animal of a given species is called the species' behavioral repertoire.

A different way of categorizing behavior is with reference to how it is organized within the animal and how it develops within the individual. Even animals with no nervous system at all, like protozoa, show behavior in that they move toward stimuli that are favorable to them and away from stimuli that are dangerous. Such a behavior is called a taxis, and is the result of bioelectrical changes in the organism's cell membrane which cause its cilia to beat in a pattern that will bring it either closer to or farther from the stimulus (Naitoh 1974). Behavior of higher organisms is organized through the nervous system, but can vary in complexity from behaviors that are reflexive (the result of a stimulus followed immediately by a response) to behaviors that are the result of learning, usually through trial and error. Reflexive behaviors are said to be "hardwired," a function of the animal's nervous system. They take the same form in every member of the species and occur in response to the same stimuli. Such behaviors are said to be innate or instinctive. Innate behaviors may be simple or complex. The distinguishing characteristic is that the behavior is performed by each animal of the species perfectly the first time without training. On the other hand, behaviors that are slowly perfected only after a period of trial and error (which may be preceded by a period of observation) are said to be learned.

Each behavior in a species' repertoire has some innate components and some learned components. For example, all mammals are born knowing instinctively how to suckle. They perform sucking motions perfectly shortly after birth. However, they don't immediately know where to suckle. The newborn infant mammal moves its sucking mouth around its environment until it encounters food (usually warm milk from its mother's teat) and then it stops searching and sucks until it is satisfied. Once the infant has learned where to suckle, its next nursing bout will be more targeted.

Some species' repertoires comprise behaviors with more innate components, while other species' repertoires comprise behaviors with more learned components. Animals with simpler nervous systems, like insects and other invertebrates, are likely to have more innate behaviors because those behaviors require fewer neural pathways. Birds and mammals tend to show behaviors with more learned components. Among species with similar nervous systems, it is thought that species that live in more predictable environments (for their mode of life) tend to have more innate behaviors than animals whose lives are less predictable. For example, among annelid worms that live in coral reefs, the worms in one group (the Christmas tree worms) live stationary lives, in calcareous tubes attached to the coral. Their behavior consists of opening up their "fan" of feeding tentacles and moving small projections on them to create a current which draws plankton into the fan and eventually into their mouth. If the area over them darkens suddenly, they rapidly withdraw their fan into their tube to wait until danger has passed. The related nereid worms are free-living predators that search the coral heads looking for smaller invertebrates to eat. Their behavioral repertoire includes searching, recognizing, capturing, subduing, and eating prey. Not only are their behaviors more varied, but they are more likely to have a greater learned component than those of their stationary relatives. Similarly, howler monkeys, which eat only leaves, show a higher proportion of innate behaviors than do spider monkeys, which eat fruit (Milton 1988). The leaves required to support a group of howler monkeys are available in a smaller area than is the fruit required to support a group of spider monkeys. Moreover, fruit is seasonal. So spider monkeys must learn not only where the various fruiting trees are within a larger area, but when each type of tree has last borne ripe fruit, and when it will again be likely to provide them with food.

HAND-REARING AND EARLY DEVELOPMENT

Early development has a critical effect on the forms some behaviors will take. Songbirds learn how to recognize their own species by seeing the characteristics of the parent birds that feed them while they are still in the nest (Marler and Tamura 1964). This nesting period is a critical period during which they imprint on the form and sound of the parent. Later, when the male bird reaches adulthood, it will match its courtship song to the song it heard during this period, and it will direct that song toward a bird that looks like the female that fed it in the nest. A female bird will be most attracted to males with songs and visual characteristics similar to those of its father. Imprinting works very well in assuring species recognition in wild birds, but what happens when birds are raised from eggs in captivity?

Some of the first whooping cranes raised in captivity became imprinted on the humans who raised them, and when they became adults they either failed to breed altogether or directed their mating dance toward humans instead of other whooping cranes. Other whooping cranes were incubated and reared by sandhill cranes, but this approach also resulted in imprinting on the wrong species and a failure of captive reared birds to reproduce. As a result, a rearing method was developed at the Patuxent Wildlife Research Center in which whooping crane eggs are incubated and reared in relative isolation from humans. Human caregivers dress in large white garments designed to disguise the human form, and use puppets in the form of the heads and necks of adult whooping cranes to feed and care for the chicks so that the chicks will imprint on these visual images rather than on people. At the same time, an adult or subadult crane of their species is housed in an adjacent enclosure within view of the developing chicks, to facilitate imprinting on their own species (Ellis et al.,1992). Similarly, people involved in the peregrine falcon rehabilitation project (Gallagher 1999) realized that if the young birds they were raising in captivity were to mate and breed with members of their own species once they reached adulthood, caregivers must be very careful not to let the birds become imprinted on humans. Although the very young hatchlings are fed by hand, by three weeks of age they are placed with peregrine foster parents that feed and care for them during the time they would normally develop in the nest (Blomqvist and Larsson, N.d.)

Many people are drawn to zookeeping because of the prospect of hand-rearing baby animals. Who would not love to cuddle a baby monkey or snow leopard kitten? But early experiences are as important for mammals as they are for birds. Although mammals do not show as clear-cut a pattern of imprinting as birds do, many are strongly affected by their experiences with humans early in life. Negative consequences can include orientation of their breeding behavior toward humans instead of toward others of their own species, and complete inability to socialize with their own species or nurture offspring of their own (Porton and Niebruegge 2006). Such extreme distortions of behavior are usually a result of an animal being reared exclusively by humans with no contact with its own species. In some species, however, parent-reared young are so fearful of humans that they do not make good exhibit animals, hiding and showing high levels of stress when humans are around.

More positive behavior patterns can result from rearing that includes both human contact and contact with the infants' own species. Wolves raised exclusively by their parents, even in captivity, are shy and secretive, difficult for zoo visitors to observe, and dangerous for zoo keepers to care for. Erich Klinghammer, at Wolf Park in Indiana, developed a protocol for hand-rearing wolves in which the pups bonded with humans but also with other wolves (Klinghammer and Goodmann 1987). Young cubs are taken from their mother at an early age (before they are 21 days old) and hand-reared, with drastically limited access to their mother until they are 16 weeks of age. At this time, they are returned to the pack and are socialized to pack life while also continuing to be socialized with humans. This early socialization with humans results in wolves that develop all the natural behaviors of wolves, socialize with wolves, establish dominance hierarchies, and breed and care for their young, but that also socialize with humans and are comfortable carrying out their natural behaviors in the presence of humans. Wolves raised in this way have no fear of humans, and thus can be dangerous to keepers who are not trained to understand their behavioral signals. On the other hand, if socialization with humans is continued throughout their lives, these wolves can be much easier to manage in a zoo setting, as they can be trained to cooperate in their own medical exams and treatment, they are less stressed in the presence of humans, and when wolf social partners are unavailable they can benefit from social interactions with humans (Goodmann 2010). Wolves raised in this way are far easier to observe than are captive wolves not socialized to humans. It is possible that such dual socialization could help other wild species that experience high levels of stress when forced to live close to humans.

SPECIES-SPECIFIC BEHAVIORAL SIGNALS

All species communicate with members of their own kind by means of behavior signals. Depending on the species' sensory system, these signals may be visual, auditory, chemical (including olfactory), or tactile. Humans are best able to detect those signals to which our own sensory system is attuned: visual and auditory signals. We can often visually observe animals using tactile cues, but can detect most chemical cues only by performing a lab test—much too slow for communication in real time. On a practical level, there is much a keeper can tell about the health, social, and reproductive status of animals by paying attention to the visual and auditory signals

Figure 38.1. Expressive behavior of wolves (based on Schenkel 1947), an ethogram of the facial expressions of wolves showing different motivational states. The wolf in the upper left figure is showing what we call a "neutral" or confident pose, typically of a dominant animal, with elements of neither aggression nor defense. The wolf at lower left shows both anxiety and subordination, which could lead to either fight or flight. The keeper should beware. The wolf in the upper right is confident and threatening, and is likely to attack. The wolf in the lower right may be most dangerous of all. It shows a defensive threat, as of an animal both threatened and subordinate but also aggressive, such as a mother protecting her young. The keeper should watch this animal closely, protect against attack, and look for ways to reduce the animal's anxiety. Illustrations by Kate Woodle, www .katewoodleillustration.com.

animals present. Perhaps of more immediate concern, the keeper can also maintain his or her own safety when caring for animals by reading their behavioral signals correctly.

Ethologists who have studied the behavior of particular species have published detailed catalogues of those behaviors, frequently with diagrams explaining what the animals' body language looks like when the animals are hungry, aggressive and likely to attack, curious, seeking social contact, sexually aroused, and so on. Such a catalog of behavior is called an ethogram. By familiarizing themselves with the behaviors described in the ethogram of a species and comparing them to the behaviors of the animals in their care, keepers can come to understand the motivational states of their animals.

By studying figure 38.1, one can see that a wolf with its ears back is expressing subordination, while a wolf with its eyes narrowed, its nose wrinkled, and its teeth bared is showing a tendency to attack. Most animals will give some warning before attacking, and a keeper who pays attention and knows the signals can avoid dangerous encounters.

MISINTERPRETATIONS OF ANIMAL BEHAVIOR

Animals read the behaviors of people too, but they interpret them according to their own ethograms rather than understanding what the people really mean. For example, a person gazing lovingly into the eyes of a rhesus monkey may be shocked when the monkey suddenly reaches out and slaps him or her. What has happened? In rhesus monkey behavior, a direct stare into the eyes is a threat and will be interpreted as such, whether the person means it that way or not. Similarly, when one feeds a dog a treat, one will sometimes hold it up above their head and have the dog jump for it. If one assumes the same posture in front of a bear, however, the bear will recognize it as a posture bears use when they are about to attack. The bear may feel threatened, and may attack the person preemptively. Wild animals can be unpredictable, but they will be much more so if the person interacting with them is unfamiliar with their ethograms.

Just as animals may misinterpret our behavior based on their ethograms, we may misinterpret their behavior based on ours. When a male baboon yawns, we may assume that he is merely sleepy. In fact, this is a threat to other male baboons (or zoo visitors), in which the male is showing how big his canines are. It could be a prelude to a chase or other aggressive behavior, so that the animal we thought was sleepy may actually be about to attack! If a monkey pulls the corners of its lips up and back, we may think it is smiling and is pleased about something. Instead, this facial expression, seen in many mammals, is a grimace; it indicates extreme discomfort, often due to the presence of a dominant individual within the boundaries of its personal space. If we approach the animal further, it is likely to defend itself by attacking.

Interpreting an animal's behavior according to the human ethogram rather than its own is called anthropomorphism—assigning human characteristics to nonhuman creatures or phenomena. Anthropomorphism can lead to inappropriate care of animals by humans, as well as to life-threatening situations. For example, many people "fall in love" with baby exotic animals, such as chimps or wild carnivores, and raise them as pets or even as their own children. The ethograms for early nurturing of many different mammals are sufficiently similar to each other so that all goes well during the early months, or even years, for animals with long developmental periods. As the animal approaches adulthood, however, the caregiver will likely fail to understand or meet the species-specific needs of the adult animal, and this may result in injury or even death, as evidenced in the 2009 mauling of a woman by "Travis," a pet male chimpanzee that had been raised by its owner from an early age (Sandoval and Schapiro 2009).

HOW STRESS IMPACTS BEHAVIOR

Stress is the activation of the animal's adrenocortical axis, which mediates the flight-or-fight response, by secretion of a number of hormones, including cortisol. The flight-or-fight response may be activated when an animal perceives itself to be threatened, when it is injured, or when its environment is not meeting some important need. The hormones that are released when this happens prepare the body to fight or flee, shutting down those systems and behaviors that are incompatible with those immediate needs. The animal usually evacuates its bowels and bladder, stops eating, and stops all reproductive activities until the crisis has passed. This reaction is adaptive in wild animals if the crisis can be resolved by the animal's running away or fighting, as it allows all of the animal's energy to be focused on escaping the dangerous situation, or on fighting off the predator or intruder.

When captive animals experience stress, the stress reaction may be far less adaptive, because the animal can neither flee nor fight its captors, and there is usually little the animal can do to change its environment. Stress may be acute (intense but of short duration) or chronic (of lower intensity, but longer duration). Symptoms of acute stress may include diarrhea, loss of appetite, extreme agitation, irritability, pacing, trembling or shaking, cowering, and attempting to hide. Unfortunately, all of these symptoms are similar to those seen with many diseases, and in some cases the animal may die before the keeper is able to figure out what is wrong and correct the situation. Chronic stress is even more difficult to detect. It may include any of the symptoms above, but usually to a lesser degree. Sometimes the animal just becomes listless. It is possible that reproductive behavior may be repressed, but most animals in zoos are in controlled breeding situations, so it may be difficult to tell if their reproductive behaviors are repressed due to stress. Perhaps the most devastating effect of chronic stress is that it suppresses the immune system, making the animal susceptible to any pathogens it may encounter. Stress can kill animals. Death may occur quickly, as in the

animal that dies from heart failure after being chased and caught for a medical examination, or it may die over months or years, like a subordinate male gazelle that wastes away due to the stress of being housed with a group of females and two dominant males it can't get away from.

What can keepers do to recognize and alleviate stress in the animals they care for? The first step is to be aware of the things that can cause stress in captive animals. Anything that seriously interferes with an animal's performance of its routine natural behaviors will cause stress. Any aversive stimulus the animal can't get away from will cause stress. Recognition of the absence of important behaviors, or of stimuli that may be aversive to an animal, requires detailed knowledge of the species' ethogram. Does a species that lives in trees have trees to climb? If it is social, do the individuals have social partners? If indivduals of the species lead a solitary life in the wild, should that situation be emulated in captivity? (Some animals that are solitary in nature, to avoid competition for food and/or mates, may do well in social situations in captivity as long as they do not have to compete for those resources.) Is the age/sex ratio of the group of animals in captivity similar to that in natural groups?

Often, the animal's behavior itself suggests that it is stressed. Does the animal race around its enclosure trying to escape? Does it cower in a corner, refusing to come out for any enticements? Does it sleep far more than the literature indicates is usual for its species? Or does it never lie down as long as visitors are present, even though the literature says that its conspecifics sleep a large percentage of the day?

Stereotypical behavior is a good indicator of stress. An animal that repeats the same behavior over and over in exactly the same place and in exactly the same way, with no apparent purpose, is showing stereotypical behavior. Pacing is an example of stereotypical behavior, particularly when it is punctuated with a head twirl as the animal turns to retrace its steps. Stereotypical behavior is believed to develop in response to an environment that is not meeting one or more of the animal's critical needs. There may be an unmet need for social partners, room to exercise, foraging opportunity, or opportunity to get out of view of the public or of other animals. Stereotypical behaviors usually develop early in an animal's development, and once started, are very difficult to eliminate. It is much more effective to monitor the animal's behavior during early development, when it is separated from its birth group, and any time it is moved to a new facility to make sure that no repetitive or aberrant behavioral habits are developing. To do this effectively, one must know what behaviors are natural for the species and what behaviors are aberrant, and one must be able to recognize stereotypical behaviors before they become too deeply ingrained. Interventions can include habitat modification, social enrichment, or various other types of environmental enrichment.

BEHAVIORAL RESEARCH IN ZOOS

Many animals that are exhibited in zoos and aquariums are difficult to study in the wild due to their rarity, their nocturnal activity patterns, or the inaccessible places in which they live (e.g., underground, or in high tree canopies). Much of what we know about the behavior of these animals has been learned by studying captive individuals, often in zoo settings. The co-operation of keepers is essential to the validity of such studies, since the keepers control the habitat and feeding schedule of animals in their care. The behavioral biologist observes the animals from the visitor side of the exhibit, and is aware of events that transpire on the keeper's side only if the keeper keeps him or her informed. Since such behavioral research can benefit other animals of the same species, all keepers should strive to learn as much as they can from researchers, and to facilitate their work as much as possible by keeping them informed of what is going on behind the scenes of the exhibit while research is in progress.

SUMMARY

A good understanding of animal behavior is critical for anyone who works directly with animals, both for their own safety and that of the animals. This brief introduction is intended to give prospective keepers an idea of the variety and types of behaviors. To provide the best care, keepers should strive to learn as much as they can about the behavior of their animals through coursework, books, journals, internet resources, and especially their own behavioral observations of the animals themselves.

REFERENCES

Blomqvist, Leif, and Christer Larsson. N.d. Restoration of a Declining Population of Peregrine Falcons in Sweden through Captive Breeding: 30 Years of Expereince. Accessed at http://centrostudinatura.it.

Ellis, David H., Glenn H. Olsen, George F. Gee, Jane M. Nicolich, Kathleen O'Malley, Meenakshi Nagendran, Scott G. Hereford, Peter Range, W. Thomas Harper, Richard P. Ingram, and Dwight G. Smith. 1992. Techniques for Rearing and Releasing Nonmigratory Cranes: Lessons from the Mississippi Sandhill Crane Program. *Proceedings of the North American Crane Workshop* 6:135–41.

Gallagher, Tim. 1999. Mission Accomplished. *The Living Bird* 10.

Goodmann, Pat. 2010. *Why Hand Raise Captive Wolves?* [newsletter]. Wolf Park 2010 [cited 13 February 2011]. Available from http://www.wolfpark.org/wolfarticles.shtml.

Klinghammer, Erich, and Patricia Ann Goodmann. 1987. Socialization and Management of Wolves in Captivity. In *Man and Wolf: Advances, Issues and Problems in Captive Wolf Research*, edited by Harry Frank. Dordecht: Junk.

Marler, Peter, and M. Tamura. 1964. Culturally Transmitted Patterns of Vocal Behavior in Sparrows. *Science* 146:4.

Milton, Katherine. 1988. Foraging Behavior and the Evolution of Primate Cognition. In *Machiavellian Intelligence: Social Expertise and the Evolution of Intellect in Monkeys, Apes and Humans*, edited by A. Whitten and R. Byrne. Oxford: Oxford University Press.

Naitoh, Yutaka. 1974. Bioelectric Basis of Behavior in Protozoa. *American Zoologist* 14(3): 11.

Porton, Ingrid, and Kelli Niebruegge. 2006. The Changing Role of Hand Rearing in Zoo-Based Primate Breeding Programs. In *Nursery Rearing of Nonhuman Primates in the 21st Century*, edited by G. P. Sackett, G. Ruppenthal and K. Elias. Devon, UK: Springer.

Pryor, Karen. 1999. *Don't Shoot the Dog!* New York: Bantam.

Sandoval, Edgar, and Rich Schapiro. 2009. 911 Tape Captures Chimpanzee Owner's Horror as 200-Pound Ape Mauls Friend. *New York Daily News*, 18 February..

Schenkel, Rudolph. 1947. Expression Studies on Wolves. *Behaviour* 1:48.

39

Animal Behavioral Concerns

Joseph C. E. Barber

INTRODUCTION

Keepers at zoos and aquariums have an enormous responsibility. It is through their actions that animals are fed, enriched, trained, sometimes medicated, and generally cared for on a day-to-day basis. The keen observational skills of keepers serve as the eyes and ears of zoo and aquarium managers and veterinarians, who rely on keeper daily reports about the health and behavior of the animals to make caring for such a wide diversity of species possible. More often than not, it is keepers who also have to educate visitors about the roles and mission of zoos and aquariums, the natural history of the species, and the specific characteristics of each individual animal that the visitors encounter. The responsibility shouldered by keepers is matched by the significant amount of time and effort they willingly invest into doing their work. Caring for captive wild animals is challenging, especially when knowledge about many species is still being gathered. This section explores some of the roles that keepers play in the behavioral management of animals in zoos and aquariums, focusing on some of the behavioral concerns that are encountered, sometimes caused, and often successfully addressed by keepers. The following topics will be covered:

- understanding animal behavior and the scientific method
- the relationship between behavioral concerns and welfare
- examples of behavioral concerns seen in zoos and aquariums
- the future role of keepers in identifying and addressing behavioral concerns.

UNDERSTANDING ANIMAL BEHAVIOR

The behavior of animals changes in response to two factors: stimuli from the external environment (e.g., temperature, the smell of conspecifics, light), and the combined effect of the physiological and neurological workings going on inside the animals (e.g., levels of reproductive hormones). Observing behavioral changes, and understanding why these changes occur, is key to understanding how an animal perceives its environment and how this affects its psychological well-being (i.e., the general sense of how animals "feel"—their species-specific, subjective emotional states). To understand behavior, one of the many attributes that all keepers need is good observational skill. This entails more than just watching animals (something that zoo and aquarium visitors do); true observation of animals allows objective information to be gathered on important behavioral changes. In some species, subtle changes in behavior can indicate significant health issues, especially in prey species in which there might be an advantage for animals to mask serious pain and distress. It is often the weakest antelope in a herd that predators will target if that weakness is obvious.

Every change in behavior seen in an animal represents some change in its internal motivational state, where motivation can be broadly defined as what an animal most wants to do in any particular situation. Observation of behavioral changes often leads to questions about why there are changes, and allows keepers to develop possible explanations of what might be causing them. These basic steps form the foundation of the "scientific method," a step-by-step process that guides the asking and answering of scientific questions. While being a keeper may not be the same as being a research scientist, keepers who can use scientific principles in their daily work are more likely to be effective at caring for animals and addressing their changing needs throughout their lives. All keepers should work hard to use this scientific approach to understand and address potential behavioral concerns. Here are the basic steps used to apply the scientific method when it comes to identifying and addressing possible behavioral concerns:

- *Observe the situation*: Time should be spent observing animals, and notes taken about any significant observations made, especially for new animals, or animals in new exhibits. Animals being introduced to one another should always be carefully observed, as these situations are often associated with periods of stress. Keepers should be very familiar with the natural history of the species. A good first step would be reading published accounts of the behavior of animals in the wild, and trying to create an ethogram (a clearly defined list of all behaviors performed for a species).

- *Ask questions*: Observations should trigger questions such as "Why is that animal doing that?"; "Is that an appropriate behavior in that context?"; "When was that behavior first seen?"; or "Do all of the animals in the group perform that behavior?"

- *Perform some background research*: If questions arise about certain behaviors—for example, about excessive aggression seen between two individuals within a social group—then animal management records for the individual animals involved can be reviewed to see if the behaviors are part of a trend (e.g., seasonal or increasing over time), and the published literature can be searched to identify any similar cases or possible explanations (e.g., have there been cases of mother-son aggression in other animals of the species at a certain age?).

- *Develop and test hypotheses*: A possible explanation of what is going on should then be developed and tested by changing different management or care approaches and observing what happens. For example, if it is thought that an animal is performing a certain behavior, such as repetitive pacing, because it is hungry, but providing more food does not change the frequency with which the behavior is performed, then this may suggest that hunger is not the cause, although it is rarely possible to be 100% certain when it comes to behavior. An alternative hypothesis might be that the sight or smell of a female leads to the pacing behavior when the male enters the breeding season. If removing the sight and smell of the female leads to a reduction in pacing, or if pacing decreases during certain months, then there is objective evidence in support of this hypothesis.

- *Draw a conclusion from the investigation*: The ultimate goal in using the scientific method is to look for objective evidence to suggest a course of action that will improve the conditions for the animals, and to be able to document these improvements in some way. While not minimizing the importance that keeper "gut feelings" can sometimes have in caring for animals, having an untested feeling about why an animal is performing a certain behavior is not as convincing to managers or veterinarians as having a more objective and supported hypothesis.

Using this scientific method, a behavioral concern, such as two llamas repetitively pacing in their petting zoo enclosure, can be clearly identified in the following way. Questions about which animals perform the behavior (e.g., both of them) and how frequently it occurs (e.g., for approximately 25% of each day) can be answered. Information can also be found showing the first occurrence of the behavior by looking at the archived daily reports for the animals, thus reinforcing the importance of collecting these detailed daily records. Different explanations for why the behavior occurs can also be tested. For example, if separating the two llamas (in terms of sight, sound, and smell) does not lead to changes in their behavior, then it may not be a social issue. If providing more food has no effect, then it may not be a hunger issue. If varying the times when food and enrichment are delivered during the day does lead to a decrease in pacing, then the hypothesis that a fixed routine minimizes the control that the animals have over their environment would receive the most support. Making an enrichment calendar to schedule different enrichment initiatives on different days, and ensuring that food and enrichment are not only delivered at 9 a.m., noon, and 3 p.m., would then be actionable items that may help to address this behavioral concern in the llamas.

ANIMAL BEHAVIOR AND WELFARE

Animal behavior is often assessed in animal welfare studies that look to see how well animals are coping with their captive conditions (Barber and Mellen 2008). North America's Association of Zoos and Aquariums (AZA) Animal Welfare Committee (AWC) has created different working definitions of animal welfare to increase understanding of this complex term and to guide the actions of the AWC in attempting to improve and measure animal welfare within the AZA community. One of the definitions states:

Animal welfare is the degree to which an animal can cope with challenges in its environment as determined by a combination of measures of health (including pre-clinical physiological responses) and measures of psychological well-being.

- Good health represents the absence of diseases or physical/physiological conditions that result (directly or indirectly) from inadequate nutrition, exercise, social groupings, or other environmental conditions to which an animal fails to cope successfully.
- Psychological well-being is dependent on there being the opportunity for animals to perform strongly motivated, species-appropriate behaviors, especially those that arise in response to aversive stimuli.
- Enhanced psychological well-being is conditional on the choices animals have to respond appropriately to variable environmental conditions, physiological states, developmental stages and social situations, and the extent to which they can develop and use their cognitive abilities through these responses (Barber and Mellen 2008, p.41).

What is important about this definition is the idea that an animal's health is closely associated with its behavior. If an animal performs behaviors that seem abnormal or out of place compared to those performed by other members of the same species in zoos or aquariums (or in the wild), then there is a strong possibility that the animal's health may also be compromised in some way.

One way to think about behavioral issues seen in zoo and aquarium animals is to picture them like rings on a tree. These rings or abnormal behaviors can accumulate throughout an animal's lifetime, providing insight into the "welfare history" of animals over time, and the types of conditions and environments they have experienced. Many of the behavioral issues seen in adult animals within zoos and aquariums can be attributed to issues experienced by the animals in their critical development period as infants and juveniles (Napolitano, De Rosa, and Sevic 2008; Rommeck, Anderson, Heagerty, Cameron, and McCowan 2009). Even the prenatal environment can affect the behavior of the yet-to-be-born animals (Kapoor, Dunn, Kostaki, Andrews, and Matthews 2006), potentially resulting in the offspring of animals stressed during pregnancy showing less exploration and play behavior, as well as impaired social and maternal behavior when they become adults (Braastad 1998). For certain captive-bred species in zoos and aquariums, this may mean that the offspring of females not given appropriate conditions during gestation are going to be at greater risk of showing behavioral issues indicative of poorer welfare. Four years after their birth, prenatally stressed monkeys showed lower levels of social interaction when faced with stressors in their environment, suggesting that they were coping less effectively with those stressors than animals that had not experienced prenatal stress (Clarke, Soto, Bergholz, and Schneider, 1996). For many species, social interaction is important as a way to cope with environmental challenges, and prenatal stress inhibited these animals' ability to use this coping mechanism effectively. Obviously, not every stressor experienced by young animals or their pregnant mothers will lead to negative behavioral consequences, and the effects will be species-specific, but it is important that keepers acknowledge and understand the influence they can have on the behavior of the animals they care for.

BEHAVIORAL CONCERNS

There is no universal list of behavioral concerns that will be relevant to every species within zoos and aquariums, but the following list provides examples of important behaviors to monitor, which may be of concern to veterinarians, managers, curators, nutritionists, and behaviorists, depending on the species in question:

FOOD-RELATED BEHAVIORS

- changes in eating habits (stopping eating, or starting to eat in excess)
- excessive drinking
- regurgitation of food
- eating of nonedible items (e.g., dirt or stones)
- appetitive behavior on nonfood items (e.g., chewing bars, licking walls).

LOCOMOTION

- increase in locomotion (e.g., constant movement in or around enclosure)
- cessation of locomotion (e.g., complete apathy)
- changes in gait (e.g., limping)
- movement of individuals that does not match that of the rest of the social group

- repetitive or exaggerated movements in or around the same area seen repeatedly.

SOCIAL BEHAVIOR

- increased aggression
- increased submission to aggressive individuals
- inappropriate parental behaviors (e.g., parental neglect or abandonment)
- lack of social interaction with conspecifics
- inappropriate sexual behaviors (e.g., courting or mounting related individuals or individuals of the same sex; self-directed sexual behaviors).

COMFORT BEHAVIOR

- overgrooming
- repetitive licking or scratching of one particular area of the body.

One of the stated goals of zoos and aquariums is to display animals in a manner that will educate zoo visitors about their physical appearance and behavior (Hutchins and Smith 2003). Ideally, the behavior of zoo animals should closely resemble the characteristics of wild animals living in their natural environments. This is important for two reasons. The first is simply that zoo visitors will learn more and have a greater understanding and appreciation of the species they are seeing if this is the case. For example, many visitors are concerned when they see animals housed by themselves within exhibits, and may think that animals such as orangutans, snow leopards, tigers, or tree kangaroos would be better off housed with companions. There are many species in the wild that are solitary for much of their lives, and it is possible that the housing together of individuals from these species (e.g., six female tigers) in a captive setting could lead to serious welfare issues (e.g., aggression and injury, chronic stress, or abnormal behavior). There are certainly exceptions to this general rule, where unnatural social groups of typically solitary animals in zoos and aquariums have been more positive than negative, but this is something that has to be carefully tested and not assumed. In appropriate captive environments that meet the behavioral needs of the animals, solitary species can still experience good welfare when housed alone.

The second reason that animals in zoos and aquariums should behave like their wild counterparts is that animals able to do so are generally better able to cope with the captive environment and its associated stressors (e.g., the presence of zoo visitors, extreme weather conditions, conflict within social groups, or changes in diet). Animals have evolved behavioral, physiological, and physical coping strategies in response to stressors that previous generations have faced in their natural environments. For example, in response to predators in the wild, a prey species may flee or hide. Species adapted to flee are athletic, agile, and usually fast (or at least as fast as the predator that usually chases them). Species adapted to hide may have cryptic coloration enabling them to be camouflaged within certain types of environments, and a useful ability to remain completely motionless even in close proximity to predators. While animals may not encounter actual predation in captive environments, they may perceive certain elements of this environment as potential predators (perhaps even the zoo or aquarium visitors themselves), and be motivated to respond accordingly (e.g., hide or flee). Animals that can perform highly motivated behaviors in much the same way as their wild counterparts will have greater control over their environment, and this is always a good goal to aim for. Animals that lack control will experience poor welfare, and there will be behavioral indicators (including some of those listed above) that indicate this negative state.

Behavioral concerns often arise when animals perform behaviors that are abnormal in type, duration, frequency, or location. One way to define abnormal in this context is a behavior performed in captivity that is not performed in the wild (e.g., self-injurious behavior) and is not adaptive from the animal's perspective. This means that the behavior would be unlikely to promote an animal's ability to survive or reproduce if performed in its natural environment. Abnormal may also mean different from other captive members of the species (e.g., a diurnal animal that is active at night while its conspecifics are sleeping). Captive animals do often perform behaviors that are slightly different from those of their wild counterparts, and that are of no concern. For example, through positive reinforcement, captive dolphins can be trained to leap in tandem to touch a hanging ball as part of a show. In this case, while the highly synchronized nature of the behavior is abnormal (compared to that of wild animals), leaping and acrobatics are very common behaviors seen in wild dolphins. Similarly, tigers in the wild do not often interact with large plastic balls (e.g., Boomer Balls), but this would not be considered an abnormal behavior because the tiger is using many of the same behavioral elements when interacting with the ball (e.g., chasing, biting, scratching, carrying) that it would use on prey in the wild.

There are some behaviors performed in captivity that clearly have negative associations, and that in no way serve to promote the survival of the animals or their ability to reproduce successfully. Examples of these types of abnormal behaviors include infanticide (e.g., Chen, et al. 2008), parental neglect (e.g., Maestripieri and Carroll 1998), self-injurious behaviors (e.g., limb biting, eye poking; Rommeck et al. 2009), and certain abnormal repetitive behaviors (e.g., Mason, Clubb, Latham and Vickery 2007) described below.

INFANTICIDE AND PARENTAL NEGLECT

In some species, adults within their natural environments may kill and sometimes consume their own young (e.g., Weber and Olsson 2008; Deloya, Setser, Pleguezuelos, Kardon, and Lazcano 2009). Animals face many extreme environmental conditions and stressors in the wild (e.g., weather, predators, and disease), and in some cases infanticide and cannibalism may be the best evolutionary strategy in a bad situation. In zoos and aquariums, these extreme conditions should be absent as a result of the daily care provided to the animals, and provision of an appropriate captive environment. The continued occurrence of infanticide-like behaviors may reveal that some aspect of the captive housing or management is inappropriate. For example, the housing together of unfamiliar female capybara can lead to females killing the

young of other females (Nogueira, Nogueira-Filho, Otta, Dias, and Carvalho 1999). Infanticide is not observed when social groups of capybara are set up using females familiar with one another since weaning (Nogueira et al. 1999). As a grouping of unfamiliar females in the wild is unlikely, it is the captive management practice that in this case should be considered abnormal, while the behavior itself could be seen as adaptive from an evolutionary perspective (females eliminating competition for resources for their offspring).

Possible causes of maternal neglect of infants in zoos and aquariums may include a lack of options for the females or both parents to create appropriate dens, nests, or rearing environments away from disturbance, human interference, excessive sound, olfactory or visual stimuli, or inappropriate social environments (Eaton 1981). Social environments in captivity that might promote maternal neglect include situations where the male is allowed access to the female in species where females nest or den alone (e.g., polar bears), or where the presence of a dominant female interferes with the subordinate female and her offspring. Keepers unfamiliar to the animals, who may want to enter the exhibit to view newborn infants, can sometimes contribute to the disturbance, so even minor changes in management and care routines can be very disruptive from the animals' perspective.

Keepers should investigate the medical and management records of animals that show parental neglect or infanticidal behaviors, as these behaviors are often habitual, and there may be evidence of previous behavioral concerns related to these same animals in their records. There have been cases where the social behavior of animals has been proactively and successfully modified using operant conditioning (Schapiro, Perlman, and Boudreau 2001). In some cases, however, animals reared in socially inappropriate environments may never show adaptive parental care for their offspring. Although successful breeding can still be accomplished in some cases by hand-rearing the offspring of animals that show poor parental care, this approach can lead to additional animals also being reared in less than optimal social conditions that lack the species-appropriate developmental conditions necessary for them to develop into effective parents. For successful captive breeding, minimizing the need for hand-rearing by providing more appropriate rearing conditions should be the primary goal. Depending on the species, steps that can be taken by keepers to achieve this goal might include allowing animals that will be bred to watch other animals rear their own offspring, assisting with the care of related offspring, or simply remaining in the company of conspecifics during their own period of rearing (Mallapur, Waran, Seaman, and Sinha 2006).

SELF-INJURIOUS BEHAVIORS

The early weaning and social isolation of group-living animals are thought to be the major contributing factors that promote the development of self-injurious behaviors in these animals (Latham and Mason 2008). These behaviors can include limb and tail biting, hair pulling, and eye poking. As these behaviors can sometimes result in serious injuries, they represent a significant behavioral concern and an obvious welfare issue. When these behaviors are seen, it is likely that the housing and management conditions are not meeting the animals' physical, social, or behavioral needs in some way. Health issues (including neurological issues) can also trigger these abnormal behaviors. An animal experiencing allergic reactions to flea bites, for example, may scratch itself repeatedly until it causes actual physical damage (e.g., Cucchi-Stefanoni, Juan-Salles, Paras, and Garner 2008). Treatment should result in a complete cessation of the self-injurious behavior. In some cases, however, the animal's current captive environment or its previous environment during development and rearing is the cause.

Self-injurious behaviors are fortunately fairly rare in zoos and aquariums, and are often more commonly seen in laboratory-reared animals (Hosey and Skyner 2007). To address these issues when they are seen, antipsychotic or antidepression medication can sometimes be used as directed by veterinarians (Mills and Luescher 2006). Laboratory researchers have had some success with basic enrichment approaches designed to provide animals with more behavioral opportunities within their restrictive environments (Bourgeois and Brent 2005). More comprehensive enrichment programs are commonplace in zoos and aquariums, and can be tailored to the needs of the species in question through a step-by-step enrichment process (Barber 2003a). Other approaches have focused on positive reinforcement training as a way to provide animals with behavioral opportunities, greater social interaction (with humans), and some control over their environment (Baker et al. 2009; Coleman and Maier 2010). Using these techniques, self-injurious behaviors can usually be reduced but not completely eliminated. Long-lasting behavioral changes can occur in animals that are reared in conditions that do not meet their physical, physiological, or behavioral needs. Keepers must be knowledgeable of these needs whenever they work with animals, so that they can contribute to providing an optimal housing and management environment.

ABNORMAL REPETITIVE BEHAVIORS

Of all of the behavioral concerns that can be found in zoos and aquariums, abnormal repetitive behaviors are usually the most common. When animals are observed performing behaviors that seem to have no obvious goal or function (e.g., pacing to and fro), and when certain behaviors are repeated time and time again, often in the same location, they may be performing abnormal repetitive behaviors (ARBs) or more specifically stereotypic behaviors or stereotypies (Swaisgood and Shepherdson 2005). A stereotypy was traditionally and broadly defined as a repetitive behavior that was invariant in its form and location and did not seem to have a purpose or to achieve anything that could be observed or measured (e.g., Mason 1991). As research has increased and become more sophisticated, this type of simple definition has been replaced with a more complex array of scientific concepts and behavioral categories (Mason and Rushen 2006). The category of behaviors that can be defined as ARBs is very broad, and many factors are linked with their performance and with complex physiological and neurological processes that lead to their development. In general terms, ARBs result from frustrated motivations—situations in which animals cannot perform a behavior they really want to perform. Some cases of ARBs are the result of neurological dysfunction, but those will not be covered here.

Despite their capacity to perform their full range of species-appropriate behaviors in the wild, there are many situations in which animals cannot perform highly motivated behaviors in their natural environments. During droughts, thirsty animals may not be able to find water despite being highly motivated to drink. Carnivores may not be able to actually catch prey, even though they may be very hungry. Prey species may be highly motivated to escape from carnivores, but many are still caught and eaten. It is only when animals are housed and managed by humans, who dictate what resources they receive and which behaviors they can perform, that their inability to perform certain highly motivated behaviors becomes a serious welfare issue. Animal welfare is concerned with the health and psychological well-being of *captive* animals.

At first glance, many ARBs resemble normal, species-appropriate behaviors. Large felids pace to and fro within their exhibits, as they might do when patrolling territories in the wild (Bashaw, Kelling, Bloomsmith, and Maple, 2007; Clubb and Mason, 2007). Giraffes lick walls and fences with their long tongues, just as they might use their tongues to feed off leaves on branches (Bashaw, Tarou, Maki, and Maple, 2001). Porcupines dig incessantly at areas within their enclosures, as they might do when digging for food or making dens. A closer look at some of these behaviors, however, reveals significant differences from normal behaviors. The pacing behavior of bears can become so fixed that some animals can wear grooves into the concrete floors of their enclosures as they place each foot in exactly the same spot every time (Wechsler 1991). Bears also pace forward and backwards, with backwards walking not commonly seen in the wild. Some species, such as llamas, develop head orbiting behaviors in which they roll their heads and necks in a large circle as they pace in their enclosure (pers. obv.). Animals may cause abrasions on their skin as they pace and twirl in areas of their enclosure where they continually rub against exhibit furniture or containment barriers. Parrots may overgroom themselves to such an extent that they remove all the feathers on parts of their bodies (Garner, Meehan, Famula, and Mench 2006). All these behaviors are similar to natural, species-appropriate behaviors, with one important difference: they do not help to maximize an animal's reproductive success or its ability to survive and thrive in its natural environment.

There are many theories associated with why ARBs develop in certain animals, and some are supported by good evidence. The theory most relevant to keepers is that ARBs develop as a result of behavioral frustration in the animals. An animal that is highly motivated to perform a behavior, but cannot do so, can become frustrated (Rushen, Lawrence, and Terlouw 1993). In the most severe case, an animal may be prevented from engaging in any aspect of a behavior that it is highly motivated to perform. A prey species housed next to a predator species may be motivated to flee or hide. A small, barren concrete enclosure may prevent either of these behaviors from occurring at all. The motivation to flee or hide then becomes a source of frustration. Frustration can also arise when animals can perform a behavior but the behavior has no functional consequence. Animals that dig burrows (e.g., for nesting, temperature regulation, or to hide) may continually

TABLE 39.1. Published accounts of behavioral concerns associated with the captive management of common species housed in zoos and aquariums, and the approaches taken to address those concerns

Species	Behavioral concern	Approach used	Reference
African wild dogs	Stereotypic pacing	Operant conditioning	Shyne and Block 2010
Amur tigers	Stereotypic pacing	Electronic feeding boxes	Jenny and Schmid 2002
Amazon parrots	Feather plucking, excessive fearfulness, aggression	Pair housing	Meehan et al. 2003
Asian elephants	Pacing, swaying, head bobbing	Auditory stimulation	Wells and Irwin 2008
Chimpanzees	Eye/ear poking, coprophagy, hair plucking	Positive reinforcement training	Pomerantz and Terkel 2009
Giraffes	Oral stereotypies (e.g., licking nonfood objects)	Puzzle feeders	Fernandez et al. 2008
Large felids	Pacing, excessive grooming	Carcass feeding	McPhee 2002
Polar bears	Stereotypic pacing	Access to indoor/outdoor space	Ross 2006
Red foxes	Stereotypic pacing	Electronic food dispensers	Kistler et al. 2009
Rhesus macaques	Locomotor stereotypy (e.g., somersaulting, pacing, bouncing)	Target training	Coleman and Maier 2010
Sea lions	Pattern swimming	Nonfood enrichment	Smith and Litchfield 2010
Vicugna	Stereotypic pacing and exaggerated head swings	Scatter feeding and browse	Parker et al. 2006

scratch at the concrete floors of their enclosures without the ability to make an actual nest or den. They go through the motions of the behavior without any functional consequences or behavioral endpoints. Animals continually frustrated in this manner may continue to perform the behaviors even though they cannot complete them. Some pacing behaviors seen in zoo animals may even be frustrated attempts to escape.

In many cases, brief periods of behavioral frustration can lead to the performance of displacement activities (if you were unable to open a door, you might start pacing up and down in front of it), and frustration-induced aggression (kicking at a stuck door will probably not help to open it, but that is often what we end up doing; Haskell, Coerse, Taylor, and McCorquodale 2004). By themselves, these responses do not necessarily indicate a welfare issue. However, prolonged frustration of highly motivated behaviors can be serious. When coupled with the stress that animals may experience as a result, it can lead to long-lasting changes to the physiological and neurological systems that control behavior. These changes are based on complex interactions of hormones and the nervous system, and are covered in more specific detail by Mason et al. (2007).

It is not only specific frustration of certain behaviors that can lead to the development of ARBs. A general lack of species-appropriate environmental stimulation and complexity has a similar effect by reducing the ability of animals to perform highly motivated behaviors. A barren environment offers few behavioral options, and when it is humans (e.g., keepers) who determine what animals eat and when, where they can go, and which conspecifics they are in close proximity to, then the animals lack control. The link between a lack of control and the development of ARBs has been clearly established (Ross 2006; Mason et al. 2007).

There can be developmental factors associated with ARBs. For example, researchers have found that animals that are weaned early (i.e., separated from their parents before the time this might normally occur in the wild) are more susceptible for developing ARBs as adults (Latham and Mason 2008), regardless of the conditions they face as adults. Early weaning does not cause the development of ARBs, but can

contribute to it. Providing a complex early rearing environment that allows animals as much control over their environment as possible may have a key role to play in "protecting" them from developing ARBs as adults. There are many studies that show that wild animals brought into captivity show fewer ARBs than captive-born animals, even if the captive-born animals are housed in the same enclosure (Lewis, Presti, Lewis, and Turner 2006). The complexity of the social or physical environment in the wild, especially during rearing periods, seems to help animals cope better with captive conditions.

There have been many attempts to address and eliminate ARBs shown by animals in zoos and aquariums. Most have focused on providing the animals with additional or more appropriate environmental enrichment, and some have used animal training programs separately or in addition to enrichment approaches. The success of these interventions is more likely when the approaches used are intensive and long-term, requiring a significant investment of time and effort by keepers. Table 39.1 lists selected studies in which enrichment and training have been used to alter the frequency of ARBs in captive animals, and where keepers have played a key role in developing and implementing these approaches.

In most cases, using enrichment and training to reduce the frequency of ARBs is not as effective as using it to prevent ARBs in the first place. Animals that show the early stages of ARBs face an immediate and acute welfare concern that should be addressed quickly. Changes to the captive environment are more likely to be successful at eliminating developing ARBs than at eliminating established ones. When performed for long periods of time, ARBs become like habits. They are hard to eliminate, and may persist in some individuals throughout their lives, despite changes to their management and care. The performance of some established ARBs may even be triggered by stimuli that would have actually prevented their development in the first place (Cooper and Nicol 1994), such as a complex, enriched environment. Those animals that have performed ARBs for many years may no longer be facing the specific environment challenges that they were failing to cope with earlier, so that the ARB may represent a "scar" from a previous environment. However, care must be taken not to assume that this is the case, as the same welfare challenge that originally caused the development of the ARB may still be present in any new environments (e.g., inappropriate social groupings or lack of foraging opportunities).

To eliminate ARBs, the focus needs to be on removing the animal's need to perform the abnormal behaviors rather than on physically preventing the animal from performing them (Mason et al. 2007). For example, foul-tasting substances can be spread on surfaces within enclosures to prevent excessive licking by giraffes (Tarou, Bashaw, and Maple 2003). Anti-pecking devices (bits) can be used in birds that overgroom to prevent them from being able to physically pull out their feathers (Savory and Hetherington 1997). Thorny plants can be placed on the paths that big cats use to pace to disrupt the pattern. These approaches may mask the performance of ARBs, but they do not address the underlying causes of the behaviors and will not improve animal welfare. Obstacles and deterrents may also contribute towards chronic frustration and stress, and potentially lead to further behavioral or health consequences (e.g., wounds resulting from animals rubbing against thorny plants as they pace). Keepers should understand that addressing behavioral concerns is complex, and that an easy or immediate "fix" is unlikely. Prevention must be the priority approach.

Keepers have a role in identifying and addressing behavioral concerns in the animals they care for. However, the frustration and lack of control that animals can experience in captive environments is often based on environmental conditions over which keepers have little control. Zoos and aquariums must address the problem of enclosures not designed for the species they contain, that are too small and without elements that would promote species-appropriate behaviors (e.g., shelters, pools, trees, shade, substrate, or browse), or which force abnormal social conditions (e.g., males and females housed together year-round, bachelor groups, or territorial species housed too close to conspecifics). The fundamental changes would require input from directors, curators, managers, keepers, veterinarians, nutritionists, researchers, and others. Even so, keepers do spend the most time with the animals, and the possibility of them affecting their welfare both positively and negatively is significant. How and when animals are fed, enriched, and trained will specifically influence their behavior, and can be a subtle source of frustration and lack of control as well. If food or enrichment is provided at set times during the day to coincide with the keepers' schedule, animals can develop anticipatory behaviors that resemble ARBs. Big cats may pace within their enclosures prior to the delivery of their food, perhaps as soon as they hear keepers in the food preparation areas. The frequency of pacing and head-orbiting in other species may also increase around the time that enrichment is normally delivered. Animals may pace and twirl by the gates of their indoor holding enclosures before being let out in the morning, and again before returning in the evening.

An animal that is highly motivated to forage for food but must wait until 9 a.m., noon, and 3 p.m. to do so lacks control. In captivity, the animal cannot make food appear if the delivery of the food is based only on the schedule of animal care staff members. There are no "free lunches" out in the wild, and many species will continue to "work" for food (e.g., forage) in captive conditions even in the presence of free food—a phenomenon known as contrafreeloading (e.g., Barber 2003b). Researchers have investigated the effects that scheduled delivery of food and enrichment have on animals, and in many cases these unchanging routines lead to the development of ARBs (Carlstead 1998; Gilbert-Norton, Leaver, and Shivik 2009). Addressing this issue may not be as simple as providing food, enrichment, and changing environmental conditions and stimuli on a random or varied basis, but these approaches seem to be part of the overall solution. Giving animals the chance to work for access to food, enrichment, social interactions, and access to resources is a clear priority, as it directly ties their behaviors to functional consequences, thus giving them greater control over their environment. The use of puzzle feeders or operant mechanisms that require time and effort from the animals for access to food is a simple approach that returns control over certain environmental factors back to the animals. The key to this approach is to keep the animals from anticipating events or stimuli over which they have little or no control.

The timing of enrichment should also be carefully monitored to ensure that it does not make abnormal behaviors more likely. Giving enrichment to animals at the very moment when they perform abnormal repetitive behaviors (i.e., as a way to address the behavioral concern) may actually reinforce those behaviors (i.e., the animals may learn that when they perform the abnormal behaviors they will receive rewards). Effective and well-thought-out enrichment should always be provided, but care should be taken that the animal does not make a connection between its own abnormal behavior and the delivery of enrichment.

THE FUTURE ROLE OF KEEPERS IN IDENTIFYING AND ADDRESSING BEHAVIORAL CONCERNS

Research has begun to validate the idea that experienced keepers have a good sense of the health and psychological well-being of the animals they care for, and that they can integrate different information from their interactions with the animals. This may include information on how much they eat, how active or social they seem, or their posture or appearance that can be used to assess their overall condition (Whitham and Wielebnowski 2009). Rather than having to rely only on long-term, complex, and sometimes expensive scientific assessments of animal welfare using behavioral and physiological studies conducted by researchers, it is possible that qualitative assessments performed by keepers could be an additional assessment tool for the future (e.g., Wemelsfelder and Lawrence 2001). These keeper assessments would still need to be validated.

The most effective way to address behavioral concerns in zoo and aquarium animals is to identify them early or, better yet, prevent them, and to understand why these concerns have arisen. By gaining knowledge of each species and becoming familiar with each individual's behavioral preferences, routines, and temperaments, keepers can identify any unusual deviations from what is considered normal behavior (Hsu and Serpell 2003; Meagher 2009). In certain cases, keepers need high levels of training and education to be effective at observing behavior objectively (Margulis and Westhus 2008)—something that should be readily available. Acquiring knowledge of species from the published literature (and gaining access to husbandry manuals or similar documents created by zoo and aquarium professionals within regional zoo associations) will also help keepers to filter out normal changes in behavior from more abnormal deviations that may be associated with poor welfare. However, it is because keepers have often been present throughout the lives of the animals they care for that they can offer such an important contribution to addressing behavioral concerns. Rarely can scientific assessments of behavior that might record an animal's discrete behavioral states every 30 seconds detect its subtle changes in posture, its interactions with conspecifics, or its energy in performing behaviors (e.g., King and Landau 2003). If keepers can detect these subtle changes and provide possible explanations for them, then they can contribute to a more integrated assessment of welfare (Whitham and Wielebnowski 2009).

This does not mean that keepers should rush to make assumptions about the welfare of their animals on the basis of overly anthropomorphic perspectives, especially since they themselves can influence the animals' behavior, perhaps even more than observers that the animals are unfamiliar with (Freeman, Schulte, and Brown 2010). By observing animals closely and thinking about their psychological well-being, keepers can offer ideas for testable hypotheses (part of the scientific method) that explain why any behavioral changes linked to negative welfare states are occurring (Scott, Nolan, Reid, and Wiseman-Orr 2007; Meagher 2009). By participating in the evaluation of these hypotheses to see whether they are based on objective and measurable criteria, keepers can also help to address the animals' welfare issues (e.g., Whitham and Wielebnowski 2009).

Keepers must be willing to play a key role in the development of future welfare assessment tools that are based on their expertise. There is growing evidence that keeper assessments can be useful, and protocols exist for developing them further (Whitham and Wielebnowski 2009). The goal is for all keepers to be able to contribute to long-term welfare assessments that are seen as part of animals' daily or routine care, and not as something extra that only researchers can do (Barber 2003c, 2009; Whitham and Wielebnowski 2009).

REFERENCES

Baker, K. C., M. Bloomsmith, K. Neu, C. Griffis, M. Maloney, B. Oettinger, et al. 2009. Positive reinforcement training moderates only high levels of abnormal behavior in singly housed Rhesus macaques. *Journal of Applied Animal Welfare Science* 12(3): 236–52.

Barber, J. 2003a. Making sense of enrichment and Auntie Joy's choice of presents. *Animal Keepers' Forum* 30(3): 106–10.

———. 2003b. Documenting and evaluating enrichment 101: Putting good science into practice. *Shape of Enrichment* 12(2): 8–12.

———. 2003c. Part II—Motivation, contrafreeloading and animal welfare: Discussion points around diet presentation. *Animal Keepers' Forum* 30(8): 344–47.

———. 2009. Programmatic approaches to assessing and improving animal welfare in zoos and aquariums. *Zoo Biology* 28(6): 519–30.

Barber, J. C., and J. Mellen. 2008. Assessing animal welfare in zoos and aquaria: Is it possible? In T. Bettinger, and J. Bielitzki, eds., *The Well-Being of Animals in Zoo and Aquarium Sponsored Research: Putting Best Practices Forward*, pp. 39–52. Greenbelt, MD: Scientists Center for Animal Welfare.

Bashaw, M. J., A. S. Kelling, M. A. Bloomsmith, and T. L. Maple. 2007. Environmental effects on the behavior of zoo-housed lions and tigers, with a case study of the effects of a visual barrier on pacing. *Journal of Applied Animal Welfare Science*, 10(2):, 95–109.

Bashaw, M., L. Tarou, T. Maki, and T. Maple. 2001. A survey assessment of variables related to stereotypy in captive giraffe and okapi. *Applied Animal Behaviour Science* 73(3): 235–47.

Bourgeois, S., and L. Brent. 2005. Modifying the behaviour of singly caged baboons: evaluating the effectiveness of four enrichment techniques. *Animal Welfare*, 14(1): 71–81.

Braastad, B. 1998. Effects of prenatal stress on behaviour of offspring of laboratory and farmed mammals. *Applied Animal Behaviour Science* 61(2): 159–80.

Carlstead, K. 1998. Determining the causes of stereotypic behaviors in zoo carnivores: Towards appropriate enrichment strategies. In D. Shepherdson, J. Mellen, and M. Hutchins, eds., *Second Nature: Environmental Enrichment for Captive Animals*, pp. 172–83. Washington, DC: Smithsonian Institute Press.

Chen, C., C. L. Gilbert, G. Yanga, Y. Guo, A. Segonds-Pichon, J. Ma, et al. (2008). Maternal infanticide in sows: Incidence and behavioural comparisons between savaging and non-savaging sows at parturition. *Applied Animal Behaviour Science* 109(2–4): 238–48.

Clarke, A., A. Soto, T. Bergholz, and M. Schneider. 1996. Maternal gestational stress alters adaptive and social behavior in adolescent rhesus monkey offspring. *Infant Behavior and Developmen*, 19(4): 451–61.

Clubb, R., and G. Mason. 2007. Natural behavioural biology as a risk factor in carnivore welfare: How analysing species differences could help zoos improve enclosures. *Applied Animal Behaviour Science* 102(3–4): 303–28.

Coleman, K., and A. Maier. 2010. The use of positive reinforcement training to reduce stereotypic behavior in rhesus macaques. *Applied Animal Behaviour Science* 124(3–4): 142–48.

Cooper, J., and C Nicol. 1994. Neighbor effects on the development of locomotor stereotypies in bank voles, *Clethrionomys glareolus*. *Animal Behaviour* 47(1): 214–16.

Cucchi-Stefanoni, K., C. Juan-Salles, A. Paras, and M. M. Garner. 2008. Fatal anemia and dermatitis in captive agoutis (*Dasyprocta mexicana*) infested with Echidnophaga fleas. *Veterinary Parasitology* 155(3–4): 336–39.

Deloya, E. M., K. Setser, J. M. Pleguezuelos, A. Kardon, and D. Lazcano. 2009. Cannibalism of nonviable offspring by postparturient Mexican lance-headed rattlesnakes, *Crotalus polystictus*. *Animal Behaviour*, 77(1): 145–50.

Eaton, R. 1981. An overview of zoo goals and exhibition principles. *International Journal for the Study of Animal Problem*, 2(6): 295–99.

Fernandez, L. T., M. J. Bashaw, R. L. Sartor, N. R. Bouwens and T. S. Maki. 2008. Tonguetwisters: Feeding enrichment to reduce oral stereotypy in giraffe. *Zoo Biology* 27(3): 200–212.

Freeman, E., B. Schulte, and J. Brown. 2010. Investigating the impact of rank and ovarian activity on the social behavior of captive female African elephants. *Zoo Biology* 29(2): 154–67.

Garner, J., C. Meehan, T. Famula, and J. Mench. 2006. Genetic, environmental, and neighbor effects on the severity of stereotypies and feather picking in Orange-winged Amazon parrots (*Amazona amazonica*): An epidemiological study. *Applied Animal Behaviour Science* 96(1–2): 153–68.

Gilbert-Norton, L., L. Leaver, and J. Shivik. 2009. The effect of randomly altering the time and location of feeding on the behaviour of captive coyotes (*Canis latrans*). *Applied Animal Behaviour Science* 120(3–4): 179–85.

Haskell, M., N. Coerse, P. Taylor, and C. McCorquodale. 2004. The effect of previous experience over control of access to food and light on the level of frustration-induced aggression in the domestic hen. *Ethology* 110(7): 501–13.

Hosey, G., and L. Skyner. 2007. Self-injurious behavior in zoo primates. *International Journal of Primatology* 28(6): 1431–37.

Hsu, Y., and J. Serpell. 2003. Development and validation of a questionnaire for measuring behavior and temperament traits in pet dogs. *Journal of the American Veterinary Medical Association* 223:1293–1300.

Hutchins, M., and B. Smith. 2003. Characteristics of a world-class zoo or aquarium in the 21st century. *International Zoo Yearbook* 38:130–41.

Jenny, S., and H. Schmid. 2002. Effect of feeding boxes on the behavior of stereotyping Amur tigers (*Panthera tigris altaica*) in the Zurich Zoo, Zurich, Switzerland. *Zoo Biology* 21(6): 573–84.

Kapoor, A., E. Dunn, A. Kostaki, M. Andrews, and S. Matthews. 2006. Fetal programming of hypothalamo-pituitary-adrenal function: prenatal stress and glucocorticoids. *Journal of Physiology—London* 572(1): 31–44.

King, J., and V. Landau. 2003. Can chimpanzee (*Pan troglodytes*) happiness be estimated by human raters? *Journal of Research in Personality* 37(1): 1–15.

Kistler, C., D. Hegglin, H. Wurbel, and B. Konig. 2009. Feeding enrichment in an opportunistic carnivore: The red fox. *Applied Animal Behaviour Science* 116(2–4): 260–65.

Latham, N., and G. Mason. 2008. Maternal deprivation and the development of stereotypic behaviour. *Applied Animal Behaviour Science* 110(1–2): 84–108.

Lewis, M., M. Presti, J. Lewis, and C. Turner. 2006. The Neurobiology of Stereotypy I: Environmental Complexity. In G. Mason and J. Rushen, eds., *Stereotypic animal behaviour: Fundamentals and Applications to Welfare,* pp. 190–226. Wallingford, UK: CAB International.

Maestripieri, D., and K. Carroll, K. 1998. Risk factors for infant abuse and neglect in group-living rhesus monkeys. *Psychological Science* 9(2): 143–45.

Mallapur, A., N. Waran, S. Seaman, and A. Sinha. 2006. Preliminary observations on the differences in reproductive behaviour between breeding and non-breeding captive lion-tailed macaques (*Macaca silenus*) housed in Indian zoos. *Applied Animal Behaviour Science* 97(2–4): 343–48.

Margulis, S., and E. Westhus. 2008. Evaluation of different observational sampling regimes for use in zoological parks. *Applied Animal Behaviour Science* 110(3–4): 363–76.

Mason, G. 1991. Stereotypies: A critical review. *Animal Behaviour* 41:1015–37.

Mason, G., and J. Rushen, eds. 2006. *Stereotypic Animal Behaviour: Fundamentals and Applications to Welfare*, 2nd ed. Wallingford, UK: CAB International.

Mason, G., R. Clubb, N. Latham, and S. Vickery. 2007. Why and how should we use environmental enrichment to tackle stereotypic behaviour? *Applied Animal Behaviour Science* 102(3–4): 163–88.

McPhee, M. E. (2002). Intact carcasses as enrichment for large felids: Effects on on- and off-exhibit behaviors. *Zoo Biology* 21(1): 37–47.

Meagher, R. 2009. Observer ratings: Validity and value as a tool for animal welfare research. *Applied Animal Behaviour Science* 119:1–14.

Meehan, C. L., J. P. Garner, and J. A. Mench. 2003. Isosexual pair housing improves the welfare of young Amazon parrots. *Applied Animal Behaviour Science* 81(1): 73–88.

Mills, D., and U. Luescher. 2006. Veterinary and pharmacological approaches to abnormal repetitive behaviour in captive animals. In G. Mason and J. Rushen, eds., *Stereotypic Behaviour in Captive Animals: Fundamentals and Applications to Welfare*, pp. 286–324. London: CAB International.

Napolitano, F., G De Rosa, and A. Sevic. 2008. Welfare implications of artificial rearing and early weaning in sheep. *Applied Animal Behaviour Science*, 110(1–2): 58–72.

Nogueira, S., S. Nogueira-Filho, E. Otta, C. Dias, and A. Carvalho. 1999. Determination of the causes of infanticide in capybara (*Hydrochaeris hydrochaeris*) groups in captivity. *Applied Animal Behaviour Science* 62(4): 351–57.

Parker, M., D. Goodwin, E. Redhead, and H. Mitchell. 2006. The effectiveness of environmental enrichment on reducing stereotypic behaviour in two captive vicugna (*Vicugna vicugna*). *Animal Welfare* 15(1): 59–62.

Pomerantz, O., and J. Terkel. 2009. Effects of positive reinforcement training techniques on the psychological welfare of zoo-housed chimpanzees (*Pan troglodytes*). *American Journal of Primatology* 71(8): 687–95.

Rommeck, I., K. Anderson, A. Heagerty, A. Cameron, and B. McCowan. 2009. Risk factors and remediation of self-injurious and self-abuse behavior in rhesus macaques. *Journal of Applied Animal Welfare Science* 12(1): 61–72.

Ross, S. 2006. Issues of choice and control in the behaviour of a pair of captive polar bears (*Ursus maritimus*). *Behavioural Processes* 73(1): 117–20.

Rushen, J., A. Lawrence, and C. Terlouw. 1993. The motivational basis of stereotypies. In A. Lawrence and J. Rushen, eds., *Stereotypic Behaviour: Fundamentals and Applications to Welfare*, pp. 41–64. Wallingford, UK: CAB International.

Savory, C., and J. Hetherington. 1997. Effects of plastic anti-pecking devices on food intake and behaviour of laying hens fed on pellets or mash. *British Poultry Science* 38(2): 125–31.

Schapiro, S., J. Perlman, and B. Boudreau. 2001. Manipulating the affiliative interactions of group-housed rhesus macaques using positive reinforcement training techniques. *American Journal of Primatology* 55(3): 137–49.

Scott, E., A. Nolan, J. Reid, and M. Wiseman-Orr. 2007. Can we really measure animal quality of life? Methodologies for measuring quality of life in people and other animals. *Animal Welfare* 16:17–24.

Shyne, A., and M. Block. 2010. The effects of husbandry training on stereotypic pacing in captive African wild dogs (*Lycaon pictus*). *Journal of Applied Animal Welfare Science* 13(1): 56–65.

Smith, B. P., and C. A. Litchfield. 2010. An empirical case study examining effectiveness of environmental enrichment in two captive Australian sea lions (*Neophoca cinerea*). *Journal of Applied Animal Welfare Science* 13(2): 103–22.

Swaisgood, R., and D. Shepherdson. 2005. Scientific approaches to enrichment and stereotypies in zoo animals: What's been done and where should we go next? *Zoo Biology* 24(6): 499–518.

Tarou, L., M. Bashaw, and T. Maple. 2003. Failure of a chemical spray to significantly reduce stereotypic licking in a captive giraffe. *Zoo Biology* 22(6): 601–7.

Weber, E., and L. Olsson. 2008. Maternal behaviour in *Mus musculus* sp.: An ethological review. *Applied Animal Behaviour Science* 114(1–2): 1–22.

Wechsler, B. 1991. Stereotypies in polar bears. *Zoo Biology* 10(2): 177–88.

Wells, D. L., and R. M. Irwin. 2008. Auditory stimulation as enrichment for zoo-housed Asian elephants (*Elephas maximus*). *Animal Welfare* 17(4): 335–40.

Wemelsfelder, F., and A. Lawrence. 2001. Qualitative assessment of animal behaviour as an on-farm welfare-monitoring tool. *Acta Agriculturae Scandinavica Section A-Animal Science, Supplement* 30:21–25.

Whitham, J., and N. Wielebnowski. 2009. Animal-based welfare monitoring: Using keeper ratings as an assessment tool. *Zoo Biology* 28(6): 545–60.

40

Enrichment

David J. Shepherdson

INTRODUCTION

The term "environmental enrichment" encompasses a wide range of practices within the zoo- and aquarium-keeping profession. These practices have in common the goal of improving the welfare of animals by changing their environment. Somewhat confusingly, even subtractions from the environment (sources of stress, for example) are sometimes described as enrichment if they are deemed to enhance the environment for the animal inhabitants.

Typically, however, "environmental enrichment" refers to the provision of environmental stimuli that increase opportunities for species-typical behaviors with the intention of enhancing animals' physical and psychological well-being (Shepherdson 1998; Newberry 1995). Most approaches to enrichment emphasize the importance of understanding the natural behavior and ecology of animals, both as a source of inspiration for ideas and as a benchmark for evaluating success (Mellen and MacPhee 2001).

The motivations for applying enrichment to zoo and aquarium environments are also varied. At its heart, enrichment is for the well-being of the animals in our care, but there are other valid reasons for it. Animals living in environments that cater to their needs and stimulate natural patterns of behavior tend to be more engaging and interesting to visitors, and thus more likely to help zoos and aquariums fulfill their mission of connecting people with wildlife. Environmental enrichment also clearly has a role to play in the conservation activities of zoos and aquariums. Endangered species are more likely to breed successfully in zoos and aquariums when enrichment is an integral part of the husbandry plan, and zoos and aquariums around the world are becoming increasingly involved in local species recovery programs that involve rearing and sometimes breeding animals for release. When we release animals into the wild from zoo and aquarium environments, it becomes our responsibility to make sure that the pre-release environment has prepared them for life

in the wild—and our success is most basically measured in terms of post-release survival (Reading et al. 2013).

Enrichment is also a field of applied scientific study, with its own guiding principles and underlying concepts. It is important that enrichment continues to be based on a sound empirical (measurable) understanding of how our enrichment activities affect the behavior and well-being of the animals in our care.

This chapter will provide a review of environmental enrichment for zoo and aquarium professionals. After studying this chapter, the reader will understand

- the history of environmental enrichment, the path of its evolution, and its significance today
- the theoretical basis of enrichment and its relationship to animal welfare
- the goals and objectives of enrichment
- the different kinds of enrichment that can be employed
- how scientific research contributes to enrichment
- how enrichment is scientifically evaluated.

HISTORY

The activity that we now know as environmental enrichment has had a long history under different names. In 1925, Robert Yerkes famously stated that animals should have their time occupied by play and work (Yerkes 1925). Heini Hediger, an influential zoo biologist in the mid-20th century, advocated strongly for the importance of understanding the natural behavior of the animals that keepers care for, and for using that information to design improved environments and husbandry (Hediger 1968). Around the same time, Desmond Morris put into practice a number of enrichment devices at the London Zoo, including a mechanical fish-feeding device to stimulate swimming in sea lions (Morris 1960).

It was Hal Markowitz, however, working in the 1980s primarily at the Oregon Zoo (then the Washington Park

Figure 40.1. Humans are a significant part of most zoo and aquarium animal environments. A dwarf mongoose is searching for food in sand substrate. Photo by Michael Durham, courtesy of the Oregon Zoo.

Zoo) in Portland, Oregon, who really focused attention on the importance of keeping zoo animals occupied and challenged with tasks that restore some element of control to their lives (Markowitz 1982). Through the force of his personality, his research, his publications, and his graduate students, he stimulated a strong and enduring interest in enrichment that continues to grow to this day. Markowitz documented a wide range of devices installed in zoo exhibits with the goal of capitalizing on the occupants' natural behaviors and motivation to stimulate their natural activity and behavior, and to restore some degree of control to their lives in the zoo environment.

For a short period of time, some of Markowitz's techniques spurred heated discussion within the zoo profession around the issue of naturalism (Hutchins and Hancocks 1984). Many of his devices were based on machines that rewarded the correct behavioral response to a stimulus with a food reward. For example, diana monkeys (*Cercopithecus diana*) at the Oregon Zoo learned to retrieve tokens from one station in response to a signal, and then used them to retrieve food rewards by inserting them at another station. In another example, white-handed gibbons (*Hylobates lar*) learned that by brachiating rapidly between two stations in response to a signal, they could obtain a food reward (Markowitz 1982). Although these were undeniably "unnatural" stimuli, the be-

haviors they stimulated were for the most part "natural." More naturalistic enrichment has predominated in the intervening years, for both conceptual and practical reasons, and with the recognition that different approaches to enrichment may be appropriate in different contexts (e.g., stark and sterile environments as opposed to complex naturalistic ones).

ENRICHMENT TODAY

Several more recent events exerted a strong influence on the shape of enrichment today. In 1992, regulations were implemented by the US Department of Agriculture (USDA) that required primate-holding facilities to maintain environmental enhancement programs (APHIS 1992). Also in 1992, the first edition of the zoo-focused quarterly newsletter *Shape of Enrichment* was published. Now a nonprofit organization, *Shape of Enrichment* has become an important resource for communication and an organizer of regional meetings and workshops. The American Association of Zoo Keepers (AAZK) has also been very active in providing information about enrichment to animal keepers and encouraging discourse and the sharing of ideas.

In 1993, the first Conference on Environmental Enrichment was held at the Oregon Zoo, in Portland. This represented the first opportunity for keepers, veterinarians, researchers,

Figure 40.2. A novel object confounds a pride of lions. Photo by Michael Durham, courtesy of the Oregon Zoo.

and other zoo professionals to meet colleagues from around the world in the zoo and aquarium profession and related professions, such as laboratory animal care researchers and laboratory animal caretakers. The conference became the first in the ongoing series of International Conferences on Environmental Enrichment, which are held every two years at locations around the world. The first conference ultimately resulted in the publication in 1998 of the first comprehensive text on the subject, "*Second Nature: Environmental Enrichment for Captive Animals*" (Shepherdson et al. 1998).

In 1999, the American Zoo and Aquarium Association (AZA) incorporated a requirement in its five-year accreditation review that zoos and aquariums provide evidence of enrichment programs with certain key elements.

As a result of all this activity, environmental enrichment is now a fully established aspect of zoo and aquarium animal husbandry and a focus of considerable research. Enrichment activities are directed at a wide range of taxa and are applied in a proactive and holistic manner rather than as a "band-aid" response to specific problems.

THEORETICAL UNDERPINNINGS AND GOALS

UNDERLYING CONCEPTS

In order for efforts to be effective and efficient, we need to have a theoretical framework based on our understanding of the ways in which enrichment might work to improve animals' well-being and stimulate their natural behaviors. This framework helps us to decide what the most appropriate actions are for a given situation, and also to evaluate our success and devise alternative strategies when our actions are not successful. Several concepts underlie our current thinking about animal welfare and enrichment.

MIMICKING NATURE

The behavior of animals has evolved to satisfy their needs in their natural habitats. So it seems intuitive that recreating natural stimuli in captivity is a good strategy. Indeed, freedom to perform natural behavior is a central tenet of much animal welfare legislation. On its own, however, it is not always helpful, for a couple of reasons. Not all "natural" behavior is indicative of good well-being; after all, fear, death, and disease are frequent experiences in the wild. But, given that one may need to choose between "good" and "bad" natural behaviors (and the circumstances from which they arise), what criteria can be used? One is forced to make similar choices when practicality makes it impossible to completely recreate in captivity the wild environment for all but the simplest organisms. Simply mimicking nature, then, is not sufficient as a guiding concept; some refinement is clearly needed.

BEHAVIORAL NEEDS

A more refined version of this concept stems from the observation that animals appear to be motivated (and therefore maybe "need") to perform behaviors independent of their outcome. Many animals will continue to eat and even hunt when satiated; similarly, some animals will continue to build nests even if presented with fully formed nests. This is the basis of the "behavioral needs hypothesis" that states that some animals may need the chance to perform certain behaviors even if they do not need the outcome. An open question with respect to this concept is the degree to which animals can meet this behavioral need through mere performance of the actions or whether those actions need a functional component. This question does help us to select the behaviors most likely to be important for animals' well-being (and, therefore, for their enrichment). For example, one can observe animals' response to being deprived of key behavioral opportunities (soil in which to dig a den, for example) and can predict on the basis of natural history which behaviors are most likely to be of short-term importance to survival (feeding and foraging and hiding or escaping from danger, for example). One can then argue that these may be the most important behaviors toward which to focus our enrichment activities.

INFORMATION SEEKING

All wild animals engage in some kind of exploratory behavior, and in a changing natural environment, information is an

Figure 40.3. Polar bear enjoys ice from an ice maker. Photo by Michael Durham, courtesy of the Oregon Zoo.

Predictability is also linked to well-being and control (Basset and Buchanan-Smith 2007). Many studies have shown that being able to predict an unpleasant event reduces the degree of stress the subject perceives (Davis and Levine 1982; Seligman and Meyer 1970), so predictability can be a good thing. However, predictable events can also result in high levels of anticipatory arousal; this itself can cause stress and has often been implicated in the formation of stereotypic behaviors (e.g., pacing in anticipation of feeding). One way to reduce predictability in a way that does not result in stress is to increase the animal's control by allowing it to react to the stimulus (e.g., food) using appropriate natural behaviors. So, for example, food can be scattered at unpredictable intervals, allowing animals to forage for it under their own control. Another way to reduce the stress associated with unpredictable arousing events is to provide a unique and reliable signal (a whistle, for example) before the event that effectively tells the animal what is about to happen and allows it to relax when the signal is absent.

One other aspect of typical zoo husbandry that might be especially stressful for zoo animals warrants discussion here. As stated above, predictable routines are preferable in many circumstances, but when animals are used to a predictable schedule and then the predicted event does not happen on time (such as when feeding is delayed by 30 minutes), this seems to be particularly stressful. The important message here seems to be that routines should be either random or predictable, but not in between, and that reliable signals and an element of control are also important factors in reducing stress.

TYPES OF ENRICHMENT

When planning to enrich a zoo and aquarium environment, it may help to break enrichment down into categories. The following five categories capture most of the diverse range of enrichment activities commonly employed.

FEEDING ENRICHMENT

This is perhaps the most commonly used category, and with good reason. The day-to-day survival of most wild animals depends on finding and eating food, and the animals have evolved finely honed, species-specific behaviors and strategies for doing so.

Feeding can be made more or less enriching by manipulating several parameters. Increasing the variety of food items offered results in more choice and a variety of feeding-related stimuli. The way in which food is prepared and offered can influence the range of behaviors required to obtain it. For example, food that is chopped requires fewer natural food processing behaviors to be employed by the forager. Food can be provided in a clean bowl or it can be distributed about the environment in a way that requires the animal to engage in natural foraging behaviors such as climbing, manipulating, smelling and searching. Food items can be dispersed in both time and space. Simply scattering food items can result in major increases in time spent foraging and consequent reductions in less desirable behaviors (Carlstead and Seidensticker 1991). Varying food items over time can maintain the novelty of some food items and combinations, and allows for the possibility of recreating natural cycles of feeding.

important key to survival. Exploration is an indicator that animals are finding useful information in their environment, and perhaps that it is "interesting" to them. Information seeking is of such fundamental importance to wild animals that many argue (Mench 1998) that animals need an environment that provides them with a constant source of new and relevant information. Enrichment can certainly be used to provide this.

CONTROL, CONTINGENCY, AND PREDICTABILITY

Wild animals are in control of their lives in the sense that their survival is dependent or contingent on their actions. For example, a hungry animal forages for food, a cold animal may make a nest, and a frightened animal may try to escape. When an animal increases the chance that it will receive what it needs through its own actions, it can be said to have control. An animal's control is typically much reduced in a captive environment, where human caretakers typically attempt to anticipate its needs. A large body of human literature and a growing body of literature on animals suggests that this kind of control is critical to an animal's well-being. Enrichment typically increases control by allowing animals to perform behaviors from their natural repertoire to obtain some of the things they want and need (food, for example).

Figure 40.4. An ocelot hunts live fish. Photo by Michael Durham, courtesy of the Oregon Zoo.

Some of the most effective feeding enrichment seems to occur when the food reward is unpredictable but the animal can still control it through natural food-finding behaviors. For example, a fishing cat, when given live fish with plenty of opportunities to hide (Shepherdson et al. 1993), spent many hours "fishing" even though the reward rate was low. As always, though, the animal's natural history must be taken into account. While intermittent food rewards may be effective and stimulating for some carnivores, they may be totally inappropriate for some other species (e.g., grazing herbivores).

SENSORY ENRICHMENT

Many of the animals in our care live in a sensory world very different from ours. It can be hard for us to appreciate this important aspect of their life, but with a little effort it can provide rewarding avenues for enrichment. As always, a sound knowledge of natural history is a good starting point.

Smells can be used to provide variety and novelty, and when combined with food rewards or other desired items it can be used to provide clues to reward investigation and exploration. Sounds can be used similarly, and Markowitz has published the results of several enrichment feeding activities that use sounds as cues for feeding rewards (Markowitz 1982). Many animals (e.g. elephants and chimpanzees) make sounds in order to communicate to others, and incorporate them into behavioral displays; this creates some interesting potential for enrichment activities (Shepherdson and Bemment 1989). Bear in mind, of course, that many animals hear well beyond the range of our own hearing.

COGNITIVE ENRICHMENT

All animals use their cognitive abilities to solve the problems and challenges of daily life. In the zoo and aquarium environment these challenges tend to be reduced, but enrichment can be used to restore the need for problem solving and thinking. Puzzle feeders are a commonly used in this capacity, and complex challenges are not difficult to set up. Chimpanzees and elephants are not the only species capable of building tools and manipulating objects in their environment to access desired items; this has been documented in ravens, sea otters, dolphins, gorillas, orangutans, and even octopuses. Indeed, the more that animal cognition is studied, the more complex are the abilities that animals have been documented to display, and the more widespread they are found to be throughout the animal kingdom.

Training is one husbandry activity that often relates directly to cognition. At least when learning a behavior for the first time, the trainee must work out what the trainer wants it to do in order for it to receive a reward. Trainers know that this is an intensely motivating activity for many animals. The act of being trained (learning) not only requires thought but also provides a certain amount of choice and control, and

Figure 40.5. Asian elephants pull up trees provided for enrichment and food. Photo by Michael Durham, courtesy of the Oregon Zoo.

results in increased activity and behavioral diversity. There can be little doubt, then, that being trained can be enriching. When thinking about training as a component of an enrichment strategy, it is important to keep in mind the specific enrichment goals. If, for example, the goal of training is to stimulate problem solving, then training for new behaviors must be a central focus, since simply reinforcing existing behaviors is unlikely to have the required effect.

SOCIAL ENRICHMENT

The animals in our care live naturally in a wide variety of social systems that cover the spectrum from being relatively solitary for most of their lives to being highly gregarious and social at all times. Creating an appropriate social context for them is one of the most critical factors with respect to their well-being. The tremendously damaging consequences of isolation or inappropriate separation are well documented for many social species, and the lessons learned can clearly be generalized to other less studied species.

Social behavior is often complex and may change through seasons and through an individual's life. Male elephants start life in matriarchal herds; they then disperse and live the rest of their lives in very different kinds of associations with both males and females. Female elephants remain in the matriarchal herds and live very different lives. Many species are only social at certain times of the year, the breeding season for example, or at certain times during their lives, such as when rearing young. Yet others may make temporary associations while migrating or searching for mates. For many species, life in captivity is necessarily different from this.

When establishing measurable behavior goals for the social animals in our care, their social behavior should be a priority, and our creative thought should be directed at trying to recreate captive social contexts analogous to those experienced by wild animals. This can include making sure that groups contain the right number of individuals with the appropriate range of age and gender. Ideally however, it would go further than this and do things to recreate some of the temporal variability seen in nature and referred to above, such as moving animals around and changing their social structures at different times of the year and at different life stages to simulate natural events. For some species we are just beginning to understand how to use the information and tools at our disposal to help give them rich and rewarding social lives.

Last, but not least, keepers have relationships with their animals and these also should be carefully considered and adapted to best suit the animals' needs. Although most scientific studies of the caretaker-animal bond have been conducted on domestic animals (Waiblinger et al. 2006) it is quite clear that the behavior of caretakers can have a very direct influence on animal welfare both positively and negatively. It may well turn out, for example, that many benefits of training are realized indirectly through the effect of training on the

Figure 40.6. Training stimulates the mind and can be considered enrichment in some circumstances. Photo by Michael Durham, courtesy of the Oregon Zoo.

human-animal relationship. This is currently an active area of research in zoos.

PHYSICAL ENRICHMENT

After feeding enrichment, this is probably the category into which most other enrichment activities fall. It includes the basic elements of exhibit design and construction, such as shape, relief (topography), microclimate, substrate, and sight lines. Also included are features such as hiding places, climbing structures, shelter, perches, and water sources. Rotating animals amongst two or more enclosures is also an enrichment option made available in a number of exhibits built fairly recently (Coe 1995). This category also includes the placement and renewal of more temporary objects such as vegetation, climbing apparatuses, pools, mud wallows, and moveable visual barriers. It includes those smaller objects that we place in exhibits every day to stimulate specific behaviors (e.g., thrashing of a branch by an elk) and provide novelty, bedding and resting areas, and things for animals to interact with in various ways. A large hollow plastic ball filled with a warm mixture of water and blood may stimulate hunting and prey

capture behavior, and a piece of PVC pipe drilled with holes can be used as a puzzle feeder to stimulate foraging behavior and problem solving. Objects placed in an animal's enclosure can also serve a function simply by virtue of their novelty, thus stimulating exploration, interest, and associated behaviors.

EVALUATING ENRICHMENT

Enrichment needs to be effective, as the animals' well-being depends on the animal care team; and since zoos and aquariums have limited resources, they must use them effectively. It is also quite possible that enrichment, though provided with the goal of improving the animals' well-being, actually reduces it. For example, in social groups the provision of food as enrichment can cause aggression and frustration. For these reasons it is imperative that any enrichment activity should have a hypothesis or goal stated in terms of measurable outcomes. For example, a novel object may be introduced into an exhibit with the goal of increasing exploratory behavior (see the chapter 41). Even if this goal is not measured objectively, its existence stated in measurable terms increases the likelihood of a realistic assessment of the enrichment's effectiveness.

EXPECTED OUTCOMES AND GOALS

The specific goals for a given enrichment activity will depend on the circumstances, the species, and the guiding concepts described above. However, since ultimately we are interested in improving animals' well-being it is worth considering how that might be measured in an environment using behavioral measures. These measures fall into two groups: those that indicate reduced well-being and those that indicate improved well-being.

Indicators of reduced well-being are the traditional measures that most will be familiar with, including abnormal behaviors and behaviors that directly indicate suffering or physiological stress. Many of these will be discussed elsewhere in this volume.

Stereotypic behavior is perhaps the best known behavioral indicator of reduced well-being. While there is no doubt that stereotypic behavior can indicate a problem, its link to well-being is not a one-to-one relationship (Mason et al. 2007), and it should never be used as the sole indicator of well-being. Other commonly accepted indicators include self-directed behaviors (e.g., overgrooming with alopecia, and self-directed [auto-] aggression) and other abnormal behaviors, such as regurgitation and reingestion and coprophagia. In addition to these easily identifiable behaviors, other indicators include lack of appetite, reduced social interaction, inappropriate social interaction, and fear associated with certain contexts. An animal care professional should always watch for these behaviors.

In some cases physiological measurements may be used to back up behavioral ones; these could include measurements of hypothalamic-pituitary-adrenal axis (HPA) activity (cortisol levels), immune function, and respiratory function (e.g., breathing and heart rate). It should be noted, however, that many of these measures are notoriously difficult to interpret with respect to a patient's well-being.

In contrast, there are few generally accepted measures of improved or good well-being in the welfare literature; the cup

Figure 40.7. African wild dogs exhibit social behavior in a naturalistic exhibit. Photo by Michael Durham, courtesy of the Oregon Zoo.

has been half full for many years! Recently however, there has been a resurgence of interest in identifying these measures (Boissy et al. 2006). Some of the most likely behavioral candidates include expressions of contentment (e.g., play, some vocalizations, exploration, affiliative actions, self-grooming, and maintenance behaviors) and indicators of interest, alertness and curiosity, and reproduction (which can be assessed both by observation and by hormonal analysis).

Patterns of behavior may also be important, and both behavioral diversity and diversity of space use have also been suggested to correlate to well-being (Shepherdson et al. 1993; Swaisgood et al. 2005).

Physiological measures of positive well-being are for the most part considered to be simply the absence of negative indicators, although some hope is held out for the potential of measuring pleasure responses by physiological indicators such as levels of dopamine, serotonin, and endorphins (Yeats and Main 2008).

TYPES OF STUDY

Various experimental designs are available to evaluate enrichment, and each has its pros and cons (Swaisgood and Shepherdson 2005; Martin and Bateson 1986). Measuring the behavior of an individual or group before providing it with enrichment, and then comparing those measurements to similar ones made afterwards is the most commonly used design, and is often referred to as a "pre-post design" or "AB." It is simple to apply, and easy to analyze and understand. It has some significant potential drawbacks, however. Since the enrichment is applied only once, it is possible that any subsequent changes in behavior may in fact be due to some other event that has coincided with the enrichment (e.g., changes in social group, weather, season, and other husbandry activities). There may also be some effects due to the mere passage of time: so-called maturational effects. It will always be true, for example, that the animals in the post-measurement phase will be older, and we know that age can have a significant effect on behavior. One way of controlling for both of these potential confounds (variables) is to use a "multiple base line design" or "ABA" (or "ABABA"). The ability to compare the enrichment condition after returning to the baseline condition multiple times will increase the chance of distinguishing effects that might be due purely to coincidence and maturation effects. However, there are practical and methodological drawbacks to this design. Withdrawing enrichment can be difficult if

Figure 40.8. Polar bear with boomer ball. Photo by Michael Durham, courtesy of the Oregon Zoo.

doing so is expensive or time-consuming, and it may also seem unethical if it appears to have resulted in alleviation of suffering. It is also possible that after animals have been exposed to enrichment, they will respond to the baseline condition (and subsequent enrichment conditions) differently; this is a sequential or experiential effect. In general, however, a multiple baseline design is much superior to a simple pre-post or AB type of study.

Yet another design is the "between subjects" or "comparative" approach. An example of this approach would be measuring the effect of scatter feeding on one or more groups of chimpanzees, and comparing it to a group or groups that has not received this form of enrichment. This kind of design lends itself well to multi-institutional studies. An inevitable drawback is that the enrichment treatment is inevitably not the only difference between the groups, so that it is always possible that a behavioral change attributed to enrichment is actually due to one of the other confounds (e.g., group size, climate, exhibit design, or diet).

A variant of the comparative experimental approach is the "correlational" study. In this design, rather than comparing the effects of the presence or absence of a specific enrichment activity, the researcher simply measures the various enrichment activities going on at each institution and then looks for correlations between the enrichment variables and the animals' behavior. With any correlational study it is difficult to definitively assign cause and effect, but such a study can be an effective way of discovering patterns of relationships between multiple variables.

SOME PROBLEMS WITH CURRENT RESEARCH

Swaisgood and Shepherdson (2005) identified several potential flaws common in the extant enrichment literature. Too many studies use the less desirable AB design, and many studies involve very small subject sample sizes which greatly reduce the ability to detect significant differences and generalize results. The effects of enrichment change through time, and when enrichment is new and novel, it may result in different behaviors than enrichment that has been used for long periods of time. Many studies take place over limited time periods, and so cannot detect the development of habituation, thus possibly overestimating the effectiveness of the measured enrichment.

One of the goals of these studies is to establish which kinds of enrichment are most effective at reaching the desired goal (e.g., reducing stereotypic behavior). Unfortunately, many studies use a "kitchen sink" approach, whereby a wide variety of different enrichment activities are employed simultaneously. Consequently it becomes difficult or impossible to determine which characteristics or kinds of enrichment are most effective. Research has clearly shown the effectiveness of enrichment; the challenge now is to learn more specifically which kinds of enrichment are most effective at achieving specific

Figure 40.9. The perfect nexus of enrichment and education! Photo by Michael Durham, courtesy of the Oregon Zoo.

goals and why. Well-designed, large-scale, multi-institutional studies and multiple baseline designs conducted over realistic time periods and combining physiological and behavioral measurements hold the promise of meeting this challenge.

EVALUATION OF ENRICHMENT

The last 15 to 20 years have seen significant growth in the number of published enrichment studies. A review of the literature (Swaisgood and Shepherdson 2005) suggests that the most commonly used measures by which enrichment is evaluated in published papers include

- reduction in stereotypic behavior
- increased time and diversity of foraging behavior
- increased diversity of space use
- increased activity
- increased play, exploration, and affiliative behavior
- increased problem-solving behavior
- increased species-typical behaviors such as digging, climbing, or swimming.

A quick perusal of this literature will be enough to convince most that enrichment can certainly improve well-being on the basis of any of these measures. However, to find out how effective enrichment is as a strategy, and whether some kinds of enrichment are more effective than others, we need

to move beyond case studies and conduct a meta-analysis of the published research literature. A meta-analysis is essentially a study of studies. Two such studies have been published, and both focus on the use of enrichment to reduce stereotypic behavior,—the most common goal of published enrichment studies, for good reason. While not all stereotypic behavior indicates current problems, there is a clear link between this behavior and reduced well-being. Indeed, Mason et al. (2007) call for "zero tolerance" of abnormal repetitive behaviors in zoo animals, due to ethical concerns about stress and poor welfare, and to the reduction in conservation value that follows from it.

In their respective studies using overlapping but different data and different statistical tools, both Swaisgood and Shepherdson (2006) and Shyne (2006) found conclusive evidence that enrichment is a highly effective strategy for reducing stereotypic behavior, with typical reductions in the region of 50 to 60% and Shyne (2006) concluding that in 90% of studies stereotypic behavior was reduced. Neither study was able to shed much light on the comparative effectiveness of different kinds of enrichment, although Shyne's results suggested that puzzle feeders were particularly effective. Both of these studies must be viewed with the caveats that in no case was a stereotypic behavior eliminated, and that many of the studies included in the analysis were of short duration and so the long-term effectiveness of the enrichment may be somewhat less. A more fundamental problem is the possibility that studies with positive outcomes are more likely to be submitted for publication than those that failed to achieve the desired outcome, thus inflating the proportion of successful studies.

SUMMARY

Enrichment is a term that covers a variety of activities including aspects of husbandry, nutrition, training, and exhibit design with the common goal of increasing behaviors indicative of well-being and reducing those that are associated with poor welfare. Scientific studies of enrichment clearly indicate that it is frequently successful at achieving this goal, especially with respect to stereotypic behavior. Enrichment is not only a desirable aspect of modern zoo and aquarium animal management; it is often a legal prerequisite for primates (e.g., in the United States, Canada, and the European Union) and may be an accreditation requirement for membership in regional zoo and aquarium associations such as the AZA and EAZA.

The most successful enrichment techniques not only reduce abnormal and undesirable behaviors but also stimulate desirable behaviors indicative of contentment or even pleasure. Successful enrichment not only improves the lives of zoo and aquarium animals but is also rewarding for keepers and makes the exhibits more rewarding and impactful for visitors. Ultimately, enrichment helps zoo and aquariums come closer to realizing their collective vision of connecting people with animals and thus creating a better future for wildlife.

By embracing enrichment, zoo and aquarium professionals are acknowledging the importance of specifically catering to the psychological needs of the animals in their care. They are also acknowledging the importance of understanding the natural behavioral biology of their charges and the need to apply knowledge, keen powers of observation, and creativity to their daily duties.

Enrichment is not a new idea within the zoo and aquarium profession, but there is still room for improvement. Looking to the future, enrichment needs to be spread to a wider diversity of species and more must be learned about which enrichment activities will result in desired behavioral changes. The principles of enrichment need to be applied holistically and preemptively to the whole lives of the animals, including birth, early development, and times spent out of the public view in quarantine and in veterinary and holding areas.

REFERENCES

Bassett, L., and H. M. Buchanan-Smith. 2007. "Effects of predictability on the welfare of captive animals." *Applied Animal Behaviour Science* 102:223–45.

Boissy, A., G. Manteuffel, M. Jensen, R. Moe, B. Spruijt, L. J. Kelling, C. Winckler, B. Forkman, I. Dimitrov, J. Langbein, M. Bakken, I. Veissier, and A. Aubert. 2007. "Assessment of positive emotions in animals to improve their welfare." *Physiology & Behavior* 92:375–97.

Brydges, N. M., M. Leach, K. Nicol, R. Wright, and M. Bateson. 2011. "Environmental enrichment induces optimistic cognitive bias in rats." *Animal Behaviour* 81:169–75.

Carlstead, K. 1991. "Husbandry of the fennec fox *Fennecus zerda*: Environmental conditions influencing stereotypic behaviour." *International Zoo Yearbook* 30:202–7.

Carlstead, K., J. Seidensticker, and R. Baldwin. 1991. "Environmental enrichment for zoo bears." *Zoo Biology* 10:3–16.

Coe, J. C. 1995. "Zoo animal rotation: New opportunities from home range to habitat theatre." *American Zoo & Aquarium Annual Proceedings*, 77–80.

Davis, H., and S. Levine. 1982. "Predictability, control and the pituitary-adrenal response in rats." *Journal of Comparative and Physiological Psychology* 96: 393–404.

Hediger, H. 1968. *Psychology and Behavior of Animals in Zoos and Circuses*. New York: Dover Publications.

Hutchins, M., D. Hancocks, and C. Crockett. 1984. "Naturalistic solutions to the behaviour problems of captive animals." *Zoological Garten* 54:28–42.

Markowitz, H. 1982. *Behavioral Enrichment in the Zoo*. New York: Van Nostrand Reinhold.

Martin, P., and P. Bateson. 1986. *Measuring Behaviour*. Cambridge: Cambridge University Press.

Mason, G. R., R. Clubb, I. N. Latham, and S. J. Vick. 2007. "Why and how should we use environmental enrichment to tackle stereotypic behaviour?" *Applied Animal Behavior Science* 102:163–88.

Mellen, J., and M. S. MacPhee. 2001. "Philosophy of environmental enrichment: Past, present, and future." *Zoo Biology* 20(3): 211–26.

Mench, J. A. 1998. "Environmental enrichment and the importance of exploratory behavior." *Second Nature: Environmental Enrichment for Captive Animals*. D. J. Shepherdson, J. D. Mellen, and M. Hutchins, eds. Washington: Smithsonian Institution Press, 30–46.

Morris, D. 1960. "Automatic seal-feeding apparatus at London Zoo." *International Zoo Yearbook* 5:70.

Newberry, R. C. 1995. "Environmental Enrichment: Increasing the Biological Relevance of Captive Environments." *Applied Animal Behaviour Science* 44(2–4): 229–43.

Reading, R. P., B. Miller, and D. Shepherdson. 2013. "The Value of Enrichment to Reintroduction Success." *Zoo Biology* doi: 10.1002/zoo.21054.

Seligman, M. E. P., and P. Meyer. 1970. "Chronic fear and ulcers in rats as a function of the unpredictability of safety." *Journal of Comparative and Physiological Psychology* 3:202–7.

Shepherdson, D. 1998. "Introduction: Tracing the path of environmental enrichment on zoos." *Second Nature: Environmental Enrichment for Captive Animals*. D. Shepherdson, J. Mellen, and M. Hutchins, eds. Washington, DC: Smithsonian Institution.

Shepherdson, D. J., K. Carlstead, J. D. Mellen, and J. Seidensticker. 1993. "The influence of food presentation on the behavior of small cats in confined environments." *Zoo Biology* 12:203–16.

Shepherdson, D. J., N. Bemment, M. Carman, and S. Reynolds. 1989. "Auditory enrichment for Lar gibbons (*Hylobates lar*) at London Zoo." *International Zoo Yearbook* 28:256–60.

Shepherdson, D. J. and J. D. Mellen, M. Hutchins, 1998. "Second Nature: Environmental enrichment for captive animals." Washington, DC: Smithsonian Institution.

Shyne, A. 2006. "Meta-analytical review of the effects of enrichment on stereotypic behavior in zoo mammals." *Zoo Biology* 25: 317–37.

Swaisgood, R. R., and D. J. Shepherdson. 2005. "Scientific approaches to enrichment and stereotypies in zoo animals: What's been done and where should we go?" *Zoo Biology* 24(6): 499–518.

———. 2006. "Environmental enrichment as a strategy for mitigating stereotypies in zoo animals: A literature review and a meta-analysis." In *Stereotypic Animal Behaviour: Fundamentals and Implications to Welfare*. G. Mason and J. Rushen, eds. Wallingford, UK: CAB International, 255–84.

Waiblinger, Susanne, X. Boivin, Vivi Pedersen, A. M. Janczak, Maria-Vittoria Tosi, R. B. Jones, and E. Kathalijne Visser. 2006. "Assessing the human-animal relationship in farmed species: A critical review." *Applied Animal Behaviour Science* 101:185–242.

Yeats, J. W., and D. C. J. Main. 2008. "Assessment of positive welfare: A review." *The Veterinary Journal* 175:293–300.

Yerkes, R. M. 1925. *Almost Human*. New York: Century.

41

Enrichment Programs

Tammy M. Root

INTRODUCTION

Enrichment programs are designed to challenge the animals physically and psychologically, allow them to exhibit their natural behaviors, and allow them some control within their environment. In a captive environment, zoo staff control when and what an animal is being fed, where it is moved to within its environment, and when it is given medical care. Enrichment offers animals an opportunity to express themselves through normal species-typical behaviors.

Throughout this chapter, the reader will become familiar with and be able to understand some key points about establishing a process for successful enrichment programs, using the framework known as SPIDER. After studying this chapter the reader will

- understand the importance of developing a formal enrichment program
- understand how to use the SPIDER framework
- understand the challenges of using SPIDER
- understand the value of SPIDER as a useful tool.

ENRICHMENT PROGRAMS

Not only do enrichment programs vary from institution to institution, but they can also vary within the institution. For example, the senior keeper of each area might develop a scheduled rotation of enrichment. Another option is that keepers could follow their institution's guidelines for enrichment without a preplanned schedule. For example, a zoo's guidelines might indicate that the sharks and rays should be offered enrichment three times a month. The keeper can then decide which days to provide this enrichment.

Although institutions differ in their approach to enrichment, it is best for them to have formalized, structured programs that give the staff time to provide enrichment as part of their daily husbandry duties. Leadership is an important part of a formalized program. An enrichment leader can be any person who assumes responsibility for the program at a facility, such as a team leader, manager, or designated enrichment coordinator. Such leaders should have a solid background in enrichment and a familiarity with the types of enrichment items that should be used for various animals. They should have strong organizational skills for structuring programs, and should also have the resources needed to successfully operate it. Enrichment programs should be proactive, cover all species within the collection, and include all staff. If a zoo or aquarium has horticulture and maintenance staff, they should also be encouraged to understand and participate in the enrichment program, as their skills and resources will often become an integral part of the process. For example, horticulture staff may be needed to cut browse for different species of herbivores. They should also be aware of tree and plant species that are nontoxic for different animals. For instance, a given species of plant could be nontoxic for a gorilla but toxic for an elephant. Before giving any browse to an animal, horticulture and animal care staff should have sufficient knowledge of plant toxicology or refer to an expert for a decision. Maintenance staff, or keepers skilled in the use of tools, may be needed to help build enrichment devices, such as large complex puzzle feeders for chimpanzees.

Zoos and aquariums are all different, and some may or may not have the resources to establish elaborate enrichment programs. In some zoos, the enrichment program may be that a keeper comes into work on a given day and decides to give an animal certain treats and/or toys to play with. Other zoos have set schedules of which enrichment item is given to which animal or group of animals. Enrichment items are normally supplied both in the holding areas and in the exhibits. Sometimes they are rotated by category. They can include the following:

- feeding/foraging items: food items that are part of the animals' daily diet that encourage or increase time spent feeding or foraging (e.g., cereal or raisins hidden in the exhibit for a primate to find)

- sensory items: generally nonfood items that encourage the use of senses, including those attuned to environmental changes such as sound and smell (e.g., radios, beepers set at different times, and scents)
- social or behavioral items: nonfood items or activities that encourage social interaction (e.g., fire hose swings or ropes for tug-o-war)
- manipulable items: non-food items that the animals can manipulate, but which are not intended to increase food intake. (e.g., a ball inside a closed plastic jar)
- environmental modification: items placed in the exhibit permanently or semipermanently (e.g., logs or plantings added to the exhibit)
- training: participation in operant conditioning training sessions (e.g., being asked to present a foot, hand, and shoulder during a specific time frame).

ADVANTAGES AND DISADVANTAGES OF SCHEDULING ENRICHMENT

The advantages of scheduling enrichment include

- lack of guesswork regarding which enrichment the animal should receive on a given day
- assurance that all categories of enrichment will be used within a given time frame

- assurance that training time is scheduled
- staff accountability for enriching the animal.

The disadvantages of scheduling enrichment include

- difficulty staying on schedule with unforeseen changes in work routines, such as new animals arriving, unscheduled media events, staff illness, or maternity leave.
- variability in the availability of enrichment items, such as seasonal fruit, foraging items, or browse.

With any enrichment program, record keeping is essential. Items that should be recorded either in a separate enrichment log or in the keeper's daily logs are

- the type of enrichment item
- where the enrichment was placed
- how the animals react or respond to the item.

Before using a new enrichment item with any given animal, it is recommended that an enrichment request form be filled out and submitted to supervisors (see table 41.1 for an example). Many zoos use this system. The form should be simple and easy to use. It is an important part of the document trail, meaning that it provides proof that the item was approved

TABLE 41.1. A behavioral enrichment request form used by one zoo. Courtesy of the Indianapolis Zoo.

Behavioral Enrichment Request Form
Indianapolis Zoo, 1200 W. Washington St, Indianapolis, IN 46222

Date Submitted: _____   Submitted By: _____

Area: _____   Species: _____   Animal: _____

Check One: ____ **Non-food Item (**indicates different route of approval)   ____ Food Item

Used for (check all that apply):   ____ Feeding/Foraging   ____ Sensory
                                         ____ Social/Behavioral   ____ Manipulable item
                                         ____ Environmental   ____ Training

Item: _____   Cost: _____

Used at other institutions?:   Yes / No   Source: _____

Where Item to be used: _____

(check one)   ____ On exhibit ____ Off exhibit   ____ Supervised ___ Unsupervised

Additional Information on Item:

Approved By: (must be routed to area manager for review)

        Yes _____ No _____   Curator: _____   Date: _____ **

        Yes _____ No _____   Nutritionist: _____   Date: _____

        Yes _____ No _____   Senior Vet: _____   Date: _____

Stipulations on Approval:

by all parties involved. When developing an enrichment request form, the choice of item requested will determine the institutional approval route it must take. If it is a food item, for example, the nutritionist or keeper responsible for making diets will have to approve the item before it is used. When completing the form, it is best to provide pictures and/or drawings, diagrams, and references (e.g., the Shape of Enrichment website, other institutions that have used the enrichment item, or reference articles) for clarity. Written descriptions alone may be unclear to those reviewing the request for approval. If everyone working with the particular enrichment item understands how it is made and what it is used for, it will be easier for everyone to approve it.

Formalized enrichment programs are used for two main reasons. One reason is to ensure that the animals are being challenged and engaged within their environment. Staff members need to understand that enrichment is used to encourage an animal's natural behavior and challenge the animal to use its mind and body. The second reason is that zoo associations, and sometimes governments, now require that an enrichment policy and program be in place. This makes the organization and staff accountable for giving enrichment, and assures that documentation of the enrichment is completed. Overall, an enrichment program is more than just throwing an item into an exhibit or a holding space; it is a process. If the animal doesn't immediately become engaged with the enrichment , that doesn't mean it is not interested in it. It may simply mean that more attempts are needed in offering that enrichment—either in the same way or in different ways to determine whether the animal will interact with the item or with the change in the environment. In general an enrichment process involves many steps, patience, and research.

Successful enrichment programs are goal oriented, self-sustaining, and integrated into the animals' daily care. For many, a successful program incorporates the concept of SPIDER (www.animalenrichment.org). The SPIDER framework was first published by Jill Mellen and Sevenich MacPhee at Disney's Animal Kingdom (Mellen and McPhee 2001). It has been applied to enrichment programs by the Association of Zoos and Aquariums (AZA) Managing Animal Enrichment and Training Programs (MAETP) course, and its name is an acronym for the generally accepted pillars of an effective enrichment program (Donald Moore, pers. comm.):

- Setting goals
- Planning
- Implementing
- Documenting
- Evaluating
- Readjusting.

SETTING GOALS

Setting goals is important to providing clear communication about where the team is headed. Writing ideas down on paper will help the team identify which behaviors are to be encouraged or discouraged. The ideas should then be listed in priority, with steps in how the goals will be achieved.

During the goal-setting process, the following information can be used to determine realistic goals for individual animals:

1. the animal's natural history
   a. its activity levels
   b. its habitat
2. the animal's individual history
   a. behavioral concerns
   b. medical issues and how they may affect the enrichment process
   c. whether the animal was parent-reared or hand-reared
3. exhibit history
   a. the role of the animal in the collection
   b. whether the exhibit fits the animal's needs
   c. the animal's degree of choice or control in the exhibit.

The desired outcome is then determined, and action steps are devised to allow the team or individual to achieve that goal. For example, if the goal is to get a penguin to play with a plastic ball, the keeper could place treats in the ball.

PLANNING

Planning is a process that involves the entire zoo or aquarium team, including keepers, curators, veterinarians, and nutritionists. This part of the process is important because it is meant to develop a plan for clear direction. It also prepares the team to implement its action steps more effectively (e.g., if penguins are to receive a new type of fish, the veterinarian or nutritionist must review the fish's quality, the curator has to review its cost, and the keeper must take the time to prepare it for the penguins). Planning helps the team identify which animals are involved, which behaviors are being considered, and what resources (instructions, materials, and schedules) are necessary. In other words, it allows the staff to determine how much time and resources it takes to perform the task. For example, if the penguins are receiving a new type of fish for enrichment, how often will the fish be offered? How will it be presented? Do the keepers have time to cut it up or place it in a free feeder device?

When a plan is created, the following questions should be asked:

- Which animals are to be enriched?
- What is the desired behavior to be encouraged or discouraged?
- What resources will be required or available?
- What is the approval process?
- How will enrichment items be acquired or constructed?

Timelines should be incorporated within the plan. They should state when each task or objective is to be completed. The plan should also be very explicit about what one should report while observing the animals and how often one should observe them. For example, if penguins are receiving a new type of fish twice a month, one should watch for overeating,

> **Good Practice Tip:** When giving an animal enrichment, take the time to watch the animal interact with it. It's simply amazing what one can learn by just taking an extra couple of minutes to watch! Don't always be in a hurry to move on to the next task.

> **Good Practice Tip:** When establishing an enrichment program using SPIDER, documentation is extremely important as it will let one know whether the enrichment was a success. Do not forget to consistently document responses to enrichment in a uniform manner.

and observe the birds for 15 minutes after the enrichment item is given to them, to watch for problems.

IMPLEMENTING

At this stage, the roles and responsibilities of the keeper, manager, and veterinarian have been established. Schedules have also been developed for everyone involved. Everyone should know whose responsibility it is to make enrichment items, to schedule the enrichment, and to offer the enrichment to the animal.

Questions about the division of duties include:

- Who will construct the enrichment?
- How often will the enrichment be given?
- Who will offer the enrichment?
- Who will monitor the animal's interactions with the enrichment?

DOCUMENTING

Documenting the results of the enrichment plan is at the center of the SPIDER process. Documentation that accurately reflects the process, including its goals and objectives, is needed in order to determine its level of success or failure.

When establishing what type of documentation is needed, one really needs to consider what is desired from the results. For example, what type and frequency of data is needed to determine whether the plan worked or not? Is a "snapshot" of the data acceptable, or are more details required?

Documentation can be as simple or complex as necessary, and may involve a variety of resources. Whatever type of documentation is chosen, it should give everyone involved a clear picture of what is happening. It should be specific, but it should also contain valuable yet objective information. Here are some examples of documentation.

SIMPLE TYPES OF DOCUMENTATION

Calendar. A calendar can be used in several different ways. One option is to post a blank calendar that staff will fill in as needed to indicate their choices as to what enrichment is given. Please see table 41.2 for an example of an enrichment calendar that was completed at the Indianapolis Zoo. It records the shark, fish, and stingray enrichment program for June 2009. The blank calendar was posted on the keeper office door, a central location, and entries were added as each keeper executed a task. Within the zoo's enrichment program, all three species of animals were given enrichment a minimum of once a month. Most of it was natural (e.g., food items that

were part of the animals' natural diet, but not a regular part of the zoo's diet due to prohibitive cost). Enrichment doesn't have to be complicated. Sometimes it takes a person relatively new to the field to come up with some great ideas. Keepers should not be afraid to ask for input from those who are new to the area, especially volunteers and interns.

Daily reports. Written communication is one of the best ways to inform several departments at once. Sometimes keepers have an area in their daily reports where they can report enrichment and what happened while they were giving it.

Other reports used could include the Zoological Information Management System (ZIMS), the Animal Record Keeping System (ARKS), photos, and video.

MORE DETAILED TYPES OF DOCUMENTATION

Numbered scales (rubrics) can be used to measure an animal's level of involvement with an enrichment item. This method of documentation is used at many zoos for evaluating enrichment, and can be found at www.animaltraining.org. The scale runs thus:

1. Animal actively avoids enrichment item.
2. Animal interacts inappropriately (e.g., dangerously or aggressively).
3. Animal interacts tentatively with item (e.g., makes brief contact with no specific behavior).
4. Animal interacts appropriately, but not with goal behaviors (e.g., it uses foraging material for nesting, which is a normal behavior for its particular species).
5. Animal interacts appropriately with intended goal (e.g., it interacts with the enrichment item for at least three out of five minutes).

Another option is the predesigned checklist. Table 41.3 is an example of such a checklist used for dolphin enrichment. The protocol simply states that the keeper should record how long the dolphins had the toy and how much the keeper interacted with the dolphins, along with the keeper's initials and the date. The keeper also needs to note whether primary reinforcement was used to encourage use of the toy. Once the toy has been given five times, it cannot be used again until all other toys have also been used five times. The scale rating runs thus:

1. No dolphins interacted with the toy.
2. Some dolphins interacted with the toy.
3. All dolphins interacted with the toy.
P. Primary reinforcement was used.
O. The toy was given to the dolphins to keep overnight.

TABLE 41.2. A fish enrichment calendar used by aquarists at the Indianapolis Zoo

Sunday	Monday	Tuesday	Wednesday	Thursday	Friday	Saturday
	1 Eel training	2	3 Eel training	4 Hand-feeding rays while diving	5	6
7	8 Offered lobster tails to eels	9	10 Shrimp to both dog shark groups	11 Rays: floating feeder with krill	12 Eel training	13 Hand-feeding rays while diving
14	15 Headless, gutless, no tails fatty herring to eels	16 Rays: floating feeder	17 Colored plastic ribbon airline to sea horse tubes	18 Rays: 1–2 lbs. herring and gel cut into krill-size pieces and scattered	19	20 Rays: beef liver
21	22 Eel training	23	24	25 Rays: squirt bottle krill feeds	26	27
28 Rays: squirt bottle feeds with fish	29	30 Beef liver to dog sharks				

TABLE 41.3. A checklist for dolphin enrichment items. The entire page must be checked before one moves on to the next. For example, every item must be given to the dolphins five times before any item can be used for a sixth time. Codes run thus: 1. No dolphins interacted with the toy. 2. Some dolphins interacted with the toy. 3. All dolphins interacted with the toy. P. Primary reinforcement was used. O. The toy was given to the dolphins to keep overnight.

Brown Boomer Ball	21 Jun 2009 2-P	26 Jun 2009 1-O			
Small hoop	9 Jun 2009 O	11 Jun 2009 O			
Green bat	6 29 Jun 2009 O	3 July 2009 1-P			
Laundry basket	22 Jun 2009 3-O	24 Jun 2009 2–0			
Trash can lid	17 Jun 2009 1-P				
Yellow stick with rings	10 Jun 2009 2-P	13 Jun 2010 2-P			

COMPLEX TYPES OF DOCUMENTATION

Complex types of documentation are records from which data can be extrapolated in graphs or charts. Examples include research (e.g., specific data on how a penguin interacts with the colony before and after molt) and personal digital assistants (PDAs), such as smartphones, PalmPilots, BlackBerry devices, and tablet computers. The following example was data collected on a PDA using the Observe Observation software published by the Chicago Zoological Society. This is a very user-friendly program that can be adapted to any animal and any behavior, and which allows collection of an unlimited amount of information within the following parameters:

- observers
- behaviors (including definitions)
- locations
- time and date (set automatically)
- visitor attendance and weather
- two extra parameters, chosen and set by the user (e.g., construction or new animal nearby).

Accurate documentation enables one to provide proof of the results, share information, show progress, and find patterns within an animal's behavior. It explains what was done, how it was done, and where it was done. It provides information about the status of the plan and what the next step should be. For example, one pattern may show that the animal interacts with enrichment prior to breeding season and then completely ignores it during that season.

EVALUATING

The next step is to evaluate what has been done. Evaluating the data will help determine whether there are any identifiable trends, such as

- the animal using the enrichment item more during cooler months than during warmer months
- the enrichment providing stimulation only before the breeding season
- the animal not interacting with enrichment during most of the year.

Evaluation also allows one to answer some important questions: Was enough enrichment provided? Did the enrichment item cause any social problems (for instance, did a youngster in the group have problems using the item due to interference by a dominant animal)? Did the enrichment item increase some of the animal's natural behavior, such as foraging, nesting, or scent marking? Did it decrease the undesirable behavior, such as the amount of time a lion spent pacing? Was there a sufficient quantity of the enrichment item for the animals (for instance, was there an item for each animal, or did a large group of animals have to share)? Was there enough interaction with the enrichment item for the desired length of time?

Tracking evaluations can be done on spreadsheets, in graphs, or in log books. One should keep in mind that success or failure is evaluated on the predetermined desired

outcomes. For example, a keeper might want to determine whether the male or the female gentoo penguin interacts more with an enrichment item. By recording the data in a spreadsheet, it can be determined that the male interacts with the item 71% of the time (table 41.4).

READJUSTING

The very important last step of SPIDER is readjustment. This clearly demonstrates that enrichment can be changed constantly to suit the animal's needs. For instance, if a colony of penguins continually stays in a small part of a relatively large exhibit, and scattering enrichment items throughout the exhibit only keeps the penguins moving around the exhibit for a short time, then it's time to readjust the plan. This may involve changing the exhibit decor or structure, changing the public viewing windows, or simply adding a snow machine and making another part of the exhibit more enjoyable for the birds.

On the basis of any enrichment plan, the following questions could be asked if readjustments are needed:

- Does the enrichment need to be provided in a different way?
- Should the enrichment item be offered for a longer duration?
- Should the enrichment be offered more frequently?
- Should the enrichment item be offered in larger or smaller quantities?
- Are there any aggression issues?
- Are there any other behavioral issues? (For example, does the enrichment item decrease one stereotypic behavior while increasing another?)

Readjusting an enrichment program allows it to be refined and improved. The team can review what has been accomplished and make changes that might make the program more effective. To do this, a change in the strategy and protocol may be needed. It is often the case that change is good both for enrichment and for the animal's overall well-being.

SUMMARY

SPIDER is simple and has many benefits. Some of these include

- showing reasonable, documented proof of what works and what doesn't
- encouraging staff to establish better relationships with the animals by spending more time with them (keepers who spend time watching their animals' reactions to enrichment will develop a better understanding of them)
- ensuring that enrichment policies are followed by all staff.

There can also be some challenges with SPIDER. One of the most common is that of developing and maintaining accurate and effective documentation. For example, proper

TABLE 41.4. Data on interest of male and female gentoo penguin in frozen fish popsicles

Date	G873 (male)	G385 (female)
11 Mar 2009	yes	no
12 Mar 2009	yes	no
13 Mar 2009	yes	no
14 Mar 2009	yes	yes
15 Mar 2009	no	yes
16 Mar 2009	yes	no
17 Mar 2009	no	no
18 Mar 2009	no	yes
19 Mar 2009	yes	no
20 Mar 2009	yes	yes
21 Mar 2009	yes	no
22 Mar 2009	no	no
23 Mar 2009	yes	no
24 Mar 2009	yes	yes

record keeping takes time. It also may not always be possible to devote the resources needed to the task (e.g., financial support or time). Sometimes a keeper may only have time to place an enrichment item in the exhibit and then leave without watching the animal interact with it. If this is the case, a trained volunteer may be able to assist in watching the animal. This volunteer can record video or take notes of what the animal is doing with the enrichment item for a specified length of time, and give a report to the keeper later when time allows. This way, there will be a better understanding of what the animal does and whether the enrichment needs to be readjusted. This in itself is a great benefit of good documentation.

Another challenge may be an individual keeper's lack of skill in documenting behavior. New keepers, for example, would need to learn this skill. The best way to learn which parameters need to be measured is to work in a team with more experienced keepers (i.e., mentors). For example, is it important to know whether an animal is interacting with the enrichment by making contact with it or by just approaching within a few feet of it.

There may also be challenges associated with evaluating the enrichment. Keepers tend to participate enthusiastically in the first three steps, S-P-I, of SPIDER. But some keepers find the final three steps, D-E-R, intimidating. For instance, data should be collected and recorded accurately. Observers need to be able to answer questions from the planning stages when evaluating data. This does not always have to be complex. Readjusting should be viewed not as a difficult task, but as a task that allows staff to become creative.

Enrichment programs can strengthen an institution's animal care and can promote a spirit of teamwork. Today's zoos and aquariums aren't just displaying animals in enclosures; they are providing for the animals' physical and psychological needs. This requires the regular use of environmental enrichment as the animals' behavior dictates for their mental stimulation, growth, and physical well-being. The use of SPIDER is a great way to understand what an animal likes or dislikes, as well as its overall behavior.

To learn more about enrichment, it is beneficial for staff to read publications and attend conferences and workshops such as those offered by the American Association of Zoo Keepers (AAZK), the Animal Behavior Management Alliance (ABMA), and the International Congress of Zoo Keeping (ICZ). For a more in-depth look at SPIDER, one can attend AZA's MAETP course, which provides students the opportunity to not only develop a SPIDER model but also practice it.

EXAMPLES OF SPIDER

Here are some examples of how SPIDER can be used in the planning and execution of enrichment for zoo animals.

EXAMPLE 1: GENTOO PENGUINS (*PYGOSCELIS PAPUA*)

Setting goals. Goal was to determine whether a blue Boomer Ball was a favorable enrichment item for the penguins. The blue Boomer Ball was placed in holding, on dry land. The penguins had never seen it.

Planning. Developed Observe Observation program on a personal digital assistant (PDA) and determined how many birds would be included in each session as well as the duration of each session and how frequently data would be collected and recorded throughout. (Observe Observation is a computer software program developed by the Brookfield Zoo near Chicago, Illinois.) The plan stated that

- data would be entered into the PDA program
- six birds would be involved in each session
- each session would be 10 minutes in length
- data would be collected daily at 11 a.m. for two weeks.
- an enrichment approval form would be prepared and submitted.

Implementing. A date was chosen and collection of observation data was begun.

- Started collecting data on 2 April 2008.
- Stopped collecting data on 16 April 2008.

Documenting. All observational data was recorded on the PDA (considered a more complex documentation level, as described above) by the following procedure:

- Set a timer to beep every 30 seconds.
- At each 30-second interval, recorded the data for each gentoo penguin.
- Trained staff to record data consistently.

Evaluating. Data was charted and reviewed (table 41.5).

- In previous enrichment attempts, contact was favorable at least 30% of the time.
- Three gentoo penguins had enrichment contact with the blue Boomer Ball.

TABLE 41.5. Amount of time each gentoo penguin spent interacting with the Boomer Ball

ID#	Gentoo penguin G873	Gentoo penguin G002	Gentoo penguin G003	Gentoo penguin G004	Gentoo penguin G006	Gentoo penguin G007
Enrichment contact	10%	0%	5%	3%	0%	0%
Looking at item	5%	0%	0%	0%	0%	3%
> 2 feet from item	35%	33%	40%	72%	68%	51%
< 2 feet from item	45%	33%	15%	28%	33%	29%
Out of view of observer	5%	33%	40%	8%	0%	18%

- Penguin G006 was never out of the keeper's view.
- Penguin G873 spent the most time closest to the blue Boomer Ball.

Readjusting. It was concluded that data should be taken again during a different time of year to determine whether the results would be different. In April the birds are just finishing molt and may not be very interested in enrichment. To get a better indication of whether the birds' interaction with the blue ball is favorable, data should also be taken in mid- to late summer, before breeding season. The final conclusion in this time frame was that the blue Boomer Ball was not a favorable enrichment item.

EXAMPLE 2: AFRICAN LION (*PANTHERA LEO*)

Setting goals. The goal was to encourage 1.3 lions to use the east side of the exhibit, given that

- the east side of the exhibit was a large, flat area of dirt
- plants and grasses had been tried unsuccessfully within that area
- during summer, that side of the exhibit received the most sun.

Planning. To persuade the lions to use the east side of the exhibit, the plan was to

- introduce enrichment items already known to be favorable into the exhibit area
- research new enrichment items used for lions at other zoos
- fill out enrichment item approval forms
- upon approval, acquire the materials needed to build the enrichment
- plan to add large rocks and construct a cave.

Implementing. For each of the above planning points, people were assigned to the task.

- All the tasks listed above under "Planning" were completed within 60 days.
- Observations of where the lions were within the exhibit began on day 61 and ended on day 69.

Documenting. Information was collected and charted using a simple chart to record the data (table 41.6).

Evaluating. The information that was evaluated included whether the enrichment encouraged the lions to use the east side of the exhibit, and whether the time of day made a difference as to whether or not there was at least one lion on the east side of the exhibit.

Readjusting. It was concluded that the lions spent more time on the east side of the exhibit when enrichment was given there. Therefore, to persuade lions to explore all parts of the exhibit on a regular basis, enrichment should be given daily in the less used areas of the exhibit.

EXAMPLE 3: GORILLA (*GORILLA GORILLA*)

Setting goals. The goal was to challenge the gorilla with a complex puzzle feeder.

Planning. A puzzle feeder was designed for the gorilla. This required determination of what the puzzle feeder would be made of, who would build it, how big it would be, whether it would hold both dry food and fresh fruits, and how often the enrichment would be given. An enrichment approval form was prepared and submitted.

Implementing. The puzzle feeder was developed with approved materials (fire hose, PVC pipe, and rope). Stainless steel bolts and hardware were used to put it together. The feeder was approximately one meter wide by one meter high (three feet wide by three feet high) and was designed to hold pieces of fruit and vegetable, each no more than one inch in length. The finished puzzle feeder was given to the gorilla at least once a week.

TABLE 41.6. Time spent by lions on newly refurnished east side of exhibit with and without various forms of enrichment

Day	Enrichment provided	Time spent by lions within three-hour watches spread throughout the day
61	None (first day, with cave and large rocks added)	1 hr.
62	Knuckle bones	2 hrs.
63	None	30 mins.
64	Scents	1 hr. 35 mins.
65	Boxes with meat	1 hr. 47 mins.
66	Knuckle bones	2 hrs. 30 mins.
67	None	1 hr. 30 mins.
68	Scattered meat	2 hrs. 40 mins.
69	Scents	2 hrs. 38 mins.

TABLE 41.7. Time needed for gorilla to remove rewards from puzzle feeder

Date	Item(s) in puzzle feeder	Time in which rewards were removed
11 Nov 2009	Cereal	4 mins.
15 Nov 2009	Apple and banana pieces	4 mins.
18 Nov 2009	Cereal	3 mins.
25 Nov 2009	Peanut butter	3 mins. 15 secs.
30 Nov 2009	Honey	2 mins.
5 Dec 2009	Cereal and yogurt	2 mins. 30 secs.
12 Dec 2009	Orange pieces	1 min.
18 Dec 2009	Applesauce	45 secs.
22 Dec 2009	Peanut butter	56 secs.
23 Dec 2009	Cereal and pumpkin paste	1 min. 10 secs.
24 Dec 2009	Banana pieces and yogurt	48 secs.

Documenting. All data was recorded from 11 November to 24 December 2009 (table 41.7). Observer comments were made after each interaction.

Evaluating. The data was reviewed and it was determined that the gorilla figured out the puzzle feeder quickly.

Readjusting. It was concluded that the next puzzle feeder that is made would have to be more challenging to keep the gorilla interested.

REFERENCES

Mellen, J., and M. Sevenich MacPhee. 2001. Philosophy of environmental enrichment: Past, present, and future. *Zoo Biology* 20(3): 211–26.

Root. 2010. www.animaltraining.org.

42

Operant Conditioning

Gary L. Wilson

INTRODUCTION

CONDITIONING

Conditioning is a process that changes an animal's behavior. Training, behavior modification, desensitization, habituation, and learning all involve conditioning. Conditioning may affect the frequency or form of a behavior, how the animal responds to specific stimuli (things perceived by the animal), or a combination of these attributes. For example, the animal might be conditioned to enter a transport crate whenever the door to the crate is opened. Because this is something the animal might do only rarely on its own, the conditioning has changed the frequency of the behavior. A new addition to the animal collection might be wary of the keeper at first. Over time, however, the animal begins to associate this keeper with the arrival of food and begins approaching the keeper whenever he appears. This conditioning, making an association between the keeper's appearance and the presentation of food, has changed the animal's responsiveness to the stimulus—that is, to the keeper.

There are different kinds of conditioning. This chapter focuses on operant conditioning (a type of learning wherein the subject makes an association between its behavior and the consequences of that behavior), but it will also touch on classical conditioning (a type of learning wherein the subject makes an association between two or more stimuli). After studying this chapter, the reader will understand

- the reasons why a keeper might need to condition the behavior of animals under human care
- the difference between classical and operant conditioning
- the basic procedures and terminology of operant conditioning
- general principles to follow in applying operant conditioning techniques
- basic guidelines to follow when addressing problems that may arise during conditioning.

REASONS TO CONDITION ANIMALS IN ZOOS AND AQUARIUMS

The reasons to modify the behavior of animals under our care fall into two broad categories: encouraging particular behaviors and reducing distress (physical or psychological stress that has detrimental effects on the animal). Animals have evolved to deal with the challenges nature has imposed upon them over the history of their species. Living in a zoo or aquarium, however, presents a different set of challenges.

An animal in the wild spends its time foraging for and consuming food, looking for a mate, and avoiding predators. In the zoo or aquarium, the animal will spend little or no time in these activities. Training can provide the animal with mental and physical stimulation as a substitute for these pursuits. The animal can be trained to display natural behaviors that can help the visitor better understand and appreciate the species.

To properly care for animals, it is often necessary to expose them to stimuli that they may find unpleasant or frightening. While occasional and irregular exposure to such stimuli is normal and animals have evolved mechanisms to deal with them, continual or frequent activation of the animal's fight-or-flight response can result in detrimental physiological effects. Since these detrimental effects can impact the animal's health, they should be avoided, and conditioning can help. For example, an animal may find it frightening to shift between holding cages or into a transport crate. It may be reluctant to approach the veterinarian for an exam, an injection, a blood draw, or administration of medication. Training can help the animal to accept these stimuli and, in many cases, even enjoy them.

ADVANTAGES OF POSITIVE REINFORCEMENT

Using positive reinforcement (an operant conditioning procedure that increases the likelihood of a behavior occurring again in the future through the presentation of a desirable stimulus) to modify an animal's behavior is preferable to other methods because it results in the animal actually seeking to do the behavior. Imagine a situation in which the keeper needs a chimpanzee (*Pan troglodytes*) to shift from one holding cage to another, but the chimpanzee is reluctant to go. The keeper might turn on a hose and direct the stream of water into the cage. To escape from this aversive stimulus, something it wants to avoid, the chimp runs into the next cage. But if, instead, the keeper trained the chimpanzee to go through the door between holding cages by using positive reinforcement and taking small, incremental steps (shaping by successive approximations; see below), the chimp would quickly shift when given the cue to do so. The chimp's motivation for shifting would be to gain a reward instead of to escape a threat. He would look for opportunities to shift on cue in order to earn the reinforcement.

Using positive reinforcement enhances the relationship between the animal and the keeper. In the example above, if the keeper uses the hose to intimidate the chimp into shifting, why would the chimp want the keeper around? The keeper would be associated with the aversive stimulus of the hose. The keeper's very presence would cause the chimp distress. On the other hand, if the keeper has trained the behavior through positive reinforcement, the chimp will associate the keeper with the reinforcers (something the animal wants) and the stimulating experience of being trained. The chimp will want the keeper around because of the rewarding things the keeper represents.

BRIEF HISTORY OF OPERANT CONDITIONING

PAVLOVIAN CONDITIONING AND THORNDIKE'S LAW OF EFFECT

Ivan Pavlov discovered the phenomenon of classical conditioning at the end of the 19th century (Schwartz 1984, 25). It is also called "Pavlovian conditioning" in his honor. It is a type of learning in which the animal makes an association between two stimuli. Stimuli are anything which the animal can perceive—for example, the sound of the keeper approaching, the temperature of the water in its pool, the texture of rocks in its enclosure, the smell and taste of its food, or the light of the sun filtering through the trees around its exhibit. Pavlov discovered that animals will associate a stimulus that has no meaning or value to the animal with another stimulus that is very meaningful. Food as a stimulus means a lot to an animal, and is of high value when the animal is hungry. The sound of a bell means nothing to the animal at first. But if the ringing of the bell always occurs before food appears, the sound of the bell becomes meaningful. It alerts the animal to the food's imminent arrival. Keepers often carry key rings with many keys that jingle. The keeper brings the animal's food. The animal hears the keys jingling, and then the food appears. Before long, the animal starts showing signs of excitement and anticipation when it hears the keys. It has made the association between the sound of the keys and the arrival of the food. This anticipatory behavior can often take the form of excited pacing. If the keeper does not want this to occur, he or she should take steps to prevent the animal from making this association by frequently jingling keys within the animal's range of hearing but without presenting food.

Pavlov thought he had discovered how animals learn everything. Edward Lee Thorndike presented animals with different problems to solve, however, and what he saw made him question Pavlov's conclusion that all learning was based on classical conditioning (Schwartz 1984, 28). Thorndike built a box from which a cat could escape to gain access to food. These puzzle boxes had doors that the cat could open if it pushed strings, levers, and latches with its nose and paws. When first put into the puzzle box, the cat tried to escape in many different ways. It would meow, scratch, and reach through the wooden slats. Eventually it would push the right lever or string, and the door would open. Over the course of repeated trials in the box, the cat's behavior changed. It stopped doing the behaviors, such as meowing, that had no effect on the door. Its behavior became focused more on the activities that resulted in the door opening. In other words, the cat learned which behaviors resulted in opening the door and allowed it to reach the food outside.

Thorndike called what he saw the Law of Effect: behavior is affected by its consequences (Thorndike 1911, 244). The cat's behavior which resulted in the door opening became more frequent. The cat's behavior which resulted in nothing became less frequent. Like classical conditioning, this is also associative learning. But instead of making an association between two stimuli, the animal makes an association between its own actions and their consequences.

SKINNER AND OPERANT CONDITIONING: PROCEDURES AND TERMINOLOGY

B. F. Skinner looked more closely at this kind of conditioning, which he called "operant" because the animal's behavior was operating on the environment. Skinner saw that consequences could result in behavior becoming more or less likely to occur. Events or procedures that make behavior more likely to occur are called reinforcement. Those that make behavior less likely to occur are called punishment.

For any behavior there are four possible consequences, two resulting in reinforcement and two resulting in punishment. Skinner used the modifiers "positive" (i.e., positive reinforcement and positive punishment) and "negative" (i.e., negative reinforcement and negative punishment) to describe these different consequences. It is important to realize that the terms "positive" and "negative" are used the same way they are used in mathematics. They don't mean good and bad. Rather, positive means "to add to." while negative means "to take away." When thinking about operant conditioning, one should first determine whether one wants a particular behavior to increase (become more frequent or more likely to occur again) or decrease (become less frequent or less likely to occur again). Then one should look at whether a stimulus is being added or taken away.

The easiest situation to understand is positive reinforcement. The animal is given something it wants when it does a behavior desired by a keeper. For example, the keeper wants a bear to shift from its outside yard into its inside holding area. The keeper opens the door between the two parts of the enclosure and, when the bear walks into the holding area, drops food onto the floor. The keeper wants the bear's behavior of coming inside to become more likely to occur.

The consequence of the bear's behavior is that a desirable stimulus (food) is added to the its environment.

A little less obvious is negative reinforcement, wherein the keeper takes away an aversive stimulus (something the animal wants to avoid) when it does the behavior desired. For example, a keeper who wants a monkey to move from its holding area into the exhibit yard brings a pole net to the keeper door of its holding area. The monkey wants to avoid the net, so it runs out into the yard. The aversive stimulus of the net is taken away when the monkey shifts into its yard.

Positive and negative reinforcement both increase the frequency of desired behavior. In the long run, positive reinforcement is generally better to use, because it doesn't involve the use of aversive stimuli. It is usually easier for a keeper to care for animals when they allow the keeper to approach them fairly closely, and this is more readily accomplished when the animals associate the keeper with desirable things rather than aversive things. If the keeper's interactions with the animals involve desirable stimuli, the animals will want the keeper to be around instead of wanting to avoid the keeper.

Positive punishment involves presenting an aversive stimulus to the animal in order to make a behavior less likely to recur. For example, a keeper wants to discourage a bear from climbing a live tree in its exhibit, so the keeper puts an electric fence wire, often called a "hot wire," around the tree trunk. When the bear climbs the tree, it touches the wire and receives a painful shock. The behavior of climbing the tree becomes less likely to recur because of the aversive consequences associated with it.

Negative punishment means taking away a desirable stimulus when the animal does an undesired behavior. For example, suppose that as the keeper is bringing food to a pair of coatis (*Nasua nasua*), one coati snarls and snaps at the other. The keeper turns and walks away with the food. The keeper punishes the aggressive behavior by taking away the desirable stimulus, the food.

The keeper should avoid using punishment in his direct interactions with the animals, because it involves aversive consequences. The use of punishment can cause frustration in the animal, resulting in aggressive behaviors directed at the keeper, and it can reduce or eliminate any trust the animal has in the keeper. Also, because punishment is less consistent in affecting behavior, it is less efficient than using reinforcement. It is important for a keeper to understand punishment to analyze animals' behavior. Even if the keeper is not using punishment to modify the behavior, the animals may experience punishment over which the keeper has no control. A keeper may also recognize that a planned procedure will have undesirable, punishing consequences that could be avoided with a different procedure.

THE BRELANDS AND APPLIED BEHAVIOR ANALYSIS

While Skinner focused on determining how operant conditioning affected behavior using laboratory animals such as rats and pigeons, two of his graduate students, Marion and Keller Breland, sought to put this science to work. Thus was born the field of applied behavior analysis, the science of controlling and predicting behavior. Two important concepts from this field are especially useful when training animals.

First, to identify the goals of training and determine its success, the keeper needs to measure behavior. A keeper who says "I want this animal to be less afraid of the veterinarian" needs to describe what "less afraid" means in terms of the animal's behavior. By describing the behaviors the animal displays when the vet shows up (running to the far side of the enclosure, defecating, panting, clinging to a cage mate, etc.), the keeper identifies what can be modified through training. Then, by measuring the frequency of these behaviors before, during, and after training, it can be determined whether the keeper has been successful in reaching the training goal.

Second, reinforcement is defined as a procedure that makes a behavior more frequent. Something isn't reinforcing just because the keeper thinks the animal should want it. Consider the example of shifting the bear into the holding area. Each time the door is opened and the bear comes in, food is dropped onto the floor for it. If this is reinforcing, one should eventually see the bear come in a little quicker or spend more time near the door, waiting for it to open. If, instead, the bear begins taking longer and longer to come in when the door is opened, it indicates that the consequences of coming through the door are not actually reinforcing. Maybe the desirability of the food does not outweigh some aversive aspect of being in the holding area.

BASIC PROCEDURES

SAFETY

Before beginning to train an animal, the keeper should spend some time considering safety issues. If food is going to be used as a reinforcer, how will it be presented to the animal? Using a rod or dowel with the food stuck on the end (a bait or feeding stick), tongs, or forceps to deliver the food reduces the risk of the keeper being bitten or grabbed by the animal. Using a feed chute (a tube, usually metal, into which the food is dropped) accomplishes the same thing and also reduces the risk of the animal damaging its teeth if it bites the bars or mesh of the enclosure where the food has been pushed through. The food's consistency should be considered as well. If it is too soft or sticky, it will cling to the keeper's fingers or feeding tool, creating a danger when the animal tries to lick this food off or otherwise pull it into the enclosure. When feeding by hand, the keeper's fingers should not break the plane of the bars or the mesh of the enclosure. Liquids can sometimes be used as effective reinforcers, but care must be taken to ensure that the animal cannot grab the tube or other container through which the liquid is being delivered.

IDENTIFYING REINFORCERS

The reinforcers that will be used must be identified. Generally, modern zoo diets include a variety of food items, ranging from whole, natural foods to manufactured biscuits or pellets and meat mixes. Not everything in the animal's diet will be equally reinforcing. Some food items are more preferred than others and therefore more reinforcing. If the keeper is not already aware of the animal's food preferences, a little time devoted to observing the animal eat will usually make such preferences obvious. The keeper may want to separate the more preferred food items from the animal's daily diet to use as reinforcers when teaching it a new behavior or a more difficult one. While the keeper might be tempted to use larger quantities of preferred food items than are specified by the animal's normal diet, such a practice can lead to the animal failing to receive a balanced diet or becoming obese. Such detrimental outcomes are not outweighed by the gains that might be achieved in training.

In general, the trainer should use the smallest piece of food that is still reinforcing to the animal. It is better for the animal to spend a minimum amount of time actually eating the food. For example, if the animal is given a large, hard-shelled nut as a reinforcer, by the time it has opened the shell and consumed the contents, it may not make the connection between its behavior and the reinforcer.

ESTABLISHING A BRIDGING STIMULUS

A useful tool for training is a bridging stimulus or bridge. For the most rapid training, the reinforcer should be presented immediately following the behavior that is being trained. If there is a delay between the performance of the behavior and the arrival of the reinforcer, the animal may do other behaviors prior to getting the reinforcer. At the very least, this will slow down the learning process; at the worst, it will cause undesirable behavior (often called superstitious behavior) to be established. Early dolphin trainers realized that it was often difficult if not impossible to get the dolphin's reinforcer of a fish to it at the moment the behavior was performed. There-fore, they started using a whistle to "bridge" the gap in time between the behavior and the reinforcer. At first the whistle has no value to the dolphin. By doing repeated trials of blow-ing the whistle and then giving the dolphin a fish, the dolphin makes the association between these two stimuli. Obviously, this is a classical conditioning scenario. We say the food is an unconditioned reinforcer or primary reinforcer because it does not require any prior experience to be reinforcing to the animal. The whistle is a conditioned reinforcer or secondary reinforcer because it only comes to have reinforcing value after it has been associated with the primary reinforcer, i.e., food. The keeper should always follow the bridging stimulus with food in order to maintain the strength of this condition-ing and the clarity of the communication between keeper and animal. While there are circumstances in which the keeper might not present food after each bridging stimulus, these are advanced training situations and beyond the scope of this introductory chapter.

While dolphin keepers most commonly use a whistle as a bridge, a clicker has been used with a wide variety of animals. Some keepers use a verbal bridge, i.e., a spoken word such as "Good." The advantage of the clicker as a bridge is that the sound it makes is distinctive and not likely to be confused with something else. The disadvantage is that it can be lost or broken. While the keeper cannot lose a verbal bridge, the animal has to learn to distinguish the verbal bridge from other words spoken by the keeper or learn to recognize verbal bridges from different keepers.

Another reason to use a clicker as a bridge is that it pro-duces a brief, discrete signal. The advantage of this relates to the bridge's function as an event marker (Pryor 2009, 4).

As the animal learns how positive reinforcement training works, the bridge provides more information than just the fact that food is on the way. It indicates the precise action of the animal that is getting reinforced. A brief sound is better at marking this event than a more drawn-out sound such as a verbal "Good."

A variety of clickers are available commercially, providing a variety of sounds. If a group of animals is being trained together, it can be advantageous to condition each animal to a different-sounding bridge to avoid confusion during the training sessions. There are electronic clickers available that can produce a variety of sounds. Some timid animals are startled by a loud clicker and respond better to a click of lower volume. A snap clip, such as many keepers use to attach their key rings to their person, can be used as a bridge.

In most cases, an auditory bridging stimulus is the most efficient and effective. In the case of an animal that cannot hear, a hand gesture or flashing light can serve as a visual bridging stimulus. The keeper's touch, with a hand or rod, can serve as a tactile bridge in situations where the animal cannot see or hear other types of bridging stimuli.

TARGET TRAINING

A useful behavior to train first is targeting. This involves train-ing the animal to touch an object with its nose. Primates often prefer to touch the object with a hand instead of the nose, but they can be trained to do both. Almost anything can be used as a target. For hoofstock, it is advantageous to use a ball on a long pole. This allows placing the target some distance from the trainer, making it easier to move the animal into the posi-tion the trainer desires. For carnivores and larger primates, it is often best to use an object that cannot pass through the mesh or bars of the enclosure. This prevents the animal from biting the target, taking it from the keeper, or ingesting it.

Targeting is trained by presenting the target and waiting for the animal to move toward it. At that point, the trainer sounds the bridge and feeds the animal. Then the trainer waits for the animal to move closer to the target before reinforc-ing. Eventually, the animal will touch the target when it is presented. Then the trainer can require the animal to hold its nose on the target for gradually longer periods before being reinforced. After this, the trainer can move the target slightly away from the animal and reinforce the animal for moving to stay in contact with the target. Following the target in this way allows teaching the animal to position itself next to a wall of its enclosure so that the trainer can get access to its side or hindquarters.

CUES

So far we have only discussed getting the animal to offer a be-havior at any time. Generally, we want the animal to perform a particular behavior at a specific time. This is where cues come in. A cue is a stimulus that indicates to the animal that reinforcement is available if it performs a particular behavior. Technically, a cue is a discriminative stimulus, because the animal has to distinguish it from other stimuli. Keepers refer to the discriminative stimulus as the S^D for short.

The S^D is often obvious to the animal. After a short period of target training, presentation of the target itself becomes the S^D for targeting. Sometimes it is advantageous to use a target that is mounted or painted on the animal's enclosure and therefore fixed in position. In such cases it is beneficial to have an S^D to tell the animal when to target, since the target is always present. Usually this is the spoken word "target." After the animal is consistently placing its nose on the target, the keeper starts saying "target" just before the animal touches the target. Then the keeper starts saying "target" earlier and earlier, waiting each time for the animal to touch the target before bridging. When the animal is responding without hesitation to the cue "target," the keeper stops reinforcing the animal for targeting on its own without the S^D being given.

In our previous example of the bear shifting into the holding area when the door was opened, the door opening was the S^D. In such a case it would be beneficial to establish another S^D, such as a bell or buzzer, for this behavior of coming into the holding area. This improves the keeper's chance of eliciting the behavior at times other than the regular time when it is usually called for. The auditory S^D gets the bear's attention at times when it might not notice the movement of the door. This would allow the keeper to shift the bear inside at any time during the day, thus allowing the placement of environmental enrichment items into the exhibit, or the retrieval of an object that has fallen into the enclosure. One zoo trained its tigers in such a behavior. This allowed the keepers to call the cats inside when a cockatoo from the zoo's bird show made an unscheduled landing inside the tiger exhibit. The bird was safely rescued (McPhee, pers. comm.).

STATIONING

Another useful basic behavior is stationing. A station is a location the animal is trained to go to and wait. Stationing is especially useful when working with groups of animals. A separate station is established for each individual. This prevents animals from interfering with each other during training sessions.

Stationing is trained in a way similar to how a dog is taught to "stay." The animal is reinforced while it is in the station location, so that the animal associates that location with reinforcement. Then the keeper gives the S^D, usually just the spoken word "station." A hand cue can be combined with the verbal cue, such as a raised open hand with the palm either facing the animal or pointing at the station. As the S^D is given, the keeper takes a small step away from the station, then immediately returns to reinforce the animal for staying on it. Over subsequent trials, the keeper gradually increases the distance he steps away from the station. It is important to always come back to the animal to reinforce it. If the animal is released from the station and then fed when it comes to the keeper, it is being reinforced for coming to the keeper, not for staying on the station.

Many animals do not appear to have the understanding of spatial relationships that humans do. Therefore, it is often helpful that the station be distinctive in appearance, giving something physical for the animal to focus on. For birds and small to medium-sized primates, a perch can serve this purpose. For larger primates, or where a perch is not practical, a carabineer clipped to the enclosure mesh can represent

the station, with the animal taught to keep a hand on the carabineer. For carnivores and hoofstock, an object placed on the ground or a shelf can be used. This is often referred to as mark training, using the terminology from the stage and film industry for a mark on the stage or ground that indicates where an actor must stand.

GENERAL PRINCIPLES

SETTING BEHAVIORAL OBJECTIVES

Beyond the basic behaviors of targeting and stationing, it is important to decide on behavioral objectives with each animal. One should be specific when setting goals, and visualize exactly what the animal is to do. Animal training can be described as a system of communicating to an animal what the keeper wants it to do. If the keeper's vision of the animal's performance of the trained behavior is clear, the keeper's communication with the animal is more likely to be understood.

SCANNING

Sometimes behaviors are trained using a method called scanning or capturing. This means that the keeper waits for the animal to do the desired behavior, and then bridges and feeds the animal. Each time the animal is reinforced for doing the behavior, the more likely it becomes that the animal will do it again. In this way, the frequency of the behavior increases. Once the animal offers the behavior frequently enough, the keeper can establish an S^D (cue) for the behavior, to gain more control over when it will be performed.

Vocalizations are frequently trained through scanning. For example, the keeper hears a sea lion bark and immediately clicks the clicker and tosses a fish to the animal. After doing this several times, the sea lion is barking more often. The keeper starts saying "speak" when it appears that the animal is about to bark, and reinforces the animal after it barks. By only reinforcing those vocalizations that occur after the cue is given, the animal learns that it is more worthwhile to bark when the keeper has said "speak."

SHAPING BY SUCCESSIVE APPROXIMATIONS

Often it is not practical or efficient to wait for the animal to do the desired behavior. Also, frequently the desired behavior is not something the animal would naturally do on its own. In these cases, the method used is called "shaping by successive approximations," or simply shaping. Another advantage of shaping over scanning can be seen when the trained behavior breaks down. If the behavior has been captured but the animal stops doing it, the keeper has to wait for the animal to offer the behavior again. But if the behavior has been trained through shaping, the keeper can go through all the training steps again to bring the behavior back. This review of the training process will generally take much less time than the initial training.

Shaping works by reinforcing small approximations toward the finished behavior. The keeper must envision what these small steps look like. The easiest way to do this is to first imagine what the completed behavior will look like.

The keeper should answer questions such as: Where in the enclosure will the animal do the behavior? What will be the animal's orientation to the keeper (facing the keeper, broadside to the keeper, or turned away from the keeper)? Will the animal be standing, sitting, lying down, walking, or jumping?

Once the keeper has a clear idea of what the final behavior should look like, he needs to think about the mechanics of the animal going from a normal position to performing the trained behavior. Breaking the animal's behavior down into the small actions that build up toward the finished behavior, the keeper can identify the steps (or approximations) to be made in the shaping process. It is often said that the keeper approximates the animal toward a particular performance standard, meaning that the keeper reinforces the animal for doing the various small steps toward the goal behavior or a significant approximation.

Imagine that we want to train a lion to place its front paws on the wire mesh of its night quarters so that the veterinarian can examine the pads of its feet. To do this behavior, the cat needs to be sitting in front of the keeper, facing the keeper. So the keeper first trains the lion to sit in front of him. The cat already has a tendency to stand facing the keeper because the keeper has been delivering food to it at this location. Now the keeper presents the target, so the cat holds its nose to the wire mesh where the target is. On each subsequent trial, the keeper presents the target an inch or two higher. As the cat raises its nose to meet the target, the keeper watches for the cat to bend its hind legs a little and lower its rump slightly before he bridges and feeds it. Next, the keeper waits for the lion to bend its legs and lower its rump farther before bridging. Then the keeper waits for the lion to sit down completely before reinforcing. The keeper could start introducing a cue for sitting at this point, but we will skip over that process since our goal behavior is presenting the paw.

When the cat is sitting, the keeper can now gradually raise the target higher on each trial. Now the keeper is looking for the lion to lift a paw slightly as it stretches its nose higher. The keeper must be sure to click the clicker when the cat lifts the paw, and not worry about the lion keeping its nose on the target. The keeper is trying in this way to get the cat to shift its focus to its paw, so that it makes the association between the movement of its paw and the sound of the bridge.

Now the keeper concentrates on reinforcing movements of the paw upwards and towards the wire mesh, gradually requiring the cat to go a little farther on each trial. Once the lion is holding its paw to the mesh, the keeper can introduce a cue for the behavior. He can also begin to fade out the use of the target. He can require the lion to hold the paw to the mesh for longer and longer periods. Eventually, the keeper will want to introduce a second person to play the role of the veterinarian who will examine the paw, first just visually and then by touch as well. Once the cat is performing this behavior well, the keeper will want to train it to present both paws.

DESENSITIZATION

Often the goal of our training is to reduce the animal's anxiety in the presence of some stimulus. This is desensitization. Sometimes all that is desired is for the animal to not react to a specific stimulus. For example, suppose a new banner is hung from a light post outside the gazelles' exhibit. The keeper notices that whenever there is even a slight breeze, the gazelles can be found pacing nervously at the back of their exhibit with their attention fixed on the fluttering banner. The movement of the banner appears to frighten the animals. The gazelles need to become habituated to the banner's presence. Habituation is one type of desensitization. By folding up the banner or tying it so that it doesn't flap in the wind, the keeper can help the gazelles become habituated to the banner. As the gazelles show less fear, the keeper can unfold the banner or loosen its ropes to allow more movement.

Another type of desensitization is counter-conditioning, a more active process in which the keeper shapes the animal's response to the stimulus. Imagine that a puma (*Puma concolor*) becomes agitated when one of the zoo trucks drives past its enclosure. It is apparent that the cat is distressed by the truck's presence. The basic principle of counter-conditioning is to present the scary stimulus at a low intensity, reinforce the animal for acting calmly, then present the stimulus again at increasingly high intensities.

First, the keeper instructs everyone not to drive the truck near the puma enclosure. Then the keeper has someone park the truck where the cat can just barely see it. The puma will notice the truck and fix its attention on it. If the puma shows signs of greater distress (pacing, panting, etc.), the truck is too close—that is, the intensity of the stimulus is too high. If the truck cannot be moved farther away and still be visible, perhaps it can be partially covered or screened from the puma's view.

The keeper waits until the cat looks away from the truck, then clicks the clicker and tosses a meatball to the cat. This is done several times. The keeper is looking for a decrease in the frequency or duration of the puma looking at the parked truck. When the cat is essentially ignoring the truck, the truck is moved a little closer and the process is repeated. The keeper may have to desensitize the cat to the sound of the truck engine as well. This is done the same way: starting the engine with the truck far away, reinforcing the puma for taking its attention away from the sound, then repeating the process with the sound closer and louder.

The goal of this counter-conditioning is to eliminate the fear the animal feels when the truck is present. With enough trials, the puma may come to make a strong association between the truck and the arrival of reinforcement, and show signs of looking forward to the truck's appearance. This example illustrates that counter-conditioning is an active process wherein the keeper shapes the animal's behavior, while habituation is a passive process wherein the keeper merely creates the environment in which habituation can take place.

RESPONDING TO INCORRECT BEHAVIOR

There is always the possibility that the trained animal will give an incorrect response. This can range from ignoring the keeper, to doing one trained behavior when cued to do another, to acting aggressively toward the keeper. Before the actual training, the keeper should plan on how to respond to each of these situations.

Ignoring the keeper is an indication that the animal lacks the motivation to interact with the keeper. This can be due to a number of reasons. The animal may not be interested in the food that is being used for reinforcement, or its appetite may be weak because of the time of day or the season. The training session may have exceeded the animal's attention span; it is important to work on a mix of behaviors to maintain the animal's interest. Also, if every training session involves an aversive stimulus, such as getting stuck with a needle, the animal may choose not to participate.

Offering the wrong behavior for the S^D presented is an indication that the animal has not learned the cues well enough or has trouble distinguishing between them. Cues should be big and obvious to help the animal discriminate between them. The perceptual systems of many animals are different from those of humans. While the keeper may believe that the cue is obvious, the animal may actually be attending to something else the keeper is doing.

Aggression can be a sign that the animal is frustrated by the training. The keeper may be asking for approximations that are too big, and the animal may not be making progress and therefore may not be getting reinforced often enough. The keeper may also have inadvertently reinforced small acts of aggression which have increased over time, and now may need to go back to asking for simpler behaviors and reinforcing for a calm, nonaggressive response. It is often instructive to have another keeper observe one or two training sessions. The observer may identify subtle errors the keeper is making. Also, by ending each training session on a high note, the keeper can make the sessions enjoyable to the animal and reduce the chance of frustration (Pryor 1999, 51).

PROBLEM SOLVING

When the keeper runs into problems with the training, it is important to keep several concepts in mind when seeking solutions.

REINFORCERS CAN TAKE MANY FORMS

Reinforcers can take a variety of forms, and it is the animal that decides what is reinforcing. For example, a female tiger at the author's facility had been taught five behaviors and then spent the summer being free-fed without being asked to perform any behaviors. A new student who took over the cat's training for the fall semester reviewed its records and familiarized herself with its behaviors and their associated cues. Then she got the tiger's diet and went to do a training session. The cat performed all five behaviors perfectly and received all of its diet for the session. The next day, the student returned for another session. This time, the tiger did only four of the five behaviors. The student tried repeatedly to get the cat to do the fifth behavior, but finally gave up. She decided to not feed the tiger the whole diet because the tiger would not work for the food. The next day, the student experienced the same scenario except that now the tiger did only three of the five behaviors. By the end of the week, the cat had stopped doing any of the behaviors. The optimism the student felt after the first session with the tiger had evaporated, and she could see her grade for the semester declining rapidly.

The student was very frustrated and believed the tiger was "testing" her, pitting its will against hers. It was pointed out that this was unlikely and that it was irrelevant to solving the problem. She was told to focus on the cat's observable behavior, not its mental state, which could only be guessed at. She was assuming that the food was the most important reinforcer to the tiger. This tiger was well fed and had been so all summer long. What the tiger had not been getting much of during the summer was attention from a keeper. With that in mind, what had happened during the week was reviewed.

The tiger had done all the behaviors easily during the first session: the student gave it all its food, and left. At the subsequent sessions the cat did not get all of its food, but the student spent more time trying to get the cat to do the behaviors it wasn't performing. Maybe the animal's failure to do the behaviors was reinforced by the attention the student was giving it.

Armed with the idea that her attention could be used as a reinforcer, the student picked up the tiger's diet and started a training session. She gave the S^D for the first behavior. When the cat just looked at her, the student took the food and walked away. She waited a couple of minutes, then returned and tried again. This time, the tiger did the behavior. The student bridged her, gave her a meatball, and then spent a minute telling her what a good girl she was. By repeating this process of reinforcing with both food and attention while leaving immediately in response to nonperformance, the student had the tiger doing all of the previously trained behaviors in just a few sessions, and started shaping new behaviors.

Sometimes the most reinforcing stimulus is for the keeper to leave. This is the case for animals that are anxious or fearful in the keeper's presence. Karen Pryor describes a situation in which a keeper could not approach a llama (*Lama glama*) without frightening the animal (Pryor 1999, 6–7). The keeper taught the animal not to run from him by using the action of going away as the reinforcer. Much as in our earlier example of counter-conditioning a puma to reduce its fear of the truck, this keeper came within sight of the llama, waited for the animal to show calm behavior, sounded the clicker, and walked away. In the next trial the keeper came a little closer before bridging the llama for calm behavior and leaving. Eventually he was able to approach the llama closely enough to offer it food. When the llama was comfortable taking food from the keeper's hand, the keeper was able to switch to using the food as the reinforcer. At this point, the food was a stronger reinforcer than the keeper's departure.

REINFORCING FOR ONE CRITERION AT A TIME

Remember that training is a system of communication. Any behavior typically can be described by multiple criteria. Each reinforcement provides the animal with a piece of information about one thing at a time. For example, if a keeper wants to train an okapi (*Okapia johnstoni*) to allow an ultrasound examination of its belly, the keeper needs to identify the various criteria that make up this behavior. The animal will have to stand parallel to a fence at a specific location while the technician reaches through the fence and touches the ultrasound probe to the okapi's abdomen. The animal will have to stand for a minimum amount of time to accomplish the exam.

The keeper might first establish the spot along the fence

where she needs the okapi to stand. She then can reinforce the animal for coming to that spot. It doesn't matter at this time whether the animal stands perpendicular or parallel to the fence, or how long it stays at this location. Once the animal consistently approaches this spot when it sees the keeper, the keeper starts shaping the okapi's orientation, reinforcing its closer and closer approximations of a standing position parallel to the fence.

Now that the okapi is coming to the proper spot at the fence and lining up parallel to it, the keeper can start working on increasing the time the animal holds this position. The keeper will gradually increase the time the animal stands in position before being bridged. If the okapi moves a little out of position during this time, it's OK. The keeper is focusing on communicating to the okapi how long it needs to remain standing in that position to be reinforced. The keeper should not bridge once for standing long enough, and then not bridge on the next trial because the animal moved slightly out of the position but stood in it long enough. The keeper can only tell the animal about one part of its performance at a time. Eventually, the okapi will be doing all parts of the behavior correctly and the keeper will only reinforce when it meets the final criteria.

MAKING THE APPROXIMATIONS SMALLER

If an animal is not progressing on a behavior, it may be because the keeper is asking too much of it. The keeper may need to reinforce for smaller approximations. For example, if an animal is being trained to go through a door from one holding area to another but bolts back through the doorway when the door is moved even slightly, the keeper needs to use smaller approximations. The keeper might reinforce the animal just for not reacting when she raises her hand toward the door handle, then for not reacting when she touches the handle, then for not reacting when she rattles the door slightly, and so on. She might even look for the situation in which the animal is on its preferred side of the doorway, and then move the door and reinforce the animal for ignoring the movement.

SUMMARY

Operant conditioning offers keepers a powerful tool for improving the lives of the animals under their care. Training can make routine husbandry procedures less stressful for both the animal and the keeper. The fear associated with medical procedures can be reduced or eliminated. The animal's well-being can be improved through mental and physical stimulation. The relationship the keeper has with the animals can be strengthened as the animals learn to trust the keeper. As both the animal and the keeper become more fluent with the communication system that positive reinforcement provides, the keeper will become more sensitive to the animal's behavior and better able to detect changes in health and better meet the animal's needs. In short, by being able to do their jobs better, keepers will find them more fulfilling and rewarding.

REFERENCES

Pryor, Karen. 1999. *Don't Shoot The Dog!* New York: Bantam Books.
———. 2009. *Reaching the Animal Mind.* New York: Scribner.
Schwartz, Barry. 1984. *Psychology of Learning and Behavior.* New York: W. W. Norton and Company.
Thorndike, Edward L. 1911. *Animal Intelligence.* New York: Macmillan.

43

Husbandry Training

Ken Ramirez

INTRODUCTION

WHAT IS HUSBANDRY TRAINING?

Although husbandry has broad definitions within the zoo and aquarium community, the term husbandry training is more specific; it refers to training animals to assist in their own care and management. The focus of this chapter will be the training of animals to cooperate in basic day-to-day care.

For many keepers and administrators, husbandry training is the single most important reason to embrace training. It provides animals with mental stimulation and physical exercise, which both have significant health benefits. But ultimately, if an animal is trained to cooperate in its own health care and day-to-day management, it makes life so much better for both the animal and the keeper. Although the initial training may require a great deal of effort, once a husbandry behavior is trained, it can facilitate tasks that would otherwise require more time and greater resources and involve some form of physical or chemical restraint. Long-term husbandry training can provide the added benefit of saving enormous time and money on individual procedures; its most important benefit is that it reduces stress for the animals and makes the procedures far safer for all involved. Ultimately it improves animal welfare and well-being.

After studying this chapter, the reader will

- appreciate the value of training for zoo and aquarium animal care
- understand basic animal behaviors that, once trained, form the foundation for more advanced behaviors
- understand some common training mistakes
- review some practical applications of husbandry training.

ASSUMPTIONS BEFORE BEGINNING

Many methods and techniques are used to train behavior; this chapter assumes that positive reinforcement will always be the tool of choice. Because husbandry training sometimes involves some degree of discomfort for the animal, success is often based on trust. Trust is built through the relationship between animal and keeper. Aversive techniques and punishers will often cause frustration, and can break down the trust that is needed for successful training of most husbandry behaviors. Throughout this chapter there will be occasional references to a bridging stimulus or marker signal (a whistle, clicker, or word such as "good") used by the trainer to indicate to an animal when it has executed a desired behavior correctly. Some trainers do not use an intentional marker signal, and everything discussed in this chapter can be accomplished without a marker signal; but because the marker or bridge is a fairly common practice in most training programs, it is included in the discussions below.

BASIC BEHAVIORS

FOUNDATIONAL TRAINING

It is important to understand that husbandry training requires a solid behavioral foundation; in other words, it requires that a great deal of preparatory work be done before starting to train difficult or uncomfortable behaviors (particularly behaviors involved in medical procedures). Often, one of the biggest stumbling blocks facing keepers is that they try to start training complex husbandry behaviors without having established the foundation needed to build strong behavior. An appropriate analogy would be trying to teach a student algebra skills without first having taught the student to count, add, subtract, multiply, or divide. Without first understanding the elementary math skills, the student will never succeed at algebra. It is the same case when training many husbandry behaviors: a variety of foundational skills must be mastered first.

STATIONING

One of the keys to many husbandry behaviors is having an animal remain stationary while a procedure takes place. This becomes much easier for an animal if it has been taught to work with the keeper from a standard, specific position or location; this is referred to as being "at station." If the animal is reinforced from the start of training for being at station often, being at station will be a comforting place to be—a definite benefit when training many behaviors. There are hundreds of different stationing locations or positions; here are just a few examples:

- Tiger: Most big cats are asked to station in a sitting position in front of the trainer, on the other side of the bars, in a back-of-house reserve area. From this position the trainer may ask for a paw, ask for a mouth open, or may cue the cat to stand and position its body in a particular position for a different behavior.
- Hawk: A perching bird is often asked to sit on a specific perch at shoulder height in front of and facing the trainer. From this position, the trainer might ask for a claw, feel the keel, request that a wing be lifted, or trim the beak.
- Sea lion: Often sea lions and fur seals are taught to sit on a specially designed pedestal or rocky outcropping. This places the sea lion in a position that gives the trainer and veterinarian access to almost every part of its body.

GATING, SHIFTING, AND KENNELING

Another critical foundational behavior is teaching an animal to shift from one location to another, which, taken a step further, includes teaching the animal to move into a kennel or restraint device. Shifting or gating is important in day-to-day management of almost all the animals in a zoo and aquarium setting, but teaching each animal to separate from its group and go into an enclosure alone can be more difficult. At first, it is usually easiest to train social animals to move through a gate or door into another enclosure as a group. When food is placed around the new enclosure, the animals are usually eager to move through the door when it is opened. As their comfort level increases, it is possible to give each individual a different cue for shifting, so that they learn to move separately. The key to succeeding at this step is to never trap an animal or close it in before it is comfortable and relaxed in its new location. Over time, the keeper can gradually increase the length of time during which the individual animal is comfortable being separated into the other enclosure. Finally, the behavior can be taken to the next step of moving the animal into a smaller cage, a kennel, or a restraint chute. Just as in the previous steps, it is important to never close the animal into this smaller space until the duration and comfort level have been gradually increased.

TARGETING

Targeting can help shape the behaviors explained above. However, it is also an essential tool and building block for many other husbandry behaviors. Targeting is more than simply teaching an animal to touch one of its body parts to an object; it is a set of behavioral skills that, when established in an animal, can form the foundation for countless procedures. Although there are hundreds of targeting variations, the following six types of targeting would give most animals a great head start at learning a larger variety of behaviors.

- Extended target: Because many procedures take time, teaching every animal to remain at a target for a lengthy period of time, usually until bridged with a clicker or whistle, helps the animal to stay focused. Frequently the animal's focus on the target will keep it from being distracted by the activity of the procedure taking place.
- Following a target: If an animal can be taught to follow a target wherever it is guided by the trainer, it can more easily be guided onto an X-ray plate, into a squeeze cage, or onto a scale.
- Different types of targets: Early in training, it is helpful if an animal is exposed to targets that come in many sizes, shapes, textures, or colors. This helps the animals to be far more accepting of unique devices touching them, whether they include a set of nail clippers, a stethoscope, an alcohol swab, or any other keeper or veterinary tool.
- Different body parts: Because medical procedures may be carried out all over an animal's body, early in training it is helpful to teach the animal to accept being touched (tactile contact) anywhere on its body by the keeper's hand, a target, or any object. If the animal learns that it must initiate the touching, the trainer has the ability to guide it more easily into many different positions: a gorilla targeting its hand or arm into a metal sleeve for insulin injections, an elephant targeting its ear through a restraint device door or window (training port) for blood taking, or a camel targeting its rump to the fence so that ointment could be applied to a lesion on its tail.
- Multiple simultaneous targets: Sometimes when an animal learns to target one body part, it can be challenging to get it to touch another body part at the same time. For many behaviors, however, it can be helpful to have the animal's head focus on one target while another body part is targeted elsewhere for a procedure: a tail for a blood draw; a leg for a hoof trim, or a side of the body for an intramuscular injection.
- "A to B" targeting: Teaching an animal to move away from the keeper can be a challenge, but using a target to make that easier can be helpful. The concept of moving from point A to point B can be useful in teaching gating and shifting as well as separation from other animals.

MANAGEMENT BEHAVIORS

Depending on which species of animal keepers work with, there are usually several additional behaviors that may be needed for their day-to-day management. These may need to be considered and trained in advance of training in husbandry behaviors. Important management behaviors vary greatly from species to species; here are just a few examples.

- Leash/harness: Animals that are walked or ridden need to be comfortable being leashed or harnessed.
- Recall: Animals being worked in large open spaces such as wild animal parks, large zoo enclosures, or free-flight bird programs need to have a signal that calls them back to the trainer, into their holding space, or to some other predetermined location. In these instances a recall signal, a cue that lets them know to go to a certain location for reinforcement, is an important management tool.

GENERAL TACTILE TRAINING

With only a few exceptions, medical behaviors require that the trainer be able to touch an animal. In fact, it is usually helpful if the animal is comfortable being touched on many parts of its body. As described above, targeting can be the start of good tactile contact with an animal. However, tactile training used in husbandry care takes time, trust, and careful use of the desensitization techniques described in the upcoming sections of this chapter. Special steps are usually required to teach an animal to accept various types of pressure, rubbing and massaging, exposure to unique devices (e.g., medical instruments), and so much more. It is important that the keeper takes time to establish safe protocols for interacting with the animal before engaging in any type of tactile contact.

MODES OF CONTACT

One way to make sure that interactions with an animal are safe is to determine the system of contact that will be used. This is not so much a training decision as it is a management decision, often based on the species and the physical setup of a particular facility. Here are the three most often-used modes of contact.

- Free contact: In this training setup, the keeper and the animal have complete access to each other. This is more common when training domestic animals such as dogs and cats, and less common in the zoo and aquarium community. However, some animals still regularly worked in free contact include sea lions, petting zoo animals, and some program animals such as dolphins, birds, reptiles, and sometimes even elephants and big cats.
- Modified contact: There is no clear line that defines where free contact ends and protected contact begins, but the area between the two is often referred to as modified contact. The decision as to where to draw the line is usually an organizational management decision. Sometimes modified contact is referred to as semifree or semiprotected contact. Some type of protection or barrier is used to facilitate management or handling, but some risk is still involved. Working with an animal on a leash or wearing a muzzle, or having the enclosure designed so that one can enter

and exit freely with the animal being unable to follow one out due to its size—these are examples of modified contact.
- Protected contact: This training system allows protection for the trainer while still allowing the trainer to touch and access various parts of the animal's body, usually through specially designed doors, windows, or training ports. In protected contact the animal has the freedom to participate or leave, but has a reduced opportunity to bite, kick, or easily injure the trainer.

DESENSITIZATION

DEFINITION

In its broadest and simplest form, desensitization is a process that helps an animal get used to new things. The process first exposes the animal to a new sensation (e.g., sound, touch, smell, or movement) at a level below the animal's nervous or fear threshold; then the trainer gradually increases the level or frequency of that new sensation.

USES AND IMPORTANCE TO HUSBANDRY

Understanding desensitization techniques is critical to a trainer's ability to successfully train medical behaviors. Certain aspects of medical treatment may be associated with some sort of pain or discomfort, and in the process of being exposed to a medical procedure the animal will come to associate various sensations with that procedure. Through proper use of desensitization, an animal can be taught to accept those sensations and participate in the important procedure.

HABITUATION

Habituation is a form of desensitization in which the animal gets used to a new stimulus through repeated exposure to it. It is often referred to as a passive form of desensitization because the trainer does not actively pair reinforcement with the new stimulus; the animal gets used to it simply through repeated exposure. If a building is painted a different color or a tree has been removed from the exhibit, these changes may make an animal nervous at first, but over time the animal will get used to these changes to its environment through natural habituation.

COUNTERCONDITIONING

Counterconditioning is a more active form of desensitization because the trainer reinforces the animal while it is exposed to a new stimulus. As the animal's comfort level increases, the trainer can increase the level, frequency, or duration of the new stimulus and continue to reinforce the animal's calm acceptance of it. Counterconditioning is frequently used in training animals for husbandry behaviors. To get an animal used to the smell of alcohol, the presence of the veterinarian, or the feel of a needle, the trainer can reinforce the animal for gradual increases in acceptance of the stimulus.

TEAM COMMUNICATION

MEETINGS AND COMMUNICATION

The relationship between the veterinarian and the keeper is key to successful training of husbandry behaviors. The secret to solving husbandry challenges can be summed up by the acronym CURE.

- Communication: The communication among keepers, managers, and veterinarians must be ongoing.
- Understanding: All parties must understand the needs of the others on their team. Keepers must fully understand the procedure for which they are training an animal, including the type of equipment to be used, the length of the procedure, the number of support staff needed, and many other details. The veterinarian must understand how the behavior will be trained, and must know which aspects of the behavior make the animal the most nervous. Managers must understand all of these things and communicate their needs and wishes as well.
- Respect: Everyone involved should respect the skill and knowledge that the others bring to the procedure. Veterinarians and keepers have complimentary skills and knowledge, which everyone should respect.
- Evaluation: Finally, at each stage of the planning, training, and execution, all involved should get together to evaluate what has worked and what might be improved.

COMMON MISTAKES

WHY MISTAKES ARE MORE PREVALENT IN MEDICAL TRAINING

No keeper can expect to go through a training career without making a mistake from time to time. The greatest number of mistakes tend to happen during the training of behaviors planned for use in medical procedures. When an animal is sick or there is an urgency to treat it, keepers, veterinarians, or managers may feel compelled to take short cuts or skip steps so that they can accelerate good care. Keepers will often make the decision to jump ahead on purpose—an animal's serious illness may be a good reason to do so—but each time they do so, it is vital that they recognize the benefits and risks involved. The list that follows is not comprehensive, but covers the most common errors.

TWELVE MOST COMMON MISTAKES IN ANIMAL MEDICAL TRAINING

1. Looking for the quick fix: It is not unusual for keepers to want to solve behavioral problems quickly. However, the search for a quick solution often leads to the use of punishers or aversive stimuli, which in turn can lead to bigger problems.
    - Example: When an animal does not shift into a kennel or enclosure right away, a keeper may feel compelled to use a net or baffle board to force the issue. This approach may get the animal to move, but it can break down the trust that has been developed. Worse, it may lead to aggression.
2. Forgetting that learning is always taking place: It is important for a keeper to remember that an animal is learning 24 hours a day, not just when a training session is taking place. When animal care staff forget this fact, they will often inadvertently shape unwanted behavior (teach the animal inappropriate behavior).
    - Example: Staff may be working near an animal's enclosure, dealing with a task that does not involve the animal in any way. If the animal has access to the area where it can see and/or hear the keeper at work, and it happens to pass through a doorway or gate just as the keeper tosses a shovel or other tool to the ground or shouts to another keeper for assistance, the unexpected sound can startle the animal and make it fearful of walking through that gate in the future.
3. Using medical behaviors before they are completely trained: When a medical concern presents itself, it can be very tempting to use a partially trained behavior before it is ready for regular use. While a very serious medical issue may warrant trying to use the unfinished behavior, less serious medical issues probably do not warrant using the behavior before completion. Although doing this may occasionally work, it will often break down the trust the trainer has worked so hard to develop and may cause long-term breakdown in the behavior, delaying completion by weeks, months, or even years.
    - Example: A keeper desires to use a restraint chute with an animal that has just learned to enter the device. The early training is progressing well, and the animal finds it very reinforcing to go into the restraint cage. A concern arises that necessitates a medical exam and the keeper suggests that the fastest and easiest way to get the animal is to use the restraint chute, since the animal will easily go into the device. The procedure is accomplished successfully, but the next time the animal is asked to go into the device, it refuses to do so. The positive association with the device was not strong enough to compensate for the negative experience of being restrained so early in the training process.
4. Not using a bridging stimulus, sometimes called a marker signal: It is important to note that a marker signal is not necessary to train or maintain behavior. When such a signal is used, however, many skilled trainers will quit using it once a behavior is completed. This is usually acceptable for basic behaviors like a sit, a target, or kenneling because these behaviors have clearly defined termination points. Medical behaviors, however, are never really completed; there are always more incremental steps (referred to as approximations or successive approximations in animal training) that can be taken, such as longer duration, new staff, new sounds, and so on. The use of a marker signal can help improve and strengthen the quality of key husbandry behaviors when these new stimuli are involved.

○ Example: Sometimes when an animal has learned to remain still for a blood sampling, if there is no signal telling it when the sampling is done, the animal will anticipate the end of the behavior. Animals have been known to pull away as they feel the veterinarian's needle being removed. Sometimes this can be dangerous to the animal, because if it moves before the needle is completely withdrawn, there can be damage to the muscle or skin. However, if a well-trained animal knows that there is a signal (like a whistle, clicker, or word) indicating completion of the procedure, it probably will be more patient and less likely to pull away early.

5. Assuming that desensitization is complete: When keepers make the assumption that desensitization for a husbandry behavior is complete, they are setting themselves up for disappointment. This is a natural extension of the previous mistake regarding the use of a bridging stimulus. No matter how much work goes into training a behavior, it is not possible to truly desensitize the animal to every possible stimulus. There will always be longer duration, new people, new distractions, new sounds, different types of touch, and so on. Desensitization to all these stimuli can never truly be complete.

6. Using too many keepers to train one task: Sometimes the urgency to get a medical behavior trained necessitates multiple keepers, but it is important to remember that when more individuals are involved, it becomes harder to maintain consistency. Lack of consistency sends mixed signals to the animal and can make training a behavior more challenging. If, due to organizational constraints, a behavior must be trained by more than one individual, extra measures must be taken to assure clear communication between keepers, and a system should be implemented to monitor the animal more closely for signs of confusion.

7. Making assumptions about what an animal likes: Keepers may believe that an animal enjoys a good rubdown, a particular toy, or a certain treat but that does not mean that those things will serve to reinforce a husbandry behavior. Some husbandry procedures can be uncomfortable for an animal, and to overcome any anxiety, fear, or uncertainty the keeper must make sure to use the appropriate reinforcer—meaning one that will actually strengthen and improve the behavior. That information can only be developed through knowing the animal and building a relationship between it and the keeper.

○ Example: It is not uncommon for keepers to observe an animal interacting enthusiastically to a toy. This will lead them to anthropomorphically assume that the animal "likes" the toy. While this assumption may in fact be true, if a husbandry procedure has been trained using a specific type of treat or food, to suddenly change the reinforcer offered and give the toy instead of the proven treat can actually frustrate the animal. Toys can be excellent reinforcers, but keepers need to have used toys in the training process and taught an animal

to accept them as reinforcers before they can be useful in a session.

8. Making approximations that are too large: Eager keepers will often try to rush to the final steps and finish a behavior quickly. Sometimes an animal will learn quickly, making it possible to skip a few steps and not make gradual incremental approximations to the finished behavior. However, it is helpful for the keeper to remember that each step forms a foundation for the finished behavior. By making sure to go through each step on the way to the finished behavior, the keeper can help ensure that the animal has learned the behavior well, and that there will be approximations to fall back on should the behavior break down.

○ Example: When a procedure requires several people and numerous pieces of equipment, it can be time-consuming to get the animal accustomed to each person and each individual piece of equipment. A keeper may bring equipment into a training area, find that the animal seems not to notice or care, and thus skip the step of desensitizing the animal to every individual stimulus. This saves time and speeds up the training; the animal may still learn the behavior and do very well for a period of time. But then, one day, without explanation, the animal notices a cord, a light, an extra person, or something else that has always been there but that it simply has never really seen or noticed. This startles the animal, and from that point forward it becomes fixated on, or afraid of, the previously unnoticed stimulus. Had the keeper taken the time to desensitize the animal to every piece of equipment during the training, this setback might not have happened.

9. Forgetting the importance of a calm response: It is critical that one criterion for most husbandry behaviors is calm acceptance of the behavior by the animal. If the animal is moving, tense, shifting, or rocking when the behavior is completed, excited keepers will often reinforce the animal because the end behavior was successfully achieved (blood was drawn, an ultrasound image was acquired, or medication was administered). If the keeper accepts this movement or tension on a regular basis, it may lead to the animal exhibiting the unwanted behavior more intensely, gradually getting to a point where humans working with the animal can no longer get a sample or complete a procedure. A calm response should be a goal for most husbandry behaviors.

10. Trying just "one more time" or pushing for a few extra seconds: If a sampling procedure is not producing the desired result, but the sample needed is particularly critical or urgent, there can be a strong desire by everyone involved to try again or push beyond the limit of the original training. While this desire is understandable, it is important that keepers, veterinarians, and managers recognize that pushing farther than the animal is used to can break the trust that has been established, and can sometimes lead to significant regression or even total loss of the

behavior. Animals can be trained to accept multiple trials or almost any length of procedure, but it is a gradual process and it requires good use of desensitization techniques to accomplish. It is also helpful to remember that if the animal does the correct behavior, but the staff fails to get the desired sample, the animal has performed correctly and should be reinforced.

11. Lack of communication: Communication can be the key to a well-executed plan. Failure to communicate and plan effectively can cause more than just husbandry behaviors to fail. The importance of communication cannot be overstated.

12. Assuming that training can be done by anyone: At first glance training can seem simple, and it is not unusual to assign complex husbandry behaviors to keepers with limited or no training experience. But it is important to remember that many husbandry behaviors can be the most difficult of behaviors to train and maintain; it is important that the keeper who is assigned to a behavior or the person who oversees the training has the experience to handle the complexity and challenges of husbandry behaviors.

PRACTICAL APPLICATIONS: SAMPLE TRAINING PLANS AND TIPS

There are many different ways to approach husbandry behaviors. Below are several shaping plans for a few of the most common medical behaviors. These are offered as examples to help stimulate the imagination and help keepers consider various options for designing their own shaping plans. Here are a few factors that will influence each shaping plan:

- the species being trained
- the location of sampling on the animal's body
- setup of the facility or area where the procedure will be conducted
- the mode of contact
- prior training and knowledge of the animal's basic behaviors

MEASURING AN ANIMAL'S WEIGHT

There are many ways to get a measurement of an animal's weight, but in most cases this is a straightforward behavior to train, and one that allows staff to monitor the animal's food intake and growth. The training plan below assumes that a walk-on scale is safely integrated in the enclosure, that it is of a size and material deemed safe and appropriate for the species, and that the keeper will be working the animal in protected contact. The plan should work for most birds, mammals, reptiles, and amphibians.

1. Allow the animal into the enclosure; reinforce if the animal enters the enclosure.
2. If the animal approaches the scale to investigate, reinforce.
3. If the animal touches or sniffs the scale, reinforce.

> **Good Practice Tip:** Cues can come in many forms: visual, tactile, audible, and so on. Pick a cue that the animal can easily perceive, and that every member of the staff can easily replicate.

4. Increased interest by the animal followed by increased reinforcement by the keeper should motivate the animal to approach the scale more frequently.
5. Reinforce any movement that causes any part of the animal's body to be over the scale.
6. Note: if the animal does not offer to move toward the scale on its own, the keeper can introduce a target to help guide it toward and onto the scale.
7. Each time the animal stands on the scale, reinforce. It is beneficial to reinforce the animal while it is on the scale. Depending on positioning of the animal's body and location of the scale, if delivery of the reinforcer cannot be done without causing the animal to leave the scale, the keeper may need to toss reinforcement onto the scale. The goal is to have the animal receive reinforcement while remaining on the scale.
8. Delay reinforcement by a few seconds so that the animal remains on the scale for a few seconds.
9. Gradually increase delay of reinforcement so that the animal remains on scale for longer and longer periods of time.
10. Because movement by the animal can cause the scale's readout to fluctuate, it is important to reinforce the animal when it is calm and remaining still.
11. Once the animal is reliably stepping onto the scale, introduce a cue just prior to it actually stepping onto the scale.

BLOOD COLLECTION

For many zoo and aquarium programs, getting a blood sample can be the single most valuable diagnostic tool in an animal's treatment. But it can also be one of the more challenging behaviors to train on a reliable basis. It is valuable to look into the successes or challenges other keepers may have experienced when training the behavior with a particular species. It is also important to discuss with the veterinarians where on the animals' body they prefer to draw blood. There can often be multiple locations where a blood draw might be possible. Training plans will vary depending on the species and the location of the blood draw. Below is a sample training plan for drawing blood from the leg of a rhinoceros being worked in protected contact, but not in a restraint device.

1. Condition the animal to follow a target along the fence or wall where blood will be drawn.
2. As the animal gets comfortable walking parallel to the fence, slow down and reinforce it for staying parallel and close to the fence.

3. Begin to reinforce the animal for stopping its forward motion and staying still.

4. Reinforce the animal for leaning against the fence or touching the fence. If the animal will not offer this on its own, use multiple targets at the head and rump so that the animal learns to target both its front and back ends against the fence.

5. Position the animal along the fence so that the opening in the fence that allows access to the foot is located within comfortable reach.

6. Practice movement in both directions, so that the animal presents both its right and left sides. Also, practice positioning so that both a front and a rear leg are located near the opening in the fence. Eventually, the animal should be comfortable with blood being drawn from all four legs.

7. Incrementally increase the length of time that the animal remains in position along the fence, gradually increasing the duration to 10 minutes (or whatever length of time is agreed upon by the team). If the animal remains still while receiving reinforcement, it should be possible to reinforce it periodically throughout the procedure. However, if the animal moves too much, reinforcement may have to wait until the procedure is completed.

8. As the animal's comfort with remaining stationary along the fence increases, desensitize it to a second person (a veterinarian, technician, keeper, or volunteer) kneeling at the opening in the fence and moving around with various blood collection supplies. This extra person can be introduced earlier in the training process if the animal appears relaxed and accepting.

9. Gradually increase the movement of the second person, to replicate movements of the actual procedure. Reinforce heavily for the calm acceptance of new stimuli.

10. Begin to briefly touch the animal's leg in the area where blood will be drawn. Reinforce well for the first calm acceptance of touch.

11. Increase the length of time that the leg is touched; gradually introduce rubbing and holding of the leg. Make sure the animal is in a position in which it cannot kick or step on the person touching it.

12. Introduce the smell of alcohol, iodine, or whatever cleaning agent the veterinary staff plans to use. Begin with a dilute concentration or a smaller amount if the animal has shown sensitivity to the smell. Gradually increase to full intensity.

13. Incorporate alcohol or iodine into the practice of touching and preparing the area of the leg for needle insertion. Always reinforce well for the acceptance of any new sensation or increase in duration.

14. Begin to introduce pressure and the sensation of a needle gradually. Try different types of pressure: begin dull pressure with finger, and move towards use of a fingernail. It can be helpful to switch items used to desensitize the animal to a needle so that the animal is comfortable with several different sensations. In addition to a fingernail, use different

> **Good Practice Tip:** For any behavior that might entail some degree of discomfort, it is wise to practice it many times without the uncomfortable element. This will set up a history of high reinforcement for the position or the procedure. For example, if blood samples are desired on a monthly basis, it can be helpful to put the animal in position for blood sampling once or twice each day without actually inserting the needle. If an animal must be closed into a kennel for any length of time every week, put the animal in the kennel for just a few seconds or minutes every day so that it does not assume that every request to kennel will be for a long duration.

types of brush bristles, paper clips, or the snap of a stretched rubber band. These different objects prepare the animal for varied sensations.

15. Once these steps have been completed, the next step should be the brief insertion of a needle without actual blood draw, so that the needle can be removed and the animal can be reinforced quickly. This step must be done in consultation with the veterinarian.

16. Wait several weeks, continuing to do all of the other steps, before doing another needle insertion, this time increasing the duration enough to see a few drops of blood drawn into the syringe or tube.

17. If the veterinarian desires to redirect or move the needle to get proper placement of the needle, additional steps will be required to desensitize the animal for that sensation. Often it is the rotation or redirection of the needle that will startle an animal unless these steps were incorporated into the training.

18. Depending on the animal's comfort level with each step, it is wise to allow time between actual needle insertions; a few weeks is recommended, but continue with the other earlier steps to make positioning along the fence a well-reinforced activity.

19. Finally, try an actual blood draw. If frequent blood draws are required or needed, it is important to train the animal to accept the level of frequency desired by taking incremental steps toward more frequent insertion of the needle.

NAIL AND BEAK TRIMS

Clipping of nails and trimming of beaks require similar training procedures. However, because beak trims are more dangerous, not every trainer or veterinarian will be comfortable working around the mouth. The training procedure requires contact with the beak or nail, and desensitization of the animal to a nail file, clippers, or a coping drill. The vibrations and discomfort caused by the filing, clipping, or grinding can take an animal quite some time to get used to. These procedures can be done in protected contact if the cage or area where the training is done is set up with appropriate access ports. Below is a sample training plan for the nail clipping of a hawk in free or modified contact (the bird can be tethered with jesses to the training perch or platform). Because many hawks are used in free flight bird shows or as program animals, the type of access described is not uncommon.

1. First, get the bird comfortable with standing on a perch that is between chest and shoulder height (see previous section on stationing).
2. Desensitize the bird to the keeper's presence close by.
3. Reinforce the bird for allowing hands to approach and touch the perch. (Note: depending upon the level of trust and comfort with the individual bird, goggles and gloves should be worn for safety.)
4. Make contact with individual talons, and reinforce each time the bird allows a talon to be touched. Safety is key, and until there is trust and the bird's reaction is understood and known, gloves should be worn.
5. Bring the clipping or trimming device towards the bird and reinforce for calm acceptance of the new device. At this point the device is not activated or making any noise.
6. Touch the device to a talon and reinforce for calm acceptance.
7. If the device makes noise, desensitize the bird to it by making the noise from a distance (i.e., by clicking the clippers or running the drill). Gradually bring the noise closer to the bird, reinforcing for calm acceptance.
8. Get the animal comfortable with a second person present. Although a skilled keeper can handle the talons, the trimming device, and the reinforcers, it might be easier to have a second person present. Reinforce the bird for calm acceptance of the closeness of the second person.
9. Reinforce the bird for allowing trainers to gradually bring all four elements together: touching and manipulating the talon, bringing the trimming device toward the talon, making noise with the device, and allowing a second person to approach.
10. Once the bird is comfortable and calm with all four elements simultaneously, begin with a very minor trim or clip on a single talon. Reinforce well for calm acceptance.
11. At this stage, do not clip or trim more than one talon in a single session. To maintain a strong reinforcement history, conduct several sessions between each minor clipping or trimming.
12. As the bird becomes comfortable and desensitized to all of this simultaneous activity, begin increasing the amount of nail clipped or the duration of filing or trimming. Always reinforce well for calm acceptance.
13. Within several months the bird should be comfortable allowing all nails on one foot to be trimmed or clipped in a single session.

MEDICATION

The administration of medication or vitamin supplements can be a very important part of daily animal care. There are

Good Practice Tip: Confidence plays an important role in successful husbandry training. When the keeper is nervous about a procedure or fearful of the animal, this change in normal demeanor may create nervousness and fear in the animal.

many methods of applying or delivering medication and vitamins, which will be described in chapter 49. However, keepers can help an animal get its medication through some basic training practices.

Oral administration. Many supplements and medications are designed to be administered by mouth. While it is possible to teach an animal to swallow pills, capsules, or tablets, such training can inadvertently teach the animal to eat other foreign objects. Therefore, disguising or hiding the medication is important. Also, some medications can have an unpleasant or unfamiliar taste, which requires significant desensitization. One way to disguise medication is to dilute or disguise it in a liquid or food with a stronger taste or smell than the medication. If using liquid, it will be helpful if the animal has already been trained to accept liquid treats of some type. A few commonly used examples of such treats are blood for big cats, sugar-free juice for primates, or fish gruel for sea otters. The animal should be conditioned to accept the liquid from a syringe, squeeze bottle, or other device. Because one cannot predict how a medication will taste, it is possible to get the animal used to new tastes by regularly mixing different veterinarian-approved flavors into the liquid, to desensitize the animal to changing tastes.

Injections. Some medications must be injected, either intramuscularly, subcutaneously, or intravenously. The steps for training this behavior can be very similar to training the blood collection behavior described above. It is important for the keeper to remember a few key aspects of training an animal for injections.

- If an injection is required daily or more frequently, the animal will start to predict that the injection is coming; and if the behavior does not have a strong reinforcement history, it will "break down."
- If an injection must be given at a new location on the animal's body, training needs to start from the beginning. It is rare that an animal will accept being injected in a location to which it has not been previously conditioned.
- Certain injectable drugs sting, so be prepared for the animal to be startled or to begin to resist cooperating with the behavior after repeated injections. To overcome this, it is important to go through the motions of an actual injection many times each day, and to reinforce every practice session heavily.

Ointment. Lotions and other topical treatments may often be needed for various reasons. The key to this is tactile preparation and training, as discussed earlier in this chapter. The application of something to the animal's body requires the

Good Practice Tip: When one is working around its head (mouth, nose, eyes, or ears), even the tamest and best-trained animal can and may bite. Even when working with an animal that is normally trained in free contact, many keepers will use protected contact to train medical behaviors involving the head, so as to increase the comfort level of the animal, keeper, and veterinarian.

animal's trust, which is developed through consistent training. The more the keeper has desensitized an animal to various types of tactile contact, the more likely that the animal will accept the application of an ointment. Here are a few helpful pointers in dealing with the application of ointments.

- Some creams and liquids have strong odors. If the keeper knows that certain animals will be exposed to creams and lotions regularly, it is wise to desensitize them to their odors on a regular basis.
- If a lotion or topical medication creates a cold or stinging sensation, the animal can become startled or frightened. To prepare the animal for the unexpected, the trainer can teach it to accept unique sensations during training. The trainer can massage the animal with a cloth dipped in hot or ice-cold water, or rub it animal with a hard- or soft-bristled brush—the more creative the trainer is during the training process ,the less likely the animal is to be surprised by a stinging or tingling sensation generated by a new medication.
- Often topical medication needs time to remain on the animal's body without the animal licking it off, rolling in the dirt, or jumping in the water. An important part of successfully applying topical medication is training the animal to remain relatively still for a lengthy period of time after the medication is applied. An animal that is trained to target or remain at station can give the medication a chance to do its job.

EYE CARE

Two basic types of eye care behavior are most useful in exotic animal husbandry: staring without blinking for a general exam, and allowing eyedrops to be placed in the eyes. The early steps of training are similar for both. The following plan assumes that the animal is being worked in protected contact.

1. Use a target or platform as a place for an animal to rest its chin or head. The platform (target) or animal should be positioned so that the animal's eye is close to the edge of the enclosure, where there is an opening wide enough to shine a light into the eye or insert a dropper.
2. The animal should be trained to position itself in both directions, so that the veterinarian can have access to either eye.
3. Acclimate the animal to staying still in that position for several minutes.

4. Gradually increase the trainer's movement, and reinforce the animal for remaining still.
5. Bring hands toward the animal's eye; get the animal comfortable with varied types of movement around the eye and toward the eye. It is helpful to make all movements slow and measured so as to never startle the animal.
6. Approximate the animal to the presence of a second person situated next to the keeper, and gradually begin making the kind of movements that the veterinarian or technician may eventually perform.
7. Reinforce the animal for maintaining its eye wide open. If the trainer uses slow movements of the hand and arm, it will cause the animal to pay attention and watch what the trainer is doing. Take advantage of that and reinforce it. Begin to reinforce for gradually increasing lengths of time during which the animal's head remains motionless and on the target. Keep in mind that fast movements can cause the animal to blink, which is counterproductive.
8. As calm, stationary time increases, begin to introduce the gradual movement of an object, such as a penlight or eyedropper, toward the eye.

For the eye exam behavior, the following pointers may be helpful.

- Begin to introduce the actual light by using a dim setting or covering the light with several layers of gauze or colored filters. Reinforce the animal for not blinking and for keeping the eye open.
- Gradually increase the intensity of the light.
- Get the animal comfortable with the light turning on and off throughout the exam.

For the eye drop behavior:

- Consult the veterinarian about the best liquid substance to use for eyedrop training. It is ideal to use a substance that will not sting the animal or cause it any discomfort. Sterile water or saline are most commonly used. It is important that the keeper be aware of the temperature of the solution so that it is comfortable for the particular species being trained.
- Once the animal's position and eye-open behavior are reliable with the eyedropper brought toward and slightly above the eye, begin to drop liquid into the eye.
- If the animal remains still and does not pull away (although it might blink), reinforce it very well.
- If the animal pulls away the first time liquid is dropped into its eye, be prepared to cue it to return to target and into position. If the animal returns immediately and is still, reinforce as soon as possible. Use of the liquid may take several weeks to desensitize.
- Various drops and medications may tingle or sting the animal's eye. Be prepared for novel reactions when using a new substance, as it may result in slight regression (steps backward).

EAR CARE

In many ways, ear care is similar to eye care (see above). The key is to get the animal to target or rest its head on a platform, next to an approved opening in the enclosure barrier that allows the keeper easy access to the ears. Approximations to touch, manipulation, and ear drops must be taken slowly. Unlike eye care, where the animal can see what the keeper is doing, ear care is out of the animal's sight. This can cause it some degree of anxiety. The keeper should take his or her time and make slow approximations, making sure to build a very strong reinforcement history with all head contact.

MOUTH CARE

Although an animal can be trained in many behaviors that could be useful for mouth care, most of them rely on getting the animal to open its mouth. This is usually followed by teaching the animal to allow an object (e.g., toothbrush, dental pick, or tweezers) to be placed in its mouth. Here are two different approaches to teaching an animal to open its mouth.

- For animals that can be worked in free contact, the "mouth open" behavior is often treated as a two-target behavior. The upper jaw is targeted to one of the trainer's hands and the lower jaw or chin is targeted to the other hand. By gradually raising the hand on the upper jaw or lowering the hand under the chin, the animal will eventually learn that it must open its mouth to maintain contact to both hands. The trainer must watch and feel for any small muscle movement that would indicate understanding of the two-target idea. Over time the animal can be approximated to opening its mouth wider, and eventually to the introduction of objects. Obviously this behavior requires a great deal of trust.
- If the animal is being worked in protected contact, teaching it the "mouth open" behavior can still be accomplished in several ways. Sticks or targets could be used in place of the trainers' hands and the technique described above would still work. Another approach is to capture the mouth-opening behavior when the animal offers it on its own. The trainer can facilitate this by moving a hand with food in it above the animal's head in a motion that looks to the animal as though the trainer is about to toss or drop food into its mouth. When the animal opens its mouth in anticipation, but before it lunges or snaps, the trainer should drop the food into the animal's mouth. This procedure, if timed well, causes the animal to open its mouth while also teaching it to freeze for a moment because the food arrives so quickly. The most important aspect of this timing is in the first two or three times that the trainer attempts it. When it is obvious that the animal is holding its mouth open, the trainer can prolong the time before food is actually offered; these approximations must be done in very small increments. If the trainer tries to increase the

> **Good Practice Tip:** Avoid predictability. If an animal learns that a procedure always happens at a specific time of day, it may do well in training but avoid the real procedure because it has learned to predict when the real procedure is about to happen. Similarly, if the veterinarian only shows up when an actual procedure is about to take place, the animal may learn that uncomfortable things happen when that person is present. This can be avoided if the veterinarian comes out for training sessions from time to time, or if the number of people conducting the procedure is increased to reduce predictability. Another example: If the person taking blood always wears a white lab coat or other predictable clothing, animals may learn to avoid a desired behavior in the presence of that piece of clothing.

time too quickly, most animals will start to lunge or snap. However, if food is delivered quickly just a brief second after the animal actually opens its mouth, the animal will learn to hold its mouth open. Once the duration has been approximated, the trainer can begin to introduce objects near the mouth and teeth. He or she must be careful not to accidentally reinforce the animal for biting at something or for grabbing onto the enclosure fence or bars.

KENNEL (CRATE), SQUEEZE CAGE, AND RESTRAINT CHUTE

Gating and kenneling were discussed briefly at the start of this chapter. Many zoos consider these to be the most important husbandry behaviors. If animals can be trained to enter the restraint device or chute without fail, most other husbandry behaviors can be accomplished under that type of restraint. However, to maintain the behavior of reliably entering into these devices, the animal must be reinforced for going into the device many times without a procedure taking place. Zoos and aquariums that only require medical exams semiannually, or even less frequently, can often walk an animal into or through a restraint device every day. If the behavior is reinforced well and the device is only used rarely and unpredictably, the behavior can be maintained. Some facilities incorporate restraint devices into animal areas as part of regularly used passageways, and this greatly aids in training and maintaining the behavior.

Even if a restraint device is used more often, it is possible to get the animal more comfortable with entering it if other aspects of its use are approximated and reinforced. For many animals, the sound and the unexpected confinement can be aversive. If those aspects are approximated gradually and calm acceptance is reinforced well, the animals' comfort levels will increase. If the trainers can also approximate other aspects of restraint—including the presence of extra staff, various kinds of tactile contact while in the device, the smell of alcohol, flashing lights, and so on—the animal will learn to associate those things with training and with a pleasant experience, as opposed to only experiencing them during an unpleasant medical exam.

SUMMARY AND FINAL THOUGHTS

CHOOSING WHAT TO TRAIN

It is not possible to train every animal for every conceivable husbandry behavior. It is important to evaluate which behaviors and which species need to be taught. A diabetic gorilla that needs a daily injection of insulin is a good candidate for training that might be high on a zoo's priority list; without training, the animal would have to be restrained and anesthetized, or the lifesaving treatment would have to be skipped—two equally undesirable alternatives. On the other hand, if the veterinarian needs to examine a gazelle only once a year, and if training the gazelle appears stressful or overly time-consuming, it is understandable that this animal might be low on a zoo's priority list; training it under these circumstances may not be in the animal's best interest. Staff members must weigh the pros and cons of each training plan to determine where best to use their resources. This decision must be made through consultation between the keepers who work with the animal every day, the veterinarians, the supervisors, the curators, and all other members of the animal management team. In addition, animal and staff safety must always be considered. The fact that an animal can be trained doesn't mean it should be trained if the training puts any person or animal at risk.

ONGOING TRAINING

Husbandry training is a process that never really ends. For husbandry behaviors to remain strong and truly useful, they must be practiced and maintained with care. There are always new stimuli to desensitize, new variations to train, new husbandry and medical developments to prepare for. Keepers cannot afford to wait until an animal is in serious need of medical care to begin a training program, because at that point the animal is not in the best frame of mind to learn, and the requisite foundational behaviors have not been established. For husbandry training to have long-term beneficial effects, zoo and aquarium professionals should begin a training program as soon in the animal's life as possible.

LEARNING IS ALWAYS TAKING PLACE

All zoo and aquarium personnel should recognize that animals are learning 24 hours per day, every day of their lives. Learning is not confined to those times when keepers are actively training. Animal behavior is affected by the public's behavior, and is changed by the activities of conspecifics. Animals learn new behavior through environmental changes such as wind, rain, cold, and heat; they learn it when maintenance is done on their exhibits. There is no end to the ways in which animal behavior is changed and learning takes place. While keepers, veterinarians, zoo managers, and curators cannot hope to control every minute of an animal's life, they can design programs and exhibits to maximize learning potential and minimize unwanted learning. Sometimes it is the unexpected learning that takes place outside of actual training sessions that causes behavior and trust to break down. An awareness of learning principles and knowledge of the necessities of good husbandry training can go a long way toward setting every zoo and aquarium program up for success.

IMPROVING TRAINING SKILLS

Just as an animal is always learning, so too is the keeper. A keeper who is interested in increasing his or her skill set should not be content with learning a few basic training concepts. Training is a very advanced and technical skill that can continually be improved. Keepers can build their knowledge in many ways.

- Training books: Many information resources are available to interested trainers today. Technical resources that look at the science of animal learning are plentiful, but often are not practical for the beginner who is just learning to train. However, once keepers have gained experience and a grasp of many of the concepts, the technical resources can be a great way for them to further their knowledge. It helps to be open-minded about the types of training books and resources that might be useful. From the numerous resources available on the training of domestic animals (dogs, horses, parrots, etc.), keepers can learn a great deal about training applications that can often translate directly from pet training to zoo and aquarium training.
- Professional conferences and seminars: More and more professional animal organizations are devoting time to training discussions. Some organizations like the Animal Behavior Management Alliance (ABMA) and the International Marine Animal Trainers Association (IMATA) are completely devoted to training and conduct conferences annually. The growth of positive reinforcement training in the world of dog training has also increased the number of seminars designed to improve one's understanding of operant conditioning.
- Chatting with other trainers: Our modern digital world enables keepers to communicate with colleagues through discussion groups, e-mail, and various other web-based portals. Trainers in today's zoo and aquarium community are very open to sharing, and the internet can get keepers connected to others who work with the same species anywhere in the world.
- Seeking out a mentor: Experienced trainers are usually very willing to share their knowledge and help others. As keepers meet other keepers and trainers at conferences or through internet chat sites, it is helpful to develop a relationship with an experienced individual who can serve as a mentor.
- Gaining practical experience: Finally, it is important to realize that there is no substitute for practical experience. Keepers who learn new skills should apply them and keep practicing. Only through constant application of training principles will a keeper continue to grow and learn.

Part Seven

Veterinary Care

44

Principles of Animal Health

Mark D. Irwin

INTRODUCTION

Animal health is a priority of zoos and aquariums; it is something that other public attractions do not have to consider, and it is at the heart of a keeper's responsibilities. It should be a constant concern of all keepers as they go about their daily work in coordination with keeper colleagues, curators, administrators, and the veterinary team. Daily animal care as conducted by a keeper provides the foundation upon which an animal's health is nurtured and maintained. It is the cumulative result of many factors and does not occur passively; a conscious effort and effective collaboration between all departments, including animal care, veterinary medicine, and administration is critical. Only through this team approach can the best animal health and general care be provided. Animal care and veterinary medicine must go hand in hand. Every zoo should have available the services of a licensed veterinarian who will work with the animal care department to promote animal health. Keepers can strengthen this relationship and its effectiveness through good communication, observation, and cooperation. Knowledge of key principles of animal health will be helpful.

The primary focus of this chapter will be on key principles of animal (physical) health and related medical principles. Psychological health will be covered in chapters 38, 39, and 40. After studying this chapter, readers should understand

- the keeper's role in promoting animal health
- basic principles of health and disease
- indications of illness in animals
- common animal-health–related terminology
- principles of infectious disease
- principles of disease transmission and biosecurity
- basic principles of immunity and disease resistance
- objectives of veterinary disease treatment and management.

THE KEEPER'S ROLE IN ANIMAL HEALTH

The role of a keeper in animal health cannot be overemphasized; even the basic foundational keeping activities are critical. Feeding, provision of fresh water, and removal of feces, if not done properly, will be detrimental to the animals' health. Poor water quality, inadequate humidity, or inappropriate lighting will similarly harm animal health, as will other deficiencies of care. The most important characteristic a keeper can have is to be attentive (i.e., thorough, with attention to detail) to the animals and to his or her own job duties; subtle changes are important. Even a highly qualified veterinarian cannot keep an animal healthy if it isn't receiving good daily care. Since the keepers have intimate contact with the animals, the veterinary team will rely on them to identify and effectively communicate animal health concerns, and to be an essential component of the disease prevention and treatment plans.

Beyond daily care, keepers will need to take an active role in other routine activities to promote and maintain animal health. This may include working with the veterinary team for regular health checks and preventive medicine procedures, providing special care to sick animals, or providing medical treatments under the direction of the veterinary team. Astute observation skills and attention to detail are important skills of keepers, and they are particularly important when it comes to animal health. A keeper who can accurately and perceptively identify subtle changes in an animal's health, whether behavioral or physical, and who can effectively communicate concerns to the rest of the animal health team will have healthier animals. This attention to subtlety is particularly important when working with wildlife species that may inherently hide signs of illness to avoid predation or social problems in the natural environment.

Veterinarians are often said to "practice the art of veterinary medicine." While veterinary medicine is based on science, its practice often involves subtle monitoring and judgment calls. Decisions are not always "black or white,"

> **Good Practice Tip:** Understanding the "why" behind a course of action or procedure can often improve success, compliance, and team effectiveness. In a zoo or aquarium, keepers with a foundational knowledge of animal health principles are better able to contribute to the health of the animals and to coordinate with curators, managers, veterinarians, nutritionists, researchers, and other team members.

and what may have worked for one patient may not work on the next. Animal care is similar to veterinary medicine in this regard, and a keeper needs to be alert to variables and observant to what is occurring if he or she is to make the best decisions regarding animal care. For example, a keeper could feed two similar animals the same diet, but one animal could become too thin. Weight loss in an animal can be due to a number of factors. Perhaps one animal is dominant and is not letting another eat, or there could be a disease present, such as dental problems or gastrointestinal parasites. Only through observation and coordination with other members of the team can the reason be identified. Sharp observation, good judgment, and communication skills are therefore essential to a keeper's ability to promote animal health.

HEALTH AND DISEASE

Health can most simply be defined as the absence of disease, whereas disease refers to abnormal, impaired, or disrupted physical or physiological function. In practice, this means that an animal that is healthy will grow at a species-typical rate, engage in species-specific behaviors (e.g., reproduction [fecundity]), and have a strong immunity to pathogens. For example, a diseased animal may appear seriously ill and ultimately die, or it could show no outward evidence of disease, yet fail to thrive. In the latter case, the animal may not be outwardly "sick" but nonetheless could be considered less healthy. A keeper's primary responsibility is to provide for the animals' needs. If an animal is well cared for, it will be healthier and the likelihood of disease (morbidity) and death (mortality) will be less.

Disease can be considered any variation from the normal functioning of the body. and may be associated with

- environment (e.g., inappropriate temperatures)
- nutrition (e.g., inappropriate diet)
- genetics (e.g., inherited hairlessness [hypotrichosis] in squirrels; Wobeser 2006, 41)
- aging or degeneration (e.g., inflammation of the joints [arthritis])
- development (e.g., congenital defects, such as a cleft palate)
- infection (e.g., bacterial infection)
- toxin (e.g., ingestion of a poisonous nightshade species [*Solanum* spp.])
- injury.

Diseases can be categorized on the basis of other criteria as well. One category of disease that is of particular importance to keepers is zoonotic disease, which can be transmitted between animals and people (some people use the term anthropozoonosis for a disease that transmits specifically from people to animals). This will be discussed in depth in chapter 46, but it is good to emphasize that keepers must be aware of their own health to be sure that they do not pick up or spread zoonotic disease. Many zoonoses are reportable (or "notifiable") diseases, which require that government or regulatory bodies be formally notified. Other, nonzoonotic diseases may be reportable if they represent a significant threat to domestic animals (especially agricultural livestock) or wildlife (e.g., chytrid disease of amphibians [the fungus *Batrachochytrium dendrobatidis*]; OIE 2011).

Disease that develops quickly (in hours or days) is said to be acute, while long-term disease is termed chronic. Often, acute disease is of short duration with more intense and sudden onset of signs. Chronic disease develops slowly, with a gradual onset of signs. Both types of disease can be serious, but chronic can be more challenging to identify due to its slow onset; reviewing records or seeking other opinions can aid the identification of chronic conditions. Clinical disease has observable (i.e., behavioral or physical) signs, while subclinical disease is not readily apparent (i.e., it is asymptomatic) and may require specific diagnostic testing to identify. For example, an animal that is clinically ill with an influenza virus may be lethargic and not eating (anorexic), whereas a subclinically ill animal would have the virus without any noticeable signs. For purposes of survival in the wild, zoo and aquarium animals are less likely to express outward signs of disease (and therefore may be more likely to have subclinical disease) than domestic animals. By the time a disease causes obvious clinical signs in a zoo animal, it may be advanced and very serious; early detection can increase the chance of recovery. For this reason, a keeper must be particularly attuned to subtle changes in the animal's appearance and behavior.

IDENTIFICATION OF ABNORMALITIES

It is impossible to identify abnormalities in animals without first understanding what represents normal. Health observations of animals should be a conscious and purposeful part of a keeper's daily routine. Only through regular and repeated observation will a keeper know the normal behavioral and physical aspects of animals. The best opportunities for observation will vary with the animal, keeper, and facility, but they should allow one to see each animal move, feed, and engage in species-specific activity. Often the provision of food or enrichment will facilitate such observation. One should observe every aspect of the animal, purposefully looking for anything that seems odd or unusual, and comparing with other animals in the group. Is the animal limping? Eating as quickly as yesterday? Interacting with conspecifics? One should look critically at every part of the anatomy: Are both eyes open? Are they tearing? Blinking frequently? Are they partly discolored? Are the shape and size of the pupils unchanged? Over time, observation skills can be developed and can improve significantly, requiring less time and gaining accuracy.

There are as many variations in the presentation of ill-

> **Good Practice Tip:** Animal observation is a fundamental responsibility of keepers, who should practice observing and describing the normal behavior and appearance of animals so that they can identify and describe abnormalities when they occur.

ness as there are diseases. Sometimes illness can be easily identified, as in the case of a large traumatic cut (laceration). At other times, only medical testing will identify an underlying illness. Keepers should be alert to signs that indicate the need for further investigation. Animals cannot describe their ailments and will often repress or hide their signs. An ill or injured animal would be targeted by predators in the wild. This is best illustrated by an animal that acts ill until the veterinarian arrives, at which time they immediately become bright, alert and responsive (BAR).

SIGNS OF ILLNESS

Behavioral signs of physical illness can encompass any variation from that animal's normal behavior. Common behavioral signs include decreased alertness, decreased movement or activity, changes in eating behavior or in amounts eaten, changes in defecation or urination behavior, and unusual social interactions with keepers or conspecifics. For example, an animal that is normally excited to see their keeper and which always rushes onto exhibit, eats its breakfast, and defecates in the same corner may, if in pain, instead exhibit aggression to the keeper, refuse to go onto exhibit, not eat (i.e., manifest anorexia), or not defecate in the normal location.

Physical signs of illness often relate directly to the disease process. For example, an injured foot may result in a limp, a parasitized animal may look thin or unthrifty (not thriving), or an infected eye may blink or tear. In cases of generalized disease the signs may be less obvious. Subtle changes such as change in posture, gait (i.e., locomotion), hair coat, or other physical characteristics may be the only signs. Posture can change when animals are weak or in discomfort. Often an animal's head will hang lower than normal and the spine may appear hunched. The ears may droop and the tail may be limp or carried between the legs. In general an animal's hair coat should be clean and shiny. This of course will vary significantly between species, and often between seasons. The animal's face should be observed for signs of problems. Is the nose unusually dry, wet, warm, or cold? Are the ears cold or hot to the touch? In cases of dehydration, the eyes may take on a sunken appearance due to loss of body water. Similarly, the skin can lose some of its elasticity with dehydration—resulting in a "tent" that doesn't bounce back to the animal when pinched and pulled out. Comparison with conspecifics can help to determine the significance of an observation. It may also be possible to compare with the animal's opposing feature for bilateral symmetry (e.g., comparing one back leg to the other, or one eye with the other). Any suspicious findings should be communicated to supervisors and the veterinary team.

INJURY

Physical injury is as common in zoo and aquarium animals as it is in people and pets, and can result from the same types of incidents. Injuries may be referred to as traumatic (e.g., lacerations or bruising [contusions]), acute (e.g., frostbite) or chronic (e.g., gradual foot problems). Some injuries that are not immediately apparent can nonetheless affect internal organs. For example, ingestion of foreign material or a foreign object (e.g., a tin can or metal nail) can block the gastrointestinal tract (GIT) or even perforate it. It is possible for the body to be injured at the cellular level as well (e.g., antifreeze, if ingested, can injure the cells of the kidneys). Tissue that is damaged or diseased is referred to as a lesion.

Keepers should practice describing injuries that they observe. It is not appropriate to state that an animal "is limping." It is important to be descriptive and describe exactly what is being observed, including the identification of the animal, the body part or location that is affected, the appearance of any lesions, any behavioral or physical characteristics (e.g., vocalization or discharge of mucus) and anything else that might be of value in assessing or treating the injury. Most importantly, keepers must assess if the problem constitutes an emergency and requires urgent care. If so, or if unsure, they should immediately contact their supervisor and/or veterinarian.

WEIGHT

One of the simplest things a keeper can do to monitor the health of animals is to routinely measure and record their weight. The feasibility of doing this will vary depending on species and facility, but the information can be invaluable for establishing baseline values (determination of "normal" weights for a species or individual) and screening for disease, and it will provide important information if action or treatment is required (e.g., chemical immobilization or dosing of medication). Frequency of weighing will vary by species and should be determined in consultation with other members of the animal care and veterinary teams. In some situations it may be advisable to weigh not only the animals but also their food intake (i.e., the amount of food left uneaten subtracted from the amount of food offered to the animals).

Of particular note is obesity (excess "body condition" or body fat content). Obesity sometimes has a medical cause, but it most often occurs as a result of overfeeding or too little activity (i.e., too little exercise). In either case, keepers must appreciate that obesity in their animals is a form of disease, and they should work with the veterinary team, nutritionist, and others to manage the problem. Keepers may need to feed animals separately to be sure that each one is fed the appropriate amount (subordinate animals may be displaced by dominant animals when fed in the same enclosure). It is important that keepers as professionals should adhere to dietary recommendations; providing excessive amounts of food is considered a form of malnutrition (bad nutrition) and constitutes inappropriate care. Treats and training rewards must be considered part of an animal's daily ration. Even if an animals wants (or begs) for extra treats, the keeper should adhere to the prescribed diet for its well-being. Obese animals are at risk for a number of health concerns (e.g., diabetes, cataracts, arthritis, and anesthetic complications). Obesity is a chronic problem, and as such it can be insidious (i.e., it can have a slow and gradual onset). Routine body condition scoring and monitoring of weight can help in comparisons with other individuals in the collection or at other zoo facilities. For more, please see chapter 16.

GROWTH AND PRODUCTION

An animal's growth or production can be another indication of health status. Slow or excessively fast growth rates among individuals or litters can be associated with health problems. For example, young carnivores that grow too fast may develop weak or deformed bones. As discussed in chapter 21, professional colleagues, publications, and growth charts, when available, are useful resources for assessing an animal's growth rate. Keepers can contribute to this database with thorough record keeping and sharing of information through the International Species Information System (ISIS) or publications. For example, routine weighing of a neonate can help a keeper know what to expect from a neonate of the same species in the future. Abnormal (i.e., unusual, decreased, or excessive) production of milk, fiber (e.g., hair and fur), or eggs can also indicate a problem.

FECES AND URINE

Urine and feces are important to monitor. Changes in amount or composition can indicate numerous health concerns. These will be discussed in chapter 45.

INFECTIOUS DISEASE

Infectious disease is particularly important in a zoo or aquarium setting because it can pose a direct threat to multiple animals, to staff, and to public health. This type of disease involves the introduction of a disease-causing organism (pathogen) into the body. Not all infectious diseases are easily transmitted (communicable) between animals, but many are. Categories of pathogens include the following.

- **Bacteria** are simple, unicellular microscopic organisms that do not have nuclei (i.e., that are prokaryotes). Many types of bacteria occur in all environments, with many serving beneficial roles (e.g., all multicellular animals have naturally occurring populations on their skin and in their gastrointestinal tracts). A common method of categorizing bacteria is based on their microscopic appearance. For example, bacilli are rod-shaped, staphylococci are round and clumped together, streptococci are rounded and connected in a string, and spirochetes are elongated and spiraled or coiled. Pathogenic bacteria may cause disease directly, through disruption of the body's normal function, or indirectly, through production of toxins or disruption of the normal bacterial populations. Many types of bacteria are common around the world and may be ubiquitous (everywhere) in a given environment;

however, some subtypes or strains may have different characteristics. For example, *Escherichia coli* (*E. coli*) is a bacterium common in the intestinal tract of animals, but some strains (e.g., *E. coli* 0157:H7) are pathogenic and can cause diarrhea and kidney disease. Antibiotics are drugs that either kill bacteria or inhibit their reproduction, thereby making it easier for the body to clear the infection. Bacteria can develop resistance to some or all antibiotics, rendering them ineffective. Keepers can take steps to minimize the risk of bacterial disease. For example, encouraging all people to wash their hands after handling animals or raw meat will help prevent salmonellosis (which is caused by *Salmonella* spp.). Some facilities may require keepers to be tested for exposure to tuberculosis (TB), especially if they will be working with certain susceptible species such as nonhuman primates or elephants (Mikota 2008, 362).

- **Viruses** are tiny organisms that are too small to be seen by light microscopes; they consist of genetic material enclosed in a protein coat. They must invade cells to reproduce. Viruses are often specialized to certain species; however, it is possible for some species-specific viruses to mutate (change their properties) and infect new species. For example, avian influenza virus, like the human influenza virus, rapidly mutates. Outbreaks of avian influenza in bird populations therefore represent a human health concern. Antiviral drugs are available, but they are generally not as effective, widely used, or available as are antibiotics for bacteria in veterinary medicine. Examples of some viruses that are important to keepers include rabies, avian influenza, elephant endotheliotropic herpesvirus (EEHV; see box 30.1), and the herpes B virus of primates.

- **Fungi** include molds and yeasts. Antifungal drugs are available for treatment of many fungal diseases. Some fungi, such as those that cause ringworm (dermatophytosis), can infect the skin of animals, while others can cause internal disease. For example, aspergillosis (due to *Aspergillus* spp.), when inhaled, can lead to respiratory problems in birds and other species of zoo or aquarium animals. *Aspergillus* spp. are common in the environment, and keepers can help to minimize the chances of disease by preventing the exposure of animals to mold (e.g., preventing growth of mold in enclosures, not feeding moldy feed).

- **Parasites** are organisms that rely upon a host organism and survive at its expense. Technically, the term parasite can be applied to the above mentioned groups (e.g. ringworm is a fungal parasite), but in practice it is often limited to pathogens that do not fit into the above categories (e.g., fleas, ticks, mites, helminths, and protozoa). In these cases the term infestation is generally used, rather than infection. The term ectoparasite refers to parasites occurring on the body (e.g., fleas or mites) and the term endoparasite refers to parasites occurring within the body (e.g., gastrointestinal nematodes or coccidia). Helminth parasites include worms such as tapeworms, roundworms, and hookworms. Protozoa are microscopic (usually single-celled) organisms that have nuclei (i.e., that are eukaryotes). Examples of diseases caused by protozoa include malaria, coccidiosis, toxoplasmosis, and amoebiasis. Please see chapter 47 for more information. A variety of antiparasitic drugs can be used, depending on the parasite and species causing the infestation.

- **Prions** are disease-causing proteins; they are therefore considered pathogens, but since they are not alive they are not considered organisms. They have been associated with a group of diseases termed "transmissible spongiform encephalopathies" (TSE), including bovine spongiform encephalopathy (BSE or "mad cow disease"), scrapie of sheep, and chronic wasting disease (CWD) of deer species. These diseases cause gradual weight loss and progressive neurological disease (e.g.. difficulty standing or moving, or low hanging head), and have no cure.

DISEASE TRANSMISSION

The term disease transmission refers to how a pathogenic microorganism is spread, which may be either direct (through contact between infected animals) or indirect (spread through other means such as vectors or contaminated substrate). The following are some common terms used to describe disease transmission.

- **aerosolization** (airborne): microscopic droplets moving through the air
- **injection:** puncture of the skin (e.g., a needle stick or insect bite)
- **ingestion:** intake with food or through contact with the mouth (e.g., when a person touches their lips while smoking)
- **fecal-oral contamination:** acquisition from feces that are ingested (e.g., from unwashed hands, contaminated food, or water)
- **horizontal transmission:** spread of disease from one animal to another
- **vertical transmission:** spread of disease from mother to offspring (e.g., during gestation or lactation)
- **vector:** an organism (usually an arthropod) that transmits a disease (e.g., when mosquitoes transmit West Nile virus)
- **fomite:** an inanimate object that can carry pathogens between animals, enclosures, or other areas within facilities (e.g., keeper footwear, a grooming tool, or a blanket).

Communicable diseases that easily transmit through direct contact (e.g. a human cold virus) are said to be contagious. Keepers can minimize many opportunities for disease transmission through their actions, and can thus decrease the likelihood of disease occurring. Knowledge of disease threats, and coordination with zoo managers and the veterinary team to mitigate them, can significantly decrease the chances of disease spread. The simplest and often best defense against most communicable disease is good hygiene (i.e., sanitation) practices such as frequent and thorough hand washing, ap-

Figure 44.1. Animal food and tools should not be located where people eat or break. Physical separation and labeling can be used to prevent cross-contamination. Photo by Mark D. Irwin, New York State Zoo at Thompson Park.

propriate use of personal protective equipment (PPE; e.g., using disposable gloves when handling dead animals, feces, or raw meat), and routine cleaning of animal food bowls, enclosures, and tools.

The term biosecurity refers to minimizing and controlling the introduction and spread of pathogens within a facility. Zoos and aquariums should have biosecurity programs designed to protect their animals, staff, and visitors from communicable disease through protocols and best practices. The program should include consideration of the facilities, animal feed, wild and feral animals, animal care staff (through training, education, safety), and management of the animal collection (e.g., through disease surveillance, sick animal procedures, and preventive medicine and quarantine; San Diego 2010). Obviously, keepers and their husbandry procedures are critical to the implementation of such a program.

Infection of the body with a pathogenic organism is often not enough to cause disease. The actual occurrence of disease in an individual animal will depend upon a number of interacting factors relating to the environment, the pathogen, and the animal (host) itself (figure 44.3).

THE ENVIRONMENT

An environment that is well suited to an animal will make it less likely for the animal to develop infectious disease. It will favor the animal (and its immune system), and disadvantage the pathogen. For example, if keepers keep a zoo enclosure clean and regularly remove feces from it, the pathogens that are transmitted through feces will be fewer in number and the animal will be less likely to encounter them. Furthermore, a smaller pathogen "load" is easier for an animal to defend against than a large load.

Figure 44.2. Use of rubber boots and foot baths may be an important part of a biosecurity program. Keepers can play an important role by encouraging compliance and changing the foot bath solution daily or as necessary. Photo by Mark D. Irwin, New York State Zoo at Thompson Park.

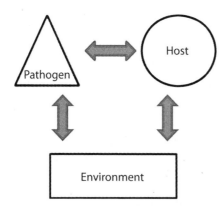

Figure 44.3. The occurrence of disease in an animal depends on factors relating to the disease-causing organism (pathogen), the host (animal), and the environment they share. Illustration by Mark D. Irwin and Kate Woodle, www.katewoodleillustration.com.

THE PATHOGEN

Some pathogens have greater ability to cause disease (i.e., greater virulence) than others. Therefore, some pathogens are more likely to cause disease than others under identical conditions. Additionally, some pathogens may be species-specific, capable of infecting only certain species but not others (e.g., canine distemper virus does not cause disease in humans).

THE HOST

Animals with strong immune systems are less likely to develop disease from a pathogen. Conversely, animals with weakened immune systems (due to age, distress, pregnancy, or other illness) will be less able to defend against a pathogen. Some species may be particularly vulnerable or resistant to certain pathogens (e.g., cheetahs [*Acinonyx jubatus*], which seem particularly vulnerable to feline infectious peritonitis [Wack 2008, 497]). Proper husbandry and care (e.g., diet, exercise, avoidance of distress), monitoring, and specific preventive measures (e.g., vaccines) can do much to improve an animal's resistance to pathogens.

IMMUNITY

Animals have many defenses against infectious disease. The body's first line of defense is nonspecific, composed of features that work against a variety of disease threats. The skin provides an impermeable barrier that most pathogens cannot penetrate. The respiratory system has cilia, and secretes mucus to protect itself and to expel foreign material such as dust or pathogens. Other first-line defenses include skin secretions, stomach acid, and tears. Secondarily, the body works to remove invaders by other nonspecific means such as fever (increase in body temperature) and inflammation. Inflammation involves an influx of defense cells and blood supply, and results in tissue that is characteristically red, swollen, warmer than normal, and often painful. The suffix "-itis" is used to identify inflammation. For example, arthritis is inflammation of the joint, and mastitis is inflammation of the mammary tissue. The body also has white blood cells that, with special proteins in the blood, can engulf and envelope foreign material (phagocytosis). Animals in which any of these nonspecific defenses are impaired will be at a greater risk of disease. This is one reason why older, sick, and young animals may be more susceptible to illness; they are more likely to have impaired defenses.

Animals also have immune systems, components throughout the body that are designed to attack, disable, and remove specific foreign material within the body (e.g. pathogens, vaccinations, foreign material, damaged cells). From the Latin word *immunitas*, meaning exemption, immunology is its own specialty field within medicine (Romich 2009, 320). An animal with immunity is considered to be protected against or exempted from a specific pathogen or foreign substance. The word antigen applies to anything that enters the body and triggers the immune system (e.g., a pathogen, vaccine, or toxin). When the body identifies an antigen, it will begin to produce antibodies. These specialized proteins are created specifically to bind to that foreign substance, and they help the body's white blood cells to remove it. Once the antigen has been removed, the body will often (but not always) have immunity, which is the ability to fight off the same specific substance if encountered again.

Antibodies can often be identified through blood samples (serology). This means that a veterinarian can test an animal's blood for antibodies to a specific antigen. For example, a wolf (*Canis* sp.) can be tested for antibodies to Lyme disease. If the wolf is "positive" (has antibodies), the veterinarian will know that it has been exposed to the pathogen that causes Lyme disease (the bacteria *Borrelia burgdorferi*). It is important to note that having antibodies to a disease pathogen only means that the animal has been *exposed* to the pathogen. It does not necessarily mean that the pathogen is still in the animal's system, or that it is causing any current disease signs. For some diseases, the veterinarian can test for the actual antigen and therefore confirm that it is still present.

To be protected from a pathogen, an animal must have an adequate quantity of antibodies (i.e., an adequate antibody level). Antibodies can be measured to evaluate whether a person or animal has a protective level of antibodies. This test is referred to as a titer. For example, some facilities may require keepers to be vaccinated against rabies. Following the initial series of vaccinations, annual titers may be done to ensure that the keepers' antibody levels are protective. If the levels are too low, development of more antibodies can be promoted through additional ("booster") vaccinations.

Sometimes the immune system identifies nonpathogenic materials (e.g., certain foods, pollen, dander, or vaccines) as foreign threats and develops immunity against them. In other words, these materials become antigenic in the individual. The individual is said to be allergic to these items, and the body will mount an immune response to these materials, which can be called allergens (antigens that cause allergic reactions). A progressively intense reaction may be seen with each exposure in some cases, and this has practical significance to keepers. Keepers shouldn't assume that because an animal has previously had only a minor reaction to a vaccine, it will therefore have the same reaction next time. In severe cases the animal may go into anaphylaxis (anaphylactic shock) as its body overreacts to the vaccine. Signs of anaphylaxis in animals are similar to those in people (see chapter 9), and may include drooling (ptyalism), rapid breathing (tachypnea), rapid heart rate (tachycardia), skin rash or hives, disorientation, coma, and death. For this reason it is important to monitor animals following vaccination to be sure that they receive immediate treatment in the case of an allergic reaction.

Without proper care, an animal cannot mount an effective immune response. For example, an animal that is fed a deficient diet or kept in an enclosure that is too cold will have a weaker immune system (and be considered immunocompromised) because it will lack the nutrients and energy required for a strong immune response. Secondary infections are common because as an animal mounts a response against one pathogen, its defenses may be weakened and body reserves lowered, thus allowing another infection to begin. It is possible that an animal fighting off a flu virus may develop a secondary bacterial respiratory, eye, or ear infection. Old (geriatric), young (neonatal), and distressed animals typically have weaker immune systems. Also, some drugs are immunosuppressive and weaken the body's ability to defend itself.

DIAGNOSIS

Determination of the cause of illness is called diagnosis (sometimes abbreviated Dx). Diagnosis should always be done by a veterinarian; however, keepers can serve an important role in achieving early and accurate diagnosis. Early detection of health concerns, when communicated precisely and accurately to managers and veterinarians, can result in the implementation of treatment before diseases are too far advanced, and can consequently result in better prognoses (outcomes).

In order for a veterinarian to make or confirm a diagnosis, he or she may need to conduct testing that requires the assistance of keepers for collection of samples (e.g., feces or urine) or restraint (physical or chemical) for physical examination. A keeper should feel comfortable in asking questions about what is required for the testing and about the diagnosis itself. A keeper who understands disease, diagnostic testing, treatment, and prognosis will have a better perspective of how he or she can best meet the needs of the animal.

TREATMENT

It is the responsibility of zoos and aquariums to treat any animals in their collections that become ill. Treatment (sometimes abbreviated as Tx) will be determined on the basis of what is in the best interest of the animal, what is most feasible, and what is the likely prognosis. The objective of treatment is not always to cure the disease. Sometimes the disease cannot be cured and must instead be managed. For example, diseases related to aging such as arthritis cannot be "cured." In such cases a veterinarian may institute palliative care to alleviate signs of the disease and to give the animal a good quality of life with minimal discomfort.

Sometimes treatment is symptomatic, designed to address the signs of a disease but not the cause. For example, if a viral disease cannot be treated with medication, the veterinarian may opt to treat signs such as dehydration and pain with intravenous fluids and pain medication. Often symptomatic treatment that supports the body will help it to mount a faster and stronger immune response.

Frequently, treatment of disease will require the use of medication, and the veterinary team may rely upon keepers to administer it. It is important that keepers precisely follow the prescribed treatment regime. A poorly administered medication may not work and could potentially harm the patient. The drug must be given for the entire prescribed period, since the absence of clinical disease does not mean that the animal is cured. In fact, in the case of infectious disease, it is possible that after the most susceptible pathogens have been destroyed, hardier pathogens remain in numbers too small to cause clinical disease. If treatment is discontinued prematurely, the hardy pathogens may reproduce to cause a clinical disease that is more difficult to treat, thus resulting in a drug-resistant pathogen population.

There are many other methods of treating disease including surgery, wound management, changes to husbandry practices (e.g., wetting hay to decrease dust), exhibit changes (e.g., increasing the ambient temperature for a reptile), and diet change (e.g., providing a vitamin supplement). Success may require creativity and knowledge of both the species and the individual animal.

SUMMARY

Disease and death are part of life and will occur regularly in every zoo and aquarium, as they do in natural settings.

However, zoo and aquarium animals will often live long past the life spans typical for their species in nature, and will still experience a good quality of life. Proper animal care and quality veterinary medicine contribute to this. The health and wellness of the zoo or aquarium's animals is the foundational priority of animal care. Keepers have a critical role in assisting the veterinary team with the prevention, recognition, diagnosis, monitoring, and treatment of animal disease. Working together and with animal care managers and other departments, they can ensure that animals receive the highest standard of care. A basic understanding of health and disease principles will help a keeper to more effectively participate as a member of this team.

REFERENCES

Mikota, Susan K. 2008. Tuberculosis in Elephants. In *Zoo and Wild Animal Medicine Current Therapy 6*, edited by Murray E. Fowler and R. Eric Miller, 355–64. Philadelphia: Saunders Elsevier.

OIE (World Organization for Animal Health). 2011. OIE Listed diseases. Accessed 24 March at http://www.oie.int/animal-health-in-the-world/oie-listed-diseases-2011/.

Romich, Janet A. 2009. *An Illustrated Guide to Veterinary Medical Terminology, Third Edition*. Clifton Park, NY: Delmar Cengage Learning.

Wack, Ray F. 2003. Felidae. In *Zoo and Wild Animal Medicine, Fifth Edition*, edited by Murray E. Fowler and R. Eric Miller, 491–501. Philadelphia: Saunders Elsevier.

Wobeser, Gary A. 2006. *Essentials of Disease in Wild Animals*. Ames, IA: Blackwell Publishing Professional.

Zoological Society of San Diego. 2010. General Biosecurity Guidelines. Updated 21 May.

45

Veterinary Care and Technology

Tracey L. Anderson

INTRODUCTION

Animal management in a zoo or aquarium encompasses many components in which the keeper must be proficient. The veterinary team works together with the keepers to monitor, maintain, and improve the health of the zoo's animals. The goal of any zoo or aquarium veterinary department program is to maintain a healthy collection. The first and foremost concern of a keeper is animal care, of which health management in concert with the veterinary and supervisory teams is paramount. Many zoos or aquariums do not have veterinary staff on site; they may only have a veterinarian "if and when" required. In some cases the department may not have veterinary assistants and keepers may, on occasion, fill some of that niche. This chapter will provide an overview of animal hospital principles within a zoo or aquarium, and help the reader to

- develop an understanding of the different staff positions and specific roles that make up the veterinary team
- learn how to assist and interact with the veterinary team during routine and emergency procedures, including collaboration for basic health care maintenance
- learn how to use the veterinary team as a resource
- know how important it is to develop strong observational skills and be able to evaluate and report a concern
- become familiar with the differences between domestic veterinary practices and zoo and aquarium veterinary medicine
- explore the importance and practical application of operant conditioning for veterinary procedures
- learn how to be prepared for veterinary procedures that may take place in the hospital or in the field
- understand proper techniques, behavior, and restrictions in the surgery room

- recognize the governing institutions for veterinary hospital management, and become familiar with hospital safety concerns.

THE VETERINARY TEAM

While job titles and staffing vary greatly between institutions, the following comprise a typical veterinary team that may be found at a zoo or aquarium.

VETERINARIAN

Zoo and aquarium veterinarians must obtain all the educational requirements that all other veterinarians meet. In addition to the educational aspect of training (US educational requirement is eight years of college: four years undergraduate, and four years of vet school), a specialized residency of one to four years is often necessary. While educational requirements and licensure will vary between countries, positions in this field will all require extensive experience working with exotic species for animal health management, diagnostics, treatment, and development and implementation of preventative medicine programs. Veterinarians must also be thoroughly trained for the additional challenges that accompany medical management of large, wild, and dangerous animals.

The zoo or aquarium veterinarian practices medicine much like a veterinarian in domestic animal practice, which includes routine examinations, diagnosis, emergency care, and surgery. In addition to the preventative medicine component and other veterinary responsibilities, they are often involved in ex situ and in situ conservation efforts, exhibit design, and collection management.

Preventative medicine (vaccinations, dental cleanings, nutrition, quarantine, necropsy, etc.) is generally the focus of zoo and aquarium animal medicine, as it is of domestic animal medicine. While the domestic animal veterinarian routinely treats sick animals, the zoo or aquarium veterinarian concentrates on keeping vast numbers of animals healthy.

The zoo or aquarium veterinarian must have the skills and knowledge to manage the health of every animal from the smallest invertebrates to some of the largest mammals in the world, along with just about everything in between. Keepers help the veterinarian by monitoring the health of the animals under their care and conveying clear, concise information to the veterinarian through professional oral and written communication. The keepers should be able to recognize subtle changes in an animal's behavior, because the veterinarian will rely on their observational skills. Wild animals tend to hide or mask signs of illness or injury—particularly in the presence of a veterinarian, who may be seen as a threat or "predator." Thus it is imperative that keepers be able to observe and accurately report descriptions of injuries, injury locations, relevant behavioral changes, perceived illnesses, and such to the veterinarian. Keepers will be responsible for providing thorough histories of the animals in their care. including their food and water intake, current medications, behaviors, feces and urine output, and any other symptoms.

VETERINARY TECHNICIAN

The veterinary technician (sometimes called "vet tech") works directly under the veterinarian's supervision. His or her primary responsibility is to provide assistance as directed by the veterinarian. The degree of this assistance will vary between facilities and the number of staff at each institution. The role of the veterinary technician is quite diverse, and includes support to the keeper staff and the entire animal collection including hospitalized and quarantined animals. Typically, the physical and technical support provided by a veterinary technician in a zoo or aquarium setting is similar to that found in domestic animal practices. The veterinary technician is a resource for the keeper and can provide supplies, medication, and advice. A veterinary technician is not allowed to diagnose illness or prescribe medication, but he or she provides assistance to the keeper staff for routine maintenance (e.g., beak and nail trims) as well as animal health evaluation. Keepers will work closely with the veterinary technician in carrying out treatments; collecting samples; monitoring and reporting treatment responses, medical procedures, and surgical recovery; assisting during procedures; and collecting data. In the event that keepers are requested to collect specimens or samples (e.g., of urine or feces) for laboratory testing, they must be diligent and adhere to the correct protocols for collection, storage, and transportation of the sample. If samples are not obtained correctly, test results may be inaccurate or inconclusive.

The veterinary technician's responsibilities may include

- laboratory procedures (e.g., conducting fecal analysis and urinalysis)
- clinical procedures (e.g., drawing blood, performing nail trims)
- pharmacy management (e.g., filling prescriptions, stocking supplies)
- medicating (e.g., giving injections)
- radiology (e.g., taking and developing radiographs)
- surgery assistance (e.g., sterilizing instruments, assisting in the surgery room, preparing patients)

- anesthesiology (e.g., maintaining and monitoring anesthesia during surgery)
- office support (e.g., filing, scheduling, reception)
- medical record keeping (e.g., entering data)
- neonatal, hospitalized, and quarantined animal management.

Countries vary in their requirements for certification as a veterinary technician, also referred to as an animal health technician, veterinary technologist, veterinary nurse, or veterinary medical assistant. "In many ways, the veterinary technician's role is broader than the nurse's role. For example, veterinary technicians (under supervision) can induce, intubate, and maintain anesthesia. Veterinary technicians must be skilled in an array of roles that are frequently performed by specialized nurses or technicians in human medicine (e.g., radiology technicians, surgical nurses, dental hygienists, and laboratory technicians)" (*Techniques* CACVT 2001). For an extensive list, visit the International Veterinary Nurses and Technicians Association website at http://ivnta.org/index.php.

In the United States, each state has its own regulations and requirements for obtaining credentials as a licensed or registered veterinary technician or technologist. The term "certification" can be applied from a variety of training, but it is not always a state-recognized credential. While the training and certification processes for "veterinary technicians" and "veterinary technologists" could be the same and generally have overlapping job duties, the terms are frequently applied according to education (two-year Associate of Science degree for "technician," or four-year Bachelor of Science degree for "technologist") and workplace roles. The technician typically works under the technologist. It has been the author's experience that the terms "technician" and "technologist" are often used interchangeably, with "technician" being the most popular. Veterinary technicians are individuals who have either received some formal training without certification or licensure, some on-the-job training, or a combination of both. Most require graduation from an accredited two-, three-, or four-year academic program, such as that offered by the American or Canadian Veterinary Medical Associations (AVMA and CVMA), which oversee the certification process. Licensing and/or registration also requires state-mandated testing and a certain number of continuing education hours each year. These hours are acquired through workshops, conferences, and other professionally-related educational experiences.

HOSPITAL KEEPER

The hospital keeper position will vary greatly between zoos and aquariums, but its primary role is usually supportive to the rest of the hospital staff and provides husbandry needs and medical care to animals in the hospital, nursery, or quarantine setting. Hospital keepers will often play a role in medical record-keeping as well as contribute to the day-to-day operations of the veterinary hospital. The hospital keeper may act as a liaison between the hospital staff and the other keepers, providing information regarding treatments, quarantined animals, and hospitalized animals. Sometimes he or she will be the only support to the veterinarian.

VETERINARY TEAMWORK

Working with animals in a zoo or aquarium presents many challenges. From the veterinary perspective, one of the major challenges is the inability to physically restrain some of the patients. If an animal becomes ill, it often requires an anesthetic or sedation procedure, which means that something as simple as an ingrown toenail treatment can be risky for both the animal and zoo staff and consume a great deal of time.

WHEN TO REPORT

There are so many different species managed in a zoo or aquarium setting that it is impossible to provide exact instruction on how to gauge when a keeper should call the vet staff for a potential emergency, or simply make a note on the daily report. While a keeper should be neither too eager nor too hesitant to make a call regarding animal health, it is always better to be overly cautious when making decisions on reporting health concerns. It is important that keepers learn species-specific differences to guide them in the urgency of the reporting. For example, a few drops of blood in an enclosure can be very serious if one is dealing with a finch, but not necessarily if one is dealing with a raptor. Comparatively, blood observed in an eagle's enclosure could be a more serious condition than an equal amount of blood observed in the enclosure of a mammal of the same size, simply because the percentage of blood in the body can differ significantly between the taxa (i.e., avian versus mammalian) of animals. With blood loss, it is important to consider how the blood is presented in the exhibit. A few drops of blood that have been smeared over a wall can look like a lot of blood, whereas a significant amount of blood can pool or be absorbed by the substrate in the enclosure and not appear to be significant. It is important for a keeper to be able to report accurately the estimated amount of blood in an enclosure. The best way to develop this skill would be through discussing scenarios or roleplaying with mentors (keepers, supervisors, and the veterinary team).

OBSERVATIONAL SKILLS AND GENERAL ASSESSMENT

Observational skills are highest on the priority list of skills that every keeper should possess. Keepers should know the individual animals under their care more thoroughly than anyone else. They must be able to recognize the most subtle change in an animal's behavior or appearance. Often, a keeper will hesitate to report a suspected problem with an animal because he or she is not sure what the exact problem is. The majority of the time, when a keeper suspects that there is something amiss with an animal, no matter how minor, there is probably something wrong and it should be reported.

A keeper must be familiar with normal behaviors (including individual "quirks") of each animal to be able to recognize any abnormal behaviors. This includes

- learning the normal stance, movement, posture, and activity patterns
- recognizing that some variation in behavior is normal
- learning to observe the entire group
- learning and understanding animal communication.

Frequent daily observations must be made to insure that an animal is healthy and normal. A keeper is the eyes and ears for the veterinary department. Observations must be made routinely, particularly while the animal is active, to visualize every part of the body and interpret its movement. One should always observe the animal in its entirety, as well as its individual body parts, from head to tail. The animal should be observed discretely and from a distance, using binoculars as necessary. The animals will develop a routine just as the keepers do. Observing that routine and its multiple components is an essential part of the keeper's day. These observations should include (but not be limited to)

- pacing, perching, and sunning habits (e.g., normal or stereotypic pacing behavior)
- eating and drinking habits (e.g., normal daily fluctuations versus anorexia; just because the food and water is disappearing, keepers should not assume the animals are consuming it)
- waste output (e.g., feces and urine)
- temperament and attitude (e.g., a normally passive or docile animal acting aggressively, or vice versa)
- activity level (e.g., a normally active animal being abnormally lethargic)
- hair, feathers, scales, and skin (e.g., excessive shedding or molting)
- eyes and nose (e.g., blood, discharge, inflammation)
- hooves, claws, and nails (e.g., cracking, overgrowth)
- weight (e.g., notation of visible changes, but also routine true measurement when possible, such as with reptiles and amphibians)
- injuries (e.g., lameness, bleeding)
- physical evidence of changing hormonal or reproductive status (e.g., rut, musth, pregnancy, lactation).

Animals are usually more "relaxed" around their keepers. They tend to regard the veterinary staff as a threat, and thus will more likely mask any signs of illness or injury in their presence. Animals recognize veterinary staff and keepers visually; they also recognize equipment (food buckets versus dart guns), voices, and smells. Keepers are more apt to identify health problems early or get a closer look at injuries, and this allows them to provide necessary details to the veterinary staff. Keepers must be able to

- convey correct and concise information to the veterinary staff
- understand clearly the information they receive from the veterinary staff
- communicate with other colleagues (keepers, supervisors).

While a keeper knows the animal better than most and will be able (and be expected) to note subtle changes, sometimes changes that occur over a period of time go unrecognized by the primary keeper and a second opinion can be extremely useful (just as parents do not see the significant growth of their children until someone who does not see the children on a daily basis points it out). A keeper should be open to second opinions from coworkers and not hesitant to ask for them.

BASIC ANIMAL HEALTH MAINTENANCE

Keepers are responsible for certain basic maintenance components for the animals under their care (e.g., beak trims, nail trims, wing clips). They must work closely with the veterinary team in obtaining the appropriate supplies to accomplish these tasks (hint: bandage scissors work well for wing clips). Occasionally, a routine maintenance procedure can turn into a medical procedure (e.g., when a nail is "quicked" or a blood feather is clipped) or even an emergency situation (e.g., in the case of hyperthermia or stress-related shock from capture and restraint). So, even though the keeper may be expected to schedule and carry out basic maintenance procedures, the veterinary team should always be alerted and their availability determined before such procedures begin. Other basic treatments may include injections, changing dressings, or topical medical applications.

OPERANT CONDITIONING

A keeper's ability to train an animal to perform desirable behaviors (or dissuade undesirable behaviors) is an important attribute in the modern zoo. Operant conditioning is an important training tool. This training can be applied to aiding in the medical management of zoo and aquarium animals. The keeper can train the animal for certain behaviors that will make veterinary care much easier to perform and less stressful for the animal. Some examples of how operant conditioning can simplify veterinary care could include training a diabetic white-cheeked gibbon (*Hylobates concolor*) to present a hip or shoulder to receive a daily insulin injection, training an American black bear (*Ursus americanus*) to open its mouth wide to allow for complete oral and dental evaluation, or training a Grevy's zebra (*Equus grevyi*) to lean into a target to enable hand injection of its annual vaccination. The list of benefits is endless, and this not only reduces stress for the animals but can improve a keeper's ability to monitor their health.

THE VETERINARY HOSPITAL

PROCEDURE PREPARATION

Significant preparation goes into a veterinary procedure, whether it will involve transporting the animal to the hospital or traveling to work on the animal in the field. The phrase "in the field" refers to any location other than the veterinary hospital. This can be a keeper service area, an animal den, or an exhibit; it can be indoors, outdoors, dry, wet, hot, cold, or any combination thereof. Working in the field eliminates many luxuries that one takes for granted when working in the veterinary hospital. What is the availability of electricity? What is the availability of water? Will building closure be necessary? What steps should be taken to insure privacy to exclude the public? All equipment and materials needed for the procedure, including items potentially needed for complications and emergencies, must be packed and transported to the site of the procedure.

The keeper must make sure that the area is prepared for the procedure. This can involve making space for necessary manual restraint and for the sedation, extra cleaning, and addition of extra bedding or padding. The keeper will also need to prepare the animal. This may include isolating the animal. If anesthesia is to be involved, then the animal will need to be fasted. For example, food and/or water may be withheld prior to anesthesia as prescribed by the veterinary team to reduce the risk of the animal aspirating regurgitated material. The duration of fasting can vary depending on the digestive system of a particular animal (e.g., nonruminants versus ruminants). If a keeper forgets to remove an animal's food and/or water the night before a procedure is planned, the procedure may have to be delayed or canceled.

Some of the best planned or most "simple" procedures can become extremely time-consuming even if all goes well. The preparation, procedure, cleanup, and recovery can take an entire day or longer. Patience is often required when working with animals in any capacity.

There are hazards to be aware of during a veterinary procedure, and a keeper should keep an eye on other team members involved in the procedure, as they may not be aware of those hazards or be paying attention to them. They may be so engrossed in the veterinary procedure that they may not notice that they are about to kneel on an uncapped syringe or knock over an open bottle of antiseptic. If a keeper is not busy actively assisting by taking vital signs or doing another assigned duty, he or she should be observing the animal, the other staff, and the surrounding area for any abnormalities or hazards.

A keeper must also be aware of the hazards associated with certain anesthetic drugs. Not every syringe or dart is safe to pick up. Residual drugs could be hazardous or even fatal if accidentally ingested, injected, or put in contact with cuts or mucous membranes. Darts may be used to introduce initial anesthetic or preanesthetic agents, and they must always be recovered and safely stored before a veterinary procedure begins.

STERILIZATION

Sterilization processes are necessary for surgical instruments and other equipment such as gloves, gowns, towels, suture material, syringes, and gauze, to name a few items. These techniques kill all present microorganisms, rendering the items sterile and reducing the risk of wound or surgical site contamination and possible infection. There are two main techniques for sterilizing items in the veterinary hospital. Primary sterilization is accomplished through the use of an autoclave. This incorporates a combination of steam and heat (under pressure) for a given amount of time. The steam penetrates items packed in the autoclave and allows the heat to kill any microorganisms. Certain delicate items can be damaged by heat and moisture, however, so for them an alternative sterilization technique can be used which involves a poisonous gas called ethylene oxide. This gas is introduced into a secured chamber that is properly ventilated to reduce staff exposure to the gas. The gas kills microorganisms and renders the surgical items sterile. Before items to be sterilized are placed in the autoclave, they are packaged in materials intended specifically for this use, such as plastic sleeves or disposable cloths. These packaging materials may be equipped with sterilization indicators, or sealed with specialized tape that includes sterilization indicators. There are also separate indicator strips that can be placed inside the package to be sterilized. When the package is opened, the color of the indicators immediately reveals whether the items inside have been successfully sterilized. The indicators used should be specific to the type of sterilization being practiced. Autoclaving (heat) indicators will visually change from no color to black when items are properly sterilized. Gas sterilization indicators will change from green to red.

SURGERY ROOM ETIQUETTE

A keeper will sometimes accompany an animal into surgery. During such a procedure it is critical that the keeper be aware of the processes involved in the surgical activity. First and foremost, the keeper must be aware of the sterile field that must be established and maintained. The area around the procedure that must be kept free of all contaminants (i.e., kept sterile) during activities such as opening of sterile packs, handing of instruments to the veterinarian, repositioning of the animal, hookup or adjustment of equipment, and monitoring of vital signs. The veterinarian and/or veterinary technician may wear surgical attire ranging from sterile gloves to a full surgical wardrobe consisting of booties, scrub suit, sterile gown, cap, mask, and sterile gloves. As a rule, only sterile items should touch other sterile items. Accidentally brushing against a surgical tray, picking up a surgical instrument in an effort to help, or touching something sterile during an attempt to monitor an animal's vital signs will compromise the sterile field; this can be detrimental to the surgery's success and the animal's recovery.

Aseptic technique is applied during any medical procedure in which the protective barrier of the animal's body is or has been compromised, whether it involves a wound being cleaned and dressed or a surgical procedure. It involves precautions that are taken to prevent or minimize the risk of contamination which could potentially lead to infection. Infection can prolong the healing process and even lead to death. Contamination occurs when microorganisms are introduced to the surgical area. These microorganisms (which may be bacterial, viral, fungal, or protozoan) can be normal to the environment or pathogenic. Unless an item has been sterilized, rendering it completely void of microorganisms, it will be contaminated to some degree. If the microorganisms become sufficiently numerous or virulent, contamination can lead to infection. Infection means that the contaminated area has established an unwanted population of microorganisms, often leading to problems such as redness, swelling, and/or drainage, which in turn can lead to a systemic infection.

Maintaining aseptic technique and establishing a sterile field begins with patient preparation. Keepers may assist with this procedure, and it is important for them to know the proper steps to be taken and the reason for them. Surgical preparation of the patient following anesthetic induction (the introduction of anesthesia resulting in an altered state of consciousness) begins with the clipping of fur or plucking of feathers as necessary. A smooth surface is the goal, because it will facilitate aseptic technique. An electric groomer's clipper equipped with a surgical clip blade (no. 40 or 50) used against the direction of hair growth will give the closest shave. It may also be necessary to use a razor after clipping most of the of hair away. One should use cream or lotion on the animal's skin before shaving it with a razor to reduce irritation that could cause the animal to chew or scratch the area after surgery. A small handheld vacuum or shop vacuum works well to remove the shaved hair from the area. The location and diameter of the area to be clipped will be predetermined by the veterinarian. Once the surgical surface is exposed, cleansing of the area with plain soap and water will remove any oils, dirt, or other detritus before the surgical scrub process begins.

Site preparation using aseptic technique insures that the location of the surgical incision is as free of microorganisms as possible, thus reducing the risk of infection. Starting this process with an antiseptic scrub (e.g., povidone iodine, or chlorhexidine with soap bubbling properties) will aid in the removal of residual skin oils. Using antiseptic scrub-saturated gauze or cotton batting, and starting in the middle of the clipped and cleansed area, one should scrub in a circular motion, working outwards toward the hairline. Scrubbing in this pattern improves the likelihood that any present microorganisms will be pushed outward and away from the surgical site. One should never work inward, double back, or zigzag back and forth, as this will recontaminate the cleansed area. Next, following the same pattern, gauze or cotton batting saturated with 70% isopropyl alcohol should be used to wipe away the antiseptic scrub. This should be done a total of three times. Finally, the clipped and scrubbed area should be sprayed with an antiseptic solution (povidone iodine, or chlorhexidine without any soap bubbling properties). The area must then be regarded as a sterile field; measures such as the use of surgical drapes can be used to protect its sterility during the process.

ANIMAL MONITORING DURING PROCEDURES

The keeper should be prepared to be more than just a passive observer during medical procedures. One of the most valuable things a keeper can to do to assist the veterinary staff during a procedure, whether it involves general anesthesia, sedation, or hand restraint, is monitor the animal's vital signs: body temperature, pulse (number of heart beats per minute), and respiration (number of complete inhale-and-exhale cycles per minute). These signs are often referred to as TPR (temperature, pulse, and respiration).

An animal's body temperature should be monitored for abnormal increase (hyperthermia) or decrease (hypothermia). Hyperthermia can be of noninfectious origin (e.g. exertion, ambient temperature, or stress) or infectious origin (e.g., virus or bacteria). An elevated temperature is called a fever. When reporting that an animal is exhibiting hyperthermia, do not say that the animal "has a temperature," as everything has a temperature. The term "fever" specifically denotes an abnormally increased body temperature. Be prepared to help lower an animal's body temperature when it is deemed necessary. Techniques can include intravenous fluids, ice packs, or the lowering of ambient temperature. A trick is to spray 70% isopropyl alcohol on the pads of the animal's feet, under its wings, on its ventral aspect, or on other safe locations on the animal's body. The alcohol's quick dissipation will facilitate evaporative cooling. Larger animals in the field can be cooled using a gentle stream of water from a hose, buckets, wet blankets, or foam pads. Animals suffering from hypothermia are at great risk, as they can no longer maintain their own body temperature. One should keep in mind that mild hypothermia can be quite common following surgery. Preventative measures should be taken for animals under sedation or anesthesia. Heating pads, hot water bottles, heat lamps, warmed towels, cage warmers, and even blow dryers can be used. Expired bags of intravenous fluids can be stored for this purpose and warmed in a microwave oven when needed; they are also helpful for positioning animals during procedures. (Note: One should be aware of the risks for each species; for example, rockhopper penguins [*Eudyptes chrysocome*] are more at risk of hyperthermia than of hypothermia during veterinary procedures). A heating or cooling component should never be placed in direct contact with an animal; it should be wrapped or covered in a towel to prevent burns or frostbite.

An increase in normal body temperature is expected during procedures due to the stress of shifting, netting, darting, restraint, and so on. It has been the author's experience that in general, many exotic species tend to exhibit greater fluctuations from their normal body temperature without ill effects than do domestic species.

There are different ways to measure body temperature. The most common method is to use a digital rectal thermometer. Glass thermometers can be used but are not recommended, due to the risk of breakage. A digital thermometer gives a quick reading, and beeps audibly when the final reading is obtained. Before insertion, the end of the thermometer should be lubricated with petroleum jelly or a water soluble lubricant. Insertion depth is very dependent on the animal's size, but for an accurate reading on a digital thermometer, the silver tip should be completely inserted. Once inserted, a rectal thermometer must never be left unattended, as it can cause injury to the surrounding tissue.

Other means of obtaining body temperature include the use of an ear thermometer; this method is not recommended by the author, as it gives unreliable and inconsistent results. Infrared thermometers can be also used, though they are an unnecessary expense, and dense hair coats and down can interfere with the accuracy of their readings. However, they can be useful in other areas of zoos to obtain temperatures of distant animals, substrate and flooring, nest sites, and so on.

The pulse measurement is generally the measure of the number of heartbeats in one minute. It can be taken manually by applying slight pressure to a blood vessel with two fingers. The vessel must have enough blood pressure for the pulse to be detected. Too much pressure applied by the fingers can stop or slow the blood flow, making it difficult if not impossible to detect the pulse. Many people have taken their own pulse using this same technique. Common locations to feel for a pulse would be the jugular, femoral, or brachial arteries. Sometimes, due to the muscle mass or excessive fat of a particular species (pulse can't be felt on most hogs), it may be necessary to use a stethoscope, pulse oximeter, endoesophageal stethoscope, or Doppler to obtain an accurate pulse or maintain continual readings during a procedure. It is important to note that electronic pulse oximeters are not designed for use with wildlife species. Due to the density of skin, fur, and feathers, as well as movement during hand

> **Good Practice Tip:** When counting heartbeats (pulse) or respirations per minute in animals with fast and regular rates, one can count the number of beats or respirations for 15 seconds and multiply the number by 4.

restraint procedures, errors in monitor readings can result. Manual monitoring of vital signs (e.g., observing respirations, and feeling for a pulse) is most reliable. A time-saving tip is to count the number of beats in 15 seconds and multiply by four.

Respiration, the act of breathing, is documented as breaths per minute, or bpm (note, however, that the abbreviation "bpm" is also used for beats per minute when documenting pulse). When monitoring an animal's respiration, the inhalation and exhalation are included together as one count. It is essential that in addition to the rate of respiration, the actions associated with the respiration are noted. These can include sounds like wheezing or clicking; behaviors such as head extension or straining; and patterns such as fast, slow, deep (excessive inflation or movement of chest and rib cage), or shallow (chest and rib cage movement barely detectable). Respirations can be counted visually, by watching the rise and fall of the chest or rib cage, by observing components on the anesthesia machine, or by feeling for the presence of air from the mouth or nostrils.

HOSPITAL REGULATIONS AND SAFETY

Safety is the responsibility of everyone! It is always in the keeper's best interest to learn about the hazards associated with the job and stay aware of current and changing risks. In the United States there are three main safety and compliance governing bodies for zoo and aquarium veterinary hospitals: the Occupational Safety and Health Administration (OSHA), the Association of Zoo and Aquariums (AZA), and the US Department of Agriculture (USDA).

While each individual is ultimately responsible for his or her own safety, OSHA was established under the Department of Labor to ensure safety guidelines for the workplace. Some potential safety hazards in the veterinary hospital can include

- bites, scratches, and other animal-induced injuries
- laboratory accidents (e.g., chemical or gas burns to skin, eyes, or lungs)
- X-ray radiation exposure (e.g., as a result of not wearing appropriate protective gear)
- waste anesthetic gas exposure (e.g., due to improper ventilation during procedures)
- excessive noise (e.g., ferruginous hawk vocalizations in a confined space during manual restraint)
- toxic chemicals (e.g., splashback from bleach or other disinfectants)
- zoonotic diseases (e.g., salmonellosis)
- dangerous drugs (e.g., carfentanil).

OSHA will enforce safety regulations for the veterinary hospital through inspections that can be unscheduled and unannounced or that can result from documented accidents or reported safety concerns. OSHA concerns itself with all workplace hazards, not just those in the veterinary hospital. However, there are many special hazards to be aware of when working in the hospital area. Through training, establishment of and adherence to protocols, proper storage and handling of chemicals and gases, maintenance of the Material Safety Data Sheet (MSDS) log, and proper ventilation, these hazards can be reduced.

Two great risks in the veterinary hospital involve exposure to radiation and to anesthetic gas. Even though there are ways to monitor radiation exposure, each state is responsible for establishing its own regulations for radiograph equipment (X-ray machines). Generally, X-ray machines should be registered with the state's department of health, and periodic inspections will occur. These inspections, which are conducted by privately hired and professionally qualified inspectors, can take place every one to three years, depending on the type of X-ray equipment being used. Equipment that passes the inspection will receive certification for continued use until the next scheduled inspection. If radiograph equipment is found to be substandard, aged, or leaking radiation, it must be repaired or retired. Failure to comply with inspection recommendations will result in denial of certification and a report to the state's board of veterinary medicine. The area of the veterinary hospital that houses the X-ray equipment should be constructed with protective walls and lead-lined glass and doors to reduce the risk of exposing other hospital employees or bystanders to radiation during a procedure. Darkrooms should also be constructed so as to protect stored radiographic film from scattered radiation, which can affect radiograph quality. Scattered radiation is the main cause of unintended exposure by humans. Keepers who assist in radiographic procedures will wear lead-lined gloves and aprons to be protected from scattered radiation. Facilities will also need to regulate radiation exposure through the use of dosimeter badges. These badges (sometimes called film badges or X-ray badges) are worn on the attire of hospital staff and assistants to monitor the level of radiation exposure. Control badges (which monitor hospital environment radiation exposure) and staff badges (which monitor the radiation exposure levels of each individual who assists in X-ray procedures) are submitted to a contracted company for analysis to ensure that exposures are kept within the acceptable safety guidelines. Gas anesthesia machines are not regulated by state or federal laws at this time. However, multiple safety precautions (e.g., proper ventilation, properly fitted anesthetic masks, and endotracheal tubes) and safety features on anesthesia machines (e.g., scavenging systems, which trap waste gas) should be in place to reduce the risk of harmful anesthetic gas exposure.

North America's Association of Zoos and Aquariums (AZA)-accredited facilities undergo an inspection for accreditation every five years. These facilities are expected to uphold the best standards between inspections. In such an inspection, the veterinary hospital will be reviewed for safe and appropriate practices in line with the veterinary care component of AZA accreditation, including

- whether the institution's animal health care program is under the direction of a licensed veterinarian
- whether the institution follows the Guidelines for

Zoo and Aquarium Veterinary Medical Programs and Veterinary Hospitals of the American Association of Zoo Veterinarians
- the animal record-keeping system
- whether medical records are up-to-date
- whether an adequate number of persons are employed in the animal health care program
- whether the extent of veterinary services provided to the collection is adequate
- USDA reports, and what is being done to correct any concerns
- quarantine procedures and their implementation
- the alarm system and emergency procedures
- drug emergency protocols
- whether the veterinarian's response time from home will be adequate in an emergency
- whether adequate policies and procedures are in place for the safe handling of venomous animals
- whether adequate serums (e.g., antivenin) are available
- whether drugs used in aquariums or aquatic exhibits comply with FDA guidelines
- whether animal foods, especially seafood products, are purchased from sustainable or well-managed sources (AZA 2011).

There are numerous zoo and aquarium associations worldwide that establish and maintain common guidelines for their countries' zoos and aquariums. These include the Canadian Association of Zoos and Aquariums (CAZA), the European Association of Zoos and Aquariums (EAZA), the British and Irish Association of Zoos and Aquariums (BIAZA), and the Japanese Association of Zoos and Aquariums (JAZA).

Agricultural and wildlife agencies will monitor and inspect certain components within the zoo and aquarium community due to the risks that exotic species may pose to the food animal industry and to native species. Zoo and aquarium guidelines and restrictions are set by agencies such as the US Department of Agriculture (USDA); inspections can occur at any time and can be either scheduled or unannounced. USDA agents can inspect an entire zoo facility or a single area within it. The USDA will monitor certain components of the veterinary hospital including controlled drugs, expired drugs, human food location restriction, porous surfaces, food thawing and handling, and caging.

A BASIC INTRODUCTION TO VETERINARY CLINIC TOOLS AND EQUIPMENT

THE MICROSCOPE

Keepers may be expected to run routine or clinical microscopic examinations or aid in the analysis of blood work. Many types of microscopes may be encountered, and it takes a great deal of practice to become efficient in microscopy. This section is intended to give a brief overview of the microscope, its parts, and the associated terminology (figure 45.1). Microscopes are used to see anything that is too small to be analyzed by the naked eye (i.e., grossly), such as endoparasites, ectoparasites, bacteria, and blood cells. They are most commonly monocular (with a single lens and eyepiece) or

Monocular Microscope

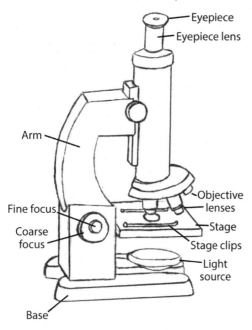

Figure 45.1. Parts of a monocular microscope. Illustration by Travis J. Pyland.

binocular (with two lenses and eyepieces). The lens at the top of the microscope is equipped with 10× magnification. The eyepiece can also house a micrometer, which allows for exact measurement in micrometers (mcm or µm) of the item(s) being examined, as may be necessary for positive identification of the items (e.g., of certain parasitic ova). A specimen will usually be presented on a microscope slide and covered with a cover glass. Its preparation may involve dilution, filtration, concentration, or staining. The slide is then placed on the stage. Once it is secured, focus of the microscope should always start with the lowest-powered objective. A light source will penetrate up through the slide into the objective and eyepiece. The intensity of this light can be controlled, as the density of specimens may vary (e.g., dark staining of blood smears will require more light to penetrate, to allow visualization of the blood cells). Any increase in magnification should occur with sequential changing of the objectives from lower to higher magnification. Objectives are the magnification components of the microscope, and most microscopes will have multiple objectives ranging from 4× to 100× magnification. The 100× objective is used only with immersion oil, a single drop of which is placed on the microscope slide before the magnification is turned to 100×. Immersion oil allows light to be concentrated into the objective for effective viewing at such a high magnification. The magnification provided by the objective is multiplied by the 10× magnification supplied by the eyepiece. For example, a 4× objective lens will actually provide 40× magnification power when combined with the eyepiece's 10× lens. There will usually be two dials for focusing, which is done by moving the stage up or down. One must be careful not to bring the stage housing the microscope slide directly into contact with the objectives; it can cause breakage of the microscope slide

and scratching of the objective lens. Initial focus is made with the coarse focus dial, used only with the lower magnifications to bring the object into initial focus. Coarse focus allows for a great amount of movement of the stage. Once the object is brought into coarse focus, the fine focus dial can be used to achieve exact clarity. Fine focus creates subtle changes in the image, and only allows for minute movements of the stage to be made.

Objective lenses and eyepieces should be cleaned only with lens paper, to prevent scratching. When immersion oil is used, the objective should be wiped clean with alcohol immediately afterward. The microscope stage should also be wiped and disinfected after every use. When not in use, a microscope should be kept covered to protect it from dust and other contaminants. It should not be scooted across a surface or transported, but when it must be moved, a supporting hand should always be placed under the base.

VETERINARY INSTRUMENTS

A keeper should be familiar with commonly used surgical instruments in order to effectively assist in surgical procedures, but some such instruments are also useful to keepers in day-to-day animal care (e.g., a hemostat or needle holder can be used to remove the broken shaft of a blood feather). Many instruments are used in veterinary medicine, each for a specific purpose. The most common instruments are listed here.

- Hemostatic forceps (i.e., hemostats) are available in many different sizes and can have straight or curved clamping ends. The inside of their "jaws" can have grooved surfaces, which provide better gripping capability, or smooth surfaces, which are less traumatic to tissues. They have handles similar to those of scissors, and a locking or ratchet mechanism that allows the grip to be secured. This feature makes it possible to maintain a grip on a bleeding vessel or blood feather when one cannot maintain hand contact with the instrument.
- Thumb forceps (i.e., forceps) are similar to tweezers but can have a variety of tips which may include "teeth" or may be blunt, pointed, smooth, or grooved. These instruments must be held closed by hand, and this allows for easy release of the item being grasped.
- Scalpels consist of two separate components, the handle and the blade. The blade is detachable and disposable. Blades are numbered for reference; the numbers reflect their size and style. They can be rounded, pointed, or curved. Scalpel handles come in two commonly used sizes, and certain blades are used with each size.

THE RADIOGRAPH MACHINE

When it comes to diagnostic imaging, the keeper may participate in restraining and positioning an animal, maintaining safety precautions, taking radiographs, and developing radiographs. The X-rays produced by radiograph machines cannot be seen, felt, smelled, or heard, and so the risk of exposure to them is easily overlooked. Scattered radiation is caused by X-ray photons that "bounce" when they hit their target instead of being absorbed into the film. Setting the machine properly can reduce the risk of scattered radiation. This section is intended to familiarize the keeper with basic X-ray machine settings and their function. Machine settings (exposure factors) are based on the size of the object being radiographed and the type of tissue being examined (e.g., bone or soft tissue). A technique chart is a "cheat sheet" of sorts that should be developed for each X-ray machine; it allows for quick and accurate guidance in setting film exposures based simply on a measurement (in centimeters) of the body part being radiographed. The three primary X-ray machine settings are kilovoltage, milliamperage, and exposure time. These are sometimes referred to as either quality or quantity factors. The kilovoltage (kV) is the setting that regulates the energy of the X-ray beam and will affect the scale of contrast seen in the radiograph. This is regarded as a quality factor. The scale of contrast is reflected by the number of shades of gray. Higher-quality radiographs contain a great variety of shades of gray. Milliamperage (mA) is regarded as a quantity factor because it regulates the amount of X-rays produced and introduced to the targeted area being radiographed. Adjustments in the milliamperage will affect the film's lightness or darkness. Exposure time means the same thing in radiograph production as it does in conventional photography, and is simply the amount of time during which the radiographic film is exposed to the X-rays produced by the machine. Because exposure time determines the number of X-rays that are introduced to the film, it is regarded as a quantity factor.

Because the film used in diagnostic imaging is light-sensitive, it is contained in a lightproof holder called a cassette. The film must be loaded into the cassette in a darkroom, and protected from scattered radiation. It is important to keep the inside and outside of the film cassette clean to prevent any false readings or misdiagnosis from the presence of hair, lint, dirt, or dust. One should always use approved cleaning solution and lint-free wipes to clean a cassette, making sure that the inside is completely dry before reloading it with a new sheet of film. The darkroom must be sealed to prevent any light from entering through, around, or under the door or elsewhere. The room can be illuminated with a special red light to allow for a small degree of visibility. It should be secured from the inside while occupied, to prevent unexpected entry, and an "in use" notification should be posted outside the room to further reduce this risk.

THE ANESTHESIA MACHINE

The anesthesia machine can be used to induce and maintain anesthesia during a veterinary procedure, or to maintain a state of sedation after injectable sedatives have been given to an animal. Gas anesthesia is considered safer for prolonged procedures than most injectable forms of anesthesia; since is quickly metabolized by the body, the animal's recovery is usually faster and less traumatic than is seen with injectable anesthetic agents. Gas anesthetic agents are referred to as inhalant anesthetics: a liquid is combined with oxygen in a machine to produce an oxygenated gas. The concentration of this gas is controlled in the machine's anesthetic vaporizer. The gas is delivered to the animal through an endotracheal tube (ET tube), a face or muzzle mask, or an induction chamber.

Anesthesia machines are usually set up as "rebreathing" or "non-rebreathing" systems. The preference in system is generally based on the size of the animal undergoing the exposure. Breathing circuits are the tubing systems that carry the anesthetic gas and oxygen to the patient, and the waste gases to the filtration unit, which is called the scavenging system. Non-rebreathing systems are recommended for use with animals under 7 kg in body weight (Bassert and McCurnin 2010, 905); they provide fresh oxygen and anesthetic gas with each breath; instead of being recycled, the waste gas is sent directly into the filtering system. These systems create less resistance, which makes for easier breathing by animals with small lung capacity. Rebreathing systems filter out the waste gas (carbon dioxide) and recirculate the anesthetic gases back to the animal, and are thus more efficient and cost-effective. These systems are also called "circle systems." Their primary advantage is that they allow for the visualization of respiration through valve components within the machine, thus making it easier to monitor the animal's condition and the depth of anesthesia.

NEEDLES AND SYRINGES

Keepers make use of syringes and needles for many things other than giving injections. Dosing oral medications, collecting samples, and measuring waste output are just a few examples. A syringe has three parts: the hub at the end, which serves as an attachment site for the needle; the barrel, which measures and contains the contents; and the plunger, which is equipped with a rubber end that is used to withdraw or inject the syringe contents (figure 45.2). Syringes come in many sizes—most that are commonly used in veterinary medicine range from 0.5 ml to 60 ml—and because of this, the gradu-

Syringe and Needle

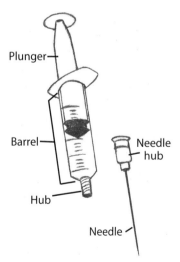

Figure 45.2. Parts of a medical needle and syringe. After injection, some contents will remain in the "dead space" of the syringe hub, needle hub, and needle. Illustration by Travis J. Pyland.

ated marks on differently sized syringes can represent very different measurements. Keepers must familiarize themselves with these differences. For example, the lines on the outside of a 3-milliliter (ml) syringe each represent a volume of 0.1 ml, whereas the lines on a 20 ml syringe would represent a volume difference of 1 ml. The volume reference for syringes and medication fluctuates with the user, and is sometimes referred to as cc (cubic centimeter) or as ml (milliliter). A cc and an ml are the same volume. A keeper should practice operating syringes, and should ideally be able to pull and push the plunger using a one-handed technique. The section of the needle and syringe that includes the hubs (of both and the needle and syringe) and the needle itself is regarded as "dead space," where contents of the syringe will always remain. Putting the needle cap back on a needle after use is not recommended. However, in the area of veterinary care there are times when there is greater risk in leaving needles exposed (e.g., during emergencies and procedures in the field). During these times, keepers should employ a one-handed recapping technique that will reduce the risk of injury or infection from the stick of a contaminated needle:

> One-Handed Needle Recapping:
> **Step 1.** Place the cap on a flat surface, such as the countertop or even the floor
> **Step 2.** Using only one hand, hold the syringe in the tips of your fingers with the needle pointing away from your body
> **Step 3.** Place your fingertips on the flat surface so that the needle and the syringe are parallel to and in line with the cap
> **Step 4.** You may then use your other hand to "seat the cap firmly" (Bassert and McCurnin 2010, 136).

It is the author's recommendation to not use one's other hand to "seat" the cap, but to instead guide the needle into the cap and in the same motion lift the plunger of the syringe, pointing the needle and cap into the surface (i.e., the countertop or floor). Then, when the syringe is in a vertical position, push the needle firmly into the cap. This eliminates any risk of a needle stick to the other hand.

Needles come in a variety of sizes, measured both by length (e.g., 1.9 cm [3/4 in.], 2.5 cm [1 in.], 3.8 cm [1-1/2 in.] and by internal diameter (ID), which is expressed as a "gauge." The larger the gauge (G) number the smaller the diameter of the needle, and vice versa (e.g., a 16G needle is much larger in diameter than a 25G needle). Gauges commonly used in veterinary medicine range from 25G to 14G.

The broad definition of a "sharp" in the veterinary field is anything that can puncture the skin. While the most common sharp exposure would come from a syringe needle, it could also be a scalpel blade, a broken microscope slide, or a dart needle. Whether a keeper is assisting in the field or in the veterinary hospital, all sharps should be accounted for, collected, and disposed of in a puncture-resistant "sharps container" that is that is properly labeled with the biohazard symbol. These containers are available through veterinary or other medical supply companies. When full, sharps containers should be incinerated (there are medical services companies that can do this), not thrown into the regular trash. These containers are designed to limit the risk of spilling, and are disposable. Sharps containers should never be emptied and reused. Sharps, as well as other veterinary instruments and collection containers, can be contaminated with biological waste; they should be handled with caution and disposed of properly. One should always remember to double-check the label of the medication vial as well as the dose before and after drawing up any medication.

SUMMARY

The extent to which the keeper will assist the veterinary team will vary with every facility. The size of the veterinary team and its members will also vary. However, it is very important for keepers to develop a basic understanding of the veterinary team, their roles, and the cooperative efforts that will be shared between the veterinary team and the animals' primary caregivers. Through obtaining a basic understanding of the medical management issues encountered in the zoo setting, and by developing a working knowledge of the applications employed during routine or emergency veterinary procedures, as well as the equipment that will be used, the keeper can better assist the veterinary team as needed. This will allow the keeper to improve communications with the veterinary team and play a more active role in veterinary care and health management. All of this will ultimately improve the quality of care that is provided.

REFERENCES

Association of Zoos and Aquariums. 2011. The Guide to Accreditation of Zoological parks and Aquariums. Accessed February 2011 at http://www.aza.org/uploadedFiles/Accreditation/Guide%20to%20Accreditation.pdf.

Bassert and McCurnin. 2010. *Clinical Textbook for Veterinary Technicians.* Saint Louis: Saunders Elsevier.

Colorado Association of Certified Veterinary Technicians. Veterinary Technician vs. Veterinary Nurse. 2001. Accessed in February at http://www.cacvt.com/files/Veterinary%20Technician%20vs%20Veterinary%20Nurse.pdf.

International Veterinary Nurses and Technicians Association. 2008. "United." Accessed in February at http://ivnta.org/index.php.

United States Department of Agriculture, Animal and Plant Health Inspection Service. 2011. Animal Welfare. Accessed in February at http://www.aphis.usda.gov/animal_welfare/.

46

Zoonotic Disease

Scott P. Terrell

INTRODUCTION

The simplest definition of a zoonotic disease is "a disease that is transmitted between animals and humans." The World Health Organization website states that more than 200 zoonotic diseases have been described. According to the Centers for Disease Control and Prevention, approximately 75% of the newly emerging infectious diseases affecting humans and 60% of all human pathogens are potentially zoonotic. The high percentage of newly emerging infectious disease being zoonotic diseases is a reflection of a variety of factors, which include domestic animal and human population expansion, intensive agricultural and farming practices, microbial adaptation, international traffic in animals and animal parts, and increased contact and interaction between humans, domestic animals, and wildlife (Daszak et al. 2001, 103–16). Zoonotic disease biology and prevention should be an important consideration for the zoo or aquarium professional, with regard to the health of zoo employees as well as visitors. For the purposes of this chapter, references to "zoos" and "zoo personnel" apply equally to those professionals who work in the aquarium industry, as all of the same concepts apply.

The goals of this chapter are to

- provide a basic understanding of disease biology, terminology, and transmission
- discuss the components of a zoo employee disease prevention program
- discuss the components of minimizing zoonotic disease risks for the public
- provide a summary of some of the more important zoonotic disease issues in the zoo environment.

ZOONOTIC DISEASE TERMINOLOGY

DISEASE BIOLOGY

A zoonotic disease or zoonosis is a disease transmitted from vertebrate animals to humans. Some authors define zoonotic diseases as those diseases common to animals and humans, but this definition may be too broad; it includes a wide variety of environmental diseases not necessarily acquired from animals. Humans are the susceptible hosts for zoonotic diseases. A susceptible host is an organism (human or animal) in which a disease agent is able to establish an infection. The susceptibility of the host depends on a variety of factors, not limited to the host's immune status, the timing of the infection, the dose of the disease agent, the route of exposure, the mode of transmission, and previous history of exposure to the disease agent. Once infected, an organism (human or animal) may exhibit active disease or may be subclinical or asymptomatic. Active disease suggests that the organism is overtly ill or abnormal. Subclinical and asymptomatic infections suggest that the organism is infected, but is not showing any outward signs of disease. Animals that are subclinical or asymptomatic may still act as carriers and may be able to transmit disease to humans.

A reservoir is the place in the environment where a disease persists naturally. Vertebrate animals are the reservoirs of zoonotic diseases. For example, the reservoir for West Nile virus in the environment is a wild bird population, such as a population of American crows (*Corvus brachyrhynchos*). Zoonotic diseases may be transmitted from the reservoir to humans through direct or indirect contact. Direct contact involves physical contact with an animal or animal bodily fluids or wastes. Indirect contact involves transmission by vectors, fomites, or environmental contact. A vector is any living object that transmits a disease agent. Insects are the most common vectors and are responsible for the transmission of a wide variety of zoonotic diseases. For example, ticks (*Ixodes* sp.) are the vector for the transmission of Lyme disease from white-tailed deer (*Odocoileus virginanus*) to humans.

455

Insects may act as mechanical vectors, whereby they simply physically carry the disease agent from the reservoir to the susceptible host. Alternatively and more commonly, insects act as biological vectors when the disease agent must undergo some developmental change inside the insect before infecting a susceptible host. A fomite is an inanimate object involved in the transmission of a disease agent. Fomites physically carry the disease agent from one location to another, such as when viral or bacterial particles adhere to the surfaces of work boots or shoes in a contaminated environment. Contaminated clothing, tools, and cleaning supplies can also act as fomites. Fomites may also participate in disease transmission by puncturing the skin and introducing pathogens, as in the case of viral infection from an accidental needle stick, or bacterial infection following a puncture wound from a thorn or splinter.

DISEASE PREVENTION IN THE WORKPLACE

Professions that bring humans and animals into close contact (farming, veterinary practice, animal care, etc.) increase the risk of exposure to a wide variety of animal pathogens, and may increase the risk for zoonotic disease transmission. Occupational hazards are working conditions that can lead to injury, illness, or death in the workplace. Thus, zoonotic disease exposure would be an occupational hazard of the animal care/zoo keeping profession. By its very nature, zookeeping encourages or requires close contact with a wide variety of species from diverse locations throughout the world. Contact is not limited to the animals themselves, as the keeper staff may also come into contact with bodily fluids, wastes, secretions, or even blood on a regular basis. Such exposure to animals, animal waste, and bodily fluids creates an opportunity for zoonotic disease exposure. Based on this fact, it is important for keepers to have at least a basic understanding of zoonotic disease biology in order to prevent occupational exposure to this disease. Zoo associations may require this training as a component of accreditation standards (e.g., the 2010 Accreditation Standards [11.1.2] of the US Association of Zoo and Aquariums [AZA] require that "training and procedures must be in place regarding zoonotic diseases"). Reports suggest that the vast majority of zoonotic disease exposures at zoos involve the public in "petting zoo" environments (Blackmore 2009, 1–15; LeJeune and Davis 2004, 1440–45; Bender and Shulman 2004, 1105–9). It is reasonable to assume that zoo personnel are also exposed to zoonotic diseases in the work environment. Disease associated with these exposures may go unreported in the wake of larger public concerns about disease, but there is no shortage of examples of zoo personnel being exposed to zoonotic disease (Hill et al. 1998, 484–88; Janssen et al. 2009, 200–201; Kiers et al. 2008, 1469–73; Stetter et al. 1995, 1618–21)

As mentioned at the beginning of this chapter, zoonotic diseases may comprise as many as 60% of all known human pathogens. Many of these diseases can have serious consequences or require intensive treatment. Thus it is important to focus on disease prevention rather than rely on treatment following exposure to or diagnosis of a serious zoonotic disease. Disease prevention can take many forms. In the workplace it may involve training to deal with specific circumstances or

risks (as in the case of human blood-borne pathogen training) or implementation of specific actions (as in the case of rabies vaccination or tuberculosis testing). Disease prevention may also involve general guidelines that workers may practice or protective equipment workers may use during their day-to-day work life. In this section we will first discuss some of the specific training and actions that should be undertaken in the work environment. These topics focus primarily on human health, and thus the final decision about how to implement training, vaccination, or testing programs should be made in consultation with a human health professional, such as a company physician, or with the public health authorities.

Discussion of zoonotic diseases in the work environment often raises questions that are best addressed by a physician. It is not uncommon for the veterinarian in a zoo environment to be looked upon as the "expert" in zoonotic disease transmission and prevention. In fact, a survey of veterinarians and physicians demonstrated that veterinarians encounter and discuss zoonotic diseases in their practice significantly more frequently than physicians (Grant and Olsen 1999, 159–63). Although a veterinarian may be highly trained and skilled in infectious disease diagnostics, the ultimate source of human disease information should be a human health professional such as an employer health service department, company doctor, public health agency, or personal physician. Veterinarians should be cautious about providing human medical advice. Information they provide regarding zoonotic diseases should be limited to the general aspects of diseases, modes of transmission, effects on the animal host, and methods of exposure prevention. A human health professional should provide specific information about human symptoms, diagnostic techniques, or treatments. Collaboration and consultation between human health professionals and veterinarians on zoonotic disease issues should be encouraged. Most states in the United States employ the services of public health veterinarians in the health department system who may facilitate this collaboration. A list of American state public health veterinarians can be found at the website for the National Association of State Public Health Veterinarians (NASPHV; www.nasphv.org). In addition, the NASPHV has developed a useful document addressing the standard precautions for zoonotic disease prevention in veterinary personnel (Elchos et al. 2008, 415–32). Many of the protocols outlined in this NASPHV document have direct application to the zoo environment.

The American Association of Zoo Veterinarians (AAZV) has created a document that specifically addresses zoo personnel health program recommendations as part of the Guidelines for Zoo and Aquarium Veterinary Medical Programs and Veterinary Hospitals (Shellabarger 1998). The document discusses the specific components of a personnel health program in the zoo environment. Topics covered include employee physical examination and medical history collection, employee health education programs, serum banking of employees, serologic screening for specific zoonotic pathogens, fecal analysis, tuberculosis testing, and immunization protocols. Implementation of a personnel health program at the zoo requires collaboration between human medical professionals and veterinarians. Many components of the above-recommended health program could be difficult to implement in today's world of privacy and liability concerns. In this chapter, we will focus primarily on four components of the zoo personnel health program: vaccination recommendations, tuberculosis testing, health education programs, and proper use of personal protective equipment and actions.

VACCINE RECOMMENDATIONS FOR ZOO PERSONNEL

Rabies vaccination. Rabies is the most serious zoonotic disease that zoo workers could potentially come into contact with. It is fatal in almost all cases unless prevented or treated in a timely fashion. Fortunately, a pre-exposure rabies vaccine exists for people at high risk for exposure to the virus. The 1998 Guidelines for Zoo and Aquarium Veterinary Medical Programs and Veterinary Hospitals prepared by the AAZV states that pre-exposure vaccination against rabies should be provided to the following employees (Shellabarger 1998):

1. those working directly with the rabies virus
2. those in contact with large animals (hoofstock and carnivores) or with susceptible animals (mammals) that are housed primarily outdoors
3. those who have direct contact with animals in quarantine
4. those who are exposed to potentially infected animal tissues, or who perform postmortem examinations on animals with histories of poorly defined neurologic disorders.
5. those with responsibility for capturing or destroying wild animals on zoo grounds.

The 2008 Centers for Disease Control (CDC) guidelines regarding pre-exposure vaccination for humans do not contain specific recommendations for zoo personnel. The general recommendation is as follows:

Pre-exposure vaccination should be offered to persons in high-risk groups, such as veterinarians and their staff, animal handlers, rabies researchers, and certain laboratory workers. Pre-exposure vaccination also should be considered for persons whose activities bring them into frequent contact with rabies virus or potentially rabid bats, raccoons, skunks, cats, dogs, or other species at risk for having rabies (Manning et al. 2008, 1–26).

Zoological institutions, in consultation with human health professionals, may choose to implement vaccine programs that follow either or both of these sets of guidelines.

Tetanus vaccination. Tetanus is not typically a zoonotic disease, but it is a serious bacterial disease that can be acquired in the workplace through any break in the skin. Employees should be encouraged to stay current on their tetanus vaccine. A booster should be given every 10 years for adults.

Hepatitis B vaccination. Hepatitis B is a blood-borne pathogen. Nonhuman primates can be infected with it, but there is little evidence to support zoonotic transmission of the virus from nonprimates to people (Acha and Szyfres 2003, 356–62). Most cases of human hepatitis B infection are acquired through contact with bodily fluids of infected humans through mucous membranes or breaks in the skin, through needle punctures (in a medical setting or through drug abuse), or by sexual contact (Hugh-Jones et. al 2000, 344). The hepatitis B vaccine should be made available to employees at high risk of contact with infected human blood or body fluids. Such exposure should be rare in a zoological environment, except perhaps for hospital staff.

Influenza vaccination. Recent outbreaks of avian (H5N1) and swine (H1N1) influenza viruses in both human and animal populations have raised awareness and concern about the transmission of zoonotic influenza viruses. Avian influenza, H5N1, is a disease primarily of domestic poultry and wild birds that has caused illness and deaths in humans in some areas of the world. The swine influenza virus, H1N1, is primarily a human disease, although populations of pigs and a few other animal species have been infected. Concerns over the possible mixing of the swine flu and human flu viruses have led to recommendations that employees who work with captive swine species be vaccinated against seasonal flu.

According to the CDC, employers should consider providing access to the seasonal influenza vaccine for animal care staff who work with swine. Vaccination of workers who care for pigs will decrease their risk of passing seasonal flu viruses to the pigs. Seasonal influenza vaccination of workers might also decrease the potential for people or pigs to become coinfected with both human and swine flu viruses. Such dual infections could result in reassortment of the two different viruses and lead to a new influenza virus (www.cdc.gov, 2010).

TUBERCULOSIS TESTING FOR ZOO PERSONNEL

There are numerous examples of zoo personnel exposed to tuberculosis through contact with zoo animals (see specific discussion of tuberculosis below). AZA standards (11.1.3) require that "a tuberculin (TB) testing/surveillance program must be established for appropriate staff in order to ensure the health of both the employees and the animal collection" (AZA 2010). Zoo personnel at greatest risk for exposure to the tuberculosis disease agents (*Mycobacterium tuberculosis, M. bovis,* or closely related bacteria) include keepers who work with nonhuman primates, elephants, seals (pinnipeds), and hoofstock, as well as veterinary personnel, quarantine staff, and immunocompromised individuals. The AAZV Guidelines for Zoo and Aquarium Veterinary Medical Programs and Veterinary Hospitals suggest that "high risk individuals have a PPD tuberculin (intradermal) skin test at least annually" (Shellabarger 1998). The specific schedule for each zoo and each type of zoo employee should be determined in consultation with human health professionals.

HEALTH EDUCATION PROGRAMS FOR ZOO PERSONNEL

Zoonotic disease training. The US AZA standards require that "training and procedures must be in place regarding zoonotic diseases" (AZA 2010). Zoonotic disease education is an essential component of a zoo employee health education program. In most zoos, a veterinarian or other member of the zoo hospital staff provides zoonotic disease training to the staff. In some instances, human medical providers, health department officials, or local academic institutions may provide this kind of training.

The goal of a zoonotic disease training program should be to introduce zoo personnel to the concepts of zoonotic diseases, modes of transmission, and methods of disease prevention in the workplace. Common and serious zoonotic disease can be used as examples to emphasize the information provided in the course. Much of the framework of this chapter comes from zoonotic disease training sessions that the author has conducted at his institution. This training is valuable not only for the zookeeper staff and hospital personnel, who work directly with animals, but also for others who may enter or work in animal holdings or exhibits. Security, maintenance, pest management, commissary, and horticulture staff may all come into contact with animal waste or other potentially infectious materials. At some institutions, all personnel who routinely work in animal exhibits or holding areas are required to complete at least a basic introductory zoonotic disease prevention training program.

At-risk training/education. One of the more important aspects of health education training for zoo personnel involves education of specific employees who are considered to be at high risk of disease exposure and infection. These high-risk employees should be encouraged to seek additional information regarding disease prevention in the workplace, which should be provided by a company physician, by a personal physician, or by local health department officials at the employees' request. High-risk individuals include elderly employees over the age of 65; pregnant women; people with chronic illnesses such as diabetes, heart, lung, liver or kidney disease; and people who are immunosuppressed due to drug treatments or infection with HIV (Grant and Olsen 1999, 159–63).

Blood-borne pathogen training. Blood-borne pathogens are human diseases that are transmitted through contact with blood or other bodily fluids. Most blood-borne pathogen training programs focus on the human immunodeficiency virus (HIV) and hepatitis viruses B and C. These pathogens are not zoonotic diseases, but it seems logical to discuss them in the context of workplace disease prevention. The US government's Occupational Safety and Health Administration (OSHA) established standards for workplace blood-borne pathogen training in 1991. The standard is referenced in title 29 of the Code of Federal Regulations (29 CFR), part 1910.1030. (www.osha.gov/pls/oshaweb/owadisp.show_document?p _table=STANDARDS&p_id=10051). All employees who can

"reasonably anticipate" contact with blood and other potentially infectious materials while performing their job duties are covered by the standard. Most keepers do not perform job functions that necessitate blood-borne pathogen training, but those who work in the zoo hospital or use needles for training or medical treatment of animals may need training. In addition, some blood-borne pathogen exposure prevention techniques may prove valuable for preventing general and zoonotic disease among general animal care staff.

PROPER USE OF PERSONAL PROTECTIVE EQUIPMENT AND ACTIONS IN THE WORKPLACE

Personal protective equipment and actions are the front line of defense against zoonotic pathogens. Proper hand washing, animal handling, and use of personal protective equipment can prevent almost all disease transmission in the workplace. The following is a summary of the personal protective equipment and actions that can and should be used in the zoo environment.

Hand washing and hygiene. Hand hygiene is the single most important disease prevention tool that can be used when working with animals or materials contaminated by animal waste or bodily fluids. (Boyce and Pittet 2002, 1–48; Larson 1995, 251–69). The term "hand hygiene" includes hand washing with soap and water as well as the use of alcohol-based products (gels, rinses, foams) that do not require water (Siegel et al. 2007). Zoo personnel should not underestimate the importance of hand hygiene in preventing a wide variety of zoonotic diseases that are transmitted primarily by fecal-oral contact (tables 46.2–46.6). Hands should be washed in between each animal contact and after any contact with animal wastes or bodily fluids. Similarly, hands should be washed following any contact with bedding, caging, tools, or other material potentially contaminated by animal wastes or bodily fluids.

Gloves. Gloves reduce the risk of disease transmission by providing a barrier for the skin, but they are not a substitute for hand washing (Olsen et al. 1993, 350–53). Two major types of gloves are used in the zoo environment: cloth work gloves and disposable gloves (typically made of latex, vinyl, or nitrile). Cloth or leather work gloves are common in traditional zoo animal care settings, and they may be used appropriately when working with healthy animals. Cloth or leather gloves provide physical protection for the hands, but do not provide an impermeable barrier to pathogens or fluids. Disposable gloves may be used in healthcare or hospital settings and are often used in nonhuman primate care; they should be used when zoo personnel are knowingly handling ill animals, materials contaminated by waste, or bodily fluids from ill animals. Disposable gloves should be removed and discarded between animal contacts and after contact with animal waste or bodily fluids. Care should be taken to avoid contact with the outer surface of gloves during removal. Hand washing or hand hygiene is recommended following removal of gloves.

Facial protection (face shields and goggles). Facial protection prevents droplet or aerosol transmission of infectious particles

> **Good Practice Tip:** Hand washing is the single most important disease prevention tool at home or at work and is an essential component of any animal contact program. Wash your hands or use a hand sanitizer product after coming into contact with any animal, animal products, or material contaminated by animal waste.

onto the mucous membranes of the eyes, nose, and mouth. The two most common pieces of equipment used for facial protection are goggles and full-face shields. To be effective, goggles must be combined with a mask or other form of respiratory protection to prevent nose and mouth exposure. Facial protection should be used in any situation where sprays or splashes may be generated (Siegel et al. 2007). The most common scenario in the zoo environment where facial protection should be used is when zoo personnel use hoses to clean areas contaminated by animal waste or bodily fluids. Splashes or sprays may also be generated in a hospital or necropsy environment, and facial protection may be appropriate.

Respiratory protection. Respiratory protection is designed to protect the wearer from exposure to disease agents (primarily viruses and bacteria) through inhalation of small particles. The need for it is limited in most animal care settings (Elchos et al. 2008, 415–32). Situations in which respiratory protection may be required include cases of bovine or human tuberculosis, chlamydiosis in birds, and human exposure to rodent fecal material in areas where hantavirus is a concern. Standard surgical masks are not designed to filter small particles, and thus do not provide adequate protection from small particle disease agents (Elchos et al. 2008, 415–32). Adequate respiratory protection requires the use of NIOSH-certified N-95 respirators designed to filter at least 95% of airborne particles (Elchos et al. 2008, 415–32). These respirators are disposable and relatively inexpensive, although their proper use requires fit testing to ensure a proper seal with the wearer's face. Fit-testing programs should be established in institutions were respiratory protection is used for disease prevention per local jurisdictional requirements (NIOSH [www.cdc.gov/niosh/topics/respirators/] and OSHA [www.osha.gov/SLTC/respiratoryprotection/index.html] standards).

Protective outerwear (coveralls, aprons, and footwear). Coveralls are designed to protect the underlying clothing from contamination by animal materials such as feces, hair, or contaminated bedding. They are not fluid-resistant and should not be used where splashing or soaking is possible. In such situations, nonpermeable aprons may provide better protection. Footwear should provide protection from traumatic injury and be impermeable to fluids to facilitate cleaning. Traditional work boots or shoes with rubber soles may be adequate in environments where contamination by fecal material, fluids, or infectious materials is unlikely. Rubber boots should be used in environments containing ill animals, in heavily contaminated areas, or in quarantine situations. Boots should be cleaned with a brush, water, and appropriate disinfectant solution in between animal areas or

after contamination. Isolation and quarantine environments may require keepers to completely change their footwear, and sometimes their clothing, before entering other areas of the zoo. Protective outwear (coveralls or footwear) worn at work in the zoo should not be worn outside the work area or laundered at home (Belkin 2001, 58–64).

Cleaning and disinfection. Work with animals, by its very nature, occurs in a contaminated work environment. Steps taken to minimize fecal and urine contamination will reduce the opportunities for contact with disease agents. Animal holding areas and exhibits should be kept clean. Appropriate cleaning techniques, tools, and chemicals should be used. Particular attention should be paid to areas holding ill animals; if possible, they should be constructed with nonpermeable materials to allow disinfection. Heavily contaminated environments should be cleaned prior to disinfection, as organic materials deactivate many disinfectant chemicals (Dwyer 2004, 531–42). Most disinfectants have established contact times and dilution rates that improve their effectiveness. Chemicals labeled as "disinfectants" are best used in hospital and quarantine environments, disease outbreak situations, ill animal holding areas, or in special circumstances such as aviary brooder rooms or nurseries.

Animal injury prevention. Traumatic injury is a type of disease. Animals are inherently unpredictable, and this increases the possibility of traumatic workplace injury due to bites, scratches, kicking, or crushing. A study of zoo veterinarians revealed that animal-related injuries were the most common occupational injuries suffered by that group (Hill et al. 1998, 484–88). Zoo personnel who work directly with animals should be trained to work safely around the various species or to use appropriate protective facilities and tools such as gloves, caging, shifts, and restraint chutes. Policies and procedures should be established to minimize the chance of injury, especially in the case of dangerous animals. In the hospital or zoo field environment, proper use of sedation and anesthesia can make specific tasks safer for human handlers. Zoo personnel who do not work directly with animals should be explicitly told to avoid any and all animal contact. Keepers and other animal husbandry staff possess skills that allow them to work safely around specific animals. People lacking such training can be severely injured or killed as a result of seemingly innocent actions or minor mistakes around animals.

DISEASE PREVENTION FOR PUBLIC AND EMPLOYEE SAFETY

Prevention of zoonotic disease exposure within the workplace or workforce represents only a portion of the possible exposure concerns in the zoo environment. A second component of zoonotic disease prevention is the prevention of disease transmission to zoo visitors and to the public. AZA accreditation standards state: "Responsible zoos should make reasonable attempts to limit the risk of spread of disease from the animals in their care to their employees and to the general public" (Miller and Janssen 1997, 96–99). Disease prevention with regard to the public may take a variety of forms, including exhibit design considerations, animal contact policies, and veterinary surveillance programs. These same measures also contribute to maintaining a safe and healthy work environment for zoo personnel.

EXHIBIT DESIGN CONSIDERATIONS

The design of an animal exhibit is the first step in preventing transmission of disease from zoo animals to the visiting public. Most zoonotic diseases require direct or close contact for transmission. The majority of zoonotic disease incidents involving the public take place in the petting zoo or education environment, where people may come into direct contact with animals or their waste (Bender and Shulman 2004, 1105–9; Blackmore 2009, 1–15). Elsewhere, disease transmission between animals and the public can be minimized by eliminating the chance of direct contact, contact with bodily waste or fluids, and droplet contact or aerosolization. Double barriers, impermeable barriers, or distance separation should all be considered, depending on the animal species in the exhibit and on specific disease concerns. Double barriers may eliminate the possibility of direct contact with an animal, but they do little to prevent droplet transmission of a disease like tuberculosis. They may also fail to prevent zoo visitors from exposure to infectious material when nonhuman primates throw their feces or elephants spray water from their trunks. Impermeable barriers may be more appropriate in instances where droplet transmission is a concern. Distance separation is a potential solution in exhibits where impermeable barriers are not practical, as with elephants or large hoofstock.

ANIMAL CONTACT POLICIES

Petting zoos and educational programs that involve the touching of animals have been responsible for the majority of zoonotic disease outbreaks among humans visiting public animal exhibits (Bender and Shulman 2004, 1105–9; Blackmore 2009, 1–15). A major focus of a zoo or aquarium's zoonotic disease prevention program should be areas where the public may come into contact with animals or animal materials. Disease prevention in animal contact settings involves proper design of the zoo and its exhibits, proper signage, control of entry and exit points, staff training, facility maintenance, and availability of hand-washing facilities. Table 46.1 summarizes the recommendations of the US National Association of State Public Health Veterinarians measures

TABLE 46.1. Recommendations to prevent disease associated with animals in public settings (from Blackmore 2009)

Design: Animal contact venues should be divided into "nonanimal areas," "transition areas," and "animal areas."

Nonanimal areas (areas where animals are not permitted, with the exception of service animals)
  Prepare, serve, and consume food and beverages only in these areas.
  Provide hand-washing facilities and display hand-washing signs where food or beverages are served.

Transition areas (entry and exit points to animal contact areas)
  Post signs or otherwise notify visitors that they are entering an animal area and that risks are associated with animal contact.
  Instruct visitors not to eat, drink, smoke, place their hands in their mouths, or use bottles or pacifiers while in the animal area.
  Do not allow strollers and related items (e.g., wagons and diaper bags) in areas where direct animal contact is encouraged. Establish storage or holding areas for these items.
  Control visitor traffic to prevent overcrowding.
  Design exits to facilitate hand washing.
  Post signs or otherwise instruct visitors to wash their hands when leaving the animal area.
  Provide accessible hand-washing stations for all visitors, including children and persons with disabilities.
  Position staff members near exits to encourage compliance with hand washing.

Animal areas (areas where animal contact is possible or encouraged)
  Remove ill or suspected ill animals from animal contact area.
  Do not allow food and beverages in animal areas.
  Do not allow toys, pacifiers, spill-proof cups, baby bottles, or strollers in animal areas.
  Prohibit smoking in animal areas.
  Supervise children closely to discourage hand-to-mouth activities (e.g., nail-biting and thumb-sucking), contact with manure, and contact with soiled bedding.
  Do not allow children to sit or play on the ground in animal areas. If their hands become soiled, supervise hand washing.
  Ensure that animal feed and water are not accessible to the public.
  Allow feeding only when contact with animals is controlled (e.g., with barriers).
  Do not provide animal feed in containers that can be eaten by humans (e.g., ice cream cones); this is to decrease the risk of children eating food that has come into contact with animals.
  Assign trained staff members to encourage appropriate human-animal interactions, identify and remove potential risks for patrons (e.g., by promptly cleaning up wastes), and process reports of injuries and exposures.
  Promptly remove manure and soiled animal bedding from animal areas.
  Store animal waste and specific tools for waste removal (e.g., shovels and pitchforks) in designated areas that are restricted from public access.
  Avoid transporting manure and soiled bedding through nonanimal areas or transition areas. If this is unavoidable, take precautions to prevent spillage.
  Where feasible, disinfect animal areas (e.g., flooring and railings) at least once daily.
  Provide adequate ventilation for both animals and humans.
  For birds in bird encounter exhibits, refer to the psittacosis compendium for recommendations regarding disease screening.
  If visitors to aquatic touch tank exhibits have open wounds, advise them not to participate. Provide hand-washing stations.
  When using animals or animal products (e.g., animal pelts, animal waste, or owl pellets) for educational purposes, use them only in designated animal areas.
  Do not bring animals or animal products into cafeterias or other food-consumption areas.

to prevent disease associated with animals in public settings (Blackmore 2009, 1–15).

The emphasis of a disease prevention program in an animal contact environment should be to reduce the chances of fecal-oral transmission of pathogens. Hand washing is the single most important prevention tool in this regard. Visitors should be encouraged by signage and zoo personnel to use hand washing stations, which should be available in sufficient numbers to handle the human traffic flow in animal contact areas. If soap and water are not available, alternative products such as alcohol-based hand sanitizers may be used. Eating and drinking should be forbidden in animal contact areas, which should also be kept as clean and free of fecal material as possible. Exhibit surfaces, barriers, and floors should be designed for easy cleaning and adequate drainage, which is important to reduce pooling of contaminated water or other fluids.

Animals participating in animal contact programs should be monitored daily for signs of illness or disease. Ill or suspected ill animals should be removed from such programs, especially if they experience diarrhea or have spontaneous abortions, and they should be kept out until a veterinarian deems them healthy. When animals are giving birth, the public should not come into contact with any fluids or materials (i.e., the placenta) associated with the birthing process. The birthing area should be thoroughly cleaned and disinfected before it is reopened to the public. Finally, dangerous animals (e.g., nonhuman primates or large carnivores), venomous animals (e.g., snakes), and animals that can be reservoirs of rabies virus (e.g., bats, raccoons, skunks) should not be used in animal contact programs.

VETERINARY SURVEILLANCE PROGRAMS

By definition, the source of a zoonotic disease is the animal reservoir; hence the importance of veterinary surveillance to detect zoonotic diseases and prevent exposure of humans. Active and passive surveillance for a variety of zoonotic diseases can further reduce the risks associated with working with, handling, or being near animals.

Active surveillance involves sampling of seemingly healthy animals for specific diseases. It helps to detect diseases where animals may be subclinical or asymptomatic. It can take the form of routine preventative health exams, specifically scheduled medical screening, or opportunistic fecal sampling. Examples of active surveillance programs in the zoo might include sampling of parrots (psittacines) for the avian chlamydia pathogen (*Chlamydophila psittaci*), yearly trunk washes of elephants for tuberculosis (*Mycobacterium tuberculosis*) screening, and routine fecal cultures of reptiles for the presence of *Salmonella* sp. bacteria. Routine preventative health examinations increase opportunities for active disease surveillance that can target a wide variety of zoonotic and nonzoonotic pathogens.

Diagnostic sampling that is done as a result of illness or disease falls into the category of passive surveillance. Passive surveillance involves sampling of overtly ill animals for specific diseases. To ensure the protection of hospital staff, zookeepers, and the public, diagnostic sampling of diseased animals should take into account the possibility of zoonotic infections. Fecal screening of animals with diarrhea for enteric pathogens can identify possible zoonotic bacterial pathogens such as *Salmonella* sp., *Shigella* sp., and *Campylobacter* sp., all of which can cause severe diarrhea and illness in humans. Similarly, fecal analysis may identify infections of *Cryptosporidium*, *Giardia*, and myriad other fecal-oral pathogens. These are just a few examples of the many diagnostic tests a zoo veterinarian may use for surveillance.

Necropsy examination is a component of a passive disease surveillance program, and it is an important aspect of zoonotic disease diagnosis and prevention. US AZA accreditation standards state that "deceased animals should be necropsied to determine the cause of death" (AZA 2010). Necropsy examination provides the opportunity for detailed diagnostic investigation of a variety of pathogens that can prove beneficial for animals remaining in the collection, their human caretakers, and the general public. Routine examination of birds dying in a zoo environment led to the diagnosis of one of the most serious human and animal disease outbreaks in recent memory, the West Nile virus outbreak in North America. There are countless other examples of the value of necropsy examination of zoo animals.

PEST AND WILDLIFE MANAGEMENT PROGRAMS

Management of pest rodents, wildlife, and insects can play an important role in disease prevention by reducing the reservoirs for disease agents on zoo grounds and by reducing the vectors that transmit those disease agents. Rodents act as reservoirs for a variety of zoonotic diseases and may deposit disease agents into the environment through fecal or urine contamination. Hantavirus is a serious viral disease in humans that is acquired through inhalation of or fecal-oral contact with rat feces or urine (Hugh-Jones 2000). Rodents may also be reservoirs for other diseases such as lymphocytic choriomeningitis virus, tularemia, bubonic plague, and leptospirosis.

Insect control programs on zoo grounds are typically targeted at reducing mosquitoes. Reduction of mosquito numbers at the zoo has the dual benefit of improved comfort for guests and zoo personnel as well as decreased opportunities for vector-borne diseases such as West Nile virus, and Eastern equine encephalitis. Flea surveillance and control, especially in areas where plague or tularemia is common, can reduce potential for exposure to these serious diseases (Dennis et al. 1999). Finally, general insect control is a component of a clean, well-maintained environment.

Management of wildlife and feral animals (e.g., cats and dogs) may be necessary on zoo grounds to reduce the potential of zoonotic disease introduction, particularly rabies virus. Wildlife and feral animal management typically involves the trapping and removal of mammals capable of transmitting the rabies virus, such as raccoons, skunks, foxes, coyotes, and bats. It should be done in accordance with local laws concerning these species.

SPECIFIC DISEASE EXAMPLES

The focus of this chapter so far has been to discuss the knowledge, policies, and techniques that enable zoo personnel to avoid and prevent zoonotic disease. It is beyond of the scope of this chapter to discuss specific details of the more than 200 known zoonotic diseases. However, a summary list organized by general taxonomic group is provided (tables 46.2–46.6), and a few diseases that bear individual discussion due to their severe consequences (rabies), their common nature (enteric pathogens, cutaneous fungal infections), or both (tuberculosis, psittacosis).

RABIES

Rabies is a fatal zoonotic disease with worldwide distribution. The virus infects the brains of animals and humans causing severe inflammation, neurologic deficits, and eventually death. An excellent review of the human and animal aspects of rabies management in the zoo environment is available (Rupprecht 1999, 136–46). Reservoirs for rabies virus in the environment include species of carnivore (order Carnivora) and bats (order Chiroptera). Canids (*Canis* sp.), mongooses (viverrids), raccoons (*Procyon lotor*), skunks (*Mephitis* sp.), and many species of bats are considered high-risk rabies vectors, but all mammals are susceptible to infection (Niezgoda et al. 2002, 163–207). Any unvaccinated mammal exhibiting neurologic disease or abnormal behavior should be considered a possible rabies risk. Rabies is a highly variable disease and can resemble a wide variety of common illnesses. Caution should be exercised around any unvaccinated mammal displaying any signs of illness.

Humans are infected with rabies almost always through the bites of infected animals. Rare cases of ingestion, mucosal, and aerosol exposure to infected saliva or cerebrospinal fluid have also been documented (Niezgoda et al. 2002, 163–207). Prevention of rabies in the zoo environment involves a four-pronged approach: pre-exposure vaccination of at-risk personnel; label and off-label vaccination of mammals; monitoring, quarantine, and/or testing of mammals involved in human bite incidents; and exhibit/facility/protocol design that reduces or prevents exposure of humans and zoo animals to wildlife. There is no treatment for rabies infection in animals.

Although it is common practice to vaccinate mammals held in zoological parks using off-label rabies vaccine, no rabies vaccine is approved for use in wild, exotic, or hybrid animals. The US Centers for Disease Control and Prevention (CDC) policy on rabies in animals maintained in exhibits and in zoological parks is as follows:

> Captive mammals that are not completely excluded from all contact with rabies vectors can become infected. Moreover, wild animals might be incubating rabies when initially captured; therefore, wild-caught animals susceptible to rabies should be quarantined for a minimum of 6 months. Employees who work with animals at such facilities should receive pre-exposure rabies vaccination. The use of pre- or post-exposure rabies vaccinations for handlers who work with animals at such facilities might reduce the need for euthanasia of captive animals that expose handlers. Carnivores and bats should be housed in a manner that precludes direct contact with the public (Manning et al. 2008, 1–26).

Pre-exposure vaccination for zoo personnel has been discussed above. Rabies vaccines licensed for use in domestic animals are considered safe and effective in dogs, cats, horses, ferrets, and livestock. Although these vaccines are not licensed for exotic animals, a number of factors combine to suggest that they should provide protection for exotic animals as well (Rupprecht 1999, 136–46). Any incident in which a human is bitten by a zoo mammal is serious and should be reported to a physician or local health department official. The physician or health department, in consultation with the zoo's veterinary staff, can determine the appropriate course of action for the human and the animal. The need for quarantine and/or euthanasia and rabies testing of an animal is dependent on a number of factors summarized by Rupprecht (Rupprecht 1999, 136–46). Lastly, exhibits and other animal holding facilities should be designed to reduce or eliminate public contact with wildlife, especially species at high risk of rabies. Zoo animals not completely excluded from rabies virus vectors can become infected. Likewise, zoo personnel should be trained to avoid high-risk rabies vectors. Protocols for safe handling, medical treatment, euthanasia, necropsy, and disposal of high-risk rabies vectors should be developed, and only specifically trained personnel should be given these jobs.

ENTERIC PATHOGENS

Enteric pathogens include a wide variety of viral, bacterial, protozoal, and parasitic pathogens that infect the gastrointestinal tract. The most common zoonotic enteric pathogens that occur in association with animal exhibits are *E. coli* 0157:H7, *Cryptosporidium* sp., and *Salmonella* sp. (LeJeune and Davis 2004, 1440–45). These pathogens are common, and may be found in most livestock and animal collections in North America and other parts of the world (LeJeune and Davis 2004, 1440–45). Many more enteric pathogens exist in the wide variety of animal species typically encountered in the zoo environment (tables 46.2–46.6). Enteric pathogens are transmitted by fecal-oral contact, and most of the animal contact policies discussed above are designed specifically to prevent accidental fecal-oral contamination. Regardless of the pathogen, enteric infections are best prevented by avoiding contact with feces and by stringently enforcing policies of hand washing and hand hygiene.

E. coli 0157:H7. The bacterial enteric pathogen *E. coli* 0157:H7 can be acquired from infected food and water sources or from contact with animals. Cattle are the primary reservoir for this disease, although the bacteria can be isolated from a large variety of animal species (Hancock et al. 1998, 11–19). *E. coli* 0157:H7 is one of the most important zoonotic pathogens due to its prevalence, the low dose required for infection, and the severity of disease it causes in infected humans (LeJeune and Davis 2004, 1440–45). Infected humans may exhibit diarrhea, severe colitis, and a severe kidney disease called hemolytic uremic syndrome (HUS; Besser et al. 1999, 355–67). The pathogen has often been associated with disease outbreaks in humans who have visited petting zoos and agricultural fairs (LeJeune and Davis 2004, 1440–45). The common risk factors for most outbreaks have been animal contact, hand-to-mouth activities (eating, drinking), and suboptimal hand washing policies (LeJeune and Davis 2004, 1440–45). The screening of healthy animals and prophylactic treatment of animals with antibiotics are not recommended as control techniques for this disease (Elchos 2008, 415–32). Animals showing signs of disease should be removed from possibility of contact with the public, and keepers should take extra precautions around them. However, the bulk of prevention techniques should focus on hand hygiene and elimination of fecal-oral contact.

Cryptosporidiosis. Cryptosporidiosis is a worldwide disease of humans and other mammals caused by infections of *Cryptosporidium* sp. (primarily *C. parvum*, a protozoan parasite). Cryptosporidium infections can be acquired through contaminated food and water sources, through contact with contaminated recreational water sources, and through contact with infected animal feces (Fayer et al. 2000, 1305–22; LeJeune and Davis 2004, 1440–45). In both animals and humans, cryptosporidiosis typically causes profuse watery diarrhea that is sometimes accompanied by nausea and vomiting. The disease is self-limiting in most cases, with the exception being immunocompromised people, in whom the disease can be fatal. Although most cases of cryptosporidiosis occur in association with contaminated food or water, *Cryptosporidium parvum* has been associated with human disease outbreaks at farms and petting zoos (Fayer et al. 2000, 1305–22; LeJeune and Davis 2004, 1440–45). As with *E. coli* 0157:H7, the risk factors for cryptosporidiosis in animal settings include contact with young sheep or calves, hand-to-mouth activities, and suboptimal hand-washing policies (LeJeune and Davis 2004, 1440–45). Treatment and eradication of *Cryptosporidium* in animals and the environment is not possible, so prevention techniques should focus on hand washing and hand hygiene.

Salmonellosis. *Salmonella* ssp. are bacterial enteric pathogens common in domestic and wild animal populations around the world (LeJeune and Davis 2004, 1440–45). Most *Salmonella* spp. infections in humans result from ingestion of contaminated food items (Schutz et al. 1999). However, salmonella outbreaks have also been associated with public contact at animal exhibits, specifically those containing small mammals, reptiles, and birds (LeJeune and Davis 2004, 1440–45). Salmonellosis is characterized by severe, sometimes bloody diarrhea, nausea, cramps, and fever. The disease is self-limiting in most cases, but serious disease and extraintestinal disease can develop even in immunocompetent hosts. As with most diseases, salmonellosis is more severe in immunocompromised individuals. Eradication of *Salmonella* from animals and the environment is impossible, as the organism is ubiquitous and many species can act as asymptomatic carriers of the bacteria (LeJeune and Davis 2004, 1440–45). Animals that are known carriers of *Salmonella* sp. (as determined by active or passive surveillance of their fecal cultures) should be removed from public contact situations. Zoo personnel should practice stringent hand washing and hand hygiene to prevent infection, and take care when cleaning fecal material (including dried feces, which can lead to inhalation of aerosolized material).

CUTANEOUS AND DEEP FUNGAL INFECTIONS

Cutaneous fungal infections are confined primarily to the skin, hair, and nails. They are among the most common zoonotic infections in the animal care environment. A study in zoo veterinarians found that "ringworm" or other superficial fungal infections were the most common zoonotic diseases acquired at work (Hill et al. 1998, 484–88). "Ringworm" or "tinea" is caused by fungi called dermatophytes, which cause the disease known as dermatophytosis. Dermatophytes include fungi in the genus *Microsporum*, *Trichophyton*, and *Epidermophyton*. Dermatophytes which infect humans are further subcategorized as "zoophilic" (Acha and Szyfres 2001, 283–99). Animals are the reservoir of zoophilic dermatophytes, and these fungi can be found associated with the skin or haircoat of a wide variety of species. The most common animals involved in zoonotic dermatophytosis are cats, dogs, horses, cattle, and rodents (Acha and Szyfres 2001, 283–99). Infection of humans is by direct skin contact with infected animals. Control of zoonotic dermatophytes is best accomplished through avoidance of infected animals or control of infection in animals by medical treatment.

Deep fungal infections involve the deep tissues of the body, such as the subcutaneous tissues, the muscles and bones, the internal organs, and even the brain. The reservoir for these fungal pathogens is the environment (soil, water, air), so they are not true zoonoses; rather, they are diseases common to humans and animals. Examples of these diseases include aspergillosis, blastomycosis, candidiasis, coccidiomycosis, cryptococcosis, histoplasmosis, mycetoma, protothecosis, rhinosporidiosis, and zygomycosis (Acha and Szyfres 2001, 283–99). Animals play no role in the transmission of these diseases to humans, with two exceptions. Animals may deposit or concentrate the infectious organisms of cryptococcosis and histoplasmosis in their fecal material. Pigeon fecal material is an important source of exposure to *Cryptococcus neoformans*, the cause of cryptococcosis (McDonough et al. 1966, 1119–23). Infection can be due to direct contact with fecal material or contact with soil enriched by pigeon feces. Similarly, bird and bat feces or soil enriched by it can be a source of human infection with *Histoplasma capsulatum*, the cause of histoplasmosis (Ajello 1964, 266–70; Lyon et al. 2004, 438–42). Prevention of these two diseases in an animal care setting involves avoidance and/or removal of large concentrations of fecal material from the environment. Cryptococcosus and histoplasmosis, as well as the other deep fungal infections, can be particularly dangerous for immunocompromised individuals (Lyon et al. 2004, 438–42).

TUBERCULOSIS AND NONTUBERCULOUS MYCOBACTERIA

Tuberculosis (TB) is a serious and potentially fatal respiratory disease caused by bacteria in a group known as the *Mycobacterium tuberculosis* complex. The human disease is most commonly caused by infection with human tuberculosis (*Mycobacterium tuberculosis*) and bovine tuberculosis (*M. bovis*). Other members of the tuberculosis complex include *M. africanum*, *M. microti*, *M. pinnipedii*, and *M. canetti*. TB is transmitted directly from one human host to another and can be transmitted directly from animals to a human host through droplet or aerosol exposure. It is a chronic disease that typically infects the lungs, causing severe pneumonia. There are numerous examples of zoonotic transmission of tuberculosis-complex bacteria to humans in zoo and aquarium environments. Zoo personnel have been exposed to tuberculosis through contact with nonhuman primates, elephants, hoofstock, and marine mammals (Dalovisio et al. 1992, 568–600; Fanning and Edwards 1991; 1253–55; Kiers et al. 2008, 1469–73; Michalak et al. 1998, 283–87; Stetter et al. 1995, 1618–21; Thompson et al. 1993, 164–67). Humans may also be a source of infection for animals with which they come into contact (Acha and Szyfres 2001, 283–99). Tuberculosis is best prevented through routine screening of susceptible animals and their human caretakers. Treatment of infected animals is difficult and involves expensive long-term antibiotic therapy. Humans that must work with infected animals (in a hospital or necropsy environment) should adhere to strict PPE policies for respiratory protection, including the use of NIOSH-certified N95 respirators.

Nontuberculous (non-TB) mycobacteria are bacteria in the genus *Mycobacterium* that rarely cause disease in healthy, immunocompetent humans. Emergence of the human immunodeficiency virus (HIV) resulted in a surge of human cases of non-TB *Mycobacterium* infections in the mid to late 1980s (Cleary and Batsakis 1995, 830–33). Examples of non-TB mycobacteria include *Mycobacterium avium*, *M. kansasii*, *M. fortuitum*, *M. marinum*, and many others. Non-TB mycobacteria are ubiquitous in the environment and have been found in a wide variety of mammals, birds, reptiles, amphibians, and fish (Lamberski 1999, 146–50). In humans the disease is typically acquired through contact with infective organisms in the environment (plants, soil, food, water). It can involve skin infections, gastrointestinal infections, and widespread systemic involvement. Zoo personnel may come into contact with non-TB mycobacteria during contact with animals or with soil or water while servicing animal holding areas. However, the risk of infection is low, and most infections are reported in immunocompromised humans (Lamberski 1999, 146–50). Good hygiene practices prevent non-TB mycobacterial infection in healthy, immunocompetent zoo personnel. Other zoo personnel who are immunosuppressed due to HIV infection or any other cause should talk with human health professionals and take additional steps to avoid infection.

PSITTACOSIS

Psittacosis (also known as "parrot fever," ornithosis, or avian chlamidiosis) is a disease acquired from birds infected by the bacterium *Chlamydophila psittaci*. Psittacosis in humans is characterized by flu-like disease, fever, conjunctivitis, and often pneumonia (Harkinezhad 2009, 68–77). Infection is acquired by droplet transmission or inhalation of dried secretions (e.g., respiratory secretions or feces) from infected birds. Although all birds are susceptible to infection, psittacines and poultry are most often involved in transmission to humans. Bird owners and pet shop employees are considered the highest risk group for infection, and psittacosis is the second most

common occupationally acquired zoonosis among zoo veterinarians (Hill et al. 1998, 484–88; Weese et al. 2002, 631–36). In 2006 the US National Association of State Public Health Veterinarians (www.nasphv.org) published a compendium of measures to control *C. psittaci* infections among humans and pet birds, including prevention and control recommendations such as (1) protecting persons at risk, (2) maintaining accurate records of all bird-related transactions for at least one year to aid in identifying the sources of infected birds and potentially exposed persons, (3) avoiding the purchase or sale of birds that have signs consistent with avian chlamydiosis, (4) isolation of newly acquired, ill, or exposed birds, (5) testing of birds before they are boarded or sold on consignment, (6) routine screening for antichlamydial antibodies of groups of birds that will have frequent public contact, (7) preventive husbandry, (8) control of the spread of the infection, and (9) disinfection measures (Harkinezhad 2009, 484–88). Hand washing and PPE are also important components of psittacosis prevention.

DISEASES BY TAXONOMIC GROUP

Tables 46.2–46.6 list zoonotic diseases by major taxonomic group (nonhuman primates, miscellaneous mammals, birds, reptiles and amphibians, and aquatic animals). These are not meant to be exhaustive lists of the more than 200 known zoonotic diseases. The tables are meant to be a quick reference guide for some of the more common, dangerous, or well known zoonotic pathogens. They may be viewed as a starting point for a zoo professional interested in zoonotic diseases.

SUMMARY

Zoos, aquariums, and the zookeeping profession, by their very nature, bring humans and animals into close contact every day. This close contact between humans, domestic animals, exotic animals, and wildlife increases the chances for zoonotic disease transmission. Personnel who work in the zoo environment should understand and respect zoonotic diseases but not fear them. The number of zoonotic diseases and the scope of associated information are immense and impossible for all but a few highly specialized health professionals to fully grasp. However, it is not necessary to be an infectious disease expert to understand and apply zoonotic disease biology. We have demonstrated in this chapter that education, actions, policies, and prevention tools make it very possible to work safely with animals and their diseases. Zoos should focus on education and prevention for their employees and visitors to ensure a safe environment. The immense educational, conservation, and emotional value of zoos and aquariums should not be overshadowed by disease concerns.

Ten golden rules of zoonotic disease biology and prevention:

1. Keepers should avoid fecal contamination of hands, clothing, and protective equipment, as this is the most common way in which zoonotic diseases are transmitted.

TABLE 46.2. Selected zoonotic diseases of nonhuman primates

Disease/agent	Transmission	Reservoir	Human symptoms	Risk of infection/disease	Prevention
Herpes B / Macacine herpesvirus 1	B/S, MM	Macaque species	Systemic	High/High	PPE, A
Hepatitis A	FO	Any	Hepatic	Low/Low	HW, PPE
Hepatitis C	B/S, MM	Chimpanzees	Hepatic	Low/Mod	PPE
Hepatitis E	FO	Any (experimental)	Hepatic	Low/Low	HW, PPE
Hemorrhagic fevers (Ebola, Marburg, etc.)	DC, unk	Unknown, fruit bats	Systemic	Low/High	A, PPE
Monkey pox	B/S, MM	Old World primates	Cutaneous	Low/Mod	PPE, A
Campylobacteriosis	FO	Any	Gastrointestinal	Mod/Mod	HW, PPE
Mycobacterium tuberculosis	A	Old World primates	Respiratory	Low/High	PPE, A
Salmonellosis	FO, MM	Any	Gastrointestinal	Mod/Mod	HW, PPE
Shigellosis	FO	Large apes, gorillas, etc.	Gastrointestinal	Mod/Mod	HW, PPE
Tularemia	I, B/S, FO, MM, C	New World primates	Cutaneous, systemic	Low/Mod	HW, PPE, PC
Yersinia enterocolitica	FO, C, DC	Any	Gastrointestinal	Low/Mod	HW, PPE
Blastocystis sp.	FO	Any	Gastrointestinal	Mod/Low	HW
Entamoeba histolytica	FO	Any	Gastrointestinal, hepatic	Low/Low	HW, PPE
Filariasis	V	Old World primates, apes	Systemic, respiratory	Low/Mod	PC, IR
Strongyloidiasis	FO, DC	Old World primates	Gastrointestinal, cutaneous, respiratory	Low/Low	HW, PPE

Transmission	Prevention
A = aerosol, respiratory, or droplet contact	V = vaccine available
FO = fecal or oral	PPE = personal protective equipment
DC = direct skin contact with contaminated materials	HW = hand washing
C = consumption of undercooked meat or milk products	A = avoidance
B/S = bite, scratch, needle stick, or skin puncture	PC = pest control
MM = mucous membrane contact	IR = insect repellents
I = insect vector	
CS = contaminated soil, bedding, or other material	

Risk of infection	Risk of disease
High = infection is common or easily acquired without proper prevention	High = consequences of infection can be severe/fatal
Mod (moderate) = infection can be acquired without proper care in certain circumstances	Mod (moderate) = disease is serious but treatable
Low = infection is rare or difficult to acquire	Low = disease is mild or easily treatable

2. Disease surveillance by veterinary professionals is an important part of a zoonotic disease prevention program.

3. Keepers should be appropriately vaccinated against high-risk diseases and trained about disease prevention techniques at their jobs.

4. Hand washing is the single most important disease prevention tool available to keepers.

5. Protective clothing and equipment can be an important disease prevention tool in some circumstances.

6. Keepers who are immunocompromised, or who have underlying health issues or any other risk factors, should seek medical advice about how to work safely around animals.

7. Reducing exposure to insects such as mosquitoes, ticks, and fleas reduces opportunity for exposure to vector-borne diseases.

8. One can avoid animal accidents and injuries by demonstrating care around all animals and by only working with animals for which one has been specifically trained.

9. Rabies is a deadly disease; great care should be taken to avoid contact with ill mammals, particularly those that are not vaccinated against rabies

10. Human health professionals and health departments can be a great source of information about zoonotic diseases and prevention techniques.

TABLE 46.3. Selected zoonotic diseases of miscellaneous mammals

Disease/agent	Transmission	Reservoir	Human symptoms	Risk of infection/disease	Prevention
Bovine papularstomatitis	DC, B/S	Cattle	Cutaneous	Mod/Low	A, PPE
Contagious ecthyma	DC	Ruminants	Cutaneous	Low/Low	HW, PPE, A
Hantavirus	A, FO, B/S	Rats, rodents	Renal, respiratory	Mod/High	PC, PPE
Influenza A	A, MM	Swine	Respiratory	Low/Mod	HW, PPE, V
Lymphocytic choriomeningitis	B/S ,MM, FO	Mice, rodents	Systemic	Mod/Mod	PC, PPE
Monkey pox, pox virus	DC, B/S	Prairie dogs	Cutaneous, systemic	Low/Low	A, PPE
Nipah/hendra virus	DC, A, C	Fruit bats, swine, horses	Neurologic, respiratory	Low/High	A, PPE
Rabies	B/S	Any mammal	Neurologic	High/High	V, A
Rotavirus	FO	Domestic ruminants, swine	Gastrointestinal	Low/Low	HW
SARS (coronavirus)	A, DC	Civets	Respiratory	Low/Mod	HW, PPE, A
Anthrax (*Bacillus anthracis*)	DC, A, C	Any, ruminants	Cutaneous, respiratory, gastrointestinal	Mod/High	A, PPE, V
Brucellosis	C, DC, A	Ruminants, swine, others	Systemic	Low/Mod	HW, PPE, A
Chlamydiophila sp.	A, DC	Many, ruminants	Respiratory	Mod/Mod	HW, PPE, A
Q fever (*Coxiella* sp.)	A, C	Sheep and goats	Systemic	Mod/Mod	PPE
E. coli (0157:H7)	FO, C	Ruminants	Gastrointestinal, renal	ModMod	HW, PPE
Erysipelas(oid)	B/S, FO, C	Swine	Cutaneous, gastrointestinal	Low/Low	HW, PPE, A
Leptospirosis	B/S, MM, A DC, C	Many mammals, rodents	Systemic, hepatic	Mod/Mod	PPE, A, PC
Listeriosis	C, DC, FO	Sheep, goats, cattle, other	Neonatal/fetal, neurologic	Mod/High	HW, PPE
Methicillin-resistant staphylococcus (MRSA)	DC	Any	Cutaneous	Mod/High	HW,PPE,A
Lyme disease (Borreliosis)	I	Deer, ticks	Systemic	Mod/Mod	A, PC, IR
Mycobacterium TB-complex (*tuberculosis, bovis, pinnipedi microti, africanum, canetti*)	A, C	Hoofed stock, pinnipeds	Respiratory	Low/High	PPE, A
Pasteurellosis	B/S, A	Any mammal, cats	Cutaneous, respiratory	Low/Low	HW, PPE
Salmonellosis	FO	Any mammal, carnivores	Gastrointestinal	High/Mod	HW, PPE
Streptobacillus and spirillium (rat bite fever)	B/S	Rats, rodents	Cutaneous, systemic	Low/Low	
Tularemia	I, B/S, FO, MM, C	Rabbits, small mammals, ticks, fleas	Cutaneous, systemic	Low/Mod	HW, PPE, PC
Typhus	I, DC	Rodents, small mammals, fleas, ticks, lice, mites	Systemic	Low/Low	A, PC
Plague (*Yersinia pestis*)	I, B/S	Rodents, prairie dogs, fleas	Cutaneous, systemic	Low/High	PPE, A, PC
Yersinia enterocolitica	FO, C, DC	Swine, other mammals	Gastrointestinal	Low/Mod	HW, PPE
Balantidium coli/suis	FO	Swine	Gastrointestinal, hepatic, respiratory	Low/Low	HW
Blastocystis sp.	FO	Any	Gastrointestinal	Low/Low	HW
Cryptosporidium sp.	FO	Ruminants	Gastrointestinal	Mod/Low	HW
Entamoeba polecki	FO	Swine	Gastrointestinal, hepatic	Low/Low	HW
Giardia sp.	FO	Any	Gastrointestinal	Mod/Low	HW
Leishmania donovani	I	Canids, small mammals	Cutaneous, systemic	Low/Mod	PC, A
Toxoplasma gondii	C, FO	Felids	Neonatal/fetal	Mod/High	A, PPE, HW
Hookworms (*Ancylostoma* sp.)	FO, DC	Canids	Gastrointestinal	Low/Low	HW
Ascarids	FO, CS	Swine	Gastrointestinal	Low/Low	HW
Baylisascaris procyonis	FO	Raccoons	Systemic	Low/Mod	HW
Echinococcosis	FO	Canids	Hepatic, respiratory, neurologic	Low/High	HW,PPE
Filariasis	V	Canids, cats	Systemic, respiratory	Low/Mod	PC,IR
Larval migrans (nematode larvae)	FO, CS	Canids, felids, others	Cutaneous, systemic	Mod/Mod	HW

TABLE 46.3. continued

Disease/agent	Transmission	Reservoir	Human symptoms	Risk of infection/disease	Prevention
Pentastomiasis	FO	Canids	Systemic	Low/Low	HW
Sarcoptic mange ("scabies")	DC	Canids, felids, pigs, other	Cutaneous	Mod/Low	A,PPE
Strongyloidiasis	FO, DC	Canids	Gastrointestinal, cutaneous, respiratory	Low/Low	HW,PPE
Trichinosis	C	Swine	Muscular	Low/Mod	A
Whipworms (*Trichuris* sp.)	FO	Canids, swine	Gastrointestinal	Low/Low	HW
Dermatophytosis/ringworm	DC	Cats, dogs, cattle, horses, rodents	Cutaneous	Mod/Low	HW,PPE,A
Sporotrichosis	DC, B/S	Cats	Cutaneous	Low/Low	A,PPE

Transmission	Prevention
A = aerosol, respiratory, or droplet contact	V = vaccine available
FO = fecal or oral	PPE = personal protective equipment
DC = direct skin contact with contaminated materials	HW = hand washing
C = consumption of undercooked meat or milk products	A = avoidance
B/S = bite, scratch, needle stick, or skin puncture	PC = pest control
MM = mucous membrane contact	IR = insect repellents
I = insect vector	
CS = contaminated soil, bedding, or other material	

Risk of infection	Risk of disease
High = infection is common or easily acquired without proper prevention	High = consequences of infection can be severe/fatal
Mod (moderate) = infection can be acquired without proper care in certain circumstances	Mod (moderate) = disease is serious but treatable
Low = infection is rare or difficult to acquire	Low = disease is mild or easily treatable

TABLE 46.4. Selected zoonotic diseases of avian species

Disease/agent	Transmission	Reservoir	Human symptoms	Risk of infection/disease	Prevention
Arboviruses (West Nile virus, EEE, others)	I	Any, raptors, crows	Neurologic, systemic	Mod/High	PC, IR
Influenza A	A	Any, poultry, ducks	Respiratory	Low/Mod	PPE
Newcastle's disease	A	Poultry	Respiratory	Low/Low	A
Chlamydiophila psittaci	A	Turkeys, psittacines, pigeons	Respiratory	Mod/Mod	PPE, A
E. coli	FO	Any	Gastrointestinal	Low/Low	HW
Erysipelas(oid)	B/S, FO, C	Any, turkeys	Cutaneous, gastrointestinal	Low/Low	HW, PPE
Listeriosis	C, DC, FO	Any	Neonatal/fetal, neurologic	Low/High	HW, PPE
Pasteurellosis	A	Any, poultry	Respiratory	Low/Low	PPE, HW, A
Salmonellosis	FO	Any, poultry	Gastrointestinal	High/Mod	HW, PPE

Transmission	Prevention
A = aerosol, respiratory, or droplet contact	V = vaccine available
FO = fecal or oral	PPE = personal protective equipment
DC = direct skin contact with contaminated materials	HW = hand washing
C = consumption of undercooked meat or milk products	A = avoidance
B/S = bite, scratch, needle stick, or skin puncture	PC = pest control
MM = mucous membrane contact	IR = insect repellents
I = insect vector	
CS = contaminated soil, bedding, or other material	

Risk of infection	Risk of disease
High = infection is common or easily acquired without proper prevention	High = consequences of infection can be severe/fatal
Mod (moderate) = infection can be acquired without proper care in certain circumstances	Mod (moderate) = disease is serious but treatable
Low = infection is rare or difficult to acquire	Low = disease is mild or easily treatable

TABLE 46.5. Selected zoonotic diseases of reptiles and amphibians

Disease/agent	Transmission	Reservoir	Human symptoms	Risk of infection/disease	Prevention
Salmonellosis	FO	Any reptile	Gastrointestinal	High/Mod	HW, PPE
Yersinia enterocolitica	FO, C, DC	Any reptile	Gastrointestinal	Low/Mod	HW, PPE
Campylobacteriosis	FO	Any reptile	Gastrointestinal	Low/Mod	HW, PPE
Armillifer (pentostomiasis)	DC, FO, C	Any reptile	Gastrointestinal, systemic	Low/Low	HW, PPE

Transmission	Prevention
A = aerosol, respiratory, or droplet contact	V = vaccine available
FO = fecal or oral	PPE = personal protective equipment
DC = direct skin contact with contaminated materials	HW = hand washing
C = consumption of undercooked meat or milk products	A = avoidance
B/S = bite, scratch, needle stick, or skin puncture	PC = pest control
MM = mucous membrane contact	IR = insect repellents
I = insect vector	
CS = contaminated soil, bedding, or other material	

Risk of infection	Risk of disease
High = infection is common or easily acquired without proper prevention	High = consequences of infection can be severe/fatal
Mod (moderate) = infection can be acquired without proper care in certain circumstances	Mod (moderate) = disease is serious but treatable
Low = infection is rare or difficult to acquire	Low ± disease is mild or easily treatable

TABLE 46.6. Selected zoonotic diseases of aquatic animals

Disease/agent	Transmission	Reservoir	Human symptoms	Risk of infection/disease	Prevention
Aeromoniasis	FO, DC, B/S	Fish	Systemic, cutaneous	Low/Low	HW, A
Brucellosis	C, DC, A	Pinnepeds, cetaceans	Systemic	Low/Mod	A, PPE, HW
Erysipelas(oid)	B/S, FO, C	Fish, pinnipeds	Cutaneous	Low/Low	HW, PPE, A
Leptospirosis	B/S, MM, A, DC, C	Sea lions, seals	Systemic, hepatic	Low/Mod	PPE, A, PC
Mycobacterium TB-complex (*bovis, pinnipedi*)	A	Pinnipeds	Respiratory	Low/High	PPE, A
Mycoplasmosis (seal finger)	DC, B/S	Pinnipeds	Cutaneous, musculoskeletal	Low/Low	PPE, A
Streptococcus inea	DC, B/S	Fish	Cutaneous, systemic	Low/Low	HW, PPE
Vibriosis	FO, DC, B/S	Fish	Systemic, cutaneous	Mod/Mod	HW, A

Transmission	Prevention
A = aerosol, respiratory, or droplet contact	V = vaccine available
FO = fecal or oral	PPE = personal protective equipment
DC = direct skin contact with contaminated materials	HW = hand washing
C = consumption of undercooked meat or milk products	A = avoidance
B/S = bite, scratch, needle stick, or skin puncture	PC = pest control
MM = mucous membrane contact	IR = insect repellents
I = insect vector	
CS = contaminated soil, bedding, or other material	

Risk of infection	Risk of disease
High = infection is common or easily acquired without proper prevention	High = consequences of infection can be severe/fatal
Mod (moderate) = infection can be acquired without proper care in certain circumstances	Mod (moderate) = disease is serious but treatable
Low = infection is rare or difficult to acquire	Low = disease is mild or easily treatable

REFERENCES

Acha, P. N., and B. Szyfres. 2001. "Zoonoses and communicable diseases common to man and animals, vol. I: Bacterioses and mycoses." Scientific and Technical Publication no. 580, Pan American Health Organization,

Acha, P. N., and B. Szyfres. 2003. "Zoonoses and communicable diseases common to man and animals, vol. II: Chlamydioses, rickettsioses, and viroses." Scientific and Technical Publication no. 580, Pan American Health Organization.

Ajello, L. 1964. "Relationship of *Histoplasma capsulatum* to avian habitats." *Public Health Reports* 79(3).

Association of Zoos and Aquariums (AZA). 2010. "Accreditation standards and related policies." Accessed April 1 at http://www.aza.org/Accreditation/.

Barr, C. 2008. "Avian Chlamydiosis." In *The Merck Veterinary Manual*, Ninth Edition, C. M. Kahn, ed. Whitehouse Station, NJ: Merck & Company.

Belkin, N. L. 2001. "Home laundering of soiled surgical scrubs: Surgical site infections and the home environment." *American Journal of Infectious Control* 29.

Bender, J. B., and S. A. Shulman. 2004. "Reports of disease outbreaks associated with animal exhibits and availability of recommendations for preventing disease transmission from animals to people in such settings." *Journal of the American Veterinary Medical Association* 224:8.

Besser, R. E, P. M. Griffin, and L. Slutsker. 1999. "*Escherichia coli* 0157:H7 gastroenteritis and the hemolytic uremic syndrome: an emerging infectious disease." *Annual Review of Medicine* 50.

Blackmore, C. 2009. "Compendium of measures to prevent disease associated with animals in public settings." *National Association of State Public Health Veterinarians Morbidity and Mortality Weekly Report* 58:RR–05.

Boyce, J. M., and D. Pittet. 2002. "Guidelines for hand hygiene in health-care settings: Recommendations of the Healthcare Infection Control Practices Advisory Committee and the HICPAC/SHEA/APIC/IDSA Hand Hygiene Task Force." *Morbidity and Mortality Weekly Report, Recommendations and Reports* 51:RR–16.

Centers for Disease Control and Prevention. 2010. "National Center for Zoonotic, Vector-borne, and Enteric Diseases." Accessed 1 April at http://origin.cdc.gov/NCZVED/.

Cleary, K. R., and J. G. Batsakis. 1995. "Mycobacterial disease of the head and neck: Current perspective." *Annals of Otology, Rhinology, and Laryngology* 104.

Dalovisio, J. R., M. Stetter, S. Mikota-Wells. 1992. "Rhinoceros' rhinorrhea: Cause of an outbreak of infection due to airborne *Mycobacterium bovis* in zookeepers." *Clinical Infectious Diseases* 15.

Daszak, P., A. A. Cunningham, and A. D. Hyatt. 2001. "Anthropogenic environmental change and the emergence of infectious disease in wildlife." *Acta Tropica* 78.

Dennis, D. T., K. L. Gage, N. Gratz, J. D. Poland, and E. Tikhomirov. 1999. *Plague Manual: Epidemiology, Distribution, Surveillance, and Control.* Geneva: World Health Organization.

Dwyer, R. M. 2004. "Environmental disinfection to control equine infectious diseases." *Veterinary Clinics of North America Equine Practice* 20.

Elchos, B. L., J. M. Scheftel, B. Cherry, E. E. DeBess, S. G. Hopkins, J. F. Levine, and C. J. Williams. 2008. "Compendium of veterinary standard precautions for zoonotic disease prevention in veterinary personnel: National Association of State Public Health Veterinarians Veterinary Infection Control Committee." *Journal of the American Veterinary Medical Association* 233:3.

Fanning, A. and S. Edwards. 1991. "*Mycobacterium bovis* infection in humans being in contact with elk (*Cervus elaphus*) in Alberta, Canada." *Lancet* 338.

Fayer, R., U. Morgan, and S. J. Upton. 2000. "Epidemiology of Cryptosporidium: Transmission, detection, and identification." *International Journal for Parasitology* 30.

Grant, S. and C. W. Olsen. 1999. "Preventing zoonotic diseases in immunocompromised persons: The role of physicians and veterinarians." *Emergent Infectious Diseases* 5(1).

Hansen, E. 2004. "An update on zoonotic disease concerns for pregnant zoo keepers and expectant fathers." *Animal Keepers Forum* 31:4.

Hancock, D. D., T. E. Besser, D. H. Rice, E. D. Ebel, D. E. Herriott and L. V. Carpenter. 1998. "Multiple sources of *E. coli* 0157 in feedlots and dairy farms in the northwestern USA." *Preventative Veterinary Medicine* 35.

Harkinezhad, T., T. Geens, and D. Vanrompay. 2009. "*Chlamydophila psittaci* infections in birds: A review with emphasis on zoonotic consequences." *Veterinary Microbiology* 135:1–2.

Hill, D. J., R. L. Langley, and W. M. Morrow. 1998. "Occupational injuries and illnesses reports by zoo veterinarians in the United States." *Proceedings of the AAZV and AAWV Joint Conference.*

Hugh-Jones, M. E., W. T. Hubbert, and H. V. Hagstad. 2000. *Zoonoses: Recognition, Control, and Prevention.* Ames, IA: Iowa State University Press, 344.

Janssen D. L., N. Lamberski, T. Donovan, D. E. Sugerman, and G. Dunne. 2009. "Methicillin-resistant *Staphylococcus aureus* infection in an African elephant (*Loxodonta africana*) calf and caretakers." *Proceedings of the AAZV AAWV Joint Conference.*

Kiers, A., A. Klarenbeek, B. Mendelts, D. Van Soolingen, G. Koeter. 2008. "Transmission of *Mycobacterium pinnipedii* to humans in a zoo with marine mammals." *International Journal of Tuberculotic Lung Disease* 12(12).

Lamberski, N. 1999. "Nontuberculous mycobacteria: Potential for zoonoses." In *Zoo and Wild Animal Medicine: Current Therapy 4*, M. Fowler and R. E. Miller, eds. Philadelphia: W. B. Saunders.

Larson. E. L. 1995. "APIC guideline for handwashing and hand antisepsis in health care settings." *American Journal of Infectious Control* 23.

Lejeune, J. T. and M. A. Davis. 2004. "Outbreaks of zoonotic enteric disease associated with animal exhibits." *Journal of the American Veterinary Medical Association* 24:9.

Lyon, G. M., A. V. Bravo, A. Espino, M. D. Lindsley, R. E. Gutierrez, I. Rodriguez, A. Corella, F. Corrilo, M. M. McNeil, D. W. Manning, S. E., C. E. Rupprecht, D. Fishbein, C. A. Hanlon, B. Lumlertdacha, M. Guerra, M. I. Meltzer, P. Dhankhar, S. A. Vaidya, S. R. Jenkins, B. Sun, and H. F. Hull. 2008. Human rabies prevention, United States, 2008: Recommendations of the advisory committee on immunization practices. *Morbidity and Mortality Weekly Report* 57, RR–3.

McDonough, E. S., A. L. Lewis, and L. A. Penn. 1966. "Relationship of *Cryptococcus neoformans* to pigeons in Milwaukee, Wisconsin." *Public Health Reports* 81(12).

Michalak, K., C. Austin, S. Diesel, J. M. Bacon, P. Zimmerman, and J. N. Maslow. 1998. "*Mycobacterium tuberculosis* infection as a zoonotic disease: Transmission betweem humans and elephants." *Emerging Infectious Diseases* 4: 2.

Miller, R. E., and D. L. Janssen. 1997. "American zoo and aquarium association draft guidelines for animal contact with the general public (revised January 25, 1997)." *Proceedings of the American Association of Zoo Veterinarians.*

National Association of State Public Health Veterinarians. 2007. "Compendium of measures to prevent disease associated with animals in public settings." Accessed 1 Apr 2010 at http://www.nasphv.org/Documents/AnimalsInPublicSettings.pdf.

National Institute of Occupational Safety and Health. 2010. "NIOSH-approved disposable particulate respirators (filtering facepieces)." Accessed April 1 at http://www.cdc.gov/niosh/npptl/top-ics/respirators/disp_part/.

Niezgoda, M., C. A. Hanlon, and C. E. Rupprecht. 2002. "Animal rabies." In *Rabies,* A. C. Jackson and W. H. Wunner, eds. San Diego: Academic Press.

Olsen, R. J., P. Lynch, M. B. Coyle, J. Cummings, T. Bokete, and W. E. Stamm. 1993. "Examination gloves as barriers to hand: Contamination in clinical practice." *Journal of the American Medical Association* 270.

Rupprecht, C. E. 1999. "Rabies: Global problem, zoonotic threat, and preventative management." In *Zoo and Wild Animal Medicine, Current Therapy 4*, M. E. Fowler and R. E. Miller, eds., 136–46. Philadelphia: W. B. Saunders.

Schutz, G. E., J. D. Sikes, R. Stefanova, and M. D. Cave. 1999. "The home environment and salmonellosis in children." *Pediatrics* 103:E1.

Shellabarger, W. C. 1996. Appendix 6: "Zoo personnel health program recommendations, American Association of Zoo Veterinarians (AAZV) infectious disease committee." In *The Guidelines for Zoo and Aquarium Veterinary Medical Programs and Veterinary Hospitals.* 1998. American Association of Zoo Veterinarians. Accessed 1 April 2010 at http://www.aazv.org/associations/6442 /files/zoo_aquarium_vet_med_guidelines.pdf.

Siegel, J. D., E. Rhinehart, M. Jackson, L. Chiarello, and the Healthcare Infection Control Practices Advisory Committee. 2007. Guideline for Isolation Precautions: Preventing Transmission of Infectious Agents in Healthcare Settings. Accessed 1 April 2010 at http://www.cdc.gov/ncidod/dhqp/pdf/isolation2007.pdf.

Stetter, M. D., S. K. Mikota, A. F. Gutter, E. R. Monterroso, J. R. Dalovisio, C. Degraw, and T. Farley. 1995. "Epizootic of *Mycobacterium bovis* in a zoological park." *Journal of American Veterinary Medical Association* 15:207(12).

Thompson, P. J., D. V. Cousins, B. L. Gow, D. M. Collins, B. H. Willamson, and H. T. Dagnia. 1993. "Seals, seal trainers, and mycobacterial infection." *American Review of Respiratory Diseases* 147.

US Department of Labor, Occupational Safety and Health Administration. 2010. Respiratory protection OSHA standards. Accessed 1 April at http://www.osha.gov/SLTC/respiratoryprotection /standards.html.

———. 2010. Code of Federal Regulations (29 CFR), Part 1910.1030. Accessed 1 April at http://www.osha.gov/pls/oshaweb/owadisp .show_document?p_table=STANDARDS&p_id=10051.

Warnock, D. W., and R. A. Hajjeh. 2004. "Histoplasmosis associated with exploring a bat-inhabited cave in Costa Rica, 1998–1999." *American Journal of Tropical Hygiene* 70(4).

Weese, J. S., A. S. Peregrine, and J. Armstrong. 2002. "Occupational health and safety in small animal veterinary practice: Part I— Nonparasitic zoonotic diseases." *Canadian Veterinary Journal* 43(8).

47

Preventive Medicine

Noha Abou-Madi

INTRODUCTION

The goals of a preventive medicine program are to prevent and control diseases and medical problems that may occur during the normal life of captive animals, and to effectively prevent the introduction of diseases into an established collection when a new animal is acquired. Furthermore, the program should address the prevention of zoonotic diseases both for the public and for the staff. Responsibilities for such a program are shared between the veterinarians and the animal keeper staff. The program and its protocols should be revised yearly and updated as necessary. Preventive medicine relies on the application of basic principles of medicine, integrating medical and natural history knowledge, judgment, medical ethics, and science-based technology for the care and treatment of the animals. Although not all problems are preventable, a comprehensive program proactively evaluates all aspects of animal husbandry and health care to limit the spread of disease and facilitate early detection of problems (Hinshaw, Amand, and Tinkelman 1996, 16–24; Whitaker 1999, 163–81; Jensen 1999, 585–99). For example, a captive coyote (*Canis latrans*) eating a soft commercial diet is likely to develop teeth problems with time (tartar buildup and periodontal disease), or may chew on hard material to compensate for the lack of proper dietary material and break some of its teeth. The dental disease may eventually lead to loss of the teeth, chronic infection of the surrounding bone, and subsequently infection in the heart, kidneys, liver, and so on. Routine cleaning of the teeth with the animal under anesthesia plays a significant role in delaying these problems. The risk of general anesthesia is outweighed by the gain of preventing and correcting dental diseases. The diet should be evaluated and modified to include specialized diets and safe chewing items that will delay the buildup of tartar. The animal's surroundings and behavior should be evaluated, and corrective measures should be taken to reduce the behavior of chewing on inappropriate materials.

Three major facets of a preventive program are often quoted; they are, however, all related. Early detection of problems through routine monitoring of the animals will allow timely and accurate intervention to prevent disease or complications. Each institution should design a program that is appropriate for the species in the collection, its facilities, and current resources, and should emphasize priorities when resources are limited. The second facet is to optimize the animal's immune system via stress reduction, optimal welfare standards, appropriate diet, and vaccination. Finally, good quarantine measures are essential to prevent or reduce the risk of introducing pathogens with the arrival of new animals or by the presence of wildlife on grounds.

This chapter will review the principles of preventive medicine as they apply to a collection of captive animals. After studying this chapter, the reader should be able to

- understand the goals of preventive medicine as they apply to the different aspects of animal care
- identify the keeper's roles in establishing and maintaining these programs
- understand the steps involved in performing a physical examination on an animal, including dentistry, blood testing, and hoof care
- understand the principles and limitations of vaccination
- understand the principles of a quarantine program and biosafety recommendations
- understand the interaction between parasites and hosts and the importance of a comprehensive parasite control program for the health of the animals
- know the correct methods for collecting and storing fecal samples
- understand the steps necessary for a comprehensive parasite control program.

MONITORING THE ANIMALS IN THE COLLECTION
TIMELINESS IN ADDRESSING HEALTH PROBLEMS

The importance of daily observation of the animals by the keepers has already been described in other chapters, but its importance is again emphasized here as it is essential for early detection of a problem. Keepers know their animals and can detect the smallest deviation in their normal behavior and routine. These small signs should be reported and monitored carefully, as they may indicate the beginning of serious problems. The signs may not require veterinary intervention, but discussion of the case should be initiated to make other animal care staff, managers, and veterinary staff aware of it. A plan can then be agreed upon in case the concern does not resolve itself in a timely manner.

ROUTINE PHYSICAL EXAMINATION AND TESTING

As an integral part of their survival, wild animals hide signs of disease and appear to function normally when sick. By the time clinical signs are detected, many maladies are already in advanced stages, thus emphasizing the importance of due diligence. Furthermore, each animal carries its own microenvironment of parasites and infectious agents. As long as the host's immune system is healthy, these agents are usually kept in check. Stress is one factor known to decrease the function of the immune system so that an animal may become sick or start shedding an infectious agent that has been dormant in its system for years. Because wild animals are adapted to hide weakness and illness, routine physical examination, radiography, and testing of blood and secretions may allow for timely detection and correction of early conditions. Our knowledge of normal physiologic values and blood parameters is still fragmented in many species; it is therefore important to establish baseline information and data on each animal while it is still healthy. If in the future the same animal or individuals from the same species are ill, values obtained at that time can be compared to the baseline information. Changes observed will help with the interpretation of disease processes (Hosey 2009, 379–426).

Monitoring and eradication of several diseases (in the United States, brucellosis, chronic wasting disease, and bovine tuberculosis) are controlled by governmental agencies. The government can mandate specific tests to check for these diseases every year or so. Additionally, cooperative science programs such as the North American Association of Zoos and Aquarium's SSP and TAG may require species-specific tests (e.g., for genetic studies). Physical restraint or general anesthesia will allow for thorough examination of the animal and collection of the required data.

RESTRAINT

Depending on the temperament, tractability, and training of an animal, the physical examination and blood collection may be performed with minimal or no use of drugs. This would be the case in many elephants, most reptiles and birds, and even some mammals (for example hoofed species habituated to people or chute systems). In most cases, however, general anesthesia or sedation is warranted to allow veterinarians to perform a complete examination while reducing the stress on the animal. The risk of general anesthesia should be weighed against the benefit of the examination and testing. Keeper staff may be called upon to help monitor the animal during and after anesthesia.

PHYSICAL EXAMINATION

Once the animal is safely restrained or under anesthesia, the veterinarian can proceed with the examination (figure 47.1). First, a stethoscope is used to listen to (auscultate) the lungs and the heart, a pulse is palpated (usually under the jaw or in the groin area), and a core body temperature is obtained. The mouth is opened and examined to ensure that there is no obstruction to breathing. To make sure that the heart is able to pump blood and therefore deliver adequate oxygen to the body, the veterinarian will look at the mucus membranes. These membranes line passages between air and the body (nose, mouth, genital opening, and anus) and contain thin blood vessels (capillaries). Changes in the blood flow and in the amount of oxygen in the blood will be reflected by a deviation from the normal pink color to a pale or bluish tinge. Also, when the lining (mucosa) is compressed with one finger, causing it to "whiten" (blanch out), and is then released, the time it takes to turn pink again is recorded. This is called the capillary refill time (CRT), and it is another indicator of how well blood is being delivered to the body. The CRT should be less than two seconds. Keepers and technicians are often called to help check the animal's condition under anesthesia, and can be asked to perform these tasks. If these parameters are normal, the physical examination continues.

The animal's identity should be confirmed by reading the tattoo, checking the microchip number (transponder), or reading the ear tag. A review of the animal's breeding history and corresponding breeding recommendations will guide the use or removal of contraceptive implants, the promotion of pregnancy, and monitoring of the health of the reproductive system.

As soon as possible, an accurate body weight should be obtained and compared to previously recorded values. A weight loss equal to or greater than 10% should be investigated. Weight gain should also be addressed. Body condition is scored with a scale. One example of the body scoring scale is a scale between 1 and 9. Five is the neutral point at which the animal is in good body condition; a score of less than 5 indicates progressively thinner body condition, and above 5 indicates various degrees of overcondition (obesity). Some veterinarians use a scale from 1 to 5 in a similar way (3 being good body condition). If possible, an animal should be weighed whenever it is handled. It is also useful to incorporate weigh scales into an animal's routine.

The eyes are examined for evidence of diseases such as opaque lenses, scars on the cornea, and other problems. The eyes are then kept moist with the application of an ointment ("eye lube"). The ears are cleaned and any material collected (ear wax, debris) is checked for ear mites or other signs of infection. As the mouth is opened and examined, the joints of the jaw (temporomandibular joints) are checked to make sure they are working normally. The mouth is checked for color; wounds (ulcers); masses; hydration; the presence of food, dirt,

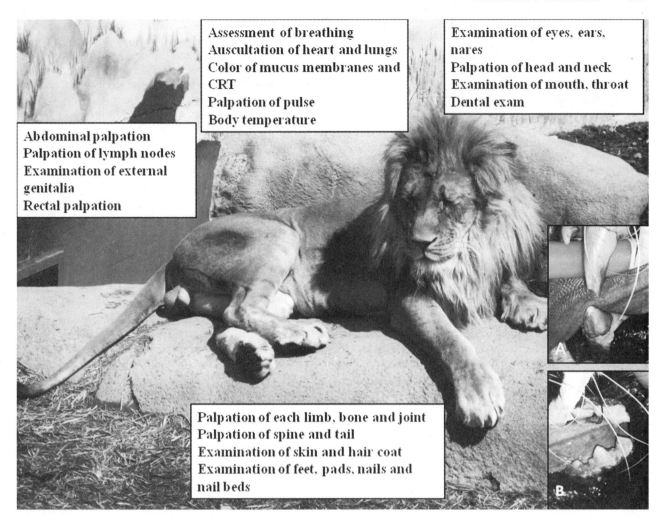

Figure 47.1. Example of the components of a physical examination. Insert A illustrates broken and worn canines. Insert B shows severe tartar build up on premolar and molar teeth. Photos by Noha Abou-Madi.

and bedding, and so on. The tongue is inspected in a similar fashion. Examination of the teeth will determine the presence of tartar buildup and gingivitis, loose teeth, abscesses, fracture, and/or abnormal wear of the teeth (see section on dental procedures, below).

The veterinarian will then continue the examination by pressing down gently on (palpating) sections of the body, feeling for normal and abnormal structures. The head, neck, and body are examined in that way to look for lumps and bumps. For example, palpation of the contents of the abdominal cavity will help detect large liver, small kidneys, irregular spleen, tumors, and the possibility of pregnancy. The veterinarian will then complete the examination of the abdomen by inserting a gloved finger (or hand, if the animal is large) into the rectum and feeling the organs located in that area (e.g., the prostate, uterus, and cervix, as well as the intestines in large animals).

Each limb and the tail will be palpated for abnormal swelling, and the flexibility (range of motion) of each joint will be evaluated. Any joint inflammation (arthritis) will be noted, and disease will be further evaluated with radiographs. Early detection of arthritis will permit early treatment and timely modification of the display in order to limit progression of the

disease. Any muscle smaller than normal (muscle atrophy) should be noted. That muscle will then compared to its counterpart on the other side. If both muscles are equally smaller than normal, the muscle atrophy is called symmetric; if one muscle is normal while the corresponding one on the other side is smaller, the muscle atrophy is called asymmetrical. Asymmetrical muscle atrophy in one leg may indicate decreased function in that limb, whereas widespread muscle wasting points towards a more generalized problem (a nerve problem involving both sides, the animal not eating enough, long-term kidney disease with muscle wasting, cancer, etc.). Nails, claws, or hooves are closely examined for abnormal growth, shape, wear, signs of infection, or other diseases. They should be trimmed as required to resemble the hoof's normal anatomy. Foot pads should be checked for abnormal wear that would indicate pacing or excessively abrasive flooring.

The quality of the hair coat, plumage, or scales will be assessed and any skin lesion, skin lumps, crusts, irritation, or hair loss will be noted. Mammary glands, if present, should be evaluated for milk production and abnormal swelling. Samples, such as pieces of tissues (biopsies) or swabs (for cultures), can be collected if indicated. If needed, treatment for parasites of the skin, hair, feathers, or scales (ectoparasites) is initiated during the procedure.

TESTING AND TREATMENTS

Blood is routinely collected and analyzed to look for evidence of anemia, infection, or inflammation (hematological parameters) and levels of electrolytes (e.g., calcium or potassium), as well as to assess how well the kidneys and liver are functioning. Additionally, previous exposure to viruses or bacteria can be checked by measuring the level of antibodies produced by the animal. Other tests can also be performed on blood, such as measuring levels of vitamins and hormones (i.e., checking thyroid hormones, reproductive hormones, insulin, etc.). Any extra blood collected may be frozen and saved for a bank of samples that will be available for future testing. Swabs of the rectum can be collected to check feces for parasites; ideally this should be done before the thermometer is used for a check of temperature, so that the feces sample is not contaminated. If urine is available, it can be collected and analyzed to evaluate kidney function and to look for signs of infection. Surveying animals for tuberculosis (or mycobacteriosis) is often mandatory in primates, elephants, and specific species of hoofed animals. The tests vary with the species to include a skin test, blood tests, radiographs of the chest, and/or cultures. Veterinarians have additional "tools" to help them "see" what is happening inside the body, such as imaging (e.g., radiography and ultrasound), electrocardiography (ECG), and endoscopy. These techniques are discussed in chapter 48.

Dental care is an essential aspect of a preventive medicine program. The presence of tartar on the teeth compromises the integrity of each tooth affected. Furthermore, the tartar harbors bacteria that can invade other tissues, cause local infection, and travel via the blood to the heart and kidneys and cause severe and sometimes fatal lesions. Thus the cleaning of the teeth is essential, and fully warrants a procedure under anesthesia if indicated. Depending on the species and the problem identified, dental procedures can vary from a simple cleaning to the removal of an infected or broken tooth, or to a root canal (Van Foreest 1993, 263–68).

Deworming, administration of vaccines, fluid supplementation, and injections of vitamins (such as vitamin E) and other supplements (including selenium) are performed as deemed necessary by the veterinarian at the time of the examination.

HOOF, CLAW, AND NAIL CARE

An abnormal wear pattern of the nails, claws, or hooves of animals is not an uncommon finding in captive animals and can affect young and older individuals. This condition can be the result of many factors (i.e., multifactorial); these factors are important aspects of husbandry that keepers should monitor and correct as soon as possible. A few examples include a diet that is either too rich (i.e., containing too much protein) or not balanced for the specific species (i.e., missing an important element); metabolic abnormalities; lack of enough movement for an animal to wear down its hooves (sedentary behavior, lack of enrichment, arthritis); substrate that is inappropriate (too moist, too soft, too abrasive, too regular, etc.); and abnormal healing of an old trauma. Abnormal growth of claws in large cats and bears can be caused by a lack of appropriate exhibit furniture that would allow these animals to wear and clean them routinely. Elongated claws can become imbedded into the digital, metatarsal, or metacarpal pads, causing pain and infection. The causes of these conditions should be investigated and corrected immediately. Pain and infection should be controlled medically and/or surgically as indicated. While some conditions (e.g., long or broken nails in birds) are minor and can easily be corrected, others, such as abnormal hooves, may cause permanent damage to the growth center of the hooves and compromise the animal's ability to walk normally. In ungulates, elongation of one or more hooves combined with asymmetrical wear is a common finding (figure 47.2). The shape of the hoof should be reestablished to allow the weight-bearing surfaces to return to the wall of the hoof. If the malformation is severe, the trimming might have to be done in stages to avoid pain and strain on the joints. It is well worth it to seek the recommendations of a farrier, both for routine and severe cases. Signs of laminitis (irregular hoof wall, thickening of the white line) indicate other underlying problems warranting radiography of the feet. Abscesses, foot rot, and sole bruising necessitate additional surgical interventions and long-term medical treatment (Fowler 1986, 549–57).

NEONATAL CARE

A successful birth can be a major event in a zoo. Protocols for proper husbandry measures and veterinary interventions should be in place ahead of time. These include setting up means of monitoring while allowing appropriate privacy for the dam and the babies, lists of emergency contact information, and "decision trees" for intervention in case of complications during the birthing process or after the animal is born. In certain species, handling should not occur after birth, but in others examination, weighing, and basic testing for passive immunity are recommended. These protocols are discussed in chapter 21.

USE OF SENTINEL ANIMALS

While not common, the use of sentinel animals is sometimes instituted in zoos to allow for early detection of serious infectious diseases (mostly viruses). The sentinel animals are often domestic animals that are resistant to a specific virus or bacteria but which can mount a detectable immune response when exposed to the agent. At the time of their introduction into the collection, they do not have antibodies against that agent. These animals subsequently will undergo routine testing for the presence of an antibody response (seroconversion) that would indicate the presence of that specific pathogen on the zoo grounds. If the animals seroconvert, the staff can take immediate measures to prevent the spread of that disease to the resident animals. When West Nile virus was spreading in North America, some zoos kept domestic chickens on their grounds and monitored their blood for signs of exposure to the virus. Similarly, opportunistic or routine testing of wild animals found on zoo grounds can

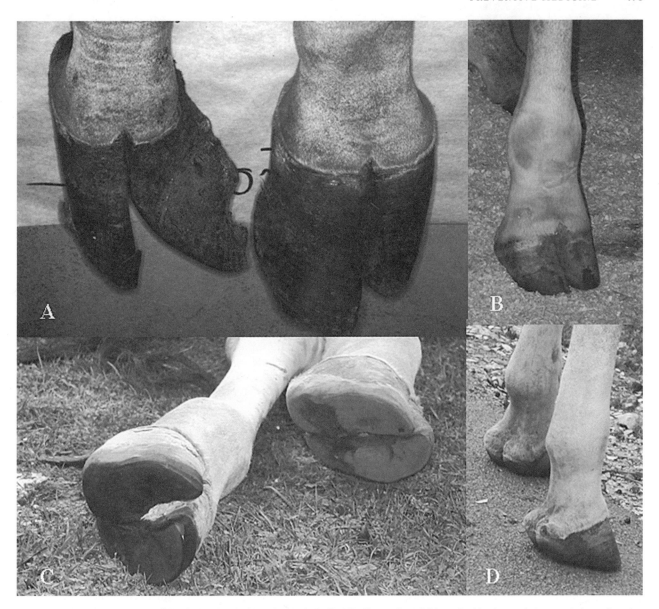

Figure 47.2. Failure to keep up with hoof trimming in this reticulated giraffe (*Giraffa camelopardalis*) resulted in abnormal shape and elongation of the keratin layer, causing pain, abnormal hoof structures underlying the keratin, arthritis, and predisposition of the animal to infection. Proper hoof trimming should restore the normal shape of the hoof (insert B) so that the weight is on the wall of the hoof (insert C) and the proportion of the hoof allows for normal angle of the joints in the rest of the leg (insert D). Photos by Noha Abou-Madi.

provide information on the presence of potential pathogens near zoo animals (e.g. in the testing of free-living amphibians for chytrid disease).

OPTIMIZING THE IMMUNE SYSTEM OF THE ANIMALS IN A ZOO

MAINTAINING PROPER HUSBANDRY MEASURES AND DECREASING STRESS

Husbandry, exhibit design, exclusion of poisonous plants, and pest control are topics covered in other chapters. They are essential elements of the preventive medicine program and should be evaluated in conjunction with general animal

welfare, the prevention of animal injuries, and the potential for the introduction and control of disease. Keeping animals in an environment where they feel safe yet challenged will greatly contribute to their overall health. They should be allowed to retreat away from the public, and their flight distance should be respected. Naturalistic and mixed-species exhibits are excellent educational opportunities; they are esthetically pleasing for the public and can be enriching for the mixed group of inhabitants. They do contribute to further health concerns and variables such as trauma, unseen stressors by dominant individuals or species, dietary problems, and cross-species disease transmission (Lowenstine 1999, 26–29). Inappropriate lighting cycles (circadian cycles) will cause chronic stress. Chronic stress may not be clinically visible, but will in the long term eventually compromise the animals' immune function.

Proper nutrition bolstered by routine food quality monitoring will ensure that each animal has the essential building blocks to maintain a healthy immune system. For example, a deficiency in vitamin A will prevent the normal maintenance of the cells lining the inside of the mouth of many species of parrots. Normally, this layer of cells constitutes an effective barrier against infectious agents crossing into the body. When this barrier is compromised it becomes ineffective, and bacteria may traverse into deeper tissues and start an infection. Similarly, improper ventilation, which leads to the buildup of noxious fumes or wastes (e.g., dust from hay, ammonia from urine) will chronically irritate the lining of the airways. This may compromise the function of the lungs and the local protective mechanisms in place to prevent the invasion of pathogens. Finally, many infectious agents (opportunistic pathogens) will cause disease only when the immune system is weakened. There may be a common condition or underlying stressor that is weakening the individual (such as overcrowding). If such a disease is seen either in one individual or in a group of animals, it should be investigated in light of a common underlying condition (Childs-Sanford 2009, 8–14).

VACCINATION

IMMUNIZATION

Many infectious diseases can be prevented by providing the animal with proper immunity against these specific diseases (immunization). With this approach, the pathogen is controlled by the animal's immune system instead of being treated after it has caused disease. An animal's immune system can be strengthened against specific diseases by the administration of a vaccine, also referred to as vaccination. A vaccine contains a specific dose of an antigen to trigger the animal's immune system without causing disease. The dose of antigen is adjusted to produce a controlled immune reaction within the body; the body responds to the antigen by producing antibodies against it. The duration of the protection is variable, and repeated vaccinations (boosters) are often required to achieve and maintain protection. When a vaccination is successful, the animal is immunized, and the resulting "memory" (antibodies or protective immune cells) will help the animal fight off the specific infectious organism during natural exposure. Some vaccines may contain antigens to protect against multiple pathogenic organisms. For example, one vaccine contains antigens to protect against Clostridium C, Clostridium D, and tetanus (*Clostridium tetani*).

The host's response can be monitored by measuring the levels of antibodies after vaccination. These levels (titers) are often lower than the levels expected to be seen after exposure to the natural disease. They typically increase with each booster and decrease with time. However, the lack of sound scientific research in this field limits our interpretation; indeed, while a low titer after vaccination may still be protective, a high titer after vaccination does not necessarily ensure protection. More species-specific research is needed in this field. Safe use of vaccines in a pregnant female before parturition will increase maternal antibodies so that a stron-ger passive immunity (antibodies) can be delivered to either the fetus across the placenta, or to the baby via ingestion of the colostrum. It is important to remember that vaccines are not 100% protective, and that vaccination should not be considered as an alternative to proper hygiene and biosecurity.

TYPES OF VACCINES

There are different types of vaccine. Some (modified-live vaccines) use live, attenuated infectious agents while others (killed vaccines) deliver inactivated pathogens or one of their constituents or products (recombinant vaccine, toxoid). It is believed that the response produced by a modified-live vaccine is stronger, and closer to that obtained following natural infection. However, in species that are very susceptible to a pathogen, modified-live vaccines have been known to cause the disease against which they are supposed to protect, and can cause an animal to shed the organism. For example, pandas and ferrets are very susceptible to the paramyxovirus responsible for canine distemper. Most vaccines formulated for domestic dogs carry modified-live forms of the distemper virus that are still very potent. If administered to pandas or ferrets, these vaccines will act as the "real" virus and cause disease. Because the sensitivity of many nondomestic species to modified-live vaccines is largely unknown, using killed vaccines is often recommended.

ADMINISTRATION OF VACCINES

In many countries, by law, a vaccination must be performed by a veterinarian or in the presence of a veterinarian. A vaccine can be administered by spraying the product in the nasal cavity (intranasally), by mouth (orally), by injection in the muscle (intramuscularly), under the skin (subcutaneously), or in the egg while the embryo is developing. Vaccines are usually dispensed in a multidose or in single-dose vials. In some cases, the vaccine is dried (lyophilized) for longer shelf life and must be mixed (reconstituted) with a diluent prior to administration. In all cases, it is essential to keep the vaccines at the temperature recommended by the manufacturer. Shipping and storage under inappropriate conditions will inactivate the vaccine. For example, many vaccines against canine distemper are lyophilized and must be stored in a refrigerator at 2 to 8°C. Once the initial series of vaccine administrations is completed, the proper frequency of subsequent administration is still open to debate. Many veterinarians will follow recommendations extrapolated from research in the species they are vaccinating, or recommendations for domestic animals. Monitoring of titers to assess antibody levels can sometimes be done to indicate when vaccination is needed.

ADVERSE REACTIONS

Adverse (allergic) reactions to the administration of a vaccine may occur, and the vaccinated animal must therefore be monitored for 30 minutes to 2 hours after injection for signs of difficulty breathing, diarrhea, vomiting, bleeding (rectal or in vomitus), facial swelling, or lethargy. If clinical signs appear they should be addressed immediately, as they can be fatal. Delayed adverse reactions have been seen, and clinical signs have been seen in ferrets up to 24 hours after vaccination with canine distemper.

QUARANTINE

GENERAL PRINCIPLES

Quarantine is one of the most important aspects of a preventive medicine program. It is not the mere isolation of an animal from the rest of the established collection. Quarantine is an active process and needs to be done under the supervision of the veterinarian. It is a period of isolation, observation, and testing prior to the introduction of a new animal to the resident population. *The main objective of a quarantine program is to protect the resident animals from diseases that may be harbored by newly acquired animals, using biosecurity measures.* Biosecurity measures are defined as preventive measures or management practices designed to reduce the risk of transmission of infectious agents. They include cleaning protocols, disposal of wastes, and human and animal traffic around a designated area such as a quarantine area. They also provide an opportune time to collect baseline health information on the new animal, start individual records, slowly introduce the new diet if required, and acclimate the animal to its new environment. Close attention to detail (such as changes in attitude, behavior, posture, appetite, and fecal and urinary output as well as subtle clinical signs of illness) and detailed reporting are essential for the success of a quarantine program (Miller 1999, 13–16; Roberts 1993, 326–30).

In the newly acquired animal the stress of shipment along with a change in environment, social group, and diet often results in a decreased immune function. Pathogens normally kept in check by the immune system of the healthy host may then start proliferating. The host can shed these pathogens, contaminating the environment and infecting other animals, and may possibly become ill. Failure of quarantine may result in the introduction of diseases with possibly disastrous consequences for the collection.

QUARANTINE FACILITIES AND DURATION

Disease agents are potentially disseminated through aerosolization via the ventilation system, in water, via urine or feces, on food and water bowls, in the overflow of water from hosing, in bedding carried out of an enclosure, and on people. For example, a virus shed by an animal can be transported on contaminated clothing or on the soles of unprotected shoes, and can be transferred to another exhibit without losing its power to infect and cause disease.

Ideally, quarantine facilities should be physically isolated from the main collection. If this is not possible, alternate areas should be designated as quarantined and should be set up to prevent the spread of disease via both direct and indirect contact with other animals (e.g., by physical contact or shared water, resources, or ventilation). Separate rooms should be allocated for birds, mammals, reptiles, and fish. Quarantine of larger mammals and marine mammals poses significant challenges. Quarantine tanks for fish should be equipped with filtration systems completely separate from the main life-support system of the resident collection. Infective microbes are a challenge because they are invisible; a constant effort must be made to remember and abide by basic rules of hygiene and biosecurity (Whitaker 1999, 163–81).

Outreach (education) animals pose a particular challenge as they routinely travel in and out of a collection, thus having a high probability of coming into contact with new pathogens. In general, facilities with outreach animals should manage them as a separate and closed collection, housing them in a facility or a room that is isolated from the rest of the collection. Permanent quarantine measures and designated keepers can be arranged for these animals.

Each country has specific legislation guiding quarantine requirements. International and local government regulations may also dictate additional procedures based on the source of newly acquired animals (from other zoos, from wild to zoo or from captivity to wild, and from approved versus nonapproved institutions), their destination (national or international transfers), and risk assessments (World Organization for Animal Health [OIE] 2007 *Terrestrial Animal Health Code* Section 1.4; Wells-Mikota 1993, 3–10).

For most species, the minimum length of quarantine is 30 days from arrival. If more than one group of animals arrives from different locations, the groups should be housed separately from each other. If there is no alternative and it is acceptable to all parties, several animals arriving on different dates may be housed in the same room; the length of quarantine is then minimally 30 days from the arrival of the last animal. Snakes and other reptiles may require longer periods (three to six months) to allow for detection of diseases (especially viral and parasitic diseases), as many of the disease processes can take longer to become clinically detectable in these species. Animals arriving with ongoing problems (e.g., parasites or poor body condition) will also require prolonged quarantine until those problems are resolved. An "all-in, all-out" policy should be respected, meaning that all of the animals arrive together and all leave quarantine only when the last animal in the quarantined group has cleared testing.

QUARANTINE PERSONNEL

To prevent the spread of diseases by humans, only a very limited number of trained staff should have access to the designated quarantine areas. Ideally, these individuals should be assigned to quarantine and should not go from a quarantine area to the main collection. If this is not possible, they should service quarantine only at the end of the day. Personnel entering the quarantine area should always wear personal protective equipment (PPE). The list of protective gear varies with the species of animal, concern, and more specifically the type of disease to be contained. Visitors should not be allowed in quarantine areas.

PREPARATION AND SETUP

The quarantine area should be set up so that the animal is comfortable. Food containers, water bowls, perches, cage furniture, and material used in quarantine must stay in that area and be used for that animal only until its release. Food and water bowls should be cleaned in the quarantine area and remain there as designated quarantine items. Food may be brought in plastic bags or disposable containers that will then be discarded. Disposable furniture can be used and discarded when the animals leave quarantine. Disposable material (including perches and hide boxes) can be sprayed with diluted bleach and discarded along with the refuse from quarantine. Providing adequate substrate that can be disinfected is sometimes problematic. The flooring should be comfortable and nonslippery but also made of a material that can be disinfected. Since pathogens survive in organic matter, direct contact with the earth should be avoided. This is often difficult to achieve, especially for hoofed species. In these cases, at the end of each quarantine period the top 20 cm of soil can be removed, limed, and replaced with fresh soil. Soil, straw, shavings, sawdust, or peat moss can be added to a cement surface. Each room, holding area, and vivarium should be equipped with appropriate lighting for daytime and dim light for nighttime. If needed, the use of timers will ensure proper lighting cycles with minimum invasion of the quarantine area. Details of the current diet should be provided prior to the animal's arrival, and are often included on the Animal Data Transfer Form, such as the one published by the American Association of Zoo Keepers (AAZK). This form details husbandry practices (diet, housing, enrichment, etc.) from the sending institution and allows the receiving institution to match the animal's routine until it is judged stable enough to make adjustments as necessary.

HUSBANDRY PROTOCOL

When entering a quarantine area, one must wear personal protective equipment. The attire used for quarantine is often disposable and should always stay in that area. In some quarantine facilities, a complete change of clothes is required and separate laundry facilities may be provided on site. It may be mandated that one must shower before and/or after servicing a quarantine area. The list of PPE should be posted for each quarantine area; it may include cap, mask, gown, coverall, rubber boots, plastic shoe covers, and gloves. Care should be taken to avoid contamination of regular clothing by the protective gear. When the PPE is removed, the following order should be respected: removal of gross organic matter from boots or shoe covers and scrubbing with recommended disinfectant, loosening of shoe cover ties (but not removal), removal of the outer layer of gloves if double gloves are used, and removal of the gown or coverall, the mask, and finally the cap. At this point one shoe cover can then be removed and that foot can step into the clean area; then the second shoe cover is removed, and the person can step completely into the clean area. The second set of gloves can then be removed and discarded. Hands should be washed immediately and a disinfecting footbath should be used upon exiting the quarantined area (figure 47.3). A person should never walk with unprotected shoes in quarantine, and should not walk outside quarantine with the protective shoe covers or boots that have been worn in quarantine. These quarantine protocols apply to everyone without exception, including curatorial staff, keepers, directors, government officials, and veterinarians.

Every animal should be checked first thing each morning for problems or distress, and fresh food and water should be provided. Servicing the area (cleaning) should be done at the

Figure 47.3. Proper preparation of a foot bath. The solution is changed daily and whenever there is contamination with organic material. The label above the bath indicates the product, instructions for dilution, and exposure to the chemicals. The towels under the bath allow excess solution to be collected in order to keep the area dry. Photo by Noha Abou-Madi.

end of the day, and further food and fresh water should be offered as necessary. Nocturnal species will require a different schedule. Although physical contact with the animal should be kept to a minimum, special attention must be provided to identify early signs of disease and any changes in behavior. To improve monitoring and follow-up, normal and abnormal behavior, attitude, food and water consumption, presence and quality of wastes, and any abnormal clinical sign (coughing, nasal discharge, lameness, etc.) must be recorded and reported to the veterinarian.

Tools, furniture, bowls, and containers should be dedicated and labeled exclusively for use in quarantine, and should never be used outside of that area. They may be cleaned with regular soap and disinfected with diluted bleach (1 part of bleach to 30 parts of water) or as otherwise prescribed in the quarantine protocol. To reduce the risk of infection and re-infestation, the cages should be cleaned daily according to the same standards as are followed for the rest of the collection (or as otherwise prescribed when developing the quarantine protocol). Any used towels or blankets should be bagged tightly and washed the same day using a separate laundry facility assigned for quarantine. Waste and bedding should be bagged in quarantine and discarded daily, with no contact with the resident animal collection's refuse. If possible, the material should be double-bagged so that the external surface of the second bag is not contaminated. This includes using a separate route or access into the manure disposal areas and dumpsters. In cases of serious disease concern, the wastes should be handled separately and can be incinerated. Upon the animal's departure the cage and any cage furniture and dishes should be thoroughly cleaned and disinfected. Any disposable furniture such as rope, wooden perches, or nest boxes should be safely discarded as described above.

RECOMMENDATIONS FOR VETERINARY PROCEDURES

Prior to the animal's arrival in quarantine, the veterinarian should review its medical record and the general medical history of the shipping institution. This will help assess the health status of this individual and its potential risk to the resident population, if any. To ensure each individual animal's health, ideally it should arrive in quarantine with a complete and updated medical record and a current health certificate, usually issued no more than 30 days prior to shipment. The health certificate is an official document, completed by the veterinarian and often endorsed by the appropriate governmental office, confirming that the veterinarian has examined the animal, has completed appropriate tests to screen for infectious and metabolic diseases, and has confirmed that the animal is healthy and free of infectious diseases. These tests often include fecal analyses for parasites, fecal cultures, testing for tuberculosis, testing for brucellosis, and other tests as deemed appropriate. Records of vaccination and other treatment, such as deworming, are included. Each animal should be identified as is appropriate for its species, and this information should be present in its medical record and on the health certificate. Updated vaccinations should be performed prior to shipment; however, if vaccinations are attempted too soon before or after shipping, the stress of the change will decrease the animal's immune function and will prevent its adequate response to the vaccine.

As soon as the animal is released from the transport container, a visual examination is performed to make sure it has not been injured during transportation, and a body weight is obtained. This can be done by weighing the transport crate with the animal inside it and then again after the animal has been transferred into the quarantine area. Monitoring of subsequent body weight will help to assess the newly arrived animal's adaptation to its new environment and also may be an indicator of its health and condition. If a problem is discovered at the time of arrival (e.g., a shipping injury), further interventions should be considered, taking into account the nature and the severity of the problem discovered. A quarantine tracking record and PPE form should be filled out and posted immediately upon the animal's arrival.

A thorough physical exam, including a second body weight, analyses of blood, and storing of blood for future testing, should be performed by the veterinarian. Some procedures may be omitted if pre-shipment testing and medical history of the animal and of the originating institution are satisfactory. Fecal samples should be collected once a week, starting within a week of the animal's admission, and should be submitted for parasite examination. Because of the length of treatment, early detection of parasites will help keep the duration of quarantine to a minimum. At least three fecal samples negative for parasites are usually required to clear an animal from quarantine. However, the number of negative fecal tests required may vary, and should be determined by the attending veterinarian. Dates of sample collections and results should be documented on the quarantine tracking records. The veterinarian will approve the time of discharge on basis of the animal's condition and the results of the laboratory work, and a third body weight should be recorded if possible at the time of departure from quarantine (Miller 1999, 13–16; AZA quarantine recommendations www.aza.org /animal-health).

During the quarantine period, complete and detailed medical records must be maintained which include daily observation of the animal's behavior, amount of food ingested, quality of the feces and presence of urine (even when deemed normal), and recording of veterinary procedures, treatments, and laboratory results. Any animal that dies in quarantine should have a complete necropsy performed, and tissues should be submitted for microscopic examination to rule out infectious diseases. Other animals that have been housed with the now-deceased individual should not be released from quarantine until results eliminate the risk of infectious or contagious disease.

PARASITOLOGY

Parasites are organisms living inside or on the surface of a host. As they depend on the host's survival to feed and propagate, they often coexist in balance with their host. However, a significant number of parasites (pathogenic parasites) may cause disease and may cause considerable losses if introduced into a population. For example, the parasite *Toxoplasma gondii* causes severe disease and death in Pallas cats (*Otocolobus manul*). Moreover, a parasite may develop without affecting its primary host, but cross-infection of this parasite to a host of a different species may cause severe disease, as seen with the raccoon roundworm (*Baylisascaris procyonis*) infecting humans, and the white-tailed deer parasite *Parelaphostrongylus tenuis* infecting llamas. Many parasites such as *Giardia*, *Cryptosporidium*, *Baylisascaris*, and *Sarcoptes* mites have well-documented zoonotic potential and present a threat to the staff and the public. Most are opportunistic pathogens, and as long as the host's immune system is strong, their development may stop (arrested stage) and they may not cause clinical illness. However, if the host is weakened by other disease or stress or is very young, these parasites will start to reproduce and proliferate, causing significant morbidity and mortality.

A complete review of veterinary parasitology is beyond the scope of this section, but the topic is an essential part of every preventive medicine program, and keepers play a critical role in control and eradication through hygiene and sound management decisions (Samuel 2001; Atkinson 2008; see table 47.1 for a simplified classification of common parasites). Morbidity and mortality related to parasitic infections have a significant impact on captive and free-living populations of animals. Ideally, new parasites must not be introduced into a collection. Potentially pathogenic parasites that are already recognized in a collection must be either eradicated

TABLE 47.1. Simplified classification of common parasites

Common name	Examples of parasites	Systems affected
Ticks	*Dermacentor, Otobius, Ixodes*	Skin; many ticks transmit diseases that can affect different organs
Mites	*Sarcoptes, Knemidokoptes, Psoroptes, Demodex, Cheyletiella*	Skin and skin structures
Flagellates	*Giardia, Trichomonas, Trypanosoma, Leishamania*	Alimentary tract,* blood, lymph, etc. (depending on species)
Amoebas	*Entamoeba histolytica, Entamoeba invadens*	Alimentary tract
Sporozoans	*Toxoplasma, Eimeria, Isospora*	Heart, muscles, brain, liver, intestinal tract, kidneys etc. (depending on species)
Ciliates	*Balantidium coli*	Alimentary tract
Trematodes (flukes)	*Fasciola hepatica, Schistosoma, Alaria*	Liver, intestinal tract, skin, blood, blood vessels, etc. (depending on species)
Cestodes (tapeworms)	*Taenia, Echynococcus* (hydatic cyst), *Moniezia*	Alimentary tract, cysts in muscles or liver (intermediate hosts)
Nematodes (roundworms, pinworms, hookworms, threadworms, whipworms)	*Strongyloides, Enterobius, Haemonchus, Ascaris, Trichuris, Capillaria, Rhabdia, Parelaphostrongylus, Nematodirus*	Alimentary tract, respiratory tract, brain, etc. (depending on species)
Acanthocephalans (thorny-headed worms)	*Moniliformis*	Alimentary tract

* The alimentary tract in these animals is analogous to the gastrointestinal tract of vertebrates.

or controlled to prevent clinical disease while stimulating the host's immune system. Young animals, individuals with decreased immune function (i.e., that are immunocompromised), and animals that are naive to a specific parasite are more susceptible to diseases (pathology) caused by these organisms. A transport host may not shed eggs or cysts for several months until conditions in the host or the environment become favorable to trigger parasite development.

LIFE CYCLE OF PARASITES

Understanding a parasite's life history is crucial for successful control and selection of appropriate management strategies. Parasites infect a host and depend on it not only for food but also for sustaining their reproductive stages. Various life cycles have been described, and some parasites (e.g., ascarids) may use more than one option. Parasites live either inside a host (i.e., they are internal parasites or endoparasites, and may live in the intestinal tract or in the lungs), or they spend part or all of their life cycle on the surface, hair or skin of an animal (i.e., they are external parasites or ectoparasites, like ticks and fleas).

DEFINITIONS

- **ovum and oocyst.** Results of sexual development of parasites. Parasitic worms (helminths) produce eggs called ova (singular ovum), which can hatch and develop into larvae; protozoa produce oocysts which can develop into cysts (i.e., they can encyst) with protective outer layers to resist environmental pressures.
- **larva.** The immature stage of a worm or insect.
- **infective stage.** A stage of the parasite which can invade a definitive host and develop in it.
- **definitive host.** A host species in which the parasite's

sexual reproduction takes place. Fertilization occurs inside the definitive host.
- **intermediate host.** A host used by a parasite to complete a specific stage of development (e.g., maturation or asexual reproduction) but not sexual reproduction. Intermediate hosts are species other than the definitive host; for example, if the definitive host is a bird, the intermediate host may be a fish, a mosquito, or a mouse.
- **aberrant host.** A host that is not usually invaded by a parasite. In an aberrant host, the parasite is unable to complete its life cycle as it would normally in a definitive host.
- **paratenic host.** A host used by a parasite for transportation and dissemination. A paratenic host does not sustain the development of the parasite.
- **vector.** A carrier, usually an animal that transports an infective agent (parasite, bacterium, virus, or fungus) and can transmit it from one host to another. In parasitology, vectors are often arthropods (e.g., mosquitoes are vectors of the malaria parasite).

LIFE CYCLES

Many parasites have direct life cycles, which require only one species of host. The host acquires the infective stage of the parasite. The parasite matures and reproduces within this host, then releases the infective stage which can reinfect the definitive host directly (figure 47.4). Classic examples of parasites with direct cycles are hookworms and coccidia. Infective eggs and oocysts are shed in the feces of the definitive host and then into the environment, after which the host becomes reinfected or a new animal becomes infected by eating contaminated feed or by ingesting the infective stage in the environment.

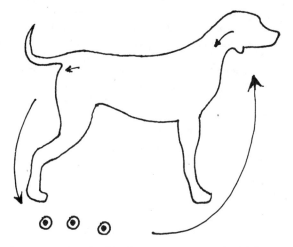

Figure 47.4. Life cycle of *Giardia*. *Giardia* has a simple and direct life cycle. The animal becomes infected after ingesting cysts from a contaminated environment. The parasite comes out of the cyst in the small intestine and starts multiplying (no sexual reproduction). Cysts are then formed in the small and large intestines and are later shed in the feces. These cysts are resistant in the environment, and can remain infective for several months under moist and cool conditions. Illustration by Noha Abou-Madi.

The indirect life cycle is more complex, as a parasite with such a life cycle needs one or more intermediate hosts to mature and reproduce (figure 47.5). The definitive host releases a stage of the parasite that is immature and cannot reinfect that animal. These forms are picked up by intermediate hosts, in which they develop into the infective stage. For the definitive host to become infected, it must eat or be bitten by the intermediate host that is carrying the infective form of the parasite. Once the infective intermediate host is ingested by the definitive host or bites the definitive host, the parasite matures and sexual reproduction completes the life cycle.

The life cycle of the meningeal worm *Parelaphostrongylus tenuis* provides a good example to illustrate the concepts of an indirect life cycle and intermediate and aberrant hosts. The adult parasites live between the membranes of the brain (meninges) and in the veins and venous sinuses of the brain of white-tailed deer (*Odocoileus virginianus*), essentially without causing disease or inflammation. The eggs are released by the females, usually in the blood stream; they stop in the lungs, where each egg hatches as a first-stage larva (L1). These larvae are coughed up, swallowed, and, surviving the passage through the gastrointestinal tract, are excreted on the

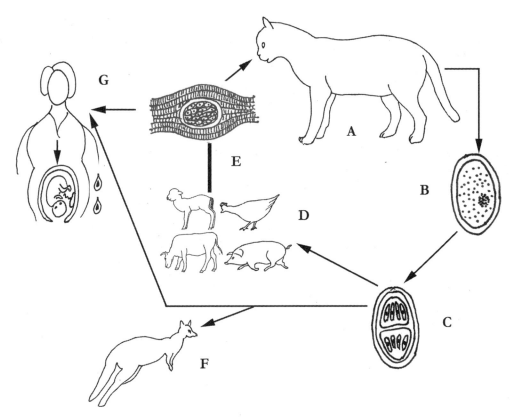

Figure 47.5. Life cycle of *Toxoplasma gondii*. *Toxoplasma* has a complex life cycle requiring intermediate hosts to complete its development. The definitive hosts (A) for *Toxoplasma* are wild and domestic cats; they become infected when ingesting infected prey or raw meat (E). The cats shed oocysts (B) in feces 3 to 18 days after ingesting the parasite. These oocysts are very resistant in the environment; they may contaminate water, food, and soil but they are not infective to other animals. Once sporulated, these oocysts (C) can infect other animals (D, F, G). Humans, other mammals, and birds are the intermediate hosts to this parasite; once ingested, the parasite will form cysts in tissues in various organs of their bodies (E). In susceptible species (F), the development of the parasite in the tissues will cause severe disease and may be fatal. In humans (G), congenital infection may cause severe disease in children, and immunodeficient individuals are susceptible to developing clinical toxoplasmosis. Illustration by Araceli Lucio-Forster.

surface of the feces in a mucous layer. Once excreted with the feces, the successful larvae will leave the fecal pellets and be ingested by, or penetrate into, terrestrial snails or slugs. These gastropods become intermediate hosts in which L1 larvae develop into second- and third-stage (L2 and L3) larvae. The third-stage larva or L3 is the infective stage. White-tailed deer become infected by accidentally ingesting the snails containing L3 while grazing. The parasite completes its cycle when the larvae exit the intestinal tract, migrate to the spinal cord and brain, become sexually mature and reproduce. If ingested by other hosts (aberrant hosts) such as llamas, moose, mule deer, elk, reindeer, sheep, or goats, the same parasite can be lethal. The larvae stay in the spinal cord and brain tissue, and cannot migrate to the veins and sinuses. Their migration causes physical damage and inflammation, resulting in neurologic signs (weakness, inability to stand, etc.), paralysis, and often death (Lankester 2001, 228–78).

TRANSMISSION

The mode of transmission depends on the type of parasite, and some organisms may employ more than one approach. The most common form of transmission is oral ingestion of the infective stage of the parasite (e.g., egg, larva, oocyst, or cyst) from contaminated water, food, or environment, or through grooming. Parasites can also be transmitted through bites from insects (e.g., mosquitoes). Alternatively, some organisms are free in the environment and can physically traverse the host's skin, such as the amphibian parasite *Rhabdias* sp., which may cause pneumonia. Staff can help spread infective forms of the parasites to other areas in a zoo facility in organic material carried on hands, footwear, service vehicles, and cleaning equipment. Since parasites prefer moist environments, runoff from one contaminated cage can be flushed to a clean pen during hosing or heavy rains.

CLINICAL SIGNS ASSOCIATED WITH PARASITIC INFECTION

Clinical signs consistent with parasitism vary according to the organ system(s) targeted by a parasite. These signs are often nonspecific, and nonparasitic causes for disease must be ruled out. Moreover, parasites can further weaken an animal's immune system and may cause it to become more susceptible to other diseases. In wild populations, animals with heavy burdens of worms often have other underlying problems.

1. Parasitized animals often appear unthrifty, weak, and thin (in poor body condition).
2. Abnormal stools are the most common clinical signs associated with gastrointestinal parasites. Keepers must collect any abnormal feces; these can vary from a sample with a soft consistency, the presence of blood, or excessive mucus around feces to mild or severe diarrhea. Tapeworm segments and sometimes roundworms can be visible in feces.
3. Anemia and hypoproteinemia caused by blood loss can manifest as weakness associated with pale mucous membranes, and subcutaneous swelling (edema) will

appear mostly under the chin (intermandibular space) and lowest parts of the chest and belly (ventrum).
4. If a parasite infects the central nervous system, neurologic deficiencies can develop and may progress from weakness, to lethargy, weakness on one or both sides of the body, paralysis, seizures, and death.
5. Animals with lungworm infections may not show symptoms at first. However, signs of pneumonia or bronchitis—such as increased respiratory effort, decreased resistance to exercise, and productive chronic coughing— may develop if the damage from the parasite is severe, or if there is a secondary bacterial infection.
6. Animals presenting with hair loss, self-induced trauma to the skin, or secondary bacterial dermatitis must be tested for ectoparasites. Many but not all ectoparasites are visible on an animal's coat or skin.
7. Severe and acute parasitic infections can also be fatal with few forewarning clinical signs. Death with few preceding clinical signs has been reported in snakes following infection with the parasitic amoeba *Entamoeba invadens*.

DIAGNOSIS

The majority of parasites produce eggs, oocysts, or larvae that are too small to be seen or identified without magnification. Monitoring for the presence of parasites in the host and the environment must be carried out regularly (as often as every three to four months) and should be intensified during quarantine and high-risk periods, such as spring, fall, and around birthing times.

COLLECTION OF SAMPLES

A representative amount of biological material (or sample) is often requested by veterinarians so that they can analyze its contents. Commonly screened materials include feces, hair and skin scrapings, urine, subcutaneous cysts, oral parasites, and blood. Additionally, during necropsy the gastro-intestinal contents and mucosal stripping, sections of muscle or liver, cerebrospinal fluid, content of eyes and cysts will be examined for parasites, along with routine histological examination of sections of the various organs.

Keepers should always wear gloves when collecting and handling samples and should wash their hands frequently, as many commonly found parasites have a zoonotic potential. Samples should not be submitted in an inverted plastic glove or plastic bag. It is best to place these samples in screw-top leakproof containers to avoid accidental spills and contamination. The containers must be labeled clearly with the species and individual identification of the animal, the date and time of collection, and the material obtained. If analysis is not performed on site, referring laboratories often request double packaging of fecal samples for shipping. Primate samples should be clearly labeled, as laboratories take extra precautions during handling of such items.

Fresh samples must be collected and analyzed within 24 to 48 hours for optimum results. It is important to avoid delays

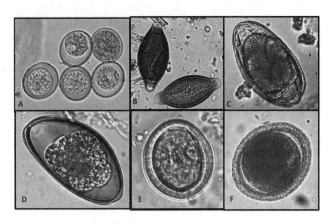

Figure 47.6. Examples of commonly seen parasite eggs: (A) coccidia, from a cardinal (*Cardinalis cardinalis*); (B) *Trichuris*, from a bison (*Bison bison*); (C) *Cruzia americana*, from a raccoon (*Procyon lotor*); (D) pinworm, from a bearded dragon (*Pogona vitticeps*); (E) taeniid egg, from a bobcat (*Lynx rufus*); (F) *Toxocara canis*. Photos by Noha Abou-Madi.

in analyzing the samples, or the parasites and eggs will decay or hatch and become harder to identify. Parasites are fragile, and a fully intact specimen is needed for accurate identification. If the samples cannot be processed or examined immediately, they must be stored under proper conditions (see below).

PROPER STORAGE OF SAMPLES

If possible, at least 10 grams of feces should be submitted. The corresponding volume of feces will vary with its consistency (moisture content), but a good approximation is about one to two tablespoons. All fecal samples should be refrigerated at 4 °C and protected from heat and dryness. Refrigeration is essential if the samples are not placed in a preservative (a preserved sample is said to be "fixed"), but it is effective only for a short time (24 to 48 hours). Fecal samples can be fixed in polyvinyl alcohol, glycine, or formalin. Refrigerators used to store fecal samples must never be used to store food items for human or animal consumption.

Free nematode worms can be stored in saline and refrigerated for one to two days maximum; otherwise they should also be placed in 70% alcohol. Free flukes and tapeworms are best stored in tap water and refrigerated for 24 to 48 hours only. Collected ectoparasites are stored in 70% alcohol, glycine, or 10% buffered formalin.

Screening for blood parasites (hemoparasites) is performed on blood smears made from blood samples collected fresh without anticoagulant or stored in ethylenediaminetetraacetate (EDTA). Blood samples from bird and reptile species that react negatively to EDTA can be stored in tubes using a different anticoagulant such as lithium heparin.

EXAMINATION OF FECES

Parasitologists, veterinarians, and veterinary technicians are trained to perform tests and interpret results to screen for the presence of parasite ova and cysts (figure 47.6). The use of a microscope is essential to identify the type of parasite present, and sometimes eggs or oocysts must be cultured until they hatch or sporulate to confirm identification (e.g., in the case of nematode and coccidian species). Other techniques can be used in specialized laboratories to find specific parasites or parasite antigens (e.g., polymerase chain reaction [PCR], or enzyme linked immunosorbent assay [ELISA]).

Selected examination techniques involve looking at a sample of feces or blood directly under the microscope with minimal processing. Other methods concentrate the eggs or parasites in order to increase yield and detection while

eliminating nonrelevant material (Foreyt 2001, 3–10). This is achieved using flotation techniques based on the difference in specific gravity of each component. Regardless of which tests are used, examination of the material should start with a description of the sample (i.e., amount, consistency, odor, etc.) and gross evaluation for the presence of blood, mucus, whole parasites, or segments of parasites.

DIRECT SMEAR OR WET MOUNT

The examination of a direct smear must be performed immediately after collection of the sample. Using this technique, water samples and fish skin scrapings can be evaluated for parasitic infection (and occasionally bacterial infections). Direct examination of feces will allow for the detection of moving organisms (protozoa, larval worms) that would otherwise be destroyed by flotation solutions. This method also allows for the detection of other parasite eggs, cysts, and larvae, and is very valuable if the sample submitted is very small (fecal samples from birds, amphibians, and small reptiles).

FLOTATION TECHNIQUES

All fecal flotation techniques rely on the difference in density (buoyancy) between the parasites and the rest of the fecal material. Most available flotation solutions will allow for the detection of nematode and cestode eggs and protozoan cysts. With some solutions, however, heavier trematode eggs fail to float, and protozoan and certain nematode larvae become distorted beyond recognition.

To circumvent this problem, many laboratories recommend using two flotation techniques, one using zinc sulfate (a salt solution) and the other a saturated sugar solution. The zinc sulfate solution is superior for the detection of *Giardia*, nematode larvae, and fragile protozoan cysts, while heavier parasite eggs will float in the sugar solution. Salt solutions require prompt reading of the slides, as those preparations tend to crystallize quickly.

CENTRIFUGAL FLOTATION TECHNIQUES

Fecal flotation techniques rely on the difference in density (buoyancy) between the parasites and the rest of the fecal material. In a saturated salt solution, most parasite eggs will float above the fecal material. The separation of the eggs in the solution is enhanced by spinning down the sample (centrifugation). These techniques are the best methods to detect parasite eggs and cysts, but they require a straight-head, swinging-bucket centrifuge. A quantitative fecal examination can be requested to provide an estimate of the number of eggs or oocysts in a sample. This information helps veterinarians determine if a treatment is effective or if drug resistance is developing, and gives an idea of the intensity of infection.

PASSIVE FLOTATION

Passive flotation techniques, which use gravity instead of centrifugation, are used widely in zoos, but their rate of parasite recovery is inferior to that of centrifugal flotation methods (figure 47.7). They have the advantage of being commercially available, with some in kits (Ovassay®, Fecalizer®, etc). They should only be used when centrifugation is not possible.

PARASITE CONTROL PROGRAM

As with all of the aspects of preventive medicine, a parasite control program must be adapted to the zoo's collection of animals, and it should never be limited to a single facet. In fact, "routine" administration of deworming drugs (anthelmintics) will cause severe resistance problems if repeated treatments are done without prior knowledge of an individual's status, without monitoring, and with no change in the environment. As a first step, an institution must identify and quantify the level of parasitism in all of the animals from fish to elephants, and for those animals with positive results the level of contamination in their environment must be estimated. The importance of each parasite to the individual, the species, and the collection must be assessed.

Often, the result of a successful program will be limited to a decrease in the morbidity and mortality associated with a parasite, since complete eradication of an organism is often impossible. This is why strict quarantine protocols that will effectively prevent the introduction of new species of parasites in a colony are so important. Maintaining a negative status is exceedingly difficult, especially in a facility with outdoor habitats in which new animals are routinely acquired. A parasite control program must include measures to improve the health of the animals so they can fight off parasitic infections, to decrease exposure to parasites, and to provide a deworming schedule appropriate for the individual animals and for the collection as a whole (Isaza and Kollias 1999, 593–96).

OPTIMIZING THE ANIMAL'S HEALTH

Animals in poor body condition or with chronic debilitating diseases will be more susceptible to parasitic diseases. With improvement in their general health status and with proper nutrition and vaccination programs, they will have stronger

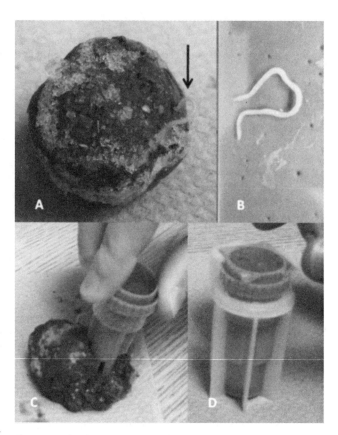

Figure 47.7. Stool sample collected from a lynx (*Lynx lynx*), showing an adult nematode (A, black arrow). The parasite is collected and submitted for identification (B). A sample of the feces is scooped into the base of the container (C). The flotation solution is added until a meniscus is created, and the microscope slide is placed over the rim of the container. Photos by Noha Abou-Madi.

resistance to parasitic infection and their immune systems will be able to mount appropriate responses to pathogen exposure. Basic hygiene measures applied by the entire staff will also have a significant impact on the animals' overall health by decreasing their exposure to pathogens (not only parasites).

CONTROL OF WILDLIFE, VERMIN, AND INSECTS

Outdoor habitats present the additional challenge of parasite contamination by wildlife. Often the zoo animal population will not have been exposed to these parasites previously (i.e., they will be naive) or will have a greater susceptibility to infection (e.g., in the infection of llamas with *P. tenuis*). Wild mice carry *Toxoplasma gondii*, among other pathogens, and may transmit the parasite to susceptible species (e.g., Pallas cats and kangaroo-related species [macropods]). Cockroaches and beetles are intermediate hosts to significant pathologic parasites (*Gongylonema*, *Moniliformis*, etc.), bugs such as kissing bugs (triatomids) transmit *Trypanosoma cruzi* (responsible for Chaga's disease), and specific species of ticks (*Ixodes scapularis* and *Ixodes pacificus*) will transmit *Borrelia burgdorgeri*, the bacteria responsible for Lyme disease. Mosquitoes transmit the avian malaria parasite (*Plasmodium*). Therefore, hygiene measures and judicious, humane, and safe control of wildlife and vermin are essential in a comprehensive parasite control program.

ANTHELMINTIC DRUGS

Anthelmintic drugs are an important tool for combating parasites in a zoo. They can be used to decrease the number of live parasites, leaving the host's defense mechanisms to cope with the secondary inflammatory response or the remaining parasites. The dose and method of administration of the drugs must be prescribed by a veterinarian, who will also recommend a treatment regimen to avoid complications related to large numbers of parasites dying simultaneously within an animal. Intestinal impactions and death have been documented in cases when large numbers of intestinal nematodes have died and were evacuated at once.

Antiparasitic agents can be administered topically, injected in the muscle or subcutaneously, or given orally. The frequency of administration depends on the drug and should be confirmed with the veterinarian. Some drugs are given once monthly, others once a day for several days. Some drugs must be repeated, while others are given only once. Some drugs target one type of parasite (e.g., imidocarb dipropinate, effective against protozoa), while others have broad spectrums of action (e.g., ivermectin, fenbendazole).

It is important to remember that antiparasitic drugs can be toxic to certain species, breeds, individuals, and young animals (e.g., organophosphates, ivermectin, fenbendazole, ionophors, levamisol). Side effects vary with the product and can include anorexia, lethargy, muscle weakness, paralysis, bone marrow depletion, necrosis of the intestinal mucosa, and death. Some drugs are available over the counter. However, because of the many variables (e.g., many different animal species) the value of the animals (many of which are endangered or threatened), drugs should be used with caution and appropriate doses should be prescribed by and used under the supervision of a knowledgeable veterinarian. The efficacy of the drugs and the changes in husbandry should be monitored with repeated fecal analyses and estimations of the number of eggs per gram of feces collected. Efficacy and confirmed adverse effects should be reported to veterinarians and zoo managers for management and possible publication.

Resistance to anthelmintic drugs is a serious complication and common cause of failure of treatment; it is caused by the repeated use of the same product and/or administration at an incorrect dosage and frequency, and possibly also by flawed hygiene measures. The parasite becomes resistant to the drug, and the drug becomes ineffective against infections.

DECREASING THE PRESSURE OF INFECTION

Eggs and cysts that are shed into the environment often have considerable resistance to external conditions, including routine disinfectants. As an animal becomes infected and sheds parasitic eggs and oocysts in a confined area, amplifying the contamination of the environment, the chance for reinfection and infection of other animals is increased. This is especially true when the substrate (e.g., soil) cannot be readily changed. Many parasites flourish in moist environments. Lush green pastures are esthetically pleasing and are sought after by zoo visitors, but they are also a great environment from a parasite's point of view!

No matter how many and how often anthelmintic drugs are used, a parasite control program will fail if an animal continues to be exposed to parasites and reinfects itself. Keepers play a crucial role in controlling the exposure and the concentration of parasites that an animal is exposed to. Decreasing the chance that an animal encounters an infective stage of a parasite is the keystone to every parasite control program.

With the parasite's life cycle in mind, a parasite control program must consider the following recommendations:

- Judicious use of anthelmintics, considering route, frequency, and ease of administration; side effects; safety in pregnant females and young animals; and treatment of a herd prior to birthing season. Anthelmintics should be used in conjunction with other measures to decrease the pressure of infection. Alternating the types of anthelmintics used will help to decrease the selection of drug-resistant parasites.
- Removal of susceptible animals (e.g., young animals) from a contaminated area, grouping of animals according to their susceptibility to a parasites, and rotation of animals from one pasture to another after treatment. These measures will decrease the pressure of infection and contamination of an area. For example, equids and ruminants should be alternated, since they share a limited number of parasites.
- Habitat modifications (e.g., changing soil, directing drainage to an area without animals, or letting a pasture dry up).
- Improved hygiene (e.g., wearing gloves, washing hands, cleaning food and water bowls once or twice daily, repositioning the bowls so that the animal does not defecate in them, promptly removing manure and contaminated material).
- Removal of intermediate hosts.
- Monitoring of the efficacy of the drugs and the parasite control program through regular fecal analyses and necropsies.

SUMMARY

In conclusion, a comprehensive preventive medicine program should be integrated into the daily care of every collection of animals. Attentive monitoring, timely intervention, and appropriate collection techniques of samples for diagnostic tests must support strategic treatment and correction of problems. Proper sanitation and effective quarantine protocols are required to prevent the introduction and the spread of pathogens. Finally, integration of sound husbandry practices with preventive medicine will ensure the welfare, good health, and reproductive success of captive animals and the safety of the staff and the public.

REFERENCES

AZA Animal Health: www.aza.org/animal-health.
Childs-Sanford, Sarah E., George V. Kollias, Noha Abou-Madi,

Michael M. Garner, and Husni O. Mohammed. 2009. "*Yersinia pseudotuberculosis* in a closed colony of Egyptian fruit bats (*Rousettus aegyptiacus*)." *J Zoo Wild Med* 40:8–14.

Foreyt, William J. 2001. "Diagnostic Parasitology." In *Veterinary Parasitology*, edited by William J. Foreyt, 3–10. Ames: Iowa State University Press.

Fowler, Murray E. 1986. "Hoof, Nail, and Claw Problems in Mammals." In *Zoo and Wild Animal Medicine*, edited by Murray E. Fowler, 549–57. Philadelphia: W. B. Saunders.

Hinshaw, Keith C., Wilbur B. Amand, and Carl L. Tinkelman. 1996. "Preventive Medicine." In *Wild Mammals in Captivity: Principles and Techniques*, edited by Devra G. Kleinman, Mary Allen, Katrina V. Thompson, and Susan Lumpkin, 16–24. Chicago: University of Chicago Press.

Hosey, Geoff, Vicky Melfi, and Sheila Pankhurst. 2009. "Health." In *Zoo Animals: Behavior, Management and Welfare*, 379–426. Oxford: Oxford University Press.

Isaza, Ramiro, and George V. Kollias. 1999. "Designing a Trichostrongyloid Parasite Control Program for Captive Exotic Ruminants." In *Zoo and Wild Animal Medicine: Current Therapy 4*, edited by Murray E. Fowler and R. Eric Miller, 593–96. Philadelphia: W. B. Saunders.

Jensen, James M. 1999. "Preventive Medicine Program for Ranched Hoofstock." In *Zoo and Wild Animal Medicine: Current Therapy 4*, edited by Murray E. Fowler and R. Eric Miller, 585–92. Philadelphia: W. B. Saunders.

Lankester, Murray W. 2001. "Extrapulmonary Lungworms of Cervids." In *Parasitic Disease of Wild Mammals*, edited by William M. Samuel, Margo J. Pybus, and A. Alan Kocan. 228–78. Ames: Iowa State University Press.

Lowenstine, Linda J. 1999. "Health Problems in Mixed Species Exhibits." In *Zoo and Wild Animal Medicine: Current Therapy 4*, edited by Murray E. Fowler and R. Eric Miller, 26–29. Philadelphia: W. B. Saunders.

Miller, E. Eric. 1999. "Quarantine: A Necessity for Zoo and Aquarium Animals." In *Zoo and Wild Animal Medicine: Current Therapy 4*, edited by Murray E. Fowler and R. Eric Miller, 13–16. Philadelphia: W. B. Saunders.

Roberts, James W. 1993. "Primate Quarantine." In *Zoo and Wild Animal Medicine: Current Therapy 3*, edited by Murray E. Fowler, 326–30. Philadelphia: W. B. Saunders.

Van Foreest, Andries W. 1993. "Veterinary Dentistry in Zoo and Wild Animals." In *Zoo and Wild Animal Medicine: Current Therapy 3*, edited by Murray E. Fowler, 263–68. Philadelphia: W. B. Saunders.

Wells-Mikota, Susan K. 1993. "Wildlife Laws, Regulations and Policies." In *Zoo and Wild Animal Medicine: Current Therapy 3*, edited by Murray E. Fowler, 3–10. Philadelphia: W. B. Saunders.

Whitaker, Brent. 1999. "Preventive Medicine Program for Fish." In *Zoo and Wild Animal Medicine: Current Therapy 4*, edited by Murray E. Fowler and R. Eric Miller, 163–81. Philadelphia: W. B. Saunders.

48

Veterinary Diagnostics

Cynthia E. Stringfield

INTRODUCTION

Obtaining diagnostic information from zoo and aquarium animals can be dangerous for humans and/or the animals, and difficult to obtain depending on the species of animal and the veterinarian's expertise with that species. Because of this, historically in zoo and aquarium medicine, animals were often treated on the basis of minimal diagnostic information. As medical and husbandry knowledge has improved, this is no longer acceptable in zoo and aquarium animal health care, as more diagnostic information can be obtained to lead the veterinarian to a diagnosis. Keepers may be best suited to collect a sample or perform a diagnostic procedure due to their relationship with the animal, or via a trained behavior. For example, blood draws may be trained in many species, trunk washes may be done voluntarily in elephants, and swabs may be taken if an animal is trained to "present" to the keeper. The keeper's ability to obtain diagnostic information can contribute greatly to the success of achieving an accurate diagnosis. Additionally, decisions for testing will be based on the animal's history, any signs of illness, and the ability to achieve diagnostic information in a safe, minimally stressful way for the animal. The keeper's knowledge of the animal and ability to relay correct and important information can lead the veterinarian down the correct diagnostic path, or, to the detriment of the animal, an incorrect diagnostic path. Understanding the value and importance of information to achieving a correct diagnosis will allow the keeper to actively participate in the diagnosis and health monitoring of the animals they care for.

This chapter will provide basic information about diagnostic procedures, emphasizing how animal keepers can assist with obtaining diagnostic information about the animals they care for. After studying this chapter, the reader will understand

- clinical pathology procedures, including how to obtain and handle blood samples, cultures, cytology slides,

biopsies, urine and other fluid samples, fecal parasitology, and skin scrapings
- diagnostic imaging types and radiation concerns
- endoscopy and exploratory surgery
- necropsy.

NEEDLE AND SYRINGE

A keeper may assist veterinary personnel, or use a hypodermic needle and/or syringe themselves, to obtain a diagnostic sample. The reader should refer to chapter 45 for more on needle and syringe usage. At its most basic, the plunger of a syringe (used with or without a needle) will be pulled back to collect a diagnostic sample. A needle will be needed if the sample must be collected through tissue. The sample, such as blood or other biological fluids, will be contained in the barrel of the syringe.

Sterility is an important concern, and a needle must remain sterile before it is introduced into an animal to prevent the introduction of possible pathogens. When desensitizing an animal during training, a needle may be reused if it is only touching the skin, but once it punctures the skin, a new sterile needle must be used each time.

The length of the needle will depend on how deep the needle needs to penetrate. The diameter will be chosen based on the size of the animal: a larger-diameter needle will collect fluid more quickly, and will also be more resistant to bending or breaking. For example, when collecting blood samples, the largest needle that can be placed in the vein successfully should be chosen. When training an animal to accept a needle, smaller-diameter needles may be used initially to minimize discomfort. The syringe size chosen will also depend on the job for which the syringe will be used. Two things must be taken into account: how much volume will need to be collected in the syringe, and how much vacuum (suction) is necessary when the plunger is pulled back. For example, too much suction from a large syringe in a small vein would collapse the vein and prevent collection of blood. Using a syringe that is too small can make it impossible to create enough suction to collect the sample, or it may not hold enough of the liquid being collected. The right gauge (diameter) and length of needle must be combined with the right size of syringe. Considering the variety of jobs that needles and syringes must do, and the variety of collection sites and sizes of patients in a zoo or aquarium, the scenarios may seem endless! However, there are combinations (called "set-ups") that are commonly used for certain sizes of animals. For example, for small dogs and cats, a 3 cc syringe with a 2.54 cm (1 in.), 22 gauge needle is a common prepackaged "set-up." This may be what the zoo or aquarium veterinarian chooses for blood collection in a serval or a raccoon. For small to medium-sized birds, reptiles, and small mammals (e.g., a turaco, kingsnake, or tamarin), however, the choice may be a 1 cc syringe with a 25 gauge needle. Conversely, an 18 gauge, 3.81 cm (1.5 in.) needle with a 12 cc syringe may be used for collecting blood from an elephant. Remember that an 18 gauge needle has a larger diameter than a 25 gauge needle, as the smaller the gauge number, the larger the diameter of the needle.

"Butterfly needles" or winged-infusion sets have the needle on the end of plastic tubing, and allow the user to manipulate the needle with greater ease; they also allow some distance between the needle and the syringe. These types of needles are very helpful when training an animal through mesh, or for an animal that is moving a little. Vacutainer needles are unique, and are used only to collect blood. They are composed of two needles, facing in opposite directions, with one needle covered in a plastic sheath. After inserting the unsheathed needle into the animal's vein, the sheath-covered needle is pushed into a blood tube (vacutainer tube). The tube's vacuum draws the blood into it. Due to the vacuum effect on the vein, this method only works with larger zoo animals. Vacutainer needles also come in a "butterfly" form.

BLOOD SAMPLES

Veterinary books and articles abound with information regarding the anatomy of peripheral veins that can be used for venipuncture (puncture of a vein) in different species to collect venous blood samples for analysis. Veins commonly used include the jugular vein on the neck, the saphenous vein on the inside or outside of the rear leg, the cephalic vein in the front leg, the tail vein, the wing or ulnar vein, and the tarsal vein of the ankle. The site to be used for a blood sample can be wetted with alcohol prior to venipuncture to improve visibility through hair or feathers, and to aid in disinfecting the site. Especially in mammals, the vein may be "held off" (occluded by applying pressure) manually or with plastic tubing tied around an appendage (figure 48.1). This occludes the vein, as it drains towards the heart. This will make the vein appear larger, as the blood that would otherwise drain to the

Figure 48.1. Keepers devised this method of obtaining blood voluntarily from this yearling giraffe for a plasma transfusion needed by a newborn. Notice the jugular vein being occluded as blood comes down the giraffe's neck toward the heart, and the needle being inserted into the vein. The hand-raised animal is sucking on a bottle of water, with his head out of a keeper observation door. Courtesy of Cynthia Stringfield.

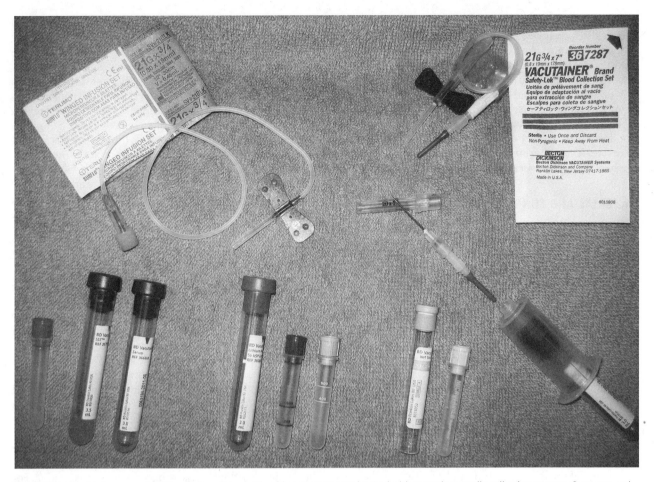

Figure 48.2. Blood tubes come in different sizes. Tubes with rubber stoppers may be used with vacutainer needle collection systems. Green-top and lavender-top tubes have anticoagulants in them. Red- or marble-top tubes (also called tiger-top tubes) are empty, for collecting samples to analyze serum. Some tubes contain gel to assist in separating the formed elements from the liquid part of blood when the tube is centrifuged. Also shown in this picture are a butterfly needle (winged infusion set) and a vacutainer needle system. Courtesy of Cynthia Stringfield.

heart backs up in the vein. Before use, the syringe plunger should be pulled slightly and replaced to "pop" the syringe vacuum, so that this does not happen when the needle is in the vein. The needle, bevel (hole) up, penetrates the skin and underlying tissue to enter the vein at an angle to allow it to just enter the vein, or to thread into the vein in a parallel fashion.

The plunger is pulled to aspirate blood into the syringe as the needle enters the vein and the proper amount is collected. Before the needle is removed, one should stop occluding the vein to decrease bleeding from the injection site. Once the needle is removed, pressure must be applied to the site to prevent bleeding until a clot is formed.

Blood samples are used for a variety of diagnostic tests, and must be immediately placed in the appropriate tube for transport, storage, and testing (figure 48.2). Some tubes contain an anticoagulant to prevent the blood from clotting (coagulating), and it is important that the blood is obtained quickly and placed in these tubes before it coagulates. These tubes must then be gently rocked back and forth (not shaken) to mix the blood with the anticoagulant, and the samples should be checked to be sure they have not clotted. Other tubes have no anticoagulant because blood clotting is de-

sired. Blood consists of formed elements (red blood cells, white blood cells, and cell fragments called platelets) that are suspended in liquid. When circulating in the body, or in a tube with anticoagulant, the liquid part of blood is called plasma. When blood has clotted, the liquid part of blood is called serum. Therefore, tubes with no anticoagulant are used to collect serum samples. These tubes do not need to be rocked back and forth.

Blood is an incredibly important diagnostic sample because of the information it can provide. The formed elements can only be analyzed in a blood sample that has not clotted. Some tests require only the serum; others require only the plasma. Thus it is very important to obtain the sample properly and in the proper tubes.

The most common of diagnostic blood tests is often called "a blood panel," which refers to a complete blood count (CBC), which looks at the formed elements, and a wide variety of other tests that analyze plasma or serum, to assess kidney, liver, and muscle function as well as glucose ("blood sugar"), protein, cholesterol, and electrolyte levels. More specialized tests can be run to evaluate certain organs or screen for certain diseases. Serologic tests (or serology) are run on serum to check antibody and antigen levels for very specific diseases such as rabies or tetanus. Measurements of antibody levels are termed antibody "titers"; they help identify exposure to a pathogen and the body's response to that exposure. They may also be used to assess response to vaccination.

Blood samples should be placed in a cool dark area—such as a refrigerator, a cooler with a cold pack, or the shade—as soon as possible after sampling, and until processing. Some blood samples are very difficult to obtain and may be irreplaceable. It is important to ensure they are not ruined by overheating or improper processing.

BACTERIAL AND FUNGAL CULTURES

Culture is the term for growing organisms such as bacteria and fungus in a laboratory. Blood can be cultured for bacteria, but the more common samples for bacterial cultures are bodily discharges, urine, and feces. Fungal cultures are often done on bodily discharges and skin samples. For culture samples, a culturette is commonly used (figure 48.3). This is a cotton swab on a long metal or plastic stick that is put into a tube filled with preservative after the sample is obtained. It is essential that the swab only touches the substance to be cultured and is then put into the sterile culturette tube.

The sterile swab must contact nothing other than what is to be cultured, avoiding things such as fingers, dirt, or fur, to avoid microorganism contamination from those sources. It is important that the only microorganisms that grow in the lab are from the target site. An example would be an animal with an abscess, or pocket of pus, that has burst. Only the pus should be touched with the swab. In a lab, the swab is applied onto a special medium (or gel) on a plate (also called a petri dish), to inoculate the gel with the microorganisms on the swab. The microorganism of concern (in this case, bacteria) will grow. Aerobic cultures expose the plate to air, and anaerobic cultures are grown without air to mimic the conditions the bacteria needs to grow. The plates are also kept at the proper temperatures to facilitate growth. Through microbiologic techniques, the bacteria can be identified. Antibiotic sensitivities are tests to see which antibiotics will kill or prevent growth of the bacteria; this is often referred to as "running a sensitivity." This is done by dropping little paper discs full of the antibiotics onto the plate before the plate is inoculated. If the bacterium is sensitive to the antibiotic, it will not grow where the disc is. Results of a bacterial culture and sensitivity usually take several days. Fungal cultures are similar in nature, but are slower growing, and will not have sensitivities run. These results usually take weeks instead of days.

CYTOLOGY

Cytology refers to the study of cells, and it occurs when cells are put on a microscope slide, stained, and examined microscopically. Samples may be obtained via a fine needle aspirate (FNA), wherein a needle is inserted into a lump or mass and a sample drawn into the syringe (aspirated). Material in the syringe is then expelled onto a slide. The author has used this test for a wolf with a soft, slow growing mass on her side (figure 48.4). The wolf was trained to permit FNA, and the cytology of the aspirate revealed fat cells leading to a diagnosis of the mass being a harmless collection of fat cells (a benign lipoma).

Urine and other body fluids and discharges may also be examined via cytology. The suffix "-centesis" is used to denote the obtaining of fluid by inserting a needle and aspirating. For example, cystocentesis refers to putting a needle into the blad-

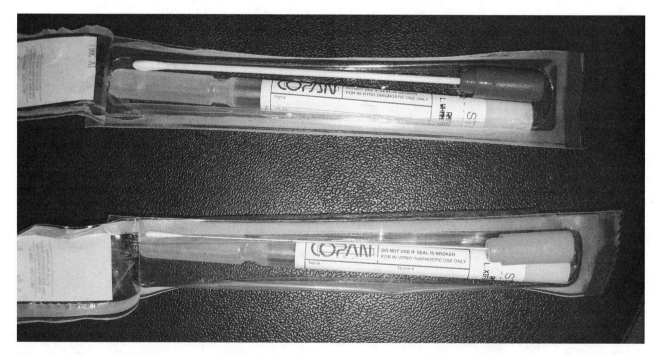

Figure 48.3. Culture swabs come in different sizes and in sterile packaging (opened in this photo). After the sample is obtained, the swab is placed in the tube of transport media. Courtesy of Cynthia Stringfield.

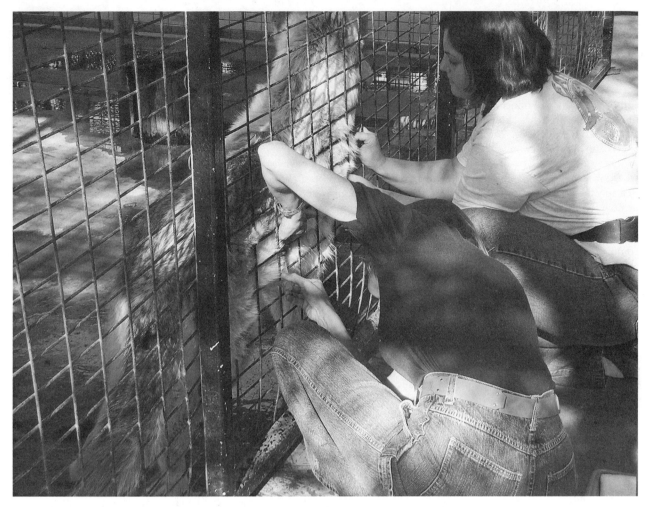

Figure 48.4. Fine needle aspirate of a fatty tumor (lipoma), obtained via a trained behavior in a wolf. Courtesy of Cynthia Stringfield.

der to obtain a urine sample; thoracocentesis, arthrocentesis, abdominocentesis refer to chest, joint, and abdomen aspirations respectively. Additionally, a swab may be smeared or rolled on a slide, or a tissue may be touched or smeared on a slide (a procedure called touch prep or impression smear), to get cells from it onto a slide. The slides do not need refrigeration. They are then stained with specific dyes to color the cells and other material on the slides. Examination of the slides will be completed by a laboratory or veterinary technician, a veterinarian, or a clinical pathologist.

BIOPSY

A biopsy is when a piece of tissue is collected and then "fixed" or preserved in formalin (formaldehyde), and then extremely thin (one cell thick) slices are placed on a microscope slide, stained, and examined. (This is done in a pathology laboratory, using a specialized piece of equipment called a microtome to slice the tissue.) The examination of these slides is termed histopathology ("patho-" referring to the abnormal, and "histo-" to tissue) or histology (the study of tissues).

A veterinary pathologist is a specialist who examines these slides. If a keeper obtains a piece of tissue for biopsy under

a veterinarian's guidance, one should drop it in formalin, taking care not to inhale the formalin fumes or get formalin on one's hands, as formalin is toxic. Formalin samples do not need refrigeration.

URINALYSIS

Urinalysis is the analysis of urine. Urine is a direct reflection of a mammal's urinary tract function, and it shows what is being excreted by the kidneys (since the kidneys filter the blood). Kidneys normally concentrate urine, since the body usually needs to conserve water. Urine will also contain any cells being sloughed off by the bladder, or blood, if bleeding is occurring somewhere in the urinary tract. As such, urinalysis can identify bladder and kidney problems as well as other problems in the body. For example, urine from an animal with a bacterial urinary tract infection may contain blood, bacteria, and white blood cells. Urinalysis results can also be extremely important in diagnosing diseases outside the urinary tract, such as diabetes and liver disease: a diabetic animal would be losing glucose in the urine, and an animal with liver disease may have bilirubin in the urine. Normal animals would not have these things in their urine. An animal that is losing kidney function may be drinking excessive amounts of water and urinating large volumes of dilute urine because the kidneys have lost their ability to concentrate the urine. Conversely, a dehydrated animal may have very concentrated urine. Protein may be lost in the urine due to kidney disease. An animal with increased minerals in its diet or a decreased ability to metabolize minerals properly may have crystals of these minerals in their urine. Stones that occlude the urinary system can then develop. Since it can be easy to obtain and analyze, urine is a very important body fluid that can provide a lot of information.

Training an animal to urinate on command, or to urinate in a specific clean location or container where the urine can be collected easily, is very valuable for its husbandry. Urine is normally sterile, and will become contaminated as it leaves the body, but if it is obtained in a clean container or location, a complete urinalysis can still be done. Only when infection is a concern and the urine needs to be cultured will a sterile sample be important. Urine can be collected in a clean container, or aspirated with a syringe and placed in a plastic laboratory urine container. This author has often seen keepers reach for whatever is handy to collect a precious urine sample when the opportunity arises unexpectedly. (A plastic toy boat was once used to collect urine from a sick beaver urinating outside its pool—a rarity!) It is a better idea to keep a syringe or plastic container in one's pocket to be prepared (a plastic syringe case works well). Samples should be refrigerated upon collection.

Other bodily fluids can be analyzed as well. One example is fluid in the abdomen (termed ascites, which can develop in liver and/or heart disease). Additionally, fluid may be introduced into an area and subsequently collected for analysis. Lavage means "wash," and a tracheal or peritoneal lavage means putting sterile fluid into the trachea or abdomen (respectively), "washing the area," and then withdrawing the fluid for analysis. A trunk lavage may be done in an elephant to culture as a means of routine tuberculosis screening. Cytology and culture is also commonly done on urine and other bodily fluids.

FECAL PARASITOLOGY

Feces can be examined by several laboratory techniques to look for gastrointestinal tract parasites. The parasites themselves and their ova (eggs) can be identified. Since feces are not sterile, they do not need to go into a clean or sterile container for this test. Only a small amount of material is needed; a pea-sized sample is commonly requested, but the veterinary team may request more. Samples should be refrigerated after collection.

SKIN SCRAPINGS

Some ectoparasites (e.g., mange mites) live in the dermal layer (the second layer from the top) of the skin; this is where blood vessels nourish the skin. A skin scraping can be performed with a scalpel blade held perpendicular to the skin to scrape away the outer layer (epidermis) and collect a sample of dermis on the blade. The skin will start to bleed slightly (like a very mild abrasion) when the dermal layer is reached. A little mineral oil on the blade will help the scrapings stick to the blade until they can be transferred to a microscope slide. The slide is immediately examined to look for parasites. Skin scrapings should be done in several locations where a skin disease symptom such as hair loss is apparent, to increase the chance of finding parasites.

DIAGNOSTIC IMAGING

RADIOLOGY

X-rays (or Röntgen rays) are nonluminous electromagnetic radiations, similar to visible light, radio, and TV signals, but of much shorter wavelength. The shorter the wavelength, the greater the energy of the X-ray beam. This allows increased penetration, so that on a radiograph (also called an X-ray film or image) black indicates air and white indicates more dense objects. "Radioopacity," or complete whiteness, occurs with metal objects. "Contrast agents" may be used to aid diagnostic imaging, as they appear radioopaque on a radiograph. For example, barium administered orally or anally will outline the digestive tract. The X-ray tube is the source of radiation. To get the proper image, the technician must calculate the proper X-ray setting of the beam and the amount of time during which the radiation is coming from the beam. These settings are recorded to form what is called a "technique chart." The X-ray film must then be developed in a darkroom using a machine called a developer.

Digital radiograph machines are similar to digital cameras. No film is used in a cassette as with traditional radiographs; instead the image is transmitted via a cord to a computer. Overall, digital radiographs require less radiation, but they still use it. The benefits of digital radiographs are numerous. No film has to be developed, and the image is seen immediately. The image can be changed on the computer screen, eliminating the need to retake films multiple times because they are "too light" or "too dark." Areas of the digital image can be magnified on the screen, and the resolution and detail far exceeds that of film images. Film images, just like film cameras, will become outdated. Expense is the primary reason why some facilities still use film radiographs.

Fluoroscopy is a moving X-ray (like a video instead of a still shot) and it uses a much larger amount of radiation because of the time involved. CT stands for computerized tomography, and is also called a CT scan or a CAT scan (for computerized axial tomography). These studies are three-dimensional X-rays: many X-rays are taken in "slices" through tissue, and a computer puts them together to give a three-dimensional X-ray view. With so many X-rays, radiation exposure is much higher with these studies.

People are exposed to radiation when in the primary beam, and also to a lesser extent when near the primary beam. This latter exposure is termed "scatter radiation" and it occurs due to lower-energy photons that have undergone a change in direction after interacting with tissue. It is important for a keeper taking part in procedures involving radiation to be properly protected and monitored, as described in other chapters. Animals can be desensitized to radiology equipment, allowing radiographs to be taken without sedation or anesthesia (figures 48.5 and 48.6). The animal must hold still very briefly, but it will not experience discomfort. Sedation

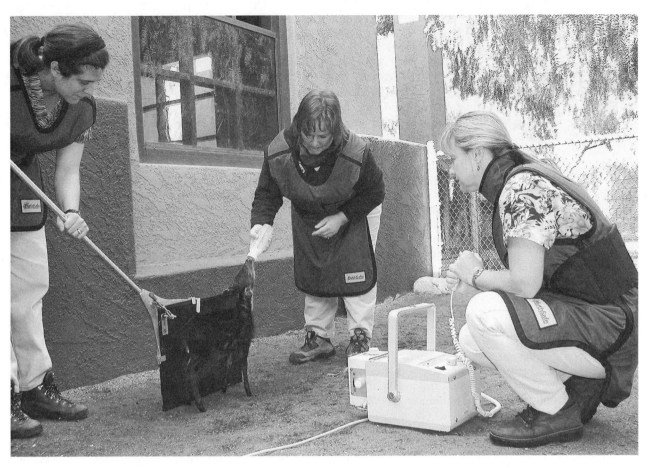

Figure 48.5 and 48.6. This young black duiker was desensitized to radiology equipment and trained to stand in front of the X-ray cassette by its keeper to prevent a risky anesthesia. Note that the keepers and veterinary technicians are wearing protective lead shielding aprons. Courtesy of Cynthia Stringfield.

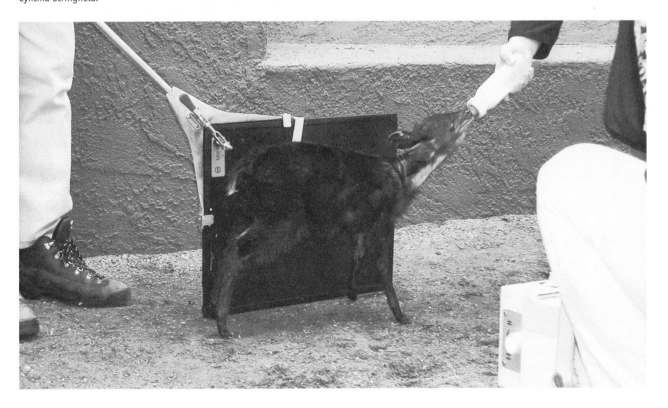

or anesthesia is needed for CT scans due to the time and machinery needed to create these images.

ULTRASOUND

Ultrasound uses sound waves of 2.5 to 12 MHz frequency (20 kHz is audible to humans, while other species are capable of hearing higher and lower frequencies). In ultrasound, a short pulse of sound is generated through a wand called a transducer. The sound waves strike the patient and return to the transducer. The strength of the returning sound determines the brightness of the image, and the time it takes to come back determines where it will appear on a screen (the transducer is attached via a cord to the ultrasound machine). This type of study is also called sonography or a sonogram. Gel is used to make the contact between the transducer and the patient's skin; there must be no air between the transducer and patient. Fur and feathers must be clipped, plucked, or parted. Newer ultrasound machines commonly use digital images on a computer screen. Older machines are larger, and look like old-fashioned television screens rather than monitor screens. Since this is a completely painless procedure, many zoo animals can be trained for voluntary ultrasound, and there are no radiation hazards related to this imaging method. The animal must be comfortable with the gel and a transducer being moved on its body, and with some slight pressure being applied through the transducer. A little movement by the animal is often acceptable. A keeper will not be allowed to desensitize the animal with a real transducer, as transducers are very expensive, so some object that looks like a real transducer should be used for training this behavior. Transducers are of various sizes and shapes, depending on the depth of the patient and the area to be "scanned."

MAGNETIC RESONANCE IMAGING

In magnetic resonance imaging (MRI), a computer reads the intensity of a radio wave signal from a tissue in which the hydrogen nuclei have been disturbed by a radio frequency pulse generated by a very powerful magnet. An animal must hold completely still and be enclosed in a tube with loud noise, so this procedure is done under anesthesia. MRI image resolution is superior to CT for some soft tissue structures, and it may show better anatomic definition. It also involves no radiation concerns. However, special anesthesia equipment is necessary due to the powerful magnet used, and no magnetic metals are allowed in the room. All metal worn or carried by humans in attendance must also be removed.

NUCLEAR MEDICINE

Nuclear medicine uses radioactive elements taken up by the body, not to generate visual images but to generate data that can be interpreted. Electromagnetic radiation is detected with a gamma camera. Due to the radioactive nature of this type of testing, it requires very specialized locations and safety equipment, and is rarely employed in zoo and aquarium medicine. New technology has created a new scan called a PET scan. PET stands for positron emission tomography, which combines nuclear medicine with CT scanning. It is leading a new wave of imaging in human medicine, since now not only anatomy but also functioning (physiology) can be imaged. The use of nuclear medicine methodology is expected to increase in the future.

THERMOGRAPHY

Digital infrared thermal imaging (DITI) is a diagnostic technique that is noninvasive and involves no exposure to potentially dangerous radiation. During an exam, a DITI camera is used to capture multicolored images called thermograms. Thermal imaging cameras detect radiation in the infrared range of the electromagnetic spectrum (roughly 9,000 to 14,000 nanometers, or 9 to 14 micrometers) and produce images of that radiation. The amount of radiation emitted by an object increases with temperature, so thermography allows one to see variations in temperature. Inflammation in the body appears as a warmer area, and this can be helpful in diagnosing areas of inflammation. Since this test can be done from a distance , use of the DITI imaging modality is becoming increasing common with zoo animals.

ENDOSCOPY

Endoscopy is a diagnostic procedure in which a scope is placed into an animal to see that part of its anatomy. Biopsies or other samples can be taken using special instrumentation with the scope. Rigid endoscopy uses a rigid metal scope or tube, and flexible endoscopy uses a bendable scope. Flexible endoscopy is commonly used to examine the inside of the gastrointestinal and respiratory systems. Rigid endoscopy is commonly used to examine the coelomic cavities of birds and reptiles, and to look at joint spaces. These rigid scopes are hooked up to light sources that shine through them; flexible scopes use a fiber optic system. These pieces of equipment are very expensive, sensitive, and fragile. An otoscope is a simple variation of this idea, using a plastic cone that goes into the animal's external ear canal and shines light through it to illuminate the ear canal. An opthalmoscope shines light into an animal's eye to enable a veterinarian to examine the eye thoroughly. Drops may be used to stain the cornea, or to dilate the pupil so the back of the eye (retina) can be examined.

EXPLORATORY SURGERY

Sometimes the quickest and best way to reach a diagnosis is to perform surgery to examine the anatomical area of concern. This is termed "exploratory surgery." Exploratory surgeries of the abdomen and joints are common. Rigid endoscopy is a recent technology that has evolved to be used surgically with additional specialized surgical instruments. Endoscopic surgery is much less invasive than traditional surgery, since the small scopes need only small incisions to enter the body, but it also requires more sophisticated technology and greater expertise.

NECROPSY

A necropsy is an "animal autopsy," or postmortem (after death) examination. Gross necropsy is examination of the animal's organs and tissues with the naked eye via dissection. Samples are collected for histopathology (similar to biopsy, except that the animal is deceased), culture, toxicology, and content analysis. Veterinarians, veterinary technicians, and pathologists perform gross necropsies. Pathologists will also examine the tissues and evaluate the information. They work with clinical veterinarians to determine a final cause of death, and identify contributing or additional conditions. Often a final diagnosis may only be found at necropsy, and this diagnostic examination is very important in order to establish a cause of death. The need for changes in husbandry practices or treatment regimes may be indicated, and infectious disease concerns identified, via this final examination. Also, and importantly, this is how zoo and aquarium veterinarians learn about disease processes in animals so that the field of zoo and aquarium medicine continues to move forward. It is not uncommon to find pathologic processes or infectious diseases "accidentally"—things that were diagnosed only after an animal's death instead of, or in addition to, other things that were suspected. A complete necropsy should be done on every animal in a zoo's collection, since it may yield potentially important information. An animal that has died should be immediately refrigerated and not frozen. The necropsy should take place as soon as possible, or at least within 24 hours of the death. It is extremely important to remove animals from heating sources or warm environments, as the heat causes autolysis (tissue breakdown) very quickly, and will destroy valuable information. For example, a reptile that dies overnight on a heat source may not be found until morning, coining the term "snake soup" at one zoo the author worked at. No diagnostic information is available when the tissues have been destroyed by heat. Conversely, if a body is frozen, ice crystals will destroy its cellular architecture, and histopathology will not be possible after it is thawed.

SUMMARY

After examining an animal and evaluating its history, a veterinarian can get more diagnostic information by obtaining samples and performing tests. It is important for the keeper to understand the purpose and importance of diagnostic testing, so that they can provide the veterinarian with support and help whenever possible. Training an animal for sample collection eliminates the stress of restraint and the risks of anesthesia, and it can also provide more accurate blood results. When an animal is approached in this manner, it voluntarily takes part in its own health care, instead of fighting it. A keeper needs to know exactly what each diagnostic procedure will entail, so they can train or desensitize the animal for it. Obtaining high-quality diagnostic information is not only critical to giving animals good medical care, but also to learning about disease processes in zoo and aquarium species so that animal care can continue to advance.

49

Medications and Dose Calculations

Mary O'Horo Loomis and Tony Beane

INTRODUCTION

Keepers have many responsibilities associated with basic animal care, such as feeding, cleaning, grooming, and training. Since keeping involves the day-to-day well-being of zoo and aquarium animals, it stands to reason that these animal caretakers need an understanding of basic drug usage and dosing. Preparing and administering medications sometimes falls to the keeper, and as one might imagine, it is very important for the animal's health and well-being that this task be carried out properly.

As a member of the animal care team, it is important that the keeper understand the serious nature of his or her role in handling and delivering therapeutic agents. All drugs are potential poisons, and care must always be taken to make sure that the proper animal is treated with the proper drug at the proper dose for the proper amount of time. Without this attention to detail, prescribed therapies may be rendered useless or poisonous (toxic). It is always important to remember that these animals depend on the keeper for their well-being; therefore, one should never become complacent when administering therapeutic medications to veterinary patients. The purpose of this chapter is to give keepers an understanding of the process and a background to properly use medications. Keepers are an important part of the animal health care team, but it is imperative that they realize that the actual diagnosing of disease and prescribing of medications is done only by a veterinarian. Keepers should never initiate treatment on their own.

This chapter will provide basic information about the use of veterinary medications, pharmacy procedures, and dose calculations. After studying this chapter, the reader will understand

- that all drugs are potential poisons and should be used with great care
- that the administration of drugs requires strict atten-

tion to detail, and prescribed directions must not be altered
- that the drug label and package insert contain valuable information about use of the drug
- that drugs are available in many forms and have many possible routes of administration
- dose intervals and the basics of calculating doses.

CLASSIFICATION

All veterinary drugs sold today are regulated by one or many government agencies, depending on the country. For example, in the United States the Food and Drug Administration (FDA) is responsible for categorizing how drugs can be sold. Over-the-counter drugs (abbreviated OTC), are just that: drugs sold on shelves in stores. Examples of OTC drugs that might be used are aspirin, triple-antibiotic ointment, and hydrocortisone cream. Drugs can be sold in this way if the FDA believes that the directions for their use of the drug can be written clearly enough so that lay people can easily understand them. Although these drugs are not without danger, they are generally considered safe to use without the direct oversight of a veterinarian.

Prescription drugs can be obtained only through a veterinarian or on the order of a veterinarian from a pharmacy. In the United States, all prescription drugs are labeled with the statement: "Caution: Federal law restricts this drug to use by or on the order of a licensed veterinarian." Prescription drugs may be abbreviated "Rx" and are labeled in this way because they may be toxic or easily misused. Labels that state "for veterinary use only" or "sold only to licensed veterinarians" do not mean that the drug is a prescription drug; they only mean that the drug was formulated for use on animals and not humans.

Prescription drugs have been registered and approved by the FDA for use on certain species of animals for certain conditions. The conditions and animals for which the drug has been approved are stated very clearly on the label or on the package insert. Veterinarians may use a drug outside of these approved conditions if they take the responsibility for the outcome of the use. When veterinarians use drugs on animals for conditions not specifically stated on the label, it is said that they use the drug in an "extra label" fashion. It is important that veterinarians have the right to do this, since most veterinary drugs are only labeled for the common domestic species (e.g., dogs, cats, cows, and horses). Very few drugs are labeled for the less common species (e.g., goats and llamas) and almost none are labeled for most zoo animals.

Another classification of drugs is that of a controlled substance. Drugs in this category may be narcotics or any other drugs with high potential for abuse by people. These drugs often may be addictive, and there is a risk that they may be used indiscreetly or stolen from pharmacies; for this reason they require special handling and must be kept in locked cabinets. In the United States, the Drug Enforcement Administration (DEA) oversees the laws regulating these substances. In order for veterinarians to purchase, administer, or dispense any substance in these controlled categories, they must first be registered with the DEA. The drugs must be carefully inventoried, and any use of them should be tracked through extensive record keeping. In the United States, drugs that are deemed to be controlled substances are placed in one of five schedules denoted with Roman numerals. A drug label that indicates a controlled substance will bear the capital letter C with a Roman numeral one through five (I through V) after it.

- Schedule I or CI drugs have a high potential for abuse and little or no accepted medical use. Drugs included in this class include LSD, marijuana, cocaine, and heroin. The ability to legally purchase these drugs is usually limited to researchers; general practitioners (human physicians or veterinarians) cannot normally purchase them.
- Schedule II or CII drugs have accepted medical use, but also a high potential for abuse. Drugs in this category include morphine, codeine, oxymorphone, and pentobarbital.
- Schedule III or CIII drugs have less potential for abuse and include ketamine, telazol, and some barbiturates.
- Schedule IV or CIV drugs have a still lower potential for abuse, and include such things as phenobarbital, butorphanol, or diazepam (Valium).
- Schedule V or CV drugs have the lowest abuse potential on the scale, and are usually mixtures that contain some narcotic, such as cough medications with codeine.

DRUG LABELS

Keepers who handle and administer medications to animals should make themselves familiar with the substances they use. Typically, the drug label or the package insert contains most of the useful and important information and should be given close attention. The label informs the user or dispenser as to what is in the bottle and what conditions or diseases the drug is to be used for. The label and insert are also important sources of information regarding possible adverse reactions or poisonings that might occur.

Care should be taken when using various bottles of drugs made by the same company, as they often will bear similar-looking labels and at a casual glance may be mistaken for one another. A good rule of thumb when preparing to medicate an animal is to read the label three times: once when the bottle is taken off the shelf, once when the needed medicine is taken out, and once when the bottle is replaced on the shelf.

When one examines the label or package insert, one of the first items seen is the drug name. Most drugs can be referred to in three ways, or by three separate names. The compound or chemical name is the drug's basic chemical formula. Examples of this would be acetylsalicylic acid, which is the chemical name of aspirin, and D alpha amino benzyl penicillin sodium salt, which is the antibiotic known as ampicillin. This compound name is important to chemists in the pharmaceutical business, but otherwise is not a name in day-to-day use. Once a compound is produced for sale, it is given a generic name such as aspirin or ampicillin (from above). This generic name will be found on the label and under the "active ingredients" list. The final name is the brand name or manufacturer's name. These are names given by individual companies for their versions of generic drugs. Advil® and Motrin® are both brand names for the generic pain reliever ibuprofen. The symbol ® refers to a trade or brand name that has been registered by the pharmaceutical company.

Labels or inserts will also contain other important pieces of information, such as the following.

Quantity of drug is how much is actually in the container, such as 50 tablets or 100 milliliters (ml). A milliliter (ml) is a standard unit of liquid measurement for most drugs. One milliliter is the same as a cubic centimeter (cc), and the two units can be used interchangeably.

Concentration is the actual amount of a drug in a specific unit of measure, such as 50 milligrams (mg) per tablet, or 20 mg per ml in the case of liquids. It is an indication of the strength of the particular formulation. This information is very important when calculating how much of a drug to give an animal.

Route of administration instructs as to how the drug should be given. Oral drugs are given by mouth. Oral administration is often abbreviated PO, referring to the Latin *per os* or "by mouth." Parenteral drug forms are drugs meant to be not taken orally but injected. They require the use of a needle and syringe. The most common routes of administration for parenteral drugs are

- intramuscular (IM) injection, into a muscle mass
- intravenous (IV) injection, into a vein
- subcutaneous (SC or SQ) injection, under the skin.

The actual locations on the animal where the shots are given will vary with the individual species.

Drugs that are applied to the skin are called topical medications. Depending on the substance, they may be rubbed into the skin or dabbed on and allowed to dry. Substances meant to be administered in the eye are called ophthalmic medications, and those meant for the ear are otic medications.

The method and location where medications are administered is dependent on the individual animal species. Since keepers work with a wide variety of very different animals, they will need instruction in restraint and therapeutic approach to each type of animal separately.

Other information on the drug label and insert may include the following.

- **Storage instructions** refer to any special environmental conditions the drug requires (e.g., "Keep refrigerated" or "Protect from sunlight").
- **Expiration date** is the date after which the medication should not be used.
- **Indications** are uses for which the drug has been approved (e.g. "For pain relief").
- **Precautions** warn of conditions in which the drug should be used carefully (e.g., if the animal is also being treated with another drug).
- **Side effects** are normally occurring effects of the drug that are not its intended use (e.g., excessive thirst).
- **Adverse reactions** are undesirable reactions to the drug that are significant enough to warrant caution (e.g., "This drug has been reported to cause anemia").
- **Contraindications** are conditions in which the drug should not be used at all.
- **Overdose information** describes the signs exhibited in the case of an overdose and may give information on how the overdose should be treated.

A keeper who wants more information on a particular drug can often go to the *Physicians' Desk Reference* (PDR), the *Veterinary Pharmaceuticals and Biologicals* (VPB), the Material Safety Data Sheets (MSDS), or veterinary drug handbooks. These sources may give added knowledge about a certain drug and how it has been used by veterinary practitioners in other species; however, it is essential that keepers follow the prescribing veterinarian's directions and ask for clarification if the directions are unclear or if they have any questions.

CLASSES OF DRUGS

Individual drugs may fit into one of several broad categories based on their effects. Some of the more common classes are as follows.

- **antipyretic:** a drug that lowers a fever
- **antitussive:** a drug that inhibits or controls coughing
- **analgesic:** a drug that relieves pain
- **antibiotic:** a drug that kills or inhibits the reproduction of bacteria (it is important to note that antibiotics have no effect on viruses)
- **antiinflammatory:** a drug that reduces the effect of inflammation, such as reducing swelling or itching
- **anesthetic:** a drug that produces unconsciousness
- **anthelmintic:** a drug used to remove internal parasites (specifically nematodes)
- **cathartic:** a drug that stimulates defecation, a laxative
- **depressant:** a drug that inhibits body function (usually an anesthetic drug)
- **emetic:** a drug that induces vomiting

- **expectorant:** a drug that increases secretions from the respiratory tract
- **sedative:** a drug that produces drowsiness, relaxation, or relief from anxiety.

PHARMACOKINETICS

HOW DRUGS WORK

Once a drug is administered to an animal, the expectation is that it will move to the desired site of action (target tissue) and cause a desired effect in the animal. It is then expected that the drug will be removed or eliminated from the body. The movement of a drug through the body is called pharmacokinetics. There are four steps to pharmacokinetics: absorption, distribution, metabolism, and excretion.

ABSORPTION

The major way in which drugs are moved through the body is by blood flow. Before a drug can travel from the site where it was administered to the organ or system where its effect is needed, it must first get into the bloodstream. The process by which a drug is taken up into the bloodstream is called absorption. Where and how drugs are administered has a great influence on the speed of absorption. Drugs administered intravenously or intra-arterially have no absorption phase and therefore exert their effects very quickly. Drugs that are injected intramuscularly are absorbed somewhat more slowly. Drugs either given orally or injected subcutaneously have an absorption phase that is slower still. The forms of some drugs also effect their absorption time, and this is particularly true of oral drugs. In most cases, liquids are absorbed faster than solids, since solid tablets must first dissolve before they can be absorbed into the bloodstream.

DISTRIBUTION

Once a drug is absorbed, it must be carried to the intended area or target tissue to have an effect. Drugs move to target cells, and once there, they bind to specific molecules on the cell surface called receptors. When the drug comes into contact with cells that have these specific receptor sites, the two bind together and some desirable changes occur within the cell. The cell might contract or secrete a product, or the destruction of a bacteria cell might occur. All cells do not have receptors for all drugs, which is why some drugs only work on nerve cells and others only work on lung cells. Many factors can affect the distribution of drugs, such as the blood supply in the target tissue, the amount of fat on the animal, or the nutritional status of the animal.

METABOLISM

Metabolism of drugs involves a chemical change. In most cases the metabolism phase alters the drug so that it becomes inactive and can be eliminated from the body. This inactivation process most often occurs as a result of enzymes found in the livers of animals.

Things that can alter the expected metabolism of a drug

are drug interactions (giving two or more drugs at the same time), the age of the animal (since very young and very old animals have decreased ability to metabolize), disease conditions (particularly liver disease), and species differences (since not all animals have the same enzyme systems).

In some cases the metabolism phase involves the activation of a drug rather than its inactivation. A few drugs are administered in an inactive form and must first pass through an animal's liver to be transformed into the active form before they are distributed. An example of this is the conversion of the corticosteroid prednisone into the active drug prednisolone after it is activated in the liver.

ELIMINATION

Once a drug has been transformed into an inactive form by the liver, it can be eliminated or excreted from the body. Most drugs leave the body by way of the kidneys, in the urine. However, other organs of excretion may be used by some drugs, such as the intestinal tract, the lungs, the skin and sweat glands, and the mammary glands. Diseases that affect an excretory organ can allow the buildup of drugs in the body and potentially lead to a toxic overdose.

By being aware of the movement of drugs through the body, one can be more aware of their possible undesirable effects. Adverse drug reactions usually are considered a problem of an individual animal being unusually sensitive to a medication. They can be as mild as a total absence of effect on an animal, or as severe as the actual death of an animal. Sometimes these undesirable reactions can be the result of a disease state in one of the organ systems discussed above (e.g., liver disease causing the drug not to be inactivated, or kidney disease causing the drug not to be eliminated). Administration of two or more drugs at the same time may result in undesirable drug interactions. Many drugs are safe to administer together and some actually increase in their effectiveness when given in this way. In some cases, however, one drug may block the effects of another, or cause additional side effects that can be dangerous and sometimes be fatal. Drugs should be given in combination only when prescribed by a veterinarian with previous knowledge of the outcome of that combination.

DRUG FORMS

The form of a drug is the consistency or physical state in which it is administered. Drug preparations may be administered orally, by injection, by inhalation, or by application to the skin.

Topical drugs are administered to the skin or body surface, with the effects of the drug meant to be delivered at that site (locally) rather than throughout the body (systemically). Some drug forms that are meant to be applied to the external body surface are as follows:

Tinctures are substances that have been dissolved in alcohol (such as tincture of iodine), and are normally meant to be dabbed on the skin. Liniments usually have an oily base and are meant to be applied to the skin by rubbing. Semisolid topical products are usually in the form of ointments or creams; these are meant to be spread across the area to be treated. It is important that the area of the body that is being treated is clean and free of debris. In some cases the hair around the affected area should be clipped so that the area can be treated more completely. Any changes in appearance of the area should be reported.

As stated previously, ophthalmic drops or ointments are meant to be administered to the eye. One should pay close attention to their administration. They are supplied in sterile containers, and their cleanliness must be maintained. The cap of the vial must remain clean and the applicator tip must not touch the eye. Either of these mishaps could result in contamination of the entire vial or tube, which in turn could lead to disease in the eye.

Drugs labeled for otic use are meant to be carefully administered in the ear canal. Again, it may be important to remove dirt or debris prior to treatment.

Oral administration is a common method of delivery for some drugs. These drugs are available in a variety of forms including powders, solids (such as tablets or capsules), liquids, and pastes. Again, the administration of drugs by the oral route is abbreviated PO (per os).

Solid forms include tablets and capsules. Tablets such as aspirin or vitamin C are made by compressing the drug into a pill form. Some tablets are scored, meaning that a line is etched into them so that they can be broken into smaller doses. Some aspirin tablets have enteric coatings, which help to protect the gastrointestinal tract from irritation.

Capsules are actual small containers that hold medications; they are most often made from gelatin, which breaks down quickly in the gastrointestinal tract. They may contain powder or liquid. For some patients they may be easier to swallow than tablets, and they hide any unpleasant taste the medication might have. A disadvantage of capsules is that they cannot be split into smaller doses. Boluses are large oblong pills usually meant for larger herbivores. They may be compressed powder, like tablets, or medicine in gelatin boluses, like capsules. They are meant to contain the appropriate amount of drug that these large animals require.

Liquid medications also take several forms, and it is important to identify types used, since they are handled in different ways. Two common forms are solutions and suspensions. Solutions are often water-based (aqueous), and are formulated so that the drug is totally dissolved in the liquid. In other words, the drug will not settle out. Suspensions contain drugs that are not completely dissolved, so that over time the drug will settle out. Suspensions must be mixed or shaken before use, to ensure that proper strength is attained. Liquid medications can be further classified on the basis of what has been added to a mixture. Emulsions are drug suspensions containing fat or oil, and are typically very thick or viscous. Syrups are mixed with a sugar and water combination to disguise the taste of a medicine. Elixirs are substances dissolved in alcohol; they are used with medications that do not readily dissolve in water.

Powders or granules are loose grains normally meant to be mixed with an animal's feed. Often these products must be mixed in such a way, or otherwise have their taste disguised, in order for the animal to eat them. Capsules or tablets are administered manually either by being placed in the animal's mouth over the back of the tongue, or with the use of a pill gun or balling gun. Liquids may be given with a dose or drenching syringe, again with care to place the liquid in the back of the mouth over the tongue. In some cases, such as when a large amount of liquid needs to be given, it may be delivered by a tube inserted into the mouth (i.e., an orogastric tube) or into the nostril (i.e., a nasogastric tube) and passed to the stomach.

Parenteral administration is the giving of a drug by way of injection. These types of drugs are supplied most commonly in vials, of both single-dose and multidose varieties; in ampules, and in large-volume bottles or bags.

Single-dose vials are small bottles sealed with rubber stoppers. As single-dose containers, they are meant to have their contents used all at once and then to be discarded. Multidose vials contain more than one dose and are used repeatedly. They often contain preservatives to ensure a longer shelf life for the drug. Ampules are glass containers that are meant to be broken in order to remove their contents; they are usually single-dose containers. Very few drugs are packaged in ampules today.

Some drugs are supplied as powder or pellets in one vial that are meant to be mixed with a fluid before they can be administered. This is known as reconstitution; often the fluid is sterile water and is supplied with the drug. Once the drug is mixed, it is meant to be used immediately; it should not be stored in the reconstituted form.

Injections are given with sterile syringes and sterile single-use needles, as described in chapter 45. It is always best to select a syringe size that is closest to the volume of medicine one is going to inject. For example; if an animal needs to receive 2 ml of medicine, one should select a 3 ml syringe, not a 12 or 20 ml. The amount of medicine drawn into the syringe will be most accurate when an appropriately sized syringe is used.

FILLING A SYRINGE WITH MEDICATION

When drawing medicine into a syringe for administration, it is important to maintain cleanliness at every step. The medication vial should be kept clean, and one should always read the label carefully to double-check that the vial being drawn from is the actual medicine that has been ordered for the animal. The rubber stopper on the vial should be wiped clean with a cotton swab moistened with 70% isopropyl alcohol (rubbing alcohol). Open the needle and syringe packages and carefully fit the needle onto the hub of the syringe. Pull back the plunger and fill the barrel with air equal to the volume of medicine you want to draw out. Insert the needle through the rubber stopper and inject this volume of air into the vial; this creates positive pressure in the vial and makes it easier to remove the medicine. (Note: It may not be recommended to inject air into the vial if there is a risk of contaminating the medication with unclean air. It is always best to check with the veterinarian when unsure of a procedure.) With the needle still in the vial, invert the bottle and needle and syringe, and slowly draw out the desired amount of medicine. Be careful to keep air bubbles to a minimum by keeping the tip of the needle below the fluid line.

Care must be taken at all times when handling needles and syringes to avoid accidentally sticking oneself with the needle.

Once an injection is given, the needle must be disposed of in an appropriate container. These containers should be made of rigid plastic that the needles cannot pierce, and they should have labels stating that they contain discarded sharp instruments. The needles are designed to be used only once and then discarded. They should never be reused. Contaminated needles may transfer diseases from one animal to another.

DOSING AND DOSE INTERVALS

The term "dose" refers to the total amount of a drug that is meant to be given at one time (for example, two tablets or 30 cc of an injectable medicine). The dose is generally calculated for individual animals on the basis of the manufacturer's recommended dosage (amount of drug given per size of animal) and the animal's weight. The drug manufacturer will recommend how often the calculated dose should be given in order to achieve the drug's desired effects. The time between administrations is known as the dose interval. Keepers who administer drugs to sick animals must pay close attention to the veterinarian's drug orders. Drugs given in incorrect dosages or at improper dose intervals may have no effect on the animal's disease or, more importantly, may have a very serious and undesirable (toxic) effect on the animal.

Dose intervals are usually written as Latin abbreviations (table 49.1). It is usually desirable to divide these time periods into equal numbers of hours; for example, something given t.i.d. or three times a day should be given roughly every eight hours. Sometimes this is also expressed with Latin abbreviations (table 49.2). It is always important that medications be administered according to directions. Giving too little of a drug may have no effect on an animal. Conversely, giving more than the recommended amount may be dangerous. Some drugs become toxic at doses only slightly higher than what is recommended. It is also important that the drug is administered for as long as the directions state. Skipping doses or stopping treatment too early will not deliver enough medicine to the animal to treat the disease. This carelessness

TABLE 49.1. Dose intervals and their commonly used Latin abbreviations

s.i.d. (*simul in die*)	Once a day
b.i.d. (*bis in die*)	Twice a day
t.i.d. (*ter in die*)	Three times a day
q.i.d. (*quater in die*)	Four times a day
p.r.n. (*pro re nata*)	As needed
ad lib. (*ad libitum*)	Freely or without restriction

TABLE 49.2. Additional dose interval abbreviations, used to indicate frequency of medication

q2h (*quaque 2 hora*)	Every 2 hours
q4h (*quaque 4 hora*)	Every 4 hours
q6h (*quaque 6 hora*)	Every 6 hours
q8h (*quaque 8 hora*)	Every 8 hours
q12h (*quaque 12 hora*)	Every 12 hours

can also lead to the development of antibiotic resistances in bacteria.

DOSING CALCULATIONS

It is important that keepers understand the basics of drug calculations, even though in most cases they will be primarily administering drug doses that have already been measured for individual animals. Veterinarians and veterinary technicians will have received significant formal training on drug administration, and will generally be responsible for calculating and dispensing medications. It is important that keepers work cooperatively with their facility's veterinary team and follow all veterinary-prescribed medication directions.

There are many ways to calculate drug doses; all require the use of some basic algebra. Also required are three basic pieces of information: the animal's weight, the concentration of the drug, and the dosage of the drug.

1. **Weight.** In most cases, drugs are dosed according to how large the animal is. An accurate measurement of the animal's weight is therefore essential. The metric system is used almost universally, so the animals should be weighed in either grams or kilograms. In the United States, however, the use of pounds and ounces is still widespread. Keepers should be able to convert from pounds to kilograms and vice versa. This can be done using a simple conversion factor:

   1 kg = 2.2 pounds (lbs.)

   If an animal weighs 44 pounds, how many kilograms does it weigh?

   44 lbs. divided by 2.2
   (remember that 1 kg = 2.2 lbs.)
   = 20 kg

   If an animal weighs 30 kilograms, how many pounds does it weigh?

   30 kg multiplied by 2.2
   (remember that 1 kg = 2.2 lbs.)
   = 66 lbs.

2. **Concentration of the drug.** This was defined earlier as the drug's basic strength. The concentration of the particular drug used will be found on the label, as well as in any package insert. It is always reported as a unit of weight (or mass) per unit of volume, for example 5 milligrams (mg) per milliliter (ml) in the case of a liquid, or 5 mg per tablet in the case of a solid medication.

3. **Dosage.** The dosage of the drug is the final piece of information needed. This is the recommended amount of drug to be given. It is usually the veterinarian's responsibility to recommend a certain dosage. He or she may get the information from the label or package insert, or from experience in using the medication. The dosage is expressed as a unit of mass

of drug (mg) per unit of animal mass (kg or lbs.): for example, 10 mg per kg or 5 ml per lb.

Once these three pieces of information are gathered, the dose of medicine for a particular animal can be calculated. Before the calculation can be made, however, all the information must be expressed in the same units. For example the measurements of weight may have to be converted to pounds or kilograms.

EXAMPLE

How many milliliters of a drug should a 20 kg animal receive, if the drug is 200 mg/ml and the dose is 100 mg/kg?

   Weight = 20 kg
   Concentration = 200 mg/ml
   Dosage = 100 mg/kg

Weight × dosage will tell us how many milligrams the animal will need (20 × 100 = 2000 mg).

Dividing this number by the concentration will give us the number of ml to administer (2,000/200 = 10 ml)

The animal should receive 10 ml of the drug.

Sometimes, instead of a liquid medicine, the drug is in a solid form such as tablets. In this case, the information needed will be the same, but the concentration of the drug will be expressed in mg per tablet. For example, a 22 lb. animal needs a drug dosed at 10 mg/kg. How many tablets would it receive each day if the concentration of the drug was 100 mg per tablet?

First, it is necessary to convert the 22 lbs. into kg by dividing by 2.2 (the conversion factor). This animal weighs 10 kg. Now that the units of weight are the same, one can find out how many milligrams of drug the animal needs.

   10 kg × 10 mg per kg = 100 mg

Now divide by the concentration 100 mg. This equals one 100 mg tablet.

What if the animal needs this medicine b.i.d.? It is known that b.i.d. means twice a day. This animal would therefore receive one tablet twice a day, or one tablet every 12 hours.

An alternative to working through these problems step by step is a method called "factor labeling." In this method, pieces of information are aligned so that numerators and denominators cancel each other out and the answer appears in the appropriate units that were asked for in the original question. This is denoted in the following equation by strikethrough of the units that have been canceled out. In this way the previous problem can be solved as follows:

22 lbs. × 1 kg/2.2 lbs. × 10 mg/kg × 1 tab/100 mg × 2 = 2 tablets

$$\frac{22\ \cancel{\text{lbs.}} \times \dfrac{1\ \cancel{\text{kg}}}{2.2\ \cancel{\text{lbs.}}} \times \dfrac{10\ \cancel{\text{mg}}}{\cancel{\text{kg}}} \times \dfrac{1\ \text{tablet}}{100\ \cancel{\text{mg}}} \times 2}{} = 2 \text{ tablets each day}$$

Some medications come in solutions, and require calculation of the amount of solution to be given. The principles are the same, but dealing with a solution of a certain percentage can confuse people. For example, a 5 % solution may be written either as "5 grams/100 cc" or as "5,000 mg/100 cc." The con-

Case Study 49.1. Calculating a Dosage for a Gorilla

An 88-pound male gorilla needs to be immobilized so that a tuberculosis test can be performed. The dose of anesthetic is 10 mg/kg and the concentration of the drug in the bottle is 100 mg/ml. How many milliliters of drug should be administered?

Weight = 88 lbs. Convert this to kilograms by dividing by 2.2 (2.2 lbs. per kg).

Weight = 40 kg.

Concentration = 100 mg/ml.

Dosage = 10 mg/kg.

Weight × dose = 40 kg × 10 mg/kg = 400 mg. This is how much of the anesthetic the gorilla will require.

Divide this by the concentration of the drug = 400 mg ÷100 mg/ml = 4 ml.

The gorilla should receive 4 ml of anesthetic.

The "factor labeling" method would be as follows:

88 lbs. × 1 kg/2.2 lbs. × 10 mg/kg × 1ml/100 mg = 4 ml.

$$88 \text{ lbs.} \times \frac{1 \text{ kg}}{2.2 \text{ lbs.}} \times \frac{10 \text{ mg}}{\text{kg}} \times \frac{1 \text{ ml}}{100 \text{ mg}} = 4 \text{ ml}$$

All the pertinent information from the question has been plugged into the equation. The animal's weight is converted to kilograms, since the dosage is in kilograms. The dosage of the drug is inserted into the equation, as is the concentration of the drug in the bottle. Since pounds are in both the numerator and the denominator, the units cancel each other. Since there are kilograms on both the top and the bottom of the equation, they also cancel each other. Finally the milligrams cancel each other, and we are left with the desired units: ml. The gorilla should receive 4 ml of anesthetic.

version factor can also be inverted: the 100 cc can appear on top as the numerator and the grams or milligrams can be put into the denominator, so that the solution is written as "100 cc/5,000 mg." This concentration can then be substituted to solve the problem.

For example, a 500 lb. animal needs a drug that is dosed at 10 mg/lbs. The drug comes in a 5% solution. How many cc of this drug does the animal need each day?

500 lbs. × 10 mg/lbs. × 100 cc/5,000 mg = 100 cc

$$500 \text{ lbs.} \times \frac{10 \text{ mg}}{1 \text{ lb.}} \times \frac{100 \text{ cc}}{5,000 \text{ mg}} = 100 \text{ cc}$$

SAFETY WITH DRUG USE

Records must always be kept of the drugs that are dispensed to an animal. Every animal should have an official record at the zoo or aquarium, and there should be daily sheets with room for information on the medications that are given.

Just writing down which medication was given to an animal is not sufficient. The records should include details of the exact amount of drug that was given, as well as the time when the medication was administered. The initials of the person giving the medication are also vital to the completeness of the records. If any follow-up information on the animal is needed, the initials readily identify the person who has given the treatments.

An inventory should be maintained of all drugs used at a zoo facility, and it may be up to the keeper to do this, although it is usually the veterinary staff that is responsible. It is important to know which drugs are used frequently so that they can be replenished when supplies are depleted. Inventories are also important so that if a theft should occur, it can be readily discovered. All drugs should be maintained in a safe, secure location at all times to avoid theft or misuse.

Keepers should be aware of any potential hazards that drugs may pose to them. Some drugs may be hazardous if inhaled or splashed into one's eyes and some may be absorbed through the skin. Keepers who assist in treating animals should always ask the veterinarian whether any personal protective equipment is recommended when administering medications. Examples of substances that warrant such protection are chemotherapy drugs, which are used to treat certain cancers. Certain drugs used to treat external parasites may also be very toxic. Keepers with special health considerations of their own should pay very close attention to what substances they come into contact with. Keepers with weakened immune systems or chronic diseases, as well as women who are pregnant, should be very cautious about handling certain drugs.

SUMMARY

Medications are powerful tools in animal health management; they have the potential to save lives. They also have the potential to take lives, and therefore must be used with care at all times. The members of the animal care team who are responsible for handling these drugs and medicating sick animals must never become complacent with them. The keeper must take the time to make sure that the proper animal is treated with the proper drug for the proper amount of time. One should never give a drug to an animal, or change treatment times in any way, without first consulting the veterinarian.

50

Chemical Restraint

Mark D. Irwin

INTRODUCTION

The use of drugs to restrain an animal—termed "chemical restraint," "sedation," "anesthesia" or "immobilization"—is frequently needed in zoos and aquariums to provide hands-on access and proper care to the animals. As it requires specialized medical training, it should always be conducted by a veterinarian with knowledge of wild animal medicine. However, keepers will often be asked to assist, and they have an important role. Since small animals can often be manually restrained or moved awake to the animal hospital where the hospital staff will conduct the procedure, this chapter will focus on situations where the keeper must assist with chemical restraint outside of the animal hospital setting (i.e., "in the field"). The chapter will explain the basic principles, procedures and tools used. After studying this chapter, the reader will

- understand the basic principles of sedation and anesthesia (chemical restraint), including when it may be and may not be appropriate
- understand the role that a keeper can take during a chemical restraint procedure
- be better able to identify safety concerns
- be familiar with the different methods of inducing chemical restraint
- become familiar with the basic procedure for conducting chemical restraint, including planning, preparation, induction, monitoring, and recovery
- identify and understand the basic use of common equipment.

CHEMICAL RESTRAINT

Sometimes it is necessary to sedate or anesthetize an animal to properly care for it. This may apply to

- a dangerous or nontractable animal that must be handled (e.g., in a routine physical exam, or for place-

ment of identification markers such as PIT chips or ear tags)
- emergencies (e.g., when the animal represents a threat to itself, conspecifics, or people)
- medical or surgical emergencies
- painful procedures
- animal movements (transportation).

Chemical restraint is not as easy as many people think; it requires a great deal of skill and training. It is a useful tool and it enables a level of care that was not possible before its development for zoo and aquarium use, but it is not without risks and drawbacks. The term "knockdown" is sometimes used when referring to chemical restraint (particularly when it is done outside of the hospital setting); however, this term is not ideal and is somewhat misleading. An anesthetic procedure should be as smooth, quick, and minimally traumatic as possible.

For the purposes of this chapter, chemical restraint will refer to the use of drugs to immobilize (stop the movement of), sedate, and anesthetize zoo and aquarium species in the field. Sedation is the induction of calmness and drowsiness. The term "tranquilizer" is often used interchangeably with "sedative," although a tranquilizer is a drug that induces a state of being tranquil (relaxed and not anxious) without drowsiness (Bassert 2010, 826), and is not necessarily an anesthetic drug, as is most often used in a zoo or aquarium. "Anesthesia" is the term for inducing a lack of sensation or awareness. Anesthetic drugs, therefore, can inhibit the sensation of pain and are essential for procedures, such as surgery, that require relief of pain (analgesia). Those drugs that result in unconsciousness are termed general anesthetic drugs, in contrast with a local anesthetic that would prevent sensation in only part of the animal.

THE KEEPER'S ROLE

Animals may feel threatened or stressed, sometimes even at the appearance of a veterinarian or new person. Keepers

> **Good Practice Tip:** Animals are often faster and stronger than humans. Keepers have the advantage of being able to plan ahead. Animals should be outthought rather than outfought.

can take steps to minimize this stress and to facilitate the procedure. Often this will involve isolating the animal to a smaller, secure area that is safer for the procedure. Keepers will have relationships and established routines with their animals. This knowledge can be used when planning logistics. In some cases the keepers may be asked to administer some initial drug (premedication) to the animal through feed or other means in order calm them. Keepers can offer insight into how an animal will respond and they should try to anticipate the needs of the veterinary team and the procedure. Their unique knowledge of the animals, both as a species and as individuals, can be a tremendous asset. For example, an antelope that is prone to panic could run into a stall wall. Keepers could prepare an induction and recovery stall that is dim and quiet, and could line the walls with bales of straw or other suitable materials for padding if necessary.

DRUGS

Different drugs will have different effects and should be carefully chosen for each situation. For example, some drugs may provide immobilization, but may not provide adequate sedative and anesthetic effects. Without adequate sedation and anesthesia, alarm responses may occur that can harm the animal, such as elevated heart rate or blood pressure. Furthermore, some drugs might permit handling of an animal for surgery, but their use could be cruel without the pain relief provided by anesthesia (Fowler 2008, p.235). For these reasons, chemical restraint of zoo and aquarium species should always include a component of sedation and/or anesthesia.

Anytime that drugs are used, there is the potential for adverse reactions. Different species and different individuals may respond differently to the same drug. Anesthesia of zoo animals is still a young field, and less is known about its use with wild species than about its use with domestic farm and pet animals. Common effects of anesthetic drugs on animals are slowing of the heart rate (bradycardia), slowing of the respiration rate (bradypnea), and lowering of the blood pressure (hypotension). These can consequently lead to low oxygen levels in the animal's blood (hypoxemia, making the animal "hypoxic"). However, some drugs can have different and even opposite effects; the veterinarian will know what to expect and can advise. The animal's ability to regulate its temperature (thermoregulate) may also be compromised.

Zoo and aquarium veterinarians will often use combinations of drugs to provide a smoother induction of anesthesia and balance the drugs' effects. For example, the drug ketamine (sold under multiple trade names including Ketaset, VetaKet, and Ketanest), which can result in excitement and muscle rigidity, may be paired with midazolam (sold under trade names including Versed, Hypnovel, and Dormicum), which causes sedation and muscle relaxation.

Some zoo and aquarium animals are very large and powerful; they require potent drugs. For example, the anesthetic drug Carfentanil (trade name Wildnil) is in the same family of drugs as morphine (opioids), but is widely considered to be approximately 10,000 times more potent than morphine (Wildlife 2011). An accidental drop in the eye or needle poke could be disastrous for a human. One should always exercise extreme caution when working around drugs, and follow the direction of the veterinary team.

Many anesthetic drugs have "reversal agents," other drugs that counteract their anesthetic effects and can result in a faster recovery. In some cases, an animal may experience "re-sedation" (or "renarcotization" with the opioid drugs); a phenomenon in which the animal seems to have recovered from anesthesia or sedation after administration of a reversal agent, but only until the effect of the reversal wears off, allowing the effects of the anesthetic drug to return. The animal may appear off-balance, uncoordinated, or poorly aware, or may manifest other side effects of the drug (Allen 1996).

METHODS OF DRUG ADMINISTRATION

In oral administration, a drug may be given in food or water or simply be sprayed into the animal's mouth. This approach is noninvasive (it does not require the penetration of the skin or healthy tissue) and it can often be done with minimal stress to the animal, but the animal must ingest enough of the drug. Unfortunately, the time for absorption will vary (depending on stomach contents, the amount ingested, etc.). Oral administration is less reliable than other approaches, as it is difficult to gauge how much of a drug will be needed for the desired effect.

Inhalation of an anesthetic gas such as isoflurane is typically used to maintain anesthesia once an animal is immobilized. It can also be used to induce anesthesia, but typically only in a small animal that can either be manually restrained, so that a mask can be held over its mouth and nose, or placed in an induction chamber, a sealed box into which inhalant is infused. Inhalant can leak around the mask or out of the chamber if it is not airtight. Also, an animal may struggle or hold its breath. Inhalation is not suitable for the induction of anesthesia in large animals in the field.

A conventional hand syringe with needle can also be used to administer the drug. This can ensure complete injection, but it requires close contact with the animal to perform the injection. Speed and accuracy of injection will be important. To extend the reach of a syringe, a pole syringe or "jab stick" may be used. It is essentially a syringe on a stick and can extend a person's reach by several feet. The pole or stick itself functions as the plunger of the syringe, so that when the pole is pushed into the animal, the drug is expelled. This approach is effective for large animals when the veterinarian can approach within a few feet (generally with a barrier between the animal and the person). Some pole syringes may be spring-loaded to add more force and more quickly eject the drug from the syringe—sometimes important, as animals seldom stand still when poked.

Remote delivery of drugs can be accomplished using blow pipes or dart guns (figure 50.1). A blow pipe is essentially a

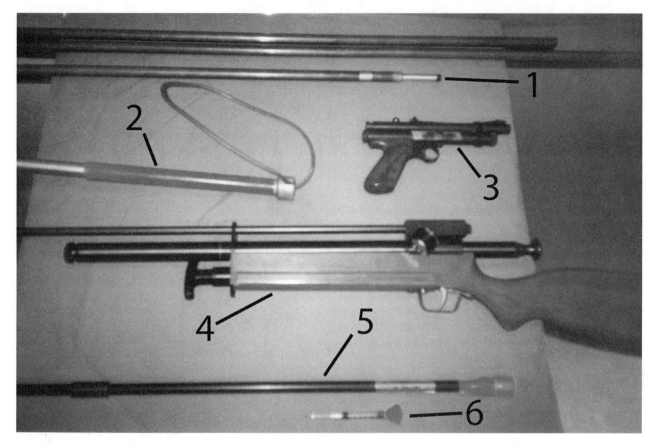

Figure 50.1. Equipment available for restraint procedures: (1) pole syringe (2) catch pole or snare), (3) dart pistol, (4) dart gun, (5) blow pipe, and (6) dart. Photo by Mark D. Irwin.

hollow tube with a mouthpiece or adapter at one end. The dart is placed into the tube and the person gives a short, strong burst of air into the tube to shoot the dart. This burst of air may be from lung power or from an air pistol (e.g., one using a CO_2 cartridge). Blow pipes are quiet and effective in close quarters; they project darts at a relatively low velocity, thereby causing minimal tissue trauma. They are limited in range and are generally not used for distances greater than 15–30 m (49–98 ft.; less if using lung power).

Dart guns are designed to project darts over longer distances. There are many different types (makes and models) of dart guns made by different companies. Dan-inject, Telinject, Cap-Chur, and Pneudart are just some of the different brands. Each brand has its own system and type of dart. "Guns" may also take different forms such as pistols, rifles, and even crossbows and are typically powered by explosive charges or compressed air. Care must be taken when using dart guns. If there is not enough power or the animal is too far away, the dart will not enter the animal with enough force. In fact, the dart may not even make it to the animal, since the size and weight of a dart can make it drop or be redirected by the wind. Worse yet, a dart can cause significant injury to the animal if too much power is used or the gun is shot from too close. Effective use of a dart gun can be very challenging when one considers that the target may be moving.

DARTS

Darts are designed to carry and inject drugs into animals. A number of commercial companies specialize in selling darting equipment, and some people have been able to make their own. A dart is typically double-chambered, with the front chamber designed to carry the drug. The back chamber will hold compressed air, gas, a spring, or an explosive charge to expel the drug. In some models, the impact will trigger an explosive charge to expel the drug. Some darts are made of metal and some are made of plastic. Dart needles are unique and adapted to their use. For example, some needles have a solid conical tip. The drug is expelled through side ports to prevent the force of the drug's ejection from pushing the needle out of the animal before the drug is injected into the animal. This reverse jet effect is analogous to a full balloon that is released before it is tied. The dart should not be forced backwards out of the animal until injection is complete. In this kind of needle, a small plastic or rubber "slider" covers the side-ports in the tip. When the needle contacts the animal the slider is pushed back, exposing the needle ports, and the compressed air in the back chamber forces the drug out through the ports and into the animal. Some needles have barbs or collars to prevent the needle and dart from falling out. Whatever the system used, a dart should be cared for and used as directed by the manufacturer, and disposed of when damaged. If in doubt, assume that the dart has drug in it and is pressurized or charged, and handle it with care.

PLANNING AND PREPARATION

The most important part of a chemical restraint procedure is the planning, which should include careful consideration of what will be done and how it will be done, as well as contingency planning. Proper planning can both smooth and speed the process, which is very important since the longer an animal is anesthetized, the greater the risk of complications and the more challenging the recovery. The principles detailed in chapter 14 apply; that chapter should be reviewed prior to participation in a chemical restraint procedure.

In short, each person should know their responsibilities and duties. Careful thought should be given to each step in the process; one should anticipate needs (supplies, equipment, extra support, etc.) and potential problems. One should consult with the veterinary team and, when necessary, with professional colleagues who have conducted such procedures before. Foresight is invaluable. For example, a keeper could preemptively train an animal to enter a small pen to be used for anesthetic events. They should work to mitigate all safety risks such as aggressive conspecifics or hazardous exhibit features (e.g., water features that could drown an animal). They should also ensure that a power supply, water hoses, and other tools are available if needed. Preparation of an induction or recovery pen with adequate padding and nonslip substrate could also be important.

In case of emergency, zoos and aquariums should have emergency drug dosages listed on a quick reference chart, and all necessary equipment ready to be used. Drills should be conducted regularly to practice emergency response. Emergency scenarios (e.g., animal escape, visitor in animal enclosure) should be reviewed to evaluate when chemical restraint is an option and when other approaches such as lethal force are more appropriate. Sometimes it is not feasible to drug an animal.

INDUCTION

Once an animal has been injected with an anesthetic drug, it will become progressively uncoordinated and less aware of its surroundings. It may become agitated or, as the drug takes effect, may stumble and fall. An animal that is extremely stressed (with elevated levels of adrenaline) may require significantly more of the drug and may struggle or hurt itself during induction. Therefore, it is best to have only required personnel present, to keep noise to a minimum, and to take other measures to avoid stress in the animal. If the animal becomes too agitated or stressed, the procedure should be rescheduled (if possible). This is a reason why chemical re-

Figure 50.2. Some animals are more calm when blindfolded. This giraffe has been blindfolded and had its ears plugged to enable its transfer into a trailer for transport. Photo by Mark D. Irwin.

straint may not be an appropriate choice in the case of an emergency: drug effects are less reliable and may need much more time to take effect in an excited animal.

Once the animal appears unconscious, caution must still be exercised because it is still very dangerous. Sometimes it may only be drowsy, and actually still be conscious and not anesthetized. Using a watch to track the time from the initial and any follow-up injections will provide a guide as to when the drug should take effect, but reactions to drugs can vary significantly. One should follow the lead of experienced animal care and veterinary staff. Typically it is wise to use a pole to poke or vigorously shake an animal from a safe distance to be sure it is unconscious and immobilized. Placement of a blindfold on the animal is recommended to protect its eyes, decrease its stimulation, and provide extra time for nearby people to respond should the animal wake up (figure 50.2). Eye ointment should also be administered, once it is safe to do, so to prevent the animal's eyes from drying.

Once an animal is safely anesthetized, an endotracheal tube may be placed into its airway (trachea). Such a tube

> **Good Practice Tip:** Use caution when approaching and handling an anesthetized or sedated animal. Animals are still dangerous, even when unconscious.

has an inflatable cuff that can be filled with air to completely fill the airway, making the animal breathe entirely through the tube; the cuff prevents air from flowing around the tube. The tube can then be connected to an anesthetic machine to deliver inhalant anesthetic or oxygen. If the animal stops breathing, the veterinary team can "breathe for the animal" by pushing air into the endotracheal tube and airway (positive pressure ventilation) through a resuscitation bag, or by other means. Additionally, the endotracheal tube will help to prevent the animal from aspirating foreign material into the lungs (see below).

MONITORING OF THE ANIMAL

An anesthetized animal must be monitored to ensure that it remains under anesthesia and that its body continues to function within healthy parameters. At its most basic, this involves monitoring the animal's "vital signs": temperature, pulse and respiration ("TPR"). It is valuable to record the vital signs throughout the procedure for the reasons mentioned above, and to provide baseline values for future anesthetic procedures involving the individual and its conspecifics. Keepers may be asked to assist with monitoring during a procedure or with record keeping. If unsure about what is expected or how to do something, a keeper should ask.

Temperature is typically measured via a digital rectal thermometer, although other locations may be used, such as the axilla (the "armpit," for an axillary temperature) or the ear (using an "aural" thermometer). The mouth should not be used, as the animal could break the thermometer or reflexively bite the person. Outdoor ("in the field") procedures should be scheduled for times of moderate temperature to avoid hyperthermia or hypothermia; one should avoid particularly hot or cold times of year and times of day when possible. A sedated or anesthetized animal will have a decreased ability to thermoregulate, and may need assistance from blankets, supplemental heat, fans, or application of cooling water.

The pulse of an animal can be felt in many locations; however, some species and some individuals may present a challenge. Locations may include the inguinal region (femoral artery), tail, feet, neck, or anywhere with adequate vasculature. For reasons of sensitivity, it is best to use a finger rather than one's thumb. One should count the pulses for 15 seconds and then multiply by four to acquire the pulse rate (heartbeats per minute). This is a good reason for a keeper to carry a timepiece as part of the work uniform. The pulse rate should match the heart rate (HR); however, in some cases there is a "pulse deficit," where the pulse is not felt. In such cases, the veterinarian will need to evaluate the animal. An experienced technician or veterinarian will additionally be able to evaluate the character of the pulse to detect some problems. If the heart rate is being measured directly, it can be felt by finger (palpated) or listened to via stethoscope (auscultated).

Respiration will be monitored to assess breathing. To determine a respiration rate (RR), one should count the number of breaths over a 15-second interval and multiply by four. This will be the number of respirations per minute. Frequency of respiration can indicate depth of anesthesia, and can indicate problems such as poor oxygenation. Character of respiration is also important to assess. One should observe the respiration to see if it is fast, slow, shallow (the chest moving only a little), deep, or labored (when the animal seems to have difficulty breathing), or if there seem to be any abnormal sounds (e.g., gurgling or crackles). Just because the animal's chest is moving, it doesn't mean enough air is being exchanged. Additional monitoring methods may be employed by the veterinary team.

Animals may also be assessed by examining their capillary refill time (CRT). Pressing an area of the gums firmly with a finger will cause the area to whiten (blanch). Once the finger is removed, the blood should return and refill the capillaries to change it back to its normal color (usually pink). The time for the normal color to return is the CRT. In general, it should take less than two seconds for the skin to return to normal. If animals have pigmented gums, other areas of the skin may be used; this will depend upon the individual and species. A CRT of greater than two seconds indicates poor perfusion of the tissue with oxygenated blood, in which case a veterinarian should be notified. Similarly, an abnormal bluish-grey color is indicative of low levels of oxygen (hypoxia).

A pulse oximeter is an electronic device designed to measure pulse rate and oxygenation (oxygen saturation) of the blood. It has a sensor that is placed onto an animal in an area of minimal pigmentation and hair (e.g., tongue, ear, digit, or tail fold; see figure 50.3). Oxygen saturation should approach 100% in a well-ventilated animal (meaning an animal that is having good, effective respirations). A pulse oximeter also commonly has audio settings that can beep with each heartbeat, enabling the veterinary team to listen to the heart rate while working. Electronic devices are helpful, but periodic confirmation by other means is recommended.

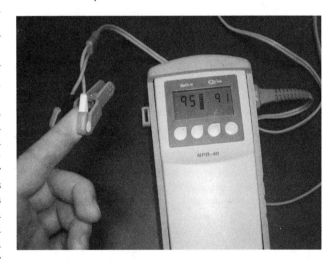

Figure 50.3. A pulse oximeter uses a light sensor to measure the oxygen saturation (SpO_2) of the blood. Values as low as 90% are common, although measurements should ideally be 95% to 100%. Many models also measure heart rate. In this photo, the person has an SpO_2 of 95% (left number) and a heart rate of 91 beats per minute (right number). Photo by Mark D. Irwin.

Depth of anesthesia refers to the extent of consciousness and sensation in an animal. An animal that is "deep" is completely unconscious and unresponsive. This state is often accompanied by slow heart and respiration rates and absent reflexes. An animal that is "light" will be close to becoming conscious. It will have some sensation to stimuli. Breathing and heart rate will typically be faster and reflexes will be present: for example, touching the corners (medial or lateral canthus) of the animal's eye will stimulate the palpebral reflex to cause blinking. Another sign that an animal is "getting light," or is experiencing a lightening of the anesthetic plane, is an increase in jaw muscle tone. The animal will begin to close its mouth and move its tongue, and might swallow. These are signs of a lightening plane of anesthesia, but some animals may not show the signs as clearly as others. The signs will also vary with the drugs used. If the animal becomes too light, the veterinarian may "top it up," meaning administering more drug to return it to the desired depth of anesthesia.

THE PROCEDURE

Once the animal is anesthetized and at an appropriate depth of anesthesia, the primary procedure can be started. As previously stated, there are a number of reasons why an animal may need to be anesthetized. The key is to be prepared so that the procedure can be done as quickly and effectively as possible. The longer the procedure, the longer the recovery can be, and the greater the associated risks.

Chemical anesthesia should be as short and infrequent as possible, to minimize stress on the animal. For this reason, during the planning phase one should be sure to consider secondary procedures that, if done during anesthesia, could eliminate or postpone the need for future restraint. For example, if a zebra is being anesthetized for transport to another enclosure, it may be wise to also give it its routine physical examination, trim its hooves, vaccinate it, or float its teeth at the same time.

RECOVERY

Recovery from anesthesia should be as smooth and as quick as possible. However, making it smooth is most important. A smooth recovery is one in which the animal regains consciousness and coordination with minimal disorientation and struggling. The animal should ideally be placed in a safe, dim, quiet location where it can gradually metabolize the drug and wake up. If it is likely to thrash or stumble, appropriate substrate and padding should be provided. All potential hazards (e.g., sharp corners, water sources [e.g., ponds, moats, or water troughs], or aggressive conspecifics) should be removed beforehand. The animal should continue to be quietly observed during its recovery until the veterinarian is satisfied. This may require hours of observation, and therefore it is often wise to begin procedures early in the day. Reintroduction of an animal into its social group should be conducted carefully, and with consideration of the species and individuals' behavior.

If the veterinarian chooses to use a reversal agent, the animal's recovery can be very quick; intravenous administration of a reversal drug can have an immediate effect. Therefore, it is important to evacuate all nonessential personnel and equipment from an enclosure before "reversing the animal." Remember that it is possible that the animal may re-sedate (renarcotize). In this case, or with any other unexpected event, consult with the veterinary team.

SPECIAL CONCERNS

Chemical restraint is a valuable tool, but it does have risks. Many of the risks are similar to those covered in chapter 15. Capture myopathy can be of particular concern if the animal is extremely agitated and has experienced muscle exertion leading up to the procedure. In such cases, the procedure should be postponed if possible.

Aspiration of foreign material is a major concern during any anesthetic procedure. While anesthetized, an animal's ability to swallow is impaired, its reflexes are diminished or absent, and some individuals may vomit. Consequently, there is a risk that stomach contents can be regurgitated and enter the airway (i.e., be aspirated). Foreign material such as stomach contents in the lungs can cause a potentially life-threatening pneumonia. Risk of aspiration pneumonia can be mitigated by use of an endotracheal tube and proper positioning of the head—keeping the neck on an incline (to prevent stomach contents from traveling up from the esophagus) and the nose or muzzle tipped down (so that any stomach juices will flow out of the mouth). Additionally, veterinarians may direct that an animal be fasted prior to an anesthetic procedure, to minimize the volume of stomach contents and therefore decrease the risk of regurgitation and aspiration. Fasting in some species is less effective than others. For example, a ruminant with its large forestomach will still have a significant volume of stomach contents with fermentative microbes. Furthermore, its copious amount of saliva presents an aspiration risk.

Ruminants are also predisposed to another significant risk: bloat. Normally, a ruminant will eructate (burp) to release the large amount of gas produced in its forestomach by fermentative microbes. When it is unable to stand or lay in sternal recumbency (lying on its chest or sternum) and when anesthetized, a ruminant cannot do this as it normally would. The gases from feed fermentation will build up and cause bloat (which can be caused by other factors as well). The buildup of gas will become apparent as one notices a visible swelling on the side of the animal. If not managed, bloat will compress the heart and lungs and can ultimately be fatal. Minimizing the duration of an anesthetic procedure, keeping the animal in sternal recumbency when possible, and minimizing its intake of fermentable feed before the procedure can help mitigate the risk of bloat becoming a problem.

SAFETY

Any anesthetic procedure carries significant risk. Animals and people may be anxious, excited, or stressed. It is important that the procedure is well planned and that each person knows their role and remains focused on it. Here are some critical points a keeper must remember during a chemical restraint procedure:

- Follow the lead of more experienced animal care staff and the veterinary team; ask for clarification if unclear.
- Treat an animal based on its potential danger. In stressful situations, the animal will act on instinct. Training, established relationships, and routine behavior patterns may change or become irrelevant.
- Anesthetized animals may still bite, kick, or scratch, sometimes by reflex. Always take care if required to work around their mouth, feet, or other dangerous areas.
- Be drug-conscious. Do not come into contact with any drug, and avoid touching objects that may have drug on or in them, including the injection site on the animal. Drugs can most easily be absorbed through mucous membranes (eyes, mouth, and nose) or through skin wounds. Wear personal protective equipment (PPE) when necessary, and follow the veterinary team's direction. If in doubt, ask for clarification.
- If picking up a dart used to deliver anesthetic drugs, be cautious. It may be charged and/or still contain drug residue.
- Stay focused on the procedure. Scan for safety hazards and opportunities to assist if requested. Do not chat with bystanders about other topics, as the unnecessary noise may stimulate the animal or distract the team.
- Always have a plan in case the animal spontaneously wakes up during the procedure. Always ensure that there is an escape route for each member of the team, and that the animal will be contained should it regain consciousness.

SUMMARY

Anesthesia enables direct contact with animals that would not be possible otherwise, and requires specialized skills and training. A number of methods and drugs are available that will be chosen by the veterinarian to fit the situation. Any chemical restraint procedure involves risk to the animal and people involved. The ultimate goal of anesthesia is to complete the required procedure in a safe, quick, and effective manner. Keepers have an important role to play in chemical restraint procedures, particularly when they are conducted at the animal's enclosure. They can contribute to a successful procedure by participating in the planning and preparation stages through anticipation of the animal and team's needs, and through assisting during and after the procedure. Their intricate knowledge of the animal and enclosure provides insight that can contribute to a successful procedure.

REFERENCES

Bassert, Joanna, and Dennis McCurnin. 2010. *McCurnin's Clinical Textbook for Veterinary Technicians*. Saint Louis: Saunders Elsevier.

Fowler, Murray. 2008. *Restraint and Handling of Wild and Domestic Animals*. Ames, IA: Wiley-Blackwell.

Wildlife Pharmaceuticals. 2011. Carfentanil from Wildlife Pharmaceuticals. Accessed 19 March at http://www.wildpharm.com /carfentanil.html.

Part Eight

Education, Outreach, and Public Interaction

51

Educating Entertainingly: Basic Interpretation

D. Andrew Saunders

INTRODUCTION

Two visitor groups enter a zoo at the same time, and then separate. Both groups appear to be more or less identical in age, gender, ethnicity, and so on. In fact, they could easily pass as one large family or neighbors from the same block, apart from the fact they arrive and tour as separate groups. One group exits the zoo an hour later, visibly tired and bored. The other visitors, however, leave three hours after the first, chatting amiably, obviously energized, and eagerly sharing their experiences with each other. Each person in this latter group seems to be trying to outdo the others in recounting some of the zoo experiences of that day. They even make references to encouraging their friends to visit the zoo.

Why the differences between these two groups? As the visitors in the first group walked around, they were mostly on their own. On four or five occasions, they did encounter keepers, but the keepers never made any effort to welcome them. In fact, the keepers were downright indifferent, even gruff and unpleasant—shaking off questions, answering in monotones, and making no attempt to prolong any interaction. The people in the second group had a vastly different experience; they were fortunate, or so they felt. They encountered about the same number of keepers during their walk, but the keepers were smiling and friendly, inviting conversations, welcoming questions, and giving no hint of any impatience. In fact, the keepers seemed to enjoy their encounters with the visitors, and volunteered interesting stories with their answers, some of which included messages about not just the care of the animals but also their ecology and conservation. The stories were fascinating, lending a behind-the-scenes quality to the second group's experience. They also provoked thought, some stories ending with tactful suggestions on practical measures for the visitors to consider: measures ensuring the quality of their own future and that of the animal they were observing. In other words, the visitors received conservation education. A few of the keepers used the group's interest to offer short, spontaneous miniprograms. The group was pleased,

delighted, and impressed with how entertaining, interesting, and informative the programs were. Some of the keepers had pictures and other objects to pass around. One keeper near the three-toed sloth exhibit brought out a radio collar to explain how radio telemetry studies had revealed the home range size and social structure of this species. No wonder the second visitor group remained at the zoo so much longer than the first one.

The purpose of this chapter is to

- briefly define and summarize environmental interpretation
- provide basic ideas and methods of environmental interpretation for improving communication with the visiting public
- offer suggestions for some of the possible benefits of implementing environmental interpretation approaches in the zoo setting.

Can keepers and zoo educators make a trip to the zoo more enjoyable and educational? The answer is a resounding yes, and the above example explains some of the reasons why this is so true. One way keepers can do this is by using methods of interpretation, and this is just what the keepers interacting with the second group were doing: they were interpreting the zoo and their work for the visitors. The zoo setting provides many opportunities for the use of interpretive methods. Adapting these methods to optimize visitor education and communication requires only modest effort and is likely to be, for lack of a better word, fun.

Few individuals in today's society have the glamour and prestige of zoo keepers. If keepers are not gods, then they must be the next best thing in the minds of many visitors, especially youngsters. Keepers can use this prestige to their advantage when talking to visitors, combining it with interpretive methods to plant the seeds of science and conservation education. At the same time and in the same way, keepers can serve as ambassadors for the animals in their care and

Figure 51.1. Keepers have a special opportunity to connect visitors with animals and nature. Photo by D. Andrew Saunders, courtesy of Rosamond Gifford Zoo, Syracuse, NY.

their wild counterparts. This, along with welcoming and educating visitors, is an important part of the keepers' job.

Keepers can also contribute other important benefits to contemporary society. They have a special opportunity to connect all visitors, especially younger ones, to nature. This opportunity should never be undervalued. Today's children who enter zoos suffer from "videophilia." Increasingly, their connections to the natural world are being traded for connections to cyberspace, and this has serious consequences for their emotional, physical, cognitive, spiritual, and evaluative well-being (Kellert 2002). In other words, today's children, parked in front of computers and televisions, are not as happy as earlier generations of children who played outdoors, especially the fortunate ones who explored nearby fields, creeks, and woods. Furthermore, because early contact with the outdoors is important for recruiting future generations of environmentalists, the siren call of cyberspace could thin the ranks of conservationists and nature enthusiasts who are willing to work to support zoos, or even seek employment within them.

Keepers possess a rich inventory of direct experience with nature, something they may take for granted but should not. They care for real, not virtual, animals every work day, and they see and learn things about their animals that most people would enjoy knowing and would find interesting, but never get to witness. The methods of interpretation provide a means for keepers to take this experience and translate it into won-

derful learning opportunities for zoo visitors. Keepers should use their experience to educate. They can provide a great service by doing so, and they may find themselves discovering new talents they did not know they possessed. They also may create from their efforts an unanticipated source of professional pride and job satisfaction.

Quite apart from the immediate benefits of forging connections with the natural world are other benefits, such as helping to maintain the financial health of the zoo and the well-being of source populations and the habitats that sustain them. First, there is the obvious benefit of successfully ensuring that visitors understand zoo etiquette and policies, and adhere to them. But keepers can do a great deal more than this. A visitor study conducted by North America's Association of Zoos and Aquariums (AZA), with the Institute of Learning Innovation and the Monterey Bay Aquarium, determined that only 35% of the visitors contacted from 7 to 11 months after their visits reported that their visits had reinforced their existing beliefs about conservation, stewardship, and love of animals. Keepers can increase this percentage by mastering a few simple approaches to visitor communication and education. Keepers who possess and use a working knowledge of contemporary interpretation methods are likely to strengthen visitor education and, by doing so, to increase the likelihood of visitor support of conservation in its broadest dimensions, both locally and globally.

Another benefit of mastering interpretation is perhaps one of the most important. By developing their interpretative skills, keepers can prosper and thrive as professionals. Increasingly, a keeper's job interview will include a live public talk in the presence of the interview panel to determine his or her ability to synthesize and use interpretive methods. Today's zoos emphasize visitor satisfaction and education.

This chapter highlights key interpretive methods. Keepers who want to delve deeper are encouraged to consult the general references listed in the bibliography as well as the ones cited in the text. It is important, too, to become familiar with the information and workshops offered by such organizations as the AZA.

MECHANISMS OF INTERPRETATION

WHAT IS INTERPRETATION?

Most of us have experienced interpretation. We have watched it from a distance, observed it up close, and even been a part of it in places like national parks, nature centers, or maybe even on a whale watch. We have attended fascinating lectures and walks, read colorful brochures, studied exhibits, and heard interesting explanations of the things we are looking at from interpreters and interpretive booklets and signs. What is the essence of this interpretation?

At first glance the methods of interpretation may seem like a loose collection of ideas for transmitting new information in a manner that makes it understandable, memorable, and inspirational. On a deeper level, however, interpretation as a process is a profound educational tool that is most effective only when someone consciously strives to share a passion and knowledge for a subject with an audience. The process is anything but superficial chitchat or careless recitation or writing. It is a means of communication based on specific principles, concepts, and methods. The US National Association for Interpretation's (NAI) latest official definition of interpretation, "a mission-based communication process that forges emotional and intellectual connections between the interests of the audience and the meanings inherent in the resource," has as its strength an emphasis on *connecting*, which always seems like a good practice. For example, a visitor who feels a connection with a red panda from visiting a zoo is more likely to be sympathetic and caring about future issues affecting the species and the ecosystems that sustain it. Visitor sympathy could extend to financially supporting red panda conservation projects, and perhaps even to changing a comfortable lifestyle that threatens global biodiversity. The definition recognizes the old saying, "What the mind does not know, the heart cannot feel"; in other words, it reflects the dual nature of our minds that operate both on emotion and on reason or rational thought. To be truly connected and committed to the causes that zoos promote, zoo visitors will have to not only *know* but also *feel* connected to them.

However, whether *resources* (in this definition, describing things of merit, e.g., natural resources such as forests or a particular plant or animal in a forest, or an historical building or battleground) really possess inherent meaning is debatable. For most keepers, it will be up to them to create and reveal the "resource meanings." Doing this with visitors who come from differing backgrounds will be challenging, but it can add spice to otherwise routine tasks.

There are as many formal definitions of interpretation as there are professional organizations employing its methods. For example, the American Association of Museums (AAM) defines interpretation not as a process, but as "the planned effort to create for the visitor an understanding of the history and significance of events, people, and objects with which the site is associated." (Alderson and Low 1985). This definition easily could be adapted to the zoo world, and often is, on a local basis within a zoo's mission statements. Sam Ham, the author of a popular and much used interpretation text, *Environmental Interpretation* (1992), crisply defines environmental interpretation as "translating the technical language of a natural science or related field into terms and ideas that people who aren't scientists can readily understand." Finally, the first formal published definition still carries a lot of weight in its utility: "an educational activity which aims to reveal meanings and relationships through the use of original objects, by firsthand experience, and by illustrative material, rather than to simply communicate factual information" (Tilden 1957). This definition defines what interpretation is and is not, and is a key statement of the process.

Each definition offers a different perspective; each has value. Comparing them stimulates thinking. From these we learn that interpretation is a planned educational effort that connects audiences to information made understandable, for example, by a context of meanings and relationships; it is not a simple statement of factual information. Finally, I would add to this mix my own definition: "Interpretation is educating entertainingly." Regardless of how it is defined, for interpretation to be effective, the keeper who offers it and the visitor who receives it must enjoy their shared experience.

ASSEMBLING THE PROCESS

This equation for interpretive excellence offers a more tangible definition of the interpretative process. It is a good place to begin when first attempting to understand and use the methods:

$$\text{Interpretive Excellence} =$$
$$\text{Goals} + \text{Audience} + \text{Methods} + \text{Subject} + \text{Site}$$

The author has been sharing this equation for the past two decades in his university courses and workshops. Students report that the equation works for them to capture and explain, or to render transparent, the major components of the interpretive process. This is important because the components must be put into play for good interpretive outcomes. To understand the components and, more importantly, to successfully use them in bouts of communication with visitors requires mastering the entire interpretive process. This is a challenge for neophytes, not so much in understanding the components, but in learning to put them all into play at once. And doing just this is the critical key to good outcomes.

Before dissecting each of these components, it may be helpful to consider the basic question of what really makes interpretation work. Why is interpretation effective in forging connections, improving communication, and, often, even in changing visitor behaviors? And how does interpretation differ from traditional teaching? Decades after Enos Mills pioneered the interpretive process while working as a nature guide up Long's Peak in the early 1900s, articulating it in his books and his trail school, Freeman Tilden described the essence of the process in a set of six core principles. These he published in his classic book *Interpreting Our Heritage* (1957). The principles summarized what he defined as "educational activities" gleaned from observing park rangers at work in US national parks. His principles can be boiled down to three key qualities or ideas that effective interpretation should possess: *relevance*, *revelation*, and *provocation*. Tilden also perceived that the park rangers understood that audiences possess different learning abilities and modalities, something the rangers mostly recognized in their interpretation for children. These qualities can be restated by explaining that for a keeper to be noticed, understood, and remembered and to initiate an ongoing interest (i.e., one that inspires more learning), the visitor must perceive something interesting and meaningful in the interpretation presented. This means that the keeper who is interpreting will need to connect new information to something familiar to the visitor and explain its significance. This is no easy feat when working with today's diverse, large, and often metropolitan audiences. The keeper will also need to provoke the visitor pleasantly, to capture and hold his or her interest. There are many ways to do this—for example, by using directed dialogue, objects, artifacts, or gadgets, but most of all, by thoughtfully shaping and constructing messages. Here keepers have the edge among the many kinds of interpreters at work throughout the world. Their ample fund of personal experiences and stories and their proximity to the zoo or aquarium's animals gives them this edge, which may include providing visitors with direct access to animals that can be safely touched or at least brought near them.

GOALS

An old adage applies here: If you don't know where you are heading, then any old road map will do. A clear sense of purpose is important in communicating with any audience, whether the purpose stems from personal goals or from institutional goals that are driven by meticulous short- and long-range planning and mission-vision statements. Without this sense of purpose, the keeper may become mired in disjointed trivia and anecdotes, perhaps even slipping into a canned recitation or a "lecture in the field." This "catalog of facts" is certain to drive visitors away. Better results come from having clear goals and restating them as realistic, even quantifiable objectives. Even when visitors are more interested in entertainment than in learning, it is often possible to initiate and capture their interest and then channel it toward the goal of education by using good interpretive techniques. After all, interpretation is educating entertainingly.

Goals need not be mutually exclusive, but they may be hierarchical. This means they may be "nested" or "layered," with each layer reinforcing a succeeding one. For example, the keeper may in one interpretive bout discuss a captive animal's needs, and the zoo's efforts to meet them. A little later, within the same conversation, the keeper can reveal more about the species' needs. The whole interpretive episode begins simply by calling attention to a physical property of the captive animal: the shape of its legs or nose, or its colors. It is also possible to mention the importance of supporting the zoo (e.g., through memberships or donations). All of these goals can be threaded into the interpretation.

AUDIENCE

Nature interpretation is "audience-sensitive" communication. Interpretation is not just communication beamed at a particular audience, it is communication tailored for the audience at hand. To work and be relevant, it must be related to something that exists within the visitor's understanding and interest. For this to happen, a keeper must know as much as possible about the visitors receiving the interpretation. If visitors are Hispanic and understand only Spanish but the keeper communicates only in English, the problem will soon be apparent. Mistakes of similar magnitude can occur, however, when keeper and visitors share a common language but the keeper makes erroneous assumptions about the visitor's level of knowledge and interest. In this instance, the lack of communication may not be so apparent.

Fortunately, communication can be enhanced and mistakes minimized by taking steps to determine visitor characteristics such as age, socioeconomic and educational background, ethnicity, profession, and hobbies. Knowing one's audience is key. Questionnaires and surveys are formal visitor assessments that are often a part of the zoo scene but are usually administered by the marketing and education staff. These documents should be read and the information shared among all zoo personnel. The staff in charge of the assessments should routinely share this information with keepers, and keepers should make a habit of reviewing the data. Keepers can take steps to collect their own information. Making a habit of chatting with visitors, especially before informal and scheduled presentations and demonstrations, and retaining the knowledge gained from those encounters is one way to do this. Another approach is to draw out this information from visitors by questioning them thoughtfully and tactfully. This becomes an interpretive strategy that can include directed dialogue, a keeper-steered conversation that is also valuable for engaging visitors and reviving their waning attention.

Are there other factors that can help a keeper to know the audience? The value of general knowledge cannot be emphasized enough as a means of improving interpretive skills and connecting with a wide range of audiences. One can build one's fund of general knowledge by listening to radio, watching television, seeing some of the movies intended for young audiences such as *The Lion King* and *Madagascar*, scanning internet listservs, and especially reading from a broad range of publications including newsmagazines and newspapers. Trade publications and journals are indispensable, but these should be supplemented with publications, such as nature magazines, that are read on a regular basis, and also with recreational reading such as novels. Taking an interest and participating in local civic matters and community organizations, as well as professional society involvement, helps to round out this picture. Collectively, all these measures build vocabularies and the facility to quickly form interpretive linkages with zoo visitors—one essential aspect of relevance.

METHODS

PERSONAL INTERPRETATION: MAKING THE MOST OF PERSONAL CONTACTS

One rule should always apply to zoo interpretation: Greet all visitors in a friendly, warm manner, and keep the interpretation pleasant and upbeat. Any message wrapped in gloom and doom will discourage visitor interest and the potential for positive action. As one experienced zoo professional said, "Conservation issues are very serious, but without a sense of hope and a positive outlook, there is not much interest among the general public for getting involved." Another way of thinking about this is to remember to accentuate the power of the positive in visitor interactions.

Keepers need to *reveal* their message and forge connections by linking the tangible to the intangible and the seen to the unseen. This intellectual excursion should be a fun and exhilarating journey for zoo visitors. To accomplish this, the information should be offered in a way that is relevant, personal, and meaningful to visitors (Ham 1992). For example, why should the visitors care about the care and conservation of Asian elephants? Why should they take an interest in international conservation projects that safeguard their future? Why should they change their consumer habits or support efforts to cope with burgeoning population pressures that negatively affect these elephants? Why should visitors be concerned with the differences between community-based and fortress-style conservation programs, the former taking into consideration indigenous residents and their needs and wisdom, the other superimposed and enforced from without (Dowie 2009)?

Visitor interest in any zoo animal is likely to begin with the mundane: with its age, size, weight, and diet, or with exclamations and questions about its observed behavior, its peculiar structure, or its shape. Explaining these qualities or traits in terms of how they adapt the animal to specific tasks and to a particular habitat and lifestyle is an easy way to make the observed understandable and meaningful—in other words, relevant.

Going beyond this to relate one's answers to the well-being of society and the preservation of biodiversity will often be appropriate. Interpreters attempt to *reveal* these ideas in a manner that provides the meanings behind the meaning, the "big picture ideas" wrapped in an understandable context. When done correctly and effectively, interpretation makes all lives more enjoyable. Keepers will not have trouble with this objective when they couple their professional interpretive methods with their solid base of knowledge and experience.

The biggest mistake any interpreter can make, and this is true for all forms of interpretation, is to assume that the audience shares the same level of interest as the interpreter. Often this is not the case. However, it is appropriate for the keeper to initiate and enlarge the visitors' interest through the use of *focusing elements* and similar methods. A focusing element sounds more complicated than it is. It is just a question, an exclamation, a command, or even a story that is used to spark visitors' interest. It is followed with information offered in a pleasing or even irresistible format. Here are five specific examples of focusing elements used by a keeper interpreting a sun bear (the focusing object) to a group of visitors:

1. What does this face remind you of?
2. Look at this face!
3. Have you ever seen a face like this before?
4. Sharp observers, like I know you all are, will tell me you have never seen such a strange face as this!
5. You know, I can remember three years ago, when I saw another mammal with a face like this, in circumstances you would appreciate. It was when. . . . (the short story follows).

These examples of focusing elements carry with them the ability to jump-start or *focus* visitors' attention on the subject or object at hand. What follows the focusing element should be a simple statement relating the interpretation's purpose, the main idea or *theme,* after which supporting or explanatory information is then *revealed* in a brief and creative presentation. Brevity is key here, as is offering the message through revelation—that is, by revealing it and bringing it out in a relevant context and an interesting way, which is the essence of interpretation, the hallmark that sets it apart from traditional education (i.e., the mere statement of information). The keeper will need to consciously connect the information or message to something within the visitors' grasp, and explain its importance to create a context of relevance. Questions such as "Why is this important to visitors?" "Why would they want to know about this?" and "Why should they know about this?" should be uppermost in the keeper's mind when communicating. Visitors have many opportunities to aid wildlife, whether it is through their choice of consumer goods or political leaders, their willingness to pay taxes, or even their modes of transportation. Many visitors have the option of modifying their yards to make them more wildlife-friendly by planting native species, and by avoiding the use of home and lawn pesticides.

Imagine for a moment the interpretive opportunity created when a zoo visitor shows an interest in fruit bats, perhaps attending a short keeper-driven demonstration about them. It would be easy in such a circumstance to first do the obvious: reveal the reasons behind the mammal's morphology after drawing attention to it and identifying it in relation to its life history. The next interpretive step, which might not be so obvious, could be to move on to the plight of fruit bats and the current attempts to relieve pressure on the species resulting from persecution by humans. An appropriate ending could include an explanation of the value of bats in general, even here in the United States, and their contributions to the well-being of ecosystems; the reasons for concern about their declining populations; and the actions that can be taken immediately to support their conservation. A final component of interpretation should be evaluation. Tying this part of the process to initial goals and objectives will permit quantification of outcomes. At the very least, informal evaluations will be helpful. Keepers who are committed to "educating entertainingly" are not likely to ignore the conspicuous signs of visitor dissatisfaction, disconnection, and discomfort. Still, formal program evaluations such as questionnaires and surveys should be used. These do not need to be lengthy. Asking colleagues such as administrators, head keepers, marketers, experienced educator staff, and even seasoned docents to join in the process is likely to produce better outcomes.

Figure 51.2. Formal program evaluations, such as questionnaires and surveys, are useful. Photo by D. Andrew Saunders, courtesy of Rosamond Gifford Zoo, Syracuse, NY.

How long should an episode of interpretation be? The length should vary with circumstances. Informal demonstrations are likely to be shorter than formal programs, especially scheduled and advertised programs, although both formats will adhere to the same general framework. *Letting visitors extend interpretive bouts with their spontaneous remarks and questions should be one of the golden rules of interpretation.* It is important to remember that visitors may not expect or need to be educated during their zoo trips—circumstances that sometimes, perhaps often, can be modified with the right approach on the part of the attentive keeper who can capitalize on an "interpretive moment" (i.e., an unexpected event, an unplanned encounter or experience with an animal that has the potential to grab the visitors mind, even if temporarily). The keeper's personal judgment should shape the length of an episode. If there is general agreement here on correct protocol, more is not better; only the insensitive keeper will prolong interactions to the point where visitors lose interest and begin walking away or shifting from foot to foot. Presentations and demonstrations are often most effective when kept to 5 or 10 minutes, but they may be extended when visitors inundate the keeper with questions.

Keepers may have other choices when engaging zoo visitors. Personal interpretation options will range from scheduled formal presentations for large audiences to small impromptu chats. Other members of the zoo team will be practicing interpretation too. Often this will include non-personal, secondhand approaches to interpretation, such as web pages, brochures, exhibits, and wayside exhibit panels.

The greatest challenge for interpreting thoughtfully is in choosing the most significant idea or message to communicate to visitors, and then preparing an effective means to reveal it. Translating the main idea into one simple sentence, a *theme* stated at the beginning and again at the conclusion, is a good approach to successful interpretation (Ham 1992). Three to five *subthemes* should be developed to support the main theme, the number approximating the limit of new ideas that most of us are capable of retaining in one learning episode. This message format is not always appropriate when interacting with visitors, but often it will be. The thoughtful keeper who is sensitive to interpretive opportunities can use it as an effective framework. The theme will become the central organizing principle of the interpretation. Visitors are most likely to remember this major idea, although it may be facts that initiate and hold their interest. The choice of a theme will depend on many factors, especially the zoo's goals and the keeper's knowledge, experience, and motivation. *A positive attitude will be most important if a keeper is to identify the best interpretive opportunities.*

The theme-and-a-few-subthemes approach is not as easy to use as it may first appear, and it is often ignored by practitioners despite its effectiveness. The power of the concept lies in the mental discipline necessary to organize, synthesize, and simplify ideas and information, and to ensure that main ideas or points are tied to the original theme. Keepers can do this by using a *thematic connector*, a statement or even a question that relates subthemes to the overall theme. Interpretive presentations can be strengthened with these connectors, and questioning visitors can draw them into the process and into the presentation itself. In this case the more interactive the interpretation, the better it will be, as long as it is done tastefully and respectfully. Whatever the approach, one should never embarrass the visitor—for example, by telling them they will be quizzed later about the information presented.

USE OF OBJECTS

Interpreters in all fields frequently use secondary objects in their interpretations—something in addition to the main subject at hand, which for zoo keepers will most likely be the animals in their care. A can of sardines displayed by the keeper while explaining the diet of fish-eating Humboldt penguins is an example. These secondary objects may be called teaching objects, props, or museum materials, but no matter what they are called, their impact is the same.

When used correctly, objects provoke in a pleasant way, enriching and enlivening communication while reviving and enhancing the audience's interest. Objects used in a presenta-

tion might even include items such as the keeper's uniform, and other standard features for first-person interpretation (Roth 1998). These items have their place in successful communication. Keepers will often dress by necessity in garb that appeals to visitors—the standard zoo uniform—and they are very likely to be aware of its power. But standard museum objects such as feathers, bones, eggshells, and skulls also have a potent capacity to improve visitors' comprehension because they bring into play the visitors' various learning modalities and styles. Engaging the senses—touch, smell, hearing, and vision—is a part of the learning process. Objects quickly transform passive education into active education They should be used whenever possible, but never without careful thought about visitor and animal safety, and about whether they reinforce the theme being presented. There is no need to use them if they are superfluous, unnecessary, or offensive to visitors. Using objects to demonstrate and compare animal adaptations such as teeth, claws, and coloration or passing around tools, photos, and graphics during a presentation is a good strategy. In many zoos the education staff uses "biofact" carts to transport materials for portable, spontaneous visitor programs on the zoo grounds. While keepers may not function in quite the same mode, it may be advantageous for them to have some of the same materials at their disposal, or to share materials used by the traditional zoo educators.

Figure 51.3. An interpreter can use "biofacts" such as this flamingo skull to make a presentation more interesting for the audience. Photo by D. Andrew Saunders, courtesy of Rosamond Gifford Zoo, Syracuse, NY.

Using technology in interpretive episodes is a great way to activate or even reactivate visitor interest. GPS units, radio transmitters, night vision goggles, gadgets used in feeding and examining animals, and remote cameras are examples of devices that can help deliver the message. Sometimes they will serve as wonderful bridges to off-site research programs and conservation projects. They also may help to reveal how animal health is monitored within the zoo.

NONPERSONAL INTERPRETATION: WHEN YOU CAN'T DO IT ALL YOURSELF

To some, the notion of keepers being involved in the production of nonpersonal interpretive products such as exhibits may seem far-fetched, but it is a common occurrence. Keepers represent a unique perspective and body of knowledge, and their input should be solicited more often than it probably is. For this matter of non-personal interpretive product development and its content, a little knowledge will ensure good outcomes. When nonpersonal approaches are appropriate, the same basic concepts of message management, such as themes and subthemes, apply. However, limiting verbiage is paramount. The sin most often committed in the development of these products is to create the "text on a stick" that zoo visitors ignore. Web pages, brochures, exhibits, and wayside panels should all honor the mandate "attractive, brief, and clear." (Sometimes the word "dynamic" is added, but its meaning should be encompassed by the word "attractive.") When interpretation is conducted by any of these textual means, brevity will often mean using no more than 220 to 240 words in a message pyramid of theme-title, subthemes, and small blocks of contrasting text, all in readable fonts. A strong focal point, good illustrations, and plenty of "air" (blank areas) complete this picture, especially in the case of exhibits and wayside interpretive panels. Finally, another suggestion is to recognize the importance of consistent design and coordinated, compatible messages in all the zoo's outreach efforts.

Nonpersonal interpretation products such as wayside panels or zoo exhibits can complement and supplement keepers' personal interpretations. When properly designed, these products will augment and reinforce the themes keepers share with visitors. For example, by directing visitors to a nearby interpretive panel near the giraffe compound, a keeper can invite visitors to learn more about the unique features of a giraffe's neck through the panel's colorful graphics and a few well-honed text blocks. An interpretive episode can be planned to include a panel as a central illustration for a subtheme: "Look at the neck of this giraffe! And we think swallowing salad is a chore! This long, skinny neck is one of many unique adaptations that equip this huge herbivore, a plant eater, to convert plant energy to giraffe power by munching on plants that its competitors cannot reach. This exhibit panel shows some of these herbivore competitors. What other structures give the giraffe its reaching power?"

SUBJECT MATTER

"No content, no process" could be the motto for interpretation. The animal and the facts about its morphology, physiology, behavior, and ecology are the currency the keeper can use to capture a visitor's interest and transmit a theme. "To educate entertainingly you must know thy animal!" This is the keeper's commandment. Visitors will always appreciate interpretation that helps them make sense of what they perceive of zoo animals. General knowledge of species-specific traits can be gleaned from many sources. The introductions to orders, families, and genera contained in survey courses and contemporary texts treating the "ologies," such as mammalogy, are good places to pick up this information. Web-based sites of reputable origin are another. A keeper's success and recognition by visitors and superiors as an interpreter will be based in part on the breadth, depth, and veracity of the knowledge he or she reveals to visitors. Knowledge is the reward for effort and persistence, but when a question cannot be answered it is OK to say, "I do not know." One can invite the visitor to submit the question by e-mail so that it can be addressed later, or point the visitor to another information source within the zoo. Redirecting questions should always be handled in a friendly, patient manner that rewards the visitor's interest and curiosity: "You know, that is a really interesting question that I cannot answer, but if you will. . . ."

Feeding schedules, the transfer of animals to different quarters, and the chief animal activity periods are elements of the zoo animal's life that are likely to interest visitors. Enrichment schemes are other important avenues for interpretation, and they have much value in demonstrating the zoo profession's concern for animals' health and well-being. Enrichment activities are "works in progress" with wonderful education potential for zoo visitors. The keeper's awareness and use of these activities greatly expands the interpretive repertoire.

THE SITE

The pressures of the job make it easy to overlook the comfort of visitors, who are likely to accommodate and enjoy the keeper's interpretation. A little forethought will help the keeper in anticipating potential interactions and trying to place the interpretive encounters at times and places most comfortable for all concerned. No one wants to stand in torrid heat staring into the glaring sunlight during even a brief a presentation. Merely asking visitors to move a few paces into the shade or, if that is not possible, just reversing positions with them so the keeper faces the sun can dramatically improve the experience and make it more pleasant. Choosing a location with an eye toward visitor comfort and safety is always a wise strategy. Anticipating crowds, background noise, and traffic pattern problems and minimizing them are aspects of this strategy. Keepers may be called upon to determine the animal-related needs and parameters of off-site programs at schools or similar locations. Effort invested in this matter will prove helpful and could save those programs. Some off-site programs may be condensed into interpretive talks embellished with objects (props) and equipment such as laptop computers or digital projectors. PowerPoint presentations that contain text, video and sound clips, and quality still images are potent communication tools when used thoughtfully. Quite apart from audience parameters and the physical space of the presentation, even adequate directions for travel and the means to undertake it, electrical outlets, extension cords, a table or tables, room darkening shades, each or all can become truly critical needs in these circumstances.

Finally, nothing should ever compromise the visitors' safety or the animals' well-being. Therefore, personal hygiene, health considerations, and safety issues are not options but musts—another important rule for keeper interpretation, and the one at the top. All institutional animal welfare guidelines must be scrupulously honored, from rules about hand sanitizing to restrictions on access to animals that are dangerous or whose well-being can be compromised by stressful experiences.

WHY DOES INTERPRETATION WORK?

Interpretation works because of its *professionalism* and *pragmatism*. In other words, interpretation works because it has been relatively free of the constraints of formal pedagogy. Interpreters seldom enjoy the luxury of teachers who work within traditional classrooms with captive audiences. If an interpreter's communication is defective and the interpretation fails, the "class" simply leaves. Both the organization providing the service and the person offering it may soon be "out of business." Trial-and-error has guided and perfected the interpretive process. Components that work have been retained; those that do not have been jettisoned. Passing grades in this case are not just options; much like the health and safety of the animals and visitors, they are a necessity. As with any personal communication, much of the effectiveness of this communcation will be found within the interpreter's personality. The passion, enthusiasm, enjoyment, and energy of the keeper will always be one of the foundations for interpretive success. Global applications of these methods applied to conservation education will be found in *Conservation Education and Outreach Techniques* (Jacobson, McDuff, and Monroe 2007); meanwhile, the US National Association for Interpretation now offers a series of modestly priced manuals and guides that treat most phases of environmental interpretation.

OTHER CONCERNS?

Professionalism means paying careful attention to many extraneous details such as appearance, personal hygiene, reliability, and punctuality. A quality sometimes overlooked in this list is congeniality, working seamlessly with the larger zoo or aquarium family of which visitors are only one part. Administrators, marketing and development personnel, formal educators, and especially the volunteers often known as docents are other branches of this family. These should all be regarded as full partners and treated accordingly. Volunteers are a case in point, and volunteerism is a looming factor in the success or failure of many organizations today that face dwindling budgets and burgeoning workloads. Volunteers reflect institutional qualities and extend them out into community and regional areas, and even beyond. Keepers who value volunteers and treat them accordingly will prosper, and those who regard volunteers with disdain or indifference will do themselves and their employer a disservice. A thoughtful articulation of the rules pertaining to zoo or aquarium docents—to their duties and limitations, with careful atten-

tion given to rewards and supervision—is important for a successful docent program. These rules must be understood and endorsed by the entire staff. The key, again, is open communication coupled with careful position descriptions and boundaries. Docents are likely to regard keepers with the same awe, envy, and appreciation as do visitors; building good will on both sides of this fence will multiply institutional productivity.

SUMMARY

Keepers should design and conduct their personal interpretations to be active and interactive, rich with sensory experience and information, incorporated within a theme and supported by three to five subthemes. They should strive to connect their interpretations to something within the interests and experiences of visitors, recognizing interpretation's importance and the challenge of doing it, given the diversity of visitors. Keepers should deliver their interpretations to meet personal, professional, and institutional goals, remembering to make provisions for evaluation and improvement; they should never overlook opportunities to connect the animals

Figure 51.4. Keepers can connect the animals that brought them into their profession with the visiting public while sharing with zoo visitors the values, opportunities, and responsibilities of stewardship. Photo by D. Andrew Saunders, courtesy of Rosamond Gifford Zoo, Syracuse, NY.

that brought them into the profession with the visiting public, and to share with zoo visitors the values, opportunities, and responsibilities of stewardship.

THE GOLDEN RULES FOR KEEPER INTERPRETATION

- Never overlook opportunities to "sell" the zoo's mission, animals, and programs to visitors on- and off-site.
- Base approaches to interpreting on a thorough understanding of the animals.
- Always greet visitors warmly and invite their curiosity and questions.
- Anticipate visitor interests and prepare for them, responding eagerly and pleasantly to their questions with thoughtful answers.
- Develop strategies to know zoo visitors in terms of their background parameters, whether through formal or informal assessments.
- Think in terms of messages to connect visitors, using the interpretive process, to the bigger picture and big ideas of the zoo world.
- Whenever possible, build each message around a theme and a few subthemes, taking care to connect the information to something familiar within the visitor's experience.
- Use focusing elements to catch and hold attention at the beginning of an interpretive bout, but begin and end by stating and restating the theme.
- Enrich each interpretive bout with objects such as biofacts and technological gadgets, to engage the visitors' senses.
- Analyze the zoo grounds and working environment to determine the safest, most comfortable places to interact with visitors, and steer them to these locations if necessary.
- Value and compliment the zoo docents, and extend the same congenial behavior to all zoo colleagues.
- Practice and promote professionalism.

REFERENCES

Beck, Larry, and Ted Cable. *Interpretation for the 21st Century, 2nd edition*. 2002. Champaign, IL: Sagamore Publishing.

Gross, Michael, Ron Zimmerman, and Jim Buchholz. 2006. *Signs, Trails, and Wayside Exhibits, Connecting People and Places, 3rd edition*. Stevens Point, WI: UW-SP Foundation Press.

Ham, Sam. 1992. *Environmental Interpretation*. Golden, CO: North American Press.

Jacobson, Susan. 2009. *Communication Skills for Conservation Professionals, 2nd edition*. Washington: Island Press.

Jacobson, Susan, Mallory McDuff, and Martha Monroe. 2007. *Conservation Education and Outreach Techniques*. New York: Oxford Press.

Kahn, P. H. Jr., and Stephen Kellert, eds. 2002. *Children and Nature*. Cambridge, MA: MIT Press.

Knudson, Douglas, Ted Cable, and Larry Beck. 2003. *Interpretation of Cultural and Natural Resources, 2nd edition*. State College, PA: Venture Publishing.

Lewis, William. 1991. *Interpreting for Park Visitors*. Yorktown, VA: Eastern Acorn Press.

Roth, Stacy. 1998. *Past into Present*. Chapel Hill: University of North Carolina Press.

Serrell, Beverly. 1996. *Exhibit Labels: An Interpretive Approach*. Walnut Creek, CA: AltaMira Press.

Tilden, Freeman. 1977. *Interpreting Our Heritage, 3rd edition*. Chapel Hill: University of North Carolina Press.

52

Public Relations in Zoos and Aquariums

Jason A. Jacobs

INTRODUCTION

Public relations (PR) is vital to promotion of the mission and work of zoos and aquariums. This chapter will provide basic information about media and public relations as it affects and is affected by the keeper position. After studying this chapter, the reader will understand

- the importance of publicity to the operation of zoos and aquariums
- the keeper's role in publicizing zoos and aquariums
- the preparation of zoo and aquarium keepers, animals, and exhibits for media opportunities
- methods of addressing proprietary and sensitive information
- the development of stories about the zoo and aquarium
- the pros and cons of anthropomorphic stories in the media
- branding in relation to the keeper's work
- keepers' use of the internet and social media.

THE BASICS OF PUBLIC RELATIONS

Many zoos and aquariums have limited budgets for paid advertising. Using the press (internet, television, radio, or newspaper) is a great way for institutions to share their stories and increase public awareness of their work. Public relations, unlike marketing and promotion, is not a paid form of advertising. A positive steady flow of publicity helps to tell the stories that take place in a zoo or aquarium, raises public awareness about the institution throughout the community, and increases its public attendance. This increased attendance will increase revenue at the institution's various concessions, rides, and gift shops, and this revenue will ultimately go back into the zoo or aquarium to support its mission. In most cases, a zoo or aquarium's publicity or PR department can measure its own value by assessing the value of the space given to its institution in newspapers or by looking at the

segment's ratings on television and comparing it to the cost of paid advertising during those times.

Another form of positive PR is the guest experience; word-of-mouth advertising is invaluable. People are far more likely to talk about a bad experience at a zoo or aquarium than about a good one. It is for this reason that all zoo and aquarium staff should have their "publicity hats" on at all times. In the age of Twitter, Facebook, and smartphones, a good or bad experience can instantly be shared by thousands of people electronically. Zoo and aquarium guests can post their own reviews of institutions on websites such as Yelp and Trip Advisor. A guest does not need to be on the zoo or aquarium grounds to instantly learn about news or events happening there. Zoos and aquariums often find themselves competing for the "family dollar" with theme parks and movie theaters, and it is important for them to realize that each and every one of their guests is a valued customer whom they want to return and support them for years to come.

Keepers are an integral part of a zoo or aquarium's publicity effort. With the proliferation of television shows about zoos, aquariums, wildlife, and natural habitats, zoo and aquarium visitors are more knowledgeable than ever about the natural world, and interest in zoos and aquariums is increasing. Like characters in a popular television show, keepers are the most popular characters at a zoo or aquarium, after the animals themselves. Since the animals in these institutions cannot speak for themselves, keepers have the job of telling their stories and those of their wild counterparts. While some zoo and aquarium staff may be offended at being compared to characters playing roles, it is important to realize that many children look up to keepers as they would firefighters or police officers.

More than 171 million people visit North America's Association of Zoos and Aquariums (AZA)-accredited zoos and aquariums each year, more than those who attend all professional sporting events in the United States combined (AZA 2010). This gives zoos and aquariums the opportunity to foster stewardship and care for the environment on a broad scale. Zoo visitors are fascinated with the work that keepers do, and keepers receive more questions from the public than any other zoo staff. For this reason, keepers need to keep themselves updated with information about not only the animals they care for, but also their facilities and their related programs. When a keeper engages the public, he or she is not only educating them but also selling them on the institution's mission and programming. When giving presentations, keepers will often be asked questions by children. It's important for them to remain patient and, if possible, adjust your dialogue to suit the situation. Studies show that most people who visit zoos bring their children; they see it as an educational experience. Keepers should remember to talk in words that children can understand. For example, in the description of an animal that is exothermic, it might be easier for a child to understand the term "cold-blooded." There have been cases where a child's conversation with a zookeeper has led to the child later becoming a zoo volunteer, pursuing a career in working with wildlife, or perhaps becoming a million-dollar donor to a zoo or aquarium.

Basic information the keeper needs to know about his or her institution includes its hours of operation, admission prices, animal and exhibit highlights, visitor's amenities, upcoming special events, programming, and ways in which patrons can help the institution. Staff should also be familiar with the history of their institution, including the year it was founded, its major births, and its conservation milestones. For instance, the Denver Zoo opened Bear Mountain, the first open-moat zoo habitat in the United States, in 1916. In 1956 the Columbus Zoo had the first-ever gorilla born in captivity. The Los Angeles Zoo and San Diego Zoo have played a major part in introducing the California condor (*Gymnogyps californianus*) back into the wild. Every institution has some notable milestones in its history that should be shared with staff and visitors. An effective way for zoos and aquariums to provide their employees with this information is with a "cheat card," which serves as a quick reference for frequently asked questions. The "cheat card" is simply a small laminated business card that contains the hours of operation, admission prices, upcoming events, and so on, and it can be updated several times a year. It would most likely be produced by the public relations and marketing personnel.

In addition to acquiring general knowledge of the institution they work for, keepers should learn as much as possible about the species they care for, to be prepared for media interviews as well as general public speaking. First and foremost, the media and the public are often interested in the animals' individual histories (e.g., names, ages, and social demographics). One should avoid anthropomorphizing when describing the animals in this context. These descriptions will vary from institution to institution. In addition, the public is often interested in what the animals eat. It's important for a keeper to know an animal's wild diet as well as its zoo or aquarium diet. Animals like lions (*Panthera leo*) obviously are not fed zebras and gazelles in zoos, so in some cases a keeper would need to be able to explain where the zoo obtains food for carnivores and the different feeding strategies it uses (e.g., fast days, enrichment). Many zoos use horse meat to feed their carnivores, or small mammals (e.g., mice and rats) to feed raptors and reptiles; one should be aware that this can cause controversy with animal rights groups, or simply upset small children. Some zoos and aquariums do not feed their animals in front of the public, while other institutions choose to use feeding behavior as an educational opportunity.

Speaking points designed to address controversial subjects are a great way for institutions to speak with a unified voice. Two examples follow.

Speaking point: Feeding rodents to animals at the zoo.
Scenario: Keepers have laid out several prekilled rats to feed to the condors in an aviary at the zoo; this angers some young children.
Response: The keeper can explain that in the wild, condors get their food by flying around and looking for dead animals. The meat people buy at the grocery market looks a lot different from an animal that has just been killed. Just as some people eat meat, these condors have to eat meat to stay healthy. These rats were bred and humanely killed to provide food for these birds. Condors are an endangered species, and both the Andean and California condors have been bred in zoos and released into the wild.

Speaking Point: Smells at the cat habitat.
Scenario: The visitors are complaining that a male tiger's zoo habitat strongly smells of urine.
Response: The keeper can thank the visitors for their concern, and explain that the male tiger often sprays his habitat because he considers it part of his territory. He is doing the same territorial marking that a tiger would do in the wild. In fact, after the keepers have cleaned his habitat, one of the first things a tiger will do is spray it again. It might smell a little strong to us, but it's a natural part of being a tiger.

Other pertinent information for the keeper to know is the taxonomy of the animal he or she is working with, examples of related species, their wild range (or, in some cases, what used to be their wild range), challenges the species faces in the wild, and the species' status in zoological collections. Conservation projects that zoos and aquariums support which keepers participate in make great stories to pitch to the media.

Many zoos and aquariums send their staff members around the world to participate in field conservation projects, and this represents a great opportunity for them to meet with local journalists months in advance to inform them of the project and how the facility and the media can communicate that work back to the journalists and, indirectly, to their readers. It is also important for keepers to be able to tell zoo visitors what they can do to support conservation or help a given species in the wild. Many zoos and aquariums support conservation programs, so if a visitor expresses a desire to support giant panda (*Ailuropoda melanoleuca*) conservation efforts, it is important that they be told the proper groups to contact. It should be remembered that while the charismatic megavertebrates often receive the lion's share of the publicity, all animals have a story to tell, and the efforts of staff working with critically endangered frogs or fish can be just as compelling. Keepers should not assume that people are only interested in the mammals.

MEDIA MISCONCEPTIONS VERSUS REALITY

Several common misconceptions about animal keeping and zoos and aquariums are important for animal care staff to remember when engaging media and the public. Many media people see keepers as trainers, and while a lot of training is done in zoos (that is, with operant conditioning), the training many journalists envision references circus-like images of keepers entering the chimpanzee's cage or taking a lion for a walk around the zoo. While there are examples of such practices still going on in today's zoos and aquariums, the vast majority of animal care professionals never enter the same physical space with dangerous animals such as great apes, bears, or big cats. Often the media are unaware of the concepts of overnight holdings (night houses), shifting of animals into containment, or behind-the-scenes service areas. This presents an opportunity for the publicity department to pitch stories to media about operant conditioning and the training done in today's zoos and aquariums for the animals' welfare and also for the keepers' safety. As a side note, one should remember that some behind-the-scenes service areas are not as aesthetically pleasing as the public sides of the animal's habitats. Some zoos and aquariums do not allow filming behind the scenes. If a zoo or aquarium does choose to bring media or guests behind the scenes, a keeper should be prepared to explain that those areas are often built primarily for functionality (cleanliness, safety, and husbandry), and not for the benefit of the viewing public.

The social lives of animals are often the subjects of misconceptions held by members of the media and the public. Many popular zoo species such as tigers, leopards, and bears are solitary by nature. Well-intentioned journalists or patrons may petition for "zoo habitat" companions for these animals because they are simply unaware of the animals' natural lifestyle. Many people still think that zoos and aquariums obtain most of their animals from the wild. It is important to know how one's institutions obtain their animals, whether through breeding program recommendations or through collaborative importation of animals between institutions to establish new bloodlines or sustainable populations. Keepers should be able to explain the various reasons for cooperative breeding programs, import consortiums, and other sources. Another misconception is that animals are bought and sold between zoos and aquariums for profit. In some cases animals are sold, but it is much different from 30 to 40 years ago, when breeding was less regulated and there were many more animal dealers. Today, most zoos and aquariums are involved in breeding programs that look at a species' overall population and the individual animals' genetics to make the best decisions for the future of the species. Media often look favorably at a cuddly tiger cub, and when it's announced that the cub is going to another zoo, some might protest; however, explaining the reasoning behind the decision is an opportunity to educate viewers and the public about the modern breeding goals of zoos.

WORKING WITH THE MEDIA

Interactions with media and the public are exciting opportunities to promote a zoo and aquarium and to inform the general public about animal care, conservation issues, and the institution's mission. Sometimes a keeper can find it difficult to do a television interview or a public talk due to nervousness, the work schedule, or the unpredictable nature of animals. In most zoos and aquariums the primary media contact is not a keeper, but either the institution's director or a member of the PR staff. Administration and PR staff will work on building relations between the institution and the media. Keepers play a vital role in solidifying this relationship by submitting potential stories and ideas for the PR department to pitch to the media. The media provide news and information to the general public about zoos and aquariums in both good times and bad. If a new exhibit opens or a baby animal is born, the local media will cover this story and help increase attendance. Likewise, if there is a major death of an animal or a crisis at the institution, the media will also cover it. It is important to remember that it is the media's job to cover all stories, both negative and positive. Every zoo and aquarium has had its share of negative stories as well as positive ones. Occasionally, news reporters will target a zoo or aquarium for controversial stories about an animal death or turmoil within the institution. In some cases they will speak with opposition groups or animal rights activists to get outsiders' opinions on the issue. A keeper should avoid offering personal opinions on these subjects and should strictly adhere to the institution's message points, or refer controversial questions to administration. For controversial subjects such as these, it is often the duty of the director or trustees to speak on behalf of the zoo or aquarium. In some scenarios the institution may not wish to comment on the story at all. In most media requests, a keeper's role in communicating with the media will be directly tied to interacting with or speaking about the animals. Administrative or PR staff will escort the media into the zoo or aquarium, or will give staff prior notice that they will be bringing them in. If a keeper is conducting a presentation in front of an exhibit and a member of the press identifies himself or herself and asks for comment, regardless of the nature of the story, the keeper should refrain from comment and direct the journalist to the zoo's administrative or PR office. It is important to forward these inquiries because media will sometimes visit the zoo unannounced, and it's important for the institution to get the proper message out about the animal or story.

It is every keeper's job to promote the institution. Promoting it by participating in a news story is a great way to tell the story of the animals in one's care. One should be a willing participant, as the last complaint an administrator wants to hear is from a keeper who does not want to do an interview because he or she is shy or sees it as an inconvenience. It is OK to be nervous about being interviewed by a journalist. Keepers know their animals better than anyone, and when asked by their administrators to speak on behalf of their animals they should see it as the perfect opportunity to tell the animals' story, the importance of the institution's commitment to the species, or the institution's involvement in in-situ or ex-situ conservation programs.

INTERVIEWS AND BEING MEDIA-READY

If keepers have any concerns before doing a media interview, they should talk to their supervisors or to the institution's publicity department, which can coach them and perhaps perform a mock interview beforehand. Larger institutions may offer media training classes in which keepers must participate before being interviewed by media. Small institutions can approach local PR professionals for assistance in training or conducting workshops for employees. A good way to prepare for a media interview is to write a few ideas on paper with bullet points on subjects that could be addressed. A keeper may be able to talk to the journalist before the interview about these points, or about others in which the journalist is interested. Animals that are conditioned to being around people can make excellent subjects for stories. Live animals are great conversation pieces during an interview, and sometimes the opportunity presents itself to use them in interactions with media and the public. It is always important to emphasize the professional care and expertise that goes into working with these animals, and that just because an animal keeper is on television with a trained cheetah, it doesn't mean that these animals make good pets. It is also a good idea to point out the animals' specialized diets, housing, and care, and how most nonprofessionals are ill-equipped to handle and maintain them. Live animals are not always available or cooperative for interviews and interactions, however, so it's a good idea to have a "biofact" or related item available as a backup. A keeper cannot bring a camera crew inside a zoo habitat with a family of gorillas, or into an aquarium's shark tank. Instead, one might invite the crew to see the gorillas' breakfast being prepared, or perhaps show them some shark teeth or shark jaws. If the interview is to take place on zoo or aquarium grounds, one should work with both the PR staff and the media to determine the best area for it. Although the media cannot film inside a hippopotamus habitat, keepers can strategically place food for the animal in a location where the animal can be comfortable but still meet the photographer's needs.

Appearing to be—or actually being—media-ready is just as important as knowing the facts about the animals. The keeper's appearance will be seen as an indication of the type of people the zoo or aquarium employs. If a staff member wears a stained shirt or has unkempt hair, it gives a sloppy look to the institution. While keeping can be a physical and sometimes dirty job, it's important to make a good impression during an interview or presentation. Perhaps it is a good idea to keep an extra zoo shirt close by or in a locker, together with some personal grooming materials such as a towel, comb, and brush along with some hair styling product and deodorant. The keeper should also wear a uniform that shows the institution's logo. Collared shirts are preferable to T-shirts for a more professional look; but the keeper should wear the uniform issued or required by the zoo or aquarium and, most importantly, remember to smile and act naturally.

Of equal importance is the look of the animals and their habitat at the zoo or aquarium. On many occasions, media interviews or feature stories have been scheduled or canceled at the last minute due to concern for the animal. If an interview is scheduled and the animal is off its feed or not looking particularly well due to a medical or husbandry issue, any concerns should be shared ahead of time with a supervisor or with PR staff. Being media-ready also applies to the animal exhibit. Viewing glass should be clean without any smudges, while leaves and debris should be removed from mesh-covered habitats. If the public viewing area is not clean or an adjacent garbage can is overfilled, it can make the institution as a whole look bad.

Media equipment such as satellite trucks, white balance screens, boom microphones, and even cameras can elicit different types of behavior from animals. The author has seen tigers react to camera crews in some cases with aggression and in other cases with ambivalence. Hoofstock and birds tend to react negatively to satellite trucks. Giraffes have been spooked by the antennas going up above their heads, while other animals such as great apes have reacted to the same thing with curiosity. If possible, it is a good idea to have the camera crew at the zoo or aquarium well ahead of time to allow animals time to adjust to the presence of the equipment. Once the journalists arrive at the zoo or aquarium, one should talk to them to discuss the camera angles they are looking for. Keepers can often strategically place food or enrichment items in desired locations so that photojournalists can obtain great film footage of animal behavior. Sometimes animals can be kept off exhibit pending a film crew's arrival to give the crew an opportunity to plan their initial shot of the habitat.

Before the interview begins, there are a few key points to remember. First, the keeper is the expert, and with some practice, staff can direct the interview by referring to knowledge of the animals and the institution and by offering appropriate "sound bites" to the reporter. If time is available, it's best to speak to the reporter before the interview begins to learn his or her expectations, find out what questions may be asked, and perhaps suggest interesting questions. One should determine the length of the segment and set the ground rules. This can also help to build some rapport and "break the ice" before the interview. If the interview will focus on an animal that the keeper is handling, the keeper should let the journalist know ahead of time where they can and cannot touch the animal, or if they shouldn't touch it at all; this prevents what could become a potentially dangerous or stressful situation for both parties. The journalist should be told that the interview may be cut short if the animal's behavior changes. On occasion some journalists can be overzealous and may attempt to reach out and hold an animal, especially a cute juvenile, while on the air. If a situation like this occurs, it's OK to smile and say, "Sorry, only zoo or aquarium staff can handle the animals."

When speaking to a reporter, one should avoid using certain words that have negative connotations. Instead of referring to the animal's home within the zoo as a cage, one can instead use such terms as "habitat," "island," or "paddock." Granted, most zoos and aquariums do not display their animals in cages or bare aquariums; however, behind-the-scenes areas often use caging, so it is may be appropriate to refer to these as "bedrooms" or "holdings." Keepers can stress that these areas offer the animals safe places to retire, eat, and in some cases birth and rear their young. Many species enjoy their holding and service areas, which are integral parts of zoos and aquariums. The word "captivity" can also sound negative, but most zoo and aquarium animals were not born in the wild; so instead of saying "animals in captivity," perhaps say "animals in zoos and aquariums" or "animals in our care." Occasionally a reporter might ask a question "off the record," but one should remember the freedom of the press and that there is no such thing as "off the record." Some journalists compete with each other for stories, and a keeper's off-the-record comment might in fact "trump" another reporter's story or reveal details that are proprietary to the zoo or aquarium. It's important to be friendly and open with the media but to know what information to offer and not offer. Finally, one should always act as if one is on air while the cameras are rolling, even if the cameras seem to be focused somewhere else. Microphones can pick up sounds even after an interview is completed, so one should be cautious about saying anything until after the journalists and their crew have left the area.

FEEL-GOOD STORIES

Positive "feel-good" stories include animal babies, exhibit openings, enrichment opportunities, and involvement in conservation projects. Zoos and aquariums offer endless opportunities for such stories. One of the tasks of a zoo or aquarium's publicity department may be to strategically plan press releases and stagger stories so as to best benefit the institution. The media cannot be expected to visit every week, and many times a story is held by the zoo or aquarium's publicity department to capitalize on its potential to increase the number of weekend visitors. If the weather forecast calls for rain or if a world or community event is going on, the zoo administration may decide to hold the story until circumstances improve. The media are very flexible, and other news events can tremendously influence the amount of coverage a zoo or aquarium receives. It is important for a keeper to be accommodating, and to smile whether one reporter or a dozen show up to cover a story.

The day-to-day care and enrichment of some species might seem routine, but in some cases these daily responsibilities of a keeper provide a fascinating look into the science of animal husbandry and the inner workings of zoos and aquariums.

Media personalities enjoy a hands-on approach to stories; one might consider creating an activity in which a journalist can help prepare enrichment for an animal and then give it to the animal. Other potential story ideas could involve the journalist helping to prepare the animals' diet or cleaning the habitat. In several media markets, journalists are tested specifically for tuberculosis so that they can meet the zoo or aquarium's bio-security requirements to report on behind-the-scenes events or participate in stories with certain animals. Baby animals are always a media and crowd pleaser. In many larger collections, animals may be born fairly frequently; it can be a challenge to ensure their safety while at the same time disseminating information about their birth. Often zoos and aquariums will have a grace period (e.g., two weeks) before they introduce a newborn animal to the media and public, to ensure first that it is healthy and being cared for. In some cases it is impossible for a camera crew to film newborn animals due to protective parents or other factors; keepers can provide invaluable assistance by working with the PR department to allow trusted staff members to take the photos or videos and have them distributed to the media. While the media enjoy the babies, this is also a great opportunity to tell the story of collaborative breeding programs, the status of this animal's wild and zoo population, and so on. While animals like baby apes and cats might draw crowds, it is just as important to the zoo or aquarium's mission to promote smaller animals that the media also find interesting, such as invertebrates and amphibians.

The staff of zoos and aquariums may use their experience from field research, conferences, and in-situ and ex-situ conservation projects to better understand the animals they care for. In many instances, the institution offers staff and expertise to help with projects that support wild animal populations. These experiences provide great media stories and truly demonstrate the good work that zoos and aquariums do to help threatened or endangered animals. Often such activities require travel, and one might consider inviting reporters along. If a reporter's organization does not have the budget to fund travel, one might work with the PR department to write an online journal or blog to document the trip or perhaps take photos that later could be submitted to local media.

Many zoos and aquariums offer varied experiences (rides, events, fundraisers) for guests—events far beyond the scope of the animal collections. Sometimes keepers (with or without their animals) are asked to participate in these activities, even when they have little to do with the animals directly. For example, many institutions hold large Halloween or holiday events. To publicize them, a zoo or aquarium will often give animals special treats, such as pumpkins, or perhaps enrichment devices shaped like a trick-or-treat bags or holiday gifts. These activities are designed to promote the event and to elicit some added media exposure, which in turn should yield increased attendance, revenue, and donations for the institution.

CONTROVERSY, CRISIS, ACTIVISTS, AND SENSITIVE ISSUES

Controversial issues may include animal deaths, escapes, and illness. While it's easy to speak about "feel-good" stories, it can be difficult to address others that are controversial. It's important to remember that every negative news story about a zoo or aquarium is also a chance to focus on positive aspects of the institution. If an animal has died at the zoo or aquarium, what was the reason? If the animal lived to an old age due to advances in husbandry, nutrition, and veterinary medicine, it is certainly a positive story, as opposed to that of a young healthy animal having died due to staff error. In regard to most controversial issues, the zoo director or PR staff will be the ones responding to the media. In many cases, however, the animal keepers will be called upon to give statements and personal reflections on their relationship with an animal. If a keeper is asked to do such an interview, he or she should remember to answer in an open and honest manner. Zoos and aquariums are public institutions, and transparency in all situations is key to their credibility with both the media and the public. Any zoo or aquarium can experience escapes, injuries, and tragedies. Many such controversial incidents result in increased media scrutiny or a change in the public's perception of the institution. Often these are isolated incidents or freak accidents. While a zoo director might talk to a local newspaper's feature editor, every zoo staff member should consider himself or herself a spokesperson who might be asked by a member of the public a question regarding a recent incident. In times of crisis it's important for staff to communicate frequently with administration and PR so that everyone within the institution acts in coordination and properly communicates the institution's message.

Another challenge facing zoos and aquariums is opposition by animal rights groups and the media coverage it receives. While these groups care about animals, some are diametrically opposed to keeping them in zoos and aquariums. Typically, these organizations are successful in using emotion and creating controversy to plan demonstrations and events that generate media coverage. It is important to remember a few key facts when dealing with media and the public under these circumstances. It is appropriate to defend the zoo or aquarium by stating facts that pertain to the care the animals receive. If an activist group lodges a specific complaint to a keeper, the keeper should discuss this with administration to develop an appropriate response. Many animal rights groups have strong ideas about zoos and aquariums, just as zoo and aquarium professionals have strong ideas and beliefs about the value and importance of their own institutions. It is very difficult to change the ideas or philosophy of those who do not believe in the mission of zoos and aquariums. While sometimes it might be possible to find value in some of their points, they are often philosophically committed to closing zoos and aquariums regardless of the facts. When responding to allegations, it is best to remain professional and calm and to respond with facts, rather than to counter-attack and start a "war of words." When giving an interview, one should remember that collectively our institutions are valuable assets to the community and serve as important resources for conservation and education. Zoos and aquariums provide many people with their only opportunity to observe wildlife firsthand. Many species of animals that zoos and aquariums care for are endangered; and others, such as the Arabian oryx, American bison, California condor, and Kihansi spray toad, would be extinct if not for the dedicated efforts of zoos and aquariums. Also, many aquariums serve as marine rescue and rehabilitation centers for species such as pelicans, sea turtles, and manatees. These are great stories to tell to the media.

ANTHROPOMORPHISM

Sometimes publicity can lead to zoos and aquariums featuring animals in anthropomorphic ways. Some zoos and aquariums do not disclose the names of their animals to the public, due to concern that it could lead to zoo visitors calling the animal by its name and potentially interfering with its training. Some institutions do share animals' names with the public, however, and sometimes those animals become local celebrities. Their celebrity status can be used to generate public awareness, media publicity, or perhaps donations. The great apes are charismatic animals that are very popular in zoos, and which often attract both visitor and media attention. Willie B., a male gorilla who lived at Zoo Atlanta for 37 years, was named after a former mayor of Atlanta and was a household name throughout the state of Georgia. For more than 20 years he lived alone in a cage in the zoo's old ape house. In 1984, *Parade* magazine named Zoo Atlanta as one of the ten worst zoos in the nation. This article motivated the community of Atlanta to renew its zoo, with Willie B. becoming its public face. A major capital campaign funded a new gorilla habitat, and several other gorillas were added to the zoo's collection. When Willie B. eventually met female gorillas and bred, bumper stickers were distributed bearing the slogan "Willie B. a Boy or a Girl." In December 2009, the Los Angeles Zoo's male orangutan, Bruno, made a prediction about the winner of the Rose Bowl football championship game. The two teams playing that year were Alabama and Texas, and in the days leading up to the game the zoo's enrichment team made papier-mâché mascots representing both teams. The first mascot that the orangutan tore up would indicate the loser of the upcoming game. Bruno failed to correctly predict the game's outcome, but the event received a massive amount of media coverage and was commented on by sports reporters from around the world. This was important, as it exposed the zoo to an audience of sports fans.

In early 2011 an Egyptian cobra escaped from its enclosure within the Bronx Zoo's reptile house. Within days, an anonymous person created a Twitter account that quickly generated more than 100,000 followers. The tweets were humorously imagined anecdotes of the cobra's supposed voyage throughout New York City. By the time the cobra was recovered by keepers within the reptile house, it had received a large deal of national press. Regardless of the anthropomorphic nature of these stories, the benefit to the institutions was that that millions of people were more aware of gorillas, orangutans, and cobras and were perhaps motivated to visit them in the zoo and learn about the challenges they faced in the wild.

BRANDING

In recent years, the marketing and branding of zoos and aquariums has focused on animals, exhibits, and events. People know what a zoo or aquarium is; these institutions do not need to spend marketing dollars to inform the public

---

Case Study 52.1. The "Lonely" Meerkat

This case study demonstrates how misconceptions about animal husbandry and communications using the internet and other media can create challenges for a zoo dealing with small mammals.

In 2008, a zoo received a considerable amount of publicity regarding a male meerkat. At the time, *Meerkat Manor* was a popular television show on the cable network Animal Planet, and several fans posted information about meerkats in zoos on Animal Planet's website forum. The zoo held an older meerkat that had recently lost its companion. The general public was unaware that as an older animal, this meerkat was missing several teeth and had arthritis, making it nearly impossible for it to be introduced to an established group. Meerkats in zoos can be extremely aggressive towards newcomers, and in some cases they can attack existing group members without warning. While meerkats are commonly kept in zoos, officials at this zoo could not just introduce this meerkat into an established group, as it could have been attacked and critically injured. Instead, the zoo began searching for a similar older male that would make a suitable companion. Complicating the process was the fact that within the zoo's region, meerkats were considered injurious wildlife and their transport and display was heavily regulated by permits. While the zoo communicated these points to the public and asked for patience, several meerkat fans contacted the media to describe how the zoo was depriving this meerkat of a social life. These fans also contacted meerkat keepers at other zoos to inquire about the structure of their meerkat groups and the possibility of introducing another male. The fan group created an on-line petition to find the meerkat a companion, all while the zoo was already doing just that. The group represented itself in some cases as being composed of "zoo volunteers." Though the zoo tried to communicate meerkat husbandry issues and the unique challenges of this situation, it received negative press while searching for a suitable meerkat companion and waiting for the proper permits. When the curator located a companion animal, an older male meerkat, at another zoo, it was transferred to the first zoo, where a habitat specially designed to meet the needs of older meerkats had been prepared. In the end, both zoos received positive press coverage from the story and the meerkat fans were happy. The moral of the story is not that meerkats are hard to find, but that the public may sometimes misinterpret a situation and that communication between zoos, the public, and the media is very important.

---

of what they are. In most cases, zoos and aquariums do not compete against each other, as usually there is only one zoo or aquarium within a given geographic region. Zoos and aquariums do, however, compete with other family destinations such as theme parks, movie theaters, and museums for the family entertainment dollar. Zoos and aquariums brand themselves by promoting the unique features and experiences they provide. A great example of branding within an aquarium is SeaWorld's "Shamu." The name "Shamu" became synonymous with orcas (*Orcinus orca*) through SeaWorld's marketing of their star orca, Shamu, and their "Shamu and the Killer Whale Family Show."

To succeed in branding, a zoo or aquarium must understand the needs and wants of its visitors. The institution does this by integrating brand strategies at every point of public contact. Branding is the sum total of the visitors' experiences and perceptions, some of which can be influenced, and some of which cannot. For example, weather and the sometimes unpredictable behavior of animals are influences that cannot be controlled by the zoo and aquarium. In most cases a rainy or snowy day may not yield as pleasant a visit to the institution as a mild or sunny day. Likewise, an animal suddenly off exhibit due to medical needs will not create as positive an experience as a group of animals enjoying their new habitat. Factors that can be controlled include public talks, interactions, and events. It is the responsibility of every employee within a zoo or aquarium to uphold the brand. Many keepers have direct contact with visitors and play a vital role in their total experience of the zoo; in effect, they are brand ambassadors. The zoo's brand resides within the hearts and minds of the visitors. A positive interaction with a keeper can change the guest's experience or perception of the venue.

KEEPER COMMUNICATION IN THE 21ST CENTURY

With the advent of the internet, media is constantly evolving. Whereas journalists once worked only in newspapers, radio, and television, today they also reach the public through web sites and social networking to talk about zoos and aquariums. Internally, release of information from zoos and aquariums is carefully coordinated and timed according to the institutions' needs. Social networking sites such as Facebook, YouTube, Flickr, and Twitter are excellent ways to promote the latest happenings. In most cases it will be an institution's administration or PR staff that uses these websites; keepers should provide them with photographs, ideas, or news pertaining to the animals. Keepers should also know what is appropriate and inappropriate to post to social networking sites. Personal photographs, behind-the-scenes video, and pictures of keepers in anthropomorphic poses with their animals (e.g., hugging or kissing them) may be considered proprietary or perhaps harmful to the institution. Even well-meant online postings— such as discussions about negative interactions with visitors, or about new animal births, acquisition, or deaths that have not been publicly announced—can inadvertently cause damage to a zoo or aquarium. Members of the media and animal rights organizations will search the internet to obtain information about individual people, animals, and organizations. A keeper should never post anything on a personal website, blog, or social networking account that he or she would not want to see on the front page of the newspaper. A zookeeper might be excited about an animal's birth and innocently post a photo of it on Facebook before it has received a clean bill of health, or before the zoo has decided to publicize it. In some such occasions, the animal has then died or the premature Facebook post has interfered with a major promotional announcement the zoo had planned around the animal.

Keepers often work in full view of zoo visitors, who can easily photograph or video-record them with their cell phones and post the pictures or video to YouTube within seconds. The internet is instant, and in some cases major media stories of events at zoos and aquariums have arisen from personal posts by zoo patrons on social media. It is best to remember that keepers, like their animals, are on display. A keeper who is wearing the uniform, even after work hours at a social event or elsewhere in public, is representing his or her institution. Many zoos and aquariums have policies relating to personal internet use by staff members and the information they can share.

Part Nine

Conservation Science

53

Conservation Biology

Gerald Dick and Markus Gusset

INTRODUCTION

Conservation biology, a relatively new stage in the application of science to conservation problems, addresses the biology of species, communities, and ecosystems that are perturbed, either directly or indirectly, by human activities or other agents (Soulé 1985). Its goal is to provide principles and tools for preserving biological diversity. Caughley (1994) identified two dominant themes in the theoretical and methodological development of conservation biology as a scientific discipline: the "small population" and "declining population" paradigms. The former focuses on the dynamics and persistence of small populations, while the latter examines the factors that reduce populations to small sizes in the first place. Conservation biology thus provides an understanding of both threatening and mitigating factors in order to propose solutions for real-world conservation challenges.

The aims of this chapter are to present an overview of

- the definition of biodiversity conservation
- global species crisis and conservation challenges
- world zoo and aquarium conservation strategies
- conservation through zoos and aquariums.

WHAT IS BIODIVERSITY CONSERVATION?

The following section is based on the introduction to "Building a Future for Wildlife: Zoos and Aquariums Committed to Biodiversity Conservation" (Dick and Gusset 2010) by Reid (2010). "Biodiversity" appears at first sight to be a fairly straightforward concept. In one context it refers to the immense global variety of wildlife, some of which is cared for in zoos and aquariums. But biodiversity can also be understood in terms of microorganisms and the complex genetic variation in animal and plant chromosomes, genes, DNA, and other biochemistry. Such intricate microscopic or molecular variation ultimately determines the uniqueness and success of individuals, species, and higher-level taxa. This contrasts with an equally complex macroscopic biodiversity, including the large range of habitats, landscapes, and ecosystems.

The focus on wildlife diversity has changed since the 18th century. Carolus Linnaeus (1707–78) described some 12,100 species in his lifetime, of which 4,400 were animals and 7,700 were plants. He confidently predicted that the complete eventual plant list would not exceed 10,000. However, the present rate of discovery of species is extremely high. At least 2,057 new vascular plants were named in 2006, along with 8,995 insects and 486 fishes. Since 2004 more than 13 new amphibian species have been recognized each month, with more than 6,000 named so far from a total list of perhaps 9,000 (Hilton-Taylor et al. 2009). Indeed, there has been an extraordinary explosion of knowledge across all major taxa, with new species being discovered daily. There is no absolutely agreed-upon projected total, nor any easy scientific means of establishing such a figure. The current global biodiversity assessment (see CBD 2010) recognizes 1.8 million described species as possibly being valid, with a final projected total of perhaps 14 million. If true, this means that less than 13% of all species have scientific names, and a staggering 12.2 million remain to be formally described.

Considering also our limited knowledge of the world's microscopic and macroscopic biodiversity, we have a heady brew of ignorance! For example, comparatively few vertebrate species have had their conservation status assessed in any scientific detail. While there are many notable species conservation successes, the general situation, as documented by the Convention on Biological Diversity (CBD), steadily gets worse with each advancing year (CBD 2010). Clearly, zoos and aquariums must do much more to conserve vertebrates, and must pay far more attention to equally imperiled invertebrates and plants.

THE ROLE OF THE WORLD'S ZOOS IN CONSERVATION

"Conservation" can be defined as action that substantially enhances the survival of species and habitats, whether conducted in or out of the natural habitat. From this, "United for Conservation" is a headline statement used to communicate the central purpose of the World Association of Zoos and Aquariums (WAZA). Certainly, a biodiversity conservation and environmental sustainability ethos pervades the vision, mission, and values of the separate membership components of WAZA, ranging from regional and national associations to individual institutions. Clearly, the 1,300 or more mainstream zoos and aquariums on our planet serve a vital and serious purpose that is not always well publicized.

This substantial conservation role exists alongside providing the visiting public a fun-filled and educational day out in a leisure context. Collectively, this creates millions of dollars of income that is applied to worthwhile conservation projects at home and abroad that are either operated by zoos and aquariums or conducted in partnership with them. Such activities are often targeted directly at practical issues in biodiversity conservation both in nature (in situ) and outside of the natural habitat (ex situ), at home and abroad. Increasingly, these two different designations are dissolving into a continuum in which threatened species are managed extensively (mainly in the wild) and intensively (mainly in captivity), with every sort of management combination in between.

To assess global zoo and aquarium attendance, the authors approached 12 national and regional zoo and aquarium associations, covering all regions of the world, to provide a figure for how many visitors their member institutions received in 2008 (Gusset and Dick 2011a). About 600 million people reportedly visited zoos and aquariums worldwide in 2008. A comparison of zoo and aquarium attendance in 1990, the year of the last global survey, and in 2008 (table 53.1) shows that those zoo and aquarium associations reporting higher numbers in the 2008 survey represent regions with established documenting structures (North America, Australasia, and Europe), thus suggesting a growing number of visits in these regions. Conversely, those associations reporting lower numbers in the current survey represent regions where obtaining comprehensive numbers is more challenging (Latin America, Africa, and Asia).

While the 2008 survey was specifically aimed at collecting documented figures from the associations' members—something that generally proved feasible for the former three regions—the 1990 survey (IUDZG/CBSG 1993) relied on the associations' estimates of zoo and aquarium attendance. This may be more appropriate for Latin America, Africa, and Asia, given the underestimates in documented figures confirmed by the associations in those regions in the current survey. Considering this variation in reporting between the two surveys, and assuming a largely unchanged number of existing zoos and aquariums, it seems legitimate to adjust the results accordingly (table 53.1), in which case zoos and aquariums worldwide receive more than 700 million visits annually. This figure, which may include multiple individual visits, is most certainly an underestimate (WAZA 2009) and is unparalleled by any other group of conservation-oriented institutions.

The authors also assessed the conservation expenditures of the world zoo and aquarium community, approaching the same 12 national and regional zoo and aquarium associations to provide a figure for how much money their member

TABLE 53.1. Annual number (in millions) of visits to zoos and aquariums worldwide in 1990 (IUDZG/CBSG 1993), 2008 (this survey), and adjusted. From Gusset and Dick 2011.

	1990	2008	Adjusted[1]
North America	106	186	186
Latin America	61	11	61
Africa	15	8	15
Australasia	6	17	17
Europe	125	142	142
Asia	308	221	308
Global total	621	585	729

[1]1990 figures for Latin America, Africa, and Asia. 2008 figures for North America, Australasia, and Europe.

institutions spent on wildlife conservation in 2008 (Gusset and Dick 2011a). (Wildlife conservation in this context encompasses in situ conservation of wild species and habitats, and also includes related ex situ work.) The world zoo and aquarium community reportedly spent about US$350 million on wildlife conservation in 2008. This amount includes the expenses of zoo-based conservation organizations, but given that only 7 of the 12 associations submitted figures on conservation expenditures, it is most certainly an underestimate. Across regions, zoos and aquariums in North America and Europe spent the most by far on wildlife conservation (97% of expenses reported). In relation to major international conservation organizations (figure 53.1), the world zoo and aquarium community is among the main providers of conservation funding.

This growing vocational conservation-focused ethos reflects the many cooperative initiatives of zoos and aquariums in supporting threatened species and habitats (e.g., Zimmermann et al. 2007; Dick and Gusset 2010; Zimmermann 2010); it may, for example, involve breeding programs, studbook management, assisted reproduction, species reintroductions or translocations, educational outreach, and benign (noninvasive and nonintrusive) scientific research. Often this now involves an integrated approach that takes in rapidly advancing disciplines such as conservation medicine, ecological restoration, and the sociology, ethnology, and psychology of "human-wildlife conflict." Sometimes this entails a need to address worldwide issues in human development and poverty alleviation, which are a major root cause of the global decline in biodiversity (Niekisch 2010). Strategy planning and stakeholder support for these exercises is often supplied by the Conservation Breeding Specialist Group (CBSG) and the Reintroduction Specialist Group (RSG) of the Species Survival Commission (SSC) of the International Union for Conservation of Nature (IUCN), while bulk electronic data handling is expertly covered through the Zoological Information Management System (ZIMS) of the International Species Information System (ISIS) organization (Althaus et al. 2010; Stuart et al. 2010).

International studbooks, kept under the auspices of WAZA, represent the highest level of global monitoring and management. They are meant to provide a valuable service to the zoological community, offering the most complete

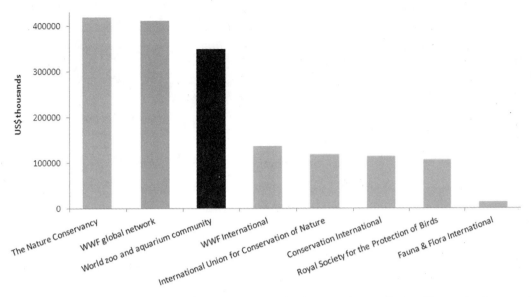

Figure 53.1. Amount of money spent on wildlife conservation by major international conservation organizations (figures taken from annual reports) and the world zoo and aquarium community (this survey) in 2008, in thousands of US dollars (Gusset and Dick 2011).

and accurate global data on the ex situ population's pedigree and demography, including husbandry and veterinary guidance where possible, thus enabling better management of the population through analysis of the data (Althaus et al. 2010). In most cases, the staff members of WAZA member institutions serve as studbook keepers. Within WAZA the international studbook program is overseen by the Committee for Population Management (CPM) and is coordinated by a nominated employee at the WAZA executive office, in collaboration with ISIS. The Zoological Society of London (ZSL) regularly publishes the updated list of current international studbooks in the *International Zoo Yearbook*. As of November 2011, there were 123 active international studbooks on 162 species or subspecies (some international studbooks cover more than one taxon).

Conservation breeding programs typically are established and administered at the level and under the auspices of the regional associations. At its 2003 annual conference, WAZA adopted a procedure for establishing interregional programs, which may involve a number of species for which international studbooks have been established. These programs, now called Global Species Management Plans (GSMPs), are those officially recognized and endorsed by WAZA. As of November 2011, there were three such global programs: one each for the Sumatran tiger (*Panthera tigris sumatrae*) and the Javan gibbon (*Hylobates moloch*), and a combined program for African and Asian elephants (*Loxodonta africana* and *Elephas maximus*). Given the emerging challenges of keeping viable captive populations regionally (Gusset and Dick 2011b), additional GSMPs will be established in the near future.

GLOBAL SPECIES CRISIS AND CONSERVATION CHALLENGES

Human beings are becoming increasingly cut off from nature. More than 50% of the world's population currently lives in cities, and 70% will live in cities by 2030 (Djoghlaf 2010). As a result, the majority of the population does not appreciate that biodiversity is their ultimate source of goods such as food, timber, and medicines, and that it provides society with irreplaceable ecosystem services like crop pollination, air and water purification, erosion control, and the renewal of soil fertility.

This estrangement from nature makes it difficult for people to see the dangers inherent in the ongoing loss of biodiversity. Human activities are currently driving species extinct at up to 1,000 times the prehuman background rate (see CBD 2010). The 2010 Living Planet Index showed that vertebrate population sizes have on average declined by almost 30% over the last 40 years (figure 53.2). In the long term, this loss will radically undermine the potential of sustainable development, exacerbating poverty and fostering conflicts over dwindling resources.

The following section is based on a review of the global species crisis and conservation challenges by Stuart et al. (2010). The rapid disappearance of species, and thus biodiversity, is often referred to as one of the world's greatest environmental concerns. The IUCN Red List of Threatened Species is the world's most comprehensive data resource on the status of species, containing information and status assessments on more than 40,000 species of animals and plants (Vié et al. 2009). As well as measuring the extinction risk faced by each species, the IUCN Red List includes detailed species-specific information on distribution, threats, conservation measures, and other relevant factors. This list is increasingly used by scientists, governments, nongovernmental organizations, businesses, and civil society for a wide variety of purposes, including zoos and aquariums that want to educate their visitors about the extinction risks faced by the exhibited species.

The IUCN Red List categories and criteria are the world's most widely used system for gauging the extinction risk faced by species. Each species assessed is assigned to one of eight different categories (figure 53.3), based on a series of quantitative criteria. Species classified as Vulnerable, Endangered, and

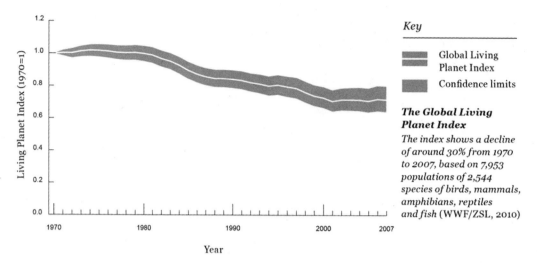

Figure 53.2. The Living Planet Index of global biodiversity, measured as the relative aggregate size of 7,953 populations of 2,544 vertebrate species across all regions of the world over time, with 1970 as the baseline (WWF International 2010).

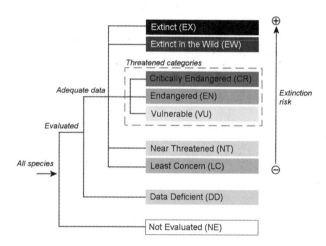

Figure 53.3. Structure of the IUCN Red List categories (Vié et al. 2009).

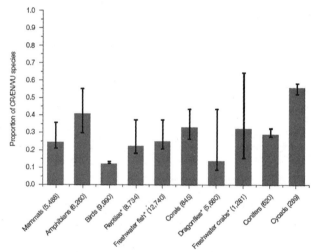

Figure 53.4. Proportion of species threatened with extinction in different taxonomic groups. The total number of described species in each group is indicated in parentheses. Error bars show minimum and maximum estimates (Hilton-Taylor et al. 2009).

Critically Endangered are all regarded as "threatened." The Red List shows that the status of the world's species is deteriorating in all regions and in all taxonomic groups. However, three major ongoing extinction crises stand out in particular: amphibians, corals, and Asian large animals. Nearly one-third of the planet's amphibians (31.1%), one-quarter of its reef-building corals (27%), and nearly one-quarter of its mammals (22.2%) are threatened or extinct (figure 53.4).

AMPHIBIAN EXTINCTION CRISIS

Amphibians, representing more than 6,200 species worldwide, are one of the most threatened major taxonomic groups on the IUCN Red List (Hilton-Taylor et al. 2009). At least 42% of amphibian species have populations that are declining, indicating that the percentage of threatened species will likely only rise in the future. In contrast, fewer than 1% of amphibian species have populations that are increasing. Overall, there is strong evidence that the pace of amphibian extinctions is increasing: of the 38 known extinctions since

the year 1500, 11 have occurred since 1980. Also, 120 species of amphibians have been listed as possibly extinct, most having not been seen since 1980. The most severe impact has been in Mesoamerica, the northern Andes, and the Greater Antilles.

Habitat loss is the greatest threat to amphibians, affecting nearly 61% of all known species and a very large percentage (87%) of those species that are threatened. However, the fungal disease chytridiomycosis has been the major driver of known and suspected amphibian extinctions over the past three decades. The disease-causing pathogen (*Batrachochytrium dendrobatidis*) was probably introduced to the affected regions (e.g., the Americas, Australasia, and Europe) by international trade in African clawed frogs (*Xenopus laevis*), which were previously used for pregnancy assays in humans (Weldon et al. 2004).

In response to the amphibian crisis, the IUCN/SSC Amphibian Specialist Group (ASG) and other partners have developed the Amphibian Conservation Action Plan (Gascon et al. 2007), which provides a comprehensive framework for combating amphibian declines and extinctions. A major priority is to protect the habitat of the many threatened amphibian species that do not occur in any protected areas. Another priority is working with zoos, aquariums, and other centers to shelter and breed amphibians threatened by chytridiomycosis, which cannot yet be treated in the wild and which can cause up to 100% mortality in certain species. The Amphibian Ark (AArk) is a global program created by IUCN/SSC and WAZA to manage threatened amphibians in captivity until it is safe to reintroduce them into the wild.

CORAL EXTINCTION CRISIS

Warm-water, reef-building corals provide essential habitat for many species of fishes and invertebrates, making them the most biologically diverse ecosystems in the ocean. All 845 known species of reef-building corals in the world have been assessed on the IUCN Red List. More than one-quarter (27%) have been listed as "threatened" or at a high risk of extinction, with an additional 20% listed as "near threatened" (Hilton-Taylor et al. 2009), meaning that they will likely join a threatened category in the future. Moreover, reef-building

corals are declining at a faster rate than any other group of species currently on the IUCN Red List. Just 15 years ago the overall level of threat to reef-building corals was very low.

The catastrophic decline in the abundance of reef-building corals stems primarily from increased bleaching (i.e., the stress-induced whitening of corals) and disease events linked to higher sea temperatures that result from global warming. Coastal development and other human activities such as coral extraction and pollution have also contributed to dramatic declines since the mid-1990s. Ocean acidification resulting from increased levels of atmospheric carbon dioxide (CO_2) is further impacting reef-building coral species by negatively affecting calcification.

The highest number of threatened species according to the IUCN Red List is in the Indo-Malay-Philippine Archipelago, or "Coral Triangle," which is the global epicenter of marine biodiversity, with the highest number of coral species. In all regions, the loss of coral ecosystems will have huge cascading effects for reef-dependent species as well as for the large number of people and nations that depend on coral reef resources for economic and food security. Ex situ conservation may be necessary for corals and other coral-dependent species (figure 53.5; see Penning 2010), since measures to reduce the level of CO_2 in the atmosphere are still a long way from having an effect.

Figure 53.5. Ex situ propagation of corals at Oceanário de Lisboa in Portugal. Photograph by Gerald Dick.

LARGE ANIMALS OF ASIA' EXTINCTION CRISIS

The third extinction crisis revealed by the IUCN Red List is for animals in Southeast Asia (Hilton-Taylor et al. 2009). Globally almost one-quarter of mammals (22.2%) are threatened with extinction, but Asian countries are showing the most impact. Of the top 20 countries with threatened mammal populations, ten are in Asia, where there have been massive population declines over the past two decades. The Indo-Malayan region shows rapid declines in both birds and mammals, driven by deforestation, habitat loss, and high rates of hunting for mammals. Large-bodied taxa throughout Asia, including mammals, birds, fishes, and reptiles (such as turtles), are being impacted by massive and largely uncontrolled overexploitation. Two mammalian extinctions have likely resulted in the last few years: the baiji or Yangtze River dolphin (*Lipotes vexillifer*) in China and the kouprey (*Bos sauveli*), a forest-dwelling large ungulate once found mainly in Cambodia but also in Laos, Vietnam, and Thailand.

There is an urgent need throughout Asia to address the overexploitation of wildlife through antipoaching measures and the control of trade in wildlife products. For example, the often unsustainable and illegal hunting of wild animals for meat—so-called bushmeat—does not only reduce populations of several species, including the great apes (Bennett et al. 2007). The consumption of great apes as bushmeat is also considered a vector of human immunodeficiency virus (HIV) and Ebola virus in Africa. Initiatives are needed that focus not only on antipoaching but also on providing alternative livelihoods for local people—allowing the root causes of poaching to be addressed through such measures as providing alternative protein sources (e.g., breeding of cane rats, *Thryonomys* spp.) and implementing capacity-building and training programs. The conversion of lowland forests for palm oil and other biofuels also needs to be urgently addressed, especially in Indonesia and Malaysia.

There is growing evidence that climate change will become one of the major drivers of species extinction in the coming years. IUCN recently completed the first phase of a project to identify species most vulnerable to climate change (Foden et al. 2009). The results showed that 52% of all amphibian species and 71% of warm-water, reef-building corals are potentially susceptible. This information will highly influence our approach to species conservation in the future. However, other more "traditional" threats, such as habitat loss, invasive species, and overharvesting, remain critically important and should not be overlooked because of the current attention being given to climate change by many of the world's leading environmental agencies and donors.

Stemming the tide of global extinctions requires urgent action by all parts of society. Species are of enormous importance to human livelihoods, and the benefits they provide will continue in perpetuity if we learn to manage biodiversity sustainably. In the next decade the tide must turn for species conservation, and society must respond to the wake-up call that the IUCN Red List represents. This is the challenge we must all address in the next decade if future generations are to have the chance not only to enjoy the full diversity of the world's species but, ultimately, to survive.

WORLD ZOO AND AQUARIUM CONSERVATION STRATEGIES

The following section is based on a review of zoo and aquarium conservation strategies by Gipps (2010). The first World Zoo Conservation Strategy was published in 1993. It was jointly produced by the International Union of Directors of Zoological Gardens (IUDZG) and the IUCN/SSC Conservation Breeding Specialist Group (CBSG).

In the foreword, HRH the Duke of Edinburgh, then president of WWF International, wrote the following: "Much can be done by establishing and managing protected areas, but there are many species whose natural habitats have already been degraded or destroyed. For these the only hope of survival is the direct stewardship and human care in zoological and botanical gardens in captive breeding centres." Note the emphasis on "human care . . . in captive breeding centers." The document itself contained eleven chapters:

- Introduction: Zoos in a changing world
- The world conservation strategy and zoos
- The global zoo network
- Education
- Zoo animal collections and their conservation
- *Ex situ* conservation of animal populations
- Capacity: Space limitations and choice of species
- Artificial reproduction and cryopreservation: Biotechnology in support of conservation
- Back to nature: Animals for reintroduction and restocking
- Knowledge and research
- The way forward: Towards a new integration

A notable feature of this list is that it concentrates almost exclusively on ways in which zoos can maintain sustainable populations within their institutions, with the notion that reintroductions or restocking are the principle [sic] objectives of the exercise.

In 2002, the council of WAZA (formerly known as IUDZG) decided that the strategy needed substantial revision. At its annual conference that year, under the auspices of its newly formed Conservation Committee, a workshop was held to discuss what a new World Zoo and Aquarium Conservation Strategy might contain (Gipps 2010). The resulting strategy document, "Building a Future for Wildlife: The World Zoo and Aquarium Conservation Strategy" (WZACS), was published by WAZA in 2005.

The chapters in that document, and the vision statements at the beginning of each chapter, are:

1. Integrating conservation: The major goal of zoos and aquariums will be to integrate all aspects of their work with conservation activities. The fundamental elements of each organization's culture will be the values of sustainability and conservation,

and social and environmental responsibility. These values will permeate all areas of their work and will be understood and promoted by all those working within the WAZA network.

2. Conservation of wild populations: Zoos and aquariums will make further contributions to conservation in the wild by providing knowledge, skills, and resources through initiatives in zoo breeding, translocations and reintroduction, wildlife health, research, training, education, and by funding field activities. Zoos and aquariums will be an important force for worldwide conservation by their employment or support of field workers active in the conservation of wild animals and their habitats.

3. Science and research: Zoos and aquariums are fully and actively integrated into the research community and into public consciousness and understanding of science, as serious, respected scientific institutions that make significant contributions and sound scientific decisions for wildlife worldwide.

4. Population management: All zoos and aquariums will be primary centers of expertise in small population management and will be involved in global or regional cooperative breeding programs. All such programs will be based on sound knowledge using the latest available data on population management, reproductive biology, genetics, behavior, physiology, nutrition, veterinary care, and husbandry.

5. Education and training: Zoos and aquariums with their unique resource of live animals, their expertise, and their links to field conservation will be recognized as leaders and mentors in formal and informal education for conservation. The educational role of zoos and aquariums will be socially, environmentally, and culturally relevant, and by influencing people's behavior and values, education will be seen as an important conservation activity. Zoos and aquariums will expand the training of their own staff and of others engaged in *in situ* and *ex situ* work.

6. Communication: Marketing and public relations: Zoos and aquariums and their national and regional associations will become highly effective in communicating conservation issues and their role in conservation. They will become better recognized as one of the major and most trusted voices speaking on behalf of wildlife and wild places.

7. Partnerships and politics: Through increased cooperation and judicious encouragement, zoos and aquariums will continue to raise standards of animal management, educate the public to act on behalf of conservation issues, and assist in field projects. Partnerships will strengthen global cooperation and help all zoos, aquariums, and other conservation organizations to improve and to achieve their conservation goals. Zoos and aquariums will be encouraged to help one another, particularly those that have fewer resources and/or expertise.

8. Sustainability: All zoos and aquariums will work towards sustainability and reduce their "environmental footprint." They will use natural resources in a way that does not lead to their decline, thus meeting the needs of the present without compromising future generations. All zoos and aquariums will serve as leaders by example, using green practices in all aspects of their operations and by demonstrating methods by which visitors can adopt sustainable lifestyles.

9. Ethics and animal welfare: All zoos and aquariums will follow ethical principles and maintain the highest standards of animal welfare in order to establish and sustain viable populations of healthy animals for conservation purposes and to convey credible conservation messages to the public.

What is particularly striking about this list of chapters is that, although there is overlap with the contents of the 1993 strategy (particularly with respect to the conservation education role of zoos), there is also a strong degree of divergence. In particular, the conservation of wild populations has become a core element of the new strategy, because it is now a core element of the conservation work (both action and research) that zoos and aquariums do.

The introduction to the 2005 strategy is littered with references to conservation in the wild. For example, the definition of conservation is given as "securing of long-term populations of species in natural ecosystems and habitats wherever possible." This definition is followed by an explanatory note that says that the words "natural ecosystems and habitats" signify that "no amount of worthy endeavor is of ultimate value if it doesn't translate into animals and plants surviving in the wild."

Apart from the change in emphasis of the 2005 strategy compared with that of 1993, the world has changed in another hugely important way; we now live in the age of the Internet. A few thousand copies of the World Zoo Conservation Strategy were printed in 1993 and basically, that was it. Most major zoos in the world bought several copies, and some copies went to government and nongovernment conservation agencies around the world. But, if you had not received a printed copy, you probably did not see it.

The contrast between the 1993 and 2005 documents is astonishing. The WZACS has now been translated into Chinese, Czech, French, German, Hungarian, Polish, Portuguese, Russian, Spanish, and Swedish and has been downloaded from the WAZA website more than 250,000 times, in addition to 10,000 printed copies of the English version plus many more thousand copies in the other languages. Clearly, this is a document that has been seen and used by a very large number of people. Indeed, the conservation strategies of many regional zoo and aquarium associations, and many of their institutional members, are based on the WZACS. A subtheme of the WZACS is the "Global Aquarium Strategy for Conservation and Sustainability," published by WAZA in 2009 (see Penning 2010).

It is believed that the WZACS has been, at least in part, instrumental in the way that zoos and aquariums have changed how they view their role in the wider world of conservation. Zoos and aquariums all over the world now believe their role in helping to conserve wild populations of animals (and plants) to be as important as conservation breeding programs and the possibilities of reintroductions. Of course, the two processes remain inextricably linked; zoos and aquariums are, after all, an astonishingly powerful and large resource for maintaining assurance populations of species threatened in the wild.

But, in the modern world, zoos and aquariums acknowledge that breeding programs and reintroductions are not, and never will be, enough. We know of no responsible zoo or aquarium director who would regard the existence of the last remnants of a species, held solely and permanently in human care, as any sort of success. Rather, the conservation world at large, which includes all those zoos and aquariums committed to conservation, would count that as a sad, and indeed abject, failure.

CONSERVATION THROUGH ZOOS AND AQUARIUMS

In the years 2000 and 2001, WAZA organized three workshops on how it should become more involved through in situ conservation. One recommendation was that with the goal of becoming an internationally recognized conservation organization, WAZA should brand suitable conservation projects. The idea behind the WAZA branding is to create a win-win situation: The brand promotes the project and also lets WAZA use it to convey what zoos and aquariums do for conservation globally. Projects or programs are branded on application, with three sets of endorsement criteria focusing on biological, operational, and institutional and partnership issues. Applications have to meet the criteria set forth by the WAZA executive office. Applications may be submitted by either the project receiving support from a member of WAZA or by a WAZA member providing support to the project. Since inception of the scheme in 2003, the number of WAZA-branded projects has steadily increased to 214 (in November 2011).

The authors of this chapter compiled and assessed conservation projects supported by the world zoo and aquarium community, focusing on 113 WAZA-branded projects (Gusset and Dick 2010a, b). The majority of projects had their main focus on species protection. They mainly focused on mammals, and mostly on charismatic primates and carnivores. Most of the projects involved working with taxa classified as globally threatened with extinction according to the 2008 IUCN Red List of Threatened Species. Amphibians and fishes were underrepresented in the number of projects relative to the percentage of threatened species described in these taxa. The projects were primarily active in the Eurasian, sub-Saharan African, or Southeast Asian regions (figure 53.6). They focused strongly on terrestrial habitats, among them chiefly tropical and subtropical forests. Terrestrial habitats at high conservation risk, especially Mediterranean woodlands and temperate grasslands, were underrepresented.

Project leaders typically applied for WAZA branding to increase their publicity or credibility or to attract support from zoos and aquariums. Later, these project leaders were generally undecided later about whether the WAZA branding had resulted in the desired outcome. Projects mainly each received support from one or two to five zoos and aquariums. Zoos and aquariums primarily became involved in projects either on their own initiative or because the projects had requested their support. The main form of project support they provided was monetary funding. Their contributions (including nonmonetary support) often covered more than half of a project's annual financial expenditures. Most of the projects would not have been viable without the support from zoos and aquariums. Their financial expenditures were typically in the range of US $10,000 to $100,000 per year, with their duration often being longer than 10 years.

In assessing these projects' conservation impact, we also asked how much the projects improved the conservation status of their target species or habitats (Gusset and Dick 2010a, b). The projects, on average, reached a self-assessed impact score of three out of four possible points. We then did an evaluation of the concurrent influence of all of the above attributes on a project's impact score. This analysis revealed that the higher a project's financial expenditures and the higher the contribution made by zoos and aquariums to these financial expenditures, the higher was the project's impact score.

In the following section, conservation success stories are presented in which zoos and aquariums were heavily involved, the focus is on reintroduction projects where a direct link between ex situ and in situ conservation was often most visible and communicable to the public. These are classic examples of how conservation breeding programs in zoos and aquariums, when effectively coordinated with protective measures in the wild, can help re-diversify life on the planet (Stanley Price and Fa 2007; Earnhardt 2010). However, as is pointed out repeatedly throughout this chapter, zoos and aquariums need to move beyond the paradigm of captive breeding and reintroduction.

The handful of iconic species in these projects includes the American bison (*Bison bison*), probably the first captive-bred animal to be reintroduced for conservation purposes back in 1907; the black-footed ferret (*Mustela nigripes*), for which removal from the wild for captive breeding started in 1985; the California condor (*Gymnogyps californianus*), which after being taken into captivity started to breed successfully in 1988; the Arabian oryx (*Oryx leucoryx*), released in Oman in 1982 using stock mainly from the United States; and the golden lion tamarin (*Leontopithecus rosalia*), which was the subject of great interzoo collaboration and successful release back into Brazil, starting from 1984.

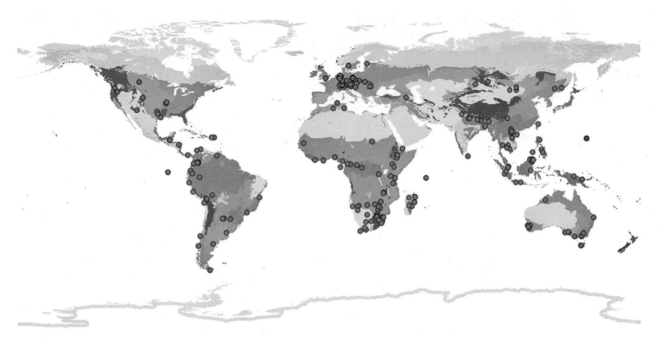

Figure 53.6. Geographic distribution of in situ conservation projects supported by the world zoo and aquarium community. The map shows major terrestrial habitat types, or biomes, of the world (Olson et al. 2001).

American bison once numbered in the tens of millions and ranged from Alaska to northern Mexico. They were then nearly eliminated by intensive hunting in the late 19th century. In 1906 only about 1,000 bison, captive and wild, remained in North America. One year later the American Bison Society, founded at New York's Bronx Zoo, started reintroducing captive-bred bison into reserves in the western United States. In addition to approximately 500,000 bison in captive commercial populations, there are now about 30,000 bison in conservation herds.

Mainly as a consequence of the degradation of the North American prairie ecosystem, only 18 black-footed ferrets remained in the wild by 1985. In the same year, a number of zoos and other organizations started collaborating on a captive breeding program with ferrets collected from the wild. After a period of prerelease conditioning, in which the ferrets lived in pens with prairie dog (*Cynomys* spp.) burrow systems, reintroduction started in 1991. About 1,000 black-footed ferrets now live in the wild.

As a consequence of habitat destruction, poaching, and lead poisoning, the California condor declined to the extent that in 1981 only 22 birds remained in the wild. After the last remaining wild specimens were brought into captivity at the Los Angeles Zoo and San Diego Zoo's Wild Animal Park in 1987, a captive breeding program was started through collaboration between zoos and various other organizations. From 1992 onwards, captive-bred condors have been reintroduced. Innovative methods include removing eggs from condor nests to encourage females to lay replacement eggs, using adult lookalike condor puppets to feed newly hatched chicks to prevent them from imprinting on humans, and aversive conditioning to avoid power lines and humans. About 170 California condors now live in the wild.

The Arabian oryx was exterminated in the wild as a con-sequence of intensive hunting, with the last animal shot in Oman in 1972. Luckily, a captive herd had been established in the early 1960s by the Phoenix Zoo. Other zoos in the United States and Europe joined the captive breeding program, and Arabian oryx have been released from 1978 onwards at several sites throughout the Arabian Peninsula. In addition to approximately 6,000 to 7,000 captive animals, about 1,100 Arabian oryx now live in the wild.

Mainly as a consequence of habitat destruction, only a few hundred golden lion tamarins remained in the early 1970s. Starting in 1972, a captive population was built up at the Smithsonian's National Zoological Park in Washington, DC, and in other zoos, in association with a primate center in Rio de Janeiro, Brazil. Before reintroduction started in 1984, the tamarins went through a prerelease training program during which the animals learned how to search for hidden food and how to use natural substrates for locomotion. The tamarins were subsequently acclimatized in enclosures built around natural vegetation at the release site, and were subjected to post-release training. About 1,000 golden lion tamarins now live in the wild.

SUMMARY

The goal of conservation biology is to provide principles and tools for preserving biological diversity. This biodiversity—the variety of genes, species, and ecosystems that constitute life on earth—is increasingly threatened by human activities. Zoos and aquariums are involved in numerous practical issues in biodiversity conservation both in nature (in situ) and outside the natural habitat (ex situ), at home and abroad. However, zoos and aquariums cannot leverage a conservation message when keeping animals in substandard conditions, and they cannot cultivate a conservation ethic in their visitors without doing business in a sustainable way. Integrated management between wild and captive populations is increasing, and we see it as a major niche on which zoos and aquariums should focus their conservation efforts, including the recovery and management of small populations and translocation biology. Zoos and aquariums, unlike other conservation organizations, enjoy the distinct advantage of having physical sites that people can visit. Can we envision a day when the zoo or aquarium becomes a conservation organization that exhibits animals as part of its mission to connect people to nature, and uses this connection to preserve species in the wild? A number of zoos and aquariums, keeping alive the flame of species conservation, are almost there.

REFERENCES

Althaus, Thomas, Laurie Bingaman-Lackey, Fiona A. Fisken, and Dave Morgan. 2010. The role of international studbooks in conservation breeding programmes. In *Building a Future for Wildlife: Zoos and Aquariums Committed to Biodiversity Conservation*, ed. Gerald Dick and Markus Gusset, 49–52. Gland: WAZA Executive Office.

Bennett, Elizabeth L., Eric Blencowe, Katrina Brandon, David Brown, Robert W. Burn, Guy Cowlishaw, Glyn Davies, Holly Dublin, John E. Fa, E. J. Milner-Gulland, John G. Robinson, J. Marcus Rowcliffe, Fiona M. Underwood, and David S. Wilkie. 2007. Hunting for consensus: Reconciling bushmeat harvest, conservation, and development policy in West and Central Africa. *Conservation Biology* 21:884–87.

Caughley, Graeme. 1994. Directions in conservation biology. *Journal of Animal Ecology* 63:215–44.

CBD. 2010. *Global biodiversity Outlook 3*. Montreal: CBD Secretariat.

Dick, Gerald, and Markus Gusset, eds. 2010. *Building a Future for Wildlife: Zoos and Aquariums Committed to Biodiversity Conservation*. Gland, Switzerland: WAZA Executive Office.

Djoghlaf, Ahmed. 2010. Convention on Biological Diversity: Conservation of Biodiversity through 2010 and Beyond. In *Building a Future for Wildlife: Zoos and Aquariums Committed to Biodiversity Conservation*, ed. Gerald Dick and Markus Gusset, 17–20. Gland, Switzerland: WAZA Executive Office.

Earnhardt, Joanne M. 2010. The role of captive populations in reintroduction programs. In *Wild Mammals in Captivity: Principles and Techniques for Zoo Management*, 2nd ed., ed. Devra G. Kleiman, Katerina V. Thompson, and Charlotte Kirk Baer, 268–80. Chicago: University of Chicago Press.

Foden, Wendy B., Georgina M. Mace, Jean-Christophe Vié, Ariadne Angulo, Stuart H. M. Butchart, Lyndon DeVantier, Holly T. Dublin, Alexander Gutsche, Simon N. Stuart, and Emre Turak. 2009. Species susceptibility to climate change impacts. In *Wildlife in a Changing World: An Analysis of the 2008 IUCN Red List of Threatened Species*, ed. Jean-Christophe Vié, Craig Hilton-Taylor, and Simon N. Stuart, 77–87. Gland, Switzerland: IUCN.

Gascon, Claude, James P. Collins, Robin D. Moore, Don R. Church, Jeanne E. McKay, and Joseph R. Mendelson III, eds. 2007. *Amphibian Conservation Action Plan*. Gland, Switzerland: IUCN.

Gipps, Jo. 2010. The world zoo and aquarium conservation strategy: Where do we go from here? In *Building a Future for Wildlife: Zoos and Aquariums Committed to Biodiversity Conservation*, ed. Gerald Dick and Markus Gusset, 35–38. Gland, Switzerland: WAZA Executive Office.

Gusset, Markus, and Gerald Dick. 2010a. 'Building a Future for Wildlife'? Evaluating the contribution of the world zoo and aquarium community to *in situ* conservation. *International Zoo Yearbook* 44:183–91.

———. 2010b. Biodiversity conservation projects supported by the world zoo and aquarium community. In *Building a Future for Wildlife: Zoos and Aquariums Committed to Biodiversity Conservation*, ed. Gerald Dick and Markus Gusset, 57–60. Gland, Switzerland: WAZA Executive Office.

———. 2011a. The global reach of zoos and aquariums in visitor numbers and conservation expenditures. *Zoo Biology* 30:566–69.

Gusset, Markus, and Gerald Dick, eds. 2011b. *WAZA Magazine 12: Towards Sustainable Population Management*. Gland, Switzerland: WAZA Executive Office.

Hilton-Taylor, Craig, Caroline M. Pollock, Janice S. Chanson, Stuart H. M. Butchart, Thomasina E. E. Oldfield, and Vineet Katariya. 2009. State of the world's species. In *Wildlife in a Changing World: An Analysis of the 2008 IUCN Red List of Threatened Species*, ed. Jean-Christophe Vié, Craig Hilton-Taylor, and Simon N. Stuart, 15–41. Gland, Switzerland: IUCN.

IUDZG/CBSG. 1993. *The World Zoo Conservation Strategy: The Role of the Zoos and Aquaria of the World in Global Conservation*. Chicago: Chicago Zoological Society.

Niekisch, Manfred. 2010. International conservation policy and the contribution of the zoo and aquarium community. In *Building a Future for Wildlife: Zoos and Aquariums Committed to Biodiversity Conservation*, ed. Gerald Dick and Markus Gusset, 45–48. Gland, Switzerland: WAZA Executive Office.

Olson, David M., Eric Dinerstein, Eric D. Wikramanayake, Neil D. Burgess, George V. N. Powell, Emma C. Underwood, Jennifer A. D'Amico, Illanga Itoua, Holly E. Strand, John C. Morrison, Colby J. Loucks, Thomas F. Allnutt, Taylor H. Ricketts, Yumiko Kura, John F. Lamoreux, Wesley W. Wettengel, Prashant Hedao, and Kenneth R. Kassem. 2001. Terrestrial ecoregions of the world: A new map of life on earth. *BioScience* 51:933–38.

Penning, Mark. 2010. Aquariums and the conservation of water-dependent species. In *Building a Future for Wildlife: Zoos and Aquariums Committed to Biodiversity Conservation*, ed. Gerald Dick and Markus Gusset, 35–38. Gland, Switzerland: WAZA Executive Office.

Reid, Gordon McGregor. 2010. Introduction. In *Building a Future for Wildlife: Zoos and Aquariums Committed to Biodiversity Conservation*, ed. Gerald Dick and Markus Gusset, 11–14. Gland, Switzerland: WAZA Executive Office.

Soulé, Michael E. 1985. What is conservation biology? *BioScience* 35:727–34.

Stanley Price, Mark R., and John E. Fa. 2007. Reintroductions from zoos: A conservation guiding light or a shooting star? In *Zoos in the 21st Century: Catalysts for Conservation?* ed. Alexandra Zimmermann, Matthew Hatchwell, Lesley Dickie, and Chris West, 155–77. Cambridge: Cambridge University Press.

Stuart, Simon N., Dena Cator, and Jane Smart. 2010. International Union for Conservation of Nature: Global species crisis and conservation challenges. In *Building a Future for Wildlife: Zoos and Aquariums Committed to Biodiversity Conservation*, ed. Gerald Dick and Markus Gusset, 21–24. Gland, Switzerland: WAZA Executive Office.

Vié, Jean-Christophe, Craig Hilton-Taylor, Caroline M. Pollock, James Ragle, Jane Smart, Simon N. Stuart, and Rashila Tong. 2009. The IUCN Red List: A key conservation tool. In *Wildlife in a Changing World: An Analysis of the 2008 IUCN Red List of Threatened Species*, ed. Jean-Christophe Vié, Craig Hilton-Taylor, and Simon N. Stuart, 1–13. Gland, Switzerland: IUCN.

Weldon, Ché, Louis H. du Preez, Alex D. Hyatt, Reinhold Muller, and Rick Speare. 2004. Origin of the amphibian chytrid fungus. *Emerging Infectious Diseases* 10:2100–2105.

WAZA. 2005. *Building a Future for Wildlife: The World Zoo and Aquarium Conservation Strategy*. Bern: WAZA Executive Office.

———. 2009. *Turning the Tide: A Global Aquarium Strategy for Conservation and Sustainability*. Bern: WAZA Executive Office.

WWF International. 2010. *Living Planet Report 2010*. Gland, Switzerland: WWF International.

Zimmermann, Alexandra. 2010. The role of zoos in contributing to *in situ* conservation. In *Wild Mammals in Captivity: Principles and Techniques for Zoo Management,* 2nd ed., ed. Devra G. Kleiman, Katerina V. Thompson, and Charlotte Kirk Baer, 281–87. Chicago: University of Chicago Press.

Zimmermann, Alexandra, Matthew Hatchwell, Lesley Dickie, and Chris West, eds. 2007. *Zoos in the 21st Century: Catalysts for Conservation?* Cambridge: Cambridge University Press.

54

Research in Zoos

Rebecca E. Spindler and Joanna Wiszniewski

THE IMPORTANCE OF ZOO RESEARCH

Zoos and aquariums provide an important avenue for increased understanding of wildlife species, their environmental needs and preferences, and their ability to adapt. This has often filled an important gap in knowledge that cannot be gained from free-ranging populations because of cryptic animal behavior, inaccessible environments, limited access to the animals, prohibitive costs of studying enough animals, and the likelihood of the study itself impacting the animals being studied. Zoo-based populations provide access to individuals on a long-term basis, providing context and life history parameters to understand the significance of samples taken at a single time point. Keepers are a key to providing this lifelong data.

Zookeepers are the principal observers and recorders of the many events that occur in animals' daily lives, some seemingly mundane and some obviously significant, but all providing an essential foundation for research investigation. Keeper observations and daily reports have provided long-term records of animals' temperament, food and environmental preferences, signs of illness and injury, response to medical treatment, interactions with individuals of their own or other species, response to enrichment and training, and daily and seasonal behavioral patterns. Such information, gained through routine management of wildlife species, has, along with targeted investigation in zoos and aquariums, increased our understanding of physiology and function in the areas of reproduction, nutrition, immunology, and behavior and improved our ability to diagnose disease and abnormalities. This data indicates the division between physiological values considered normal (the average of values in animals known to be of good health and function) and abnormal. Further, this knowledge has improved the management of other captive individuals of the same or closely related species, and has provided a foundation for the assessment and management of free-ranging populations.

In addition to the passive collection of valuable data, zoos and aquariums have the capacity to create controlled conditions, so specific questions can be isolated and investigated with few confounding factors (factors that add variability not related to the treatment). Using this kind of experimentation, zoos and aquariums have increased our understanding of wildlife species' environmental requirements, specific disease treatments, contraception, and cognition, as well as of the effectiveness of treatments for their general health maintenance.

Zoos and aquariums have also provided essential avenues for expanding and improving the data obtainable from free-ranging populations. For example, samples collected noninvasively from free-ranging populations (e.g., fecal and hair samples) are now routinely analyzed using techniques that could only have been developed using captive populations. To validate these tests and understand their limits, samples from individuals of known identity, age, health status, nutrition, reproductive history, and so on have been analyzed to determine a reliable correlation between test results and the individual's functionality. This data could then be used to estimate the biology and health of a free-ranging animal from one or a few samples collected in the wild.

CURRENT STATUS OF ZOO RESEARCH

The following summaries are not exhaustive, but rather are a brief indication of the research topics published using zoo-based populations of wildlife species. Additional summaries of zoo or aquarium-based research can be found on the websites and in the publications of the regional zoo and aquarium associations (AZA, EAZA, BIAZA, ZAA, etc.). as well as the regional keeper associations (AAZK, ASZK, ABWAK, etc.) Excellent reviews of zoo and aquarium-based issues and research may be found in publications such as *Wild Mammals in Captivity*, 2nd ed. (Kleiman, Thompson, and Kirk Baer 2010), *After the Ark* (Mazur 2001) and *Zoos in the 21st Century* (Zimmermann et al. 2007).

BEHAVIORAL BIOLOGY

Behavioral observations provide keepers and other staff with vital information on the activity and welfare of animals in their care. For example, behavioral cues can be used to examine social and spatial relationships among individuals (Freernan, Schulte, and Brown 2010), personality (McDougall et al. 2006), interactions with local wildlife (Ross, Holmes, and Lonsdorf 2009), interactions with visitors (Mallapur, Sinha, and Waran 2005; Shen-Jin et al. 2010) and vocal communication among individuals (Dooley and Judge 2007; Stoeger-Horwath et al. 2007). Behavioral observations in collaboration with other disciplines such as medical research (Terio, Marker, and Munson 2004) and endocrinology (Terio, Citino, and Brown 1999; Wielebnowski et al. 2002) have provided increased understanding of stressors and their impact on the well-being and health of animals. Such studies provide essential information in the management of captive populations for best animal welfare.

Environmental enrichment may include altering feeding protocols and food types, increasing the complexity of enclosures or interactions, and regularly providing novel equipment for play (Videan and Fritz 2007; Smith and Litchfield 2010). Enrichment devices can stimulate normal species-specific behaviors; create opportunities for exploration, foraging, and social interactions; and reduce stereotypical patterns of behavior. As a result, zoo- and aquarium-based individuals in an appropriate enrichment program may exhibit reduced aggression and abnormal behavior as well as improved reproduction (Fuller et al. 2010; Carlstead and Shepherdson 1994; Hunter et al. 2002).

An increased understanding of the behavioral biology, learning ability, and motivation of a species and of individuals within the species is key to a successful training program. Training can be used by keepers as a type of cognitive enrichment, or to elicit a specific behavior from a collection animal (Mellen and MacPhee 2010). Operant conditioning is the most popular method of training, whereby animals are rewarded for the requested behavior. In many species this training has decreased or eliminated the need for anesthesia in completing medical assessments, providing treatment, and collecting samples for research.

The capacity of zoological institutions to conduct behavioral research on captive animals can contribute to our understanding of the evolution of social behavior and the environmental factors influencing it. For example, simple but systematic observations of the social interactions and patterns of play of some primate species have demonstrated their high cognitive ability and complex social structure (Koyama, Caws, and Aureli 2006; Tanner and Byrne 2010).

REPRODUCTIVE BIOLOGY AND ASSISTED REPRODUCTION

Keepers' notes and long-term behavioral observations have increased knowledge of life history characteristics—including age at sexual maturity, average litter size, correlations between age and reproductive fitness, and offspring survival—of a diverse range of species. Data of this kind has been important in species breeding and conservation plan-ning, with the use of population-modeling software programs such as Vortex (Lacy 1993).

Increasingly, zoos and aquariums require detailed reproductive monitoring of collection animals to enhance or control reproduction. Hormone monitoring can provide keepers and animal managers with a way to accurately detect reproductive cycling and seasonality (Weissenbock, Schwammer, and Ruf 2009), synchronize estrus (Brown et al. 2004), and time artificial insemination (Hodges, Brown, and Heistermann 2010). Noninvasive hormone analysis using fecal samples also allows the assessment of adrenal activity without causing the animal distress (Wielebnowski et al. 2002; Wasser et al. 2000). This data must be interpreted carefully, as increased adrenal activity does not necessarily indicate distress (Wielebnowski 2003).

Increased knowledge of gamete biology and development of effective cryopreservation techniques can significantly improve the success of artificial insemination and in vitro fertilization (Howard et al. 1991; Herrick et al. 2010). Long-term cryopreservation of sperm from wild animals, especially, can facilitate zoo-based breeding programs by replenishing genetic material without the associated cost and anxiety of animal translocations (Wildt 1997). Increasingly sophisticated methods of assisted reproductive technologies are being developed, such as embryo and oocyte transfer, gender selection, and cryopreservation of embryos (Hermes et al. 2009).

As reproductive success and veterinary care in zoos and aquariums have increased the fecundity and longevity of captive wildlife populations, contraceptive techniques are being sought and developed (Asa and Porton 2004). Contraception is employed to maintain group dynamics for social species, and to manage the genetics of captive populations (e.g., for equalizing offspring across founders and minimizing inbreeding). Recent studies have focused on testing the effects of long-term use of contraceptives (Munson et al. 2002; Moresco, Munson, and Gardner 2009) and devising new, safer techniques that are applicable to multiple species.

GENETIC MANAGEMENT OF CAPTIVE BREEDING PROGRAMS

The knowledge of a threatened species' reproductive biology, combined with effective genetic management of captive populations, can often determine the success of captive breeding and reintroduction programs (Frankham, Ballou, and Briscoe 2002). The primary focus of captive breeding programs is to maximize the retention of genetic diversity and minimize the occurrence of inbreeding. This can be achieved by increasing the numbers of founders and decreasing the levels of genetic relatedness among the potential breeders (Ballou and Foose 1996). In many captive breeding programs, however, the number of founders is restricted and the genetic relationships among individuals is unknown (Zeoli, Sayler, and Wielgus 2008). Genetic and analytical techniques are now being used to assess relatedness among captive animals, as well as to reconstruct multigenerational pedigrees if the ancestry of captive individuals is unknown—for example, in the black-footed ferret, parma wallaby, Persian wild ass, and giant panda (Ballou and Foose 1996; Ivy et al. 2009; Li et al. 2010; Nielsen, Pertoldi, and Loeschcke 2007).

Research is also being undertaken to minimize genetic adaptation of individuals to captivity, thereby increasing the success of future reintroduction attempts. Genetic changes in individuals as a result of unnatural environmental conditions and selection pressures have been documented in fish, insects, amphibians, and mammals (Woodworth et al. 2002), but they can be alleviated by manipulating the environment and breeding opportunities (Frankham 2008).

NUTRITION

Nutrition plays a fundamental role in animal health, behavior, and reproduction. The difficulty of obtaining direct data on the complete nutritional range of free-ranging wildlife, as well as a limited ability to replicate the wild diet in captivity, has prompted research into the effect of various diets and feeding protocols on animal health. This requires an initial understanding of the digestive physiology of the different types of animals, as well as information on potential diet-related diseases (Kirk Baer et al. 2010). Careful optimization of diets has been facilitated by routinely assessing the health of captive animals, investigating the causes of deficiencies and toxicities, and by identifying species-specific nutritional requirements through comparative analysis with domestic animals (Vester et al. 2010; Lavin, Chen, and Abrams 2010; Gutzmann, Hill, and Koutsos 2009).

Primary areas of research currently focus on novel ways of providing natural diets in nonnative species, the impact of supplementation with micronutrients, and opportunities for enrichment using appropriate food items.

ANIMAL HEALTH AND DISEASE SURVEILLANCE

Veterinary medicine has been a major focus area of zoo and aquarium research for many decades. *Zoo and Wild Animal Medicine Current Therapy* (Fowler and Miller 2007) is an excellent reference source for information on wildlife medicine.

In addition to providing daily medical care for the zoo or aquarium's animal collection, veterinarians play a critical role in zoo- and aquarium-based conservation programs as well as in field conservation. Specifically, zoo veterinarians have contributed to our knowledge of species biology, wildlife disease and illness, and treatment options. They often perform this work in close collaboration with keepers.

Information obtained during routine health assessments, veterinary procedures, and necropsies, combined with detailed nutritional and behavioral records, has been applied to the development of improved treatment options and husbandry practices (Müller et al. 2010; Napier et al. 2009). Zoos increasingly have been implementing novel methods to prevent, diagnose, and treat diseases in captive populations (Abril et al. 2008; Zelepsky and Harrison 2010). For example, a highly sensitive and rapid detection of tuberculosis in elephants provides an opportunity to limit the spread of the disease, and has increased treatment success of infected individuals (Greenwald et al. 2009). Understanding the risk factors that increase an individual's susceptibility has also reduced the

incidence and spread of infectious disease within and between captive populations (Witte et al. 2010). For example, closely monitoring aged and younger animals, minimizing stressors and inbreeding, and providing enclosures of sufficient size and complexity may significantly reduce an animal's susceptibility to disease. Further, captive studies have provided a unique resource of information about the effectiveness of a broad range of treatment options for various infections and diseases (Aruji et al. 2004; Hunter and Isaza 2002).

Veterinary staff often contribute significantly to the assessment of wild populations through attending field excursions, or by developing techniques in captivity to be used in the field. For example, the optimization of noninvasive techniques for collecting saliva from captive gorillas to establish accurate diagnostic techniques has enabled researchers to assess the prevalence of infectious diseases in wild gorilla populations (Smiley et al. 2010). Further, the detection of diseases in free-ranging animals brought to zoos and rehabilitation centers for diagnosis, rehabilitation, or treatment has helped to indicate the incidence and threat of emerging diseases in wildlife populations (Nemeth et al. 2007; Sleeman 2008).

ECOLOGY AND WILDLIFE CONSERVATION

Biodiversity is currently being lost at an overwhelming rate (Brooks et al. 2002). So, the value of zoo- and aquarium-based "insurance populations" and multidisciplinary research has increased significantly for wildlife conservation. Captive animal collections have been used for a number of applications, including reintroduction programs (when threatening processes have been addressed), to preserve genetic diversity within threatened species, to increase public awareness of conservation issues and instigate change in human behavior, and to obtain baseline information on species biology that can be used in wild population models (Hutchins, Foose, and Seal 1991; Osofsky, Karesh, and Deem 2000). In particular, technical advancements in the methods used to study the reproduction and health of zoo and aquarium-based animals without capturing or immobilizing them has contributed significantly to wildlife conservation (Lesley and Kirkpatrick 1991; Kirkpatrick et al. 1991). For example, the development and validation of a noninvasive technique to monitor adrenal activity using feces from captive African wild dogs (*Lycaon pictus*) has enabled researchers to evaluate how ecological conditions and conservation strategies (such as the use of radio collars) affect the health of free-ranging animals (Monfort et al. 1998).

Local community participation and knowledge from zoo and aquarium-based studies has also greatly helped in the development of effective field conservation initiatives. For example, to assist in reintroduction of the endangered Egyptian tortoise, members of a local community were involved in the study of activity patterns and diet in captivity. Equipped with this knowledge and training, the local community monitored the effectiveness of the reintroduction program in the wild. This research subsequently progressed into a broader study addressing more conservation-relevant concerns about movement patterns, habitat selection, and population dynamics (Attum et al. 2007). Knowledge of behavioral budgets and nutritional needs and preferences has greatly helped reintroduction programs and small-population management by facilitating the capacity for free-ranging population monitoring. Keepers have a unique opportunity to provide specialist expertise to field-based research projects, given institutional support.

EVALUATION OF EDUCATION AND CONSERVATION INITIATIVES

Motivational factors behind people visiting zoos and aquariums is an area of great interest and can include desires for education, enjoyment, and/or developing a connection to wildlife (Falk 2006). Investigating how visitors interpret animal and educational encounters is an important area of research, and it leads directly to continual improvement of zoos and aquariums from the visitor's perspective (Balmford et al. 2008). For example, factors such as the presence of children and level of social interaction among visitors can influence the proportion of time spent on various zoo-based activities (e.g., watching display animals, engaging with interpretive displays, and reading educational signs; Ross and Gillespie 2009).

One of the primary influences on visitor experience and conservation attitude is the effectiveness of animal enclosures and animal behavior, as well as the visitor's attitude towards zoo and aquarium husbandry practices (Anderson et al. 2003; Blaney and Wells 2004; Swanagan 2000). Some studies have shown that knowledge of conservation issues does not always improve following a zoo or aquarium visit (Balmford et al. 2008; Lindemann-Matthies and Kamer 2006). However, a visitor's concern for a species or conservation issue can be dependent on the knowledge gained from the zoo or aquarium experience, the initial motivation to attend the zoo, and the degree of connection that the visitor felt with the animals (Clayton, Fraser, and Saunders 2009; Falk and Adelman 2003). Ultimately, determining whether zoo and aquarium-based educational experiences instigate changes in visitors' conservation values and action towards sustainability is still a critical research area (Miller et al. 2004).

HOW TO STRUCTURE AND PERFORM ZOO RESEARCH

Research is always initiated by an observation that generates a question. Important research questions on wildlife biology often arise as a result of the close relationship that keepers share with the animals under their care. Consequently, keepers often initiate research, perform research, and/or assist with the training of animals or the collection of samples for research projects. Many projects are made possible only through the contribution of enthusiastic and knowledgeable keepers. Many keepers will perform their own research, and collaboration with established scientists may facilitate research planning and original designs. Reading articles published in peer-reviewed journals and popular science magazines will also greatly help keepers in developing appropriate questions for investigation. Regardless of who performs the research, zoo and aquarium experimental design will often benefit from early collaboration with keepers.

INSTITUTIONAL SUPPORT

Once a question has been generated, the keeper and the managers must determine whether the study is of value to the management, understanding, or conservation of the species. Institutional investment is essential to the success of the study, as it ensures that research is conducted with sufficient support and consistency. Withdrawing support from an unfinished project prevents an evidence-based conclusion from being reached. This wastes resources and reduces the perception of the value of science to provide clear answers. In a research proposal written to gain support from peers or managers, it is extremely important that the writing is clear, free of emotional language, and unbiased. The following elements should be covered.

1. What is the problem that generated the question?
2. What is the specific research question?
3. Why is the research important to the facility?
4. How will the research be performed?
5. What will it cost, and how will the funds be secured?
6. Who will assist with the project and what experience do they have?
7. What impact will the results have, depending on what they might be?
8. How will the results be published or presented?
9. What will be the overall benefits to the zoo or aquarium (including in staff development)?

EXPERIMENTAL DESIGN

Researchers must have an important and testable question with a good experimental design in order to proceed. Specific methods are determined by the research question and are too diverse to explore individually here. The project team should review the literature and seek expert advice when designing project methods. The following are some general elements that can assist in the development of a robust experimental design. A hypothetical example will be used to illustrate these points. In this scenario, the organization would like to investigate the potential of adding an early-morning behind-the-scenes tour of the tigers.

1. The research question must be clearly defined and kept in mind throughout the development of the design ("Does the implementation of the trial tour change the behavior of tigers during the day?").
2. The method to be used must be known to accurately measure the behavior or trait being studied. For example, new analytical tests should be run in parallel with tests known to work in similar species and with similar sample types (e.g., blood, feces, or urine). With respect to our tiger example, a species-specific ethogram (chart that lists common species-specific behaviors) and an accepted method to accurately determine tiger position in the enclosure (an exhibit plan divided into horizontal and vertical sections) should be used.
3. A statistical power analysis should be performed to

determine the minimum number of study animals and samples required to detect any changes beyond normal variability. The tigers in our example should be observed before any changes are made, to determine their normal or "baseline" behavior. The time they spend performing each behavior and the variability over time and among individual animals should be noted specifically.

4. The required animals must be available for study during the entire study period. Any tigers that are part of a breeding program, or which otherwise may not be available for the entire time, should not be included in the study. Any confounding (unrelated) factors that may influence the results should be identified and controlled throughout the experimental period. For example, animals should be age- and gender-matched across treatments and have similar environments and diet where possible. The introduction of the trial tours should not coincide with other changes in the animals' environment such as season or diet.

5. Measurement of any treatment's effects should be consistent to avoid decreasing the significance of results. The tigers should be observed at standard times and from similar positions before and after the addition of the tours.

6. To be sure that any detected behavioral change is due to the treatment, the original conditions should be restored and observations continued to monitor the effect on behavior. The tigers in our experiment should be examined, before, and during a trial run of the tours, and again after the tours have stopped.

SELECTING THE RESEARCH TEAM

The research team should consist of people who have generated the question and those who have experience in the study's sample collection and analysis methods. Keepers may provide valuable advice for a behavior project but may not make appropriate observers, because the animals' behavior is likely to change in their presence. The research team should agree upon and designate a project manager at the project's onset. This person must ensure that each team member is clear about their responsibilities, and must ensure that the agreed-upon methods are adhered to strictly.

FUNDING RESEARCH

Many funding sources are available for the wide diversity of research conducted around the world. Government sources will often require affiliation with a recognized academic institution, such as a university. These grants may also involve long and complicated application processes, and may only suit larger projects with significant costs.

Pilot projects and low-cost projects can often be funded through foundation grants. Many countries will produce voluntary foundation registers, while others mandate registration of charities that meet certain criteria (e.g., regarding their activities and income levels). For example, charities are registered with the Charities Directorate of the Canada Revenue Agency, The Charity Commission for England and Wales, the Office of the Scottish Charity Regulator, the European Foundation Centre, and Philanthropy Australia. There are multiple avenues for discovering foundations in the United States. Foundations relevant to funding a given project can also be found by searching for similar projects online that declare their funding sources.

The guidelines for writing a foundation grant may include factors similar to those of the report to gain institutional support for research (see above). It is critical to a funding application's success that the grant's philosophy, restrictions, and goals match those of the project. For example, some but not all foundations aimed at conservation will fund research. Foundations may also have geographic restrictions or areas of interest (e.g., work locations, specific habitats), species or taxa preferences, or requirements for education components. Foundations may also exclude funding for specific activities such as travel, captive work, or staff costs. Most foundations will provide guides to their funding ranges, or at least their maximum grant funding levels. Examining their lists of projects already funded can provide valuable information about their preferences. If in doubt, contact the foundation's administrator to avoid writing an application with little or no chance of success.

Corporate sponsorships may also be an option, but they should be coordinated with ongoing fundraising programs at the zoo or aquarium, so that potential sponsors are not bombarded with requests that seem unrelated. Similarly, efforts to match a research project with a funding organization's philosophy are important to success.

THE IMPORTANCE OF PUBLICATION

Even small-scale studies that provide better understanding of wildlife biology and management strategies should be published. This is necessary to avoid duplication of research effort by others in the same profession, to provide evidence-based guidelines on management options, to increase understanding in the wider community as to the complexities of wildlife management, and to provide evidence of the value of research to managers and peers. Results of studies that do not provide an improved approach will often not be submitted for publication, but if the science has been conducted well and the answer is clear, the results of a method that has not produced the goal outcome should still be published to avoid its duplication by other professionals.

Peer-reviewed journals relevant to the zoo and aquarium industry include, but are not limited to, *Zoo Biology*, the *Journal of Zoo and Wildlife Medicine*, the *International Zoo Yearbook*, the *Journal of Zoology*, and *International Zoo Educator's Journal*. An article should be submitted to journals with the best fit for its subject, and a journal's instructions to authors should be adhered to closely to avoid rejection on trivial grounds. Discipline-specific journals focusing on behavior, reproduction, wildlife medicine, ecology, genetics or nutrition, or specific taxa may be more appropriate for publication of certain studies. Some publications are more widely read and have articles that contain more citations; these data are

> **Good Practice Tip:** Google Scholar is a good start for finding general information on a topic and relevant journal articles. For reading full articles, access to a university search engine is ideal, but must be negotiated with the university.

used to calculate a journal's impact factor. These should not be the primary reasons for choosing a given journal, but they will provide some indication of its quality and may affect an author's future funding opportunities.

Journal articles will generally include the following sections:

1. Title: The title should be succinct and specific to your project, and should provide interest by indicating the result. It should be written last.
2. Abstract: This should be one paragraph of approximately 200 to 400 words that clearly summarizes each section (introduction, methods, results, and discussion). It should be written second to last.
3. Introduction: This section should include a description of the issue that generated the question, the current knowledge around that problem (a literature review of relevant work), the specific question at the study's focus, and the project's aims and hypotheses.
4. Methods: This should include specific detail about the location, number, and environmental conditions of the animals studied, the time of year and days on which the study was undertaken, the source of materials, and a description of exactly what was carried out in enough detail so that someone else can repeat the study exactly.
5. Results: This is a pure description of the results without interpretation. For example, it is appropriate to describe a decrease in aggressive behavior between male oryx in the months of December through February, but not to attribute it to seasonal reproduction of the males in question or the neighboring females. That attribution should be included in the discussion as a possible explanation of the observed results.
6. Discussion: The meaning of the results. This section should also include a discussion of whether the results met the study expectations—that is, whether the hypothesis is to be accepted. If it is not to be accepted, the discussion should explore the possible reasons and describe the relevance of the information that was gained. Finally, the impact of the results on the original problem should be explained, and suggestions for further study given.

NON-PEER-REVIEWED JOURNALS

Publication in industry-specific journals published by the regional keeper associations (e.g., *Animal Keepers Forum*, published by the AAZK; *Thylacinus*, published by the ASZK; *Ratel*, published by ABWAK) may be most appropriate for zoo-based studies where the primary audience is zookeepers and other staff. The format may differ from the one described above, but it should contain the same basic elements.

Publications in popular journals or in the media have a greater reach to the public than scientific journals, and advance public understanding of the zoo and aquarium industry. Such articles will likely employ a more discursive and informal style of writing, but when the article discusses science, the content must remain accurate and unbiased. Some details may be left out for the sake of clarity, but not if that will change the reader's interpretation of the results. The author's credibility will often be bolstered by references to original peer-reviewed articles.

AUTHORSHIP

While convention is different in different countries, the Vancouver Protocol is generally accepted. This states that all authors must have had significant input into the conception or design, interpretation, or analysis of the results and into writing or revising the article significantly for intellectual input, and all must approve the final version for publication. The order in which the authors' names are listed is determined by the amount of effort contributed by each. The person who has performed the most work (both in performing the research and in writing the article) is the primary author, with the other authors listed in the order of decreasing amounts of effort contributed. In some countries, the leader of the group who has provided significant intellectual input and approved or funded the research is often placed at the end of the author list—thus, this position has become recognized as significant.

RESEARCH OPPORTUNITIES AND LIMITATIONS

INFRASTRUCTURE

Zoos and aquariums have a unique opportunity to present a range of environmental conditions to test animals' preferences and needs or basic biology. This will increasingly be an interesting and important area of study. However, the cost and logistics of altering key elements of enclosures is often prohibitive. Refurbishment and rebuilding provide excellent opportunities to engineer novelty into exhibits or create ways to consistently provide individual animals with choices. Communication within a zoo or aquarium should include discussions between building project managers and keepers or researchers. An ongoing, up-to-date list of potential research projects that require building or maintenance assistance may help in incorporating these needs into the building plan.

LIMITED NUMBERS

Because most zoos and aquariums have the capacity to hold only a few individuals of each species, few rigorous studies can be conducted in a single institution. Regional associations and cooperative breeding programs mentioned earlier in this volume have created excellent opportunities and channels of communication for the development of multi-institutional studies. Performing research across multiple institutions does introduce some variability in environmental

Good Practice Tip: Communication is the key to good research. First, making sure that everyone who will be involved in the project is included in its design can help ensure that the study receives support to the end. Second, presenting and publishing the results will help staff in other zoos learn from the research work.

Good Practice Tip: Search out and read as many papers as you can on the topic you want to research to find out what has already been done, how best to go about your study, and who might be helpful contacts.

conditions, and this must be taken into consideration when designing and performing the research. Open range zoos and wildlife parks provide an excellent opportunity for study of a large number of animals of the same species under similar, generally natural conditions. A possible limitation with these populations is limited access and the staff resources needed to collect samples. With institutional support, this can be overcome.

ETHICAL CONSIDERATIONS

Zoos and aquariums are facing greater scrutiny over all aspects of operation. This is an opportunity for keepers and other staff to review internal operating procedures, perform rigorous research, and either make management changes or provide evidence of the suitability of current practices. All research that alters the environment of the animal being studied should be approved by an animal ethics committee licensed by the relevant government body before starting. Some zoos and aquariums have their own animal ethics committees, while others may need to link with university or government departments to ensure that research is reviewed appropriately. These committees weigh the degree and longevity of any stress or discomfort that may be experienced by individual animals as a result of the research against the overall benefit to the population or species. The benefit to the organization should not be considered here.

RESOURCES

The skills and knowledge of keepers are invaluable to research, but spending these resources on research rather than on operational needs is often difficult to accommodate. External funding can help if the research projects require a significant amount of the keeper's time. Individual institutions must decide how much research can be incorporated into core business, and should manage the expectations of those wishing to contribute to this area.

THE VALUE OF PARTNERSHIPS

Crossdivisional cooperation within the zoo or aquarium is an important avenue for the development of novel ideas and appropriate execution of the methods. For example, studying the impact of enrichment items on animal well-being can require input from the keeping, maintenance, food preparation, veterinary, and behavioral staff. Early recognition of the potential benefits of research and its incorporation into the design of a new exhibit can ensure that small additions (e.g., attachments in an enclosure for enrichment or camera

setup) or changes (e.g., in configuration flexibility or diet) can be incorporated with less effort and cost.

Communication and collaboration across zoos is essential to the performance of rigorous research, but it can be limited by a sense of competition. Fortunately, this competitive character is diminishing as zoos and aquariums work more cooperatively to constantly improve their practices and the public perception of well-managed zoo organizations in general. In cases where this competition persists, it is rarely at the level of keepers and other frontline staff. Communication with peers can generate new ideas, improve project design, and streamline work methods. Regional zoo and aquarium and keeper associations can promote research communication and collaboration among facilities by disseminating and coordinating information on specific projects. Presentation at association conferences and publication in industry-specific journals can improve understanding of issues, increase sample size, improve project design, and provide feedback on results.

Collaboration with university partners can provide a valuable source of advice (particularly on statistics), journal articles, and equipment for research projects. Training postgraduate students at zoos or aquariums and having staff spend time at universities while completing projects or degrees can benefit zoos or aquariums by providing insights into different audiences, building relationships, and broadening understanding. In many cases, this can also lead to impartial revue and support from unbiased, respected sources in the face of criticism.

Many zoos and aquariums are managed by federal, state, municipal, or local governments, and will therefore have mandates to provide services to those governments. Unfortunately, research is often not considered a key element of such services. Early and consistent communication with government environmental departments can advance the planning of research that uses captive animal populations to inform upcoming management decisions pertinent to free-ranging populations. Often a government department lacks sufficient data or resources to perform needed experiments, but does not think to turn to local zoos or aquariums for assistance. Partnership with nongovernmental organizations can also provide important information for wildlife managers and assist with institutional efforts toward continuous improvement, as long as the relationships are fruitful and respectful.

THE FUTURE OF ZOO-BASED RESEARCH

While not all zoos and aquariums have dedicated research staff, almost all cooperate in collaborative research, provide samples for research, or undertake projects of their own design. Although often not formalized, keepers will continue to undertake research by altering individual animals' environments to ensure their best welfare. Increased engagement of enthusiastic keepers in the generation of project ideas, performance of research, and publication of species knowledge is an area of great opportunity for improved management of our captive populations. The keepers' intimate knowledge of animal preferences and tolerance limits could also provide vital species information and increase our understanding of the environmental requirements of zoo- and aquarium-based species, to be applied to their management in the wild.

With the advent and wide use of global data organization and amalgamation systems such as the Zoological Information Management System (ZIMS), the zoo community will be able to provide this type of data. By holding individuals of key species throughout the world, zoos expose their collection animals to a broad range of environmental conditions such as temperatures, humidity, and food sources. Combining data on animal function and preferences in each of these conditions will provide a valuable resource for determining the likely responses of species to a changing environment. Issues such as data consistency, ownership, and organizational sensitivities should be considered throughout the process of combining data. In many cases, specific experiments will have to be performed to eliminate confounding factors and to ensure that questions and species are prioritized to provide the most useful information to ecosystem modelers and managers. This is clearly important for improving the management of wildlife species in captivity, but it will also be increasingly important for predicting and alleviating the biological impact of environmental change, including climate change.

In the face of a changing environment, populations in the wild will either persist as they are, adapt over generations while remaining in their original areas, migrate to preferred conditions, or go extinct (Hewitt and Nichols 2005; Midgley, Thuiller, and Higgins 2005). Each of these reactions is likely to affect a species' interactions with others around it, and therefore to affect the functionality of the ecosystem as a whole.

For example, in response to increasing temperatures, a species may adapt by foraging and hunting at a different time of day. It may persist in its traditional range and be found in similar numbers on a census, but the species it interacts with and the role it plays in the ecosystem may change along with its behavior. Further, over generations, species may adapt genetically to change their metabolic rates or take on different coat colors—changes that would affect food source preferences and predator-prey relationships respectively. Perhaps more obviously, species that migrate or go extinct leave gaps in the ecosystem function, which may in turn affect the services (clean air, clean water, etc.) the ecosystem provides for every species in the catchment area.

Currently, there is very little evidence on species' preferences, tolerance ranges, and functions under altered conditions. Without this data, it is impossible to predict how resilient each species will be to a changed environment, and how species interaction and therefore ecosystem function is likely to change, or to know which actions will be most helpful to maintaining ecosystem function. A good resource for searching data relevant to biodiversity and climate change is www.bioclimate.org, hosted by the Zoological Society of London.

SUGGESTED AREAS OF RESEARCH ATTENTION

WILDLIFE HEALTH

A significant gap in current knowledge of wildlife health includes the early behavioral signs of disease in wildlife. Clearly, wildlife species are well adapted to show few signs of illness as a predator avoidance strategy, so the early warning signs are difficult to detect. Keepers are the first line of inquiry into the health of individual animals, and although this works well within zoos and aquariums, specific observations of behavioral changes are not standardized across institutions. Standardized daily keeper notes should be entered into a common database that can be analyzed to determine common early warning signs, and published to improve management of wildlife species globally.

Medication and delivery procedures have generally been developed for domestic animals, and they clearly create issues for the repeated medication of animals that are not tractable or which present human health and safety concerns. The development of novel, noninvasive techniques for the delivery of medications to wildlife species would be of great value to both veterinarians and keepers. For example, the stress of forced ingestion of medications can be avoided by smearing palatable oral medication on the patient's coat or skin and allowing the animal to lick it off. This avoids stress to the animal and promotes a healthy animal-keeper relationship.

Research into alternate methods of reward for species that may not be food-oriented, such as ungulates, could be conducted by keepers who observe those animals' preferences and understand their biology.

ECOLOGY

Captive individuals can provide better understanding of species' nutrient requirements, heat tolerance, or disease susceptibility as a result of changes in environmental conditions such as humidity. By providing choices it is also possible to determine their preferences in food source species, temperature, and so on. This kind of experiment is often prohibited by a lack of infrastructure, but over time the problem could be overcome with the development of flexible exhibits with a range of environmental states. This would provide avenues for research as well as enrichment, ongoing choice, and optimal comfort for the animals in our care.

Many keepers and other zoo or aquarium staff are involved in field research programs specifically because they are knowledgeable about the handling and biology of the animals being studied. This rare skill should be considered when surveying the range of skills within an institution. As the wild becomes more intensively managed, skills such as those developed in zoos and aquariums will be called upon more and more.

POPULATION DYNAMICS AND VIABILITY

Keeper participation is essential for the collection of reproductive and genetic samples and of behavioural and environmental data that can be correlated with physiology.

Anesthesia is often required to collect samples from wildlife unless alternate methods can be employed reliably and safely. Enthusiastic keepers with in-depth understanding of the animals in their care may generate and test novel ways of collecting specimens. For example, hair samples for genetic or hormone analysis can be captured using field techniques (e.g., sticky pads or brushes) as an animal walks through a tight hallway or doorway.

BEHAVIOR

Assessment of behavior is intrinsic to many of the areas discussed above. Assessments of species' cognitive abilities, behavioral adaptability to changing conditions, and approaches to novelty in known environments would also be of great interest.

Targeted analysis of the effect of enrichment on the behavior and welfare of wildlife species is not undertaken as often or extensively as it should be. Keepers could help identify target behaviors to elicit through enrichment and stereotypic behaviors to be reduced. They could also help design novel enrichment devices and assist in assessing a research program's impact.

NUTRITION

Food source preferences and resistance to food source changes would provide a great deal of information on the adaptability limits of species in the wild. Nutrition combined with the provision of novel enrichment and medication is also an area of great interest and relevance to managing captive populations.

An area of nutritional physiology that could also use zoo populations to improve free-ranging wildlife management is the development of safe, food-based deterrents for wildlife control in areas that border human communities and agricultural lands (Osborn 2002).

EDUCATION OPTIMIZATION

The high regard in which society holds zoo and aquarium staff is important to our ability to convey messages about wildlife, the primary threats to their survival, and the action needed for their conservation. Determining how to transmit this information most effectively (e.g., through signs, public talks, one-on-one communications, or digital media) and how to convey it through animal demonstrations (e.g., with the animal in a natural exhibit close by or in the hands of the keeper, available to be touched) would help to guide best practices in zoo and aquarium education.

SUMMARY

Zoo and aquarium professionals have used research to improve animal health, welfare, husbandry, nutrition, and population management for centuries. A strong foundation of these changes has been the daily interaction between keepers and the animals in their care. With appropriate resources and collaboration, this research can now provide data that is invaluable for predictive models and optimal management of free-ranging animal populations as well as zoo-based wildlife. Further, the application of scientific

principles and knowledge to the daily management of captive populations can also lead to continual improvement in zoos. For this to happen efficiently and without duplication of effort, research findings must be published and accessible to people across institutions in all countries. So one of the great challenges is to increase the impact of research through effective communication of experimental findings in zoos and aquariums. Keepers already play an essential role in research, and each keeper has the potential to apply and expand his or her knowledge through many exciting future research opportunities.

REFERENCES

Abril, C., H. Nimmervoll, P. Pilo, I, Brodard, B. Korczak, M. Seiler, R. Miserez, and J. Frey. 2008. "Rapid Diagnosis and Quantification of Francisella Tularensis in Organs of Naturally Infected Common Squirrel Monkeys (*Saimiri sciureus*)." *Veterinary Microbiology* 127:203–8.

Anderson, U. S., A. S. Kelling, R. Pressley-Keough, M. A. Bloomsmith, and T. L. Maple. 2003. "Enhancing the Zoo Visitor's Experience by Public Animal Training and Oral Interpretation at an Otter Exhibit." *Environment and Behaviour* 35:826–41.

Aruji, Y., K. Tamura, S. Sugita, and Y. Adachi. 2004. "Intestinal Microflora in 45 Crows in Ueno Zoo and the In Vitro Susceptibilities of 29 *Escherichia coli* Isolates to 14 Antimicrobial Agents." *Journal of Veterinary Medical Science* 66:1283–86.

Asa, C. S., and I. J. Porton, eds. 2004. *Wildlife Contraception: Issues, Methods and Applications*. Baltimore: Johns Hopkins University Press.

Attum, O., M. B. El Din, S. B. El Din, and S. Habinan. 2007. "Egyptian Tortoise Conservation: A Community-Based Field Research Program Developed From a Study on a Captive Population." *Zoo Biology* 26:397–406.

Ballou, J. D., and T. J. Foose. 1996. "Demographic and Genetic Management of Captive Populations." In *Wild Mammals in Captivity: Principles and Techniques for Zoo Management*, edited by D. G. Kleiman, K. V. Thompson, and C. Kirk Baer. Chicago: University of Chicago Press.

Balmford, A., N. Leader-Williams, G. Mace, O. Manica, C. West, and A. Cimmerman. 2008. "Message Received? Quantifying the Impact of Informal Conservation Education on Adults Visiting UK Zoos." In *Zoos in the 21st Century: Catalysts for Conservation?*, edited by A. Zimmermann, M. Hatchwell, L. A. Dickie, and C. West, Cambridge: Cambridge University Press.

Blaney, E. C., and D. L. Wells. 2004. "The Influence of a Camouflage Net Barrier on the Behaviour, Welfare, and Public Perceptions of Zoo-Housed Gorillas." *Animal Welfare* 13:111–18.

Brooks, T. M., R. A. Mittermeier, C. G. Mittermeier, G. A. B. Da Fonseca, A. B. Rylands, W. R. Konstant, P. Flick, J. Pilgrim, S. Oldfield, G. Magin, and C. Hilton-Taylor. 2002. "Habitat Loss and Extinction in the Hotspots of Biodiversity." *Conservation Biology* 16:909–23.

Brown, J. L., F. Goritz, N. Pratt-Hawkes, R. Hermes, M. Galloway, L. H. Graham, C. Gray, S. L. Walker, A. Gomez, R. Moreland, S. Murray, D. L. Schmitt, J. Howard, J. Lehnhardt, B. Beck, A. Bellem, R. Montali, and T. B. Hildebrandt. 2004. "Successful Artificial Insemination of an Asian Elephant at the National Zoological Park." *Zoo Biology* 23:45–63.

Carlstead, K, and D. Shepherdson. 1994. "Effects of environmental enrichment on reproduction." *Zoo Biology* 13:447–58.

Clayton, S., J. Fraser, and C. D. Saunders. 2009. "Zoo Experiences: Conversations, Connections, and Concern for Animals." *Zoo Biology* 28(5): 377–97.

Dooley, H., and D. Judge. 2007. "Vocal Responses of Captive Gibbon Groups to a Mate Change in a Pair of White-Cheeked Gibbons (*Nomascus leucogenys*)." *Folia Primatologica* 78(4): 228–39.

Falk, J. H. 2006. "An Identity-Centred Approach to Understanding Museum Learning." *Curator* 49:151–66.

Falk, J. H., and L. M. Adelman. 2003. "Investigating the Impact of Prior Knowledge and Interest on Aquarium Visitor Learning." *Journal of Research in Science Teaching* 40:163–76.

Fowler, M. E, and R. E. Miller, eds. 2007. *Zoo and Wild Animal Medicine Current Therapy*, 6th edition. Saint Louis: Elsevier Science.

Frankham, R. 2008. "Genetic Adaptation to Captivity in Species Conservation Programs." *Molecular Ecology* 17(1): 325–33.

Frankham, R., J. D. Ballou, and D. A. Briscoe. 2002. *Introduction to Conservation Genetics*. Cambridge: Cambridge University Press.

Freernan, E. W., B. A. Schulte, and J. L. Brown. 2010. "Using Behavioral Observations and Keeper Questionnaires to Assess Social Relationships among Captive Female African Elephants." *Zoo Biology* 29(2): 140–53.

Fuller, G., L. Sadowski, C. Cassella, and K. E. Lukas. 2010. "Examining Deep Litter as Environmental Enrichment for a Family Group of Wolf's Guenons, *Cercopithecus wolfi*." *Zoo Biology* 29(5): 626–32.

Greenwald, R., O. Lyashchenko, J. Esfandiari, M. Miller, S. Mikota, J. H. Olsen, R. Ball, G. Dumonceaux, D. Schmitt, T. Moller, J. B. Payeur, B. Harris, D. Sofranko, W. R. Waters, and K. P. Lyashchenko. 2009. "Highly Accurate Antibody Assays for Early and Rapid Detection of Tuberculosis in African and Asian Elephants." *Clinical and Vaccine Immunology* 16(5): 605–12.

Gutzmann, L. D, H. K. Hill, and E. A. Koutsos. 2009. "Biochemical and Physiological Observations in Meerkats (*Suricata suricatta*) at Two Zoos during a Dietary Transition to a Diet Designed for Insectivores." *Zoo Biology* 28:307–18.

Hermes, R., B. Behr, T. B. Hildebrandt, S. Blottner, B. Sieg, A. Frenzel, A. Knieriem, J. Saragusty, and D. Rath. 2009. "Sperm Sex-Sorting in the Asian Elephant (*Elephas maximus*)." *Animal Reproduction Science* 112(3–4): 390–96.

Herrick, J. R., M. Campbell, G. Levens, T. Moore, K. Benson, J. D'Agostino, G. West, D. M. Okeson, R. Coke, S. C. Portacio, K. Leiske, C. Kreider, P. J. Polumbo, and W. F. Swanson. 2010. "In Vitro Fertilization and Sperm Cryopreservation in the Black-Footed Cat (*Felis nigripes*) and Sand Cat (*Felis margarita*)." *Biology of Reproduction* 82(3): 552–62.

Hewitt, G. M., and R. A. Nichols. 2005. "Genetic and Evolutionary Impacts of Climate Change." In *Climate Change and Biodiversity*, edited by T. E. Lovejoy and L. Hannah. New Haven and London: Yale University Press.

Hodges, K., J. Brown, and M. Heistermann. 2010. "Endocrine Monitoring of Reproduction and Stress." In *Wild Mammals in Captivity: Principles and Techniques for Zoo Management*, edited by D. G. Kleiman, K. V. Thompson, and C. Kirk Baer. Chicago: University of Chicago Press.

Howard, J. G., M. Bush, C. Morton, F. Morton, K. Wentzel, and D. E. Wildt. 1991. "Comparative Semen Cryopreservation in Ferrets (*Mustela putorius furo*) and Pregnancies after Laparoscopic Intrauterine Insemination with Frozen-Thawed Spermatozoa." *Journal of Reproduction and Fertility* 92:109–18.

Hunter, R. P, and R. Isaza. 2002. "Zoological Pharmacology: Current Status, Issues, and Potential." *Advanced Drug Delivery Reviews* 54:787–93.

Hunter, S. A., M. S. Bay, M. L. Martin, and J. S. Hatfield. 2002. "Behavioral Effects of Environmental Enrichment on Harbor Seals (*Phoca vitulina concolor*) and Gray Seals (*Halichoerus grypus*)." *Zoo Biology* 21:375–87.

Hutchins, M., T. Foose, and U. S. Seal. 1991. "The Role of Veterinary Medicine in Endangered Species Conservation." *Journal of Zoo and Wildlife Medicine* 22(3): 277–81.

Ivy, J. A., A. Miller, R. C. Lacy, and J. A. DeWoody. 2009. "Methods and Prospects for Using Molecular Data in Captive Breeding Programs: An Empirical Example Using Parma Wallabies (*Macropus parma*)." *Journal of Heredity* 100(4): 441–54.

Kirk Baer, C., D. E. Ullrey, M. L. Schlegel, G. Agoramoorthy, and D. J. Baer. 2010. "Contemporary Topics in Wild Mammal Nutrition." In *Wild Mammals in Captivity: Principles and Techniques for Zoo Management*, edited by D. G. Kleiman, K. V. Thompson, and C. Kirk Baer. Chicago: University of Chicago Press.

Kirkpatrick, J. F., V. Kincy, K. Bancroft, S. E. Shideler, and B. L. Lasley. 1991. "Oestrous Cycle of the North American Bison (*Bison bison*) Characterized by Urinary Pregnanediol-3-glucuronide." *Journal of Reproduction and Fertility* 93:541–47.

Kleiman, D. G., K. V. Thompson, and C. Kirk Baer, eds. 2010. *Wild Mammals in Captivity: Principles and Techniques for Zoo Management*. 2nd edition. Chicago: University of Chicago Press.

Koyama, N. F., C. Caws, and F. Aureli. 2006. "Interchange of Grooming and Agonistic Support in Chimpanzees." *International Journal of Primatology* 27(5): 1293–1309.

Lacy, R. C. 1993. "VORTEX: A Computer Simulation Model for Population Viability Analysis." *Wildlife Research* 20:45–65.

Lavin, S. R., Z. S. Chen, and S. A. Abrams. 2010. "Effect of Tannic Acid on Iron Absorption in Straw-Coloured Fruit Bats (*Eidolon helvum*)." *Zoo Biology* 29:335–43.

Lesley, B. L., and J. F. Kirkpatrick. 1991. "Monitoring Ovarian Function in Captive and Free-Ranging Wildlife by Means of Urinary and Fecal Steroids." *Journal of Zoo and Wildlife Medicine* 22:23–31.

Li, Y.Z., X. Xu, F. J. Shen, W. P. Zhang, Z. H. Zhang, R. Hou, and B. S. Yue. 2010. "Development of New Tetranucleotide Microsatellite Loci and Assessment of Genetic Variation of Giant Panda in Two Largest Panda Captive Breeding Populations." *Journal of Zoology* 282:39–46.

Lindemann-Matthies, P, and T. Kamer. 2006. "The Influence of an Interactive Educational Approach to Visitor's Learning in a Swiss Zoo." *Science Education* 90:296–315.

Mallapur, A, A. Sinha, and N. Waran. 2005. "Influence of Visitor Presence on the Behaviour of Captive Lion-Tailed Macaques (*Macaca silenus*) Housed in Indian Zoos." *Applied Animal Behaviour Science* 94:341–52.

Mazur, N. A. 2001. *After the Ark*. Melbourne: Melbourne University Press.

McDougall, P. T., D. Reale, D. Sol, and S. M. Reader. 2006. "Wildlife Conservation and Animal Temperament: Causes and Consequences of Evolutionary Change for Captive, Reintroduced, and Wild Populations." *Animal Conservation* 9(1): 39–48.

Mellen, J., and M. MacPhee. 2010. "Animal Learning and Husbandry Training for Management." In *Wild Mammals in Captivity: Principles and Techniques for Zoo Management*, edited by D. G. Kleiman, K. V. Thompson, and C. Kirk Baer. Chicago: University of Chicago Press.

Midgley, G. F., W. Thuiller, and S. I. Higgins. 2005. "Plant Species Migration as a Key Uncertainty in Predicting Future Impacts of Climate Change on Ecosystems: Progress and Challenges." In *Terrestrial Ecosystems in a Changing World*, edited by J. G. Canadell, D. E. Pataki, and P. L. F. Berlin, Heidelberg, New York: Springer.

Miller, B., W. Conway, R. P. Reading, C. Wemmer, D. Wildt, D. G. Kleiman, S. L. Monfort, A. Rabinowitz, B. Armstrong, and M. Hutchins. 2004. "Evaluating the Conservation Mission of Zoos, Aquariums, Botanical Gardens, and Natural History Museums." *Conservation Biology* 18:86–93.

Monfort, S. L., K. L. Mashburn, B. A. Brewer, and S. R. Creel. 1998. "Evaluating Adrenal Activity in African Wild Dogs (*Lycaon pictus*) by Fecal Corticosteroid Analysis." *Journal of Zoo and Wildlife Medicine* 29:129–33.

Moresco, A., L. Munson, and I. A. Gardner. 2009. "Naturally Occurring and Melengestrol Acetate-Associated Reproductive Tract Lesions in Zoo Canids." *Veterinary Pathology* 46(6): 1117–28.

Müller, D. W., L. B. Lackey, W. J. Streich, J. M. Hatt, and M. Clauss. 2010. "Relevance of Management and Feeding Regimens on Life Expectancy in Captive Deer." *American Journal of Veterinary Research* 71(3): 275–80.

Munson, L, A. Gardner, R. J. Mason, L. M. Chassy, and U. S. Seal. 2002. "Endometrial Hyperplasia and Mineralization in Zoo Felids Treated with Melengestrol Acetate Contraceptives." *Veterinary Pathology* 39:419–27.

Napier, J. E., S. H. Hinrichs, F. Lampen, P. C. Iwen, R. S. Wickert, J. L. Garrett, T. A. Aden, E. Restis, T. G. Curro, L. G. Simmons, and D. L. Armstrong. 2009. "An Outbreak of Avian Mycobacteriosis Caused by *Mycobacterium intracellulare* in Little Blue Penguins (*Eudyptula Minor*)." *Journal of Zoo and Wildlife Medicine* 40(4): 680–86.

Nemeth, N., G. Kratz, E. Edwards, J. Scherplz, R. Brown, and N. Komar. 2007. "Surveillance for West Nile Virus in Clinic-Admitted Raptors, Colorado." *Emerging Infectious Diseases* 13(2): 305–7.

Nielsen, R. K., C. Pertoldi, and V. Loeschcke. 2007. "Genetic Evaluation of the Captive Breeding Program of the Persian Wild Ass." *Journal of Zoology* 272(4): 349–57.

Osborn, F. V. 2002. "Capsicum Resin as an Elephant Repellent: Field Trials in the Communal Lands of Zimbabwe." *Journal of Wildlife Management* 66:674–77.

Osofsky, S. A, W. B. Karesh, and S. L. Deem. 2000. "Conservation Medicine: A Veterinary Perspective." *Conservation Biology* 14:336–37.

Ross, S. R., and K. L. Gillespie. 2009. "Influences on Visitor Behavior at a Modern Immersive Zoo Exhibit." *Zoo Biology* 28(5): 462–72.

Ross, S. R., A. N. Holmes, and E. V. Lonsdorf. 2009. "Interactions between Zoo-Housed Great Apes and Local Wildlife." *American Journal of Primatology* 71(6): 458–65.

Shen-Jin, L., P. A. Todd, Y. Yan, Y. Lin, F. Hongmei, and W. Wan-Hong. 2010. "The Effects of Visitor Density on Sika Deer (*Cervus nippon*) Behaviour in Zhu-Yu-Wan Park, China." *Animal Welfare* 19(1): 61–65.

Sleeman, J. M. 2008. "Use of Wildlife Rehabilitation Centres as Monitors of Ecosystem Health." In *Zoo and Wild Animal Medicine Current Therapy*, edited by M. E. Fowler and R. E. Miller. Saint Louis: Elsevier Science.

Smiley, T., L. Spelman, M. Lukasik-Braum, J. Mukherjee, G. Kaufman, D. E. Akiyoshi, and M. Cranfield. 2010. "Noninvasive Saliva Collection Techniques for Free-Ranging Mountain Gorillas and Captive Eastern Gorillas." *Journal of Zoo and Wildlife Medicine* 41(2): 201–9.

Smith, B. P., and C. A. Litchfield. 2010. "An Empirical Case Study Examining Effectiveness of Environmental Enrichment in Two Captive Australian Sea Lions (*Neophoca cinerea*)." *Journal of Applied Animal Welfare Science* 13(2): 103–22.

Stoeger-Horwath, A. S., S. Stoeger, H. M. Schwammer, and H. Kratochvil. 2007. "Call Repertoire of Infant African Elephants: First Insights into the Early Vocal Ontogeny." *Journal of the Acoustical Society of America* 121(6): 3922–31.

Swanagan, J. S. 2000. "Factors Influences Zoo Visitor's Conservation Attitudes and Behavior." *Journal of Environmental Education* 31(26–31).

Tanner, J. E., and R. W. Byrne. 2010. "Triadic and Collaborative Play by Gorillas in Social Games with Objects." *Animal Cognition* 13(4): 591–607.

Terio, K. A., S. B. Citino, and J. L. Brown. 1999. "Fecal Cortisol Metabolite Analysis for Noninvasive Monitoring of Adrenocortical Function in the Cheetah (*Acinonyx jubatus*)." *Journal of Zoo and Wildlife Medicine* 30(4): 484–91.

Terio, K.A., L. Marker, and L. Munson. 2004. "Evidence for Chronic Stress in Captive but not Free-ranging Cheetahs (*Acinonyx jubatus*) Based on Adrenal Morphology and Function." *Journal of Wildlife Diseases* 40:259–66.

Vester, B. M., A. N. Beloshapka, I. S Middelbos, S. L Burke, C. L Dikeman, L. G. Simmons, and K. S. Swanson. 2010. "Evaluation of Nutrient Digestibility and Fecal Characteristics of Exotic Felids Fed Horse- or Feed-Based Diets: Use of the Domestic cat as a Model for Exotic Felids." *Zoo Biology* 29:432–48.

Videan, E.N, and J. Fritz. 2007. "Effects of Short- and Long-Term Changes in Spatial Density on the Social Behaviour of Captive Chimpanzees (*Pan troglodytes*)." *Applied Animal Behaviour Science* 102:95–105.

Wasser, S. K., K. E. Hunt, J. L. Brown, C. Crockett, U. Bechert, J. Millspaugh, S. Larson, and S. L. Monfort. 2000. "A Generalized Fecal Glucocorticoid Assay for Use in a Diverse Array of Nondomestic Mammalian and Avian Species." *General Comparative Endocrinology* 120:260–75.

Weissenbock, N. M, H. M. Schwammer, and T. Ruf. 2009. "Estrous Synchrony in a Group of African Elephants (*Loxodonta africana*) under Human Care." *Animal Reproduction Science* 113:322–27.

Wielebnowski, N. 2003. "Stress and Distress: Evaluating Their Impact for the Well-being of Zoo Animals." *Journal of the American Veterinary Medical Association* 223:973–77.

Wielebnowski, N. C., N. Fletchall, K. Carlstead, J. M. Busso, and J. L. Brown. 2002. "Noninvasive Assessment of Adrenal Activity Associated with Husbandry and Behavioral factors in the North American clouded Leopard Population." *Zoo Biology* 21:77–98.

———. 2002. "Noninvasive Assessment of Adrenal Activity Associated with Husbandry and Behavioral Factors in the North American Clouded Leopard Population." *Zoo Biology* 21(1): 77–98.

Wildt, D. E. 1997. "Genome Resource Banking: Impact on Biotic Conservation and Society." In *Tissue Banking in Reproductive Biology*, edited by A. M. Karow and J. Critser. New York: Academic Press.

Witte, C. L., L. L. Hungerford, R. Papendick, I. H. Stalis, and B. A. Rideout. 2010. "Investigation of Factors Predicting Disease among Zoo Birds Exposed to Avian Mycobacteriosis." *Javma-Journal of the American Veterinary Medical Association* 236(2): 211–18.

Woodworth, L. M., M. E. Montgomery, D. A. Briscoe, and R. Frankham. 2002. "Rapid Genetic Deterioration in Captive Populations: Causes and Conservation Implications." *Conservation Genetics* 3:277–88.

Zelepsky, J., and T. M. Harrison. 2010. "Surveillance of Rabies Prevalence and Bite Protocols in Captive Mammals in American Zoos." *Journal of Zoo and Wildlife Medicine* 41(3): 474–79.

Zeoli, L. F., R. D. Sayler, and R. Wielgus. 2008. "Population Viability Analysis for Captive Breeding and Reintroduction of the Endangered Columbia Basin Pygmy Rabbit." *Animal Conservation* 11(6): 504–12.

Zimmermann, A., M. Hatchwell, L. A. Dickie, and C. West, eds. 2007. *Zoos in the 21st Century.* Cambridge: Cambridge University Press.

55

Cooperative Management Programs

Candice Dorsey, Debborah E. Luke, and Paul Boyle

INTRODUCTION

Accredited zoos and aquariums serve as conservation centers that are concerned about ecosystem health. They take responsibility for species survival; contribute to research, conservation, and education; and provide society with the opportunity to develop personal connections with the animals in their care. Zoo- and aquarium-managed populations, often referred to as ex situ populations, are usually small compared to their wild or in situ counterparts. Accredited zoos and aquariums provide invaluable opportunities for both in situ and ex situ research to advance scientific knowledge of the animals in their care and to help in the conservation of wild populations. They also have served as important genetic and demographic reservoirs to help reestablish threatened or endangered wild populations.

Ex situ populations must be carefully managed to ensure their long-term viability as well as their genetic health and integrity. Protocols, tools, and software have been developed for animal record keeping, collection planning, and population management that serve to facilitate exceptional animal management techniques. Small population management must take several variables into account, including animal husbandry expertise, welfare, genetics, demographics, space availability, and inter- and intranational regulations and policies. Addressing each of these variables individually is pivotal to achieving exceptional management of animals within and among zoos and aquariums. It is also vital for ensuring that healthy, genetically diverse populations are sustained so that visitors can appreciate these species for many generations to come.

This chapter will provide an overview of cooperative animal management programs. After reading this chapter, the reader will understand

- the theory and concepts surrounding cooperative regional and global animal management

- the roles and responsibilities of the people who lead and participate in animal management programs
- the history of the Association of Zoos and Aquarium's (AZA's) animal management programs
- the AZA's emphasis on incorporating long-term population sustainability into animal management administration
- the structure and functions of AZA's cooperative Animal Programs, Scientific Advisory Groups, and committees, and their respective roles within the AZA.

REGIONAL MANAGEMENT

Regional zoological associations around the world provide support and member services to a portion of the zoos and aquariums located within a particular region (table 55.1). These associations are responsible, to varying degrees, for the oversight of their member institutions, as well as for the management of their associated animal programs. A variety of animal programs have been established to enhance collaboration within and between regional zoological associations, their program leaders, and member institutions to meet animal care and management goals.

Animal management is most successful when it is carefully planned and coordinated between all holding institutions and executed with the input of all stakeholders. These types of cooperatively managed animal programs have been established in several zoo and aquarium regional associations, and their efficacy is constantly evolving and improving. Managing small populations in hopes of preserving their genetic and demographic diversity requires effective management of animal transfers between institutions, breeding success, and husbandry expertise. Without science-based, coordinated management, ex situ populations could suffer from inbreeding depression. The animals within such a population could experience decreased breeding success with fewer offspring and decreased fitness (i.e., lower survival rates and

TABLE 55.1. WAZA-member regional zoo and aquarium associations (www.waza.org).

Acronym	WAZA member association
ACOPAZOA	Colombian Association of Zoos and Aquariums
AFDPZ	Association française des parcs zoologiques
AIZA	Iberian Association of Zoos and Aquaria
ALPZA	Latin American Zoo and Aquarium Association
AMACZOOA	Mesoamerican and Caribbean Zoos and Aquaria Association
AZA	Association of Zoos and Aquariums
AZCARM	Asociacion de Zoologicos, Criaderos y Acuarios de Mexico AC
BIAZA	British and Irish Association of Zoos and Aquariums
CAZA	Canadian Association of Zoological Parks and Aquariums
DAZA	Danish Zoological Gardens and Aquaria
DTG	Deutsche Tierpark-Gesellschaft e.V.
DWV	Deutscher-Wildgehege-Verband e.V.
EARAZA	Eurasian Regional Association of Zoos and Aquariums
EAZA	European Association of Zoos and Aquaria
JAZA	Japanese Association of Zoos and Aquariums
PAAZAB	African Association of Zoos and Aquaria
SAZA-SDF	Swedish Association of Zoological Parks and Aquaria
SAZARC	South Asian Zoo Association for Regional Cooperation
SEAZA	South East Asian Zoo Association
SNDPZ	Société national des parcs zoologiques
UCSZ	Union of Czech and Slovak Zoological Gardens
UIZA	Italian Union of Zoos and Aquaria
VDZ	German Federation of Zoo Directors
ZAA	Zoo and Aquarium Association (Australasia)

less adaptability in the face of changing environments and new diseases).

STUDBOOKS

Regional studbooks document the pedigree and entire demographic history of each individual in a population of species. These collective histories are known as the population's genetic and demographic identity, and they are valuable tools for tracking and managing each individual in an ex situ population. They are compiled and maintained by a regional studbook keeper. Thousands of regional studbooks represent the animals within a single regional association. However, there are times when it is more appropriate to track managed species globally, through international studbooks. The Committee on Population Management (CPM), which reports to the World Association of Zoos and Aquariums (WAZA), oversees all of the international studbooks. These studbooks are recognized and endorsed by both WAZA and the World Conservation Union/Species Survival Commission (IUCN/SSC). International studbooks provide a valuable service to the zoological community by offering the most complete and accurate global data on the ex situ population's pedigree and demography, and frequently include husbandry, behavioral, and veterinary guidance (International Studbook Keeper Manual 2010). Through the CPM, WAZA approves new international studbooks and international studbook keepers as needed. The international studbook keeper is responsible for communicating with all of the regional studbook keepers, requesting studbook data updates and compiling these data into an international studbook, which in turn is available to all of the regional zoological associations.

TAXON ADVISORY GROUPS

The animal programs of many regional associations (e.g., North America's AZA, the European Association of Zoos and Aquaria [EAZA], and Australasia's Zoo and Aquarium Association [ZAA]) are overseen by Taxon Advisory Groups (TAGs), which examine the management and conservation needs of entire taxa, promote sustainability, develop recommendations for population management through collection plans based upon the needs of the species and member institutions, and oversee the creation of guidelines for husbandry or animal care. TAGs may also play a large role in education programs, conservation initiatives, and scientific research within a regional association. They oversee any number of specific animal programs (e.g., AZA's Species Survival Plans [SSPs] program, EAZA's European Endangered Species Programmes [EEPs], and the ZAA's Australasian Species Management Programs [ASMP]) that focus on the intensive population management of particular species among member institutions. These programs are generally led by a species coordinator who is assisted by a management committee and a regional studbook keeper. The species coordinator uses the regional studbook data to perform demographic and genetic analyses which result in a management plan that identifies the population's long-term goals and makes recommendations to ensure the sustainability of a healthy, genetically diverse, and demographically varied population. More specifically, the management plan recommends which animals should or should not breed, and which animals should be transferred to other facilities for breeding purposes.

SCIENTIFIC ADVISORY GROUPS

In addition to genetics and demographics, successful animal management programs require detailed knowledge of a species' biology. To address these needs, many zoo and aquarium associations have invested extensively in scientific research to ensure superior animal care and to develop and advance technologies surrounding managed propagation. Scientific Advisory Groups (SAGs) are designed to help coordinate, facilitate, and monitor the cooperative animal management programs in a specific discipline of interest. SAGs play a pivotal role in animal programs by acting as advisors in their specialty disciplines, which may include reproduction, contraception, behavior, and nutrition. They are encouraged to develop cooperative relationships with and glean knowledge from experts from other regional associations, appropriate scientific societies, and other expert collaborators from outside the zoo and aquarium community.

GLOBAL MANAGEMENT

For some animal program populations, management at the regional level (solely within one regional association such as AZA or EAZA) may be sufficient to achieve the program's goals. For other animal programs, however, the desired population size for maintaining optimal gene diversity is greater than the current carrying capacity (maximum available space) within one regional association. In these instances, global management provides an opportunity to combine several small and potentially unsustainable regional populations into a metapopulation, thus improving the entire captive population's genetic and demographic management potential by increasing its size and potential carrying capacity. For certain populations, these additional resources may markedly increase their long-term management success. Global Species Management Plans (GSMPs), overseen by WAZA's CPM, are composed of animal programs from two or more regional associations, which are then managed in coordination. TAGs and species coordinators from the participating regional associations communicate and collaborate with each other to develop a GSMP for their species. These global programs encourage effective collaboration at a multiregional level and ensure appropriate consultation with all stakeholders. Successful GSMPs and efficient global population management rely on current international studbook data, detailed record keeping by all regional species coordinators, a thorough understanding of species biology, reliable husbandry practices, and excellent communication among all of the stakeholders.

AZA COOPERATIVE ANIMAL MANAGEMENT PROGRAMS

AZA'S FOCUS ON POPULATION SUSTAINABILITY

In the late 1970s, a recognition that wildlife populations were declining in the wild inspired a group of visionary zoologists to create the SSP concept as a cooperative breeding and conservation program. In 1981 AZA formally established its first SSP programs, which included the Sumatran tiger and the Aruba island rattlesnake. These animal programs functioned by managing each individual of a species cared for within AZA member institutions (AZA accredited zoos and aquariums, certified related facilities, and approved non-AZA members) as a member of a single population for breeding purposes. The SSPs developed a Breeding and Transfer Plan which details "breeding" or "do-not-breed" recommendations in order to maintain demographically stable populations with the greatest possible genetic diversity for a healthy and sustainable long-term future. Population sustainability is dependent on several variables, including gene diversity, demographic stability, breeding success, level of husbandry expertise, and amount of available space within zoos or aquariums dedicated to the species.

In 1994, AZA published *Species Survival Plans: Strategies for Wildlife Conservation*, which stated: "The SSP Program was originally conceived to provide a blueprint for cooperative captive breeding programs in North America, but more recently the concept has also evolved to include field conservation efforts." The AZA Conservation Department and the AZA Wildlife Conservation Management Committee (WCMC) initiated a variety of processes designed to sustain zoo and aquarium populations and wild species. Population Management Plans (PMP) were established to genetically and demographically manage additional zoo and aquarium species that did not require the intense management parameters of SSP programs, and thus the mandatory participation of AZA member institutions. The TAGs became responsible for creating and maintaining Regional Collection Plans (RCPs), which recommended that species for cooperative management within AZA member institutions be given finite space and resource availability.

In 2000, the AZA Population Management Center (PMC), hosted by the Lincoln Park Zoo in Chicago, became responsible for using the data derived from studbooks and RCPs to identify science-based breeding recommendations within each SSP and PMP program. The AZA Wildlife Contraception Center (WCC), hosted by the St. Louis Zoo, became responsible for providing information on safe, effective, and reversible contraceptive products to the AZA community and for helping zoo professionals make informed decisions on how to manage their animal populations.

A 2009 publication entitled *Sustaining the Ark: The Challenges Faced by Zoos in Maintaining Viable Populations* (Lees and Wilckens 2009), highlighted a concern that many have increasingly recognized over the past several years. "Over the last decade . . . ark-related activity [i.e., maintaining sustainable ex situ populations] has declined as zoos have diversified their conservation activities, re-directing efforts into other areas, such as conservation education, fund-raising and other support for in situ projects. . . . Zoo populations are not achieving the conditions for sustainability." The declining sustainability of zoo populations likely results from a variety of factors including insufficient animal holding and breeding space, low breeding success, need for more advanced husbandry techniques, or, occasionally, lack of success in completing breeding recommendations. The AZA board of directors quickly formed its Task Force on the Sustainability of Zoo-Based Populations.

The task force obtained input from the diverse conservation community as it reviewed the mission, goals, and limits of AZA's cooperatively managed animal programs. It determined which factors were having the greatest impact on the sustainability of zoo populations, and it assessed which additional resources were most needed to successfully sustain zoo-based or aquarium-based populations. The results of the task force's research and analyses include a restructured management scheme that emphasizes long-term population sustainability, and improvements in succession planning and training. The new management scheme emphasizes those population parameters most affecting long-term sustainability as the criteria used to designate animal programs.

AZA ANIMAL PROGRAMS

In 2011 AZA's new board-approved animal program management strategy was implemented, eliminating PMP programs as a management type. The cooperatively managed animal programs that have resulted from this change continue to

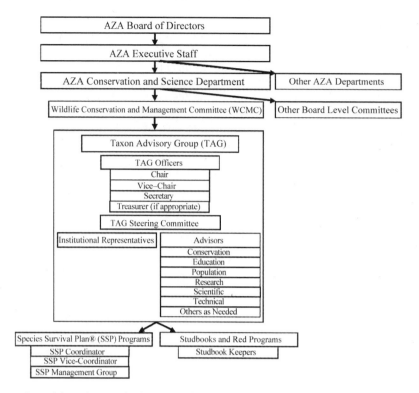

Figure 55.1. Structure of oversight and communication among the AZA Conservation and Science Department and its animal programs. Illustration courtesy of AZA.

ensure that each species population is managed to maximize its genetic potential, and that each individual animal within the population is cared for in the best possible way. The administrative structure of AZA animal programs has grown and evolved over the decades.

AZA program leaders are dedicated employees of AZA member institutions who volunteer their time to oversee the conservation and management of an ex situ population within the region, whether of an entire taxon (as in the case of a TAG chair) or a specific species (as in the case of SSP coordinators or studbook keepers). An AZA animal program may be managed by a program leader who works independently for smaller populations, or by a program leader and a management group for larger populations. These program leaders and management groups coordinate with up to hundreds of active stakeholders who cooperate to achieve the program's goals.

The program leader communicates all program-related matters to participating institutions via institutional representatives (IRs). Each zoo or aquarium participating in the program designates an IR to act as the primary contact between that institution and the program leader. The IR's primary responsibilities include communicating with and disseminating information between the animal program and the zoo or aquarium's director and staff. This open communication is fundamental to successful animal management. The IR responds to critical inquiries from the program leader during the population planning process, reviews draft and final TAG RCPs and SSP breeding and transfer Plans, and communicates the zoo institution's comments and concerns to the program leader.

Supporting and advising these animal programs are SAGs, AZA board-level committees, and the AZA conservation department, all of which collectively offer guidance to AZA animal programs and their management, research, welfare, and conservation initiatives. All animal programs are managed by the WCMC (figure 55.1).

TAXON ADVISORY GROUPS (TAGS)

In 2011 there were 46 AZA TAGs. Each TAG is led by a chair, a vice-chair, and a steering committee and is supported by several expert advisors from within and outside the zoological community. The officers and steering committee members are elected from the TAG's group of IRs, and they form the voting body for the animal program. A TAG's primary responsibilities include the development of an RCP and appropriate animal care manuals, oversight of the SSP and studbook programs within its purview, and establishment of management, research, and conservation priorities (table 55.2).

REGIONAL COLLECTION PLANS (RCPS)

A TAG's RCP identifies a list of species recommended for cooperative management among AZA zoos and aquariums (i.e., studbooks and SSP programs) and, with the help of the PMC and other approved population advisors, designates the manner in which each population should be managed, based upon the needs of the species and those of AZA's member institutions. The RCP explains how each recommendation was developed, defines the long-term goals of each animal program, describes the available space for each population,

TABLE 55.2. AZA Taxon Advisory Groups, and examples of Species Survival Plan programs within their purview

Taxon Advisory Group (TAG)	Sample SSP Programs
Amphibian	Puerto Rican crested toad, Houston toad
Anseriformes	Marbled teal, white-winged wood duck
Antelope and Giraffe	Addax, okapi
Ape	Chimpanzee, lowland gorilla
Aquatic Invertebrate	—
Bat	Large flying fox, Rodrigues fruit bat
Bear	Polar bear, sloth bear
Bison, Buffalo, Cattle	Lowland anoa
Canid and Hyaenid	African wild dog, swift fox
Caprid	Sichuan takin, Tadjik markhor
Charadriiformes	Atlantic puffin, Inca tern
Chelonian	Coahuilan box turtle, radiated tortoise
Ciconiiformes/ Phoenicopteriformes	African spoonbill, Chilean flamingo
Columbiformes	Marianas fruit dove, Nicobar pigeon
Coraciiformes	Laughing kookaburra, rhinoceros hornbill
Crocodilian TAG	Chinese alligator
Deer (Cervid/Tragulid)	Barbary stag, white-lipped deer
Elephant	African elephant, Asian elephant
Equid	Asian wild horse, Grevy's zebra
Felid	African lion, caracal
Freshwater Fishes	Lake Victoria cichlid, river stingray
Galliformes	Attwater's prairie chicken, Malayan argus
Gruiformes	Kori bustard, red-crowned crane
Lizard	Chinese crocodile lizard, Komodo dragon
Marine Fishes	Lined seahorse, zebra shark
Marine Mammal	Harbor seal, bottlenosed dolphin
Marsupial and Monotreme	Brush-tailed bettong, Parma wallaby
New World Primate	Golden lion tamarin, white-faced saki
Old World Monkey	Japanese macaque, mandrill
PACCT* (Passerines)	Bali mynah, tawny frogmouth
Pangolin, Aardvark, Xenarthra	Linne's two-toed sloth, giant anteater
Parrot	Hyacinth macaw, palm cockatoo
Pelecaniformes	—
Penguin	African penguin, chinstrap penguin
Piciformes	Green aracari, Toco toucan
Prosimian	Coquerel's sifaka, ring-tailed lemur
Raptor	Burrowing owl, California condor
Ratite/Tinamiformes	Elegant crested tinamou, greater rhea
Rhinoceros	White rhinoceros, Indian rhinoceros
Rodent, Insectivore, Lagomorph	Capybara, Prevost's squirrel
Small Carnivore	Black-footed ferret, red panda
Snake	Aruba island rattlesnake, bushmaster
Tapir	Malayan tapir
Terrestrial Invertebrate	American burying beetle, Partula snail
Turaco/Cuckoo	Greater roadrunner, Lady Ross' turaco
Wild Pig, Peccary, Hippo	Chacoan peccary, river hippopotamus

*Passeriformes, apodiformes, coliiformes, caprimulgiformes, and trogoniformes

and evaluates how much additional space is needed for the population to achieve its goals. RCPs also assist AZA member institutions during their institutional collection planning processes by identifying which species require more resources (e.g., space, optimal social structure) and detailing species' projected population sustainability.

Since 2011, the RCP has identified the management level designated to each recommended species, on the basis of the population's sustainability criteria, which include its size and its sustainability score. A population's size determines whether the recommended species is managed as an SSP program and therefore requires a regional studbook and formal population planning. Populations with 50 or more individuals are automatically designated as SSP programs. Populations with fewer than 50 individuals are automatically designated as red programs, which are not SSP programs and are managed only through AZA regional studbooks.

Applying the sustainability score to an SSP program determines whether the population is designated as a green or yellow SSP program. In general, the criterion for defining a population's sustainability score is the projected gene diversity (% GD) at 100 years or 10 generations. Green SSP programs are populations projected to retain at least 90.0% GD for at least 100 years, or 10 generations. Yellow SSP programs are projected to retain less than 90.0% GD at 100 years or 10 generations, and therefore may require additional resources and/or alternate management strategies to increase their sustainability. Some TAGs, however, may find it more appropriate to use other criteria (e.g., reproductive rates or population size) to develop the sustainability scores for certain taxa (e.g., species that live in herds, flocks, and harems). An SSP program's designation may change in accordance with the population becoming more or less sustainable over the course of time.

SPECIES SURVIVAL PLAN PROGRAMS

There are currently more than 250 AZA SSP programs; however the number of animal programs changes throughout the years as the needs of species and institutions evolve and population sustainabilities change. Every SSP program is led by a SSP coordinator, a vice coordinator, a studbook keeper, and, if deemed appropriate, a management group. SSP coordinators and studbook keepers are selected by the overseeing TAG's steering committee, and the management group members are elected from the SSP program's IRs. The SSP program may also be supported by several expert advisors from within and outside the zoological community. The primary function of an SSP program is to use its studbook to develop a Breeding and Transfer Plan in coordination with the PMC, or an approved population advisor.

All SSP programs are required to go through formal population planning processes at least once every three years—although many SSP programs do so more often, as they deem appropriate for their species' life history traits. The SSP coordinator, the population advisor, and IRs from each participating institution work together to develop the SSP program's Breeding and Transfer Plans. Each breeding and transfer plan summarizes the current demographic and genetic status of an ex situ population, and identifies breeding or nonbreeding recommendations with consideration given to each animal's social and biological needs, as well as the feasibility of moving each animal to another facility in a practical, safe, and healthy way. All recommendations are designed to maintain or increase a healthy, genetically diverse, and demographically stable population.

The AZA manages its animal programs in three distinct management levels: green , yellow, and red. All AZA member institutions and animal programs, regardless of management designation, must adhere to the AZA acquisition and disposition policy and code of professional ethics. Green SSP programs must also adhere to the AZA policy for full participation in the SSP program and approved nonmember participant policies.

GREEN SSP PROGRAMS

Populations that are presently sustainable, and which can thus retain 90% gene diversity (GD) for at least 100 years or 10 generations, are designated as green SSP programs. Consistent, collaborative management is critical to maintaining the long-term sustainability of green SSP programs. Therefore, these programs must adhere to the AZA Policy for Full Participation in the SSP Program, used to ensure that AZA-accredited institutions have input into the SSP planning process, and that they fully comprehend, agree to, and follow the final Breeding and Transfer Plan recommendations.

Some Green SSP programs benefit from partnering with facilities outside the AZA community, such as international zoos and aquariums that are members of other regional associations, US zoos and aquariums that are not AZA-accredited, animal ranches, and private individuals. These non-AZA entities may offer much-needed resources such as space, founder animals, or husbandry expertise that may increase the particular animal program's population sustainability, but they must apply and become approved nonmember participants (through the WCMC) to participate in Green SSP programs.

YELLOW SSP PROGRAMS

Populations that currently have 50 or more individuals but cannot retain 90% GD for 100 years or 10 generations are designated as yellow SSP programs. These programs manage populations that require additional considerations, resources, and flexible efforts to increase their sustainability. Yellow SSP program sustainability may be affected by too few individuals, too little space, low gene diversity, poor demographics, and reduced levels of husbandry and/or breeding expertise and predictability. Due to these issues, Yellow SSP programs have more flexibility in their administration and may be more creative in their management strategies. Yellow SSP programs' adherence to AZA's policy for full participation in the SSP program is voluntary, and they can form partnerships with non-AZA members without going through the formal AZA nonmember application and approval process. However, cooperation is still fundamental to management success, and each institution is encouraged to participate in and focus its attention on these yellow SSP programs to increase their populations' sustainability.

RED PROGRAMS

Populations recommended for management by the TAG that have fewer than 50 individuals are not designated as SSP program on the basis of collection sustainability criteria, and are thus instead designated Red Programs. Red programs maintain official AZA regional studbooks and, while population planning is encouraged if resources allow, formal planning on a regularly scheduled basis is not required.

The Red Program's adherence to AZA's policy for full participation in the SSP program is voluntary, and red programs can form partnerships with non-AZA members without going through the AZA nonmember application and approval process. Red program designation may serve as a strong call to action, indicating that the population is in critical need of additional assistance. Some start-up programs (e.g., importations, rescue populations) may also fall into this designation. It is important to note that AZA animal program designations can change as the appropriate sustainability criteria are reached, and that this may function as an incentive for program leaders to increase their populations' long-term sustainability. For example, if a red program succeeds in raising its population size to 50 or more individuals, it can be considered for designation as a Yellow or Green program.

AZA SCIENTIFIC ADVISORY GROUPS

Successful AZA animal programs require detailed knowledge of a species' ecology, biology, genetics, behavior, nutrition, and diseases. For this reason, AZA-accredited zoos and aquariums have invested extensively in scientific research to develop effective methods of managed propagation and animal care, as well as to gain knowledge or develop technology that has direct applications towards the conservation of in situ populations.

SAGs are composed of experts in their field: typically scientists, veterinarians, animal care staff, and researchers. SAGs also help to coordinate, facilitate, and monitor the cooperative scientific programs among AZA's conservation and science committees. AZA animal programs and related committees benefit from the cooperative relationships that SAGs develop with appropriate scientific societies and expert collaborators, as well as from the technical advice they may provide. SAG members frequently serve as advisors to animal programs and assist in writing and reviewing their chapters in the animal program's animal care manuals. AZA SAGs include the following groups.

- The AZA Avian SAG effectively represents the 16 AZA avian TAGs. Many of the issues concerning husbandry, management, and conservation pertain to multiple bird taxa. This group works cohesively to address these issues through collaborative research and management initiatives.
- The AZA Behavior SAG assists AZA members in addressing animal management issues that involve behavior or husbandry, and in identifying and encouraging the development of animal welfare and enrichment studies. This group provides information

and guidance to enhance behavioral monitoring and record keeping techniques, as well as expertise and assistance in animal husbandry and research training.

- The AZA Biomaterials Banking SAG develops policies and protocols relating to the collection, storage, transfer, and distribution of biomaterials (e.g., tissues, blood, and semen) within AZA-accredited zoos and aquariums. This may also include reviewing criteria for the selection of partners such as tissue repositories, natural history museums, and researchers.
- The AZA Contraception SAG organizes the comprehensive pathologic examinations on reproductive tracts, initiates and coordinates research trials of new contraception methods, monitors research trials, provides data to the US Food and Drug Administration (FDA), and works with commercial partners to make contraceptives available and affordable.
- The AZA Endocrinology SAG provides endocrine-related advice to institutions to ensure superior health, reproductive management, and welfare; to support the development of assisted reproductive technologies; to monitor adrenal, thyroid, and other endocrine functions; and to help zoos and aquariums assess reproductive status, diagnose and monitor pregnancy, and treat infertility in their animals.
- The AZA Institutional Data Management SAG promotes effective data management including data collection, storage, analysis, and retrieval. The group includes data managers, registrars, population biologists, and curators.
- The AZA Nutrition SAG provides advice to AZA institutions and animal programs on species' nutrition and dietary needs. This includes developing species-specific diets and supplements for infants, juveniles, pregnant females, and geriatric animals as well as evaluating pellet and browse components.
- The AZA Reproduction SAG encourages integrative research and multidisciplinary studies that encompass and address wildlife reproductive management problems. It also works with animal programs to identify and prioritize research needs and advise them on appropriate reproductive technologies. This SAG works closely with the AZA Biomaterials Banking SAG on developing policies and procedures for the storage of gametes.
- The AZA Small Population Management SAG is composed of population biologists who strive to advance the science of applied small population biology, to develop tools for use by approved population advisors, to provide training for population managers in North America and other regions, and to advise AZA animal programs on the genetic and demographic management of their species.
- The AZA Veterinary SAG is comprised of veterinarians who work to identify and review medical topics for specific species and taxa and to recommend preventative health monitoring, surveillance programs, diagnostic evaluation tools, and procedures for AZA members.

AZA BOARD-APPOINTED COMMITTEES

AZA animal programs are advised, guided, and supported by several board-appointed committees. Each committee represents a diverse subset of AZA's membership that works together to address the association's mission and management priorities.

The WCMC. The WCMC facilitates the professional, cooperative, and scientific management of AZA-accredited institutions' animal populations. The committee develops, oversees, promotes, evaluates, and supports AZA's animal programs and their conservation and scientific initiatives. The committee is responsible for the formulation and communication of the various guidelines and protocols essential to all AZA animal programs and serves as a conflict mediation and reconciliation body for issues arising within those programs when the parties directly involved cannot resolve their concerns. The WCMC also acts as the approving body for TAG RCPs, program leader appointments, and non-AZA partnerships in Green SSP programs.

The Animal Health Committee. The Animal Health Committee (AHC) ensures that AZA institutions have the highest-quality and most comprehensive animal health care, including disease management and veterinary medical care. This committee drafts and reviews health-related regulatory and legislative guidelines and policies and develops protocols, guidelines, and recommendations to help AZA animal programs and member institutions respond to emerging disease issues. The AHC works with the AZA Accreditation Commission to develop and review health-related accreditation standards, and acts as a liaison between AZA, the American Association of Zoo Veterinarians (AAZV), and animal health regulatory agencies such as US Department of Agriculture (USDA) and Centers for Disease Control (CDC).

The Animal Welfare Committee The Animal Welfare Committee (AWC) ensures that the welfare of animals is a central tenet of all AZA animal programs by continuously defining and increasing the common understanding of animal welfare, identifying and encouraging the development of animal welfare research projects and assessment tools, educating and engaging AZA member institutions in the use and application of welfare assessment tools, proactively identifying and addressing internal and external animal welfare issues, and understanding and influencing public perceptions about animal welfare in AZA-accredited zoos and aquariums. The AWC played a guiding role in the conception of AZA animal care manuals, and continues to manage their development.

The Research and Technology Committee. The Research and Technology Committee (RTC) advances and promotes the use of science and technology throughout the zoo and aquarium profession, builds relationships with the scientific community, and facilitates collaboration to identify the best scientific practices to improve wildlife survival. The RTC works closely with AZA member institutions, committees, SAGs, and animal Programs to facilitate coordinated approaches

Box 55.1. European Association of Zoos and Aquaria (EAZA) cooperative breeding programmes, European Endangered Species Programmes (EEP) and European studbooks (ESB)

Danny de Man

The animals, are of course, the absolute key element in each zoo or aquarium. Modern and well-managed zoos and aquariums select the species they keep very carefully and for specific reasons. EAZA member institutions have 41 established Taxon Advisory Groups (TAGs) covering all the species of animals that are held in zoos and aquariums, from aquatic invertebrates to sheep and goats. Indeed, one of the TAGs' main tasks is to develop Regional Collection Plans that describe which species are recommended to be kept, why they should be kept, and how they should be managed. The Regional Collection Plans also identify which species need to be managed in European Endangered Species Programmes (EEPs) and European studbooks (ESBs).

Like TAGs in other regions, the EAZA TAGs consider a variety of criteria before deciding on the RCP recommendation of a certain species. These include space, IUCN Red List status and (ex situ) conservation needs, educational value, husbandry expertise, demographic and genetic parameters of the current population in EAZA, research needs, and priorities set by other regional zoo associations. Ex situ conservation of species native to Europe has priority in the collection planning process. EAZA's breeding programs generally strive to maintain 90% of genetic diversity in a period of 100 years. At the moment, however, EAZA TAGs are committed to setting more program-specific goals for each EEP and ESB. These goals will give a clearer picture of what each program aims to achieve, what potential problems need to be solved, what actions need to be done, and so on. This, furthermore, enables evaluation of each breeding program against its goals, and the monitoring of its progress.

The EAZA EEP Committee oversees the breeding program structure within EAZA (41 TAGs, 178 EEPs, 178 ESBs). Around 75% of the species in EAZA breeding programs are listed as threatened (vulnerable [VU], endangered [EN], or critically endangered [CR]) on the IUCN Red List (2010). EAZA breeding programs receive support and advice from various committees and working groups, such as the EAZA Population Management Advisory Group, the EAZA Research Committee, the EAZA Nutrition Working Group, the EAZA Conservation Committee, the EAZA Veterinary Committee, the Transport Working Group, and the EAZA Legislation Committee. EEPs are evaluated once every five years by the EEP committee. TAGs are responsible for the process of evaluating each program. The EEP's coordinator, species committee, participants, and TAG and the EAZA executive office all provide input to the evaluation process through the use of standardized questionnaires. The TAG compiles the questionnaires and prepares a summary report that includes an overall assessment for discussion by the EEP Committee. Normally the EEP Committee follows the advice of the TAG and ends the evaluation; but sometimes additional requirements are made to improve the program's functioning, or support is offered to help solve problems.

The collection, coordination, and conservation department of the EAZA Executive Office, based in Amsterdam, also supports and assists the TAGs, EEPs, and ESBs as well as the EEP Committee. One recent development is ensuring that information on EAZA's breeding programs (e.g. husbandry guidelines, studbooks, and annual reports) are available on the EAZA website (www.eaza.net). Together with ISIS, the EAZA Executive Office has worked on updating and improving the quality of Single Population Analysis and Records Keeping System (SPARKS) datasets in preparation for the Zoological Information Management System ZIMS.

The TAG chairs, EEP coordinators, and ESB keepers are based at EAZA member institutions.

---

and collaborative advancement of research, science, and technology in zoos and aquariums.

The Animal Data Information System Committee. The Animal Data Information System Committee (ADISC) works with AZA member institutions, other regional zoological associations, and ISIS to promote the development of initiatives to improve the collection, storage, analysis, retrieval, and use of animal records data in zoos and aquariums.

SUMMARY

Effective cooperative management of animal programs ensures that healthy, diverse populations are available for zoo and aquarium visitors to appreciate and personally connect with for many generations to come. Successful animal program management requires cooperation and collaboration among numerous stakeholders. As zoos and aquariums focus more on maintaining healthy, self-sustaining populations for the long term, we must turn our focus toward global management and fostering those relationships among regional zoological associations, while maintaining the utmost standards in animal care.

REFERENCES

———. 2011. *Taxon Advisory Group (TAG) Handbook.* Silver Spring, MD: Association of Zoos and Aquariums,.

Lees, C. M., and J. Wilcken. 2009. Sustaining the ark: The challenges faced by zoos in maintaining viable populations. *International Zoo Yearbook* 43:6–18.

Wiese, Robert J., and Michael Hutchins. 1994. *Species Survival Plans: Strategies for Wildlife Conservation.* Silver Spring, MD: American Zoo and Aquarium Association.

World Association of Zoos and Aquariums, Committee on Population Management. 2010. Resource Manual for International Studbook Keepers. Accessed online.

56

Going "Green" in the Workplace

Beth Posta and Michelle E. S. Parker

WHAT DOES "GREEN" MEAN?

The trend to be "green" has been growing for many years. For many, being green means recycling or turning off lights when leaving a room, whereas for others it means much more, from energy and water conservation to green building. Today, earth-friendly practices have become more prominent in many communities and have become integrated in mainstream business practices. What started as a recycling movement has grown into a culture of the three R's: reduce, reuse, and recycle; although now, the phrase has evolved into reduce, reuse, *and then* recycle. While many people begin their green activities by recycling, they should look to first reduce their consumption of resources. What is used should be reused as much as possible before finally being recycled.

This chapter will provide the reader with insight into the many green practices available to today's zoos and aquariums. After studying this chapter, the reader will have an understanding of

- the importance of living sustainably and examples of sustainable practices
- types of green energy and the uses of each to limit CO_2 emissions
- methods of reducing the negative impact of humans on the planet through daily practices such as cleaning, purchasing, and waste removal
- techniques for developing green teams and green practices within zoos and aquariums.

THE MANY SHADES OF GREEN

There are various degrees of green practices, often based on several parameters. New research is continuously leading to improved green technology. An examination of recycling programs in zoos and aquariums reveals that these programs have grown from recycling aluminum cans to include everything from paper and cardboard to light bulbs, batteries, and cooking oil. When designing new animal exhibits and human spaces, zoos and aquariums are now faced with making decisions about low-flow water faucets and toilets, recycled carpet, and walls made from compressed straw. Deciding which choices are most appropriate can be a daunting task, as there is an often overwhelming amount of information on green practices. Conversely, while green technologies improve, other, less environmentally-friendly practices also continue to spread. For example, in many developing nations, where residents are learning new ways to earn a living and improve their way of life, natural habitat is being cut down to build roads and houses and to extend electricity to outlying areas. However, these technologies come at an environmental cost of habitat loss and higher use of coal, water, and other resources. In many nations, the use of petroleum-based products has led to a great reliance on oil for plastics, synthetic fibers, food coloring, and many cosmetics. Keepers can help reduce human dependence on these resources by limiting their use of plastics and recycling those they do use, by ordering supplies that come responsibly packaged, by turning off lights, and by using vehicles wisely. For example, when transporting animals or equipment or simply using a vehicle to travel to other areas of the zoo, it is wise to combine trips and not leave vehicles idling, which uses excess gasoline and sends toxic emissions into the air. With energy consumption rising considerably throughout the world, the use of resources has increased, and with it emissions and other pollutants. Therefore, the choice to be green comes with a great number of decisions. Many of these choices are based on current practices and require little forethought, whereas others may be dependent on a number of factors such as personnel resources, knowledge of green practices, staff commitment, and financial resources. Each of these considerations will play a role in the degree to which individuals and organizations commit to green practices.

The decision to "go green" involves input from all levels of the organization, from keepers to the zoo director to the board of directors. Each department will have a different perspective on how to promote environmental conservation.

Box 56.1. Going "Green" in the Workplace

Being green means different things to different people:

- reducing, reusing, recycling
- carpooling
- turning off lights
- composting
- reducing junk mail and paper use
- decreasing water and electricity use
- using environmentally friendly or sustainable materials
- reducing use of chemicals
- sustainability
- green building
- energy conservation and use of alternative (solar, wind, or geothermal) energy
- storm water management.

For example, horticulturists might choose to use a rain barrel water collection system, or to harness sources of grey water, such as water rerouted from sinks and washing machines for irrigation, while the grounds crew might choose to rake leaves instead of using a gas-powered leaf blower. The public relations staff might choose to e-mail its press releases rather than send paper facsimiles, while the membership department might e-mail its membership renewals and updates. So where do keepers fit in? Keepers can find ways to minimize their water use when cleaning exhibits, and can help design those exhibits with conservation in mind. They can also reuse many household items as animal enrichment, use green cleaners to clean the exhibits, and recycle not only the plastic cleaner bottles and cardboard boxes but also the baling wire that holds straw and hay bales together. In response to habitat destruction and the possibility of orangutans becoming extinct, keepers at several zoos have led initiatives to help save the rain forests by using only products containing certified orangutan-friendly palm oil.

PERSONNEL RESOURCES

Many zoos and aquariums today have staff members within their conservation departments who assist with green practices. Some zoos have obtained green certifications, such as the International Organization for Standardization's ISO14001, which is designed to identify and control the environmental impact of the organization's activities, products, and services (see www.iso.org). Others have adopted green policies and constructed green buildings. More common within zoos, however, is the formation of a green team comprised of representatives from various animal and nonanimal departments who are interested in advancing an institution's green practices and commitment to environmental conservation. The number of staff members who participate and the tasks assigned to them can have a profound effect on the zoo or aquarium's green progress. Staffers at facilities that include green duties in their job descriptions may have a greater ability to advance green practices where other facilities may depend more on gaining individual staff member support (buy-in).

KNOWLEDGE OF GREEN PRACTICES

With different job positions and duties come a variety of knowledge and expertise. For example, employees in a construction department will likely be well versed in green building and exhibit design, but they may have less expertise and knowledge of other issues such as the palm oil crisis in Asia. Collectively, the green team will likely have a broad range of knowledge based on the skills and expertise of its members. When developing a green team it is beneficial to include representatives from several diverse departments and to include both management and nonmanagement staff. Zoo volunteers typically come from a wide range of backgrounds and can be valuable resources to the green team as well.

STAFF COMMITMENT

The environmental commitment of an organization will be influenced by several factors including funding sources, the availability of staff to research green practices, local support of environmental initiatives, and especially staff acceptance of and enthusiasm for the program. As zoos and aquariums present themselves as conservation organizations, they must remember that wildlife conservation begins in their own backyards. For some, gaining staff support is fairly simple, yet for others it seems like an uphill battle. Green practices can take hold from the bottom up, as generated by front-line staff, or from the top down, from upper management. The most effective method of gaining the staff's participation, however, is to find a compelling reason for them to go green, whether it be a financial incentive or the conservation of a species the staff is passionate about. The palm oil crisis in Borneo and Sumatra, where the rain forest is being cleared to make room for palm oil plantations (figure 56.1), has moved several zoo green teams to reduce their use of palm oil and purchase products containing only certified orangutan-friendly palm oil. Palm oil is found in many foods, such as candy, cookies and crackers, as well as in lotions, shampoos, and other personal products. The clear cutting and burning of the rain forest to make way for palm oil plantations is destroying critical habitat for orangutans and other species, such that orangutans may be extinct by 2020 if nothing is done to counter the habitat loss. The Roundtable on Sustainable Palm Oil (RSPO, 2010), an international organization that promotes sustainable palm oil farming and global standards for palm oil products, is working to develop a certification process for sustainable palm oil. In the meantime, a number of zoos have agreed to eliminate palm oil from their gift shop items and concessions, and many in the United States have committed to giving out palm oil–free candy during their Halloween trick-or-treat events.

FINANCIAL RESOURCES

In the past, the cost of being green was often considerably higher than using conventional products and resources that were more established. However, many green items have decreased in cost as they become more mainstream and technology improves. For example, the cost of bamboo flooring can be less expensive than a less environmentally friendly hardwood counterpart, because bamboo can be harvested for

Figure 56.1. In countries such as Borneo, much of the rainforest has been clear cut to make way for palm oil plantations. Courtesy of Cheyenne Mountain Zoo.

flooring at three to four years of age while a hardwood tree must mature 100 years to reach the size that can be used for flooring. However, when deciding on which green practices to employ, it is important to look at the initial cost as well as the lifecycle cost—that is, the cost of the item throughout its entire life. Some green technologies, such as wind energy, solar panels, and geothermal heating and cooling, can be very expensive up front but cost-saving over time. Because it often takes years to realize the cost savings, they must be planned and incorporated into the budget. Careful consideration is needed when deciding upon green practices to ensure that the institution can accommodate the initial costs in its operating budget and that the net outcome over time will be worth the investment. For example, solar energy during the winter may not be as efficient as during the longer summer days. Wind energy is effective only when there is ample wind. The institution should research each technology and ensure that it is appropriate for the region and make sure to have contingency plans in place for green practices that rely on environmental conditions.

SUSTAINABILITY

There is an old Native American saying: "Treat the earth well; we do not inherit the Earth from our ancestors; we borrow it from our children." Sustainability is "the ability to meet the needs of the present without compromising the ability of future generations to meet their own needs" (United Nations World Commission on Environment and Development 1987), including those related to biodiversity, habitat preservation,

and natural resources. Sustainable living can help ensure that future generations live in a world that is as healthy as it is today. This can be challenging, given the environmental consequences of daily human life. The rise of greenhouse gases in our atmosphere from burning fossil fuels, such as gasoline in automobiles or coal in power plants, has contributed to global climate change. The changing climate is causing such events as the rapid melting of glaciers and polar ice, which threatens polar bear populations. The resulting rise in sea levels, another climate change issue, threatens coastlines and the wildlife and human life associated with them. Carbon dioxide (CO_2) is readily absorbed by the oceans, and the increase of CO_2 is causing the ocean pH to decrease and the oceans to become more acidic (ocean acidification). This decrease in pH leads to a higher level of bicarbonate in the water, thus negatively affecting the growth of stony corals and shellfish, which are key to the marine ecosystem and food chain. Combined with pollution of our waterways—such as oil, garbage, and agricultural runoff, as well as air and land pollution—these challenges create a need for people to take action today and in the future to ensure that healthy resources remain for future generations of all life to not only survive but thrive.

WHY IS SUSTAINABILITY IMPORTANT?

The United States currently uses approximately 25% of the planet's energy resources, but houses only 5% of the world's population. As with other resources, there are a number of actions people can take to ensure a healthy planet for future generations. For example, according to the US Environmental

Protection Agency, if each home in the US replaced one incandescent light bulb with a compact fluorescent lamp (CFL), enough energy could be saved to light three million homes for a year and save the same amount of greenhouse gases from entering the atmosphere as would be produced by 800,000 cars on the road (EPA 2010). This practice would help reduce the need for coal and thus reduce mercury emissions created by coal-fired power plants, which generate most of the mercury emissions in the air. Mercury is a neurotoxin that can be especially dangerous to children. It takes 13.6 mg of mercury to power an incandescent lightbulb, but only 3.3 mg of mercury to power a CFL for the same amount of time (EPA 2010). By reducing the use of electricity with CFLs, humans can decrease their reliance on coal and the amount of mercury in the environment. However, it is important that CFLs be disposed of properly and recycled rather than being thrown in the garbage where they can break, releasing mercury, into the environment. Many stores that sell CFLs will take old ones back for proper disposal.

There are a number of other actions one can take, such as turning a car off when stopped rather than letting it idle. In fact, leaving a car to idle for 10 minutes per day can add more than one-half kilogram of toxic gases into the air as well as waste almost $200 per year in gasoline (Clean Air Partnership 2010). Making small changes can have a large effect for future generations. Several zoos have anti-idling policies, and a number of cities and local governments have banned vehicle idling.

THE POLAR BEAR CRISIS

The polar bear was listed as threatened under the US Endangered Species Act in 2008. While many species on the list are endangered due to habitat destruction or other specific human activities, the polar bear is in danger due to climate change. As the earth's temperature has warmed, the polar ice cap has been melting, creating more distance between arctic ice floes. As a result, polar bears are unable to swim the long distances between floes and are dying from drowning and starvation. If the world does not take measures to reduce the emission of greenhouse gases, polar bear populations could disappear completely, leaving a void in the natural ecology. Zoos and aquariums are well poised to help spread the message of the effects of climate change on our ecosystems. Keepers are especially well positioned to educate zoo visitors on the impacts of daily living on our environment. Many zoos offer keeper talks aimed to educate and inspire visitors to adopt positive habits that help preserve the environment and make it safer and healthier for both humans and animals. Keeper-initiated programs have been successful in promoting a better understanding of our impact on wildlife. For example, campaigns that encourage people to turn off lights when leaving a room or to switch from incandescent light bulbs to CFLs can be linked to energy conservation and climate change initiatives.

HOW CAN WE HELP?

Zoo staff can reduce their impact on the environment by reducing their use of gasoline powered equipment, such as lawn mowers, trimmers, and leaf blowers; by purchasing items that are packaged responsibly; and by reusing items rather than sending them to landfills. There are a number of resources for buying, selling, and giving away used items. Many cities now have free online classifieds or groups where items can be offered for free or requested from members. These are useful ways to keep unwanted items out of the landfills and reduce the need for new construction of items that use physical resources and require energy to construct and ship. Buying local can also have a large impact on the environment. Most local produce is picked four to seven days prior to shipping, whereas food from farther away must be picked earlier and shipped on average 937 km (1,500 miles) to its destination (Pirig and Benjamin 2009). Choosing locally grown or locally produced foods for humans and animals ensures fresher food, reduces the need for shipping, and supports local businesses. In addition, when items are purchased from locally owned businesses, more money stays within the local economy rather than going into the pockets of larger corporations.

CARBON FOOTPRINTS

A carbon footprint is the sum total of all the greenhouse gases emitted by an individual, organization, or building. Greenhouse gases are those that trap heat within the atmosphere, including carbon dioxide, methane, and nitrous oxide. These gases are emitted through different means of energy use, including gasoline-powered vehicles, agriculture, manufacturing, and many other activities. Humans are responsible for producing greenhouse gases that contribute to global warming, or our carbon footprint. Numerous online carbon footprint calculators are available with which one can determine one's impact on the environment (see www.epa.gov for one example). Carbon footprints can be decreased through responsible energy use such as using CFLs or LED lights, changing temperature settings on heating or air conditioning a few degrees, and using alternative energy. We can further reduce our environmental impact by driving hybrid or other fuel-efficient vehicles or using alternate transportation, buying local (especially used) goods, using only what we need, and recycling what we don't.

As environmental and conservation organizations, many zoos have reduced their carbon footprints through responsible practices. Numerous zoos have recycling programs for staff and the public to keep not only plastic, glass, aluminum, paper, and cardboard out of landfills, but also batteries, cell phones, and print cartridges that contain environmentally hazardous components. Several zoos have adopted alternative energy programs, such as geothermal energy, solar panels, and wind turbines.

ENERGY

Since the onset of the Industrial Revolution in the 18th century, developed countries have depended on fossil fuels such as coal, oil, and natural gas for energy and power. Many of these power sources are not renewable: they are consumed faster than they are replaced. In addition, most fossil fuels release pollutants such as carbon dioxide, sulfur dioxide, and heavy metals into the air. People are now beginning to recognize the need for cleaner, renewable energy sources that emit fewer toxins into the environment and don't require clear-cutting forests or destroying habitat. Green energy now takes several forms, such as solar, wind, hydroelectric, geothermal, biomass, and landfill energy, and several zoos have embraced some of these technologies. Solar power can take the form of solar water heaters, photovoltaic solar cells that directly convert the sunlight into electricity, or the use of building positioning, skylights, and solar tubes that reduce the need for indoor daytime lighting. Wind energy can be harnessed through the use of wind turbines, both on a small residential scale and on a large industrial scale. Wind passes over the turbine blades, causing them to turn a rotor that powers an electric generator. More than two-thirds of the United States has adequate wind to use wind turbines as a green energy source. According to the American Wind Energy Association, if 20% of the electricity used in the United States was supplied by wind energy, coal plant emissions could be reduced by one-third, leading to a significant reduction in air pollution (American Wind Energy Association 2010). Geothermal energy is used throughout the world to create clean energy from the heat of the earth's magma, as hot water under the earth's surface is captured as steam which propels turbines to generate electricity. Geothermal heat pumps move water through an exchanger where it can provide heating and cooling for buildings. The temperature of the ground under the earth's surface stays at approximately 10 °C (50 °F). Air or water is pumped through pipes or wells underground, where it is cooled or warmed through heat exchange with the underground temperature. In summer, for example, water is pumped underground, cooled, and then returned aboveground as natural air conditioning for buildings. In winter water is transported through the same system, where it is warmed by the underground temperature, and then returned aboveground as a natural heat source for buildings, so that the buildings' temperature remains at approximately 10 °C (50 °F), and less extra heating is required to achieve a comfortable temperature. The Toronto Zoo recently renovated its lion-tailed macaque exhibit using geothermal energy to replace less energy-efficient and more expensive radiant heating. Geothermal pumps can be used not only to heat and cool buildings, but also to heat and cool rocks for zoo animals to provide them with comfortable resting areas during hot and cold months.

Several zoos have integrated different types of alternative energy into their operations. For example, the Toledo Zoo in Ohio investigated several options for alternative energy that could be used throughout the zoo's campus. The results of this study included the installation of a power monitoring system on several buildings, and an energy conservation plan that is used during peak times. For example, on the hottest days, low-priority equipment is systematically shut down when electricity usage reaches a certain level. Also, in all new construction projects, heating, ventilation, and air conditioning (HVAC) equipment is connected to an automated system that monitors and controls it. By monitoring energy use, the zoo can diagnose the areas of high usage and develop methods to reduce it.

The zoo's annual display of more than one million holiday lights contains more than 70% LED lights, which use 10% of the energy of regular lights. Most of the exterior site lighting, including the holiday display, is connected to a software-based management system that allows the zoo to have lights turn on and off at specific times to reduce energy usage. New construction includes solar tubes and skylights to allow natural lighting in the buildings.

Further integrating alternative energy, a geothermal test well was installed to aid in calculating the lifecycle cost and savings for geothermal systems at the zoo. Later, a geothermal system was installed to support the aquarium and nearby conservatory, and a wind turbine was installed which, in combination with three solar panels, now powers the parking lot ticket booths. The same zoo recently installed a solar walkway comprising more than 1,400 solar panels (figure 56.2). The energy produced by the SolarWalk panels enables the zoo to reduce its annual CO_2 output by 75 metric tons—the same amount of CO_2 as 15 medium-sized cars would have put out in one year.

But what if a zoo or aquarium can't afford the initial cost of solar panels, wind turbines, or geothermal wells? There are still many ways to reduce dependence on coal-powered electricity, such as by switching to CFLs and motion-sensored lights, programming computers to go to sleep mode rather than use energy-sapping screen savers, and using programmable thermostats. In addition, green power can be purchased from local utility companies that use some of these technologies.

PURCHASING

OFFICE SUPPLIES

Green office supplies, such as paper made from 100% recycled fiber and refillable pens, are widely available but are often more expensive than their non–eco-friendly counterparts. Many office supply companies are willing to discount the greener products if an organization agrees to make the company its exclusive supplier. Organizations can further save money if they commit to certain products with their suppliers and restrict their purchasing to those discounted items. For example, rather than allowing staff to choose their own preferred file folders from a supply catalog, a zoo can negotiate a reduced price for a recycled option and eliminate the other choices. Beyond purchasing greener office supplies, the better course of action is to reduce the amount of new office supplies altogether. Office supply swaps and supply closets are great ways to reduce purchasing: unnecessary purchasing of new items can be avoided by having designated areas for unwanted supplies or by hosting events at which staff can donate excess supplies.

FOOD

The food outlets at a zoo or aquarium are managed either by the institution's organization or by an outsourced company. Typical zoo and aquarium fare consists of quick, child-friendly options such as hamburgers, hot dogs, and pizza. Zoo and aquarium visitors have recently come to expect and appreciate greener options such as fair-trade coffee, organic snacks, vegetarian options, and local produce. Cost can be a challenge, but some guests are willing to pay more for options perceived to be more healthful, such as organic and vegetarian foods. Costs can also be curbed by minimizing

Figure 56.2. Solar paneling and wind turbine. Photograph courtesy of Toledo Zoo.

waste: for example, by offering condiments in bulk, using washable serviceware, removing items such as straws and cup lids, and recycling cooking oil. Food service offers a great opportunity to encourage green behavior by guests, whether through signage asking to minimize the number of napkins taken or by providing information on the green wisdom behind certain practices. Behind the scenes, waste can be minimized and resources saved through some simple operational changes such as insuring that water is not left running, composting food prep scraps, and taking care to recycle material and food packaging.

GIFT SHOPS

Like the food outlets, gift shops at a zoo or aquarium are managed by the institution's organization or by an outside company. The overall goal of a gift shop is to provide additional revenue for the organization. Greening a gift shop can be profitable and easy, if it is approached with sensitivity to the shop's business needs. The first step is to review what the shop sells to identify which products support the institution's mission, which might be in conflict with the mission, and which can be considered "green." Most zoos and aquariums have policies against selling live animals and products made from animal parts, including shells and fur. Consideration of other possible contradictions to a zoo's mission or conser-

vation strategy is important for overall gift shop greening. For example, are guests informed of the impact that palm oil products have on orangutan populations but then offered palm oil products in the gift shop? Does the zoo or aquarium have a sustainable seafood program but sell a cookbook featuring recipes that call for shark meat? It is helpful to make note of opportunities to align the mission with the inventory, and of ways to offer more green products such as those made from organic cotton and recycled materials. Many gift shops already offer a robust selection of these products and find them popular among guests. Often a simple conversation with the person responsible for the gift shops is enough to get the green ball rolling. One prominent gift shop company created an entire green line of products after an aquarium employee inquired about the possibility of adding an organic cotton T-shirt to the apparel collection! When making a gift shop request, it is important to focus on the positive aspects of offering greener products, such as their wide availability and popularity among guests. Other zoos, aquariums, and museums can be contacted for sales information on their green products. Also, it is essential to be reasonable in any request for greening, keeping in mind that the gift shop exists to bring in revenue: key chains, magnets, and knickknacks are often big sellers. Rather than asking for these to be removed, it might be more reasonable to ask for organic cotton T-shirts or corn-based plastic travel mugs to be added to the collection.

ANIMAL SUPPLIES

Many animal foods can be purchased from local suppliers such as meat and produce farmers. Buying local can reduce the use of petroleum and related emissions in long-distance shipping. Farming cooperatives are becoming more popular, and they might be a helpful alternative for some zoos and aquariums and local keeper associations as sources of fresh, locally grown produce. It is important to keep in mind that many farms use harmful pesticides and herbicides on their produce, some of which might be harmful to animals that eat it. Livestock farmers might also include hormones and antibiotics or other chemicals in their animal feed to help boost growth and decrease the incidence of disease. Therefore, it is important to check with farmers regarding their use of chemicals or hormones before committing to purchasing food from them. Armed with this information, the zoo is well positioned to make appropriate feed choices. If space allows, some zoos can grow some of their own produce and browse for their animals, or can partner with local gardening organizations to maintain organic gardens. Partnerships with local grocers and farmers can also serve the animals, since local businesses are often willing to donate excess animal food supplies to zoos and aquariums.

Animal enrichment items can be created from many sources. Many items from the home can be reused in the zoo. Cardboard tubes and boxes can be fashioned into puzzle-type feeders for various animals, or used to create shelters and hide boxes. Old sheets and towels can be used as animal bedding when deemed safe. Yogurt cups and other plastic containers can be used to create frozen treats. Other items can be rotated among exhibits, a practice that provides the animals with variety and saves money and environmental resources. It is important, however, to check with the facility's veterinarian before moving an item from one exhibit to another; some items will need to be disinfected before reuse. Many enrichment items can be purchased secondhand at yard sales and thrift stores, and many, such as cardboard and paper items, can be recycled again after being used at the zoo. When purchasing new items, it is wise to be aware of packaging and to avoid items that come in excessive wrapping.

GREEN CLEANING

Cleaning today involves new responsibilities. With the number of chemicals introduced in our environment, there is an increased risk for asthma and other diseases. Reducing the use of chemicals can lead to improved human and animal health, as well as a healthier environment. Cleaning a zoo or aquarium requires a number of different types of cleaners. Animal areas require grease cutters, surface cleaners, disinfectants, and window cleaners. In choosing the best cleaning product, one must consider animal safety as well as human safety. Every chemical will have an accompanying material safety data sheet (MSDS) that provides information about the product as well as recommendations for first aid in case of exposure. Green cleaners come in different forms. Some can be made from household ingredients, such as baking soda and vinegar, while others are certified green by Green Seal and have passed a rigorous evaluation of their design and manufacturing. In all, green cleaners should meet numerous criteria: in undiluted form they should be nontoxic to all animals (including aquatic life), should not produce skin sensitivity, and should be biodegradable and low in air pollutants. In addition, green cleaners should not have been tested on animals, and should be sold in recyclable and responsible packaging. Many green cleaners today are plant-based and come in concentrated quantities. When diluting them for use, it is important to follow the manufacturer's instructions to obtain the best results.

Green cleaning goes beyond using environmentally friendly chemicals. A keeper's day is filled with opportunities to reduce impact on the environment. Instead of hosing an enclosure, one can save water by sweeping it first. Food dishes and pans can be washed in large sinks rather than under running water, thus saving gallons with every cleaning. Food scraps can be composted rather than thrown in the garbage or landfill. Cleaning with reusable sponges and rags greatly reduces waste. When paper towels are necessary, unbleached options not only are less toxic to produce but are often compostable as well. When using cleaning machines such as clothes or dish washers, running full loads and using cold water when appropriate saves substantial water and energy.

WASTE COMPOSTING

Composting allows for nutrients in food to be returned to the soil rather than being trapped in a landfill or burned in an incinerator. At a time when soil integrity is of utmost concern, composting is more important than ever. Many opportunities exist for composting at zoos and aquariums, including food scraps and horticulture and animal waste. Many zoos have space available for onsite composting, and can return the composted waste directly to their gardens and grounds. Organizations without onsite composters may rely on outside contractors, which can be challenging to find and fund, given that some jurisdictions hesitate to allow commercial composting. Zoos and aquariums also need to be cognizant of regulations and concerns regarding disease transmission via animal waste. Vermiculture, composting with worms, is an option for small-scale composting at organizations that have limited access to other options. For less than US$25, a zoo's business department can compost its daily food waste quietly and without odor or insects through vermiculture. (See http://www.sheddaquarium.org/pdf/Shedd_Worm_Brochure.pdf for information on setting up a worm compost bin.)

EXHIBIT DESIGN

Exhibit design, whether renovation to an existing exhibit or construction of a new one, should be based on the animals' needs and the public experience. That said, exhibits can be constructed in a way that leaves minimal negative impact on the environment.

WHAT IS LEED?

Much green building today is according to Leadership in Environmental Energy and Design (LEED) standards. Coordinated by the US Green Building Council (www.usgbc.org), LEED provides a framework for the design and construction of new or renovated buildings, beginning with the land they sit on. With each green practice in areas such as energy conservation, water savings, emissions, improvements to indoor air quality, and stewardship of good environmental practices, an organization earns points toward a level of LEED certification ranging from certified to platinum. Green construction has become popular in Europe as well, with many companies following LEED criteria in construction and renovation. In Australia many organizations look to the Green Building Council Australia for guidelines and "Green Star" ratings. A number of other zoos and related organizations have also committed to green building practices. For example, the International Union for Conservation of Nature (IUCN) constructed what is said to be one of the greenest office buildings in Europe, while in the United States the Happy Hollow Zoo in California has achieved gold level LEED certification for its entire facility. The Vine Street Village Entrance Complex at the Cincinnati Zoo in Ohio and the Lee H. Brown Family Conservation Learning Center at Arizona's Reed Park Zoo both achieved LEED platinum certification in 2009. Applying for LEED certification can be expensive and time-consuming, but some zoos and aquariums that do not want the expense of certification will still follow LEED standards in exhibit design and construction.

GREEN ELEMENTS IN EXHIBIT DESIGN

From the floor to the ceiling, there is a green alternative for almost every construction element in exhibit design. A few examples are listed here, but more information can be found at the websites of the US Green Building Council (USGBC 2010) or the Green Building Council Australia (GBCA 2010).

The exhibit building can be placed to optimize natural lighting and cooling. When the exhibit site is being cleared, concrete or asphalt can be ground and reused in new paving. Walls can be made from Agriboard, and wall boards made from compressed straw, a readily renewable resource, rather than wood. Floors can be designed from renewable resources such as cork or bamboo, or from recycled elements such as carpet made from recycled plastic or environmentally friendly linoleum with nontoxic adhesive. Wall insulation made from recycled denim is also widely available.

Within the buildings much can be done to promote sustainability and reduce the carbon footprint. Natural lighting through windows, sky lights and solar tubes is optimal and, when artificial light is needed, CFLs use 75% less energy than incandescent lightbulbs. Motion sensors can be used to control lighting so that lights turn off when rooms are empty. Motion sensors can also be used for water fixtures such as sinks and low-flow toilets in restrooms, thus reducing water waste.

Given the diverse needs of different animal species, the design requirements of animal living areas can vary. Some animals need certain lights, temperatures, or nesting or denning areas, among other things. When designing these elements, one can look to green architects and designers to address eco-friendly options.

Outdoor exhibit and public areas are typically landscaped to reflect a theme. When designing these areas, one might consider rain gardens that collect rainwater and filter it through the soil instead of allowing it to run off into storm sewers with pollutants in tow, or bioswale planting areas that also divert rainwater from the storm drains (figure 56.3). Bioswales are sloped and graded areas that create drainage systems to slow or divert storm water away from buildings and infrastructure. Sometimes planted with grass or other plantings, they help to trap silt and pollutants from traveling into storm sewers or watersheds and often are used as aesthetic elements of construction design. Permeable paver stones have a similar effect and can be incorporated into public pathways. All of these methods help to filter rainwater that might otherwise collect harmful fertilizers, pesticides, and oils that would eventually contaminate the water supply.

Like all construction, exhibit construction and renovation results in waste. Much of this construction waste can be diverted from landfills by being reused, if possible, or by being donated to building suppliers and other organizations. Materials like steel and many carpets can be recycled, while concrete can be ground into gravel for future use. These practices can also lower costs by eliminating dumping fees. There is a growing trend to require that specific amounts of construction waste be diverted from landfills, and to place a large part of the responsibility for this diversion on construction contractors.

ENGAGING STAFF IN GOING GREEN

GAINING LEADERSHIP SUPPORT

It is important to consider the business side of a zoo or aquarium when employing green operations. Zoo and aquarium leadership is tasked with the financial health of the organization; therefore, gaining its support requires a strong business case for greening. Whether green project ideas promise resource reduction, operational efficiencies, or added value to the overall organization, they should be discussed with leadership so that the impact to the facility's business operations can be considered. Because cross-departmental buy-in is essential to the success of such projects, strategic planning for sustainability should be a team effort with representatives from all levels of the organization.

GREEN TEAMS

Green teams are great for initial staff engagement and the identification and implementation of quick and easy changes. Direct leadership support and a certain amount of decision-making authority are vital to a green team's success. It is also important to ensure that the team tackles appropriate projects. For example, green team activities might include auditing waste in office areas, organizing office supply swaps, or running special recycling drives for staff. Asking a green team to collaborate with responsible departments in compiling a list of greening opportunities to present to the zoo and aquarium leadership would also be appropriate. Conflict may arise when a green team feels authorized to make changes that are not desired by departments or that are outside its own members' expertise. This can lead to frustration for the green team members as well as for the staff and departments that are being pressured to change their practices. Clearly articulating the expectations and authority of a green team at

Figure 56.3. A rain garden outside of the Toledo Zoo's Butterfly Conservation Center. Photograph courtesy of Toledo Zoo.

its inception can help curtail these conflicts. At some point, a green team should be transitioned into a wider staff culture in which sustainability is everyone's responsibility. This can help to avoid the "us-versus-them" mentality, in which the green team's agenda becomes everyone else's burden. Ideally, a green team promotes green practices until they are so embedded in the organizational culture that the green team itself becomes obsolete.

SHIFTING STAFF CULTURE

A common assumption is that everyone who works at a zoo or aquarium holds strong environmental values. This may not be the case, however. Recognizing that being eco-friendly is not intuitive or attractive to everyone is an important first step toward gaining staff support for green efforts. Everyone employed by a zoo or aquarium is compensated for fulfilling specific responsibilities. The more that green practices and green thinking can be aligned with the achievement of organizational goals and the responsibilities of individual staff members, the easier it becomes to advance sustainability across the organization. For example, rather than forcing desktop composting onto a public relations department because "it's

the right thing to do," the task can be approached as a unique PR story that caters to green audiences. Rather than trying to guilt the human resources staff into going paperless to save trees, the green team can discuss with that staff the growing importance of demonstrated environmental responsibility by employers to employees. For example, a recent study of MBA graduates showed that the average graduate would give up $14,000 in salary to work for a socially or environmentally responsible company (Montgomery and Ramus 2007). It is important to remember that one person's agenda (e.g., a green team's desire to reduce paper waste) can be another person's burden (e.g., a staffer's annoyance at the slow hand driers that replace paper towels). Once being green is on everyone's agenda, innovative improvements become the norm.

STAFF TRAINING

One of the best ways to engage staff in greening is to offer them educational opportunities that are designed to build their interest, knowledge, and investment in sustainable practices. These opportunities should not focus only on the environmental case behind greening, however; the business case and personal connections should also be highlighted.

Examples of these events might include dedicating part of a new employee orientation to establishing a green culture, providing opportunities for staff to learn about new processes and the wisdom behind them, and sessions designed to build eco-literacy. The more a zoo or aquarium can helps its staff view the environment as something comfortable and relatable, the easier greening will be.

REFERENCES

American Wind Energy Association. 2010. Accessed in August at http://www.awea.org/la_utilitywind.cfm.

Clean Air Partnership. 2010. Accessed September 18 at http://www.cleanairpartnership.org/idle_free.

Green Building Council of Australia. Accessed August at http://www.gbca.org.au/.

Green Energy Ohio. 2010 Accessed in August at http://www.greenenergyohio.org.

Green Seal. 2010. Accessed in August at http://www.greenseal.org.

John G. Shedd Aquarium. 2010. "Care and Keeping of Worms." Last modified in March at http://www.sheddaquarium.org/pdf/Shedd_Worm_Brochure.pdf.

Montgomery, D. B., and C. A. Ramus. 2007. "Including Corporate Social Responsibility, Environmental Sustainability and Ethics in Calibrating MBA Job Preferences." Research Paper Series, Stanford University Graduate School of Business, no.1981.

Occupational Health and Safety Administration. 2010. Accessed in August at http://www.osha.gov.

Pirog, Rich, and Andrew Benjamin. 2003. "Checking the food odometer: Comparing food miles for local versus conventional produce sales to Iowa institutions." Leopold Center for Sustainable Agriculture, Iowa State University. Last modified in July at http://www.leopold.iastate.edu/pubs/staff/files/food_travel072103.pdf.

Polar Bears International. 2010. Accessed in August at http://www.polarbearsinternational.org.

Roundtable on Sustainable Palm Oil. 2010. Accessed in August at http://www.rspo.org.

US Environmental Protection Agency. 2010. Accessed in August at http://www.epa.gov.

57

Wildlife Rehabilitation

Erica Miller and Sandra Woltman

INTRODUCTION

Some people say that wildlife rehabilitation has been happening for as long as humans and wild animals have roamed the planet together; many humans feel a need or a responsibility of stewardship to assist injured animals. It's even been said that had there been wildlife rehabilitators, the dinosaurs would not have become extinct!

Early efforts to treat or raise native wildlife resulted in varying degrees of success, as individuals attempted to apply domestic or zoo animal husbandry techniques and found they were not always appropriate (Thrune 1997). The concept of wildlife rehabilitation as a specialty separate from other animal care specialties began to grow in the 1960s and 1970s. The first wildlife rehabilitation center in the United States was established in the early 1950s and the first large-scale wildlife center in Great Britain was established in the late 1970s. So wildlife rehabilitation is actually a relatively new field.

CHAPTER OBJECTIVES

This chapter provides an introduction to the field of wildlife rehabilitation, particularly as it relates to zoos and aquariums and the participation of keepers. After studying this chapter, the reader will understand

- what wildlife rehabilitation is
- how it differs from the general care of captive wildlife
- methods for keeping wildlife "wild"
- the necessity for proper permits and licenses to conduct wildlife rehabilitation
- some ethical considerations relating to the practice of wildlife rehabilitation
- who wildlife rehabilitators are
- the roles that zoos, aquariums, and keepers can have in wildlife rehabilitation
- the importance of proper record keeping on wildlife received for rehabilitation

- the importance of behavioral enrichment to wildlife during rehabilitation
- the role of nonreleasable native wildlife as educational ambassadors
- disaster-related wildlife rehabilitation, such as that done after oil spills.

DEFINITION AND PURPOSE

In addition to satisfying the human need to assist injured or abandoned animals, wildlife rehabilitation serves several purposes. To understand these purposes, it helps to understand exactly what is meant by wildlife rehabilitation. It is officially defined as the treatment and temporary care of injured, diseased, and displaced indigenous animals, and the subsequent release of healthy animals to appropriate habitats in the wild (Miller 2000).

Herein lies the difference between wildlife rehabilitation and zoo and aquarium work. While both disciplines involve the care of wild animals in captive settings, wildlife rehabilitation has the goal of returning animals to full-functioning lives in the wild. Animals undergoing rehabilitation, as the name suggests, are in need of medical care or are too young to survive on their own. Not only must they be hand-reared, but many may need to be taught to forage and fend for themselves in the wild. All of these animals must be not only physically healthy but also behaviorally normal and wild (e.g., not imprinted or too strongly socialized with humans).

Under normal circumstances, wild animals that are rehabilitated and released do not make a significant impact on the overall population of that species—with a few exceptions. Twenty years ago, the return of even a single bald eagle (*Haliaeetus leucocephalus*) to the northeast United States was a significant contribution to the survival of the species; the return of every California condor (*Gymnogyps californianus*) or mountain gorilla (*Gorilla beringei beringei*) to its natural habitat may be vital for the survival and genetic diversity of that species as a whole. In the case of large numbers of a single animal species, such as the 38,500 African penguins (*Spheniscus demersus*) affected by the *Treasure* oil spill in 2000 (Barham et al. 2006), the successful return of that many animals to the wild certainly affects the population.

So why are efforts made to rehabilitate wildlife if the results have such little impact on the populations? When an individual wild animal is injured, ill, or for some reason separated from its parents, its only chance for survival may be the care it receives from humans. In such a situation, wildlife rehabilitation matters very much to that individual animal; it also means very much for the person who has found the animal and wants to ensure that it receives help. In this manner, wildlife rehabilitators not only help the animals they treat, but also provide a service to the public. Furthermore, when a person brings a wild animal to a rehabilitator, the rehabilitator will often take time to explain to the person how that injury or illness could have been prevented, or offer some other natural history information regarding the species, thereby providing an element of public education. Other reasons for rehabilitating wildlife—such as opportunities for learning about the species, gathering data, and learning about the effect of humans on wildlife—can be found in Frink 1998.

LAWS AND REQUIRED PERMITS

In most countries, native wildlife is protected by federal or national, state or provincial, and local laws. These laws differ in every country and even from state to state. In the United States, Canada, Japan, Mexico, and Russia, all migratory birds are protected by the Migratory Bird Treaty Act of 1918. This law prohibits anyone from possessing a migratory bird without proper permits. Additional permits for wildlife rehabilitation may be required for the possession and treatment of threatened and endangered species as well as marine mammals.

Permits for wildlife rehabilitation may be difficult to obtain. Rehabilitators are required to have adequate animal caging, to provide professional medical treatment for the animals, and to provide appropriate release sites. Before starting a wildlife rehabilitation program of any kind, it is important to know the local regulations regarding the possession of wildlife in captivity. Some local laws may be restrictive, perhaps not allowing the care of predatory animals, while other laws may restrict caring for any injured wildlife in captivity within a town's limits.

If a zoonotic disease is locally common (endemic) within a species, rehabilitation of that species may be limited or excluded from a permit, or special training may be required for rehabilitators of that species. Rabies is an example of a zoonotic disease for which common vector species are regulated differently in different areas of the United States. Many states require an individual wishing to rehabilitate "rabies vector species" (RVS) to have pre-exposure vaccinations, special training, and successful completion of an examination; other states prohibit RVS rehabilitation entirely, while some states have no RVS restrictions or requirements for rehabilitators. Again, familiarization with and routine review of local laws is the best way to keep informed about any changes to regional regulations regarding wildlife care and possession.

It is necessary for wildlife rehabilitators to maintain accurate and up-to-date records for the animals in their care (tables 57.1–57.6). In many countries, they are required to report their activity to government officials at the end of each year. This information often includes the number of animals received, their conditions upon arrival, and their disposition (i.e., the outcome). Some government agencies require rehabilitators to contact them within a certain period of time if they receive threatened or endangered species or if an animal has been purposely harmed by people, including being shot, poisoned, or abused.

Good patient records may also include pathology reports, genetic sampling, daily treatment schedules and efficacy, and the numbers on any federal or permanent identification bands that have been placed on the animal prior to release. It is recommended to write on the animal's chart any data or information about it that may be useful at a later date. This information may be used in connection with studies conducted by researchers, geneticists, biologists, and wildlife rehabilitators. Detailed record keeping may also indicate certain trends in a given location, including multiple cases of disease seen in the same area, or an increased number of illegally shot animals in a neighborhood.

ETHICAL CONCERNS

Wildlife rehabilitation can be quite controversial among the public, government officials, and even wildlife biologists. People often question whether rehabilitators should be spending thousands of dollars each year to save common species that in some areas are considered a nuisance. They also question whether rehabilitators should be caring for and releasing potentially invasive species whose behavior may be reducing the population of native species. Wildlife rehabilitators need to consider these views before accepting nonindigenous or "nuisance" wildlife for treatment in their facilities.

Some wildlife rehabilitators choose to treat and release common and/or nonindigenous species. In some areas this practice is illegal, and in many places it is seen by wildlife biologists as unethical. Should rehabilitators be releasing animals whose populations are already exploding in numbers and depleting appropriate habitats for native species? Is the release of nonnative wildlife that competes for prime nesting or breeding habitats of indigenous species a responsible choice for the environment and the native species in a region? These are all questions that wildlife rehabilitators face each and every day. Although many wildlife rehabilitators are dedicated to nondiscriminatory care for all wildlife in need, that care should extend to the environment and should consider the survival of indigenous species.

In order for a rehabilitated animal to be released responsibly into the wild, it must demonstrate that it can survive on its own. It must be healed of all injuries and free of disease, must exhibit behaviors appropriate for its species, must be self-feeding, and must possess vision adequate for its species. A full list of release criteria can be found in *Minimum Standards for Wildlife Rehabilitation, Fourth Edition* (Miller 2012). If the animal exhibits signs of permanent injury that prevent it from living a normal, healthy life in the wild, the animal is often euthanized. Some injuries, such as complete blindness or, in the case of birds, full amputation of a wing, make it obvious that an animal cannot survive on its own.

TABLE 57.1. Sample patient admission form for wild avian rehabilitation cases. Courtesy of Tri-State Bird Rescue.

PATIENT ADMISSION FORM

DATE__/__/__   SPECIES_____   CASE #: _____ - _____

Time _____   Admitted by _____   Transported by _____

PLEASE COMPLETE THE FOLLOWING INFORMATION TO THE BEST OF YOUR KNOWLEDGE:

Name (person who found bird): Mr. Mrs. Ms. _____

Address _____ City/State/Zip _____

E-mail address _____ Phone (___) _____

Date/time you first saw bird(s) _____ Date/time captured _____

Where specifically it was found (in yard, by window, etc.) _____

Address where found _____ City/State/Zip _____

County_____ Nearest intersection _____

Did you feed the bird? _____ If yes, what and how? _____

What else did you do to help the bird? _____

Please circle any of the following that pertain to this bird and its capture:

Easy/hard to catch	Parents/nestmates dead	Hit by vehicle	Trapped/entangled
Can't stand/fly	Attacked by cat/dog/bird	Found in/beside road	Fishing line/hook
Hopping on ground	Hit window	Oiled or sticky	Shot or bleeding
Nest destroyed	Found by window	Wet or cold	Exposed to chemical/toxin

Additional remarks: _____

Wild birds admitted to this facility become the sole responsibility of the organization.
We are licensed by the state and federal governments to provide care for wild animals.

YES! I support the work of this organization with my tax deductible gift of $ _____
If you are leaving a cash donation and would like to be acknowledged, please put it in one of the envelopes provided and clearly
print your name on it before placing it in the donation box.

FOR OFFICE USE ONLY:
Check if referral _____ or readmission _____

Assumed Cause of Injury   Attack by cat/dog/other   Contaminant/oiled   Dangerous site (describe)

Disease (FES/pox/WNV/other)   Electrocution/burns   Entrapment (specify)   Fell from nest

Gunshot   Hit by vehicle   Human interference (describe)   Impact (specify)   Nest destroyed

Nutritional/developmental   Orphaned   Toxin (botulism/lead/other)   Undetermined (acute/chronic)

Types of Injury   Primary   Behavioral   Contaminant   Feather damage   Gen'l debilitation

Neurologic   Ophthalmic   Orthopedic   Parasites   Respiratory   Soft tissue   No appt. injury

Other/Comment: _____

Secondary   Behavioral   Contaminant   Feather damage   General debilitation   Neurologic

Ophthalmic   Orthopedic   Parasites   Respiratory   Soft tissue   Other/comment: _____

Disposition: Date _____ Initials _____   **Final Cause of Injury** _____

DBA*   Died   EOA**   Euth   Rel   Release/Return to Nest   Rel/placed   Transf **by** _____

Rel Location _____·_____   Band # _____ - _____

*Deceased by arrival. **Euthanized on arrival.

However, some other injuries are not so obvious, and the decision to release an animal with such an injury can be quite controversial. Some such questionable scenarios can include the release of a raptor with vision in only one eye, of a predatory mammal missing one leg, or of an animal that is habituated to humans but physically normal. Again, the animal must meet all the criteria to be considered for release.

Wild animals that are to be released into the wild must also be cared for and housed in settings that do not allow them to become habituated to or imprinted on humans. Such behavioral changes can be detrimental to an animal, greatly decreasing its chance of survival after release. To pre-vent such habituation, all wildlife undergoing rehabilitation should have minimal visual and audible contact with humans. This can easily be achieved by housing wildlife in areas as far away from human activities as possible. Visual barriers such as solid walls, sheets or other fabric, or natural foliage can be erected in or around cages to decrease an animal's visual contact with people (figure 57.1). The elimination or minimizing of talking around wildlife in captivity is also important for reducing an animal's stress, and for preventing it from becoming accustomed to human voices. Young animals are especially vulnerable to habituation and human imprinting, so extra care must be exercised to prohibit such

TABLE 57.2. Sample physical examination form for wild avian rehabilitation cases. Courtesy of Tri-State Bird Rescue.

PHYSICAL EXAM

DATE: ___/___/___   SPECIES: _____   CASE #: ___ - _____

TIME: _____ INITIALS: _____ WEIGHT: _____ TAG #: _____

BODY CONDITION:		emaciated	underweight	normal	overweight	AGE/SEX: _____
HYDRATION:		good	fair	poor		TEMP: _____
ATTITUDE:	BAR	Remarks: _____				
NARES:	Clear	Remarks: _____				
BEAK/MOUTH:	WNL	Remarks: _____				
RESPIRATION:	WNL	Remarks: _____				
CROP:	full empty	Remarks: _____				
GI TRACT/ABDM:	WNL	Remarks: _____				
DROPPINGS:	WNL none	Remarks: _____				
EYES:	WNL	Remarks: _____				
EARS:	WNL	Remarks: _____				
FEATHERS:	WNL	Remarks: _____				
ECTOPARASITES:	none	Remarks: _____				
SKIN:	WNL	Remarks: _____				
FEET:	WNL	Remarks: _____				
NERVOUS SYSTEM:	WNL	Remarks: _____				
MUSCULOSKELETAL:	WNL	Remarks: _____				
INJURIES/PROBLEMS (wounds, etc.):		Remarks: _____				

On Entry:

D2.5LRS _____
abx _____
PO _____
Fecal _____
Other _____

PCV: _____%
BC: _____%
TS: _____g/d

Initial

Location

Note: WNL = within normal limits. BAR = bright, alert, responsive. D2.5LRS = dextrose 2.5% lactated ringers solution. abx = antibiotics. PO = per os (provided by mouth). PCV = packed cell volume. BC = blood count. TS = total solids.

Please mark cause and type of injury on front of form.

outcomes. Wildlife rehabilitators often use puppets, mirrors, costumes, foster adults, recorded animal sounds, and other techniques to reduce the chance of these young animals becoming imprinted on the wildlife rehabilitator. These tools help to reduce or eliminate visual and audible contact with caretakers, and allow the animal to imprint on conspecifics rather than humans.

ECOTOURISM AND WILDLIFE REHABILITATION

In some areas, wildlife rehabilitation may play an important role in ecotourism, attracting individuals who might otherwise not even be aware of the concept of wildlife rehabilitation. Wildlife rehabilitation facilities that participate in ecotourism—such as the Alaskan Raptor Center, which provides tours to vacationers on cruise ships along the Alaskan seacoast—help to spread a message of environmental conservation. These tours are often the first exposure people have to wildlife rehabilitation; upon returning home, they often seek out local rehabilitation centers they can visit and support through donations or volunteering. In this way tourists become engaged with wildlife while rehabilitation centers receive support. The Karanambu Trust in Guyana is an excellent example of ecotourism not only helping to support a wildlife rehabilitation program, but also helping to maintain the local economy. Those involved with the program say that for many years the only reason people went to Karanambu was to see the hand-reared giant river otters (*Pteronura brasiliensis*) that live there. Thus the rehabilitation program served to increase conservation awareness among local, national, and international visitors to Guyana (McTurk and Spellman 2005).

Ecotourism, however, can also have some detrimental effects on wildlife and wildlife rehabilitation. For example,

TABLE 57.3. Sample patient admission form for wild turtle rehabilitation cases. Courtesy of Turtle Rescue of NJ.

	Case number:

TURTLE ADMITTANCE RECORD: NATIVE SPECIES

Date: _____ E-mail _____

Finder's name: _____

Address: _____

_____

Phone: _____

Where turtle was found: Street: _____

Town: _____

County: _____

Length of time in finder's care: _____

Hit by car ☐	Caught by dog ☐	Fishhook ☐	Ear abscess ☐

Other: _____

_____

_____

Results

DOA: _____ Euthanized: _____

Died within 24 hours _____ Died: _____

Transferred: _____

Released: _____

Place of release: _____

while tourism has done much to help fund the mountain gorilla project in Rwanda and to educate people about environmental conservation both in that area and globally, the extensive exposure to humans and human diseases has been fatal to some gorillas, and has also prevented the release of most gorilla orphans (Spellman, pers. comm.).

WILDLIFE REHABILITATORS

So who are wildlife rehabilitators? Most are simply concerned people who care for injured native wildlife in their homes. According to a member survey conducted in 2008 by the National Wildlife Rehabilitators Association (a not-for-profit organization of rehabilitators based in the United States), 78% of wildlife rehabilitation is done in private homes (Thrune 2009). These rehabilitators have a wide range of backgrounds: they may be nurses, wildlife biologists, teachers, veterinarians, or housewives. Most begin with little formal education or training for wildlife care, and most are funded solely by donations from the public and their own savings. Wildlife rehabilitators obtain their skills in several main ways:

1. trial and error, often at the expense of the animal they are treating

2. training by other rehabilitators via formal apprenticeships, internships, or volunteer work
3. books and articles published by other rehabilitators and veterinarians (see appendix 1 for sample sources)
4. formal training sessions offered by some of the larger wildlife rehabilitation centers, or conferences offered by state, regional, and national wildlife rehabilitation organizations.

A few colleges in North America now offer courses in wildlife rehabilitation. For example, Lees-McRae College offers a Bachelor of Science degree in biology with a concentration on wildlife rehabilitation, University of Wisconsin–Stevens Point offers a minor in captive wildlife management that includes wildlife rehabilitation, Coastal Carolina Community College offers classes in wildlife rehabilitation, and Northern College in Ontario offers a veterinary technology and wildlife rehabilitation combined degree program. In addition, many of the larger wildlife rehabilitation centers in several countries offer training programs, and most wildlife rehabilitation organizations offer annual conferences which provide both beginning courses and continuing education for wildlife rehabilitators.

Rehabilitators are not typically paid for the work they do. In fact, in the United States it is illegal to charge a fee for

TABLE 57.4. Sample physical examination form for wild turtle rehabilitation cases. Courtesy of Turtle Rescue of NJ.

TURTLE EXAMINATION FORM

Condition: alert   lethargic   unconscious

Hydration: normal   dehydrated (slight/moderate/severe)

Respiration: normal   open mouth   labored   noise   other _____

Neurological: normal   head tilt   circling   other _____

Head:	Normal	Trauma	Fracture	Blood	Fly eggs	Maggots
Skull						
Upper beak						
Lower beak						

Mouth: normal   blood   discharge   fly eggs   maggots

Glottis: clear   debris   blood   mucous

Color: normal   pale   red   other

Notes:

Ears: normal   trauma   swelling

discharge   blood   fly eggs   maggots

Notes:

Eyes: normal   clear   open   closed

trauma   swelling   sunken

discharge   blood   fly eggs   maggots

Notes:

Nose: normal   discharge   blood   trauma

fracture   pigmentation   erosion

Notes:

Shell:	Normal	Trauma	Fracture	Fly eggs	Maggots	Blood	Discharge	Odor	Discoloration
Carapace									
Plastron									
Bridge									

Notes:

	Normal	Trauma	Fracture	Fly eggs	Maggots	Missing
Legs						
Feet						
Toes						

Notes:

Skin: normal   dry/flaky   discoloration   punctures   cuts   abrasions   fly eggs   maggots

Cloaca: normal   trauma   swelling   discharge   fly eggs   maggots

Weight at admittance:

Observations:

caring for injured or orphaned wildlife. Some rehabilitation facilities are managed by local government agencies and receive local tax dollars to offset their expenses, while many are not-for-profit organizations funded solely through various grants and donations. However, even the government-run facilities must rely extensively on donations from the public. Wildlife rehabilitation is also done at domestic animal shelters, zoos, aquariums, and veterinary hospitals, but typically on a much smaller scale as it is usually not the mission or goal of these organizations.

TABLE 57.5. Sample avian necropsy form. Courtesy of Tri-State Bird Rescue.

AVIAN NECROPSY FORM

Case no. _____   Date/time of death _____
Species _____   Storage after death _____
Sex _____ Age _____   Date posted _____
                              DOA   EOA   Died   DBA   Euth

Location found: State _____ Town _____ County _____
History/clinical diagnosis _____
_____
Radiographic findings _____
_____

Weight _____g Tarsus _____mm   Beak depth _____mm   Beak width _____mm
                                             Culmen length _____mm Wing chord _____cm

Mark N for no gross lesions, O for not examined, X for examined and lesions found

External exam
__Carcass condition (autolyzed) 1 2 3 (fresh)
__Body condition (emaciated) 1 2 3 (good) _____
__Body openings _____
__Feather condition (poor) 1 2 3 (good) _____
__Molting yes no _____
__Skin _____
__Subcutaneous fat (poor) 1 2 3 (good) _____
__Feet (poor) 1 2 3 (good) _____
__External parasites _____

Internal exam
GI system
__Peritoneum, mesentery _____
__Abdominal fat (none) 1 2 3 (lots) _____
__Esophagus _____
__Proventriculus _____
__Gizzard (ventriculus) _____
__Stomach contents _____(organics) _____(rocks) _____(lead) _____
__Small intestine _____
__Cecae _____
__Large intestine _____
__Pancreas _____
__Liver _____
__Gall bladder _____
__Spleen _____(longest length) _____mm(widest) width _____mm
__Parasites (none) 1 2 3 (lots)__type: _____

Cardiovascular system
__Heart _____
__Pericardium _____
__Major vessels _____
__Coronary fat (none) 1 2 3 (lots) _____

Respiratory system
__Nasal cavity _____
__Trachea _____
__Syrinx _____
__Lungs _____
__Air sacs _____

Urogenital systems
__Kidneys _____
__Cloaca _____
__Ovary _____mm_____
__Testis, left _____ mm, right _____ mm _____
__Eggs _____

Endocrine system
__Adrenals _____longest length_____mm widest width_____mm
__Thyroid _____longest length_____mm widest width_____mm
__Bursa _____longest length_____mm widest width_____mm

Musculoskeletal system
__Pectoral muscles (little) 1 2 3 (lots) _____
__Bones _____
__Joints _____
__Skull _____

(continued)

TABLE 57.5. continued

Nervous system
__Brain _____
__Spinal cord _____
__Eyes _____
__Other nerves _____

Samples saved
Histopathology _____
Parasites or foreign bodies _____
Photographs _____
Toxicology _____
Virology _____
Bacteriology/mycology _____

Summary of gross findings
_____
_____
_____
_____
_____

Gross diagnosis _____
Prosector _____ Reviewer _____

WILDLIFE REHABILITATION AT HOME

Most wildlife rehabilitators are people who love animals and want to help them. These people may not live near established wildlife rehabilitation facilities, and may thus find it necessary to create places for the public to bring injured and orphaned animals for treatment. All see a need in their communities to help local wildlife, and all strongly desire to help. With determination and passion, their homes slowly become wildlife hospitals for injured animals.

This may sound like fun, but setting up a facility for injured wildlife in one's home can be quite challenging. For example, proper quarantine areas, cages, and cleaning areas are necessary, and should be kept separate from a family's living quarters to ensure minimal stress on the wildlife patients and a reduced risk of disease transmission to the family. Since a lot of extra room is not always available, rehabilitators must often sacrifice living space in their homes and make modifications to accommodate these requirements.

WILDLIFE REHABILITATION IN ZOOS AND AQUARIUMS

Wildlife rehabilitation can sometimes take place within an established zoo or aquarium. However, most zoos and aquariums need the help of separate rehabilitation facilities to complete the rehabilitation process (e.g., to exercise and condition the animals for release), since they lack the space, caging, and staff time to commit to this purpose. Most zoos and aquariums will not accept injured, orphaned, or diseased wildlife for care, but this doesn't mean that the public will not bring the injured wildlife to them anyway. These facilities should, and many do, maintain current lists of local wildlife rehabilitators that will care for native animals. They should also work with the public to ensure that those animals are transferred appropriately.

Zoos and aquariums usually do not have adequate quarantine areas separate from their animal collections for po-

tentially diseased wild animals coming into their facilities. They do not want to risk the possibility of disease entering their facilities and infecting their populations, especially since most are working with critically endangered species. However, zoos and aquariums can often provide critical care for wildlife in need, even if only temporarily. Most have the necessary equipment and knowledgeable staff to initially stabilize an animal before transferring it to a licensed wildlife rehabilitator for ongoing care.

Some zoos and aquariums have established wildlife rehabilitation facilities within their own confines. These zoos and aquariums may have keepers who work strictly with the injured wildlife, while others may have staff who alternate between working with resident animals and with injured wildlife. In order for a zoo or aquarium to successfully rehabilitate wildlife within its confines, it must be able to quarantine the wildlife properly, with separate caging to prevent transmission of diseases to zoo animals, and large outdoor cages to condition the injured animals before releasing them back into the wild. Furthermore, the rehabilitation facilities need to be in off-exhibit areas, as it is imperative that the injured animals maintain their wild nature so that they can survive after release.

Wildlife in captivity is stressed from being injured, captured, and then housed in unfamiliar settings. If zoos and aquariums participate in wildlife rehabilitation, it is critical that they provide caging in areas that are quiet and out of public view. Most wild species, especially adults, are very sensitive to sudden loud noises, human voices, and the sight of humans and predators. It is extremely important to their recovery that all measures be taken to reduce their stress during their treatment.

BEHAVIORAL ENRICHMENT

Providing wild animals in captivity with behavioral enrichment helps to improve their psychological and physiologi-

TABLE 57.6. Sample daily care sheet for wild avian rehabilitation cases. Note abbreviations: CP = changed papers in cage; H2O = provided fresh water; ID = initials of individual providing treatment or husbandry care. Courtesy of Tri-State Bird Rescue.

DAILY CARE SHEET (INSIDE)
USE FOR RAPTORS, GULLS, WATERFOWL, ETC.

Date	Time	CP	Chngd perch	H2O	Food in	Food out	Swim/ spritz	Meds	Notes	ID

cal well-being by providing them with stimulus. Whether animals are in captivity for a few days or for several weeks, they should all receive some type of behavioral enrichment. It can help prevent them from self-mutilating or overgrooming, reduce their stereotypic behaviors, increase their activity, and help them develop and maintain normal behaviors (Woltman 2005). It also reduces their stress, which in turn allows them to heal from injuries more quickly. Enrichment can be as simple as providing an animal's daily food in different loca-

tions within its enclosure, thus requiring it to forage around its cage. It can also take the form of an assortment of toys, various natural scents, rearrangement of cage furniture, provision of nesting materials during breeding season, or a window in the cage that provides a view of natural surroundings.

When providing an animal with any type of behavioral enrichment, keepers must keep its safety in mind. This includes checking a toy's construction to ensure that it will not injure the animal. Keepers must also be certain that the

Figure 57.1. Rehabilitating a yellow-crowned night heron (*Nyctanassa violacea*) in a cage with natural foliage to provide hide areas (for stress reduction) and obstacles and perches (for enrichment). Photo courtesy of Erica A. Miller.

animal cannot become trapped or entangled in the enrichment that is provided. The enrichment item should either be disposable or easy to disinfect, to prevent the spread of disease and bacteria to other animals. Keepers should also be mindful of the animal's injury, and should place food rewards where the animal can reach them without becoming frustrated.

USE OF NONRELEASABLE WILDLIFE IN EDUCATION

Wildlife rehabilitators routinely receive animals with severe injuries that are difficult or impossible to fully repair. With new technology and improved veterinary care, these animals are often saved from death. However, many of the injuries cannot be repaired well enough to allow the animal to resume normal activity and be returned to the wild. Unfortunately, many such animals are euthanized because of the lack of permanent homes for them. An increasing number of zoos and aquariums use these permanently disabled animals in exhibits and educational programs that focus on native wildlife. Many zoos and aquariums thus are able to educate the public about native wildlife and incorporate wildlife rehabilitation into the message of their outreach programs.

Rehabilitated animals give zoos and aquariums the opportunity to teach the public how humans can affect the wildlife around them, and how they can help wildlife and the environment by changing their own actions. An example of such a case is an opossum whose foreleg was strangulated by the plastic rings used to hold beverage cans together (figure 57.2). The plastic rings had wrapped around the animal's neck and foreleg and deeply severed the axilla ("armpit") of its foreleg, resulting in permanent damage to nerves and blood vessels. The leg was amputated, rendering the opossum nonreleasable to the wild. The animal, however, was kept on display and used in educational programs for many years, serving as an "ambassador" to teach people about the harm that rubbish may have on wildlife, and how cutting plastic beverage rings before disposing of them can prevent injury to animals (figure 57.3).

Some wildlife rehabilitation facilities not only treat injured and orphaned wildlife, but they also have public displays of nonreleasable wildlife. Many of the animals on display have been directly or indirectly injured by humans: hit by cars, shot illegally, attacked by unleashed pets, poisoned by insecticides or lead fishing sinkers, concussed by flying into windows, or raised improperly by well-meaning humans.

Like zoos and aquariums that rehabilitate wildlife, wildlife rehabilitation centers caring for both injured wildlife and animals on permanent display face several challenges. They must provide separate caging for animals in rehabilitation and for those on permanent display. Wildlife rehabilitation

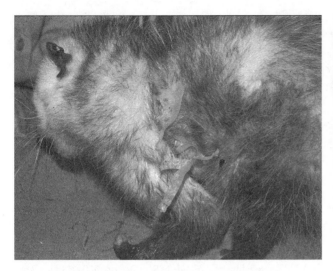

Figure 57.2. Virginia opossum (*Didelphis virginiana*) presented for rehabilitation with its left forelimb constricted by a six-pack ring. This leg had permanent nerve damage and was later amputated. Photo courtesy of Erica A. Miller.

Figure 57.3. Same opossum as in figure 57.2, in its new role of educational ambassador to teach people about the importance of proper trash disposal to prevent wildlife injuries. Straw placed regularly in this animal's cage provided excellent enrichment, as the opossum used its prehensile tail to carry the straw for fresh bedding. Photo courtesy of Sandra Woltman.

centers typically have the same staff caring for animals in both categories. When the caseload of injured and orphaned wildlife is high, this can be very draining on the staff. Animals on permanent display may be overlooked and their routine care, including routine beak and nail trims and the regular replacement of cage furniture, may be rushed or delayed.

With increased environmental awareness, education of the public regarding local wildlife has become increasingly important. The use of nonreleasable wildlife for this purpose allows for a "win-win" situation for both nature centers and wildlife rehabilitators. Nature centers can educate the community about local wildlife, and the animals are given a second life as ambassadors. Nonreleasable animals may also be good candidates for breeding programs at zoos, aquariums, nature centers, or even wildlife rehabilitation centers that maintain permanent residents.

DISASTER RESPONSE AND REHABILITATION OF OILED WILDLIFE

Oil spill response is a specialized type of disaster-related wildlife rehabilitation. Many types of oils, from petroleum oils to edible (vegetable and animal-based) oils, may have detrimental effects on wildlife through simple physical contact, inhalation, ingestion, and transcutaneous absorption. These effects include but are not limited to contamination of feathers, fur, and skin and damage to vital organ systems, including the lungs, air sacs (in birds), kidneys, liver, heart, blood, and gastrointestinal tract (Leighton 1991; Pierce 1991; Langenberg 1983).

Rehabilitation of oiled wildlife is complex, crisis-oriented work that requires experienced staff with medical, technical, and crisis-management skills. It requires hazmat (hazardous material) training to minimize petroleum exposure and potential harm to the workers, and in many areas it also requires special permits issued by the government. The many toxic effects that different oils can have on an animal's skin,

eyes, feathers or fur, central nervous system, and internal organs—as well as the secondary effects of hypothermia, aspiration pneumonia, and immune suppression—require unique medical treatments. Finally, a single oil spill can result in many and diverse animals all needing care at once. The number of animals can quickly overwhelm a wildlife rehabilitation facility that is not prepared for such a response.

Several organizations specialize in caring for oiled wildlife with trained and experienced staff, specially designed washing facilities and equipment, housing for species most commonly affected by oil spills, and skills and protocols for managing large numbers of animals at one time. However, none has enough staff to fully respond to a large oil spill and provide adequate care for all of the animals that might be affected. For this reason, these organizations rely heavily upon individuals with experience in handling and treating the affected species. Zookeepers often provide excellent support during oil spill responses, contributing knowledge about the husbandry requirements and behavioral attributes of the species involved. During the *Treasure* oil spill off the coast of South Africa, staff from 59 different zoos and aquariums in 14 countries assisted in the cleaning and rehabilitation of approximately 19,500 penguins, eventually returning 95% of them to the wild (IBRRC 2011).

Rehabilitation of oiled wildlife focuses primarily on the adverse physiological effects of oil on individual animals. These effects, while complex, can often be successfully counteracted through the cooperation of experienced veterinarians, biologists, and rehabilitators. Working together under the management of an organization classified as a Qualified Oiled Wildlife Responder, such a team can quickly turn an empty warehouse or other large space into a response facility for the treatment of a large number of affected animals. There the animals can be examined, stabilized through the administration of oral and intravenous fluids, sampled for legal evidence of contamination (usually photos, feather or fur samples, and swabs of oil are collected to prove that these animals were contaminated by the spilled product), and given appropriate care so that their body temperatures can normalize. Several hours or days may elapse before the animals are strong enough to withstand the stress of being washed; until then, proper hydration and nutrition must be provided. Once the animals are maintaining their body temperature, weight, and hydration, they are washed and rinsed thoroughly with detergents at temperatures necessary for removing the oil without causing the animals more harm. The rehabilitative care of the animals then continues for several days or weeks to ensure that their waterproofing, body condition, and overall health are sufficient for survival in the wild. Release of the animals is usually coordinated with the overseeing wildlife authorities (such as the US Fish and Wildlife Service).

Other large-scale wildlife disasters may be handled similarly. Such disasters may include wildfires, floods, or earthquakes, as well as botulism and other toxin-related events. In addition to contributing skills and labor during oil spills or other wildlife disasters, zoos and aquariums and their staff can also help by procuring or providing supplies, by stabilizing animals before their transportation to designated wildlife response facilities, by providing freezer space for the large amount of food needed to feed the many affected animals, by providing temporary housing for clean animals, and by coordinating the placement of animals determined to be nonreleasable. An example of such a response occurred in 2007 when sudden drops in temperature resulted in large numbers of brown pelicans (*Pelecanus occidentalis*) developing frostbite. The sudden influx of almost 30 of these large birds overwhelmed the local rehabilitation center. Local zoos were able to assist by providing knowledgeable zookeepers, medical supplies, and freezer space to store the large amounts of fish needed to feed these birds.

SUMMARY

Wildlife rehabilitation shares many similarities with the keeping and care of permanently captive wildlife. Since its goal is the eventual release of animals back into the wild, however, it differs from zookeeping in some important ways. Despite

those differences, keepers and other individuals with zoo and aquarium training and experience can play important roles in wildlife rehabilitation. Whether they share their knowledge regarding the husbandry and care of a particular species, donate supplies to local rehabilitation organizations, work together to use nonreleasable rehabilitation animals as education ambassadors, or volunteer their time and expertise to assist with disaster response, keepers can contribute tremendously to the success of wildlife rehabilitation.

REFERENCES

Barham, Peter, Robert Crawford, Les Underhill, et al. 2006. "Return to Robben Island of African Penguins That Were Rehabilitated, Relocated or Reared in Captivity following the *Treasure* Oil Spill of 2000." *Ostrich* 77 (3/4): 202–9.
Frink, Lynne. 1998. "Philosophy of Wildlife Rehabilitation: The Wildlife Rehabilitator's Guide to Self-Defense." *NWRA Quarterly Journal* (Spring 1998): 13–14.
Langenberg, Julie, and F. Joshua Dein. 1983. "Pathology of Ruddy Ducks (*Oxyura jamaicensis*) Contaminated with Spilled #6 Fuel Oil." In *The Effects of Oil on Birds: A Multi-discipline Symposium.* D. Rosie and S. N. Barnes, eds., 139–42. Wilmington, DE: Tristate Bird Rescue and Research.
Leighton, Frederick A. 1991. "The Toxicity of Petroleum Oils to Birds: An Overview." *The Effects of Oil on Wildlife: Research, Rehabilitation and General Concerns.* J. White and L. Frink, eds., 43–57. Suisun City, CA: International Wildlife Rehabilitation Council.
McTurk, Diane, and Lucy Spelman. 2005. "Hand-Rearing and Rehabilitation of Orphaned Wild Giant Otters, *Pteronura brasiliensis*, on the Rupununi River, Guyana, South America." *Zoo Biology* 24:153–67.
Miller, Erica, ed. 2012. *Minimum Standards for Wildlife Rehabilitation, Fourth Edition.* Saint Cloud, MN: National Wildlife Rehabilitators Association.
Pierce, Virginia. 1991. "Pathology of Wildlife following #2 Fuel Oil Spills." In *The Effects of Oil on Wildlife: Research, Rehabilitation and General Concerns.* J. White and L. Frink, eds., 78–94. Suisun City, CA: International Wildlife Rehabilitation Council.
Thrune, Elaine. 1997. *Wildlife Rehabilitation: A History and Perspective.* In *Principles of Wildlife Rehabilitation*, Adele Moore and Sally Joosten, eds. St. Cloud, MN: National Wildlife Rehabilitators Association.
———. 2009. "NWRA 2008 Member Survey Report, Part 1." *The Wildlife Rehabilitator* 9, no. 1.
"Treasure Report: 20,000 Patient Penguins." International Bird Rescue Research Center. Accessed 12 March 2011 at http://www.ibrrc.org/treasure_report_1.html.
Wolf, L. A. 1993. *NWRA Quick Reference.* Saint Cloud, MN: National Wildlife Rehabilitators Association.
Woltman, Sandy Heyn. 2005. "Behavioral Enrichment for Mammals." *Wildlife Rehabilitation Bulletin* 23(2): 32–35.

Part Ten

Government and Legislation

58

Introduction to Regulation of Zoos and Aquariums

John B. Stoner, Mark D. Irwin, and Aaron M. Cobaugh

INTRODUCTION

Regulation of zoos and aquariums varies a great deal around the world. This section will provide an overview of some regions' zoo and aquarium legislation. As the reader progresses through the following chapters, about regulations in Africa, Asia, Australia, Canada, Europe, New Zealand, and the United States, they will begin to appreciate the complexities of the bureaucracies that have developed to safeguard the wildlife that zoos, aquariums, and other agencies care for throughout the world. Some of the regulations, especially in countries that rely heavily on agriculture, have been developed to protect that industry from disease. In other cases, especially island nations, there is a focus on preventing the importation of invasive species that could severely impact a unique natural environment. For example, the United Kingdom and Ireland have an additional six-month quarantine on rabies-susceptible species to prevent the introduction of that disease. Each piece of legislation has its own unique background and is often a reflection of cultural differences between countries and regions.

It is impossible to cover all regulations in every region within these chapters. In some of the following sections a single country will be profiled. Others will provide a very broad overview of an entire region. For example, the reader will be given an overview of Asia that includes the Indian subcontinent and its extraordinary diversity of wildlife legislation. The chapter will begin with an introduction to two sources of international regulation, the Air Transport Association (IATA) and the Convention on the International Trade in Endangered Species (CITES).

After reading the next several chapters, the reader will:

- have a basic understanding of some regional and international legislation
- have a broad background in the diversity of zoo and aquarium regulation
- be more sympathetic to the constraints placed upon the staff responsible for shipping and receiving animals
- understand why time constraints and attention to detail are important for animal transportation, record keeping, licensing, and permitting
- appreciate why it is important for zoos and aquariums to be members of national and international associations
- comprehend the need for these organizations to be actively involved in the legislative process.

59

CITES and IATA

Andrea Drost

THE CONVENTION ON INTERNATIONAL TRADE IN ENDANGERED SPECIES OF WILD FAUNA AND FLORA

The Convention on International Trade in Endangered Species of Wild Fauna and Flora (CITES) is an agreement between countries to monitor and regulate the international trade of animals, plants, and their products. It is a means of ensuring the continued survival of specific animals and plants by safeguarding them from overexploitation caused by international trade. In 1963 the International Union for Conservation of Nature (IUCN) began the process of what would become CITES, and in 1975 CITES was initiated with at least 80 countries agreeing to the content of the convention. To date, just under 200 countries have become parties to CITES.

Countries that join CITES becomes bound by its rules, but it is up to each individual country's government to enforce these rules. They do this by establishing national laws and assigning the responsibility of implementing CITES to a management authority. In Canada, Environment Canada (EC) has this oversight, while in the United States CITES is administered by the US Fish and Wildlife Service (USFWS). The Conference of the Parties (CoP), the decision-making body of CITES, comprises representatives from all member countries

CITES categorizes species into three appendixes, according to their degree of endangerment and therefore their need of protection:

- Appendix I: Species viewed by the CoP as "currently threatened with extinction" are listed as Appendix I. In order to trade an Appendix I species (or any by-product of that species, e.g., ivory tusks, pelts, or blood) between countries, the exporting country normally issues a CITES I export permit. In the case of a specimen undergoing repeated exports from multiple countries, a CITES I re-export permit would be issued. The importing country would normally issue a CITES I import permit.
- Appendix II: Species listed as Appendix II are viewed by the CoP as "currently not threatened with extinc-

tion but vulnerable to trade which could cause the species status to be elevated to threatened with extinction." The permit requirements for an Appendix II species is normally a CITES II export (or re-export) permit from the exporting country.
- Appendix III: Decision-making for Appendix III species differs from that for species in Appendixes I and II. The individual member party can decide whether trade of a species in its particular country needs to be regulated. The CoP no longer decides if the species is threatened with extinction; instead, the individual country does. When moved out of country, an Appendix III species normally requires a CITES III export (or re-export) permit or a certificate of origin issued by the management authority of the exporting country.

INTERNATIONAL AIR TRANSPORT ASSOCIATION LIVE ANIMAL REGULATIONS

The International Air Transport Association (IATA) Live Animals Regulations (LAR) are the global standard and essential guide to transporting animals by air in a safe, humane, and cost-effective manner (IATA 2010). The LAR is applicable to airlines that abide by the IATA Multilateral Interline Traffic Agreement for Cargo. This simply means that anyone involved with shipping an animal, accepting an animal as air cargo, or loading an animal as air cargo must be familiar with the specific handling requirements for the individual species, regardless of whether the animal is being moved domestically or internationally. This will ensure that animals always travel in safe, healthy, and humane conditions. These regulations have been accepted by the Convention on International Trade in Endangered Species of Wild Fauna and Flora (CITES) and by the Office international des épizooties (OIE; World Organisation for Animal Health) as guidelines for animals being shipped either by air or by land. Federal inspection officers of most countries will look closely at live animal crates to ensure they have met the conditions set out by LAR for that particular species.

60

Government and Legislation in Africa

Dave Morgan

Africa is the world's second largest and second most populous continent, after Asia. At about 30.2 million km² including adjacent islands, it covers 6% of the Earth's total surface area and 20.4% of the total land area. It encompasses some 53 nation states and, as a consequence, legislation pertaining to zoo operation within the continent ranges from diverse to nonexistent. However, Africa continues to boast perhaps the world's largest densities of terrestrial free-ranging wild animal populations, and conceivably one of the greatest in diversity. As a result, some degree of wildlife legislation per se exists in most African countries.

From this perspective, zoo legislation is then influenced by

- common law pertaining to wildlife
- prevailing wildlife-specific legislation
- prevailing agricultural livestock legislation
- prevailing animal welfare legislation
- international conventions.

All of these will have an impact upon the initiation of zoos and also their operation. Within countries themselves, zoo legislation can also be national or federal and provincial or state. Countries may have both, neither, or indeed contradictory sets of both national and provincial legislation.

In this treatment, however, it simply is not practical to review such legislation country by country. Indeed, the African Association of Zoos and Aquaria (PAAZAB) is only tracking approximately 250 zoos in 48 countries within Africa. Instead, we will have a look at the two African countries with the highest gross domestic products (GDP) on the continent: South Africa and Nigeria. As such, these countries represent both extremes of zoo legislation on the continent, with other countries falling somewhere in between.

SOUTH AFRICA

As in a number of African countries, South African wildlife legislation is predicated on the common-law concept of res

nullius. As of March 2011, the encyclopedia website Wikipedia defined res nullius as "a Latin term derived from Roman law whereby res (objects in the legal sense, anything that can be owned, even slaves, but not subjects in law such as citizens) are not yet the object of rights of any specific subject. Such items are considered ownerless property and are usually free to be owned. Examples of res nullius in the socio-economic sphere are wild animals or abandoned property." This means, then, that private ownership of wild animals is legal, and that licensing authorities can only attach conditions to that ownership. This in turn allows the existence of private commercial zoos.

In South Africa, there are currently in excess of 80 "zoo-like" facilities and probably more than 100 other private nature reserves, game farms, and hunting operations that include captive breeding facilities. This situation poses two questions: "What exactly is a zoo, anyway?" and "How should legislation pertaining to zoos differ from that pertaining to other captive animal operations?" In themselves these questions have become significant challenges to developing specific zoo legislation in the country, as two separate government departments have a say in this regard. The South African Department of Water and Environmental Affairs has promulgated the National Environmental Management: Biodiversity Act (NEMBA), whereas the South African Department of Agriculture has the Animals Protection Act and the Performing Animals Act.

NEMBA regulates the registration of all facilities that have anything whatsoever to do with wildlife. It also defines them; zoos are defined as "commercial exhibit facilities" irrespective of the size or scope of their operation. It also defines sanctuaries and animal rehabilitation facilities as being two separate and distinct operations. Finally the act lists prohibited activities for certain species of indigenous wildlife. All of which of course affect zoo operation.

The Animal Protection Act primarily regulates draught animal use and animal fighting for sport. It also mandates a magistrate to determine a contravention of the act according to the degree of offense caused to the community. This effects zoo operation in that it consequently prohibits the feeding of live animals—as in live rodents to reptiles, for example. Also, if a local community takes offense to anything a zoo does, it theoretically may be convicted under the Animals Protection Act.

The Performing Animals Act was designed to regulate the use of animals in public display, with special emphasis on circuses. However, it is inconsistently applied throughout the country, with some legally mandated animal welfare organizations insisting upon its application to nongovernmental zoos. The act requires a magistrate to issue a license authorizing the facility to operate under the act. The magistrate can be influenced by animal welfare organizations. Zoos out of favor with their local animal welfare organizations might find their Performing Animals Act license barred as a result.

Both the Animals Protection Act and the Performing Animals Act were promulgated in the 1950s, and are actually both inappropriate and irrelevant to modern zoo operation, as their primary intent was to prohibit the use of animals for fighting and to control their use for entertainment purposes. Indeed, the Performing Animals Act specifically excludes "public zoos." However, both acts may be used by animal rights or welfare organizations against zoo operation, as South Africa lacks a clear-cut legal definition of what constitutes a "public zoo."

The Department of Animal Health also operates under the South African Department of Agriculture. Certain animal health legislation affects zoo operation in terms of quarantine, locale, and transportation restrictions on wild animals. It is worth noting, too, that such biosecurity regulations can, and usually do, overrule almost all elements of wildlife legislation.

This legislation is of course at a national or federal level. South Africa consists of nine provinces or states, each of which has its own nature conservation department and its own zoo and wildlife legislation. The most stringent legislation is to be found in the province of KwaZulu-Natal, where the provincial wildlife authority, eVimvelo KZN Wildlife, has promulgated a policy for the management of ex situ wild animals. This policy requires all zoos in the province to establish awareness programs and conservation research and breeding programs. The policy also states species-specific minimum requirements in terms of space and enrichment.

In 2000 PAAZAB, in collaboration with the South African Bureau of Standards and the National Council for the Societies for the Prevention of Cruelty to Animals, developed the South African National Code for Modern Zoo and Aquarium Practice. This document was based upon the British Secretaries of State Standard of Modern Zoo and Aquarium Practice, and it covers all elements of zoo operation, from animal welfare to visitor services. The South African code was officially gazetted in the South African Government Gazette in 2004 as "SANS 10379; the South African Code for Zoo and Aquarium Practice." However, compliance with the standard is entirely voluntary, and to date not a single South African zoo has applied for certification against SANS 10379.

NIGERIA

Nigeria has very different regulations. PAAZAB speculates that there may be in excess of 100 zoos or zoo-type facilities in Nigeria. However, no official figures exist. National wildlife legislation is restricted to the National Fauna Conservation Law, which makes no reference to zoos or captive animal facilities. Indeed, no federal/national zoo or animal welfare legislation appears to exist.

Nigeria consists of 36 states, and each has some form of wildlife legislation. In the case of Kano state (located in northern Nigeria, on the edge of the Sahel), the state arm is the Kano State Zoological and Wildlife Management Agency, mandated by the Zoological and Wildlife Agency Law. However, the law solely refers to the establishments and regulation of the agency itself. It does not regulate zoo operation or zoo initiation.

AFRICAN ZOO LEGISLATION: POINTS TO CONSIDER

In the African context, wildlife legislation is more often than not merely a list of prohibited activities regarding animals. While such legislation may influence the initiation of zoos, it rarely prescribes zoo standards or animal welfare. Also, national agricultural legislation will influence zoo activities to a degree, and not always positively.

The fact remains that the absence of appropriate or at least adequate zoo-specific legislation—not just in African countries—is problematic for the global zoo community. That community is judged by its weakest member, and the absence of zoo legislation allows for the proliferation of zoos of a very poor standard.

PAAZAB has set mandatory operational standards for its institutional members. These standards were taken directly from the South African National Code of Zoo and Aquarium Practice, SANS 10379, partially as a means of getting them enacted, and also to give the association operational quality parameters for its membership. The "Ops Std," as it is referred to by PAAZAB, covers all aspects of zoo operation including welfare, enrichment, veterinary health, nutrition, education, conservation, research, staff training, and visitor services, and is very similar to the EAZA and AZA standards of accreditation. PAAZAB members are thus required under the association's constitution to demonstrate compliance with the standard in an audit conducted by association personnel. However, this requirement only pertains to PAAZAB members.

The Pan African Sanctuary Alliance (PASA) similarly governs its primate sanctuary members, while the Animal Keepers Association of Africa (AKAA) espouse core animal welfare values to which its individual members are required to sign.

61

Government and Legislation in Asia

Sally R. Walker

"Asia" and the topic "legislation" have many commonalities: both entities are large, dissimilar and similar, subtle, uncertain, complex, and even contradictory. Asia, with 43 countries and a few protectorates in five regions, is the most diverse of the world's seven continents. Zoo legislation, or even environmental legislation referring to zoos, is diverse where it exists, reflecting Asian countries' widely varied cultures, attitudes, politics, and economies.

Asia is the largest continent in both area and population. Zoo legislation is patchy in each region and in most of the countries. Unlike other large land areas in North America, Europe, and Australia/New Zealand, each region in Asia contains from 5 to 17 distinct countries. Each has its own government, economy, developmental status, religions, and traditions which affect its concept of a zoo and its administration.

Therefore, many zoos in Asia look good on the outside, with their dramatic size and displays, but are not keeping up on the inside, where ethics, interest, and technical zoo animal management skills are required. Moreover, zoo legislation as such is meager in Asia, and this contributes to the imbalance.

LEGISLATION

It would appear that Asian zoo legislation is sparse. Many Asian countries now have wildlife and welfare legislation, which sometimes names zoos generally if any laws in the document apply. That is better than no legislation, but governments should know that specific standards of zoo exhibition and care are of paramount importance, and should include them. These standards are conspicuously absent in many of the Asian countries, whereas they are of primary importance in many other parts of the world. Western countries legislate standards of care and carry out inspections to ensure that zoos are following the law. Zoo associations work together with governments on accreditation or recognition. In Asia, few countries include standards of care, welfare, or display, and none have taxon-specific legislation that details precise

values for different animal groups. The United Kingdom, Europe, and Australia have these standards in governmental legislation and national and or regional accreditation, while the American Zoo Association (AZA) covers it in their accreditation program.

It is noteworthy that most Asian national wildlife legislation rarely refers to zoos as part of the conservation effort. Ironically, most Asian forest and wildlife legislation includes strengthening, restocking, or reintroduction for declining wildlife populations, and also the leisure aspect of wildlife without linking zoos. It is as if they cannot conceive of zoos being helpful to wildlife and environmental managers. Another interesting feature of Asian zoos is the paucity of regional or national zoo associations, "Friends of the Zoo," and other nongovernmental organizations that are so productive, helpful, and appreciated in Western countries. Asian governments seem reticent to recognize the few extant organizations unless they are themselves part of government.

SOUTH ASIA

South Asia includes eight countries: Afghanistan, Bangladesh, Bhutan, India, Maldives, Nepal, Pakistan, Sri Lanka. Two of these countries have zoo legislation, and two more have drafts of legislation that are plodding through various governmental offices until their passage. Other South Asian countries will most likely opt for it in the coming years. The South Asian Zoo Association for Regional Cooperation (SAZARC) was founded to encourage zoo legislation and attention to conservation, welfare, and standards.

Sri Lanka was the first country in Asia to pass legislation specifically for zoos and not as part of other legislation. The National Zoological Gardens Act (No. 42 of 1982) provided for administration and management of the National Zoo, a zoo fund, a description of the zoo director's duties and responsibilities, standards of behavior, and penalties for noncompliant visitors.

When **India** became independent there was great interest

Case Study 61.1. Government and Zoo Legislation in India

India, the largest country in South Asia, probably has done the most of any country in the world legislatively to improve the quality and control the quantity of its once insufficiently managed but now dramatically proliferating zoos. In the decades after Indian independence (1947) there was official recognition that the existing zoos were not keeping up with the rest of the world, and many well-meant actions were taken to improve them, including establishment of a zoo wing in the Indian Board for Wildlife, the establishment of an Indian expert committee, a visit of American zoo experts, inclusion of zoos in the Wildlife Action Plan, a meeting of Indian zoo directors, and other recommendations which largely stayed on paper.

A seminal event took place in 1987: A list of 187 zoos was published in a zoo magazine (Walker 1987) that challenged the official ministry list of 44. This demonstrated that zoos were springing up faster than government could keep up with. In 1988, the Department of Environment created the post of joint director for zoo affairs in the ministry, and from this point on, regular progress took place. Serious steps had been taken during these four decades to improve existing zoos, but in the end it was determined that the way forward was to develop serious and stringent national legislation. An enthusiastic minister gave the order for ministry officials to create the framework for the new Central Zoo Authority (CZA) and operational principles for implementing zoo legislation. An act establishing CZA was passed in 1991, and another one establishing norms and standards was passed in 1992.

According to law, all zoos were then (as now) required to register with the government. When the registration process was complete, the number of zoos was about 450. According to these new laws, all zoos had to be inspected and given time to improve. The zoos were also given funds by the CZA to do the work in a 50% matching scheme in which their respective states also contributed funds and did the work, after which they would be reinspected and either given recognition or be ordered to close.

Over the next decade, CZA gave recognition to almost 200 zoos, while closing more than 200 wild animal facilities and distributing those animals to recognized zoos. Also, a vast number of projects were designed and carried out to bring Indian zoos up to international standard and improve animal welfare. Those projects included training for all levels of zoo personnel, high-level committees (e.g., zoo design, education, and technical guidance), zoo associations (e.g., for curators, keepers, and educators), institution of small grants and fellowships for zoo personnel to conduct research projects, and establishment of a high-level DNA research facility.

Over the next 15 years there were a number of amendments to the act. Whenever CZA decided an existing statute was insufficient or inappropriate, it would improve the statute as an ongoing process. Later it was decided that CZA would set up regional offices that will have closer relations with zoos of north, northeast, south, west, and central India. These regional offices help insure that needs are met and work completed more effectively and efficiently for the further improvement of zoos.

in all wildlife, and also in zoos. Decades later it was realized that zoos had proliferated too rapidly and were out of control. Only legislation could bring about restraint. India's story of addressing this and other problems is so remarkable that a case study has been included in this book.

Afghanistan and **Bhutan,** with only one zoo and one Takin Centre respectively, have yet to see the need for zoo legislation, although both countries plan other wild animal facilities. Current and future zoos would benefit a great deal by having the structure and incentive provided by standards and guidelines.

Bangladesh has about 10 zoos currently managed by different governmental authorities. The government has drafted zoo legislation that various ministries must examine. The South Asian Zoo Association for Regional Cooperation (SAZARC) is a strong encouraging force, but transfer of senior ministry officials who could push legislation is an obstacle to its timely passage.

In **Nepal,** concerned officials identified a need for captive animal legislation and have acted on it. Now, the Nepalese government has tracked the number of captive facilities or their agencies and has drawn up legislation which is shy of one signature to become law. In addition, the Central Zoo, Kathmandu, which took the lead in convincing government of the need for legislation, has also created a network which brings all zoos in Nepal together for meetings.

There is no central zoo legislation for captive wild ani-

mals in **Pakistan**, but there is wildlife legislation in the four provinces, some of which can be applied to zoos. There are no standards for exhibition of animals or inspections, but animal welfare legislation can be used in some situations. Pakistan has a number of zoos, some of which date from the 19th century and others which are recently constructed.

SOUTHEAST ASIA

Southeast Asia includes ten countries: Singapore, Brunei, Malaysia, Indonesia, Thailand, Philippines, Laos, Cambodia, Myanmar, and Vietnam. The zoos in this region have some specific or implied legislated requirements within wildlife legislation. The "connect" between zoos and conservation of wildlife is lacking in these laws, but the regional association, South East Asian Zoos Association (SEAZA), connects more than 100 zoos and parks and offers training in conservation, helpful inspections, and advice.

Three zoos in **Brunei** are under some governmental authority: Tamburong Zoo, Louis Mini Zoo, and Hassanal Bookiah Aquarium, all of which are of good repute. All but one of the royal family's private collections, four zoos and an aquarium, are open to the public. His Majesty the Sultan of Brunei is to make rules regarding control of ownership and of import or export of any wildlife species.

Currently the organization of zoos in **Thailand** is on two levels: official zoo endeavors via the Zoological Parks Organization (ZPO) of the government of Thailand, and independent spurious facilities that are unregulated. In 1992 a Wildlife Preservation and Protection Act (BE 2535) was passed that applied to ex situ conservation in public zoos, as defined in chapter V.29 of the act. It includes rules for permission, establishment of zoos, compliance with notifications, and documenting of species.

Captive facilities in the **Philippines** include zoos, breeding centers, and rescue centers. Although strengthened with relatively recent legislation, enforcement is not effective. The Department of Environment instituted an accreditation process for zoos and wildlife facilities to control trade and improve standards, but implementation is scant.

Laos (Lao People's Democratic Republic) is a rural country devastated by recent wars. The primary zoo in the country is the Vientiane Zoo, founded in 1992 and located in Vientiane, the national capital. There are other zoos associated with hotels and private menageries. Lao's zoo legislation was passed in 2007 and is the major piece of legislation with application to captive facilities, covering wild capture, use of wildlife for public benefit (zoos, aquariums, etc.), need for veterinary units, and so on.

Wars destroyed the modern zoos established in the cities of colonial **Cambodia**. Today the Phnom Tamao Zoo, located outside Phnom Penh, founded in 1995 and managed by Department of Forestry and Wildlife, seems the only major zoo in the country. It is large and includes carnivore, bird, and herbivore parks, with native species donated or confiscated locally. Draft circulars have been issued by the State Wildlife Department that refer to zoos, but they are not available.

Currently there are two main zoos in **South Vietnam**, along with several small ones owned by businesses and private individuals. There are also several breeding and rescue centers for conserving endangered species and reintroducing animals back into the wild. Environmental legislation in South Vietnam is written in Vietnamese. There are translations of Vietnamese laws that include the management of captive bear facilities (due to the bear bile industry) up to the most detailed biodiversity legislation. Also available online are lists of decisions, circulars, and decrees from **North Vietnam** relating to captive management of bears, wild plants and animals, and management of import and export, including CITES regulations and penalties (Education for Nature-Vietnam: www.envietnam.org/library/law-library .html).

Of importance in **Singapore** is the Wild Animals and Birds Act, passed in 1965, which has had frequent revisions, most recently in 2002. Many other laws, such as the Animal Welfare Law, apply to captive animals, including licensing for commercial animals (e.g., those for sale and for exhibition), and the Endangered Species Act. Taken together, the requirements that apply to zoos in Singapore are stringent, and Singapore's animal facilities reflect this.

There are three zoos in **Myanmar**: Yangon Zoo, Mandalay Zoo, and the relatively new Nay Pyi Taw Garden Project. Myanmar's only apparent zoo legislation is a regulation that provides for the importation of wild animals going to zoos, issued under auspices of the Ministry of Livestock. This requires a valid import license from the Myanmar authority, disease certification, certification of year-long disease-free area of provenance, and quarantine on arrival for a minimum of 14 days.

In **Malaysia** the Association of Zoological Parks and Aquaria, founded in 1996, is an aid to that nation's zoos. The Malaysian government also encourages good management and maintenance of zoos through its Wildlife Protection Act of 1973. A national zoo policy provides guidelines with minimum standards for safety, animal welfare, veterinary medicine, and enclosure design. Separate territories in Malaysia each have their own wildlife departments and regulations that apply to captive facilities.

Indonesia has made substantial efforts to organize and raise the standards of its zoos. The Indonesian Zoological Parks Association is a partner of the Ministry of Forestry and the Indonesian Scientific Institution. There are a diversity of zoos on three islands, including traditional zoological gardens, aquariums, bird parks, reptile parks, and butterfly parks. The most recent legislation for Indonesia is Ministerial Decree No. 479/Kpts-11/1998, which describes the functions of zoos and special fauna parks.

EAST ASIA

Of the six East Asian countries—China, Japan, South Korea, Taiwan, North Korea, and Mongolia—only Mongolia does not have a zoo or any similar captive animal facility. All other countries in the region have two to hundreds of zoos. None has a trace of zoo legislation as such, but the ways in which the different countries mind their zoos is varied and interesting.

China has as many as 700 zoos, but no specific zoo legislation. It does have a number of laws, acts, and decrees focused upon wild animal protection and management that can be brought to bear on zoos, if there is sufficient coordination between different agencies. The State Forestry Administration (SFA) has demonstrated its ability to issue and implement controls in zoos. The Chinese Association of Zoological Gardens (CAZG), itself a government organization, can, with its connections to other government ministries, put pressure on zoos that are out of line.

Japan lacks specific legislation for zoos but other laws cover some of the need for zoo regulation. The Act on Welfare and Management covers exhibition standards and breeding permits, although it places zoos and pet shops on the same level. The Japanese Association of Zoos and Aquariums (JAZA), with 89 member zoos, has brought out a husbandry manual with standards for care and exhibition and also acts as a monitor for quality and conditions.

Taiwan has at least 14 zoos or zoolike facilities displaying animals for public viewing. The Taiwan Aquarium and Zoological Parks Association (TAZA) currently has 14 institutional members, including zoos, aquariums, bird parks, museums, and rescue centers. There are two major pieces of legislation, the Taiwan Conservation Law and the Taiwan Animal Protection Law, that have to be followed by zoos, aquariums, and animal ranches, although these are not specifically zoo legislation.

Hong Kong is a special administrative region of the People's Republic of China (PRC) and since 1871 has had a public zoo, the Hong Kong Zoological and Botanical Gardens. There are other zoos in Hong Kong now, most notably Ocean

Park Hong Kong. There is no legislation devoted specifically to zoos, but certain portions of other legislation, such as the Amendments to the Protection of Endangered Species of Animals and Plants Ordinance, apply to zoos.

North Korea has one zoo, Pyongyang Central Zoo, with 5,000 individual animals and 600 species. There is no indication of zoo legislation, zoo ethics, or standards of care, as information about policy is not easily forthcoming. Certain animal species have been designated as "natural monuments" by the Cultural Properties Preservation Law.

South Korea can boast about the Seoul Grand Park Zoo, which is the 10th largest zoo in the world. There are also three other zoos in the country. There is no zoo legislation, but it may not be required in this disciplined country. The Animal Protection Act (1991, revised 2004) describes very simple standards of animal care and provides for wild animals to go to a zoo. The Protection of Wild Fauna and Flora Act (2004, revised 2008) purports to prevent extinctions by protecting and managing species in their habitat as well as in "ex-habitat conservation agencies."

WEST ASIA (MIDDLE EAST)

West Asia includes two groups: the federation of seven countries known as the United Arab Emirates (UAE)—Abu Dhabi, Dubai, Sharjah, Ajman, Umm al-Qaiwain, Ras al-Khaimah and Fujaira—and the other countries of the region: Bahrain, Georgia, Iran, Iraq, Israel, Jordan, Kuwait, Lebanon, Oman, Qatar, Saudi Arabia, Syria, and Yemen. Nearly all these countries have zoos, but none have zoo legislation as such. Some include zoos in their wildlife or environmental legislation, and some have no regulation at all. However, in December 2012 a well-attended meeting of Arab zoo personnel was held to discuss the need for an Arab regional zoo association, backed by Al Ain Zoo and its general director, Ghanim Al Hajeri. Representatives from the World Association of Zoos and Aquariums (WAZA) and other regional zoo associations assisted in discussions, and it was decided overwhelmingly to undertake this important step.

Zoos in the **United Arab Emirates** are privately owned and "governed" by the wealthy sheikhs who own them. This is also the case in some non-Emirates countries, as some zoos are owned by wealthy individuals and others are not. Many zoos not under the guidance of wealthy patrons are poorly organized and dysfunctional. In some countries, CITES legislation permits the registration of "breeding centers," and zoos that are so recognized can get import/export permits for wild animals. Animal welfare legislation in the Emirates covers captive and wild animals and stipulates adequate diet, safe transport, and medical treatment for them. The Emirates have adopted the IATA Live Animals Regulations (LAR), and were the first in the Middle East to do so.

Of the non-Emirates countries in this region, Georgia, Jordan, and Israel each have an administrative setup that seems to have direction and controls. The **Georgia** Law on Wildlife, Article 21, covers captivity, ex situ conservation, zoological collection, and the creation and filling of zoos. It includes some standards other than "appropriate care," such as dimensions of fences and cages. **Jordan,** with its Directive No. Z/44(2003), establishes technical and sanitary require-

ments for zoos and other animal facilities with articles on definitions; a Committee for Licensing Zoos; steps and conditions for licensing, transportation, fences, and cages; penalties for offenses. In **Israel** the Wildlife Protection Law 5715-1955 empowers the Minister of Agriculture to implement it and to make regulations for wildlife in zoos and other animal facilities. There are 15 known zoos and aquariums, all with different themes and venues, some with unique specialties and objectives. Several Israeli zoos have nongovernmental organizations (NGOs) contributing to their quality and care in various ways.

Saudi Arabia has several small zoos; two private facilities that display and breed animals, Assayd Nature Reserve and Riyadh Zoo (a municipal zoo); and two research centers managed by the National Commission for Wildlife Conservation. There is no obvious zoo legislation in the country, perhaps due to the sheikhs' interest and involvement. In **Bahrain,** environmental law is pending under pressure from animal welfare groups to pass legislation for general animal welfare protection. Al Areen Nature Reserve protects a large community of local birds and mammals, and is divided into two parts, for researchers and other visitors. The only public zoological facility in **Oman** is the Oman Mammal Breeding Centre, of Arabian Oryx fame. Oman is credited by United Nations Environment Programme (UNEP) with establishing one of the world's best records in environmental conservation, along with one of the world's most seriously "green" governments. There is a plethora of legislation for wildlife and wildlife protected areas, in the form of ministerial decrees and laws.

The only zoo in **Kuwait** was destroyed in the 1991 war with Iraq and later rebuilt with volunteer help, but it has no apparent controls or legislation. **Iraq** has Baghdad Zoo in Baghdad, which was nearly destroyed in the Gulf War, and bits of it were salvaged with help from zoos and welfare NGOs. Zoo legislation may be a long time coming to Iraq. **Iran** has two significant zoos in Tehran and Mashhad, and a few small zoos, but it still has no obvious zoo laws. The Department of the Environment runs the Pardisan Nature Park and also maintains an animal orphanage with no apparent legal controls. In **Qatar,** Doha Zoo seems to be the only significant public zoo and is a major tourist attraction in the country, but it is not without problems. **Syria**'s only functioning zoo is Duma Zoo, which is not of a good standard. Another facility, called Zoo Damascus, is under construction and looks very promising, with a sophisticated theme and two very professional architectural firms involved.

CENTRAL ASIA

The nations of Central Asia (Kazakhstan, Kyrgyzstan, Tajikistan, Turkmenistan, and Uzbekistan) were once parts of the Soviet Union, and in those days they each had one or more zoos. This region has passed through very difficult times during its civil wars, and its zoos suffered losses of animals, infrastructure, and finance. The new countries are still in the early stages of establishing their various governmental agencies.

Kazakhstan has legislation in place that covers zoos reasonably well, although it is not called zoo legislation and may not apply to the two extant zoos, Almaty Zoo and Karaganda

Zoo, which are not without problems. The legislation is part of a greater government conservation project that includes the setting up of elaborate state zoological parks.

Turkmenistan's Environmental Law covers some aspects of zoos, such as State Protected Natural Areas (1992) and Protection and Rational use of Fauna (1997). In this country the government has recently opened the National Wildlife Museum Zoo, but a lack of standards is already affecting the animals, which suffer from a shortage of shade and water in 30 °C heat. **Tajikistan** and **Uzbekistan** each have at least one public zoo, but no legislation and no effective controls by any agency. Uzbekistan's only official zoo was set up by the Cabinet of Ministers in 1997 and is a member of the European Zoo Association, which has a Code of Ethics and Standards that may compensate for lack of government-imposed standards. **Kyrgyzstan**'s only zoo was private and belonged to the country's now deposed ruler. It is now closed and the animals have been moved outside the country.

REFERENCES

Anon. 2009. *Zoos in India: Legislation, Policy, Guidelines and Strategies.* New Delhi: Central Zoo Authority.

Bell, C., ed. 2001. *Encyclopedia of the World's Zoos*, Vols. I, II, III. Chicago: Fitzroy Dearborn.

Indian Zoo Yearbook 3:126–97. New Delhi: Indian Zoo Directors Association and Central Zoo Authority,.

Kisling, V. N., Jr., ed. 200. *Zoo and Aquarium History: Ancient Animal Collections to Zoological Gardens.* Boca Raton, FL: CRC Press.

Spitsin, Vladimir, ed. 2009. *Information Issue of Eurasian Regional Association of Zoos and Aquariums* 28, no. 28. Moscow: Department of Culture.

Walker, Sally. 1987. "How Many Zoos?" *Zoo's Print: Journal of Zoo Outreach Organisation* 2:7–10.

62

Government and Legislation in Australia

Sara F. K. Brice

INTRODUCTION

Legislation is the act of making or enacting laws which are enforceable by a management authority; penalties may apply if breaches of legislation occur. Legislation provides the framework under which zoos and aquariums in Australia can operate; the resulting regulations influence species selection and the way animals are displayed, transferred between institutions, and collected from the wild. It also determines how animal health and welfare standards are maintained.

The Commonwealth of Australia comprises six states (Queensland, New South Wales, Victoria, Tasmania, South Australia, and Western Australia) and two territories (Australian Capital Territory and Northern Territory). The diversity of zoo and wildlife parks within Australia includes 58 institutions that are full and associate members of the Zoo Aquarium Association, Australasia (ZAA). These include zoos that are government-owned (14), not-for-profit trusts (4), and privately owned institutions (40). Most of the institutions are located in the eastern side of the country, including in Queensland (22), New South Wales (15), and Victoria (7). There are also at least 100 small wildlife parks that are not ZAA members, most of which display only Australian native species and domesticated animals. Holders of native and exotic animals are licensed under state law; in New South Wales, for example, zoos and aquariums are licensed under the Exhibited Animals Protection Act 1986 (EAPA) and therefore need to comply with state regulations.

Australian zoos and zoo-related activities are regulated by legislation on the Commonwealth (national) and state levels. Commonwealth legislation is developed to deal with issues which need a consistent national approach, such as quarantine risks, customs, and international treaties (e.g., CITES). State laws govern more localized concerns, such as the protection of native flora and fauna within the state, protection from the incursion of pest species across state borders,

and licensing of zoos and aquariums, wildlife parks, and sanctuaries. State laws may have similarities across the states, but they operate separately from each other; therefore, there can be substantial disparity between the legislation of different states. This chapter describes commonwealth legislation, state legislation, and other standards that affect Australian zoo operations; see table 62.1.

COMMONWEALTH LEGISLATION

QUARANTINE ACT 1908

The Quarantine Act 1908 is probably one of the oldest laws that specifically refer to zoos and aquariums in Australia. Australian zoos and aquariums have been, and are still widely perceived as, potential agents for the introduction of exotic pests and diseases into Australia by the importation and holding of exotic and nondomesticated animals. Australia currently uses a process of import risk analysis (IRA) to provide a policy enabling the importation of products and to manage their potential quarantine risk. The IRA process can take at least two years to complete. Previous and current controls have been effective in controlling major disease outbreaks; Australia is free from rabies, foot-and-mouth disease (FMD), and bovine spongiform encephalopathy (BSE).

The IRA can have a direct impact on Australian zoo operations by restricting zoo animal imports. For example, an IRA for Bovidae was suspended in 2001 following outbreaks of BSE and FMD in Europe, the United Kingdom, and Japan (Animal Health Australia website, last modified May 22, 2009). As a result, some Australian captive populations of exotic hoofstock are facing the real possibility of a demographic and genetic crisis (Barlow and Hibbard 2006). A similar ban was placed on rodents in 2002 pending a review of blood parasites and the possibility of the introduction of insect-borne diseases. This ban was lifted in 2009 when importation of a limited number of zoo-resident rodent

TABLE 62.1. Legislation and standards relating to Australian zoos

Level	Legislation/ standards	Management authority (current)	Scope	Impact on zoo operations
Commonwealth (national)	Quarantine Act (1908)	Australian Quarantine and Inspection Service (AQIS)	Protects Australia's international borders from incursions by exotic pests and diseases.	• Protects Australian zoos from disease risk. • Requires comprehensive assessment for quarantine approved premises. • Places restrictions of imports of certain orders of species, resulting in reduction of species and genetic diversity within zoos.
Commonwealth (national)	Environment Protection and Biodiversity Conservation Act (1999)	Department of Sustainability, Environment, Water, Population, and Communities (DSEWPaC)	Provides framework to protect national biodiversity, and regulates the import and export of Australian native fauna and CITES listed species.	• Administers the transfer of CITES species internationally. • Sanctions the import and export of Australian native species. • Requires significant periods of time to acquire permits.
State (New South Wales)	Exhibited Animals Protection Act (1986)	Department of Primary Industries New South Wales (DPI NSW)	Provides prescribed standards for the exhibition of all vertebrate animals, irrespective of whether they are native, exotic, or domestic. Each exhibitor must have a license to operate and also a license to exhibit specific species.	• Specifies minimum enclosure size, design, and interpretation. • Renders enclosure construction and transfer of specimens subject to interpretation, which in turn affects how standards are applied. • Requires permanent method of identification for animal specimens.
State (New South Wales)	Non-Indigenous Animals Act (1987)	Industry and Investment New South Wales (I & I NSW)	Provides a model to protect the state from the release and spread of harmful exotic species.	• Requires extra time for animal transfers, to allow for processing time.
State (New South Wales)	Threatened Species Conservation Act (1995) (NB: There is comparable legislation in all other states and territories.)	Office of Environment and Heritage New South Wales (OEH NSW)	Aims to protect terrestrial threatened species, populations, and ecological communities within New South Wales. Lists threatened species and regulates their use by licensing.	• Requires zoos to obtain licenses to collect from the wild for display and research.
State (New South Wales)	National Parks and Wildlife Act (1974)	Office of Environment and Heritage New South Wales (OEH NSW)	Aims to protect native flora and fauna species.	• Requires zoos to obtain licenses to transfer native species over state borders.
State (New South Wales)	New South Wales Occupational Health and Safety Act (2000)	Work Cover (New South Wales)	Protects staff in workplaces against risks to health or safety arising out of work-related activities.	• Requires zoo staff to take reasonable care to prevent work-related injuries to themselves and others.
State (Western Australia)	Agriculture and Related Resources Protection Act (1976)	Department of Agriculture and Food (Western Australia)	Provides framework to protect the state from introduced pests and diseases.	• Protects the state from introduction of pest species. • Monitors animal movements between states.
Nonstatutory	IATA Live Animal Regulations Standards	International Air Transport Association	Provides minimum standards for transporting animals in containers.	• Defines animal crate design and size.
Nonstatutory	Vertebrate Pest Committee (national)	Representatives from all state and territory pest control agencies	Provides coordinated response to vertebrate pest management in Australia.	• Can place bans on exotic species. • Requires extra time for animal transfers to allow for processing.
Nonstatutory	ZAA (Australasia) Code of Practice	Zoo Aquarium Association (Australasia)	Promotes high standards of animal welfare and exhibition by all institutional members.	• Promotes best practices within zoos belonging to ZAA. • Provides guidelines for animal euthanasia, regional collection planning and species management, animal transactions, and animal record keeping. • Requires permanent method of identification for animal specimens.
Nonstatutory	National standards: under development Australian Animal Welfare Strategy (AAWS)	Department of Agriculture, Fisheries, and Forestry	Develops new, nationally consistent policies and standards that will be implemented in all Australian states and territories.	• Will provide prescriptive standards for the display and welfare of animals kept in captivity.

species (capybara [*Hydrochoerus hydrochaeris*], Brazilian agoutis [*Dasyprocta leporina*], Patagonian maras [*Dolichotis patagonum*], and Cape porcupines [*Hystrix africaeaustralis*] from Canada, New Zealand, the United States, and Europe) was permitted (www.daff.gov.au 2008). There is also a current ban on the importation of parrot species, which is likely to continue until more knowledge and testing methods for a range of diseases hosted by parrots become available (www .daff.gov.au 1999).

The Quarantine Act is administered by the Australian Quarantine Inspection Service (AQIS). This federal government body manages the operational side of animal imports. They are responsible for accrediting quarantine-approved premises (QAP), which can be located at zoos and at animal inspection areas of airports.

Importing an animal into Australia is conditional on compliance with a number of health and travel stipulations associated with an AQIS import permit. Strict health requirements, such as comprehensive screening for specific diseases and mandatory isolation periods, can lead to delays in obtaining new animals. On occasion, animals are unable to arrive because the sending institutions are unable to comply with Australia's import conditions.

ENVIRONMENT PROTECTION AND BIODIVERSITY CONSERVATION ACT 1999

The Environment Protection and Biodiversity Conservation Act 1999 (EPBC Act), administered by the Department of Sustainability, Environment, Water, Population and Communities (DSEWPaC), is the mechanism which allows for the protection of native species and communities on a national and international level; for example, management of world heritage properties, places of national significance, and migratory and national marine species. The act is also the legislative basis for meeting Australia's responsibilities under CITES, and it allows for the import and export of CITES-listed and Australian native species subject to the act. With few exceptions, each time a zoo in Australia wishes to move an animal internationally, it seeks a permit from DSEWPaC. Most often animals are transferred for exhibition purposes, and therefore an Australian import/export permit requires a comprehensive description of the facility (usually provided by a keeper, who completes a specific application form) where the animal is to be housed, and an explanation of how the organization intends to display information about the cultural, scientific, or conservation value of the species.

The level of information required for a permit can lead to lengthy delays in its issue; some such permits have taken more than a year to obtain. The following specific provisions of the act associated with the import and export of animals goes some way to explain the length of time required to obtain the Australian permit:

I. Only animals belonging to the "list of species approved for live import" under the EPBC Act will be issued with import permits. The species included in the list are those that were imported from 1984 and those subsequently added under the current act. If a species is not included in the list, there is a process for adding it to the approved register, but it can take more than a year.

II. For the live export of a koala, platypus, wombat, or Tasmanian devil or an animal of an eligible listed threatened species, an Ambassador Agreement must be made between DSEWPAC, the exporter, and the importer regarding the treatment and disposal of the animal and any of its progeny. The ambassador agreement applies not only to the specimen exported but also to those specimens belonging to the same species that the zoo already holds in its collection.

III. A CITES I–listed species is permitted to travel in or out of Australia only for the purposes of conservation breeding and propagation. It therefore has to be part of a cooperative conservation program (CCP). Development of a CCP is a complex process, and it requires the collaboration of all zoos that hold the species within the Australasian region and within the region where the species is to be transferred.

Apart from the expense of transferring animals internationally, the above stipulations can be a barrier to a zoo importing new specimens for its collection or providing animals to overseas zoos. These decisions are usually determined by the curator or director of a zoo or aquarium, but it is valuable for the keeper to understand the complexities of moving animals to or from Australia.

VERTEBRATE PEST COMMITTEE AND NON-INDIGENOUS ANIMALS ACT 1987

Australia is an island continent, and the majority of its species are endemic to the continent. Since European settlement, many introduced and native animals have become pests, establishing large populations across Australia and causing significant pressures on its environment and agricultural production. Australia is very determined to protect its environment and agricultural sectors from vertebrate pests on both national and state levels. Each state has legislation and policies designed to manage the 800-plus nonindigenous vertebrate species currently included in the VPC List July 2007, documented on the Feral website (www.feral.org.au)—for example, the Non-Indigenous Animals Act 1987 (in New South Wales) and the Agriculture and Related Resources Protection Act 1976 (in Western Australia). Each state has the legislative power to prohibit nonindigenous species from entering its borders and to dictate how and where they are held. For example, certain exotic species are allowed only if they are held in statutory zoos, which are considered to be highly secure in comparison to smaller wildlife parks which only exhibit native fauna.

Nationally, the role of coordinating a pest control policy falls to the Vertebrate Pest Committee (VPC). The VPC is made up from members of each state or territory as well as from other relevant government bodies. It determines the potential threat an exotic species may have to the environment, primary agricultural production, and human safety. Each species is assessed on the risk of its endangering people, damaging agriculture, and establishing itself in the wild. The

VPC is a policy-level group that lacks the power to change legislation but generally looks to harmonize regulation across jurisdictions.

STATE LEGISLATION

STATE WILDLIFE ACTS

All Australian states have independent legislation to protect their unique native wildlife. State wildlife laws concentrate on areas of state and local significance, such as management of local animal populations and licensing of zoos, aquariums, and wildlife parks. They require zoos, aquariums, and individuals to obtain permits to legally possess Australian native fauna and to transfer it across state borders. These permits are designed as a mechanism to track the movement of fauna and to ensure that the animals have been obtained and traded legally.

Each state has an individual list of threatened native fauna and flora which are protected by the legislation. This means that an Australian native animal can have three different threat statuses attached to it. For example, the brush-tailed rock wallaby (*Petrogale penicillata*) is near threatened at the international level (per IUCN), vulnerable at the national level (per the EPBC Act) and endangered at the state level (per the NSW National Parks and Wildlife Act 1974).

A keeper needs to be aware that there are differences between the wildlife legislation of one state and another. Some states do not allow imports of particular species; for example, Tasmania prohibits the entry of 25 species of birds (e.g., the gang gang cockatoo [*Callocephalon fimbriatum*], the red-whiskered bulbul [*Pycnonotus jocosus*], and the chukar partridge [*Alectoris chukar*]) from other states. Similarly, the Northern Territory does not allow ferrets, frogs, or turtles to enter from other states.

THE EXHIBITED ANIMALS PROTECTION ACT 1986

The strongest zoo animal welfare and display legislation in Australia is in New South Wales. The Exhibited Animals Protection Act 1986 (EAPA) has no equivalent in any other state or territory. It was developed in response to public unease with the conditions of captive animals in some circuses and fauna parks (www.dpi.nsw.gov.au). Any zoo or "animal display establishment" wishing to operate in NSW has to be licensed under the EAPA. This act has a wide range of effects on the operation of zoos and aquariums in New South Wales, and it covers all vertebrate species, both native and non-indigenous to Australia.

The regulations ensure that

- exhibits meet high and exacting standards (minimum requirements for size and containment type, exhibit furniture arrangements, pest control, educational signs, keeper experience, and record keeping)
- comprehensive records are kept for all species and provided to Department of Primary Industry New South Wales annually, as a requirement of zoo licensing
- specific individual approvals (permits) are obtained to

exhibit a wide range of animal species, especially those not endemic to Australia
- approvals are sought before new exhibits are designed or old exhibits are modified to ensure compliance with current standards
- transfer approvals are obtained every time an animal is transferred from or to another zoo either within New South Wales, nationally or internationally.

Under the EAPA, animal facilities can be inspected at any time to ensure that standards are maintained, and also when new exhibits have been completed. This long list of conditions is meant to ensure that the essential environmental and welfare needs of captive animals are always met. There are attempts to apply similar standards nationally (see below).

NEW SOUTH WALES OCCUPATIONAL HEALTH AND SAFETY ACT 2000

Every Australian state and territory has an Occupational Health and Safety Act (OHS), whose regulations require a zoo or wildlife park to identify, assess, and control work hazards. At Taronga Zoo in New South Wales, OHS policies cover activities such as working with dangerous animals, in high-ultraviolet conditions and weather extremes, in confined spaces, and with chemicals, as well as managing duties that involve manual handling.

NONSTATUTORY STANDARDS AND PRACTICES

ZOO AND AQUARIUM ASSOCIATION ACCREDITATION AND CODE OF PRACTICE

The Zoo and Aquarium Association (ZAA) is the equivalent organization in Australasia to the Association of Zoos and Aquariums (AZA) in the United States. The ZAA has recently developed an accreditation program which provides operating standards in the areas of general operations, species collection management, animal husbandry, animal health care, education, and conservation. The program includes review of the institution's policies and procedures and a site inspection by an Association-trained welfare officer. Accreditation provides confidence that animals transferred between institutions will receive the same high standard of care and maintenance at both. Some zoos, such as the Taronga Conservation Society Australia and Zoos Victoria, also undertake site inspections of smaller animal parks that are not ZAA members before transferring animals to them. One stipulation for accreditation is that the zoo must meet all government legislative requirements.

All zoos that belong to the ZAA adopt its code of practices and by default comply with its guidelines concerning policy and practice, which can be accessed on the ZAA website (www.zooaquarium.org.au):

- animal management guidelines
- guidelines on animal euthanasia
- guidelines on regional collection planning and species management
- guidelines on animal transactions
- guidelines on animal records keeping.

FUTURE DIRECTIONS

NATIONAL ZOO AND AQUARIUM INDUSTRY WELFARE STANDARD

In terms of legislation governing animal welfare, all zoos and aquariums in Australia operate independently of each other. There are no regulatory standards that apply across state borders. For example, it is easier to exhibit species in South Australia than in New South Wales, since zoos and aquariums in New South Wales have to abide by specific exhibitory standards stipulated by Department of Primary Industries New South Wales that do not exist in South Australia. The ZAA zoo accreditation program provides a framework to establish general operating standards amongst member zoos and aquariums, and there is a regional approach to the management of some species (supervised by the Australasian Species Management Program, the association's species management arm)—but compliance is voluntary, only relevant to ZAA members, and zoos and aquariums can still operate outside the auspices of the ZAA.

Recently there has been a move to establish a national standard for animal welfare. The Australian Animal Welfare Strategy was established in 2004 and developed over five years by the Australian government. The strategy covers an extraordinary range of animal groups including livestock; animals in work, recreation, and display; wild animals; aquatic animals; and animals used in research and teaching. In 2010, ZAA was asked to establish working groups within the zoo and aquarium community to provide overarching general standards and guidelines, as well as more specific associated taxon standards for all animals used for exhibition purposes. The proposed national standards (in draft) are very similar to the existing welfare and display standards of New South Wales.

The taxon-specific standards are still in the early stages of development, but will eventually cover 20 groups including anurans (frogs), crocodilians, koalas, wombats, macropods (kangaroos and wallabies), ratites (emus and ostriches), reptiles, carnivores, ungulates, primates, dasyurids (marsupial mice, Tasmanian devils, and quolls), chiropterans (bats), raptors (birds of prey), pinnipeds (seals), dolphins, monotremes (platypus and echidnas), possums, gliders, rodents, bandicoots, and birds.

SUMMARY OF THE IMPACT OF LEGISLATION ON KEEPERS

The legislation

- places restrictions on animal imports which limit species and genetic diversity in zoo collections
- places restrictions on animal movements between states
- requires accurate and up-to-date records for all vertebrate animals
- requires animal exhibits to meet particular standards including appropriate staff experience, specific exhibit sizes, and naturalistic settings
- requires lengthy periods for the conduct of animal imports and exports and obtaining necessary permits
- requires permanent animal identification
- occasionally calls for specific health and isolation requirements
- occasionally requires keeper accompaniment for animal exports
- requires specific crates for international animal transport.

REFERENCES

Animal Health Australia website, last modified 22 May 2009. http://www.animalhealthaustralia.com.au/programs/adsp/tsefap/tse_quarantine.cfm.

AQIS website. 1999. "Importation of Psittacine Birds into Australia." Technical issues paper, last modified December 1999. http://www.daff.gov.au/__data/assets/pdf_file/0013/12046/99-090a.pdf.

Barlow, S. C., and C. Hibbard. 2006. Going, Going, Gone: A Zoo Without Exotic Mammals. Paper presented at the Annual Australasian Regional Association of Zoological Parks and Aquariums [ARAZPA] conference, Perth.

Biosecurity Australia website. 2008. Importation of Rodents into Australian Zoos: Draft Policy. Biosecurity Australia Advice 2008/10, last modified 26 March 2008. Accessed at http://www.daff.gov.au/__data/assets/pdf_file/0020/600464/2008_10.pdf.

Feral website. 2007. List of Exotic Vertebrate Animals in Australia. Last modified 27 July. Accessed at http://www.feral.org.au/list-of-exotic-vertebrate-animals-in-australia.

NSW DPI website. 2004. A Guide to the Exhibition of Animals in New South Wales. Last modified 13 July 2004. Accessed at www.dpi.nsw.gov.au/__data/ . . . /exhibition-of-animals-guide.pdf.

ZAA Policies and Guidelines. 2010. Accessed 1 September 2010 at http://(www.zooaquarium.org.au)/Policies-of-Arazpa/default.aspx.

63

Government and Legislation in Canada

Andrea Drost and William A. Rapley

INTRODUCTION

Zoos and aquariums in Canada must adhere to both federal and provincial or territorial regulations. The government of Canada has three main federal offices that have direct impact on Canadian zoos and aquariums: Environment Canada, the Canadian Food Inspection Agency, and the Department of Fisheries and Oceans. Canada is comprised of 10 provinces and 3 territories, each with its own provincial or territorial governmental regulations that zoos and aquariums must be familiar with. Provincial and territorial regulations for Canadian zoos and aquariums vary greatly. For instance, in Ontario the most influential regulatory bodies are the Ontario Ministry of Natural Resources (OMNR) and the Ontario Society for the Prevention of Cruelty to Animals (OSPCA). The government of Alberta, however, has the Government of Alberta Standards for Zoos in Alberta, which has a very direct impact on zoos and aquariums operating in that province.

ENVIRONMENT CANADA AND THE WILD ANIMAL AND PLANT PROTECTION AND REGULATION OF INTERNATIONAL AND INTERPROVINCIAL TRADE ACT

The responsibility for coordinating environmental policies and programs falls under the responsibility of Environment Canada (EC), which is also charged with preserving and enhancing the natural environment and renewable resources. The two branches of EC that most influence Canadian zoos are (1) the Enforcement Branch, specifically Wildlife Enforcement, and (2) the Environmental Stewardship Branch, specifically the Canadian Wildlife Service (CWS). CWS is the Canadian management authority whose mandate it is to fulfill the purpose of the Wild Animal and Plant Protection and Regulation of International and Interprovincial Trade Act (WAPPRIITA). WAPPRIITA's purpose is "to protect certain species of animals and plants, particularly by implementing CITES and regulating international and interprovincial trade in animals and plants" (Canadian Ministry of Justice 1992).

WAPPRIITA applies to the following animal and plant species: (a) species on the CITES control list; (b) foreign species whose capture, possession, and export are prohibited or regulated by laws in their country of origin; (c) Canadian species whose capture, possession, and transportation are regulated by provincial or territorial laws; and (d) species whose introduction into Canadian ecosystems could endanger Canadian species. Wildlife Enforcement is the body that ensures that CITES regulations are maintained, while WAPPRIITA is the act that forbids the import, export, and interprovincial transportation of these species unless the specimens are accompanied by the appropriate licenses and permits. In all cases the act applies to the plant or animal, alive or dead, as well as to its parts and any derived products. The CWS issues the CITES permits as required for the international movement of live animals that are CITES-listed. Wildlife Enforcement will become involved with an international CITES-listed live animal transfer if the appropriate permits have not been issued, and it has the authority to confiscate CITES-listed animals and/or CITES-listed animal products upon entry or exit into or out of Canada.

ENVIRONMENT CANADA SPECIES AT RISK ACT

The Species at Risk Act (SARA), which became law in 2002, is in place primarily to prevent Canadian indigenous species, subspecies, and distinct populations of animals and plants from becoming extirpated or extinct. It also provides for the recovery of endangered or threatened species. The act is meant to encourage the management of other species to prevent them from becoming "at risk." SARA-listed species can be held for exhibit at Canadian zoos. Some already have breeding and release programs in place. The black-footed ferret (*Mustela nigripes*) is such an animal. In 2009 Canada had its first ever black-footed ferret release into the wild on Canadian soil, and SARA played a key role in the release. SARA categorizes mammalian, bird, reptile, and amphibian species thus:

TABLE 63.1. Legislation and standards affecting Canadian zoos

Level	Legislation/ standards	Management authority (as of 2013)	Scope	Impact on zoo operations
Federal (international)	Convention on International Trade in Endangered Species of Wild Fauna and Flora (CITES)	Environment Canada (EC)	• Aims to ensure that international trade in specimens of wild animals and plants does not threaten their survival (international agreement among governments, adhered to voluntarily).	• Pending CITES listing, requires import and/or export permits for species. • Requires some countries to demonstrate financial support for in situ conservation efforts of a particular species proposed for import or export.
Federal	Species at Risk Act (2002)	Environment Canada (EC)	• Aims to prevent wildlife species from becoming extinct and secure the necessary actions for their recovery. • Provides for the legal protection of wildlife species and the conservation of their biological diversity.	• Regulates activities in which animals listed as species at risk can be involved.
Federal	Migratory Birds Convention Act (1994)	Environment Canada (EC)	• Aims to protect and conserve migratory birds.	• Requires a permit be issued to exhibit or display any migratory bird. • Prohibits the killing of migratory birds unless under the direction of a permit.
Federal	Health of Animals Act (1990)	Canadian Food Inspection Agency (CFIA)	• Provides the legislative and regulatory authority for the animal health import program. • Minimizes and manages risk by protecting against previously known disease outbreaks in Canada, and the entry of new diseases currently not existing in Canada	• Protects livestock economy and industry, as well as the public, from disease risk. • Requires comprehensive assessment of quarantine-approved premises. • Regulates the importation of live animals.
Federal	Fisheries Act (1985)	Department of Fisheries and Oceans (DFO)	• Delivers programs and services that support sustainable use and development of Canada's waterways and aquatic resources.	• Applies if any zoo or aquarium houses marine mammals that are native to Canada.
Provincial (Ontario)	Fish and Wildlife Conservation Act (1997)	Ontario Ministry of Natural Resources (OMNR)	• Regulates fishing, hunting, and trapping. • Is enforced by conservation officers.	• Requires zoos to obtain "License to Keep Specially Protected and Game Wildlife in a Zoo" in order to exhibit and/or display animals native to Ontario that are listed in the act.
Provincial (Ontario)	An act to amend the Ontario Society for the Prevention of Cruelty to Animals Act (2008): Bill 50	Ontario Society for the Prevention of Cruelty to Animals (OSPCA)	• Facilitates a more proactive enforcement program in Ontario with respect to animal welfare legislation (amendment to the Ontario SPCA Act).	• Prohibits animal cruelty. • Imposes fines and bans on animal ownership for violations. • Gives inspection powers to the OSPCA. • Expands standards of care to be applicable to all animals.
Provincial (Alberta)	Government of Alberta Standards for Zoos in Alberta (2005)	Government of Alberta	• Ensures that all provincial zoos meet the same standards as set by the provincial government.	• Requires suitable environments for animal collections and visitors. • Requires emphasis on education. • Ensures that off-site displays pose minimum risk to free-ranging wildlife, domestic animals, and people.
Provincial (British Columbia)	Provincial Animal Welfare Act (2008)	British Columbia Society for the Prevention of Cruelty to Animals (BCSPCA)	• Recognizes the responsibility to protect animals and manage dangerous exotic species.	• Investigates any complaints lodged against zoos or aquariums in the province.
Municipal (Example: Toronto, Ontario)	Municipal Act (1990): City of Toronto Municipal Code and By-laws	City of Toronto, Toronto Animal Services (TAS)	• Promotes and supports a harmonious environment in which humans and animals can coexist free from conditions that adversely affect their health and safety.	• Prohibits the keeping of exotic animals in many municipalities (accredited facilities are normally exempt).

Level	Legislation/ standards	Management authority (as of 2013)	Scope	Impact on zoo operations
Nonstatutory	Accreditation of zoos and aquariums in Canada through evaluation and assessment based on specific animal care and program standards	Canadian Association of Zoos and Aquariums (CAZA)	• Aims to promote animal welfare by encouraging the advancement of animal care, safety, conservation, education, and science. • Seeks to raise awareness and motivate support for animals and threatened environments.	• Requires members to adhere to a strict code of accreditation standards. • Offers benefits including position statements, position policies, grants, and awards.
International (nonstatutory)	International Air Transport Association – Live Animal Regulations (IATA-LAR)	International Air Transport Association (IATA)	• Provides minimum standards for transporting animals in containers by land, sea, or air.	• Dictates animal crate design and size.
Nonstatutory	Standards of research, animal care, housing and transport	Canadian Council on Animal Care (CCAC)	• Sets and maintains standards for the care and use of animals in science in Canada.	• Requires facilities conducting animal research to be inspected every three years. • Provides certificate of GAP (Good Animal Practice). • Provides minimum standards for the housing, care, and transport of animals.
If a zoo or aquarium is involved in scientific research and has the CCAC's Certificate of GAP (Good Animal Practice), then the federal Animals for Research Act (1990) applies.				
Provincial	Animals for Research Act (1990)	Ontario Ministry of Agriculture, Food and Rural Affairs (OMAFRA).	Provides minimum standards for the housing, care, and transport of animals.	

1. Extirpated species: As of October 2010, 23 species of animals and plants are listed as extirpated in Canada. Example: black-footed ferret (*Mustela nigripes*).
2. Endangered Species: As of October 2010, 262 species are listed as endangered in Canada. Example: swift fox (*Vulpes velox*).
3. Threatened species. As of October 2010, 151 species are listed as threatened in Canada. Example: wood bison (*Bison bison athabascae*).
4. Special concern species: As of October 2010, 166 species are listed as of special concern in Canada. Example: sea otter (*Enhydra lutris*).

A full list of all species in the above categories can be found in the Canadian Wildlife Species at Risk summary tables (COSEWIC 2010).

ENVIRONMENT CANADA MIGRATORY BIRDS CONVENTION ACT

Environment Canada also has oversight of the Migratory Bird Convention Act (MBCA). This act is meant to ensure the conservation of migratory bird populations by regulating potentially harmful human activities. A permit must be issued for all activities affecting migratory birds. Most Canadian zoos display and exhibit one or more birds that are listed under the MBCA, and therefore must comply with the act's

regulations. The following categories comprise upwards of 500 species and are controlled under the MBCA.

1. Migratory game birds.
   - As of November 2010, 8 families are listed as migratory game birds in Canada.
2. Migratory insectivorous birds
   - As of November 2010, 29 families are listed as migratory insectivorous birds in Canada.
3. Other migratory nongame birds
   - As of November 2010, 8 families are listed as other migratory nongame birds in Canada.

A full list of all species found in the above categories can be found in Environment Canada's list of birds protected in Canada under the Migratory Birds Convention Act (Environment Canada 2010). Birds not falling under federal jurisdiction within Canada include grouse, quail, pheasants, ptarmigan, hawks, owls, eagles, falcons, cormorants, pelicans, crows, jays, and kingfishers. These birds are protected under provincial or territorial jurisdictions within Canada.

CANADIAN FOOD INSPECTION AGENCY HEALTH OF ANIMALS ACT

The Canadian Food Inspection Agency (CFIA) is the federal agency that regulates the humane treatment of animals, disease control, and the care, handling, and transportation of animals in Canada. It operates under the authority of the Health of Animals Act. CFIA's plans and priorities link directly to the government of Canada's priorities for bolstering economic prosperity, strengthening security at the border, increasing the safety of the food supply, protecting the environment, and contributing to the health of Canadians (CFIA 2010). CFIA's strategic aims are (a) to ensure that public health risks associated with the food supply and transmission of animal disease to humans are minimized and managed, (b) to ensure a safe and sustainable plant and animal resource base, and (c) to contribute to consumer protection and market access based on the application of science and standards. To address these aims, the CFIA has both an export program and a National Import Service Centre. The objective of the export program is to ensure that only healthy animals and animal products and by-products which meet the import health requirements of an importing country are exported from Canada, and that live animals are transported humanely. An excellent source of information pertaining to the humane treatment of animals during transit can be found on the World Organization for Animal Health (OIE) website (www.oie.int) under the Terrestrial Animal Health Code 2010, chapters 7.2 ("Transport of Animals by Sea"), 7.3 ("Transport of Animals by Land"), and 7.4 ("Transport of Animals by Air"). A large number of species exported from Canada must be accompanied by a health certificate issued or endorsed by a CFIA veterinary inspector. The health certificate is a legal document which confirms that the health requirements of an importing country have been complied with.

The National Import Service Centre (NISC) is operated in cooperation with CFIA and the Canada Border Services Agency (CBSA), Canada's national customs agency. NISC processes import request documentation and/or data sent by the importers across Canada. Staff will review the information and return the decision electronically to the CBSA, which then relays it to the client or the client's broker or importer as the case may be. CFIA also regulates the health requirements of live animals presented for import into Canada. Many import reference documents can be accessed easily through the CFIA's website. Some examples that could have a direct influence on a Canadian zoo include:

- AHPD-DSAE-IE-2006-2-5 "New Import Measures for Live Birds to Prevent the Introduction of Avian Influenza in Domestic Birds," which encompasses birds proposed for import from any exporting country
- AHPD-DSAE-IE-2009-1-2 "Requirements for Non-Human Primates Imported into Canada"
- AHPD-DSAE-IE-2009-5-1 "Requirements for Wild Ruminants Imported from the United States to Canada."

Once the animal presented for import into Canada has been identified, CFIA will issue the appropriate CFIA import permit, which outlines the conditions and requirements for import found in the import reference documents. Depending on the species of animal and the country of export, a CFIA veterinary inspector may be required to inspect the live animal at the first port of entry into Canada. CFIA also regulates specific deer (cervid) species movements within Canada. For instance, a Toronto Zoo reindeer (*Rangifer tarandus tarandus*) used for an offsite Christmas holiday event in Toronto requires a cervid movement permit issued by CFIA. This permit certifies that the animal has been tested and found negative for both brucellosis and bovine tuberculosis within a specified time period.

DEPARTMENT OF FISHERIES AND OCEANS; THE FISHERIES ACT

The Department of Fisheries and Oceans (DFO) is primarily responsible for developing and implementing policies and programs in support of Canada's economic, ecological, and scientific interests in oceans and inland waters, and it operates under the authority of the Fisheries Act. The Fisheries Act primarily protects saltwater fishing interests in Canada and has little consequence for Canadian zoos unless a zoo wishes to acquire marine mammals from Canada's oceans. If that is the case, then the Marine Mammal Regulations would have to be adhered to. Any activities involving marine mammals would have to be authorized by the DFO, any collection and transportation would have to be licensed by the DFO, and the collector vessel would require licensing as well.

ONTARIO MINISTRY OF NATURAL RESOURCES FISH AND WILDLIFE CONSERVATION ACT

The Ontario Ministry of Natural Resources (OMNR) operates under the authority of the provincial Fish and Wildlife Conservation Act, which dictates that any zoo wishing to hold and exhibit any animals native to the province must maintain a "License to Keep Specially Protected and Game Wildlife in a Zoo," and can only do so if it is for educational and public display purposes. This license must be renewed annually; the renewal process requires the zoo to outline exactly the holding facilities for each species of native wildlife it holds, and provide information on each of the native wildlife specimens, such as date of acquisition, source of acquisition, and so on. The OMNR can and does occasionally inspect the premises of the permit holders, and will perform a thorough examination of the permit holder's record keeping of those animals covered by the permit.

ONTARIO SOCIETY FOR THE PREVENTION OF CRUELTY TO ANIMALS; AN ACT TO AMEND THE ONTARIO SOCIETY FOR THE PREVENTION OF CRUELTY TO ANIMALS ACT—BILL 50

An Act to amend the Ontario Society for the Prevention of Cruelty to Animals Act—Bill 50 is the amendment to the OSPCA Act. It is the most comprehensive amendment adopted since the provincial animal welfare legislation's inception in 1919. Bill 50 has established new provincial laws against animal cruelty. It has also provided judges in Ontario with greater flexibility to impose stiffer penalties, including jail time, fines of up to $60,000 CAD and a potential lifetime ban on animal ownership. Bill 50 gives inspection powers to the OSPCA, whose investigators can now inspect premises where animals are kept for the purposes of exhibition, entertainment, boarding, sale, or hire. This bill also allows for the OSPCA to apply for custody of an animal victim while a case is still before the courts. Under Bill 50, veterinarians are now required to report suspected animal abuse or neglect to the OSPCA with protection under the law. Overall, this bill contains a number of very significant and beneficial changes to the OSPCA Act that have facilitated a more proactive enforcement program within the province of Ontario. It can be noted, however, that if a zoo or aquarium is registered with the Animals for Research Act, that facility is exempt from the OSPCA enforcement program. In such a case, a representative of the Animals for Research Act would investigate an animal welfare complaint instead of an OSPCA officer.

GOVERNMENT OF ALBERTA STANDARDS FOR ZOOS IN ALBERTA

All zoos in the province of Alberta have to adhere to the Government of Alberta Standards for Zoos in Alberta. As outlined in the standards, "the general purpose is to ensure that facilities requiring an Alberta Zoo Permit meet acceptable standards that provide: a suitable environment for the animal collections and visitors; an environment that emphasizes education; where off-site display occurs, there is minimum risk to surrounding free-ranging wildlife, domestic animals and people, and in some cases provide an opportunity for scientific research and animal propagation to support wildlife conservation programs" (Government of Alberta, Canada 2005). The standards that Alberta zoos must meet are divided between those included within the mandate of the Alberta Sustainable Resource Development Wildlife Act and Regulations, and those that are related to the Animal Protection Act. The Wildlife Act and Regulations standards have been developed as they relate to (1) animal collection management, (2) record keeping, (3) wildlife and controlled animal transportation, (4) wildlife and controlled animal containment, (5) off-site display, (6) species conservation, (7) conservation education, (8) public and staff safety, and (9) staff experience and training. The Animal Protection Act standards have been developed as they relate to (1) protocol development requirements for animal care, (2) animal exhibition, (3) animal health care, (4) animal behavior husbandry, and (5) general animal care. The Alberta Zoo Permit must be renewed by each zoo annually.

BRITISH COLUMBIA SOCIETY FOR THE PREVENTION OF CRUELTY TO ANIMALS; PROVINCIAL ANIMAL WELFARE ACT

The British Columbia Society for the Prevention of Cruelty to Animals (BCSPCA) operates under the authority of the Provincial Animal Welfare Act. If the BCSPCA receives animal welfare complaints from the public about zoo animals, it is its responsibility to respond to these complaints by physically going to the zoo to carry out an investigation. Not all provincial SPCAs will perform this function. The BCSPCA has ongoing involvement and input in British Columbia's legislation as it applies to wild animals kept in captivity. The BCSPCA acts in the role of advisor to the provincial authority responsible for the permitting and/or licensing that zoos in British Columbia must comply with. For instance, in 2007 the BC SPCA played a role in assisting with the creation of the Wildlife Act's new Controlled Alien Species (CAS) Regulation. This new regulation controls the possession, breeding, and shipping of animals that are not native to British Columbia and that pose serious risks to human safety should they come into direct contact with people. Animals regulated by CAS include large felids, primates, and venomous reptiles and amphibians. The CAS mandate is to ensure that animals such as these are exhibited at zoos and aquariums in British Columbia accredited by the Canadian Association of Zoos and Aquariums (CAZA), and are prohibited as pets .

MUNICIPAL BYLAWS

Municipalities can exercise influence over zoos. For example, the City of Toronto Municipal Code and Bylaws contain a section entitled "Keeping of Certain Animals Prohibited," which prohibits the holding of most exotic animals. The City of Toronto, however, has also added a very specific exception stating that the section does not apply to the premises of the city zoo (Toronto Zoo) or to the premises of facilities accredited by CAZA. The Toronto Animal Services (TAS) are responsible for enforcing the municipal bylaws, but their involvement with zoos is usually limited to requesting assistance in the capture of escaped exotic pets from members of the public.

CANADIAN ASSOCIATION OF ZOOS AND AQUARIUMS; ACCREDITATION

The Canadian Association of Zoos and Aquariums (CAZA) is a nonprofit organization that promotes the welfare of animals and encourages the advancement and improvement of zoological parks in the areas of recreation, conservation, education and science. CAZA's mission statement commits the organization to "unite the Canadian zoo and aquarium community in connecting people to nature through demonstrating dedication to conservation and excellence in animal care" (CAZA 2013). Once a Canadian zoo or aquarium has gone through the process of peer inspection and review to become accredited, it is then officially recognized as an approved facility. This accreditation carries many benefits, the primary one being that the zoo or aquarium is publicly recognized as an institution that meets or exceeds the professional standards set out by CAZA. Other benefits, as outlined in CAZA's official website (www.caza.ca), include "eligibility for grants (makes institutions eligible for consideration for funding and grants from certain foundations, corporations and other sources), cuts red tape (exempts institutions from certain government requirements), allows organizations to learn from other institutions and better understand the importance of accreditation through participating in training and subsequent participation as accreditation inspectors, fosters staff and community pride, and significantly improves the organizations ability to attract and retain a high quality, professional staff."

CANADIAN COUNCIL ON ANIMAL CARE; ZOOLOGICAL INSTITUTIONS ACTIVE IN ANIMAL RESEARCH

The Canadian Council on Animal Care (CCAC) sets standards and guidelines for the care and use of experimental animals in Canada, and is a national peer review organization founded in Ottawa in 1968. Its mandate is "to work for the improvement of animal care and use on a Canada-wide basis." The CCAC not only ensures that high standards are met for the ethical use and care of animals used in research but also emphasizes the educational value of such research. It requires every institution in Canada conducting animal-based research to establish an animal care committee. The Toronto Zoo, for example, has an active research program and has established its Animal Care Research Committee (ACRC) to meet this CCAC requirement. Animal welfare is very important in the CCAC regulations and the welfare of animals in the province of Ontario is governed by the Ministry of Agriculture, Food and Rural Affairs (OMAFRA), which operates under the authority of the Animals for Research Act. If a zoo is not active in any animal-based research, then CCAC (and subsequently OMAFRA) are not involved with regulating the zoo.

REFERENCES

Canadian Food Inspection Agency. 2010. About the CFIA. Last modified May 4. http://www.inspection.gc.ca/english/agen /agene.shtml.

Canadian Ministry of Justice. 1992. Wild Animal and Plant Protection and Regulation of International and Interprovincial Trade Act. Last modified February 26. http://laws.justice.gc.ca/PDF /Statute/W/W-8.5.pdf.

CAZA. 2013. "Vision and Mission Statement." Accessed April 21 at http://www.caza.ca/en/about_caza/mission_statement/index.php.

Committee on the Status of Endangered Wildlife in Canada. 2010. "Canadian Wildlife Species at Risk Summary Tables." Last modified in October. http://www.cosewic.gc.ca/eng/sct0/rpt /dsp_booklet_e.htm.

Environment Canada. 2010. "Migratory Birds Protected in Canada." Last modified 8 November. http://www.ec.gc.ca/nature/default .asp?lang=En&n=00F8E609–1.

Government of Alberta, Canada. 2005. Standards for Zoos in Alberta. Last modified 30 September. http://www.srd.alberta.ca /BioDiversityStewardship/ZooStandards/documents/Alberta _Govt_Standards_for_Zoos_in_Alberta_Sept2008.pdf.

International Air Transport Association. 2010. "IATA Live Animal Regulations." Accessed 22 December at http://www.iata.org/ps /publications/Pages/live-animals.aspx.

64

Government and Legislation in Europe

Lesley A. Dickie and Miranda F. Stevenson

INTRODUCTION TO EUROPE AND THE EUROPEAN UNION

Europe comprises the westernmost peninsula of Eurasia, bounded by the Ural Mountains to the east and the Mediterranean Sea to the south, encompassing a total of 50 countries. Yet this short geographic description belies the complex legal structure of the European Union (EU), which dominates government and legislation in this region and in particular in the 27 member states of the union. Founded in 1957, with the signing of the Treaty of Rome, the EU today is a powerhouse of economic and legislative integration and a major force in the world. Zoos within the union and outside its borders are significantly affected both positively and negatively by the legislation enacted from its headquarters in Brussels, Belgium.

The EU can be a bewildering behemoth to the uninitiated, with confusion as to its many constituent parts; however, there are four main bodies which came into being via the Treaty of Rome (Leonard 2010):

- the European Parliament, with 736 elected "members of the European Parliament" (MEPs) from the 27 member states, who represent not national political parties but cross-national political alliances
- the European Commission (EC), the body of permanent staff that implements decisions taken by the Council of Ministers and, to an increasing extent, by the European Parliament
- the Council of Ministers, the supreme legislative arm of the union and the representatives of the national governments of the member states
- the European Court of Justice, which interprets EU legislation in the case of disputes.

The EU has three main legislative vehicles: directives, regulations, and decisions. Directives, such as the Zoos Directive (1999/22/EC), require member states to reach a final outcome without imposing strict requirements about the national legislative framework that should be in place to achieve the results. Directives must therefore go through a process of being transposed into national law, which allows substantial flexibility. The more exacting regulations are legally binding in all 27 member states, without further transposition. Decisions are binding through already enacted directives or regulations, and can be directed toward a government, another body, or an individual.

ZOOS IN EUROPE

The European Association of Zoos and Aquaria (EAZA), with 322 members, is the largest professional zoo and aquarium association in the world today. Although it is represented by membership in 36 countries, the bulk of its membership is located within the EU, and EAZA members are located in 25 of the 27 member states. Therefore, EAZA as an association must be aware of EU legislation, changes to such legislation, and its impact on EAZA members. It should also be noted that significant numbers of institutions are not members of EAZA. In addition, EAZA has as members a number of national zoo and aquariums associations, such as the British and Irish Association of Zoos and Aquariums (BIAZA), which work with the transposed directives of the EU. As a nongovernmental organization (NGO) based within the EU, EAZA has the opportunity to give input to the legislative framework and its implementation via working with MEPs at the parliament and with commission officials. In addition, national associations such as BIAZA work via their national government bodies and parliamentarians. The remainder of this chapter will describe in more detail some of the legislative framework within which European zoos and aquariums operate. It will also detail requirements under global conventions and how these relate to European legislation. However, it cannot cover all relevant legislation; rather, it focuses on that legislation most pertinent to the role and function of zoos.

TREATIES, CONVENTIONS, AND THE EUROPEAN UNION

As the international community has no legislature capable of formulating binding laws, treaties and conventions have developed as a means by which EU member states wishing to cooperate on a certain course of action can establish mutual legal obligations. A state will sign a treaty, but this does not impose a legal obligation on it to ratify the treaty at a later stage. Ratification puts the treaty into enactment in the state.

In the EU, treaties and international agreements form primary legislation. Secondary legislation comprises binding legal instruments which take the form of the regulations, directives, and decisions previously described.

CONVENTION ON BIOLOGICAL DIVERSITY

The Convention on Biological Diversity (CBD) was signed at the Rio "earth" summit in 1992 by 168 countries, and entered into force in 1993. The CBD seeks to conserve global biological resources for the fair benefit of all of humanity (CBD 2010). The EU ratified the CBD in 1993 (European Commission 2006). Thereafter, it put in place a number of legislative mechanisms to strengthen those already in place, in order to implement the convention. Article 8 of the CBD requires contracting parties to, among other things, establish a system of protected areas or areas where special measures need to be taken to conserve biological diversity. This is implemented at EU level by a number of mechanisms. The Birds Directive (1979) substantially predates the CBD and was designed to provide protection for birds across the union, particularly those that cross national boundaries. This directive was in fact the first EU legislation to protect nature in situ. It was joined in 1992 by the Habitats Directive, which obliged member states to manage Special Areas of Conservation (SACs). This was bolstered, after the CBD Convention of the Parties in 2004, by the Natura 2000 network. The Natura 2000 program designates areas of significant biological diversity and importance that require higher levels of protection. It is envisaged that between 15 and 20% of the total EU terrestrial area should eventually be designated as Natura 2000 sites. It is also being extended to cover important marine EU habitats. While this is legislation that primarily impacts in situ environments, zoos have been important partners in various aspects of its implementation, from running awareness-raising campaigns of the Natura 2000 network for zoo visitors to managing Natura 2000 sites or even being designated as Natura 2000 areas, like the Nuremburg Zoo in Germany, with its reintroductions of native amphibians (WAZA 2011).

Article 9 of the CBD focuses on ex situ conservation. While some aspects of agricultural policy within the EU have ex situ components, the main instrument for the application of Article 9 in the EU is found within the Zoos Directive (1999/22/EC). It is also the only piece of EU legislation that specifically focuses on zoos.

ZOOS DIRECTIVE

The Zoos Directive was introduced in 1999 and has been implemented into national legislation throughout the EU, although the extent of implementation varies substantially from country to country. The directive specifically calls for the following action from zoos:

Participating in research from which conservation benefits accrue to the species, and/or training in relevant conservation skills, and/or exchange of information relating to species conservation, and/or where appropriate captive breeding, repopulation or reintroduction of species into the wild;

Promoting public education and awareness in relation to the conservation of biodiversity, particularly by providing information about the species exhibited and their natural habitats;

Accommodating their animals under conditions which aim to satisfy the biological and conservation requirements of the individual species, inter alia (among other things), by providing species specific enrichment of the enclosures; and maintaining a high standard of animal husbandry with a developed programme of preventive and curative veterinary care and nutrition;

Preventing the escape of animals in order to avoid possible ecological threats to indigenous species and preventing intrusion of outside pests and vermin;

Keeping of up-to-date records of the zoo's collection appropriate to the species recorded.

It is notable that the directive does not include significant details about any of the above, and this has been the subject of some criticism of its text. In a recent evaluation by the Directorate General for Health and Consumers (DG SANCO, 2010) of the EU Animal Welfare Policy, lack of implementation of the Zoos Directive in some member states and the absence of more detailed welfare requirements has been highlighted. The directive makes specific mention of EAZA, recognizing its role as the preeminent zoo and aquarium association in Europe:

Whereas a number of organisations such as the European Association of Zoos and Aquaria have produced guidelines for the care and accommodation of animals in zoos which could, where appropriate, assist in the development and adoption of national standards.

In relation to the above, EAZA has suggested to the European Commission that the EAZA Minimum Standards for the Care and Accommodation of Animals (compliance with which is a condition of EAZA membership) be added as an annex to the directive (February 2011), which would help strengthen its animal welfare component (see the case study of how the directive is implemented in practice in the United Kingdom).

CONVENTION ON WETLANDS

Known as the Ramsar Convention, after the town in Iran where it was signed in 1971, this global treaty seeks to provide protection for a specific ecosystem, wetlands, and is the only such global treaty to do so. The Ramsar Convention has 160 contracting parties and covers an area of 187,055,551 hectares (as of February 2011). It specifically highlights the concept of "wise use" of such designated wetland areas.

CONVENTION ON THE CONSERVATION OF MIGRATORY SPECIES

The Convention on the Conservation of Migratory Species of Wild Animals (also known as CMS or the Bonn Convention; http://www.cms.int/) aims to conserve terrestrial, marine, and avian migratory species throughout their range. Since the convention came into force in June 1979, its membership stands at 114 signatory (member) parties as of January 2010. Parties or member states strive towards strictly protecting threatened migratory species, conserving or restoring the places where they live, mitigating obstacles to their migration, and controlling other factors that might endanger them.

The World Association of Zoos and Aquariums (WAZA) has established working agreements with international bodies, such as the International Union for Conservation of Nature (IUCN), CMS, CBD, and Ramsar. These partnerships help the international zoo and aquarium community to use synergies in international endeavors like the conservation of gorillas (e.g., in the Year of the Gorilla 2009) or the preservation of biodiversity (within the United Nations Decade on Biodiversity 2011–2020). Being a more visible partner of international conservation bodies opens up new contacts, such as the Ramsar focal points for communication, education, and public awareness (CEPA), and increases awareness of the conservation work of zoos and aquariums. Materials and international events of the international conservation community, such as World Wetlands Day or Biodiversity Day, can be used and celebrated as well.

CONVENTION ON THE CONSERVATION OF EUROPEAN WILDLIFE AND NATURAL HABITATS

The Convention on the Conservation of European Wildlife and Natural Habitats, also known as the Bern Convention, was adopted in September 1979 in Bern, Switzerland, and came into force on 1 June 1982. As of 2010, it has 40 contracting parties, including 35 member states of the Council of Europe as well as the EU.

The aims of the convention are "to conserve wild flora and fauna and their natural habitats, especially those species and habitats whose conservation requires the co-operation of several states, and to promote such cooperation. Particular emphasis is given to endangered and vulnerable species, including endangered and vulnerable migratory species." The convention lists protected species in four appendixes. This convention has been transposed into national law; for example, in England and Wales it has been transposed into the Wildlife and Countryside Act 1981. This act makes illegal the taking from the wild of native species and the release of nonnative species; zoos and aquariums must take care to abide by it.

ANIMAL HEALTH LEGISLATION

OFFICE INTERNATIONAL DES ÉPIZOOTIES: WORLD ORGANIZATION FOR ANIMAL HEALTH

The Office international des épizooties or World Organization for Animal Health (OIE) was founded in 1924 as a global response to counter the effects of the spread of animal disease (www.oie.int). It is an intergovernmental organization of 176 member countries (in 2010). The OIE produces lists of notifiable animal diseases; if a member country has a case of a disease on one of these lists, it *must* notify the OIE. This also means that the member country must have domestic legislation in place to ensure that diseases on OIE lists are notifiable within its own borders.

In the EU the majority of animal health legislation is implemented through regulations, directives, and decisions. However, most legislation is aimed at the agricultural and farming industries and zoos can inadvertently be negatively affected by restrictions which are not actually intended for the exotic species that they keep. Vigilance is therefore required to ensure that this does not occur, and that the necessary movement of zoo animals for conservation and managed breeding programs can still take place. EAZA and the national associations spend much time and effort checking relevant legislation and working with the EU commissions and national government departments. It is also important to carry out research on how some diseases affect exotic species; for example, a recent study on the bluetongue virus in European zoos has demonstrated that African species of antelope are not susceptible, but that many non-African species are. Zoos must initiate research such as this and keep databases so that their disease information is evidence-based. The European Association of Zoo and Wildlife Vets (EAZWV) works closely with EAZA on veterinary issues in zoos. EAZWV has for several years produced a comprehensive document entitled the *Transmissible Diseases Handbook* (www.eaza.net), now in its fourth edition. This provides a review of diseases that can affect both common and domestic stock, and advice on how best to identify their occurrence.

Zoos may be forced to close, and therefore lose vital revenue, if they are situated within certain disease outbreak zones. For example, many zoos in the UK keeping hoofstock had to close during the foot-and-mouth outbreak in 2001, not only because some were within disease control zones, but also because it was necessary to reduce the risk of transmission.

BALAI DIRECTIVE

One EU directive that is designed to help zoos move animals is the Balai Directive (EC 92/65/EEC). *Balai* means "broom" in French, and the directive was designed to "sweep together" and make the movement of exotic animals between EU member states easier. Under the Balai Directive a zoo becomes approved though an inspection process; this approval means that the zoo has to have certain controls and biosecurity measures in place, including an animal disease surveillance plan. This focuses on the notifiable diseases listed by the OIE and relevant diseases listed in annexes to the directive. Animals can be moved between approved zoos with health certificates, which may remove the need for some pre- and post-movement tests, due to the implementation of the surveillance plan. However, for this plan to work effectively, all member states have to put the directive into operation, and although it came into effect in 1992, this still has not occurred. Full details can be found at www.defra .gov.uk.

Some EU member states may have additional health restrictions. For example, the United Kingdom and Ireland have additional quarantine restrictions on rabies-susceptible species.

ANIMAL BY-PRODUCTS REGULATION

A more recent regulation that has negatively affected zoos is Regulation (EC) no. 1 (1774/2002), which lays down health rules concerning animal by-products not intended for human consumption. This regulation came in response to animal health crises such as the bovine spongiform encephalopathy (BSE) outbreak in the early 1990s, and it is intended to protect animal and human health. Animal by-products are defined as the bodies or body parts of animals or products of animal origin not intended for human consumption, including ova, embryos, and sperm. Animals and their by-products are placed into three different categories, and zoo animals are placed in category 1. Under the initial regulation, category 1 animals could not be fed to other category 1 animals. In practice, for some EU member states this meant that if rabbits culled from domestic education or from children's farms within zoos have been on public display, they cannot be fed to carnivores within the zoo. In addition, the use of culled hoofstock, often fed to carnivores in zoos to make the most sustainable use of resources, would be restricted by such legislation. Some EU member states had to employ a flexible and pragmatic approach to this issue. Recent amendments (derogations) have partially revoked the law relating to zoos and now allow for the feeding of zoo herbivores to zoo carnivores with certain controls and checks. The Animal By-Products Regulation is a good example of zoos having to abide by legislation that largely focuses on agriculture.

HORSE PASSPORT REGULATION

The Horse Passport Regulation (EC) no. 504/2008, which came into force on 1 July 2009, requires the identification of horses, donkeys, and other species of the genus *Equus*. It requires that all zoo equids receive a paper "passport" and electronic chip within six months of birth. The passport requires that distinctive visual features of the specimen are required to be identified and documented on the passport. In practice this should not pose too many compliance problems for zoos although, as with all legislation fundamentally designed for domestic stock, it should be monitored carefully as to how it affects zoos.

PROTECTION OF ANIMALS DURING TRANSPORT REGULATION

As with the regulation above, (EC) no. 1/2005 was primarily intended for the tens of millions of agricultural animals that move around Europe annually. In comparison, EAZA has estimated that within the EAZA community approximately 25,000 animals are moved in the same time period, often for very short distances without crossing member state boundaries. EAZA abides by all international guidelines (such as those of the Animal Transportation Association

[ATA] and the International Air Transport Association [IATA]) in relation to animal moves, generally supports a "self-regulatory" approach, and as such has defined its own EAZA Guidelines for Animal Transport (2010). The regulation is currently under review and EAZA has issued a position statement (EAZA 2010) asking the EC to ensure that any revisions take into account the specific needs of zoos animals and arguing that an approach combining EAZA's own guidelines, the AATA and IATA guidelines, and CITES's Animal Committee Guidelines (currently in preparation) would be the best approach to dealing with animal transport for EAZA members. Currently, Regulation (EC) no.1/2005 is interpreted differently between member states, and this can delay or hamper breeding program movements.

DISEASE CONTROLS FOR IMPORT

Disease control in the EU is, as in all such integrated economic groupings, an area of grave concern. The Directorate-General for Health and Consumers (DG SANCO), the EC directorate that monitors food, health, and agriculture, must specifically provide a legislative framework that protects the agricultural herds and flocks of the EU, their economic importance, and the safety of consumer food products. Zoos and aquariums must therefore bear in mind that DG SANCO operates within an enormous remit, and that while legislation is not designed to hamper zoos, they inevitably are not the first priority. Association zoos, such as those belonging to EAZA and national associations, also must bear in mind that legislation must be applied to all entities that fall under their purview—in this case all licensed zoos and aquariums in the EU—and not just to the most progressive and responsible members of the zoo community.

As an example, Council Directive 2004/68/EC governs the import into the EU of ungulates from "third countries" (those outside member states) and therefore affects zoo movements. This also relates specifically to the health surveillance conditions laid down in the Balai Directive (EC/92/65/EEC). Imports are allowed from listed countries, but only in a relatively small number, and many imports are not possible (though the details differ from one member state to another). An example is that the import of a rhinoceros from an AZA-accredited zoo in the United States to an EAZA zoo in the United Kingdom, even if recommended under an international breeding program, would be prohibited at the time of drafting this chapter (March 2011), despite risk to agricultural stock being extremely small. The EC is aware, however, that while the legislation is designed to protect economically important agriculture stock, it does hamper legitimate movement of nondomestic stock in some cases. The balancing act for the EC is to continue to protect agricultural animals while enabling nondomestic movement of importance for conservation aims. Again, zoos must recognize this balancing act and they should seek to work productively with the commission to meet such challenges. This legislation is currently under review, and zoos have given a stakeholder's perspective on the proposed changes. A productive dialogue with legislative bodies is essential if zoos are to operate most efficiently.

Case Study 64.1. Zoo Legislation in the United Kingdom

The Zoo Licensing Act 1981 came into force in 1984 and regulated zoos in the United Kingdom (UK) until 1999. In 1999 the act was devolved to Wales, Scotland, and Northern Ireland with the UK government retaining its central authority under the act in England only.

In 2002 the existing act was amended to incorporate the Zoos Directive (1999/22/EC). The amending legislation for England is the Zoo Licensing Act 1981 as amended by the Zoo Licensing Act 1981 Amendment. England and Wales Regulations 2002 and came into force on 8 January 2003. Scotland, Wales, and Northern Ireland also introduced their own equivalent amending regulations in 2003 to ensure that the full provisions of the directive were implemented in their respective countries.

The Zoo Licensing Act (ZLA) is administered at the local government level (except in Northern Ireland, where the Northern Ireland executive undertakes that role) and requires that a detailed and regular inspection process is carried out to ensure that acceptable standards are maintained. The act requires that a system of formal inspections is carried out every three years. Zoo licenses are issued for a period of six years, except in the case of a first or original license, which is issued for four years. Within a six-year period, a zoo will have a formal inspection in the third year and again no later than six months before the end of the sixth year of the license period.

Although most of the act's requirements are administered at local government level except in Northern Ireland, central government does have a role. The ZLA requires the relevant central government or administration (i.e., the UK government, Scottish executive, Welsh assembly government, or Northern Ireland executive) to compile and manage a two-part list of persons suitable to inspect zoos under the ZLA. Inspectors retained on the list are expected to meet certain criteria, which include being able to demonstrate continued professional development in the zoo field. In the case of the UK government, the list of zoo inspectors is currently maintained by Animal Health (an executive agency of Defra). Part 1 of the list names veterinary surgeons with experience of wild animal medicine, and part 2 names persons with experience in managing zoos, including the care and welfare of animals.

The act also requires the central government or administration to specify, from time to time, standards of modern zoo practice that zoos are expected to meet and which zoo inspectors should refer to when carrying out inspections. These standards are called the Secretary of State's Standards of Modern Zoo Practice, and they are updated as knowledge develops and zoo practice evolves. The standards cover animal welfare, conservation, public education, and public safety. The UK government also has an independent advisory body, the Zoos Forum, to advise it on matters pertaining to zoos. The Zoos Forum's remit can extend to provide advice to the devolved administrations. Among other things, the Zoos Expert Committee has produced a handbook (Defra 2002) which is designed to aid implementation of good standards and legislation. This handbook has chapters added from time to time, and it is a useful resource document for the production of standards.

Central government also has powers under the ZLA to issue dispensations for smaller zoos, which may result in it directing that a zoo be exempt from the requirements of the act or that a zoo's formal inspections are undertaken by a smaller number of inspectors than would ordinarily be required under the act.

LEGISLATION THE BASE, ASSOCIATIONS THE STANDARDS

Along with all of the legislative elements identified above, there is also a specific layer of nonlegislative accreditation, codes, standards, and guidelines imposed via membership in national and regional associations, such as BIAZA and EAZA. Legislation, such as the Zoos Directive, forms the baseline to which EU member states' zoos and aquariums must reach. However, associations can build on this baseline and demand, as a condition of membership, more stringent and defined standards of behavior in relation to all aspects of zoo and aquarium operation. Previously in this chapter we have mentioned the EAZA Minimum Standards for the Care and Accommodation of Animals (see www.eaza.net) and their role in defining good welfare standards for EAZA zoos and aquariums. EAZA has a defined Membership and Ethics Committee that investigates any reported breaches in its codes and standards, and which has terminated membership of some institutions and imposed additional sanctions on individual zoos to ensure improvements in their performance.

ADDITIONAL NATIONAL LEGISLATION

Apart from animal health and wildlife conservation legislation and specific legislation pertaining to zoo licensing, zoos may also be affected by general animal welfare legislation. For example, in England, Scotland, and Wales additional animal welfare legislation imposes a "duty of care" on all those who keep vertebrate animals in the "custody or control of man." There is also legislation covering performing animals, which may apply to zoos that feature animals in demonstrations or shows.

SUMMARY

Zoos and aquariums are subject to a plethora of legislation that affects almost all aspects of their work. In a culturally diverse community such as the EU, legislation can be controversial and can take many years to draft and thereafter implement. This chapter has detailed only a small selection of the legislation that covers zoos and aquariums within the EU, and has not covered any of the laws regarding health and safety, employment, and so on. The most important role for

Case Study 64.2. A Zoo Inspection in England for a Large Zoo with a Full Inspection Team

In this case the inspection team will comprise two government inspectors (one from each part of the list), a local authority–appointed vet and up to two more inspectors that the local authority has the option of appointing. In the case of a large zoo, the inspection may take two or three days. All arrangements for the inspection are organized by the local authority. The date of the inspection is prearranged and agreed upon, and a preinspection audit form is completed and sent to inspectors beforehand. This provides the team with up-to-date information about the zoo, and it is accompanied by background information such as animal stock lists. This makes the inspection more efficient.

Most inspectors start by reviewing the audit forms and accompanying paperwork, and then discuss any emerging issues. This is followed by an on-site inspection of the zoo. Inspectors are issued with blank pro forma inspection reports to complete,

and this helps to give all inspections a reasonable level of consistency. Towards the end of an inspection it is standard practice for the inspector to let the zoo management know, in broad terms, what will be in the inspectors' report. This provides an opportunity for contentious issues to be discussed, and avoids factual errors in the final report.

The final report contains two categories of comment from the inspection team. *Recommendations* are suggestions for improvement, but they are not intended to be enforced. Sometimes a recommendation may be turned into a condition at a future inspection. *Conditions* are the inspection team's requirements for changes or improvements, usually within a given time scale, that must be enforced by the local authority. Failure to comply with a condition may threaten a zoo's license. Inspectors will acknowledge excellence, good practice, and improvements in their reports.

zoos is in making sure that they abide by existing legislation, but also in making sure their voices are heard as an important stakeholder group when legislative change is envisaged. The professional and progressive zoos must also identify themselves as organizations that can *assist* the EU in formulating legislation that works in practice, and can be trusted partners in this process.

REFERENCES

Barber, J, D. Lewis, G. Agoramoorthy, and M. Stevenson. 2010. "Setting Standards for Evaluation of Captive Facilities." In *Wild Mammals in Captivity: Principles and Techniques for Zoo Management*, Second Edition, edited by D. Kleiman, K. Thompson, and C. Kirk Baer. Chicago: University of Chicago Press.

Cooper, M. E., and A. M. Rosser. 2002. "International Regulation of Wildlife Trade: Relevant Legislation and Organisations." *Rev. Sci. Tech. Off. Int. Epiz.* 21:103–23.

Department for Environment, Food, and Rural Affairs (Defra). 2012. *The Zoos Forum Handbook.* Bristol: Defra. Accessed at http://www.gov.uk/government/publications/zoos-expert-committee-handbook.

Department of the Environment, Transport, and the Regions (DETR). 2000. *Secretary of State's Standards of Modern Zoo Practice.* London: DETR. Accessed at https://www.gov.uk/government/publications/secretary-of-state-s-standards-of-modern-zoo-practice.

EU Council 99 Council Directive 1999/22/EC of 29 March 1999, relating to the keeping of wild animals in zoos.

European Commission. 2006. *The Convention on Biological Diversity: Implementation in the European Union.* Luxembourg: Office for Official Publications of the European Communities.

Kirkwood, James K. 2001a. "United Kingdom: Legislation." In *Encyclopedia of the World's Zoos*, edited by C. E. Bell, 1281–83. Chicago and London: Fitzroy Dearborn.

———. 2001b. "United Kingdom: Licensing." In *Encyclopedia of the World's Zoos*, edited by C. E. Bell, 1284–85. Chicago and London: Fitzroy Dearborn.

Leonard, Dick. 2010. *Guide to the European Union: The Definitive Guide to all Aspects of the EU.* Tenth Edition. London: Profile Books.

Transmissible Diseases Handbook. Accessed at http://www.eaza.net/activities/Pages/Transmissible%20Diseases%20Handbook.aspx.

WAZA. 2005. *Building a Future for Wildlife: The World Zoo and Aquarium Conservation Strategy.* Bern, Switzerland: WAZA Executive Office. Accessed at http://www.waza.org/en/site/conservation/conservation-strategies.

———. 2011. http://www.waza.org/en/site/conservation/waza-conservation-projects/amphibian-reintroduction.

The Zoo Licensing Act (Amendment) (England and Wales) Regulations 2002. London: Stationery Office. Accessed at http://www.gov.uk/government/publications/zoo-licensing-act-1981-guide-to-the-act-s-provisions.

65

Government and Legislation in New Zealand

Tineke Joustra

INTRODUCTION

The earliest attempt at wildlife legislation in New Zealand was in 1864, when some protection was given to native species with the declaration of a closed season for native ducks and pigeons. The first significant legislation relating to wildlife was in 1867 where the introduction of "any fox, venomous reptile, hawk, vulture or other bird of prey" became forbidden. Not until 1895 was it necessary to obtain written consent to introduce "any animal or bird whatsoever" (McLintock 2009).

It was not until the late 19th century that the belief that native birds and plants would inevitably disappear due to introduced species was recognized. The common belief was that while a few species were perhaps doomed to extinction for various reasons (e.g., huia [*Heteralocha acutirostris*], which became extinct in 1907), it was possible that most native fauna might be preserved. The earliest legislation was almost entirely concerned with introduced species, and even included protection of introduced species from poaching (McLintock 2009). In 1886, power was granted to the governor of New Zealand to give absolute protection to any indigenous bird; this power was exercised with the passing of the Animals Protection Act 1907. In the Animals Protection and Game Act 1921–22, the principle of absolute protection for the majority of native birds and a few land animals (two species of bats, the tuatara [*Sphenodon* sp.], and native frogs) was a major step in the conservation of wildlife in New Zealand (McLintock 2009).

Protection of native fauna was further extended with the passing of the Wildlife Act 1953. This act provided some form of protection to all native birds, excluding the great black cormorant (*Phalacrocorax carbo*), the harrier hawk (*Circus approximans*), and the kea (*Nestor notabilis*), but including regular migrants such as the godwit. No introduced mammals or birds, except for seven species of game birds and the mute swan (*Cygnus olor*), were protected. The passing of this act reflected an increased concern for New Zealand's native species (McLintock 2009).

MAJOR LEGISLATION

The major New Zealand laws and regulations currently in place that relate to zoo and aquarium operation are

- the Conservation Act 1987, which promotes the conservation of New Zealand's natural and historic resources and to establish for that purpose a Department of Conservation
- the Conservation Law Reform Act 1990, which amends the law relating to conservation organizations, freshwater fish and game, conservation management planning, and marginal strips (strips of land 20 meters wide extending along and abutting the landward margins of foreshores, or beds from other natural water bodies)
- the Trade in Endangered Species Act 1989, which furthers the protection and conservation of endangered species of wild fauna and flora by regulating the export and import of such species and any product derived from them
- the Wildlife Act 1953, which consolidates and amends the law relating to the protection and control of wild animals and birds and the management of game, sets out the different levels of protection for wildlife, and provides for the administration of permits
- the Marine Mammals Protection Act 1978, which provides for the protection, conservation, and management of marine mammals within New Zealand and its fisheries' waters
- the Native Plants Protection Act 1934, which provides for the protection of native plants
- the Animal Welfare Act 1999, which reforms the law relating to the welfare of animals and the prevention of their ill treatment (PCO 2010)
- the Animal Welfare (Zoos) Code of Welfare 2004, which has minimum standards for the care of zoo animals (Wassilieff 2009)

613

- the Hazardous Substances and New Organisms Act 1996, which sets the standards for containment of animals in zoos, and outlines the regulations for importing new species into the country (Wassilieff 2009).

A number of organizations are responsible for the administration of the above acts, including the Department of Conservation (DOC), established under the Conservation Act 1987. Under this act the DOC has a number of functions, including (DOC 2010)

- management for conservation purposes of all land and natural and historic resources held under the Conservation Act
- preservation of indigenous freshwater fisheries (so far as is practicable)
- protection of recreational freshwater fisheries and freshwater fish habitats
- conservation advocacy
- promotion of the benefits of international cooperation on conservation matters
- promotion of the benefits of the conservation of natural and historic resources in New Zealand, the subantarctic islands, the Ross Dependency, and Antarctica
- provision of educational and promotional conservation information
- allowance of tourism and recreation on conservation land, provided that its use is consistent with the conservation of the resource
- provision of advice to the conservation minister.

The Ministry of Primary Industries (MPI) is a descendant of the old Department of Agriculture, which was founded in 1892 from the amalgamation of the Stock and Agriculture Branches of the Department of Crown Lands. MPI's function in the past was to provide farmers with expert scientific advice for improving the quality and quantity of their production. Since then the ministry has undergone a series of major restructurings which have changed its role and key functions. The mission of MPI today is to enhance New Zealand's natural environment. The ministry does this by encouraging high-performing sectors (e.g., agriculture), developing safe and free trade, ensuring healthy New Zealanders, and protecting the country's natural resources for the benefit of future generations (MAF 2010). In relation to the zoo industry, MPI Biosecurity New Zealand is the division of MPI that is charged with leadership of the New Zealand Biosecurity system. It encompasses facilitating international trade, protecting the health of New Zealanders, and ensuring the welfare of its environment, flora and fauna, marine life, and Maori resources (MAF 2010). It achieves this by providing zoo codes and standards such as the MPI Containment Standard 154.03.04 and the MPI Code of Recommendations for the Welfare of Exhibited Animals 1999, both of which are comparable to the Standards of Modern Zoo Practice in the United Kingdom.

The Environmental Risk Management Authority (ERMA) was established under the Hazardous Substances and New Organisms (HSNO) Act 1996. Its main role is to make decisions on applications to import, develop, or "field test" new organisms, or to import or manufacture hazardous substances in New Zealand (ERMA 2010). Zoos and aquariums must have ERMA approval when importing animals into the country. These animals may carry unwanted parasites currently not found in New Zealand.

SUMMARY

New Zealand legislation affects zookeepers in the following ways.

- It places restrictions on animal imports, which limits the range of species and genetic diversity that can appear in zoo collections.
- It requires that accurate and up-to-date records be maintained for all zoo animals.
- It requires that an annual animal inventory and mortality analysis be completed on 30 June of each year.
- It requires animal exhibits to meet containment standards, for which they must be inspected before they are used, and annually thereafter.
- It requires lengthy time periods for animal imports and exports to be approved nationally and internationally.
- It requires that each animal be identified with methods that are permanent.
- It requires specific health and isolation (quarantine) measures.
- It requires specific crates for international animal transport.

REFERENCES

Department of Conservation (DOC). 2010. "Wildlife Act 1953." Accessed 1 April at http://www.doc.govt.nz/about-doc/role /legislation/wildlife-act/.

Environmental Risk Management Authority (ERMA). 2010. "About ERMA New Zealand." Accessed 1 April at http://www.ermanz .govt.nz/.

Ministry of Agriculture and Forestry (MAF). 2010. "Ministry of Agriculture & Forestry, New Zealand." Accessed 1 April at http:// www.maf.govt.nz.

Ministry of Agriculture and Forestry (MAF). 2010. "MAF Biosecurity New Zealand." Accessed 1 April at http://www.bio security.govt.nz.

McLintock. 2009. "Wildlife Legislation." In *An Encyclopaedia of New Zealand*, edited by A. H. McLintock, originally published in 1966, updated 22 April. Accessed 1 April 2010 at http://www.TeAra .govt.nz/en/1966/acclimatisation-of-animals/6.

Parliamentary Counsel Office (PCO). 2010. "Conservation Act 1987." Accessed 1 April at http://www.legislation.govt.nz/act /public/1987/0065/latest/DLM103610.html.

Wassilieff, Maggy. 2009. "Zoos and Aquariums: Zoos in the 21st Century." In *Te Ara: The Encyclopedia of New Zealand*, updated 1 March. Accessed 1 April 2010 at http://www.TeAra.govt.nz/en /zoos-and-aquariums/4.

66

Government and Legislation in the United States

Steve Olson

It is important for zoo and aquarium professionals to be knowledgeable about the legislative and regulatory activities around them—to be familiar with the government actions that could impact their work environment and their day-to-day activities. For zoos and aquariums, it is essential for persons involved in the capture, shipment, receipt, sale, transportation, or display of wildlife to be familiar with existing local, state, national, and international wildlife laws. Ignorance of the laws and assumptions of compliance will not withstand legal challenge. Measures spelled out in much of the legislation affecting zoos and aquariums make this clear. This chapter will also examine some of the key US laws and regulations which affect zoos and aquariums and how these measures directly impact directors, curators, and keepers alike.

OVERVIEW OF US LEGISLATIVE AND REGULATORY PROCESS

On average, about 5,000 to 6,000 bills are introduced during every congressional cycle (a congressional cycle lasts for two years and is divided into two one-year sessions). Only about 500 to 600 bills become US law in each two-year congressional cycle, very few of which impact zoos and aquariums. If a bill fails to become enacted during the congressional cycle, it has to be reintroduced in the next cycle and go through the same clearance processes all over again to become a law.

Once a bill is signed into law, the appropriate agency or department within the federal government then implements the provisions of each law through the development or promulgation of pertinent rules, regulations, or policies. For example, the Animal Welfare Act is administered by the Animal and Plant Health Inspection Service (APHIS) within the US Department of Agriculture (USDA). Various federal agencies have jurisdiction over the rules and regulations that impact zoos and aquariums. Chief among these are

- the US Fish and Wildlife Service (permits under the Endangered Species Act, Marine Mammal Protection

Act, Lacey Act, Wild Bird Conservation Act, and Migratory Bird Act)
- the National Marine Fisheries Service (permits under the Marine Mammal Protection Act and Magnuson-Stevens Fishery Management Act)
- the US Department of Agriculture (USDA's Animal and Plant Health Inspection Service [APHIS]; Animal Welfare Act and animal care standards)
- the US Centers for Disease Control (regulations governing the import of certain animals such as bats, tortoises, and nonhuman primates)
- the US Transportation Security Administration (regulations governing animal transportation by air)
- the Occupational Safety and Health Administration (OSHA; keeper safety).

In total, US zoos and aquariums are monitored and regulated by no fewer than 12 federal agencies.

KEY LAWS AFFECTING ZOOS AND AQUARIUMS

THE USDA/APHIS AUTHORIZATION ACT

In the late 1950s, the US Department of Agriculture (USDA), under the authority of the Animal and Plant Health Inspection Service (APHIS) Authorization Act, initiated the first major restriction by a federal agency to affect zoos. The APHIS Authorization Act, implemented by the USDA, was designed to protect animals in the United States against infectious or contagious diseases by regulating the import, export, and quarantine of birds and poultry, horses, ruminants, swine, animal semen, animal blood, and blood serum.

Initially the USDA had planned to ban the importation of all ruminants into the United States, but a small group of zoo professionals convinced the agency that zoological displays must be allowed to import ruminants to keep their gene pools viable and to ensure future public display of ruminants in the United States. The regulations were structured to allow entry of ruminants into the United States under a Permanent Post Entry Quarantine (PPEQ) procedure. Approval of a facility is based on inspection of its operation by APHIS personnel. Requirements include satisfactory pens, cages, or enclosures. Disposition of waste, sewage, and specimens that die must be within the zoological park or at preapproved burial sites. All PPEQ animal carcasses must be incinerated. As the act stands now, public display facilities must meet APHIS quarantine requirements, in addition to any other permit or authorization, to import any ruminants, swine, horses, birds, or animal semen.

Shipments into the United States are restricted to APHIS-designated ports—Los Angeles, Miami, Honolulu, or New York City—and quarantine is required. Ruminants and swine must be quarantined for at least fifteen (15) days. APHIS has separate regulations for elephants, hippopotamuses, rhinoceros, and tapirs which include APHIS permits, health certificates, ectoparasite inspections, and shipping requirements.

Upon entering the United States and passing a period in quarantine, ruminants must go directly to a PPEQ facility. If the animals do not show clinical evidence of communicable disease after a year in this facility, they can be moved to other PPEQ facilities, to facilities accredited by the Association of Zoos and Aquariums (AZA), or to facilities with disease prevention procedures equivalent to those of AZA. However, progeny of PPEQ animals can be moved without restriction.

ANIMAL WELFARE ACT

In 1970, Congress passed the Animal Welfare Act (AWA) to regulate animals used in research facilities and for exhibition purposes, to ensure that they are provided with humane care and treatment. It was also enacted to address problems with pet theft and the sale of dogs and cats for research. The AWA regulates aspects of transportation, purchase, sale, housing, care, handling, and treatment. The USDA's Animal and Plant Health Inspection Service (APHIS) is the agency responsible for administering the AWA.

Under the AWA, US zoological parks and aquariums that display certain mammals must be licensed as exhibitors by APHIS. Current AWA regulations cover dogs, cats, monkeys, guinea pigs, hamsters, rabbits, and most other mammals. The regulations do not apply to cold-blooded animals, rats and mice bred for use in research, fish, horses, or farm animals. However, as a result of language inserted into the 2002 Farm Bill, APHIS is mandated to develop regulations that would cover birds not bred for use in research, which would include birds used for public display. The proposed regulations governing the inspection of birds had not been developed by USDA/APHIS at the time of writing.

All US zoological parks and aquariums that display animals must be licensed as exhibitors with APHIS. Applications for licenses must be submitted to the APHIS veterinarian in charge in the state where the applicant operates. The applicant must pay a licensing fee and APHIS will inspect the facility to ensure it meets minimum standards. APHIS usually inspects facilities annually. A facility must submit an annual report to APHIS identifying the number of individuals of each species and number of species that it maintains in its collection. Also, animals that have been bought or sold (including those transported, traded, or donated) must be identified.

The AWA regulations include minimum standards for handling, housing, feeding, watering, transportation, sanitation, ventilation, shelter from extremes of weather and temperature, adequate veterinary care, and separation of incompatible animals. Facilities must also meet minimum standards for structural strength of enclosures, sources of water and power, storage for food and bedding, and control of ambient temperature and ventilation.

Regulations pertaining to exotic animals are divided into three categories:

- nonhuman primates (including psychological well-being);
- marine mammals;
- other warm-blooded animals.

The AWA was amended by Congress in 1985 to add a requirement for facilities holding nonhuman primates. The amendment makes it mandatory for such facilities to develop and follow a plan for environmental enrichment to promote the psychological well being of nonhuman primates.

MARINE MAMMAL PROTECTION ACT

By 1972, the plight of marine mammals was generating considerable interest in the United States and abroad as the public began learning about the incidental take of marine mammals by commercial fisheries, particularly the take of dolphins by tuna fishermen. Expensive, highly emotional lobbying campaigns took place. Consequently, the Marine Mammal Protection Act of 1972 (MMPA) was enacted to protect all species of whales, dolphins, seals, polar bears, walruses, manatees, and sea otters. The basic premise of the MMPA is a moratorium against the taking of marine mammals in US waters. The MMPA makes it illegal to take or import any marine mammal without a permit, illegal to import a marine mammal taken in violation of the MMPA or similar foreign law, and illegal to use, possess, transport, or sell an illegally taken marine mammal. However, the MMPA does provide for the issuance of public display and scientific research permits authorizing the taking or importation of marine mammals. This represents a specific exemption to the moratorium against taking marine mammals. Public display permits cannot be issued for species designated as depleted (i.e., below their optimum sustainable population) or listed as either threatened or endangered under the Endangered Species Act.

Both the Department of the Interior's Fish and Wildlife Service (FWS) and the Department of Commerce's National Marine Fisheries Service (NMFS) administer the MMPA. Polar bears, sea otters, walrus, dugongs, and manatees are under the jurisdiction of the FWS. Cetaceans (whales and porpoises) and pinnipeds (other than walruses but including seals and sea lions) are under the jurisdiction of NMFS. USDA/APHIS has jurisdiction over the care and maintenance standards for marine mammals in captivity, including public display.

Facilities desiring a public display permit must (1) be open to the public on a regularly scheduled basis, with access to such facilities not limited or restricted other than by an admission fee; (2) be registered or licensed by APHIS; and (3) offer education or conservation programs based on professionally recognized standards of the public display community, such as the AZA in the United States.

Each permit is evaluated by the Marine Mammal Commission and the Committee of Scientific Advisors. The permit is also published in the *Federal Register* for 30 days for review and comment. A hearing may be held if requested.

ENDANGERED SPECIES ACT

The Endangered Species Act of 1973 (ESA) was enacted into law to provide a means for the conservation of threatened and endangered species and the ecosystems upon which they depend. The ESA places restrictions on a wide range of activities involving endangered and threatened animals and plants to help ensure their continued survival. The ESA prohibits the taking of these species without a permit. It defines an "endangered" species as any species listed by federal regulations as being in imminent danger of extinction. A "threatened" species is defined as any species listed by regulations that is likely to become endangered in the foreseeable future. The list of endangered and threatened species includes both native and foreign species, though only native species have provisions for critical habitat designations and species conservation and recovery plans.

Under the ESA, the term "take" is very broadly defined. "Take" means to harass, hunt, capture, or kill an endangered species, or to attempt to do so. Without a permit it is unlawful for any person subject to US jurisdiction to conduct the following activities with endangered or threatened species: import or export; deliver, receive, carry, transport, or ship in interstate or foreign commerce in the course of a commercial activity; sell or offer for sale in interstate or foreign commerce; take, possess, ship, deliver, carry, transport, or receive unlawfully taken wildlife.

In the case of foreign or interstate commerce, "commerce" is defined and interpreted as an exchange for a gain. For example, an exchange of rhinos between two institutions is not commerce, whereas an exchange of a rhino for a giraffe would be considered such. Similarly, a loan of an animal to an institution which includes a donation to the lending institution's conservation fund would be viewed as commerce. Finally, a loan of an animal which includes a stud fee paid to the lending institution is again viewed as commerce.

Certain situations are also exempt from the ESA. These include

- intrastate commerce (activities by residents of a state that occur entirely within that state)
- offering for sale (endangered and threatened species may be advertised for sale, but a permit must first be obtained)
- loans and gifts (endangered and threatened species may be shipped interstate as bona fide gifts or loans, but there may be no barter, credit, or other form of compensation, nor any intent to profit or gain)
- breeding loan agreements where offspring are returned to lenders
- hybrids (offspring of two species, one of which is protected by the act; hybrids are not protected by the ESA).

The secretary of the interior or the FWS may permit any act otherwise prohibited by the ESA for scientific purposes or to enhance the propagation or survival of the affected species. The latter "enhancement" provision is one commonly used in the zoo and aquarium community for importing and exporting ESA-listed species. In that situation, the institution must show that the proposed activity has direct/tangible benefit to the survival of the species in the wild. This would include such factors as

- showing the institution's role in a conservation program, such as the Species Survival Program (SSP), for the species in question
- explaining the in situ conservation programs of the SSP
- explaining how these programs assist the wild population
- demonstrating the adequacy of purpose to justify removal from the wild (if necessary)
- showing the probable effect on wild populations
- offering opinions of others with expertise on the species
- demonstrating the adequacy of expertise, facilities, and resources to accomplish specific objectives.

Under the authority of the ESA, the FWS has also issued regulations to encourage responsible breeding through captive bred wildlife (CBW) permits. The regulations (1) require registration with the FWS; (2) are restricted to nonnative, captive-born endangered, or threatened species; (3) are used for commercial transactions (purchase, sale, or exchange), not for loans or donations; and (4) require that both the sender and the recipient are registered. These permits are only good for responsible breeding initiatives, and cannot be used to move endangered species around the country for educational or public display purposes alone.

LACEY ACT

The oldest of all US wildlife laws, the Lacey Act was enacted in 1900 to protect both plants and animals by creating civil and criminal penalties for a wide variety of violations—including the trade of plants and animals that had been illegally taken, transported, or sold. In 1981 several large Washington-based animal welfare and environmental organizations were urging the Department of Interior to propose stricter laws regarding the importation of wildlife into the United States. As a result, the Department proposed a far-reaching amendment to the Lacey Act which prohibited the importation, exportation, transportation, sale, receipt, acquisition, or purchase of any fish or wildlife taken or possessed in violation of any law, treaty, or regulation of the United States or any US state, any Native American tribe, or any foreign country. The Lacey Act also was amended to regulate the humane and healthful transport of wild mammals and birds to the United States. (The Animal Welfare Act regulates shipment of wildlife within the United States.)

Another provision of the Lacey Act prohibits the import, acquisition, or transport of injurious wildlife except for zoological, educational, medical, or scientific purposes.

TABLE 66.1. Key legislation relating to United States zoos and aquariums

Level	Legislation	Regulatory authority	Scope	Impact on zoo and aquarium operations
National	USDA/APHIS Authorization Act	Animal and Plant Health Inspection Service (APHIS),US Department of Agriculture	Protects animals in the United States against infectious or contagious diseases by regulating the import, export, and quarantine of certain animals and parts of animals.	• Develops permanent post-entry quarantine (PPEQ) procedures. • Requires public display facilities to meet APHIS quarantine standards to import any ruminants, swine, horses, birds, or animal semen. • Enforces separate regulations for import of elephants, hippopotamuses, rhinoceroses, and tapirs.
National	Animal Welfare Act	Animal and Plant Health Inspection Service (APHIS),US Department of Agriculture	Regulates animals used in research facilities and for exhibition purposes, to ensure that they are provided with humane care and treatment.	• Requires US zoological parks and aquariums that display certain mammals to be licensed as exhibitors. • Inspects such facilities to ensure that hey meet minimum standards. • Requires facilities to submit annual reports identifying the number of individuals of each species and number of species they maintain in their collections. • Requires facilities holding nonhuman primates to develop and follow plans for environmental enrichment to promote their psychological well-being. • Regulates the humane and healthful transport of wild mammals and birds within the United States.
National	Marine Mammal Protection Act	US Fish and Wildlife Service,-US Department of the Interior National Marine Fisheries Service, US Department of Commerce Animal and Plant Health Inspection Service (APHIS),US Department of Agriculture	Protects all species of whales, dolphins, seals, polar bears, walrus, manatees, and sea otters by establishing a moratorium against the taking of marine mammals in US waters.	• Provides for the issuance of public display and scientific research permits authorizing the taking or importation of marine mammals. • Prohibits the issuance of public display permits for species designated as depleted or listed as either threatened or endangered under the Endangered Species Act. • Requires facilities desiring a public display permit to (1) be open to the public on a regularly scheduled basis, (2) be registered or licensed by APHIS, and (3) offer an education or conservation program based on professionally recognized standards of the US public display community.
National	Endangered Species Act	US Fish and Wildlife Service, US Department of the Interior National Marine Fisheries Service, US Department of Commerce	Provides a means for the conservation of threatened and endangered species and the ecosystems upon which they depend by placing restrictions on a wide range of activities involving endangered and threatened animals.	• Requires a permit for any person subject to US jurisdiction to import or export, or to transport or ship endangered or threatened species in interstate or foreign commerce in the course of a commercial activity. • May permit any act otherwise prohibited by the ESA to proceed for scientific purposes or to enhance the propagation or survival of the affected species (the "enhancement" provision is commonly used in the zoo and aquarium community for importing and exporting ESA-listed species). • Requires an institution to show that the proposed activity has direct or tangible benefit to the survival of the species in the wild. • Encourages responsible breeding through captive bred wildlife permits.
National	Lacey Act	US Fish and Wildlife Service, US Department of the Interior	Prohibits the importation, exportation, transportation, sale, receipt, acquisition, or purchase of any fish or wildlife taken in violation of any law, treaty, or regulation of the United States or any foreign country.	• Regulates the humane and healthful transport of mammals and birds to the United States. • Prohibits the import, acquisition, or transport of injurious wildlife except for zoological, educational, medical, or scientific purposes. • Requires facilities to be inspected to ensure the protection of the public.

TABLE 66.1. continued

Level	Legislation	Regulatory authority	Scope	Impact on zoo and aquarium operations
National	Wild Bird Conservation Act	US Fish and Wildlife Service, US Department of the Interior	Promotes the conservation of wild exotic birds by ensuring that all trade in exotic bird species involving the United States is biologically sustainable and not detrimental to the species.	• Prohibits the importation of wild-caught birds (with exceptions for zoological breeding or display programs and scientific research).
National	Migratory Bird Treaty Act	US Fish and Wildlife Service, US Department of the Interior	Prohibits taking, possession, import, export, transport, sale, purchase, barter, or offer for sale of any migratory bird, or the nests or eggs of such a bird.	• Authorizes exceptions by a valid permits (for public zoological parks, accredited institutional members of AZA, and public, scientific, or educational institutions).
National	Public Health Service Act	Centers for Disease Control and Prevention (CDC), US Department of Health and Human Services	Regulates imports to prevent the introduction, transmission, or spread of communicable diseases from foreign countries into the United States.	• Covers turtles, tortoises, terrapins (excluding sea turtles), nonhuman primates, and "vectors." (with exceptions for exhibition or scientific purposes, subject to a permit process).

The importation or transportation of live injurious wildlife is deemed to be a threat to the health and welfare of human beings, the interests of forestry, agriculture, and horticulture and the welfare or survival of the United States wildlife. Zoological institutions can receive zoological purpose permits. Such a facility must be inspected to ensure protection of the public. Some examples of injurious wildlife include, but are not limited to, any species of flying bat, mongoose, meerkat, or snakehead (fish), and any species of European rabbit or Indian wild dog. Recently conservation organizations have petitioned the US Department of the Interior to add additional species to the injurious wildlife list, such as certain constrictor snakes and amphibians exhibiting the chytrid fungus, but no definitive decision has been rendered on these proposals.

WILD BIRD CONSERVATION ACT

In the late 1980s the international wild bird trade began taking an unprecedented toll on wild bird populations worldwide. Importation into the United States alone contributed to the decline of many species in the wild. In the early 1990s about 3.5 million wild birds, mostly songbirds and psittacines (parrots, macaws, and cockatoos), were involved in the international commercial trade each year. Concerned groups recognized this as a serious conservation problem. The AZA participated in a coalition of conservation, scientific, animal welfare, pet industry, and bird breeder organizations to discuss protection for wild-caught birds. The coalition met for more than two years to develop draft legislation to curb the problems of the bird trade.

The Wild Bird Conservation Act was subsequently enacted in 1992 to promote the conservation of wild exotic birds by assisting conservation and management programs in wild bird countries of origin, and by ensuring that all trade in exotic birds involving the United States is biologically sustainable and is not detrimental to the species. The law prohibits the importation of wild-caught birds. There are four exceptions to the prohibition: zoological breeding or display programs,

scientific research, cooperative breeding, and importation of previously owned pet birds. The FWS implements the law.

MIGRATORY BIRD TREATY ACT

The Migratory Bird Treaty Act, first enacted in 1918, implements four separate treaties to which the United States is a party along with Great Britain (on behalf of Canada), Mexico, Russia, and Japan. The act states that no person shall take, possess, import, export, transport, sell, purchase, barter, or offer for sale any migratory bird, or the nests or eggs of such birds, except as authorized by a valid permit. The FWS implements the Migratory Bird Treaty Act; its regulations outline general exceptions to permit requirements. Those exceptions include state game departments, municipal game farms or parks, public museums, public zoological parks, accredited institutional members of the AZA, and public scientific or educational institutions.

PUBLIC HEALTH SERVICE ACT

The Public Health Service Act (PHSA), first enacted in 1944 and implemented by the US Department of Health and Human Services Centers for Disease Control, regulates imports to prevent the introduction, transmission, or spread of communicable diseases from foreign countries into the United States. The PHSA covers turtles, tortoises, terrapins (excluding sea turtles), nonhuman primates, and "vectors," which are defined as any animal (including insects) that may be able to convey an infectious disease to other animals or people, such as bats or rodents. Exceptions allow the importation of prohibited animals and insects for exhibition or scientific purposes, subject to a permit process.

Live nonhuman primates may be imported into the United States and sold, resold, and distributed for bona fide scientific, educational, or exhibition purposes only. Importers must register with the Secretary of Health and Human Services. At press time, the CDC is in the process of developing regulations for the quarantine and transportation of nonhuman primates.

Acknowledgments

This book has been a true team effort and would not have reached completion without the contributions (and in some cases, sacrifices) of many people. Of course, at the heart of this project have been the contributing authors and artists. We have listed their names in the contributors list. Their time, patience, dedication, and hard work is much appreciated.

First, we need to acknowledge the patience and support of our friends and family, and also those of each contributor. This project has required passion (with hearty doses of obsession), time and dedication. Thank you for sharing us!

We would also like to expressly thank the employers of our contributors; please see the contributors list for details. Without the support of their institutions (zoos, aquariums, colleges, etc.)—whether through the direct donation of resources, or indirectly through the general support of professional development, continuing education workshops, conferences, flexibility, and so on—many contributors would not have been able to participate in this project.. In tough economic times it can be difficult to invest in the future, but such investment is essential if zoos and aquariums are to fulfill all aspects of their mission, including science, education, recreation, and conservation.

We would like to acknowledge the support of the American Association of Zoo Keepers; its board has provided input, guidance, and encouragement from the beginning. Also, many professional members and staff from organizations such as the Association of Zoos and Aquariums (AZA), the European Association of Zoos and Aquaria (EAZA), and the World Association of Zoos and Aquariums (WAZA) have contributed and been supportive.

As this is a practical book, we wanted to include visuals as much as possible; it is not as easy as it sounds! Thank you to the contributing artists who provided custom artwork:

Lia Brands, Amy Burgess, Anthony Galván III, Kim Lovich, Lisa McLaughlin, and Travis Pyland. Kate Woodle provided artistic editorial support far above and beyond the contribution of her illustrations.

A special thank you to the team at the State University of New York's Jefferson Community College (JCC). Without their support and their grant of sabbatical time, this project may never have started. Appreciation goes to Diane Chepko-Sade, Linda Dittrich, Jack Donato, Kate Fenlon, Frank Florence, Pete Gaskin, Ed Knapp, Monica LeClerc, Carole McCoy, John Penrose, Kelly Rusho-Boyer, Todd Vincent, Vicki White, Deltra Willis, and the Jefferson Board of Trustees. We also extend a heartfelt thank you to the AMG student keepers for their patience, enthusiasm, and support. Thanks to Gulos, Red Pandas, Timber Wolves, Black Mambas, Wombats, Bandicoots, Ocelots, Grizzlies, Hyenas, Fossa, Flying Foxes, and Kinkajous!

Many other people have also offered advice, support, input and guidance. These include David Barney, David Blasko, Paul Boyle, Hans Christoffersen, Bob Cisneros, Jeff Cole, Peter Dickinson, Candice Dorsey, Murray Fowler, Shane Good, Devra Kleiman, Debborah E. Luke, Kay Mehren, Deborah Olson, William Rapley, Kevin Robinson, Kyle Stevenson, Sally Walker, Jinelle Webb, Adrienne Whiteley, and Glenn Zugehar. During "crunch time," Brie Foltz provided organizational assistance and Nicholas Ladd assisted with editorial tasks.

Finally we would like to thank Christie Henry, Abby Collier, Renaldo Migaldi, and the team at the University of Chicago Press. We appreciate their support and shared vision for this project.

With so many people involved, we have probably missed someone who has made a very important contribution to this book. Thank you!

621

Appendix 1

Further Readings by Chapter

PART ONE: PROFESSIONAL ZOOKEEPING

1. THE PROFESSION OF ZOOKEEPER

Animal Keepers Forum (AKF). Magazine of the American Association of Zookeepers

International Zoo News (IZN). Subscription publication, 8 × per year

International Zoo Yearbook (IZY). London: Zoological Society of London

PART THREE: WORKPLACE SAFETY AND EMERGENCIES

7. WORKPLACE SAFETY

Hediger, Heini. 1950. *Wild Animals in Captivity*. London: Butterworth.

Resources for Crisis Management in Zoos and Other Animal Care Facilities. 1999. American Association of Zoo Keepers.

8. EMERGENCY READINESS AND CRISIS MANAGEMENT

www.osha.gov/SLTC/etools/evacuation/index.html. OSHA Evacuation Plans and Procedures. This online resource helps small businesses like zoos develop and implement Emergency Action Plans (EAPs) and comply with OSHA's emergency standards.

www.osha.gov/SLTC/firesafety/index.html. OSHA's online "Safety and Health Topics: Fire Safety."

PART FOUR: ZOO ANIMAL MANAGEMENT

11. TAXONOMY

Mortenson, Philip B. 2004. *This is Not a Weasel: A Close Look at Nature's Most Confusing Terms*. Hoboken, NJ: John Wiley & Sons.

Yoon, Carol Kaesuk. 2009. *Naming Nature: The Clash Between Instinct and Science*. New York: W.W. Norton.

15. PHYSICAL RESTRAINT AND HANDLING

Fowler, M. E. 2008. *Restraint and Handling of Wild and Domestic Animals*, 3rd ed. Ames, IA: Wiley-Blackwell.

Lewbart, G. A., ed. 2006. *Invertebrate Medicine*, Ames, IA: Wiley-Blackwell.

Moberg, G. P., ed. 1985. *Animal Stress*. Bethesda, MD: American Physiological Society.

Ramirez, K. 1999. *Animal Training: Successful Animal Management through Positive Reinforcement*. Chicago: Shedd Aquarium.

Sonsthagen, T. F. 1991, *Restraint of Domestic Animals*. Goleta, CA: American Veterinary Publications.

16. NUTRITION

Kleiman, D., M. Allen, K.Thompson, and S. Lumpin, eds. 1996. *Wild Animals in Captivity: Principles and Techniques*. Chicago: University of Chicago Press.

Maynard, L., J. Loosli, H. Hintz, and R. Werner. 1979. *Animal Nutrition*. McGraw-Hill.

McDowell, L.R. 1989. *Vitamins in Animal Nutrition: Comparative Aspects to Human Nutrition*. Academic Press.

———. 1991. *Minerals in Animal and Human Nutrition*. Academy Press.

Stevens, E., and D. Hume. 1995. *Comparative Physiology of the Vertebrate Digestive System*. Second Edition. Cambridge University Press.

Schmidt, D., and R. Barbiers. The Giraffe Nutrition Conference Proceedings, May 25–26, 2005. Chicago: Lincoln Park Zoo.

Schmidt, D., D. Travis, and J. Williams. 2006. Guidelines for Creating a Food Safety HACCP Program in Zoos or Aquaria. *Zoo Biology* 25:125–35.

Zoo Animal Nutrition. Proceedings of the European Zoo Nutrition Conferences, vols. 1 (2000), 2 (2003), and 3 (2006). Filander Verlag Furth.

17. RECORDKEEPING

Alley, Michael. 1987. *The Craft of Scientific Writing*. Englewood Cliffs, NJ: Prentice-Hall.

Miller, Jean, and Block, Judith. 1992 (revised 2004). *Animal Records Keeping*. Available at http://www.buffalozoo.org/E-manual_short_form_Word_format.pdf.

18. IDENTIFICATION

Anon. 2009. Toe-Clipping in Amphibians. *Journal of Herpetological Medicine and Surgery* 19:38–41.

Elasmobranch Husbandry Manual. 2004. Available at http://www .elasmobranchhusbandry.org/.

Lander, Michelle E., Andrew J. Westgate, Robert K. Bonde, and Michael J. Murray. 2001. Tagging and Tracking. In *CRC Handbook of Marine Mammal Medicine Second Edition*, edited by Leslie A. Dierauf and Frances M.. Gulland, 851–80. Boca Raton, FL: CRC Press.

Loomis, Michael. 1993. Identification of Animals in Zoos. In *Zoo and Wild Animal Medicine, Current Therapy 3*, edited by Murray E. Fowler, 21–23. Philadelphia: W. B. Saunders.

Transponder Placement (microchips) AZA Recommendations. Available at www.aza.org/uploadedFiles/Animal_Care_and_Management /Animal_Management/Animal_Data_and_Recordkeeping /IDMAG_Documents/TransponderStatement2010.pdf.

19. REPRODUCTION

Burton, Michael, and Ed Ramsey. 1986. Cryptorchidism in maned wolves. *Journal of Zoo and Animal Medicine* 17:133–35.

Rowell, T. E. 1972. Female Reproduction Cycles and Social Behavior in Primates. In *Advances in the Study of Behavior*, Vol. 4, edited by D. S. Lehrman. New York: Elsevier, Academic Press.

21. MANAGEMENT OF NEONATAL MAMMALS

Alldredge, A.W., R. Deblinger, and J. Peterson. 1991. Birth and Fawn Bed Site Selection by Pronghorns in a Sagebrush-Steppe Community. *Journal of Wildlife Management* 55, no. 2.

Anon. 2000. Special Issues: Duiker Management and Conservation. *Animal Keepers' Forum* 27, no. 11.

Berman, C. 1984. Variation in Mother-Infant Relationships: Traditional and Nontraditional Factors. In *Female Primates: Studies by Women Primatologists*, 17–36. New York: Alan R. Liss.

Brown, M. T. 2006. Birth and Development of a La Plata Three-Banded Armadillo." *Animal Keepers' Forum* 33, no. 2: 73–78.

Burch, Catherine. 1999. "Oral Re-hydration Therapy and Its Effect on Curd Formation in Cow's Milk. *Zoological Society of San Diego Neonatal Symposium Paper Presentations.*

Clutton-Brock, T. H. 1993. *The Evolution of Parental Care: Monographs in Behavior and Ecology*. Princeton, NJ: Princeton University Press.

Columbus Children's Zoo Staff. 1993. Columbus Zoological Gardens Gorilla Nursery Protocol. *Animal Keepers' Forum* 20, no. 2.

Deem, S., S. Citino, and L. Cree. 2002. Parenteral Nutrition in a Neonatal Reticulated Giraffe (*Giraffa camelopardalis reticulata*). *Proceedings of the American Association of Zoo Veterinarians.*

Edwards, M. S. 1998. Preliminary Observations of a Revised Formula for Hand-rearing a Malayan Sun Bear *(Helarctos m. malayanus)*: Days 0–75. Zoological Society of San Diego Neonatal Symposium Paper Presentations.

Edwards, M.S., C. D. Burch, and M. Diehl. 1995. Evaluation of 3 Milk Replacers for Hand-Rearing Indian Nilgai (*Boselaphus tragocamelus*). Zoological Society of San Diego Neonatal Symposium Paper Presentations.

Edwards, M. S., and K. J. Lisi. 2004. Observations of Growth and Caloric Intake of Two Polar Bears (*Ursus maritimus*) Hand Raised Using a Revised Protocol. International Polar Bear Husbandry Conference, San Diego, CA.

Edwards, M. S., K. J. Lisi, K. Lang, and L. Ware. 2007. Evaluation of a Formula for Hand-Rearing Red Pandas (*Ailurus fulgens*).

Annual Conference Proceedings of American Association of Zoo and Wildlife Veterinarians, Knoxville, TN.

Espinoza-Bylin, Debra L. 1992. Digestion and Immunity in the Ungulate Neonate. AAZK National Conference Workshop.

———. 1994. Plasma Transfusion Program: A Retrospective Look at the Records. Zoological Society of San Diego Neonatal Symposium Paper Presentations.

Forbes, S., R. Sealy, H. Frazier, N. Farnham, L. Moneymaker, and G. Toffic. 2006. Breaking the Cycle: Hand-Rearing and Early Reintroduction as a Step towards Appropriate Behavioral Development and Successful Family Bonding in Two Species of Callitrichids. *AAZK Conference Proceedings.*

Fowler, M., and R. E. Miller. 1978, 1986, 1993, 1999, 2003, 2008. *Zoo and Wild Animal Medicine Current Therapy*, vols. 1–6, 1978, 86, 93, 99, 2003, Saunders Elsevier.

Godson, D., S. Acres, D. Haines, 2003. Failure of Passive Transfer and Effective Colostrum Management in Calves. *Large Animal Veterinary Rounds*, 3, no. 10.

Graham, C., and J. Bowen. Clinical Management of Infant Great Apes." *Monographs in Primatology* 5.

Graham, C. and J. Bowen, eds. 1982. *Clinical Management of Infant Great Apes*. New York: Alan R. Liss.

Gunn-Moore, Danielle. 2006. Neonatal Resuscitation. University of Edinburgh. World Small Animal Veterinary Association World Congress Proceedings.

Hnida, John. 1985. Mother-Infant and Infant-Infant Interactions in Captive Sable Antelope: Evidence for Behavioral Plasticity in a Hider Species. *Zoo Biology* 4:339–49.

Hoskins, J. 1999. Pediatrics: Puppies and Kittens. *The Veterinary Clinics of North America, Small Animal Practice*. Philadelphia:W. B. Saunders.

Hu, D. and G. Zhang. 2001. Some Aspects of Hand-rearing Giant Panda Cubs at the China Research and Conservation Center for the Giant Panda in Wolong. *Zoological Society of San Diego Neonatal Symposium Paper Presentations.*

Ingram, Kathryn. 1990. Use of Equine Plasma Transfusions for an Immuno-deficient Brazilian Tapir Neonate. Zoological Society of San Diego Neonatal Symposium Paper Presentations.

Izard, M. K., and E. Simons. 1986. Isolation of Females Prior to Parturition Reduces Neonatal Mortality in *Galago*." *American Journal of Primatology* 10:249–55.

Janssen, D. L., B. R. Rideout, and M. S. Edwards. 1999. Tapir Medicine. In *Zoo and Wildlife Medicine*, edited by M. E. Fowler and R. E. Miller. Philadephia: W. B. Saunders.

Joseph, D., and A. Reed. 2001. Aspiration Pneumonia. *Veterinary Technician* 22, no. 9.

Keith, M. 1994. Weight Gain of a Giraffe After Being Orphaned at 3.5 Months of Age. *Animal Keepers' Forum* 21, no. 5.

Kennedy, C. 1992. The Early Introduction of a Hand-Reared Orangutan Infant to a Surrogate Mother. Proceedings of the 19th National Conference of the American Association of Zoo Keepers.

Kinzley, Colleen, ed. 1997. *The Elephant Hand Raising Notebook*. Oakland, CA: Oakland Zoo.

Kirkwood, James K., and Katherine Stathatos. 1992. *Biology, Rearing, and Care of Young Primates*. New York: Oxford University Press.

Levy, J., et al. 2001. Use of Adult Cat Serum to Correct Failure of Passive Transfer in Kittens. *JAVMA* 219, no 10.

Lincoln Park Zoological Society. 1993. *Maned Wolf Diaries: Growth and Development of Hand-Reared Pups* Video.

Lopate, C. 2009. The Critical Neonate: Under 4 Weeks of Age. *NAVC Clinician's Brief, Patient Support.*

Lopez Schaller, P. 1999. One Unsuccessful African Wild Dog (*Lycaon pictus*) Litter. *Animal Keepers' Forum* 26, no. 8.

Markham, R. 1990. Breeding Orangutans at Perth Zoo: Twenty Years of Appropriate Husbandry. *Zoo Biology* 9:171–82.

McCauley, D. 2003. *Macropods: Their Care, Breeding, and the Rearing of Their Young.* San Antonio: Lithopress.

McCracken, H., et al. 1994. Two Successive Cesarean Deliveries in an Elderly Primigravid Lowland Gorilla (*Gorilla gorilla*) and Neonatal Care of the Offspring. *Proceedings American Association of Zoo Veterinarians.*

Mikota, S., E. Sargent, G. S. Ranglack. 1994. *Medical Management of the Elephant.* West Bloomfield, MI: Indira Publishing House.

Miller-Edge, Michele. 1994. Neonatal Immunology. *Zoological Society of San Diego Neonatal Symposium Paper Presentations.*

Mulnix, P. D. Collello, and M. Baeyens. 2002. Flat Puppy Syndrome in Maned Wolves Corrected Through Physical Therapy. *Animal Keepers' Forum* 30, no. 7.

Price, E. 1992. The Benefits of Helpers: Effects of Group and Littler Size on Infant Care in Tamarins (*Saguinus oedipus*). *American Journal of Primatology* 26:179–90.

Quavillon, M. D. 2009. Methods of Recognizing and Recuperating an Underweight.

Reason, R. 1999. Successful Rearing of a 10 1/2-Week-Old Orphaned Giraffe (*Giraffa camelopardalis*) Calf at Brookfield Zoo." *Animal Keepers' Forum* 26, no. 2.

Reed-Smith, Janice, ed. 1995. *North American River Otter (Lontra Canadensis) Husbandry Notebook.* Grand Rapids: John Ball Zoological Garden.

Robinson, Edward. 1997. Immunodeficiencies of Foals. *Current Therapy in Equine Medicine,* chap. 13, "The Foal."

Roussel, A., and P. Woods. 1999. Colostrum and Passive Immunity. *Current Veterinary Therapy: Food Animal Practice,* Fourth Edition.

Ruppenthal, G. and G Sackett. 1992. *Research Protocol and Technicians Manual: A Guide to the Care, Feeding, and Evaluation of Infant Monkeys.* Seattle: Infant Primate Research Laboratory, University of Washington.

Sodaro, C., and N. Greenblatt. 1998. Training Caregiving Behaviors in Orangutans (*Pongo pygmaeus pygmaeus*). *The Behavior Bridge* 1, no. 3.

Staker, L. A. 2009. Handrearing of Developing Macropods and Macropod Husbandry. *Macropod Husbandry Specialist, Queensland, Australia. ICZ/AAZK 2009 Proceedings,* Seattle, WA.

Sutherland-Smith, M. 1992. Diarrhea in Neonatal Hoofstock: What's the Poop?*12th Annual Proceedings of the Association of Zoo Veterinary Technicians,* San Diego.

Swanson, W. 1999. Toxoplasmosis and Neonatal Mortality in Pallas' Cats: A Survey of North American Zoological Institutions. *Proceedings of the American Association of Zoo Veterinarians.*

Wagner, D. C., and M. S. Edwards. 1999. Hand-Rearing Black and White Rhinoceroses: A Comparison. *Proceedings of the 26th National Conference of the American Association of Zoo Keepers,* Portland, OR.

Wanders, K. 2009. Conditioning of a Red-Flanked Duiker (*Cephalophus rufilatus*) to Determine and Monitor Pregnancy. *ICZ/AAZK 2009 Proceedings,* Seattle, WA.

Weaning Juvenile Giant Anteater (*Myrmecophaga tridactyla*). *ICZ/AAZK 2009 Proceedings.* Seattle, WA.

Whitman, Kim. Raising Jesse: Notes on Hand-rearing a Sloth Bear Cub and Its Subsequent Introduction to a Brown Bear Cub. Associate Curator of Large Mammals, Philadelphia Zoological Gardens.

Zuba, Jeffery R. 1991. Factors Influencing Neonatal Infections. *Zoological Society of San Diego Neonatal Symposium Paper Presentations.*

22. MANAGEMENT OF GERIATRIC ANIMALS

Animal Keeper's Forum Special Issue: The Care and Management of Geriatric Animals in Zoos April/May 2009: http://aazk.org.

PART FIVE: ZOO ANIMAL HUSBANDRY AND CARE

26. HUSBANDRY AND CARE OF SMALL MAMMALS

Bertram, B. C. R. 1986. Breeding Small Mammals in Zoos, In *International Zoo Yearbook, Vol 24/25,* edited by C. Jarvis. London: Zoological Society of London.

Crandall, L. S. 1964. *Management of Wild Mammals in Captivity.* Chicago: University of Chicago Press.

Eisenberg, J. F. and D. G. Kleiman, eds. 1983. *Advances in the Study of Mammalian Behavior.* Special Publication #7, American Society of Mammalogists, Lawrence, KS.

Gittleman, J. L., ed. 1996. *Carnivore Behavior, Ecology and Evolution,* Vols. 1 and 2. Ithaca, NY, and London: Comstock Press.

Jarvis, C. 1968. *International Zoo Yearbook, Vol. 8: Canids and Felids.* London: Zoological Society of London.

Jarvis, C., and D. Morris. 1962 (reprinted 1971). *International Zoo Yearbook,* Vol. III: Special Section, "Small Mammals." London: Zoological Society of London.

Kessler, D. S., K. Hope, M. Maslanka. In L Hess. 2009. *Veterinary Clinics of North America,* Vol. 12, no. 2: Exotic Animal Practice. Saint Louis: Saunders Elsevier.

Kleiman, D. G., M. E. Allen, K. V. Thompson, and S. Lumpkin, eds. 1997. *Wild Mammals in Captivity: Principles and Techniques.* Chicago: University of Chicago Press.

MacDonald, D., ed. 1985. *The Encyclopedia of Mammals.* New York: Facts on File Publications.

Olney, P. J. S., and P. Ellis. 1992. *International Zoo Yearbook,* Vol. 31, section 1: Australasian Fauna. London: Zoological Society of London.

Wharton, D. C. 1986. Management Procedures for the Successful Breeding of the Striped Grass Mouse. In *International Zoo Yearbook, Vol. 24/25,* edited by C. Jarvis. London: Zoological Society of London.

Williams, M. 1990. Beaver Country: A North American Beaver Exhibit at Drusillas Zoo Park. *International Zoo Yearbook* 29. (This exhibit won the 1989 UFAW Animal Welfare Award.)

27. HUSBANDRY AND CARE OF HOOFSTOCK

Bergeron, R., A,. J. Badnell-Waters, S. Lambton, and G. Mason. 2006. Stereotypic Oral Behaviour in Captive Ungulates: Foraging, Diet, and Gastrointestinal Function. In *Stereotypic Animal Behaviour:Fundamentals and Applications to Welfare,* edited by Georgia Mason and Jeffrey Rushen, 19–57. Trowbridge, UK: Cromwell Press.

Burgess, A., ed. 2004. *The Giraffe Husbandry Resource Manual.* Silver Spring, MD: Association of Zoos and Aquariums Antelope/Giraffe Taxon Advisory Group.

Fischer, Martha, ed. 2002. *Husbandry Guidelines for the Babirusa* (Babyrousa babyrussa) *Species Survival Plan.* American Zoo and Aquarium Association Babirusa SSP.

Gage, Laurie J., ed. 2002. *Hand-rearing Wild and Domestic Mammals.* Ames: Iowa State University Press (Blackwell Publishing).

Gilbert, Tania, and Tim Woodfine, ed. 2004. *The Biology, Husbandry, and Conservation of Scimitar-horned Oryx* (Oryx dammah). Winchester, UK: Marwell Preservation Trust. Available online at http://www.marwell.org.uk/downloads/biology_husbandry_conservation_SHO_2004.pdf

Joseph, Sharon, ed. 2004. *Husbandry Manual for Wild Cattle Species.* American Zoo and Aquarium Association Bison, Buffalo, and Cattle Advisory Group.

Junge, Randall E. 2007. Basic Medical Management of Exotic Hoofstock. *Proceedings of the 79th Western Veterinary Conference* Vol. 540. Available online at http://wvc.omnibooksonline.com/data/papers/2007_V540.pdf.

Kleiman, Devra G., Mary E. Allen, Katerina V. Thompson, and Susan

Lumpkin, eds. 1996. *Wild Mammals in Captivity: Principles and Techniques.* Chicago: University of Chicago Press.

Melcher, Carol R. 1975. *Elements of Stable Management.* Cranbury, NJ: A. S. Barnes.

Morrow, C. J., L. M. Penfold, and B. A. Wolfe. 2009. Artificial Insemination in Deer and Non-Domestic Bovids. *Theriogenology* 71(1): 149–65.

Nowak, Ronald M., ed. 1991. *Walker's Mammals of the World, Fifth Edition.* Baltimore: Johns Hopkins University Press.

Rietschel, W. 2002. Plant Poisoning of Zoo Animals or an Unsuspicious Method of Population Control in Zoo Collections. *European Association of Zoo and Wildlife Veterinarians (EAZWV) 4th Scientific Meeting, May 8–12, 2002, Heidelberg, Germany.* 109–13.

Thompson, Valerie D. 1989. Behavioral Response of 12 Ungulate Species in Captivity to the Presence of Humans. *Zoo Biology* 8(3): 275–97.

Ullrey, Duane E. 1997. *Hay Quality Evaluation.* AZA Nutrition Advisory Group Fact Sheet 001. Available online at http://www.nagonline.net/Technical%20Papers/technical_papers.htm.

Young, E., ed. 1975. *The Capture and Care of Wild Animals.* Hollywood, FL: Ralph Curtis

HOOFSTOCK TRAINING ARTICLES

Barrios, S. 2000. Operant Conditioning of Lowland Tapirs for Purpose of Blood Collection. *Proceedings of the 27th National Conference of the American Association of Zoo Keepers, Inc., Columbus, Ohio, October 8–12, 2000,* 1–3.

Bouwens, Nichole, Mollye Nardi, and Lisa Smith. 2008. Training a Yellow-Backed Duiker for Radiographs. *Animal Keepers Forum* 35(6): 226–28.

Crump, J. P. Jr., and J. W. Crump. 1994. Manual Semen Collection from a Grevy's Zebra Stallion (*Equus grevyi*), Onset of Sperm Production, Semen Characteristics, and Cryopreservation of Semen, with a Comparison to the Sperm Production from a Grant's Zebra Stallion (*Equus burchelli boehmi*). *Theriogenology* 41(5): 1011–21.

Dumonceaux, G. A., M. S. Burton, R. L. Ball, and A. Demuth. 1998. Veterinary Procedures Facilitated by Behavioral Conditioning and Desensitization in Reticulated Giraffe (*Giraffa camelopardalis*) and Nile Hippopotamus (*Hippopotamus amphibius*). *Proceedings of the 1998 Annual Conference of the American Association of Zoo Veterinarians,* 388–91.

Houston, E. William, Patricia K. Hagberg, Martha T. Fischer, Melissa E. Miller, and Cheryl S. Asa. 2001. Monitoring Pregnancy in Babirusa (*Babyrousa babyrussa*) with Transabdominal Ultrasonography. *Journal of Zoo and Wildlife Medicine* 32(3): 366–72.

Phillips, Megan, Temple Grandin, Wendy Graffam, Nancy A. Irlbeck, and Richard C. Cambre. 1998. Crate Conditioning of Bongo (*Tragelaphus eurycerus*) for Veterinary and Husbandry Procedures at the Denver Zoological Gardens. *Zoo Biology* 17(1): 25–32.

Robertia, J., J. Sauceda, and R. Willison. 2000. Conditioning Three Species of Aridland Antelopes for Weight Collection: A Case Study on Hippotraginae. *Animal Keepers Forum* 27(5): 214–23.

Watkins, V., and J. Gregory. 1997. Conditioning a Greater One-Horned Rhino (*Rhinoceros unicornis*) to Accept Foot Treatment without Anesthetic. *Animal Keepers Forum* 24:250–56.

30. HUSBANDRY AND CARE OF ELEPHANTS

Adams, J. 1081. *Wild Elephants in Captivity.*

AZA Elephant Management Guidelines. 2010.

AZA School. Principles of Elephant Management.

Blond, G. 1961. *The Elephants.*

Carrington, R. 1959. *The Elephants.*

Crandall, L. 1964. *Management of Wild Mammals in Captivity.*

Hediger, H. 1964. *Wild Animals in Captivity.*

Olson, Deborah. *Elephant Husbandry Resource Guide.*

Pryor, Karen. *Don't Shoot the Dog!* New York: Random House.

Schmidt, M. 1978. "Elephants." In Zoo and Wild Animal Medicine, edited by M. Fowler, 709–52.

Sillar, F. C., and R. M. Meyler. 1968. *Elephants Ancient and Modern.*

Ramirez, Ken, 1999. *Animal Training.*

Sikes, S. 1971. *Natural History of the African Elephant.*

31. HUSBANDRY AND CARE OF MARINE MAMMALS

Bonner, W. Nigel. 1994. *Seals and Sea Lions of the World.* New York: Facts on File.

Dierauf, L. A., and F. M. D. Gulland. 2001. *CRC Handbook of Marine Mammal Medicine Second Edition.* Boca Raton, FL: CRC Press.

King. J. E. 1983. *Seals of the World.* London: British Museum Natural History.

Klinowska. 1991. *Dolphins, Porpoises and Whales of the World: The IUCN Red Data Book,* IUCN, Gland. Switzerland, and Cambridge: IUCN.

32. HUSBANDRY AND CARE OF BIRDS

Beebe, C. William. *The Bird: Its Form and Function.*

Handbook of the Birds of the World. Vols. 1–16. Barcelona: Lynx Editions.

34. AMPHIBIAN HUSBANDRY AND CARE

Browne, Robert K., et al. 2007. Facility Design and Associated Services for the Study of Amphibians. *ILAR Journal* 48(3): 188–202.

Browne, Robert K., and Kevin Zippel. 2007. Reproduction and Larval Rearing of Amphibians. *ILAR Journal* 48(3): 214–34.

Burger, R. Michael, et al. 2007. *Evaluation of UVB Reduction by Materials Commonly Used in Reptile Husbandry.* Zoo Biology 26:417–23.

Gehrmann, William H. 1987. Ultraviolet Irradiances of Various Lamps Used in Animal Husbandry. *Zoo Biology* 6:117–27.

Pough, F. Harvey. 2007. Amphibian Biology and Husbandry. *ILAR Journal* 48(3): 203–13.

35. AQUARIUM SCIENCE: HUSBANDRY AND CARE OF FISHES AND AQUATIC INVERTEBRATES

Creswell, LeRoy. 1993. *Aquaculture Desk Reference,* edited by Frank Hoff. Dade City: Florida Aqua Farms.

Noga, Edward J. 2000. *Fish Disease: Diagnosis and Treatment.* Ames, Iowa State University Press.

Lewbart, Gregory A. 2006. *Invertebrate Medicine.* West Sussex, UK: Blackwell Publishing.

36. HUSBANDRY AND CARE OF TERRESTRIAL INVERTEBRATES

Proceedings of the International Invertebrates in Captivity Conference 1993-2007. SASI.

Proceedings of the Invertebrates in Education & Conservation Conference 2008-2012.

PART SIX: ANIMAL BEHAVIOR, ENRICHMENT, AND TRAINING

37. INTRODUCTION TO ANIMAL BEHAVIOR

Bekoff, M., ed. 2009. *Encyclopedia of Animal Behavior.* Westport, CT: Greenwood Press.

Breed, M.. and J. Moore, eds. 2010. *Encyclopedia of Animal Behavior.* Elsevier.

40. ENRICHMENT

Kleiman, Devra G., Katerina V. Thompson, and Charlotte K. Baer. 2010. *Wild Mammals in Captivity: Principles and Techniques.* Chicago: University of Chicago Press.

Shepherdson, D. J., J. D. Mellen, and M. Hutchins, M. 1998. *Second Nature: Environmental Enrichment for Captive Animals.* Washington: Smithsonian Institution.

Young, Robert. 2003. *Environmental Enrichment for Captive Animals.* Oxford: Blackwell Science.

41. ENRICHMENT PROGRAMS

Mellen J., and J. Barber. 2002. Animal Enrichment Framework. Disney's Animal Kingdom, available at www.animalenrichment.org.

Mellen J., and M. Sevenich MacPhee. 2002. Animal Training Framework. Disney's Animal Kingdom, available at www.animaltraining.org.

43. HUSBANDRY TRAINING

Chance, Paul. 2009. *Learning and Behavior, 6th Edition.* Belmont, CA: Wadsworth, Cengage Learning.

Fowler, Murray E., and Miller, Eric R. 2003. *Zoo and Wild Animal Medicine, 5th Edition.* Philadelphia: W. B. Saunders.

Kazdin, Alan E. 2001. *Behavior Modification in Applied Settings, 6th Edition.* Long Grove, IL: Waveland Press.

Kleiman, Devra G., Katerina V. Thompson, and Charlotte Kirk Baer. 2010. *Wild Mammals in Captivity: Principles and Techniques for Zoo Management, 2nd Edition.* Chicago: University of Chicago Press.

Ramirez, Ken (1999). *Animal Training: Practical Animal Management through Positive Reinforcement.* Chicago: Shedd Aquarium Press.

Pryor, Karen (1999). *Don't Shoot the Dog!* New York: Random House.

PART SEVEN: VETERINARY CARE

44. PRINCIPLES OF ANIMAL HEALTH

Ballard, Bonnie and Ryan Cheek. 2003. *Exotic Animal Medicine for the Veterinary Technician.* Ames: Blackwell Publishing.

Fowler, Murray. 2008. *Restraint and Handling of Wild and Domestic Animals, Third Edition.* Ames: Blackwell Publishing.

Fowler, Murray E., and R. Eric Miller. 2003. *Zoo and Wild Animal Medicine, Fifth Edition.* Philadelphia: Saunders Elsevier.

Hosey, Geoff, Vicky Melfi, and Sheila Pankhurst. 2009. *Zoo Animals: Behaviour, Management, and Welfare.* New York: Oxford University Press.

Kleiman, Devra G., Katerina V. Thompson, and Charlotte Kirk Baer. 2010. *Wild Mammals in Captivity: Principles and Techniques for Zoo Management.* Chicago: University of Chicago Press.

Romich, Janet A. 2009. *An Illustrated Guide to Veterinary Medical Terminology, Third Edition.* Clifton Park, NY: Delmar Cengage Learning.

46. ZOONOTIC DISEASE

Acha, P. N., and B. Szyfres. 2001. *Zoonoses and communicable diseases common to man and animals, Vol. I. Bacterioses and Mycoses.* Pan American Health Organization, Scientific and Technical Publication no. 580.

———. 2003. *Zoonoses and communicable diseases common to man and animals, Vol. II. Chlamydioses, Rickettsioses, and Viroses.* Pan American Health Organization, Scientific and Technical Publication no. 580.

———. 2003. *Zoonoses and communicable diseases common to man and animals, vol. III. Parasitoses.* Pan American Health Organization, Scientific and Technical Publication no. 580.

Flammer, K. 1999. Zoonoses acquired from birds. In *Zoo and Wild Animal Medicine, Current Therapy 4,* edited by M. E. Fowler and R. E. Miller, 151–156. Philadelphia: W. B. Saunders.

Grant, S., and C. W. Olsen. 1999. Preventing Zoonotic Diseases in immunocompromised persons: The role of physicians and veterinarians. *Emerg Inf Dis* 5(1): 159–63.

Hugh-Jones, M. E., W. T. Hubbert, and H. V. Hagstad. 2000. *Zoonoses: Recognition, Control, and Prevention.* Ames, IA: Iowa State University Press. 344.

Ott-Joslin, J. E. 1993. Zoonotic Diseases of Non-human Primates. In *Zoo and Wild Animal Medicine, Current Therapy 3,* edited by M. E. Fowler, 358–73. Philadelphia: W. B. Saunders.

Rupprecht, C.E. 1999. Rabies: Global problem, zoonotic threat, and preventative management. In *Zoo and Wild Animal Medicine, Current Therapy 4,* edited by M. E. Fowler and R. E. Miller, 136–46. Philadelphia: W. B. Saunders.

47. PREVENTIVE MEDICINE

Atkinson Carter, T., Nancy J. Thomas, and D. Bruce Hunter. 2008. *Parasitic Diseases of Wild Birds.* Ames: Wiley-Blackwell.

Bowman, Dwight D. 2009. *Georgi's Parasitology for Veterinarians.* St. Louis: Saunders Elsevier.

Foreyt, William J. 2001. *Veterinary Parasitology.* Ames: Iowa State University Press.

Fowler, Murray E. 1993. *Zoo and Wild Animal Medicine: Current Therapy 3.* Philadelphia: W. B. Saunders.

Fowler, Murray E., and R. Eric Miller. 1999. *Zoo and Wild Animal Medicine: Current Therapy 4.* Philadelphia: W. B. Saunders.

———. 2003. *Zoo and Wild Animal Medicine.* St. Louis: Saunders Elsevier.

———. 2008. *Zoo and Wild Animal Medicine: Current Therapy 6.* St. Louis: Saunders Elsevier.

Hosey Geoff, Vicky Melfi, and Sheila Pankhurst. 2009. *Zoo Animals: Behavior, Management and Welfare.* Oxford: Oxford University Press.

Kleiman, Devra G., Mary Allen, Katrina V. Thompson, and Susan Lumpkin. 1996. *Wild Mammals in Captivity: Principles and Techniques.* Chicago: University of Chicago Press.

Samuel, William M., Margo J. Pybus, and A. Alan Kocan. 2001. *Parasitic Diseases of Wild Mammals.* Ames: Iowa State Press.

Thomas, Nancy J., D. Bruce Hunter, and Carter T. Atkinson Carter. 2007. *Infectious Diseases of Wild Birds.* Ames: Blackwell Publishing.

Williams, Elizabeth S., and Ian K. Barker. 2001. *Infectious Diseases of Wild Mammals.* Ames: Iowa State University Press.

49. MEDICATIONS AND DOSE CALCULATIONS

Bassert, Joanna M., and Dennis M. McCurnin. 2010. *McCurnin's Clinical Textbook for Veterinary Technicians, Seventh Edition.* Saint Louis: Saunders Elsevier.

Bill, Robert L. 2006. *Clinical Pharmacology and Therapeutics for the Veterinary Technician, Third Edition.* Saint Louis: Mosby Elsevier.

Wanamaker, Boyce P., and Kathy Lockett Massey. 2009. *Applied Pharmacology for Veterinary Technicians, Fourth Edition.* Saint Louis: Saunders Elsevier.

50. CHEMICAL RESTRAINT

Fowler, Murray. 2008. *Restraint and Handling of Wild and Domestic Animals, Third Edition.* Ames: Blackwell Publishing.

Hosey, Geoff, Vicky Melfi, and Sheila Pankhurst. 2009. *Zoo Ani-*

mals: Behaviour, Management, and Welfare. New York: Oxford University Press.

Kleiman, Devra G., Katerina V. Thompson, and Charlotte Kirk Baer. 2010. *Wild Mammals in Captivity: Principles and Techniques for Zoo Management.* Chicago: University of Chicago Press.

Kock, M. D., D. Meltzer, and R. Burroughs. 2006. *Chemical and Physical Restraint: A Training Field Manual.* Greyton, South Africa: International Wildlife Veterinary Services.

Kreeger, Terry. 2007. *Handbook of Wildlife Chemical Immobilization, Third Edition.*

West, Gary, Darryl Heard, and Nigel Caulkett, eds. 2007. *Zoo Animal and Wildlife Immobilization and Anesthesia.* Ames, IA: Wiley-Blackwell.

PART NINE: CONSERVATION SCIENCE

53. CONSERVATION BIOLOGY

Dick, Gerald, and Markus Gusset, eds. 2010. *Building a Future for Wildlife: Zoos and Aquariums Committed to Biodiversity Conservation.* Gland: WAZA Executive Office.

Hosey, Geoff, Vicky Melfi, and Sheila Pankhurst. 2009. Conservation. In *Zoo Animals: Behaviour, Management, and Welfare.* Oxford: Oxford University Press.

WAZA. 2005. *Building a Future for Wildlife: The World Zoo and Aquarium Conservation Strategy.* Bern: WAZA Executive Office.

WAZA. 2009. *Turning the Tide: A Global Aquarium Strategy for Conservation and Sustainability.* Bern: WAZA Executive Office.

Zimmermann, Alexandra, Matthew Hatchwell, Lesley Dickie, and Chris West, eds. 2007. *Zoos in the 21st Century: Catalysts for Conservation?* Cambridge: Cambridge University Press.

54. RESEARCH IN ZOOS

Asa, C. S., and I. J. Porton, eds. 2004. *Wildlife Contraception: Issues, Methods and Applications.* Baltimore: Johns Hopkins University Press.

Edenhard, T. 1995. Conservation breeding as a tool for saving animal species from extinction. *Trends in Ecology and Evolution* 2:279–86.

Frankham, R., J. D. Ballou, and D. A. Briscoe. 2002. *Introduction to Conservation Genetics.* Cambridge: Cambridge University Press.

Fowler, M. E, and R. E. Miller, eds. 2007. *Zoo and Wild Animal Medicine Current Therapy.* Sixth Edition. St. Louis: Elsevier Science.

Hutchins, M., T. Foose, and U. S. Seal. 1991. The role of veterinary medicine in endangered species conservation. *Journal of Zoo and Wildlife Medicine* 22(3) :277–81.

Karow, A. M., and J. Critser, eds. 2010. *Tissue Banking in Reproductive Biology,* edited by A. M. Karow and J. Critser. New York: Academic Press, Inc.

Kleiman, D. G., K. V. Thompson, and C. Kirk Baer, eds. 2010. *Wild Mammals in Captivity: Principles and Techniques for Zoo Management.* Second Edition. Chicago: University of Chicago Press.

Mazur, N.A. 2001. *After the Ark.* Melbourne: Melbourne University Press.

Miller, B., W. Conway, R. P. Reading, C. Wemmer, D. Wildt, D. G. Kleiman, S. L. Monfort, A. Rabinowitz, B. Armstrong, and M. Hutchins. 2004. Evaluating the conservation mission of zoos, aquariums, botanical Gardens, and Natural History Museums. *Conservation Biology* 18:86–93.

Osofsky, S. A., W. B. Karesh, and S. L. Deem. 2000. Conservation medicine: A veterinary perspective. *Conservation Biology* 14: 336–37.

Wielebnowski, N. 2003. Stress and Distress: Evaluating their impact for the well-being of zoo animals. *Journal of the American Veterinary Medical Association* 223:973–77.

Woodworth, L. M., M. E. Montgomery, D. A. Briscoe, and R. Frankham. 2002. Rapid genetic deterioration in captive populations: Causes and conservation implications. *Conservation Genetics* 3:277–88.

Zimmermann, A., M. Hatchwell, L. A. Dickie, and C. West, eds. 2007. *Zoos in the 21st Century.* Cambridge: Cambridge University Press.

57. WILDLIFE REHABILITATION

Gage, Laurie, ed. 2002. *Hand-Rearing Wild and Domestic Mammals.* Ames, IA: Blackwell Publishing.

Gage, Laurie, and Rebecca Duerr, eds. 2007. *Hand-Rearing Birds.* Ames, IA: Wiley-Blackwell.

Moore, Adele, and Sally Joosten, eds. 1997. *Principles of Wildlife Rehabilitation.* Saint Cloud, MN: NWRA.

Stocker, Les. 2000. *Practical Wildlife Care.* Ames, IA: Blackwell.

Appendix 2

Recommended Web Links

PART 1: ZOOKEEPING

www.australasianzookeeping.org
Australasian Zoo Keeping

www.aza.org
American Association of Zoos and Aquariums

www.biaza.org.uk
British and Irish Association of Zoos and Aquariums

www.caza.ca
Canadian Association of Zoos and Aquariums

www.eaza.net
European Association of Zoos and Aquaria (EAZA)

www.eresumes.com
eResumes and Resume Writing Services

www.glassdoor.com
Glassdoor: An Inside Look at Jobs and Companies

www.iczoo.org
International Congress of Zookeepers

www.jobsearch.about.com
About.com job searching

www.nextstep.direct.gov.uk
Next Step career advice

www.videojug.com
Video Jug

www.waza.org
World Association of Zoos and Aquariums. Site includes a list of regional zoo association members

www.zooaquarium.org.au
Australasian Zoo Aquarium Association (ZAA)

www.zoosafrica.com
African Association of Zoos and Aquaria (PAAZAB)

PART 3: WORKPLACE SAFETY AND EMERGENCIES

www.aza.org
American Association of Zoos and Aquariums

www.bls.gov
US Bureau of Labor Statistics

www.ccohs.ca
Canadian Centre for Occupational Health and Safety

www.cdc.org
US Centers for Disease Control

www.iczoo.org
International Congress of Zookeeping

www.iso.org
International Organization for Standardization

www.niosh.org
National Institute for Occupational Safety and Health

www.osha.gov
Occupational Safety and Health Administration

www.waza.org
World Association of Zoos and Aquariums

www.who.int
World Health Organization

www.zaa.org
Zoo and Aquarium Association

PART 4: ZOO ANIMAL MANAGEMENT

www.aazk.org
American Association of Zoo Keepers

www.aazv.org
American Association of Zoo Veterinarians

www.ai.uga.edu/mc/latinpro.pdf
Latin Pronunciation Demystified

www.aphis.usda.gov/
US Department of Agriculture Animal and Plant Health
Inspection Service

www.arma.org
ARMA International (Association of Records Managers
and Administrators)

www.australasianzookeeping.org
Wonderful resource for husbandry information and other
resources

www.avidid.com/
AVID microchips

www.aza.org/
Association of Zoos and Aquariums

**www.aza.org/uploadedFiles/Animal_Care_and
_Management/Animal_Management/Animal
_Data_and_Recordkeeping/IDMAG_Documents
/TransponderStatement2010.pdf**
AZA recommendations for transponder (microchip)
placement

www.azh.org
Association of Zoological Horticulture

www.biolac.com.au
Biolac, a resource for marsupial milk formulas

**www.biology.ualberta.ca/courses.hp/zool250/Roots
/RootsMain.htm**
Zool 250 Latin and Greek Roots

www.bio-serv.com
Resource for primate and rodents milk formulas

www.bvzs.org
British Veterinary Zoological Society

www.catalogueoflife.org/dynamic-checklist/search.php
Species 2000 and ITIS Catalogue of Life: Dynamic
Checklist

www.cbsg.org/cbsg/
Conservation Breeding Specialist Group

www.cdc.gov/ncezid/
US Centers for Disease Control and Prevention / National
Center for Emerging and Zoonotic Infectious Diseases

www.cnsweb.org/
Comparative Nutrition Society

www.derm.qld.gov.au
Department of Environment of the Queensland
government, with a section on caring for wildlife

www.destronfearing.com
Destron Fearing

www.eazwv.org
European Association of Zoo and Wildlife Veterinarians

www.elasmobranchhusbandry.org/
Elasmobranch Husbandry Manual

www.fourthcrossingwildlife.com
Information on Australian native animals, including hand-
rearing information

www.fws.gov/
US Fish and Wildlife Service

www.isis.org
International Species Information System

www.lyndastaker.hainsnet.com
Kangaroo care site

www.marsupialsociety.org
Marsupial Society of Australia

www.nagonline.net
AZA Nutrition Advisory Group

www.nicuvet.com
NICU Vet, web page of the Large Animal Neonatal
Intensive Care program at the Graham French Neonatal
Section of the Connelly Intensive Care Unit, New Bolton
Center, University of Pennsylvania

www.nmt.us/
Northwest Marine Technology

**www.nwhc.usgs.gov/publications/amphibian_research
_procedures/toe_clipping.jsp**
Amphibian Toe Clipping

www.organismnames.com
Index to Organism Names

www.petag.com
Replacement formulas: Pet Ag and Zoologic

**www.stlzoo.org/animals/scienceresearch
/contraceptioncenter/**
AZA Wildlife Contraception Center

www.trovan.com/index.html
Trovan microchips

www.waza.org/
World Association of Zoos and Aquariums

www.wildcare.org.au
Wildcare Australia

www.wombaroo.com.au
Nutrition and health care for birds and animals

www.zooborns.com
News alerts on latest births within zoos

www.zoolex.org
Site established to help improve exhibition of wild animals
in zoos by publishing and disseminating information
related to exhibit design

www.zooregistrars.org/
Zoological Registrars Association

PART 5: ZOO ANIMAL HUSBANDRY AND CARE

www.aalso.org
Aquatic Animal Life Support Operators

www.aazv.org
American Association of Zoo Veterinarians

www.afsc.noaa.gov/nmml
National Marine Mammal Laboratory

www.allaboutbirds.org
Cornell Lab of Ornithology

www.amphibianark.org
Ex situ components of the Amphibian Conservation
Action Plan (ACAP); includes link to AZA's Amphibian
Husbandry Manual

www.amphibiancare.com
Information about amphibians

www.amphibiaweb.org
Information on amphibian declines, conservation, natural
history, and taxonomy

www.Animaldiversity.ummz.umich.edu/site/index.html
Animal Diversity web

www.animalenrichment.org/MammalsArticles.pdf
A comprehensive list of enrichment articles

www.ansci.cornell.edu/plants
Cornell University poisonous plants page

www.antelopetag.com/
AZA Antelope and Giraffe TAG

www.asianwildcattle.org
IUCN Asian Wild Cattle Specialist Group

www.australasianzookeeping.org/
Australasian zookeeping site

www.aza.org/animal-care-manuals/
Animal care manuals published by the Association of Zoos
and Aquariums (AZA)

www.azh.org
Association of Zoological Horticulture

www.awic.nal.usda.gov
Search to find informational bibliography on companion
animal welfare

www.birds.cornell.edu
Cornell Lab of Ornithology

www.caudata.org
Science-based information about newts and salamanders,
with an emphasis on their maintenance in captivity

**www.csfs.colostate.edu/cowood/library/01_White_Fir
_Shavings.pdf**
One of the few scientifically rigorous studies about bedding
types and small mammals

www.cwd-info.org/
Chronic Wasting Disease Alliance

www.dairylandhoofcare.com
Even-toed ungulate hoof trimming resource

www.data.iucn.org/themes/ssc/sgs/gecs/
IUCN South American Camelid Specialist Group

www.data.iucn.org/themes/ssc/sgs/pphsg/home.htm
IUCN Pigs, Peccaries, and Hippos Specialist Group

www.data.iucn.org/themes/ssc/sgs/equid/
IUCN Equid Specialist Group

www.dels.nas.edu/ilar
Links to ILAR journal volume 48(3), *Use of Amphibians in
the Research, Laboratory, or Classroom Setting*

www.eaam.org
European Association for Aquatic Mammals

www.eekma.org

**www.elephant.se/eekma.php?open=Elephant%20
organizations**
European Elephant Keeper and Manager Association

www.elasmo.org
American Elasmobranch Society

www.elephantconservation.org
Elephant conservation and education organization

www.enrichment.org/
The Shape of Enrichment

www.fishbase.org
A comprehensive source of information about fish species
worldwide

www.foragerssource.org/
The Forager's Source: Plant nutrition database (under
development)

www.fws.gov
US Fish and Wildlife Service

www.gerbils.co.uk/gerbils/burrow.htm
Covers gerbil burrowing stereotypies and solutions

www.globalamphibians.org
Global Amphibian Assessment

www.grandin.com/behaviour/transport.html
Dr. Temple Grandin's resource on the behavior of
ungulates during handling and transport

www.horseshoes.com
The Farrier and Hoofcare Resource Center

www.iibce.edu.uy/DEER/
IUCN Deer Specialist Group

www.imata.org
International Marine Animal Trainer's Association

www.iucn.org
International Union for the Conservation of Nature

www.iucnredlist.org/initiatives/amphibians
Amphibians on the IUCN Red List

www.johnes.org
Johne's disease information center

www.moray.ml.duke.edu/projects/hippos/
IUCN Hippo Specialist Subgroup

www.nagonline.net
AZA Nutrition Advisory Group

www.nmfd.noaa.gov/index.html
National Oceanic and Atmospheric Administration

www.pages.usherbrooke.ca/mfesta/iucnwork.htm
IUCN Caprinae Specialist Group

www.phoenixzoo.org/learn/animals/behavioral_enrichment.shtml
Practical enrichment options

www.rawconference.org
Regional Aquatics Workshop, an annual symposium of public aquarium professionals

www.rhinoresourcecenter.com
Rhino Resource Center

www.rhinos-irf.org
International Rhino Foundation

www.stateofthebirds.org
US Fish and Wildlife Lab of Ornithology

www.stlzoo.org/contraception
AZA Wildlife Contraception Center

www.tapirs.org
IUCN Tapir Specialist Group

www.torontozoo.com/meet_Animals/enrichment/ungulate_enrichment.htm
Practical enrichment options

www.ultimateungulate.com
The Ultimate Ungulate page

www.vetmed.wsu.edu/mcf/
Malignant catarrhal fever resource center

www.wildcattleconservation.org
Wild Cattle Conservation

www.who.int/zoonoses/
World Health Organization: Zoonoses and veterinary public health

www.zooplants.net/
BIAZA Zoo Plant Wiki (requires free registration)

PART 6: ANIMAL BEHAVIOR AND BEHAVIORAL HUSBANDRY

www.animalbehaviorsociety.org
Animal Behavior Society

www.animalenrichment.org
Disney's Animal Kingdom enrichment program "SPIDER"

www.animaltraining.org
Disney's Animal Programs Animal Training website

www.enrichment.org/index.php
The Shape of Enrichment

www.ethograms.org
EthoSearch, a database of peer-reviewed ethograms

www.honoluluzoo.org/enrichment_activities.htm
Honolulu Zoo enrichment list

www.iaate.org
International Association of Avian Educators and Educators

www.imata.org
International Marine Animal Trainers Association

www.nal.usda.gov/awic/pubs/Primates2009/primates.shtml
USDA nonhuman primate enrichment guide

www.oregonzoo.org/Cards/Enrichment/conserv_enrichment.htm
Oregon Zoo practical enrichment options

www.theabma.org
Animal Behavior Management Alliance

PART 7: VETERINARY CARE

www.aazv.org
American Association of Zoo Veterinarians

www.avma.org/public_health/default.asp
American Veterinary Medical Association public health page

www.aza.org/Accreditation/
Association of Zoos and Aquariums Accreditation Standards and Guidelines

www.aza.org/animal-health
Association of Zoos and Aquariums

www.cdc.gov/mmwr/preview/mmwrhtml/rr5805a1.htm
Compendium of Measures to Prevent Disease Associated with Animals in Public Settings, 2009

www.cdc.gov/nczved/
Centers for Disease Control and Prevention National Center for Zoonotic, Vector-borne, and Enteric Diseases

www.eazwv.org
European Association of Zoo and Wildlife Veterinarians

www.isis.org
Link to International Zoo Vet Forum for veterinarians

www.merckvetmanual.com
Merck Veterinary Manual

www.nasphv.org/
National Association of State Public Health Veterinarians

www.oie.int
World Organization for Animal Health

www.who.int/en/
World Health Organization

PART 9: CONSERVATION SCIENCE

www.aazk.org
American Association of Zoo Keepers

www.abwak.org
Association of British Wild Animal Keepers

www.amphibianark.org
Amphibian Ark

www.aszk.org.au
Australasian Society of Zoo Keeping

www.aza.org
Association of Zoos and Aquariums

www.aza.org/animal-programs
AZA Animal Programs

www.biaza.org.uk
British and Irish Association of Zoos and Aquariums

www.bioclimate.org
Combined biodiversity and climate change–related data

www.cbd.int
Convention on Biological Diversity

www.eaza.net
European Association of Zoos and Aquaria

www.edgeofexistence.org
Evolutionary Distinct and Globally Endangered

www.enrichment.org
The Shape of Enrichment

www.fws.gov
US Fish and Wildlife Service

www.icmje.org
International Committee of Medical Journal Editors

www.isis.org
International Species Information System

www.iucn.org
International Union for Conservation of Nature

www.iucnredlist.org
IUCN Red List of Threatened Species

www.iwrc-online.org
International Wildlife Rehabilitators Council

www.nwrawildlife.org
National Wildlife Rehabilitators Association

www.scholar.google.com
Google Scholar for searching articles

www.unitedforconservation.org
WAZA Conservation and Sustainability Resource Center

www.waza.org
World Association of Zoos and Aquariums

www.zooaquarium.org.au
Australasian Zoo and Aquarium Association

PART 10: GOVERNMENT AND LEGISLATION

INTERNATIONAL

www.cites.org
Convention on International Trade of Endangered Species of Wild Fauna and Flora

www.iata.org
International Air Transport Association

AUSTRALIA

www.daff.gov.au/ba/ira
Australian Import Risk Analysis Website

www.daffa.gov.au/aqis/import
Australian (AQIS) requirements to Import Animals

www.environment.gov.au/biodiversity/trade-use/index.html
Australian (DSEWPAC) requirements to import and export CITES-listed and native animals

www.environment.gov.au/biodiversity/trade-use/lists/import/pubs/live-import-list.pdf
Australian (DSEWPAC) approved animal species for live import

www.environment.gov.au/biodiversity/trade-use/sources/non-commercial/ccp/index.html
Australian (DSEWPAC) guidelines for cooperative conservation programs

www.feral.org.au/
Guidelines for the import and keeping of exotic vertebrates in Australia

www.zooaquarium.org.au/Policies-of-Arazpa/default.aspx
ZAA policies and guidelines (refer to "Accreditation General Standards")

AFRICA

www.nda.agric.za
South African Department of Agriculture

www.akaafrica.com
Animal Keepers Association of Africa

www.environment.gov.za
South African Department of Water and Environmental Affairs

www.nspca.co.za
National Council of SPCAs, South Africa

www.paazab.com
African Association of Zoos and Aquaria

www.pasaprimates.org
Pan African Sanctuary Alliance

ASIA

www.cza.nic.in/
India's Central Zoo Authority

www.eaza.net
European Association of Zoos and Aquaria

www.ecolex.org
Short description of legislation with link to actual law

www.jazga.or.jp/english/index.html
Japanese Association of Zoos and Aquariums

www.zooreach.org/
Zoo legislation (South, Southeast, and East Asia, and Australia)

www.zooreach.org/SAZARC/SAZARC.htm
South Asian Zoo Association for Regional Cooperation

NEW ZEALAND

www.doc.govt.nz
New Zealand Department of Conservation

www.pa.govt.nz/
New Zealand Environmental Protection Authority

www.isis.org
International Species Information System

www.iucn.org
International Union for the Conservation of Nature

www.mapi.govt.nz
New Zealand Ministry for Primary Industries

www.redlist.org
IUCN Redlist

www.waza.org
World Association of Zoos and Aquaria

www.zooaquarium.org.au
Zoo Aquarium Association (Australasian Regional Association)

Appendix 3

Professional Colleges and Universities in the United States

COLLEGES AND UNIVERSITIES OFFERING TWO-YEAR ZOOKEEPING-RELATED DEGREES

Alamance Community College
AAS degree in animal care and management technology
P.O. box 8000
Graham, NC 27253
www.alamancecc.edu

Davidson County Community College
AAS degree in zoo and aquarium science
P.O. box 1287
Lexington, NC 27293
www.davidsonccc.edu

Jefferson Community College (SUNY)*
AAS degree in zoo technology
1220 Coffeen St.
Watertown, NY 13601
www.zookeeping.com
www.sunyjefferson.edu

Moorpark College*
AS degree or certificate of achievement in exotic animal training and management (EATM)
7075 Campus Rd.
Moorpark, CA 93021
www.moorparkcollege.edu

Niagara County Community College (SUNY)*
AAS degree in animal management
3111 Saunders Settlement Rd.
Sanborn, NY 14132
www.niagaracc.suny.edu

Pikes Peak Community College*
AAS degree in zookeeping technology
5675 S. Academy Blvd.
Colorado Springs, CO 80906
www.ppcc.edu

Portland Community College
AAS degree in biology and management of zoo animals
Rock Creek Campus
17705 N.W. Springville Rd.
Portland, OR 97229
www.pcc.edu

Santa Fe Community College
AS degree in zoo technology
3000 N.W. 83rd St.
Gainesville, FL 32606
www.sfcollege.edu

COLLEGES AND UNIVERSITIES OFFERING FOUR-YEAR OR GRADUATE ZOOKEEPING-RELATED DEGREES

Canisius College*
BS degree in biology with a minor in zoo biology
2001 Main St.
Buffalo, NY 14208
www.canisius.edu

Delaware Valley College
BS degree in animal biotechnology and conservation with a major in zoo science, small animal science, or conservation and wildlife management
700 East Butler Ave.
Doylestown, PA, 18901
www.delval.edu

Friends University
BS degree in zoo science and MS programs with emphasis in zoo science, zoo management, and zoo education
2100 W. University St.
Wichita, KS 67213
www.friends.edu

George Mason University

MA degree in interdisciplinary studies (MAIS) in zoo and aquarium leadership with specializations in collections management, administration, and conservation education
4400 University Dr.
Fairfax, VA 22030
www.gmu.edu

Michigan State University

BS degree in zoology with concentration in zoo and aquarium science and proMSc degree in zoo and aquarium science management
203 Natural Sciences Bldg.
East Lansing, MI 48824
www.msu.edu

Unity College

BS degree in captive wildlife care and education
90 Quaker Hill Rd.
Unity, ME 04988
www.unity.edu

University of West Florida

BS degree in interdisciplinary science with specialization in zoo science
11000 University Pkwy.
Pensacola, FL 32514
www.uwf.edu

Western Illinois University, School of Graduate Studies

Post-baccalaureate certificate in zoo and aquarium studies
Sherman Hall 116
Macomb, IL 61455
www.wiu.edu

COLLEGES AND UNIVERSITIES OFFERING AQUARIUM SCIENCE DEGREES

Brunswick Community College

AAS degree, diploma, and certificate programs in aquaculture technology
PO box 30
Supply, NC 28462
www.brunswickcc.edu

Carteret Community College

AAS degree, diploma, and certificate programs in aquaculture technology
3505 Arendell St.
Morehead City, NC 28557
www.carteret.edu

Gadsden State Community College (ACCS)

AS degree in aquatic biology and aquaculture technician certificate
1001 George Wallace Dr.
Gadsden, AL 35903
www.gadsdenstate.edu

Hillsborough Community College

AS degree in aquaculture
10414 E. Columbus Dr.
Tampa, FL 33619
www.hccfl.edu

Indian River State College

AAS in agriculture production technology with a specialization in aquaculture
3209 Virginia Ave.
Fort Pierce, FL 34981
www.irse.edu

Oregon Coast Community College*

AAS degree and completion certificate in aquarium science
332 S.W. Coast Hwy.
Newport, OR 97365
www.occc.cc.or.us/aquarium

Trinidad State Junior College

AAS degree in aquaculture
Alamosa, CO
www.cccs.edu

University of New England

BS degree in aquaculture and aquarium sciences (AQS)
11 Hills Beach Rd.
Biddeford, ME 04005
www.une.edu

*Contributor to this volume; see list of contributors.

Glossary

Much of the terminology listed here will have more than one usage. Definitions relate to the most common usage in the zoo and aquarium industry.

abnormal repetitive behavior (ARB). A behavior that seems to have no obvious goal or function (e.g., pacing to and fro) and which is repeated time and time again, often in the same location

accession number. A string of numbers and/or letters assigned by an institution to an individual (individual identification number) or group (group identification number) of animals and used to identify that animal or group in the institution's files, and which may or may not be the same as the ISIS number

acid detergent fiber (ADF). The least digestible components of plant feed, including lignin, silica, and cellulose. Hemicellulose is not included in this category. The higher the ADF, the lower the digestibility of the feedstuff.

acrodonts. The teeth attached to the crest of the jawbone, as seen in many lizards (e.g., chameleons and bearded dragons) and fish

actinic. (1.) Demonstrating chemical changes as a result of radiation (e.g., from light) (2.) Blue-spectrum light used to stimulate light-sensitive species and promote photosynthesis, especially in marine aquariums

acute. Having developed quickly over a short time period; the opposite of **chronic**

ad libitum ("ad lib"). Latin for "to one's pleasure," usually pertaining to the provision of as much food as will be consumed. Also called "free choice."

adrenal cortex. The outer layer of the adrenal endocrine gland (usually closely associated with the kidneys), concerned with the production of cortisol

adrenocortical axis. The release and collective interactions of hormones in response to stress (the "fight-or-flight" or alarm response); may also be referred to as the hypothalamic-pituitary-adrenal (HPA) axis, since it involves all of these glands.

AED. See **automated external defibrillator**

aerobic. Pertaining to the presence of or need for oxygen

aerosolization / droplet contact. A mode of disease transmission that involves any mechanism whereby infectious agents are suspended in the air with moist droplets that can come into contact with the mucous membranes that line the eyes, mouth, nose, trachea, or lungs

aestivation. A period of reduced metabolism and activity (i.e., dormancy) in response to a combination of high temperature and extreme dryness to prevent desiccation

aged water. Municipally treated water that has been left to stand after being drawn from a tap so that gases (specifically chlorine, but also carbon dioxide) can dissipate, thereby rendering it safer for sensitive aquatic species

agonistic. Pertaining to competitive, combative, or aggressive behavior

air cell (or **air space**). A pocket of air located in the broad end of a bird's egg and situated between the inner and outer shell membranes. Before the egg hatches, the air cell increases in size, giving the chick air to breathe during the hatching process.

albumin. A category of protein that is water-soluble and is naturally found in blood plasma and other biological fluids such as the liquid ("white" or "albumen") portion of eggs. It is a commonly measured component of blood, since decreased or elevated levels of albumin can indicate disease.

algae turf scrubber. An aquarium filtration component used for growing algae to remove excess nutrients from water

alkalinity. The quantitative capacity of water to neutralize an acid (i.e., the buffering capacity of the water due to its mineral content) and therefore resist a change in pH when acid is added

allantois. A membranous sac that develops from the posterior part of the alimentary canal in the embryos of mammals, birds, and reptiles, and which is important in the formation of the umbilical cord and placenta in mammals. In egg-laying animals it connects the developing embryo to the yolk and facilitates oxygenation of the embryo.

allergy. An exaggerated sensitivity and immune response to a substance (allergen) such as pet dander, pollens, or microorganisms

all-in, all-out. Movement of animals together as a group into and out of a situation or location; often applied to quarantine conditions, since moving new individuals into a group would potentially contaminate it and require the group to restart the quarantine period

allogrooming. An affiliative behavior in which one individual grooms another of the same species. Allogrooming serves a hygienic purpose, although it functions primarily as a social bonding mechanism. Grooming frequency among individuals can be used as an indicator of an individual's social status within a group.

altricial. Requiring significant parental care to move or feed following birth or hatching (contrasts with **precocial**); a characteristic of a species

amino acids. Organic molecules that contain carbon, hydrogen, oxygen, and nitrogen and serve as the subunits of protein. There are many different types of protein, but only 20 amino acids from which they are formed. Some amino acids are essential to life and must be ingested if a species cannot create them through metabolic processes.

amniotic fluid. The fluid contained in the amniotic sac of a pregnant mammal, which surrounds the developing fetus. Ingested amniotic fluid contributes to the **meconium.**

amplexus. The mating clasp of amphibians, in which the male clings to the back of the female with its front legs to initiate and assist with egg fertilization, which is often external

analgesia. The relief of pain

anaphylactic shock. A life-threatening allergic reaction characterized by dilation of the blood vessels with a sharp drop in blood pressure, bronchial spasm, and shortness of breath

anesthesia. Inducement of a controlled and reversible lack of sensation or state of unconsciousness through the use of pharmaceutical agents (anesthetics) such as gas or injectable drugs

anestrus. The state of not being in estrus or not ovulating; a normal part of some animals' reproductive cycles

ankus (guide). A hand tool used to guide and direct elephants

anorexia. A failure to eat, or a refusal of food.

anoxia. The state of being without oxygen

anthelmintic. Any drug used to kill internal parasitic worms

anthropomorphism. The attribution of human behaviors, thought processes, and responses to nonhuman animals

antibiotic. A category of drugs that either kill bacteria or inhibit their reproduction, thereby helping an animal's body to clear such infections. Some bacteria can become resistant to the effects of one or more antibiotic drug.

antibody. A specialized protein created to bind to specific foreign substances (**antigens**) and help the body's white blood cells to remove them

antigen. A substance that, when inside the body, will stimulate an immune response and the production of antibodies

antiseptic. A substance that kills microorganisms (particularly bacteria) on living tissue, and is often used prior to surgery

antivenin (or **antivenom** or **antivenene**). A product containing antibodies to a particular species' venom, used in the treatment of venomous bites and stings

antler. One of a pair of deciduous bony growths (often shed and regrown annually) which grow from the top of the head of most male deer (family Cervidae) and are not regularly seen in females except in reindeer (*Rangifer tarandus*). Antlers are generally branched and may be simple or complex in structure.

aposematic. Colored or constructed in a special way to indicate special defense abilities

applied behavioral analysis. The field of science focused on controlling and predicting behavior

approximation or successive approximation. A progressive step or stage in behavior training that leads to a desired final outcome, such as a more complex behavior

arboreal. Living in or among trees

arboriculture. The cultivation of trees and shrubs, especially for ornamental purposes

asexual reproduction. The production of new, genetically identical offspring from a single parent. Forms of asexual reproduction include budding (e.g., in sponges and many corals), in which a new individual begins by developing from a small part of the parent organism and may or may not remain attached to the parent; and regeneration from fragments of the parent animal. See also **parthenogenesis.**

as-fed basis. Evaluation of a feed in its natural state, when it has not been modified or dried (i.e., when it includes water content). It is often preferable and more accurate to compare feed analyses on a dry matter basis.

aspiration. (1.) Inhalation of fluid, food particles, or other material into the lungs. (2.) The practice of drawing a sample of cells or fluid through a needle, as in fine-needle aspiration.

asymptomatic. Lacking readily apparent signs of disease. An asymptomatic animal may be disease-free or have **subclinical disease.**

ataxia. A failure of muscle coordination; muscular incoordination that impairs the ability to move normally

atrophy. Wasting away or diminishing in the size of an organ or tissue

auditory. Pertaining to hearing, the organs of hearing, or the sense of hearing

auscult (auscultate). To listen to the internal sounds of the body (heart, lungs, gastrointestinal tract, etc.), often with the aid of a stethoscope

autoclave. A device used to sterilize items using a combination of steam and heat under pressure

autogrooming. Grooming of an individual animal by itself. If excessive, autogrooming behavior can become self-mutilation.

automated external defibrillator (AED). A portable device used in diagnosing and treating life-threatening cardiac arrhythmias of ventricular fibrillation and ventricular tachycardia

autonomic nervous system (ANS). The part of the nervous system concerned with automatic regulation of the function of the heart, lungs, intestines, and pupils without conscious effort by the animal

aversive stimuli. Sensory input that the animal wants to avoid

backwashing. The process of reversing the flow of water through a filter and dislodging trapped particles

baculum. A bone inside the penis of most mammals that helps to stiffen it to aid in intercourse, slid into place from the abdomen through specific muscles; also known as an os penis

baseline. A standard guideline that approximates normal; often used with measurements such as weight, dimensions, and blood values

behavioral budget. A representation, in percentages, of time spent in different behavioral states; also called an activity budget

bezoar. An abnormal mass, typically occurring in the gastrointestinal tract. Bezoars are named on the basis of their composition. Lactobezoars result from intestinal milk bolus obstruction, phytobezoars are composed of nondigestible plant material, and trichobezoars are composed of hair.

biodiversity (biological diversity). The variety of plants and animals occurring within a system. As animal populations diminish and species become extinct, global biodiversity will decrease.

biofact. Material from a once living organism. In zoos and aquariums, the term is sometimes loosely applied to any object used for biology interpretation purposes by keepers, educators, docents or others; examples include bones, pelts, scat, used enrichment items, and samples of enclosure furniture.

biofiltration. The use of specific, naturally occurring bacteria species to oxidize nitrogen-based wastes (specifically ammonia and nitrites) generated by fish and other aquatic organisms, thus rendering them less harmful

biological vector. An organism (usually an arthropod) in which a disease agent (pathogen) undergoes some developmental change and is then physically transferred to infect another susceptible host animal

biosecurity. Prevention of the introduction and spread of disease organisms within an animal collection, often involving a formal program with a set of procedural guidelines

bioswale. A graded area that diverts storm water from buildings and infrastructure. It may be planted with grass or other plantings, is often used as an aesthetic element of construction design, and helps to trap silt and pollutants from traveling into storm sewers, drainage areas, or bodies of water.

blastocyst. An early stage in the development of an embryo; basically a hollow ball of developing cells. It is at this stage of development that implantation may be delayed in some species.

bloat. A condition in which the stomach becomes overly distended with gas or foam, and which can be serious and potentially fatal if not treated appropriately

blood feather. An immature feather that still has a significant blood supply, and which if damaged may require management to stop the bleeding

body condition scoring (BCS). Assignment of a numeric score to assess the amount of body fat and muscling an animal has. BCS systems vary between species and may be on a nine-point, five-point, or three-point scale. Evaluation may incorporate photograph evaluation, palpation, body weight comparisons, or other means.

brachiation. A form of arboreal locomotion in which primates swing from tree limbs or enclosure furniture using their arms; from the Latin word *brachium*, meaning arm. Primate shoulder joints allow a high degree of movement, thus accommodating this behavior.

brackish water. Slightly briny or salty water that results from fresh water mixing with seawater at natural interfaces such as river mouths and estuaries

bradycardia. A slower-than-normal heart rate. Cholinergic bradycardia is associated with stimulation of the parasympathetic autonomic nervous system.

bradypnea. A slower than normal rate of respiration

brood. To sit on or incubate eggs

browse. Parts of trees or shrubs used as a food source (usually leafy branches)

brumation. Period of winter dormancy in reptiles marked by reduced activity and metabolism, often triggered by lower temperature and reduced daylight, and needed to trigger reproductive behavior in many species

bushmeat. Meat of wild animals, killed for subsistence or commercial purposes throughout the tropics of many areas of the world

calorie (or joule). A unit of energy. One calorie equals 4.184 joules (J); one joule equals 0.239 calories. Energy in food is often expressed as calories (cal) or kilocalories (kcal) per gram of food or per day.

capture myopathy. A syndrome (also called **exertional myopathy**) that can develop following restraint of a wild animal, associated with a buildup of lactic acid in muscle that can cause either sudden death or muscle stiffness and lameness

carapace. The dorsal exoskeleton of many groups of animals, such as turtles and tortoises (order Testudines) and arthropods such as some crustaceans and insects

carbohydrate. An organic compound composed of carbon, oxygen, and hydrogen that can be complex in form (e.g., starch, glycogen, or cellulose) or simple (e.g., monosaccharides, like glucose; or disaccharides, like sucrose). Carbohydrates are commonly used as fuel for metabolism, meaning that they are used as an energy source or for storing energy in other compounds (biosynthesis).

carbon footprint. The sum total of all greenhouse gases emitted by an individual or company

carnivore. An organism adapted to eat flesh and other whole or parts of animals

carrying capacity. The maximum population (as of an animal species) that an area can support without experiencing damage or deterioration

catchpole. A long-handled snare used to capture and restrain animals (also called a snare or a dog noose)

cell mediation. Resistance to an infectious disease at the cellular level by special defensive cells. See also **humoral immunity.**

cellulose. A complex carbohydrate which is found as a component of the cell wall in the fibrous portion of plants, and is low in digestibility.

cerebral cortex. The major part of the brain, concerned with the control of all higher functions.

chalazae. Thick strands of egg white, attached to egg membranes at the ends of the yolk in birds and reptiles, which keep the yolk suspended in the middle of an egg

chitin. The main structural component (polysaccharide) of the exoskeletons of arthropods such as crustaceans and insects, which is similar to cellulose in its chemical structure

choanal slit. The slit located in the roof of a bird's mouth which connects the trachea to the nares

chronic. Persistent over a long term; the opposite of **acute**

chute. A narrow confined space in which an animal can be moved, restrained, or manipulated

chytridiomycosis. An infectious fungal disease of amphibians, causing death and population declines in many parts of the world, caused by the chytrid fungus *Batrachochytrium dendrobatidis*

cladistics. Method of classifying organisms into clades (groups of members sharing features derived from a common ancestor); also called phylogenetic systematics

claspers. Specialized appendages developed to deliver sperm in male sharks and rays, formed from modified pelvic fins that are inserted into the female **cloaca** to transport the sperm. Claspers may contain cartilaginous hooks that "clasp" the inside of the female's oviduct.

classical conditioning. A type of learning in which an animal makes an association between two stimuli; also called Pavlovian conditioning

clinical disease. Disease with observable signs

cloaca. A posteriorly located body opening which serves as the terminal part of the digestive, urinary, and reproductive systems in birds, reptiles, amphibians, fish, and a few primitive mammals

cognitive. Pertaining to conscious reasoning, judging, memory, creativity, or learning

colic. A condition of gastrointestinal pain or discomfort, which can have one of many different causes

colormetric testing. Measurement of the concentration of a known constituent of a solution by comparison with colors of standard solutions of that constituent, using a specialized meter or a relatively inexpensive, commercially available aquarium test kit

colostrum. The first milk produced following birth, which is typically rich in **immunoglobulins** (**antibodies**) and often provides important immunological support to the newborn

communicable disease. A disease that can be transmitted between animals

concentrate. A feed that is high in nutrients and typically low in fiber; generally, the processed pellet or cereal grain portion of a herbivore's diet, which can be energy-dense or protein-dense

concentrate selectors. Ruminant animals that prefer to browse selectively on leaves, buds, and other high-protein, lower-fiber plant parts

conformation. The body posture and bone structure of an animal.

congenital condition. A condition present at birth, caused by heredity or other effects on the fetus

consignee. The zoo or individual to whom a shipment is to be delivered

consignor. The zoo or individual responsible for sending a shipment to be delivered to a destination or **consignee**

conspecific. Another member of the same species

contrafreeload. The behavior of an animal that will work hard for access to a resource (e.g., food) even when it is already freely available.

convergent evolution. The evolution of similar characteristics in unrelated species

copulatory organ. Any anatomical part of an animal that is involved in the act of sexual reproduction

cortisol. A steroid hormone (glucocorticoid) produced by the adrenal gland, whose secretion increases in times of stress. Among its functions, it will suppress the immune system and inflammation, and elevate levels of blood glucose.

creep. A protected part of an enclosure which permits entry by some but not all of its inhabitants, thereby providing refuge

crepuscular. Active at twilight (dawn and dusk)

cryopreservation. Storage (preservation) of substances at very low temperatures

cryptorchid. Having one or more undescended or "hidden" testes

cursorial. Adapted to run, typically for long distances at high speeds

cytology. The study of the origin, structure, and function of cells, which helps in pathology and the diagnosis of disease, as well as in the understanding of genetics and reproductive activities

dam. Female parent of an animal

deadfall. Dead trees or shrubs, typically used as exhibit furniture

deciduous. Inclined to fall off or be shed, either seasonally, in reaction to environmental conditions (as with plant foliage) or at a certain stage of development (as with antlers or "baby teeth" in some animals)

defense spray. A tool used by a keeper who may be exposed to an animal attack; alternatively called pepper spray, bear spray, or OC (*Oleoresin capsicum*), and available in various-sized pressurized containers

de-gas. To remove excess dissolved gases (e.g., nitrogen, carbon dioxide, or ozone) from water

delayed implantation. In some placental (eutherian) mammal species, intimate attachment of the early embryo to the uterine wall that is postponed after fertilization, with embryonic development halted until external conditions are favorable and the embryo attaches to the uterus. Also called embryonic diapause.

demography. The scientific study of populations, especially with regard to characteristics such as the size of the population, sex ratio, births, and deaths. A balanced demographic population would include all age groups, an appropriate balance of genders, and an average birth rate that equals the average death rate.

denitrification. The chemical process of converting nitrate to nitrogen gas, accomplished by anaerobic bacterial action

dermal denticles. Small dermal teeth which cover the bodies of the cartilaginous fishes (sharks, rays, and chimaeras) and are structurally similar to the teeth of vertebrate animals. Sometimes called placoid scales.

desensitize. To accustom an animal to a potentially aversive stimulus by habituation or counter-conditioning

desiccation. The process of drying and losing moisture

detritus. Material and debris, such as excreta, mucus, uneaten food, and shed skin, that accumulates in an enclosure

detrivore. An organism that feeds on detritus or organic waste

diet drift. Unplanned change from a zoo or aquarium's prescribed animal diet (e.g., provision of training treats that have not been factored into the diet)

diorama. A life-size display representing a scene from nature or human history that uses real artifacts such as preserved wildlife, biofacts, or wax figures in front of a painted or drawn mural background

direct transmission. Spread of disease through direct contact with another infected animal (e.g., bites, scratches, or bare skin contact)

disinfection. Use of a chemical solution to kill microorganisms such as bacteria, viruses, or fungi on nonliving surfaces. Disinfection of an object does not necessarily sterilize it.

distress. Stress that is prolonged and intense and which induces detrimental (e.g., behavioral or physiological) responses

diurnal. Happening or active during the day

docent. A trained volunteer; usually one who provides educational and interpretive programs within a zoo or aquarium

domesticate. To accustom a species to living in close proximity with people through generations of selective breeding

double clutch. Removal of a clutch of eggs to promote the laying of a second clutch, thus maximizing offspring production

dry matter (DM). The content of a feed with water removed, important to consider when considering the feed's actual nutrient content

dyspnea. Difficulty breathing, or shortness of breath

dystocia. Difficulty during parturition (giving birth)

ecdysis. Shedding of the outer skin layer (epidermis) in amphibians and reptiles, or of the cuticle layer in arthropods

ecology. The study of the interactions of organisms with each other and with the environment

ectoparasite. A parasite (such as a flea) that lives on the exterior of another organism

ectotherm. An organism that cannot generate metabolic heat and is dependent on environmental heat sources; colloquially referred to as "cold-blooded"

egg tooth. A temporary toothlike structure at the tip of the beak in birds, and on the snout in reptiles, that a young animal uses in breaking out of the egg in which it has developed

embryo. The earliest multicellular form of an animal following fertilization of egg and sperm

emesis. The forceful ejection of stomach contents up the esophagus and out of the mouth; also called vomiting

endocrinology. The study of glands and hormones and how they affect the behavior and physiology of animals

endoparasite. A parasite, such as a tapeworm or fluke, that lives inside another organism

endotherm. An organism that controls its own body temperature through internal metabolic means and is therefore not dependent on environmental heat sources; colloquially referred to as "warm-blooded"

endotracheal (ET) tube. A plastic or rubber tube designed for placement into the airway (trachea) to facilitate breathing or the administration of gas (oxygen or anesthetic agent)

ergonomics. The study of repetitive motion, lifting techniques, posture, and statistical tendencies towards injury or illness. Much of the study relates to the development of equipment and techniques to mitigate related workplace risks.

eructation. Release of gas from the digestive tract through the mouth; also known as belching or burping

esophageal groove. A feature of young ruminants that diverts ingested milk directly to the abomasum, bypassing the undeveloped rumen. As the animal begins to eat solid food, the rumen will develop and the esophageal groove will no longer divert ingesta from entering it.

essential amino acids. Those amino acids which an animal needs and cannot create in adequate amounts within the body to use as part of proteins or for other purposes

essential fatty acids. Those fatty acids that an animal needs and cannot create in adequate amounts within the body to use as part of fat molecules or for other purposes

estrus. A period of sexual receptivity in females that can result in ovulation; referred to in some animals as "heat"

estuarine. Pertaining to things found in estuaries, which are partly enclosed bodies of coastal waters into which rivers or streams flow

ethogram. A catalogue of the discrete behaviors that make up the typical behavioral repertoire of a species. These behaviors are sufficiently consistent (stereotyped) so that an observer can record the number of such behaviors or the amount of time engaged in them.

ethology. The study of animal behavior

ex situ. In a zoo, aquarium, or other setting that is not the typical or natural habitat. Contrasts with **in situ.**

extinction vortex. The tendency of a small population to decline towards extinction.

facultative. Characterized by a behavior that is optional or not required. A facultative carnivore is an animal (e.g., a wolf) that eats meat but can also consume some plant material.

farrier. A professional hoof specialist, typically one who trims or places shoes on hooves

fecal-oral contact. A mode of disease transmission in which fecal particles from an organism are transmitted into the oral cavity of the same or another organism

fecundity. An animal's potential reproductive capacity

feral. Pertaining to a domestic animal that has reverted to a wild (untamed) state

fetus. The stage of development occurring between that of an embryo and birth. At this stage the major structures are present and provide some resemblance to the adult animal.

fiber. Complex carbohydrates found within plant cells. Fiber (e.g., cellulose) is generally indigestible by mammals although some, such as ruminants, have developed means to access some of the nutrients. Feeds high in fiber are often termed roughages, and are important dietary components in many species.

fission-fusion. A social grouping structure in which size and composition changes periodically with activities (e.g., hunting), varying environmental fluctuations (e.g., fruiting season), and reproductive cycles (females in estrus). Individuals enter and leave communities from time to time.

fledge. To leave the nest once sufficient plumage and musculature has developed (in birds)

flehmen. A behavioral reaction in which the upper lip rolls upward making the animal appear to grimace; associated with investigation of scents and urine tasting in many different species, and particularly prominent in many species of cat and hoofstock

flight distance. A measure of how close an animal will let a potential threat approach before fleeing

float. To remove overgrown enamel ridges on teeth by rasping; frequently associated with horses

flocculation. A chemical process that occurs when a clarifying agent such as ferric sulphate or aluminum sulphate is injected into water. These chemicals bind particles which would normally pass through the filter bed, causing them to clump together and become trapped in the filter.

follicle, A small bodily cavity or sac. In mammalian reproduction, the oocyte develops within an ovarian follicle.

fomite. An inanimate object or substance (e.g., clothing, bedding, or grooming tool) capable of carrying infectious organisms from one individual to another

forage. (1.) the act of searching for food; (2.) the fibrous portion of an ungulate's diet; (3.) plants fed to herbivores

foregut. Generally, the length of the digestive tract from the mouth to the beginning of the small intestine, including the stomach

fossorial. Adapted to living underground

founder. (1.) Sinking or rotation of the bones within the hoof as a result of laminitis, a common and serious cause of lameness in horses; (2.) an individual that is assumed to be genetically unrelated to others in the population and which provides a genetic contribution to the population's future

frass. A fine powdery material that insects pass after digesting their food; insect feces

furniture. Objects added to a zoo or aquarium exhibit to improve the welfare of the animals or to make the exhibit more aesthetically pleasing to visitors

fuzzy. An immature mouse or rat that has only begun to develop fur (and thus feels fuzzy); often used as food for carnivores

gamete. A mature reproductive cell, such as a spermatozoon or ovum

gametogenesis. Production of different cells (gametes) for the purpose of reproduction, as in the case of oocytes and sperm

gavage feeding. Supplying of liquified food directly into the stomach through a tube or feeding needle (gastric gavage)

gestation (period). The duration of pregnancy

glucocorticoids. A class of steroids produced by the adrenal glands that is responsible for regulating glucose metabolism. Levels of these hormones (particularly cortisol) can increase in response to stress.

glutaraldehyde agglutination test. A commercial test that can detect failure of passive transfer in some types of neonatal animal. Immunoglobulins (antibodies), if present, will result in clotting (agglutination) in the sample. This is one of many types of tests to evaluate the success of passive transfer of immunity from the female to the offspring.

goiter. A condition characterized by swelling of the thyroid gland and resulting in swelling of the neck and larynx, commonly caused by iodine deficiency

gonochoric. Having separate sexes

gout. A condition usually characterized by recurrent episodes of acute inflammatory arthritis; a red, tender, hot, swollen joint

grain. Seeds of plants such as corn, wheat, oats, rice, or millet, which are high in carbohydrates such as starch and are often used in commercial diets. Pelleted feed may often contain grain.

gravid. Pregnant or carrying eggs

grazer. An animal that feeds on grass and other ground-level vegetation

gut-loading. The practice of feeding prey species (often invertebrates) nutrient-rich food before feeding them to a carnivore as supplemental nutrition

"ha-ha." A physical barrier that incorporates a trench or moat to permit an unobstructed view of an exhibit. Typically it conceals at least one side from view; it may include a retaining wall that is not visible to visitors, and its dimensions are chosen so as to prevent escape by animals or trespass by visitors.

habituate. To accustom to a stimulus through repeated or prolonged exposure

hallux. The great toe or inner toe of the foot, which is opposable in primates

hard-standing surface. An enclosure surface with a solid, rough, and/or abrasive substrate

harrow. A piece of farm equipment comprising a heavy frame to which are attached discs or spikes. The whole apparatus is pulled across a paddock to break up and level the surface, and it can also be used to incorporate manure as a fertilizer into the substrate.

hay. Forage material provided to herbivorous animals, created by drying grass (such as Bermuda grass) or legumes (such as alfalfa or clover) to increase stability and longevity of the feed. Grass hays are generally lower in protein and energy than legume hay.

heliotherm. An animal (e.g., a reptile) that basks in the sun to absorb heat

helminths. Parasitic worms, often of the digestive tract (e.g., tapeworms, roundworms, and hookworms)

hematopoiesis. Production of the cellular components of blood, including red and white blood cells

hemicellulose. A major component of plant fiber and an important component of fibrous feeds such as roughages and agricultural by-products (soy hulls, oat hulls, etc.) Hemicellulose is important for proper rumen function, and while low in digestibility for most animals, it is more digestible than cellulose. It is also a component of cell walls in plants.

hemipenes. In snakes and lizards, paired eversible (which can be turned inside out) structures within the cloaca of the male, used for internal fertilization of the female

hemolymph. Fluid that circulates through and around the tissues of many invertebrates, serving many of the functions of blood in vertebrates

hemorrhage. Bleeding

herbaceous plant. A soft-tissued plant that dies back to ground level at the end of the growing season with no woody stem persisting aboveground

herbivore. An organism that consumes and digests plant-based foods, and which may have special adaptations to aid digestion (as in ruminants)

herpetology. The branch of zoology responsible for the scientific study of amphibians and reptiles

heterodonty. Possession of more than one type of tooth morphology, as in mammalian species

hibernation. A period of seasonal dormancy in which metabolic rate (respiration, heart rate, and body temperature) decreases

hierarchy. A specific ordered arrangement of animals within a group, based in many species on social dominance and subordination

hindgut. Generally, the colon, cecum, and rectum in most animals

hindgut fermentation. Microbial fermentation that occurs somewhere in the colon or cecum

histology. The microscopic study of plant and animal cells; known as **histopathology** when used to diagnose and study disease

holding area. The area where an animal sleeps, eats, or otherwise spends time when not on display to the public

homeostasis. The ability of an animal to maintain internal equilibrium by adjusting its physiological processes

hopper. A small, young mouse or rat that has developed a full hair coat and is mobile; often used as food for carnivores

hormone. A chemical messenger within the body which is produced by specialized cells and elicits specific changes in other (target) cells

horn. One of a pair of permanent, bony-cored, keratin-covered growths on the heads of ungulates in the family Bovidae. Horns may have ridges and sharp points, and are never branched.

hotwire. An electrified fence used to restrict access to an area or prevent escape

humoral immunity. Resistance of an infectious disease by circulating antibodies in the blood. See also **cell mediation.**

hydrometer. An instrument that is used to measure the specific gravity or density of water, and which must be accurately calibrated

hydroregulation.—A method used by animals, especially amphibians, to regulate a proper water balance within their bodies

hygrometer. An instrument used to measure the moisture content of air, and often used to measure humidity in terrestrial animal enclosures

hyper-. Prefix meaning elevated or in excess

hyperglycemia. High blood sugar level, which if chronic can cause organ damage

hyperthermia. Excessive elevation of the core body temperature

hypo-. Prefix meaning low or deficient

hypoglycemia. Abnormally low levels of blood glucose

hypothalmus. An area in the bottom of the brain that has a regulatory effect on many basic body functions and which serves as a switchboard, directing nerve impulses and regulatory chemicals to other areas of the body

hypothermia. A state characterized by subnormal core body temperature

hypoxemia. Abnormally low levels of oxygen in the blood (the term **hypoxia** refers to a general state of having low oxygen levels, but is not specific to the blood)

identifier. A natural marking, physical characteristic, or artificial device that makes an animal unique. Natural identifiers include coat or skin markings and defects such as broken horns or kinked tails. Artificial identifiers include ear tags, leg bands, transponders (microchips, passive integrated transponder[PIT] tags), tattoos, and ear notches.

immune system. The system of special cells, proteins, tissues, and organs that protects the body from foreign materials, including pathogens, and abnormal cells

immunity. A body's ability to resist a specific disease or pathogen

immunization. The introduction of an antigen to the body to trigger the production of antibodies which will provide a level of protection against the actual disease pathogen

immunocompromise. A condition in which the immune system's effectiveness is decreased, thus leaving an organism more vulnerable to opportunistic infections

immunoglobulin. A protein created by the body to fight off infection as part of the immune system; also called an **antibody** and primarily found in the blood

imprint. To create a long-lasting association learned by a young animal in a species-specific critical period during which it must establish proper social attachment and identification to avoid abnormalities later in life

in situ. In the wild or natural setting. Contrasts with **ex situ**.

indigenous. Native; naturally occurring

indirect contact. A mode of disease transmission involving a fomite or contact with the environment

inflammation. A response of tissue to injury or infection in which an influx of bodily defense cells and blood supply results in tissue that is characteristically red, swollen, warmer than normal, and often painful

innate. Present in an individual from birth

inoculate. To introduce microorganisms into a new setting

inorganic nutrient. A non–carbon-based nutrient, such as a macro- or micromineral.

insectivore. A carnivorous animal that specializes in feeding on insects and other similar small invertebrates

inter-. Prefix meaning between

intra-. Prefix meaning within

intraosseous (IO). Within the bone (as in osseous tissue)

intraperitoneal (IP). Within the peritoneal or body cavity

intravenous (IV). Introduced directly into a vein, as in the case of a drug or other fluid

invasive species. An unnaturally occurring species within an environment that becomes endemic and has a negative impact upon the naturally occurring species

joule. See **calorie**.

keratinized. Covered in keratin, a fibrous protein that is present in skin and scales and is important to the structure of hair, horns, claws, and nails. Also known as **cornified**.

knockdown. Slang for an anesthetic or immobilization procedure

kriesel. A specially designed enclosure that provides a continuous circular flow of water to support successful jellyfish culture

laceration. A jagged tissue cut or wound

laminitis. A hoof condition characterized by inflammation of the laminae (the structures that connect the outer keratinized layer of the hoof), and which in severe cases may result in **founder**

lateral. Located away from an animal's midline, and toward its right or left side.

legume. Any plant or fruit of the family Leguminosae; in zoos, usually a clover-like plant such as alfalfa that is fed either fresh or in the form of hays (e.g., alfalfa/lucerne, or clover hay).

lesion. Abnormal tissue

ligate. To tie or bind with a ligature or suture

lipids. A broad category of organic compounds known for high energy density (e.g., fats like triglycerides) for their roles in cell membrane structure (e.g., phospholipids and cholesterol), as signaling pathways (e.g., compounds like estrogen and cortisol), and as essential nutrients (e.g., fat-soluble vitamins and essential fatty acids)

live rock. Calcerous (calcareous) substrate, usually compacted coral skeletons, that has been colonized by living marine organisms

livestock. Domesticated hoofstock

lordosis. The downward (in quadrupeds) or inward (in orthograde or bipedal animals) arching of the spine; sometimes called "saddleback" or "swayback." In some small cat and rodent species it is a reproductive behavior seen in females ready to mate.

macrominerals. Nonorganic chemical elements used and stored in large quantities within the body, including calcium (Ca), phosphorous (P), and magnesium (Mg); usually expressed in percentage units

mechanical vector. An organism, usually an arthropod, that physically carries a pathogen from one organism to another

meconium. The first feces produced by a newborn. It is comprised of material ingested during gestation, and is generally dark (tarlike) and without odor.

medial. Being towards the midline or medial plane

menarche. A female's first menstrual cycle

metabolism. The essential chemical reactions which take place within an organism to permit life, and which can be divided into two categories: catabolism and anabolism. Catabolism is the breaking down of molecules, as in digestion and respiration. Anabolism is the building of molecules, as is essential for functions such as growth and reproduction.

metamorphosis. Change; the rapid transformation of a larva into an adult (e.g., a tadpole into a frog)

metapopulation. A collection of separate but connected local species populations. For example, some insect populations may exist in a small area but occasionally have new individuals enter from the larger metapopulation.

microminerals. Nonorganic chemical elements required by animals in minute concentrations for normal development and growth. These include iron (Fe), copper (Cu), zinc (Zn), molybdenum (Mo), selenium (Se), cobalt (Co), manganese (Mn), and many others. Micromineral content is expressed as parts per million (ppm or mg/kg) or parts per billion (ppb).

mineralocorticoids. A class of steroids produced by the adrenal glands that are important in body water and salt regulation

minerals. Naturally occurring solid chemical substances formed through geological processes, many of which are used by animals for normal body function and can be classified as either macrominerals or microminerals

monoestrus. Having one estrus cycle per year

monogamous. Having a single, exclusive mate

monogastric. Having a single (simple) stomach

morbidity. The incidence or prevalence of disease

morphology. In biology, the study of form and structure of organisms

mortality. The condition of being subject to death, or the measurement of death within a population over a period of time

mucous membranes. The moist, absorptive linings of body openings and canals such as the nostrils, eye, mouth, lips, throat, anus, and genitals

mucus. The thick (viscous) fluid produced and secreted by glands lining mucous membranes to prevent drying, provide lubrication, aid with some absorption, and protect the body from pathogens and other foreign material

multiparous. Having given birth on multiple occasions

musth. A periodic condition of heightened aggression in adult male elephants, associated with reproductive activity and usually accompanied by draining of an oily substance from the temporal glands

myopathy. Any disease of the muscles

natal. Pertaining to birth

necropsy. The postmortem (after death) examination of an animal. An autopsy is a necropsy performed on a human being.

necrosis. The death of tissue

negative punishment. The removal of a desirable stimulus when the animal does an undesired behavior, to make the behavior less likely to occur again. For example, an animal that wants its keeper's attention will be less likely to bite its keeper once it learns that biting (undesired behavior) will make the keeper leave (loss of desired stimulus).

negative reinforcement. In operant conditioning, the removal of an aversive stimulus to make a behavior more likely to occur. For example, a horse will be more likely to move as directed (a desired behavior) if doing so will remove the aversive stimulus of the rider pulling on the reins

neonate. A newly born infant; typically up to one week to one month old, depending on the species

neoteny. Retention of juvenile or larval morphological features by the adults of a species

neutral detergent fiber (NDF). Less digestible components of plant feed, including lignin, silica, cellulose, and hemicellulose. The NDF level is an estimate of the feed's total fiber content; with higher NDF levels, animals typically consume less feed.

nidiculous animals. Young animals (typically birds) that stay around the nest or den site and rely on the parents for food and care for a period of time after birth or hatching

nidifugous (nidicolous) animals. Young animals (typically birds) that leave the nest shortly after hatching, and which may remain with their parents but find their own food

nitrate (NO_3). A compound resulting from the breakdown of nitrite by beneficial bacteria, usually considered harmless in trace to moderate amounts

nitrification. The series of oxidation processes by which ammonia is converted first to nitrite and then to nitrate by aerobic (nitrifying) bacteria

nitrite (NO_2). A toxic compound resulting from the breakdown of ammonia by beneficial bacteria, and which is less toxic than ammonia and more toxic than nitrate

nocturnal. Active at night

noninvasive. Pertaining to a procedure or action that does not negatively affect an animal. Information can sometimes be collected noninvasively from a distance without any direct interaction with an animal

nulliparous. Pertaining to a female that has never given birth

nutrient. A chemical substance the animal body needs as a source of energy or as part of its metabolic machinery, and which is essential for life

obesity. A medical condition in which excess body fat accumulates to the extent that it may have an adverse effect on health, leading to reduced life expectancy and/or increased health problems

ocean acidification. The gradual decrease in pH of the world's oceans through absorption of carbon dioxide, making the oceans more acidic and causing the decline of coral reefs and other calcium-dependent ocean dwellers

olfactory. Pertaining to the sense of smell, and to the detection of chemicals dissolved in air or water

omnivore. An organism adapted to eat both animal and plant material

oocyte. The female gametocyte or germ cell involved in reproduction; an immature ovum or egg cell

operant conditioning. A method of learning in which the subject makes an association between its behavior and the consequences of that behavior

ophthalmic medication. A drug that is administered into the eye

organic nutrients. Carbon-based nutrients including protein, fat, carbohydrates, and vitamins

ossification. The process in which the body lays down new bone

otic medication. A drug that is administered into the ear canal

ovariohysterectomy. The surgical removal of one or both ovaries and the uterus

ovary. A female reproductive organ, usually found in pairs, which contains the female gametes, the ova (singular: **ovum**). Unlike the testes, which in mammals continually produce spermatozoa, the female mammal's ovary at birth contains a finite number of ova that will be steadily depleted during her lifetime.

overseeding. The addition of seeds to an enclosure to rejuvenate it with new and healthy plant growth as required. The seeds may be of commercial grasses or other types of ground cover.

oviparous. Pertaining to animals that lay eggs

ovoviviparous. Pertaining to animals that give birth to live young hatched from eggs that develop internally

ovulation. Release of the ovum ("egg") from the ovary

ovulator, induced. A species in which the female releases the ovum or ova in response to the stimulation of a period of mating

ovulator, spontaneous. A species in which an ovum or ova are released regularly from the ovary, irrespective of whether the female has mated

ovum. (plural: **ova.**) The female gamete that fuses with the male's sperm during fertilization; sometimes called an "egg"

ozone (O$_3$). A highly reactive molecule that is used as an oxidative chemical filter or disinfectant

paddock. A large outdoor hoofstock exhibit, often containing grass or other vegetation

palatable. Pleasant to taste

palisade. A wall comprised of vertical posts, often wooden

palliative care. Specialized care which focuses on relief and prevention of suffering rather than on curing the ailment

palm oil crisis. The spread of palm oil plantations into the rain forests of Borneo and Sumatra, threatening populations of orangutans and other rainforest inhabitants through habitat destruction and conflict with humans

palpate. To explore by touch using the hands and fingers, a method used to aid in diagnosis of illness or to assess body condition

parasite. An organism living on or in another organism (the host), from which it derives nourishment

parasympathetic system. One of the two main divisions of the autonomic nervous system, responsible for the body when it is at rest, ensuring that essential functions (such as secretion and smooth muscle function) continue

parenteral medication. A drug administered in a way other than by mouth (e.g., by injection)

paresis. Slight or incomplete paralysis

parthenogenesis. Development of young without fertilization. See **asexual reproduction.**

parturition. The birth process

passive integrated transponder (PIT). A small (typically rice-sized) electronic device that is injected into an animal to aid animal identification. When scanned using a handheld reader, the PIT will provide an individual identification number. Often referred to as a chip, tag, or microchip.

passive transfer. The means by which maternal antibodies in colostrum are absorbed by a neonate and provide immunity to disease until the neonate can develop its own immunity. Some species depend on passive transfer of antibodies through colostrum and can only absorb them in the hours immediately following birth

pathogen. Any agent that can cause disease

pathogenesis. The pathway for development of a disease in an animal

pedipalps. Leglike appendages situated between the chelicera and the first set of legs on spiders. Depending on the species, they may be used for catching and holding prey. Male spiders have specialized structures on the "palps" used for transferring sperm to females.

periodontal. Pertaining to the tissues which surround and support the teeth

peristaltic movement. The contraction and relaxation of the smooth muscles of the gastrointestinal tract, which propagates a wave along the muscular tube and propels digesta with it

personal protective equipment (PPE). Equipment, such as gloves or safety glasses, normally provided to employees by the employer to be worn as protection from injury and disease

phagocytosis. The process whereby some cells, such as certain white blood cells, can engulf and envelope foreign material such as bacteria (literally, "cell eating")

phallodeum. A penis-like intromittent organ present in the cloaca of male caecilians and used for internal fertilization of the female in a process unique to this group of amphibians

phallus. The male reproductive organ, also called the penis

pheromone. A chemical secreted by an animal that influences the behavior or development of others of the same species

photophobia. Intolerance to or avoidance of light

photosynthesis. The process in which carbon is fixated by plants and which results in the release of oxygen and creation of organic compounds such as sugars, using the energy of the sun

physiological stages. The stages of life that can alter the needs of metabolism, including growth, lactation, gestation, and old age

phytotoxin. A toxin manufactured by or found in plants

pica. An abnormal craving to ingest nonfood substances such as clay, dirt, or hair

piloerection. Reflexive raising of the hair that occurs in response to perceived threats or cold temperatures

Pillstrom tongs. Long forceps used for grasping reptiles, often referred to as "reptile tongs"

pinioning. A surgical process which removes the end of a bird's wing at the joint farthest from the body, usually carried out on only one wing to create an imbalance that prevents flight

pinky. An immature mouse or rat that has not yet begun to develop fur; often used as food for carnivores

piscivore. A carnivorous animal that primarily eats fish

pituitary gland. An endocrine gland located at the base of the brain, often called the "master gland" because it regulates the activity of all other endocrine glands (adrenals, thyroid, pancreas, ovaries, testicles)

placenta. The organ that connects the fetus with the uterine wall and provides it with nutrients and oxygen

plasma. The clear, yellowish fluid portion of whole (unclotted) blood

plastron. The ventral shell of the turtles and tortoises (order Testudines)

pleurodont. An animal in which teeth are attached to the inner (labial) wall of the jaw with an eroded lingual aspect, as seen in snakes and some lizards such as iguanas

polyandry. A breeding system in which animals live in groups each consisting of a single adult breeding female, more than one adult male, and their immature offspring

polyestrus. Having multiple estrus cycles within a year

polygyny. A breeding system in which animals live in groups each consisting of a single adult breeding male, more than one adult female, and their dependent offspring

porcine zona pellucida (PZP). The outer membrane (zona pellucida) of a pig's ovum (egg), injected into an individual of another species to trigger an immune reaction that will prevent fertilization

positive punishment. Presentation of an aversive stimulus to the animal to make a behavior less likely to recur: for example, administration of an undesired squirt of water (aversive stimulus) to a cat that has jumped onto the kitchen counter (undesired behavior)

positive reinforcement. In operant conditioning, the provision of a desired stimulus to make a behavior more likely to occur: for example, provision of a food treat (desired stimulus) to an animal that has come when called by name (desired behavior)

precocial. Pertaining to a young animal that is highly developed at a young age, and for whose species such high development at a young age is typical

pregnancy resorption. The resolution of an unsustained pregnancy, in which the embryo or fetus dissolves and is absorbed by the dam

prenatal. Before birth

primary containment. The main barrier of an animal enclosure, whose purpose is to ensure that the animal does not escape

primiparous. Having given birth once (for the first time). A first-time mother might be referred to as a **primiparousa.**

prion. The smallest and simplest disease agent composed of misfolded proteins that cause other proteins to assume the same nonfunctional shape, destroying tissue, usually of the brain

prognosis. The likely outcome of a medical condition

prophylactic. Pertaining to that which protects from disease or infection

prosimian. Mammals defined as primates but not as monkeys or apes. The group includes lemurs, lorises, and tarsiers.

proteins. Organic nutrients consisting of carbon, nitrogen, and hydrogen arranged into folded chains of amino acids, which serve many functional, structural, and mechanical purposes within the body. For example, proteins make up myosin (in muscles), enzymes, antibodies, and keratin (in nails and horns).

pseudopregnancy. A condition, common in carnivore species, in which the animal exhibits signs of pregnancy but is not pregnant. The period of pseudopregnancy usually represents the lifespan of the corpus luteum and is often 50 to 66% of the normal gestation length. Also sometimes called false pregnancy.

psychological. Pertaining to, dealing with, or affecting the mind

pulse oximeter. A device used to detect pulse and the oxygen saturation levels in the blood, and which can be used to aid in the monitoring of anesthesia

punishment. In operant conditioning, something that makes a behavior less likely to occur

quarantine. A period of containment and isolation for purposes of disease control. Often quarantine lasts 30 to 60 days, but its duration will depend upon the species, the region, and disease concerns.

radiograph. An X-ray image of the bones and internal organs. X-rays may also be called Röntgen rays.

rain garden. A landscape feature constructed in such a location and in such a way so as to act as a filter to minimize precipitation runoff into natural waterways via storm drains

ram ventilation. The respiratory process in some fish in which water flows through the mouth and across the gills as they swim; used by fish with little or no ability to pump water buccally, such as sharks

reflex. An automatic response to the stimulation of a nerve

refractometer. An instrument used to measure the concentration of dissolved solids in water, and thereby indicate its specific gravity

refugia. A shelter or protected area

regurgitate. To move undigested stomach contents back up the esophagus from the stomach; sometimes called "vomiting," although vomiting (emesis) is typically more forceful and unpleasant

reinforcement. In operant conditioning, something that makes a behavior more likely to occur. A reinforcer, therefore, causes the frequency of a behavior to increase

relative humidity. Usually stated as a percentage, the amount of water vapor in air (relative to the level of water vapor saturation, which changes with temperature). A high relative humidity (near 100%) will decrease the ability of animals to lose heat through evaporation (e.g. perspiration) and will increase the time needed for objects to air dry. A low relative humidity (approaching 0%) will speed drying (desiccation) and evaporation.

renarcotization. Sedative effects that occur after initial recovery from an immobilization (technically involving a narcotic drug); more generally referred to as resedation

reservoir. (1.) A place in the environment where a disease naturally resides (e.g., water) (2.) A containment for water storage in both natural and man-made environments

respirator. A device worn most often over the nose and mouth, to protect the respiratory system by filtering out potentially toxic or injurious substances, chemicals, and organisms

reverse osmosis (RO) water. Ultrapure fresh water that has passed through a reverse osmosis membrane filter, which has openings so small that only water molecules can pass through under pressure

roughage. Animal feed from plant sources that contain high levels of dietary fiber (e.g., browse, hay, and straw) and takes the form of browse, hay, straw, and other agricultural products. Caution must be exercised when using some of these products as animal bedding, as they may be ingested by the animals in place of other parts of the diet, thus causing potential nutritional concerns

ruminant. A member of the suborder Ruminantia, defined by the presence of a four-chambered stomach and the habit of regurgitating food to aid digestion (also termed "chewing cud" or ruminating). This group includes the chevrotains, musk deer, true deer (cervids), giraffes, pronghorns, and bovids (cattle, antelopes, sheep, etc.). Ruminants are considered foregut fermenters, since they rely on microbial fermentation within their stomach chambers. The largest of the stomach compartments is called the **rumen.**

rut. The breeding season of some ungulate species (such as deer and camels), typically marked by heightened aggression in males

sanitation. The use of a soap and water solution to clean

sanitize. To remove dirt, pathogens, or other unwanted material from; to make clean or sanitary

satiation. The state of being satisfied and not desiring more; may be applied to food, heat, enrichment, etc.

scientific method. A procedure involving the formulation and evaluation of a testable hypothesis through systematic experimentation, measurement, and observation

scuba. Originally an acronym for "self-contained underwater breathing apparatus"; now considered a word on its own

scutes. Horny, chitinous, or bony outer plates covering the shells of most turtles and tortoises, and also seen on crocodilians and the feet of birds

secondary barrier. An additional enclosure barrier that serves as a fail-safe backup to the primary barrier, that complements the primary barrier (e.g., electric fencing that discourages animals from contacting the primary barrier), or that acts as an additional barrier serving a different purpose (e.g., a public safety barrier on the outside of an enclosure that prevents visitors from getting too close to the primary barrier)

secondary infection. An additional infection that begins during an existing period of illness or shortly after, often because the body's immune system has already been compromised by the primary disease

sedation. The induction of calmness and drowsiness. The term "tranquilizer" is often used interchangeably with "sedative," although a tranquilizer does not characteristically cause drowsiness

senescence. The state of having grown old

sentience. A state ascribed to animals considered to be aware of their environment by means of sensations and feelings, especially feelings resulting from the experience of subjective emotional states (e.g., happiness or sadness).

seroconversion. The development of antibodies in response to the administration of a vaccine or exposure to an infectious disease

serum. The clear fluid portion of blood after clotting (coagulation); contrasts with **plasma**, the liquid portion of unclotted blood

service area. An area, accessible to keepers but not to zoo visitors or animals, where keepers store their tools and equipment, and

which is connected by service doors to the animals' enclosure or holding area

sexual dimorphism. Differences in physical appearance between males and females of a given species

sharps container. A puncture-resistant container with appropriate hazardous materials labeling (e.g., National Fire Protection Agency in the United States), used to safely house syringes, needles, and other items that can puncture the skin, until they can be appropriately disposed of

shock. A physiological state in which the body is unable to supply the vital organs with enough blood

sire. Male parent of an animal

species. A classification grouping of organisms that share common characteristics and can reproduce; they cannot produce fertile offspring with individuals of other species. Taxonomically, species is ranked below genus and is written in italics with the genus name capitalized and the species name in lowercase: e.g., *Loxodonta africana* (African elephant).

spermatozoon. A sperm cell capable of moving independently. A **spermatophore** is a capsule or mass containing **spermatozoa** (pl.), which is transferred to the female animal, as seen in some amphibians and invertebrates.

splay. Spreading the limbs apart, past their normal degree of movement

starch. A complex carbohydrate composed of large numbers of simple sugar molecules joined together. Starch is the carbohydrate found in many grains, tubers, and rhizomes (roots) used in feeds. It stores energy and can be digested down to its sugar molecules in most animals far more easily than the fibrous components of plants. It provides the greatest amount of energy for animal needs of any of the carbohydrates in nature. Excess starch can cause obesity in some animals and affect normal rumen function.

stereotypical behavior. Persistent, repetitive behavior considered abnormal or unnatural

sterile. Rendered completely void of any living microorganisms

sternal. Pertaining to the sternum (breast bone). Sternal recumbency is a position in which an animal lies on its sternum; it is often preferred for an animal under or recovering from anesthesia.

stethoscope. An instrument with two earpieces that is used to hear internal sounds such as those produced by the heart, lungs, and intestines

stimulus. Something in an animal's external or internal environment that it can detect

straw. The dry stems of various grains (e.g., oats or barley), often used for bedding

stress. The cumulative response of an animal to interaction with its environment via receptors. Any stimulus that elicits a biological response when perceived by the animal is termed a **stressor**. Stress is sometimes categorized; **eustress** refers to stress that produces beneficial consequences, while **distress** is detrimental to the animal.

studbook. A database used to maintain the records of the parentage of all specimens of a single species held in a particular region (in the case of a regional studbook) or held globally (in the case of a global studbook). The information is used to make recommendations for the genetic management of the species. Each individual animal within the studbook population is assigned its own studbook number.

studbook keeper. The person responsible for gathering information and updating and maintaining the studbook

subclinical disease. Disease without apparent signs. Some diseases may progress from subclinical to clinical, or from clinical to subclinical.

subcutaneous. Beneath the skin; in the area of connective tissues and adipose tissues that lie between the skin and above the muscles

sump pump. A small water pump designed for moving water from low areas of a building where water pools

surrogate. A person, animal, or object that substitutes for another in a social or family role; also, a keeper or animal that functions as a parent for an orphaned animal

symbiosis. A close relationship between two different organisms which is beneficial to both (e.g., between a ruminant and the microbes in its stomach that digest plant fiber for mutual benefit)

sympathetic. One of the two main divisions of the autonomic nervous system, responsible for the "fight-or-flight" response (including increased heart rate, decreased secretion, and decreased contractility of smooth muscle), among other things

syndrome. A group of signs characteristic of a specific disease

tactile. Pertaining to the sense of touch

tartar. A hard, yellowish deposit that forms on teeth, made up of organic secretions and food particles deposited in various salts such as calcium carbonate

terrapin. A turtle with webbed digits and a streamlined carapace that spends much of its life in fresh or brackish water

terrestrial. Living in the ground or on land

thalamus. A nerve center in the brain that relays impulses to the cortex and is involved in sensory perception and the regulation of motor functions

thecodont. An animal that has teeth embedded within a deep bony socket that lacks a periodontal ligament, as seen in crocodilians

thermoregulation. The ability of an organism to maintain its core body temperature within certain parameters

thigmothermic. Pertaining to animals, particularly reptiles, that lie on warm surfaces to absorb heat through direct contact

titer. The measure of the concentration of a substance in fluids or tissues at a given point in time. For example, antibody titers refer to the concentration of antibodies present in blood at the time the blood was collected.

top-dressing. Spreading of a layer of soil, compost, or other material over the surface of an enclosure to promote healthy groundcover growth

topical medication. A drug applied to the surface of the skin

tortoise. A member of the order Testudines (turtles). These reptiles are primarily herbivorous land dwellers with heavy domed carapaces.

TPR. Abbreviation for temperature, pulse, and respiration; vital signs commonly monitored during physical examination and anesthesia

tractable. Easily managed or controlled

trauma. Injury, either physical or psychological

troglodytic. Adapted to living in caves

tubal ligation. A generally permanent method of birth control in which the oviducts (fallopian tubes) are cut, tied, or occluded to stop the passage of ovum or sperm and prevent reproduction

Tyndall effect. The formation of colors via the scattering of white light by particles in its path

ubiquitous. Existing everywhere within a given environment

udder. The baglike structure containing the mammary (milk-producing) glands of many ungulates

ultraviolet radiation (UV). High-energy, short-wavelength light, used for many functions in zoo and aquarium animal care, including disinfection of water in aquatic enclosures and promotion of vitamin D synthesis in many terrestrial species

umbilicus. The neonatal remnant of the umbilical cord that provided nutrients to the fetus in the uterus

ungulate. A hoofed mammal from the orders Perissodactyla (odd-toed ungulates: horses, tapirs, rhinoceroses) or Artiodactyla (even-toed ungulates: cattle, deer, pigs, hippopotamuses). Taxonomically the term Cetartiodactyla may be used in place of Artiodactyla. Technically, ungulates also include Tubulidentata (aardvarks), Cetaceans (whales), Sirenia (manatees) and Proboscidea (elephants).

unken reflex—Arching defense posture used by some salamanders and frogs that exposes warning coloration (i.e., aposomatic coloration) of potential toxicity if consumed, although not all species that exhibit the behavior are toxic

urachus. A canal that connects the bladder of a fetus with the allantois, and contributes to the formation of the umbilicus

urolithiasis. A disease in which stones or calculi form in the urinary tract and can lead to obstruction and sometimes death

urticating hair. Stinging or irritating hairs used in self-defense by some plants and animals, such as New World tarantulas and some lepidopteran caterpillars

vaccine. A biological preparation that promotes immunity to a specific disease, typically containing an antigen that resembles a disease-causing agent

vasectomy. A generally permanent method of birth control which involves severing the male's vasa deferentia and tying off the cut ends, thus preventing the release of sperm

vector. A living organism that transmits a pathogen. For example, a mosquito can be considered a vector of the West Nile virus.

venipuncture. The procedure of inserting a needle or other surgical instrument into a vein for the purpose of collecting blood or for intravenous therapy

venturi. A device that infuses microfine bubbles into a steam of water to remove undesired charged particles, such as dissolved organic compounds within a protein skimmer

vermiculture. A form of composting in which worms fed organic waste turn it into compost soil through their castings

vernix. A white, cheesy substance that covers and protects the skin of the fetus and is present at birth. Also known as **vernix caseosa.**

vertical transmission. Disease transmission from a mother to offspring

virulence. The relative ability of a pathogen to cause disease

viscera. The internal organs of the body (e.g., lungs, stomach, intestines, liver, spleen)

vital signs. Indications of physical function that can be detected to assess health status. Most commonly measured signs include body temperature (T); pulse rate and character, expressed in number of heart beats per minute (P), respiration rate, expressed in complete inhale-and-exhale cycles per minute (R) and blood pressure. Measurement of temperature, pulse, and respiration is often abbreviated as TPR.

vitamin. An organic molecule required for normal metabolism and needed in small quantities

viviparous. Pertaining to animals that give birth to live young rather than laying eggs

volatile fatty acid. A short chain fatty acid (e.g., acetate, propionate, or butyrate) that is a major product of microbial fermentation and breakdown and is used by animals such as ruminants for energy

vomeronasal organ (or Jacobson's organ). An auxiliary olfactory sense organ found in many animals, used for detecting pheromones and other chemical substances in the environment

woody plant. A plant that has wood as its structural tissue and which supports continued vegetative growth above ground from year to year; usually a tree, shrub, or vine

xeric. Characterized by an extremely dry environment

yolk sac. The part of the egg which supplies the developing embryo with nutrients

zeolite. A natural compound used in freshwater systems to remove ammonia

zoonotic disease. Disease that can be transmitted between animals and people. Sometimes the term **anthropozoonosis** is used to refer to a disease that is transmitted specifically from people to animals.

zooxanthellae. Symbiotic algae, present in the cells of many corals and some marine clams, that produce oxygen and sugars through photosynthesis to the benefit of the host tissue

Contributors

Noha Abou-Madi
Cornell University
Ithaca, NY 14853
USA

Tracey L. Anderson
Pikes Peak Community College
Colorado Springs, CO 80906
USA

Joseph C. E. Barber
Hunter College (CUNY)
New York, NY 10065
USA

Tony Beane
State University of New York at Canton
Canton, NY 13617
USA

Andrew Allan Birr
Toledo Zoo
Toledo, OH 43614
USA

Jacqueline J. Blessington
American Association of Zoo Keepers
Kansas City, MO 64131
USA

Paul Boyle
Association of Zoos and Aquariums
Silver Spring, MD 20910
USA

Sara F. K. Brice
Taronga Zoo
Mosman, NSW
Australia

B. Diane Chepko-Sade
SUNY Jefferson
Mexico, NY 13114
USA

Aaron M. Cobaugh
Niagara County Community College (SUNY)
Sanborn, NY 14132
USA

Adrienne E. Crosier
Smithsonian's National Zoo
Smithsonian Conservation Biology Institute
Front Royal, VA 22630
USA

Erika K. (Travis) Crook
Utah's Hogle Zoo
Salt Lake City, UT 84108
USA

Gerald Dick
World Association of Zoos and Aquariums
Gland
Switzerland

Lesley Dickie
European Association of Zoos and Aquaria
Amsterdam
The Netherlands

Candice Dorsey
Association of Zoos and Aquariums
Silver Spring, MD 20910
USA

Chuck Doyle
Rosamond Gifford Zoo
Syracuse, NY 13204
USA

Andrea L. Drost
Toronto Zoo
Toronto, ON
Canada

Michelle R. Farmerie
Pittsburgh Zoo and PPG Aquarium
Pittsburgh, PA 15206
USA

Murray E. Fowler
University of California, Davis (retired)
Davis, CA 95616
USA

Ted Fox
Rosamond Gifford Zoo
Syracuse, NY 13204
USA

Harmony B. Frazier
Woodland Park Zoo
Seattle, WA 98103
USA

Markus Gusset
World Association of Zoos and Aquariums
Gland
Switzerland

Ed Hansen
American Association of Zoo Keepers, Inc.
Topeka, KS 66614
USA

Janet Hawes
Zoological Society of San Diego
San Diego, CA 92112
USA

Daryl Hoffman
Houston Zoo
Houston, TX 77030
USA

Brent A. Huffman
Toronto Zoo
www.ultimateungulate.com
Toronto, ON
Canada

Mark D. Irwin
SUNY Jefferson
Watertown, NY 13601
USA

Jason A. Jacobs
Reid Park Zoo
Tucson, AZ 85716
USA

Tineke Joustra
Auckland Zoo
Auckland
New Zealand

Ken Kawata
Staten Island Zoo (retired)
Staten Island, NY 10314
USA

Bruce Koike
Oregon Coast Community College
Newport, OR 97366
USA

Josef H. Lindholm III
Tulsa Zoo
Tulsa, OK 74115
USA

Andrew M. Lentini
Toronto Zoo
Toronto, ON
Canada

Mary O'Horo Loomis
State University of New York at Canton
Canton, NY 13617
USA

Debborah E. Luke
Association of Zoos and Aquariums
Silver Spring, MD 20910
USA

Danny de Man
European Association of Zoos and Aquaria
Amsterdam
The Netherlands

Michael T. Maslanka
Smithsonian's National Zoo
Smithsonian Conservation Biology Institute
Washington, DC 20013
USA

Thomas R. Mason
Toronto Zoo
Toronto, ON
Canada

Colleen McCann
Bronx Zoo
Wildlife Conservation Society
Bronx, NY 10460
USA

Gordon McGregor Reid
North of England Zoological Society (Chester Zoo)
University of Liverpool
British Museum of Natural History
Cheshire
UK

Gerard H. Meijer
Ouwehands Dierenpark
Rhenen
The Netherlands

Jill D. Mellen
Disney's Animal Kingdom
Lake Buena Vista, FL 32830
USA

Karla J. Michelson
San Diego Zoo's Safari Park
Escondido, CA 92027
USA

Erica A. Miller
Tri-State Bird Rescue and Research, Inc.
Newark, DE 19711
USA

Jean D. Miller
Buffalo Zoo
Buffalo, NY 14214
USA

Donald E. Moore III
Smithsonian's National Zoo
Smithsonian Conservation Biology Institute
Washington, DC 20008
USA

Dave Morgan
Wild Welfare
North Carolina Zoological Park
Asheboro, NC
USA

Michael Noonan
Canisius College
Buffalo, NY 14208
USA

Steve Olson
Association of Zoos and Aquariums
Silver Spring, MD 20910
USA

Michelle Parker
Daniel P. Haerther Center for Conservation and Research
John G. Shedd Aquarium
Chicago, IL 60605
USA

Linda M. Penfold
South-East Zoo Alliance for Reproduction and Conservation
Yulee, FL 32097
USA

Beth A. Posta
Toledo Zoo
Toledo, OH 43614
USA

Ken Ramirez
John G. Shedd Aquarium
Chicago, IL 60605
USA

William A. Rapley
Toronto Zoo
Toronto, ON
Canada

Tammy M. Root
Indianapolis Zoo
Indianapolis, IN 46222
USA

Jay H. Ross
Tulsa Zoo
Association of Zoological Horticulture
Allison Park, PA 15102
USA

D. Andrew Saunders
SUNY-ESF
Syracuse, NY 13210
USA

David J. Shepherdson
Oregon Zoo
Portland, OR 97221
USA

Rebecca E. Spindler
Taronga Conservation Society Australia
Mosman, NSW
Australia

Judie Steenberg
Woodland Park Zoo (retired)
Maplewood, MN 55109
USA

Martina Stevens
Houston Zoo
Houston, TX 77030
USA

Miranda F. Stevenson
British and Irish Association of Zoos and Aquariums
London
United Kingdom

John B. Stoner
Toronto Zoo (retired)
Bethany, ON
Canada

Cynthia E. Stringfield
Moorpark College
Moorpark, CA 93021
USA

Scott P. Terrell
Disney's Animal Kingdom
Bay Lake, FL 32837
USA

Erika K. Travis
Utah's Hogle Zoo
Salt Lake City, UT 84108
USA

Patrick Thomas
Bronx Zoo
Wildlife Conservation Society
Bronx, NY 10460
USA

Eduardo V. Valdes
Disney's Animal Kingdom
Bay Lake, FL 32830
USA

Sally R. Walker
Zoo Outreach Organization
Coimbatore, Tamil Nadu
India

Adrienne Whiteley
Rosamond Gifford Zoo
Syracuse, NY 13204
USA

Douglas P. Whiteside
Calgary Zoo
University of Calgary, Faculty of Veterinary Medicine
Calgary, Alberta
Canada

Gary L. Wilson
Moorpark College
Moorpark, CA 93021
USA

Joanna Wiszniewski
Taronga Conservation Society Australia
Mosman, NSW
Australia

Sandra Woltman
National Wildlife Rehabilitators Association
Glen Ellyn, IL 60137
USA

Index

655

newts, handling of, 147
New Zealand, 613–14
niacin, dietary, 164–65
Nigeria, 590
Night Safari exhibit, 237
nightshade (*Solanum* spp.), 268
nipah/hendra virus, 466t
nipples, types of, 211, 212f
Nitrococcus, 351
nitrogen (N₂), 348, 350–52
nitrogen cycle, 351f
nitroglycerine, 90
Nitrosomonas, 351
nocturnal species, 278, 478
noisemaking devices, 103
nongovernmental conservation organizations
  (NGOs), 49
Non-Indigenous Animals Act, 1987, 598
norepinephrine, 127
North American beaver (*Castor canadensis*),
  259
North American opossum (*Didelphis virgin-iana*), 142, 144
North American porcupine (*Erethizon dorsatum*), 144
North American red tree mice (*Phenacomys longicaudus*), 259
North Korea, 594
Northwest Marine Technology, 188
nose, observation of, 447
notebooks, 174
nuclear medicine, 494
nudibranch (*Berghia verrucicornis*), 360
Nuremburg Zoo, Germany, 608
nursery areas, 210
nutrients, 151–52
nutrition, 151–70; amphibians, 340–41;
  aquarium specimens, 358–59; body condition scoring (BCS), 160; challenges, 151–52; dams, 209; disease and, 438; geriatric animals, 225–26; hoofstock, 270–71; immune systems and, 476; infant care, 211–12; mammal, 262–63; reptile requirements, 330–31; research into, 54, 546, 552; role of keepers in, 162; small mammal environment, 262–63; supplementation, 341. *See also* diets; feed
Nutrition Advisory Group (NAG), 158
nutritional diseases, 161
Nutrition SAG, 561
nyala (*Tragelaphus angasii*), 37

obesity, 161, 439–40
observation: animal welfare and, 379–80; bird, 318; dams and neonates, 209; keeper responsibilities, 439; skills, 66; small mammal keepers, 260–61; understanding behavior, 387–89
OC (*Oleoresin capsicum*), 74–76
Occupational Safety and Health (OSH) Act, 65
Occupational Safety and Health Administration (OSHA): hospital safety and, 450–51; jurisdiction of, 615; workplace, 458
ocean sunfish (*Mola mola*), 358
ocelot (*Leopardus pardalis*), 198, 400f
octopi, 361
Odobenidae (walruses), 305
Odontoceti, 305, 306

Office international des épizooties (OIE), 588,
  609–10
Office of the Scottish Charity Regulator, 548
office supplies, 567
oil spills, 574, 583–84
ointments, 499
okapi (*Okapia johnstoni*), 193, 422
olfaction: amphibian, 128; fish, 128; reptile, 128
olive baboons (*Papio anubis*), 195
olive ridley turtle (*Lepidochelys olivacea*), 193
Oman, 594
ommochromes, 111
omnivores, 330
Ontario Ministry of Natural Resources
  (OMNR), Canada, 601, 604
Ontario Society for the Prevention of Cruelty to
  Animals (OSPCA), 601, 604–5
oocysts, parasite, 480
oocytes, transfer of, 545
operant conditioning, 416–23; amphibians, 341; basic health maintenance, 448; history of, 417–18; positive reinforcement and, 416–17; small mammals, 264–65
ophthalmic drugs, 497–98
Opiliones, 369
opisthosoma, 112
ophthalmic drops, 499
oral administration, 499–500, 504
orcas (*Orcinus orca*), 529
orchard grass hay, 164
order, classification, 108
organ failure, 224–25
Oriental water shrews (*Chimarrogale* sp.), 258
orientation, facility-related, 97–98
ornate horned frogs (*Ceratophrys ornata*), 123
ornithosis. *See* psittacosis
orogastric tubes, 500
orphaned animals, 582
Orthopteroids, 369–71
oryx antelope (*Oryx* spp.), 269
osprey (*Pandion haliaetus*), 322
ostrich (*Struthio camelus*), 144, 145, 321
otic medications, 497, 499
otter (*Lutra* spp.), 144, 259
outdoor habitats, 484–85
outerwear, protective, 459
ova, 191, 480, 483f
ovaries, 191
ovariohysterectomy, 196
overcrowding, 133
overdose information, 498
overfeeding, 215
over-the-counter (OTC) drugs, 496
oviparity, 332
ovulation, 192–93
oxygenation, 507
oxytocin, 127
ozone disinfection, 310, 355
ozone generators, 355

pacemakers, 89
pachyderms, 118
Pacific white-sided dolphin (*Lagenorhynchus obliquidens*), 306
pacing, 447
padlocks, 72f, 73–74

paging systems, 70
pain management, 224
Pakistan, 592
Pallas cats (*Otocolobus manul*), 192, 215, 479
palm cockatoos (*Probosciger aterrimus*), 195
Palm Desert Zoo, 46
palm oil crisis, 564, 565f
Pan African Sanctuary Alliance (PASA), 590
Panama Amazon parrot (*Amazona ochrocephala panamensis*), 46
Panamanian golden frog (*Atelopus zeteki*), 49f
pancreas, 123, 127
panda. *See* giant panda
pandas, paramyxovirus in, 476
panleukopenia, 284
paramecium, cultured, 358
paramyxovirus, 476
paraphyly, 106
parasites: amphibians, 345; bird, 322; classification of, 480t; control programs, 484; diagnosis, 482; disease and, 440–41; habitat contamination by, 484–85; life cycles, 480–82; microenvironment of, 472; pressure of infection, 485; reptile care, 333; sample collection, 482–83; signs of infection by, 482; transmission, 482
parasitology, 479–85, 491
paratenic hosts, 480
parathyroid glands, 127
parathyroid hormone, 127
PAR bulbs, 327
Parc Zoologique de Cleres, 37
Parelaphostrongylus tenuis, 479, 481
parental neglect, 390
parenteral administration, 500
parent rearing, birds, 322
parrot fever. *See* psittacosis
parrots: repetitive behaviors, 391; restraint of, 145
parthenogenesis, 190, 332
partnerships, 47–51, 550
Partula snails, 40
parturition: captive animals, 380; pregnancy and, 193–95; prelabor/labor assessment checklist, 208f; primates, 293
parvovirus, 284
passenger pigeon (*Ectopistes migratorius*), 39
passive identification, 179–81
pasteurellosis, 466t, 467t
Patagonian cavies (*Dolichotus patagonum*), 260
Patagonian sea lion (*Otaria flavescens*), 305
pathogens: aerosolization, 477; diseases and, 442; parasites, 479–85
pathologists, 491
patient records: admission forms, 575f; physical exam forms, 576f; turtle admission records, 577f; wildlife rehabilitation and, 574. *See also* documentation
Patuxent Wildlife Research Center, 383
paver stones, 570
Pavlov, Ivan, 417
Pavlovian conditioning, 417
peafowls, 118f
Peale Museum, Philadelphia, 35
pedicel, 112
pedipalps, 112